Studies in Logic

Mathematical Logic and Foundations

Volume 98

Classification Theory

Second Edition with a New Introduction

Volume 91
A View of Connexive Logics
Nissim Francez

Volume 92
Aristotle's Syllogistic Underlying Logic: His Model with his Proofs of Soundness and Completeness
George Boger

Volume 93
Truth and Knowledge
Karl Schlechta

Volume 94
A Lambda Calculus Satellite
Henk Barendregt and Giulio Manzonetto

Volume 95
Transparent Intensional Logic. Selected Recent Essays
Marie Duží, Daniela Glavaničová, Bjørn Jespersen and Miloš Kosterec, eds

Volume 96
BCK Algebras versus m-BCK Algebras. Foundations
Afrodita Iorgulescu

Volume 97
The Logic of Knowledge Bases, Second Edition
Hector Levesque and Gerhard Lakemeyer

Volume 98
Classification Theory. Second Edition
Saharon Shelah

Studies in Logic Series Editor
Dov Gabbay dov.gabbay@kcl.ac.uk

Classification Theory

Second Edition with a New Introduction

Saharon Shelah

© Individual author and College Publications, 2023
All rights reserved.

ISBN 978-1-84890-423-1

College Publications
Scientific Director: Dov Gabbay
Managing Director: Jane Spurr

http://www.collegepublications.co.uk

All rights reserved. No part of this publication may be reproduced, stored in a retrieval system or transmitted in any form, or by any means, electronic, mechanical, photocopying, recording or otherwise without prior permission, in writing, from the publisher.

Introduction to the new edition of Classification Theory

Saharon Shelah
Einstein Institute of Mathematics, Edmond J. Safra Campus, Givat Ram, The Hebrew University of Jerusalem, Jerusalem, 9190401, Israel
and
Department of Mathematics, Hill Center - Busch Campus, Rutgers, The State University of New Jersey, 110 Frelinghuysen Road, Piscataway, NJ 08854-8019 USA
shelah@math.huji.ac.il

Introduction

I like to believe this book (i.e. the 1978, 1990 editions) has been influential, not just by honors the author received, but because there has been much work on stable theories as well as on simple theories, dependent theories, applications of this framework, and other works continuing Chapters II,II,V and the generalizations to AEC. In fact there has been so much it seem to me pointless to elaborate on it (See [2]). I will rather try two things. First, to comment on facets which while not so popular, I believe will lead to fruitful research, primarily on further directions where I have worked.

Second, I will try to point out what look to me to be outstanding open problems relating to each chapter, some of which the author has started to work on, and generally pointing out topics for people

The author thanks Alice Leonhardt for the beautiful typing. First typed July 11, 2018. The reader should note that the version in my website is usually more updated than the one in the mathematical archive.

to work on and I will comment on some of my work. Not incidentally, the book was called "Classification theory and the number of non-isomorphic models" and not "stability and the number of non-isomorphic models". Classification was my focus from the beginning.

The major thesis of the book was the dividing lines thesis. Truly the main result, the main gap theorem, characterize first order countable theories for which we can classify their models by generalized cardinal invariant but this is a secondary meaning of the word classification. See more on dividing lines in [Sh:E53]. The main gap was fulfilling the program for one test question, a major question, both intrinsically and historically important but in no way unique. Also, as expected, the properties investigated were not interesting just toward this end but in themselves. E.g. an elementary class is stable iff it has a unique resplendent model in λ for some $\lambda = (\mu^{|T|})^+$ iff this occurs in every such λ (see [25]). To repeat the discussion of [Sh:E53] , the aim of the book has been to classify elementary classes (i.e. first order theories) in the sense of biology–finding natural dividing lines. The point is the belief that:

1. There are such natural dividing lines.

2. Though not all interesting properties of first order theories are dividing lines, there are many such interesting ones; and being a dividing line is a major point in favour of a property.

3. This should eventually lead also to aiding the analysis of specific classes; this was peripheral motivation for me; but I believed it would occur and is certainly important.

4. Outside test problems are excellent ways to develop such a program. The function $I(\lambda, T)$, which counts the number of non-isomorphic models of T of cardinality λ was an excellent test problem. But this is just "the first brother, not unique in any way".

Applying and expanding the classification of theories in specific contexts has been very successful; see in particular the works of Hrushovski.

Chapters I,II,III (forking), V §1,§2 (which dealt with orthogonality of infinite indiscernibility) call for investigating families other than the stable ones, in particular, for the notion of forking and its generalizations. This has been very successful for simple theories (see [12], a later text, e.g. [3]), and dependent classes (see e.g. introduction to [Sh:950]); and lately on NPT_2 and more see, e.g. [5], [4], [7], [6]. Not included in the book were parallel theorems on non-elementary classes which exist already in the seventies; also this direction has been very successful, see e.g. the introduction to [Sh:h] and Baldwin's [1] though very few read [31], [32].

Chapter V, §3, §4 deal with weight, p-weight were very successfully continued in Hrushovski works. See more in [Sh:863, §(5G),§(5F)] and Shuus-Usvyatsov [10]. On more general non-first order see Part I of the work [Sh:839] (still in preparation).

Concerning Chapter V, §5 see [28], [14], [26, Ch.III].

Chapter VI, on Keisler order ($\triangleleft$) and saturation of ulrapowers, received little attention for many years; but during the last decade this dramatically changed, e.g. see Keisler [Ke17]. I had hopes for the StOP (strict order property), i.e. some $\varphi(\overline{x}_n, \overline{y}_n) \in \mathbb{L}(\tau_T)$ defines on $M \models T$ a partial order with arbitrarily long finite chains. It seemed to be very pervasive, but with no evidence, still see Kikyo-Shelah [17]. Related are the SOP_n ($n \geq 3$) ([500]), which suffered a similar fate.

The StOP had seemed a good candidate for maximality for $\triangleleft$ but also SOP_3 suffices [29], [39]. However, by [9], [39] if T is NSOP_2 then T is not $\triangleleft^*$-maximal. (Well, assuming an instance of GCH; $\triangleleft^*$ is a relative of $\triangleleft$ from [29]) the proof seems like a beginning of a positive theory.

More recently [20, 7.14] proves the other direction: if T is SOP_2, then it is $\triangleleft$-maximal and even $\triangleleft^*$-maximal and deal more with a beginning of a positive theory; this is a further suggestion of a positive theory. Together, we conclude that SOP_2 is a real dividing line.

Interestingly enough, the results above were preceded by a breakthrough [19] telling us a weaker result: $SOP_3 \Rightarrow \triangleleft$-maximal and also, out of the blue, that $\mathfrak{p} = \mathfrak{t}$, i.e. that two classical cardinal characteristics of the continuum are equal. This is another example of how investigating "good" dividing lines may lead to interesting and unexpected discoveries in other areas of mathematics. So SOP_2 is provably a dividing line having equivalent characterization of internal and external conditions.

Lately there was considerable work on SOP_1, see [39], [15], [18] and references there.

Another direction is 'half-dependent'; see [37], [35].

Lately there had been works on exact saturation; see [16], [15].

Recently, it seems to me that the universality spectrum would be a good test problem. We may consider the $\leq_{\text{univ}}$-maximal ones or just the so called almost-$\leq_{\text{univ}}$-maximal ones. The second seems to indicate a dividing line below SOP_4 but including some SOP_3 theories, see [8], [36], [22] and references there.

Concerning Chapters VII, VIII they have received little attention from model theorists, but one off-shoot, the black box construction, has been very successful among Abelian group theorists, see the books Eklof-Mekler [EM02] and Goebel-Trlifaj [11], and see [23], [34], [38] and references there.

The author's non-book on non-structure continuing it has not materialized, but most parts are available in mathematical arXive and my web site. We may mention early [Sh:262], Baldwin [Bal89] then [24] which deal more abstractly with constructions, it's third section improve [27, Ch.VII,§4]. Black boxes are dealt with in [23]. In [27, Ch.VIII] it is proved that under the relevant conditions (unstable, unsuperstable, DOP, OTOP) we get the maximal number of non-isomorphic models for $\lambda > |T|$. But not so fully (also for singular cardinals) for the maximal number of models of cardinality λ, no one elementarily embeddable in another; this is particularly interesting as this number behaves better for the superstable NDOP NOTOP countable first order T. This is done in [Sh:331], [Sh:309], [Sh:511].

The rest, in particular, the book's main theorem, the main gap has, except for the refinement in [13], has not been continued much.

In particular the $\mathscr{P}(n)$-diagrams have not been extensively developed; this notion is generalized in [33, §12], [31]. A related project is classification over a predicate which hopefully is now in progress and on game logics, see [30], Palacin-Shelah [21].

References

[1] John Baldwin. *Categoricity*, volume 50 of *University Lecture Series*. American Mathematical Society, Providence, RI, 2009.

[2] John T. Baldwin. *Model theory and the philosophy of mathematical practice*. Cambridge University Press, Cambridge, 2018. Formalization without foundationalism.

[3] Enrique Casanovas. *Simple theories and hyperimaginaries*, volume 39 of *Lecture Notes in Logic*. Association for Symbolic Logic, Chicago, IL; Cambridge University Press, Cambridge, 2011.

[4] Artem Chernikov. Theories without the tree property of the second kind. *Ann. Pure Appl. Logic*, 165(2):695–723, 2014.

[5] Artem Chernikov and Itay Kaplan. Forking and dividing in NTP_2 theories. *J. Symbolic Logic*, 77(1):1–20, 2012.

[6] Artem Chernikov, Itay Kaplan, and Pierre Simon. Groups and fields with NTP_2. *Proc. Amer. Math. Soc.*, 143(1):395–406, 2015.

[7] Artem Chernikov and Nicholas Ramsey. On model-theoretic tree properties. *J. Math. Log.*, 16(2):1650009, 41, 2016.

[8] Mirna Džamonja. Club guessing and the universal models. *Notre Dame J. Formal Logic*, 46:283–300, 2005.

[9] Mirna Džamonja and Saharon Shelah. On $\triangleleft^*$-maximality. *Ann. Pure Appl. Logic*, 125(1-3):119–158, 2004. arXiv: math/0009087.

[10] Darío García, Alf Onshuus, and Alexander Usvyatsov. Generic stability, forking, and thorn-forking. *Trans. Amer. Math. Soc.*, 365(1):1–22, 2013.

[11] Rüdiger Göbel and Jan Trlifaj. *Approximations and endomorphism algebras of modules*, volume 41 of *de Gruyter Expositions in Mathematics*. Walter de Gruyter, Berlin, 2006.

[12] Rami Grossberg, Jose Iovino, and Olivier Lessmann. A primer of simple theories. *Archive for Mathematical Logic*, 41:541–580, 2002.

[13] Bradd Hart, Ehud Hrushovski, and Michael C. Laskowski. The uncountable spectra of countable theories. *Annals of Mathematics*, 152:207–257, 2000.

[14] A. Hernandez. *On ω_1–saturated models of stable theories.* PhD thesis, Univ. of Calif. Berkeley, 1992. Advisor: Leo Harrington.

[15] Itay Kaplan, Nicholas Ramsey, and Saharon Shelah. Local character of Kim-independence. *Proc. Amer. Math. Soc.*, 147(4):1719–1732, 2019. arXiv: 1707.02902.

[16] Itay Kaplan, Saharon Shelah, and Pierre Simon. Exact saturation in simple and NIP theories. *J. Math. Log.*, 17(1):1750001, 18, 2017. arXiv: 1510.02741.

[17] Hirotaka Kikyo and Saharon Shelah. The strict order property and generic automorphisms. *J. Symbolic Logic*, 67(1):214–216, 2002. arXiv: math/0010306.

[18] Alex Kruckman and Nicholas Ramsey. Generic expansion and Skolemization in NSOP_1 theories. *Ann. Pure Appl. Logic*, 169(8):755–774, 2018.

[19] Maryanthe Malliaris and Saharon Shelah. Cofinality spectrum theorems in model theory, set theory, and general topology. *J. Amer. Math. Soc.*, 29(1):237–297, 2016. arXiv: 1208.5424.

[20] Maryanthe Malliaris and Saharon Shelah. Model-theoretic applications of cofinality spectrum problems. *Israel J. Math.*, 220(2):947–1014, 2017. arXiv: 1503.08338.

[21] Daniel Palacín and Saharon Shelah. On the class of flat stable theories. *Ann. Pure Appl. Logic*, 169(8):835–849, 2018. arXiv: 1801.01438.

[22] S. Shelah et al. Tba. In preparation. Preliminary number: Sh:F1808.

[23] Saharon Shelah. Black Boxes. arXiv: 0812.0656 Ch. IV of The Non-Structure Theory" book [Sh:e].

[24] Saharon Shelah. General non-structure theory and constructing from linear orders. arXiv: 1011.3576 Ch. III of The Non-Structure Theory" book [Sh:e].

[25] Saharon Shelah. On spectrum of κ-resplendent models. arXiv: 1105.3774 Ch. V of [Sh:e].

[26] Saharon Shelah. Stable frames and weights.

[27] Saharon Shelah. *Classification theory and the number of nonisomorphic models*, volume 92 of *Studies in Logic and the Foundations of Mathematics*. North-Holland Publishing Co., Amsterdam, 2nd edition, 1990. Revised edition of [Sh:a].

[28] Saharon Shelah. Multi-dimensionality. *Israel J. Math.*, 74(2-3):281–288, 1991.

[29] Saharon Shelah. Toward classifying unstable theories. *Ann. Pure Appl. Logic*, 80(3):229–255, 1996. arXiv: math/9508205.

[30] Saharon Shelah. Theories with Ehrenfeucht-Fraïssé equivalent non-isomorphic models. *Tbil. Math. J.*, 1:133–164, 2008. arXiv: math/0703477.

[31] Saharon Shelah. *Classification theory for abstract elementary classes*, volume 18 of *Studies in Logic (London)*. College Publications, London, 2009.

[32] Saharon Shelah. *Classification theory for abstract elementary classes. Vol. 2*, volume 20 of *Studies in Logic (London)*. College Publications, London, 2009.

[33] Saharon Shelah. *Toward classification theory of good λ frames and abstract elementary classes.* 2009. arXiv: math/0404272 Ch. III of [Sh:h].

[34] Saharon Shelah. Pcf and abelian groups. *Forum Math.*, 25(5):967–1038, 2013. arXiv: 0710.0157.

[35] Saharon Shelah. A.E.C. with not too many models. In A. Hirvonen, M. Kesala, J. Kontinen, R. Kossak, and A. Villaveces, editors, *Logic Without Borders: Essays on Set Theory, Model Theory, Philosophical Logic and Philosophy of Mathematics*, volume Ontos Mathematical Logic, vol. 5, pages 367–402. Berlin, Boston: DeGruyter, 2015. arXiv: 1302.4841.

[36] Saharon Shelah. No universal group in a cardinal. *Forum Math.*, 28(3):573–585, 2016. arXiv: 1311.4997.

[37] Saharon Shelah. Definable groups for dependent and 2-dependent theories. *Sarajevo J. Math.*, 13(25)(1):3–25, 2017. arXiv: math/0703045.

[38] Saharon Shelah. Quite free complicated Abelian groups, pcf and black boxes. *Israel J. Math.*, 240(1):1–64, 2020. arXiv: 1404.2775.

[39] Saharon Shelah and Alexander Usvyatsov. More on SOP_1 and SOP_2. *Ann. Pure Appl. Logic*, 155(1):16–31, 2008. arXiv: math/0404178.

CONTENTS

ACKNOWLEDGEMENTS

My research in this area started with my thesis, and I thank M. O. Rabin for his kind guidance. I thank the Israel Academy of Science for partially supporting my research, as well as the United States–Israel Binational Science Foundation (grant 1110) and the NSF (grants 144–H747 and MCS–08479). Much of the proofreading and checking was done industriously by M. Abramsky who was my research assistant from 1973 to 1976, and for this I am grateful to him, as well as to L. Marcus who did a similar work previously. They are responsible for the few paragraphs which are in good English. I owe a special debt of gratitude to D. Ehrman, J. Alon and mainly D. Sharon for typing, correcting and recorrecting the manuscript. Lastly, I thank J. Baldwin, G. Cherlin, S. Koppleberg, D. Monk, M. Rubin and P. Schmidt for detecting various errors and inaccuracies.

INTRODUCTION

The aim of this book is to represent works of the author on classification and related topics. The author, in a moment of insanity, believed this would be the easiest way to represent his work.

There is no point in trying to convince you that I think the book is important; since otherwise I would not have spent my time writing it. Anyhow, I have said it in the introduction to [Sh 75c], which is supposed to be a propaganda for this book. There is also no point in explaining how the subject evolved because I say almost everything in the introduction to [Sh 74]. Let us note only that the founding stone is [Mo 65] (there are historical notes at the end of the book).

So we shall now explain how to read the book. The right way is to put it on your desk in the day, below your pillow at night, devoting yourself to the reading, and solving the exercises till you know it by heart. Unfortunately, I suspect the reader is looking for advice on how not to read, i.e. what to skip, and even better, how to read only some isolated highlights.

If you are generally not interested in model-theory but are interested in ultrafilters and ultraproducts, you can read Chapter IV, Sections 1, 2 and 3. There, many classical results are represented but also some new ones. We prove that there is a λ-good not λ^+-good ultrafilter iff λ is regular (improving Keisler, cf. [Ke 65]); show that $\mu \le 2^\lambda \wedge \mu^{\aleph_0} = \mu$ iff there is an ultrafilter D over λ and $n_i < \omega$ such that $\prod n_i/D = \mu$, thus answering a question of Keisler (cf. [Ke 67a]). We also answer one of the questions from [CK 73] (due to Keisler): there is a regular ultrafilter D over $\lambda > \aleph_0$, D not good, but $\prod n_i/D \ge \aleph_0 \Rightarrow \prod n_i/D = 2^\lambda$. Also an $\aleph_1$-incomplete D over I is λ-good iff for every order J, J^I/D is λ-saturated.

Also Chapter VII is quite isolated. The first section deals with a generalization of λ-saturated, so that such models exist in more cardinalities. The second section presents a generalization of Ehrenfeucht–Mostowski models, and assume only the knowledge of Skolem functions. The third section uses the second, and some definitions from previous chapters, and pretends to show the generalization of Section 2 is useful, so e.g., we have a model generated by a tree of indiscernibles and this is important for unsuperstable theories. The fourth section again requires no prerequisite, and there we prove that for *any* theory, and $\mu \geq 2^{|T|}$, it has μ-universal, μ-stable models of arbitrarily high cardinalities. In Section 5 we present the theorems on Hanf numbers of omitting types.

Chapter VIII is devoted to the number of non-isomorphic models, but for this we use only some definitions (or rather equivalent conditions explicitly stated and used as alternative definitions). Also Chapter VII, Section 3 is used in Chapter VIII, Sections 2 and 3. Most of Chapter VIII, Section 1 is quite easy (and has real results) and there is not much dependence.

In Chapter VIII, Section 2 we have somewhat heavier material; you can start by reading Theorem 2.2(1): if T is unsuperstable, $\lambda > |T|$ is regular, then T has 2^λ non-isomorphic models of power λ. Chapter VIII, Sections 2 and 3 are aimed at proving the same for any $\lambda \geq |T| + \aleph_1$; but Chapter VIII, Section 3 and the later parts of Section 2 are, in their present form, suitable for your too arrogant students (but no guarantee).

Section 7 of Chapter III is independent, and contains results on the possible $K(\lambda) = \sup\{|S(A)|: |A| \leq \lambda\}$ and related problems.

Also the combinatorial appendix needs no prerequisite, but it is not so interesting. If you arrive here I hope you understand that you should really start to read from Chapter I.

Now Chapter I, Section 1 is a restatement of some classical theorems (compactness and Lowenheim–Skolem which are proved later, elementary chains, saturated models, etc.). Now the reader should know what are first order theories, formulas and satisfaction; and no more; but I would be very interested to meet a successful reader who has not read before a significant part of [CK 73], or [Sa 72]. We here also make some conventions.

Chapter I, Section 2 contains central definitions (stability, indiscernibility) and some theorems on existence of indiscernibles and the connection between unstability and order.

Chapter II, Sections 1 and 2 are easy and central. Here ranks are introduced and stability of formulas and theories are really introduced, i.e., we give many equivalent conditions.

Chapter II, Sections 3 and 4 are quite peripheral and independent. In Chapter II, Section 3 we investigate more deeply ranks, and connect them with superstability. Only Claim 3.12 of Chapter II is used in Chapter II, Section 4.

In Chapter II, Section 4 we investigate some more properties of formulas and theories: the finite cover property, the strict order property and the independence property. Almost the only use later is in the investigation of Keisler orders and saturation of ultrapowers, in Chapter VI, Sections 4 and 5.

In Chapter III, Sections 1, 2, 3, 4 and 6 are very necessary for what follows; Sections 1–4 can be read only successively, Section 6 depends mainly on Chapter II, Sections 1 and 2. Unlike Chapter II we concentrate here on complete types and investigate forking, and in Sections 1–5 deal with stable T only. Forking is introduced in Section 1, investigated in Section 2, used in Section 3 to (almost) give the stability spectrum theorem, and in Section 5 really to give it. In Section 4 we see the connection between ranks and forking, use them to prove results on both sides. In Section 6 we extend our models so as to have names for equivalence classes, this is important when we look for canonical forms. In Section 7 we investigate various properties, like in Chapter II, Section 4, mostly for unstable T.

Chapter IV deals with prime models. As we have to deal with five kinds of prime models, we give an axiomatic setting (in Section 1) find which examples satisfy which axioms (in Section 2, the results sum up in a table at the end). In Section 3 we prove theorems in this axiomatic setting (prime models exists, they realize only isolated types, etc.). Section 3 uses only Chapter IV, Section 1, but in Chapter IV, Section 2 extensive use of Chapter III, Section 4 is made. The fourth section is somewhat heavier; we here concentrate on stable theories, and give characterization of prime models (implying uniqueness). Only a few lemmas from it are used later.

The fifth section contains scattered results, most of them do not require much knowledge. The main theorem says, e.g., that for countable stable T, if a primary model (i.e. a prime model constructed step by step) exists, the prime model is unique. We also prove a theorem on the existence of a model of T omitting $< 2^{|T|}$ complete types.

Chapter V contains some of the deepest results of the book. It uses Chapter IV but usually not in a deep sense, the reader would better go to its detailed introduction.

Now in Chapter VI we turn to ultraproducts, Sections 1, 2 and 3 do not refer to anything. In Sections 4, 5 and 6 we use Chapter II. The first half was already reviewed. In the second half we investigate

Keisler order on theories ([Ke 67]) and get quite a complete picture (to complete the picture we should know more on unstable theories without the strict order property), and find, quite accurately, how saturated are ultraproducts and ultralimits.

Chapters VII and VIII were already reviewed. Chapter VIII, Section 4 is to a large extent a summary of results, and has some not so hard theorems.

In Chapter IX, Section 1 we prove the categoricity theorems, relying on various previous results (including Chapters V and VIII). We also prove some results on the number of models in elementary classes (in Chapter VIII we concentrate on pseudo-elementary classes), and when a model can be represented as a union of strictly increasing chain of length δ.

In Chapter IX, Section 2 we deal with the number of countable models of a superstable T; and what can be the number of $\mathbf{F}^a_{\aleph_0}$-models of T in $\aleph_\alpha$ (it is 1, $|\alpha|^\mu$ ($\mu \leq 2^{|T|}$), or $\geq 2^{|\alpha|}$), and just models when T is totally-transcendental.

The Appendix contains the combinatorial results we need.

Note. (1) Each chapter has its introduction, it may be wise to look at it again during the reading.

(2) Exercises are scattered randomly among the sections. Some are the result of the public pressure due to the prejudice that examples clarify notions, others are variants of theorems I want to mention, or remnant of obsolete proof. And some were the result of my preferring to give an exercise with a generous hint rather than a theorem with a thin proof.

(3) The change in the name of the book is not incidental, but a change in point of view during the years in which the book was written, explained in [Sh 75c].

The reader may also want to know which of my papers becomes obsolete by this book. So this is the fate of [Sh 69a], [Sh 70a], [Sh 70b] [Sh 71], [Sh 71b], [Sh 71d], [Sh 72a], [Sh 74] and [Sh 74a]. Also [Sh72] except Section 2 which deals with the uncountable case, [Sh 72b] except the purely combinatorial part and the theorem on $L_{\infty,\omega}$. But in [Sh 71b], the proof of the last theorem is not covered.

We did not deal here with the results on non-elementary classes ([Sh 69], Section 6; [Sh 69c], [Sh 70], [Sh 75a] and [Sh 75c]).

INTRODUCTION TO THE REVISED EDITION

In this edition four new chapters (X, XI, XII and XIII) have been added. (In addition, many corrections have been made, including many which previously were in the "Added in Proof" part of the First Edition.)

The additional chapters present the solution to countable first order T of what the author sees as the main test of the theory (XII, 6.1).

In Chapter X we introduce the dop (dimensional order property) and show that it is a very meaningful dividing line for superstable theories: if it holds, T has many $\aleph_\epsilon$-saturated ($= \mathbf{F}^a_{\aleph_0}$-saturated) models; if it fails, we have a structure theory: every $\aleph_\epsilon$-saturated model M of T is $\mathbf{F}^a_{\aleph_0}$-prime over a non-forking tree of models $\langle M_\eta : \eta \in I \subseteq {}^{\omega >}\lambda \rangle$ (each M_η $\mathbf{F}^a_{\aleph_0}$-prime over $\emptyset$). If the tree I were to have an infinite branch, the theory is deep and still has many models. Otherwise, we can compute the number of $\aleph_\epsilon$-saturated models. This chapter is an improved version of [Sh 83a], [Sh 84b]. Mainly the *IE* case is explicated and the facts on trivial (regular) types are gathered in one section.

In Chapter XI we prove the needed decomposition theorems assuming

(∗) there are prime atomic models over $M_1 \cup M_2$ when $M_0 \prec M_1$, M_2 $\{M_1, M_2\}$ is independent over M_0.

Chapter XII is the crux of the matter. We prove that the negation of (∗) above implies that in models of T we can define a relation which order a large subset of ${}^m|M|$; the definition being $\bar{a}R\bar{b}$ iff a type $p(\bar{x}, \bar{a}, \bar{b})$ is omitted.

Chapter XIII is intended to exemplify that Theorem XII, 6.1 fulfills its aim. For this we consider several questions which are solved using its partition to cases. In Section 1 we deal with "when $L_{\infty,\lambda}$-equivalent models of cardinality λ are isomorphic".

In Section 2 we deal with computing $IE(\lambda, T)$. Considering what was done in Chapter X, the main problem is to show $IE(\lambda, T) = 2^\lambda$ when T has otop. Usually, the reader thinks this is immediate, forgetting that elementary embedding does not necessarily preserve the omission of types.

In Chapter XII, §3, we prove the Morley conjecture: for countable first order T, $\aleph_0 < \lambda < \mu \Rightarrow I(\lambda, T) \leqslant I(\mu, T)$ except when T is categorical in $\aleph_1$ (so $(\forall\lambda > \aleph_0) I(\lambda, T) = 1$) but $I(\aleph_0, T) = \aleph_0$.

In Chapter XIII, §4, we compute $I(\lambda, T)$ for λ large enough (though we do not know if a relevant invariant can be $> \aleph_0$ but $< 2^{\aleph_0}$).

For a discussion on the significance of the new chapters, see [Sh 85b].

We thank Leo Harrington for hearing the proofs in Chapter XII while they were generated, and the reader may thank him for persuading me to prove the otop there using finite sequences rather than some considerably more complicated and infinite versions.

We thank Saffe and Hrushovski for pointing out the need for corrections, Grossberg for proofreading VIII, and the *Israel Journal of Mathematics* for allowing us to use portions of [Sh 83] and [Sh 83a] which appear there. Last, but not least, we thank Damit Sharon and Alice Leonhardt for typing the new material and corrections.

S. SHELAH
November 1987

OPEN PROBLEMS

A. Completions

Here we deal with problems which just complete theorems in the book. We feel existing methods will suffice.

(1) Complete the computation of the possible function $I(\lambda, T)$ ($\lambda \geqslant \aleph_0, T$ countable complete first order theory). See XIII, §§3 and 4.

Though some think this was "the problem", I could not make myself excited about it. Still it would be nice to know.

The most appealing part is:

(1a) What can $SND(T)$ be? If it is a limit cardinal $> \aleph_0$, what can $k(T), l(T)$ be?

Hopefully $SND(T) \in (\omega+1) \cup \{\aleph_1, (2^{\aleph_0})^+\}$: even so, we shall need another cardinal invariant to compute $I(\lambda, T)$ for $\aleph_\omega \leqslant \lambda < \aleph_{\beth_1}$.

If we are interested in $I(\aleph_0, T)$, too, we should ask

(1b) Does the assumption $I(\aleph_0, T) = \aleph_0$, T superstable, bound the depth of T? [$(\omega+1)$ is possible.]

(2) What can λ be from Theorem V, 5.8?

(3) Is $DP(T) > |T|^+$ possible?

Remember $DP(T) < \delta(|T|)$, all successor $< |T|^+$ are possible; however, cf. $|T| > \aleph_0$ imply $\delta(|T|) > |T|^+$.

(4) Suppose $\lambda \geqslant \kappa_r(T)$ (or even $\lambda = \kappa_r(T)$), M is $\mathbf{F}^a_\lambda$-atomic over A, and for every $\boldsymbol{I} \subseteq M$ indiscernible over A, $\dim(\boldsymbol{I}, A, M) \leqslant \lambda$. Then, is M necessarily $\mathbf{F}^a_\lambda$-prime over A?

Even various weakenings are open. You can assume $\lambda > |T|$ [is regular] or it can be to prove for $\mathfrak{C}^{eq}$ (this strengthens the hypothesis — we have more $\boldsymbol{I}$'s), or try the parallel for $\mathbf{F}^t_\lambda$. Note that this is an attempt to improve the characterization theorem IV, 4.14; note also that for T superstable this is true (see IV, 4.18).

(5) Suppose T is stable $|D(T)| > |T|^{<\kappa}$, $(\forall \chi < \mu)[\chi^{<\kappa} < |D(T)|]$ κ

regular $\geqslant \kappa(T)$. Prove that for every $\lambda = \lambda^{<\kappa} \geqslant |T|$ T has at least $\min\{2^\lambda, 2^\mu\}$ non-isomorphic κ-compact models of power λ.

See IX, 1.20.

(6) For a shallow superstable without the dop variety the depth is $\leqslant \omega$.

(This means that if every completion of an equational theory is shallow superstable without the dop, then every such completion has finite depth.) See [Sh 86].

(7) Is every unidimensional stable T, superstable?

(8) See questions in Baldwin and Shelah [BSh 85] and [Sh 86a].

B. Classifying further unstable theories

Clearly not all unstable theories are equally complicated, and we may want finer divisions: Theorem II, 4.7(1) suggest the strict order property and independence property, and IV, §7 the tree property. We shall deal here with some problems that look like reasonable approaches.

(1) What can $\{(\lambda, \mu)$ be: every model of T of power λ has a μ-compact extension of power $\lambda\}$?

The interesting case is $|T| < \mu < \lambda, \chi^{<\mu} < \lambda < 2^\chi$ (see Ex. VIII, 4.5, which is solved in [Sh 81, Th 1, 18]), and we can assume T is unstable but simple (i.e. without the tree property, since otherwise we have an answer). By [Sh 80], a positive non-trivial answer is consistent; by [Sh 81] if we change the problem a little (restricting the μ-compactness to a predicate) this cannot be proved in ZFC. If the idea can be transferred to the usual case, we may still hope for $\omega+1$ cases (with [Sh 81]'s example being for 3 and ω's one are those like T_{ind}; the needed analysis may look like [Sh 83a]).

(2) Continue the investigation of Keisler order.

See VI and [Sh 72, §1]. Note that the countable theories left are those unstable T without the strict order property, and that the simple partition with the tree property seems meaningful, so [Sh 80] seems relevant. It seemed that the inner partition of unstable simple theories (if exists) is different than in (1). Also note that this involves building special ultrafilters, so we may choose another variant of the order (e.g. limit ultrapower $M_D^I \mid W$ with W $\aleph_1$-complete; see Keisler [Ke 63]).

(3) The following definition was suggested by Grossberg and the author.

DEFINITION 3a: Let $\lambda \Rightarrow (\mu)^{<\omega}_{T,\chi}$ mean that if $\boldsymbol{I}, A \subseteq M, M \models T, |A| \leqslant \chi$, $|\boldsymbol{I}| = \lambda$ then some $\mathbf{J} \subseteq \boldsymbol{I}$ is an indiscernible sequence over A of power μ.

We can ask,

CONJECTURE 3b: Suppose T is without the independence property, then for some α, $\beth_\alpha(\mu + |T| + \chi) \to (\mu)^{<\omega}_{T,\chi}$.

This is confirmed in [Sh 86a] for theories that have no monadic expansions with the independence property. Note that we get complete negative results if T has the (**)-independence property, where

DEFINITION 3c: (i) T has the *-independence property, if for some $\varphi(\bar{x}_0, \bar{x}_1, \bar{x}_2, \bar{y})$ for every $k < \omega$ and $w \subseteq k$ there are $\bar{b}$ and $\bar{a}_i$ $(i < k)$ such that for every distinct $i_0 < i_1 < i_2 < k$

$$\models \varphi[\bar{a}_{i_0}, \bar{a}_{i_1}, \bar{a}_{i_2}, \bar{b}] \quad \text{iff } \langle i_0, i_1, i_2 \rangle \in w.$$

(ii) T has the (**)-independence property if there are formulas $\varphi_n(\bar{x}_0, \ldots, \bar{x}_n, \bar{y}_n)$ $(\ell(\bar{x}_\ell)$ constant) such that for every $W_n \subseteq {}^n\omega (n < \omega)$ there are $\bar{b}_n (n < \omega)$ $\bar{a}_i (i < \omega)$ such that for $n < \omega$ and $i_0 < \ldots < i_{n-1}$

$$\models \varphi_n[\bar{a}_{i_0}, \ldots, \bar{a}_{i_{n-1}}, \bar{b}] \quad \text{iff } \langle i_0, \ldots, i_{n-1} \rangle \in W_n.$$

We may weaken (ii) and still obtain the result.

(4) Find what $K = \{\lambda : T$ has a universal model of power $\lambda\}$ can be.

Note that for "saturated" we know the answer (see VIII, 4.8). Also, for a strong limit $\lambda > |T|$, $\lambda \in K$, and if T has a saturated model of power λ, then $\lambda \in K$. Surprisingly, some independence results were found: for T_{ord} see [Sh 80a], and for T_{ind} [Sh 84].

C. Problems parallel (to the book)

(1) Prove the main gap (etc.) for uncountable theories. [Here and elsewhere we concentrate on the main gap, but expect the parallel to XIII too.]

Note that almost everywhere in the book the countability of the theory is not particularly important. In fact, there is just one point (but a crucial one) in which we use countability: in XII, §4 (the otop/existence dichotomy).

It seems we should try to deal with inevitable types (instead of isolated ones) (see proof of IV, 5.9) or even with just prime (but not necessarily primary) models over suitably chosen non-forking trees of models.

We may concentrate on unidimensional superstable T (this suffices to prove either $I(\lambda, T) \leqslant 2^{2|T|}$ or $I(\aleph_\alpha, T) \geqslant |\alpha|$). Note that if $\varphi(x, \bar{a})$ is weakly minimal, $\bar{a} \cup \varphi(M, \bar{a}) \subseteq A \subseteq M$, then M is $\mathbf{F}^{\ell}_{\aleph_0}$-atomic over A.

(2) Prove the main gap for the class of $|T|^+$-saturated models.

Again only "one point" is missing: the existence of a regular type in X, §3's proof. Note the analysis of non-multidimensional T (in V, §5) and the dimensional continuity property (see X, §2) (which take care of the cases that an increasing union of saturated models do not). We can look also at other kinds of saturation.

(3) Prove the main gap (etc.) for the class

$$K_T = \bigcap\{PC(T_1, T) : T \subseteq T_1, |T| = 2^{\aleph_0}\}$$

for T countable.

See VI, 5.3, 5.4, 5.5 and 5.6.

A priori we considered the order of difficulty in increasing order as: Ch. X (i.e. the main gap for $\mathbf{F}^{a}_{\aleph_0}$-saturated models), question (2) above, the present question C3 and Theorem XII, 6.1. The fact that XII, 6.1 was solved before C2 and C3 is due to our greater interest, but we may have been wrong.

A relevant theorem is [Sh 75]. Note that a hidden order property for those classes is:

(Pr) In $\mathfrak{C}^{\mathrm{eq}}$, for some formula $\varphi(x, \bar{y}, \bar{z})$ for every $k, n < \omega$, there are $\bar{a}_\ell$, ($\ell < n$) such that for every $\ell_1 < \ell_2 < n$, $m_1 < m_2 < n$,

$$|\varphi(\mathfrak{C}^{eq}, \bar{a}_{\ell_1}, \bar{a}_{\ell_2}|^k \leqslant |\varphi(\mathfrak{C}^{eq}, \bar{a}_{m_2}, \bar{a}_{m_1})|,$$

$$k \leqslant |\varphi(\mathfrak{C}, \bar{a}_{\ell_1}, \bar{a}_{\ell_2}| < \aleph_0.$$

By [Sh 75] if T has (Pr), $I(\lambda, K_T) = 2^\lambda$ for $\lambda > 2^{\aleph_0}$ (using cardinality quantifiers).

(4) Prove the main gap for strongly λ-saturated models (i.e. for $\mathbf{F}^a_\lambda$-saturated models strongly λ-homogeneous, i.e. if $\langle a_i : i < \alpha < \lambda \rangle \subseteq M \langle b_i : i < \alpha < \lambda \rangle \subseteq M$ realizes the same type, then some automorphism of the model takes one to the other).

The following problem should be easier to solve.

(5) Prove the main gap (etc.), for the class of $\aleph_0$-saturated models (T countable).

(6) Classify when we replace "isomorphic" by "$L_{\infty,\omega}$-equivalent".

(6a) Classify the possible Scott heights of models of T.

See the works of Nadel and [Sh 71a]. This does, of course, have some connection to the Vaught conjecture.

(7) We know that $\{\lambda : T$ has a rigid model of power $\lambda\}$ may be "bad" (see [Sh 76c]). But what for $|T|^+$-saturated models? What about the number of non-isomorphic ones when at least one exists? See [Sh 83b] on very partial positive results.

(8) Vaught conjecture, i.e. (for complete countable T) $I(\aleph_0, T) > \aleph_0 \Rightarrow I(\aleph_0, T) = 2^{\aleph_0}$. We can look also at variants of it: the number of minimal, models or rigid models.

Morley [Mo] proves $I(\aleph_0, T) > \aleph_1 \Rightarrow I(\aleph_0, T) = 2^{\aleph_0}$, and the proof applies to many other cases (but this really belongs to description set theory, where stronger theorems have since been proven, e.g. that of Burgess; see [Sh 84b]).

Some people think this is the most important question in model theory as its solution will give us an understanding of countable models which is the most important kind of models. We disagree with all those three statements.

D. Further reaching problem

(1) Define "T has a structure theory over a class K" and classify accordingly.

It seems reasonable to look at specific K's. A reasonable interpretation is: for some κ, λ for every model M of T there is a model $I \in K$, and functions $f: M \to {}^{\kappa}I$, $g: M \to \lambda$ such that if $a_i^{\ell} \in M (\ell = 1, 2, i < n)$, $g(a_i^1) = g(a_i^2)$ and $\langle f(a_i^1) : i < n \rangle$, $\langle f(a_i^2) : i < n \rangle$ realize the same atomic type in I, then $\langle a_i^1 : i < n \rangle$, $\langle a_i^2 : i < n \rangle$ realizes the same type in M; moreover, if $\bar{c}, \bar{d} \in {}^{\kappa}I$, $\bar{\alpha} < \lambda$ and $\bar{c}, \bar{d}$ realizes the same atomic type in I, then

$$(\exists \bar{a} \subseteq M)[f(\bar{a}) = \bar{c} \wedge g(\bar{a}) = \bar{\alpha}] \quad \text{iff } (\exists \bar{a} \subseteq M)[f(\bar{a}) = \bar{d} \wedge g(\bar{a}) = \bar{\alpha}].$$

So the main gap says which T has a structure theory over $K = \{I : I$ a tree with $\leqslant \omega$ levels and depth $\leqslant \alpha\}$, and the decomposition theorem (XIII, 5.1) says which T has a structure over the class of trees with ω levels.

The natural next step is to look at the class of linear order or trees.

(2) Classification over a predicate.

We assume T has a designated predicate P, and we want to know how much $M \models T$ is determined by $M \restriction P$. See [PSh 85] (and previous references therein). The author's recent works are [Sh 85d], [Sh 86b].

Related problems are

(2a) What can be

$$f(\lambda) = \operatorname{Sup}\{f(\lambda, N) : \|N\| = \lambda_1\},$$

$$f(\lambda, N) = \{(M, a)_{a \in N} / \cong : M \models T, \|M\| = \lambda_1, M \restriction P^M = N\}.$$

(2b) Replace $(M, a)_{a \in N}$ by M in (2a) above.

(2c) When over $M \restriction P^M$ there is a prime model.

(3) Classify for $\psi \in L_{\omega_1, \omega}$ or $\psi \in L_{\lambda^+, \omega}$ or for abstract elementary classes see [Sh 85a].

For example:

(3a) If $\psi \in L_{\lambda^+, \omega}$ is categorical in one $\mu \geqslant \beth_{\delta(\lambda)}$, then it is categorical in every $\mu \geqslant \beth_{\delta}(\lambda)$.

(3b) If $\psi \in L_{\lambda^+, \omega}$, Φ as in VII, §2, Φ is such that $EM(I, \Phi) \models \psi$ for every I, then either for every $\lambda \geqslant \beth_{\omega}(|\psi|)$ $|\{EM(I, \Phi)/\cong : |I| = \lambda\}| = 2^{\lambda}$, or (possibly changing Φ) for every $\lambda |\{EM(I, \Phi)/\cong |I| = \lambda\} = 1$.

See [Sh 75], [Sh 83a], [Sh 85a].

ADDED IN PROOF

In order to make the process of writing converge, the author "closed" it to innovations in the summer of 1983 (except that Chapter X, Section 5, was expanded drastically to include the proof for *IE*). So let us comment on some of the relevant improvements and advances.

1. *Uniqueness of prime models (IV, 5.6)*

This is old, but still note [Sh 79a] while reading Theorem IV, 5.6 (which says that if T is countable and stable, and there is a primary model over A, *then* the prime model over A is unique). By [Sh 79a] the assumptions "countable" and "stable" cannot be omitted and an alternative proof of the theorem is given.

2. *Regular, semi-regular and simple types (Chapter V)*

See several works of S. Buechler, the thesis and some works of E. Hrushovski and Buechler and Shelah [BuSh] and Hrushovski and Shelah [HSh]. More on the dichotomy of multi-dimensionality for stable theories (strengthening it) will appear, relevant to problem (12).

3. *Chapter VI*

S. Koppleberg continues the work on ultraproducts of finite cardinals (and gets quite general sets).

4. *Numbers of non-isomorphic/pairwise non-embeddable models (Chapter VIII)*

There are improvements and (more) independence results and

counterexamples restricting possible improvements, see [Sh 80a], [Sh 84], [Sh 85e, Ch III], [Sh 88b] and [Sh 89].

The improvements are:

(a) We improve the proof of VIII, 3.2 (give new results for non-first order cases). See [Sh 85e, Ch. III].

(b) Improving VII, 1.8: if $T \subseteq T_1$ are complete, countable T, $\aleph_0$-unstable not stable, then $IE(\lambda, T_1, T) \geqslant \mathrm{Min}\{2^\lambda, \beth_2\}$ for $\lambda > \aleph_0$ by [Sh 88b, Theorems 1 and 2].

(c) Improving VIII, 1.2 if $T \subseteq T_1$, $|D(T)| > |T_1|^+$, then $IE(\lambda, T_1, T) \geqslant |D(T)|^{\aleph_0}$ for $\lambda > |T_1|^+$. See [Sh 88b, Fact 3].

(d) If T is unsuperstable, $\lambda > |T_1| + \aleph_0$, $T \subseteq T_1$, then $IE(\lambda, T_1, T) = 2^\lambda$ (and general principles), which implies this is proved (see [Sh 83] for many cases and some principles, and [Sh 89] for all).

For independent results:

(A) It is consistent that for some unsuperstable countable T, and T_1, $T \subseteq T_1$, $|T_1| = \aleph_1$ and $I(\aleph_1, T_1, T) = 1$ (by [Sh 80a]).

(B) It is consistent that for $T = T_{\mathrm{ord}}$, the theory of dense linear order, or T^*_{ind}, for some T_1, $\lambda = |T_1| > \aleph_0$, $IE(\lambda, T_1, T) = 1$ (by [Sh 80a] $T = T_{\mathrm{ord}}$, $\lambda = \aleph_1$, by [Sh 84] $T = T^*_{\mathrm{ind}}$, $\lambda = \aleph_1$, or one of many others). Mekler generalizes the results on graphs to classes of universal theories

(C) For a quite comprehensive list of restricting independence results $|D(T)| \geqslant |T_1|$, see [Sh 88b].

5. *Decomposition theorem (Chapter XI)*

We here prove [by XII, XI 2.4, for $(\mathbf{T}_t, \subseteq_a)$]

(*) if T is countable superstable without dop and otop, *then* for every M there is a non-forking tree $\langle N_\eta : \eta \in I \rangle$, M primary over $\bigcup N_\eta$, $\|N_\eta\| \leqslant 2^{\aleph_0}$. Moreover, any suitable tree of this form by (i.e. using $(\mathbf{T}^t_{\aleph_0}, \subseteq_a)$) can be extended such that the model M is prime over it.

By [BuSh 89] this missing axiom for, e.g., $(\mathbf{T}^t_{\aleph_0}, \subseteq_e)$ holds (in the cases we need) so we can apply this pair in Chapter XIII to improve the cardinals in the results ($\|N_\eta\| \leqslant \aleph_0$).

Earlier results do this too.

Saffe (in Trento 7/1984) suggests using the notion of elementary submodels used in IX, 1.1, 1.5 used also in V, §6, i.e.

$A \subseteq^*_d B$ *if* for every $\bar{b} \in B$, finite Δ and $\bar{a} \in A$ and φ such that $\models\varphi[\bar{b},\bar{a}]$ there is $\bar{b}'\in B$, satisfying $\models\varphi[\bar{b}',\bar{a}]$ and $R^m(\text{tp}(\bar{b},\bar{a}),\Delta) = R^m(\text{tp}(\bar{b}',\bar{a}),\Delta)$,

and prove (*) for $(\mathbf{T}_t, \subseteq_a)$. Hence, we can get there $\|N_\eta\| \leqslant \aleph_0$. But he does not prove Ax(C1). Somewhat later, Harrington proves it and the author notes that since trivially $[A \subseteq_a B \Rightarrow A \subseteq_e B]$ it follows that the other axioms which hold for $(\mathbf{T}^t_{\aleph_0}, \subseteq_e)$ hold for $(\mathbf{T}^t_{\aleph_0}, \subseteq_a)$ [see XI, 3.17(3)] [only for Ax(A5) use (C1)], hence X, 2.7, X, 2.15, X, 2.16 apply. Thus, showing that very little need to be added to XI to get the above improvement, as well as replacing XIII, 1.1, 1.4, §2 "$\lambda > |T| + 2^{\aleph_0}$" by "$\lambda > |T|$}, "$(2^{\aleph_0})^+$" by "$\aleph_1$".

Let us turn to the list of open problems.

6. *Problem A7*

The problem was: "Is every unidimensional, stable T, superstable?" Hrushovski answers positively, using the interpretation of groups.

7. *Problem C4*

The problem was: "Prove the main gap from a F^a_λ-saturated strongly λ-homogeneous model". For $\lambda = \aleph_0$, we compute the possible spectrum functions [Sh 88a]. Also, the related problem of a λ-resplendent model, $\lambda > |T|$, was solved.

8. *Problem D2* (Classification over a predicate)

This is essentially solved (but using independence results for non-structure, and much is not yet written, see [PSh 85], [Sh 85d] and [Sh 86b]).

9. *Problem D3*

The classification of a universal class (i.e. $M \in K$ if and only if every finitely generated substructure of M belongs to K) is essentially solved see [Sh 85e] and [Sh 89], but not completely written.

On the categoricity of $\psi \in L_{\lambda^+,\omega}$ see a work by Grossberg and the author (on the categoricity of ω successive cardinals) in preparation; and a work by Makkai and the author (and more in preparation).

10. *Chapter III, Section 5 (on instability and D(T))*

Newelski had shown that we cannot in general improve the results there.

11. *SND*

E. Hrushovski has proved that for countable first order T, SND(T) satisfies CH. We report below on other relevant works of the author.

12. *Better invariants*

We prove that if T is superstable without dop, for characterizing up to isomorphism $\aleph_\varepsilon$-saturated models we need only finitary invariants; more specifically we can use the theory of the model in the logic with quantification over sets which are the algebraic closure of finite sets [in $\mathfrak{C}^{eq}$] and has quantifiers on dimension.

13. *Multidimensionality*

As we say previously the results of V, §5 can be improved. In fact we have now a typescript [Sh 89] proving the following. Assume T is stable (first order complete) and it is multidimensional (see V, §5), then there is a (stationary) type orthogonal to the empty set [hence: if T is stable in $\aleph_\beta$ and $\alpha < \beta$ then T has $\geqslant 2^{|\beta-\alpha|}$ pairwise non isomorphic $\mathbf{F}^a_{\aleph_\alpha}$-saturated models of cardinality $\aleph_\beta$].

14. *Isomorphic ultrapowers*

We proved that it is consistent with ZFC, that there are countable models M, N which are elementary equivalent, but for every ultrafilter D on ω, M^ω/D and N^ω/D are not isomorphic, moreover $\mathrm{Th}(M) = \mathrm{Th}(N)$ does not have the strict order property and even for different ultrafilters we do not get isomorphic ultrapower. This will probably appear in the proceedings of the MSRI 10/89 symposium in set theory. It also seems that, e.g., in some models of set theory, some ultrafilter will not give isomorphic ultrapower in the use of Ax Kochen.

15. *Universal models*

In a work in preparation by Kojman and the author more is done on characterizing the possible classes $\{\lambda : T$ has a universal model of power $\lambda\}$ for T first order, e.g., T is countable with the strict order property or unsuperstable and $\aleph_3 < 2^{\aleph_0}$, then T has no universal model in $\aleph_3$.

16. *Resplendency*

We note that the following are equivalent for a first order T: (a) T is stable, (b) if M is a $|T|^+$-saturated model of T satisfying the following then it is saturated: for every $c_i \in M$ for $i < |T|$, if T_1 is a consistent theory extending $\mathrm{Th}(M, c_i)_{i<|T|}$ of power $\leqslant |T|$ then $(M, c_i)_{i<|T|}$ can be expanded to a model of T_1.

17. *Omitting types*

We forgot to mention a work of Hrushovski and the author, to appear in the *Israel J. of Math.*, characterizing quite fully the possible sets $\{h(T,p) : p$ a type$\}$ where $h(T,p)$ is $\sup\{\|M\| : M$ a model of T omitting $p\}$ for T (countable first order) stable and for T superstable.

18. *Universal classes*

The appearance of [Sh 89] has now been delayed for several years. In addition to the four chapters which appeared (in [Sh 85e]), a reasonable amount of material has been available in the form of a preprint before those four were sent for printing (i.e., Chs. V, VI, and III, §6). Meanwhile Ch. VII has been lengthened and rewritten and more material has been added (mainly III, §7). We shall summarize this below. Note also that works of Baldwin and Shelah on the primal framework are beginning to appear (it is a more general framework of course).

19. *Chapter III*

Sections 1–5 have been revised (mainly the later parts of §3, e.g., a complete proof that if in $\mathrm{EM}(I)$ we can define (in any logic) order on $\{\langle F_i(x_t): i<\mu\rangle: t\in I\}$ then for any $\lambda \geqslant \mu^{++}+|\tau^1|$ we have 2^λ pairwise non isomorphic models of the form $\mathrm{EM}(I)$.

Section 6, a survey and results on black boxes, is mainly in more finished form (but there is the knowledge that, e.g., a club of $\delta<\omega_1$ $\mathrm{pp}(\beth_\delta)=2^{\beth_\delta}$ helps to get better black boxes on $(\beth_{\delta+1})^+$).

We add section 7, done later, where it is proved that, e.g., $\mathrm{IE}(\lambda, T_1, T)=2^\lambda$ when $\lambda>|T_1|$ (even if λ is singular). Some of the cases depend on "guessing clubs in ZFC" which has meanwhile found other applications.

20. *Chapter IV*

This has been revised but nothing novel added; here and later we work mainly on Axiom Framework 1.

21. *Chapter V*

In Section 1 we prove that if $\mathbf{K}$ is $(\leqslant\mu, \leqslant\mu)$-smooth and $(\leqslant\mu^+,\mu)$-based and satisfies $\mathrm{LSP}(\mu)$, then $\mathbf{K}$ satisfies smoothness, is μ^*-based and satisfies $\mathrm{LSP}(\mu^*)$ for every $\mu^*\geqslant\mu$. For $\mu=\chi_{\mathbf{K}}$, if the assumption fails we have strong non structure results, hence we assume the conclusion. Having smoothness, we can allow ourselves to work inside the monster model $\mathfrak{C}$; $\mathrm{tp}(\bar a, A)$ is now defined as $\{F(\bar a): F$ an automorphism of $\mathfrak{C}$ over $A\}$, $(\mathbf{D}_{\mathbf{K}},\mu)$-homogeneous is the replacement of μ-saturated, $\mathrm{NF}(M_0,M_1,M_2,M_3)$ is the replacement of: M_i are algebraically closed sets (not necessarily universes of models of the fix first order T) $M_0\subseteq M_1\subset M_3$, $M_0\subseteq M_2\subseteq M_3$ and $\mathrm{tp}(M_1,M_2)$ does not fork over M_0. Now Sections 2–7 are concerned with "all that we know on stable theories for $|T|^+$-saturated models holds in this context". Well, some definitions have to be adapted – we deal only with types of the form $\mathrm{tp}(N,M)$, where $\mathrm{NF}(N\cap M,N,M,\mathfrak{C})$, usually $\|N\|\leqslant\chi_{\mathbf{K}}$. The proof of "every type p, $\mathrm{dom}(p)$ of power $\leqslant\mu$ has a μ^+-isolated extension" is somewhat more complicated (we have to add variables to the type during the extensions, in order to continue to deal with types of the allowed form only, i.e., of models).

22. *Chapter VI*

Here we generalize the superstable theory to our context. But first we have to be able to deal with types $\mathrm{tp}(a,M)$, i.e., types of elements.

We say tp(a,N) does not fork over $M \leqslant N$ if for some N_1 we have $M \leqslant N_1$ and $a \in N_1$ and NF$(M,N,N_1,\mathfrak{C})$. We define stationarization and parallelism too, and prove the expected results (though, e.g., the uniqueness of stationarization is no longer totally trivial). Next $\kappa(\mathbf{K})$ is no longer a cardinal but a set of regular cardinals $\leqslant \chi_{\mathbf{K}}$, $\kappa \in \kappa(\mathbf{K})$ iff for some $\langle M_i, i \leqslant \kappa \rangle$ increasing continuous sequence of $M_i < \mathfrak{C}$, and some $a \in \mathfrak{C}$ we have: tp(a, M_κ) forks over M_i for every $i < \kappa$. $\mathbf{K}$ is called superstable iff $\kappa(\mathbf{K})$ is empty. Under those definitions the theory generalizes – e.g., non superstability implies strong non structure theorems, the union of an $<_{\mathbf{K}}$-increasing chain of $(\mathbf{D}_{\mathbf{K}}, \mu)$-homogeneous models of length δ, cf(δ) not in $\kappa(\mathbf{K})$ is $(\mathbf{D}_{\mathbf{K}}, \mu)$-homogeneous, also the theory of regular types. [Also generalizing Chapter X here is straightforward, and we can prove the categoricity theorem, i.e., like Los conjecture.]

23. *Chapter VIII*

The aim of this chapter is as follows. We strengthen the notion of elementary submodels to: $M \leqslant^* N$ iff $M \leqslant N$ and for every type over $M_1 \leqslant M$ with $\leqslant \kappa$ variables with $\|M_1\| < \lambda$ which is realized in N, is realized in M too. Our aim is to show that all the good properties still hold (i.e., Axiomatic Framework 1). More exactly, we show it except when the class has some non structure property. Unfortunately one of these gives λ (usually) λ^+ non isomorphic models (rather than 2^λ). Essentially the point is that we do not know to discard classes which behave like the class of linear well ordering. One of our problems is to get non structure from order. The problem is that the relevant order is not preserved when we pass to Ehrenfeucht Mostowski models. The solution is a theorem on the existence on indiscernibles which is novel also in the first order case (and is explained below), and then use universality. Subsequently we prove non structure when unions are not O.K.

Maybe it is the right place to say again what the point is of the indiscernibility theorem in Chapter I, last section, of [Sh 85e]. For a first order T, to get an indiscernible subset $\mathbf{J}$ of $\mathbf{I}$ of power λ, we need a large cardinal (indiscernible set rather than just n-indiscernible set). Of course we get the theorem assuming stability of T, i.e., no order on n-tuples for every n. There we say that if $\mathbf{I}$ consists of n-tuples it suffices to assume there is no long order on $2n$-tuples (and if we phrased it rightly – on n-tuples). What we have to say on order, stability and existence of indiscernibles, for any model – not necessarily of a stable theory, is said here in Chapter I, 2.12, 2.10 [the first says there are indiscernibles in the presence of stability and lack of long order, the second says that the lack of long orders implies stability; the case of order of length ω is spelled out in 2.11, and remember the trivial $|S(A)| \leqslant \prod_\varphi |S_\varphi(A)|$ (or see II, 2.15)].

The indiscernibility theorem we use in ([Sh 89]) Chapter VII says, e.g. for first order stable T, that if $\langle A_t : t \in \mathscr{P}_{\leqslant 2}(\lambda)\rangle$ is given, $|A_t| \in \kappa$, $\kappa(T) \leqslant \kappa$ (where $\mathscr{P}_{\leqslant 2}(\lambda) = \{t : t \subseteq \lambda, |t| \leqslant 2\}$) then we can find $Y \subseteq \lambda$, $|Y| = \mu^+$ and $\langle B_t : t \in \mathscr{P}_{\leqslant 2}(Y)\rangle$ such that: (a) $\langle B_t : t \in \mathscr{P}_{\leqslant 2}(Y)\rangle$ is independent, i.e., for every t, $\mathrm{tp}(B_t, \bigcup\{B_s : s \in \mathscr{P}_{\leqslant 2}(Y), t$ is not a subset of $s\})$ does not fork over $\{B_s : s$ a proper subset of $t\}$, (b) $A_t \subseteq B_t$, $|B_t| \leqslant \kappa$, provided that $\lambda > \beth_1(\mu^\kappa)$.

Similar theorems hold for the structures $\{\rho : \rho$ a finite sequence of ordinals $< \lambda\}$ and for the structures $\{\rho : \rho$ a strictly decreasing (finite) sequence of ordinals $< \lambda\}$.

NOTATION

We use $i, j, \alpha, \beta, \gamma, \xi, \zeta$ for ordinals; k, l, m, n for natural numbers (i.e., finite ordinals); $\kappa, \lambda, \mu, \chi$ for cardinals (i.e., initial ordinals, infinite if not stated otherwise). δ is reserved for limit ordinals. $i < \alpha, \kappa < \lambda$ mean $i \in \alpha, \kappa \in \lambda$, respectively.

A sequence is a function $\bar{s}$ whose domain, Dom $\bar{s}$, is an ordinal, which is also called the length of $\bar{s}$ and denoted by $l(\bar{s})$. $\bar{s}[i]$ denotes the image of i under $\bar{s}$, i.e., the ith member of the sequence. $\bar{s} \upharpoonright \alpha$ (or sometimes $\bar{s} \mid \alpha$) is $\{\langle i, \bar{s}[i]\rangle : i \in \alpha \cap l(\bar{s})\}$. $\bar{s} \lessdot \bar{t}$ means $\bar{s} = \bar{t} \upharpoonright l(\bar{s})$; we write $\bar{s}$ is an initial segment of $\bar{t}$. For any set A, ${}^{\alpha}A = \{\bar{s} : l(\bar{s}) = \alpha, \text{Range } \bar{s} \subseteq A\}$, ${}^{\alpha>}A = \bigcup_{\beta<\alpha} {}^{\beta}A$. We sometimes write $A^{\alpha}, A^{<\alpha}$ instead of ${}^{\alpha}A, {}^{\alpha>}A$, respectively. Sequences of ordinals are denoted by η, ν, ρ, σ (σ is also used to denote permutations). The empty sequence is denoted by $\langle\ \rangle$.

$\aleph_\alpha$ is the αth infinite cardinal. $\beth_\alpha(\lambda) = \beth(\lambda, \alpha)$ is defined inductively by $\beth_0(\lambda) = \lambda$, and for $\alpha > 0$, $\beth(\lambda, \alpha) = \sum_{\beta<\alpha} 2^{\beth(\lambda,\beta)}$; $\beth_\alpha(\aleph_0)$ is written $\beth_\alpha$. λ^+ is the cardinal successor of λ; $\lambda^{<\kappa} = \sum_{0 \le \mu < \kappa} \lambda^\mu$. The cardinality of a set A is denoted by $|A|$.

L denotes a first order language with equality (unless L is defined explicitly, without mention of the equality predicate). Variables are denoted by x, y, z, finite sequences of variables by $\bar{x}, \bar{y}, \bar{z}$, predicate symbols by P, Q, R, and E (for equivalence relations), and function symbols by F. Terms of L are denoted by τ and formulas of L (L-formulas) by φ, ψ, θ. $\varphi(\bar{x})$ denotes the pair $\langle\varphi, \bar{x}\rangle$ where all the free variables of φ belong to Range $\bar{x}$. We identify φ and $\neg\neg\varphi$.

We also use L to denote the set of L-formulas or the set of L-formulas of the form $\varphi(\bar{z})$ so always $|\text{L}| \ge \aleph_0$. We consider L to have $\aleph_0$ variables for cardinality considerations but we often add λ new variables to L for compactness purposes while regarding $|\text{L}|$ as unchanged.

T denotes a complete theory with infinite models in a first order language $\text{L} = \text{L}(T)$; $|T| = |\text{L}(T)|$.

An L-model M is $\langle |M|, \ldots, R^M, \ldots, \ldots, F^M, \ldots, \rangle_{R \in \text{L}, F \in \text{L}}$ where $|M|$

is a non-empty set (the "universe" of the model), and if R is an n-place predicate symbol of L then R^M, the interpretation of R, is an n-place relation over $|M|$, and if F is an m-place function symbol of L then F^M, the interpretation of F, is an m-place function from $|M|$ to $|M|$. Since $|M|$ is the universe of M, $\|M\|$ is its cardinality and it is called the cardinality of M.

L(M) is the language such that M is an L-model. Models are denoted by M, N and occasionally by $\mathfrak{B}$. A, B and C denote sets contained in some $|M|$; a, b and c denote elements of some $|M|$ and $\bar{a}$, $\bar{b}$ and $\bar{c}$ finite sequences of elements; $\bar{a}^\frown\bar{b}$ denotes the concatenation of $\bar{a}$ and $\bar{b}$. We write $\bar{a} \in A$ instead of $\bar{a} \in {}^{\omega>}A$ and sometimes write $\bar{a} \in M$ instead of $\bar{a} \in |M|$. We often write just $\bar{a}$ instead of Range $\bar{a}$, e.g., we write $A \cup \bar{a}$ instead of $A \cup$ Range $\bar{a}$ and if I is a set of finite sequences we write $\bigcup I$ or sometimes just I for $\bigcup \{\text{Range } \bar{a} : \bar{a} \in I\}$.

If $\varphi(\bar{x})$ is an L(M)-formula $\bar{a} \in |M|$ $l(\bar{a}) = l(\bar{x})$ we write $M \models \varphi[\bar{a}]$ to mean $\varphi[\bar{a}]$ is true in M. Th(M) is the set of all sentences true in M. M is elementarily equivalent to N, written $M \equiv N$, if Th(M) = Th(N).

If P is a unary predicate symbol of L(M) we often write $P(M)$ for P^M (note that $P(M) \subseteq |M|$), and if $\varphi(\bar{x}; \bar{y})$ is an L(M)-formula, $\bar{a} \in |M|$, $l(\bar{a}) = l(\bar{y})$, we write $\varphi(M; \bar{a})$ for $\{\bar{b} : \bar{b} \in |M|, l(\bar{b}) = l(\bar{x}), M \models \varphi[\bar{b}; \bar{a}]\}$.

M is a submodel of N, written $M \subseteq N$, if $\mathrm{L} = \mathrm{L}(M) \subseteq \mathrm{L}(N)$, $|M| \subseteq |N|$, for every $R \in \mathrm{L}$ and $\bar{a} \in |M|$, $\bar{a} \in R^M \Leftrightarrow \bar{a} \in R^N$, and for every $F \in \mathrm{L}$ and $\bar{a} \in |M|$, $F^M(\bar{a}) = F^N(\bar{a})$. If $M_i \subseteq M_j$ for all $i < j < \alpha$ then the union of the M_i, $\bigcup_{i<\alpha} M_i$, is the model M such that $|M| = \bigcup_{i<\alpha} |M_i|$, $M_i \subseteq M$ for $i < \alpha$, and $\mathrm{L}(M) = \bigcup_{i<\alpha} \mathrm{L}(M_i)$. M is the L-reduct of N, written $M = N \upharpoonright \mathrm{L}$, if $\mathrm{L}(M) = \mathrm{L} \subseteq \mathrm{L}(N)$, $|M| = |N|$, for $R \in \mathrm{L}$, $R^M = R^N$, and for $F \in \mathrm{L}$, $F^M = F^N$. If M is the L(M)-reduct of N, then N is called an expansion of M.

Δ denotes a set of formulas of the form $\varphi(\bar{x}; \bar{y})$. An m-formula, or φ-m-formula, is a formula of the form $\varphi(\bar{x}; \bar{y})$ or $\varphi(\bar{x}; \bar{a})$ where $l(\bar{x}) = m$ and we regard $\bar{y}$ as a sequence of parameters for which we will usually substitute some $\bar{a}$ and obtain $\varphi(\bar{x}; \bar{a})$. We sometimes consider the same formula as an m-formula for several different values of m. A Δ-m-formula is a φ-m-formula for some $\varphi \in \Delta$.

p is a Δ-m-type over A in M if:

(1) p is a set of formulas of the form $\varphi(\bar{x}; \bar{a})$ where $\bar{a} \in A \subseteq |M|$, $\bar{x} = \langle x_0, \ldots, x_{m-1} \rangle$, and $\varphi(\bar{x}; \bar{y})$ or $\neg\varphi(\bar{x}; \bar{y})$ belongs to Δ. (Note that we *identify* φ and $\neg\neg\varphi$ as mentioned below.)

(2) p is consistent with M, i.e., p is finitely satisfiable in M: for every finite $q \subseteq p$, $M \models (\exists \bar{x}) \bigwedge_{\varphi \in q} \varphi$.

If we are not interested in A, we just write p is a Δ-m-type in M. Types are denoted by p, q and r. We write m-type instead of L-m-type, Δ-type instead of Δ-1-type, φ-m-type instead of $\{\varphi\}$-m-type, and Δ-$(<\aleph_0)$-type instead of Δ-m-type for some $m < \omega$. But from III, §3 on type [Δ-type] means $(<\aleph_0)$-type [Δ-$(<\aleph_0)$-type].

p is a complete Δ-m-type over A in M if it is maximal, that is, if $\bar{a} \in A$, $\varphi(\bar{x}; \bar{y}) \in \Delta$, $l(\bar{a}) = l(\bar{y})$, then $\varphi(\bar{x}; \bar{a}) \in p$ or $\neg\varphi(\bar{x}; \bar{a}) \in p$. $\bar{c}$ realizes p in M if $\varphi(\bar{x}; \bar{a}) \in p \Rightarrow M \models \varphi[\bar{c}; \bar{a}]$. We identify $\neg\neg\varphi$ with φ.

p restricted to A, written $p \restriction A$, is $\{\varphi(\bar{x}; \bar{a}) \in p : \bar{a} \in A\}$; p restricted to Δ, written $p \restriction \Delta$, is $\{\varphi(\bar{x}; \bar{a}) \in p : \varphi(\bar{x}; \bar{y}) \in \Delta$ or $\neg\varphi(\bar{x}, \bar{y}) \in \Delta\}$; p restricted positively to Δ, written $p \restriction^+ \Delta$, is $\{\varphi(\bar{x}; \bar{a}) \in p : \varphi(\bar{x}; \bar{y} \in \Delta\}$. If $\Delta = \{\varphi\}$ we write $p \restriction \varphi$ and $p \restriction^+ \varphi$ instead of $p \restriction \{\varphi\}$ and $p \restriction^+ \{\varphi\}$ respectively. The domain of p, Dom p, is the smallest set over which p is a type. We sometimes consider types with an infinite number of free variables.

Let Δ be a set of m-formulas, i.e., $\Delta = \{\varphi_i(\bar{x}; \bar{y}_i) : l(\bar{x}) = m,\ i < |\Delta|\}$ $\mathrm{cl}_1(\Delta)$ denotes the closure of Δ under negation, i.e.,

$$\begin{aligned}\mathrm{cl}_1(\Delta) &= \{\varphi_i(\bar{x}; \bar{y}_i) : i < |\Delta|\} \cup \{\neg\varphi_i(\bar{x}; \bar{y}_i) : i < |\Delta|\}\\ &= \{\psi_i(\bar{x}; \bar{y}_i) : i < |\mathrm{cl}_1(\Delta)|\}.\end{aligned}$$

$\mathrm{cl}_2^m(\Delta)$ denotes the closure of $\mathrm{cl}_1(\Delta)$ under conjunction, i.e.,

$$\mathrm{cl}_2^m(\Delta) = \left\{\bigwedge_{i \in w} \psi_i(\bar{x}; \bar{y}_i) : w \subseteq |\mathrm{cl}_1(\Delta)|,\ |w| < \aleph_0\right\}.$$

$\mathrm{cl}_3^m(\Delta)$ denotes the Boolean closure of Δ, i.e.,

$$\mathrm{cl}_3^m(\Delta) = \left\{\bigvee_{i \in w} \bigwedge_{j \in v_i} \psi_{ij}(\bar{x}; \bar{y}_{ij}) : \psi_{ij}(\bar{x}; \bar{y}_{ij}) \in \mathrm{cl}_1(\Delta),\ |w|, |v_i| < \aleph_0\right\}.$$

$\mathrm{cl}_4(\Delta)$ denotes the closure of Δ under negation, conjunction and permutation of variables, i.e., $\mathrm{cl}_4(\Delta) = \mathrm{cl}_3^0(\Delta')$ where

$$\Delta' = \{\varphi_i(\sigma(\bar{z}_i)) : \bar{z}_i = \bar{x}^\frown\bar{y}_i,\ \sigma \text{ a permutation of } \bar{z}_i \text{ (strictly speaking of Range } \bar{z}_i),\ i < |\Delta|\}.$$

We use mainly $\mathrm{cl}_3^m(\Delta)$.

We define φ^i as φ if $i = 0$ and $\neg\varphi$ if $i = 1$, $\varphi^{\mathrm{if}(\cdots)}$ is φ if "..." is true and $\neg\varphi$ if "..." is false.

Sometimes when we do not want to write $\varphi(\bar{x}; \bar{a})$ out in full too many times we write "let $\varphi = \varphi(\bar{x}; \bar{a})$" in an abuse of notation. When we say "a formula" we may mean φ, $\varphi(\bar{x})$, $\varphi(\bar{x}; \bar{y})$ or $\varphi(\bar{x}; \bar{a})$.

Let f be a function. For a set A, $f(A) = \{f(a) : a \in A\}$; for a sequence $\bar{s}$,

Range $\bar{s} \subseteq \operatorname{Dom} f$, $f(\bar{s})$ is a sequence $\bar{t}$, $l(\bar{t}) = l(\bar{s})$, $\bar{t}[i] = f(\bar{s}[i])$; for a type p, $f(p) = \{\varphi(\bar{x}; f(\bar{a})) : \varphi(\bar{x}; \bar{a}) \in p\}$. (In practice there will be no inconsistencies.)

I, J denote orders or index sets. $\boldsymbol{I}, \boldsymbol{J}$ denote sets of finite sequences; this has been reversed in Chapters I–V and IX.

Theorems, lemmas, claims, conclusions and corollaries are numbered together and are often referred to just by their number, e.g., II, 4.16 means Theorem 16 of Section 4 of Chapter II. Within Chapter II itself the theorem would be referred to as 4.16. Exercises, problems, questions and conjectures are numbered together; they and definitions are referred to always with name attached, e.g., Definition VII, 3.2 means Definition 2 of Section 3 of Chapter VII. Within the chapter itself, the chapter number is not given. However in Chapters X–XIII definitions are numbered together with theorems.

When we wish to write two almost identical statements we often write only one and insert the variations from the other statement at the appropriate places in round or square brackets, see e.g., Lemma I, 1.12 and Definition I, 2.2(2).

CHAPTER I

PRELIMINARIES

I.0. Introduction

We here introduce our notation, which is quite standard, then, in Section 1, give some of the classical theorems of model theory (compactness, Lowenheim–Skolem, and the existence and uniqueness of saturated models). They are included mainly for the sake of completeness. Some of them are proved later (compactness after VI, 1.3, Lowenheim–Skolem in VII, 1.1–1.3).

In Section 1 we also introduce $\mathfrak{C}$—a saturated model such that we restrict ourselves to elementary submodels of it of cardinality $< \|\mathfrak{C}\|$. By this way we do not lose generality, and hopefully improve presentation.

In Section 2 we deal with some problems concerning stability of models rather than of theories (on which we concentrate in the rest of the book). We define $\mathrm{tp}_\Delta(\bar{a}, A, M)$, $S^m_\Delta(A, M)$, stability, indiscernibility and splitting; and then investigate the connections between stability, order and the existence of indiscernible sequences: By 2.5 if, e.g., $p_i = \mathrm{tp}(\bar{a}_i, A \cup \bigcup \{\bar{a}_j : j < i\})$ does not split over A, $j < i \Rightarrow p_j \subseteq p_i$ for $i < \alpha$, then $\{a_i : i < \alpha\}$ is an indiscernible sequence over A. (This will be the way by which we construct indiscernible sets.)

We prove in 2.8 that if $\mathrm{Th}(M)$ is stable in λ, $I \subseteq M$, $|I| > \lambda$, $|A| \leq \lambda$, $A \subseteq M$, then there is a $J \subseteq I$ indiscernible over A, $|J| > \lambda$. The method here is to find $p \in S^m(C, M)$ and $B \subseteq C$, such that for every C', $C \subseteq C' \subseteq |M|$, p has a unique extension in $S^m(C, M)$ not splitting over B, and when $|C'| \leq \lambda$, this type is realized by some $\bar{c} \in I$.

In 2.9 we prove that unstability implies the existence of ordered I (remember that by II, 2.13, T is unstable iff it has the order property).

I.1. Preliminaries and saturation

We shall use freely the:

COMPACTNESS THEOREM 1.1: *If T_1 is a set of sentences, and every finite subset of T_1 has a model, then T_1 has a model.*

Proof. See Chapter VI after Lemma 1.3.

DEFINITION 1.1: (1) M is an *elementary submodel* of N, $M \prec N$ if $M \subseteq N$ and for every $\bar{a} \in |M|$, $\varphi \in L(M)$, $M \vDash \varphi[\bar{a}] \Leftrightarrow N \vDash \varphi[\bar{a}]$. (If not mentioned otherwise, we assume $L(M) = L(N)$.)

(2) M_i, $i < \alpha$ is an *elementary chain* if $i < j < \alpha \Rightarrow M_i \prec M_j$.

(3) M_i, $i < \alpha$ is a *continuous elementary chain* if it is an elementary chain and $\delta < \alpha \Rightarrow M_\delta = \bigcup_{j<\delta} M_j$.

(4) If M is an elementary submodel of N, N is called an *elementary extension* of M.

LEMMA 1.2 (Tarski–Vaught Test): *M is an elementary submodel of N iff M is a submodel of N, and when $\bar{a} \in |M|$, $b \in |N|$, $\varphi \in L(M)$, $N \vDash \varphi[b, \bar{a}]$, then there is a $b' \in |M|$ for which $N \vDash \varphi[b', \bar{a}]$.*

Proof. The "only if" is clear. So suppose the second condition is satisfied. We let $\bar{a} \in |M|$, $\varphi(\bar{x}) \in L$ and we shall prove by induction on φ that $M \vDash \varphi[\bar{a}] \Leftrightarrow N \vDash \varphi[\bar{a}]$.

If φ is atomic, this holds as M is a submodel of N. If we get φ by connectives, the result follows by using truth tables.

So we are left with the case $\varphi = (\exists y)\psi(y, \bar{x})$. If $M \vDash \varphi[\bar{a}]$, then for some $b \in |M|$, $M \vDash \psi[b, \bar{a}]$, so by the induction hypothesis $N \vDash \psi[b, \bar{a}]$ thus $N \vDash (\exists y)\psi(y, \bar{a})$, i.e., $N \vDash \varphi[\bar{a}]$. If $N \vDash \varphi[\bar{a}]$, then for some $b \in |N|$, $N \vDash \psi[b, \bar{a}]$, so by assumption for some $b' \in |M|$, $N \vDash \psi[b', \bar{a}]$, so by the induction hypothesis $M \vDash \psi[b', \bar{a}]$ thus $M \vDash \varphi[\bar{a}]$. We have shown $M \vDash \varphi[\bar{a}] \Leftrightarrow N \vDash \varphi[\bar{a}]$ and thus have finished the induction and the proof of the theorem.

THE ELEMENTARY CHAIN LEMMA 1.3: (1) *If M_i, $i < \alpha$ is an elementary chain, then $M_i \prec \bigcup_{i<\alpha} M_i$.*

(2) *$\prec$ is a partial order; and if $M \prec N$, then p is an m-type in M iff p is an m-type over $|M|$ in N.*

If $M_l \prec N$, $l = 1, 2$ and $M_1 \subseteq M_2$, then $M_1 \prec M_2$, so if $M_i \prec N$ and $M_i \subseteq M_j$ for $i < j < \alpha$, then $\bigcup_{i<\alpha} M_i \prec N$.

Proof. (1) Let $M = \bigcup_{i<\alpha} M_i$; so by definition, $M_i \subseteq M$ for $i < \alpha$. We shall prove that for every $i < \alpha$, $\bar{a} \in |M_i|$, $\varphi(\bar{x}) \in L(M_i)$, $M \vDash \varphi[\bar{a}] \Leftrightarrow M_i \vDash \varphi[\bar{a}]$; by induction on φ.

The only non-trivial case is $\varphi = (\exists y)\psi(y, \bar{x})$. If $M_i \vDash \varphi[\bar{a}]$, then for some $b \in |M_i|$, $M_i \vDash \psi[b, \bar{a}]$ hence by the induction hypothesis $M \vDash \psi[b, \bar{a}]$ so $M \vDash \varphi[\bar{a}]$. If $M \vDash \varphi[\bar{a}]$, then for some $b \in |M|$, $M \vDash \psi[b, \bar{a}]$; but as $|M| = \bigcup_{i<\alpha} |M_i|$ for some j, $i \leq j < \alpha$, $b \in |M_j|$. By the induction hypothesis $M_j \vDash \psi[b, \bar{a}]$, hence $M_j \vDash \varphi[\bar{a}]$, so, as $M_i \prec M_j$, $M_i \vDash \varphi[\bar{a}]$.

(2) Immediate.

DOWNWARD LOWENHEIM–SKOLEM THEOREM 1.4: *If* $A \subseteq |M|$, *then* M *has an elementary submodel* N $(\mathrm{L}(N) = \mathrm{L}(M))$, $A \subseteq |N|$, $\|N\| = |A| + |\mathrm{L}|$.

Proof. See VII, 1.1, 1.2 and 1.3.

LEMMA 1.5: (1) *If* $a_i \in |M|$ *for* $i < \alpha$, *then* $(M, \ldots, a_i, \ldots)_{i<\alpha} \prec (N, \ldots, a_i, \ldots)_{i<\alpha}$ *iff* $M \prec N$.

(2) *If* $M \subseteq N$, $\mathrm{Th}(M) \subseteq \mathrm{Th}(N)$ *and in* M *every element is an individual constant, then* $M \prec N$.

Proof. Immediate.

LEMMA 1.6: *If* p *is a type in* M, *then in some elementary extension* N *of* M, p *is realized and* $\|N\| \leq \|M\| + |\mathrm{L}(M)|$.

Proof. By 1.5(1) we can assume every element of M is an individual constant (so we can assume p is over $\emptyset$). Let c be a new individual constant and

$$T_1 = \mathrm{Th}(M) \cup \{\varphi(c) : \varphi(x) \in p\}.$$

As p is finitely satisfiable in M, every finite subset of T_1 has a model. Hence by the compactness theorem, T_1 has a model N, and by the downward Lowenheim–Skolem theorem we can assume $\|N\| \leq |T_1| = \|M\| + |\mathrm{L}(M)|$. As N is a model of $\mathrm{Th}(M)$, if for every individual constant $c_1 \in \mathrm{L}(M)$ we identify c_1^M and c_1^N, then M will be a submodel of N. By 1.5(2), $M \prec N$ and by the definition of T_1, c^N realizes p.

DEFINITION 1.2: (1) A model M is λ-*saturated* if every type in M over some $A \subseteq |M|$, $|A| < \lambda$, is realized in M.

(2) A model M is λ-*compact* if every type in M of cardinality $< \lambda$ is realized in M.

(3) A model M is *saturated* (*compact*) if it is $\|M\|$-saturated (-compact).

THEOREM 1.7: (1) *Suppose* $\|M\| + |L(M)| \leq \lambda = \lambda^{<\kappa}$, $|M|$ *infinite, then* M *has a* κ*-compact elementary extension of cardinality* λ.

(2) *Suppose* $\|M\| + |L(M)| \leq \lambda = \lambda^{<\kappa}$, $L_0 \subseteq L(M)$, M *infinite and for every* m *there are* $\leq \lambda$ *complete* L_0*-m-types over* $\emptyset$ *in* M. *Then* M *has an elementary extension of power* λ, *whose* L_0*-reduct is* κ*-saturated.*

Proof. (1) We shall define by induction on $\alpha \leq \lambda$ models M_α such that:

(i) $M_0 = M$ and for limit $\delta \leq \lambda$, $M_\delta = \bigcup_{\alpha<\delta} M_\alpha$;

(ii) $|M_\alpha| \neq |M_{\alpha+1}|$, $M_\alpha \prec M_{\alpha+1}$, $\|M_{\alpha+1}\| \leq \lambda + \|M_\alpha\|$;

(iii) every type over M_α of cardinality $< \kappa$ is realized in $M_{\alpha+1}$.

We can easily prove by induction on $\alpha \leq \lambda$, that $\beta < \alpha \Rightarrow M_\beta \prec M_\alpha$ (by 1.3) and that $\|M_\alpha\| \leq \lambda$; as $|M_\alpha| \neq |M_{\alpha+1}|$ clearly $\|M_\lambda\| = \lambda$. Now if p is a type in M_λ, $|p| < \kappa$, then for some $\alpha < \lambda$, p is over $|M_\alpha|$ (as $|p| < \kappa \leq \operatorname{cf} \lambda$ because $\lambda^{<\kappa} = \lambda$). As $M_\alpha \prec M_\lambda$, p is in M_α, hence realized in $M_{\alpha+1}$ so (as $M_\alpha \prec M_{\alpha+1} \prec M_\lambda$) p is realized in M_λ. So M_λ is κ-compact, $\|M_\lambda\| = \lambda$ and $M \prec M_\lambda$ (as $M = M_0 \prec M_\lambda$). So it suffices to define the M_α's; so let $M_0 = M$ and $M_\delta = \bigcup_{\alpha<\delta} M_\alpha$.

Now suppose M_α is defined, and we shall define $M_{\alpha+1}$. Let $\{p_i : i < \lambda\}$ be a list of all types in M_α of cardinality $< \kappa$ (as $\|M\| + |L(M)| \leq \lambda = \lambda^{<\kappa}$, their number is $\leq \lambda$; by allowing repetitions we get $i < \lambda$). We define an elementary chain M_α^i, $i \leq \lambda$, such that $M_\alpha^0 = M_\alpha$ $\|M_\alpha^i\| \leq \lambda$, $M_\alpha^\delta = \bigcup_{i<\delta} M_\alpha^i$ (for limit $\delta \leq \lambda$) and M_α^{i+1} realizes p_i (possible by 1.6 and 1.3). As $|M|$ is infinite, also $|M_\alpha^\lambda|$ is infinite, hence $q_\alpha = \{x \neq a : a \in |M_\alpha^\lambda|\}$ is finitely satisfiable in M_α^λ; hence by 1.6, M_α^λ has an elementary extension of cardinality $\leq \lambda$ which realizes q_α, and we call it $M_{\alpha+1}$. So clearly $M_\alpha \prec M_\alpha^\lambda \prec M_{\alpha+1}$, $|M_{\alpha+1}| \neq |M_\alpha|$, and every type in M_α of cardinality $< \kappa$ is realized in $M_{\alpha+1}$. We have finished the definition of the M_α's, thus the proof.

(2) A similar proof.

DEFINITION 1.3: Let $L(M) \subseteq L(N)$. An (M, N)*-elementary* mapping is a (one-to-one) function f, $\operatorname{Dom} f \subseteq |M|$, $\operatorname{Range} f \subseteq |N|$, such that for every $a_1, \ldots, a_n \in \operatorname{Dom} f$ and formula $\varphi(x_1, \ldots, x_n) \in L(M)$, $M \models \varphi[a_1, \ldots, a_n] \Leftrightarrow N \models \varphi[f(a_1), \ldots, f(a_n)]$. When no confusion can arise we omit "(M, N)".

Remark. Note that if $M \equiv N$, then the empty function is an (M, N)-elementary mapping. We denote elementary mappings by f, F and g.

DEFINITION 1.4: (1) M is *isomorphic* to N, written $M \cong N$, if $\mathrm{L}(M) = \mathrm{L}(N)$ and there is an (M, N)-elementary mapping f with $\mathrm{Dom}\, f = |M|$, $\mathrm{Range}\, f = |N|$. f is called an *isomorphism* from M onto N.

(2) f is an *automorphism* of M if f is an isomorphism from M onto M.

DEFINITION 1.5: (1) M is *κ-homogeneous* if for every (M, M)-elementary mapping f, $|\mathrm{Dom}\, f| < \kappa$, and every $a \in |M|$, there is an (M, M)-elementary mapping g which extends f, $\mathrm{Dom}\, g = \{a\} \cup \mathrm{Dom}\, f$.

(2) M is *strongly κ-homogeneous* if for every (M, M)-elementary mapping f, $|\mathrm{Dom}\, f| < \kappa$, there is an automorphism of M extending f.

(3) M is *strongly κ-saturated* if it is κ-saturated and strongly κ-homogeneous.

LEMMA 1.8: *Suppose* $\|M\| + |\mathrm{L}(M)| \le \lambda = \lambda^{<\kappa}$; *then M has an elementary extension of cardinality λ which is κ-homogeneous (in fact strongly κ-homogeneous).*

Proof. The proof is, in fact, identical to the proof of 1.7, provided we prove the analog to 1.6:

(*) If f is an (M, M)-elementary mapping, $a \in |M|$, then there is an elementary extension N of M, and an extension g of f, which is an (N, N)-elementary mapping and $\mathrm{Dom}\, g = \{a\} \cup \mathrm{Dom}\, f$.

Proof of (*). Let $\mathrm{Dom}\, f = \{a_i : i < \alpha\}$, and $f(a_i) = b_i$, and

$$p = \{\varphi(x, b_{i(1)}, \ldots, b_{i(n)}) : i(1), \ldots < \alpha,\ M \models \varphi[a, a_{i(1)}, \ldots, a_{i(n)}]\}.$$

We want to prove that p is consistent, i.e., finitely satisfiable. Clearly p is closed under conjunctions, so it suffices to prove, for $\varphi(x, b_{i(1)}, \ldots) \in p$ that $M \models (\exists x)\varphi(x, b_{i(1)}, \ldots)$. By the definition of p, $M \models \varphi[a, a_{i(1)}, \ldots]$, hence $M \models (\exists x)\varphi(x, a_{i(1)}, \ldots)$; so as f is an elementary mapping, $M \models (\exists x)\varphi(x, b_{i(1)}, \ldots)$. So p is consistent, hence by 1.6, p is realized in some elementary extension N of M; say by $b \in |N|$. Let us extend f to g by letting $g(a) = b$. It is easy to check that N, g satisfy our demands.

DEFINITION 1.6: (1) M is *λ-universal* if for every N elementarily equivalent to M and $A \subseteq |N|$, $|A| \le \lambda$ there is an (N, M)-elementary mapping f, $\mathrm{Dom}\, f = A$. M is *universal* if it is $\|M\|$-universal.

(2) M is *$(<\lambda)$-universal* if M is μ-universal for every $\mu < \lambda$.

THEOREM 1.9: (1) *Every λ-saturated model is λ-compact.*

(2) *If $\lambda > |L(M)|$, then M is λ-saturated iff M is λ-compact.*

(3) *Every λ-saturated model is λ-homogeneous and λ-universal.*

(4) *Every $(<\aleph_0)$-universal, λ-homogeneous model is λ-saturated.*

(5) *If $\mu \leq \lambda$, then every λ-saturated model is μ-saturated, every λ-compact model is μ-compact, and every λ-universal model is μ-universal.*

(6) *Let $\lambda > \aleph_0$ be a limit cardinal. If M is μ-saturated (μ-homogeneous) for every $\mu < \lambda$, then M is λ-saturated (λ-homogeneous). (This does not hold for universality but holds for compactness.)*

(7) *If M is λ-saturated (λ-compact) and $L_0 \subseteq L(M)$, then $M \restriction L_0$ is λ-saturated (λ-compact).*

(8) *Every strongly λ-homogeneous model is λ-homogeneous.*

Proof. (1) Immediate.

(2) Immediate.

(3) For this it suffices to prove:

CLAIM 1.10: *Let $A \subseteq B \subseteq |M|$, $|A| < \lambda$, $|B| \leq \lambda$, f an (M, N)-elementary mapping,* $\operatorname{Dom} f = A$. *If N is λ-saturated or $N = M$ is λ-homogeneous, then we can extend f to an (M, N)-elementary mapping g,* $\operatorname{Dom} g = B$.

Proof of 1.10. Let $B = \{b_i : i < |B|\}$, $B_\alpha = A \cup \{b_i : i < \alpha\}$ (so $|B_\alpha| < \lambda$ for $\alpha < |B|$) and we shall define by induction on $\alpha \leq |B|$ (M, N)-elementary mappings f_α; $f_0 = f$, $f_\delta = \bigcup_{i<\delta} f_i$, $f_{\alpha+1}$ extends f_α and $\operatorname{Dom} f_\alpha = B_\alpha$, so clearly $g = f_{|B|}$ is the desired mapping. For $\alpha = 0$, α limit the definition is trivial. If f_α is defined, $N = M$ is λ-homogeneous; the existence of a suitable $f_{\alpha+1}$ follows by the definition of homogeneity. If f_α is defined, N is λ-saturated, the existence of $f_{\alpha+1}$ is proved like $(*)$ in the proof of 1.8.

Proof of 1.9 (continued). (4) Let N be $(<\aleph_0)$-universal and λ-homogeneous; and we shall prove it is λ-saturated. We prove this by induction on λ. So let $A \subseteq |N|$, $|A| < \lambda$, p a type over A in N and we shall prove that p is realized in N. By 1.6 there are M and a such that $N \prec M$ $a \in |M|$ and a realizes p. There is an (M, N)-elementary mapping f, $\operatorname{Dom} f = A \cup \{a\}$ [if A is finite—by the hypothesis; if A is infinite—N is $|A|$-saturated by the induction hypothesis, hence N is $|A|$-universal by 1.9(3)]. Then $g = f \restriction A$ is an (N, N)-elementary mapping, and so $g_1 = g^{-1}$ is an (N, N)-elementary mapping. As N is λ-homogeneous,

$|A| < \lambda$, some (N, N)-elementary mapping g_2 extends g_1 and $\text{Dom}\, g_2 = \text{Dom}\, g_1 \cup \{f(a)\}$. It is easy to check that $g_2(f(a))$ realizes p.

(5) Immediate.

(6) Immediate.

(7) Immediate.

(8) Immediate.

THE UNIQUENESS THEOREM 1.11: *If $M \equiv N$, M, N are saturated and have the same power, then M, N are isomorphic. Moreover, any (M, N)-elementary mapping f with domain of cardinality $< \|M\|$ can be extended to an isomorphism from M onto N.*

Proof. Let $\lambda = \|M\| = \|N\|$, $|M| = \{a_i : i < \lambda\}$, $|N| = \{b_i : i < \lambda\}$. Now we define by induction on $\alpha < \lambda$, (M, N)-elementary mappings f_α, such that: $f_0 = f$ (or the empty function), $f_\delta = \bigcup_{i<\delta} f_i$, $|\text{Dom}\, f_{\alpha+1} - \text{Dom}\, f_\alpha| \leq 1$, $|\text{Dom}\, f_\alpha| \leq |\text{Dom}\, f_0| + |\alpha| < \lambda$; and $a_\alpha \in \text{Dom}\, f_{2\alpha+1}$, $b_\alpha \in \text{Range}\, f_{2\alpha+2}$. This is possible by 1.10; so f_λ is the required isomorphism.

LEMMA 1.12: *Let M be a λ-compact [λ-saturated] model. If p is a $(< \aleph_0)$-type (or even a type with $\leq \lambda$ free variables) in M, $|p| < \lambda$ [$|\text{Dom}\, p| < \lambda$], then p is realized in M.*

Proof. Easy, so left to the reader.

Remark. We use Lemma 1.12 very frequently but without explicit mention.

Conventions. We work within ZFC set theory in the standard way, i.e., all the symbols and formulas of a language L are sets of hereditary power $< |\text{L}|$.

We assume also that there exists an inaccessible cardinal $\bar{\kappa}$ (i.e., a regular cardinal $\bar{\kappa}$ such that $(\forall \lambda < \bar{\kappa})(2^\lambda < \bar{\kappa})$) bigger than all the cardinalities we shall deal with. Note that $\langle R(\bar{\kappa}), \epsilon \rangle$ is a model of ZFC set theory ($R(\bar{\kappa})$ is the set of all sets of hereditary power $< \bar{\kappa}$).

Every theory T will have a saturated model $\mathfrak{C} = \mathfrak{C}(T)$ of cardinality $\bar{\kappa}$, (exists by 1.7) and, unless otherwise stated M, N will be elementary submodels of $\mathfrak{C}$ of cardinality $< \bar{\kappa}$; A, B, C subsets of $|\mathfrak{C}|$ of cardinality $< \bar{\kappa}$; a, b, c, d elements of $|\mathfrak{C}|$; and "type" will mean "type in $\mathfrak{C}$". We will not distinguish between individual constants and elements of $\mathfrak{C}$.

These conventions are convenient as now all formulas and sets of formulas of L are elements of $\langle R(\bar{\kappa}), \epsilon \rangle$, and $M \subseteq N \Leftrightarrow M \prec N$, and in writing $M \models \varphi[\bar{a}]$ we can omit M. We shall deal with a fix complete T (unless stated otherwise).

All our important properties (such as rank, forking etc.) are preserved under automorphisms of $\mathfrak{C}$ and therefore also under elementary mappings as an elementary mapping is a $(\mathfrak{C}, \mathfrak{C})$-elementary mapping and can be extended to an automorphism of $\mathfrak{C}$ by 1.11. We do not state and prove these preservation theorems as they are trivial.

The assumption on $\bar{\kappa}$ is justified as

(1) It does not, in fact, add any extra axiom of set theory as a hypothesis to our theorems.

(2) Any model of T of cardinality $< \bar{\kappa}$ is isomorphic to some $M \prec \mathfrak{C}$.

(3) If $M \prec \mathfrak{B}$, $\|\mathfrak{B}\| < \bar{\kappa}$, then $\mathfrak{B}$ is isomorphic over $|M|$ to some $N \prec \mathfrak{C}$ ("over $|M|$" means an isomorphism which is the identity on $|M|$).

If we assume in addition GC (the axiom of global choice) we can take $\mathfrak{C}$ to be a saturated class—the union of an elementary chain $\{M_\lambda: \lambda$ a cardinal, M_λ is λ-saturated$\}$ of models of T. It is known that ZFC + GC is equiconsistent with ZFC. See [Fe 71] and [Ga 75].

EXERCISE 1.1: Prove that every finite model is λ-saturated (for any λ).

EXERCISE 1.2: Give an example of a $(< \lambda)$-universal model which is not λ-universal for every λ.

CONCLUSION 1.13: *If M is a saturated model, $T = \mathrm{Th}(M) \subseteq T_1$, T_1 consistent, $|T_1| \le \|M\| = \lambda$, then M can be expanded to a model of T_1.*

Proof. We can assume T_1 is complete. If T_1 has a saturated model of cardinality λ then its restriction to $\mathrm{L} = \mathrm{L}(T)$ is also saturated by 1.9(7) and the theorem follows by 1.11. In general let N_1 be a λ^+-saturated model of T_1, N the L-reduct of N_1. Define inductively on $\alpha < \lambda$, N^0_α, N^1_α such that: $N^1_\alpha \prec N_1$, $N^0_\alpha \prec N$, $\|N^1_\alpha\| = \lambda$, $\beta < \alpha$ implies $|N^1_\beta| \subseteq |N^0_\beta| \subseteq |N^1_\alpha|$, and N^0_α is saturated. Now $N^0 = \bigcup_{\alpha<\lambda} N^0_\alpha$ has cardinality λ; and when λ is regular is trivially saturated. For singular λ by VIII, 4.7 T is stable, $\kappa(T) \le \mathrm{cf}\,\lambda$, so III, 3.11 implies N^0 is saturated (no vicious circle arises in this proof).

EXERCISE 1.3: Prove 1.13 directly.

EXERCISE 1.4: Call M P-saturated, P a one-place predicate, if every type p over $A \subseteq |M|$, $|A| < \|M\|$, such that $P(x) \in p$, is realized in M.

(1) Prove that if M is saturated, N is P-saturated, $\mathrm{L}(N) \subseteq \mathrm{L}(M)$, $\|M\| = \|N\|$, $M \restriction \mathrm{L}(N) \equiv N$, then there is an (N, M)-elementary mapping $f: |N| \to |M|$, $f(P^N) = P^M$.

(2) Suppose L(T) has the one place predicate P and the two place predicate ϵ, and for every model M of T and $a_1, \dots, a_n \in P^M$ there is $b \in P^M$ such that $M \models (\forall x)(x \in b \equiv \bigvee_{i=1}^{n} x = a_i)$. Prove that if for $\alpha < \delta < \lambda^+$, $\|M_\alpha\| = \lambda$, $P(M_\alpha) = P(M_0)$, M_α is P-saturated, then $\bigcup_{\alpha<\delta} M_\alpha$ is P-saturated.

(3) Prove that if a theory T has a model M, $\|M\| > |P^M| \geq \aleph_0$ $\lambda = \lambda^{<\lambda} > |T| + \aleph_0$, then T has a model N, $\|N\| = \lambda^+$, $|P(N)| = \lambda$. (Show first that we can assume w.l.o.g. L(T) has a predicate ϵ as in (2), and then define inductively saturated models M_α ($a < \lambda^+$) of T, $\|M_\alpha\| = \lambda$, $P(M_\alpha) = P(M_0)$. For $\alpha = 0, 1$ use the assumption, for limit δ use (2) and (1) and for $\alpha = \beta + 1$ remember $M_\beta \cong M_0$.)

I.2. Order, stability and indiscernibles

DEFINITION 2.1: Let $A \subseteq |M|$.

(1) $\mathrm{tp}_\Delta(\bar{b}, A, M) = \{\varphi(\bar{x}; \bar{a})^{\mathbf{t}} : \bar{a} \in A, \mathbf{t} \in \{0, 1\}, \varphi \in \Delta, M \models \varphi[\bar{b}; \bar{a}]^{\mathbf{t}}\}$.

(2) $S_\Delta^m(A, M) = \{\mathrm{tp}_\Delta(\bar{b}, A, M) : \bar{b} \in {}^m|M|\}$.

(3) If $\Delta = \mathrm{L}$ we omit it, if $m = 1$ we omit it, if $M = \mathfrak{C}$ we omit it, and if $A = \emptyset$, $M = \mathfrak{C}$, we omit them.

Remark. If $\mathrm{tp}(\bar{a}, A) = \mathrm{tp}(\bar{b}, A)$ and $\bar{c} \in A$, then $\mathrm{tp}(\bar{a}^\frown\bar{c}, A) = \mathrm{tp}(\bar{b}^\frown\bar{c}, A)$.

DEFINITION 2.2: (1) The model M is *stable in* (λ, Δ) if for all $A \subseteq |M|$, $|A| \leq \lambda$ and $m < \omega$, $|S_\Delta^m(A, M)| \leq \lambda$.

(2) The theory T is *stable in* (λ, Δ) if every model of T is. We sometimes say "$M[T]$ is (λ, Δ)-stable", instead of "$M[T]$ is stable in (λ, Δ)".

(3) If $\Delta = \mathrm{L}$ we omit it.

(4) T is *stable* if there exists a λ in which T is stable.

(5) T is *superstable* if there exists a λ such that T is stable in all $\mu \geq \lambda$.

LEMMA 2.1: *If λ is regular, $A \subseteq |M|$, $|A| < \lambda$, but $|S^m(A, M)| \geq \lambda$, then there is a finite $B \subseteq |M|$, $|B| < m$, such that $|S(A \cup B)| \geq \lambda$.*

Proof. By induction on m. For $m = 1$, trivial. Now assume the claim for m and look at $S^{m+1}(A, M)$. If $|S^m(A, M)| \geq \lambda$ we are through by the induction hypothesis. So assume $|S^m(A, M)| < \lambda$. For all $q \in S^{m+1}(A, M)$ define

$$q^* = \{(\exists x_m)\psi(x_0, \dots, x_m; \bar{a}) : \psi(x_0, \dots, x_m; \bar{a}) \in q\}.$$

Clearly q^* is a (consistent) m-type over A and has a unique extension $q^+ \in S^m(A, M)$. Now $|\{q^+ : q \in S^{m+1}(A, M)\}| \leq |S^m(A, M)| < \lambda$, so by the regularity of λ, for some $p \in S^m(A, M)$ $|\{q \in S^{m+1}(A, M): q^+ = p\}| \geq \lambda$. In other words p has $\geq \lambda$ extensions in $S^{m+1}(A, M)$. Thus for any $\langle b_0, \ldots, b_{m-1}\rangle$ realizing p in M $|S(A \cup \{b_0, \ldots, b_{m-1}\})| \geq \lambda$.

Remark. If M is $(|A|^+ + \aleph_0)$-homogeneous we can strengthen the above conclusion to $|S(A \cup B, M)| \geq \lambda$.

COROLLARY 2.2: *T is stable in μ iff for every A, $|A| \leq \mu \Rightarrow |S(A)| \leq \mu$.*

Proof. μ^+ is a regular cardinal.

DEFINITION 2.3: (1) $\{\bar{a}^i: i < \alpha\}$ is a *Δ-n-indiscernible sequence over A* if for every $i^0 < \cdots < i^{n-1} < \alpha$, $j^0 < \cdots < j^{n-1} < \alpha$, and permutation σ of $\{0, \ldots, n-1\}$,

$$\mathrm{tp}_\Delta(\bar{a}^{i^{\sigma(0)}} \frown \cdots \frown \bar{a}^{i^{\sigma(n-1)}}, A) = \mathrm{tp}_\Delta(\bar{a}^{j^{\sigma(0)}} \frown \cdots \frown a^{j^{\sigma(n-1)}}, A).$$

If $i < j < \alpha \Rightarrow \bar{a}^i \neq \bar{a}^j$ we call the sequence non-trivial.

(2) If $\Delta = L$ we omit it; if the sequence is Δ-n-indiscernible for all n we say it is *Δ-indiscernible*; if $A = \emptyset$ we omit it; if $\Delta = \{\varphi\}$ we write φ instead of Δ. For simplicity we treat $I = \{\bar{a}^i: i < \alpha\}$ as if it were ordered by $<$. We always assume $l(\bar{a}^i) = l(\bar{a}^0) = m$. I is Δ-$(<n)$-indiscernible if it is Δ-m-indiscernible for all $m < n$. We sometimes write (Δ, n)-indiscernible instead of Δ-n-indiscernible.

DEFINITION 2.4: (1) $\{\bar{a}^i: i < \alpha\}$ is a *Δ-n-indiscernible set over A* if for any two sets of n distinct ordinals $<\alpha$, $\{i^0, \ldots, i^{n-1}\}$, $\{j^0, \ldots, j^{n-1}\}$,

$$\mathrm{tp}_\Delta(\bar{a}^{i^0} \frown \cdots \frown \bar{a}^{i^{n-1}}, A) = \mathrm{tp}_\Delta(\bar{a}^{j^0} \frown \cdots \frown \bar{a}^{j^{n-1}}, A).$$

If $i < j < \alpha \Rightarrow \bar{a}^i \neq \bar{a}^j$ we call the set non-trivial.

(2) We adopt the same conventions as in Definition 2.3(2).

LEMMA 2.3: (1) *If $\Delta \subset \Delta_1$ and $I = \{\bar{a}^i: i < \alpha\}$ is a Δ_1-n-indiscernible set [sequence] over A, then I is a Δ-n-indiscernible set [sequence] over A.*

(2) *If I is a φ-n-indiscernible set [sequence] over A for all $\varphi \in \Delta$, then I is a Δ-n-indiscernible set [sequence] over A.*

(3) *Let $m < \omega$ and for all $k < m$ let Δ_k be a finite set of formulas and $n_k < \omega$. Let $\alpha \geq n = \max\{n_k: k < m\}$ $[\alpha > n]$. Then there is a finite Δ such that for any $I = \{\bar{a}^i: i < \alpha\}$, I is a Δ-n-indiscernible set [sequence]*

iff for all $k < m$, I is a Δ_k-n_k-indiscernible set [sequence]. (Δ also depends on $m = l(\bar{a}^i)$.)

(4) *If $\{\bar{a}^i: i < \alpha\}$, $\alpha \geq \omega$ [α a limit ordinal] is a Δ-n-indiscernible set [sequence] over A and $\beta > \alpha$, then we can define $\bar{a}^i$ for $\alpha \leq i < \beta$ such that $\{\bar{a}^i: i < \beta\}$ is also a Δ-n-indiscernible set [sequence] over A.*

(5) *If $(\bar{x} = \bar{y}) \in \Delta$, then a trivial indiscernible set or sequence has only one element.*

Note. The requirement in (4) that α be a limit ordinal is necessary: take $\alpha = \omega + 1$, $\Delta = \{\varphi(x, y)\}$; and let $\mathfrak{C} \models \varphi[a^i, a^j]$ iff $i < j < \omega + 1$, $\mathfrak{C} \models (\neg\exists y)\varphi(a^\omega, y)$; then $\{a^i: i < \omega + 1\}$ is a Δ-2-indiscernible sequence which cannot be extended.

The proof is left to the reader.

DEFINITION 2.5: Let I be an infinite set of finite sequences, all of the same length, $\varphi(\bar{x}^0, \ldots, \bar{x}^{n-1}; \bar{y})$ a formula, and $\bar{c}$ a sequence. $\varphi(\bar{x}^0, \ldots, \bar{x}^{n-1}; \bar{c})$ is *connected* and *antisymmetric* over I if for any n distinct sequences $\bar{a}^0, \ldots, \bar{a}^{n-1}$ from I there is a permutation σ of n such that $\models\varphi[\bar{a}^{\sigma(0)}, \ldots, \bar{a}^{\sigma(n-1)}; \bar{c}]$ and there is a permutation σ of n such that $\models\neg\varphi[\bar{a}^{\sigma(0)}, \ldots, \bar{a}^{\sigma(n-1)}; \bar{c}]$.

THEOREM 2.4: (1) *If I is an infinite set of sequences of the same length, A, Δ and n are finite, then I has an infinite subset $\{\bar{a}^i: i < \omega\}$ which is a Δ-n-indiscernible sequence over A.*

(2) *If A, Δ, n are finite, then for every $k < \omega$ there is $l = l(|A|, \Delta, n, k) < \omega$ such that every set of sequences (of the same length) of cardinality $\geq l$ has a subset of cardinality $\geq k$ which is a Δ-n-indiscernible sequence over A.*

Proof. (1) follows from Ramsey's theorem and (2) from the finite analog of Ramsey's theorem. (See 2.1 of the Appendix.)

DEFINITION 2.6: (1) The type p (Δ_1, Δ_2)-*splits* over the set A if there are $\bar{b}, \bar{c}$ such that $\mathrm{tp}_{\Delta_1}(\bar{b}, A) = \mathrm{tp}_{\Delta_1}(\bar{c}, A)$ but there is $\varphi \in \Delta_2$ such that $\varphi(\bar{x}; \bar{b}), \neg\varphi(\bar{x}; \bar{c}) \in p$.

(2) p *splits* over A if p (L, L)-splits over A.

Remark. Clearly if p splits over $A \supseteq B$, then p splits over B.

LEMMA 2.5: *For $1 \leq n < \gamma \leq \omega$ let Δ_n be sets of formulas closed under permutations of variables. Let $I = \{\bar{a}^i: i < \alpha\}$ and $A_i = \bigcup\{\bar{a}^j: j < i\}$*

$\cup A$. *Assume that for all* $i < \alpha$, $2 \le n + 1 < \gamma$, $p_i = \mathrm{tp}(\bar{a}^i, A_i)$ *does not* (Δ_n, Δ_{n+1})*-split over* A *and that* $p_j \upharpoonright \Delta_n \subseteq p_i \upharpoonright \Delta_n$ *for all* $j < i < \alpha$, $1 \le n < \gamma$. *Then* I *is a* Δ_n*-*n*-indiscernible sequence over* A *for* $1 \le n < \gamma$.

Proof. Let $i(0) < i(1) < \cdots < i(n-1) < \alpha$, $j(0) < j(1) < \cdots < j(n-1) < \alpha$, $\varphi = \varphi(\bar{x}^0, \ldots, \bar{x}^{n-1}; \bar{y}) \in \Delta_n$, and $\bar{c} \in A$. We must show $\models\varphi[\bar{a}^{i(0)}, \ldots, \bar{a}^{i(n-1)}; \bar{c}]$ iff $\models\varphi[\bar{a}^{j(0)}, \ldots, \bar{a}^{j(n-1)}; \bar{c}]$. The proof is by induction on n. For $n = 1$, $p_0 \upharpoonright \Delta_1 \subseteq p_{j(0)} \upharpoonright \Delta_1$, $p_{i(0)} \upharpoonright \Delta_1$ so the result is obvious. Now assume the result for n. Let $\beta = \max\{i(n), j(n)\}$. By the induction hypothesis $\bar{a}^{i(0)\frown} \ldots {}^\frown\bar{a}^{i(n-1)}$ and $\bar{a}^{j(0)\frown} \ldots {}^\frown\bar{a}^{j(n-1)}$ realize the same Δ_n-type over A. Now p_β does not (Δ_n, Δ_{n+1})-split over A so if $\varphi \in \Delta_{n+1}$, then

$$\varphi(\bar{a}^{i(0)}, \ldots, \bar{a}^{i(n-1)}, \bar{x}, \bar{c}) \in p_\beta \Leftrightarrow \varphi(\bar{a}^{j(0)}, \ldots, \bar{a}^{j(n-1)}, \bar{x}, \bar{c}) \in p_\beta$$

(by the remark after Definition 2.1). Thus since $p_{j(n)} \upharpoonright \Delta_{n+1}$, $p_{i(n)} \upharpoonright \Delta_{n+1} \subseteq p_\beta \upharpoonright \Delta_{n+1}$ we get

$$\begin{aligned}
\models\varphi[\bar{a}^{i(0)}, \ldots, \bar{a}^{i(n)}, \bar{c}] &\Leftrightarrow \varphi(\bar{a}^{i(0)}, \ldots, \bar{a}^{i(n-1)}, \bar{x}, \bar{c}) \in p_{i(n)} \\
&\Leftrightarrow \varphi(\bar{a}^{i(0)}, \ldots, \bar{a}^{i(n-1)}, \bar{x}, \bar{c}) \in p_\beta \\
&\Leftrightarrow \varphi(\bar{a}^{j(0)}, \ldots, \bar{a}^{j(n-1)}, \bar{x}, \bar{c}) \in p_\beta \\
&\Leftrightarrow \varphi(\bar{a}^{j(0)}, \ldots, \bar{a}^{j(n-1)}, \bar{x}, \bar{c}) \in p_{j(n)} \\
&\Leftrightarrow \models\varphi[\bar{a}^{j(0)}, \ldots, \bar{a}^{j(n)}, \bar{c}].
\end{aligned}$$

LEMMA 2.6: *Let* M *be stable in* λ *and assume that in* M*:*

(*) *There is no* $n < \omega$ *and increasing sequence* $\{A_i\}_{i \le \lambda}$, *and* $p \in S^n(A_\lambda, M)$ *such that* $p \upharpoonright A_{i+1}$ *splits over* A_i *for all* $i < \lambda$.

If $A \subseteq |M|$ *and* $I \subseteq {}^m|M|$, $|I| > \lambda \ge |A|$ *then there is* $I' \subseteq I$, $|I'| > \lambda$ *such that* I' *is an indiscernible sequence over* A.

Proof. First we shall show:

(**) There exist B, C, $A \subseteq B \subseteq C \subseteq |M|$, $|C| \le \lambda$, and $p \in S^m(C, M)$ such that

(1) for all C', $C \subseteq C' \subseteq |M|$, $|C'| \le \lambda$ p has an extension $p' \in S^m(C')$ realized in $I - {}^mC'$ which does not split over B (in particular p does not split over B), and

(2) for all $\bar{c}' \in |M|$ there is $\bar{c} \in C$ such that $\mathrm{tp}(\bar{c}, B) = \mathrm{tp}(\bar{c}', B)$.

By way of contradiction assume $\neg(**)$. We define by induction an increasing sequence $\{B_i\}_{i\le\lambda}$ such that $B_i \subseteq |M|$, $|B_i| \le \lambda$ as follows: $B_0 = A$, $B_\delta = \bigcup_{i<\delta} B_i$ for limit δ. Assume B_i is defined. Let $C_i \supseteq B_i$, $C_i \subseteq |M|$, $|C_i| \le \lambda$ be such that for all $\bar{c}' \in |M|$ there is $\bar{c} \in C_i$ with $\text{tp}(\bar{c}, B_i) = \text{tp}(\bar{c}', B_i)$ (this by the λ-stability of M). Since $\neg(**)$ for every $p \in S^m(C_i, M)$ which does not split over B_i there is C'_p contradicting $(**)(1)$ for $C = C_i$, $B = B_i$. That is, $C_i \subseteq C'_p \subseteq |M|$, $|C'_p| \le \lambda$, and p has no extension in $S^m(C'_p)$ which is realized in $I - {}^m(C'_p)$ and does not split over B_i. Define $B_{i+1} = \bigcup \{C'_p : p \in S^m(C_i, M), p \text{ does not split over } B_i\} \cup C_i$.

Now let $\bar{c} \in I - {}^m B_\lambda$, and $p = \text{tp}(\bar{c}, B_\lambda)$. We shall show that $p \restriction B_{i+1}$ splits over B_i, in contradiction to $(*)$. If not, then certainly $q = p \restriction C_i$ does not split over B_i; $q \in S^m(C_i, M)$; thus C'_q is defined and $C'_q \subseteq B_{i+1}$. So $p \restriction C'_q$ does not split over B_i. But $p \restriction C'_q$ is realized by $\bar{c} \in I - {}^m(C'_q)$, contradicting the definition of C'_q. So $p \restriction B_{i+1}$ does split over B_i, a contradiction. This proves $(**)$.

Now define by induction $\bar{c}^i \in I$ for $i < \lambda^+$: If $\bar{c}^j$ is defined for $j < i$ let $C^i = \bigcup \{\bar{c}^j : j < i\} \cup C$ and let $p_i \in S^m(C^i)$ be an extension of p which is realized by $\bar{c}^i \in I - {}^m(C^i)$ and does not split over B ($|C^i| \le \lambda$). By Lemma 2.5 we just have to show that $p_j \subseteq p_i$ for $j < i$. Let $\varphi(\bar{x}; \bar{b}) \in p_j$. Let $\bar{b}' \in C$ be such that $\text{tp}(\bar{b}', B) = \text{tp}(\bar{b}, B)$. Since p_j does not split over B, $\varphi(\bar{x}; \bar{b}) \in p_j \Rightarrow \varphi(\bar{x}; \bar{b}') \in p_j$. Also p_j extends p so $\varphi(\bar{x}; \bar{b}') \in p$. But so does p_i, hence $\varphi(\bar{x}; \bar{b}') \in p_i$. p_i does not split over B so $\varphi(\bar{x}; \bar{b}) \in p_i$ and the proof is complete.

LEMMA 2.7: *If T is stable in λ, then every model of T satisfies $(*)$ from Lemma 2.6.*

Proof. By way of contradiction assume that $\{A_i\}_{i\le\lambda}$ is an increasing sequence, $p \in S^m(A_\lambda)$, and $p \restriction A_{i+1}$ splits over A_i. Choose $\bar{b}^i, \bar{c}^i \in A_{i+1}$ such that $\text{tp}(\bar{b}^i, A_i) = \text{tp}(\bar{c}^i, A_i)$ and $\varphi_i(\bar{x}; \bar{b}^i)$, $\neg\varphi_i(\bar{x}; \bar{c}^i) \in p$ for some formula φ_i. Let $\bar{a}$ realize p. Let $\mu = \min\{\mu : 2^\mu > \lambda\}$. Define by induction on $l(\eta)$ elementary mappings F_η, $\eta \in {}^{\mu\ge}2$, Dom $F_\eta = A_{l(\eta)}$: If $l(\eta) = 0$ F_η is the identity on A_0. If $l(\eta) = \delta$ then $F_\eta = \bigcup_{\alpha<\delta} F_{\eta|\alpha}$. If $l(\eta) = i + 1$ let $F_{\eta\frown\langle 0\rangle}$ be an arbitrary extension of F_η to A_{i+1}. $F_{\eta\frown\langle 1\rangle}$ will be defined as follows: Let G be such that $G \restriction A_i = F_\eta$ and $G(\bar{b}^i) = F_{\eta\frown\langle 0\rangle}(\bar{c}^i)$. G is an elementary mapping and $F_{\eta\frown\langle 1\rangle}$ will be an arbitrary extension of G to A_{i+1}. Let $B = \bigcup \{F_\eta(\bar{b}^i \frown \bar{c}^i) : \eta \in {}^{i+1}2, i < \mu\}$. Clearly $|B| \le \aleph_0 \cdot \sum_{\alpha<\mu} |{}^\alpha 2| \le \lambda$. For every $\eta \in {}^\mu 2$ extend F_η to a mapping F'_η with domain $A_\mu \cup \{\bar{a}\}$. Let $p_\eta = \text{tp}(F'_\eta(\bar{a}), B)$. Now assume $\eta \ne \nu \in {}^\mu 2$. Let $\alpha = \min\{\beta : \eta[\beta] \ne \nu[\beta]\}$. W.l.o.g. $\eta[\alpha] = 0$. Thus $\varphi_\alpha(\bar{x}; \bar{b}^\alpha) \in p$ and $\vDash \varphi_\alpha[\bar{a}; \bar{b}^\alpha]$.

Thus $\vDash\varphi_\alpha[F'_\nu(\bar{a}); F'_\nu(\bar{b}^\alpha)]$, and $\varphi_\alpha(\bar{x}; F'_\nu(\bar{b}^\alpha)) \in p_\nu$. Similarly $\neg\varphi_\alpha(\bar{x}; F'_\eta(\bar{c}^\alpha)) \in p_\eta$. But $F'_\nu(\bar{b}^\alpha) = F'_\eta(\bar{c}^\alpha)$; thus $\neg\varphi_\alpha(\bar{x}; F'_\nu(\bar{c}^\alpha)) \in p_\eta$. In other words $p_\eta \neq p_\nu$. Hence $|S^m(B)| \geq |\{p_\eta: \eta \in {}^\mu 2\}| = 2^\mu > \lambda \geq |B|$.

So T is not stable in λ, a contradiction.

THEOREM 2.8: *If T is stable in λ, $|I| > \lambda \geq |A|$, then there is $J \subseteq I$, $|J| > \lambda$, which is an indiscernible sequence over A.*

Proof. Immediate by the two preceding lemmas.

Remark 1. J is in fact an indiscernible set over A, see 2.9 and II, 2.13(1) and (9).

Remark 2. If κ is regular, and for all $|A| < \kappa$ and $m < \omega$, $|S^m(A)| < \kappa$, and there is $\lambda < \kappa$ for which 2.6(*) holds, then for all A and I with $|A| < \kappa \leq |I|$, there is $J \subseteq I$, $|J| \geq \kappa$, J an indiscernible sequence over A. The proof is similar.

LEMMA 2.9: *If I is a Δ-indiscernible sequence over A but not a Δ-indiscernible set over A then there are $n < \omega$, $\varphi(\bar{x}^0, \ldots, \bar{x}^{n-1}; \bar{y}) \in \Delta$, and $\bar{a} \in A$ such that $\varphi(\bar{x}^0, \ldots, \bar{x}^{n-1}; \bar{a})$ is connected and antisymmetric over I.*

Proof. Immediate.

THEOREM 2.10: *Suppose $\varphi = \varphi(\bar{x}; \bar{y})$, $l(\bar{x}) = m$, $l(\bar{y}) = n$, $|S^m_\varphi(A, M)| > \sum_{0 \leq \mu < \lambda} (|A|^\mu + 2^{2^\mu})$, $\lambda \geq \aleph_0$.*

(1) *Let $\theta(\bar{x}_1, \bar{x}_2, \bar{x}_3; \bar{y}_1, \bar{y}_2, \bar{y}_3) = [\varphi(\bar{x}_1; \bar{y}_2) \equiv \varphi(\bar{x}_1; \bar{y}_3)]$, $l(\bar{x}_i) = l(\bar{y}_i)$, $i = 1, 2, 3$. Then there are in $|M|$ sequences $\bar{a}^i$ $i < \lambda$ such that $M \vDash \theta[\bar{a}^i; \bar{a}^j] \Leftrightarrow i < j$.*

(2) *Suppose $\lambda \to (\chi)^2_2$ [e.g., $\lambda = \chi = \aleph_0$ or $\chi = \kappa_0^+$, $\lambda = (2^{\kappa_0})^+$, see Appendix, Definitions 2.1 and 2.5] then there are in $|M|$ sequences $\bar{a}^i$, $\bar{c}^i$, $i < \chi$ and $\mathfrak{t} \in \{0, 1\}$ such that $M \vDash \varphi[\bar{c}^i; \bar{a}^j]^{\mathfrak{t}} \Leftrightarrow i < j$.*

Remark. For a finite A, see a stronger result: II, 4.10(4), Appendix 1.7(2).

Proof. Let $\kappa = \sum_{0 \leq \mu < \lambda} (|A|^\mu + 2^{2^\mu})$ and $\bar{c}^i$, $i < \kappa^+$ be sequences from $|M|$ such that the types $\mathrm{tp}_\varphi(\bar{c}^i, A, M)$ are distinct and let $\psi = \psi(\bar{y}; \bar{x}) = \varphi(\bar{x}; \bar{y})$. Now define by induction on $\alpha \leq \lambda$ sets $A_\alpha \subseteq |M|$ such that

$A_0 = A$; for limit δ, $A_\delta = \bigcup_{\alpha<\delta} A_\alpha$; and for every $p \in S^m_\varphi(A_\alpha, M) \cup S^n_\psi(A_\alpha, M)$ and $B \subseteq A_\alpha$, $|B| < |\alpha|^+ + \aleph_0$, $p \upharpoonright B$ is realized by a sequence from $A_{\alpha+1}$, $A_{\alpha+1} \supseteq A_\alpha$, and $|A_\alpha| \leq |A|^{|\alpha+1|}$. So $\alpha < \lambda \Rightarrow |A_\alpha| \leqslant \sum_{\beta<\alpha} mn(|A|+2)^{mn|\beta|}$ hence $|A_\lambda| \leqslant \kappa$.

Now we shall prove:

(*) There is $i < \kappa^+$ such that for every $\alpha < \lambda$ and every $B \subseteq A_\alpha$, $|B| < |\alpha|^+ + \aleph_0$, $\mathrm{tp}_\varphi(\bar{c}^i, A_{\alpha+1}, M)$ (ψ, φ)-splits over B.

Suppose not, then for every $i < \kappa^+$ there are $\alpha_i < \lambda$, $B_i \subseteq A_{\alpha_i}$ which contradict (*). Clearly we can replace $\{\bar{c}^i : i < \kappa^+\}$ by any subset of the same cardinality. As $\lambda \leq \kappa$ we can assume $\alpha_i = \alpha$ for every i. Then as $\alpha < \lambda$ the number of subsets of A_α of cardinality $<|\alpha|^+ + \aleph_0$ is $\leq \sum_{0\leq\mu<\lambda} |A_\alpha|^\mu \leq \sum_{0\leq\mu<\lambda} |A|^{|\alpha+1|\mu} \leq \kappa$ so we can assume $B_i = B$ for every i. Let $|B|^m = \mu < |\alpha|^+ + \aleph_0 \leq \lambda$. By the requirements on $A_{\alpha+1}$ there is $B \subseteq C \subseteq A_{\alpha+1}$. $|C| \leq (n+1)2^\mu$ in which every $p \in S^n_\psi(B, M)$ is realized. Clearly $|S^m_\varphi(C, M)| \leq 2^{|C|^n} \leq 2^{((n+1)\cdot 2^\mu)^n} \leq \kappa$ so we can assume that for every $i < \kappa^+$, $\mathrm{tp}_\varphi(\bar{c}^i, C, M) = p$. Now as $\mathrm{tp}_\varphi(\bar{c}^0, A, M) \neq \mathrm{tp}_\varphi(\bar{c}^1, A, M)$ there is $\bar{a} \in A$ such that $M \models \varphi[\bar{c}^0; \bar{a}] \equiv \neg\varphi[\bar{c}^1; \bar{a}]$. Choose $\bar{a}' \in C$ such that $\mathrm{tp}_\psi(\bar{a}, B, M) = \mathrm{tp}_\psi(\bar{a}', B, M)$ (possible by the definition of C). As $\mathrm{tp}_\varphi(\bar{c}^l, A_{\alpha+1}, M)$ does not (ψ, φ)-split over B, for $l = 0, 1$, $M \models \varphi[\bar{c}^l; \bar{a}] \equiv \varphi[\bar{c}^l; \bar{a}']$. So $M \models \varphi[\bar{c}^0; \bar{a}'] \equiv \neg\varphi[\bar{c}^1; \bar{a}']$ but this contradicts $\mathrm{tp}_\varphi(\bar{c}^i, C, M) = p$. So we have proved (*) and now we prove:

(1) Define by induction on $\alpha < \lambda$, sequences $\bar{a}_\alpha, \bar{b}_\alpha, \bar{c}_\alpha \in A_{2\alpha+2}$ as follows. Suppose we have defined them for every $\beta < \alpha$. Let $B_\alpha = \bigcup_{\beta<\alpha} \bar{a}_\beta{}^\frown\bar{b}_\beta{}^\frown\bar{c}_\beta$, so by (*) $\mathrm{tp}_\varphi(\bar{c}^i, A_{2\alpha+1}, M)$ (ψ, φ)-splits over B_α, so there are $\bar{a}_\alpha, \bar{b}_\alpha \in A_{2\alpha+1}$, $\mathrm{tp}_\psi(\bar{a}_\alpha, B_\alpha, M) = \mathrm{tp}_\psi(\bar{b}_\alpha; B_\alpha, M)$ but $M \models \varphi[\bar{c}^i; \bar{a}_\alpha] \wedge \neg\varphi[\bar{c}^i; \bar{b}_\alpha]$. Now choose $\bar{c}_\alpha \in A_{2\alpha+2}$ which realizes $\mathrm{tp}_\varphi(\bar{c}^i, B_\alpha \cup \bar{a}_\alpha \cup \bar{b}_\alpha, M)$.

So clearly:

(i) If $\beta \leq \alpha$, then $M \models \varphi[\bar{c}^i; \bar{a}_\beta] \wedge \neg\varphi[\bar{c}^i; \bar{b}_\beta]$ and by the definition of $\bar{c}_\alpha$,

$$M \models \varphi[\bar{c}^i; \bar{a}_\beta] \equiv \varphi[\bar{c}_\alpha; \bar{a}_\beta], \qquad M \models \varphi[\bar{c}^i; \bar{b}_\beta] \equiv \varphi[\bar{c}_\alpha; \bar{b}_\beta];$$

hence $M \models \varphi[\bar{c}_\alpha; \bar{a}_\beta] \wedge \neg\varphi[\bar{c}_\alpha; \bar{b}_\beta]$.

(ii) If $\alpha < \beta$, then $\mathrm{tp}_\psi(\bar{a}_\beta, B_\beta, M) = \mathrm{tp}_\psi(\bar{b}_\beta, B_\beta, M)$, but $\bar{c}_\alpha \in B_\beta$ hence

$$M \models \psi[\bar{a}_\beta; \bar{c}_\alpha] \equiv \psi[\bar{b}_\beta; \bar{c}_\alpha] \quad \text{or} \quad M \models \varphi[\bar{c}_\alpha; \bar{a}_\beta] \equiv \varphi[\bar{c}_\alpha; \bar{b}_\beta].$$

So if we let $\bar{a}^\alpha = \bar{c}_\alpha{}^\frown\bar{b}_\alpha{}^\frown\bar{a}_\alpha$, we prove (1).

(2) Let $\bar{a}_\alpha, \bar{b}_\alpha, \bar{c}_\alpha$ be as in the proof of (1). For the definition of $\lambda \to (\chi)^2_2$; and why $\aleph_0 \to (\aleph_0)^2_2$; see Appendix, Definitions 2.1 and 2.1. Here we define the *colouring* f of λ: if $\alpha < \beta < \lambda$, $f(\{\alpha, \beta\}) = 0$ when $M \models \varphi[\bar{c}_\alpha; \bar{a}_\beta]$, and $f(\{\alpha, \beta\}) = 1$ otherwise. As $\lambda \to (\chi)^2_2$, f is constant on (all pairs from) a subset of λ of cardinality χ, so, by renaming, on χ. If $M \models \neg\varphi[\bar{c}_0; \bar{a}_1]$, then for $\alpha, \beta < \chi$, $M \models \varphi[\bar{c}_\alpha; \bar{a}_\beta] \Leftrightarrow \alpha \geq \beta$, so let $\bar{a}^\alpha = \bar{a}_\alpha$, $\bar{c}^\alpha = \bar{c}_\alpha$ for $\alpha < \chi$ and $\mathfrak{t} = 1$. If $M \models \varphi[\bar{c}_0; \bar{a}_1]$, then for $\alpha, \beta < \chi$, $M \models \varphi[\bar{c}_\alpha; \bar{b}_\beta] \Leftrightarrow \alpha < \beta$ so let $\bar{a}^\alpha = \bar{b}_\alpha$, $\bar{c}^\alpha = \bar{c}_\alpha$ for $\alpha < \chi$ and $\mathfrak{t} = 0$.

CONCLUSION 2.11: *If $|S^m_\varphi(A, M)| > |A| + \aleph_0$, then there are sequences $\bar{a}_n, \bar{c}_n$ in $|M|$ ($n < \omega$) and $\mathfrak{t} \in \{0, 1\}$ such that $M \models \varphi[\bar{c}_l; \bar{a}_n]^{\mathfrak{t}} \Leftrightarrow l < n$.*

THEOREM 2.12: *Suppose M is stable in λ, and in it there is no ordered (by some formula of $L(M)$) set of sequences of length μ, and $\mu = \lambda$, $|L(M)| < \mathrm{cf}\,\lambda$ or $\mu = \lambda^+$. If $I \subseteq |M|$, $|I| > \mu$, then there is an indiscernible set $J \subseteq I$ of length λ^+.*

Proof. Use the ideas of the proof of 2.6 and 2.10, and use II, 2.16.

EXERCISE 2.1: Suppose $\lambda_0 < \lambda_1 < \cdots < \lambda_n$, $I \subseteq |M|$, $|I| > \lambda_n^+$ and M is stable in λ_i for $i \leq n$, then there is an $(n + 2)$-indiscernible sequence $J \subseteq I$, $|J| = \lambda_0^+$.

EXERCISE 2.2: (1) Find a model of power $\aleph_1$, which is stable in $\aleph_0$, but there is no indiscernible sequence of length $\aleph_1$ in it.

(2) Moreover, for some symmetric $\varphi = \varphi(x, y)$, there is no φ-2-indiscernible sequence of power $\aleph_1$ in it. (Hint: Use a dense Specker order for (1), and add a well-ordering to it for (2) (see e.g. [GS 73]).)

(3) In (1) and (2), show that we can choose a model with a countable, $\aleph_0$-categorical theory.

EXERCISE 2.3: Suppose $A \subseteq B \subseteq C$ and for every $\bar{c} \in C$ there is a $\bar{b} \in B$ such that $\mathrm{tp}(\bar{c}, A) = \mathrm{tp}(\bar{b}, A)$. If $p \in S^m(B)$ does not split over A, then p has a unique extension in $S^m(C)$ which does not split over A.

EXERCISE 2.4: Generalize 2.12 to (λ, Δ)-stability.

EXERCISE 2.5: Suppose λ is regular, M a model, and for every $A \subseteq |M|$, $|A| < \lambda$ implies $|S^m(A, M)| < \lambda$, and in M there is no ordered set of sequences of length μ for some $\mu < \lambda$. Then for every $I \subseteq |M|$, $|I| \geq \lambda$ there is an indiscernible set $J \subseteq I$, $|J| = \lambda$.

EXERCISE 2.6: In 2.12 (cf. Exercise 2.5) add an $A \subseteq |M|$, $|A| <$ cf $\mu(|A| < \lambda)$ and demand that J is indiscernible over A.

EXERCISE 2.7: (1) Suppose $S \subseteq S^m_\varphi(A)$, $\varphi = \varphi(\bar{x}; \bar{y})$, $|S| > |A|$. Prove that there are $\bar{a}_n \in A$, $p_k \in S$ $(n, k < \omega)$, and $\mathbf{t} \in \{0, 1\}$, such that $\varphi(\bar{x}; \bar{a}_n)^{\mathbf{t}} \in p_k$ iff $n < k$ (iff $\neg\varphi(\bar{x}; \bar{a}_n)^{\mathbf{t}} \notin p_k$, of course). (Hint: Generalize 2.10(1) and 2.11.)

(2) Generalize 2.10(2) similarly.

EXERCISE 2.8: Let $\varphi = \varphi(\bar{x}; \bar{y})$,

$$\varphi_n = \bigwedge_{i<n} \varphi(x_{mi}, x_{mi+1}, \ldots, x_{mi+m-1}; \bar{y}_i) = \varphi_n(\bar{x}^n; \bar{y}^n),$$

λ be regular, $\bar{c} \in A \subseteq |M|$, $|A| < \lambda$. Show that if $\bar{c}^n = \bar{c}^\frown\bar{c}^\frown\cdots{}^\frown\bar{c}$, $|\{p \in S^{mn}(A, M)\colon \varphi_n(\bar{x}^n; \bar{c}^n) \in p\}| \geq \lambda$, then for some $B \subseteq |M|$, $|B| \leq m(n-1)$, $|\{p \in S^m(A \cup B)\colon \varphi(\bar{x}; \bar{c}) \in p\}| \geq \lambda$.

EXERCISE 2.9: (1) Suppose T is stable, λ regular, $|A| < \lambda \Rightarrow |S(A)| < \lambda$, and $\mu < \lambda \Rightarrow \mu^\kappa < \lambda$. For any $|A_0| < \lambda$ and elements a^i_α $(i < \kappa, \alpha < \lambda)$ there is a set $S \subseteq \lambda$, $|S| = \lambda$, such that for any $i(0) < \cdots < i(n) < \kappa$, $\{\langle a^{i(0)}_\alpha, \ldots, a^{i(n)}_\alpha\rangle\colon \alpha \in S\}$ is indiscernible over A_0. (Hint: You can assume T is stable in some $\mu < \lambda$ by II, 2.13 or see III, 4.23.)

(2) Generalize (1) when we replace the condition on T by a condition on M.

EXERCISE 2.10: Let M be an infinite κ^+-saturated model, and define the model M^κ as follows:

$$|M^\kappa| = |M| \cup {}^\kappa|M|,$$

$$F^{M^\kappa}_\alpha(a) = \begin{cases} a & \text{if } a \in M, \\ a[\alpha] & \text{if } a \in {}^\kappa|M|, \end{cases}$$

$$P^{M^\kappa} = |M| \quad \text{and} \quad M \subseteq M^\kappa.$$

Show that if M is λ-stable, $\lambda = \lambda^\kappa$, then M^κ is λ-stable; and if Th(M) is λ-stable, $\lambda = \lambda^\kappa$, then Th(M^κ) is λ-stable; but if $\lambda < \lambda^\kappa$ or M is not λ-stable, then M^κ is not λ-stable; and if $\lambda < \lambda^\kappa$, Th(M^κ) is not λ-stable.

Also show that Th(M^κ) depends only on Th(M) and κ.

Chapter II

RANKS AND INCOMPLETE TYPES

II.0. Introduction

Section 2 (and Section 1 on which it relies) are the cornerstones of the book, but they are not difficult.

This chapter has essentially two interlocked aims; the first is to investigate the family of formulas $\varphi(\bar{x}; \bar{a})(\bar{a} \in \mathfrak{C})(\varphi = \varphi(\bar{x}; \bar{y})$ fixed); or other words the family of the sets $\varphi(\mathfrak{C}, \bar{a})(\bar{a} \in \mathfrak{C})$. We try to classify them by complexity, in Sections 2 and 4. Clearly this has various implications for the theory T. The second aim is to investigate various notions of ranks (e.g., possible values, equality between distinct ranks), we deal with it in Section 1, somewhat in Section 2, and mainly in Section 3.

The properties we shall deal with are the order property ($\equiv$instability), the f.c.p. (the finite cover property) the independence property and the strict order property. We say T has such a property if some $\varphi(x; \bar{y})$ has the property (or, equivalently, by our theorems, some $\varphi(\bar{x}; \bar{y})$ has the property). Essentially we prove that the order property implies the f.c.p. and the order property is equivalent to the disjunction of the independence property and the strict order property. Other properties of this sort are suggested in Chapter III, Section 7, ($\kappa_{\mathrm{cdt}}(T) = \infty$, $\kappa_{\mathrm{inp}}(T) = \infty$, essentially). The most important of these properties is stability ($\equiv$not the order property) with which we shall deal in Section 2. The main theorem (2.2) lists many properties equivalent to unstability of a formula, some of them are helpful in proving assertions about stable formulas, and some are helpful in proving assertions about unstable formulas. For unstable formulas the important properties are: the order property (i.e. for some $\bar{a}_n, \bar{b}_n (n < \omega) \vDash \varphi[\bar{a}_l, \bar{b}_k]$ iff $l > k$) (this shows the similarity to the "classical" case—the theory of dense linear order); and for every λ there is an A, $|A| \leq \lambda < |S^m_\varphi(A)|$. Among the properties important for stable φ are: $|S^m_\varphi(A)| \leq |A| + \aleph_0$ (=stability in every λ); $R^m(p, \varphi, \infty) \leq R^m(\bar{x} = \bar{x}, \varphi, 2) < \omega$; and every φ-m-type over A is definable by a formula over A, of a fixed form.

For theories we get a similar theorem (2.13): T is stable in every $\lambda = \lambda^{|T|}$ iff T is stable in some λ iff every $\varphi(x; \bar{y})$ is stable iff every $\varphi(\bar{x}; \bar{y})$ is stable iff T does not have the order property (i.e. no formula orders an infinite set of sequences), iff every infinite indiscernible sequence is an indiscernible set.

We also prove that if $|I| \geq \lambda > |A|$, λ regular and Δ is finite, then some $J \subseteq I$ is Δ-indiscernible over A, $|J| = \lambda$.

In Section 4 we shall deal with the other properties. We say $\varphi(\bar{x}; \bar{y})$ does not have the f.c.p. if for some $k_\varphi < \aleph_0$ we can strengthen the compactness properties of $\mathfrak{C}$ to: if p is a set of φ-m-formulas, and every subset of p of cardinality $\leq k_\varphi$ is realized, then p is realized. The "classical" example of a theory with the f.c.p. is the theory of one equivalence relation such that for every $n < \omega$ there are (infinitely many) E-equivalence classes of n elements. It appears that for stable T, the possession of the f.c.p. is equivalent to several natural properties (see the f.c.p. Theorem 4.4). In particular if T has the f.c.p., then there are a formula $E(x, y; \bar{z})$ and sequences $\bar{c}_n$ such that $E(x, y, \bar{c}_n)$ is an equivalence relation with k_n equivalence classes and $n \leq k_n < \aleph_0$ (this shows the affinity of any stable T with the f.c.p. to the classical example, and is helpful in proving assertions about such theories). On the other hand if T does not have the f.c.p., then $R^m[\theta(\bar{x}; \bar{a}), \varphi, \lambda] = k$ is an elementary property of $\bar{a}$; and there is $k^1_{\theta,\varphi} < \aleph_0$ such that for each $\bar{a}$, for all $\lambda \geq k^1_{\theta,\varphi}$ the value of $R^m[\theta(\bar{x}; \bar{a}), \varphi, \lambda]$ is fixed; and the value of $\mathrm{Mlt}[\theta(\bar{x}; \bar{a}), \varphi]$ is bounded by some $k^2_{\theta,\varphi} < \aleph_0$; and for each finite Δ, there is $k_\Delta < \aleph_0$ such that any set of (Δ, m)-formulas p is realized iff each $q \subseteq p$, $|q| < k_\Delta$ is realized; and every (Δ, m)-type has a finite subtype of *bounded* cardinality which has the same rank $R^m(-, \Delta, \lambda)$.

Let $\mathrm{Kr}^m_\varphi(\lambda)$ be the first regular cardinality μ such that $|A| \leq \lambda \Rightarrow |S^m_\varphi(A)| < \mu$. It appears that the number of possible functions is small: for stable $\varphi = \varphi(\bar{x}; \bar{y})$: n $(1 < n < \aleph_0)$, and λ^+. For unstable $\varphi = \varphi(\bar{x}; \bar{y})$: $\mathrm{Ded}_r\, \lambda$ and $(2^\lambda)^+$; when they are distinct, φ satisfies the second case iff $\varphi(\bar{x}; \bar{y})$ has the independence property. The strict order property of T is equivalent to the existence of a formula $\varphi(\bar{x}; \bar{y})$ defining a quasi-order (i.e., a transitive, reflexive, anti-symmetric relation) with infinite chains. In Section 4 we also start to investigate dimensions.

Ranks were invented and investigated for their use in stability. Several kinds of ranks were used, and most of them are particular cases of $R^m(p, \Delta, \lambda)$, on which we concentrate. We investigate them also when there is no apparent application; more information is obtained in Chapter III, Section 4, and Chapter 5, Section 7.

$R^m(p, \Delta, \lambda)$ is interesting mainly for $\lambda = 2, \aleph_0, \infty$ and $\Delta = \mathrm{L}$ or Δ finite. What is the meaning of the rank $R^m(p, \Delta, \lambda)$? For finite p, we can say that it measures the complexity of the family of sets $\{\bar{a}: \bar{a}$ realizes $p \cup \{\varphi(\bar{x}; \bar{b})\}\}$ for $\varphi \in \Delta$, $\bar{b} \in \mathfrak{C}$.

The basic properties are proved in Section 1. The rank is monotonic in the parameters (decreasing for p and λ, increasing in Δ), and every type has a finite subtype of the same rank. When $\lambda \geq \aleph_0$ we can extend p while preserving the rank, i.e., if p is over A, there is a complete $q \supseteq p$, $R^m(p, \Delta, \lambda) = R^m(q, \Delta, \lambda)$; when $\lambda = 2$, $\varphi \in \Delta$ there is no $\bar{a}$ such that

$$R^m[p \cup \{\varphi(\bar{x}; \bar{a})^{\mathbf{t}}\}, \Delta, \lambda] = R^m(p, \Delta, \lambda) \quad \text{for } \mathbf{t} \in \{0, 1\};$$

so for $\Delta = \mathrm{L}$, p over A, p has at most one extension to a complete type over A of the same rank. For every λ, the maximal number of such complete extensions is, essentially denoted by $\mathrm{Mlt}^2(p, \Delta, \lambda)$.

In Section 1 we also prove that for limit λ, $R^m(p, \Delta, \lambda) = R^m(p, \Delta, \lambda^+)$.

In Section 2 we prove that for finite Δ, $R^m(p, \Delta, 2) < \omega$ or $R^m(p, \Delta, \infty) = \infty$; and $R^m[\theta(\bar{x}; \bar{a}), \Delta, k] = l$ is an elementary property of $\bar{a}$.

In Section 3 we complete the investigation on the stability spectrum for countable theories. If $|T| < 2^{\aleph_0}$, then T is stable in $|T|$ iff T is stable in some $\lambda < 2^{\aleph_0}$ iff T is stable in every $\lambda \geq |T|$ iff $R(x = x, \mathrm{L}, 2) < \infty$. Also T is superstable iff T is stable in every $\lambda \geq 2^{|T|}$ iff T is stable in some $\lambda < \lambda^{\aleph_0}$ iff $R(x = x, \mathrm{L}, \infty) < \infty$. But the aim of Section 3 is the investigation of R and D.

We prove that $R^m(p, \Delta, \lambda)$ is $< |\Delta|^+ + \aleph_0$ or ∞ usually (i.e., $\lambda = \infty$, or $\lambda \leq \aleph_0$). Also $\alpha_\lambda = R^m(p, \Delta, \lambda)$ (except for being decreasing) is essentially arbitrary for $\aleph_0 \leq \lambda \leq |T|^+$, has restricted changes for $\lambda \leq \aleph_0$, is fixed for $\lambda > (2^{|T|})^+$. In fact it is fixed for $\lambda > |\Delta|^+ + \aleph_0$ except when for some μ, $|\Delta|^+ \leq \mu \leq (2^\lambda)^+$, $|\Delta|^+ < \lambda < \mu$ implies $\alpha_\lambda = \infty$, and $\lambda > \mu^+$ implies $\alpha_\lambda = \alpha_\infty < \infty$; and $\alpha_\mu = \infty$ except possibly when μ is a limit cardinal (remember that for limit μ, $\alpha_\mu = \alpha_{\mu^+}$) so α_λ ($\lambda > |\Delta|^+ + \aleph_0$) has at most three values (see Theorem 3.13 and Exercise 3.16).

Moreover for $\Delta = \mathrm{L}$, $\lambda \leq \aleph_0$ we can compute $R^m(p, \Delta, \lambda)$ from $R^m(p, \Delta, 2)$ (Exercise 3.21); and $R^m(p, \Delta, \infty)$, for finite p, is usually directly characterized as the maximal α for which $\Gamma^*_{|T|^+}(\bigwedge p, h)$ is consistent for some (Δ, α)-function h. We deal much with $D^m(p, \Delta, \lambda)$, which is a version of $R^m(p, \Delta, \lambda)$ (essentially we require that the many contradictory extensions are finite) and we use it to prove assertions about R for $\lambda > |\Delta|^+ + \aleph_0$ and prove that for $\lambda > |\Delta|^+ + \aleph_0$ they are usually equal. We also give a method of constructing counterexamples (Exercise 3.20) and many exercises.

II.1. Ranks of types

DEFINITION 1.1: Let p be an m-type, Δ a set of m-formulas, λ a cardinal (possibly finite), or $\lambda = \infty$. We define the *rank* $R^m(p, \Delta, \lambda)$ or $R^m[p, \Delta, \lambda]$ by defining inductively when $R^m(p, \Delta, \lambda) \geq \alpha$, α an ordinal.

(1) $R^m(p, \Delta, \lambda) \geq 0$ when p is a (consistent) type. (When p is inconsistent we stipulate $R^m(p, \Delta, \lambda) = -1$.)

(2) $R^m(p, \Delta, \lambda) \geq \delta$ for δ a limit ordinal if $R^m(p, \Delta, \lambda) \geq \alpha$ for all $\alpha < \delta$.

(3) $R^m(p, \Delta, \lambda) \geq \alpha + 1$ if for all $\mu < \lambda$ and all finite $q \subseteq p$ there are types $\{q_i\}_{i \leq \mu}$ which are Δ-m-types (i.e., m-types whose formulas are all of the form $\varphi(\bar{x}; \bar{a})$ or $\neg\varphi(\bar{x}; \bar{a})$ where $\varphi(\bar{x}; \bar{y})$ is in Δ) such that:

(i) for $i \neq j$ there is a formula φ such that $\varphi \in q_i$, $\neg\varphi \in q_j$ (or vice versa). In this case we say that q_i and q_j are explicitly contradictory.

(ii) $R^m(q \cup q_i, \Delta, \lambda) \geq \alpha$ for all $i \leq \mu$.

(4) Now if $R^m(p, \Delta, \lambda) \geq \alpha$ but not $R^m(p, \Delta, \lambda) \geq \alpha + 1$ we say $R^m(p, \Delta, \lambda) = \alpha$. (It is easy to see by induction on α that $R^m(p, \Delta, \lambda) \geq \alpha$ implies $R^m(p, \Delta, \lambda) \geq \beta$ for all $\beta \leq \alpha$.) If $R^m(p, \Delta, \lambda) \geq \alpha$ for all α we define $R^m(p, \Delta, \lambda) = \infty$.

DEFINITION 1.2: If $R^m(p, \Delta, \lambda) = \alpha \neq \infty$, define the *multiplicity*:

(1) $\mathrm{Mlt}^1(p, \Delta, \lambda)$ is the first power μ_0 such that there is a finite $q \subseteq p$ such that q has no more than μ_0 Δ-m-types q_i satisfying 3(i), (ii) from Definition 1.1 (μ_0 may be finite).

(2) $\mathrm{Mlt}^2(p, \Delta, \lambda)$ is the first power μ_0 such that there are no more than μ_0 Δ-m-types q_i satisfying 3(i), (ii) from Definition 1.1 for $p = q$. If $R^m(p, \Delta, \lambda) = \infty$, define $\mathrm{Mlt}^1(p, \Delta, \lambda) = \mathrm{Mlt}^2(p, \Delta, \lambda) = \infty$.

If $\mathrm{Mlt}^1(p, \Delta, \lambda) = \mathrm{Mlt}^2(p, \Delta, \lambda)$ we shall just write $\mathrm{Mlt}(p, \Delta, \lambda)$. If a statement is true for both $\mathrm{Mlt}^1(p, \Delta, \lambda)$ and $\mathrm{Mlt}^2(p, \Delta, \lambda)$ we shall write $\mathrm{Mlt}^l(p, \Delta, \lambda)$. If $\lambda = \aleph_0$ we may omit it.

Remark. Clearly if $R^m(p, \Delta, \lambda) \neq \infty$, then $0 < \mathrm{Mlt}^2(p, \Delta, \lambda) \leq \mathrm{Mlt}^1(p, \Delta, \lambda) < \lambda$ and if p is finite, then $\mathrm{Mlt}^2(p, \Delta, \lambda) = \mathrm{Mlt}^1(p, \Delta, \lambda)$. In fact $\mathrm{Mlt}^1(p, \Delta, \lambda)^+ < \lambda$. See Lemma 1.9.

Notation. If $p = \{\theta(\bar{x}; \bar{a})\}$ we may write $R^m[\theta(\bar{x}; \bar{a}), \Delta, \lambda]$ instead of $R^m[\{\theta(\bar{x}; \bar{a})\}, \Delta, \lambda]$. The same convention applies to Mlt and D (see Definition 3.2).

Special Cases. $R^m(p, \Delta, \lambda) \geq 1$ if for all $\mu < \lambda$ and all finite $q \subseteq p$ there are Δ-m-types $\{q_i\}_{i \leq \mu}$ each of which is consistent with q, but every two are explicitly contradictory. $R^m(p, \Delta, \lambda) \geq 2$ if for every $\mu < \lambda$ and every finite $q \subseteq p$ there are Δ-m-types $\{q_i\}_{i \leq \mu}$ which are pairwise explicitly contradictory and $R^m(q \cup q_i, \Delta, \lambda) \geq 1$ for all i.

Example. Let M be a model with equivalence relations $\{E_\alpha: \alpha < \alpha_0\}$ such that $\alpha < \beta \Rightarrow E_\alpha \subseteq E_\beta$, E_α divides every equivalence class of E_β into infinitely many classes, every equivalence class of E_0 is infinite, and if $\alpha_0 = \gamma + 1$, then E_γ has infinitely many equivalence classes.

Let $T^{\alpha^*} = \mathrm{Th}(M)$ and let $\mathfrak{C} = \mathfrak{C}(T^{\alpha^*})$ be the saturated model of T^{α^*} of cardinality $\bar{\kappa}$. Then for every $a \in |\mathfrak{C}|$:

(1) $R^1[\{x = a\}, \mathrm{L}, \infty] = 0$,

(2) $R^1[\{xE_0a\}, \mathrm{L}, \infty] = 1$,

(3) $R^1[\{xE_\alpha a\}, \mathrm{L}, \infty] = 1 + \alpha$.

Remark. T^{α^*} admits elimination of quantifiers and is totally transcendental (see Definition 3.1).

Proof. (1) Clearly R^1 $[\{x = a\}, \mathrm{L}, \infty] \geq 0$, and there are no q_0, q_1 such that $q_0 \cup \{x = a\}$ and $q_1 \cup \{x = a\}$ are consistent while q_0, q_1 are explicitly contradictory, otherwise a would realize q_0 and q_1. Thus $R^1[\{x = a\}, \mathrm{L}, \lambda] \ngeq 1$ for $\lambda \geq 2$, in particular for $\lambda = \infty$.

(2) First we show $R^1[\{xE_0a\}, \mathrm{L}, \infty] \geq 1$. Let μ be any cardinal. Choose elements a_i, $i \leq \mu$, in the E_0 equivalence class of a and let $r_i = \{x = a_i\} \cup \{x \neq a_j : i \neq j \leq \mu\}$. Clearly they are pairwise explicitly contradictory, but a_i realizes $\{xE_0a\} \cup r_i$. In order that $R^1[\{xE_0a\}, \mathrm{L}, \infty] \geq 2$ there must be two explicitly contradictory types p_0, p_1 such that for $i = 0, 1$ $R^1[\{xE_0a\} \cup p_i, \mathrm{L}, \infty] \geq 1$. In particular $\{xE_0a\} \cup p_i = q_i$ is consistent, $i = 0, 1$. By Theorem 1.6, extend q_0, q_1, to complete types q_0^*, q_1^* over the same sets and of the same rank. But $E_0 \subseteq E_\alpha$ for all $\alpha < \alpha_0$ so q_0^*, q_1^* can be explicitly contradictory only if there is a_0 such that $[x = a_0] \in q_0^*$, $[x \neq a_0] \in q_1^*$ or vice versa. But then from the proof of (1) it is clear that $R^1(q_0^*, \mathrm{L}, \infty) \ngeq 1$. So $R^1[\{xE_0a\}, \mathrm{L}, \infty] \ngeq 2$.

(3) We leave this as an exercise for the reader.

DEFINITION 1.3: $p \vdash q$ when every sequence realizing p realizes q. If $p \vdash q$ and $q \vdash p$ we write $p \equiv q$: p is equivalent to q.

Remark. For finite types, $p \vdash q$ is equivalent to $(\forall \bar{x})[\bigwedge p \to \bigwedge q]$, and for infinite types to the condition that for every finite $q' \subseteq q$ there is a finite $p' \subseteq p$ such that $p' \vdash q'$.

THEOREM 1.1: (1) *If* $p_1 \vdash p_2$, *then* $R^m(p_1, \Delta, \lambda) \leq R^m(p_2, \Delta, \lambda)$. *In particular this is the case when* $p_2 \subseteq p_1$.

(2) *If equality holds in* (1), *then* $\mathrm{Mlt}^l(p_1, \Delta, \lambda) \leq \mathrm{Mlt}^l(p_2, \Delta, \lambda)$.

Proof. (1) We prove by induction on α that $R^m(p_1, \Delta, \lambda) \geq \alpha \Rightarrow R^m(p_2, \Delta, \lambda) \geq \alpha$. For $\alpha = 0$ or limit there is no problem. Assume for α and prove for $\alpha + 1$. Let $R^m(p_1, \Delta, \lambda) \geq \alpha + 1$ and let $q^2 \subseteq p_2$ be finite, $\mu < \lambda$. We shall find types $\{q_i\}_{i \leq \mu}$ as required in the definition. Since $p_1 \vdash p_2$, clearly $p_1 \vdash q^2$ so there is a finite $q^1 \subseteq p_1$ such that $q^1 \vdash q^2$. Choose pairwise explicitly contradictory Δ-m-types $\{q_i\}_{i \leq \mu}$ such that $R^m(q^1 \cup q_i, \Delta, \lambda) \geq \alpha$ for all $i \leq \mu$. By the induction hypothesis, $R^m(q^2 \cup q_i, \Delta, \lambda) \geq \alpha$ for all $i \leq \mu$. Thus $R^m(p_2, \Delta, \lambda) \geq \alpha + 1$.

(2) Similar proof.

THEOREM 1.2: *For all* p, Δ, λ, m *there is a finite* $q \subseteq p$ *such that* $R^m(p, \Delta, \lambda) = R^m(q, \Delta, \lambda)$, *and* $\mathrm{Mlt}^1(p, \Delta, \lambda) = \mathrm{Mlt}^1(q, \Delta, \lambda)$.

Proof. If $q \subseteq p$, then $R^m(p, \Delta, \lambda) \leq R^m(q, \Delta, \lambda)$ by the previous theorem. So if the left-hand side is ∞, we may take q arbitrarily, for example $q = \emptyset$. If the left side is $\alpha \neq \infty$, then $R^m(p, \Delta, \lambda) \ngeq \alpha + 1$ and thus there is a finite $q \subseteq p$ and there is $\mu < \lambda$ such that there are no pairwise explicitly contradictory Δ-m-types $\{q_i\}_{i \leq \mu}$ satisfying $R^m(q \cup q_i, \Delta, \lambda) \geq \alpha$. But that means that $R^m(q, \Delta, \lambda) \ngeq \alpha + 1$. By 1.1(2) $\mathrm{Mlt}^1(p, \Delta, \lambda) \leq \mathrm{Mlt}^1(q, \Delta, \lambda) = \mu_q$, say.

Among the pairs q, μ_q satisfying the above mentioned condition, choose one with minimal μ_q. Thus $R^m(q, \Delta, \lambda) = \alpha$, $\mathrm{Mlt}^1(q, \Delta, \lambda) = \mathrm{Mlt}^1(p, \Delta, \lambda)$.

LEMMA 1.3: (1) *If* $\Delta_1 \subseteq \Delta_2$, $\lambda \geq \kappa$, *then* $R^m(p, \Delta_1, \lambda) \leq R^m(p, \Delta_2, \kappa)$.

(2) *If there is equality in* (1), *then* $\mathrm{Mlt}^l(p, \Delta_1, \lambda) \leq \mathrm{Mlt}^l(p, \Delta_2, \kappa)$.

Proof. (1) We prove by induction on α that $R^m(p, \Delta_1, \lambda) \geq \alpha \Rightarrow R^m(p, \Delta_2, \kappa) \geq \alpha$. The proof is immediate.

(2) Likewise.

LEMMA 1.4: (1) *If* $R^m(p, \Delta, 2) = \alpha \neq \infty$, *then there is no* $\varphi(\bar{x}; \bar{y}) \in \Delta$ *and sequence* $\bar{a}$ *such that* $R^m[p \cup \{\varphi(\bar{x}; \bar{a})^{\mathfrak{t}}\}, \Delta, 2] = \alpha$, *for* $\mathfrak{t} = 0, 1$.

(2) *If* $R^m(p, \Delta, \lambda) = \alpha \neq \infty$ *and* $\mathrm{Mlt}^l(p, \Delta, \lambda) = 1$ *then the parallel conclusion still holds.*

(3) *If* $R^m(p, \Delta, \lambda) = \alpha \neq \infty$ *and* $\mathrm{Mlt}^l(p, \Delta, \lambda) < \aleph_0$, *then there is no* $\varphi(\bar{x}; \bar{y}) \in \Delta$ *and sequence* $\bar{a}$ *such that* $R^m[p \cup \{\varphi(\bar{x}; \bar{a})^{\mathfrak{t}}\}, \Delta, \lambda] = \alpha$ *and* $\mathrm{Mlt}^l[p \cup \{\varphi(\bar{x}; \bar{a})^{\mathfrak{t}}\}, \Delta, \lambda] = \mathrm{Mlt}^l(p, \Delta, \lambda)$ *for* $\mathfrak{t} = 0, 1$.

Proof. (1) If there were such a φ we would choose $q_0 = \{\varphi(\bar{x}, \bar{a})\}$, $q_1 = \{\neg\varphi(\bar{x}, \bar{a})\}$ and get $R^m(p, \Delta, 2) \geq \alpha + 1$.

(2), (3) The proof is similar.

Remark. If $q \subseteq p$, $p \restriction \Delta \in S^m_\Delta(A)$, $R^m(q, \Delta, 2) = R^m(p, \Delta, 2) = \alpha \neq \infty$, then $p \restriction \Delta$ may be defined in terms of q in the following sense: $\varphi(\bar{x}; \bar{a}) \in p \restriction \Delta \Leftrightarrow \varphi(\bar{x}; \bar{y}) \in \mathrm{cl}_1(\Delta), \bar{a} \in A$, and $R^m[q \cup \{\varphi(\bar{x}; \bar{a})\}, \Delta, 2] = \alpha$; for, if $\varphi(\bar{x}; \bar{a}) \in p \restriction \Delta$, then $q \subseteq q \cup \{\varphi(\bar{x}; \bar{a})\} \subseteq p$ and the rank of $q \cup \{\varphi(\bar{x}; \bar{a})\}$ is between the rank of q and the rank of p, and thus is α. And if $\varphi(\bar{x}; \bar{y}) \in \mathrm{cl}_1(\Delta)$, $\bar{a} \in A$, $\varphi(\bar{x}; \bar{a}) \notin p$, then $\neg\varphi(\bar{x}; \bar{a}) \in p$, since $p \restriction \Delta \in S^m_\Delta(A)$. Thus the rank of $q \cup \{\neg\varphi(\bar{x}; \bar{a})\}$ is α and by the previous lemma the rank of $q \cup \{\varphi(\bar{x}; \bar{a})\}$ is not α. (Compare Theorem 2.12.)

COROLLARY 1.5: *If* $q \subseteq p$, $p \in S^m_\Delta(A)$, $R^m(p, \Delta, 2) = R^m(q, \Delta, 2) = \alpha \neq \infty$, *then* p *is the unique type in* $S^m_\Delta(A)$ *extending* q *with rank* α.

THEOREM 1.6: *If* p *is an m-type over* A, $\lambda \geq \aleph_0$, $R^m(p, \Delta, \lambda) = \alpha$, *then there is* $q \in S^m(A)$, $p \subseteq q$, $R^m(q, \Delta, \lambda) = \alpha$.

Proof. Denote $\Gamma = \{\neg\psi(\bar{x}; \bar{a})\colon R^m[\{\psi(\bar{x}; \bar{a})\}, \Delta, \lambda] < \alpha,\ \bar{a} \in A\}$. It is sufficient to show that $p \cup \Gamma$ is consistent since then we can find $q \in S^m(A)$ extending $p \cup \Gamma$. Thus $R^m(q, \Delta, \lambda) \leq R^m(p, \Delta, \lambda) = \alpha$ but there is no finite $q_1 \subseteq q$ such that $R^m(q_1, \Delta, \lambda) < \alpha$ (since then we would have $R^m[\{\bigwedge q_1\}, \Delta, \lambda] < \alpha$ and $(\neg\bigwedge q_1) \in \Gamma \subseteq q$, thus making q contradictory). So by Theorem 1.2, $R^m(q, \Delta, \lambda) = \alpha$.

Now we show that $p \cup \Gamma$ is consistent. Otherwise there is a finite $r \subseteq p$ of equal rank and there are ψ_i, $\bar{a}^i \in A$, $i = 1, \ldots, n$, such that $R^m[\{\psi_i(\bar{x}, \bar{a}^i)\}, \Delta, \lambda] < \alpha$ and $r \cup \{\neg\psi_i(\bar{x}, \bar{a}^i)\colon 1 \leq i \leq n\}$ is contradictory. Thus $r \vdash \psi_1(\bar{x}, \bar{a}^1) \vee \cdots \vee \psi_n(\bar{x}, \bar{a}^n)$. Now $R^m(r, \Delta, \lambda) = \alpha > \max_i R^m[\{\psi_i(\bar{x}, \bar{a}^i)\}, \Delta, \lambda]$. On the other hand, $R^m(r, \Delta, \lambda) \leq$

$R^m[\{\bigvee_i \psi_i(\bar{x}; \bar{a}^i)\}, \Delta, \lambda]$. Thus

$$R^m[\{\bigvee_i \psi_i(\bar{x}; \bar{a}^i)\}, \Delta, \lambda] > \max_i R^m[\{\psi_i(\bar{x}; \bar{a}^i)\}, \Delta, \lambda].$$

But we shall prove:

CLAIM 1.7: *For all* r, $\lambda \geq \aleph_0$,

$$R^m\left[\left\{\bigvee_{i=1}^{n} \psi_i(\bar{x}; \bar{a}^i)\right\} \cup r, \Delta, \lambda\right] = \max_{1 \leq i \leq n} R^m[\{\psi_i(\bar{x}; \bar{a}^i)\} \cup r, \Delta, \lambda].$$

Proof. Let $\psi_i = \psi_i(\bar{x}; \bar{a}_i)$; note that for $1 \leq i \leq n$ $\{\psi_i\} \cup r \vdash \{\bigvee_{i=1}^n \psi_i\} \cup r$, and so by Theorem 1.1(1) $R^m[\{\psi_i\} \cup r, \Delta, \lambda] \leq R^m[\{\bigvee_{i=1}^n \psi_i\} \cup r, \Delta, \lambda]$. Thus $\max_{1 \leq i \leq n} R^m[\{\psi_i\} \cup r, \Delta, \lambda] \leq R^m[\{\bigvee_{i=1}^n \psi_i\} \cup r, \Delta, \lambda]$.

For the other direction we prove

$$(*) \qquad R^m\left[\left\{\bigvee_i \psi_i\right\} \cup r, \Delta, \lambda\right] \geq \beta \Rightarrow \max_i R^m[\{\psi_i\} \cup r, \Delta, \lambda] \geq \beta.$$

The proof is by induction on β. For $\beta = 0$ (*) says that if $\{\bigvee_i \psi_i\} \cup r$ is consistent, then there is i such that $\{\psi_i\} \cup r$ is consistent, and this is correct. When β is a limit ordinal the claim follows directly from the induction hypothesis. We must now show the passage from β to $\beta + 1$. Assume by way of contradiction that $R^m[\{\bigvee_i \psi_i\} \cup r, \Delta, \lambda] \geq \beta + 1$ but $\max_i R^m[\{\psi_i\} \cup r, \Delta, \lambda] < \beta + 1$.

Since $R^m[\{\psi_i\} \cup r, \Delta, \lambda] < \beta + 1$ for every $i = 1, \ldots, n$, there are $\mu_i < \lambda$ and finite $q_i \subseteq r$ such that there are no $\{r_j\}_{j \leq \mu_i}$ satisfying the demands of Definition 1.1(3). Let $q = \bigcup_{i=1}^n q_i$, $\mu = (n + 1) \max_i \mu_i$. λ is infinite by assumption, so $\mu < \lambda$. Thus there is a finite $q \subseteq r$ and there are $\mu_1, \ldots, \mu_n < \lambda$ such that for all i there do not exist pairwise explicitly contradictory types $\{r_j\}_{j \leq \mu_i}$ such that $R^m[\{\psi_i\} \cup q \cup r_j, \Delta, \lambda] \geq \beta$ for all $j \leq \mu_i$. On the other hand, since $R^m[\{\bigvee_i \psi_i\} \cup r, \Delta, \lambda] \geq \beta + 1$, and $\mu < \lambda$, for all finite $q \subseteq r$, there are pairwise explicitly contradictory $\{r_j\}_{j \leq \mu}$ such that $R^m[\{\bigvee \psi_i\} \cup q \cup r_j, \Delta, \lambda] \geq \beta$ for all $j \leq \mu$. By the induction hypothesis with $q \cup r_j$ in place of r, we get that $\max_i R^m[\{\psi_i\} \cup q \cup r_j, \Delta, \lambda] \geq \beta$ for all j. Thus for all $j \leq \mu$ there is $i(j)$, $1 \leq i(j) \leq n$, such that $R^m[\{\psi_{i(j)}\} \cup q \cup r_j, \Delta, \lambda] \geq \beta$. There are $\mu + 1$ j's and n i's, so there is i_0 equal to $i(j)$ for at least $\max_i \mu_{i_0} + 1$ j's (by the definition of μ). Then w.l.o.g. $i_0 = i(j)$ *for all* $j \leq \mu_{i_0}$, and so $R^m[\{\psi_{i_0}\} \cup q \cup r_j, \Delta, \lambda] \geq \beta$ for all $j \leq \mu_{i_0}$. This contradicts what we obtained previously by the choice of the μ_i. Thus we have proved (*), Claim 1.7 and Theorem 1.6.

COROLLARY 1.8: (1) *If* $\lambda \geq \aleph_0$, *in Definitions 1.1 and 1.2 we can replace "explicitly contradictory" by "contradictory"; or by "*$q \cup q_i \cup q_j$

implies a contradiction for all i, j" *and even by* "$R^m(q \cup q_i \cup q_j, \Delta, \lambda) < \alpha$ *for* $i < j \le \mu$".

(2) *If* Δ *is a set of m-formulas and* $\Delta_1 = \mathrm{cl}_3^m(\Delta)$, *then for every m-type* p *and infinite cardinality* λ, $R^m(p, \Delta_1, \lambda) = R^m(p, \Delta, \lambda)$.

LEMMA 1.9: (1) *If* $R^m(p, \Delta, \lambda) = \alpha \neq \infty$, $\lambda \geqslant \aleph_0$, *then the value* $\mu_0 = \mathrm{Mlt}^l(p, \Delta, \lambda)$ *is achieved, that is, for every finite* $q \subseteq p$ [*for* $q = p$] *there are* Δ*-m-types* $\{q_i\}_{i<\mu_0}$ *satisfying Definition* 1.1(3).

(2) $R^m(p, \Delta, \lambda) = R^m(p, \Delta, \lambda^+)$ *for* λ *a limit cardinal.*

Proof. (1) The proof is the same for both multiplicities so we just prove for Mlt^1.

If $\mu_0 < \aleph_0$ or μ_0 is a successor cardinal the claim is obvious; so assume $\mu_0 \ge \aleph_0$, μ_0 a limit cardinal. Clearly for every finite $q \subseteq p$ for every $\mu < \mu_0$ there are pairwise contradictory Δ*-m-types* $\{q_i^\mu\}_{i<\mu}$ such that $R^m(q \cup q_i^\mu, \Delta, \lambda) \ge \alpha$. Let A be a set such that for all $i < \mu < \mu_0$, q_i^μ and q are types over A. By Theorem 1.6 for all $i < \mu < \mu_0$ there are types $r_i^\mu \in S_\Delta^m(A)$ such that $q_i^\mu \subseteq r_i^\mu$ and $R^m(q \cup r_i^\mu, \Delta, \lambda) \ge \alpha$. Clearly for every $\mu < \mu_0$, $i \neq j < \mu$ implies $r_i^\mu \neq r_j^\mu$ since q_i^μ, q_j^μ are contradictory and so $|\{r_i^\mu : i < \mu < \mu_0\}| = \mu_0$ since μ_0 is a limit cardinal and $|\{r_i^\mu : i < \mu\}| = \mu$ for every $\mu < \mu_0$. So omitting duplicates and relabelling, for every finite $q \subseteq p$ there are pairwise explicitly contradictory Δ-*m*-types $\{q_i\}_{i \le \mu_0}$ such that $R^m(q \cup q_i, \Delta, \lambda) \ge \alpha$.

(2) By 1.3(1), $R^m(p, \Delta, \lambda^+) \le R^m(p, \Delta, \lambda)$ so it is sufficient to show by induction on α, that $R^m(p, \Delta, \lambda) \ge \alpha \Rightarrow R^m(p, \Delta, \lambda^+) \ge \alpha$. For $\alpha = 0$ or α a limit ordinal this is obvious. So suppose $\alpha = \beta + 1$, $R^m(p, \Delta, \lambda) \ge \alpha$, then for every finite $q \subseteq p$, for all $\mu < \lambda$ there are pairwise contradictory Δ-*m*-types $\{q_i^\mu\}_{i<\mu}$ such that $R^m(q \cup q_i^\mu, \Delta, \lambda) \ge \beta$. But λ is a limit cardinal so exactly as in part (1) for every finite $q \subseteq p$ there are pairwise explicitly contradictory Δ-*m*-types $\{q_i\}_{i \le \lambda}$ such that $R^m(q \cup q_i, \Delta, \lambda) \ge \beta$. By the induction hypothesis, $R^m(q \cup q_i, \Delta, \lambda^+) \ge \beta$ so by Definition 1.1(3), $R^m(p, \Delta, \lambda^+) \ge \beta + 1 = \alpha$.

LEMMA 1.10: *Assume* $R^m(p, \Delta, \lambda) = \alpha \neq \infty$, $\mathrm{Mlt}^l(p, \Delta, \lambda) = \mu_0$.

(1) *There is a set* A, $|A| < \lambda$, *such that for all* $q \in S_\Delta^m(A)$, *if* $R^m(p \cup q, \Delta, \lambda) = \alpha$, *then* $\mathrm{Mlt}^2(p \cup q, \Delta, \lambda) = 1$. (*This will also be true for all* $A' \supseteq A$.)

(2) *If* $\aleph_0 \le \lambda$ *and* $R^m[p \cup \{\neg\varphi\}, \Delta, \lambda] < \alpha$, *then* $R^m[p \cup \{\varphi\}, \Delta, \lambda] = \alpha$, $\mathrm{Mlt}^l[p \cup \{\varphi\}, \Delta, \lambda] = \mu_0$.

(3) *If* $R^m[p \cup \{\varphi\}, \Delta, \lambda] = R^m[p \cup \{\neg\varphi\}, \Delta, \lambda] = \alpha$ *and* $\lambda \ge \aleph_0$, *then* $\mathrm{Mlt}^l(p, \Delta, \lambda) \le \mathrm{Mlt}^l[p \cup \{\varphi\}, \Delta, \lambda] + \mathrm{Mlt}^l[p \cup \{\neg\varphi\}, \Delta, \lambda]$, *and equality holds if* $\varphi \in \Delta$.

(4) $\mathrm{Mlt}^1(q, \Delta) = \mathrm{Mlt}^2(q, \Delta)$ (*i.e.*, $\mathrm{Mlt}^1(q, \Delta, \aleph_0) = \mathrm{Mlt}^2(q, \Delta, \aleph_0)$).

Proof. (1), (2) and (3) are left as an exercise.

(4) If $R^m(q, \Delta, \aleph_0) = \infty$ this holds by Definition 1.2. Otherwise let $\alpha = R^m(q, \Delta, \aleph_0)$, $k = \mathrm{Mlt}^1(q, \Delta)$ so by the remark after Definition 1.2, $k < \aleph_0$ and $k \geq \mathrm{Mlt}^2(q, \Delta)$. For proving the other inequality, choose finite $p \subseteq q$ such that $\alpha = R^m(p, \Delta, \aleph_0)$, $\mathrm{Mlt}^1(p, \Delta) = \mathrm{Mlt}^1(q, \Delta)$ (by Theorem 1.2). By Definition 1.2 there are pairwise explicitly contradictory Δ-m-types r_i $(i < k)$ such that $\alpha = R^m(p \cup r_i, \Delta, \aleph_0)$. If for each $i < k$, $\alpha = R^m(q \cup r_i, \Delta, \aleph_0)$, then $\mathrm{Mlt}^2(q, \Delta) \geq k$ so we finish. Otherwise for some $j < k$, $R^m(q \cup r_j, \Delta, \aleph_0) < \alpha$ hence for some finite p_1, $p \subseteq p_1 \subseteq q$, $R^m(p_1 \cup r_j, \Delta, \aleph_0) < \alpha$. As $p \subseteq p_1 \subseteq q$ clearly $\alpha = R^m(p_1, \Delta, \aleph_0)$, $k = \mathrm{Mlt}^1(p_1, \Delta)$ so as p_1 is finite, there are pairwise explicitly contradictory Δ-m-types r_i^1 $(i < k)$ such that $R^m(p_1 \cup r_i^1, \Delta, \aleph_0) = \alpha$. Choose A such that r_i, r_i^1 $(i < k)$ and q are over A, and for $i < k$, choose r_i^2 such that $r_i^1 \subseteq r_i^2 \in S_\Delta^m(A)$, $R^m(p_1 \cup r_i^2, \Delta, \aleph_0) = \alpha$ (by 1.6) and similarly r_k^2, such that $r_j \subseteq r_k^2 \in S_\Delta^m(A)$, $\alpha = R^m(p \cup r_k^2, \Delta, \aleph_0)$. The r_i^2 $(i \leq k)$ are distinct (for $i(1) \neq i(2) < k \Rightarrow r_{i(1)}^1, r_{i(2)}^1$ are explicitly contradictory; and if $i < k$, $r_i^2 = r_k^2$, then $\alpha \leq R^m(p_1 \cup r_k^2, \Delta, \aleph_0)$ hence $\alpha \leq R^m(p_1 \cup r_j, \Delta, \aleph_0)$ contradicting the choice of p_1, r_j). As $r_i^2 \in S_\Delta^m(A)$, the r_i^2 $(i \leq k)$ are pairwise explicitly contradictory, and $\alpha \leq R^m(p \cup r_i^2, \Delta, \aleph_0)$ hence $\mathrm{Mlt}^1(p, \Delta) \geq k + 1$, a contradiction.

LEMMA 1.11: *There is λ_T depending on T only such that for $\lambda \geq \lambda_T$ $R^m(p, \Delta, \lambda) = R^m(p, \Delta, \infty)$ for every p and Δ.*

Proof. For every $\theta(\bar{x}; \bar{a})$ and Δ, $R^m[\theta(\bar{x}; \bar{a}), \Delta, \lambda]$ is a decreasing function of λ, hence there is $\lambda = \lambda(\theta(\bar{x}; \bar{a}), \Delta)$ such that for all $\mu \geq \lambda$ $R^m[\theta(\bar{x}; \bar{a}), \Delta, \mu] = R^m[\theta(\bar{x}; \bar{a}), \Delta, \lambda]$. It is clear that $\lambda(\theta(\bar{x}; \bar{a}), \Delta)$ depends on $\theta(\bar{x}; \bar{y})$, $\mathrm{tp}(\bar{a}, \emptyset)$, Δ only (as rank is preserved by automorphisms of $\mathfrak{C}$). As the number of such triples is $\leq 2^{|T|}$, there is $\lambda_T = \sup\{\lambda(\theta(\bar{x}; \bar{a}), \Delta): \theta(\bar{x}, \bar{a}), \Delta\} < \infty$. As for finite p, $R^m(p, \Delta, \lambda) = R^m(\bigwedge p, \Delta, \lambda)$, and every type has a finite subtype of the same rank (see 1.1(1), 1.2), for every p and Δ, $\lambda \geq \lambda_T$ implies $R^m(p, \Delta, \lambda) = R^m(p, \Delta, \lambda_T)$. Now we can easily prove by induction on α that for $\lambda \geq \lambda_T$, $R^m(p, \Delta, \lambda) \geq \alpha$ implies $R^m(p, \Delta, \infty) \geq \alpha$. Hence for $\lambda \geq \lambda_T$, $R^m(p, \Delta, \infty) \geq R^m(p, \Delta, \lambda)$, and 1.3 implies equality.

EXERCISE 1.1: Let f be an automorphism of $\mathfrak{C}$. Show that $R^m(p, \Delta, \lambda) = R^m[f(p), \Delta, \lambda]$ for every p, Δ, λ; where $f(p) = \{\varphi(\bar{x}; f(\bar{a})): \varphi(\bar{x}; \bar{a}) \in p\}$.

EXERCISE 1.2: Show that for every Δ, λ, that if for no p does $R^m(p, \Delta, \lambda) = \alpha_0$, then $R^m(p, \Delta, \lambda) \geq \alpha_0 \Rightarrow R^m(p, \Delta, \lambda) = \infty$, for every m-type p.

EXERCISE 1.3: Show that for all $\lambda \geq \aleph_0$, $R^m(p, \mathrm{L}, \lambda) = 0$ iff p is (consistent and) algebraic (i.e. realized by only a finite number of sequences).

EXERCISE 1.4: Show that if T is the theory of $\aleph_0$ independent unary predicates, then for all p, $R^m(p, \mathrm{L}, \infty) = 1 \Leftrightarrow p$ is not algebraic and for p not algebraic $\mathrm{Mlt}^1(p, \mathrm{L}, \infty) = 2^{\aleph_0}$. (So $T = \{(\exists x) \bigwedge_{i \in w} P_i(x)^{\text{if } (i \in u)}$: $u \subseteq w \subseteq \omega$, w finite$\}$.)

EXERCISE 1.5: (1) Show that if δ is a limit ordinal (or 0), then for all p,

$$R^m\left[p \cup \left\{\bigvee_{i<k} \varphi_i\right\}, \Delta, n\right] \geq \delta + lk^* \Rightarrow (\exists i) R^m(p \cup \{\varphi_i\}, \Delta, n) \geq \delta + l,$$

where $k, l < \omega$ and $k^* = \min\{j\colon n^j \geq nk\}$.

(2) Generalize Theorem 1.6 accordingly. (Hint: take $l = 0$.)

EXERCISE 1.6: Let k, l be natural numbers, $k \geq 2$, $l \geq 1$. Show that for all α, p and Δ ((3) and (4) are rephrasings of (1) and (2), respectively):

(1) $R^m(p, \Delta, k) \geq l \cdot \alpha \Rightarrow R^m(p, \Delta, k^l) \geq \alpha$,

(2) $R^m(p, \Delta, k) \geq \omega \cdot \alpha \Rightarrow R^m(p, \Delta, \aleph_0) \geq \alpha$,

(3) $R^m(p, \Delta, k) \leq l \cdot R^m(p, \Delta, k^l) + (l - 1)$,

(4) $R^m(p, \Delta, k) < \omega \cdot R^m(p, \Delta, \aleph_0) + \omega$.

(See also Exercise 3.22(1).)

EXERCISE 1.7: Show that if any one of p, Δ, or λ is finite, then $\mathrm{Mlt}^1(p, \Delta, \lambda) = \mathrm{Mlt}^2(p, \Delta, \lambda)$. Show that in general equality does not hold.

EXERCISE 1.8: Prove that 1.10(3) may fail for $\lambda < \aleph_0$.

EXERCISE 1.9: Prove that in 1.10(3) even for finite λ, $\varphi \in \Delta$

$$\mathrm{Mlt}^l(p, \Delta, \lambda) \geq \mathrm{Mlt}^l[p \cup \{\varphi\}, \Delta, \lambda] + \mathrm{Mlt}^l[p \cup \{\neg\varphi\}, \Delta, \lambda].$$

EXERCISE 1.10: Show that if $R^m(p, \Delta, \lambda) \geq n$, then for some finite $\Delta^1 \subseteq \Delta$, $R^m(p, \Delta^1, \lambda) \geq n$ for p finite, λ finite or ∞. Show that those restrictions are necessary.

II.2. Stability, ranks and definability

Now we show that there is no essential difference between Δ-m-types for finite Δ and φ-m-types.

LEMMA 2.1: *Let* $\Delta = \{\varphi_k(\bar{x}; \bar{y}_k): k < l < \omega\}$, $l(\bar{x}) = m$ *and*

$$\psi_\Delta = \psi_\Delta(\bar{x}; \bar{y}_0, \ldots, \bar{y}_{2l-1}; z, z_0, \ldots, z_{2l-1})$$

$$= \bigwedge_{k=0}^{l-1} (z = z_k \to \varphi_k(\bar{x}; \bar{y}_k)) \wedge \bigwedge_{k=l}^{2l-1} (z = z_k \to \neg\varphi_{k-l}(\bar{x}; \bar{y}_k))$$

$$\wedge \bigvee_{k<2l} z = z_k \wedge \bigwedge_{k<n<2l} \neg(z = z_k \wedge z = z_n).$$

Then:

(1) *For every type* p *and power* λ, $R^m(p, \Delta, \lambda) = R^m(p, \psi_\Delta, \lambda)$.

(2) *For every formula* $\varphi(\bar{x}; \bar{a})$ *(where* $\varphi \in \Delta$ *or* $\neg\varphi \in \Delta$*) and* $c_0 \neq c_1$ *there is* $\bar{a}'$, $\bar{a}' \subseteq \bar{a} \cup \{c_0, c_1\}$ *such that* $\varphi(\bar{x}; \bar{a}) \equiv \psi_\Delta(\bar{x}; \bar{a}')$ *(more exactly,* $\vDash(\forall\bar{x})(\varphi(\bar{x}; \bar{a}) \equiv \psi_\Delta(\bar{x}; \bar{a}'))$*). Hence for every* Δ-m-*type* p *over* A, $|A| \geq 2$, *there is a* ψ_Δ-*type* q *over* A *equivalent to it (i.e.,* $p \equiv q$*). Also if* $p \in S^m_\Delta(A)$, *then* $q \in S^m_{\psi_\Delta}(A)$.

(3) *For every* $\bar{b}$ *such that* $\vDash(\exists\bar{x})\psi_\Delta(\bar{x}; \bar{b})$ *there is* $\bar{a} \subseteq \bar{b}$ *and* φ *(*$\varphi \in \Delta$ *or* $\neg\varphi \in \Delta$*) for which* $\varphi(\bar{x}; \bar{a}) \equiv \psi_\Delta(\bar{x}; \bar{b})$. *Hence for every* ψ_Δ-m-*type* q *over* A *there is a* Δ-m-*type* p *over* A *equivalent to it. Also* $q \in S^m_{\psi_\Delta}(A)$ *implies* $p \in S^m_\Delta(A)$.

Proof. Let $c_0 \neq c_1 \in A$. (2) and (3) may be proved as follows:

$$\varphi_k(\bar{x}; \bar{a}) \equiv \psi_\Delta(\bar{x}; \underbrace{c_0, \ldots, c_0}_{\bar{y}_0}, \underbrace{c_0, \ldots, c_0}_{\bar{y}_1}, \ldots, \underbrace{\bar{a}}_{\bar{y}_k}, \ldots, \underbrace{c_0, \ldots, c_0}_{\bar{y}_{2l-1}};$$

$$\underset{z}{c_1}, \underset{z_0}{c_0}, \ldots, \underset{z_k}{c_1}, \underset{z_{k+1}}{c_0}, \ldots, \underset{z_{2l-1}}{c_0}).$$

In other words, in ψ_Δ we substituted $\langle c_0, \ldots, c_0 \rangle$ for $\bar{y}_n$, $n \neq k$, $\bar{a}$ for $\bar{y}_k$, c_1 for z and z_k, c_0 for z_n, $n \neq k$. A similar trick works for $\neg\varphi_k(\bar{x}; \bar{a})$. And on the other hand,

$$\psi_\Delta(\bar{x}; \bar{a}_0, \ldots, \bar{a}_{2l-1}; c_k, c_0, \ldots, c_{2l-1}) \equiv \varphi_k(\bar{x}, \bar{a}_k) \quad \text{or} \quad \neg\varphi_{k-l}(\bar{x}, \bar{a}_k)$$

or is false (the c_i's are not necessarily distinct). So in this way q may be defined from p and vice versa. Now for the case where $p[q]$ is complete. If $p[q]$ is a complete Δ-type (ψ_Δ-type) over A, $|A| \geq 2$, then

(i) for every k, $\bar{a}_k \in A$,

$$q \vdash \varphi_k(\bar{x}, \bar{a}_k) \quad \text{or} \quad q \vdash \neg\varphi_k(\bar{x}, \bar{a}_k).$$

(ii) For every c, c_k, $\bar{a}_k \in A$,

$$p \vdash \psi_\Delta(\bar{x}; \bar{a}_0, \ldots; c, c_0, \ldots)$$

or

$$p \vdash \neg\psi_\Delta(\bar{x}; \bar{a}_0, \ldots; c, c_0, \ldots).$$

The proof is straightforward.

Proof of (1). By the above it is sufficient to show:

Claim 2.1(4): If Δ_1, Δ; are sets of m-formulas which are closed under $\neg$, and:

(iii) For every $\varphi(\bar{x}; \bar{y}) \in \Delta_1$ and $\bar{a}$ there is a finite Δ_2-m-type $p^\varphi_{\bar{a}}$ such that $\varphi(\bar{x}; \bar{a}) \equiv \bigwedge p^\varphi_{\bar{a}}$, and $p^\varphi_{\bar{a}}$ and $p^{\neg\varphi}_{\bar{a}}$ are explicitly contradictory, or $\models \neg(\exists\bar{x})\varphi(\bar{x}; \bar{a})$.

(iv) For every $\varphi(\bar{x}; \bar{y}) \in \Delta_2$ and $\bar{a}$ there is a finite Δ_1-m-type $q^\varphi_{\bar{a}}$ such that $\varphi(\bar{x}; \bar{a}) \equiv \bigwedge q^\varphi_{\bar{a}}$ and $q^\varphi_{\bar{a}}$ and $q^{\neg\varphi}_{\bar{a}}$ are explicitly contradictory, or $\models \neg(\exists\bar{x})\varphi(\bar{x}; \bar{a})$.

Then for every type r and cardinal λ, $R^m(r, \Delta_1, \lambda) = R^m(r, \Delta_2, \lambda)$.

Remark. If we assume (iii) alone we get $R^m(r, \Delta_1, \lambda) \leq R^m(r, \Delta_2, \lambda)$.

Proof. By the symmetry of the assumptions it is enough to prove for finite r that $R^m(r, \Delta_1, \lambda) \geq \beta \Rightarrow R^m(r, \Delta_2, \lambda) \geq \beta$. For $\beta = 0$ or limit this is immediate. So assume it for α and let $\beta = \alpha + 1$. Let $\mu < \lambda$. Since $R^m(r, \Delta_1, \lambda) \geq \alpha + 1$ there are Δ_1-m-types $\{r_i\}_{i \leq \mu}$ which are pairwise explicitly contradictory and $R^m(r \cup r_i, \Delta_1, \lambda) \geq \alpha$. By (iii) we can define $r_i^* = \bigcup \{q^\varphi_{\bar{a}}: \varphi(\bar{x}; \bar{a}) \in r_i\}$, and r_i^*, r_j^* are explicitly contradictory for $i \neq j$. Also $r_i \equiv r_i^*$. Thus $R^m(r \cup r_i^*, \Delta_1, \lambda) = R^m(r \cup r_i, \Delta_1, \lambda) \geq \alpha$. So by the induction hypothesis and 1.2, $R^m(r \cup r_i^*, \Delta_2, \lambda) \geq \alpha$. Thus $R^m(r, \Delta_2, \lambda) \geq \alpha + 1 = \beta$. This completes the proof of the claim and the lemma.

Remark. We will use Lemma 2.1 without explicit mention.

In the following theorem we show that the unstability of $\varphi(\bar{x}; \bar{y})$ is equivalent to many conditions.

THE UNSTABLE FORMULA THEOREM 2.2: *The following properties of* $\varphi = \varphi(\bar{x}; \bar{y})$, $m = l(\bar{x})$, *are equivalent (relative to a given theory* T*).*

(1) φ *is* unstable (*in every infinite power*); *i.e., for every* $\lambda \geq \aleph_0$ *there is* A *such that* $|S^m_\varphi(A)| > \lambda \geq |A|$.

(2) *φ is unstable in at least one power $\lambda \geq \aleph_0$.*

(3) *φ has the* order property; *i.e., there are $\bar{a}^n$, $n < \omega$ such that for every $k < \omega$ $\{\varphi(\bar{x}; \bar{a}^n)^{\text{if } (k \leq n)}: n < \omega\}$ is consistent.*

(4) *For every $n < \omega$, $\Gamma(\varphi, m, n)$ is consistent, where*

$$\Gamma(\varphi, m, \alpha) = \{\varphi(\bar{x}_\eta; \bar{y}_{\eta|\beta})^{\eta[\beta]}: \eta \in {}^\alpha 2, \beta < \alpha\}$$

(usually we omit the m).

(5) *$\Gamma(\varphi, \alpha)$ is consistent for every ordinal α.*

(6) *$R^m(\bar{x} = \bar{x}, \varphi, \infty) = \infty$.*

(7) *$R^m(\bar{x} = \bar{x}, \varphi, 2) \geq \omega$.*

(8) *There are $A, p, p \in S^m_\varphi(A)$, such that p is not A-*definable *(see Definition 2.1).*

(9) *There is no ψ such that for every A, $|A| \geq 2$, and $p \in S^m_\varphi(A)$, p is (ψ, A)-*definable *(see Definition 2.1).*

DEFINITION 2.1: (1) The φ-m-type p is $\psi(\bar{y}; \bar{c})$-*defined* if $\varphi(\bar{x}; \bar{a}) \in p \Rightarrow \vDash \psi[\bar{a}; \bar{c}]$ and $\neg\varphi(\bar{x}; \bar{a}) \in p \Rightarrow \vDash \neg\psi[\bar{a}; \bar{c}]$.

(2) The φ-m-type p is $(\psi(\bar{y}; \bar{z}), A)$-*definable* if there is a $\bar{c} \in A$ such that p is $\psi(\bar{y}; \bar{c})$-defined.

(3) The φ-m-type p is A-*definable* if there is a ψ for which p is (ψ, A)-definable.

(4) The m-type p is A-definable if $p \restriction \varphi$ is A-definable for every φ.

Remark. If p is a complete φ-m-type over A, then the arrows in (1) go both ways for $\bar{a} \in A$. We sometimes write "definable over A" instead of "A-definable".

The proof of Theorem 2.2 will be broken down into a series of lemmas and theorems and completed with Theorem 2.12. From now till then all notations will be as in Theorem 2.2.

Obviously, (1) $\Rightarrow$ (2).

LEMMA 2.3: (2) $\Rightarrow$ (3).

Proof. Let $|S^m_\varphi(A)| > \lambda \geq |A| \geq \aleph_0$. By Conclusion I, 2.11 there are sequences $\bar{a}^n$, $n < \omega$ such that either

(i) for every $k < \omega$, $\{\varphi(\bar{x}; \bar{a}^l)^{\text{if } (l<k)}: l < \omega\}$ is consistent, or

(ii) for every $k < \omega$, $\{\varphi(\bar{x}; \bar{a}^l)^{\text{if } (k<l)}: l < \omega\}$ is consistent.

In case (ii) clearly φ has the order property. In case (i) in order to prove (3) we use the compactness theorem, replacing the $\bar{a}^l$ by parameters $\bar{y}^l$ and reversing the order of the first $n < \omega$ of them.

LEMMA 2.4: (3) ⇒ (4).

Proof. Define an order $<$ on ${}^{\omega\geq}2$ as follows: if $\eta \restriction k = \nu \restriction k$, $\eta[k] = 0$, $\nu[k] = 1$, then $\eta < \nu$; if $\eta \restriction k = \nu \restriction k$, $l(\eta) = k$, $\nu[k] = 1$ then $\eta < \nu$; if $\eta \restriction k = \nu \restriction k$, $l(\eta) = k$, $\nu[k] = 1$, then $\nu < \eta$. It is easily seen that $<$ is a total order. So by the compactness theorem and the fact that φ has the order property we see that the set

$$T \cup \{\varphi(\bar{x}_\eta; \bar{y}_\nu)^{\text{if}(\eta<\nu)}: l(\eta) = n, l(\nu) < n\}$$

is consistent for all $n < \omega$.

So we get that

$$\Gamma(\varphi, n) = \{\varphi(\bar{x}_\eta; \bar{y}_{\eta|k})^{\eta[k]}: \eta \in {}^n2, k < n\}$$

is consistent for every $n < \omega$.

LEMMA 2.5: (4) ⇒ (5).

Proof. This is immediate since every finite subset of $\Gamma(\varphi, \alpha)$ is equal to a subset of some $\Gamma(\varphi, n)$, after changing names of variables.

LEMMA 2.6: (5) ⇒ (1).

Proof. Let λ be infinite; we must show that φ is unstable in λ. Let $\mu = \inf\{\mu: 2^\mu > \lambda\}$ and let M be a model of $\Gamma(\varphi, \mu)$ where $\bar{a}_\nu$ realizes $\bar{y}_\nu$ and $\bar{c}_\eta$ realizes $\bar{x}_\eta$. Let $A = \bigcup\{\bar{a}_\nu: l(\nu) < \mu\}$. Clearly $|A| \leq (\sum_{\alpha<\mu} |{}^\alpha 2|)\cdot\aleph_0 \leq \mu\cdot\lambda\cdot\aleph_0 = \lambda$ (since $\lambda < 2^\lambda$, we have $\mu \leq \lambda$). For $\eta \in {}^\mu 2$, let p_η be the φ-m-type $\bar{c}_\eta$ realizes over A. If $l(\eta) = l(\nu) = \mu$, $\eta \neq \nu$, then $p_\eta \neq p_\nu$, since if α is the first ordinal such that $\eta[\alpha] \neq \nu[\alpha]$, then $\varphi(\bar{x}; \bar{a}_{\eta|\alpha})^{\eta[\alpha]} \in p_\eta$ and $\varphi(\bar{x}; \bar{a}_{\nu|\alpha})^{\nu[\alpha]} = \neg\varphi(\bar{x}; \bar{a}_{\eta|\alpha})^{\eta[\alpha]} \in p_\nu$. Thus

$$|S^m_\varphi(A)| \geq |\{p_\eta: l(\eta) = \mu\}| = |\{\eta: l(\eta) = \mu\}| = 2^\mu > \lambda \geq |A|.$$

This completes the proof of the equivalence of properties (1) through (5).

LEMMA 2.7: (5) ⇒ (6).

Proof. Let $\bar{a}^\alpha_\rho$ realize $\bar{y}_\rho$ and $\bar{c}^\alpha_\eta$ realize $\bar{x}_\eta$ for all α, $\rho \in {}^{\alpha>}2$, $\eta \in {}^\alpha 2$. Let λ be any cardinal. For all $\eta \in {}^{\lambda>}2$ let $p_\eta = \{\varphi(\bar{x}; \bar{a}^\lambda_{\eta|\beta})^{\eta[\beta]}: \beta < l(\eta)\}$; p_η is a (consistent) φ-m-type. Let p_ν be such a p_η with minimal $R^m(p_\eta, \varphi, \lambda)$, say $R^m(p_\nu, \varphi, \lambda) = \alpha$.

Suppose $\alpha \neq \infty$. Let $\bar{0}^\gamma$, $\gamma < \lambda$, be a sequence of zeros of length γ and $p^\gamma = p_{\nu\frown\bar{0}\frown\langle 1\rangle}$. The p^γ's are pairwise explicitly contradictory, $p_\nu \subseteq p^\gamma$, and $R^m(p^\gamma, \varphi, \lambda) \geq \alpha$, by the minimality of α. So $R^m(p_\nu, \varphi, \lambda) \geq \alpha + 1$, a contradiction. Hence $R^m(p_\nu, \varphi, \lambda) = \infty$, so $R^m(\bar{x} = \bar{x}, \varphi, \lambda) = \infty$. As λ was arbitrary, by 1.11 $R^m(\bar{x} = \bar{x}, \varphi, \infty) = \infty$.

LEMMA 2.8: (6) $\Rightarrow$ (7).

Proof. Exercise.

Notation. From now on in this section, $R(p)$ or $R^m(p)$ denotes $R^m(p, \varphi, 2)$.

LEMMA 2.9: (1) *Let p be an arbitrary m-type. $R(p) \geq n$ iff*

$$\Gamma_p(\varphi, n) = \{\psi(\bar{x}_\eta; \bar{a}): \psi(\bar{x}; \bar{a}) \in p,\ \eta \in {}^n 2\}$$
$$\cup \{\varphi(\bar{x}_\eta; \bar{y}_{\eta|k})^{\eta[k]}: \eta \in {}^n 2,\ k < n\}$$

is consistent. (*Note that* $\Gamma_\emptyset(\varphi, n) = \Gamma(\varphi, n) \equiv \Gamma_{\{\bar{x}=\bar{x}\}}(\varphi, n)$.)

(2) *For any m-type p, for all $k, n < \omega$, $R^m(p, \varphi, k) \geq n + 1$ iff*

$$\{\psi(\bar{x}_\nu; \bar{a}): \nu \in {}^{n+1}k,\ \psi(\bar{x}; \bar{a}) \in p\}$$
$$\cup \{\varphi(\bar{x}_\eta; \bar{z}_\nu^{i,j}) \equiv \neg\varphi(\bar{x}_\rho; \bar{z}_\nu^{i,j}): \eta, \rho \in {}^{n+1}k,\ \nu \in {}^{n\geq}k,\ l(\nu) = m < n + 1,$$
$$\nu = \eta \upharpoonright m = \rho \upharpoonright m,\ i = \eta[m],$$
$$j = \rho[m],\ i \neq j\}$$

is consistent. $R^m(p, \varphi, k) = n$, $\mathrm{Mlt}(p, \varphi, k) \geq k_0 < k$ *iff the set of formulas given above is inconsistent, but if the further conditions* $\eta[0] < k_0$, $\rho[0] < k_0$ *are added, the set is consistent.*

(3) *For any m-type p for all $n < \omega$, $R^m(p, \varphi, \aleph_0) \geq n + 1$ iff*

$$\{\psi(\bar{x}_\nu; \bar{a}): \nu \in {}^{n+1}\omega,\ \psi(\bar{x}; \bar{a}) \in p\}$$
$$\cup \{\varphi(\bar{x}_\eta; \bar{z}_\nu^{i,j}) \equiv \neg\varphi(\bar{x}_\rho; \bar{z}_\nu^{i,j}): \eta, \rho \in {}^{n+1}\omega,\ \nu \in {}^{n\geq}\omega,\ l(\nu) = m < n + 1,$$
$$\nu = \eta \upharpoonright m = \rho \upharpoonright m,\ i = \eta[m],$$
$$j = \sigma[m],\ i \neq j\}$$

is consistent. $R^m(p, \varphi, \aleph_0) = n$, $\mathrm{Mlt}(p, \varphi, \aleph_0) \geq k_0 < \aleph_0$ *iff the set of formulas given above is inconsistent, but if the further conditions* $\eta[0] < k_0$, $\rho[0] < k_0$ *are added the set is consistent.*

Proof. (1) First suppose $R(p) \geq n$. By the compactness theorem it is sufficient to show that $\Gamma_p(\varphi, n)$ is consistent when p is finite. Now we define $\bar{a}_\rho$ for $\sigma \in {}^{k>}2$, by induction on $k \leq n$, such that for all η, $\eta \in {}^{k}2$, $R(p_\eta) \geq n - k$ where $p_\eta = p \cup \{\varphi(\bar{x}; \bar{a}_{\eta|j})^{\eta[j]} : j < k\}$. For $k = 0$, $p_{\langle\rangle}$ is just p; so $R(p_{\langle\rangle}) \geq n$. Suppose the definition for $k < n$ is completed, and look at $k + 1$. Let $l(\eta) = k$; then $R(p_\eta) \geq n - k$. Thus there is a sequence $\bar{a}_\eta$ such that if $p_{\eta\frown\langle 0\rangle} = p_\eta \cup \{\varphi(\bar{x}; \bar{a}_\eta)\}$ and $p_{\eta\frown\langle 1\rangle} = p_\eta \cup \{\neg\varphi(\bar{x}; \bar{a}_\eta)\}$, then $R(p_{\eta\frown\langle i\rangle}) \geq n - k - 1$ for $i = 0, 1$. And this is the condition for $k + 1$. Thus when $l(\eta) = n$ we get $R(p_\eta) \geq 0$. Thus p_η is consistent. Let $\bar{c}_\eta$ realize p_η. Taking $\bar{c}_\eta$ for $\bar{x}_\eta$ and $\bar{a}_\eta$ for $\bar{y}_\eta$ we see that $\Gamma_p(\varphi, n)$ is consistent.

The proof of the other direction is similar and is left to the reader (now the induction is downward).

(2) and (3) have a similar proof.

COROLLARY 2.10: (1) (7) $\Rightarrow$ (4).

(2) *$R(p) \geq \omega$ iff $R(p) = \infty$ iff $R^m(p, \varphi, \infty) = \infty$ iff there are $\bar{a}_n$, $n < \omega$, such that for every $k < \omega$ $p \cup \{\varphi(\bar{x}; \bar{a}_n)^{\text{if}\,(k \leq n)} : n < \omega\}$ is consistent.*

Proof. (1) This is because $\Gamma_{\{\bar{x}=\bar{x}\}}(\varphi, n)$ is equivalent to $\Gamma(\varphi, n)$.

This completes the proof of the equivalence of (1) through (7).

(2) $R(p) \geq \omega$ implies $\Gamma_p(\varphi, n)$ is consistent for every n, hence $\Gamma_p(\varphi, \alpha)$ is consistent for every α, so by the same proof as that of 2.7, $R^m(p, \varphi, \infty) = \infty$. Trivially $R^m(p, \varphi, \infty) = \infty$ implies $R^m(p) = \infty$ implies $R^m(p) \geq \omega$. The last "iff" is proved as in 2.3 and 2.4 for finite p, and then by compactness.

LEMMA 2.11: (1) $\Rightarrow$ (8).

Proof. There is an A such that $|S^m_\varphi(A)| > |T| \geq |A|$. But the number of $p \in S^m_\varphi(A)$ which are A-definable is

$$\leq |\{\psi(\bar{y}; \bar{c}) : \psi \in \mathrm{L}(T), \bar{c} \in A\}| \leq |T| \cdot \sum_{n<\omega} |A|^n \leq |T| < |S^m_\varphi(A)|.$$

So some $p \in S^m_\varphi(A)$ is not A-definable.

THEOREM 2.12: (1) *Assume $R(\bar{x} = \bar{x}) < \omega$. Then there is a formula $\psi(\bar{y}; \bar{z})$ such that for every A ($|A| \geq 2$), every $p \in S^m_\varphi(A)$ is (ψ, A)-definable.*

(2) *For every $\theta(\bar{x}; \bar{y})$, $l(\bar{x}) = m$, finite Δ, k, and n there is a $\psi(\bar{y})$ such that: for all $\bar{b}$, $R^m[\theta(\bar{x}; \bar{b}), \Delta, k] \geq n$ iff $\vDash \psi[\bar{b}]$.*

(3) *For every finite* Δ, n, $\varphi(\bar{x}; \bar{y}) \in \Delta$, *and* $\theta(\bar{x}; \bar{z})$ *there is a* $\psi(\bar{y}; \bar{z})$ *such that: if* $\theta(\bar{x}; \bar{c}) \in p$, $R^m(p, \Delta, 2) = R^m[\theta(\bar{x}; \bar{c}), \Delta, 2] = n$, *then* $p \restriction \varphi$ *is* $\psi(\bar{y}; \bar{c})$*-definable.*

Remark. Theorem 2.12(1) proves (9) ⇒ (7). It is obvious that (8) ⇒ (9) so we have (1) ⇒ (8) ⇒ (9) ⇒ (7) and thus (1) through (9) are equivalent, completing the proof of Theorem 2.2.

Proof. (1) Let A be any set, $p \in S^m_\varphi(A)$. We shall show first that p is (ψ, A)-defined for some ψ. By Theorem 1.2 there is a finite $q \subseteq p$ such that $R(q) = R(p) = k$ $(<\omega)$. Assume $q = \{\varphi(\bar{x}; \bar{a}^l)^{\eta[l]}: l < n\}$ where η is an appropriate 0–1 sequence. For every $\bar{a} \in A$ let $q(\bar{a}) = q \cup \{\varphi(\bar{x}; \bar{a})\}$. If $\varphi(\bar{x}; \bar{a}) \in p$, then $q(\bar{a}) \subseteq p$ and so $R[q(\bar{a})] \geq R(p)$. But $q \subseteq q(\bar{a})$ so $R(q) \geq R[q(\bar{a})]$. Thus $R[q(\bar{a})] = k$. On the other hand if $\neg\varphi(\bar{x}; \bar{a}) \in p$, then $R[q \cup \{\neg\varphi(\bar{x}; \bar{a})\}] = k$ and so by Lemma 1.4(1) $R[q(\bar{a})] < k$. Thus we have shown that for any $\bar{a} \in A$, $\varphi(\bar{x}; \bar{a}) \in p$ iff $R[q(\bar{a})] \geq k$. By Lemma 2.9(1), $R[q(\bar{a})] \geq k$ iff $\Gamma_{q(\bar{a})}(\varphi, k)$ is consistent. Let $\theta(\bar{x}; \bar{c}) = \bigwedge q$. Clearly $\Gamma_{q(\bar{a})}(\varphi, k)$ is consistent iff $\vDash\psi[\bar{a}; \bar{c}]$, where

$$\psi[\bar{a}; \bar{c}] = (\exists y_\rho \cdots)_{\rho \in {}^{k>}2} \bigwedge_{\eta \in {}^k 2} (\exists \bar{x})\left[\theta(\bar{x}; \bar{c}) \wedge \varphi(\bar{x}; \bar{a}) \wedge \bigwedge_{l<k} \varphi(\bar{x}; \bar{y}_{\eta|l})^{\eta[l]}\right].$$

Hence $\varphi(\bar{x}; \bar{a}) \in p$ iff $\vDash\psi[\bar{a}; \bar{c}]$, $\bar{c} \in A$, and p is (ψ, A)-definable. Now we show that the choice of ψ depends only on φ, and not on p. In fact it suffices to prove that there is a finite Δ such that every $p \in S^m_\varphi(A)$ is (ψ, A)-definable for some $\psi \in \Delta$. For if $\Delta = \{\psi_i(\bar{y}; \bar{z}^i): i < n\}$ then

$$\begin{aligned}\psi(\bar{y}; \bar{z}) &= \psi(\bar{y}; \bar{z}^0, \ldots, \bar{z}^{n-1}; z, z_0, \ldots, z_{n-1}) \\ &= \bigwedge_{l<n} [z \neq z_l \rightarrow \psi_l(\bar{y}; \bar{z}^l)]\end{aligned}$$

is the required single formula.

By way of contradiction assume there is no such finite Δ. Let P be a new one-place predicate and $b_0, \ldots, b_{m-1}$ new individual constants. Remember $l(\bar{x}) = m$ and let $l(\bar{y}) = n$ in $\varphi(\bar{x}; \bar{y})$. For any (not necessarily finite) set Δ of formulas $\psi(\bar{y}; \bar{z}_\psi)$, let

$$T_\Delta = T \cup \left\{\neg(\exists z_0, \ldots, z_{l(\bar{z}_\psi)-1})\left[\bigwedge_{l=0}^{l(\bar{z}_\psi)-1} P(z_l) \wedge (\forall y_0, \ldots, y_{n-1})\left[\bigwedge_{k=0}^{n-1} P(y_k) \rightarrow \varphi(\bar{b}; \bar{y}) \equiv \psi(\bar{y}; \bar{z}_\psi)\right]\right]: \psi \in \Delta\right\}$$

(where $\bar{b} = \langle b_0, \ldots, b_{m-1}\rangle$).

Now T_Δ is consistent for any finite Δ, since by assumption there are $A, p, p \in S^m_\varphi(A)$ and p not (ψ, A)-definable for any $\psi \in \Delta$. So letting $M \models T$, $A \subseteq |M|$, and $\bar{b}$ realize p we get $(M, A, b_0, \ldots, b_{m-1}) \models T_\Delta$.

Thus T_{Δ_0} is consistent where Δ_0 is the set of all formulas of L of the form $\psi(\bar{y}; \bar{z}_\psi)$ with $l(\bar{y}) = n$. Let $(M, A, b_0, \ldots, b_{m-1}) \models T_{\Delta_0}$ where $A = P^M$. If p is the φ-type $\bar{b}$ realizes over A, then for no ψ is p (ψ, A)-definable, contradiction. This proves (1).

(2) is clear by Lemma 2.9(2) using Lemma 2.1.

(3) is left as an exercise.

THEOREM 2.13: *The following properties of T are equivalent:*

(1) *T is unstable.*

(2) *T is unstable in at least one λ, $\lambda^{|T|} = \lambda$.*

(3) *Some formula $\varphi(x; \bar{y})$ is unstable.*

$(4)_m$ *Some formula $\varphi(\bar{x}; \bar{y})$ is unstable, $l(\bar{x}) = m$.*

(5) *There is a formula $\varphi(\bar{x}; \bar{y})$ and sequences $\{\bar{a}^n : n < \omega\}$, $l(\bar{x}) = l(\bar{y}) = l(\bar{a}^n)$, such that for all $n, k < \omega$, $\models \varphi[\bar{a}^n; \bar{a}^k] \Leftrightarrow n < k$.*

(6) *There is a formula $\varphi = \varphi(\bar{x}^0, \ldots, \bar{x}^{n-1}; \bar{a})$ and an infinite set I of sequences of length $l(\bar{x}^0) = \cdots = l(\bar{x}^{n-1})$ such that φ is connected and antisymmetric over I (see Definition I, 2.5).*

(7) *There is an infinite indiscernible sequence (of sequences) which is not an indiscernible set.*

(8) *There is $n < \omega$, a finite Δ, and A such that for all $n_0 < \omega$ there is a Δ-n-indiscernible sequence I over A which is not a Δ-n-indiscernible set over A, and $|I| \geq n_0$.*

(9) *There is a formula $\varphi = \varphi(\bar{x}^0, \ldots, \bar{x}^{n-1}; \bar{y})$ such that for all $n_0 < \omega$ there is a set I of $\geq n_0$ sequences of length $l(\bar{x}^0) = \cdots = l(\bar{x}^{n-1})$ and a sequence $\bar{a}$ such that $\varphi(\bar{x}^0, \ldots, \bar{x}^{n-1}; \bar{a})$ is connected and antisymmetric over I.*

The proof will be delayed until after Lemma 2.16.

LEMMA 2.14: *For every A, φ, $|S^m(A)| \geq |S^m_\varphi(A)|$. If $\varphi(\bar{x}; \bar{y})$ is unstable in λ, then T is unstable in λ.*

Proof. Exercise.

LEMMA 2.15: (1) *Assume $|S^m_\Delta(A, M)| \geq \lambda$ where $\lambda_i < \lambda$ for $i < |\Delta| \Rightarrow \prod_{i<|\Delta|} \lambda_i < \lambda$. Then there is $\varphi \in \Delta$ such that $|S^m_\varphi(A, M)| \geq \lambda$.*

(2) *If $|S^m_\Delta(A, M)| > \mu^{|\Delta|}$, then there is $\varphi \in \Delta$ such that $|S^m_\varphi(A, M)| > \mu^{|\Delta|}$.*

Proof. (1) Let $\Delta = \{\varphi_i(\bar{x}; \bar{y}^i): i < |\Delta|\}$. The mapping $g: S^m_\Delta(A, M) \to \prod_{i<|\Delta|} S^m_{\varphi_i}(A, M)$ defined by $g(p) = \langle \ldots, p \restriction \varphi_i, \ldots \rangle_{i<|\Delta|}$ is one-one, since if $p \neq q \in S^m_\Delta(A, M)$, then there is $i < |\Delta|$ and $\bar{c} \in A$ such that $\varphi_i(\bar{x}; \bar{c}) \in p \Leftrightarrow \varphi_i(\bar{x}; \bar{c}) \notin q$; thus $g(p) \neq g(q)$. Thus

$$\lambda \leq |S^m_\Delta(A, M)| \leq \Big| \prod_{i<|\Delta|} S^m_{\varphi_i}(A, M) \Big| = \prod_{i<|\Delta|} |S^m_{\varphi_i}(A, M)|.$$

Let $\lambda_i = |S^m_{\varphi_i}(A, M)|$. So $\lambda \leq \prod_{i<|\Delta|} \lambda_i$. By the assumption on λ we can find an i for which $\lambda_i \geq \lambda$, and that is exactly what had to be proved.

(2) Follows from (1) by taking $\lambda = (\mu^{|\Delta|})^+$.

LEMMA 2.16: *Let $I = \{\bar{a}^i: i < i_0\}$ be a Δ-n-indiscernible sequence over A which is not a Δ-n-indiscernible set over A. Then there is $\psi(\bar{x}; \bar{y}, \bar{z}) \in \Delta$, $\bar{c} \in A$ such that for $0 < \alpha \neq \beta$, $\models \psi[\bar{b}^\alpha, \bar{b}^\beta, \bar{b}^0]$ iff $0 < \alpha < \beta$, where $\bar{b}^\alpha = \bar{c}^\frown \bar{a}^{n\alpha \frown} \bar{a}^{n\alpha+1\frown} \ldots {}^\frown \bar{a}^{n\alpha+n-1}$ (and $n(\beta + 1) \leq i_0$).*

Proof. By I, 2.9 there is $\varphi \in \Delta$ and $\bar{c} \in A$ such that $\varphi(\bar{x}^1, \ldots, \bar{x}^n; \bar{c})$ is connected and antisymmetric over I. So by 3.9 of the Appendix it follows that there is such a ψ.

Proof of Theorem 2.13. By definition, (1) $\Rightarrow$ (2).

By Corollary I, 2.2, (2) implies that there is an A for which $|S(A)| > \lambda(=\lambda^{|T|}) \geq |A|$. So (2) $\Rightarrow$ (3) by Lemma 2.15(2). By adding dummy variables, (3) $\Rightarrow$ $(4)_m$. By Lemma 2.14, $(4)_m \Rightarrow$ (1). So (1) through $(4)_m$ are equivalent.

Now we show (3) $\Rightarrow$ (5): (3) implies that φ has the order property, by Theorem 2.2. Hence there are $\bar{a}^n$, $n < \omega$, such that for every $k < \omega$ $p_k = \{\varphi(x; \bar{a}^n)^{\text{if}\,(k \leq n)}: n < \omega\}$ is consistent. Let b_k realize p_k, and set $\bar{c}^n = \langle b_n \rangle^\frown \bar{a}^n$. Then $\models \varphi[b_k, \bar{a}^n]$ iff $k \leq n$. Let $\psi = \psi(x_1, \bar{y}^1; x_2, \bar{y}^2) = \varphi(x_1; \bar{y}^2) \wedge x_1 \neq x_2$. Then $\models \psi[\bar{c}^k; \bar{c}^n]$ iff $k < n$, and this is (5). (5) $\Rightarrow$ $(4)_m$ (for some m): Assume (5) and let $\bar{c}^n = \bar{a}^{2n+1}$. Then for every k $\{\varphi(\bar{x}; \bar{c}^n)^{\text{if}\,(k \leq n)}: n < \omega\}$ is consistent (since it is realized by $\bar{a}^{2k}$). So φ has the order property and thus is unstable by Theorem 2.2.

(5) $\Rightarrow$ (6) is clear.

(6) $\Rightarrow$ (7): By Theorem I, 2.4, for every finite Δ, $n(\varphi \in \Delta)$, the infinite set mentioned in (6) has a subset $\{\bar{a}^i: i < \omega\}$ which is a Δ-n-indiscernible sequence over $\bar{a}$. This is of course not a Δ-n-indiscernible *set* over $\bar{a}$ (by (6)). Thus by compactness we get (7).

(7) $\Rightarrow$ (5) by Lemma 2.16.

(6) ⇒ (9) and (7) ⇒ (8) are obvious and (9) ⇒ (6) and (8) ⇒ (7) follow by compactness. This completes the proof of the theorem.

THEOREM 2.17: *Let T be stable and Δ be finite. Then there is a finite Δ^* such that if for $\beta < \alpha$ p_β is the Δ^*-m-type that $\bar{a}^\beta$ realizes over $A_\beta = A \cup \bigcup \{\bar{a}^\gamma: \gamma < \beta\}$, $p_0 \subset p_\beta$, and for every $\varphi \in \Delta^*$ $R^m(p_\beta \restriction \varphi) = R^m(p_0 \restriction \varphi)$ $(m = l(\bar{a}^\beta))$ and p_β is $(\psi_\varphi(\bar{y}, \bar{z}); A)$-definable, ψ_φ from 2.12; then $\{\bar{a}^\beta: \beta < \alpha\}$ is a Δ-indiscernible sequence over A.*

Proof. Since Δ is finite, there is an $n < \omega$ such that any sequence $\{\bar{b}^\gamma: \gamma < \beta\}$ is Δ-indiscernible over a set B iff it is Δ-n_1-indiscernible over B for all $n_1 \le n$.

Now define a sequence of finite sets $\Delta_n, \Delta_{n-1}, \ldots, \Delta_0$ by downward induction such that (1) $\Delta \subseteq \Delta_i$, (2) each Δ_i is closed under permutations of variables, and (3) if $\varphi = \varphi(\bar{x}^0, \ldots, \bar{x}^{i-1}; \bar{y}) \in \Delta_i$, $l(\bar{x}^0) = \cdots = l(\bar{x}^{i-1}) = m$, then every φ-m-type p such that $R^m(p) = R^m(p \restriction A)$ is (ψ, A)-definable for some $\psi = \psi(\bar{x}^1, \ldots, \bar{x}^{i-1}, \bar{y}; \bar{z}) \in \Delta_{i-1}$. (3) is accomplished by Theorem 2.12(1)–(3). Now choose $\Delta^* = \bigcup_{i \le n} \Delta_i$. So in order to prove the conclusion of the theorem it is sufficient to show that $\{\bar{a}^\beta: \beta < \alpha\}$ is Δ_i-i-indiscernible over A for all $i \le n$, since $\Delta \subseteq \Delta_i$. Now if $\gamma < \beta < \alpha$, then for every $\varphi \in \Delta^*$, $p_0 \restriction \varphi \subset p_\gamma \restriction \varphi$, $p_\beta \restriction \varphi$ and $R(p_\gamma \restriction \varphi) = R(p_0 \restriction \varphi) = R(p_\beta \restriction \varphi)$. By Corollary 1.5, $p_0 \restriction \varphi$ has a unique extension in $S^m_\varphi(A_\gamma)$ of equal rank, so since $p_\gamma \restriction \varphi, p_\beta \restriction \varphi \restriction A_\gamma \in S^m_\varphi(A_\gamma)$ and $R(p_0 \restriction \varphi) \ge R[(p_\beta \restriction \varphi) \restriction A_\gamma] \ge R(p_\beta \restriction \varphi) = R(p_0 \restriction \varphi)$ we have $p_\gamma \restriction \varphi = (p_\beta \restriction \varphi) \restriction A_\gamma$ for every $\varphi \in \Delta^*$, $\gamma < \beta < \alpha$. So by the definition of Δ^*, $\mathrm{tp}(\bar{a}^\gamma, A_\gamma) \restriction \Delta_i \subseteq \mathrm{tp}(\bar{a}^\beta, A_\beta) \restriction \Delta_i$ for all $\gamma < \beta < \alpha$, $i \le n$. It also follows from (3) that $\mathrm{tp}(\bar{a}^\beta, A_\beta)$ does not (Δ_i, Δ_{i+1})-split over A for $i < n$, $\beta < \alpha$ so the result follows by Lemma I, 2.5.

COROLLARY 2.18: *If T is stable and for every $\beta < \alpha \ge \omega$ p_β is the m-type $\bar{a}^\beta$ realizes over $A_\beta = A \cup \bigcup \{\bar{a}^\gamma: \gamma < \beta\}$, for every φ $R^m(p_\beta \restriction \varphi) = R^m(p_0 \restriction \varphi)$, and $p_0 \subseteq p_\beta$, then $\{\bar{a}^\beta: \beta < \alpha\}$ is an indiscernible set over A.*

Remark. We can use many other parallel conditions in place of "for every φ, $R^m(p_\beta \restriction \varphi) = R^m(p_0 \restriction \varphi)$", e.g., $R^m(p_\beta, \mathrm{L}, 2) = R^m(p_0, \mathrm{L}, 2)$.

Proof. By the previous theorem it is an indiscernible sequence over A. So by Theorem 2.13(7) it is an indiscernible set over A.

THEOREM 2.19: *Assume T is stable, Δ finite, λ regular, I a set of sequences of length m, $|A| < \lambda \le |I|$. Then there is $J \subseteq I$, $|J| = \lambda$, which is a Δ-indiscernible set over A.*

Proof. Let Δ^* be as in Theorem 2.17. Among the Δ^*-m-types p, $|p| < \lambda$, which are realized by at least λ sequences from I, choose one p_0 with minimal $\langle R^m(p, \varphi_j, 2)\colon j < n_0\rangle$ (ordered lexicographically, where $\Delta^* = \{\varphi_j\colon j < n_0\}$). We can assume that p_0 is a type over $A_0 \supseteq A$, $|A_0| < \lambda$. Now we shall define $\bar{a}^i \in I$ for $i < \lambda$. Let $\bar{a}^j$, $j < i < \lambda$ be defined, $A_i = A_0 \cup \bigcup \{\bar{a}^j\colon j < i\}$. $|S^m_{\Delta^*}(A_i)| < \lambda$ since T is stable and $|A_i| < \lambda$. Thus there is $p_i \in S^m_{\Delta^*}(A_i)$, $p_0 \subseteq p_i$, and p_i is realized by λ sequences from I. Clearly for $j < n_0$, $R^m(p_i, \varphi_j, 2) = R^m(p_0, \varphi_j, 2)$. Choose $\bar{a}^i \in I - {}^m(A_i)$ realizing p_i.

By 2.17 and 2.13(8), $J = \{\bar{a}^i\colon i < \lambda\}$ is a Δ-indiscernible set.

If $\lambda = \aleph_0$, the claim follows from Theorem I, 2.4(1) and the fact that as Δ is finite every set of sequences is a Δ-n-indiscernible set for all sufficiently large n.

LEMMA 2.20: *Suppose $\varphi(\bar{x}; \bar{y})$ is stable. Then there are finite Δ, n such that if I is a Δ-n-indiscernible set of sequences of length $l(\bar{y})$ and $\bar{a}$ is any sequence of length $l(\bar{x})$, then*

$$|\{\bar{c} \in I\colon \vDash \varphi[\bar{a}; \bar{c}]\}| < n \quad \text{or} \quad |\{\bar{c} \in I\colon \vDash \neg\varphi[\bar{a}; \bar{c}]\}| < n.$$

Proof. By contradiction. Assume that the conclusion is false. Then by compactness there are an $\bar{a}$ and an indiscernible set I such that $\{\bar{c} \in I\colon \vDash \varphi[\bar{a}; \bar{c}]^{t}\}$ is infinite for $t = 0, 1$. Let $\{\bar{c}^n\}_{n<\omega} \subseteq I$. For every $w \subseteq \omega$, $\{\varphi(\bar{x}; \bar{c}^n)^{\text{if}\,(n \in w)}\colon n < \omega\}$ is consistent (since I is indiscernible). Thus

$$\left| S^m_\varphi\Big(\bigcup_{n<\omega} \bar{c}^n\Big)\right| \geq 2^{\aleph_0} > \aleph_0 = \left|\bigcup_{n<\omega} \bar{c}^n\right| \quad (m = l(\bar{x})).$$

This contradicts the stability of φ.

EXERCISE 2.1: Show that in Lemma 2.20, for big enough Δ,

(1) we can choose $n = R^m(\bar{x} = \bar{x}, \varphi, 2) + 1$,

(2) we can choose $n = 2^{k+1}$ where $k = R^m(\bar{x} = \bar{x}, \psi, 2)$ and $\psi(\bar{x}; \bar{y}) = \varphi(\bar{y}; \bar{x})$,

(3) when I is infinite we can choose $n = R^m(\bar{x} = \bar{x}, \varphi, \infty)$.

EXERCISE 2.2: Show that Exercise 2.1 cannot be essentially improved. That is for some stable (even $\aleph_0$-stable) theory T, infinite indiscernible set I, φ and $\bar{c}$,

$$|\{\bar{a} \in I\colon \vDash \varphi[\bar{c}; \bar{a}]^{t}\}| \geq n \quad \text{for } t \in \{0, 1\},$$

but $R^m(\bar{x} = \bar{x}, \varphi, \lambda) \le n$ for every $\lambda \ge 2$ and $R^m(\bar{x} = \bar{x}, \psi, 2) = [\log_2 n]$, $R^m(\bar{x} = \bar{x}, \psi, \lambda) = 1$ for $\lambda \ge \aleph_0$. (Hint: Let $T = \mathrm{Th}(M)$, $|M| = \omega \cup \{w \subseteq \omega : |w| = n\}$, $P = \omega$, $R^M = \{\langle n, w\rangle : n \in w\}$ and $\varphi(x, y) = R(x, y)$, $I = \omega$.)

EXERCISE 2.3: Let $(I, \le)$ be an order; $G^I = \{E_s : s \in I\}$. We define the L^I model M^I as follows:

$$|M^I| = \{f : f : I \to \omega\}; \qquad E_s^{M^I} = \{\langle f, g\rangle : (\forall t \le s) f(t) = g(t)\};$$
$$T^I = \mathrm{Th}(M^I).$$

(1) Show that T^I admits elimination of quantifiers.

(2) Let $\kappa(I) = \inf\{\kappa : I$ contains no increasing sequence of length $\kappa\}$. Show that T^I is stable in λ iff $\lambda = |I| + \lambda^{<\kappa}$.

Compare this with the example after Definition 1.2. There, $(I, <)$ is α^*, i.e., $(\alpha, \ni)$.

EXERCISE 2.4: Suppose M is κ-compact, p is an m-type in M, $|p| < \kappa$, $\bar{a}_l \in |M|$, and $\varphi_l \in L(M)$, $l = 1, 2$. If for all $\bar{c} \in |M|$ realizing p, $\models \varphi_1[\bar{c}; \bar{a}_1] \Leftrightarrow \models \varphi_2[\bar{c}; \bar{a}_2]$, then $\models \varphi_1[\bar{c}; \bar{a}_1] \Leftrightarrow \models \varphi_2[\bar{c}; \bar{a}_2]$ for all $\bar{c} \in \mathfrak{C}$ realizing p.

EXERCISE 2.5: Relativize Theorem 2.2 to a formula $\theta(\bar{x}; \bar{a})$ (i.e., in (1) and (2), replace "φ stable in λ" by

"$|A| \le \lambda$ implies $|\{p \in S_\varphi^m(A) : p \cup \{\theta(\bar{x}; \bar{a})$ is consistent$\}| \le \lambda$",

in (3), add $\{\theta(\bar{x}; \bar{a})\}$; in (4) and (5), add $\{\theta(\bar{x}_\eta; \bar{a}) : \eta \in {}^{\alpha}2\}$ to Γ; in (6) and (7), replace $\bar{x} = \bar{x}$ by $\theta(\bar{x}; \bar{a})$; and in (8) and (9) demand $\theta(\bar{x}; \bar{a}) \in p$.

EXERCISE 2.6: Do the same with Theorem 2.13.

EXERCISE 2.7: For any finite Δ, n, and $\theta(\bar{x}; \bar{y})$ there is an $l(\bar{y})$-type q, such that for every $\bar{c}$

$$R^m[\theta(\bar{x}; \bar{c}), \Delta, \aleph_0] \ge n \quad \text{iff} \quad \bar{c} \text{ realizes } q.$$

(Hint: Use 2.9(3).)

EXERCISE 2.8: Let $\varphi(\bar{x}; \bar{y}) = \psi(\bar{y}; \bar{x})$; and show that $\varphi(\bar{x}; \bar{y})$ is stable iff $\psi(\bar{y}; \bar{x})$ is stable. (Hint: Use 2.2(3).)

EXERCISE 2.9: If T is stable, $p \in S_\Delta^m(|M|)$, $M \subseteq A$, then there is a unique $p \subseteq q \in S_\Delta^m(A)$ such that for every φ, $R(p \restriction \varphi) = R(q \restriction \varphi)$, and if p is $\psi(x; \bar{c})$-defined, $\bar{c} \in |M|$, then so is q.

II.3. Ranks, degrees and superstability

THEOREM 3.1: *For a finite m-type p the following are equivalent:*

(1) $R^m(p, \Delta, 2) \geq |\Delta|^+ + \aleph_0$,

(2) $R^m(p, \Delta, \lambda) = \infty$ *where* $\lambda = (2^{\aleph_0})^+$,

(3) *There are* $\varphi_\eta(\bar{x}; \bar{y}_\eta) \in \Delta$ *and* $\bar{a}_\eta$ *for* $\eta \in {}^{\omega>}2$ *such that for every* $\eta \in {}^{\omega\geq}2$, $p \cup p_\eta$ $[p_\eta = \{\varphi_{\eta|n}(\bar{x}; \bar{a}_{\eta|n})^{\eta[n]}: n < l(\eta)\}]$ *is consistent.*

Remark. From 3.1 it follows that (1) and (2) are equivalent also for infinite p (by 1.2). If in (2) we replace λ by $\aleph_0$, the equivalence is still correct, but the claim is stronger with λ. If we replace λ by $\lambda_1 > \lambda$, the implication (1) $\Rightarrow$ (2) is not correct.

DEFINITION 3.1. If $R^m(\bar{x} = \bar{x}, \mathrm{L}, 2) < \infty$, T is called *totally transcendental*.

Proof of 3.1. (2) $\Rightarrow$ (1) is trivial by 1.3(1).

(3) $\Rightarrow$ (2) We shall prove by induction on α that $R^m(p \cup p_\eta, \Delta, \lambda) \geq \alpha$ for every $\eta \in {}^{\omega>}2$. As $p_{\langle\rangle} = p$, this suffices. For $\alpha = 0$ it follows by the consistency of $p_\eta \cup p$; and for α a limit ordinal from the induction hypothesis. So let $\alpha = \beta + 1$, $\eta \in {}^{\omega>}2$; then $\{p_\nu: \eta \lessdot \nu \in {}^{\omega}2\}$ is a family of $2^{\aleph_0}$ Δ-m-types which are pairwise explicitly contradictory. By 1.1(1) and the induction hypothesis $R^m(p_\nu \cup p, \Delta, \lambda) \geq \beta$ hence $R^m(p_\eta \cup p, \Delta, \lambda) \geq \beta + 1 = \alpha$.

(1) $\Rightarrow$ (3) We shall define by induction on $n < \omega$ subsets U_n of $|\Delta|^+ + \aleph_0$ of cardinality $|\Delta|^+ + \aleph_0$, formulas $\varphi_\eta \in \Delta$ for $\eta \in {}^{n>}2$ and sequences $\bar{a}_\eta^{i,n}$ for $\eta \in {}^{n>}2$, $i \in U_n$, such that $R^m(p_\eta^i, \Delta, 2) \geq i$ for $i \in U_n$, $\eta \in {}^n2$ where $p_\eta^i = p \cup \{\varphi_{\eta|k}(\bar{x}; \bar{a}_{\eta|k}^{i,n})^{\eta[k]}: k < l(\eta)\}$.

For $n = 0$ let $U_0 = \{\alpha: \alpha < |\Delta|^+ + \aleph_0\}$ and $p_{\langle\rangle}^i = p$, so by (1) the induction hypothesis holds. For $n + 1$, for every $i < |\Delta|^+ + \aleph_0$ choose $j = j(i)$, $i < j \in U_n$, so $R^m(p_\eta^j, \Delta, 2) > i$ for $\eta \in {}^n2$, hence for some $\varphi_\eta^i \in \Delta$ and $\bar{a}_\eta^i$, $R^m[p_\eta^j \cup \{\varphi_\eta^i(\bar{x}; \bar{a}_\eta^i)^t\}, \Delta, 2] \geq i$, $t \in \{0, 1\}$. Hence for some $U_{n+1} \subseteq \{\alpha: \alpha < |\Delta|^+ + \aleph_0\}$, $|U_{n+1}| = |\Delta|^+ + \aleph_0$, for every $\alpha, \beta \in U_{n+1}$, $\eta \in {}^n2$, $\varphi_\eta^\alpha = \varphi_\eta^\beta$. Let for $\eta \in {}^n2$, $\varphi_\eta = \varphi_\eta^\alpha$ (for any $\alpha \in U_{n+1}$) and for $i \in U_{n+1}$, $\bar{a}_\eta^{i,n+1}$ is $\bar{a}_\eta^{j(i),n}$ when $\eta \in {}^{n>}2$ and $\bar{a}_\eta^i$ when $\eta \in {}^n2$. Hence

$$\{\varphi(\bar{x}_\eta; \bar{a}): \varphi(\bar{x}; \bar{a}) \in p,\ \eta \in {}^{\omega}2\} \cup \{\varphi_{\eta|n}(\bar{x}_\eta; \bar{y}_{\eta|n})^{\eta[n]}: \eta \in {}^{\omega}2,\ n < \omega\}$$

is consistent so some assignment $\bar{x}_\eta \mapsto \bar{b}_\eta$, $\bar{y}_\nu \mapsto \bar{a}_\nu$ realizes it. These $\bar{a}_\nu$ show that (3) holds.

THEOREM 3.2: *For* $|\Delta| < 2^{\aleph_0}$, $m < \omega$ *the following are equivalent* (*for* $|\Delta| \geq 2^{\aleph_0}$ *still* $(2)_m \Rightarrow (1)_m \Rightarrow (3)_m$):

$(1)_m$ $R^m[\bar{x} = \bar{x}, \Delta, 2] = \infty$,

$(2)_m$ *for some* A, $|S^m_\Delta(A)| > |A| + |\Delta| + \aleph_0$,

$(3)_m$ *for some* A, $|A| \leq \aleph_0$, $|S^m_\Delta(A)| \geq 2^{\aleph_0}$.

Remark. For $\Delta = L$ these conditions are equivalent for different m's by I, 2.1, i.e., $(1)_l \Leftrightarrow (2)_m \Leftrightarrow (3)_n$ for all l, m, n.

Proof. (3) $\Rightarrow$ (2) trivially, by the same A.

(2) $\Rightarrow$ (1) Suppose not (1). For every $p \in S^m_\Delta(A)$ there is a finite $q_p \subseteq p$ such that $R^m(p, \Delta, 2) = R^m(q_p, \Delta, 2) < \infty$ (by not (1)). By 1.5, $p(1) \neq p(2) \Rightarrow q_{p(1)} \neq q_{p(2)}$ hence

$$|S^m_\Delta(A)| \leq |\{q_p : p \in S^m_\Delta(A)\}| \leq |\{q : q \text{ a finite } \Delta\text{-}m\text{-type over } A\}|$$
$$\leq |A| + |\Delta| + \aleph_0,$$

a contradiction.

(1) $\Rightarrow$ (3) By 3.1 we know that (1) implies condition (3) from 3.1. Letting $A = \bigcup \{\bar{a}_\eta : \eta \in {}^{\omega>}2\}$, and q_η be a type in $S^m_\Delta(A)$ extending p_η [we use the notation of 3.1(3)] we see

$$|S^m_\Delta(A)| \geq |\{q_\eta : \eta \in {}^\omega 2\}| = 2^{\aleph_0}$$

because if $\eta \neq \nu \in {}^\omega 2$, let $n = \min\{k : \eta[k] \neq \nu[k]\}$. Then

$$\varphi_{\eta|k}(\bar{x}; \bar{a}_{\eta|k})^{\eta[k]} \in p_\eta, \qquad \neg\varphi_{\eta|k}(\bar{x}; \bar{a}_{\eta|k})^{\eta[k]} = \varphi_{\nu|k}(\bar{x}; \bar{a}_{\nu|k})^{\nu[k]} \in q_\nu.$$

On the other hand, $|A| \leq |{}^{\omega>}2| \cdot \aleph_0 = \aleph_0$.

CONCLUSION 3.3: (1) *If* $|T| < 2^{\aleph_0}$, T *is stable in* $|T|$ (*or in some* $\lambda < 2^{\aleph_0}$), *then* T *is stable in every cardinality* $\mu \geq |T|$.

(2) *A totally transcendental* T *is stable in every* $\lambda \geq |T|$.

DEFINITION 3.2: We define the *degree* $D^m(p, \Delta, \lambda)$ (p, Δ, λ are as in Definition 1.1) by defining inductively when $D^m(p, \Delta, \lambda) \geq \alpha$ (α an ordinal):

(1) $D^m(p, \Delta, \lambda) \geq 0$ when p is a (consistent) type (when p is an inconsistent set of m-formulas, we let the degree be -1).

(2) $D^m(p, \Delta, \lambda) \geq \delta$ (δ a limit ordinal) if $D^m(p, \Delta, \lambda) \geq \alpha$ for all $\alpha < \delta$.

(3) $D^m(p, \Delta, \lambda) \geq \alpha + 1$ if for all $\mu < \lambda$ and all finite $r \subseteq p$ there are, a finite $q \supseteq r$, $n < \omega$, $\psi(\bar{x}; \bar{y}) \in \Delta$, and sequences $\bar{a}_i$, $i \leq \mu$ such that:

(i) $D^m(q \cup \{\psi(\bar{x}; \bar{a}_i\}, \Delta, \lambda) \geq \alpha$,

(ii) $\{\psi(\bar{x}; \bar{a}_i): i \leq \mu\}$ is n-contradictory (or n-inconsistent) over q, i.e., for every $w \subseteq \mu$, $|w| = n$, $\vdash \neg(\exists \bar{x})(\bigwedge_{i \in w} \psi(\bar{x}; \bar{a}_i) \wedge \bigwedge q)$.

(4) Exactly as (4) of Definition 1.1.

Remark. We ignore the degree for $\lambda < \aleph_1$, and will be mainly interested in the case $\lambda > |T|^+$. We can always replace q by $\bigwedge q$.

LEMMA 3.4: (1) *For every p, Δ, λ there is a finite $q \subseteq p$ such that $D^m(p, \Delta, \lambda) = D^m(q, \Delta, \lambda)$.*

(2) *If $p_1 \vdash p_2, \Delta_1 \subseteq \Delta_2, \lambda_1 \geq \lambda_2$, then $D^m(p_1, \Delta_1, \lambda_1) \leq D^m(p_2, \Delta_2, \lambda_2)$.*

(3) *When $\lambda \geq \aleph_0$,*

$$D^m\Big(p \cup \Big\{\bigvee_{i<n} \varphi_i(\bar{x}; \bar{a}_i)\Big\}, \Delta, \lambda\Big) = \max_{i<n} D^m(p \cup \{\varphi_i(\bar{x}; \bar{a}_i)\}, \Delta, \lambda).$$

(4) *If p is an m-type over A, $\lambda \geq \aleph_0$, Δ is given, then there is $q, p \subseteq q \in S^m(A)$ such that*

$$D^m(p, \Delta, \lambda) = D^m(q, \Delta, \lambda).$$

Proof. The proof of (2) is like 1.1(1), 1.3(1); the proof of (1) is like 1.2; the proof of (3) is like 1.7 and the proof of (4) like 1.6.

LEMMA 3.5: (1) *If $X \in \{R, D\}$, p a finite type and $\infty > X^m(p, \Delta, \lambda) \geq \alpha$, then for some Δ-m-type q, $X^m(p \cup q, \Delta, \lambda) = \alpha$.*

(2) *If $X \in \{R, D\}$ $\alpha = (2^{|T|})^+$, $X^m(p, \Delta, \lambda) \geq \alpha$, then $X^m(p, \Delta, \lambda) = \infty$. If $X = R$ we need only $\alpha = (2^{|\Delta| + \aleph_0})^+$.*

Proof. (1) We suppose there is no such q, and prove by induction on $\beta \geq \alpha$ that $q = p \cup q_1$, q_1 a Δ-m-type, $X^m(q, \Delta, \lambda) \geq \alpha$ implies $X^m(q, \Delta, \lambda) \geq \beta$.

For $\beta = \alpha$ this is the assumption, for β a limit ordinal it follows by induction. For $\beta = \gamma + 1 > \alpha$, and e.g., $X = D$, let $r_1 \subseteq q$ be finite and $\mu < \lambda$. Then there are a finite $r \supseteq r_1$ and formulas $\psi(\bar{x}; \bar{a}_i)$ ($i \leq \mu$, $\psi \in \Delta$) n-contradictory over $p \cup r$ such that

$$D^m[p \cup r \cup \{\psi(\bar{x}; \bar{a}_i)\}, \Delta, \lambda] \geq \alpha$$

(they exist as $D^m(p \cup q, \Delta, \lambda) \geq \alpha$ implies $D^m(p \cup q, \Delta, \lambda) \geq \alpha + 1$). So by the induction hypothesis

$$D^m[p \cup r \cup \{\psi(\bar{x}; \bar{a}_i)\}, \Delta, \lambda] \geq \gamma.$$

As this holds for any finite r_1, $D^m(q, \Delta, \lambda) \geq \beta$. Thus $D^m(p, \Delta, \lambda) = \infty$, a contradiction. The proof for $X = R$ is similar.

(2) By 1.2 and 3.4(1) we can assume p is finite; if $X = R$ we can also assume $|T| = |\Delta| + \aleph_0$. If (2) does not hold, then by part (1) there are finite p_i ($i < \alpha$) such that $X^m(p_i, \Delta, \lambda) = i$. By 1.1(1) and 3.4(2) we can assume $p_i = \{\theta_i(\bar{x}; \bar{a}_i)\}$. By the definition of α, for some $i < j < \alpha$, $\theta_i = \theta_j$ and $\mathrm{tp}(\bar{a}_i) = \mathrm{tp}(\bar{a}_j)$; but $X^m(p_i, \Delta, \lambda) = i < j = X^m(p_j, \Delta, \lambda)$, a contradiction.

DEFINITION 3.3: Let $ds(\alpha)$ be the set of strictly descending nonempty sequences of ordinals $< \alpha$. A (Δ, α)-*function* is a function

$$h : ds(\alpha) \to \{\langle \varphi(\bar{x}, \bar{z}), \psi(\bar{x}; \bar{y}), n \rangle : \varphi \in \mathrm{L}, \psi \in \Delta, n < \omega, l(\bar{x}) = m\};$$

say $h(\eta) = \langle \varphi_\eta, \psi_\eta, n_\eta \rangle$ (for some fixed m).

Let $\theta = \theta(\bar{x}; \bar{a})$ be an m-formula, μ a cardinal, h a (Δ, α)-function and $U \subseteq ds(\alpha)$. Define $\Gamma^U_\mu(\theta, h)$ to be the following set of formulas containing free variables of the form $\bar{y}_{\eta,\nu}, \bar{z}_{\eta,\nu}$:

$$\Gamma^U_\mu(\theta, h) = \Big\{(\exists \bar{x})\Big[\theta(\bar{x}; \bar{a}) \wedge \bigwedge_{0<l\leq l(\eta)} (\psi_{\eta|l}(\bar{x}; \bar{y}_{\eta|l,\nu|l}) \wedge \varphi_{\eta|l}(\bar{x}; \bar{z}_{\eta|l,\nu|(l-1)}))\Big] :$$
$$\eta = \langle\ \rangle \text{ or } \eta \in U; \text{ and } \nu \in {}^k\mu \quad \text{where } k = l(\eta)\Big\}$$
$$\cup \Big\{\neg(\exists \bar{x})\Big[\bigwedge_{i\in w} \psi_\eta(\bar{x}; \bar{y}_{\eta,\nu\frown\langle i\rangle}) \wedge \varphi_\eta(\bar{x}; \bar{z}_{\eta,\nu})\Big] :$$
$$\eta \in U, w \subseteq \mu, |w| = n_\eta, \nu \in {}^k\mu \quad \text{where } k = l(\eta) - 1\Big\}.$$

If $U = ds(\alpha)$, we write $\Gamma^*_\mu(\theta, h)$.

THEOREM 3.6: *Let* $|T| + |\alpha| < \mathrm{cf}\,\mu$. *Then* $D^m[\theta(\bar{x}; \bar{a}), \Delta, \mu^+] \geq \alpha$ *iff there is a* (Δ, α)-*function* h *such that* $\Gamma^*_\mu(\theta(\bar{x}; \bar{a}), h)$ *is consistent.*

Remark. For the if part no restriction on μ is needed.

Proof. ($\Leftarrow$) Suppose $\Gamma^*_\mu(\theta(\bar{x}; \bar{a}), h)$ is consistent, and let it be realized by the assignment $\bar{y}_{\eta,\nu} \mapsto \bar{b}_{\eta,\nu}$, $\bar{z}_{\eta,\nu} \mapsto \bar{c}_{\eta,\nu}$. For $\eta \in ds(\alpha) \cup \{\langle\ \rangle\}$ and $\nu \in {}^k\mu$, where $k = l(\eta)$, let

$$p_{\eta,\nu} = \{\theta(\bar{x}; \bar{a})\} \cup \{\psi_{\eta|l}(\bar{x}; \bar{b}_{\eta|l,\nu|l}) : 0 < l \leq l(\eta)\}$$
$$\cup \{\varphi_{\eta|l}(\bar{x}; \bar{c}_{\eta|l,\nu|(l-1)}) : 0 < l \leq l(\eta)\}.$$

We can easily prove by induction on β that $D^m(p_{\eta,\nu}, \Delta, \mu^+) \geq \beta$ where $\beta \leq \eta[l(\eta) - 1]$ for $l(\eta) > 0$ and $D^m(p_{\langle\rangle,\langle\rangle}, \Delta, \mu^+) \geq \alpha$. As $p_{\langle\rangle,\langle\rangle} = \{\theta(\bar{x}; \bar{a})\}$, this proves one direction.

($\Rightarrow$) We prove by induction on α, for all $\theta(\bar{x}; \bar{a})$ at once.

$\alpha = 0$ h is the empty function.

$\alpha = \delta$ (a limit ordinal). So by the induction hypothesis, for every $\beta < \alpha$ there is a (Δ, β)-function h_β for which $\Gamma^*_\mu(\theta(\bar{x}; \bar{a}), h_\beta)$ is consistent.

For $\nu \in ds(\alpha)$ define $h(\nu) = h_{\nu[0]+1}(\nu)$. It is easy to check that a combination of the assignments realizing the $\Gamma^*_\mu(\theta(\bar{x}; \bar{a}), h_\beta)$ realizes $\Gamma^*_\mu(\theta(\bar{x}; \bar{a}), h)$.

$\alpha = \beta + 1$ By the definition there are $\varphi(\bar{x}; \bar{c})$, $n < \omega$ and Δ-m-formulas $\psi(\bar{x}; \bar{b}_i)$ $(i \leq \mu)$ n-contradictory over $\varphi(\bar{x}; \bar{c})$ such that $D^m[\varphi(\bar{x}; \bar{c}) \wedge \psi(\bar{x}; \bar{b}_i), \Delta, \mu^+] \geq \beta$ and $\varphi(\bar{x}; \bar{c}) \vdash \theta(\bar{x}; \bar{a})$. Hence, by the induction hypothesis for every $i \leq \mu$ there is a (Δ, β)-function h_i such that $\Gamma^*_\mu[\varphi(\bar{x}; \bar{c}) \wedge \psi(\bar{x}; \bar{b}_i), h_i]$ is consistent. Define (Δ, α)-functions h^i (for $i \leq \mu$) as follows: If $\eta \in ds(\beta)$, $h^i(\eta) = h_i(\eta)$, if $\eta = \langle\beta\rangle$, $h^i(\eta) = \langle\varphi, \psi, n\rangle$ and if $\eta = \langle\beta\rangle^\frown\nu$, $\nu \in ds(\beta)$, then $h^i(\eta) = h_i(\nu)$ (clearly this exhausts all possibilities).

If for some $i \leq \mu$, $\Gamma^*_\mu(\theta(\bar{x}; \bar{a}), h^i)$ is consistent, we are through. Otherwise, for each i some finite subset of it is inconsistent, hence for some finite $u(i) \subseteq ds(\alpha)$ $\Gamma^{u(i)}_\mu(\theta(\bar{x}; \bar{a}), h^i)$ is inconsistent. The number of possible $u(i)$ is $\leq |\alpha| + \aleph_0$, and for each finite u the number of possible $h^i \restriction u$ is $\leq |T|$. As cf $\mu > |T| + |\alpha|$, there is $V \subseteq \mu$, $|V| = \mu$ such that for all $i \in V$, $u(i) = u$, $h^i \restriction u = h^* \restriction u$. W.l.o.g. u is closed undertaking non-empty initial segments. Now using the $\psi(\bar{x}; \bar{b}_i)$ $(i \in V)$ and $\varphi(\bar{x}; \bar{c})$ we can easily show that $\Gamma^u_\mu(\theta(\bar{x}; \bar{a}), h^*)$ is realized, a contradiction.

LEMMA 3.7: (1) *If* $\mu = |T|^+$ *and* $D^m(p, \Delta, \mu^+) \geq \mu$, *then* $D^m(p, \Delta, \mu^+) = \infty$.

(2) *If, in addition,* p *is finite, then there are* φ_k, ψ_k, n_k $(0 < k < \omega)$ *such that for all* α, $\Gamma^*_\mu(\bigwedge p, h_\alpha)$ *is consistent, where* $h_\alpha(\eta) = \langle\varphi_{l(\eta)}, \psi_{l(\eta)}, n_{l(\eta)}\rangle$, h_α *a* (Δ, α)*-function.*

(3) *For finite* p, $D^m(p, \Delta, \mu^+) \geq \mu$ *iff there are* φ_k, ψ_k, n_k $(0 < k < \omega)$ *and* $\bar{a}_\eta, \bar{c}_\eta$ $(\eta \in {}^{\omega>}\mu)$ *such that* $\{\psi_k(\bar{x}; \bar{a}_{\eta^\frown\langle i\rangle}): i < \mu,\ l(\eta) = k - 1\}$ *is* n_k*-contradictory over* $\varphi_k(\bar{x}; \bar{c}_\eta)$ *and*

$$p \cup \{\psi_k(\bar{x}; \bar{a}_{\eta|k}) \wedge \varphi_k(\bar{x}; \bar{c}_{\eta|k-1}): 0 < k < \omega\}$$

is consistent for $\eta \in {}^\omega\mu$.

Proof. (1) It suffices to prove (1) for finite p, so by the remark to 3.6 it suffices to prove (2). Notice that 3.6 holds when $|T| < \mathrm{cf}\,\mu = \alpha$.

(2), (3) By 3.6, for some (Δ, μ)-function h, $\Gamma^*_\mu(\bigwedge p, h)$ is consistent. Now we define for $i < \mu$, $0 < k < \omega$; $\varphi_k, \psi_k, n_k, \eta^i_k$ such that $\eta^i_k \in ds(\mu)$, $l(\eta^i_k) = k$, $\eta^i_k[k-1] > i$, $h(\eta^i_k \restriction l) = \langle \varphi_l, \psi_l, n_l \rangle$ for $1 \le l \le k$. For $k = 1$ there are φ_1, ψ_1, n_1, such that $|\{\alpha < \mu: h(\langle\alpha\rangle) = \langle\varphi_1, \psi_1, n_1\rangle\}| = \mu$, and for $i < \mu$ choose $\eta^i_1 = \langle\alpha\rangle$ such that $\alpha > i$, and $h(\langle\alpha\rangle) = \langle\varphi_1, \psi_1, n_1\rangle$. If we have defined for k, there are $\varphi_{k+1}, \psi_{k+1}, n_{k+1}$ such that

$$|\{\alpha < \mu: h(\eta^\alpha_k{}^\frown\langle\alpha\rangle) = \langle\varphi_{k+1}, \psi_{k+1}, n_{k+1}\rangle\}| = \mu;$$

and for $i < \mu$ choose $\eta^i_{k+1} = \eta^\alpha_k{}^\frown\langle\alpha\rangle$ where $\alpha > i$, $h(\eta^\alpha_k{}^\frown\langle\alpha\rangle) = \langle\varphi_{k+1}, \psi_{k+1}, n_{k+1}\rangle$. Clearly these φ_k, ψ_k, n_k prove (2), (3).

THEOREM 3.8: *If* $\lambda > |T|^+$, *then* $D^m(p, \Delta, \lambda) = D^m(p, \Delta, \infty)$.

Remark. If you wonder why we do not require only $\lambda > (|\Delta| + \aleph_0)^+$, then see Exercises 3.10 and 3.14.

Proof. Let $\mu = |T|^+$, and w.l.o.g. p is $\{\theta(\bar{x}; \bar{a})\}$. If $D^m(p, \Delta, \mu^+) \ge \alpha$, then by 3.6 (when $\alpha < \mu$) and 3.7 (when $\alpha \ge \mu$) there is a (Δ, α)-function h such that $\Gamma^*_\mu(\theta(\bar{x}; \bar{a}), h)$ is consistent. Hence for every λ, $\Gamma^*_\lambda(\theta(\bar{x}; \bar{a}), h)$ is consistent; hence $D^m[\theta(\bar{x}; \bar{a}), \Delta, \lambda] \ge \alpha$. By the monotonicity of $D^m(\theta(\bar{x}; \bar{a}), \Delta, \lambda)$ in λ (by 3.4(2)) it follows that for every $\lambda > \mu$, $D^m[\theta(\bar{x}; \bar{a}), \Delta, \lambda] = D^m[\theta(\bar{x}; \bar{a}), \Delta, \mu^+]$. Now we can easily finish the proof by proving by induction on α that $D^m[\theta(\bar{x}; \bar{a}), \Delta, \mu^+] \ge \alpha$ implies $D^m[\theta(\bar{x}; \bar{a}), \Delta, \infty] \ge \alpha$ using what we have just proved (for every p).

THEOREM 3.9: *If p is finite*, $D^m(p, \mathrm{L}, \mu^+) \ge \mu$, $\mu = |T|^+$, *then there are* φ_n $(0 < n < \omega)$ *and* $\bar{a}_\eta$, $\eta \in {}^{\omega>}\mu$, *such that*

(i) *for every* $\eta \in {}^\omega\mu$, $p_\eta = p \cup \{\varphi_n(\bar{x}; \bar{a}_{\eta|n}): 0 < n < \omega\}$ *is consistent*,

(ii) *for every* $\eta \in {}^n\mu$, $n < \omega$, $i < j < \mu$, $\vdash \neg(\exists\bar{x})[\varphi_{n+1}(\bar{x}; \bar{a}_{\eta^\frown\langle i\rangle}) \wedge \varphi_{n+1}(\bar{x}; \bar{a}_{\eta^\frown\langle j\rangle})]$,

(iii) *for every* $\eta \in {}^{n+1}\mu$, $0 < n < \omega$, $\vdash (\forall\bar{x})[\varphi_{n+1}(\bar{x}; \bar{a}_\eta) \to \varphi_n(\bar{x}; \bar{a}_{\eta|n})]$.

Proof. We shall define, by induction on $k < \omega$, formulas φ^k_n and natural numbers m^k_n such that

(1) for every k, there are $\bar{a}_\eta$, $\eta \in {}^{\omega>}\mu$, such that

(i) for every $\eta \in {}^\omega\mu$, $p_\eta = p \cup \{\varphi^k_n(\bar{x}; \bar{a}_{\eta|n}): 0 < n < \omega\}$ is consistent,

(ii) for every $\eta \in {}^n\mu$, $n < \omega$, $w \subseteq \mu$, $|w| \ge m^k_{n+1}$, $\{\varphi^k_{n+1}(\bar{x}; \bar{a}_{\eta^\frown\langle i\rangle}): i \in w\}$ is inconsistent.

Let $f(k) = \min\{n: m_n^k > 2\} \cup \{\omega\}$.

(2) For every k, if $f(k) < \omega$, then $f(k+1) > f(k)$ or $l = f(k) = f(k+1)$, and $m_l^k > m_l^{k+1}$,

(3) if $n < f(k)$, then $\varphi_n^k = \varphi_n^{k+1}$.

If we succeed in defining them, and define φ_n' as φ_n^k for large enough $k(f(k) > n)$, then clearly 3.9 is satisfied by $\varphi_n = \bigwedge_{1 \le l \le n} \varphi_l'(\bar{x}; \bar{y}^l) \wedge \bigwedge p$. It is also clear that for $k = 0$ there are such φ_n^0, m_n^0 (by 3.7(3)). So it suffices to prove, that if φ_n^l, m_n^l are defined for $l \le k$, $0 < n < \omega$, then we can define φ_n^{k+1}, m_n^{k+1}, $0 < n < \omega$.

Now we can assume there are $\bar{a}_\eta$, $\eta \in {}^{\omega>}\mu$, such that (1) is satisfied by φ_n^k, m_n^k, $(0 < n < \omega)$, $\bar{a}_\eta$; and

(iii) if $\eta \in {}^{\omega>}\mu$, $l < \omega$, $i_1 \le i_2 \le \cdots \le i_l < \mu$, $j_1 \le j_2 \le \cdots \le j_l < \mu$ (where $i_\alpha = i_{\alpha+1} \Leftrightarrow j_\alpha = j_{\alpha+1}$ for $\alpha = 1, \ldots, l-1$) and $\nu_1, \ldots, \nu_l \in {}^{\omega>}\mu$ then the two sequences

$$\bar{a}_{\eta\frown\langle i_1\rangle\frown\nu_1}\frown\cdots\frown\bar{a}_{\eta\frown\langle i_l\rangle\frown\nu_l},$$

$$\bar{a}_{\eta\frown\langle j_1\rangle\frown\nu_1}\frown\cdots\frown\bar{a}_{\eta\frown\langle j_l\rangle\frown\nu_l}$$

realize the same type over $\emptyset$, which depend on $l(\eta)$ only and

$$A_\eta = \bigcup \{\bar{a}_\nu : l(\nu) \le l(\eta) \quad \text{or} \quad \nu \restriction l(\eta) \ne \eta;\ \nu \in {}^{\omega>}\mu\} \cup \operatorname{Dom} p.$$

Remark. (1) We can choose the ν's as void sequences. Hence in particular, $\langle \bar{a}_{\eta\frown\langle\alpha\rangle} : \alpha < \mu\rangle$ is an indiscernible sequence over A_η.

(2) This is possible by 2.6 of the Appendix and compactness. See VII, 3.5. So we have φ_n^k, m_n^k $(0 < n < \omega)$, and $\bar{a}_\eta$, $\eta \in {}^{\omega>}\mu$, such that (i) and (ii) from (1) hold, and also (iii). We must define φ_n^{k+1}, m_n^{k+1}.

If $f(k) = \omega$, clearly 3.9 holds. So let $f(k) < \omega$ and $\eta \in {}^{f(k)-1}\mu$, and let 1_n be a sequence of n ones. Suppose first that

$$p^* = p \cup \{\varphi_n^k(\bar{x}; \bar{a}_{\eta|n}) : n < f(k)\}$$
$$\cup \{\varphi_{f(k)+n}^k(\bar{x}; \bar{a}_{\eta\frown 1_n}) : n < \omega\} \cup \{\varphi_{f(k)}^k(\bar{x}; \bar{a}_{\eta\frown\langle 0\rangle})\}$$

is consistent. Then define

$$\varphi_n^{k+1} = \varphi_n^k \quad \text{for } n \ne f(k),$$

$$\varphi_{f(k)}^{k+1} = \varphi_{f(k)}^{k+1}(\bar{x}; \bar{y}^1, \bar{y}^2) = \varphi_{f(k)}^k(\bar{x}; \bar{y}^1) \wedge \varphi_{f(k)}^k(\bar{x}; \bar{y}^2),$$

$$m_n^{k+1} = m_n^k \quad \text{for } n \ne f(k)$$

$$m_{f(k)}^{k+1} = m_{f(k)}^k - 1 \quad (\text{or, in fact } [(m_{f(k)}^k + 1)/2].$$

Clearly this definition satisfies our demands. So suppose that p^* is inconsistent. Hence it has an inconsistent finite subtype, which we can assume is

$$p^1 = p \cup \{\varphi_n^k(\bar{x}; \bar{a}_{\eta|n}) : n < f(k)\}$$
$$\cup \{\varphi_{f(k)+n}^k(\bar{x}; \bar{a}_{\eta^\frown 1_n}) : n < n_0\} \cup \{\varphi_{f(k)}^k(\bar{x}; \bar{a}_{\eta^\frown\langle 0\rangle})\}.$$

Let us define

$$\varphi_n^{k+1} = \varphi_n^k \quad \text{for } n < f(k),$$
$$\varphi_{f(k)}^{k+1} = \varphi_{f(k)}^{k+1}(\bar{x}; \bar{y}^0, \dots) = \bigwedge_{i < f(k)+n_0} \varphi_i^k(\bar{x}; \bar{y}^i) \wedge \bigwedge p,$$
$$\varphi_n^{k+1} = \varphi_{n+n_0}^k \quad \text{for } n > f(k),$$
$$m_n^{k+1} = m_n^k \quad \text{for } n < f(k),$$
$$m_{f(k)}^{k+1} = 2,$$
$$m_n^{k+1} = m_{n+n_0}^k \quad \text{for } n > f(k).$$

Clearly this definition satisfies our demands.

LEMMA 3.10: *For* $\lambda \geq \aleph_1$, $R^m(p, \Delta, \lambda) \geq D^m(p, \Delta, \lambda)$.

Proof. We prove by induction on α that $D^m(p, \Delta, \lambda) \geq \alpha$ implies $R^m(p, \Delta, \lambda) \geq \alpha$. For $\alpha = 0$, or α a limit ordinal it is immediate. Let $\alpha = \beta + 1$, and w.l.o.g. p is finite. So by definition for any $\mu < \lambda$ there are $\psi(\bar{x}; \bar{b}_i)$ $(i \leqslant \mu)$, n-contradictory over some finite $r \supseteq p$, with $D^m[r \cup \{\psi(\bar{x}; \bar{b}_i)\}, \Delta, \lambda] \geq \beta$, hence, by the induction hypothesis, $R^m[r \cup \{\psi(\bar{x}; \bar{b}_i)\}, \Delta, \lambda] \geq \beta$. By 1.6 there are $p_i \in S^m(A)$, $r \cup \{\psi(\bar{x}; \bar{b}_i)\} \subseteq p_i$, and $R^m(p_i, \Delta, \lambda) \geq \beta$, where $A = \bigcup_{i \leq \mu} \bar{b}_i \cup \operatorname{Dom} p$. As the $\psi(\bar{x}; \bar{b}_i)$ are n-contradictory over r, for every $q \in S_\Delta^m(A)$, $|\{i : p_i \restriction \Delta = q\}| < n$, hence, as μ was arbitrary, $R^m(p, \Delta, \lambda) \geq \beta + 1 = \alpha$.

THEOREM 3.11: $D^m(p, \Delta, \lambda) = R^m(p, \Delta, \lambda)$ *provided that:* $\Delta = \mathrm{cl}_2^m(\Delta_1)$, $\lambda = \mu^+$, μ *is regular;* $\mu \geq |T|^+$ *or* $\mu = \aleph_0 > |\Delta_1|$; *and* $R^m(p, \Delta, \lambda) < \infty$ or T *is stable and* $\mu_i < \mu$ $(i < |\Delta_1|) \Rightarrow \prod_{i<|\Delta_1|} \mu_i < \mu$.

Proof. By 1.2, 1.1(1) and 3.4(1), (2) we can assume $p = \{\theta(\bar{x}; \bar{a})\}$, and by 3.10 it suffices to prove by induction on α that $R^m(p, \Delta, \lambda) \geq \alpha \Rightarrow D^m(p, \Delta, \lambda) \geq \alpha$. For $\alpha = 0$ or α a limit ordinal this is immediate, so let $\alpha = \beta + 1$. If $R^m(p, \Delta, \lambda) < \infty$, by 3.5(1) we can assume $R^m(p, \Delta, \lambda)$

$= \alpha$. By 1.8(2) as p is finite and $R^m(p, \Delta, \lambda) \geq \beta + 1$, there are Δ_1-m-types p_i $(i \leq \mu)$ pairwise explicitly contradictory, much that $R^m(p \cup p_i, \Delta, \lambda) \geq \beta$. As $R^m(p, \Delta, \lambda)$ is not $> \beta + 1$, $|\{i: R^m(p \cup p_i, \Delta, \lambda) \geq \beta + 1\}| < \mu$, so w.l.o.g. $R^m(p \cup p_i, \Delta, \lambda) = \beta$ for every $i \leq \mu$. By 1.6 we can assume that $p_i \in S^m_{\Delta_1}(A)$ $(i \leq \mu)$ for some A, and of course $i \neq j \Rightarrow p_i \neq p_j$. Notice that by 1.3, for every $\varphi \in \Delta$, $R^m(p, \varphi, \infty) \leq R^m(p, \varphi, \lambda) < \infty$ hence by 2.10(2) $n_0(\varphi) =_{\text{def}} R^m(p, \varphi, 2) < \omega$ and there is $n_1(\varphi) < \omega$ such that for no $\bar{d}_l$ $(l < n_1(\varphi))$ does

$$\bigwedge_{l < n_1(\varphi)} (\exists \bar{x}) \Big[\bigwedge_{k < n_1(\varphi)} \varphi(\bar{x}; \bar{d}_k)^{\text{if}(k \geq l)} \wedge \theta(\bar{x}; \bar{a}) \Big]$$

hold.

If T is stable, we can choose A and distinct $p_i \in S^m_{\Delta_1}(A)(i \leq \mu)$ such that $R^m(p \cup p_i, \Delta, \lambda) \geq \beta$. As

$$\mu_j < \mu(j < |\Delta_1|) \Rightarrow \prod_{j < |\Delta_1|} \mu_j < \mu,$$

for some $\varphi \in \Delta_1$ there are μ distinct $p_i \restriction \varphi$ and w.l.o.g. $i < j \leq \mu \Rightarrow p_i \restriction \varphi \neq p_j \restriction \varphi$. Now $R^m(p, \varphi, 2) < \omega$ by the stability of T (2.13, 2.2) so we can proceed as before so $n_0(\varphi)$, $n_1(\varphi) < \omega$ exist.

Now the proof splits into two cases:

Case I: $\mu = \aleph_0 > |\Delta_1|$. We can find φ' such that for each $n < \omega$, $\{\varphi'(\bar{x}; \bar{b}_l)^{\text{if}(l=n)}: l \leq n\} \subseteq p_{i(n)}$, where $i(n) < \mu$. (If T is stable φ' is φ or $\neg\varphi$ and if $R^m(p, \Delta, \lambda) < \infty$, then by 1.6 of the Appendix such a φ', $\varphi' \in \text{cl}_1(\Delta_1)$ exists.) Let $n_1 = n_1(\varphi')$ and for $n \leq n_1$ let

$$\psi_n(\bar{y}_0, \ldots, \bar{y}_{n_1}; \bar{z}) = (\exists \bar{x}) \Big[\theta(\bar{x}; \bar{z}) \wedge \bigwedge_{k \leq n_1} \varphi'(\bar{x}; \bar{y}_k)^{\text{if}(k \geq n)} \Big].$$

By Theorem I, 2.4 we can assume that $\{\bar{b}_l: l < \omega\}$ is a ψ_n-n_1-indiscernible sequence over $\bar{a}$ for each $n \leq n_1$. For every $l < \omega$ let

$$\psi(\bar{x}; \bar{c}_l) = \bigwedge_{k < n_1} \neg\varphi'(\bar{x}; \bar{b}_k) \wedge \varphi'(\bar{x}; \bar{b}_{n_1 + l}).$$

It suffices to prove that $\{\theta(\bar{x}; \bar{a}) \wedge \psi(\bar{x}; \bar{c}_l): l < \omega\}$ is n_1-contradictory, because by our choice $p_{i(n_1 + l)} \vdash \psi(\bar{x}; \bar{c}_l)$, hence $R^m[\theta(\bar{x}; \bar{a}) \wedge \psi(\bar{x}; \bar{c}_l), \Delta, \lambda] \geq \beta$, hence by the induction hypothesis $D^m[\theta(\bar{x}; \bar{a}) \wedge \psi(\bar{x}; \bar{c}_l), \Delta, \lambda] \geq \beta$, hence by definition $D^m[\theta(\bar{x}; \bar{a}), \Delta, \lambda] \geq \beta + 1$ (as $\lambda = \mu^+$, $\psi \in \Delta$).

Now if $l(k) < \omega$ $(k < n_1)$ are distinct and

$$\vDash (\exists \bar{x}) \bigwedge_{k < n_1} [\theta(\bar{x}; \bar{a}) \wedge \psi(\bar{x}; \bar{c}_{l(k)})],$$

then

$$\vDash (\exists \bar{x})\Big[\theta(\bar{x};\bar{a}) \wedge \bigwedge_{k<n_1} \neg\varphi'(\bar{x};\bar{b}_k) \wedge \bigwedge_{k<n_1} \varphi'(\bar{x};\bar{b}_{n_1+l(k)}\Big]$$

hence for every $n \le n_1$,

$$\vDash (\exists \bar{x})\Big[\theta(\bar{x};\bar{a}) \wedge \bigwedge_{k<n} \neg\varphi'(\bar{x};\bar{b}_k) \wedge \bigwedge_{k=n}^{n_1-1} \varphi'(\bar{x};\bar{b}_{n_1+l(k)})\Big]$$

hence by the ψ_n-n_1-indiscernibility of $\bar{b}_l$ $(l < \omega)$ over $\bar{a}$,

$$\vDash \bigwedge_{n} (\exists \bar{x})\Big[\theta(\bar{x};\bar{a}) \wedge \bigwedge_{k<n} \neg\varphi'(\bar{x};\bar{b}_k) \wedge \bigwedge_{k=n}^{n_1-1} \varphi'(\bar{x};\bar{b}_k)\Big],$$

a contradiction.

Case II: $\mu > \aleph_0$, so we can assume $\Delta = \Delta_1$. If T is stable and $\beta \ge |T|^+$, then by the induction hypothesis, $D^m(p, \Delta, \lambda) \ge \beta$ so by 3.7 and 3.8, $D^m(p, \Delta, \lambda) = \infty$; so $D^m(p, \Delta, \lambda) = R^m(p, \Delta, \lambda)$.

Let $R^m(p, \Delta, \lambda) < \infty$, then clearly by the above $R^m(p \cup p_i, \Delta, \lambda) = \beta < |T|^+$. Now choose for each $i \leqslant \mu$ a finite $q_i \subseteq p_i$ such that $R^m(p \cup q_i, \Delta, \lambda) = \beta$. Hence for each $i \leqslant \mu$, $|\{j \leqslant \mu : q_i \subseteq p_j\}| < \mu$ so we can assume that $i < j < \mu$ implies $q_i \nsubseteq p_j$. As Δ is closed under conjunction we can replace q_i by $\bigwedge q_i = \varphi_i(\bar{x}; \bar{b}_i)$ and as the number of possible φ_i is $|\Delta| < \mu$ and μ is regular, we can assume that $\varphi_i = \varphi$ for every $i < \mu$. As $p_i \in S^m_{\Delta_1}(A)$ for $i < \mu$, clearly for $i < j < \mu$, $\neg\varphi(\bar{x}; \bar{b}_i) \in p_j$, so the $p_i \restriction \varphi$ are distinct.

So whether $R^m(p, \Delta, \lambda) < \infty$ or T is stable and $\mu_i < \mu$ $(i < |\Delta_1|) \Rightarrow \prod_{i<|\Delta_1|} \mu_i < \mu$, we can assume $\beta < |T|^+$ and there is $\varphi \in \Delta$ such that

$$\Phi = \{q \in S^m_\varphi(A) \colon R^m(p \cup q, \Delta, \lambda) \ge \beta\}$$

has cardinality $\ge \mu$. We now need:

CLAIM 3.12: *Suppose* $\Phi \subseteq \{p \in S^m_\varphi(A) \colon p \cup \{\theta(\bar{x};\bar{a})\}$ *is consistent*$\}$ *and* $R^m[\theta(\bar{x};\bar{a}), \varphi, 2] = n_0 < \omega$. *Let* κ *be* $|\Phi|$ *when* $|\Phi|$ *is infinite, and* $\log_2 \log_2 |\Phi|$ *if* $|\Phi|$ *is finite. Then we can find* r_i $(i < \kappa)$ *such that:*

(1) r_i *is a* φ-m-*type over* A,

(2) $|r_i| = n_0 + 2$,

(3) *for each* i, *there is a unique* $p_i \in \Phi$ *such that* $r_i \subseteq p_i$,

(4) $i \ne j$ *implies* $p_i \ne p_j$.

We shall prove this claim later. Now apply Claim 3.12 to our Φ, and we get r_i $(i < \mu)$ as mentioned there and let $\bigwedge r_i = \varphi^*(\bar{x}; \bar{c}_i)$. So $R^m[\theta(\bar{x};\bar{a}) \wedge \varphi^*(\bar{x};\bar{c}_i), \Delta, \lambda] \ge \beta$, but for $i \ne j$,

$$R^m[\theta(\bar{x};\bar{a}) \wedge \varphi^*(\bar{x};\bar{c}_i) \wedge \varphi^*(\bar{x};\bar{c}_j), \Delta, \lambda] < \beta.$$

For any sequence $\bar{\imath}$, $\bar{\imath} = \langle i(0), \ldots, i(n_1)\rangle$, $i(0) < \cdots < i(n_1) < \mu$, $n_1 = n_1(\varphi)$, let

$$\psi(\bar{x}; \bar{c}^{\bar{\imath}}) = \bigwedge_{l<n_1} \neg\varphi^*(\bar{x}; \bar{c}_{i(l)}) \wedge \varphi^*(\bar{x}; \bar{c}_{i(n_1)}).$$

Clearly $R^m[\theta(\bar{x}; \bar{a}) \wedge \psi(\bar{x}; \bar{c}^{\bar{\imath}}), \Delta, \lambda] \geq \beta$, hence by the induction hypothesis and 3.6, for some (Δ, β)-function $h_{\bar{\imath}}$, $\Gamma^*_\mu[\theta(\bar{x}; \bar{a}) \wedge \psi(\bar{x}; \bar{c}^{\bar{\imath}}), h^{\bar{\imath}}]$ is consistent. For each such $\bar{\imath}$ define a (Δ, α)-function $h^{\bar{\imath}}$ as follows:

If $\eta \in ds(\beta)$, then $h^{\bar{\imath}}(\eta) = h_{\bar{\imath}}(\eta)$,
if $\eta = \langle\beta\rangle$, then $h^{\bar{\imath}}(\eta) = \langle\theta, \psi, n_0\rangle$,
if $\eta = \langle\beta\rangle^\frown\nu$, $\nu \in ds(\beta)$, then $h^{\bar{\imath}}(\eta) = h_{\bar{\imath}}(\nu)$.

If for a suitable $\bar{\imath}$, $\Gamma^*_{\aleph_0}[\theta(\bar{x}; \bar{a}), h^{\bar{\imath}}]$ is consistent, then by 3.6, $D^m[\theta(\bar{x}; \bar{a}), \Delta, \lambda] \geq \alpha$, so we finish. Hence assume that for each $\bar{\imath}$, for some finite $u(\bar{\imath}) \subset ds(\alpha)$, $\Gamma^{u(\bar{\imath})}_{\aleph_0}[\theta(\bar{x}; \bar{a}), h^{\bar{\imath}}]$ is inconsistent. Clearly the function F defined by $F(\bar{\imath}) = \langle u(\bar{\imath}), h^{\bar{\imath}} \restriction u(\bar{\imath})\rangle$, has $\leq |T|$ possible values. As $\mu > |T|$, μ is regular, there is $\gamma < \mu$ such that for any $j_1 \geq \gamma$, $l_0 < \omega$, $i_l(0) < \cdots < i_l(n_1 - 1) < \gamma$ (for $l < l_0$) there is $j_2 < \gamma$ such that for any $l < l_0$, $i_l(n_1 - 1) < j_2$ and

$$F(\langle i_l(0), \ldots, i_l(n_1 - 1), j_1\rangle) = F\langle i_l(0), \ldots, i_l(n_1 - 1), j_2\rangle)$$

(this is a variant of the Downward Lowenheim–Skolem theorem, see VII, 1.4). So we can define $\gamma(l) < \gamma$ $(l < \omega)$ such that $\gamma(l) < \gamma(l + 1)$, and if $l(0) < \cdots < l(n_1 - 1) < k < \omega$, then

$$F(\langle\gamma(l(0)), \ldots, \gamma(l(n_1 - 1)), \gamma(k)\rangle) = F(\langle\gamma(l(0)), \ldots, \gamma(l(n_1 - 1)), \gamma\rangle).$$

As in Case I we can assume that the formulas

$$\theta(\bar{x}; \bar{a}) \wedge \psi(\bar{x}; \bar{c}^{\langle\gamma(0), \ldots, \gamma(n_1-1), \gamma(n_1+l)\rangle}) \quad (l < \omega)$$

are n_1-contradictory, but of course $F(\langle\gamma(0), \ldots; \gamma(n_1 - 1), \gamma(n_1 + l)\rangle) = F(\langle\gamma(0), \ldots, \gamma(n_1 - 1), \gamma\rangle)$ and let it be $\langle u, h \restriction u\rangle$ so clearly $\Gamma^u_{\aleph_0}(\theta(\bar{x}; \bar{a}), h)$ is consistent, a contradiction.

Proof of Claim 3.12. We define r_i, p_i by induction on i, so suppose $i < \kappa$, and we have defined r_j, p_j for every $j < i$.

Define an equivalence relation E_i on the φ-m-formulas over A: $\varphi(\bar{x}; \bar{a}_1)E_i\varphi(\bar{x}; \bar{a}_2)$ iff for every $j < i$, $\varphi(\bar{x}; \bar{a}_1) \in p_j \Leftrightarrow \varphi(\bar{x}; \bar{a}_2) \in p_j$. If $|\Phi|$ is finite, the number of E_i-equivalence classes is $\leq 2^i$; hence

$$|\{p \in \Phi \colon \varphi(\bar{x}; \bar{a}_1)E_i\varphi(\bar{x}; \bar{a}_2) \Rightarrow [\varphi(\bar{x}; \bar{a}_1) \in p \Leftrightarrow \varphi(\bar{x}; \bar{a}_2) \in p]\}| \leq 2^{2^i},$$

so there are $p^0 \in \Phi$, $\varphi(\bar{x}; \bar{a}^i_0) \in p^0$, $\neg\varphi(\bar{x}; \bar{a}^i_1) \in p^0$, such that $\varphi(\bar{x}; \bar{a}^i_0)$, $\varphi(\bar{x}; \bar{a}^i_1)$ are E_i-equivalent.

We can choose such $p^0, \bar{a}^l_0, \bar{a}^l_1$ also for infinite Φ. E_l has $<\kappa$ equivalence classes—otherwise choose representatives $\varphi(\bar{x}; \bar{a}_j)$ $(j < \kappa)$, and let $\bar{b}_j$ $(j < i)$ realize p_j. Then by Exercise I, 2.7 and compactness there are $\alpha(n) < \kappa$, $\beta(n) < i$ such that $\vdash\varphi[\bar{b}_{\beta(l)}; \bar{a}_{\alpha(k)}]^{\text{if}(k \geq l)}$ for $k, l \leq \omega$; contradicting 2.10(2). In the same way we can prove that

$$|\{p \in \Phi: \varphi(\bar{x}; \bar{a}_1)E_l\varphi(\bar{x}; \bar{a}_2) \Rightarrow [\varphi(\bar{x}; \bar{a}_1) \in p \Leftrightarrow \varphi(\bar{x}; \bar{a}_2) \in p]\}| < \kappa;$$

so necessarily suitable $p^0, \bar{a}^l_0, \bar{a}^l_1$ exist.

Now we define by induction on $l \leq n_0$, p^l, $\bar{a}^l_{l+1}$, $\mathfrak{t}(l) \in \{0, 1\}$ so that

(i) $r^i_l = \{\varphi(\bar{x}; \bar{a}^l_0), \varphi(\bar{x}; \bar{a}^l_1)^{\mathfrak{t}(0)}, \ldots, \varphi(\bar{x}; \bar{a}^l_{l+1})^{\mathfrak{t}(l)}\} \subseteq p^l \in \Phi$,

(ii) $R^m[\{\theta(\bar{x}; \bar{a})\} \cup r^i_l, \varphi, 2] \leq n_0 - l$ or $|\{p \in \Phi: r^i_l \subseteq p\}| = 1$.

For $l = 0$, p^0, $\bar{a}^l_{l+1}$ are defined and let $\mathfrak{t}(0) = 1$, so clearly (i) and (ii) hold. If we have defined for l, and $|\{p \in \Phi: r^i_l \subseteq p\}| > 1$, then for some $\varphi = \varphi(\bar{x}; \bar{a}^l_{l+1})$, $r^i \cup \{\varphi^{\mathfrak{t}}\} \subseteq p^{l,\mathfrak{t}} \in \Phi$ for $\mathfrak{t} \in \{0, 1\}$. Clearly there is $\mathfrak{t}(l + 1) \in \{0, 1\}$ so that

$$R^m[\{\theta(\bar{x}; \bar{a})\} \cup r^i_l \cup \{\varphi^{\mathfrak{t}(l+1)}\}, \varphi, 2] < R^m[\{\theta(\bar{x}; \bar{a})\} \cup r^i_l, \varphi, 2] \leq n_0 - l$$

and let $p^{l+1} = p^{l,\mathfrak{t}(l+1)}$.

If we have defined for l but $|\{p \in \Phi; r^i_l \subseteq p\}| = 1$ let $\bar{a}^l_{l+1} = \bar{a}^l_l$, $\mathfrak{t}(l + 1) = \mathfrak{t}(l)$.

Now let $r^i = r^i_{n_0+1}$. It is easy to check that all the conditions are satisfied.

THEOREM 3.13: (1) (A) $R^m(p, \Delta, \lambda) = R^m(p, \Delta, \infty)$, *and*

(B) $R^m(p, \Delta, \lambda) \geq |\Delta|^+ + \aleph_0$ *implies* $R^m(p, \Delta, \lambda) = \infty$, *provided that:*

(i) $\lambda \geq (2^{|\Delta|})^{++} + \aleph_0$, *or*

(ii) $\lambda > |\Delta|^+ + \aleph_0$, $R^m(p, \Delta, \lambda) < \infty$ λ *not singular nor successor of singular.*

(2) *Theorem 3.11 holds for all* $\lambda > \mu > \aleph_0$ *(μ as in 3.11).*

Proof. (1) W.l.o.g. we can assume p is finite. If Δ is finite, (A) follows from 2.10(2) if $R^m(p, \Delta, \lambda) \geq \omega$ and from 2.9(3) (using compactness) if $R^m(p, \Delta, \lambda) < \omega$, and (B) follows from 2.10(2).

So w.l.o.g. Δ is infinite, $|\Delta| = |T|$, and by 1.8(2) we can assume $\Delta = \mathrm{cl}^m_3(\Delta)$.

If (ii) holds, let $\lambda = \mu^+$, μ regular (if λ is regular but not a successor by 1.9(2) we can replace it by λ^+). By 3.11, $R^m(p, \Delta, \lambda) = D^m(p, \Delta, \lambda)$ and by 3.8, $D^m(p, \Delta, \lambda) = D^m(p, \Delta, \infty)$. By 3.10, $R^m(p, \Delta, \infty) \geq D^m(p, \Delta, \infty)$ hence $R^m(p, \Delta, \infty) \leq R^m(p, \Delta, \lambda)$ but by 1.9(2), $R^m(p, \Delta, \infty) \leq R^m(p, \Delta, \lambda)$ hence (A) holds. (B) follows similarly from 3.7(1).

Let $\theta(\bar{x};\bar{a}) = \bigwedge p$. We can assume that (i) holds. By 1.3(1) it suffices to prove (A) and (B) for $\lambda = (2^{|\Delta|})^{++}$. If $R(p, \Delta, \lambda) = \infty$, (B) is trivial, and otherwise 3.11 applies and we can finish as above. So we need only prove (A) on the assumption that $R^m(p, \Delta, \lambda) = \infty$. So $R^m(p, \Delta, \lambda) > (2^{|T|})^+$, hence there are Δ-m-types p_i $(i < (2^{|\Delta|})^+)$, pairwise explicitly contradictory such that $R^m(p \cup p_i, \Delta, \lambda) \geqslant (2^{|T|})^+$, hence $R^m(p \cup p_i, \Delta, \lambda) = \infty$. By 1.6 we can assume that for some A, $p_i \in S^m_\Delta(A)$ for every i. As

$$(2^{|\Delta|})^+ = |\{p_i: i < (2^{|\Delta|})^+\}| \le \prod_{\varphi\in\Delta} |\{p_i \restriction \varphi: i < (2^{|\Delta|})^+\}|$$

$$\le \prod_{\varphi\in\Delta} |\{q \in S^m_\varphi(A): R^m(p \cup q, \Delta, \lambda) = \infty\}|,$$

for some $\varphi \in \Delta$ there are distinct $q_i \in S^m_\varphi(A)$ $(i < (2^{|\Delta|})^+)$ such that $R^m(p \cup q_i, \Delta, \lambda) = \infty$. If $R^m(p, \varphi, 2) \ge \omega$, our conclusion follows by 2.10(2) and 1.3 hence we can assume $R^m(p, \varphi, 2) < \omega$. So we can apply Claim 3.12 to $\Phi = \{q \in S^m_\varphi(A): R^m(p \cup q, \Delta, \lambda) = \infty\}$, and get φ-m-types r_i $(i < |\Delta|^+)$, $|r_i| \le 2 + R^m(p, \varphi, 2) =_{\mathrm{def}} n_{\langle\rangle}$ such that $R^m(p \cup r_i, \Delta, \lambda) = \infty$, $R^m(p \cup r_i \cup r_j, \Delta, \lambda) < \infty$ (for $i \neq j < |\Delta|^+$). We can apply this construction to $p \cup r_i$, etc. Hence we can define for $\eta \in {}^{\omega >}(|\Delta|^+)$ Δ-m-types r_η, and natural numbers n_η so that:

(0) $r_{\eta|l} \subseteq r_\eta$ when $l < l(\eta)$,

(1) $R^m(p \cup r_\eta, \Delta, \lambda) = \infty$ for each $\eta \in {}^{\omega >}(|\Delta|^+)$,

(2) $|r_{\eta\frown\langle i\rangle}| = n_\eta$, $r_{\langle\rangle} = \emptyset$,

(3) $R^m(p \cup r_{\eta\frown\langle i\rangle} \cup r_{\eta\frown\langle j\rangle}, \Delta, \lambda) < \infty$ $(\eta \in {}^{\omega >}(|\Delta|^+)$, $i \neq j < |\Delta|^+$.

Now for $\eta \in {}^{\omega}(|\Delta|^+)$ let $r_\eta = \bigcup_{l<\omega} r_{\eta|l}$, so clearly $R^m(p \cup r_\eta, \Delta, \lambda) = \infty$. Let p, r_η be over A, so by 1.6 there are $r^\eta \in S^m_\Delta(A)$, $r_\eta \subseteq r^\eta$, $R^m(p \cup r^\eta, \Delta, \lambda) = \infty$ for $\eta \in {}^{\omega}(|\Delta|^+)$ hence if $\eta \restriction l \neq \nu \in {}^{l}(|\Delta|^+)$, then $r_\nu \nsubseteq r^\eta$. Let $\bigwedge r_n = \psi_n(\bar{x}; \bar{a}_n)$, now we can prove that for some $\psi_n(\bar{x}; \bar{y}_n)$ and $\kappa \le |\Delta|^+$,

$$\Gamma_\kappa = \{\theta(\bar{x}_\nu;\bar{a}): \nu \in {}^{\omega}\kappa\} \cup \{\psi_n(\bar{x}_\nu;\bar{y}_\eta)^{\mathrm{if}\,(\nu|n=\eta)}: \nu \in {}^{\omega}\kappa, \eta \in {}^{n}\kappa, 0 < n < \omega\}$$

is consistent; and that this implies the consistency of Γ_μ for each μ, and this implies $R^m(p, \Delta, \infty) = \infty$. As the proofs are similar to 3.6 and 3.8 we leave the details to the reader. Instead of 3.6 we can use 1.6 of the Appendix for $m = 1$.

(2) Easy by 3.11, 3.8 and 3.13(1).

THEOREM 3.14: *The following are equivalent.*

(1) *T is superstable, i.e., stable in every large enough λ (in fact $\lambda \ge 2^{|T|}$).*

(2) *T is stable in some λ for which $\lambda^{\aleph_0} > \lambda$.*

(3) $R^1[x = x, \mathrm{L}, (2^{|T|})^{++}] < |T|^+$.

(4) *For some* $m < \omega$, $R^m(\bar{x} = \bar{x}, \mathrm{L}, \infty) < \infty$.
(5) T *is stable and* $D^1(x = x, \mathrm{L}, |T|^{++}) < |T|^+$.
(6) T *is stable and for some* $m < \omega$, $D^m(\bar{x} = \bar{x}, \mathrm{L}, \infty) < \infty$.

Proof. (1) $\Rightarrow$ (2) Immediate.

(2) $\Rightarrow$ (5) By 3.9 (for $p = \emptyset$) and compactness, if (5) fails, but T is stable we can find $\bar{a}_\eta$, $\eta \in {}^{\omega>}\lambda$, satisfying (i) and (ii) from 3.9, and if $A = \bigcup \{\bar{a}_\eta : \eta \in {}^{\omega>}\lambda\}$, $q_\eta \in S(A)$, $p_\eta \subseteq q_\eta$, then $|S(A)| \geq |\{q_\eta : \eta \in {}^{\omega}\lambda\}| = \lambda^{\aleph_0}$, as the q_η are pairwise contradictory by (ii); but $|A| \leq \lambda$, a contradiction.

(5) $\Rightarrow$ (6) Follows by 3.4(2)

(6) $\Rightarrow$ (4) By 3.13(2).

(4) $\Rightarrow$ (1) For every $\varphi = \varphi(\bar{x}; \bar{y})$, $l(\bar{x}) = m$, by 1.3(1), $R^m(\bar{x} = \bar{x}, \varphi, \infty) \leq R^m(\bar{x} = \bar{x}, \mathrm{L}, \infty) < \infty$, hence, by 2.2, $\varphi(\bar{x}; \bar{y})$ is stable. As this holds for every such φ, T is stable. So by 3.13(1), $R^m(\bar{x} = \bar{x}, \mathrm{L}, \lambda) < \infty$ where $\lambda = (2^{|T|})^{++}$. Now for every A, for each $p \in S^m(A)$, choose finite $q_p \subseteq p$, $R^m(q_p, \mathrm{L}, \lambda) = R^m(p, \mathrm{L}, \lambda)$ which is $< \infty$. By definition of rank, $|\{p \in S^m(A) : q_p = q\}| \leq 2^{|T|}$ so

$$|S^m(A)| \leq |\{q_p : p \in S^m(A)\}| \cdot 2^{|T|} = |A| + |T| + 2^{|T|}.$$

So T is stable in every $\lambda \geq 2^{|T|}$.

(3) $\Rightarrow$ (4) Immediate, by 1.3.

(5) $\Rightarrow$ (3) By 3.11.

LEMMA 3.15: *If* $|S^m_\Delta(A)| > |A| + |T|$, *then for some countable* $B \subseteq A$, $|S_\Delta(A)| \geq |S_\Delta(B)| \geq 2^{\aleph_0}$, *and there are* $\bar{a}_\eta \in A$, $\varphi_\eta \in \Delta$ *such that for* $k < \omega$, $\eta \in {}^k2$:

$$\vDash (\exists x) \bigwedge_{l<n} \varphi_{\eta|l}(x, \bar{a}_{\eta|l})^{\eta[l]}.$$

Proof. Choose, by induction on $l < \omega$, formulas $\varphi_\eta(x; \bar{a}_\eta)$ for $\eta \in {}^{l>}2$ such that $\bar{a}_\eta \in A$ and for every $\nu \in {}^l2$,

$$\{p : p \in S_\Delta(A), \varphi_{\nu|k}(x; \bar{a}_{\nu|k})^{\nu[k]} \in p \text{ for every } k < l\}$$

has cardinality $> |A| + |T|$. Then we let $B = \bigcup \{\bar{a}_\eta : \eta \in {}^{\omega>}2\}$, so $|B| \leq \aleph_0$, $B \subseteq A$, $|S_\Delta(A)| \geq |S_\Delta(B)| \geq 2^{\aleph_0}$ because the types $\{\varphi_{\nu|k}(x; \bar{a}_{\nu|k})^{\nu[k]} : k < \omega\}$ are consistent and pairwise contradictory for $\nu \in {}^{\omega}2$.

EXERCISE 3.1: Prove that if $\mu \geq |\Delta| + \aleph_0$, $\mu^+ \geq \lambda$, and $R^m(p, \Delta, \lambda) \geq \delta(\mu)$, then $R^m(p, \Delta, \lambda) = \infty$. [On $\delta(\mu)$ see Definition VII, 5.1(3) and Theorem VII, 5.5.]

EXERCISE 3.2: Prove that if $\mu^+ \geq \lambda$, $X = R$ *or* $X = D$ and T is stable, $X^m(p, \Delta, \lambda) \geq [(\aleph_0 + |\Delta|)^\mu]^+$, then $X^m(p, \Delta, \lambda) = \infty$ (for $X = D$ use Exercise 3.10).

EXERCISE 3.3: For a formula $\theta = \theta(\bar{x}; \bar{a})$ define the model $\mathfrak{C}^0$. Its universe is $\{\bar{b} \in \mathfrak{C}: \mathfrak{C} \models \theta[\bar{b}; \bar{a}]\}$, and its relations are

$$R_\varphi = \{\langle \bar{c}_1, \ldots, \bar{c}_k \rangle : \mathfrak{C} \models \varphi[\bar{c}_1, \ldots, \bar{c}_k, \bar{a}]\}$$

for every $\varphi \in \mathrm{L}(T)$. Assume T is stable or at least that for every φ, $R^m(\theta(\bar{x}; \bar{a}), \varphi, \infty) < \infty$.

Prove that $R^m[\theta(\bar{x}, \bar{a}), \mathrm{L}, \lambda]$ computed in $\mathfrak{C}$ is equal to $R(x = x, \mathrm{L}^0, \lambda)$ computed in $\mathfrak{C}^0$. Moreover, for each $\Delta \subset \mathrm{L}$ there is $\Delta^0 \subset \mathrm{L}^0$ (and vice versa) such that $|\Delta| = |\Delta^0|$ and $R^m[\theta(\bar{x}; \bar{a}), \Delta, \lambda]$ computed in $\mathfrak{C}$ is equal to $R^m(\bar{x} = \bar{x}, \Delta^0, \lambda)$ computed in $\mathfrak{C}^0$.

Moreover we can assume that for every $\varphi(\bar{x}; \bar{y}) \in \Delta$ there is $\varphi^0(\bar{x}; \bar{y}^0) \in \Delta^0$, and vice versa, such that

(i) for every $\bar{b} \in \mathfrak{C}$ there is $\bar{b}^0 \in \mathfrak{C}^0$ such that for every $\bar{c} \in \mathfrak{C}^0$, $\bar{c} = \langle \bar{c}_0, \ldots \rangle$,

$$\mathfrak{C} \models \varphi[\bar{c}_0, \ldots; \bar{b}] \wedge \bigwedge_l \theta[\bar{c}_l; \bar{a}] \quad \text{iff} \quad \mathfrak{C}^0 \models \varphi^0[\bar{c}; \bar{b}^0],$$

(ii) for every $\bar{b}^0 \in \mathfrak{C}^0$, $\mathfrak{C}^0 \models (\forall \bar{x}) \neg \varphi^0(\bar{x}; \bar{b}^0)$ or there is $\bar{b}$ as in (i).

EXERCISE 3.4: Give an example of a theory T such that $D(x = x, \mathrm{L}, \infty) = 1$, $R(x = x, \mathrm{L}, \infty) = \infty$. (Hint: See 4.8(2).)

EXERCISE 3.5: Give an example of a theory T_α, $|T_\alpha| = |\alpha| + \aleph_0$ such that $D(x = x, \mathrm{L}, \infty) = \alpha$, $R(x = x, \mathrm{L}, \infty) = \infty$.

EXERCISE 3.6: Give an example of a theory T, $|T| = \lambda$, such that $R(x = x, \mathrm{L}, \mu) < \infty$ iff $\mu > (2^\lambda)^+$. (Hint: It has λ one-place predicates only.)

EXERCISE 3.7: Prove that if Δ is finite, and for every $k < \omega$, $R^m(p, \Delta, k) \geq n$, then $R^m(p, \Delta, \aleph_0) \geq n$.

DEFINITION 3.4: (1) Define $\mathrm{CR}(p, \lambda)$ as an ordinal or ∞, as follows: $\mathrm{CR}(p, \lambda) \geq 0$ iff p is a complete type, and $\mathrm{CR}(p, \lambda) \geq \alpha + 1$ iff p is a complete type and for every $\mu < \lambda$ and finite $B \subseteq \mathrm{Dom}\, p$, there are pairwise contradictory complete types q_i $(i \leq \mu)$, $p \restriction B \subseteq q_i$, and $\mathrm{CR}(q_i, \lambda) \geq \alpha$.

(2) T is transcendental if $\mathrm{CR}(p, 2) < \infty$ for every type p.

EXERCISE 3.8: Prove for CR(p, λ) theorems parallel to 1.1(1), 1.2, 1.3(1), 3.1 and 3.2.

EXERCISE 3.9: Investigate CR(p, λ). (Hint: See V, Section 7.)

EXERCISE 3.10: (1) If T is stable, show that in (3) of the definition of $D^m(p, \Delta, \lambda)$ we can restrict q to the union of r and a finite $\mathrm{cl}_3^m(\Delta)$-m-type.

(2) If in addition $\Delta = \mathrm{cl}_3^m(\Delta)$ we can replace "$\{\psi(\bar{x}; \bar{a}_i); i \le \mu\}$ is n-contradictory over q" by "$\{\psi(\bar{x}; \bar{a}_i); i \le \mu\}$ is n-contradictory" (i.e., n-contradictory over the empty set) and thus omit q completely and use just r.

(3) Hence show that if T is stable and $\Delta = \mathrm{cl}_3^m(\Delta)$ it is sufficient to define a (Δ, α)-function as a function

$$h: ds(\alpha) \to \{\langle \psi(\bar{x}; \bar{y}), n\rangle : \psi \in \Delta, n < \omega, l(\bar{x}) = m\}$$

and we can then replace $|T|$ by $(|\Delta| + \aleph_0)$ in 3.6, 3.7, 3.8 and 3.11.

EXERCISE 3.11: Show that in the previous exercise the stability is indeed necessary.

EXERCISE 3,12: Let p be a finite m-type, $\Delta \subset$ L; prove that the following conditions are equivalent when $|\Delta| < 2^{\aleph_0}$, and always (2) $\Rightarrow$ (1) $\Rightarrow$ (3).

(1) $R^m(p, \Delta, 2) = \infty$,

(2) for some A, $|\{q \in S_\Delta^m(A): p \subset q\}| > |A| + |\Delta| + \aleph_0$,

(3) for some A, $|A| \le \aleph_0$, $|\{q \in S_\Delta^m(A): p \subset q\}| \ge 2^{\aleph_0}$.

EXERCISE 3.13: Relativize 3.3(1) and 3.14 to $\theta(\bar{x}; \bar{a})$ (as in Exercise 2.5) and check the other theorems to see if they can be relativized similarly.

EXERCISE 3.14: (1) Show it is possible that $D^m(\bar{x} = \bar{x}, \psi, \aleph_1) = \infty$ but $D^m(\bar{x} = \bar{x}, \psi, \lambda) = 2$ for every $\lambda \ge \aleph_2$. (Compare with 3.8.)

(2) Show that in (1) we can replace two by any α.

(3) Show that in any example of (1), T is necessarily unstable.

[Hint: Let T_1 be the theory saying: P, Q form a partition of the universe; $P_n (0 < n < \omega)$ are disjoint subsets of P; $xRy \to P(x) \wedge Q(y)$ and $S(x) = \{y: xRy\}$ (i.e., we look at P as a family of subsets of Q); the sets $S(x)$ for $x \in P_1$ are pairwise disjoint; $P_1(c_{\langle\rangle})$, $P_n(c_\eta)$ for $\eta \in {}^n\omega$, and the c_η's are distinct; and let h be a one-to-one function from ${}^{\omega>}\omega$ into ω, for every $\eta \in {}^n\omega$, $n < \omega$ and distinct

$$x_l \in P_{n+1}, \qquad \bigcap_{k \le l(\eta)} S(c_{\eta|k}) \cap \bigcap_{l < h(\eta)} S(x_l) = \emptyset.$$

Now T_1 has a model completion with elimination of quantifiers satisfying (1) for $\psi = xRy$.]

EXERCISE 3.15: Show that in Theorem 3.11, we can replace "μ is regular" by "cf $\mu > |T|$" and if T is stable we can replace "$\mu \ge |T|^+$, $\mu_i < \mu$ $(i < |\Delta_1|) \Rightarrow \prod_{i<|\Delta|_1} \mu_i < \mu$" by "$\mu \ge (2^{|\Delta|})^+$". (Hint: See 3.13(2) and Exercise 3.10.) For 3.13(1)(B), $\lambda \ge |\Delta|^+ + \aleph_0$ suffices.

EXERCISE 3.16: Show that Exercise 3.15 is best possible, i.e., in 3.11 we cannot replace "$\mu \ge (2^{|\Delta|})^+$" by "cf $\mu \ge |\Delta|^+$".

[Hint: Assume $2^{\aleph_n} = \aleph_{\omega+n+1}$ for $n < \omega$. M is the model with universe $\aleph_\omega$ and the one-place relations $P^M_{n,\alpha}(\alpha < \aleph_n, n < \omega)$ such that $P^M_{n,\alpha} \subseteq \{i: \aleph_n < i < \aleph_{n+1}\}$, and $\{P^M_{n,\alpha}: \alpha < \aleph_n\}$ is independent for each $n < \omega$. Let $T = \mathrm{Th}(M)$, so $|T| = \aleph_\omega$, but

$$R(x = x, \mathrm{L}, \lambda) = \begin{cases} 1 & \text{if } \lambda > \aleph_{\omega+\omega+1}, \\ 2 & \text{if } \lambda = \aleph_{\omega+\omega} \text{ or } \aleph_{\omega+\omega+1}, \\ \infty & \text{if } \lambda < \aleph_{\omega+\omega}. \end{cases}$$

For each $\alpha < \aleph_{\omega+1}$, or $\alpha = \infty$ we can expand M to M_1 by one-place predicates such that 2 is replaced by α, $|D(T_1)| = \aleph_{\omega+\omega}$, $|T_1| = \aleph_\omega$ where $T_1 = \mathrm{Th}(M_1)$.]

We can use Exercise 3.20 for more general examples.

EXERCISE 3.17: Prove that in Claim 3.12, when Φ is finite, we can replace "$\kappa = \log_2 \log_2 |\Phi|$" by "$\kappa^{2n_0} \le |\Phi|$" (see 1.7 of the Appendix).

EXERCISE 3.18: (1) Prove that the following conditions on $\lambda \le \chi$ are equivalent:

(A)$_m$ For some T, $|T| \le \lambda$, $|D_m(T)| = \chi$.

(B) For some Boolean algebra $\mathbb{B}$, $\|\mathbb{B}\| \le \lambda$ and on $\mathbb{B}$ there are χ ultrafilters.

(2) If λ, χ satisfy (B) for a 1-homogeneous $\mathbb{B}$, show that in Exercise 3.6 we can replace $(2^\lambda)^+$ by χ^+.

EXERCISE 3.19: Generalize 3.9 to $D^m(p, \Delta, \mu^+)$:

(1) for stable T,

(2) in general.

EXERCISE 3.20: Let $R(\lambda, T) = R^1(x = x, \mathrm{L}, \lambda)$, $\mathrm{Mlt}(\lambda, T) = \mathrm{Mlt}(x = x, \mathrm{L}, \lambda)$. Suppose T_i is a complete theory in L_i ($i < \kappa$), L_i pairwise disjoint, $(\forall x)P_i(x) \in T_i$. Let M_i be a model of T_i, $M = \sum_{i<\kappa} M_i$ (i.e., $|M| = \bigcup_{i<\kappa} |M_i|$, $\mathrm{L}(M) = \bigcup_{i<\kappa} \mathrm{L}_i$, and if $R \in \mathrm{L}_i$ is a predicate, $R^M = R^{M_i}$; for simplicity only, assume L has no function symbols) and let $T = \mathrm{Th}(M)$. Then

(1) $R(\lambda; T) \geq \sup_{i<\kappa} R(\lambda, T_i)$.

(2) If $\lambda \geq \aleph_0$, $\beta = \sup_{i<\kappa} R(\lambda, T_i)$, then $R(\lambda, T) = \beta$ or $R(\lambda, T) = \beta + 1$. Also $R(\lambda, T) = \beta + 1$ iff $\lambda \leq \mu^+$ where

$$\mu = \sum \{\mathrm{Mlt}(\lambda, T_i): i < \kappa, R(\lambda, T_i) = \beta\}.$$

If $R(\lambda, T) = \beta + 1$, $\mathrm{Mlt}(\lambda, T) = 1$. If $R(\lambda, T) = \beta$, $\mathrm{Mlt}(\lambda, T) = \sum \{\mathrm{Mlt}(\lambda, T_i): R(\lambda, T_i) = \beta\}$.

(3) Let $\lambda < \aleph_0$, $\beta = \sup_{i<\kappa} R(\lambda, T_i)$, $\beta = \delta + k$, where $k < \aleph_0$ δ is limit or zero. Then

(i) $\beta \leq R(\lambda, T) \leq \delta + \omega$,

(ii) if $\{i < \kappa: R(\lambda, T_i) \geq \delta\}$ is infinite, then $R(\lambda, T) = \delta + \omega$, $\mathrm{Mlt}(\lambda, T) = 1$,

(iii) if $\{i < \kappa: R(\lambda, T_i) \geq \delta\}$ is finite, then $R(\lambda, T) < \delta + \omega$; and also if $k_0 = \sum \{\mathrm{Mlt}(\aleph_0, T_i): R(\lambda, T_i) \geq \delta\}$, $\delta + l = R(\lambda, T)$, then $k_0 = \max\{k: \lambda^k \leq l\}$ (see the next exercise).

EXERCISE 3.21: If $\lambda < \aleph_0$, $\alpha = R^m(p, \mathrm{L}, \lambda)$, $\beta = R^m(p, \mathrm{L}, \aleph_0)$, then $\omega\beta \leq \alpha < \omega\beta + \omega$, and if $k_1 = \alpha - \omega\beta$, $k_2 = \mathrm{Mlt}(p, \mathrm{L}, \lambda)$, $k = \mathrm{Mlt}(p, \mathrm{L}, \aleph_0)$, then $\lambda^{k_1} \leq k < \lambda^{k_1+1}$, and $k_2 = \max\{l: l \cdot \lambda^{k_1} \leq k\}$. Also k is the maximal number of pairwise contradictory extensions q of p satisfying $R^m(q, \mathrm{L}, \lambda) \geq \omega\beta$ (see Exercise 1.6).

Show that we can replace "L" by "Δ" when $\Delta = \mathrm{cl}_3^m(\Delta)$.

EXERCISE 3.22: Show that the results of Exercises 1.5 and 1.6 cannot be improved. In particular show that:

(i) For every α, and $\omega\alpha \leq \beta < \omega\alpha + \omega$ there is a totally transcendental theory T, $|T| = |\alpha| + \aleph_0$, $R(2, T) = \beta$, $R(\aleph_0, T) = \alpha$. (Hint: By induction on β using Exercise 3.20.)

(ii) Suppose $n < \omega$, $\aleph_0 = \lambda_1 < \lambda_2 < \cdots < \lambda_n = \infty$, $\alpha_1 > \alpha_2 > \cdots > \alpha_n$ where for $q < l < n$, λ_l is not the successor of a limit cardinal. Then there is a theory T, $|T| \leq |\alpha_1| + \mu$ (where $\mu^+ \geq \lambda_{n-1}$) such that $R(\lambda, T) = \alpha_l$ when $\lambda_l \leq \lambda < \lambda_{l+1}$. (Hint: Use N_μ, $|N_\mu| = {}^{\omega>}\mu$, $P_\eta^{N_\mu} = \{\nu \in N_\mu: \eta \lessdot \nu\}$ for $\eta \in {}^{\omega>}\mu$.)

QUESTION 3.23: Investigate the connection between $R^1(x = x, \mathrm{L}, \lambda)$ and $R^m(\bar{x} = \bar{x}, \mathrm{L}, \lambda)$. (Hint: See V, Section 7.)

EXERCISE 3.24: Let $\mathfrak{C}'$ be an expansion of $\mathfrak{C}$ by $\kappa < \bar{\kappa}$ individual constants. Show that

$$R^m(p, \mathrm{L}, \lambda) = R^m(p, \mathrm{L}', \lambda), \qquad D^m(p, \mathrm{L}', \lambda) = D^m(p, \mathrm{L}', \lambda).$$

EXERCISE 3.25: Show that if $R^m(p, \Delta, \aleph_0) \geq \beta + 1$, $q \subseteq p$ is finite, $\Delta = \mathrm{cl}_2^m(\Delta)$, then there are Δ-m-formulas $\varphi_n(\bar{x}; \bar{a}_n)$, pairwise contradictory, such that

$$R^m(q \cup \{\varphi_n(\bar{x}; \bar{a}_n)\}, \Delta, \aleph_0) \geq \beta \quad \text{for } n < \omega.$$

(Hint: Show there are distinct $q_n \in S^m(\Delta)$, $R^m(q \cup q_n, \Delta, \aleph_0) \geq \beta$ (by 1.6 and 1.9(2)) and then use 1.6 of the Appendix.)

EXERCISE 3.26: Let μ be a strong limit cardinal of cofinality $\aleph_0$ (i.e., $(\forall \chi < \mu) 2^\chi < \mu$ and $\mathrm{cf}\, \mu = \aleph_0$).

(1) Show that if p is an m-type over $\mathfrak{C}$,

$$\chi = |\{q \in S_\Delta^m(A) : p \cup q \text{ consistent}\}| > |A| + |\Delta| + \mu,$$

then $\chi \geq 2^\mu$; so, e.g., $|S^m(A)| > |A| + |\Delta| + \mu$ implies $|S^m(A)| \geq 2^\mu$. In particular, there are Δ-m-formulas $\varphi_\eta(\bar{x}; \bar{a}_\eta)$ over A for $\eta \in {}^{\omega >}\mu$, such that for every $\eta \in {}^\omega\mu$,

$$p \cup \{\varphi_{\eta \restriction l}(\bar{x}; \bar{a}_{\eta \restriction l}) \wedge \neg\varphi_\nu(\bar{x}; \bar{a}_\nu) : l < \omega, \nu = (\eta \restriction l)^\frown\langle i \rangle, i < \eta[l]\}$$

is consistent.

(2) Prove that for finite p, $R^m[p, \Delta, (2^\mu)^+] = \infty$ iff there are $\mathrm{cl}_1(\Delta)$-m-formulas $\varphi_\eta = \varphi_\eta(\bar{x}; \bar{a}_\eta)$ ($\eta \in {}^{\omega >}\mu$) such that for any $\eta \in {}^\omega\mu$

$$p \cup \{\varphi_{\eta \restriction l} \wedge \varphi_{\eta \restriction (l-1)^\frown\langle \alpha \rangle} : 0 < l < \omega, \alpha < \eta[l-1]\}$$

is consistent iff for some $\Delta_1 \subseteq \Delta$ and A, $|\Delta_1| + |A| = \mu$ and $|\{q \in S_{\Delta_1}^m(A) : p \cup q \text{ is consistent}\}| > \mu$ iff there are μ_n ($n < \omega$) such that $\sum_{n<\omega} \mu_n = \mu$, and $\varphi_\eta^\gamma(\bar{x}; \bar{a}_\eta^\gamma)$ for

$$\eta = \langle\langle \alpha_0, \beta_0 \rangle, \ldots, \langle \alpha_{m-1}, \beta_{m-1} \rangle\rangle, \qquad \alpha_l < \mu_l, \quad \gamma < \mu_m,$$

such that for any $\alpha_l < \beta_l < \mu_l$ ($0 < l < \omega$)

$$p \cup \{\neg\varphi_{\eta_l}^{\alpha_l}(\bar{x}; \bar{a}_{\eta_l}^{\alpha_l}) \wedge \varphi_{\eta_l}^{\beta_l}(\bar{x}; \bar{a}_{\eta_l}^{\beta_l}) : l < \omega\}$$

is consistent where $\eta_l = \langle\langle \alpha_0, \beta_0 \rangle, \ldots, \langle \alpha_{l-1}, \beta_{l-1} \rangle\rangle$.

(3) If the μ_n are as in (2) and for every $n < \omega$, for some $k < \omega$, $\mathrm{AD}(|\Delta|^+, |\Delta|, \mu_k, \mu_n)$ fails (see Definition VII, 1.11; e.g., if $|\Delta| = \mu$ or $\mu = \aleph_{\alpha+\mu}$), then in (2) we can add "iff $R^m(p, \Delta, \mu) \geq |\Delta|^+$".

EXERCISE 3.27: If T is stable, $D^m[\theta(\bar{x};\bar{a}),\Delta,\infty] \geq \alpha + 1$, then there are $n < \omega$, a formula $\psi(\bar{x};\bar{y})$ and an indiscernible set I based on $\bar{a}$ (see Definition III, 1.8) such that $\{\psi(\bar{x};\bar{b}): \bar{b} \in I\}$ is n-contradictory but $D^m[\theta(\bar{x};\bar{a}) \wedge \psi(\bar{x};\bar{b}),\Delta,\infty] \geq \alpha$ for $\bar{b} \in I$.

EXERCISE 3.28: $D^m(p,\Delta,\lambda) = D^m[p,\mathrm{cl}_1(\Delta),\lambda]$ for every p, λ; if T is stable and for each $\psi(\bar{x};\bar{y}) \in \Delta$, $n < \omega$, $\psi_n(\bar{x};\bar{y}_0,\ldots) = \bigwedge_{l<n}\psi(\bar{x};\bar{y}_l) \in \Delta$.

EXERCISE 3.29: Show that in Exercise 3.28 both conditions are necessary.

EXERCISE 3.30: Let p,Δ,φ be fixed, and for $k < \omega$, $R^m(p,\Delta,k) = \alpha^k$, $R^m[p \cup \{\varphi^{\mathrm{t}}\},\Delta,k] = \alpha^k_{\mathrm{t}}$.

(1) $\alpha^k = \delta + n^k$ and $\alpha^k_{\mathrm{t}} < \delta$ imply $\alpha^k_{1-\mathrm{t}} = \alpha^k$.

(2) If $\Delta = \mathrm{cl}^m_{3!}(\Delta)$, then for some $\mathrm{t} \in \{0,1\}$, $\alpha^k \leq \alpha^k_{\mathrm{t}} + 1$.

(3) We cannot improve (2) (to $\alpha^k \leq \alpha^k_{\mathrm{t}}$).

(4) If $\alpha^k_{\mathrm{t}} = \delta + n^k_{\mathrm{t}}$ for $\mathrm{t} \in \{0,1\}$, then $n^k \leq n^k_0 + n^k_1$.

[Hint: (1) Use Exercise 1.5: (4) We can assume p is finite and define finite Δ-m-types p_η ($\eta \in {}^{n\geq}k$, $n = n^k$) such that

(i) $p_{\eta^\frown\langle i\rangle}$ ($i < k$) are explicitly contradictory,

(ii) $R^m(p^\eta,\Delta,k) \geq \delta$ for $\eta \in {}^{n\geq}k$, where $p^\eta = p \cup \bigcup\{p_{\eta|l}: l \leq l(\eta)\}$.

We prove by downward induction on $l(\eta)$ that

(α) for some t, $R^m[p^\eta \cup \{\varphi^{\mathrm{t}}\},\Delta,k] = \alpha^k$, or ($\beta$) let

$$R^m[p^\eta \cup \{\varphi^{\mathrm{t}}\},\Delta,k] = \delta + n^{\eta,k}_{\mathrm{t}}, \qquad \mathrm{Mlt}(p^\eta \cup \{\varphi^{\mathrm{t}}\},\Delta,k) = l^{\eta,k}_{\mathrm{t}}$$

then

$$n - l(\eta) \leq n^{\eta,k}_0 + n^{\eta,k}_1 \quad \text{or} \quad n - l(\eta) = n^{\eta,k}_0 + n^{\eta,k}_1 + 1,$$

$$l^{\eta,k}_0 + l^{\eta,k}_1 \geq k.$$

QUESTION 3.31: Is Exercise 3.30(4) best possible? (This is a problem in finite combinatorics.)

EXERCISE 3.32: Suppose $l(\bar{x}) = l(\bar{y}) = m$ and $\theta(\bar{x},\bar{y};\bar{c}) \vdash \varphi(\bar{x};\bar{a}) \wedge \psi(\bar{y};\bar{b})$, $\varphi(\bar{x};\bar{a}) \vdash (\exists\bar{y})\theta(\bar{x},\bar{y};\bar{c})$, and for every $\bar{d}$, $\theta(\bar{x};\bar{d};\bar{c})$ is algebraic. Prove that for every infinite λ, $R^m[\varphi(\bar{x};\bar{a}),\mathrm{L},\lambda] \leq R^m[\psi(\bar{x};\bar{b}),\mathrm{L},\lambda]$.

DISCUSSION 3.16: The following generalizes Exercise 3.3, and shows how the work of Poizat can be put in our context.

Let p be a type over a set A. Define $\mathfrak{C}_A = (\mathfrak{C}, a)_{a \in A}$; $\mathfrak{C}_p$ is a model with universe $\{a \in C: a$ realizes $p\}$, and relation $R_\varphi = \{\bar{c} \in \mathfrak{C}_p: \bar{c}$ satisfies $\varphi\}$, for any $\varphi = \varphi(\bar{x}, \bar{a})$, $\bar{a} \in A$.

In II, Exercise 3.2 we deal with the case p a finite type, then $\mathfrak{C}_p$ is a saturated model, and so we can translate results on the stable case to the non-stable case. Here $\mathfrak{C}_p$ is not necessarily saturated, but it is like the $\mathfrak{C}$ when we deal with kind II in [Sh 75c]. As said there, all results on ranks, degrees and also *forking* goes through, and some cases are discussed in detail.

Let me elaborate a little more:

CLAIM 1: *$\mathfrak{C}_p$ is a $\bar{\kappa}$-homogeneous; moreover, if $\langle a_i: i < \alpha\rangle$, $\langle b_i: i < \alpha\rangle$ realizes the same quantifier-free formulas in $\mathfrak{C}_p$ (equivalently, the same formulas with parameters from A in $\mathfrak{C}$), then there is an automorphism f of $\mathfrak{C}_p$ which takes $\langle a_i: i < \alpha\rangle$ to $\langle b_i: i < \alpha\rangle$ (and, in fact, is a restriction of an automorphism of $\mathfrak{C}$ which is the identity on A).*

Proof. Find the required automorphism of $\mathfrak{C}$.

Remark. This would hold also for $|\mathfrak{C}'| = \{a: \text{tp}(a, A) \in \Gamma\}$, $\Gamma \subseteq S(A)$. Easily:

CLAIM 2: (1) *$\mathfrak{C}_p$ is $\bar{\kappa}$-compact for quantifier free formulas,*

(2) *$\mathfrak{C}_p$ satisfies the assumption of kind II, hence all results on forking and ranks generalized.*

But we may still be interested in the connection between what occurs in $\mathfrak{C}$ and in $\mathfrak{C}_p$.

ASSERTION 3: (1) *A set $I \subseteq |\mathfrak{C}_p|$ is discernible in $\mathfrak{C}_p$ iff it is indiscernible in $\mathfrak{C}$ over A.*

(2) *Two such infinite sets are equivalent in $\mathfrak{C}_p$ iff they are equivalent in $\mathfrak{C}$.*

(3) *For $B \subseteq C \subseteq |\mathfrak{C}_p|$, and $\bar{a} \in |\mathfrak{C}_p|$, $\text{tp}(\bar{a}, C, \mathfrak{C}_p)$ fork over B iff $\text{tp}(\bar{a}, C \cup A, \mathfrak{C})$ fork over $B \cup A$.*

(4) *Similar equivalence holds for stability in a power.*

Proof. E.g. (3) $p = \text{tp}(\bar{a}, C, \mathfrak{C}_p)$ does not fork over B iff some infinite indiscernible set I over C of sequences realizing p has, under automorphism of $\mathfrak{C}_p$ which are the identity over B, a bound number of images up to equivalence, so we finish.

For usual rank there is no nice translation, but if in the definition we do not take a finite subtype but deal with the same type, or deal with CP (and $A = \emptyset$ for simplicity) the situation will be like II, Exercise 3.3.

II.4. The f.c.p., the independence property and the strict order property

DEFINITION 4.1: (1) $\varphi(\bar{x}; \bar{y})$ has the *finite cover property* (f.c.p.) if for arbitrarily large natural numbers n there are $\bar{a}^0, \ldots, \bar{a}^{n-1}$ such that

$$\vDash \neg(\exists \bar{x}) \bigwedge_{k<n} \varphi(\bar{x}; \bar{a}^k)$$

but for every $l < n$,

$$\vDash (\exists \bar{x}) \bigwedge_{k<n,\, k \neq l} \varphi(\bar{x}; \bar{a}^k).$$

(2) T has the f.c.p. if there exists a formula $\varphi(x; \bar{y})$ which has the f.c.p.

Note. Here x is a single variable, see also Theorem 4.4.

LEMMA 4.1: *The formula $\varphi(\bar{x}; \bar{y})$ does not have the* f.c.p. *iff there is a natural number n such that: If Γ is a set of φ-m-formulas and every subset of Γ of cardinality $< n$ is consistent, then Γ is consistent.*

Proof. Immediate, by the definition.

THEOREM 4.2: (1) *If T does not have the* f.c.p., *then T is stable. (In other words, every unstable theory has the* f.c.p.)

(2) *If $\varphi(\bar{x}; \bar{y})$ is unstable, then the formula*

$$\psi(\bar{x}; \bar{z}) = \psi(\bar{x}; \bar{y}^1, \bar{y}^2, \bar{y}^3, \bar{y}^4)$$
$$= [\varphi(\bar{x}; \bar{y}^1) \equiv \neg\varphi(\bar{x}; \bar{y}^2)] \wedge [\varphi(\bar{x}; \bar{y}^3) \equiv \varphi(\bar{x}; \bar{y}^4)]$$

has the f.c.p.

Proof. (1) By Theorem 2.13, as T is unstable, some $\varphi(x; \bar{y})$ is unstable, hence (1) follows from (2).

(2) As $\varphi(\bar{x}; \bar{y})$ is unstable, by Theorem 2.2(1), (3) it has the order property, i.e., there are $\bar{a}^0, \ldots, \bar{a}^l, \ldots, l < \omega$ such that for every $k < \omega$,

$$p_k = \{\varphi(\bar{x}; \bar{a}^l)^{\text{if } (k \leq l)} : l < \omega\}$$

is consistent.

Let n be any natural number. For any $k < n$ define $\bar{c}^k = \bar{a}^0 {}^\frown \bar{a}^n {}^\frown \bar{a}^k {}^\frown \bar{a}^{k+1}$. We claim that $q = \{\psi(\bar{x}; \bar{c}^k) : k < n\}$ is inconsistent, but for every

$l < n$, $q_l = \{\psi(\bar{x}; \bar{c}^k): k < n, k \neq l\}$ is consistent. As n is arbitrary, by Lemma 4.1, this will prove the theorem.

Let us first prove that q is inconsistent. Suppose $\bar{b}$ realizes q. Then as $\models\psi[\bar{b}; \bar{c}^0]$, also $\models\varphi[\bar{b}; \bar{a}^0] \equiv \neg\varphi[\bar{b}; \bar{a}^n]$. On the other hand for every $k < n$, as $\models\psi[\bar{b}; \bar{c}^k]$, clearly $\models\varphi[\bar{b}; \bar{a}^k] \equiv \varphi[\bar{b}; \bar{a}^{k+1}]$.

Hence

$$\models\varphi[\bar{b}; \bar{a}^0] \Leftrightarrow \models\varphi[\bar{b}; \bar{a}^1] \Leftrightarrow \models\varphi[\bar{b}; \bar{a}^2] \Leftrightarrow \cdots \Leftrightarrow \models\varphi[\bar{b}; \bar{a}^n]$$
$$\Leftrightarrow \models\neg\varphi[\bar{b}; \bar{a}^0],$$

a contradiction. Therefore q is inconsistent.

Now if $l < n$, then clearly a sequence realizing p_{l+1} realizes q_l; hence q_l is consistent.

THEOREM 4.3: (1) *There is a stable theory with the* f.c.p.*; and there is a stable theory without the* f.c.p.

(2) *There are stable theories which are*

(i) *superstable and without the* f.c.p., *which are also stable in* $\aleph_0$,

(ii) *superstable with the* f.c.p., *which are also stable in* $\aleph_0$,

(iii) *unsuperstable without the* f.c.p.,

(iv) *unsuperstable with the* f.c.p.

Proof. Clearly (2) implies (1), so we prove (2) only.

(i) The theory of a model whose only relation is equality.

(ii) Let M be a model with the equality relation, and an equivalence relation, such that for every n there is an equivalence class of cardinality n. Clearly its theory satisfies our demands.

(iii) Let M be a model such that $|M| = {}^{\omega}\omega$; the relations of M are equality and for every n the equivalence relation E_n defined as

$$E_n = \{\langle\eta, \nu\rangle: \eta, \nu \in {}^{\omega}\omega, \eta \restriction n = \nu \restriction n\}.$$

The theory of M is the required theory; see Exercise 2.3.

(iv) By combining the two previous examples we can easily construct such a theory.

THE f.c.p. THEOREM 4.4: *Let T be stable. Then the following assertions are equivalent (where $1 \leq m < \omega$, $2 \leq \lambda \leq \infty$):*

(1) *T has the* f.c.p., *i.e., some $\varphi(x; \bar{y})$ has the* f.c.p.

$(2)_m$ *Some $\varphi(\bar{x}; \bar{y})$ has the* f.c.p., *$l(\bar{x}) = m$.*

$(3)_m$ *Not for every finite Δ, is there a $k < \omega$ such that a set p of Δ-m-formulas is realized iff every $q \subseteq p$, $|q| < k$ is realized.*

$(4)_m$ *For some finite Δ, k, and formula $\theta(\bar{x}; \bar{y})$, $R^m[\theta(\bar{x}; \bar{a}), \Delta, \aleph_0] = k$*

is not an elementary property of $\bar{a}$ *(that is, there is no formula* $\psi(\bar{y})$ *such that* $R^m[\theta(\bar{x}; \bar{a}), \Delta, \aleph_0] = k \Leftrightarrow \models \psi[\bar{a}]$).

$(5)_m$ *There are finite* Δ, k *and a formula* θ *such that for no* $l < \omega$ $R^m[\theta(\bar{x}; \bar{a}), \Delta, \aleph_0] = k \Rightarrow \mathrm{Mlt}[\theta(\bar{x}; \bar{a}), \Delta] \leq l$.

$(6)_{\lambda,m}$ *For some finite* Δ, *for no* $k < \omega$ *does the following hold: for every set* p *of* Δ*-m-formulas, there is* $q \subseteq p$, $|q| < k$, *such that* $R^m(p, \Delta, \lambda) = R^m(q, \Delta, \lambda)$.

$(7)_m$ *For some finite* Δ, *for no* $k < \omega$ *does the following hold: for every* Δ*-m-type* p, $R^m(p, \Delta, k) = R^m(p, \Delta, \infty)$.

$(8)_m$ *There is a formula* $\varphi(\bar{x}, \bar{y}; \bar{z})$ *such that* $l(\bar{x}) = l(\bar{y}) = m$ *and*

(A) *for every* $\bar{c}$, $\varphi(\bar{x}, \bar{y}; \bar{c})$ *is an equivalence relation.*

(B) *for every* n *and for some* $\bar{c}_n$, $\varphi(\bar{x}, \bar{y}; \bar{c}_n)$ *has* $\geq n$ *but only finitely many equivalent classes.*

Proof. We shall prove $(1) \Rightarrow (2)_m \Rightarrow (1)$, $(2)_m \Rightarrow (3)_m \Rightarrow (5)_m \Rightarrow (4)_m \Rightarrow (8)_m \Rightarrow (2)_m$ and $(3)_m \Rightarrow (6)_{\lambda,m}$, $(5)_m \Rightarrow (7)_m$, $(\forall m)$[not $(3)_m$ and not $(5)_m$] $\Rightarrow$ [not $(6)_{\lambda,m}$ and not $(7)_m$]. Clearly this is sufficient (as (1) does not depend on m and λ).

$(1) \Rightarrow (2)_m$ Trivial.

$(2)_m \Rightarrow (1)$ We shall prove it by induction on $m = l(\bar{x})$. For $m = 1$ this is true by definition. Hence assume we have proved it for m and we shall prove it for $m + 1$. Suppose it is not true. Then there is $\varphi = \varphi(\bar{x}; \bar{y}) = \varphi(x_0, \ldots, x_m; \bar{y})$ and for every $n < \omega$ a set Γ_n of formulas of the form $\varphi(\bar{x}; \bar{a})$ such that Γ_n is inconsistent but every subset of Γ_n of cardinality $< n$ is consistent. By Lemma 4.1 there is $l < \omega$ such that if Γ is a set of formulas of the form $\varphi(x_0; c_1, \ldots, c_m, \bar{a})$, and every subset of Γ of cardinality $< l$ is consistent, then Γ is consistent. Let

$$\psi(x_1, \ldots, x_m; \bar{y}^0, \ldots, \bar{y}^{l-1}) = (\exists x_0)\Big[\bigwedge_{k<l} \varphi(x_0, x_1, \ldots, x_m; \bar{y}^k)\Big]$$

and let for every n

$\Gamma_n^* = \{\psi(x_1, \ldots, x_m; \bar{a}^0, \ldots, \bar{a}^{l-1})$: for $\varphi(\bar{x}; \bar{a}^0) \in \Gamma_n, \ldots, \varphi(\bar{x}; \bar{a}^{l-1}) \in \Gamma_n\}$.

Clearly every subset of Γ_n^* of cardinality $< n/l$ is consistent. Hence by the induction hypothesis, for some sufficiently large n, Γ_n^* is consistent. Hence there are $c_1, \ldots, c_m$ which realize Γ_n^*. Let

$$\Gamma = \{\varphi(x_0; c_1, \ldots, c_m, \bar{a}): \varphi(x_0, \ldots, x_m; \bar{a}) \in \Gamma_n\}.$$

Clearly every subset of Γ of cardinality $\leq l$ is consistent, and so by the definition of l, Γ is consistent; hence there is c_0 which realizes it. So $\langle c_0, c_1, \ldots, c_m \rangle$ realizes Γ_n, a contradiction. So the implication is proved.

$(2)_m \Rightarrow (3)_m$ Trivial, $\Delta = \{\varphi(\bar{x}; \bar{y})\}$.

$(3)_m \Rightarrow (5)_m$ Suppose Δ exemplifies $(3)_m$; so for every $k < \omega$ there is a set p_k of Δ-m-formulas, which is not realized, but every $q \subseteq p_k$, $|q| < k$, is realized. W.l.o.g. p_k has minimal cardinality; so p_k is finite, $|p_k| \geq k$, and every proper subset of p_k is realized.

Let $n(0) = R^m(\bar{x} = \bar{x}, \Delta, 2)$, so $R^m(p, \Delta, \lambda) \leq n(0)$ for every p and λ. Now for each k, we define by induction on $l \leq n(0) + 1$ sets $p_k^l \subseteq p_k$. Let $p_k^0 = \emptyset$, and p_k^{l+1} be a subset of p_k of minimal cardinality such that $p_k^l \subseteq p_k^{l+1}$ and $R^m(p_k^{l+1}, \Delta, \aleph_0) \leq n(0) - l - 1$ (as $R^m(p_k, \Delta, \aleph_0) = -1$, p_k^l exists).

There is a maximal $l(0)$ such that for some $n < \omega$, $W = \{k < \omega\colon |p_k^{l(0)}| \leq n\}$ is infinite. So for some $n(1)$, $n(2) < \omega, \theta$, and infinite W_1, for each $k \in W_1$, $|p_k^{l(0)}| = n(1)$ and $\bigwedge p_k^{l(0)} = \theta(\bar{x}; \bar{c}_k)$ and $R^m[p_k^{l(0)}, \Delta, \aleph_0] = n(2)$; necessarily $n(2) = n(0) - l(0)$.

($n(2) \leq n(0) - l(0)$ by the definition of the p_k^l and $n(2) \geq n(0) - l(0)$ by the choice of $l(0)$) and $n(2) > 0$ ($n(2) \geq 0$ for we can choose $k > n(1)$ and $n(2) \neq 0$ by the choice of $l(0)$).

Clearly $|p_k^{l(0)+1}|$ tends to infinity with k; and for proving $(5)_m$ it suffices to prove that $\mathrm{Mlt}[\theta(\bar{x}; \bar{c}_k), \Delta] = \mathrm{Mlt}[p_k^{l(0)}, \Delta]$ tends to infinity with k. We prove that, more specifically, $\mathrm{Mlt}(p_k^{l(0)}, \Delta) \geq |p_k^{l(0)+1}| - |p_k^{l(0)}|$. Let A be such that all p_k are over A. For each $\varphi \in p_k^{l(0)+1} - p_k^{l(0)}$, by the choice of $p_k^{l(0)+1}$, $R^m[p_k^{l(0)+1} - \{\varphi\}, \Delta, \aleph_0] \geq n(0) - l(0)$, so there is $q_\varphi \in S^m(A)$, $p_k^{l(0)+1} - \{\varphi\} \subseteq q_\varphi$ and $R^m(q_\varphi, \Delta, \aleph_0) \geq n(2)$. But $R^m(p_k^{l(0)+1}, \Delta, \aleph_0) \leq n(0) - l(0) - 1$ so $\neg\varphi \in q$. Clearly the q_φ's show that $\mathrm{Mlt}(p_k^{l(0)}, \Delta) \geq |p_k^{l(0)+1}| - |p_k^{l(0)}|$.

$(5)_m \Rightarrow (4)_m$ Suppose Δ, k, θ exemplify $(5)_m$; so there are $\bar{a}_l (l < \omega)$ such that $R^m[\theta(\bar{x}; \bar{a}_l), \Delta, \aleph_0] = k$ and $\mathrm{Mlt}[\theta(\bar{x}; \bar{a}_l), \Delta] \geq l$. So for some A_l $|\{q \in S_\Delta^m(A_l)\colon R^m[\{\theta(\bar{x}; \bar{a}_l)\} \cup q, \Delta, \aleph_0] = k\}| \geq l$, so if $l \geq 2^{4n}$, by 1.6 of the Appendix there are φ_i^l, $\bar{b}_i^l$ $(i < n)$ such that $R^m(p_i^l, \Delta, \aleph_0) = k$,

$$p_i^l = \{\varphi_j^l(\bar{x}; \bar{b}_j^l)^{\mathrm{if}\,(i=j)}\colon j \leq i\} \cup \{\theta(\bar{x}; \bar{a}_l)\}$$

and $\varphi_i^l \in \mathrm{cl}_1(\Delta)$. So we can find $l(n+1) < \omega$ and φ_i such that for $i < n < \omega$ $\varphi_i^{l(n)} = \varphi_i$, and let for $i < n$, $\bigwedge p_{l(n)}^l = \theta_i(\bar{x}; \bar{c}_i^k)$. If $(4)_m$ is not true, there are ψ_i, ψ such that: for every $\bar{c}$, $R^m[\theta_i(\bar{x}; \bar{c}), \Delta, \aleph_0] = k \Leftrightarrow \vDash\psi_i[\bar{c}]$ and similarly for θ, ψ.

By a compactness argument we can find $\bar{a}$, $\bar{b}_j$ $(j < \omega)$ such that:

$$R^m[\theta(\bar{x}; \bar{a}), \Delta, \aleph_0] = k,$$

$$R^m[\{\theta(\bar{x}; \bar{a})\} \cup \{\varphi_i(\bar{x}; \bar{b}_j)^{\mathrm{if}\,(j=i)}\colon j \leq i\}, \Delta, \aleph_0] = k$$

contradicting the definition of the rank.

$(4)_m \Rightarrow (8)_m$ We choose Δ, k, θ exemplifying $(4)_m$, so that k is minimal. Clearly for every θ_1, the property $R^m[\theta_1(\bar{x}; \bar{a}_1), \Delta, \aleph_0] \geq k$ is elementary (as it is equivalent to $\neg\bigvee_{l<k} R^m[\theta_1(\bar{x};\bar{a}_1),\Delta,\aleph_0] = l$, and by the minimality of k). Moreover also the property "$R^m[\theta(\bar{x};\bar{a}),\Delta,\aleph_0] > k$ or $R^m[\theta(\bar{x};\bar{a}),\Delta,\aleph_0] \geq k$ and $\mathrm{Mlt}[\theta(\bar{x};\bar{a}),\Delta] \geq l$" is elementary (it is equivalent to the existence of $\varphi_{i,j} \in \Delta$ ($i \neq j < l$) and $\bar{b}_{i,j}$ such that

$$\bigwedge_{j<l} R^m[\{\theta(\bar{x};\bar{a})\} \cup \{\varphi_{i,j}(\bar{x};\bar{b}_{i,j}) : i < l\},\Delta,\aleph_0] \geq k$$

where $\varphi_{i,j}(\bar{x};\bar{b}_{i,j}) = \neg\varphi_{j,i}(\bar{x};\bar{b}_{j,i})$).

If $R^m[\theta(\bar{x};\bar{a}),\Delta,\aleph_0] = k \Rightarrow \mathrm{Mlt}[\theta(\bar{x};\bar{a}),\Delta] \leq l$ for some l (not depending on $\bar{a}$) it will follow that also $R^m[\theta(\bar{x};\bar{a}),\Delta,\aleph_0] = k$ is elementary, contradicting our hypothesis ($(4)_m$).

So for every $l < \omega$ there is $\bar{a}_l$ such that $R^m[\theta(\bar{x};\bar{a}_l),\Delta,\aleph_0] = k$, $\mathrm{Mlt}[\theta(\bar{x};\bar{a}_l),\Delta,\aleph_0] \geq l$. So for some A_l (by 1.10(1) and 1.10(4)).

$$\Phi_l = \{q \in S^m_\Delta(A_l) : R^m[\{\theta(\bar{x};\bar{a}_l)\} \cup q,\Delta,\aleph_0] = k\}$$

has cardinality $\mathrm{Mlt}[\theta(\bar{x};\bar{a}_l),\Delta]$ and $q \in \Phi_l$ implies $\mathrm{Mlt}[\{\theta(\bar{x};\bar{a})\} \cup q,\Delta] = 1$. By Claim 3.12, for each $j < \omega$, there is $l(j) < \omega$ (e.g., $2^{2^j}+1$) and distinct $q^j_i \in \Phi_{l(j)}$ ($i < j$) and $r^j_i \subseteq q^j_i$, $|r^j_i| \leq R^m(\bar{x}=\bar{x},\Delta,2)+2$, and $r^j_i \subseteq q \in \Phi_{l(j)} \Rightarrow q = q^j_i$.

We can find θ^* and $\bar{b}^j_i$ ($i < j$, $j < \omega$) such that for every $i < j$, $j < \omega$, $\theta(\bar{x};\bar{a}_{l(j)}) \wedge \bigwedge r^j_i = \theta^*(\bar{x};\bar{b}^j_i,\bar{a}_{l(j)})$, hence

$$R^m[\theta^*(\bar{x};\bar{b}^j_i,\bar{a}_{l(j)}),\Delta,\aleph_0] = k, \qquad \mathrm{Mlt}[\theta^*(\bar{x};\bar{b}^j_i,\bar{a}_{l(j)}),\Delta] = 1.$$

Clearly for every $\varphi \in \mathrm{cl}_1(\Delta)$ there is a formula ψ_φ, depending on Δ, m, k and θ^* only, such that:

$$R^m[\theta^*(\bar{x};\bar{b}^j_i,\bar{a}_{l(j)}) \wedge \varphi(\bar{x};\bar{c}),\Delta,\aleph_0] \geq k \Leftrightarrow \vDash \psi_\varphi[\bar{c},\bar{b}^j_i,\bar{a}_{l(j)}],$$

and for each $\varphi(\bar{x};\bar{c})$ exactly one of $\psi_\varphi[\bar{c},\bar{b}^j_i,\bar{a}_{l(j)}]$, $\psi_{\neg\varphi}[\bar{c},\bar{b}^j_i,\bar{a}_{l(j)}]$ holds.

We define an equivalence relation $E(\bar{y}_1,\bar{y}_2;\bar{a})$: $\bar{b}_1$, $\bar{b}_2$ are equivalent if:

(i) $R^m[\theta^*(\bar{x};\bar{b}_1,\bar{a}),\Delta,\aleph_0] = k \wedge \mathrm{Mlt}[\theta^*(\bar{x};\bar{b}_1,\bar{a}),\Delta] = 1$ iff

$$R^m[\theta^*(\bar{x};\bar{b}_2,\bar{a}),\Delta,\aleph_0] = k \wedge \mathrm{Mlt}[\theta^*(\bar{x};\bar{b}_2,\bar{a}),\Delta] = 1.$$

(ii) if the condition from (i) holds, then for each $\varphi \in \Delta$

$$\vDash (\forall \bar{z})[\psi_\varphi(\bar{z},\bar{b}_1,\bar{a}) \equiv \psi_\varphi(\bar{z},\bar{b}_2,\bar{a})].$$

It is easy to check that $E(\bar{y}_1,\bar{y}_2;\bar{a})$ is an equivalence relation, expressible by a first order formula (for (i) look at the beginning of the proof of $(4)_m \Rightarrow (8)_m$), and $E(\bar{y}_1,\bar{y}_2;\bar{a}_{l(j)})$ has $<\aleph_0$ but $\geq j$ equivalence classes.

$(8)_m \Rightarrow (2)_m$ Let $\neg\varphi(\bar{x},\bar{y};\bar{z})$ be as in $(8)_m$, and we shall prove that $\neg\varphi(\bar{x};\bar{y},\bar{z})$ has the f.c.p. For each n let $\bar{c}_n$ be such that $\varphi(\bar{x},\bar{y};\bar{c}_n)$ has $\geq n$ but $<\aleph_0$ equivalence classes; let $\bar{a}_1,\ldots,\bar{a}_l$ ($l \geq n$) be a set of representatives, so $p = \{\neg\varphi(\bar{x}; \bar{a}_i, \bar{c}_n): 1 \le i \le l\}$ is inconsistent but every $p' \subseteq p$, $p' \ne p$ is realized.

$(3)_m \Rightarrow (6)_{\lambda,m}$ Trivial (as p is realized iff $R^m(p, \Delta, \lambda) \ne -1$).

$(5)_m \Rightarrow (7)_m$ Immediate (use a minimal k). We finish by:

CLAIM 4.5: *Suppose not* $(3)_m$ *and not* $(5)_m$ *for every* m. *Then*

(A) *not* $(6)_{\lambda,m}$,

(B) *not* $(7)_m$.

Proof. (A) By (B) it suffices to prove (A) for any fixed $\lambda < \aleph_0$. By Theorem 2.2(7) and Theorems 1.1(1) and 1.3(1), for all p, $R^m(p, \Delta, \lambda) \le R^m(\bar{x} = \bar{x}, \Delta, 2) = n_0 < \omega$. By 2.1, w.l.o.g. $\Delta = \{\varphi\}$, so it suffices to prove that for every $k \le n_0$ there is $n_1(k, \lambda) < \omega$ such that every φ-m-type of rank $<k$ has a subtype q of cardinality $\le n_1(k, \lambda)$ such that $R^m(q, \varphi, \lambda) < k$.

Define for $\mathbf{t} \in \{0, 1\}$, $\psi^{\mathbf{t}}(\ldots, \bar{x}_\eta, \ldots; \bar{y}) = \bigwedge_{\eta \in {}^k\lambda} \varphi(\bar{x}_\eta; \bar{y})^{\mathbf{t}}$ and let

$$\begin{aligned}\psi(\ldots, \bar{x}_\eta, \ldots; \ldots, \bar{z}^{i,j}_\nu, \ldots)_{\eta \in {}^k\lambda, \nu \in {}^{k>}\lambda; i,j<\lambda} &= \bigwedge \{\varphi(\bar{x}_\eta; \bar{z}^{i,j}_\nu) \equiv \neg\varphi(\bar{x}_\rho; \bar{z}^{i,j}_\nu): \nu \in {}^{k>}\lambda; \eta, \rho \in {}^k\lambda;\\ & k > n = l(\nu), \nu = \eta \restriction n = \rho \restriction n, i = \eta[n], j = \rho[n]\}.\end{aligned}$$

For notational simplicity let $\psi^{\mathbf{t}} = \psi^{\mathbf{t}}(\bar{x}^*; \bar{y})$, $\psi = \psi(\bar{x}^*; \bar{z})$. By Lemma 2.9(2) it is clear that for every φ-m-type q $R^m(q, \varphi, \lambda) \ge k$ iff $\{\psi(\bar{x}^*; \bar{z})\} \cup \{\psi^{\mathbf{t}}(\bar{x}^*; \bar{a}): \varphi(\bar{x}; \bar{a})^{\mathbf{t}} \in q\}$ is consistent. Taking in Theorem 4.4(3) $\Delta = \{\psi, \psi^0, \psi^1\}$, $m' = l(\bar{x}^{*\frown}\bar{z})$ the result follows.

(B) Since for every $\lambda \ge 2$, and p:

(i) $R^m(p, \varphi, \lambda) \le n_0 =^{\text{def}} R^m(\bar{x} = \bar{x}, \varphi, 2) < \omega$,

(ii) $R^m(p, \varphi, n) = R^m(p, \varphi, \infty)$ $(n \le \lambda)$ implies $R^m(p, \varphi, n) = R^m(p, \varphi, \lambda)$;

it suffices to prove:

(*) for every n there are $\alpha(n), \beta(n) < \omega$ such that for any set of φ-m-formulas, p, if for every $q \subseteq p$, $|q| \le \alpha(n)$, $R^m[q, \varphi, \beta(n)] \ge n$, then $R^m(p, \varphi, \infty) \ge n$.

(Take $n_2 = \max\{\beta(n): n \le n_0\}$, so for (A) $k = \max\{\alpha(n): n \le n_0\}$.)

We prove it by induction on n. For $n = 0$, $R^m(p, \varphi, \lambda) \ge 0$ iff p is consistent; hence by "not $(3)_m$" our assertion holds. Suppose we have proved for n, and we shall prove for $n + 1$. By Theorem 2.12(2) for every $\eta \in {}^{\omega>}2$, $k, l < \omega$ there is a formula $\psi^{k,l}_\eta$ such that

$$R^m[\{\varphi(\bar{x}; \bar{a}_i)^{\eta[i]}: i < l(\eta)\}, \varphi, k] \ge l \quad \text{iff} \quad \models \psi^{k,l}_\eta[\bar{a}_0, \ldots, \bar{a}_{l(\eta)-1}].$$

Let $m_0 = l(\bar{y})$ where $\varphi = \varphi(\bar{x}; \bar{y})$; and $\Delta_n = \{\psi^{\beta(n),n}_\eta: \eta \in {}^{\alpha(n) \ge}2\}$.

Let $k(n) < \omega$ be such that

(i) every inconsistent set of Δ_n-m_0-formulas Γ has an inconsistent subset of power $< k(n)$, and

(ii) there are no $\bar{a}_i, \bar{b}_i$ such that $\vdash \varphi[\bar{b}_i; \bar{a}_j] \equiv \neg\varphi[\bar{b}_j; \bar{a}_i]$ for $i \neq j < k(n)$ [$k(n)$ exists by "not $(3)_m$" and Theorem 2.2 and 4.3].

Now let $\beta(n+1) = 2^{4k(n)} + \beta(n)$, and $\alpha(n+1)$ be such that if Γ is a set of φ-m-formulas, and for every $\Gamma' \subseteq \Gamma$ of power $< \alpha(n+1)$ $R^m[\Gamma', \varphi, \beta(n+1)] \geq n+1$, then $R^m[\Gamma, \varphi, \beta(n+1)] \geq n+1$ (exists by what we have already proved for part (A) of this claim) so $\alpha(n+1)$ depends on $\beta(n+1)$.

So it suffices to prove that if p is finite and $R^m[p, \varphi, \beta(n+1)] \geq n+1$, then $R^m(p, \varphi, \infty) \geq n+1$. As $R^m[p, \varphi, \beta(n+1)] \geq n+1$ there are φ-m-types $q_i \supseteq p$ ($i < \beta(n+1)$) such that $R^m[q_i, \varphi, \beta(n+1)] \geq n$, and they are explicitly contradictory in pairs. As $\beta(n+1) \geq \beta(n)$, by the induction hypothesis $R^m(q_i, \varphi, \infty) \geq n$; so by 1.6, we can assume all the q_i belong to some $S^m_\varphi(A)$. So, by 1.6 of the Appendix, by renaming, there are $\varphi(\bar{x}; \bar{a}_i)$ ($i < k(n)$) such that $\neg\varphi(\bar{x}; \bar{a}_i) \in q_j$ for $i < j < k(n)$, $\varphi(\bar{x}; \bar{a}_i) \in q_i$, and $\varphi(\bar{x}; \bar{a}_i)^t \in q_l$ for all $i < l < k(n)$, where t depends on n only. By (ii) of the definition of $k(n)$, for $i, j < k(n)$ $\varphi(\bar{x}; \bar{a}_i) \in q_j \Leftrightarrow i = j$. Now define $\bar{b}_i$ by induction: for $i < k(n)$ $\bar{b}_i = \bar{a}_i$; and for $i \geq k(n)$ define $\bar{b}_i$ so that for every $j \leq i$, $R^m[q^j_i, \varphi, \beta(n)] \geq n$ where $q^j_i = p \cup \{\varphi(\bar{x}; \bar{b}_\gamma)^{\text{if}(\gamma=j)}: \gamma \leq i\}$. By part (i) of the definition of $k(n)$ this is possible. By the induction hypothesis, $R^m(q^j_i, \varphi, \infty) \geq n$, hence $R^m(p, \varphi, \infty) \geq n+1$. So we have finished the induction, hence the proof of 4.5(B).

THEOREM 4.6: (1) *If T does not have the* f.c.p. *and Δ is finite, then there is $n_1(\Delta) < \omega$ such that: if $\{\bar{a}^\gamma: \gamma < \alpha \geq n_1(\Delta)\}$ is a Δ-n-indiscernible set over A, $\beta > \alpha$, then we can define $\bar{a}^\gamma$ for $\alpha \leq \gamma < \beta$ so that $\{\bar{a}: \gamma < \beta\}$ is a Δ-n-indiscernible set over A.*

(2) *If $A \subset |M|$, $\bar{a}^\gamma \in |M|$ for $\gamma < \alpha$ and M is $(|A|^+ + |\Delta|^+ + |\beta| + |\alpha|^+)$-compact, then we can choose $\bar{a}^\gamma \in |M|$ for $\alpha \leq \gamma < \beta$ (as in* (1)).

(3) *If A, Δ are finite $\alpha < \beta = \omega$, $A \subset |M|$, $\bar{a}^\gamma \in |M|$ for $\gamma < \alpha$ then we can choose $\bar{a}^\gamma \in |M|$ for $\alpha \leqslant \gamma < \beta$ (under the assumption of* (1)).

Proof. (3) follows from (2), and (2) will be clear from the proof of (1). So we shall prove (1) only.

If n is too large, every set of distinct $\bar{a}$'s is a Δ-n-indiscernible set. Similarly if $m = l(\bar{a}^0)$ is too large. So we can prove the theorem for fixed n, m, and still get an $n_1(\Delta)$ which is good for all n, m. Let

$$\Delta^* = \{\varphi(\bar{x}^0, \ldots, \bar{x}^{n-1}; \bar{y}): \varphi(\bar{x}^0, \ldots, \bar{x}^{n-1}; \bar{y}) = \psi(\bar{x}^{\sigma(0)}, \ldots, \bar{x}^{\sigma(n-1)}; \bar{y}) \text{ for some } \psi(\bar{x}^0, \ldots, \bar{x}^{n-1}; \bar{y}) \in \Delta \text{ and permutation } \sigma \text{ of } n\}.$$

As T does not have the f.c.p., by Theorem 4.4(3), there is a natural number $n^1 = n^1(\Delta)$ such that:

If Γ is a set of formulas of the form $\varphi(\bar{x}^0; \bar{a}^1, \ldots, \bar{a}^{n-1}, \bar{c})$, $\varphi \in \mathrm{cl}_1(\Delta^*)$ and every subset of Γ of cardinality $< n^1$ is consistent then Γ is consistent.

It is clearly sufficient to prove that we can define $\bar{a}^\alpha$ such that $\{\bar{a}^\gamma: \gamma < \alpha + 1\}$ will be a Δ-n-indiscernible set over A. For this it is clearly sufficient to find $\bar{a}^\alpha$ which realizes

$$p_\alpha = \{\varphi(\bar{x}; \bar{a}^{\gamma^{n-2}}, \ldots, \bar{a}^{\gamma^0}, \bar{c})^t: \bar{c} \in A, \gamma^0 < \cdots < \gamma^{n-2} < \alpha, t \in \{0, 1\},$$
$$\varphi \in \Delta^* \text{ and } \vDash \varphi[\bar{a}^{n-1}, \bar{a}^{n-2}, \ldots, \bar{a}^0, \bar{c}]^t\}.$$

For this, it is clearly sufficient to prove that p_α is consistent. By the definition of n^1, it suffices to prove that every subset of p_α of cardinality $< n^1$ is consistent. Let q be a subset of p_α, $|q| < n^1$. Clearly in q appear $\leq (n-1)(n^1-1)$ $\bar{a}^\gamma$'s. So if $\alpha > (n-1)(n^1-1)$, then there is $\bar{a}^\gamma$ which does not appear in any of the formulas of q, hence $\bar{a}^\gamma$ realizes q, so q is consistent. So p_α is consistent; hence we can define $\bar{a}^\alpha$, hence we can define $\bar{a}^\gamma$ for $\alpha \leq \gamma < \beta$, by induction, as required.

DEFINITION 4.2: (1) A formula $\varphi(\bar{x}; \bar{y})$ has the *independence property* if for every $n < \omega$ there are sequences $\bar{a}_l$ $(l < n)$ such that for every $w \subseteq n$,

$$\vDash (\exists \bar{x})\left[\bigwedge_{l<n} \varphi(\bar{x}; \bar{a}_l)^{\text{if } (l \in w)}\right].$$

(2) T has the *independence property* if some formula $\varphi(x; \bar{y})$ has the independence property.

DEFINITION 4.3: (1) A formula $\varphi(\bar{x}; \bar{y})$ has the *strict order property* if for every n there are $\bar{a}_l$ $(l < n)$ such that for any $k, l < n$

$$\vDash (\exists \bar{x})[\neg\varphi(\bar{x}; \bar{a}_k) \wedge \varphi(\bar{x}; \bar{a}_l)] \Leftrightarrow k < l.$$

(2) T has the *strict order property* if some $\varphi(x; \bar{y})$ has the strict order property.

Remark. (1) As in Definition 4.1(2), we can also in Definitions 4.2(2) and 4.3(2) replace $\varphi(x; \bar{y})$ by $\varphi(\bar{x}; \bar{y})$; in this case we use Theorems 4.11 and 4.16.

(2) We shall show that a theory is unstable iff it has the independence property or the strict order property; but neither of these properties implies the other (see 4.7 and 4.8).

THEOREM 4.7: (1) *T is unstable iff T has the independence property or the strict order property.*

(2) *$\varphi(\bar{x}; \bar{y})$ is unstable iff it has the independence property, or for some $n < \omega$, $\eta \in {}^n2$,*

$$\psi_\eta(\bar{x}; \bar{y}_0, \ldots, \bar{y}_{n-1}) = \bigwedge_{l<n} \varphi(\bar{x}; \bar{y}_l)^{\eta[l]}$$

has the strict order property.

Proof. (1) This follows from (2). It T is unstable, some $\varphi(x; \bar{y})$ is unstable (by 2.13); hence by (2) $\varphi(x; \bar{y})$ has the independence property or $\psi_\eta(x; \bar{y}_0, \ldots, \bar{y}_{n-1})$ has the strict order property (for some n, η). Thus in the first case T has the independence property, and in the second case the strict order property.

Suppose on the other hand that T, and thus some $\varphi(x; \bar{y})$, has one of these properties. If $\varphi(x; \bar{y})$ has the independence property, then by (2), $\varphi(x; \bar{y})$ is unstable and by Theorem 2.13, T is unstable. If $\varphi(x; \bar{y})$ has the strict order property, then so does ψ_η for $\eta = \langle 0 \rangle$ so by (2) again, $\varphi(x; \bar{y})$ is unstable and so is T.

(2) Suppose $\varphi(\bar{x}; \bar{y})$ has the independence property. Then clearly $\varphi(\bar{x}; \bar{y})$ has the order property. So by Theorem 2.2 $\varphi(\bar{x}; \bar{y})$ is unstable. Now assume that for some η ψ_η has the strict order property. Then ψ_η has the order property and thus ψ_η is unstable. So there is an A such that $|S^m_{\psi_\eta}(A)| > |A| \geq \aleph_0$, where $l(\bar{x}) = m$. For every $q \in S^m_{\psi_\eta}(A)$ let $\bar{a}_q$ realize q, and define

$$q^* = \{\varphi(\bar{x}; \bar{a})^{\mathfrak{t}} : \bar{a} \in A,\ \mathfrak{t} \in \{0, 1\},\ \models \varphi[\bar{a}_q; \bar{a}]^{\mathfrak{t}}\},$$

i.e., the φ-m-type $\bar{a}_q$ realizes over A.

Clearly $q^* \in S^m_\varphi(A)$; and for $p, q \in S^m_{\psi_\eta}(A)$, $q^* = p^*$ implies $q = p$; hence $q \neq p$ implies $q^* \neq p^*$, so

$$|S^m_\varphi(A)| \geq |\{q^* : q \in S^m_{\psi_\eta}(A)| > |A| \geq \aleph_0.$$

This means that $\varphi(\bar{x}; \bar{y})$ is unstable. So the strict order property of ψ_η implies the unstability of φ; and also the independence property of $\varphi(\bar{x}; \bar{y})$ implies $\varphi(\bar{x}; \bar{y})$ is unstable.

So it remains to be proved that if $\varphi = \varphi(\bar{x}; \bar{y})$ is unstable, then it has the independence property or for some η, ψ_η has the strict order property. By Theorem 2.2, $\varphi(\bar{x}; \bar{y})$ has the order property, so there are sequences $\bar{a}^0, \ldots, \bar{a}^n, \ldots$ such that for every $k < \omega$ there exists $\bar{c}^k$ such that $\models \varphi[\bar{c}^k; \bar{a}^n]$ iff $k \leq n$. By the compactness theorem and Theorem I,

2.4, we can assume that $\langle \bar{a}^n : n < \omega \rangle$ is an indiscernible sequence. If for every $n < \omega$, $w \subseteq n$,

$$\models (\exists \bar{x}) \Big[\bigwedge_{k<n} \varphi(\bar{x}; \bar{a}^k)^{\text{if } (k \in w)} \Big],$$

then clearly $\varphi(\bar{x}; \bar{y})$ has the independence property and so are finished. So we assume that there are $n < \omega$, $w \subseteq n$ such that

$$\models \neg (\exists \bar{x}) \Big[\bigwedge_{k<n} \varphi(\bar{x}; \bar{a}^k)^{\text{if } (k \in w)} \Big].$$

Let $|w| = n_0$. There is an $\alpha < \omega$ such that we can define w_l for $l \le \alpha$ such that:

(1) $w_0 = w$, $w_\alpha = \{n - n_0, n - n_0 + 1, \ldots, n - 1\}$,

(2) for every $l \le \alpha$, $|w_l| = n_0$, $w_l \subseteq n$,

(3) for every $l < \alpha$ there is $k_l < n$ such that $w_{l+1} = w_l \cup \{k_l + 1\} - \{k_l\}$ (and so $k_l \in w_l$, $k_l \notin w_{l+1}$, $k_l + 1 \notin w_l$, $k_l + 1 \in w_{l+1}$) (we step by step raise $w = w_0$ to w_α).

We are assuming $\models \neg (\exists \bar{x}) [\bigwedge_{k<n} \varphi(\bar{x}; \bar{a}^k)^{\text{if } (k \in w_0)}]$. On the other hand by the definition of the $\bar{a}^k$'s and w_α,

$$\models (\exists \bar{x}) \Big[\bigwedge_{k<n} \varphi(\bar{x}; \bar{a}^k)^{\text{if } (k \in w_\alpha)} \Big].$$

Hence there is $l < \alpha$ such that

$$\models \neg (\exists \bar{x}) \Big[\bigwedge_{k<n} \varphi(\bar{x}; \bar{a}^k)^{\text{if } (k \in s)} \Big], \quad \text{where } s = w_l,$$

and

$$\models (\exists \bar{x}) \Big[\bigwedge_{k<n} \varphi(\bar{x}; \bar{a}^k)^{\text{if } (k \in t)} \Big], \quad \text{where } t = w_{l+1}.$$

Let $\beta = k_l$, and

$$\psi = \psi(\bar{x}; \bar{y}, \bar{y}^0, \ldots, \bar{y}^{\beta-1}, \bar{y}^{\beta+2}, \ldots, \bar{y}^{n-1})$$
$$= \bigwedge_{\substack{k<n \\ k \ne \beta, \beta+1}} \varphi(\bar{x}; \bar{y}^k)^{\text{if } (k \in s)} \wedge \varphi(\bar{x}; \bar{y}).$$

We shall prove that ψ has the strict order property. Let $\gamma < \omega$, and define $\bar{b} = \bar{a}^0 \frown \ldots \frown \bar{a}^{\beta-1} \frown \bar{a}^{\beta+2+\gamma} \frown \ldots \frown \bar{a}^{n-1+\gamma}$. By the indiscernibility of $\langle \bar{a}^i : i < \omega \rangle$, and the definition of ψ, if $\beta \le k < l < \beta + 2 + \gamma$, then (as this holds for $k = \beta$, $l = \beta + 1$, $\gamma = 0$),

$$\models (\exists \bar{x}) [\psi(\bar{x}; \bar{a}^l, \bar{b}) \wedge \neg \varphi(\bar{x}; \bar{a}^k)];$$

but (by the same argument)

$$\vDash \neg(\exists \bar{x})[\psi(\bar{x}; \bar{a}^k, \bar{b}) \wedge \neg\varphi(\bar{x}; \bar{a}^l)].$$

Hence, observing again the definition of ψ,

$$\vDash (\exists \bar{x})[\psi(\bar{x}; \bar{a}^l, \bar{b}) \wedge \neg\psi(\bar{x}; \bar{a}^k, \bar{b})],$$

$$\vDash \neg(\exists \bar{x})[\psi(\bar{x}; \bar{a}^k, \bar{b}) \wedge \neg\psi(\bar{x}; \bar{a}^l, \bar{b})].$$

As this is true for every $\gamma < \omega$, clearly ψ has the strict order property, and it is also clear that ψ is of the required form. As this was proved under the assumption that $\varphi(\bar{x}; \bar{y})$ does not have the independence property, we have proved the theorem.

THEOREM 4.8: (1) *There is a theory T_{ord} with the strict order property and without the independence property.*

(2) *There is a theory T_{ind} with the independence property and without the strict order property. Moreover, there is no formula $\varphi(x_1, \ldots, x_n; \bar{a})$ which is connected and antisymmetric over an infinite set.*

Remark. Cf. Definition I, 2.5. Of course by 2.13 there are $m < \omega$ and a formula $\psi = \psi(\bar{x}^0, \ldots, \bar{x}^{n-1}; \bar{b})$, $l(\bar{x}^0) = \cdots = l(\bar{x}^{n-1}) = m$ such that ψ is connected and antisymmetric over an infinite set of sequences of length m.

Proof. (1) Let T_{ord} be the theory of the rational order, i.e., dense order without first or last element. Clearly $\varphi(x; y) = [x < y]$ has the strict order property. The checking that it does not have the independence property is left to the reader as an exercise.

(2) In the language of T_{ind} there will be only the equality sign, a one place predicate $P(x)$, and a two place predicate xEy. Its axioms will be:

(1) xEy implies $\neg P(x)$, $P(y)$; that is, $(\forall xy)[xEy \to \neg P(x) \wedge P(y)]$.

(2) if $P(y)$, then y is uniquely determined by $\{x: xEy\}$, and conversely; i.e.,

$$(\forall y_1 y_2)[P(y_1) \wedge P(y_2) \wedge (\forall x)(xEy_1 \equiv xEy_2) \to y_1 = y_2]$$

$$(\forall x_1 x_2)[\neg P(x_1) \wedge \neg P(x_2) \wedge (\forall y)(x_1 Ey \equiv x_2 Ey) \to x_1 = x_2].$$

$(3)_n$ For every $2n$ different elements in $\neg P, x_1, \ldots, x_n, x^1, \ldots, x^n$ there is a y such that $x_1Ey, \ldots, x_nEy, \neg x^1Ey, \ldots, \neg x^nEy$. That is,

$$(\forall x_1, \ldots, x_n)(\forall x^1, \ldots, x^n)\Bigg[\bigwedge_{\substack{0<k\le n\\ 0<l\le n}} [x_k \neq x^l \wedge \neg P(x_k) \wedge \neg P(x^l)] \rightarrow (\exists y) \bigwedge_{0<k\le n} (x_kEy \wedge \neg x^kEy)\Bigg].$$

$(4)_n$ The same as $(3)_n$, interchanging P and $\neg P$, xEy and yEx. That is,

$$(\forall y_1, \ldots, y_n)(\forall y^1, \ldots, y^n)\Bigg[\bigwedge_{\substack{0<k\le n\\ 0<l\le n}} [y_k \neq y^l \wedge P(y_k) \wedge P(y^l)] \rightarrow (\exists \bar{x}) \bigwedge_{0<k\le n} (xEy_k \wedge \neg xEy^k)\Bigg].$$

(5) All models of T_{ind} are infinite, i.e.,

$$(\exists z_0, \ldots, z_n) \bigwedge_{i<j\le n} z_i \neq z_j \quad \text{for every } n > 0.$$

It is not hard to prove that T_{ind} is consistent, by building a model for it. It is also easy and standard to prove it has elimination of quantifiers, and is complete. By this it can be shown that no formula $\varphi = \varphi(\bar{x}; \bar{y})$ has the strict order property. On the other hand, clearly $\varphi(x; y) = xEy$ has the independence property (hence the order property).

THEOREM 4.9: *If* $\varphi(\bar{x}; \bar{y})$ *is unstable,* $\lambda < \kappa < \text{Ded}\,\lambda$, *then there is an* A *such that* $|A| \le \lambda$, $|S^m_\varphi(A)| \ge \kappa$. *(See Definition 1.4 of the Appendix.)*

Proof. By Exercise 1.4 of the Appendix, there is an ordered set $\boldsymbol{J}$, $|\boldsymbol{J}| \ge \kappa$, with a dense subset $\boldsymbol{I}$, $|\boldsymbol{I}| = \lambda$. As $\varphi(\bar{x}; \bar{y})$ is unstable, by Theorem 2.2 φ has the order property. Hence by the compactness theorem there are $\bar{a}_s$, $s \in \boldsymbol{J}$ such that: for every $t \in \boldsymbol{J}$, $\{\varphi(\bar{x}; \bar{a}_s)^{\text{if } (t \le s)}: s \in \boldsymbol{J}\}$ is consistent.

Let $A = \bigcup \{\bar{a}_s: s \in \boldsymbol{I}\}$. Clearly $|A| \le |\boldsymbol{I}| \cdot \aleph_0 = \lambda$. For every $t \in \boldsymbol{J}$ let $p_t = \{\varphi(\bar{x}; \bar{a}_s)^{\text{if } (t<s)}: s \in \boldsymbol{I}\}$. Clearly p_t is a consistent φ-m-type over A. Let $p_t \subset q_t \in S^m_\varphi(A)$. Now if $s_1, s_2 \in \boldsymbol{J}$, $s_1 < s_2$, then there is $t \in \boldsymbol{I}$, $s_1 < t < s_2$, so $\varphi(\bar{x}; \bar{a}_t) \in q_{s_1}$, $\neg\varphi(\bar{x}; \bar{a}_t) \in q_{s_2}$. Hence $s_1 \neq s_2$ implies $q_{s_1} \neq q_{s_2}$. So

$$|S^m_\varphi(A)| \ge |\{q_s: s \in \boldsymbol{J}\}| = |\boldsymbol{J}| = \kappa > \lambda = |A|.$$

THEOREM 4.10: (1) *If* $\varphi = \varphi(\bar{x}; \bar{y})$ *has the independence property, then for every* λ *there is an* A *such that* $|A| = \lambda$, $|S^m_\varphi(A)| = 2^\lambda$.

(2) *If for some infinite* A, $|S^m_\varphi(A)| \geq \mathrm{Ded}_r|A|$, *then* $\varphi(\bar{x}; \bar{y})$ *has the independence property.* (Ded_r—*see Definition* 1.4 *of the Appendix.*)

(3) *If* $\varphi(\bar{x}; \bar{y})$ *has the independence property and* $l(\bar{y}) = k$, *then for every* $n < \omega$ *there is an* A, $|A| \leq nk$, $|S^m_\varphi(A)| \geq 2^n$.

(4) *If for every* $n < \omega$ *there is a finite* A, $|A| \geq 2$, *such that* $|S^m_\varphi(A)| \geq |A|^n$, *then* $\varphi(\bar{x}; \bar{y})$ *has the independence property.*

Proof. (1) Let

$$\Gamma = \left\{ (\exists \bar{x}) \left[\bigwedge_{i \in w} \varphi(\bar{x}; \bar{y}_i) \wedge \bigwedge_{i \in u} \neg\varphi(\bar{x}; \bar{y}_i) \right] : w \subseteq \lambda,\ u \subseteq \lambda;\ w \cap u = \emptyset;\ w, u \text{ are finite.} \right\}$$

As φ has the independence property, $\Gamma \cup T$ is consistent, hence has a model M. Let $\bar{a}_i$ realize $\bar{y}_i$ and $A = \bigcup_{i<\lambda} \bar{a}_i$. Clearly $|A| = \lambda$. For every (not necessarily finite) $w \subseteq \lambda$ let $p_w = \{\varphi(\bar{x}; \bar{a}_i)^{\mathrm{if}\,(i \in w)} : i < \lambda\}$. By the definition of Γ, p_w is consistent, hence there is $q_w \in S^m_\varphi(A)$, $p_w \subseteq q_w$. Clearly $w \neq u$ implies $q_w \neq q_u$, so

$$|S^m_\varphi(A)| \geq |\{q_w : w \subseteq \lambda\}| = |\{w : w \subseteq \lambda\}| = 2^\lambda$$

thus proving (1).

(2) Let $U = \{\varphi(\bar{x}; \bar{a}) : \bar{a} \in A\}$. The mapping $p \to p \cap U$ from $S^m_\varphi(A)$ to subsets of U is one-to-one. Hence $\{p \cap U : p \in S^m_\varphi(A)\}$ is a family of power $\geq \mathrm{Ded}_r|A| = \mathrm{Ded}_r|U|$. So by Theorem 1.7(1) of the Appendix for every $n < \omega$ there are $\varphi(\bar{x}; \bar{a}^i) \in U$, $i < n$, such that for every $w \subseteq n$ there is $p_w \in S^m_\varphi(A)$ such that for $i < n$ $\varphi(\bar{x}; \bar{a}^i) \in p_w \Leftrightarrow i \in w$. This implies that $\varphi(\bar{x}; \bar{y})$ has the independence property.

(3) The proof is similar to that of (1).

(4) The proof is similar to that of (2), this time using 1.7(2) of the Appendix.

THEOREM 4.11: *The following statements about a theory* T *are equivalent.*

(1) *Some formula* $\varphi(x; \bar{y})$ *has the independence property, i.e.,* T *has the independence property.*

$(2)_m$ *Some formula* $\varphi(\bar{x}; \bar{y})$ $(l(\bar{x}) = m)$ *has the independence property.*

(3) *There are* $\varphi(x; \bar{y})$ *and* $n < \omega$ *such that for every* $k < \omega$ *there is an* A, $|A| = k$, $|S_\varphi(A)| \geq 2^{k/n}$.

$(4)_m$ *There is* $\varphi(\bar{x}; \bar{y})$, $m = l(\bar{x})$, *such that for every* n *there is a finite* A, $|A| \geq 2$, $|S^m_\varphi(A)| \geq |A|^n$.

If for some λ, $(\forall\mu)[\mu < \mathrm{Ded}_r\,\lambda \rightarrow \mu^{|T|} < \mathrm{Ded}_r\,\lambda]$, *the following statements are equivalent as well.*

(5) *For every* μ *there is an* A, $|A| = \mu$, $|S(A)| \geq 2^{\mu}$.

$(6)_m$ *For some* A, $|S^m(A)| \geq \mathrm{Ded}_r|A|$, *and* $(\forall\mu)[\mu < \mathrm{Ded}_r|A| \rightarrow \mu^{|T|} < \mathrm{Ded}_r|A|]$.

Proof. Clearly (1) implies $(2)_m$ (by adding dummy variables to φ). By Theorem 4.10(3), $(2)_m$ implies $(4)_m$ and by 4.10(1), $(2)_m$ implies $(6)_m$, when assuming the existence of the suitable λ.

Suppose $(6)_m$ holds, and let $\lambda = |A|$; $\chi = \mathrm{Ded}_r\,\lambda$; so by Lemma I, 2.1 for some B, $|B| = |A|$, $|S(B)| \geq \chi$. As χ is regular, clearly by Lemma 2.15(1) for some φ, $|S_\varphi(B)| \geq \chi$. Hence by 4.10(2), $\varphi = \varphi(x; \bar{y})$ has the independence property. So $(6)_m$ implies (1). By 4.10(4) $(4)_m$ implies $(2)_m$.

By 4.10(1), (1) implies (5), and by 4.10(3), (1) implies (3), and trivially (3) implies $(4)_m$, (5) implies $(6)_m$. So we have (1) $\Rightarrow (2)_m \Rightarrow (4)_m \Rightarrow (2)_m$, (1) $\Rightarrow$ (3) $\Rightarrow (4)_m$; and when suitable λ exists (1) $\Rightarrow (2)_m \Rightarrow (6)_m \Rightarrow$ (1), (1) $\Rightarrow$ (5) $\Rightarrow (6)_m$. So the theorem is proved, by the consistency results of Baumgartner (see [Ba 76]).

Definition 4.4: (1) $K^m_\Delta(\lambda, T)$ is the first cardinal μ such that $|A| \leq \lambda$ implies $|S^m_\Delta(A)| < \mu$.

(2) $Kr^m_\Delta(\lambda, T)$ is the first regular or finite cardinal which is $\geq K^m_\Delta(\lambda, T)$.

(3) Usually we omit T. If $\Delta = L$ we omit it, if $\Delta = \{\varphi\}$ we write just φ and if $m = 1$ we omit it.

THEOREM 4.12: *$Kr^m_\varphi(\lambda)$ can be only one of the following functions:*

$$\text{constant } n \ (1 < n < \omega), \quad \lambda^+, \quad \mathrm{Ded}_r\,\lambda, \quad (2^\lambda)^+.$$

Moreover, each of these functions is realized (by some φ, T*). In fact,* φ *is unstable iff* $Kr^m_\varphi(\lambda) = \mathrm{Ded}_r\,\lambda$ *for all* λ *or* $Kr^m_\varphi(\lambda) = (2^\lambda)^+$ *for all* λ.

Proof. First suppose that $\varphi(\bar{x}; \bar{y})$ is unstable. Then by Theorem 4.9, $\lambda < \kappa < \mathrm{Ded}\,\lambda$ implies there is an A, $|A| \leq \lambda$, $|S^m_\varphi(A)| \geq \kappa$. Hence $K^m_\varphi(\lambda) \geq \mathrm{Ded}\,\lambda$. If for every λ, $Kr^m_\varphi(\lambda) = \mathrm{Ded}_r\,\lambda$, the conclusion of the theorem holds. So suppose for at least one λ, $Kr^m_\varphi(\lambda) > \mathrm{Ded}_r\,\lambda$. So by Definition 4.4, there is A, $|A| \leq \lambda$, $|S^m_\varphi(A)| \geq \mathrm{Ded}_r\,\lambda$. As $\mathrm{Ded}_r\,\lambda$ is regular, by Theorem 4.10(2) $\varphi(\bar{x}; \bar{y})$ has the independence property. So by Theorem 4.10(1) for every λ there is an A, $|A| \leq \lambda$, $|S^m_\varphi(A)| \geq 2^\lambda$, hence $Kr^m_\varphi(\lambda) > 2^\lambda$. But always $|S^m_\varphi(A)| \le 2^{|A|+\aleph_0}$, hence $Kr^m_\varphi(\lambda) \le (2^\lambda)^+$. So if φ is unstable, $Kr^m_\varphi(\lambda) = \text{Ded}_r\,\lambda$ (for every λ) or $Kr^m_\varphi(\lambda) = (2^\lambda)^+$ (for every λ).

So suppose $\varphi(\bar{x}; \bar{y})$ is stable. For every n, if for one λ, $Kr^m_\varphi(\lambda) > n$, then there is A, $|A| \le \lambda$, $|S^m_\varphi(A)| \ge n$. It is easy to find $B \subseteq A$, $|B| \le \aleph_0$ such that $|S^m_\varphi(B)| \ge n$; hence for every μ, as $\mu \ge \aleph_0 \ge |B|$, $Kr^m_\varphi(\mu) > n$. Hence if for some λ, $Kr^m_\varphi(\lambda) = n$, then for every μ, $Kr^m_\varphi(\mu) = n$. As for every A, and $\bar{a}$, $\bar{a}$ realizes a type in $S^m_\varphi(A)$, clearly $|S^m_\varphi(A)| \ge 1$, so $n > 1$.

So suppose $Kr^m_\varphi(\lambda) \ge \aleph_0$ for some λ. Then for any n there is A_n, $|S^m_\varphi(A_n)| \ge 2^n$. We can define by induction on $k < n$, $\bar{a}^k$, $\eta[k]$ such that

$$|\{p \in S^m_\varphi(A_n): \varphi(\bar{x}; \bar{a}^l)^{\eta[l]} \in p \text{ for every } l < k\}| \ge 2^{n-k}$$

and

$$\{\varphi(\bar{x}; \bar{a}^l)^{\eta[l]}: l < k-1\} \cup \{\varphi(\bar{x}; \bar{a}^{k-1})^{1-\eta[k-1]}\}$$

is consistent. For simplicity suppose every $\eta[l]$ is zero. Hence $\{\varphi(\bar{x}^n; \bar{y}^l)^{\text{if}\,(l<k)}: l \le k < n\}$ is consistent for every n. So for every λ, $\{\varphi(\bar{x}^\alpha; \bar{y}^\beta)^{\text{if}\,(\beta<\alpha)}: \beta \le \alpha < \lambda\}$ is consistent. Let $\bar{a}^\beta$ realize $\bar{y}^\beta$ and $\bar{c}^\alpha$ realize $\bar{x}^\alpha$, $A = \bigcup \{\bar{a}^\beta: \beta < \lambda\}$, and p_α be the φ-m-type $\bar{c}^\alpha$ realizes over A. Clearly,

$$|A| = \lambda, \qquad |S^m_\varphi(A)| \ge |\{p_\alpha: \alpha < \lambda\}| = \lambda.$$

Hence for every λ, $K^m_\varphi(\lambda) \ge \lambda^+$. As $\varphi(\bar{x}; \bar{y})$ is stable, $|A| \le \lambda$ implies $|S^m_\varphi(A)| \le \lambda$, so $K^m_\varphi(\lambda) \le \lambda^+$. So $K^m_\varphi(\lambda) = \lambda^+$. Clearly in the cases $K^m_\varphi(\lambda)$ is $(2^\lambda)^+$, λ^+, n; $Kr^m_\varphi(\lambda) = K^m_\varphi(\lambda)$.

Let us prove that each of the mentioned functions is $Kr^1_\varphi(\lambda)$ for some φ and T.

(1) By Theorem 4.8(2) there are T and $\varphi(x; y)$ such that $\varphi(x; y)$ has the independence property. Hence $K^1_\varphi(\lambda) = (2^\lambda)^+$.

(2) Let T be the theory of the order of the rationals, and $\varphi(x; y) = x < y$. Clearly $K^1_\varphi(\lambda) = \text{Ded}\,\lambda$.

(3) If T is the theory of equality, $\varphi(x; y) = [x = y]$, then $K^1_\varphi(\lambda) = \lambda^+$.

(4) Let T be a theory with an equivalence relation E with $n \ge 1$ equivalence classes, and $\varphi(x; y) = [xEy]$. Clearly $K^1_\varphi(\lambda) = n + 1$.

THEOREM 4.13: *Suppose $\psi(\bar{y}; \bar{x}) = \varphi(\bar{x}; \bar{y})$ and $\psi(\bar{y}; \bar{x})$ does not have the independence property. Let*

$$\Delta_n = \left\{(\exists \bar{y}) \bigwedge_{l<n} \psi(\bar{y}; \bar{x}^l)^{\eta[l]}: \eta \in {}^n2\right\}, \quad n > 0.$$

Then there is $n = n(\varphi) < \omega$ such that

(1) *if $\{\bar{a}^\beta : \beta < \alpha\}$ is a Δ_n-n-indiscernible set, $\bar{c}$ a sequence then either*

$$|\{\beta < \alpha : \models \varphi[\bar{a}^\beta; \bar{c}]\}| < n$$

or

$$|\{\beta < \alpha : \models \neg\varphi[\bar{a}^\beta; \bar{c}]\}| < n;$$

(2) *if $\{\bar{a}^\beta : \beta < \alpha\}$ is a Δ_n-n-indiscernible sequence and $\bar{c}$ is a sequence, then there are $0 = \alpha_0 \leq \alpha_1 \leq \cdots \leq \alpha_n = \alpha$ such that: if $i < n$, $\alpha_i \leq \beta^1$, $\beta^2 < \alpha_{i+1}$, then $\models \varphi[\bar{a}^{\beta^1}; \bar{c}]$ iff $\models \varphi[\bar{a}^{\beta^2}; \bar{c}]$.*

Proof. (1) As $\psi(\bar{y}; \bar{x})$ does not have the independence property there is $n_0 < \omega$ such that

$$\Gamma = \left\{ (\exists \bar{y}^w) \bigwedge_{l < n_0} \psi(\bar{y}^w; \bar{x}^l)^{\text{if } (l \in w)} : w \subseteq n_0 \right\}.$$

is inconsistent. Let $n = n_0$.

Now suppose our conclusion is incorrect. Then there are distinct ordinals $\beta_0, \ldots, \beta_{n-1}, \gamma_0, \ldots, \gamma_{n-1} < \alpha$ such that

$$\models \varphi[\bar{a}^{\beta_0}; \bar{c}], \ldots, \models \varphi[\bar{a}^{\beta_{n-1}}; \bar{c}],$$

$$\models \neg\varphi[\bar{a}^{\gamma_0}; \bar{c}], \ldots, \models \neg\varphi[\bar{a}^{\gamma_{n-1}}; \bar{c}].$$

Remembering that $\{\bar{a}^\beta : \beta < \alpha\}$ is a Δ_n-n-indiscernible set, we can see that taking $\bar{a}^l$ for $\bar{x}^l$ for $l < n$, Γ is satisfied. Hence Γ is consistent, contradiction.

(2) The proof is essentially the same.

DEFINITION 4.5: (1) Let I be a Δ-n-indiscernible set in M (i.e., $\bar{a} \in I \Rightarrow \bar{a} \in |M|$). Then $\dim(I, \Delta, n, M)$ is the first cardinal μ such that there exists a maximal Δ-n-indiscernible set I_1, in M, $I \subset I_1$, $|I_1| = \mu$ (I_1 being maximal means that there is no Δ-n-indiscernible set I_2 in M, $I_1 \subset I_2$, $I_1 \neq I_2$).

(2) Similarly we define $\dim(I, \Delta, M)$, $\dim(I, n, M)$, with Δ-indiscernibility, n-indiscernibility instead of Δ-n-indiscernibility; and the same with $\dim(I, \Delta, <n, M)$, but *not* $\dim(I, M)$.

THEOREM 4.14: *Suppose T does not have the independence property, Δ is finite. Then there are a natural number $n^* = n^*(\Delta)$ and a finite Δ^* such that: for any n; if I_1 is a Δ-n-indiscernible set in M, I_2 is a Δ^*-n^*-indiscernible set in M, $|I_1 \cap I_2| \geq n^*$, then*

$$n^*[\dim(I_1, \Delta, n, M)]^{n-1} \geq \dim(I_2, \Delta^*, n^*, M).$$

So if one of those dimensions is infinite, then

$$\dim(I_1, \Delta, n, M) \geq \dim(I_2, \Delta^*, n^*, M).$$

Proof. Remember that by Lemma I, 2.3, for any finite k; $\Delta^1, \ldots, \Delta^k$; $n^1, \ldots, n^k$ there are finite Δ, n such that I is Δ-n-indiscernible iff it is Δ^i-n^i-indiscernible for every i, $1 \leq i \leq k$. Remember also that for sufficiently large n, every set of sequences (of length m) is Δ-n-indiscernible. Thus it suffices to prove the theorem for a fixed n. Let

$$\Delta = \{\varphi_k(\bar{x}^0, \ldots, \bar{x}^{n-1}): k < k_0 < \omega\}$$

where $l(\bar{x}^0) = \cdots = l(\bar{x}^{n-1}) = m$. Define, for $k < k_0$, σ a permutation of n

$$\psi_{k,\sigma}(\bar{x}^1, \ldots, \bar{x}^{n-1}; \bar{x}^0) = \varphi_k(\bar{x}^{\sigma(0)}, \ldots, x^{\sigma(n-1)})$$

and let $\bar{z}^k = \bar{x}^1 \frown \cdots \frown \bar{x}^{n-1}$, so $\psi_{k,\sigma} = \psi_{k,\sigma}(\bar{z}^k; \bar{x}^0)$.

Now no $\psi_{k,\sigma}(\bar{z}^k; \bar{x}^0)$ has the independence property (by Theorem 4.11, as T does not have the independence property). So there is $i = i(k, \sigma) < \omega$ such that

$$\left\{(\exists \bar{z}^k) \bigwedge_{l<i} \psi_{k,\sigma}(\bar{z}^k; \bar{x}^{0,l})^{\text{if } (l \in w)}: w \subseteq i = i(k, \sigma)\right\}$$

is inconsistent. Define

$$n^* = \max\{i(k, \sigma) + n: k < k_0, \sigma \text{ a permutation of } n\}$$

$$\Delta^1 = \left\{(\exists \bar{z}^k) \bigwedge_{l<i} \psi_{k,\sigma}(\bar{z}^k; \bar{x}^{0,l})^{\text{if } (l \in w)}: w \subseteq i = i(k, \sigma), k < k_0, \right.$$
$$\left. \sigma \text{ is a permutation of } n\right\}.$$

$$\Delta^2 = \{\varphi_{k,\sigma}(\bar{x}^0; \bar{z}^k): k < k_0, \sigma \text{ a permutation of } n,$$
$$\varphi_{k,\sigma}(\bar{x}^0; \bar{z}^k) = \psi_{k,\sigma}(\bar{z}^k; \bar{x}^0)\} \cup \{\bar{x}^0 \neq \bar{y}\}, l(\bar{y}) = m.$$

We shall now show: If I_1 is a Δ-n-indiscernible set in M, I_2 a Δ^1-$(<n^*)$-indiscernible set in M, $|I_1 \cap I_2| \geq n^*$, then

$$n^*[\dim(I_1, \Delta, n, M)]^{n-1} \geq \dim(I_2, \Delta^1, <n^*, M).$$

Then we can find the required Δ^* by Lemma I, 2.3(3).

By the definition of dimension we can assume I_1 is a maximal Δ-n-indiscernible set in M of cardinality $\dim(I_1, \Delta, n, M)$; and similarly for I_2. So it suffice to prove only that $n^*|I_1|^{n-1} \geq |I_2|$.

Now there is a Δ^2-m-type q over $A = \bigcup \{\bar{a}: \bar{a} \in I_1\}$ such that $\bar{c}$ realizes q iff $I_1 \cup \{\bar{c}\}$ is a Δ-n-indiscernible set, $c \notin I_1$. We can assume that q is such that every formula in q contains elements from $\leq n - 1$

sequences of I_1. So every formula $\varphi(\bar{x}; \bar{b})$ in q is realized by every $\bar{c} \in I_1 \cap I_2$ except at most $n - 1$. Hence

$$|\{\bar{c} \in I_2: \models\varphi[\bar{c}; \bar{b}]\}| \geq |I_1 \cap I_2| - (n-1) \geq n^* - (n-1) > i(k, \sigma)$$

where φ is $\varphi_{k,\sigma}$ or $\neg\varphi_{k,\sigma}$. By the definition of Δ^1 and $i(k, \sigma)$, clearly $|\{\bar{c} \in I_2: \models\neg\varphi[\bar{c}; \bar{b}]\}| < i(k, \sigma) < n^*$ (by 4.13).

We may assume the number of formulas $\theta(\bar{x}; \bar{y})$ appearing in q is $\leq |\Delta^2|$.

Hence if $|I_2| > n^*|I|^{n-1}|\Delta^2| = n^{**}|I_1|^{n-1}$, then there is a $\bar{c} \in I_2 - I_1$ which satisfies every formula in q, hence $I_1 \cup \{\bar{c}\}$ is a Δ-n-indiscernible set, and as $\bar{c} \in I_2$, $\bar{c} \in |M|$. This contradicts the choice of I_1 as a maximal Δ-n-indiscernible set in M. So $|I_2| \leq n^{**}|I_1|^{n-1}$ so by replacing n^* by n^{**} we finish the proof.

THEOREM 4.15: *If T does not have the independence property and $|I_1 \cap I_2| \geq \aleph_0$; I_1, I_2 are indiscernible sets, then*

$$\dim(I_1, \Delta, M) + |T| = \dim(I_2, \Delta, M) + |T|.$$

Proof. The same as the previous theorem.

THEOREM 4.16: *If some $\varphi(\bar{x}; \bar{z})$ has the strict order property then T has the strict order property.*

Proof. We prove it by induction on $m = l(\bar{x})$. If $m = 1$, there is nothing to prove. Suppose we have proved for m and we shall prove for $m + 1$; for notational simplicity let $\varphi = \varphi(x, \bar{y}; \bar{z})$, $l(\bar{y}) = m$.

Let $\{\bar{a}_l: l < \omega\}$ be an indiscernible sequence such that

$$\models(\exists x, \bar{y})[\neg\varphi(x, \bar{y}; \bar{a}_l) \wedge \varphi(x, \bar{y}; \bar{a}_k)]$$

iff $l < k$ (exists by compactness and Theorem I, 2.4). If there is b such that

$$\models(\exists\bar{y})[\neg\varphi(b, \bar{y}; \bar{a}_{2l}) \wedge \varphi(b, \bar{y}; \bar{a}_{2l+1})]$$

for every $l < \omega$, then $\theta(\bar{y}; x, \bar{z}) = \varphi(x, \bar{y}; \bar{z})$ has the strict order property (this is exemplified by $\langle b\rangle^\frown\bar{a}_{2l}$, $l < \omega$) hence we finish by the induction hypothesis. Otherwise, by compactness there is a minimal $k < \omega$ such that for no b, $\models(\exists\bar{y})[\neg\varphi(b, \bar{y}; \bar{a}_{2l}) \wedge \varphi(b, \bar{y}; \bar{a}_{2l+1})]$ for every $l < k + 1$. Clearly $k \geq 1$. Let

$$\psi(x; \bar{z}_0, \ldots, \bar{z}_{2k-1}) = \bigwedge_{l<k} (\exists\bar{y})[\neg\varphi(x, \bar{y}; \bar{z}_{2l}) \wedge \varphi(x, \bar{y}; \bar{z}_{2l+1})].$$

Define for $l < \omega$,

$$\bar{b}_l = \bar{a}_0{}^\frown\cdots{}^\frown\bar{a}_{2k-3}{}^\frown\bar{a}_{2k-2}{}^\frown\bar{a}_{2k+2l}.$$

Then

(i) $\models(\forall x)[\psi(x;\bar{b}_l) \to \psi(x;\bar{b}_{l+1})]$ (for $l < \omega$) because if $b, \bar{c}$ satisfy $\models\neg\varphi[b,\bar{c};\bar{a}_{2k-2}] \wedge \varphi[b,\bar{c};\bar{a}_{2k+2l}]$, then

$$\models\neg\varphi[b,\bar{c};\bar{a}_{2k-2}] \wedge \varphi[b,\bar{c};\bar{a}_{2k+2l+2}].$$

(ii) $\models(\exists x)[\neg\psi(x;\bar{b}_l) \wedge \psi(x;\bar{b}_{l+1})]$ because

$$\models(\exists x)\psi(x;\bar{a}_0,\ldots,\bar{a}_{2k-3},\bar{a}_{2k+2l+1},\bar{a}_{2k+2l+2})$$

(by the minimality of k and the indiscernibility of $\{\bar{a}_l : l < \omega\}$) and the same x shows that (ii) holds.

By this and compactness, it follows that $\psi(x;\bar{z}_0,\ldots,\bar{z}_{2k-1})$ has the strict order property, hence we have finished the proof.

QUESTION 4.1: Suppose $|S^m_\varphi(A)| \geq \operatorname{Ded}|A|$; does φ necessarily have the independence property? (See 4.10(2).)

EXERCISE 4.2: Show that 4.6 may fail when T is stable but does not have the f.c.p.

EXERCISE 4.3: Show that we can add to 4.4 the conditions:

$(9)_m$ There is a finite Δ such that for every $l < \omega$ there is a Δ-indiscernible set I_l of sequences of length m, such that $l \leq \dim(I_l, \Delta, \mathfrak{C}) < \aleph_0$.

$(10)_m$ For some finite Δ_1, Δ_2 for no $k < \omega$ does the following hold: for every Δ_1-m-type p, $R^m(p,\Delta_2,k) = R^m(p,\Delta_2,\infty)$.

$(11)_{\lambda,m}$ For some finite Δ_1, Δ_2 for no $k < \omega$ does the following hold: for every set p of Δ_1-m-formulas, there is $q \subseteq p$, $|q| < k$ such that $R^m(p,\Delta_2,\lambda) = R^m(q,\Delta_2,\lambda)$.

EXERCISE 4.4: Show that T has the strict order property iff some $\varphi(\bar{x};\bar{y})$ is a partial non-trivial order, i.e., it is transitive $[(\forall\bar{x},\bar{y},\bar{z})(\varphi(\bar{x};\bar{y}) \wedge \varphi(\bar{y};\bar{z}) \to \varphi(\bar{x};\bar{z}))]$, irreflexive $[(\forall\bar{x})\neg\varphi(\bar{x};\bar{x})]$, asymmetric $[(\forall\bar{x},\bar{y})(\varphi(\bar{x};\bar{y}) \to \neg\varphi(\bar{y};\bar{x})]$ and with arbitrarily large finite chains [for every n, $(\exists\bar{x}_0,\ldots,\bar{x}_n)\bigwedge_{l<n}\varphi(\bar{x}_l;\bar{x}_{l+1})$].

EXERCISE 4.5: Let T^*_{ind} be the theory with the following axioms:

$$(\forall xy)(xRy \equiv yRx), \qquad (\forall x)\neg xRx, \tag{1}$$

$$(\forall x_1,\ldots,x_n;y_1,\ldots,y_n)\Big[\bigwedge_{i,j} x_i \neq y_j \to (\exists z)\Big(\bigwedge_i x_iRz \wedge \bigwedge_j \neg y_jRz\Big)\Big]. \tag{$(2)_n$}$$

Show that also T^*_{ind} satisfies the demands in Theorem 4.8(2).

EXERCISE 4.6: For stable T, regular λ, and a model M the following conditions are equivalent.

(1) There are finite Δ, m and a Δ-m-type p over $|M|$, $|p| = \lambda$, such that every $q \subseteq p$, $|q| < \lambda$, is realized in M, but p is not realized in M.

(2) There are finite Δ, m, Δ-m-types r, p over $|M|$, $|p| = \lambda$, $|r| \le \lambda$, such that for every $q \subseteq p$, $|q| < \lambda$, $r \cup q$ is realized in M, but $r \cup p$ is not realized in M.

(3) There are finite Δ, I, n such that $\lambda = \dim(I, \Delta, n, M)$.

(4) There are Δ, m and $p \in S^m_\Delta(|M|)$ such that M realizes every $q \subseteq p$, $|q| < \lambda$, and some $r \subseteq p$, $|r| = \lambda$, but not p itself.

[Hint: For (2) $\Rightarrow$ (3) let $p = \{\varphi_i(\bar{x}; \bar{a}_i): i < \lambda\}$, and let $\bar{b}_i$ realize $r \cup \{\varphi_j(\bar{x}; \bar{a}_j): j < i\}$. Now use 2.18 and 4.13(1). For (3) $\Rightarrow$ (4) use

$$p' = \{\varphi(\bar{x}; \bar{a}): \bar{a} \in M,\ \varphi \in \mathrm{cl}_1(\Delta) \text{ and } |\{\bar{c} \in J: \vDash \varphi[\bar{c}; \bar{a}]\}| \ge \aleph_0\}$$

for any J, $I \subseteq J' \subseteq |M|$, $|J| = \lambda$, $J \subseteq J'$ enough indiscernible.]

EXERCISE 4.7: Suppose T does not have the f.c.p. and M omits some Δ-m-type of cardinality λ, and Δ is finite. Then for some $E(\bar{x}, \bar{y}; \bar{z}) \in \mathrm{L}$ and $\bar{c} \in |M|$, $E(\bar{x}, \bar{y}; \bar{c})$ is an equivalence relation, and in M it has $\le \lambda$ but $\ge \aleph_0$ equivalence classes.

[Hint: By 2.1 we can assume $p = p \restriction^+ \varphi$ and by 4.2(1) T is stable, hence by 2.13 and 2.2 $n_1 = R^m(p, \varphi, \aleph_0) < \omega$. We can assume p has a unique extension q in $S^m_\varphi(|M|)$. We prove by induction on n_1. Choose a finite $p_1 \subseteq p$, $n_1 = R^m(p_1, \varphi, \aleph_0)$, let $\psi(\bar{y}; \bar{z})$ be as in 2.12(1) and let n^* be maximal such that

$$\{r \in S^m_\varphi(|M|): p_1 \subseteq r,\ n^* = R^m(r, \varphi, \aleph_0)\}$$

is infinite, so $n^* < n_1$. Let p_2 be finite, $p_1 \subseteq p_2 \subseteq q$, such that $p_1 \subseteq r \in S^m_\varphi(|M|)$, $R^m(r, \varphi, \aleph_0) > n^*$ and $r \ne q$ implies $p_2 \nsubseteq r$. Let $\theta(\bar{z}; \bar{c})$ ($\bar{c} \in |M|$) say: $\{\varphi(\bar{x}; \bar{y}_0): \psi(\bar{y}_0; \bar{z})\}$ is consistent, has rank ($R^m(-, \varphi, \aleph_0)$) $= n^*$, and extends p_2 (θ, $\bar{c}$ exist by Theorem 4.4). We could have chosen p_1 such that n^* is minimal. Then

$$E(\bar{z}_1, \bar{z}_2; \bar{c}) = \theta(\bar{z}_1; \bar{c}) \wedge \theta(\bar{z}_2; \bar{c}) \wedge (\forall \bar{y})[\psi(\bar{y}; \bar{z}_1) \equiv \psi(\bar{y}; \bar{z}_2)]$$

is as required.]

QUESTION 4.8: Suppose the condition of Exercise 4.6 holds. Is there an equivalence relation over M with exactly λ equivalence classes?

Chapter III

GLOBAL THEORY

III.0. Introduction

The main interest and motivation in this chapter is in uncountable theories. The results of Chapter II are sufficient to find the stability spectrum of T (i.e., the class of cardinals in which T is stable) when T is countable. We shall prove:

THE STABILITY SPECTRUM THEOREM: *For any stable T, there is a cardinal $\kappa(T) \leq |T|^+$ such that T is stable in λ iff $\lambda = \lambda^{<\kappa(T)} + |D(T)|$ or T is stable in λ iff $\lambda = \lambda^{<\kappa(T)} + 2^{\aleph_0}$.*

The main idea behind the theorem, is that if T is unstable in λ, $\lambda \geq 2^{|T|}$, then there is a sequence of equivalence relations (in T) E_i ($i < \delta$) each E_i partitions many equivalence class of $\bigwedge_{j<i} E_j$, into infinitely many parts, and $\lambda^{|\delta|} > \lambda$. (This is almost true, as by 6.13 if $\aleph_0 < \kappa < \kappa(T)$, κ is regular, then there are such E_i ($i < \kappa$).)

For proving the stability spectrum theorem, we seek a notion, which we call "p forks over A" and which satisfies:

(0) $\kappa(T)$ is the minimal cardinality κ; such that for every $p \in S^m(A)$ there is $B \subseteq A$, $|B| < \kappa(T)$ such that p does not fork over B (in fact, $\kappa(T) \leq |T|^+$).

(1) For every $B \subseteq A$, $|B| < \kappa(T)$, $\{p \in S^m(A): p \text{ does not fork over } B\}$ has small cardinality (in fact $\leq 2^{|T|}$).

(2) If $p \in S^m(A)$ forks over B, then for some finite $C \subset A$, $p \restriction (B \cup C)$ forks over B.

(3) If $p \in S^m(A_\delta)$, A_i ($i \leq \delta$) is increasing, and $p \restriction A_{i+1}$ forks over A_i then T is unstable in λ whenever $\lambda^{|\delta|} > \lambda$.

Clearly properties (0), (1) enable us to prove stability results, whereas (2), (3) enable us to prove instability results.

The notion of splitting is almost suitable, but it satisfies (3) only for

$\lambda < 2^{|\phi|}$. This was the motivation for defining: p strongly splits over B if $\varphi(\bar{x}, \bar{a}), \neg\varphi(\bar{x}, \bar{b}) \in p$ where $\bar{a}, \bar{b}$ belong to an infinite set indiscernible over B. This is sufficient to prove the stability spectrum theorem for $\lambda \geq 2^{|T|}$. However, for finer considerations it is not good enough, as there is an unjustified dependence on the parameters appearing in the type, and as it does not necessarily satisfy:

(4) If p does not fork over B, $\operatorname{Dom} p \subseteq A$ then there is q, $p \subseteq q \in S^m(A)$, which does not fork over B.

All this leads to defining "p forks over B" by: for some A, $\operatorname{Dom} p \subseteq A$, each extension of p in $S^m(A)$ splits strongly over B. Notice that for $B \subseteq A$, $p \in S^m(A)$, "p does not fork over B" is a good substitute for "p and $p \restriction B$ have the same rank" and it looks more natural (see Section 4 for the connection).

In Section 1 we define forking, and prove its simplest properties. It is interesting to note that "$p \in S^m(A)$ does not fork over A" is relatively problematic, and in its proof we assume T is stable and use the ranks $R^m(p, \Delta, \aleph_0)$ for finite Δ's. Hence in Sections 1–5 we deal with stable T only. Note that some of the lemmas in Section 1 are needed only to get better results in Section 2.

We say p is stationary over B, if for every A, $\operatorname{Dom} p \subseteq A$, there is one and only one $q \in S^m(A)$ extending p which does not fork over B. It is interesting to know how much information p should contain in order to be stationary over B (assuming p does not fork over B); and the finite equivalence relation theorem answers it. A formula $\varphi(\bar{x}; \bar{a})$ is almost over B if some $\psi(\bar{x}, \bar{y}, \bar{b})$ ($\bar{b} \in B$) is an equivalence relation with finitely many equivalence classes, and $\psi(\bar{x}; \bar{y}; \bar{b}) \vdash \varphi(\bar{x}; \bar{a}) \equiv \varphi(\bar{y}; \bar{a})$. So $p \in S^m(A)$ ($B \subseteq A$) is stationary over B iff it does not fork over B, and it "decides" all formulas almost over B. Hence each $p \in S(|M|)$ is stationary over $|M|$ and we can prove property (1) of forking.

In Section 3 we prove the stability spectrum theorem for $\lambda \geq 2^{|T|}$. We also define the dimension $\dim(I, A, M)$ (Definition 3.3) and show that it is well behaved when it is $\geq \kappa_r(T)$. We prove that a $\kappa_r(T)$-saturated model is not λ^+-saturated iff some I has dimension $\leq \lambda$. We also prove that λ-saturation is preserved under unions of length δ if $\operatorname{cf} \delta \geq \kappa(T)$, and that a λ-stable T has a saturated model of cardinality λ.

In Section 4 we give a quite complete picture of the connection between ranks and forking, being stationary and the multiplicity; and show that complete types over $|M|$ have usually multiplicity 1. But more important are the results:

THEOREM 0.1: (1) (*symmetry*) $\mathrm{tp}(\bar{a}, A \cup \bar{b})$ *does not fork over* A *iff* $\mathrm{tp}(\bar{b}, A \cup \bar{a})$ *does not fork over* A (*see* 4.13).

(2) (*transitivity*) *if* $A \subseteq B \subseteq C$, *and* $\mathrm{tp}(\bar{a}, C)$ *does not fork over* B, $\mathrm{tp}(\bar{a}, B)$ *does not fork over* A *then* $\mathrm{tp}(\bar{a}, C)$ *does not fork over* A (*see* 4.4).

(3) *Let* $B \subseteq A$, *then* $\mathrm{tp}(\bar{b}, A \cup \bar{a})$, $\mathrm{tp}(\bar{a}, A)$ *do not fork over* $B \cup \bar{a}$, B *resp. iff* $\mathrm{tp}(\bar{a}^\frown \bar{b}, A)$ *does not fork over* B (4.14, 4.15).

(4) *The type* p *does not fork over* $|M|$ *iff* p *is finitely satisfiable in* M (*see* 4.10).

This theorem will enable us to introduce in Chapter IV $\mathbf{F}^f_\lambda$-constructions.

In Section 5 we complete the stability spectrum theorem for $\lambda \leq 2^{|T|}$, by proving in the appropriate cases, that there are many independent formulas (over φ) using very specific dependence relations.

The idea of Section 6 is that sometimes equivalence classes are not less real than elements, hence when we seek a canonical form, sometimes it is an equivalence class. So we introduce $\mathfrak{C}^{eq}$, which is just like $\mathfrak{C}$ except that we have elements for equivalence classes. We get, e.g., that for stable T, for each $p \in S^m(|M|)$ there is a canonical $B \subseteq |M|$, over which p does not fork, B is the algebraic closure of some C, $|C| < \kappa(T)$.

In Section 7 we are interested mainly in unstable theories, and get various results. Some theorems are generalizations of the stability spectrum theorem. We find $\kappa_{\mathrm{ird}}(T) \leq |T|^+$ such that $\kappa < \kappa_{\mathrm{ird}}(T)$ iff in T there are κ independent orders *iff* for some A, $|S^m(A)| > \lambda^\kappa$ for every $\lambda < \mathrm{Ded}|A|$. (Thus we get a complete characterization of $Kr(\lambda)$ for countable T, and a better picture in general). We also find when there are $> \lambda$ pairwise contradictory types of cardinality $< \kappa$ over a set A, $|A| = \lambda$.

Those proofs indicate an alternative proof to the stability spectrum theorem for $\lambda \geq 2^{|T|}$ without using forking or ranks (see Exercise 7.8).

In Sections 1–5, T will be stable and Δ finite in all unstarred theorems and lemmas. No unstated assumptions are made on T or Δ in starred theorems and lemmas.

III.1. Forking

Notation. For an index set U let $\bar{x}_U = \{x_u : u \in U\}$ and for a sequence $\bar{u} = \langle u_i \rangle_{i<i_0}$ let $\bar{x}_{\bar{u}} = \langle x_{u_i} \rangle_{i<i_0}$. A *type in* $\bar{x}_U$ *over* B is a consistent set of formulas using only variables from $\bar{x}_U$ and parameters from B. The

type is complete if it is maximal among the types in $\bar{x}_U$ over B. For an m-type p and a set I of sequences of length m let $p(I) = \{\bar{a} \in I: \bar{a}$ realizes $p\}$. Similarly we define $\varphi(I; \bar{b})$ for a formula $\varphi(x; \bar{b})$. Instead of ${}^nA$ we sometimes write just A.

DEFINITION 1.1: $\mathrm{tp}_*(B, A) = \{\varphi(\bar{x}_{\bar{b}}; \bar{a}): \bar{b} \in B, \bar{a} \in A, \vDash\varphi[\bar{b}; \bar{a}]\}$, $\mathrm{tp}_*(\bar{b}, A) = \{\varphi(\bar{x}_{\bar{c}}; \bar{a}): \bar{a} \in A, \bar{c} \subseteq \bar{b}, \vDash \varphi[\bar{c}; \bar{a}]\}$.

DEFINITION 1.2: p *splits strongly* over A if there is an indiscernible sequence $I = \{\bar{a}^n: n < \omega\}$ over A and a formula φ such that $\varphi(\bar{x}; \bar{a}^0)$, $\neg\varphi(\bar{x}; \bar{a}^1) \in p$. (Compare Definition I, 2.6.)

Remark. Clearly if p splits strongly over A then p splits over A.

DEFINITION 1.3: The formula $\varphi(\bar{x}; \bar{a})$ *divides* over A if there are $n < \omega$ and sequences $\bar{a}^l$, $l < \omega$, such that

(1) $\mathrm{tp}(\bar{a}, A = \mathrm{tp}(\bar{a}^l, A)$,

(2) $\{\varphi(\bar{x}; \bar{a}^l); l < \omega\}$ is n-inconsistent (see Definition II, 3.2(3)(ii)).

DEFINITION 1.4: The type p in $\bar{x}_U$ *forks over* A if there are formulas $\varphi^0(\bar{x}^0; \bar{a}^0), \ldots, \varphi^{n-1}(\bar{x}^{n-1}; \bar{a}^{n-1})$, with $\bar{x}^k \subseteq \bar{x}_U$ for $k < n$, such that

(1) $p \vdash \bigvee_{k<n} \varphi^k(\bar{x}^k; \bar{a}^k)$,

(2) $\varphi^k(\bar{x}^k; \bar{a}^k)$ divides over A for all $k < n$.

Instead of writing "$\{\varphi(\bar{x}; \bar{a})\}$ forks" we write "$\varphi(\bar{x}; \bar{a})$ forks".

Note. p is not in general a type over A; see Corollary 1.3.

LEMMA 1.1*: (1) *If* $\bar{x} \subseteq \bar{y}$, $\vdash\psi(\bar{y}; \bar{c}) \to \varphi(\bar{x}; \bar{c})$ *and* $\varphi(\bar{x}; \bar{c})$ *divides over* A, *then* $\psi(\bar{y}; \bar{c})$ *divides over* A. (*In particular this holds when* ψ *is obtained by adding dummy variables to* φ.)

(2) *If* p *is a type in* $\bar{x}_{\bar{u}}$, ($\bar{u}$ *finite*) *then* p *forks over* A *iff there are formulas* $\varphi^0(\bar{x}_{\bar{u}}; \bar{a}^0), \ldots, \varphi^{n-1}(\bar{x}_{\bar{u}}; \bar{a}^{n-1})$ *such that* $p \vdash \bigvee_{k<n} \varphi^k(\bar{x}_{\bar{u}}; \bar{a}^k)$, *and* $\varphi^k(\bar{x}_{\bar{u}}; \bar{a}^k)$ *divides over* A *for* $k < n$.

(3) $\varphi(\bar{x}; \bar{a})$ *divides over* A *iff there are* $n < \omega$ *and a sequence* $I = \{\bar{a}^l: l < \omega\}$ *indiscernible over* A *such that* $\bar{a} = \bar{a}^0$ *and* $\{\varphi(\bar{x}; \bar{a}^l): l < \omega\}$ *is* n-*inconsistent*.

(4) *If* $\varphi(\bar{x}; \bar{a})$ *divides over* A *then* $\varphi(\bar{x}; \bar{a})$ *forks over* A.

(5) p *forks over* A *iff some finite subtype* q *of* p *forks over* A (*so* q *is a type in* $\bar{x}_{\bar{u}}$ *for some finite* $\bar{u}$).

(6) $\mathrm{tp}_*(C, B)$ *forks over* A *iff for some* $\bar{c} \in C$, $\mathrm{tp}_*(\bar{c}, B)$ (*or equivalently* $\mathrm{tp}(\bar{c}, B)$) *forks over* A.

(7) *If* $p_i \subseteq p_j$ *for* $i < j < \delta$ *and no* p_i *forks over* A, *then* $\bigcup_{i<\delta} p_i$ *does not fork over* A.

(8) *If* $A \subseteq B$, q *and* p *are types in the same variables*, $q \vdash p$ $[p \subseteq q]$ *and* q *does not fork* [*split strongly*] *over* A *then* p *does not fork* [*split strongly*] *over* B.

(9) *If* $p \cup \{\psi^l(\bar{x}; \bar{a}^l)\}$ *forks over* A *for* $l < n < \omega$, *then* $p \cup \{\bigvee_{l<n} \psi^l(\bar{x}; \bar{a}^l)\}$ *forks over* A.

Proof. (1) is immediate and (2) follows from the parenthetical remark in (1).

(3) If the stated condition holds then clearly $\varphi(\bar{x}; \bar{a})$ divides over A. Now suppose $\varphi(\bar{x}; \bar{a})$ divides over A. Then there are $n < \omega$ and sequences $\bar{a}^l$, $l < \omega$ such that $\text{tp}(\bar{a}, A) = \text{tp}(\bar{a}^l, A)$, and $\{\varphi(\bar{x}; \bar{a}^l): l < \omega\}$ is n-inconsistent. By Ramsey's theorem and compactness there are $\bar{b}^l$, $l < \omega$ such that $\{\bar{b}^l: l < \omega\}$ is an indiscernible sequence over A, $\text{tp}(\bar{b}^l, A) = \text{tp}(\bar{a}, A)$ and $\{\varphi(\bar{x}; \bar{b}^l): l < \omega\}$ is n-inconsistent. So there is an automorphism F of $\mathfrak{C}$ such that $F \restriction A$ is the identity and $F(\bar{b}^0) = \bar{a}$. So $\{F(\bar{b}^l): l < \omega\}$ is the required sequence.

(4) Immediate. (A converse is Exercise 4.15.)

(5) Since if $p \vdash \bigvee_{i<n} \varphi^i$ then for some finite $q \subseteq p$, $q \vdash \bigvee_{i<n} \varphi^i$.

The rest are immediate.

LEMMA 1.2: *Let* p, q *be* m-*types*, p *is over* A.

(1) *If* $\varphi(\bar{x}; \bar{y}) \in \Delta$, $l(\bar{x}) = m$, $\varphi(\bar{x}; \bar{a})$ *divides over* A, *then* $R^m(p, \Delta, \aleph_0) > R^m[p \cup \{\varphi(\bar{x}; \bar{a})\}, \Delta, \aleph_0]$.

(2) *If* $q \supseteq p$, q *forks over* A, *then there is a finite* Δ_0 *such that for every finite* $\Delta \supseteq \Delta_0$, $R^m(q, \Delta, \aleph_0) < R^m(p, \Delta, \aleph_0)$.

(3)* *If* $\varphi(\bar{x}; \bar{y}) \in \Delta$, $\varphi(\bar{x}; \bar{a})$ *divides over* A, $\lambda \geq 2$, *then*

$$R^m(p, \Delta, \lambda) < \infty \Rightarrow R^m[p \cup \{\varphi(\bar{x}; \bar{a})\}, \Delta, \lambda] < R^m(p, \Delta, \lambda).$$

$$D^m(p, \Delta, \lambda) < \infty \Rightarrow D^m[p \cup \{\varphi(\bar{x}; \bar{a})\}, \Delta, \lambda] < D^m(p, \Delta, \lambda).$$

(4)* *If* $q \supseteq p$, q *forks over* A, $\lambda \geq \aleph_0$, *then there is a finite* Δ_0 *such that for every* $\Delta \supseteq \Delta_0$,

$$R^m(p, \Delta, \lambda) < \infty \Rightarrow R^m(q, \Delta, \lambda) < R^m(p, \Delta, \lambda)$$

$$D^m(p, \Delta, \lambda) < \infty \Rightarrow D^m(q, \Delta, \lambda) < D^m(p, \Delta, \lambda).$$

(5)* *If in* (4) *we assume only that* q *splits over* A, *then the conclusion is valid for* $\lambda = 2$.

Proof. (1) By Lemma 1.1(3) there is $I = \{\bar{a}^l: l < \omega\}$ which is an indiscernible sequence over A, $\bar{a} = \bar{a}^0$, and $\{\varphi(\bar{x}; \bar{a}^l): l < \omega\}$ is n-inconsistent. Let $p_l = p \cup \{\varphi(\bar{x}; \bar{a}^l)\}$ for $l < \omega$, $\alpha = R^m(p, \Delta, \aleph_0)$.

Clearly $\alpha \geq R^m(p_l, \Delta, \aleph_0) = R^m(p_0, \Delta, \aleph_0)$, by Theorem II, 1.1(1) and Exercise II, 1.1 and $\alpha < \infty$ as T is stable.

Suppose $\alpha = R^m(p_l, \Delta, \aleph_0)$ by way of contradiction. Let $B = A \cup \bigcup I$. By Theorem II, 1.6 there are $q_l \in S^m_\Delta(B)$ such that $R^m(p_l \cup q_l, \Delta, \aleph_0) = \alpha$. Notice that $p \cup q_l = p_l \cup q_l$. Now $|\{l < \omega: \varphi(\bar{x}; \bar{a}^l) \in q_k\}| < n$ so $|\{q_l: q_l = q_k\}| < n$. Thus there are infinitely many distinct q_k, which are clearly pairwise explicitly contradictory. Since $R^m(p \cup q_k, \Delta, \aleph_0) = \alpha$, $R^m(p, \Delta, \aleph_0) > \alpha$, contradiction.

(2) Since q forks over A, let $q \vdash \bigvee_{k<n} \varphi^k(\bar{x}; \bar{a}^k)$ as in Lemma 1.1(2) where $\varphi^k(\bar{x}; \bar{a}^k)$ divides over A. Take $\Delta_0 = \{\varphi^k(\bar{x}; \bar{y}^k): k < n\}$ and let $\Delta \supseteq \Delta_0$ be finite. Clearly $q \vdash p \cup \{\bigvee_{k<n} \varphi^k(\bar{x}; \bar{a}^k)\} \vdash p$. Thus

$$R^m(q, \Delta, \aleph_0) \leq R^m\left[p \cup \left\{\bigvee_{k<n} \varphi^k(\bar{x}; \bar{a}^k)\right\}, \Delta, \aleph_0\right]$$

$$= \max_{k<n} R^m[p \cup \{\varphi^k(\bar{x}; \bar{a}^k)\}, \Delta, \aleph_0],$$

this by Claim II, 1.7. By part (1) just proved, $R^m[p \cup \{\varphi^k(\bar{x}; \bar{a}^k)\}, \Delta, \aleph_0] < R^m(p, \Delta, \aleph_0)$. Since $n < \omega$ we get

$$R^m(q, \Delta, \aleph_0) \leq \max_{k<n} R^m[p \cup \{\varphi^k(\bar{x}; \bar{a}^k)\}, \Delta, \aleph_0] < R^m(p, \Delta, \aleph_0).$$

(3), (4), (5) are proved similarly.

COROLLARY 1.3: *If p is a type in $\bar{x}_U$ over A then p does not fork over A.*

Proof. Suppose p does fork over A. By 1.1(5), (1) we can assume p is an m-type. Taking $q = p$ in part (2) of the above lemma $R^m(p, \Delta, \aleph_0) < R^m(p, \Delta, \aleph_0)$: contradiction.

Remark. Clearly if p is a type over A then p does not split strongly over A. This also follows from Theorem 1.6.

THEOREM 1.4*: *If p is a type in $\bar{x}_U$ over B which does not fork over A, then there is a complete type $q \supseteq p$ in $\bar{x}_U$ over B which does not fork over A.*

Proof. Let $\Gamma = \{\psi(\bar{x}_u; \bar{a}): \bar{u} \in U, \bar{a} \in B, \psi(\bar{x}_u; \bar{a})$ forks over $A\}$ and define $q' = p \cup \{\neg\psi(\bar{x}_u; \bar{a}): \psi(\bar{x}_u; \bar{a}) \in \Gamma\}$. We now show that q' is consistent. If not, then it has a finite inconsistent subset

$$p' \cup \{\neg\psi^k(\bar{x}_{u_k}; \bar{a}^k): k < n < \omega\},$$

where $p' \subseteq p$, $\psi^k \in \Gamma$. Thus $p \vdash \bigvee_{k<n} \psi^k$. Now ψ^k forks over A so $\bigvee_{k<n} \psi^k$ forks over A by Lemma 1.1(9) so p forks over A by Lemma

1.1(8), contradiction. Thus q' is consistent, q' is a type in $\bar{x}_U$ over B. Let $q \supseteq q'$ be any complete type in $\bar{x}_U$ over B.

We now show that q does not fork over A. If it does, then $q \vdash \bigvee_{k<n} \varphi^k(\bar{x}_{\bar{a}_k}; \bar{a}^k)$, $\bar{u}_k \in U$, where φ^k divides over A. Let $q^* \subseteq q$ be a finite subtype such that $q^* \vdash \bigvee_{k<n} \varphi^k$ and let $\theta(\bar{x}_a; \bar{b}) = \bigwedge q^*$ ($\bar{u} \in U$, $\bar{b} \in B$). q is complete so that $\theta(\bar{x}_a; \bar{b}) \in q$. Clearly $\theta(\bar{x}_a; \bar{b})$ forks over A so $\theta(\bar{x}; \bar{b}_a) \in \Gamma$. So $\neg\theta \in q' \subseteq q$, contradiction. Thus q does not fork over A.

COROLLARY 1.5*: *Let $A_1 \subseteq B_1$, $A_2 \subseteq B_2$, C_1, C_2 be any sets, F an elementary mapping from B_1 onto B_2; $F(A_1) = A_2$, $\operatorname{tp}_*(C_1, B_1)$ does not fork over A_1. Then we can extend F to an elementary mapping F',* Dom $F' = B_1 \cup C_1$ *such that* $\operatorname{tp}_*(F'(C_1), B_2 \cup C_2)$ *does not fork over A_2.*

Proof. Directly from the theorem.

THEOREM 1.6: (1) *If p strongly splits over A, then p forks over A.*

(2) *p forks over A iff there is a set B such that p is a type over B and for every $q \supseteq p$ (in the same variables) which is complete over B, q strongly splits over A.*

Remark. B may be chosen such that B − Dom p is finite.

Proof. Without loss of generality we may assume that p is an m-type.

(1) By definition there are a set $I = \{\bar{a}^n : n < \omega\}$ indiscernible over A and a formula φ such that $\varphi(\bar{x}; \bar{a}^0), \neg\varphi(\bar{x}; \bar{a}^1) \in p$. Now by Lemma II, 2.20 there is $n(\varphi) < \omega$ such that for all $\bar{c}$, $|\{n < \omega : \models\varphi[\bar{c}; \bar{a}^n]\}| < n(\varphi)$ or $|\{n < \omega : \models\neg\varphi[\bar{c}; \bar{a}^n]\}| < n(\varphi)$. Define $\bar{b}^n = \bar{a}^{2n}\frown\bar{a}^{2n+1}$ and $\psi(\bar{x}; \bar{b}^n) = \varphi(\bar{x}; \bar{a}^{2n}) \wedge \neg\varphi(\bar{x}; \bar{a}^{2n+1})$. Clearly $\{\bar{b}^n : n < \omega\}$ is indiscernible over A. If $w \subseteq \omega$, $|w| = n(\varphi)$, then $\{\psi(\bar{x}; \bar{b}^n) : n \in w\}$ is inconsistent; otherwise, letting $\bar{c}$ realize this set, $|\{n < \omega, \models\varphi[\bar{c}; \bar{a}^n]\}| \geq |\{2n : n \in w\}| = n(\varphi)$ and $|\{n < \omega : \models\neg\varphi[\bar{c}; \bar{a}^n]\}| \geq |\{2n+1 : n \in w\}| = n(\varphi)$, contradiction. Thus by Lemma 1.1(3) $\psi(\bar{x}; \bar{b}^0)$ divides over A. In addition $p \vdash \psi(\bar{x}; \bar{b}^0)$ since $\varphi(\bar{x}; \bar{a}^0), \neg\varphi(\bar{x}; \bar{a}^1) \in p$. Hence p forks over A.

(2) $\Rightarrow$ By definition, $p \vdash \bigvee_{k<n} \varphi^k(\bar{x}; \bar{a}^k)$ where $\varphi^k(\bar{x}; \bar{a}^k)$ divides over A. So by Lemma 1.1(3) there are sequences (or sets; it is the same by stability) $\{\bar{a}_i^k : i < \omega\}$ indiscernible over A, $\bar{a}_0^k = \bar{a}^k$, and there are $n(k)$ such that $\{\varphi^k(\bar{x}; \bar{a}_i^k) : i < \omega\}$ is $n(k)$-inconsistent. Let $B = \operatorname{Dom} p \cup \{\bar{a}_i^k : k < n, i < n(k)\}$ and let $q \in S^m(B)$ extend p. Since $p \vdash \bigvee_{k<n} \varphi^k(\bar{x}; \bar{a}^k)$ we have a $k < n$ such that $\varphi^k(\bar{x}; \bar{a}^k) \in q$ (remember $\bar{a}_0^k = \bar{a}^k$). Now $|\{i < \omega : \varphi^k(\bar{x}; \bar{a}_i^k) \in q\}| < n(k)$, so there is $i_0 (\leq n(k))$ such that $\neg\varphi^k(\bar{x}; \bar{a}_{i_0}^k) \in q$. Thus q strongly splits over A.

$\Leftarrow$ Now assume p is a type over B and every q, such that $p \subseteq q \in S^m(B)$, strongly splits over A. By Theorem 1.4 if p does not fork over A there is $q \in S^m(B)$ extending p which does not fork over A, contradiction by (1).

DEFINITION 1.5: Let I be an infinite indiscernible set and A any set. $\mathrm{Av}_\Delta(I, A) = \{\varphi(\bar{x}; \bar{a})^{\mathbf{t}}: \varphi(\bar{x}; \bar{y}) \in \Delta, \mathbf{t} \in \{0, 1\}, \bar{a} \in A$, and for all but finitely many $\bar{c} \in I$, $\models \varphi[\bar{c}; \bar{a}]^{\mathbf{t}}\}$. If $\Delta = \mathrm{L}$ we omit it.

DEFINITION 1.6: The infinite indiscernible sets I_1, I_2 are *equivalent* if there is an infinite J such that $I_1 \cup J$ and $I_2 \cup J$ are still indiscernible.

LEMMA 1.7: (1) *For every infinite indiscernible set* I, $\mathrm{Av}(I, A)$ *is a complete type over* A. (2) *If* I *is an infinite indiscernible set over* A *then* $\bar{c}$ *realizes* $\mathrm{Av}(I, A \cup \bigcup I)$ *iff* $I \cup \{\bar{c}\}$ *is indiscernible over* A.

Proof. (1) By Lemma II, 2.20 for every $\bar{a} \in A$ and formula φ $|\varphi(I; \bar{a})| < \aleph_0$ or $|\neg\varphi(I; \bar{a})| < \aleph_0$; thus $\neg\varphi(\bar{x}; \bar{a}) \in \mathrm{Av}(I, A)$ or $\varphi(\bar{x}; \bar{a}) \in \mathrm{Av}(I, A)$. In other words, $\mathrm{Av}(I, A)$ is complete. In order to see that $\mathrm{Av}(I, A)$ is consistent, notice that every finite subset of $\mathrm{Av}(I, A)$ is realized by all but finitely many members of I.

(2) Left as an exercise.

LEMMA 1.8: *The following conditions on the infinite indiscernible sets* I_1, I_2 *are equivalent.*

(1) I_1, I_2 *are equivalent.*

(2) *For any set* A $\mathrm{Av}(I_1, A) = \mathrm{Av}(I_2, A)$.

(3) *There is a model* M, $|M| \supseteq I_1, I_2$ *such that*

$$\mathrm{Av}(I_1, |M|) = \mathrm{Av}(I_2, |M|).$$

Proof. (1) $\Rightarrow$ (2) Let J be an infinite set such that $I_1 \cup J, I_2 \cup J$ are indiscernible. By definition $\mathrm{Av}(I_1, A) \supseteq \mathrm{Av}(I_1 \cup J, A)$, but since both are complete types over A, they are equal. Likewise, $\mathrm{Av}(I_2, A) = \mathrm{Av}(I_2 \cup J, A) = \mathrm{Av}(J, A) = \mathrm{Av}(I_1 \cup J, A)$. So $\mathrm{Av}(I_1, A) = \mathrm{Av}(I_2, A)$.

(2) $\Rightarrow$ (3) Trivial.

(3) $\Rightarrow$ (2) If (2) does not hold then there are A and a formula $\varphi(\bar{x}; \bar{a})$ such that $\varphi(\bar{x}; \bar{a}) \in \mathrm{Av}(I_1, A)$ and $\neg\varphi(\bar{x}; \bar{a}) \in \mathrm{Av}(I_2, A)$. By Lemma II, 2.20 there is $n(\varphi) < \omega$ such that for any indiscernible set I^* and sequence $\bar{a}^*$, $|\varphi(I^*; \bar{a}^*)| < n(\varphi)$ or $|\neg\varphi(I^*; \bar{a}^*)| < n(\varphi)$. Choose $\bar{b}^1, \ldots, \bar{b}^{n(\varphi)} \in I_1$ such that $\models \varphi[\bar{b}^i; \bar{a}]$ $(1 \leq i \leq n(\varphi))$ and $\bar{c}^1, \ldots, \bar{c}^{n(\varphi)} \in I_2$ such that

$\models \neg\varphi[\bar{c}^i; \bar{a}]$ $(1 \leq i \leq n(\varphi))$. Thus $\models (\exists \bar{x}) \bigwedge_{i=1}^{n(\varphi)} (\varphi(\bar{b}^i; \bar{x}) \wedge \neg\varphi(\bar{c}^i; \bar{x}))$. Since M is a model, $\bar{b}^i, \bar{c}^i \in |M|$ there is $\bar{a}^* \in |M|$ such that

$$\models \bigwedge_{i=1}^{n(\varphi)} (\varphi[\bar{b}^i; \bar{a}^*] \wedge \neg\varphi[\bar{c}^i; \bar{a}^*]).$$

Thus $|\varphi(I_1; \bar{a}^*)|, |\neg\varphi(I_2; \bar{a}^*)| \geq n(\varphi)$. Hence $\varphi(\bar{x}; \bar{a}^*) \in \mathrm{Av}(I_1, |M|)$ and $\neg\varphi(\bar{x}; \bar{a}^*) \in \mathrm{Av}(I_2, |M|)$, contradiction.

(2) $\Rightarrow$ (1) Define by induction on $n < \omega$ sequences $\bar{c}^n$ which realize $\mathrm{Av}(I_1, C_n) = \mathrm{Av}(I_2, C_n)$ where $C_n = \bigcup I_1 \cup \bigcup I_2 \cup \bigcup \{\bar{c}^i : i < n\}$. The sets $I_1 \cup \{\bar{c}^n : n < \omega\}$ and $I_2 \cup \{\bar{c}^n : n < \omega\}$ are indiscernible by Lemma 1.7(2) and so I_1 and I_2 are equivalent.

DEFINITION 1.7: p is *stationary over* A if (1) p does not fork over A and (2) p has no two contradictory extensions which do not fork over A.

Remark. By Theorem 1.4 an equivalent condition is that for all B such that p is a type over B, p has a unique extension to a complete type (in the same variables) over B which does not fork over A.

LEMMA 1.9*: *If* $A \subseteq B$, *p does not fork over* A, $p \restriction B$ *is stationary over* A, *then p does not split over* B.

Proof. By way of contradiction assume $\mathrm{tp}(\bar{b}, B) = \mathrm{tp}(\bar{c}, B)$ and $\varphi(\bar{x}; \bar{b}), \neg\varphi(\bar{x}; \bar{c}) \in p$. Let F be an elementary mapping such that $F \restriction B$ is the identity and $F(\bar{c}) = \bar{b}$. $(p \restriction B) \cup \{\varphi(\bar{x}; \bar{b})\}$ does not fork over A, since it is a subtype of p. Likewise $(p \restriction B) \cup \{\neg\varphi(\bar{x}; \bar{c})\}$ does not fork over A. Thus $F[(p \restriction B) \cup \{\neg\varphi(\bar{x}; \bar{c})\}] = p \restriction B \cup \{\neg\varphi(\bar{x}; \bar{b})\}$ does not fork over A (since $F \restriction A$ is also the identity). But then $p \restriction B$ has two contradictory extensions which do not fork over A, contradiction.

DEFINITION 1.8: The infinite indiscernible set I is *based on* A if for every set B, $\mathrm{Av}(I, B)$ does not fork over A. (The interesting case is when $B \supseteq A$, and this gives an equivalent definition.)

Remark. I is based on $\bigcup I$.

LEMMA 1.10: (1) *If p is an m-type over* $B \supseteq A$ *which is stationary over* A, $B_j = B \cup \bigcup_{i<j} \bar{a}^i$, $p \subseteq p_j = \mathrm{tp}(\bar{a}^j, B_j)$, *and p_j does not fork over* A, *then* $I = \{\bar{a}^i : i < i_0 \geq \omega\}$ *is an indiscernible set over B based on* A.

(2) *In the notation of* (1), *for all* $C \supseteq B$, $\mathrm{Av}(I, C)$ *is the unique extension of p in* $S^m(C)$ *which does not fork over* A.

Proof. (1) By Lemma 1.9 p_j does not split over B and for all $i < j < i_0$, $p_i \subseteq p_j$ because p is stationary over A. So by Lemma I, 2.5, stability, and Theorem II, 2.13(7) I is an indiscernible set over B.

Now we show that I is based on A. Let B' be any set, and define $C = B \cup B' \cup \bigcup_{i<i_0} \bar{a}^i$. Inductively define $\bar{a}^j$, for $i_0 \leq j < i_0 + \omega$, such that $p_j = \text{tp}(\bar{a}^j, C \cup \bigcup_{i<j} \bar{a}^i)$ extends p and does not fork over A. This is achieved by 1.4.

Clearly $p_j \upharpoonright B_j$ also satisfies the above conditions. Thus, as above, $\{\bar{a}^j : j < i_0 + \omega\}$ is an indiscernible set over B. Now

$$\begin{aligned}\text{Av}(I, B') \subseteq \text{Av}(I, C) = \text{Av}(\{\bar{a}^j : j < i_0\}, C) &= \text{Av}(\{\bar{a}^j : j < i_0 + \omega\}, C)\\ &= \text{Av}(\{\bar{a}^j : i_0 \leq j < i_0 + \omega\}, C) = p_{i_0},\end{aligned}$$

since $p_{i_0} = \text{tp}(\bar{a}^j, C)$ for $i_0 \leq j < i_0 + \omega$. But p_{i_0} does not fork over A so neither does $\text{Av}(I, B')$.

(2) Follows from (1).

LEMMA 1.11: (1) *If $p \in S^m(|M|)$ does not fork over $A \subseteq |M|$, and M is $((|A| + 2)^{|T|})^+$-saturated, then p is stationary over A.*

(2) *If $A \subseteq B$ and $\Gamma_1 = \{p \in S^m(B)$: p does not fork over $A\}$, then $|\Gamma_1| \leq (|A| + 2)^{|T|}$ for any B.*

(3) *If $A \subseteq B$ and $\Gamma_2 = \{p \in S^m(B)$: p does not strongly split over $A\}$, then $|\Gamma_2| \leq (|A| + 2)^{|T|}$.*

(4) *There are at most $(|A| + 2)^{|T|}$ non-equivalent infinite indiscernible sets based on A.*

Proof. (1) Assume p is not stationary over A. Then there is $B \supseteq |M|$ and $p \subseteq q_1, q_2 \in S^m(B)$, $q_1 \neq q_2$, and q_1, q_2 do not fork over A. Let $\bar{b} \in B$ be such that $\varphi(\bar{x}; \bar{b}) \in q_1$, $\neg\varphi(\bar{x}; \bar{b}) \in q_2$. Now we define an increasing sequence $A_i \subseteq |M|$, $|A_i| \leq (|A| + 2)^{|T|}$, sequences $\bar{b}_i \in |M|$ and types $p_i \in S^m(|M|)$, for $i < ((|A| + 2)^{|T|})^+$. Let $A_0 = A$, and if δ is a limit ordinal $A_\delta = \bigcup \{A_j : j < \delta\}$. Assume A_i is defined; we shall define $\bar{b}_i$, p_i and A_{i+1}. Let $\bar{b}_i \in |M|$ be such that $\text{tp}(\bar{b}_i, A_i) = \text{tp}(\bar{b}, A_i)$; there is such a $\bar{b}_i$ by saturativity. By the assumption on q_1, q_2 we have that $p \upharpoonright A_i \cup \{\varphi(\bar{x}; \bar{b}_i)\}$, $p \upharpoonright A_i \cup \{\neg\varphi(\bar{x}; \bar{b}_i)\}$ do not fork over A. Without loss of generality $\varphi(\bar{x}; \bar{b}_i) \in p$. Let p_i be such that $p \upharpoonright A_i \cup \{\neg\varphi(\bar{x}; \bar{b}_i)\} \subseteq p_i \in S^m(|M|)$, and p_i does not fork over A (by Theorem 1.4). Take $A_{i+1} = A_i \cup \bar{b}_i$.

So we have defined A_i, p_i and $\bar{b}_i$. Now from $\{\bar{b}_i : i < ((|A| + 2)^{|T|})^+\}$ choose an (infinite) indiscernible sequence over A. If $i_0 < i_1$ and $\bar{b}_{i_0}$, $\bar{b}_{i_1}$ are in the indiscernible sequence over A then $\varphi(\bar{x}; \bar{b}_{i_0})$, $\neg\varphi(\bar{x}; \bar{b}_{i_1}) \in p_{i_1}$

so p_{i_1} splits strongly over A and thus forks over A (by Theorem 1.6); contradiction.

(2) Follows from (3) since strong splitting implies forking.

(3) Define an equivalence relation $\sim$ between finite sequences from B as follows: $\bar{b} \sim \bar{c}$ if there are $\bar{a}_0, \ldots, \bar{a}_n$, $\bar{a}_0 = \bar{b}$, $\bar{a}_n = \bar{c}$, and infinite indiscernible sets $I_0, \ldots, I_{n-1}$ over A such that $\bar{a}_l, \bar{a}_{l+1} \in I_l$ for all $0 \le l < n$. Let $B' \subseteq B$ be such that every finite sequence in B has an $\sim$-equivalent sequence in B'. The mapping $p \mapsto p \restriction B'$ is a one-one mapping from Γ_2 into $S^m(B')$. (Assume $p_1 \neq p_2 \in \Gamma_2, p_1 \restriction B' = p_2 \restriction B'$. Let $\varphi(\bar{x}; \bar{b}) \in p_1$, $\neg\varphi(\bar{x}; \bar{b}) \in p_2$ and $\bar{b} \sim \bar{c} \in B'$. If $\varphi(\bar{x}; \bar{c}) \in p_1$ then $\varphi(\bar{x}; \bar{c}) \in p_2$ and p_2 splits strongly over A, contradiction; if $\neg\varphi(\bar{x}; \bar{c}) \in p_1$, then p_1 splits strongly over A, contradiction.)

By Theorem II, 2.13 T is stable in $(|A| + 2)^{|T|}$ so (3) will follow if we show that $\sim$ has $\le (|A| + 2)^{|T|}$ equivalence classes (since then B' may be chosen such that $|S^m(B')| \le (|A| + 2)^{|T|}$). If not then there is $I = \{\bar{b}_i : i < ((|A| + 2)^{|T|})^+\}$, $\bar{b}_i \nsim \bar{b}_j$ for all $i \neq j$, and we may assume $l(\bar{b}_i) = l(\bar{b}_j)$, for all i, j. By Theorem I, 2.8 I has a subsequence of the same power which is indiscernible over A. This is a contradiction to the inequivalence of the $\bar{b}_i$. This proves (3).

(4) Assume there are $\{I_j : j < ((|A| + 2)^{|T|})^+\}$ pairwise non-equivalent indiscernible sets based on A. Choose M such that $|M| \supseteq A \cup \bigcup\{\bigcup I_j : j < ((|A| + 2)^{|T|})^+\}$ and M is $((|A| + 2)^{|T|})^+$-saturated. Let $p_j = \mathrm{Av}(I_j, |M|)$. By Lemma 1.8 the inequivalence of the I_j implies that $p_i \neq p_j$ for all $i \neq j$. But by definition the fact that I_j is based on A implies that p_j does not fork over A. So there are $> (|A| + 2)^{|T|}$ complete types over $|M|$ which do not fork over A, in contradiction to (2).

This completes the proof of Lemma 1.11.

CONCLUSION 1.12: *For every sequence $\bar{b}$ and set A there is an indiscernible set I over A and based on A, such that $\bar{b} \in I$. If M is a $(|A| + \aleph_0)^+$-saturated model such that $A \cup \bar{b} \subseteq |M|$ then we may take $I \subseteq |M|$. If* $\mathrm{tp}(\bar{b}, A)$ *does not fork over $B \subseteq A$ then we can choose I so that I is based on B.*

Proof. Let M_1 be a $((|A| + 2)^{|T|})^+$-saturated model such that $|M_1| \supseteq A \cup \bar{b}$. $\mathrm{tp}(\bar{b}, A)$ does not fork over A by Corollary 1.3. Thus by Theorem 1.4 there is a complete type $r \supseteq \mathrm{tp}(\bar{b}, A)$ over $|M_1|$ which does not fork over A. By Lemma 1.11(1) r is stationary over A. Now by repeated use of Theorem 1.4 and by Lemma 1.10(1) we may build an indiscernible set (over $|M_1|$) of sequences realizing r based on A. Call this set $\{\bar{c}_i: i < \omega\}$. Now $\mathrm{tp}(\bar{b}, A) = \mathrm{tp}(\bar{c}_0, A)$ so there is an automorphism F of $\mathfrak{C}$ which is the identity on A and such that $F(\bar{c}_0) = \bar{b}$. Without loss of generality $F(\bar{c}_i) \in |M|$ for all $i < \omega$, so we may take $I = \{F(\bar{c}_i): i < \omega\}$.

If $\mathrm{tp}(\bar{b}, A)$ does not fork over $B \subseteq A$ then we can choose r so that r does not fork over B and then I is based on B.

Remark. See 6.3(5).

EXERCISE 1.1: Show that in Lemma 1.7(1) it is possible to replace the assumption that T is stable by the (weaker) assumption that T does not have the independence property.

Exercise 1.1A. Another example is: $T = \mathrm{Th}(N), N = (|N|, E, c, P_l)_{l<n}$ where E is an equivalence relation over N with infinitely many equivalence classes, each infinite, $P_l \subseteq c/E$, and every Boolean combination of the P_l has infinite intersection with c/E. Let $A = \emptyset, p = \emptyset, q = \{xEc\}$ and $\Delta = \{P_l(x): l < n\} \cup \{E(x; y)\}$. Clearly $p \cup q$ fork over A, but $R(p, \Delta, 2) = R(p \cup q, \Delta, 2) = n$ when $n > 0$. For $2 \leq \lambda < \aleph_0$, we use $N = (|N|, E, c, P_{l,m})_{l<n,m<\lambda}$ where N is as above but $\langle P_{l,m}: m < \lambda\rangle$ is a partition of c/E, and $\{P_{l,\eta(l)}(x): l < n\} \cup \{E(x, c)\}$ is not algebraic for every $\eta \in {}^{\lambda}n$.

EXERCISE 1.2: For an indiscernible sequence $I = \{\bar{a}_i: i < \delta\}$, δ a limit ordinal, define the average of I over A as $\{\varphi(\bar{x}; \bar{b}): \bar{b} \in A$, for all sufficiently large i, $\models\varphi[\bar{a}_i; \bar{b}]\}$. Check the theorems corresponding to the ones proved here when T does not have the independence property but is not necessarily stable.

EXERCISE 1.3: Show that Corollary 1.3 is not in general true for unstable T.

Example 1. M contains two one-place predicates P, R, and one 3-place predicate E such that every element a of P is defined by $E(x, y, a)$ which is an equivalence relation over R, and all possible equivalence relations over R appear precisely once in the list $\{E(x, y; a): a \in P\}$. Let $\{x: E(x; b_l, a_l)\}$ have infinitely many $E(x, y; a_{1-l})$-equivalence classes for $l = 0, 1$ and $R(x) \vdash \bigvee_{l<2} E(x, b_l; a_l)$.

Example 2 (without the independence property). Let G_1 be a connected graph without circuits in which every vertex has ∞ neighbours. $E(x, y, z) =$ "the paths (x, z), (y, z) have a point in common apart from z.

Remark. Even if in Definition 1.3 we demand, "$\{\bar{a}_n : n < \omega\}$ is an indiscernible set", the example falsifies 1.3.

III.2. The finite equivalence relation theorem

Notation. $\mathrm{FE}^m(A)$ will denote the set of formulas $\varphi = \varphi(\bar{x}; \bar{y}; \bar{a})$ such that $l(\bar{x}) = l(\bar{y}) = m$ and $\bar{a} \in A$, and φ is an equivalence relation with a finite number of equivalence classes, $\mathrm{FE}(A) = \bigcup_{m<\omega} \mathrm{FE}^m(A)$. Elements of $\mathrm{FE}(A)$ will also be denoted by $E(\bar{x}; \bar{y})$; $n(E)$ will be the number of equivalence classes of E.

DEFINITION 2.1: (1) The formula $\varphi(\bar{x}; \bar{b})$ is *almost over* A if there is $E(\bar{x}; \bar{y}) \in \mathrm{FE}(A)$ such that $(\forall \bar{x}\bar{y})[E(\bar{x}; \bar{y}) \to (\varphi(\bar{x}; \bar{b}) \equiv \varphi(\bar{y}; \bar{b}))]$. In this case φ is said to *depend* on E. Notice that if $\bar{b} \in A$ then $\varphi(\bar{x}; \bar{b})$ is almost over A.

(2) The type p is almost over A if every formula in p is almost over A.

(3) $\mathrm{stp}(\bar{a}, A)$ = the strong type of $\bar{a}$ over A = $\{E(\bar{x}; \bar{a}) : E(\bar{x}; \bar{y}) \in \mathrm{FE}^m(A)\}$. Similarly define $\mathrm{stp}_*(C, B)$ for a set C.

Remark. $\bar{a}$ realizes $\mathrm{stp}(\bar{a}, A)$ and $\mathrm{stp}(\bar{a}, A) \vdash \mathrm{tp}(\bar{a}, A)$. Notice also that $\mathrm{stp}(\bar{a}, A)$ is almost over A, and if $\mathrm{stp}(\bar{a}, A) \equiv \mathrm{stp}(\bar{b}, A)$ and $\varphi(\bar{x}; \bar{c})$ is almost over A then $\vdash \varphi[\bar{a}; \bar{c}] \equiv \varphi[\bar{b}; \bar{c}]$. If $\bar{a}$ realizes p, p is almost over A, then $\mathrm{stp}(\bar{a}, A) \vdash p$.

LEMMA 2.1*: (1) $\mathrm{FE}^m(A)$ *is closed under conjunction, i.e., if* $\varphi_i(\bar{x}; \bar{y}; \bar{a}_i) \in \mathrm{FE}^m(A)$, $i = 1, \ldots, n$ *then* $\varphi(\bar{x}; \bar{y}; \bar{a}) =_{\mathrm{def}} \bigwedge_{i=1}^n \varphi_i(\bar{x}; \bar{y}; \bar{a}_i) \in \mathrm{FE}^m(A)$. *Hence* $\mathrm{stp}(\bar{a}, A)$ *is closed under conjunctions.*

(2) *If* $E(\bar{x}; \bar{y}) \in \mathrm{FE}^m(A)$ *and* $E'(\bar{x}^\frown\bar{x}'; \bar{y}^\frown\bar{y}') =_{\mathrm{def}} E(\bar{x}; \bar{y})$, *then* $E' \in \mathrm{FE}^n(A)$ (*where* $n = m + l(x')$, $l(\bar{x}') = l(\bar{y}')$).

(3) *If* $\bar{x} = \langle x_0, \ldots, x_{m-1}\rangle$, $\bar{y} = \langle y_0, \ldots, y_{m-1}\rangle$, σ *is a permutation of* $0, \ldots, m-1$ and $E(\bar{x}; \bar{y}) \in \mathrm{FE}^m(A)$ then

$$E'(\bar{x}; \bar{y}) =_{\mathrm{def}} E(x_{\sigma(0)}, \ldots, x_{\sigma(m-1)}; y_{\sigma(0)}, \ldots, y_{\sigma(m-1)}) \in \mathrm{FE}^m(A).$$

Proof. Immediate.

LEMMA 2.2*: (1) *The set of formulas which are almost over A is closed under all the connectives and quantifiers, under the addition of dummy variables, and under permutation of variables.*

(2) *The number of formulas which are almost over A is, up to logical*

equivalence, at most $|A| + |T|$. *Thus the number of non-equivalent types which are almost over* A *is* $\leq 2^{|A|+|T|}$.

(3) *If* $\operatorname{stp}(\bar{a}, A) \equiv \operatorname{stp}(\bar{b}, A)$ *then there is an automorphism* F *of* $\mathfrak{C}$ *such that* $F(\bar{a}) = \bar{b}$, $F \restriction A$ *is the identity, and* F *preserves the formulas which are almost over* A. *We can replace* $\bar{a}, \bar{b}$ *by sets.*

Proof. (1) The cases $\wedge$ and $\neg$ are easy to see ($\wedge$ by the preceding lemma), as are dummy variables and permutations of variables. Now let $\varphi(x_0, \bar{x}; \bar{b})$ be almost over A; i.e., there is $E(x_0, \bar{x}; y_0, \bar{y}) \in \mathrm{FE}(A)$ such that $E(x_0, \bar{x}; y_0, \bar{y}) \vdash (\varphi(x_0, \bar{x}; \bar{b}) \equiv \varphi(y_0, \bar{y}; \bar{b}))$. Let $\varphi^*(\bar{x}; \bar{b}) = (\exists x_0)\varphi(x_0, \bar{x}; \bar{b})$. Define

$$E^*(\bar{x}; \bar{y}) = (\forall x_0)(\exists y_0)E(x_0, \bar{x}; y_0, \bar{y}) \wedge (\forall y_0)(\exists x_0)E(x_0, \bar{x}; y_0, \bar{y}).$$

It is not hard to see that $E^* \in \mathrm{FE}(A)$ and that φ^* depends on E^*; in fact $n(E^*) \leq 2^{n(E)}$. Note that the above proof works even when $\bar{x} = \bar{y}$ is the empty sequence.

(2) and (3) immediate.

LEMMA 2.3*: (1) *The formula* $\varphi(\bar{x}; \bar{a})$ *is almost over* A *iff* $\Gamma = \{\varphi(\bar{x}; F(\bar{a}))$: F *is an automorphism of* $\mathfrak{C}$ *which is the identity on* $A\}$ *contains only a finite number of nonequivalent formulas iff* Γ *contains* $< \|\mathfrak{C}\|$ *non-equivalent formulas.*

(2) $\varphi(\bar{x}; \bar{a})$ *is equivalent to a formula over* A *iff all the formulas of* Γ *above are equivalent.*

Proof. (1) The third statement implies the second by compactness and the inverse implication is immediate. Also the direction $\Rightarrow$ in the first "iff" is clear by definition. Now for the other direction, if Γ has only a finite number of non-equivalent formulas, then there is n such that the set $\Gamma_1 = \{\theta(\bar{y}_i; \bar{c})\colon \theta(\bar{y}_i; \bar{c}) \in \operatorname{tp}(\bar{a}), A, i < n\} \cup \{\neg(\forall \bar{x})[\varphi(\bar{x}; \bar{y}_i) \equiv \varphi(\bar{x}; \bar{y}_j)]\colon i < j < n\}$ has a contradiction. Let n be minimal. Then we can assume that there is a contradiction in

$$\Gamma' = \{\theta(\bar{y}_i; \bar{c})\colon i < n\} \cup \{\neg(\forall \bar{x})[\varphi(\bar{x}; \bar{y}_i) \equiv \varphi(\bar{x}; \bar{y}_j)]\colon i < j < n\}$$

where $\vDash \theta[\bar{a}; \bar{c}]$. Remember $\bar{c} \in A$. Since n is minimal

$$\vDash (\exists \bar{y}_1 \cdots \bar{y}_{n-1})\Big[\bigwedge_{i=1}^{n-1} \theta(\bar{y}_i; \bar{c}) \wedge \bigwedge_{i \neq j} \neg(\forall \bar{x})[\varphi(\bar{x}; \bar{y}_i) \equiv \varphi(\bar{x}; \bar{y}_j)]$$
$$\wedge (\forall \bar{y})\Big[\theta(\bar{y}; \bar{c}) \to \bigvee_i (\forall \bar{x})(\varphi(\bar{x}; \bar{y}_i) \equiv \varphi(\bar{x}; \bar{y}))\Big]\Big].$$

Now $E(\bar{x}_1; \bar{x}_2) = (\forall \bar{y})[\theta(\bar{y}; \bar{c}) \rightarrow (\varphi(\bar{x}_1; \bar{y}) \equiv \varphi(\bar{x}_2; \bar{y}))] \in \mathrm{FE}(A)$, $n(E) \leq 2^{n-1}$, and $\varphi(\bar{x}; \bar{a})$ depends on E.

(2) Similar.

LEMMA 2.4*: *If $E(\bar{x}; \bar{y})$ is an equivalence relation almost over A, then there is an equivalence relation E' over A which refines E; i.e., $(\forall \bar{x}\bar{y})(E'(\bar{x}; \bar{y}) \rightarrow E(\bar{x}; \bar{y}))$, such that if E has finitely many equivalence classes, then so does E'.*

Remark. This lemma says it is pointless to replace "E over A" by "E almost over A" in Definition 2.1(1). See Exercise 2.3.

Proof. By Lemma 2.3(1) there are E_i, $i < n$, equivalence relations such that every automorphism F of $\mathfrak{C}$, such that $F \restriction A$ is the identity, takes E to one of the E_i, and for every E_i there is such an F. Thus every automorphism of $\mathfrak{C}$ which is the identity on A yields a permutation of the E_i. Thus $\bigwedge_{i<n} E_i(\bar{x}; \bar{y})$ is preserved (up to equivalence) under all automorphisms of $\mathfrak{C}$ which are the identity on A. So by Lemma 2.3(2) $E'(\bar{x}; \bar{y}) = \bigwedge_{i<n} E_i(\bar{x}; \bar{y})$ is over A, and certainly refines E. If E has k equivalence classes so does each E_i, hence E' has at most $k^n < \aleph_0$ equivalence classes.

LEMMA 2.5: *Let $I = \{\bar{a}_l : l < \omega\}$ be an indiscernible set which is based on A. Let $\varphi(\bar{x}; \bar{y})$ be a formula and for all $n < \omega$ define $\varphi_n(\bar{x}; \bar{a}^n) = \bigvee_{w \subseteq 2n-1, |w| = n} \bigwedge_{l \in w} \varphi(\bar{x}; \bar{a}_l)$. Then for all n large enough, φ_n is almost over A. (The size of n depends on φ only.)*

Proof. By Lemma II, 2.20 there is n such that for any indiscernible set I, and for all $\bar{a}$ of length $l(\bar{x})$, $|\varphi(\bar{a}, I)| < n$ or $|\neg\varphi(\bar{a}, I)| < n$. Thus for any $\bar{b}$, $\models \varphi_n[\bar{b}; \bar{a}^n] \Rightarrow$ there is $w \subseteq 2n - 1$, $|w| = n$ such that $\models \varphi[\bar{b}; \bar{a}_i]$ for all $i \in w \Rightarrow |\varphi(\bar{b}, I)| \geq n \Rightarrow |\neg\varphi(\bar{b}, I)| < n \Rightarrow \varphi(\bar{b}; \bar{y}) \in \mathrm{Av}(I, \bar{b}) \Rightarrow |\varphi(\bar{b}, I)| \geq \aleph_0 \Rightarrow |\neg\varphi(\bar{b}, I)| < n \Rightarrow$ there is $w \subseteq 2n - 1$, $|w| = n$ such that $\models \varphi[\bar{b}; \bar{a}_i]$ for all $i \in w \Rightarrow \models \varphi_n[\bar{b}; \bar{a}^n]$. So $\models \varphi_n(\bar{b}; \bar{a}^n) \Leftrightarrow \varphi(\bar{b}; \bar{y}) \in \mathrm{Av}(I, \bar{b})$.

Now by Lemma 2.3(1) if $\varphi_n(\bar{x}; \bar{a}^n)$ is not almost over A, then there are $\{F_i : i < ((|A| + 2)^{|T|})^+\}$, $F_i \restriction A$ is the identity, $F_i(\bar{a}^n) = \bar{a}_i^n$, $\mathrm{tp}(\bar{a}_i^n, A) = \mathrm{tp}(\bar{a}^n, A)$, and such that $\varphi_n(\bar{x}; \bar{a}_i^n)$ are pairwise non-equivalent. Thus for all $i < j < ((|A| + 2)^{|T|})^+$ there is $\bar{b}_{i,j}$ such that $\models \varphi_n[\bar{b}_{i,j}; \bar{a}_i^n] \equiv \neg\varphi_n[\bar{b}_{i,j}; \bar{a}_j^n]$. Thus, by the above, $\varphi(\bar{b}_{i,j}; \bar{y}) \in \mathrm{Av}(F_i(I), \bar{b}_{i,j}) \Leftrightarrow \neg\varphi(\bar{b}_{i,j}; \bar{y}) \in \mathrm{Av}(F_j(I), \bar{b}_{i,j})$. Thus $F_i(I)$, $F_j(I)$ are not equivalent (see Lemma 1.8). Since I is based on A and $F_i \restriction A$ is the identity we have that each $F_i(I)$ is also based on A.

Thus we have $((|A| + 2)^{|T|})^+$ non-equivalent indiscernible sets based on A, in contradiction to Lemma 1.11(4).

LEMMA 2.6: (1)* *If* $p \in S^m(B)$ *does not fork over* $A \subseteq B$ *and* $q \supseteq p$ *is a type (in the same variables) which is almost over* B, *then* q *does not fork over* A. *Also* $R^m(q, \Delta, \aleph_0) = R^m(p, \Delta, \aleph_0)$ *for all finite* Δ.

(2)* *If* $p \in S^m(B)$ *does not fork over* $A \subseteq B$, $E(\bar{x}; \bar{y}) \in \mathrm{FE}^m(B)$, *and* $q = p \cup \{E(\bar{x}; \bar{a})\}$ *is consistent, then* q *does not fork over* A. *Also* $R^m(q, \Delta, \aleph_0) = R^m(p, \Delta, \aleph_0)$ *for all (finite)* Δ.

(3) *For all* A, B, C *(with* $A \subseteq B$*) there is an elementary mapping* F, $\mathrm{Dom}\, F = A \cup C$, $F \restriction A$ *is the identity, such that for all* $\bar{c} \in C$, $\mathrm{tp}(F(\bar{c}), B)$ *does not fork over* A, *and* $\mathrm{stp}(\bar{c}, A) \equiv \mathrm{stp}(F(\bar{c}), A)$.

Proof. (1) Assume that q *does* fork over A; then by Lemma 1.1(5) q has a finite subtype q' which forks over A. Every formula in q' is almost over B, so by Lemma 2.2(1) $\varphi(\bar{x}; \bar{b}) = \bigwedge q'$ is almost over B. Thus φ depends on some $E(\bar{x}; \bar{y}) \in \mathrm{FE}^m(B)$. Let $\bar{a}$ realize q. Clearly $\vdash (\forall \bar{x})[E(\bar{x}; \bar{a}) \to \varphi(\bar{x}; \bar{b})]$, so if $\varphi(\bar{x}; \bar{b})$ forks over A so does $p \cup \{E(\bar{x}; \bar{a})\}$. Thus the first part of (1) follows from the first part of (2).

The second clause (about $R^m(q, \Delta, \aleph_0)$) similarly follows from the proof of $R^m[p \cup \{E(\bar{x}; \bar{a})\}, \Delta, \aleph_0] = R^m(p, \Delta, \aleph_0)$.

(2) Notice $\bar{a}$ does not necessarily realize p.

Choose a maximal set $\bar{a}^1, \ldots, \bar{a}^n$ such that $\mathrm{tp}(\bar{a}^k, B) = \mathrm{tp}(\bar{a}, B)$ but $\vdash \neg E(\bar{a}^l; \bar{a}^k)$, for all $1 \le l \ne k \le n$. ($n < \omega$ since $n(E) < \aleph_0$.) Thus $\mathrm{tp}(\bar{a}, B) \cup \{\neg E(\bar{x}; \bar{a}^k) : 1 \le k \le n\}$ is inconsistent. $\mathrm{tp}(\bar{a}, B)$ is closed under conjunctions so there is $\varphi(\bar{x}; \bar{c}) \in \mathrm{tp}(\bar{a}, B)$ such that $\varphi(\bar{x}; \bar{c}) \vdash \bigvee_{k=1}^n E(\bar{x}; \bar{a}^k)$.

Assume $p \cup \{E(\bar{x}; \bar{a})\}$ forks over A. Then for all k, $p \cup \{E(\bar{x}; \bar{a}^k)\}$ forks over A. So by Lemma 1.1(9) $p \cup \{\bigvee_{k=1}^n E(\bar{x}; \bar{a}^k)\}$ forks over A.

$p \cup \{E(\bar{x}; \bar{a})\}$ is consistent so $(\exists \bar{y})[\varphi(\bar{y}; \bar{c}) \wedge E(\bar{x}; \bar{y})] \in p$. Thus if $\bar{b}$ realizes p there is $\bar{a}^*$ satisfying $\varphi(\bar{y}; \bar{c}) \wedge E(\bar{b}; \bar{y})$, hence $\bigvee_{k=1}^n E(\bar{a}^*; \bar{a}^k)$. Thus also $\bigvee_{k=1}^n E(\bar{b}; \bar{a}^k)$. In other words $p \vdash \bigvee_{k=1}^n E(\bar{x}; \bar{a}^k)$, or $p \vdash p \cup \{\bigvee_{k=1}^n E(\bar{x}; \bar{a}^k)\}$. Thus p also forks over A, in contradiction to the hypothesis. The second clause is proved using Claim II, 1.7 and Exercise II, 1.1.

(3) Clear by Corollary 1.3, Theorem 1.4 and (1).

COROLLARY 2.7: *If* p *is almost over* A *then* p *does not fork over* A.

Proof. Let $q \supseteq p$ be a complete m-type over $A \cup \mathrm{Dom}\, p$. $q \restriction A \in S^m(A)$ and so does not fork over A by Corollary 1.3, and clearly $q \restriction A \cup p$ is a (consistent) type almost over A. Thus $q \restriction A \cup p$ does not fork over A by Lemma 2.6(1), and so neither does p.

THEOREM 2.8 (The Finite Equivalence Relation Theorem): *Let* $p, q \in S^m(|M|)$, $p \neq q$, $A \subseteq |M|$, p, q *do not fork over* A, *and* M *is* $((|A| + 2)^{|T|})^+$*-saturated. Then there is* $E(\bar{x}; \bar{y}) \in \mathrm{FE}^m(A)$ *such that* $p(\bar{x}) \cup q(\bar{y}) \vdash \neg E(\bar{x}; \bar{y})$.

Proof. Since $p \neq q$ and both are complete over $|M|$ there is $\varphi(\bar{x}; \bar{b})$, $\bar{b} \in |M|$, such that $\varphi(\bar{x}; \bar{b}) \in p$, $\neg\varphi(\bar{x}; \bar{b}) \in q$.

By Conclusion 1.12 there is a set $I = \{\bar{b}_i: i < \omega\} \subseteq |M|$ indiscernible over A and based on A such that $\bar{b}_0 = \bar{b}$. p, q do not fork over A, thus they do not strongly split over A, so $\varphi(\bar{x}; \bar{b}_i) \in p$ and $\neg\varphi(\bar{x}; \bar{b}_i) \in q$ for all i.

By Lemma 2.5 (and using its notation) there is $n < \omega$ such that $\varphi_n(\bar{x}; \bar{b}^n)$ is almost over A. Let $E(\bar{x}; \bar{y}) \in \mathrm{FE}(A)$ be such that φ_n depends on E: $(\forall \bar{x}\bar{y})[E(\bar{x}; \bar{y}) \rightarrow (\varphi_n(\bar{x}; \bar{b}^n) \equiv \varphi_n(\bar{y}; \bar{b}^n))]$. Notice $\varphi_n(\bar{x}; \bar{b}^n) \in p$, $\neg\varphi_n(\bar{x}; \bar{b}^n) \in q$. Thus $p(\bar{x}) \cup q(\bar{y}) \vdash (\varphi_n(\bar{x}; \bar{b}^n) \wedge \neg\varphi_n(\bar{y}; \bar{b}^n)) \vdash \neg E(\bar{x}; \bar{y})$.

COROLLARY 2.9: (1) *For all* $\bar{a}$ and A, $\mathrm{stp}(\bar{a}, A)$ *is stationary over* A.

(2) *If* $p, q \in S^m(B)$ *do not fork over* $A \subseteq B$, $p \neq q$, *then there is* $E(\bar{x}; \bar{y}) \in \mathrm{FE}^m(A)$ *such that* $p(\bar{x}) \cup q(\bar{y}) \vdash \neg E(\bar{x}; \bar{y})$.

Remark. This improves Lemma 1.11(1), 2.8.

Proof. (1) Let $p = \mathrm{stp}(\bar{a}, A)$. p is almost over A, so by Corollary 2.7 p does not fork over A. Now if p were not stationary over A, then p would have two contradictory extensions p_1, p_2 which do not fork over A. Let M be a $((|A| + 2)^{|T|})^+$-saturated model such that p_1, p_2 are over $|M|$. So $A \subseteq |M|$. Let p^1, p^2 be extensions of p_1, p_2, respectively, in $S^m(|M|)$ which do not fork over A. Let $E(\bar{x}; \bar{y}) \in \mathrm{FE}^m(A)$ be such that $p^1(\bar{x}) \cup p^2(\bar{y}) \vdash \neg E(\bar{x}; \bar{y})$. But $p \subseteq p^1, p^2$, $E(\bar{x}; \bar{a}) \in p$, so that $p^1(x) \cup p^2(y) \vdash \neg E(\bar{x}; \bar{y}) \wedge E(\bar{x}; \bar{a}) \wedge E(\bar{y}; \bar{a})$, a contradiction.

(2) Assume that there is no such E. Let $\Gamma = p(\bar{x}) \cup q(\bar{y}) \cup \{E(\bar{x}; \bar{y}): E \in \mathrm{FE}^m(A)\}$, Γ is consistent (since $\mathrm{FE}^m(A)$ is closed under conjunctions). Let $\bar{x} \mapsto \bar{a}$, $\bar{y} \mapsto \bar{b}$ realize Γ. Thus $\mathrm{stp}(\bar{a}, A) \equiv \mathrm{stp}(\bar{b}, A)$, $\bar{a}$ realizes p and $\bar{b}$ realizes q. So $\mathrm{stp}(\bar{a}, A)$ is not stationary over A, in contradiction to (1).

LEMMA 2.10*: *If I is an infinite indiscernible set over A and $\varphi(\bar{x}_1, \ldots, \bar{x}_n; \bar{c})$ is almost over A, then there is $\mathfrak{t} \in \{0, 1\}$ such that for all distinct $\bar{a}_1, \ldots, \bar{a}_n \in I$, $\models \varphi[\bar{a}_1, \ldots, \bar{a}_n; \bar{c}]^{\mathfrak{t}}$.*

Proof. Assume that there are $a_1^{\mathfrak{t}}, \ldots, \bar{a}_n^{\mathfrak{t}} \in I$ for $\mathfrak{t} = 0, 1$ such that $\models \varphi[\bar{a}_1^{\mathfrak{t}}, \ldots, \bar{a}_n^{\mathfrak{t}}; \bar{c}]^{\mathfrak{t}}$. Without loss of generality $\{\bar{a}_1^0, \ldots, \bar{a}_n^0\} \cap \{\bar{a}_1^1, \ldots, \bar{a}_n^1\} = \emptyset$ (if not, choose a set $\{\bar{b}_1, \ldots, \bar{b}_n\}$ disjoint from both of them and replace one of them by it). Since $\text{stp}(\bar{a}_1^0{}^\frown \cdots {}^\frown a_n^0, A) \not\equiv \text{stp}(\bar{a}_1^1{}^\frown \cdots {}^\frown a_n^1, A)$, there is $E \in \text{FE}(A)$ such that $\models \neg E(\bar{a}_1^0{}^\frown \cdots {}^\frown \bar{a}_n^0; \bar{a}_1^1{}^\frown \cdots {}^\frown \bar{a}_n^1)$. Now since I is indiscernible, and $\{\bar{a}_1^0, \ldots, \bar{a}_n^0\}$, $\{\bar{a}_1^1, \ldots, \bar{a}_n^1\}$ are disjoint, it can be shown that E has infinitely many equivalence classes, contradiction.

COROLLARY 2.11: *Let I be an infinite indiscernible set over $A \subseteq B$.*

(1) *If for all $\bar{c} \in \bigcup I$, $\text{tp}(\bar{c}, B)$ does not fork over A then I is indiscernible over B.*

(2)* *If for all $\bar{b} \in B$, $\text{tp}(\bar{b}, A \cup \bigcup I)$ does not strongly split over A, then I is indiscernible over B.*

Proof. (1) Let $\{\bar{a}_1, \ldots, \bar{a}_n\}$, $\{\bar{b}_1, \ldots, \bar{b}_n\}$ be sets of n sequences in I. We must show that $\text{tp}(\bar{a}_1{}^\frown \cdots {}^\frown \bar{a}_n, B) = \text{tp}(\bar{b}_1{}^\frown \cdots {}^\frown \bar{b}_n, B)$. By Lemma 2.10 $\text{stp}(\bar{a}_1{}^\frown \cdots {}^\frown a_n, A) \equiv \text{stp}(\bar{b}_1{}^\frown \cdots {}^\frown \bar{b}_n, A)$, and $\text{tp}(\bar{a}_1{}^\frown \cdots {}^\frown \bar{a}_n, B)$, $\text{tp}(\bar{b}_1{}^\frown \cdots {}^\frown \bar{b}_n, B)$ do not fork over A, so by Corollary 2.9(2) they are the same.

(2) Immediate by direct check.

LEMMA 2.12: *Let $A \subseteq B$, $p \in S^m(B)$ not fork over A. If B contains sequences $\bar{a}_n \in B$, $n < \omega$, such that $\bar{a}_n$ realizes $p_n = p \upharpoonright (A \cup \bigcup \{\bar{a}_i : i < n\})$, then p is stationary over A.*

Proof. Let $\bar{a}$ realize p. By Corollary 2.9(1) it is sufficient to prove $p \vdash \text{stp}(\bar{a}, A)$, in other words $p \vdash E(\bar{x}; \bar{a})$ for all $E \in \text{FE}^m(A)$. Thus it suffices to show $p \vdash E(\bar{x}; \bar{a}_n)$ for some $n < \omega$. Assume to the contrary that $\neg E(\bar{x}; \bar{a}_n) \in p$ for all n, and hence $\neg E(\bar{x}; \bar{a}_n) \in p_l$ for all $n < l < \omega$; thus $\models \neg E(\bar{a}_l; \bar{a}_n)$. But then E has infinitely many equivalence classes, contradiction.

COROLLARY 2.13: *Let $A \subseteq |M|$, M is λ-saturated, $|A| < \lambda$, $\aleph_0 < \lambda$. Then:*

(1) *Every m-type which is almost over A is realized in M.*

(2) *If J is an infinite indiscernible set based on A, then there is an equivalent indiscernible set $I \subseteq |M|$ (I is also indiscernible over A). See Definition 1.6.*

Remark. Notice that when $|T|$ is small there is a simple direct proof. The interesting case is when $|T|$ is very large in relation to λ.

Proof. Let p be an m-type almost over A. By Corollary 2.7, p does not fork over A, so there is a complete m-type $q \supseteq p$, over $|M| \cup \operatorname{Dom} p$ which does not fork over A; so $q \upharpoonright |M| \in S^m(|M|)$. Clearly $q \upharpoonright |M|$ also does not fork over A. Now define a sequence $\{\bar{a}_i: i < \omega + \omega\}$ in M such that $\bar{a}_i$ realizes $q \upharpoonright (A \cup \bigcup \{\bar{a}_j: j < i\})$ for all $i < \omega + \omega$. This is possible by saturativity. So by Lemma 2.12 $q \upharpoonright (A \cup \{\bar{a}_j: j < \omega\}$ is stationary over A. By Lemma 1.10 $I = \{\bar{a}_i: \omega \leq i < \omega + \omega\}$ is an indiscernible set over A, whose average is q. And so by Lemma 2.10 we see that $\bar{a}_\omega$ realizes p. This proves (1). For part (2), choose distinct $\bar{a}_l \in J$, and $\bar{b}_0 \in |M|$ realizing $\operatorname{stp}(\bar{a}_0, A)$ and $\bar{b}_n \in |M|$ such that $\operatorname{tp}(\bar{a}_0{}^\frown \cdots {}^\frown a_n, A) = \operatorname{tp}(\bar{b}_0{}^\frown \cdots {}^\frown \bar{b}_n, A)$ and let $I = \{b_n: n < \omega\}$.

LEMMA 2.14*: *If $p = \operatorname{tp}(\bar{a}, A)$ is stationary over $B \subseteq A$, then $p \vdash \operatorname{stp}(\bar{a}, A)$, and thus for every φ which is almost over A, $p \vdash \varphi$ or $p \vdash \neg\varphi$.*

Proof. Let $E \in \mathrm{FE}^m(A)$ where $m = l(\bar{a})$. We must show that $p \vdash E(\bar{x}; \bar{a})$. If not, then there is $\bar{b}$, not E-equivalent to $\bar{a}$, such that $q = p \cup \{E(\bar{x}; \bar{b})\}$ is consistent. Then by Lemma 2.6(2) q and $p \cup \{E(\bar{x}; \bar{a})\}$ are two contradictory extensions of p which do not fork over B, contradiction to the fact that p is stationary over B.

LEMMA 2.15: (1)* *If $\varphi(\bar{x}; \bar{b})$ is almost over $|M|$ then for some $\bar{a} \in |M|$ and formula $\psi(\bar{x}; \bar{y})$, $\vdash \varphi(\bar{x}; \bar{b}) \equiv \psi(\bar{x}; \bar{a})$.*

(2) *If $p \in S^m(|M|)$ does not fork over $A \subseteq |M|$ then p is stationary over A. If $\bar{a}$ realizes p then $p \equiv \operatorname{stp}(\bar{a}, |M|)$.*

Proof. (1) Let $\varphi(\bar{x}; \bar{b})$ depend on $E \in \mathrm{FE}(|M|)$ and let $n(E) = n < \omega$. Since M is a model there are representatives $\bar{a}_0, \ldots, \bar{a}_{n-1} \in |M|$ of all the equivalence classes of E. We have $\vDash (\forall \bar{x}) \bigvee_{i<n} E(\bar{x}; \bar{a}_i)$ and $\vDash (\forall \bar{x})(E(\bar{x}; \bar{a}_i) \rightarrow \varphi(\bar{x}; \bar{b}))$ or $\vDash (\forall \bar{x})(E(\bar{x}; \bar{a}_i) \rightarrow \neg\varphi(\bar{x}; \bar{b}))$ for $i < n$. Without loss of generality we can assume $\vDash (\forall \bar{x})(E(\bar{x}; \bar{a}_i) \rightarrow \varphi(\bar{x}; \bar{b}))$ for

$i < m \le n$ and $\models (\forall x)(E(\bar{x}; \bar{a}_i) \to \neg\varphi(\bar{x}, \bar{b}))$ for $m \le i < n$. So $\vdash \varphi(\bar{x}; \bar{b}) \equiv \bigvee_{i<m} E(\bar{x}; \bar{a}_i)$ if $m > 0$ and $\vdash \neg\varphi(\bar{x}; \bar{b})$ if $m = 0$.

(2) Let $\bar{a}$ realize p. By Corollary 2.9(1) it is sufficient to prove that $p \vdash \operatorname{stp}(\bar{a}, A)$, i.e., for all $E \in \mathrm{FE}^m(A)$, $p \vdash E(\bar{x}; \bar{a})$. So let $E \in \mathrm{FE}^m(A)$. By definition $n = n(E) < \aleph_0$, and so $\models (\exists \bar{x}_1 \cdots \bar{x}_n)(\forall \bar{y}) \bigvee_{i=1}^n E(\bar{y}; \bar{x}_i)$. Now since the parameters of E are from $A \subseteq |M|$, and M is a model, there are $\bar{a}_1, \ldots, \bar{a}_n \in |M|$ such that $\models (\forall \bar{y}) \bigvee_{i=1}^n E(\bar{y}; \bar{a}_i)$. Let $\models E(\bar{a}; \bar{a}_i)$. Thus $p \vdash E(\bar{x}; \bar{a}_i)$ and so $p \vdash E(\bar{x}; \bar{a})$. The second clause follows from the first clause by Lemma 1.1(8) and the fact that $\operatorname{stp}(\bar{a}, |M|) \vdash \operatorname{tp}(\bar{a}, |M|)$.

EXERCISE 2.1: Is Lemma 2.15(2) still true when T is not stable?

EXERCISE 2.2: Show that if $A \subseteq B$, $p = \operatorname{tp}(\bar{a}, B)$ does not fork over A then p is stationary over A iff $p \vdash \operatorname{stp}(\bar{a}, A)$ iff $p \vdash \operatorname{stp}(\bar{a}, B)$.

EXERCISE 2.3: The E' we get in 2.4 is the coarsest equivalence relation over A refining E.

EXERCISE 2.4: If I_1, I_2 are (infinite) indiscernible sets based on B, and for some $\bar{a}_l \in I_l$, $\operatorname{stp}(\bar{a}_1, B) \equiv \operatorname{stp}(\bar{a}_2, B)$ then I_1, I_2 are equivalent.

EXERCISE 2.5: If $\operatorname{stp}(\bar{a}, A) \equiv \operatorname{stp}(\bar{b}, A)$, then there is an indiscernible set I based on A such that $I \cup \{\bar{a}\}$, $I \cup \{\bar{b}\}$ are indiscernible over A.

EXERCISE 2.6: If p does not fork over A, $p \in S^m(B)$, $\operatorname{stp}(\bar{a}, A) \equiv \operatorname{stp}(\bar{b}, A)$, $\bar{a}, \bar{b} \in B$ then for every φ, $\varphi(\bar{x}; \bar{a}) \in p \Leftrightarrow \varphi(\bar{x}; \bar{b}) \in p$.

EXERCISE 2.7: If T is stable in λ, $|A| \le \lambda$, then the number of non-equivalent types almost over A of the form $\operatorname{stp}(\bar{a}, A)$ is $\le \lambda$.

EXERCISE 2.8: Show that if p is stationary over A but $p \cup q$ forks over A then there is $\varphi \in q$ such that $p \cup \{\varphi\}$ forks over A.

III.3. The instability spectrum

THEOREM 3.1: *The following conditions on κ, T, m are equivalent:*

(1) *There is an increasing sequence A_i, $i \le \kappa$, and $p \in S^m(A_\kappa)$ such that for all $i < \kappa$, $p \restriction A_{i+1}$ forks over A_i.*

(2) *Like* (1) *but now $p \restriction A_{i+1}$ splits strongly over A_i.*

(3)$_\lambda$ *There exists an increasing sequence* A_i, $i \le \kappa$, *a type* $q \in S^m(A_\kappa)$, *sets* $I^i = \{\bar{a}^i_j : j < \lambda\} \subseteq A_{i+1}$ *indiscernible over* A_i, *and* $\varphi_i(\bar{x}; \bar{a}^i_0) \in q$ *such that* $j > 0 \Rightarrow \neg\varphi_i(\bar{x}; \bar{a}^i_j) \in q$, *and there are* $m_i < \omega$ *such that* $\{\varphi_i(\bar{x}; \bar{a}^i_j): j < \lambda\}$ *is* m_i*-inconsistent. (So* $q \restriction A_{i+1}$ *splits strongly over* A_i *for all* $i < \kappa$.)

(4)$_\lambda$ *Like* (3)$_\lambda$ *but where* $A_i = |M_i|$, M_{i+1} *being* $\|M_i\|^+$*-saturated.*

DEFINITION 3.1: $\kappa^m(T)$ = the first infinite κ for which (1) above fails; $\kappa(T) = \sup_m \kappa^m(T)$. The first regular $\kappa \ge \kappa(T)$ is denoted by $\kappa_r(T)$, or $\kappa r(T)$ (so it is $\kappa(T)$ or $\kappa(T)^+$). For unstable T we stipulate $\kappa^m(T) = \kappa_r(T) = \infty$.

Proof of Theorem 3.1. Trivially (4)$_\lambda \Rightarrow$ (3)$_\lambda$ and by definition (3)$_\lambda \Rightarrow$ (2) (for any λ). By Theorem 1.6(1), (2) $\Rightarrow$ (1). So it remains to prove (1) $\Rightarrow$ (4)$_\lambda$.

Since $p \restriction A_{i+1}$ forks over A_i we have $p \restriction A_{i+1} \vdash \bigvee_{k<n_i} \varphi^i_k(\bar{x}; \bar{a}^i_k)$ where each $\varphi^i_k(\bar{x}; \bar{a}^i_k)$, $k < n_i$, divides over A_i. So by Lemma 1.1(3) there are $\{\bar{a}^i_{k,l}: l < \omega\}$ indiscernible over A_i, $\bar{a}^i_k = \bar{a}^i_{k,0}$ and $\{\varphi^i_k(\bar{x}; \bar{a}^i_{k,l}): l \in \omega\}$ is m^i_k-inconsistent. Without loss of generality we may assume $A_\delta = \bigcup_{i<\delta} A_i$, δ a limit ordinal (otherwise define $A'_i = A_i$, for i a successor, and $A'_\delta = \bigcup_{i<\delta} A_i$; then $A'_\delta \subseteq A_\delta$ and $p \restriction A_{\delta+1} = p \restriction A'_{\delta+1}$ forks over A_δ, so it forks over A'_δ, and thus the A'_i's satisfy (1)).

Now let $B_i = \bigcup_{k,l} \bar{a}^i_{k,l}$. Define by induction elementary mappings F_i and models M_i such that Dom $F_\delta = A_\delta$, $M_\delta = \bigcup_{i<\delta} M_i$, Dom $F_{i+1} = A_{i+1} \cup B_i$, M_0 is $(\lambda + |A_\kappa|)^+$-saturated, M_{i+1} is $\|M_i\|^+$-saturated, $i < j \Rightarrow F_j \restriction A_i = F_i \restriction A_i$, $i < j \Rightarrow M_i \subseteq M_j$, Range $F_i \subseteq |M_i|$ and for any $\bar{a} \in A_{i+1} \cup B_i$, tp$(F_{i+1}(\bar{a}), |M_i|)$ does not fork over $F_i(A_i)$.

For $i = 0$ take F_0 as the identity on A_0, and take any appropriate M_0. For $i = \delta$ let $F_\delta = \bigcup_{i<\delta} F_i \restriction A_i$, $M_\delta = \bigcup_{i<\delta} M_i$. By Corollary 1.5 we may define F_{i+1}: $A_{i+1} \cup B_i \to |M_i| \cup F_{i+1}(A_{i+1} \cup B_i)$ with the desired property, and then extend $|M_i| \cup F_{i+1}(A_{i+1} \cup B_i)$ to an $\|M_i\|^+$-saturated model M_{i+1}. This completes the definition. Now choose $q \in S^m(|M_\kappa|)$ such that $F_\kappa(p) \subseteq q$. For every i, $p \restriction A_{i+1} \vdash \bigvee_{k<n_i} \varphi^i_k(\bar{x}; \bar{a}^i_k)$ so $F_{i+1}(p \restriction A_{i+1}) \vdash \bigvee_{k<n_i} \varphi^i_k(\bar{x}; F_{i+1}(\bar{a}^i_k))$. $F_\kappa \restriction A_{i+1} = F_{i+1} \restriction A_{i+1}$ so $F_\kappa(p \restriction A_{i+1}) \vdash \bigvee_{k<n_i} \varphi^i_k(\bar{x}; F_{i+1}(\bar{a}^i_k))$, so $q \vdash \bigvee_{k<n_i} \varphi^i_k(\bar{x}; F_{i+1}(\bar{a}^i_k))$. q is complete so for each $i < \kappa$ there is $k(i) < n_i$ such that $\varphi^i_{k(i)}(\bar{x}; F_{i+1}(\bar{a}^i_{k(i)})) \in q$. $\{\bar{a}^i_{k(i),l}: l < \omega\}$ is indiscernible over A_i so $\{F_{i+1}(\bar{a}^i_{k(i),l}): l < \omega\}$ is indiscernible over $F_{i+1}(A_i) = F_i(A_i)$, and if $\bar{a} \in B_i \cup A_{i+1}$, tp$(F_{i+1}(\bar{a}), |M_i|)$ does not fork over $F_i(A_i)$. Hence by Corollary 2.11(1) $\{F_{i+1}(\bar{a}^i_{k(i),l}): l < \omega\}$ is indiscernible over $|M_i|$. Define $\bar{b}^i_l = F_{i+1}(\bar{a}^i_{k(i),l})$. Since M_{i+1} is $\|M_i\|^+$-saturated, and $\|M_i\|^+ > \lambda$ we can define $\bar{b}^i_l \in |M_{i+1}|$ for

$\omega \le j < \lambda$ such that $\{\bar{b}^i_j : j < \lambda\}$ is indiscernible over $|M_i|$. Now $\{\varphi^i_{k(i)}(\bar{x}; \bar{b}^i_j) : j \in \omega\}$ is $m^i_{k(i)}$-inconsistent. Hence $\{j < \lambda : \varphi^i_{k(i)}(\bar{x}; \bar{b}^i_j) \in q\}$ is finite. We may in fact assume (by omitting if necessary) that only $\varphi^i_{k(i)}(\bar{x}; \bar{b}^i_0) \in q$, and $j > 0$ implies $\neg\varphi^i_{k(i)}(\bar{x}; \bar{b}^i_j) \in q$. Let $I^i = \{\bar{b}^i_j : j < \lambda\}$. The conclusion holds with this I^i, $\varphi^i_{k(i)}$, M_i, q.

COROLLARY 3.2: *For all $p \in S^m(B)$ there is $A \subseteq B$, $|A| < \kappa^m(T)$ such that p does not fork over A.*

Proof. By way of contradiction, assume there is no such A. We shall define A_i: $i \le \kappa^m(T)$ such that $A_i \subseteq B$ and $|A_i| < |i|^+ + \aleph_0$ for $i < \kappa^m(T)$. Take $A_0 = \emptyset$, $A_\delta = \bigcup_{i<\delta} A_i$, $A_{\kappa^m(T)} = B$. Assume A_i is defined. By assumption p forks over A_i. By Lemma 1.1(5) p has a finite subtype p' which forks over A_i; let $A_{i+1} = A_i \cup \operatorname{Dom} p'$. So by Theorem 3.1(1) $\kappa^m(T) < \kappa^m(T)$, contradiction.

COROLLARY 3.3: $\kappa(T) \le |T|^+$.

Proof. If not, then by Theorem 3.1(1) there are an increasing sequence A_i, $i \le |T|^+$, $m < \omega$, and $p \in S^m(A_{|T|^+})$ such that for all $i < |T|^+$ $p \restriction A_{i+1}$ forks over A_i. Also, by Theorem II, 1.2, for every Δ there is $\alpha(\Delta) < |T|^+$ such that $R^m(p, \Delta, \aleph_0) = R^m[p \restriction A_{\alpha(\Delta)}, \Delta, \aleph_0]$. The number of possible finite Δ is $|T|$ so $\alpha = \sup_{\Delta \text{finite}} \alpha(\Delta) < |T|^+$. But p forks over A_α and so by Lemma 1.2(2) there is Δ_0 such that $R^m(p, \Delta_0, \aleph_0) < R^m(p \restriction A_\alpha, \Delta_0, \aleph_0)$, contradiction.

LEMMA 3.4: *If J is an indiscernible set over A, $\bar{b}$ any sequence, $l(\bar{b}) = m$, then there is $I \subseteq J$ with $|I| < \kappa^m(T)$ such that $J - I$ is indiscernible over $A \cup \bar{b} \cup \bigcup I$.*

Proof. By Corollary 3.2 there is $B \subseteq A \cup \bigcup J$, $|B| < \kappa^m(T)$, such that $\operatorname{tp}(\bar{b}, A \cup \bigcup J)$ does not fork over B. Let $I \subseteq J$, $|I| < \kappa^m(T)$, be such that $B \subseteq A \cup \bigcup I$. Clearly $J - I$ is indiscernible over $A \cup \bigcup I$, and thus by Corollary 2.11(2), 1.6(1) $J - I$ is indiscernible over $A \cup \bar{b} \cup \bigcup I$.

Remark. If T does not have the independence property (but is not necessarily stable) a similar conclusion holds, using $\kappa_{\text{inp}}(T)$, see Definition 7.3.

COROLLARY 3.5: *If J is an indiscernible set over A, B is any set, then there is $I \subseteq J$ such that $J - I$ is indiscernible over $A \cup B \cup \bigcup I$, and*

(1) $|I| \le \kappa(T) + |B|$.

(2) *If* $|B| < \mathrm{cf}(\kappa(T))$ *then* $|I| < \kappa(T)$. *(The interesting case is where $|J|$ is large enough in relation to $|B|$.)*

LEMMA 3.6: *Assume* $\mu = \mu^{<\kappa^1(T)} (= \sum_{\kappa < \kappa^1(T)} \mu^{\kappa})$. *If* $\mu \ge 2^{|T|}$ *or there is* μ_0 *such that* $|T| \le \mu_0 \le \mu$ *and T is μ_0-stable, then T is μ-stable.*

Proof. By way of contradiction assume there are A, m such that $|A| \le \mu < |S^m(A)|$. By Lemma I, 2.1 we can assume $m = 1$. By Corollary 3.2 for every $p \in S(A)$ there is $B_p \subseteq A$, $|B_p| < \kappa^1(T)$ such that p does not fork over B_p. The number of possible B_p's is $\le \mu^{<\kappa^1(T)}$ so there is a B for which $|\{p \in S(A): B_p = B\}| > \mu$. By Corollary 3.3 $\kappa(T) \le |T|^+$ and so $|B| \le |T|$. Let $B \subseteq |M|$, $\|M\| = |T|$. Then $\mu < |\{p \in S(A): B_p = B\}| \le |\{p \in S(A): p$ does not fork over $B\}| \le |\{p \in S(|M| \cup A): p$ does not fork over $B\}| = |\{p \in S(|M|): p$ does not fork over $B\}|$ (this by Lemma 2.15(2)) $\le |S(|M|)|$.

But clearly $|S(|M|)| \le 2^{|T|}$, contradicting $\mu \ge 2^{|T|}$. And if T is stable in μ_0 and $\|M\| = |T| \le \mu_0$, then $|S(|M|)| \le \mu_0$ contradicting $\mu_0 \le \mu$.

Remark. Another alternative assumption is that $\mu = \mu^{<\kappa^1(T)}$ and $\mu \ge \sup\{|S(A)|: |A| \le |T|\}$.

THEOREM 3.7: *If* $\mu < \mu^{<\kappa(T)}$ *then T is not stable in μ.*

Proof. Since $\mu < \mu^{<\kappa(T)}$ there is $\kappa < \kappa(T)$ such that $\mu^{\kappa} > \mu$. Let κ be the minimal one. Thus $\mu^{<\kappa} = \mu$. Now since $\kappa < \kappa(T)$, for some m $\kappa < \kappa^m(T)$ so by Theorem 3.1, $(3)_\mu$ we have an increasing sequence A_i, $i \le \kappa$, a type $p \in S^m(A_\kappa)$, sets $I^i = \{\bar{a}^i_j: j < \mu\}$ indiscernible over A_i, $I^i \subseteq A_{i+1}$, and $\varphi_i(\bar{x}; \bar{a}^i_0) \in p$ such that $j > 0 \Rightarrow \neg\varphi_i(\bar{x}; \bar{a}^i_j) \in p$.

Now for $\eta \in {}^{\kappa \ge}\mu$ define an elementary mapping F_η such that

(1) $\mathrm{Dom}\, F_\eta = \bigcup \{\bar{a}^i_j: j < \mu, i < l(\eta)\}$,

(2) $\eta = \rho \restriction i \Rightarrow F_\rho$ extends F_η,

(3) if $l(\eta) = i$ then for every j:

(A) $F_{\eta^\frown\langle j\rangle}(\bar{a}^i_0) = F_{\eta^\frown\langle 0\rangle}(\bar{a}^i_j)$,

(B) $F_{\eta^\frown\langle j\rangle}(\bar{a}^i_j) = F_{\eta^\frown\langle 0\rangle}(\bar{a}^i_0)$,

(C) for every $\alpha \ne 0, j$ $F_{\eta^\frown\langle j\rangle}(\bar{a}^i_\alpha) = F_{\eta^\frown\langle 0\rangle}(\bar{a}^i_\alpha)$.

The definition proceeds by induction on $l(\eta)$. For $l(\eta) = 0$, F_η is the empty mapping, and for $l(\eta)$ limit take $F_\eta = \bigcup_{\gamma < l(\eta)} F_{\eta \restriction \gamma}$. If $l(\eta) = \alpha + 1$, $F_{\eta\restriction\alpha^\frown\langle 0\rangle}$ is any extension of $F_{\eta\restriction\alpha}$ with the right domain and $F_{\eta\restriction\alpha^\frown\langle\eta[\alpha]\rangle}$ is defined by (3).

Now F_η is elementary (and well-defined) since $\{\bar{a}_j^i : j < \mu\}$ is indiscernible over $A_i \supseteq \operatorname{Dom} F_{\eta\restriction i}$.

Let $B = \bigcup \{F_\eta(\bar{a}_j^i): \eta \in {}^{\kappa >}\mu,\ i < l(\eta),\ j < \mu\}$. Then $|B| \le \mu \cdot \kappa \cdot \sum_{\gamma < \kappa} \mu^\gamma \le \mu$.

For $\eta \in {}^\kappa\mu$, $F_\eta(p \restriction \operatorname{Dom} F_\eta)$ is consistent since F_η is elementary. Let $F_\eta(p \restriction \operatorname{Dom} F_\eta) \subseteq q_\eta \in S^m(B)$. We shall show that for $\eta \ne \rho \in {}^\kappa\mu$, $q_\eta \ne q_\rho$.

Let α be the first ordinal such that $\eta[\alpha] \ne \rho[\alpha]$; let $\eta \restriction \alpha = \rho \restriction \alpha = \nu$. Assume $\eta \restriction (\alpha + 1) = \nu^\frown\langle j_1\rangle$, $\rho \restriction (\alpha + 1) = \nu^\frown\langle j_2\rangle$, $j_1 \ne j_2$. We may assume without loss of generality that $j_2 \ne 0$.

If also $j_1 \ne 0$, then $F_\rho(\bar{a}_{j_1}^\alpha) = F_{\nu^\frown\langle j_2\rangle}(\bar{a}_{j_1}^\alpha) = F_{\nu^\frown\langle 0\rangle}(\bar{a}_{j_1}^\alpha)$; $F_\eta(\bar{a}_0^\alpha) = F_{\nu^\frown\langle j_1\rangle}(\bar{a}_0^\alpha) = F_{\nu^\frown\langle 0\rangle}(\bar{a}_{j_1}^\alpha)$. Thus $F_\rho(\bar{a}_{j_1}^\alpha) = F_\eta(\bar{a}_0^\alpha)$, $\varphi_\alpha(\bar{x}; \bar{a}_0^\alpha) \in p$ so $\varphi_\alpha(\bar{x}; F_\eta(\bar{a}_0^\alpha)) \in q_\eta$. $\neg\varphi_\alpha(\bar{x}; \bar{a}_{j_1}^\alpha) \in p$ since $j_1 \ne 0$ so $\neg\varphi_\alpha(\bar{x}; F_\rho(\bar{a}_{j_1}^\alpha)) \in q_\rho$, thus $\neg\varphi_\alpha(\bar{x}; F_\eta(\bar{a}_0^\alpha)) \in q_\rho$, and $q_\eta \ne q_\rho$.

Similarly if $j_1 = 0$, $F_\rho(\bar{a}_0^\alpha) = F_\eta(\bar{a}_{j_2}^\alpha)$ but $\varphi_\alpha(\bar{x}; F_\rho(\bar{a}_0^\alpha)) \in q_\rho$ and $\neg\varphi_\alpha(\bar{x}; F_\eta(\bar{a}_{j_2}^\alpha)) \in q_\eta$. Again $q_\eta \ne q_\rho$.

Thus $|S^m(B)| \ge |\{q_\eta : \eta \in {}^\kappa\mu\}| = \mu^\kappa > \mu$ and T is not stable in μ.

COROLLARY 3.8: (1) *If μ_0 is the first cardinal $\ge |T|$ in which T is stable, then $\mu_0 \le 2^{|T|}$. T is stable in μ iff $\mu = \mu_0 + \mu^{<\kappa(T)}$. In fact $\mu_0 = \sup\{|S(A)|: A$ of cardinality $\le |T|\}$.*

(2) *T is superstable iff $\kappa(T) = \aleph_0$.*

(3) *Either* (i) *$\kappa^m(T) = \kappa(T)$ for every m or* (ii) *for some m_0 $m < m_0$ iff $\kappa^m(T) = \kappa^1(T)$ iff $\kappa^m(T) \ne \kappa(T)$. In the latter case $\kappa^1(T)$ is singular and $\kappa(T) = (\kappa^1(T))^+$.*

Remark. See Exercise 3.3.

Proof. (1) Follows from Lemma 3.6 and Theorem 3.7, and the remark to 3.6.

(2) If $\kappa(T) = \aleph_0$ then $\mu = \mu^{<\kappa(T)}$ for all (infinite) μ. So T is stable in μ for $\mu \ge 2^{|T|}$, thus T is superstable. On the other hand if $\kappa(T) > \aleph_0$ then for every cardinal $\aleph_\alpha$, $\operatorname{cf}(\aleph_{\alpha+\omega}) = \omega$ so $\aleph_{\alpha+\omega}^{\aleph_0} > \aleph_{\alpha+\omega}$. Thus $\aleph_{\alpha+\omega}^{<\kappa(T)} > \aleph_{\alpha+\omega}$, and T is not stable in $\aleph_{\alpha+\omega}$ so T is not superstable.

(3) Clearly $\kappa^m(T) \le \kappa^{m+1}(T)$ (by extending the m-type in Theorem 3.1 to an $(m + 1)$-type). For every regular κ there is a strong limit cardinal $\lambda > 2^{|T|}$ of cofinality κ, so $\lambda^{<\kappa} = \lambda$, $\lambda^\kappa > \lambda$. If $\kappa < \kappa(T)$ then

T is not stable in λ by Theorem 3.7, and if $\kappa \geq \kappa^1(T)$ then T is stable in λ by Lemma 3.6. So for regular κ, $\kappa < \kappa(T)$ iff $\kappa < \kappa^1(T)$, so we finish.

DEFINITION 3.2: I is a *maximal* indiscernible set over A in M if I is indiscernible over A, $I \subseteq |M|$, and there is no $J \supsetneqq I$, $J \subseteq |M|$, J indiscernible over A.

DEFINITION 3.3: $\dim(I, A, M) = \min\{|J|: J$ is equivalent to I, and J is a maximal indiscernible set over A in $M\}$. If $A = \emptyset$ we just write $\dim(I, M)$. If for all J as above $\dim(I, A, M) = |J|$, we say that the dimension is *true*.

Remark. This definition differs technically from Definition II, 4.5, but no confusion will arise.

LEMMA 3.9: *If I is a maximal indiscernible set over A in M, then $|I| + \kappa(T) = \dim(I, A, M) + \kappa(T)$ and if $\dim(I, A, M) \geq \kappa(T)$, then the dimension is true.*

Proof. Let J be a maximal indiscernible set over A in M, equivalent to I, and for which $|J| = \dim(I, A, M)$. Clearly $|I| \geq |J|$, so that $|I| + \kappa(T) \geq |J| + \kappa(T)$. Thus it is sufficient to prove that $|I| + \kappa(T) \leq |J| + \kappa(T)$ or in fact $|I| \leq |J| + \kappa(T)$.

By way of contradiction assume $|I| > |J| + \kappa(T)$. From Corollary 3.5 we have $I_1 \subseteq I$, $|I_1| \leq |J| + \kappa(T)$ such that $I - I_1$ is indiscernible over $A \cup \bigcup I_1 \cup \bigcup J$. From the equivalence of I and J we can find an infinite I^* such that $(I - I_1) \cup I^*$ is indiscernible over $A \cup \bigcup J$ and $J \cup I^*$ is indiscernible over A (take $I^* = \{\bar{a}_n: n < \omega\}$ where the type $\bar{a}_n$ realizes over $A \cup \bigcup I \cup \bigcup J \cup \bigcup \{\bar{a}_l: l < n\}$ is the average of I over the same set).

Now with the aid of I^* we shall show that $(I - I_1) \cup J$ is indiscernible over A. Let $\bar{a}_1, \ldots, \bar{a}_n$ be distinct sequences in $(I - I_1) \cup J$, $\bar{c} \in A$, $\varphi(\bar{x}_1, \ldots, \bar{x}_n; \bar{c})$ a formula. Let $\bar{b}_1, \ldots, \bar{b}_n$ be any distinct sequences in I^*. Define $\bar{a}_i'$ as follows:

$$\bar{a}_i' = \begin{cases} \bar{a}_i & \text{for } \bar{a}_i \in J, \\ \bar{b}_i & \text{for } \bar{a}_i \in I - I_1. \end{cases}$$

Now $\models \varphi[\bar{a}_1, \ldots, \bar{a}_n; \bar{c}] \Leftrightarrow \models \varphi[\bar{a}_1', \ldots, \bar{a}_n'; \bar{c}]$ since $(I - I_1) \cup I^*$ is indiscernible over $A \cup \bigcup J$, and $\models \varphi[\bar{a}_1', \ldots, \bar{a}_n'; \bar{c}] \Leftrightarrow \models \varphi[\bar{b}_1, \ldots, \bar{b}_n; \bar{c}]$ since $J \cup I^*$ is indiscernible over A. But then the truth of $\varphi(\bar{a}_1, \ldots, \bar{a}_n; \bar{c})$ is not dependent on the particular $\bar{a}_i$. Thus $(I - I_1) \cup J$ is indiscernible over A, in contradiction to J's maximality in M.

LEMMA 3.10: (1) *Let M be an $(\aleph_1 + \kappa(T))$-saturated model. Then M is λ-saturated iff for every infinite indiscernible $I \subseteq |M|$,* $\dim(I, M) \geq \lambda$.

(2) *We can replace the assumption "$(\aleph_1 + \kappa(T))$-saturated" by "every type which is almost over some $A \subseteq |M|$, $|A| < \kappa(T)$, is realized in M" (i.e., by "$\mathbf{F}^a_{\kappa(T)}$-saturated". See Definitions IV, 1.1 and IV, 2.1).*

Proof. (1) If M is λ-saturated then clearly there is no indiscernible set of power $< \lambda$ which is maximal in M. So one direction is clear.

Now assume M is not λ-saturated, $\lambda > \aleph_1 + \kappa(T)$, and let $p \in S^m(A)$, $A \subseteq |M|$, $|A| < \lambda$, be a type omitted in M. Extend p to $q \in S^m(|M|)$ and choose $B \subseteq |M|$, $|B| < \kappa(T)$, so that q does not fork over B (by Corollary 3.2). By 1.12 and 2.13(2) since M is $(\aleph_1 + \kappa(T))$-saturated there is $I \subseteq |M|$ which is an infinite indiscernible set such that $q = \mathrm{Av}(I, |M|)$. Let $J \supseteq I$ be an indiscernible set maximal in M. If $\dim(I, M) < \lambda$, we are through. Otherwise $|J| \geq \dim(I, M) \geq \lambda$. By Corollary 3.5 there is $J' \subseteq J$, $|J'| \leq |A| + \kappa(T)$ such that $J - J'$ is indiscernible over A ($\kappa(T) < \lambda$, $|A| < \lambda$ so $|J'| < \lambda$ and thus $J - J' \neq \emptyset$). But then since $\mathrm{Av}(J, A) = \mathrm{Av}(I, A) = q \restriction A = p$ we have that all the sequences in $J - J'$ realize p, contradiction.

(2) Similar proof.

THEOREM 3.11: *If $\{M_i\}_{i<\delta}$ is an increasing sequence of λ-saturated models, and $\kappa(T) \leq \mathrm{cf}\, \delta$, then $M = \bigcup_{i<\delta} M_i$ is λ-saturated.*

Proof. Clearly if $\lambda \leq \mathrm{cf}(\delta)$ then M is λ-saturated. So we may assume $\lambda > \mathrm{cf}(\delta)$, and in particular $\lambda > \kappa(T)$. First assume $\mathrm{cf}\, \delta > \aleph_0$ so we can use 3.10(1). Let $I \subseteq |M|$ be an infinite indiscernible set. Let $B \subseteq |M|$, $|B| < \kappa(T)$, be such that $\mathrm{Av}(I, |M|)$ does not fork over B. Since $\kappa(T) \leq \mathrm{cf}\, \delta$ and $\mathrm{cf}\, \delta$ is regular, there is $j < \delta$ such that $B \subseteq |M_j|$. By Corollary 2.13(2) there is $J \subseteq |M_j|$ equivalent to I and since M_j is λ-saturated there is $J' \supseteq J$, $J' \subseteq |M_j|$, indiscernible and $|J'| \geq \lambda$. Thus by Lemma 3.9 $\dim(I, M) = \dim(J', M) \geq \lambda$ and by Lemma 3.10(1) M is λ-saturated. If $\mathrm{cf}\, \delta = \aleph_0$ we can use 3.10(2) instead of 3.10(1) (by 2.13(1)) and easily reprove a suitable version of 2.13(2) (by Exercise 2.4).

THEOREM 3.12: *If T is λ-stable then T has a λ-saturated model of power λ.*

Proof. We define an increasing sequence $\{M_i\}_{i \le \lambda}$ such that $\|M_i\| = \lambda$, $M_\delta = \bigcup_{i<\delta} M_i$ and every $p \in S(|M_i|)$ is realized in M_{i+1}. It is easy to see that M_δ is cf(δ)-saturated. Thus if λ is regular then M_λ is λ-saturated.

If λ is singular then $\lambda = \sum_{i<\mathrm{cf}(\lambda)} \lambda_i$, $\mathrm{cf}(\lambda) < \lambda_i < \lambda_j$ for all $i < j < \mathrm{cf}(\lambda)$. So for all $i < \mathrm{cf}(\lambda)$ $M_\lambda = \bigcup_{j<\lambda} M_{j+\lambda_i^+}$; $\mathrm{cf}(j + \lambda_i^+) > \lambda_i$ so $M_{j+\lambda_i^+}$ is λ_i-saturated. Now if $\kappa(T) \le \mathrm{cf}(\lambda)$ then by Theorem 3.11 we have that M_λ is λ_i-saturated, for all i. Thus M_λ is λ-saturated.

If $\kappa(T) > \mathrm{cf}(\lambda)$ then $\lambda^{<\kappa(T)} \ge \lambda^{\mathrm{cf}(\lambda)} > \lambda$. So by Theorem 3.7 T is not stable in λ, contradiction.

EXERCISE 3.1: In Lemma 3.4 prove that we have $|I| < \kappa^1(T)$.

EXERCISE 3.2: Give an example in which $\kappa(T)$ is singular.

EXERCISE 3.3: Prove $\kappa^1(T) = \kappa(T)$. [Hint: Suppose A_i ($i < \kappa$) is increasing, $\mathrm{tp}(\bar{a}{}^\frown\bar{b}, A_{i+1})$ forks over A_i for $i < \kappa$. By 4.15 $\mathrm{tp}(\bar{a}, A_{i+1})$ forks over A_i or $\mathrm{tp}(\bar{b}, A_{i+1} \cup \bar{a})$ forks over $A_i \cup \bar{a}$.]

EXERCISE 3.4: Prove that in Lemma 3.4, when J is infinite, there is a minimal $I = I_0$ among the possible I's, i.e., I_0 is contained in every possible I.

EXERCISE 3.5: Prove the parallel of 3.11 for λ-compactness.

III.4. Further properties of forking

THEOREM 4.1: *Let $p \in S^m(B)$, $A \subseteq B$, then p does not fork over A iff for all finite Δ, $R^m(p, \Delta, \aleph_0) = R^m(p \restriction A, \Delta, \aleph_0)$.*

The analogous theorem for stationary types is:

THEOREM 4.2: *Let $p \in S^m(B)$, $A \subseteq B$, then*

(1) *p is stationary over A iff for all finite Δ, $R^m(p, \Delta, \aleph_0) = R^m(p \restriction A, \Delta, \aleph_0)$ and $\mathrm{Mlt}(p, \Delta, \aleph_0) = 1$. If p is stationary over A and the m-type $q \supseteq p$, then q does not fork over A iff for all finite Δ and all λ, $2 \le \lambda \le \aleph_0$, $R^m(p, \Delta, \lambda) = R^m(q, \Delta, \lambda)$.*

(2) *p is stationary over A iff p is stationary over B and does not fork over A.*

(3) *In (1) we can restrict ourselves to all large enough Δ.*

Proof of Theorem 4.1. The direction $\Leftarrow$ follows from 1.2(2). Now assume p does not fork over A. Let $\bar{a}$ realize p. Let $q = \mathrm{stp}(\bar{a}, A)$, $r = \mathrm{stp}(\bar{a}, B)$. By 2.6(1) (and since $q \vdash p \restriction A$, $r \vdash p$), for every finite Δ, $R^m(r, \Delta, \aleph_0) = R^m(p \cup r, \Delta, \aleph_0) = R^m(p, \Delta, \aleph_0)$; $R^m(q, \Delta, \aleph_0) = R^m(q \cup p \restriction A, \Delta, \aleph_0) = R^m(p \restriction A, \Delta, \aleph_0)$. So it is sufficient to prove Lemma 4.3.

LEMMA 4.3: *Let* $q = \mathrm{stp}(\bar{a}, A)$, *$p$ a type realized by $\bar{a}$ which does not fork over A, and $q \subseteq p$. Then for every finite Δ, $2 \le \lambda \le \aleph_0$,*

(1) $R^m(q, \Delta, \lambda) = R^m(p, \Delta, \lambda)$.

(2) $\mathrm{Mlt}(q, \Delta, \lambda) = \mathrm{Mlt}(p, \Delta, \lambda)$.

Proof. First we prove (1) for $\lambda < \aleph_0$.

Let p be over B, where $A \subseteq B$.

By II, 1.1 $R^m(q, \Delta, \lambda) \ge R^m(p, \Delta, \lambda)$. By II, 2.13 and II, 2.2 $R^m(q, \Delta, \lambda) = n < \omega$ for some n. By II, 2.9(2) there are $\varphi_\eta^{i,j}(\in \Delta)$, $\bar{a}_\nu^{i,j}$ for $\eta \in {}^{n>}\lambda$, $\nu \in {}^{n<}\lambda$, $i < \lambda$, $j < \lambda$, such that for $\eta \in {}^n\lambda$

$$q \cup \{\varphi_{\eta|l}^{i,j}(\bar{x}; \bar{a}_{\eta|l}^{i,j})^{\mathbf{t}}: l < n,\ i < \lambda,\ j < \lambda,\ i \ne j;$$
$$i = \eta[l] \text{ and } \mathbf{t} = 0 \text{ or } j = \eta[l] \text{ and } \mathbf{t} = 1\}$$

is consistent; let $\bar{a}_\eta$ realize this set.

Define $\bar{b}_\eta$ $(\eta \in {}^n\lambda)$ $\bar{b}_\eta^{i,j}$ $(\eta \in {}^{n>}\lambda, i < \lambda, j < \lambda)$ by 2.6(3) such that

(i) $\mathrm{stp}(\bar{b}_\eta, A) \equiv \mathrm{stp}(\bar{a}_\eta, A)$, $\eta \in {}^n\lambda$
$\mathrm{stp}(\bar{b}_\eta^{i,j}, A) \equiv \mathrm{stp}(\bar{a}_\eta^{i,j}, A)$, $\eta \in {}^{n>}\lambda$, $i < \lambda$, $j < \lambda$.

(ii) $\mathrm{tp}_*(\bigcup \{\bar{b}_\eta: \eta \in {}^n\lambda\} \cup \bigcup \{\bar{b}_\eta^{i,j}: \eta \in {}^{n>}\lambda, i < \lambda, j < \lambda\}, B)$ does not fork over A, and extend the corresponding type of the $\bar{a}_\eta$'s.

By 2.9(1) $\bar{b}_\eta$ $(\eta \in {}^n\lambda)$ realizes p. Clearly $\vDash \varphi_{\eta|l}[\bar{b}_\eta; \bar{b}_{\eta|l}^{i,j}]^{\mathbf{t}}$ when $l < n$, $\eta \in {}^n\lambda$, $i \ne j < \lambda$; and $i = \eta[l]$, $\mathbf{t} = 0$ or $j = \eta[l]$, $\mathbf{t} = 1$. So by II, 2.9(2) (now using the other direction) $R^m(p, \Delta, \lambda) \ge n = R^m(q, \Delta, \lambda)$. So we prove the equality.

The proof of (1) for $\lambda = \aleph_0$, and of (2) is similar using suitable parts of II, 2.9.

Now we return to

Proof of Theorem 4.2. (1) Suppose p is stationary over A. So it does not fork over A, hence by 4.1 for every finite Δ, $R^m(p, \Delta, \aleph_0) = R^m(p \restriction A, \Delta, \aleph_0)$.

Let M be a $(|B|^+ + |T|^+)$-saturated model, $B \subseteq |M|$. Let $r \in S^m(|M|)$ $p \subseteq r$, r does not fork over A, and let $\bar{a}$ realize r. By 2.14 $p \vdash q =_{\mathrm{def}} \mathrm{stp}(\bar{a}, A)$, hence by 4.3 $\mathrm{Mlt}(p, \Delta, \lambda) = \mathrm{Mlt}(q, \Delta, \lambda) = \mathrm{Mlt}(r, \Delta, \lambda)$ for every finite Δ, and $2 \le \lambda \le \aleph_0$. But by II, 1.10(1) $\mathrm{Mlt}(r, \Delta, \lambda) = 1$ so $\mathrm{Mlt}(p, \Delta, \lambda) = 1$ and we finish.

Now suppose p is not stationary over A. If it forks over A, then by 4.1 for some finite Δ, $R^m(p, \Delta, \aleph_0) \neq R^m(p \restriction A, \Delta, \aleph_0)$. So assume p does not fork over A. By 2.9(1), if $p = \mathrm{tp}(\bar{a}_1, B)$, $q = \mathrm{stp}(\bar{a}_1, A)$ then $p \nvdash q$. So for some $E \in \mathrm{FE}^m(A)$ $p \cup \{\neg E(\bar{x}; \bar{a}_1)\}$ is consistent; let $\bar{a}_2$ realize it. By 2.6(1) $R^m[p \cup \{E(\bar{x}; \bar{a}_l)\}, \Delta, \aleph_0] = R^m(p, \Delta, \aleph_0)$ for $l = 1, 2$ (where $\Delta = \{E\}$). But $E(\bar{x}; \bar{a}_1) \vdash \neg E(\bar{x}; \bar{a}_2)$. Hence $\mathrm{Mlt}(p, \Delta, \aleph_0) \ge 2$. The second sentence is easy, so we have finished the proof of 4.2(1).

(2) The proof is immediate by part (1).

From 4.1 we see

COROLLARY 4.4: *Let $A \subseteq B \subseteq C$ and $p \in S^m(C)$. p does not fork over A iff p does not fork over B and $p \restriction B$ does not fork over A.*

THEOREM 4.5: *Let p be an m-type over A, q a Δ-m-type, Δ not necessarily finite, such that $p \cup q$ (is consistent and) forks over A. Then for $\lambda \ge \aleph_0$: $R^m(p, \Delta, \lambda) < \infty \Rightarrow R^m(p \cup q, \Delta, \lambda) < R^m(p, \Delta, \lambda)$.*

Remark. (1) This improves 1.2. Note that in 1.2(4) we have some unspecified Δ_0, and in 1.2(3) we have a formula which divides.

(2) In the course of proving this theorem we shall show:

(*) If p forks over A, $A \subseteq B$, then there is an elementary mapping F, $F \restriction A =$ the identity and $F(p)$ forks over B.

Proof. Let M be a μ^+-saturated model ($\mu = |A| + \lambda + |T|$) such that $A \subseteq |M|$ and q is a type over $|M|$. Assume that $\alpha = R^m(p, \Delta, \lambda) = R^m(p \cup q, \Delta, \lambda) < \infty$ and we will get a contradiction.

Since $\lambda \ge \aleph_0$, by II, 1.6 and II, 1.1 there is $q' \in S^m_\Delta(|M|)$ such that $R^m(p \cup q', \Delta, \lambda) = \alpha$. By the definition of rank there are $< \lambda$ such q'. Let them be $\{q_j: j < j_0 < \lambda\}$ and let $\bar{a}_j$ realize $p \cup q_j$. Choose $A_j \subseteq |M|$, $|A_j| \le |T|$ such that $\mathrm{tp}(\bar{a}_j, |M|)$ does not fork over A_j and let $B = \bigcup_{j<j_0} A_j \cup A$. Then $|B| \le \mu$.

By II, 1.8(2) we may assume without loss of generality that Δ is closed under conjunction and negation and by II, 1.1 that p and q are closed under conjunction. So there is $\psi(\bar{x}; \bar{a}) \in p (\bar{a} \in A)$ and $\theta(\bar{x}; \bar{b}) \in q$ such that $\psi(\bar{x}; \bar{a}) \wedge \theta(\bar{x}; \bar{b})$ forks over A. So there are $\{\bar{c}_{l,i}: i < \omega\}_{l<l_0<\omega}$ such that $\{\bar{c}_{l,i}: i < \omega\}$ is indiscernible over A, $\{\varphi_l(\bar{x}; \bar{c}_{l,i}): i < \omega\}$ is m_l-

contradictory, $\psi \wedge \theta \vdash \bigvee_l \varphi_l(\bar{x}; \bar{c}_{l,0})$. Since M is a model we can choose (by 2.6(3)) $\bar{b}', \bar{c}'_{l,i} \in |M|$, $l < l_0$, $i < \omega$, such that

$$\begin{aligned} &\operatorname{stp}_*(\bar{b} \cup \bigcup \{\bar{c}_{l,i}: l < l_0, i < \omega\}, A) \\ &\quad \equiv \operatorname{stp}_*(\bar{b}' \cup \bigcup \{\bar{c}'_{l,i}: l < l_0, i < \omega\}, A) \end{aligned}$$

and $\operatorname{tp}_*(\bar{b}' \cup \bigcup \{\bar{c}'_{l,i}: l < l_0, i < \omega\}, B)$ does not fork over A. Thus also $\psi(\bar{x}; \bar{a}) \wedge \theta(\bar{x}; \bar{b}') \vdash \bigvee_l \varphi_l(\bar{x}; \bar{c}'_{l,0})$, $\alpha = R^m[p \cup \{\psi(\bar{x}; \bar{a}) \wedge \theta(\bar{x}; \bar{b}')\}, \Delta, \lambda]$ and $\{\varphi_l(\bar{x}; \bar{c}'_{l,i}): i < \omega\}$ is m_l-contradictory, $l < l_0$. But $\operatorname{stp}(\bar{c}_{l,0}, A) \equiv \operatorname{stp}(\bar{c}_{l,i}, A)$, $l < l_0$, $i < \omega$ (by the indiscernibility of $\{\bar{c}_{l,i}: i < \omega\}$ over A) so by 2.9(1) $\operatorname{tp}(\bar{c}'_{l,i}, B) = \operatorname{tp}(\bar{c}'_{l,0}, B)$, $l < l_0$, $i < \omega$. Hence $\psi(\bar{x}; \bar{a}) \wedge \theta(\bar{x}; \bar{b}')$ forks over B. By II, 1.6 we may extend $p_1 = p \cup \{\psi(\bar{x}; \bar{a}) \wedge \theta(\bar{x}; \bar{b}')\}$ to a type $p \cup q'$, $q' \in S^m_\Delta(|M|)$ such that $R^m(p \cup q', \Delta, \lambda) = \alpha$ (because $R^m(p_1, \Delta, \lambda) = \alpha$ by Exercise II, 1.1).

Thus there is a $j < j_0$ such that $\theta(\bar{x}; \bar{b}') \in q_j$ (since $\theta(\bar{x}; \bar{b}')$ is a Δ-formula). Since $\bar{a}_j$ realizes $q_j \cup p$ and $\operatorname{tp}(\bar{a}_j, |M|)$ does not fork over B, the formula $\psi(\bar{x}; \bar{a}) \wedge \theta(\bar{x}; \bar{b}')$ does not fork over B. Contradiction.

EXERCISE 4.1: Show that in 4.5 the condition $\lambda \geq \aleph_0$ cannot be omitted.

Remark. See Exercise 4.18.

Hint: Let $M = (B, R)$ be a finite model, R a two place relation, such that for any $a, b \in B$ there is an automorphism f of M, $f(a) = b$. Let us define a model N:

$$\begin{aligned} |N| &= \omega \times B \times \{0, 1\}, \\ R^N &= \{\langle\langle n, a, 1\rangle, \langle n, b, 1\rangle\rangle: M \models R[a, b]\} \\ S^N &= \{\langle\langle n, a, 0\rangle, \langle m, b, 1\rangle\rangle: n = m \text{ or } a \neq b\} \\ P^N &= \{\langle n, a, 1\rangle: n < \omega, a \in A\}. \end{aligned}$$

Let $T = \operatorname{Th}(N)$; it is $\aleph_0$-stable; let $A = \emptyset$, $p = \{P(x)\}$, $q = \{\langle 0, a, 0\rangle Sx: a \in B\}$, $\Delta = \{S, R\}$.

COROLLARY 4.6: *Let* $p = \operatorname{stp}(\bar{a}, A)$, $\lambda \geq \aleph_0$, $l(\bar{a}) = m$ *and* Δ *is not necessarily finite.*

(1) *There do not exist two contradictory* Δ-m-*types* q_1, q_2 *such that* $R^m(p, \Delta, \lambda) = R^m(p \cup q_1, \Delta, \lambda) = R^m(p \cup q_2, \Delta, \lambda) < \infty$.

(2) *If* $R^m(p, \Delta, \lambda) < \infty$ *then* $\operatorname{Mlt}^2(p, \Delta, \lambda) = 1$.

(3) *If* $q \in S^m(|M|)$, $R^m(q, \Delta, \lambda) < \infty$ *then* $\operatorname{Mlt}^2(q, \Delta, \lambda) = 1$ *(see* 2.15(2)).

LEMMA 4.7: *Let* $q = \mathrm{stp}(\bar{a}, A)$, $l(\bar{a}) = m$, p *be a* Δ-m-*type such that* $p \cup q$ *does not fork over* A (Δ *not necessarily finite*). *Then for* $\lambda \geq \aleph_0$, $R^m(q, \Delta, \lambda) = \alpha < \infty \Rightarrow R^m(q \cup p, \Delta, \lambda) = \alpha$.

Proof. Let $B \supseteq A \cup \bar{a} \cup \mathrm{Dom}\, p$. Choose $r \in S^m(B)$ extending $p \cup q$ such that r does not fork over A. Then $r \vdash q \cup p \vdash q$ and thus $R^m(r, \Delta, \lambda) \leq R^m(p \cup q, \Delta, \lambda) \leq R^m(q, \Delta, \lambda) = \alpha$. On the other hand by II, 1.6 there is $p' \in S^m_\Delta(B)$, such that $R^m(q \cup p', \Delta, \lambda) = R^m(q, \Delta, \lambda)$. By 4.5 $q \cup p'$ does not fork over A and thus by 2.9(1) $p' \subseteq r$ hence $p \subseteq p'$.

LEMMA 4.8: (1) *If* p *does not fork over* A *then* p *is "definable almost over* A". *This means that for all* φ *there is a formula* $\psi_\varphi(\bar{x}; \bar{c}_\varphi)$ *which is almost over* A, *such that* $p \restriction \varphi$ *is definable by* $\psi_\varphi(\bar{x}; \bar{c}_\varphi)$; *i.e.*, $\varphi(\bar{x}; \bar{a}) \in p \Rightarrow \models \psi_\varphi[\bar{a}; \bar{c}_\varphi]$ *and* $\neg\varphi(\bar{x}; \bar{a}) \in p \Rightarrow \models \neg\psi_\varphi[\bar{a}; \bar{c}_\varphi]$.

(2) *If* $A \subseteq B \subseteq C$, $p \in S^m(C)$, $p \restriction B$ *is stationary over* A *and* p *does not fork over* A, *then* p *is definable over* B.

Proof. (1) Immediate by 4.3 (for $\lambda = 2$) and II, 2.12(3).

(2) As of (1), noticing that by 2.14, if $\bar{a}$ realizes p, $p \restriction B \vdash \mathrm{stp}(\bar{a}, A)$.

LEMMA 4.9: *Let* $p \in S^m(|M|)$, Δ *finite. For every type* $q \in S^m(B)$ *extending* p, *which does not fork over* $|M|$, $q \restriction \Delta = \mathrm{Av}_\Delta(I, B)$ *for some infinite* Δ-*indiscernible set* $I \subseteq |M|$.

Proof. By Lemma II, 2.20 there is a finite Δ_1 such that for all $\varphi(\bar{x}; \bar{y}) \in \Delta$, $\bar{a}$, and any infinite Δ_1-indiscernible set J, $|\varphi(J, \bar{a})| < \aleph_0$ or $|\neg\varphi(J, \bar{a})| < \aleph_0$.

Using Theorem II, 2.17 we can find a finite Δ^* such that if $p_k = \mathrm{tp}_{\Delta^*}(\bar{a}_k, A_k)$, $A_k = A \cup \bigcup \{\bar{a}_l : l < k\}$, and for all $\varphi \in \Delta^*$ and $k < \omega$ $R^m(p_k \restriction \varphi, \varphi, 2) = R^m(p_0 \restriction \varphi, \varphi, 2)$ (where $l(\bar{a}_k) = m$), then $\{\bar{a}_k : k < \omega\}$ is a Δ_1-indiscernible set over A, if p_k satisfies the condition from II, 2.17.

Choose a finite $A \subseteq |M|$ such that $p \restriction \varphi$ is (ψ_φ, A)-definable, for every $\varphi \in \Delta^*$ (φ_φ from II, 2.12) and define $\bar{a}_k$ ($k < \omega$) accordingly. As in 2.13 the result follows (by defining $\{\bar{a}_i : \omega \leq i < \omega + \omega\}$ and Exercise II, 2.9).

EXERCISE 4.2: Show that I does not depend on q but only on p and Δ.

COROLLARY 4.10: (1) *The formula* $\varphi(\bar{x};\bar{a})$ *does not fork over* $|M|$ *iff there is* $\bar{b} \in |M|$ *such that* $\models \varphi[\bar{b};\bar{a}]$.

(2)* *If there is* $\bar{b} \in A$ *such that* $\models \varphi[\bar{b};\bar{a}]$ *then* $\varphi(\bar{x};\bar{a})$ *does not fork over* A.

Proof. (1) Assume $\varphi(\bar{x};\bar{a})$ does not fork over $|M|$. Then there is $q \in S^m(|M| \cup \bar{a})$ which does not fork over $|M|$, $\varphi(\bar{x};\bar{a}) \in q$ (by Theorem 1.4). Let $p = q \restriction |M|$. By the previous lemma there is $I \subseteq |M|$ such that $q \restriction \varphi = \mathrm{Av}_\varphi(I, |M| \cup \bar{a})$. Thus $\varphi(\bar{x};\bar{a})$ is realized in M. This proves one direction, and the other follows from (2).

(2) If $\varphi(\bar{x};\bar{a})$ forks over A then $\varphi(\bar{x};\bar{a}) \vdash \bigvee_{i<n} \varphi_i(\bar{x};\bar{a}_i)$ where $\varphi_i(\bar{x};\bar{a}_i)$ divides over A. But if $\varphi(\bar{x};\bar{a})$ is realized in A then there is i such that $\varphi_i(\bar{x};\bar{a}_i)$ is realized in A. By the definition of "divides" there are $n_i < \omega$ and sequences $\bar{a}_i^j$, $j < \omega$, such that $\mathrm{tp}(\bar{a}_i^j, A) = \mathrm{tp}(\bar{a}_i, A)$ and for every $i < n$, $\{\varphi_i(\bar{x};\bar{a}_i^j) : j < \omega\}$ is n_i-contradictory. But the sequence in A which realizes $\varphi_i(\bar{x};\bar{a}_i)$ realizes also all the $\varphi_i(\bar{x};\bar{a}_i^j)$, contradiction.

An alternative proof of Corollary 4.10. Choose $q \in S^m(|M|)$ such that $q \cup \{\varphi(\bar{x};a)\}$ does not fork over $|M|$, and $n(\varphi) < \omega$ such that for any indiscernible set I, $|\varphi(I;\bar{a})| < n(\varphi)$ or $|\neg\varphi(I,\bar{a})| < n(\varphi)$. Let $I = \{\bar{b}_n : n < \omega\}$ be an indiscernible set based on q, so $\varphi(\bar{x};\bar{a}) \in \mathrm{Av}(I;\bar{a})$ hence

$$\bigvee_{\substack{w \subseteq 2n-1 \\ |w| = n}} \bigwedge_{l \in w} \varphi(\bar{b}_n, \bar{a}).$$

Now only $\mathrm{tp}_*(\bigcup I, |M|)$ was important, hence

$$\mathrm{tp}(\bar{b}_0{}^\frown \cdots {}^\frown \bar{b}_{2n-1}, |M|) \vdash \bigvee_{\substack{w \subseteq 2n-1 \\ |w| = n}} \bigwedge_{l \in w} \varphi(\bar{x}_n, \bar{a}).$$

So for some $\psi(\bar{x}_0, \ldots, x_{2n-1}, \bar{c}) \in \mathrm{tp}(\bar{b}_0 \cdots \bar{b}_{2n-1}, |M|)$,

$$\psi(\bar{x}_0, \ldots, \bar{x}_{2n-1}, \bar{c}) \vdash \bigvee_{\substack{w \subseteq 2n-1 \\ |w| = n}} \bigwedge_{l \in w} \varphi(\bar{x}_n, \bar{a}).$$

But there are $\bar{b}'_0, \ldots, b'_{2n-1} \in |M|$, $M \models \psi[b'_0, \ldots, \bar{b}'_{2n-1}, \bar{c}]$, so for some l $\models \varphi[\bar{b}'_l, \bar{c}]$, as required.

A converse to 4.8 is

LEMMA 4.11: *If* $p \in S^m_\Delta(|M|)$ *(Δ is not necessarily finite) is definable almost over A where $A \subseteq |M|$ (see 4.8(1)) then p does not fork over A.*

Proof. Without loss of generality M is $|A|^+$-saturated, since we may extend the given M to a $|A|^+$-saturated model and then extend p to a complete Δ-m-type over the new model which is still definable almost over A. The implication about forking will then hold for p. Also we may assume that Δ is closed under conjunction and negation. By way of contradiction assume p forks over A. Thus by Lemma 1.1(5) we get a $\varphi(\bar{x}; \bar{c}) \in p$ which forks over A, and by definition there is $\psi(\bar{y}; \bar{a})$ almost over A which defines $p \restriction \varphi$. Without loss of generality $\bar{a} \in |M|$. So if $\bar{c}' \in |M|$ and $\operatorname{stp}(\bar{c}', A) \equiv \operatorname{stp}(\bar{c}, A)$ then $\vDash \psi[\bar{c}; \bar{a}] \equiv \psi[\bar{c}'; \bar{a}]$ and thus $\varphi(\bar{x}; \bar{c}') \in p$.

Now define sequences $\bar{c}_n \in |M|$, $n < \omega$ such that $\operatorname{tp}(\bar{c}_n, A \cup \bigcup_{l<n} \bar{c}_l)$ does not fork over A, and each $\bar{c}_n$ realizes $\operatorname{stp}(\bar{c}, A)$. By Lemma 1.10(1) $\{\bar{c}_n : n < \omega\}$ is an indiscernible set based on A. Clearly $\vDash \psi[\bar{c}_n; \bar{a}]$; thus $\varphi(\bar{x}; \bar{c}_n) \in p$; thus for every n, $\psi_n = \bigvee_{w \subseteq 2n-1, |w|=n} \bigwedge_{l \in w} \varphi(\bar{x}; \bar{c}_l) \in p$ and [by 1.1(9)] it forks over A. This contradicts 2.5 (because of 2.7).

EXERCISE 4.3: (1) Find an example of $p \in S^m_\Delta(B)$, p is definable over $A \subseteq B$, but p forks over A.

(2) In 4.11 let q be an m-type over A, $p \cup q$ consistent; show that $p \cup q$ does not fork over A.

COROLLARY 4.12: *Let $p \in S^m(|M|)$, $A \subseteq |M|$, p does not fork over A iff p is definable almost over A iff for all φ, $R^m(p, \varphi, \aleph_0) = R^m(p \restriction A, \varphi, \aleph_0)$.*

Proof. By 4.8, 4.11 and 4.1.

THEOREM 4.13 (The symmetry theorem): *If $A \subseteq C$, $B \subseteq C$,* $\operatorname{tp}(\bar{a}, C)$ *does not fork over A,* $\operatorname{tp}(\bar{b}, C)$ *does not fork over B then* $\operatorname{tp}(\bar{a}, C \cup \bar{b})$ *forks over A iff* $\operatorname{tp}(\bar{b}, C \cup \bar{a})$ *forks over B.*

Proof. Suppose not; so by the symmetry of the hypothesis on $\bar{a}, \bar{b}$; we can assume $\operatorname{tp}(\bar{a}, C \cup \bar{b})$ forks over A, but $\operatorname{tp}(\bar{b}, C \cup \bar{a})$ does not fork over B. Hence there is $\bar{c} \in C$ and a formula $\varphi(\bar{x}; \bar{b}, \bar{c})$ which forks over A, and $\vDash \varphi[\bar{a}, \bar{b}, \bar{c}]$. Now define by induction on $n < \omega$ sequences $\bar{a}_n$, $\bar{b}_n$ such that

(i) $\operatorname{stp}(\bar{a}_n, C) \equiv \operatorname{stp}(\bar{a}, C)$

(ii) $\operatorname{stp}(\bar{b}_n, C) \equiv \operatorname{stp}(\bar{b}, C)$

(iii) $p_n = \operatorname{tp}(\bar{a}_n, \bar{a} \cup \bar{b} \cup C \cup \bigcup \{\bar{a}_i : i < n\} \cup \bigcup \{\bar{b}_i : i < n\})$ does not fork over A.

(iv) $q_n = \operatorname{tp}(\bar{b}_n, \bar{a} \cup \bar{b} \cup C \cup \bigcup \{\bar{a}_i : i \le n\} \cup \bigcup \{\bar{b}_i : i < n\})$ does not fork over B (this is possible by 2.6(3)).

Now $\vDash \varphi[\bar{a}, \bar{b}_n, \bar{c}]$ [as by 2.9(1) $\operatorname{stp}(\bar{b}, A)$ is stationary over A, we have assumed $\operatorname{tp}(\bar{b}, C \cup \bar{a})$ does not fork over B, and by (iv), (ii)] hence

$\vDash\varphi[\bar{a}_k, \bar{b}_n, \bar{c}]$ for $k \le n$ [q_n does not fork over B, hence, by 4.8(1) is definable almost over B; but by (i) $\text{stp}(\bar{a}_k, C) \equiv \text{stp}(\bar{a}, C)$].

On the other hand if $k > n$, as $\varphi(\bar{x}; \bar{b}, \bar{c})$ forks over A, by (ii) also $\varphi(\bar{x}; \bar{b}_n, \bar{c})$ forks over A, but by (iii) p_k does not fork over A, hence $\neg\varphi(\bar{x}; \bar{b}_n, \bar{c}) \in p_k$. So $\vDash\neg\varphi[\bar{a}_k, \bar{b}_n, \bar{c}]$.

So $\vDash\varphi[\bar{a}_k, \bar{b}_n, \bar{c}] \Leftrightarrow k \le n$, hence by II, 2.13(5) T is unstable. Contradiction.

THEOREM 4.14: *If* $A \subseteq B$, $p = \text{tp}(\bar{a}^\frown\bar{b}, B)$ *does not fork over* A, *then* $q = \text{tp}(\bar{a}, B \cup \bar{b})$ *does not fork over* $A \cup \bar{b}$.

Proof. Suppose the hypothesis holds, but the conclusion fails. So for some $\bar{c} \in B$, $\vDash\varphi[\bar{a}, \bar{b}, \bar{c}]$ but $\varphi(\bar{x}; \bar{b}, \bar{c})$ forks over $A \cup \bar{b}$. Now define by induction on n sequences $\bar{c}_n, \bar{a}_n$ such that

(i) $\text{stp}(\bar{a}_n, A \cup \bar{b}) \equiv \text{stp}(\bar{a}, A \cup \bar{b})$.

(ii) $\text{stp}(\bar{c}_n, A) \equiv \text{stp}(\bar{c}, A)$.

(iii) $p_n = \text{tp}(\bar{a}_n, A \cup \bar{a} \cup \bar{b} \cup \bar{c} \cup \bigcup\{\bar{a}_i: i < n\} \cup \bigcup\{\bar{c}_i: i < n\})$ does not fork over $A \cup \bar{b}$.

(iv) $r_n = \text{tp}(\bar{c}_n, A \cup \bar{a} \cup \bar{b} \cup \bar{c} \cup \bigcup\{a_i: i \le n\} \cup \bigcup\{\bar{c}_i: i < n\})$ does not fork over A (this is possible by 2.6(3)).

As $\bar{c} \in B$, $\text{tp}(\bar{a}^\frown\bar{b}, A \cup \bar{c})$ does not fork over A; hence by the previous theorem 4.13 (and by 1.3), $\text{tp}(\bar{c}, A \cup \bar{a} \cup \bar{b})$ does not fork over A, hence by (ii), (iv) and 2.9(1), $\text{tp}(\bar{c}_n, A \cup \bar{a} \cup \bar{b}) = \text{tp}(\bar{c}, A \cup \bar{a} \cup \bar{b})$, hence $\vDash\varphi[\bar{a}, \bar{b}, \bar{c}_n]$.

But by (i) $\text{stp}(\bar{a}_n, A \cup \bar{b}) \equiv \text{stp}(\bar{a}, A \cup \bar{b})$, so by (iv), $\varphi(\bar{a}_n, \bar{b}; \bar{x}) \equiv \varphi(\bar{a}, \bar{b}; \bar{x}) \in r_k$, when $n \le k$. Combining we get $\vDash\varphi[\bar{a}_n, \bar{b}, \bar{c}_k]$ when $n \le k$.

Suppose now $n > k$, by (ii), (iv) $\text{tp}(\bar{b}^\frown\bar{c}, A) = \text{tp}(\bar{b}^\frown\bar{c}_k, A)$ hence $\varphi(\bar{x}; \bar{b}, \bar{c}_k)$ forks over A, hence by (iii) $\neg\varphi(\bar{x}; \bar{b}, \bar{c}_k) \in p_n$; hence $\vDash\neg\varphi[\bar{a}_n, \bar{b}, \bar{c}_k]$. Combining we get $\vDash\varphi[\bar{a}_n, \bar{b}, \bar{c}_k]$ iff $n \le k$, so by II, 2.13(5) T is unstable, contradiction.

LEMMA 4.15: (1) *If* $\text{tp}(\bar{a}, B)$ *does not fork over* $A \subseteq B$, *and* $\text{tp}(\bar{b}, B \cup \bar{a})$ *does not fork over* $A \cup \bar{a}$, *then* $\text{tp}(\bar{a}^\frown\bar{b}, B)$ *does not fork over* A.

(2) *If the first two types are stationary over* A, *the third is also.*

(3) *In* (1) *we can use an infinite* $\bar{a}$.

Proof. (1) Let $\{\bar{b}_i: i < i_0\}$ be a maximal sequence with the property that for all $i \ne j$, $\text{stp}(\bar{b}_i, A \cup \bar{a}) \not\equiv \text{stp}(\bar{b}_j, A \cup \bar{a})$. By Lemma 2.2(2) i_0 is certainly $\le 2^{|A|+|T|}$. Now by Lemma 2.6(3) there is an elementary mapping F with $\text{Dom}\, F = A \cup \bar{a} \cup \bigcup\{\bar{b}_i: i < i_0\}$ such that $F \restriction A$ is the identity, and for all $\bar{c} \in \mathrm{Dom}\, F$, $\mathrm{tp}(F(\bar{c}), B)$ does not fork over A, and $\mathrm{stp}(\bar{c}, A) \equiv \mathrm{stp}(F(\bar{c}), A)$.

Thus, in particular, $\mathrm{stp}(\bar{a}, A) \equiv \mathrm{stp}(F(\bar{a}), A)$ and $\mathrm{tp}(\bar{a}, B), \mathrm{tp}(F(\bar{a}), B) \in S^{l(\bar{a})}(B)$ do not fork over A. But by Corollary 2.9(1) $\mathrm{stp}(\bar{a}, A)$ is stationary over A, and thus $\mathrm{tp}(\bar{a}, B) = \mathrm{tp}(F(\bar{a}), B)$.

Thus we can find an elementary mapping G with $\mathrm{Dom}\, G = B \cup \mathrm{Range}\, F$ such that $G \restriction B$ is the identity and $G(F(\bar{a})) = \bar{a}$. Now let $\bar{b}_i^* = G(F(\bar{b}_i))$. By the maximality of the $\bar{b}_i$ there is no $\bar{b}'$ such that $\mathrm{stp}(\bar{b}', A \cup \bar{a}) \not\equiv \mathrm{stp}(\bar{b}_i^*, A \cup \bar{a})$ for all $i < i_0$. Thus there is an $i < i_0$ such that $\mathrm{stp}(\bar{b}, A \cup \bar{a}) \equiv \mathrm{stp}(\bar{b}_i^*, A \cup \bar{a})$ ($\bar{b}$ is from the statement of the lemma).

Now $\mathrm{tp}(\bar{b}_i^* {}^\frown \bar{a}, B)$ does not fork over A since $\bar{a}^\frown \bar{b}_i^* \in G(\mathrm{Range}\, F)$. Thus $\mathrm{tp}(\bar{b}_i^*, B \cup \bar{a})$ does not fork over $A \cup \bar{a}$, by Theorem 4.14. But we have also assumed that $\mathrm{tp}(\bar{b}, B \cup \bar{a})$ does not fork over $A \cup \bar{a}$, so by Corollary 2.9(2) $\mathrm{tp}(\bar{b}_i^*, B \cup \bar{a}) = \mathrm{tp}(b, B \cup \bar{a})$. Thus $\mathrm{tp}(\bar{a}^\frown \bar{b}, B) = \mathrm{tp}(\bar{a}^\frown \bar{b}_i^*, B)$, and $\mathrm{tp}(\bar{a}^\frown \bar{b}, B)$ does not fork over A.

(2) Immediate by 2.14.

(3) Same proof as (1).

EXERCISE 4.4: Prove that if $p \neq q \in S^m(B)$ $p \restriction A = q \restriction A = r$ and $R^m(p, \mathrm{L}, \lambda) = R^m(q, \mathrm{L}, \lambda) = R^m(r, \mathrm{L}, \lambda) < \infty$, then there is $E \in \mathrm{FE}^m(A)$ such that $p(\bar{x}) \cup q(\bar{y}) \vdash \neg E(\bar{x}; \bar{y})$.

EXERCISE 4.5: Prove that if $p \in S^m(B)$, $A \subseteq B$ and $R^m(p \restriction A, \mathrm{L}, \lambda) = R^m(p, \mathrm{L}, \lambda) < \infty$ then p is definable almost over A.

EXERCISE 4.6: If in addition to the assumptions of Exercise 4.5 $\mathrm{Mlt}(p \restriction A, \mathrm{L}, \lambda) = 1$, then p is definable over A.

EXERCISE 4.7; Try to generalize Exercise 4.4 and Exercise 4.5 to other infinite Δ's, where $S_\Delta^m(B)$ would replace $S^m(B)$.

LEMMA 4.16: *Let $p \in S^m(|M|)$, where $A \subseteq |M|$ and M is $|A|^+$-saturated.*

(1) *p forks over A iff p strongly splits over A.*

(2) *[$p \restriction A$ is stationary over A and p does not fork over A] iff p does not split over A.*

Proof. (1) One direction is just Theorem 1.6(1), and the other direction is left as an exercise.

(2) One direction is Lemma 1.9, and the other direction is left as an exercise.

LEMMA 4.17: *Let I, J be infinite indiscernible sets.*

(1) *If* $\mathrm{Av}(I, \bigcup I) = \mathrm{Av}(J, \bigcup I)$ *and* $\mathrm{Av}(I, \bigcup J) = \mathrm{Av}(J, \bigcup J)$ *then* I *and* J *are equivalent.*

(2) *If for all distinct* $\bar{a}_1, \ldots, \bar{a}_n \in I$ *and distinct* $\bar{b}_1, \ldots, \bar{b}_n \in J$ $\mathrm{tp}(\bar{a}_1 \frown \cdots \frown \bar{a}_n) = \mathrm{tp}(\bar{b}_1 \frown \cdots \frown \bar{b}_n)$, *then* $\mathrm{Av}(I, \bigcup I) = \mathrm{Av}(J, \bigcup I) \Rightarrow I, J$ *are equivalent.*

(3) $\mathrm{Av}(I, \bigcup I)$ *is stationary over* $\bigcup I$.

(4) $\mathrm{Av}(I, \bigcup I_1)$ *is stationary over* $\bigcup I_1$ *for some* $I_1 \subseteq I$, $|I_1| < \min\{\aleph_1, \kappa(T)\}$, *and* $\mathrm{Av}(I, \bigcup I)$ *does not fork over* $\bigcup I_1$.

Proof. Let M be a $(|I|^+ + |J|^+)$-saturated model such that $\bigcup I \cup \bigcup J \subseteq |M|$. Let $p = \mathrm{Av}(I, |M|)$, $q = \mathrm{Av}(J, |M|)$. Clearly p and q do not split over $\bigcup I$ and $\bigcup J$ respectively. So by Lemma 4.16, $p \restriction \bigcup I$ is stationary over $\bigcup I$ and p does not fork over $\bigcup I$; a similar claim holds for q. This proves (3).

Now from $\mathrm{Av}(I, \bigcup I) = \mathrm{Av}(J, \bigcup I)$ we get that $p \restriction \bigcup I = q \restriction \bigcup I$ and thus (by Theorem 4.2(1)) for all finite Δ, $R^m(p, \Delta, 2) = R^m(p \restriction \bigcup I, \Delta, 2) = R^m(q \restriction \bigcup I, \Delta, 2) \geq R^m(q, \Delta, 2)$. If we assume that $\mathrm{tp}(\bar{a}_1 \frown \cdots \frown \bar{a}_n) = \mathrm{tp}(\bar{b}_1 \frown \cdots \frown \bar{b}_n)$ as in (2), equality holds; thus $p = q$ so (2) holds. Also in case (1), by the symmetry in the assumptions, again equality holds, thus (1) holds.

Let us prove (4). If $I_1 \subseteq I$, $|I_1| = \aleph_0$ then by (3) $\mathrm{Av}(I_1, \bigcup I_1) = \mathrm{Av}(I, \bigcup I_1)$ stationary over $\bigcup I_1$, and as $\mathrm{Av}(I, |M|)$ does not fork over $\bigcup I_1$ (as it does not split over $\bigcup I_1$) $\mathrm{Av}(I, \bigcup I)$ does not fork over $\bigcup I_1$. If $\aleph_0 < \kappa(T)$, we finish. If $\kappa(T) = \aleph_0$, for some finite $I_0 \subseteq I$, $\mathrm{Av}(I, \bigcup I)$ does not fork over $\bigcup I_0$. Choose $\bar{a} \in I - I_0$; let $I_1 = I_0 \cup \{\bar{a}\}$. As $\mathrm{Av}(I, \bigcup I_1) \vdash \mathrm{stp}(\bar{a}, \bigcup I_0)$ we finish.

DEFINITION 4.1: An m-type p is *stationary* if for every A, $\mathrm{Dom}\, p \subseteq A$, there is a (unique) type $q \in S^m(A)$ such that

(i) $p \subseteq q$.

(ii) For every finite Δ $R^m(p, \Delta, 2) = R^m(q, \Delta, 2)$.

Remark. The uniqueness follows by II, 1.4(1).

DEFINITION 4.2: (1) The stationary m-types p_1, p_2 are called *parallel* if for every A, $\mathrm{Dom}\, p_1 \cup \mathrm{Dom}\, p_2 \subseteq A$, the unique types q_1, q_2 from the previous definition are equal.

(2) We call p a stationarization of q if q is stationary p parallel to q *or* q is complete over some A, and for some $\bar{c}$ realizing q, p is parallel to $\mathrm{stp}(\bar{c}, A)$. A stationarization of p over A is any stationarization $q \in S^m(A)$ of p.

Remark. In (2), in any case p is stationary, and the two cases are consistent.

LEMMA 4.18: (1) *Let* $p \in S^m(B)$, *then* p *is stationary over* B *iff* p *is stationary.*

(2)* *In Definition 4.2(1) it suffices to say that this holds for some* A.

(3)* *Parallelism is an equivalence relation.*

(4)* *If* p, q *are stationary and for every finite* Δ, $R^m(p, \Delta, 2) = R^m(p \cup q, \Delta, 2) = R^m(q, \Delta, 2)$ *then* p, q *are parallel.*

(5)* *If* p *is a stationary m-type,* $\operatorname{Dom} p \subseteq A$, *there is a unique* $q \in S^m(A)$ *parallel to* p.

Proof. (1) By 4.2.

(2) Immediate.

(3) Immediate, by the definition and (2).

(4) Immediate.

(5) Immediate by the definitions.

DEFINITION 4.3: An (infinite) indiscernible set I *is based on a stationary type* p if $\operatorname{Av}(I, \bigcup I)$ is parallel to p.

DEFINITION 4.4: (1) A set I of sequences is *independent over* (A, B) where $A \subseteq B$ if for every $\bar{a} \in I$, $\operatorname{tp}(\bar{a}, B \cup \bigcup (I - \bar{a}))$ does not fork over A. When $A = B$ we write A instead of (A, A).

(2) We say I is *independent over* p, if $p \in S^m(A)$, I is independent over A, and each $\bar{a} \in I$ realizes p.

LEMMA 4.19: (1) *If* I *is independent over* p, $p \in S^m(A)$, p *is stationary, then* I *is an indiscernible set over* A *based on* p.

(2) *If* I *is an indiscernible set over* A, *based on* $p \in S^m(A)$, *then* I *is independent over* A *and over* p. *If* $C \subseteq A$, p *does not fork over* C, *then* I *is independent over* (C, A).

(3) *If* I *is independent over* (A, B), J *independent over* $(A, B \cup \bigcup I)$ *then* $I \cup J$ *is independent over* (A, B).

(4) *If* I *is independent over* (A, B), $A \subseteq C \subseteq B_1 \subseteq B$, $J \subseteq I$, *then* $I - J$ *is independent over* $(C, B_1 \cup \bigcup J)$.

(5) *If* $I = \{\bar{a}_i : i < \alpha\}$, $\operatorname{tp}(\bar{a}_i, B \cup \bigcup \{\bar{a}_j : j < i\})$ *does not fork over* $A \subseteq B$ *then* I *is independent over* (A, B).

Proof. (1) By 1.10(1).

(2) Easy by 1.7(2).

(3), (5) Immediate, by 4.13, 4.14 and 4.15.

(4) Immediate.

DEFINITION 4.5: Let $A \subseteq B$, J a set of sequences of length m.

(1) If $p \in S^m(A)$, $\dim(p, B, J)$ is $\min\{|I|: I \subseteq J,\ I$ a maximal independent set over (A, B) of sequences realizing $p\}$.

(2) If p is a stationary type, and there is $q \in S^m(B)$ parallel to it. $\dim(p, B, J) = \min\{|I|: I$ a maximal set indiscernible over B, of sequences in J realizing $q\}$. See Lemma 4.20.

(3) The dimensions in (1), (2) are called true if for every such maximal I, $|I|$ is equal to the dimension.

(4) If $J = {}^mA$, we write A instead of J. If $B = \operatorname{Dom} p$, we omit it.

LEMMA 4.20: (1) *Definitions 4.5(1) and 4.5(2) are consistent, i.e., when both apply the resulting dimensions are equal.*

(2) *If I is based on p (p stationary, I an (infinite) indiscernible set) then* $\dim(I, A, M) = \dim(p, A, M)$.

Proof. Immediate (use 4.19, see Definition 3.3).

THEOREM 4.21: (1) *If I is independent over (A, B), $\kappa \geq \kappa(T)$,* $\operatorname{cf}\kappa > |C|$, *then for some $J \subseteq I$, $|J| < \kappa$, $I - J$ is independent over $(A, B \cup C \cup J)$. Moreover there is a minimal such J.*

(2) *If* $\dim(p, A, J)$ *is $\geq \kappa(T)$ then it is true; and if the dimension is not true, for every suitable maximal J (from Definition 4.5) $|J| < \kappa_r(T)$.*

Proof. Similar to 3.5, 3.9, Exercise 3.4.

EXERCISE 4.8: Show that for some unstable T, there is $p \in S^m(|\mathfrak{C}|)$, $A \subseteq |\mathfrak{C}|$, such that p does not split over A but is not definable over A.

EXERCISE 4.9: Show no stable T satisfies Exercise 4.8.

EXERCISE 4.10: Let $A \subseteq B$, $\bar{a}$ algebraic over A. Then $\operatorname{tp}(\bar{a}, B)$ does not fork over A (see Definition 6.1).

CLAIM 4.22: (1) *Let $A \subseteq B \subseteq C$ be such that for all $\bar{b} \in C$ there is $\bar{c} \in C$ such that* $\operatorname{tp}(\bar{c}, B)$ *does not fork over A and* $\operatorname{tp}(\bar{c}, A) = \operatorname{tp}(\bar{b}, A)$. *If $\bar{a}$ is any sequence such that* $\operatorname{tp}(\bar{a}, A)$ *is stationary and* $\operatorname{tp}(\bar{a}, C)$ *does not fork over A then:* $\operatorname{tp}(\bar{a}, B) \vdash \operatorname{tp}(\bar{a}, C)$ *iff* $\operatorname{tp}(\bar{a}, A) \vdash \operatorname{tp}(\bar{a}, C)$.

(2) *We can waive the stationarity of* $\operatorname{tp}(\bar{a}, A)$, *if* $\operatorname{tp}(\bar{c}, A)$ *is stationary for every* $\bar{c} \in C$; *and then:*

$$\operatorname{stp}(\bar{a}, B) \vdash \operatorname{tp}(\bar{a}, C) \text{ iff } \operatorname{tp}(\bar{a}, A) \vdash \operatorname{tp}(\bar{a}, C).$$

Proof. (1) $\Leftarrow$ is immediate.

$\Rightarrow$ Let $\varphi(\bar{x}; \bar{b}) \in \operatorname{tp}(\bar{a}, C)$, $\bar{b} \in C$; we shall show that $\operatorname{tp}(\bar{a}, A) \vdash \varphi(\bar{x}; \bar{b})$. Let $\bar{c}$ be as above. By Lemma 1.9 $\operatorname{tp}(\bar{a}, C)$ does not split over A, and $\varphi(\bar{x}; \bar{b}) \in \operatorname{tp}(\bar{a}, C)$, so $\varphi(\bar{x}; \bar{c}) \in \operatorname{tp}(\bar{a}, C)$. Since $\operatorname{tp}(\bar{a}, B) \vdash \operatorname{tp}(\bar{a}, C)$ we have $\operatorname{tp}(\bar{a}, B) \vdash \varphi(\bar{x}; \bar{c})$. Thus there is $\bar{d} \in B$ and $\psi(\bar{x}; \bar{d})$ such that $\vdash \psi[\bar{a}; \bar{d}]$ and $\vdash \psi(\bar{x}; \bar{d}) \rightarrow \varphi(\bar{x}; \bar{c})$. Clearly $\operatorname{tp}(\bar{c}, A \cup \bar{d})$ does not fork over A so (by 4.13) $\operatorname{tp}(\bar{d}, A \cup \bar{c})$ does not fork over A. So as before, we can define $\{\bar{d}_n : n < \omega\}$ to be an indiscernible set over $A \cup \bar{c}$ based on A such that $\operatorname{stp}(\bar{d}_n, A) \equiv \operatorname{stp}(\bar{d}, A)$. By Lemma 2.5 there is n for which

$$\psi_n = \bigvee_{\substack{w \subseteq 2n-1 \\ |w| = n}} \bigwedge_{i \in w} \psi(\bar{x}; \bar{d}_i)$$

is almost over A. Of course, $\psi(\bar{x}; \bar{d}_n) \vdash \varphi(\bar{x}; \bar{c})$ for all n and so $\psi_n \vdash \varphi(\bar{x}; \bar{c})$. So by Lemma 2.14, since ψ_n is almost over A, $\operatorname{tp}(\bar{a}, A) \vdash \psi_n$ (or $\operatorname{tp}(\bar{a}, A) \vdash \neg\psi_n$, let $q \in S^m(C \cup \bigcup_{n<\omega} \bar{d}_n)$ be a stationarization of $\operatorname{tp}(\bar{a}, C)$, so $\psi(\bar{x}, \bar{d}_n) \in q$ hence $\psi_n \in q$, hence $\operatorname{tp}(\bar{a}, A) \cup \{\psi_n\}$ is consistent; contradiction). Thus $\operatorname{tp}(\bar{a}, A) \vdash \varphi(\bar{x}; \bar{c})$; but $\operatorname{tp}(\bar{c}, A) = \operatorname{tp}(\bar{b}, A)$ so $\operatorname{tp}(\bar{a}, A) \vdash \varphi(\bar{x}; \bar{b})$.

(2) Similar proof.

EXERCISE 4.11: Suppose Δ is finite, $p \in S^m(B)$ and r_i, $i < \lambda$, are Δ-m-types such that the types $p \cup r_i$ are m_i-contradictory, $m_i < \omega$, $i < \lambda$, and $|B| + |T| < \lambda$.

(1) Show that for some i, $p \cup r_i$ forks over B.

(2) Show that if $|B| + |T|^\kappa < \lambda$ we can relax the finiteness condition on Δ to $|\Delta| < \kappa$. In fact, $\lambda \geq \aleph_0$, $\lambda > 2^{|\Delta|}$ suffice.

QUESTION 4.12: Generalize the properties of the rank D such as symmetry, monotonicity, etc.) to the case where T is unstable but $D(p, \mathrm{L}, \infty) < \infty$ for every type p.

THEOREM 4.23: *Suppose* $\lambda > \aleph_0$ *is regular and* $(\forall \mu < \lambda)(\mu^{<\kappa(T)} < \lambda)$. *If* $|I| \geq \lambda > |A|$ *then there are* J, $I_0 \subseteq I$, $|J| = \lambda$, $|I_0| < \kappa(T)$, *such that* J *is independent over* $A \cup \bigcup I_0$ *(hence if* $\operatorname{stp}(\bar{a}, A \cup \bigcup I_0)$, $\bar{a} \in J$, *is fixed,* J *is indiscernible over* $A \cup \bigcup I_0$*).*

Remark. We give this theorem partly to show an alternative proof of Theorem I, 2.8: the existence of large indiscernible sets.

Proof. Let $\bar{a}_i$, $i < \lambda$, be distinct members of I, and let $A_i = A \cup \bigcup_{j<i} \bar{a}_j$. For each i let $B_i \subseteq A_i$, $|B_i| < \kappa(T)$ be such that $\text{tp}(\bar{a}_i, A_i)$ does not fork over B_i and let $f(i) = \min\{j: B_i \subseteq A_j\}$ so $\text{cf}\, \delta \geq \kappa(T) \Rightarrow f(i) < i$.

As $(\forall \mu < \lambda)(\mu^{<\kappa(T)} < \lambda)$, $\lambda > \kappa_r(T)$ hence by 1.3(3) of the Appendix, $S_0 = \{\delta < \lambda: \text{cf}\, \delta \geq \kappa(T)\}$ is stationary. Hence by 1.3(1) of the Appendix, there is a stationary set $S_1 \subseteq S_0$ on which f is constant, $i(0)$ say. As

$$|\{B \subseteq A_{i(0)}: |B| < \kappa(T)\}| \leq |A_{i(0)}|^{<\kappa(T)} < \lambda,$$

by 1.3(1) of the Appendix there are a stationary set $S_2 \subseteq S_1$ and $B \subseteq A_{i(0)}$ such that $i \in S_2 \Rightarrow B_i = B$. Let $I_0 \subseteq I$ be minimal such that $B \subseteq A \cup \bigcup I_0$ and let $J = \{\bar{a}_i: i \in S_2\}$. Clearly the conclusion holds.

EXERCISE 4.13: Given an increasing sequence of sets A_α $(\alpha < \zeta)$, for every ξ we can find elementary mappings F_η, $\eta \in {}^{\zeta>}\xi$, such that:

(i) $\eta \lessdot \nu$ implies F_ν extends F_η.

(ii) $\text{Dom}\, F_\eta = A_{l(\eta)}$.

(iii) $\text{tp}_*(\text{Range}\, F_\eta, \bigcup \{\text{Range}\, F_\nu: \nu \in {}^{\zeta>}\xi, \text{ not } \eta \lessdot \nu\})$ does not fork over $\bigcup_{\alpha < l(\eta)} \text{Range}\, F_{\eta|\alpha}$.

If $\text{tp}_*(A_\alpha, \bigcup_{i<\alpha} A_i)$ is stationary, this construction is unique and for any $\bar{a}_i \in A_{i+1}$, we call $\{F_\eta(\bar{a}_i): \eta \in {}^{i+1}\xi, i < \zeta\}$ an indiscernible tree, as it satisfies Definition VII, 2.4, and VII, 3.1.

EXERCISE 4.14: Suppose $\kappa < \kappa(T)$ and φ_α, $\alpha < \kappa$, are as in Theorem 3.1. Find sequences a_η, $\eta \in {}^{\kappa \geq}\lambda$ such that:

(i) $\{\bar{a}_\eta: \eta \in {}^{\kappa\geq}\lambda\}$ is an indiscernible tree.

(ii) $\models \varphi_\alpha[\bar{a}_\eta; \bar{a}_\nu]^{\text{if}\,(\nu \lessdot \eta)}$ where $\eta \in {}^{\kappa}\lambda$, $\nu \in {}^{\alpha+1}\lambda$.

(iii) $\{\varphi_\alpha(\bar{x}; \bar{a}_{\nu\frown\langle i\rangle}): i < \lambda\}$ is m_i-contradictory for $\nu \in {}^{\alpha}\lambda$, some $m_i < \omega$.

EXERCISE 4.15: Show that if T is stable then we can take $n = 1$ in Definition 1.4 of forking. Hence a formula divides iff it forks.

EXERCISE 4.16: Show that in 4.23, if we weaken the conclusion "$|I_0| < \kappa(T)$" to "$|I_0| < \lambda$" we can weaken the hypothesis on λ to "$\lambda > \kappa_r(T)$ is regular".

EXERCISE 4.17: (1) Suppose M is κ-compact, p an m-type, $|p| < \kappa$. *Then* p does not fork over $|M|$ iff p is finitely satisfiable in M iff p is realized in M.

(2) Suppose M is κ-saturated, $p \in S^m(A)$, $\kappa_r(T) + |A|^+ \leq \kappa$. *Then* p does not fork over $|M|$ iff p is finitely satisfiable in M iff p is realized in M.

EXERCISE 4.18: Show that in 4.5 we can omit $\lambda \geq \aleph_0$ if we add $\Delta = \mathrm{cl}_2^m(\Delta)$ or p is stationary over A. [Hint: We can assume that q is finite, so in either case that $q = \{\varphi(\bar{x}; \bar{a})\}$, so we can assume p is finite. Let M be a $(\lambda + |T| + |A|)^+$-saturated model, $A \cup \bar{a} \subseteq |M|$. Let f_α $(\alpha < \lambda)$ be elementary maps, $f_\alpha \restriction A =$ the id., $f_0 \restriction |M| =$ the id., $\mathrm{tp}_*(f_\alpha(|M|), \bigcup_{\beta<\alpha} f_\beta(|M|))$ does not fork over A. Then note:

(A) $q_j = \{\varphi(\bar{x}; f_\alpha(\bar{a}))^{\mathrm{if}\,(\alpha = j)}\colon \alpha < \lambda\}$ are pairwise explicitly contradictory, and of equal rank, hence $R^m(p, \Delta, \lambda) > R^m(p \cup q_0, \Delta, \lambda)$. .

(B) No element of $|M|$ realizes $\varphi(\bar{x}; f_\alpha(\bar{a})) \wedge \bigwedge p$ $(\alpha > 0)$ (by 4.10, Exercise 4.19), hence $R^m(p \cup q, \Delta, \lambda) = R^m(p \cup q_0, \Delta, \lambda)$ (by Exercise 4.20).]

EXERCISE 4.19: Suppose $A \subseteq B, C$, $\mathrm{tp}_*(C, B)$ does not fork over A, and q is a type over B forking over A. Then q forks over C. [Hint: Otherwise let $\mathrm{tp}(\bar{d}, B \cup C)$ extend q and not fork over C. By the Symmetry Theorem 4.13 $\mathrm{tp}_*(B, C \cup \bar{d})$ does not fork over C. But (by symmetry) also $\mathrm{tp}_*(B, C)$ does not fork over A, hence by transitivity (4.4) $\mathrm{tp}_*(B, C \cup \bar{d})$ does not fork over A, so by monotonicity $\mathrm{tp}_*(B, A \cup \bar{d})$ does not fork over A. So by symmetry, $\mathrm{tp}(\bar{d}, B)$ does not fork over A, but $q \subseteq \mathrm{tp}(\bar{d}, B)$ and q forks over A, contradiction.]

EXERCISE 4.20: Suppose p is an m-type over $|M|$, $\mu \geq |p|^+ + \lambda$, M is μ-saturated, $p \cup \{\varphi(\bar{x}; \bar{a})\}$ is not realized in M; and $p_1 = p \cup \{\neg\varphi(\bar{x}; \bar{a})\}$. Then $R^m(p, \Delta, \lambda) = R^m(p_1, \Delta, \lambda)$. [Hint: Trivially $R^m(p, \Delta, \lambda) \geq R^m(p_1, \Delta, \lambda)$, so we prove by induction on α that $R^m(p, \Delta, \lambda) \geq \alpha \Rightarrow R^m(p_1, \Delta, \lambda) \geq \alpha$. The point is that the types "witnessing" for the rank can be chosen over $|M|$ and if r is an m-type over $|M|$, $|r| < \mu$, then $p \cup r$ is consistent iff $p_1 \cup r$ is consistent. Use Exercise 4.17(1).] In fact, M μ^+-compact is sufficient.

III.5. The first stability cardinal

From now on, until after Lemma 5.11 in this section let $B \subseteq |M|$, $p \in S^m(B)$, and $\mathscr{P} = \{q \in S^m(|M|)\colon p \subseteq q$ and q does not fork over $B\}$. By Lemma 2.15(2) every $q \in \mathscr{P}$ is stationary over B, and by Corollary 2.9(1) for every $\bar{a}$ realizing p there is a unique $q_{\bar{a}} \in \mathscr{P}$ such that $q_{\bar{a}} \vdash \mathrm{stp}(\bar{a}, B)$ so if $\bar{a}$ realizes $q \in \mathscr{P}$ then $q_{\bar{a}} = q$.

LEMMA 5.1: *Choose a fixed* $\bar{a}^*$ *which realizes* p. *Then there are* λ^* *and* $E_i^*(\bar{x}; \bar{y}) \in \mathrm{FE}^m(B)$ *for* $i < \lambda^*$ *such that:*

(1) $p(\bar{x}) \cup \{E_i^*(\bar{x}; \bar{a}^*): i < \lambda^*\}$ *is stationary over* B.

(2) *For all* $j < \lambda^*$, $p(\bar{x}) \cup \{E_i^*(\bar{x}; \bar{a}^*): i < j\} \cup \{\neg E_j^*(\bar{x}; \bar{a}^*)\}$ *is consistent.*

Proof. By Corollary 2.9(1) $p(\bar{x}) \cup \{E(\bar{x}; \bar{a}^*): E \in \mathrm{FE}^m(B)\}$ is stationary over B. So there is a set Φ_1 of minimal power λ^* such that $p_{\Phi_1}^{\bar{a}^*} = p(\bar{x}) \cup \{E(\bar{x}; \bar{a}^*): E \in \Phi_1\}$ is stationary over B. Let $\Phi_1 = \{E_i(\bar{x}; \bar{y}): i < \lambda^*\}$. Let U be the set of all $j < \lambda^*$ such that $p(\bar{x}) \cup \{E_i(\bar{x}; \bar{a}^*): i < j\} \cup \{\neg E_j(\bar{x}; \bar{a}^*)\}$ is consistent, and let $\Phi = \{E_i(\bar{x}; \bar{y}): i \in U\}$.

We claim that $p_{\Phi}^{\bar{a}^*} \vdash p_{\Phi_1}^{\bar{a}^*}$. We shall prove this by proving by induction on i that $p_{\Phi}^{\bar{a}^*} \vdash E_i(\bar{x}; \bar{a}^*)$. Assume this is true for $i < j$. Now if $j \in U$ the result is immediate. If $j \notin U$ then by the induction hypothesis $p_{\Phi}^{\bar{a}^*} \vdash p(\bar{x}) \cup \{E_i(\bar{x}; \bar{a}^*): i < j\}$ and as $j \notin U$ $p(\bar{x}) \cup \{E_i(\bar{x}; \bar{a}^*): i < j\} \vdash E_j(\bar{x}; \bar{a}^*)$, so $p_{\Phi}^{\bar{a}^*} \vdash E_j(\bar{x}; \bar{a}^*)$. This proves the claim, and in particular $p_{\Phi}^{\bar{a}^*}$ is stationary over B. Thus $|\Phi| = \lambda^*$. Since $p(\bar{x}) \cup \{E_i(\bar{x}; \bar{a}^*): i < j, i \in U\} \cup \{\neg E_j(\bar{x}; \bar{a}^*)\}$ is consistent for each $j \in U$, by changing the indices we get exactly (2).

LEMMA 5.2: (1) *For all* $\bar{a}_i$, $i < \lambda^*$, *if* $p' = p(\bar{x}) \cup \{E_i^*(\bar{x}; \bar{a}_i): i < \lambda^*\}$ *is consistent* (λ^* *and* E_i^* *are from the previous lemma*), *then* p' *is stationary over* B.

(2) *For all* $\bar{a}_i$, $i < \lambda^*$, *if* $p^j = p(\bar{x}) \cup \{E_i^*(\bar{x}; \bar{a}_i): i < j\}$ *is consistent then* $p^j \cup \{\neg E_j^*(\bar{x}; \bar{a}_j)\}$ *is consistent.*

Proof. (1) Let $\bar{a}'$ realize p' and set $r = p(\bar{x}) \cup \{E_i^*(\bar{x}; \bar{a}'): i < \lambda^*\}$.

Now by Lemma 5.1(1) $p(\bar{x}) \cup \{E_i^*(\bar{x}; \bar{a}^*): i < \lambda^*\}$ is stationary over B, and $\bar{a}'$ realizes the same type over B as $\bar{a}^*(p)$; so r is also stationary over B. It is not hard to see that $p' \vdash r \vdash p'$; thus p' is also stationary over B.

(2) Now let $\bar{a}'$ realize p^j and take $r = p(\bar{x}) \cup \{E_i^*(\bar{x}; \bar{a}'): i < j\}$. Since $\bar{a}^*$ and $\bar{a}'$ realize the same type over B, $r \cup \{\neg E_j^*(\bar{x}; \bar{a}')\}$ is consistent (by Lemma 5.1(2)), and of course $r \cup \{E_j^*(\bar{x}; \bar{a}')\}$ is consistent. As above $p^j \vdash r \vdash p^j$, so that $p^j \cup \{E_j^*(\bar{x}; \bar{a}')\}$ and $p^j \cup \{\neg E_j^*(\bar{x}; \bar{a}')\}$ are consistent. If $\models E_j^*[\bar{a}_j; \bar{a}']$ then $p^j \cup \{\neg E_j^*(\bar{x}; \bar{a}')\} \vdash p^j \cup \{\neg E_j^*(\bar{x}; \bar{a}_j)\}$ and if $\models \neg E_j^*[\bar{a}_j; \bar{a}']$ then $p^j \cup \{E_j^*(\bar{x}; \bar{a}')\} \vdash p^j \cup \{\neg E_j^*(\bar{x}; \bar{a}_j)\}$. In any case (2) is proved.

Remark. This shows that λ^* is in fact *not* dependent on the choice of $\bar{a}^*$ as assumed in Lemma 5.1. Clearly it does not depend on M either, so

DEFINITION 5.1: $\lambda^*(p)$ is the λ^* defined in 5.1.

LEMMA 5.3: (1) *If* $\lambda^* \geq \aleph_0$ *then* $|\mathscr{P}| = 2^{\lambda^*}$ *and if* $\lambda^* < \aleph_0$ *then* $|\mathscr{P}| < \aleph_0$.

(2) *There is* $C \subseteq |M|$, *where* $|C| = \lambda^*$ *or both* C *and* λ^* *are finite, such that* $q_1 \upharpoonright C \neq q_2 \upharpoonright C$ *for all* $q_1 \neq q_2 \in \mathscr{P}$.

Proof. (1) For all $\eta \in \bigcup_{i<\lambda^*} {}^{i+1}2$ define $\bar{a}_\eta$ by induction on $l(\eta)$ such that $p_\eta = p \cup \{E_i^*(\bar{x}; \bar{a}_{\eta|(i+1)}): i < l(\eta)\}$ will be consistent and $\vDash \neg E_i^*[\bar{a}_{\eta\frown\langle 0\rangle}; \bar{a}_{\eta\frown\langle 1\rangle}]$ for $i = l(\eta)$. Assuming p_η is defined for $l(\eta) = i$, choose $\bar{a}_{\eta\frown\langle 0\rangle}$ such that $p_\eta \cup \{E_i^*(\bar{x}; \bar{a}_{\eta\frown\langle 0\rangle})\}$ is consistent. By Lemma 5.2(2), $p_\eta \cup \{\neg E_i^*(\bar{x}; \bar{a}_{\eta\frown\langle 0\rangle})\}$ is also consistent, so any sequence realizing it is a suitable $\bar{a}_{\eta\frown\langle 1\rangle}$.

For all $\eta \in {}^{\lambda^*}2$ p_η is stationary over B, and so p_η has a unique stationarization in $\mathscr{P}$. Thus $|\mathscr{P}| \geq 2^{\lambda^*}$.

On the other hand let $\bar{a}_i^l$, $l < n(E_i^*)$ be representatives $\in |M|$ of the different equivalence classes of E_i^*, and $\Gamma = \{E_i^*(\bar{x}; \bar{a}_i^l): i < \lambda^*, l < n(E_i^*)\}$. Then by Lemma 5.2(1) $q_1 \cap \Gamma \neq q_2 \cap \Gamma$ for $q_1 \neq q_2 \in \mathscr{P}$. So it is clear that $|\mathscr{P}| \leq 2^{|\Gamma|} = 2^{\lambda^*}$ for $\aleph_0 \leq \lambda^*$ and $|\mathscr{P}| \leq 2^{|\Gamma|} < \aleph_0$ for $\lambda^* < \aleph_0$.

(2) $C = \bigcup \{\bar{a}_i^l: i < \lambda^*, l < n(E_i^*)\}$ is a set of power λ^* for $\lambda^* \geq \aleph_0$ and is finite if λ^* is finite. It is readily seen that C satisfies the requirements stated in the lemma.

DEFINITION 5.2: Let $E^0 \in \mathrm{FE}^m(B)$, $\Phi \subseteq \mathrm{FE}^m(B)$. $\mathrm{ECN}(E^0, \Phi)$ is the maximal n such that for some $\bar{c}$ realizing p, $\Gamma(\Phi, \bar{c}, n) = \bigcup_{i=1}^n p(\bar{x}_i) \cup \{E(\bar{x}_i; \bar{c}): E \in \Phi, 1 \leq i \leq n\} \cup \{\neg E^0(\bar{x}_i; \bar{x}_j): 1 \leq i < j \leq n\}$ is consistent. (ECN stands for class equivalence number.)

Remark. Clearly $\mathrm{ECN}(E^0, \Phi) \leq n(E^0)$.

LEMMA 5.4*: (1) *In the above definition "some* $\bar{c}$*" may be replaced by "any* $\bar{c}$*".*

(2) *If* $\Phi_1 \subseteq \Phi_2$ *then* $\mathrm{ECN}(E^0, \Phi_2) \leq \mathrm{ECN}(E^0, \Phi_1)$.

Proof. (1) is true because p is a complete type over B so for every $\bar{c}, \bar{c}'$ realizing p there is an elementary mapping F such that $F \upharpoonright B$ is the identity and $F(\bar{c}) = \bar{c}'$. (2) is immediate.

DEFINITION 5.3: Let $\{E^0\}, \Phi, \Psi \subseteq \mathrm{FE}^m(B)$. Then E^0 *depends* on Φ (mod Ψ) if $\mathrm{ECN}(E^0, \Psi) > \mathrm{ECN}(E^0, \Phi \cup \Psi)$.

DEFINITION 5.4: Let $\Pi, \Phi, \Psi \subseteq \mathrm{FE}^m(B)$. Π is *independent* over Φ (mod Ψ) if for all $E \in \Pi$, E does not depend on $\Phi \cup (\Pi - \{E\})$ (mod Ψ).

LEMMA 5.5*: (1) *For every* Φ, E *there is a finite* $\Psi \subseteq \Phi$ *such that* $\mathrm{ECN}(E, \Phi) = \mathrm{ECN}(E, \Psi)$.

(2) *If* E *depends on* Φ (mod Ψ) *then for some finite* $\Pi \subseteq \Phi$, E *depends on* Π (mod Ψ).

Proof. (1) Let $n = \mathrm{ECN}(E, \Phi)$. Then $\Gamma(\Phi, \bar{c}, n + 1)$ is inconsistent. Let Γ' be a finite inconsistent subset and let Ψ be the set of $E \in \mathrm{FE}^m(B)$ appearing in Γ', $\Gamma' \subseteq \Gamma(\Psi, \bar{c}, n + 1)$ so $\mathrm{ECN}(E, \Psi) \leq n$. But by Lemma 5.4(2) $\mathrm{ECN}(E, \Psi) \geq n$, so we have equality.

(2) Immediate by part (1).

LEMMA 5.6*: *If* E^1 *depends on* $\Phi \cup \{E^2\}$ (mod Ψ), *but* E^1 *does not depend on* Φ (mod Ψ), *then* E^2 *depends on* $\Phi \cup \{E^1\}$ (mod Ψ) *and even on* $\{E^1\}$ (mod $\Psi \cup \Phi$).

Proof. Let $\bar{a}$ realize p and denote $p^{\bar{a}}_{\Pi} = p \cup \{E(\bar{x}; \bar{a}) \colon E \in \Pi\}$. Let $\bar{a}^1_1, \ldots, \bar{a}^1_{n_1}$ $(\bar{a}^2_1, \ldots, \bar{a}^2_{n_2})$ be representatives of the equivalence classes of E^1 (E^2) which contain sequences realizing p.

Let $S^1 = \{k\colon 1 \leq k \leq n_1, p^{\bar{a}}_{\Phi\cup\Psi} \cup \{E^1(\bar{x}; \bar{a}^1_k)\}$ is consistent$\}$. Since E^1 depends on $\Phi \cup \{E^2\}$ (mod Ψ) but not on Φ (mod Ψ) there is $k \in S^1$ such that $p^{\bar{a}}_{\Phi\cup\Psi} \cup \{E^2(\bar{x}; \bar{a}), E^1(\bar{x}; \bar{a}^1_k)\}$ is inconsistent. Let $\bar{a}^*$ realize $p^{\bar{a}}_{\Phi\cup\Psi} \cup \{E^1(\bar{x}; \bar{a}^1_k)\}$. Thus for $E \in \Phi \cup \Psi$, $\models E[\bar{a}^*; \bar{a}] \wedge E^1[\bar{a}^*; \bar{a}^1_k]$.

Also for some $m \leq n_2$ $\models E^2[\bar{a}; \bar{a}^2_m]$ so $p^{\bar{a}^*}_{\Phi\cup\Psi} \cup \{E^2(\bar{x}; \bar{a}^2_m)\}$ is consistent (it is realized by $\bar{a}$). But $p^{\bar{a}^*}_{\Phi\cup\Psi\cup\{E^1\}} \cup \{E^2(\bar{x}; \bar{a}^2_m)\}$ is inconsistent as every sequence which realizes it also realizes $p^{\bar{a}}_{\Phi\cup\Psi} \cup \{E^2(\bar{x}; \bar{a}), E^1(\bar{x}; \bar{a}^1_k)\}$.

Thus $\mathrm{ECN}(E^2, \Phi \cup \Psi \cup \{E^1\}) < \mathrm{ECN}(E^2, \Phi \cup \Psi) \leq \mathrm{ECN}(E^2, \Psi)$, and the conclusion follows.

LEMMA 5.7*: *If for all* $i < \alpha$ E^i *does not depend on* $\Phi \cup \{E^j\colon j < i\}$ (mod Ψ), then $\{E^i\colon i < \alpha\}$ *is independent over* Φ (mod Ψ).

Proof. By Lemma 5.5 it suffices to prove this for finite α. For $\alpha = 0$ or $\alpha = 1$ it is trivial. Now suppose the claim is true for α but fails for $\alpha + 1$. For some $i < \alpha$ E^i depends on $\Pi = \Phi \cup \{E^j\colon j < \alpha + 1, j \neq i\}$ (mod Ψ), but E^i does not depend on $\Pi - \{E^\alpha\}$ (mod Ψ). So by Lemma 5.6, with E^i for E^1, E^α for E^2, $\Pi - \{E^\alpha\}$ for Φ, and Ψ for Ψ, we get a contradiction.

LEMMA 5.8*: *Let Π be independent over Φ* (mod Ψ). *Then:*

(1) *for every E there is a finite $\Pi^* \subseteq \Pi \cup \Phi$ such that $\Pi - \Pi^*$ is independent over $\Phi \cup \{E\}$* (mod $\Pi^* \cup \Psi$).

(2) *For every Φ_1 there is $\Pi_1 \subseteq \Pi \cup \Phi$, $|\Pi_1| < |\Phi_1|^+ + \aleph_0$, such that $\Pi - \Pi_1$ is independent over $\Phi \cup \Phi_1$* (mod $\Psi \cup \Pi_1$).

Proof. (1) By Lemma 5.5(1) there is a finite $\Pi^* \subseteq \Pi \cup \Phi$ such that $\mathrm{ECN}(E, \Psi \cup \Phi \cup \Pi) = \mathrm{ECN}(E, \Psi \cup \Pi^*)$. Clearly by 5.4(2) $\Pi - \Pi^*$ is independent over Φ (mod $\Psi \cup \Pi^*$). Suppose that $\Pi - \Pi^*$ is not independent over $\Phi \cup \{E\}$ (mod $\Psi \cup \Pi^*$). Then for some $E^* \in \Pi - \Pi^*$, E^* depends on $\Phi \cup \{E\} \cup (\Pi - \Pi^* - \{E^*\})$ (mod $\Psi \cup \Pi^*$). But E^* does not depend on $\Phi \cup (\Pi - \Pi^* - \{E^*\})$ (mod $\Psi \cup \Pi^*$) (by 5.4(2): the monotonicity properties of dependency). So by Lemma 5.6 E depends on $\Phi \cup (\Pi - \Pi^*)$ (mod $\Psi \cup \Pi^*$), hence $\mathrm{ECN}(E, \Psi \cup \Pi^*) > \mathrm{ECN}(E, \Pi^* \cup \Psi \cup \Phi \cup (\Pi - \Pi^*))$, contradiction to the definition of Π^*. Now the claim follows by Lemma 5.4(2).

(2) Similar.

LEMMA 5.9*: *If $\aleph_0 < \lambda \le \lambda^*$ (see Lemma 5.1) and λ is regular, then there is a finite $\Psi \subseteq \{E_i^* : i < \lambda\}$ and $S \subseteq \lambda$, $|S| = \lambda$, such that $\Pi = \{E_i^* : i \in S\}$ is independent over the empty set* (mod Ψ) *and* $\mathrm{ECN}(E_i^*, \Psi \cup (\Pi - \{E_i^*\})) \ge 2$ *for $i \in S$.*

Proof. Let $\Pi_j = \{E_i^* : i < j\}$; by Lemma 5.1 $\mathrm{ECN}(E_j^*, \Pi_j) \ge 2$. If δ is a limit ordinal less than λ then by Lemma 5.5(1) there is $f(\delta) < \delta$ such that $\mathrm{ECN}(E_\delta^*, \Pi_\delta) = \mathrm{ECN}(E_\delta^*, \Pi_{f(\delta)})$.

By the regularity of λ and 1.3(1) of the Appendix there is $i_0 < \lambda$ such that $S_0 = \{\delta < \lambda : f(\delta) = i_0\}$ is of cardinality λ. For each $\delta \in S_0$ there is a finite $\Pi^\delta \subseteq \Pi_{i_0}$ such that $\mathrm{ECN}(E_\delta^*, \Pi_\delta) = \mathrm{ECN}(E_\delta^*, \Pi_{i_0}) = \mathrm{ECN}(E_\delta^*, \Pi^\delta)$. As $|\Pi_{i_0}| < \lambda$, $|S_0| = \lambda$, λ is regular, and $\Pi_{i_0})$ has $< \lambda$ finite subsets, for some finite Π, $S = \{\delta : f(\delta) = i_0, \Pi^\delta = \Pi\}$ is of cardinality λ. The result now follows by 5.7.

LEMMA 5.10: *Let $\lambda_0 \le \lambda^*$ be singular. Then there is $\Psi \subseteq \{E_i^* : i < \lambda_0\}$, $|\Psi| = \mathrm{cf}\,\lambda_0$, and $S \subseteq \lambda_0$, $|S| = \lambda_0$, such that $\{E_i^* : i \in S\}$ is independent over the empty set* (mod Ψ) *and* $\mathrm{ECN}(E_i^*, \Psi) \ge 2$ *for $i \in S$.*

Proof. Let $\lambda_0 = \sum_{0 < \alpha < \mathrm{cf}\lambda_0} \lambda_\alpha$ where λ_α is regular and for $0 < \alpha < \beta < \mathrm{cf}\,\lambda_0$, $\lambda_\beta > \lambda_\alpha > \mathrm{cf}\,\lambda_0$. Let Π_j be defined as in the previous lemma.

Now define by induction finite $\Psi_\alpha \subseteq \Pi_{\lambda_{\alpha+1}}$ and $S_\alpha \subseteq \lambda_{\alpha+1}$, $|S_\alpha| = \lambda_{\alpha+1}$, such that $\{E_i^*: i \in S_\alpha\}$ is independent over $\bigcup_{\beta<\alpha} \Psi_\beta \cup \Pi_{\lambda_\alpha}$ (mod Ψ_α), and without loss of generality $i \in S_\beta, j \in S_\alpha, \beta < \alpha, \Rightarrow i < j$. The proof is as in Lemma 5.9. We define by induction on $n < \omega$, $V_\alpha^n \subseteq S_\alpha$ for every $\alpha < \operatorname{cf} \lambda_0$, such that:

(1) $V_\alpha^0 = S_\alpha$.

(2) $V_\alpha^{n+1} \subset V_\alpha^n$, $|S_\alpha - V_\alpha^{n+1}| \leq \operatorname{cf} \lambda_0$.

(3) $\{E_i: i \in V_\alpha^{n+1}\}$ is independent over

$$\Pi_{\lambda_\alpha} \bmod\Big[\bigcap_{\beta<\operatorname{cf}\lambda_0} \Psi_\beta \cup \bigcup \{E_j: j \in S_\alpha - V_\alpha^n\}\Big]$$

(by 5.8(2)). Let $V_\alpha = \bigcap_{n<\omega} V_\alpha^n$, then $|S_\alpha - V_\alpha| \leq \operatorname{cf} \lambda_0$ and $\{E_i: i \in V_\alpha, \alpha < \operatorname{cf} \lambda_0\}$ is independent over

$$\emptyset \bmod\Big[\bigcup_{\alpha<\operatorname{cf}\lambda_0} \Psi_\alpha \cup \bigcup \{E_i: i \in S_\alpha - V_\alpha, \alpha < \operatorname{cf} \lambda_0\}\Big].$$

LEMMA 5.11*: *Let Π be independent over Φ (mod Ψ) and* $\operatorname{ECN}(E, \Psi) \geq 2$ *for $E \in \Pi$. Then for every $\Pi' \subseteq \Pi$,*

$$\Gamma_{\Pi'} = p(\bar{x}) \cup p(\bar{y}) \cup \{E(\bar{x}; \bar{y}): E \in \Phi \cup \Psi \cup \Pi'\} \cup \{\neg E(\bar{x}; \bar{y}): E \in \Pi - \Pi'\}$$

is consistent.

Proof. By the compactness theorem it suffices to prove for finite Π. We shall prove it by induction on $|\Pi - \Pi'|$. If $|\Pi - \Pi'| = 0$ there is nothing to prove. So assume it is proven for n, and let $|\Pi - \Pi'| = n + 1$ and $E^* \in \Pi - \Pi'$. Thus $\Gamma_{\Pi' \cup \{E^*\}}$ is consistent; so let $\bar{x} \mapsto \bar{a}$, $\bar{y} \mapsto \bar{b}$ realize it: so $\vDash E^*[\bar{a}; \bar{b}]$. Now we know that

$$\operatorname{ECN}(E^*, \Psi \cup \Phi \cup (\Pi - \{E^*\})) = \operatorname{ECN}(E^*, \Psi) \geq 2.$$

So $\Gamma = p(\bar{z}) \cup \{E(\bar{z}; \bar{b}): E(\bar{x}; \bar{y}) \in \Psi \cup \Phi \cup (\Pi - \{E^*\})\} \cup \{\neg E^*(\bar{z}; \bar{b})\}$ is consistent. Let $\bar{z} \mapsto \bar{c}$ realize it. Then $\bar{x} \mapsto \bar{a}$, $\bar{y} \mapsto \bar{c}$ realize $\Gamma_{\Pi'}$.

QUESTION 5.1: There is a countable $C \subseteq |M|$ such that for $q \neq r \in \mathscr{P}$ $q \restriction C \neq r \restriction C$ (assume M is $|B|^+$-saturated). See Lemma 5.3.

Now we abandon our fixed B, M, p.

DEFINITION 5.5: $D_m(T) = S^m(\emptyset)$, $D(T) = \bigcup_m D_m(T)$.

LEMMA 5.12*: *There is a countable A such that $|S^m(A)| \geq |D(T)|$ for some $m < \omega$.*

Remark. (1) We can take $m = 1$ unless $|(D(T)|$ is singular of cofinality $> \omega$. (2) If $\mathrm{cf}|D(T)| > \aleph_0$ we can choose an empty A.

Proof. Let $\mu = |D(T)|$. For all $m < \omega$ and $\lambda \leq |D_m(T)|$, λ regular, there is a set $A_{\lambda,m}$ of power $< m$ such that $|S^1(A_{\lambda,m})| \geq \lambda$ (this by Lemma I, 2.1).

Case 0. If $\mu \leq \aleph_0$ there is nothing to prove.

Case 1. If $\mu > \aleph_0$ is regular then $\mu = |\bigcup_m D_m(T)| \leq \sum |D_m(T)|$, so there exists m for which $\mu = |D_m(T)|$ and thus $|S^1(A_{\mu,m})| \geq \mu$ and $|A_{\mu,m}| < m < \aleph_0$.

Case 2. If $\mu > \aleph_0$ is singular of cofinality $> \aleph_0$, then again there is m for which $|D_m(T)| = \mu$ and so $|S^m(\emptyset)| \geq \mu$.

Case 3. If $\mu > \aleph_0$ is singular of cofinality $\aleph_0$, let $\mu = \sum_{n<\omega} \mu_n$ and $A = \bigcup_{m,n} A_{\mu_n^+,m}$. Then $|A| \leq \aleph_0$ and $|S(A)| \geq |D(T)|$.

LEMMA 5.13: *Suppose* $|S(A)| \geq \lambda > |A|^{<\kappa(T)}$, λ *regular. Then one of the following holds:*

(1) *For some* B, $|B| < \kappa(T)$, $|S(B)| \geq \lambda$ *and hence* $|D(T)|^{<\kappa(T)} \geq \lambda$.
(2) $\lambda \leq 2^{\aleph_0}$.

Proof. For every $p \in S(A)$ there is $B_p \subseteq A$, $|B_p| < \kappa(T)$, such that p does not fork over B_p. $|S(A)| \geq \lambda > |A|^{<\kappa(T)}$ so for some B^0 $|\{p \in S(A) \colon B_p = B^0\}| \geq \lambda$. Now if $|S(B^0)| \geq \lambda$ then (1) holds. Otherwise $|S(B^0)| < \lambda$. Thus for some p^0, $\mathscr{P}' = \{p \in S(A) \colon p \restriction B^0 = p^0, B_p = B^0\}$ is of cardinality $\geq \lambda$. If $\lambda^*(p^0) \leq \aleph_0$ (from 5.1) then by Lemma 5.3, $2^{\aleph_0} \geq 2^{\lambda^*(p^0)} \geq |\mathscr{P}'| \geq \lambda$. Thus (2) holds.

If $\lambda^*(p^0) > \aleph_0$, by Lemmas 5.9, 5.10, 5.11, $|S^2(B^0)| \geq 2^{\lambda^*(p^0)} \geq |\{p \in S(A) \colon p \restriction B^0 = p^0, B_p = B^0\}| \geq \lambda$. So for some B', $|S(B')| \geq \lambda$ and $|B'| = |B| + 1 < \kappa(T)$.

THEOREM 5.14*: *If* $\lambda = |D(T)| < |T|$ *then* L *has a sublanguage* L* *of power* λ *such that for every predicate* R *of* L *there is a formula (in fact a predicate)* R^* *of* L* *such that* $(\forall \bar{x})\,(R(\bar{x}) \equiv R^*(\bar{x})) \in T$. *(We say in this case that* T *is a definitional extension of* $T \cap \mathrm{L}^*$.)

Proof. For all m and all $p \neq q \in D^m(T)$ choose a formula $\varphi = \varphi_{p,q}(\bar{x})$ such that $\varphi \in p$, $\neg\varphi \in q$. Let $\Phi_m = \{\varphi_{p,q}(\bar{x}) \colon p \neq q \in D^m(T)\}$, and let Ψ_m

be the closure of Φ_m under the connectives. *Claim:* For every formula $\psi(\bar{x})$ ($l(\bar{x}) = m$) there is an equivalent formula in Ψ_m.

Proof. By the definition of the $\varphi_{p,q}$, the set $\Gamma = \{\varphi(\bar{x}) \equiv \varphi(\bar{y}) : \varphi \in \Phi_m\} \cup \{\psi(\bar{x}) \equiv \neg\psi(\bar{y})\}$ is inconsistent.

Thus there is a finite $\Phi \subseteq \Phi_m$ such that $\{\varphi(\bar{x}) \equiv \varphi(\bar{y}): \varphi \in \Phi\} \cup \{\psi(\bar{x}) \equiv \neg\psi(\bar{y})\}$ is inconsistent. Let $\Phi = \{\varphi_l : l < n\}$. Thus for every $\eta \in {}^n2$ there is $\mathfrak{t}(\eta)$ such that $\bigwedge_{l<n} \varphi_l(\bar{x})^{\eta[l]} \vdash \psi(\bar{x})^{\mathfrak{t}(\eta)}$. Thus

$$\psi(\bar{x}) \equiv \bigvee_{\substack{\eta \in {}^n2 \\ \mathfrak{t}(\eta)=0}} \bigwedge_{l<n} \varphi_l(\bar{x})^{\eta[l]}.$$

This proves the above claim.

Now we define L*. For every m and $\varphi(\bar{x}) \in \Psi_m$, if there is a predicate R such that $(\forall \bar{x})(\varphi(\bar{x}) \equiv R(\bar{x})) \in T$ let R_φ be any such R. If there is no such R, let R_φ be an arbitrary predicate. Then $L^* = \{R_\varphi : \varphi \in \bigcup_m \Psi_m\}$ is the required sublanguage.

Remark. Since $|D(T)|$ is always $\geq \aleph_0$ we can assume by the above theorem that $|D(T)| \geq |T|$.

Now we come to the main theorem of this section.

THEOREM 5.15: *Let λ designate an infinite cardinal. Then every stable theory T satisfies one of the following:*

(1) *T is stable in λ iff $\lambda = |D(T)| + \lambda^{<\kappa(T)}$.*

(2) *T is stable in λ iff $\lambda = 2^{\aleph_0} + \lambda^{<\kappa(T)}$; and also $|D(T)| = |T| < 2^{\aleph_0}$.*

Proof. If T is stable in λ then $|S^m(\emptyset)| \leq \lambda$ and so $|D(T)| \leq \lambda$. Also by Theorem 3.7, $\lambda = \lambda^{<\kappa(T)}$; so $\lambda = |D(T)| + \lambda^{<\kappa(T)}$. If there is a countable A such that $|S(A)| \geq 2^{\aleph_0}$ then we also get $\lambda = 2^{\aleph_0} + \lambda^{<\kappa(T)}$.

Now assume that (1) does not hold. Then there is λ such that T is not stable in λ but still $\lambda = |D(T)| + \lambda^{<\kappa(T)}$. Now λ^+ is regular and there is A such that $|S(A)| \geq \lambda^+ > \lambda \geq |A|$. Thus the assumptions of Lemma 5.13 hold for λ^+ in place of λ, but 5.13(1) does not hold (otherwise $|D(T)|^{<\kappa(T)} \geq \lambda^+$; by our hypothesis $|D(T)| \leq \lambda$ and thus $\lambda^{<\kappa(T)} > \lambda$, a contradiction). Thus 5.13(2) holds, i.e., $2^{\aleph_0} \geq \lambda^+$. In other words $\lambda < 2^{\aleph_0}$; so $|D(T)| < 2^{\aleph_0}$ and thus $|T| = |D(T)|$ by Lemma II, 3.15. Of course there is a countable A for which $|S(A)| \geq 2^{\aleph_0}$ (by Theorem II, 3.2).

Thus T is not stable in any power $< 2^{\aleph_0}$. But by our hypothesis if $\lambda \geq 2^{\aleph_0}$, $\lambda = \lambda^{<\kappa(T)}$ then T is stable in λ. Thus (2) holds.

COROLLARY 5.16: *If T is superstable then $|D(T)|$ or $|D(T)| + 2^{\aleph_0}$ is the first power in which T is stable and it is $|S(A)|$ for some countable A.*

LEMMA 5.17: *If T is countable and superstable, but not stable in $\aleph_0$, then there is m such that $|D_m(T)| \geq \aleph_0$. (Thus T is not $\aleph_0$-categorical by IX, 1.6.)*

Proof. Similar to the proof of Lemma 5.13.

III.6. Imaginary elements

DEFINITION 6.1: (1) The sequence $\bar{b}$ is defined by the formula $\varphi(\bar{x}; \bar{a})$ if $\varphi(\mathfrak{C}; \bar{a}) = \{\bar{b}\}$. It is defined by the type p if $\bar{b}$ is the unique sequence which realizes p. It is definable over A if $\mathrm{tp}(\bar{b}, A)$ defines it.

(2) The formula $\varphi(\bar{x}; \bar{a})$ is algebraic if $\varphi(\mathfrak{C}; \bar{a})$ is finite. The type p is algebraic if it is realized by finitely many sequences only. The sequence $\bar{b}$ is algebraic over A if $\mathrm{tp}(\bar{b}, A)$ is algebraic.

(3) The definable closure of A, $\mathrm{dcl}\, A$ is $\{b: b$ definable over $A\}$.

(4) The algebraic closure of A, $\mathrm{acl}\, A$ is $\{b: b$ algebraic over $A\}$. If $A = \mathrm{acl}\, A$ we say A is algebraically closed.

(5) When $\mathrm{dcl}\, A = \mathrm{acl}\, A$, $\mathrm{cl}\, A$ will be their common value.

LEMMA 6.1: (1) *The type p defines a sequence iff for some finite $q \subseteq p$, $\bigwedge q$ defines that sequence.*

(2) *The type p is algebraic iff for some finite $q \subseteq p$, $\bigwedge q$ is algebraic (moreover, we can choose a q such that p and q are realized by exactly the same sequences).*

Proof. Immediate.

LEMMA 6.2: (1) $A \subseteq \mathrm{dcl}\, A \subseteq \mathrm{acl}\, A$.

(2) *When* $A \subseteq B$, $\mathrm{dcl}\, A \subseteq \mathrm{dcl}\, B$ *and* $\mathrm{acl}\, A \subseteq \mathrm{acl}\, B$.

(3) *If* $B = \mathrm{dcl}\, A$ *then* $B = \mathrm{dcl}\, B$.

(4) *If* $B = \mathrm{acl}\, A$ *then* $B = \mathrm{acl}\, B$.

(5) $\bar{b}$ *is definable [algebraic] over A iff* $\bar{b} \in \mathrm{dcl}\, A$ [$\bar{b} \in \mathrm{acl}\, A$].

Proof. Easy, e.g., (4) If $c \in \mathrm{acl}\, B$ then for some $b_1, \ldots, b_k \in B$, $n < \omega$ and formula φ, $\vDash (\exists^{\leq n} x)\varphi(x, b_1, \ldots, b_k) \wedge \varphi[c, b_1, \ldots, b_k]$. As $b_i \in B =$

acl A there are $\bar{a}_i \in A$, $n(i) < \omega$ and ψ_i such that $\vDash (\exists^{\le n(i)} x)\psi_i(x; \bar{a}_i) \wedge \psi_i[b_i; \bar{a}_i]$. Define

$$\theta(x; \bar{a}_1, \ldots, \bar{a}_k) = (\exists y_1 \ \ldots \ y_k)\left[\bigwedge_{i=1}^{k} \psi_i(y_i; \bar{a}_i) \wedge (\exists^{\le n} z)\varphi(z, y_1, \ldots, y_k) \right.$$
$$\left. \wedge\ \varphi(x, y_1, \ldots, y_k)\right];$$

so $\vDash \theta[c; \bar{a}_1, \ldots, \bar{a}_k]$ (where $y_i \mapsto b_i$). But the number of possible $\langle y_1, \ldots, y_k\rangle$ is $\le \prod_{i=1}^{k} n(i)$, and for each such $\langle y_1, \ldots, y_k\rangle$ we can find $\le n$ x's. Hence $|\theta(\mathfrak{C}; \bar{a}_1, \ldots, \bar{a}_k)| \le n \cdot \prod_{i=1}^{k} n(i) < \aleph_0$.

LEMMA 6.3: (1) *If* $p \in S^m(A)$ *then* p *has a unique extension* $q \in S^m(\operatorname{dcl} A)$, *in fact* $p \equiv q$ *and every formula over* dcl A *is equivalent to a formula over* A.

(2) *If* $\bar{a} \in \operatorname{acl} A$, *then* $\varphi(\bar{x}; \bar{a})$ *is almost over* A. *Moreover if* $\theta(\bar{x}; \bar{b})$ *is almost over* acl A *then it is almost over* A.

(3) *If* $q \in S^m(\operatorname{acl} A)$, $B \subseteq A$, *then* q *forks over* B *iff* $q \restriction A$ *forks over* B.

(4) $\operatorname{stp}(\bar{a}, A) \equiv \operatorname{stp}(\bar{a}, \operatorname{acl} A)$, $\operatorname{tp}(\bar{a}, \operatorname{acl} A) \vdash \operatorname{tp}(\bar{a}, A)$.

(5) *Let* T *be stable,* I *an (infinite) indiscernible set based on* A. *Then* I *is trivial (all of* I*'s elements are equal) iff* $\operatorname{Av}(I, A)$ *is algebraic.*

Proof. (1) is trivial, and (3), (4) follow from (2) by 2.6(1) (and 1.1(8) of course). So let us prove (2); so by 6.2(5) assume $\psi(\bar{y}; \bar{c})$ is algebraic, $\bar{c} \in A$, $\vDash \psi[\bar{a}; \bar{c}]$. Define $E(\bar{x}; \bar{y}; \bar{c}) = (\forall \bar{z})[\psi(\bar{z}; \bar{c}) \rightarrow \varphi(\bar{x}; \bar{z}) \equiv \varphi(\bar{y}; \bar{z})]$. Clearly $E = E(\bar{x}; \bar{y})$ is an equivalence relation, and if $|\psi(\mathfrak{C}; \bar{c})| = n$, it has $\le 2^n$ equivalence classes. So $E \in \operatorname{FE}(A)$. As $\vDash \psi[\bar{a}; \bar{c}]$, $E(\bar{x}; \bar{y}) \rightarrow \varphi(\bar{x}; \bar{a}) \equiv \varphi(\bar{y}; \bar{a})$, i.e., $\varphi(\bar{x}; \bar{a})$ is almost over A. The assertion about $\theta(\bar{x}; \bar{b})$ follows by 2.4.

Now (5) is easy.

Now we define the model $\mathfrak{C}^{eq}$.

DEFINITION 6.2: (1) The universe of $\mathfrak{C}^{eq}$ is $\{\bar{a}/E: m < \omega, \bar{a} \in {}^m|\mathfrak{C}|, E = E(\bar{x}, \bar{y}) \in \mathrm{L}$ is an equivalence relation (without parameters)$\}$ (we mean that $\bar{a}_1/E_1 = \bar{a}_2/E_2$ iff $E_1 = E_2$, $\vDash E_1(\bar{a}_1, \bar{a}_2)$). We identify $a/=$ with a.

(2) The relations of $\mathfrak{C}^{eq}$ will be equality and

(i) $P_E = \{\bar{a}/E: \bar{a} \in |\mathfrak{C}|\}$,

(ii) a function F_E from $P_=$ onto P_E defined by $F_E(\bar{a}) = \bar{a}/E$ (so F_E is a partial function, i.e., a relation),

(iii) for every formula $\varphi(\bar{x}) \in \mathrm{L}$ $R_\varphi = \{\langle a_1/=, \ldots\rangle: \vDash \varphi[a_1, \ldots]\}$.

(3) The language of $\mathfrak{C}^{eq}$ is L^{eq}, and its theory T^{eq}.

(4) $\mathfrak{C}^{eq}_*$ will denote a submodel of $\mathfrak{C}^{eq}$ whose universe is $P_= \cup P_{E_1} \cup \cdots \cup P_{E_n}$ for some P_{E_i} and some $n < \omega$. So $\mathfrak{C}^{eq}_*$ is not uniquely defined.

LEMMA 6.4: *For every formula $\varphi(\bar{x})$ in L^{eq} there is an equivalent formula which is a Boolean combination of formulas of the following forms:*

(i) *a propositional constant (true or false),*

(ii) $x = y$,

(iii) $P_E(x)$,

(iv) $\bigwedge_{i=1}^n P_{E_i}(x_i) \to (\forall \bar{y}_1, \ldots, \bar{y}_n)[\bigwedge_{i=1}^n F_{E_i}(\bar{y}_i) = x_i \to R_\psi(\bar{y}_1, \ldots, \bar{y}_n)]$.

Proof. Left as an exercise for the reader (equivalence is relative to T).

CONCLUSION 6.5: *$\mathfrak{C}^{eq}$ is not saturated; but if we add to it $\|\mathfrak{C}\|$ more elements, without extending the relations (so a new element $\notin \bigcup_E P_E$) we get a saturated model.*

DEFINITION 6.3: (1) If p is a set of m-formulas in $\mathfrak{C}$, then $p^{eq} = \{R_\varphi(\bar{x}; \bar{a}): \varphi(\bar{x}; \bar{a}) \in p\} \cup \{P_=(x_i): i < l(\bar{x})\}$.

(2) If $M \prec \mathfrak{C}$ we let M^{eq} be the submodel of $\mathfrak{C}^{eq}$, whose universe is $\{\bar{a}/E: \bar{a} \in |M|\}$. See Lemma 6.6.

(3) In attributing properties we shall not distinguish strictly between M and M^{eq}, p and p^{eq} (this is justified by 6.6).

LEMMA 6.6: (1) *p is consistent iff p^{eq} is consistent. If p is an m-type in $\mathfrak{C}^{eq}$, $p \vdash \bigwedge_{l<n} P_=(x_l)$ then for some m-type q in $\mathfrak{C}$, $q^{eq} \equiv p$.*

(2) *When $M \prec \mathfrak{C}$, $M^{eq} \prec \mathfrak{C}^{eq}$.*

(3) *For every $M' \prec \mathfrak{C}^{eq}$, there is a unique $M \prec \mathfrak{C}$ such that $M^{eq} = M'$.*

Proof. Easy.

DEFINITION 6.4: (1) The *kind* θ_a of $\bar{a} = \langle a_0, \ldots, a_{m-1}\rangle$ is the unique formula $\theta(x_0, \ldots, x_{m-1}) = \bigwedge_{i<m} P_{E_i}(x_i)$ such that $\mathfrak{C}^{eq} \models \theta[\bar{a}]$.

(2) By a $(<\aleph_0)$-type in $\mathfrak{C}^{eq}$ we shall mean a type p such that $\theta_a \in p$ for any $\bar{a}$ realizing p.

LEMMA 6.7: (1) *For $\lambda \geq |L|$, T is stable in λ iff T^{eq} is stable in λ. Also $\kappa(T) = \kappa(T^{eq})$.*

(2) *If p is a type in $\mathfrak{C}$, $A \subseteq |\mathfrak{C}|$ then: p forks over A (in $\mathfrak{C}$) iff p^{eq} forks over A (in $\mathfrak{C}^{eq}$) and p is stationary iff p^{eq} is stationary.*

(3) *If $\bar{a} \in \mathfrak{C}$, $A \subseteq |\mathfrak{C}|$ then* $[\text{tp}(\bar{a}, A, \mathfrak{C})]^{eq} \equiv \text{tp}(\bar{a}, A, \mathfrak{C}^{eq})$.

(4) *If $B \subseteq A$, $\bar{a} \in \text{acl}\, B$, then $\text{tp}(\bar{a}, A)$ does not fork over B.*

(5) *If* $B \subseteq A$, $\bar{b} \in \mathrm{acl}(B \cup \bar{a})$, $\bar{b} \notin \mathrm{acl}\, A$, and $\mathrm{tp}(\bar{b}, A)$ *forks over* B, *then* $\mathrm{tp}(\bar{a}, A)$ *forks over* B (*holds in* $\mathfrak{C}$ *and in* $\mathfrak{C}^{eq}$, *in particular for* $b = F_E(\bar{a})$).

Proof. (1) By 6.5; and the second phrase by (2), (5).

(2) By 6.4.

(3) Easy.

(4) Trivial.

(5) Easy by (4) and 4.15.

Following Definition 6.4(2), the theory we have developed for $\mathfrak{C}$ *holds for* $\mathfrak{C}^{eq}$. *We shall use this freely.*

LEMMA 6.8: *For any equivalence relation in* $\mathfrak{C}^{eq}$, $E = E(\bar{x}, \bar{y}) = E(\bar{x}, \bar{y}; \bar{c})$ $(l(\bar{x}) = l(\bar{y}) = m)$ *between sequences of kind* θ (*so* $E(\bar{x}, \bar{y}) \vdash \theta(\bar{x}) \wedge \theta(\bar{y})$), *we can choose a relation* P_E, *and partial function* F_E (*both definable by formulas in* L^{eq} *with* $\bar{c}$ *as parameter; but the formulas themselves do not depend on* $\bar{c}$) *which satisfy:*

F_E *is a function from sequences of length* m *of kind* θ *to* P_E, *such that* $\models \theta[\bar{a}^1] \wedge \theta[\bar{a}^2]$, *implies* $F_E(\bar{a}_1) = F_E(\bar{a}_2) \Leftrightarrow \models E[\bar{a}_1, \bar{a}_2]$.

Proof. Technical, so we leave it to the reader.

Remark. We shall not distinguish strictly between the F_E, P_E from 6.8 and those from Definition 6.2. So we write $\bar{a}/E$ instead of $F_E(\bar{a})$.

DEFINITION 6.5: For a formula $\varphi(\bar{x}; \bar{y})$ let $E_\varphi(\bar{y}, \bar{z}) = (\forall \bar{x})[\varphi(\bar{x}; \bar{y}) \equiv \varphi(\bar{x}; \bar{z})]$ (in $\mathfrak{C}^{eq}$ this is done for fixed kinds of $\bar{y}, \bar{z}$.) Clearly E_φ is an equivalence relation, and $\varphi(\bar{x}; \bar{a})$ is equivalent to a formula over $\bar{a}/E_\varphi$.

LEMMA 6.9: *In* $\mathfrak{C}^{eq}$, *for algebraically closed* A:

(1) *Every* $p \in S^m(A)$ *is stationary over* A (*for stable* T).

(2) *Every formula* $\varphi(\bar{x}; \bar{a})$ *which is almost over* A, *is equivalent to a formula over* A.

Proof. (1) By 2.9(1), (1) follows from (2).

(2) As $\varphi(\bar{x}; \bar{a})$ is almost over A, it depends on some $E \in \mathrm{FE}^m(A)$. Let $\bar{b}_1, \ldots, \bar{b}_n$ be representatives of the equivalence classes of E. Clearly for all i $F_E(\bar{b}_i)$ is algebraic over A [as $P_E(F_E(\bar{b}_i))$ and $\vdash P_E(x) \equiv \bigvee_{i=1}^n x = F_E(\bar{b}_i)$]. So $F_E(\bar{b}_i) \in A$. As $\varphi(\bar{x}; \bar{a})$ depends on E, for some w, $\vdash \varphi(\bar{x}; \bar{a}) \equiv \bigvee_{i \in w} E(\bar{x}; \bar{b}_i)$. But $\vdash E(\bar{x}; \bar{b}_i) \equiv [F_E(\bar{b}_i) = F_E(\bar{x})]$ hence $\vdash \varphi(\bar{x}; \bar{a}) \equiv [\bigvee_{i \in w} F_E(\bar{x}) = F_E(\bar{b}_i)]$. The second formula is over A, so we finish.

THEOREM 6.10 (The Canonicity Theorem for Types): *Let T be stable. In $\mathfrak{C}^{eq}$, for every stationary m-type p there is a set $A = \mathrm{Cb}(p)$ (the canonical base of p) and a stationary type $q \in S^m(A)$, $q = \mathrm{Ctp}(p)$ (= the canonical type of p) such that:*

(1) p and q are parallel.

(2) Any automorphism F of $\mathfrak{C}^{eq}$ takes p to a parallel type iff $F \restriction A$ = identity.

(3) For any $a \in \mathfrak{C}^{eq}$, $a \in A$ iff every automorphism F of $\mathfrak{C}^{eq}$ which takes p to a parallel type necessarily satisfies $F(a) = a$.

(4) There is $B \subseteq A$, $|B| < \kappa(T)$ over which no type parallel to p forks.

(5) If a type in $S^m(C)$ is parallel to p and does not fork over $B \subseteq C$ then $A \subseteq \mathrm{acl}\, B$.

(6) $A \subseteq \mathrm{dcl}(\mathrm{Dom}\, p)$.

Proof. Let $r \in S^m(\mathfrak{C}^{eq})$ be parallel to p. By II, 2.2 for every $\varphi(\bar{x}; \bar{y})$ there is a formula $\psi_\varphi(\bar{y}; \bar{c}_\varphi)$ which defines $r \restriction \varphi$, i.e.,

$$\vDash \psi_\varphi[\bar{a}; \bar{c}_\varphi] \Leftrightarrow \varphi(\bar{x}; \bar{a}) \in r.$$

Let

$$A = \mathrm{dcl}\{\bar{c}_\varphi/E_{\psi_\varphi} : \varphi \in \mathrm{L}^{eq}\},$$
$$q = r \restriction A.$$

Clearly r is definable over A, hence does not fork over A by 4.11. So q is stationary and q, r are parallel, hence p, q are parallel (this is (1)).

Clearly an automorphism F of $\mathfrak{C}^{eq}$ takes p to a parallel type iff $F(r) = r$ iff for every φ, $\vdash \psi_\varphi(\bar{y}; \bar{c}_\varphi) \equiv \psi_\varphi(\bar{y}; F(\bar{c}_\varphi))$ iff for every φ, $\bar{c}_\varphi/E_{\psi_\varphi} = F(\bar{c}_\varphi)/E_{\psi_\varphi}$ iff $F \restriction A$ = identity (this is (2)). If $a \notin A$ then as $A = \mathrm{dcl}\, A$ there is $a' \neq a$, $\mathrm{tp}(a, A) = \mathrm{tp}(a', A)$ hence there is an automorphism F of $\mathfrak{C}^{eq}$, $F \restriction A$ = identity and $F(a) = a'$; so $F(r) = r$, $F(a) \neq a'$ (this is the missing part in (3)). From (3) it is clear that A does not depend on the particular choice of the ψ_φ: Also A depends only on r, i.e., on the equivalence class of p under parallelism. The same is true for q, hence the notation $A = \mathrm{Cb}(p)$, $q = \mathrm{Ctp}(p)$ is justified.

As we could have chosen $\bar{c}_\varphi \in \mathrm{Dom}\, p$ (by II, 2.12) and as $\bar{c}_\varphi/E_{\psi_\varphi} \in \mathrm{dcl}(\bar{c}_\varphi)$ $\{\bar{c}_\varphi/E_{\psi_\varphi} : \varphi \in \mathrm{L}^{eq}\} \subseteq \mathrm{dcl}(\mathrm{Dom}\, p)$. Hence by 6.2(2), (3) $A \subseteq \mathrm{dcl}(\mathrm{Dom}\, p)$, so (6) holds.

As for (4), for some $B \subseteq A$, $|B| < \kappa(T)$, q does not fork over B, but r does not fork over A, hence by 4.4 r does not fork over B. As any type parallel to p is $\subseteq r$ also (4) holds.

So we are left with (5). By 4.4; r does not fork over B, hence is definable almost over B, hence by 6.9(2) definable over $\mathrm{acl}\, B$, so we could have chosen $\bar{c}_\varphi \in \mathrm{acl}\, B$ hence $A \subseteq \mathrm{acl}\, B$.

CONCLUSION 6.11: *When T is stable, for every stationary type p in $\mathfrak{C}$ there is a parallel type q, such that for every automorphism F of $\mathfrak{C}$:*

$$F(p) \text{ is parallel to } p \text{ iff } F(q) \equiv q.$$

LEMMA 6.12: *In $\mathfrak{C}^{eq}$, if I is an indiscernible set based on the stationary type p (i.e., $\mathrm{Av}(I, \bigcup I)$, and p are parallel) then I is indiscernible over* $\mathrm{Cb}(p)$.

Proof. Any permutation of I is induced by some automorphism F of $\mathfrak{C}^{eq}$ which maps $\mathrm{Av}(I, \bigcup I)$ onto itself, hence p to a parallel type hence $F \restriction \mathrm{Cb}(p) =$ identity. So the result is clear by 6.10(2).

The following theorem shows that if T is stable but $\kappa(T)$ is large, then T contains an underlying structure of equivalence relations. This corroborates our intuition that the natural example is in fact quite general.

THEOREM 6.13: *In $\mathfrak{C}$, if $\aleph_0 < \kappa < \kappa(T) < \infty$, κ is regular, then there are $m < \omega$ and formulas $\varphi_i(\bar{x}, \bar{y}) \in \mathrm{L}$ $i < \kappa$, $l(\bar{x}) = l(\bar{y}) = m$ such that:*

(i) *$\varphi_i(\bar{x}, \bar{y})$ defines an equivalence relation.*

(ii) *For every λ there are $\bar{a}_\eta$ ($\eta \in {}^\kappa\lambda$) such that $\vDash \varphi_i[\bar{a}_\eta, \bar{a}_\nu]$ iff $\nu \restriction (i + 1) = \eta \restriction (i + 1)$.*

Proof. Let M_i ($i \le \kappa$) be an increasing sequence of models, $M_i \prec \mathfrak{C}$, $M_\kappa = \bigcup_{i<\kappa} M_i$, and $p \in S(|M_\kappa|)$, $p \restriction |M_{i+1}|$ forks over $|M_i|$ and M_{i+1} is $\|M_i\|^+$-saturated. (This is by 3.1, 4.16(1).)

Choose a_n ($n < \omega$) such that $\mathrm{tp}(a_n, |M_\kappa| \cup \bigcup_{l<n} a_l)$ extends p and does not fork over $|M_\kappa|$. Hence $I = \{a_n : n < \omega\}$ is indiscernible over $|M_\kappa|$ and based on it.

Now work in $\mathfrak{C}^{eq}$. Clearly I is based on p hence $\mathrm{Cb}(p) \subseteq \mathrm{acl}(\bigcup I)$. On the other hand $\mathrm{Cb}(p) \subseteq |M_\kappa^{eq}|$ but for $i < \lambda$, $\mathrm{Cb}(p) \nsubseteq |M_i^{eq}|$ (all by 6.10). Hence we can define, by induction on i, an increasing sequence $\alpha(i) < \kappa$ ($i < \kappa$) and $c_i \in \mathrm{Cb}(p)$, $c_i \in |M^{eq}_{\alpha(i)}|$, $c_i \in |M^{eq}_{\alpha(i+1)}|$. (We use the regularity of κ.) So for every i $c_i \in \mathrm{acl}(\bigcup I)$ hence for some $m(i)$, $k(i)$, ψ_i

$$\vDash \psi_i[c_i, a_0, \ldots, a_{m(i)}], \qquad \vDash (\exists^{\le k(i)} x)\psi_i(x, a_0, \ldots, a_{m(i)})$$

and

$$\psi_i(x; a_0, \ldots, a_{m(i)}) \vdash \mathrm{tp}(c_i, \bigcup I)$$

(see Exercise 6.1) and w.l.o.g. $\psi_i(|\mathfrak{C}^{eq}|; a_0, \ldots, a_{m(i)}) \cap |M_\kappa| \subseteq |M_{\alpha(i+1)}|$.

Clearly if $\mathrm{tp}(c^1, \bigcup I) = \mathrm{tp}(c^2, \bigcup I)$ then $c^1 \in \mathrm{Cb}(p) \Leftrightarrow c^2 \in \mathrm{Cb}(p)$ hence $\psi_i(|\mathfrak{C}^{eq}|; a_0, \ldots, a_{m(i)}) \subseteq |M^{eq}_{\alpha(i+1)}|$.

As $\kappa > \aleph_0$ we can assume $m(i) = m$, $k(i) = k$, $l_1 = l_1(i) = |\psi_i(|\mathfrak{C}^{eq}|; a_0, \ldots, a_m)|$, $l_2 = l_2(i) = |\psi_i(M_{\alpha(i)}; a_0, \ldots, a_m)|$ (for every i). Let

$$\varphi_i(\bar{x}, \bar{y}) = \varphi_i(x_0, \ldots, x_m; y_0, \ldots, y_m)$$
$$= (\forall z)[\psi_i(z; \bar{x}) \equiv \psi_i(z; \bar{y})] \wedge \bigwedge_{l=0}^{m} (P_=(x_l) \wedge P_=(y_l)).$$

Trivially φ_i is an equivalence relation and as $\psi_i(|\mathfrak{C}^{eq}|; a_0, \ldots, a_m) \subseteq |M^{eq}_{\alpha(i+1)}|$, $\nsubseteq |M^{eq}_{\alpha(i)}|$ we can find $\bar{a}_\eta$ satisfying (ii) (as in Theorem 3.1). By 6.4 we can go back to $\mathfrak{C}$.

EXERCISE 6.1: Prove that if $\bar{b} \in \operatorname{acl} A$, then for some $\bar{a} \in A$ and φ, $\varphi(\bar{x}; \bar{a}) \equiv \operatorname{tp}(\bar{b}, A)$.

EXERCISE 6.2: Suppose $c_i \in \operatorname{acl}(A \cup \bar{a})$ for some $\bar{a}$ but $c_i \notin \operatorname{acl}(A \cup \{c_j : j < i\})$, for $i < \kappa$. Prove $\kappa < \kappa(T)$.

EXERCISE 6.3: Prove that in 6.13 the restriction on κ cannot in general be weakened.

QUESTION 6.4: Can we in 6.13 replace "$\kappa > \aleph_0$ regular" by "cf $\kappa > \aleph_0$"?

CLAIM 6.14: *Let T be stable.* (1) *(in $\mathfrak{C}^{eq}$) If $\bar{c} \in \operatorname{acl}(A \cup B)$, $\operatorname{tp}_*(A, B \cup C)$ does not fork over B, then*

$$\operatorname{tp}(\bar{c}, A \cup \operatorname{acl} B) \vdash \operatorname{tp}(\bar{c}, A \cup B \cup C).$$

(2) *Suppose $B \subseteq C$ and*
(i) $\operatorname{stp}(\bar{b}, A \cup B) \vdash \operatorname{tp}(\bar{b}, A \cup C)$,
(ii) $\operatorname{tp}_*(\bar{b} \cup A, C)$ *does not fork over B,*
(iii) $\operatorname{tp}_*(\bar{b} \cup A, B)$ *is stationary,*
then $\operatorname{tp}(\bar{b}, A \cup B) \vdash \operatorname{tp}(\bar{b}, A \cup C)$.

Proof. (1) As $\operatorname{tp}_*(A, B \cup C)$ does not fork over B, by 6.7(5), $\operatorname{tp}_*(A \cup \bar{c}, B \cup C)$ does not fork over B; so by 6.3(3) $\operatorname{tp}_*(A \cup \bar{c}, \operatorname{acl} B \cup C)$ does not fork over B, hence over $\operatorname{acl} B$. So $\operatorname{tp}_*(C, \operatorname{acl} B \cup A \cup \bar{c})$ does not fork over $\operatorname{acl} B$. But, by 6.9 $\operatorname{tp}_*(C, \operatorname{acl} B)$ is stationary hence by Exercise 2.2 and Lemma 6.3(4),

$$\operatorname{tp}_*(C, \operatorname{acl} B \cup A) \vdash \operatorname{tp}_*(C, \operatorname{acl}(B \cup A)) \vdash \operatorname{tp}_*(C, \operatorname{acl} B \cup A \cup \bar{c})$$

hence $\operatorname{tp}(\bar{c}, A \cup \operatorname{acl} B) \vdash \operatorname{tp}(\bar{c}, A \cup \operatorname{acl} B \cup C)$ (see IV, Ax(V.1) for $\mathbf{F}^s_\sim$; IV, 2.8).

(2) We work in $\mathfrak{C}^{eq}$, so by 6.3(4), Exercise 2.2, $\text{tp}(\bar{b}, \text{acl}(A \cup B)) \vdash \text{stp}(\bar{b}, A \cup B)$ so by (i) (α) $\text{tp}(\bar{b}, \text{acl}(A \cup B)) \vdash \text{tp}(\bar{b}, A \cup C)$. By 6.14(1) and (ii) for every $\bar{c} \in \text{acl}(A \cup B)$, $\text{tp}(\bar{c}, A \cup \text{acl}\, B) \vdash \text{tp}(\bar{c}, A \cup C)$, hence: ($\beta$) $\text{tp}_*(\text{acl}(A \cup B), A \cup \text{acl}\, B) \vdash \text{tp}_*(\text{acl}(A \cup B), A \cup C)$. Now ($\alpha$) and ($\beta$) imply: ($\gamma$) $\text{tp}(\bar{b}, A \cup \text{acl}\, B) \vdash \text{tp}(\bar{b}, A \cup C)$. [If $\bar{b}_1$ realizes $\text{tp}(\bar{b}, A \cup \text{acl}\, B)$ let f_1 be an elementary mapping, $f_1 \restriction (A \cup \text{acl}\, B)$ is the identity, $f(\bar{b}) = \bar{b}_1$. Now let f extend f_1, $\text{Dom}\, f = \bar{b} \cup \text{acl}(A \cup B)$, hence $\text{Range}\, f = \bar{b}_1 \cup \text{acl}(A \cup B)$. Let g be the identity mapping on C. Clearly $(f \restriction (A \cup \text{acl}\, B)) \cup g$ is an elementary mapping. Hence by (β) $(f \restriction \text{acl}(A \cup B)) \cup g$ is an elementary mapping, hence by (α) $f \cup g$ is an elementary mapping, so $\bar{b}_1$ realizes $\text{tp}(\bar{b}, A \cup C)$.]

As $\text{tp}(\bar{b} \cup A, B)$ is stationary, by 6.3(4) $\text{tp}_*(\bar{b} \cup A, B) \vdash \text{tp}(\bar{b} \cup A, \text{acl}\, B)$ hence (δ) $\text{tp}(\bar{b}, A \cup B) \vdash \text{tp}(\bar{b}, A \cup \text{acl}\, B)$. By ($\gamma$) and ($\delta$) $\text{tp}(\bar{b}, A \cup B) \vdash \text{tp}(\bar{b}, A \cup C)$, which is our desired conclusion.

III.7. Instability

DEFINITION 7.1: (1) Let $\kappa^m_{\text{ird}}(T)$ be the first infinite cardinal κ for which T does not have κ independent orders by m-formulas: i.e., there are no formulas $\varphi_i(\bar{x}; \bar{y}_i)$ $(i < \kappa)$ $l(\bar{x}) = m$ and $\bar{a}^i_j$ $(j < \omega)$ such that for every $\eta \in {}^\kappa\omega$,

$$\{\varphi_i(\bar{x}; \bar{a}^i_j)^{\text{if}\,(\eta(i) \geq j)} : j < \omega, i < \kappa\}$$

is consistent.

(2) $\kappa_{\text{ird}}(T) = \sup_m \kappa^m_{\text{ird}}(T)$.

Remark. The expression "ird" is an abbreviation for "independent orders".

LEMMA 7.1: (1) *T has the independence property iff $\kappa^m_{\text{ird}}(T) = \infty$ iff $\kappa^m_{\text{ird}}(T) > |T|^+$ (for any m).*

(2) *For stable T $\kappa^m_{\text{ird}}(T) = \aleph_0$.*

(3) $\kappa^m_{\text{ird}}(T) \leq \kappa^{m+1}_{\text{ird}}(T)$.

Proof. Immediate.

THEOREM 7.2: $K^m(\lambda) \geq \sup\{(\mu^\kappa)^+ : \mu < \text{Ded}\, \lambda, \kappa < \kappa^m_{\text{ird}}(T)\}$ *(we assume $\lambda \geq \kappa_{\text{ird}}(T)$, of course) (see Definition II, 4.4).*

Proof. Let $\mu < \operatorname{Ded} \lambda$, and I an order $|I| = \mu$, with a dense suborder J, $|J| = \lambda$. Let $\kappa < \kappa^m_{\text{ird}}(T)$ and let φ_i $(i < \kappa)$ be the formulas from Definition 7.1. By compactness define $\bar{a}^i_s$ $(s \in J)$ so that for every $\eta \in {}^{\kappa}I$,

$$p_\eta = \{\varphi_i(\bar{x}; \bar{a}^i_s)^{\text{if } (\eta[i] \geq s)}: s \in J, i < \kappa\}$$

is consistent. Let $A = \bigcup \{\bar{a}^i_s: i < \kappa, s \in J\}$, and $p_\eta \subseteq q_\eta \in S^m(A)$. Then $|A| \leq \lambda$ and

$$|S^m(A)| \geq |\{q_\eta: \eta \in {}^{\kappa}I\}| = |I|^\kappa = \mu^\kappa.$$

THEOREM 7.3: *Let Δ and A be infinite sets such that:*

(i) $|S^m_\Delta(A)| \geq \lambda > 2^{|\Delta|} + |A|$, *where λ is regular.*

(ii) $|S^m_{\Delta_1}(A)| < \lambda$ *for all $\Delta_1 \subseteq \Delta$, $|\Delta_1| < |\Delta|$.*

(iii) *There is no $S \subseteq S^m_\Delta(A)$ such that $|S| \geq \lambda$ and $|\{p \restriction \varphi: p \in S\}| \leq |A|$ for every $\varphi \in \Delta$.*

Then $|\Delta| < \kappa^m_{\text{ird}}(T)$.

Proof. Let $\Delta = \{\varphi_i(\bar{x}; \bar{y}): i < |\Delta|\}$, and $\Delta_i = \{\varphi_j: j < i\}$. We define by induction on α a function F_α from $S^m_\Delta(A)$ to subsets of $|\Delta|$:

$F_\alpha(p) = \{\zeta < |\Delta|$: for every finite $\Delta^* \subseteq \Delta_\zeta$ there are $> |A|$ types $q \in S^m_\Delta(A)$ with distinct $q \restriction \varphi_\zeta$ such that $p \restriction \Delta^* = q \restriction \Delta^*$ and $(\forall \beta < \alpha)[F_\beta(p) = F_\beta(q)]\}$.

In the definition α appears only once: in $(\forall \beta < \alpha)$; hence for $\gamma < \alpha$ $F_\gamma(p) \supseteq F_\alpha(p)$.

Now for some $p \in S^m_\Delta(A)$, $|\bigcap_{n<\omega} F_n(p)| = |\Delta|$ (in fact this holds for α's much higher than ω). For suppose there is no such p. As $|S^m_\Delta(A)| \geq \lambda$, $\lambda > 2^{|\Delta|}$, λ regular, for some $w_n \subseteq |\Delta|$, $n < \omega$, $S_1 = \{p \in S^m_\Delta(A): F_n(p) = w_n$ for every $n < \omega\}$ has cardinality $\geq \lambda$. Let $\Delta_* = \{\varphi_i: i \in \bigcap_{n<\omega} w_n\}$, so by assumption $|\Delta_*| < |\Delta|$, hence by (ii) $|S^m_{\Delta_*}(A)| < \lambda$. As λ is regular, for some $p_0 \in S^m_{\Delta_*}(A)$,

$$S_2 = \{p \in S_1: p \restriction \Delta_* = p_0\}$$

has cardinality $\geq \lambda$.

Now for every $\zeta < |\Delta|$, $|\{p \restriction \varphi_\zeta: p \in S_2\}| \leq |A|$. We prove this by induction on ζ; if $\zeta \in \bigcap_{n<\omega} w_n$, $p \in S_2$, then $\varphi_\zeta \in \Delta_*$ and $p \restriction \varphi_\zeta =$

$p_0 \upharpoonright \varphi_\zeta$. If for some n $\zeta \notin w_n$ then by the definition of F_n, and as $(\forall p, q \in S_2)(\forall l < \omega)[F_l(p) = F_l(q)]$ we get (letting $S_{\Delta^*} = \{p \upharpoonright \Delta^*: p \in S_2\}$):

$$\begin{aligned}
&|\{p \upharpoonright \varphi_\zeta: p \in S_2\}| \\
&\quad \le \Big| \bigcup_{\substack{\Delta^* \subseteq \Delta_\xi \\ |\Delta^*| < \aleph_0}} \{p \upharpoonright \varphi_\zeta: p \in S_2 \text{ and the number of } q \upharpoonright \varphi_\zeta, \text{ for } q \in S_2, \\
&\qquad\qquad q \upharpoonright \Delta^* = p \upharpoonright \Delta^* \text{ is } \le |A|\}\Big| \\
&\quad \le \sum_{\substack{\Delta^* \subseteq \Delta_\xi \\ |\Delta^*| < \aleph_0}} \sum_{r \in S_{\Delta^*}} |\{p \upharpoonright \varphi_\zeta: p \in S_2, p \upharpoonright \Delta^* = r \text{ and the number of} \\
&\qquad\qquad q \upharpoonright \varphi_\zeta, \text{ for } q \in S_2,\ q \upharpoonright \Delta^* = r \text{ is } \le |A|\}| \\
&\quad \le |\Delta| \cdot |S_{\Delta^*}| \cdot |A| \le |A|
\end{aligned}$$

as

$$|S_{\Delta^*}| \le \prod_{\varphi_l \in \Delta^*} |S_{\varphi_l}| \le |A|^{|\Delta^*|} = |A|$$

by the induction hypothesis, and $|\Delta| < |A|$ as

$$2^{|\Delta|} < \lambda \le |S_\Delta^m(A)| \le 2^{|\Delta| + |A|}.$$

So by S_2 we get a contradiction to hypothesis (iii) of the theorem. Hence for some $p^* \in S_\Delta^m(A)$, $|w| = |\Delta|$ where $w = \bigcap_{n<\omega} F_n(p^*)$.

By compactness it suffices to prove that for any $i(0) < i(1) < i(2) < \cdots < i(n-1) < |\Delta|$, $i(l) \in w$, $l < n$ and $k < \omega$ there are $\bar{a}_j^l$ $(l < n, j < k)$ such that for any $\eta \in {}^n k$, $\{\varphi_{i(l)}(\bar{x}; \bar{a}_j^l)^{\text{if}(\eta[l] \ge j)}: l < n, j < k\}$ is consistent.

Denote $\Delta^l = \{\varphi_{i(\alpha)}: \alpha < l\}$.

Now we define by induction on $l \le n$ for every $\eta \in {}^l(|A|^+)$ a type $q_\eta \in S_\Delta^m(A)$ such that (i) $i(0), \ldots, i(n-1) \in F_{n+1-l}(q_\eta)$; (ii) when ν is an initial segment of η, $q_\nu \upharpoonright \Delta^{l(\nu)} \subseteq q_\eta$; (iii) $i \ne j \Rightarrow q_{\nu\frown\langle i\rangle} \upharpoonright \Delta^l \ne q_{\nu\frown\langle j\rangle} \upharpoonright \Delta^l$ when $l = l(\nu) + 1$.

We let $q_{\langle\rangle} = p^*$; and if q_η is defined, $l(\eta) = l$ we can define $q_{\eta\frown\langle i\rangle}(i < |A|^+)$ by the definition of F_{n+1-l}.

Let $r_\eta = q_\eta \upharpoonright \Delta^{l(\eta)+1}$. For every $\eta \in {}^{n-1}(|A|^+)$, by Exercise I, 2.7(1) there are $\bar{a}_{j,\eta}^{n-1}$ $(j < k)$ such that for every $l < k$ there is an α such that $r_\eta \cup \{\varphi_{i(n-1)}(\bar{x}; \bar{a}_{j,\eta}^{n-1})^{\text{if}(j \ge l)}: j < k\} \subseteq q_{\eta\frown\langle\alpha\rangle}$. By 2.6 of the Appendix (the "pigeonhole" principle for n-trees) and renaming we can assume $\bar{a}_{j,\eta}^{n-1} = \bar{a}_j^{n-1}$ for every $\eta \in {}^{n-1}(|A|^+)$ (note that the number of possible $\bar{a}_{j,\eta}^{n-1} \frown \cdots \frown \bar{a}_{k-1,\eta}^{n-1}$ is $\le |A| < |A|^+$). Now we choose similarly $\bar{a}_{j,\eta}^{n-2}[j < k, \eta \in {}^{n-2}(|A|^+)]$, and continuing by downward induction we define all the $\bar{a}_j^l$ $(l < n, j < k)$ and clearly they satisfy our demands, hence we finish.

CONCLUSION 7.4: *If for some* A, $|S(A)| \geq \lambda$, λ *regular*, $\lambda > 2^{|T|} + |A|^{|T|}$ *and for* $\mu < \mathrm{Ded}_r|A|$, $\mu^{<\kappa} < \lambda$ *then* $\kappa^1_{\mathrm{ird}}(T) > \kappa$.

Proof. By 7.1 we can assume T does not have the independence property Choose a Δ with minimal cardinality for which $|S_\Delta(A)| \geq \lambda$. By II, 4.10(2) when $|\Delta_1| < \kappa$, $|S_{\Delta_1}(A)| \leq \prod_{\varphi\in\Delta_1} |S_\varphi(A)| < \lambda$ hence $|\Delta| \geq \kappa$. So our conclusion follows by 7.3 as (i), (ii) hold by the choice of Δ and the existence of S as mentioned in (iii) contradicts $|A|^{|T|} < \lambda$.

THEOREM 7.5: *Suppose T does not have the independence property,* $A \subseteq B$, $p \in S^m(A)$. *Then there is* $C \subseteq B$, $|C| \leq |T|$ *and* $q \in S^m(B)$, *such that* $p \subseteq q$, *and q does not split over* $A \cup C$.

Proof. For every $C \subseteq B$ let

$$r(C) = \{\varphi(\bar{x}; \bar{a}) \equiv \varphi(\bar{x}; \bar{b}) : \bar{a}, \bar{b} \in B, \mathrm{tp}(\bar{a}, A \cup C) = \mathrm{tp}(\bar{b}, A \cup C)\}.$$

It is clear that $\mathrm{tp}(\bar{c}, B)$ does not split over $A \cup C$ iff $\bar{c}$ realizes $r(C)$. So it suffices to prove that for some C, $C \subseteq B$, $|C| \leq |T|$ and $p \cup r(C)$ is consistent. Suppose there is no such C, and define by induction on $\alpha < |T|^+$ sets $C_\alpha \subseteq B$, $|C_\alpha| \leq |T|$. Let $C_0 = \emptyset$, and $C_\delta = \bigcup_{i<\delta} C_i$. Suppose C_α is defined; so $p \cup r(C_\alpha)$ is inconsistent, hence there are $n(\alpha) < \omega$, $\bar{a}^\alpha_l, \bar{b}^\alpha_l \in B$ and formulas φ^α_l for $l < n(\alpha)$ such that

(i) $\mathrm{tp}(\bar{a}^\alpha_l, A \cup C_\alpha) = \mathrm{tp}(\bar{b}^\alpha_l, A \cup C_\alpha)$,

(ii) $p \vdash \bigvee_{l<n(\alpha)} \varphi^\alpha_l(\bar{x}; \bar{a}^\alpha_l) \equiv \neg\varphi^\alpha_l(\bar{x}; \bar{b}^\alpha_l)$.

Now let $C_{\alpha+1} = \bigcup \{\bar{a}^\alpha_l {}^\frown \bar{b}^\alpha_l : l < n(\alpha)\} \cup C_\alpha$. So C_α is defined for every $\alpha < |T|^+$.

As the number of possible $\langle n(\alpha), \ldots, \varphi^\alpha_l, \ldots\rangle_{l<n(\alpha)}$ is $\leq |T|$, there are $n < \omega$ and $\varphi_l (l < n)$ such that

$$|\{\alpha : \alpha < |T|^+, n(\alpha) = n, \varphi^\alpha_l = \varphi_l\}| = |T|^+.$$

So assume that $n(k) = n$, $\varphi^k_l = \varphi_l$ for every $k < \omega$ (by renaming). Let $\Delta = \{\varphi_l : l < n\}$, and define $p_\eta \in S^m(A \cup C_{l(\eta)})$ for $\eta \in {}^{\omega>}2$ by induction on $l(\eta)$. Let $p_{\langle\rangle} = p$, and if p_η is defined $l(\eta) = k$, let $p_{\eta^\frown\langle 0\rangle}$ be any extension of it in $S^m(A \cup C_{k+1})$. So by (ii) for some $l < n$, $\varphi_l(\bar{x}; \bar{a}^k_l) \equiv \neg\varphi_l(\bar{x}; \bar{b}^k_l) \in p_{\eta^\frown\langle 0\rangle}$. Let $\varphi_l(\bar{x}; \bar{a}^k_l)^{\mathbf{t}} \in p_{\eta^\frown\langle 0\rangle}$, then $p_\eta \cup \{\neg\varphi_l(\bar{x}; \bar{a}^k_l)^{\mathbf{t}}\}$ is consistent (by (i)) hence there is $p_{\eta^\frown\langle 1\rangle} \in S^m(A \cup C_{k+1})$, $p_\eta \subseteq p_{\eta^\frown\langle 1\rangle}$, $\neg\varphi_l(\bar{x}; \bar{a}^k_l)^{\mathbf{t}} \in p_{\eta^\frown\langle 1\rangle}$. Now for every $k < \omega$ let $A_k = \bigcup \{\bar{a}^j_l : j < k, l < n\}$; for $\eta \neq \nu \in {}^k2$, by the definition $(p_\eta \restriction \Delta) \restriction A_k \neq (p_\nu \restriction \Delta) \restriction A_k$, hence $|S^m_\Delta(A_k)| \geq 2^k$, and $|A_k| \leq k \cdot m_0$, where $m_0 = \sum_{l<n} l(\bar{a}^0_l)$. So by II, 2.1

and II, 4.11 some $\varphi(\bar{x};\bar{y}) \in \Delta$ has the independence property, a contradiction.

DEFINITION 7.2: (1) $\kappa r^m_{\mathrm{cdt}}(T)$ is the first regular cardinal κ such that: there are no formulas $\varphi_i(\bar{x};\bar{y}_i)$ ($l(\bar{x}) = m$, $i < \kappa$, i successor) numbers $n_i < \omega$ and sequences $\bar{a}_\eta$, $\eta \in {}^{\kappa>}\omega$, for which

(i) $p_\eta = \{\varphi_i(\bar{x};\bar{a}_{\eta|i}): i \text{ successor}, i < \kappa\}$ is consistent for $\eta \in {}^{\kappa}\omega$,

(ii) $\{\varphi_i(\bar{x}, \bar{a}_{\eta\frown\langle\alpha\rangle}): \alpha < \omega, \eta \in {}^{\kappa>}\omega, i = l(\eta) + 1\}$ is n_i-contradictory, i.e., for $w \subseteq \omega$, $|w| = n_i$, $\vdash \neg(\exists\bar{x})[\bigwedge_{\alpha\in w} \varphi_i(\bar{x}, \bar{a}_{\eta\frown\langle\alpha\rangle})]$.

(2) $\kappa r_{\mathrm{cdt}}(T) = \sup_m \kappa r^m_{\mathrm{cdt}}(T)$.

Remark. The expression "cdt" stands for "contradictory types".

LEMMA 7.6: (1) *If T has the strict order property then* $\kappa r^1_{\mathrm{cdt}}(T) = \infty$.

(2) $\kappa r^m_{\mathrm{cdt}}(T) > |T|^+$ *implies* $\kappa r^m_{\mathrm{cdt}}(T) = \infty$.

(3) $\kappa r^{m+1}_{\mathrm{cdt}}(T) \geq \kappa r^m_{\mathrm{cdt}}(T)$.

(4) $\kappa r^m_{\mathrm{cdt}}(T) > \aleph_0$ *iff* $D^m(\bar{x} = \bar{x}, \mathrm{L}, \infty) = \infty$.

(5) *For stable T* $\kappa r_{\mathrm{cdt}}(T) = \kappa_r(T)$.

Proof. (1) Let $\varphi(x;\bar{y})$ have the strict order property. Choose $\varphi_i = \varphi(x;\bar{y}_1) \wedge \neg\varphi(x;\bar{y}_2)$, $n_i = 2$.

(2) Because we can choose all the φ_i's equal.

(3) By adding dummy variables.

(4) By II, 3.5.

(5) Clearly $\kappa r_{\mathrm{cdt}}(T) \leq \kappa_r(T)$; the converse holds by (4) and by 6.13.

THEOREM 7.7: *The following conditions on T, $m < \omega$ and a regular cardinal κ are equivalent.*

(1) $\kappa < \kappa r^m_{\mathrm{cdt}}(T)$.

(2) $\kappa < \kappa r^m_{\mathrm{cdt}}(T)$, *and moreover, in Definition 7.2(1) we can have $n_i = 2$ for every i.*

(3) *For every λ, $\lambda^{<\kappa} = \lambda$, there is a set A, $|A| = \lambda$, and a set S of m-types over A which are contradictory in pairs, $|S| = \lambda^\kappa$, where $p \in S \Rightarrow |p| = \kappa$.*

(4) *There is a set A, a set S of m-types over A and a cardinal χ such that:*

(i) $|S| > |A|^{<\kappa} + 2^{|T|+\chi}$,

(ii) *no sequence realizes $>\chi$ types from S,*

(iii) *for every $p \in S$, $|p| \leq \chi$.*

Proof. (2) ⇒ (1) Immediate.

(1) ⇒ (4), (2) ⇒ (3) Easy, by compactness.

(3) ⇒ (4) Choose in (3) a λ which is a strong limit cardinal $> 2^{|T|+\kappa}$, which has cofinality κ. Then the A and S from (3) will satisfy (4), for $\chi = \kappa$.

(4) ⇒ (2) Let w.l.o.g. $|S| = \lambda^+$, where $\lambda = |A|^{<\kappa} + 2^{|T|+\chi}$, and $S = \{p_i : i < \lambda^+\}$. We can assume the types in S are pairwise contradictory, for define by induction $u_i \subseteq \lambda^+$ as follows: u_i is a maximal subset of λ^+, for which $j < i \Rightarrow u_i \cap u_j = \emptyset$, and $\bigcup \{p_\alpha : \alpha \in u_i\}$ is consistent. By (4)(ii) $|u_i| \le \chi$; so letting $p_i' = \bigcup \{p_\alpha : \alpha \in u_i\}$ we get what we want.

So now let $p_i = \{\varphi_\alpha^i(\bar{x}; \bar{a}_\alpha^i) : \alpha < \chi\}$. As the number of sequences $\langle \varphi_\alpha^i : \alpha < \chi \rangle$ is $\le |T|^\chi \le 2^{|T|+\chi} \le \lambda$ we can assume $\varphi_\alpha^i = \varphi_\alpha$; and similarly that $\operatorname{tp}_*(A^i)$ is fixed where $A^i = \bigcup_\alpha \bar{a}_\alpha^i$. For $i < \lambda^+$ we define finite subsets v_α^i of χ by induction on α. If v_β^i is defined for $\beta < \alpha$ and there is a finite $v \subseteq \chi$ and there are distinct $\zeta(n) < \lambda^+ (n < \omega)$ such that $\bar{a}_j^{\zeta(n)} = \bar{a}_j^i$ for $j \in \bigcup_{\beta<\alpha} v_\beta^i$, $\bigwedge_{\beta<\alpha} v_\beta^{\zeta(n)} = v_\beta^i$; and the formulas $\bigwedge_{\gamma \in v} \varphi_\gamma(\bar{x}; \bar{a}_\gamma^{\zeta(n)})$ are pairwise contradictory then we let $v_\alpha^i = v$. The first α for which we cannot define v_α^i will be denoted by $\alpha(i)$. Clearly $\alpha(i) \le \chi$; and it suffices to prove that for some $\xi < \lambda^+$, $\alpha(\xi) \ge \kappa$. (Then the φ_i of Definition 7.2 will be $\bigwedge \{\varphi_j : j \in v_i^\xi\}$ remembering that $\operatorname{tp}_*(A_i)$ is fixed.)

So suppose for every $\xi < \lambda^+$, $\alpha(\xi) < \kappa$, and we shall get a contradiction. As there are $\chi^{<\kappa} \le 2^\chi \le \lambda$ possible sequences $\langle v_\alpha^i : \alpha < \alpha(i) \rangle$ (as $|A|^{<\kappa} \le \lambda < |S| \le |A|^\chi$, $\kappa \le \chi$) for some $S_1 \subseteq S$, $|S_1| = \lambda^+$, and $p_i \in S_1$ implies $\alpha(i) = \alpha^*$, and $v_\alpha^i = v_\alpha$ for $\alpha < \alpha^*$. As $|A|^{<\kappa} \le \lambda$, there is $S_2 \subseteq S_1$ such that $|S_2| = \lambda^+$, and $p_i \in S_2$ implies $\bar{a}_j^i = \bar{a}_j$ for $j \in \bigcup_{\alpha<\alpha^*} v_\alpha^i$.

Now let $S_2 = \{\zeta(i) : i < \lambda^+\}$. For $i < j < \lambda^+$, $p_{\zeta(i)} \cup p_{\zeta(j)}$ is contradictory (we noted it at the beginning of the proof), so there is a finite subset $h(i, j)$ of χ such that

$$\{\varphi_\alpha(\bar{x}; \bar{a}_\alpha^i) : \alpha \in h(i, j)\} \cup \{\varphi_\beta(\bar{x}; \bar{a}_\beta^j) : \beta \in h(i, j)\}$$

is contradictory. But $\lambda^+ \ge (2^\chi)^+$, hence by 2.5 of the Appendix for some $S_3 \subseteq S_2$, $|S_3| = \chi$, h has a fixed value, v. This shows that for any $p_i \in S_3$ we could have defined $v_{\alpha^*}^i = v$; a contradiction.

THEOREM 7.8: *There is a theory T, without the strict order property, for which $\kappa r_{\text{cdt}}^1(T) = \infty$.*

Proof. The set of axioms of T will be as follows:

(1) E is an equivalence relation with infinitely many equivalence classes

$$(\forall x)(xEx) \quad (\forall xy)(xEy \equiv yEx),$$
$$(\forall xyz)(xEy \wedge yEz \rightarrow xEz),$$
$$(\exists x_0, \ldots, x_n) \bigwedge_{i<j\leq n} \neg x_i E x_j \text{ for every } n > 0.$$

(2) For two distinct E-equivalence classes x_1/E, x_2/E, the family of sets

$$\{z\colon zEx_1, zRy\} \text{ for } y \in x_2/E$$

forms a partition of x_1/E

$$(\forall xy)(xRy \rightarrow \neg xEy),$$
$$(\forall xy)(\neg xEy \rightarrow \exists z(zEx \wedge yRz),$$
$$(\forall xyz)(zRy \wedge zRx \wedge xEy \rightarrow y = x).$$

(3) These partitions are independent, moreover, for $m \leq n < w$,

$$\begin{aligned}(\forall x_1, \ldots, x_n)(\forall y_1, \ldots, y_n)(\forall z)\Bigg[\Bigg(\bigwedge_{l=1}^{n} \neg zEy_l \wedge \bigwedge_{l=1}^{n} \neg zEx_l \\ \wedge \bigwedge_{1\leq l<k\leq n} \neg y_l E y_k \\ \wedge \bigwedge_{l<m\leq k} \neg x_l = x_k\Bigg) \\ \rightarrow (\exists z^*)\Bigg(z^*Ez \wedge \bigwedge_{1\leq l\leq n} z^*Ry_l \\ \wedge \bigwedge_{l=1}^{m-1} x_l R z^* \\ \wedge \bigwedge_{l=m}^{n} \neg x_l R z^*\Bigg)\Bigg].\end{aligned}$$

We leave to the reader to prove that T admits elimination of quantifiers is complete and satisfies our requirements.

THEOREM 7.9: *There is a theory T, and a type $p \in S(\emptyset)$, such that for every cardinal λ there is a set A, $|A| = \lambda$, which satisfies: for every $q \in S(A)$, $p \subseteq q$, and $B \subseteq A$, $|B| < |A|$, q splits strongly over B.*

Remark. This shows the limitations on generalizing 1.3 to unstable theories.

Proof. A set of axioms for T will be:

(1) E is an equivalence relation, on $\{x: P(x)\}$; each equivalence class has exactly two elements

$$(\forall x)[xEx \equiv P(x)],$$

$$(\forall xy)[xEy \equiv yEx],$$

$$(\forall xyz)[xEy \wedge yEz \rightarrow xEz],$$

$$(\forall x)[P(x) \rightarrow (\exists! y)(x \neq y \wedge xEy)].$$

(2) When $\neg P(x)$, $P(y)$ then R chooses an element from y/E,

$$(\forall xy)[P(x) \vee \neg P(y) \rightarrow \neg xRy],$$

$$(\forall xy)[\neg P(x) \wedge P(y) \rightarrow (\exists! z)[yEz \wedge xRz]].$$

(3) The choices in (2) are independent, i.e., for every $w(1), \ldots, w(n) \subseteq \{1, \ldots, n\}$

$$(\forall x_1, \ldots, x_n)(\exists y_1, \ldots, y_n)\Big[\bigwedge_{i<j} x_i \neq x_j \wedge \bigwedge_i \neg P(x_i) \rightarrow \bigwedge_{i=1}^{n} P(y_i)$$
$$\wedge \bigwedge_{i<j} \neg y_i E y_j$$
$$\wedge \bigwedge_{i,j} (x_i R y_j)^{\text{if } (i \in w(j))}\Big],$$

$$(\forall y_1, \ldots, y_n)(\exists x_1, \ldots, x_n)\Big[\bigwedge_{i<j} \neg y_i E y_j \wedge \bigwedge_i P(y_i) \rightarrow \bigwedge_{i=1}^{n} \neg P(x_i)$$
$$\wedge \bigwedge_{i<j} x_i \neq x_j$$
$$\wedge \bigwedge_{i,j} (x_i R y_j)^{\text{if } [i \in w(j)]}\Big].$$

It is easy to see that T has elimination of quantifiers, and when M is a model of T, $A = P(M)$, $p = \{\neg P(x)\}$ satisfies our demands.

THEOREM 7.10: (1) $\kappa r_{\text{ird}}(T) = \kappa r^1_{\text{ird}}(T) = \kappa r^m_{\text{ird}}(T)$.

(2) $\kappa r_{\text{cdt}}(T) = \kappa r^1_{\text{cdt}}(T) = \kappa r^m_{\text{cdt}}(T)$.

Proof. (1) We can use 7.4 and the independence proofs of set theory or prove like II, 4.16.

(2) As $\kappa r^1_{cdt}(T) \le \kappa r^m_{cdt}(T)$, it suffices to prove by induction on m that $\kappa r^m_{cdt}(T) \le \kappa r^1_{cdt}(T)$. For $m = 1$ it is trivial, so suppose we have proved it for m and $\kappa < \kappa r^{m+1}_{cdt}(T)$ is regular; we shall prove $\kappa < \kappa r^1_{cdt}(T)$. So there are formulas $\varphi_\alpha(y, \bar{x}; \bar{z}_\alpha)$ $(l(\bar{x}) = m)$ $(0 < \alpha < \kappa$, α successor) and $\bar{a}_\eta$ $(\eta \in {}^\kappa\lambda)$ as in Definition 7.2; and $n_i = 2$, and $\{\bar{a}_\eta\colon \eta \in {}^{\kappa>}\lambda\}$ is an indiscernible tree (see Definitions VII, 2.4, VII, 3.1, VII, 3.2, Theorems VII, 3.5, VII, 3.6) and $\lambda > |T|$ is a strong limit cardinal of cofinality κ. Choose a maximal set $w \subseteq \kappa$ such that Γ_w is consistent, where $\Gamma_w = \{\varphi_\alpha(y_\eta, \bar{x}; \bar{a}_{\eta|\alpha})$ $0 < \alpha < \kappa$ a successor, $\eta \in {}^\kappa\omega$, and $\gamma + 1 \notin w \Rightarrow \eta[\gamma] = 0\}$ (note that if $w(i)(i < \delta)$ is increasing, then $\Gamma_{w(i)}$ is increasing and $\Gamma_{\bigcup_i w(i)}$ is consistent iff $\bigcup_i \Gamma_{w(i)}$ is). If $|w| = \kappa$ clearly $\{\varphi_\alpha(y; \bar{x}, \bar{z}_\alpha)\colon \alpha \in w\}$ proves our assertion. Otherwise let $\beta_0 < \kappa$, $(\forall \alpha \in w)(\alpha < \beta_0)$, and for every $\eta \in {}^\kappa\lambda$ let $(\bar{0} = \langle 0, \ldots \rangle_{i<\alpha_0})$

$$\Gamma^\eta = \{\varphi_\alpha(y_{\nu\frown\eta}, \bar{x}; \bar{a}_{(\nu\frown\eta)|\alpha})\colon 0 < \alpha < \kappa,\ \alpha \text{ successor},\ \nu \in {}^\beta\omega,$$
$$[\gamma + 1 \notin w \text{ and } \gamma < \beta_0] \Rightarrow \nu[\gamma] = 0\},$$
$$r^\eta = \{(\exists y_{\nu_1\frown\eta_1}, \ldots) \bigwedge p\colon p \subseteq \Gamma^\eta \text{ is finite}\}.$$

Now no sequence $\bar{c}$ realizes $> 2^{|T|}$ of the types r^η. [Otherwise by 2.8 of the Appendix there are η_l $(l < \omega)$ and $\gamma < \kappa$ such that $\eta_l \upharpoonright \gamma = \eta_0 \upharpoonright \gamma$, and the $\eta_l[\gamma]$ are distinct. Then $w \cup \{\gamma + 1\}$ contradicts the maximality of w.]

DEFINITION 7.3: (1) $\kappa^m_{inp}(T)$ is the first cardinal κ, such that there are no formulas $\varphi_i(\bar{x}; \bar{y}_i)$ $(i < \kappa)$ $(l(\bar{x}) = m)$ and natural numbers n_i, and sequences $\bar{a}^i_\alpha(\alpha < \omega)$ such that

(i) for any $\eta \in {}^\kappa\omega$, $\{\varphi_i(\bar{x}; \bar{a}^i_{\eta[i]})\colon i < \kappa\}$ is consistent,

(ii) for all $i < \kappa$ $\{\varphi_i(\bar{x}; \bar{a}^i_\alpha)\colon \alpha < \omega\}$ is n_i-contradictory.

(2) $\kappa_{inp}(T) = \sup_m \kappa^m_{inp}(T)$.

(3) $\kappa r^m_{inp}(T)$ $[\kappa r_{inp}(T)]$ is the first regular cardinal $\ge \kappa^m_{inp}(T)$ $[\kappa_{inp}(T)]$.

Remark. The notation "inp" stands for "independent partitions".

DEFINITION 7.4: (1) $\kappa^m_{srd}(T)$ is the first cardinal κ such that there are no formulas $\varphi_i(\bar{x}; \bar{y}_i)$ $(i < \kappa)$ $(l(x) = m)$ and sequences $\bar{a}^i_\alpha(\alpha < \omega)$ such that

(i) for $\eta \in {}^\kappa\omega$ $\{\varphi_i(\bar{x}; \bar{a}^i_\alpha)^{\text{if } (\eta[i] \ge \alpha)}\colon i < \kappa,\ \alpha < \omega\}$ is consistent,

(ii) for $i < \kappa$, $\alpha < \omega$, $\vDash (\forall \bar{x})[\varphi_i(\bar{x}; \bar{a}^i_\alpha) \to \varphi_i(\bar{x}; \bar{a}^i_{\alpha+1})]$.

(2) $\kappa_{srd}(T) = \sup_m \kappa^m_{srd}(T)$.

Remark. The notation "srd" means "strict independent order".

THEOREM 7.11: *Suppose $\kappa_{\text{cdt}}(T) = \infty$, $\kappa_{\text{inp}}(T) < \infty$ then there is a formula $\varphi(x;\bar{y})$ and sequences $\bar{a}_\eta$ ($\eta \in {}^{\omega>}\omega$) such that* (i) $\{\varphi(x;\bar{a}_{\eta|n}): n < \omega\}$ *is consistent for each $\eta \in {}^{\omega}\omega$* (ii) *if $\eta, \nu \in {}^{\omega>}\omega$ are $\triangleleft$-incomparable, then* $\vDash\neg(\exists x)[\varphi(x;\bar{a}_\eta) \wedge \varphi(x;\bar{a}_\nu)]$.

Proof. By the assumption $\kappa_{\text{cdt}}(T) = \infty$, the proof of 7.6(2), 7.10(2), and 7.7(2), there are a formula $\varphi(x;\bar{y})$ and sequences $\bar{a}_\eta$ ($\eta \in {}^{\omega>}\omega$) such that

(α) $\{\varphi(x;\bar{a}_{\eta|n}): n < \omega\}$ is consistent for each $\eta \in {}^{\omega}\omega$,

(β) $\vDash\neg(\exists x)[\varphi(x,\bar{a}_{\eta\frown\langle l\rangle}) \wedge \varphi(x,\bar{a}_{\eta\frown\langle k\rangle})]$ for $k \neq l < \omega$, $\eta \in {}^{\omega>}\omega$. By Definitions VII, 2.4, VII, 3.1 and Theorems VII, 3.5, VII, 3.6, we can assume

(γ) $\{\bar{a}_\eta: \eta \in {}^{\omega>}\omega\}$ is an indiscernible tree, i.e., $\text{tp}(\bar{a}_{\eta_1}\frown\cdots\frown\bar{a}_{\eta_k}, \emptyset)$ depends on the lexicographic order among the η_i's, and $l(\eta_i)$, $h(\eta_i,\eta_j) = \max\{l: \eta_i \restriction l = \eta_j \restriction l\}$. By Ramsey's theorem, the fact that we can take any subset of the levels and compactness we can assume

(δ) in (γ) only the order relations among the $l(\eta_i)$, $h(\eta_i,\eta_j)$ matter.

For $0,\ldots,0$ (k times) we write 0_k. Now $\{\varphi(x;\bar{a}_{\langle l,0_{l+1}\rangle}): l < \omega\}$ cannot be consistent (as then $\{\{\varphi(x;\bar{a}_{\langle l,0_l,k\rangle}): k < \omega\}: l < \omega\}$ proves $\kappa_{\text{inp}}(T) = \infty$). Hence there is a maximal n for which

$$\{\varphi(x;\bar{a}_{\langle l,0_{l+1}\rangle}): l < n\} \cup \{\varphi(x;\bar{a}_{\langle n,0_k\rangle}): k > n\}$$

is consistent. Hence there is $m < \omega$ such that

$$\{\varphi(x;\bar{a}_{\langle l,0_{l+1}\rangle}): l < n\} \cup \{\varphi(x,\bar{a}_{\langle n,0_{n+1}\rangle})\}$$
$$\cup\, \{\varphi(x;\bar{a}_{\langle n+1,0_k\rangle}): n+2 \le k < n+m+2\}$$

is inconsistent. Now define, for every $\eta \in {}^{\omega>}\omega$

$$h(\eta) = \langle n, 0_{n+1}, \eta[0], 0_m, \eta[1], 0_m, \ldots, \eta[l(\eta)-1], 0_m\rangle,$$
$$\varphi^*(x;\bar{a}^*_\eta) = \bigwedge_{l<n} \varphi(x;\bar{a}_{\langle l,0_{l+1}\rangle}) \wedge \bigwedge_{l<m} \varphi(x;\bar{a}_{h(\eta)\frown 0_l}).$$

It is easy to check that:

(1) For every $\eta \in {}^{\omega>}\omega$, $\{\varphi^*(x;\bar{a}^*_{\eta|l}): l < \omega\}$ is consistent (by the choice of n).

(2) If $\eta, \nu \in {}^{\omega>}\omega$, $l(\eta) < l(\nu)$, $\eta <^{lx} \nu$ (lexicographic order) but not $\eta \triangleleft \nu$, then $\vDash\neg(\exists x)[\varphi^*(x;\bar{a}^*_\eta) \wedge \varphi^*(x;\bar{a}^*_\nu)]$ (remember that the indiscernibility applies to levels too, and the definitions of m and n).

So the $\varphi^*(x;\bar{a}^*_\eta)$ satisfy "half" of our demand (ii) (and (i), of course). By compactness we can define $\bar{a}^1_\eta$ ($\eta \in {}^{\omega>}\omega$) which satisfy ($\alpha$), ($\beta$), ($\gamma$), ($\delta$) and when $\eta, \nu \in {}^{\omega>}\omega$, $l(\eta) < l(\nu)$, $\nu <^{lx} \eta$ but not $\eta \triangleleft \nu$ then $\vDash\neg(\exists x)[\varphi^*(x;\bar{a}^1_\eta) \wedge \varphi^*(x;\bar{a}^1_\nu)]$ (by reversing order).

Now repeat the process and we get φ^{**} as required.

THEOREM 7.12: *If T does not have the independence property, then $\kappa^m_{\text{srd}}(T) = \kappa^m_{\text{ird}}(T)$.*

Proof. By definition, always $\kappa^m_{\text{srd}}(T) \le \kappa^m_{\text{ird}}(T)$, so it suffices to prove $\kappa < \kappa^m_{\text{srd}}(T)$ assuming $\kappa < \kappa^m_{\text{ird}}(T)$, and let $\varphi_i(\bar{x}; \bar{z}_i)$ $(i < \kappa)$ exemplify the latter. We define by induction on ζ formulas $\psi_i(\bar{x}, \bar{y}_i)$ $(i < \zeta)$ such that, letting

$$\theta_i(x; \bar{y}_i) = \begin{cases} \psi_i(\bar{x}; \bar{y}_i) & \text{if } i < \zeta, \\ \varphi_i(\bar{x}; \bar{z}_i) & \text{if } i \ge \zeta, \end{cases}$$

(*) there are $\bar{a}^i_\alpha$ $(\alpha < \omega, i < \kappa)$ such that:

(i) for every $\eta \in {}^\kappa\omega$, $\bigcup_{i<\kappa} p^i_{\eta[i]}$ is consistent where

$$p^i_\beta = \{\theta_i(\bar{x}; \bar{a}^i_\alpha)^{\text{if } (\beta \ge \alpha)} : \alpha < \omega\},$$

(ii) $\psi_i(\bar{x}; \bar{a}^i_\alpha) \vdash \psi_i(\bar{x}; \bar{a}^i_{\alpha+1})$ for $i < \zeta$,

(iii) $\{\bar{a}^i_\alpha : \alpha < \omega\}$ is an indiscernible sequence over $\bigcup_{j\ne i} \bigcup_{\beta<\omega} \bar{a}^j_\beta$.

Clearly $(*)_0$ holds, and $(*)_\kappa$ suffice and $(*)_\delta$ follows from $(*)_\zeta$ $(\zeta < \delta)$. So suppose $(*)_\zeta$, and we shall prove $(*)_{\zeta+1}$. As T does not have the independence property, for some $n < \omega$ and finite $w \subseteq n$, and $\eta \in {}^\kappa\omega$,

$$\bigcup_{\substack{i<\kappa \\ i\ne\zeta}} p^i_{\eta[i]} \cup \{\varphi_i(\bar{x}; \bar{a}^i_\alpha)^{\text{if } (\alpha\in w)} : \alpha < n\}$$

is inconsistent. So as in the proof of II, 4.7(2), there are $u, v \subseteq n$, $k < n$ such that $(k + 1) \in u - v$, $k \in v - u$, $u - \{k + 1\} = v - \{k\}$, and for every $\eta \in {}^\kappa\omega$

$$\bigcup_{\substack{i<\kappa \\ i\ne\zeta}} p^i_{\eta[i]} \cup \{\varphi_i(\bar{x}; \bar{a}^i_\alpha)^{\text{if } (\alpha\in v)} : \alpha < n\}$$

is consistent but for some $\rho \in {}^\kappa\omega$,

$$\bigcup_{\substack{i<\kappa \\ i\ne\zeta}} p^i_{\rho[i]} \cup \{\varphi_i(x; a^i_\alpha)^{\text{if } (\alpha\in u)} : \alpha < \omega\}$$

is inconsistent and we can replace $\bigcup_{i<\kappa, i\ne\zeta} p^i_{\rho[i]}$ by a finite subset q. We continue as in II-4.7(2), conjuncting ψ with $\bigwedge q$. We let $\psi_\zeta = \psi$, and $(*)_{\zeta+1}$ is satisfied because by the indiscernibility assumption for any $k < \omega$, we can assume

$$q \subseteq \{\varphi_i(x, \bar{a}^i_\alpha)^{\text{if } (\rho(i)\ge\alpha)} : i \ne \zeta, \alpha < \omega; \text{ but not } l(i) \le \alpha < l(i) + k\}.$$

EXERCISE 7.1: Prove that $\kappa^m_{\text{inp}}(T) \le \kappa^m(T)$, $\kappa^m_{\text{inp}}(T) \le \kappa r^m_{\text{cdt}}(T)$. Show that even for stable T equality need not hold.

EXERCISE 7.2: Prove that when T does not have the independence property, $\kappa^m_{\text{inp}}(T) \ge \kappa^m_{\text{srd}}(T)$.

EXERCISE 7.3: Suppose T does not have the independence property; and there are κ φ_i, $\bar{a}^i_\alpha$, $i < \kappa$, $\alpha < \omega$ such that (where $l(\bar{x}) = m$) for any $\eta \in {}^\kappa\omega$, $\{\varphi_i(\bar{x}; \bar{a}^i_\alpha)^{\text{if } (\alpha = \eta(i))}: i < \kappa, \alpha < \omega\}$ is consistent. Prove $\kappa < \kappa^m_{\text{inp}}(T)$.

EXERCISE 7.4: Prove that in Definition 7.3 we can w.l.o.g. take $n_i = 2$.

QUESTION 7.5: Prove that $\kappa_{\text{inp}}(T) = \kappa^m_{\text{inp}}(T) = \kappa^1_{\text{inp}}(T)$.

QUESTION 7.6: Is $K^m(\lambda) = K^1(\lambda)$?

EXERCISE 7.7: Suppose $A \subseteq B$, $p \in S^m(A)$. T does not have the independence property. Then there are $q \in S^m(B)$, $q \supseteq p$ and $C \subseteq B$, $|C| < \kappa^m_{\text{inp}}(T)$ such that: if $\{\bar{a}_n : n < \omega\}$ is an indiscernible sequence over $A \cup C$, and $\bar{a}_n \in B$, then $\varphi(\bar{x}; \bar{a}_0) \in q \Leftrightarrow \varphi(\bar{x}; \bar{a}_1) \in q$.

Remark. Instead of "T does not have the independence property" we can add "where $\{\varphi(\bar{x}; \bar{a}_n) \equiv \neg\varphi(\bar{x}; \bar{a}_{n+1}) : n < \omega\}$ is k-contradictory for some $k < \omega$".

EXERCISE 7.8: Prove Theorem 3.8 by the method of the proof of 7.3 (changing the definition of F_α).

QUESTION 7.9: Investigate $\kappa^m_{\text{srd}}(T)$. If T does not have the independence property is $\kappa_{\text{srd}}(T) = \kappa_{\text{ird}}(T)$, at least when both are regular cardinals?

PROBLEM 7.10: Characterize the possible functions $K(\lambda)$, and $K(\lambda, \kappa) = \sup\{|S|^+ : S$ is a set of 1-types, contradictory in pairs; $p \in S \Rightarrow |p| < \kappa$; and the types in S are over some A, $|A| \le \lambda\}$. (Possibility: Maybe the function $h(\lambda, \bar{w})$ should be used, where $\bar{w} = \langle w_i : i < \kappa\rangle$, w_i a subset of $\{j : j < i\}$ and $h(\lambda, \bar{w}) = \sup\{|I \cap {}^\kappa\lambda|^+ : I \subseteq {}^{\kappa \ge}\lambda$; I closed under initial segments; and for $i < \kappa$, $|\{\eta \restriction w_i : \eta \in I\}| \le \lambda\}$ (remember η is a function with domain $\{j : j < i\}$).)

EXERCISE 7.11: Suppose in $\mathfrak{C}^{eq}$, in $\operatorname{acl}(A \cup \bar{a})$ there is I independent over A, assuming T is stable (see Definition 4.4). Prove $\kappa_{\rm inp}(T) > |I|$.

EXERCISE 7.12: There is a theory T without the strict order property, $\kappa_{\rm cdt}(T) = \infty$, $\kappa_{\rm inp}(T) < \infty$. Hint: Let T consist of:

(1) P, Q form a partition. $(\forall x)(P(x) \equiv \neg Q(x))$.

(2) R is a symmetric relation on Q, $\in$ on $P \times Q$: $\forall x(\neg xRx)$, $(\forall xy)(xRy \equiv yRx)$, $(\forall xy)(xRy \to Q(x) \wedge Q(y))$, $(\forall xy)(x \in y \to P(x) \wedge Q(y))$.

(3) xRy implies x, y are disjoint: $(\forall xyz)(xRy \wedge z \in x \to \neg z \in y)$.

(4) $(\forall xyz)(xRy \wedge Q(z) \to xRz \vee yRz)$.

(5) There is an x satisfying a quantifier free formula, except when (1)–(4) prevent it, e.g.,

$$\left[(\exists z)\left(\bigwedge_{l<n} (z \in x_l)^{\text{if } (l \in w)} \wedge \bigwedge_{l<m} z \neq y_l\right)\right] \equiv \left[\bigwedge_{l} Q(x_l) \wedge \bigwedge_{l,n \in w} \neg x_l R x_n\right].$$

QUESTION 7.13: Can we in 7.10(1) replace κr by κ?

DEFINITION 7.5: Let $\kappa^m_{\rm sct}(T)$ be the first cardinal κ for which we cannot find an m-formula $\varphi_\alpha(\bar{x}; \bar{y}_\alpha)$ $(\alpha < \kappa)$ and sequences $\bar{a}_\eta$ $(\eta \in {}^{\kappa >}\omega)$ such that:

(i) For every $\eta \in {}^\kappa\omega$ $\{\varphi_\alpha(\bar{x}; \bar{a}_{\eta|\alpha}): 0 < \alpha < \kappa,\ \alpha \text{ a successor}\}$ is consistent.

(ii) If $\eta \in {}^\alpha\omega$, $\nu \in {}^\beta\omega$, $\alpha, \beta < \kappa$ are successors, and η, ν are $\triangleleft$-incomparable then $\vDash \neg(\exists \bar{x})(\varphi_\alpha(\bar{x}; \bar{a}_\eta) \wedge \varphi_\beta(\bar{x}; \bar{a}_\nu))$.

We let $\kappa_{\rm sct}(t) = \sup_m \kappa^m_{\rm sct}(T)$, $\kappa{\rm r}^m_{\rm sct}(T)$ $[\kappa{\rm r}_{\rm sct}(T)]$ be the first regular cardinal $\geq \kappa^m_{\rm sct}(T)$ $[\kappa_{\rm sct}(T)]$.

QUESTION 7.14: Investigate $\kappa^m_{\rm sct}(T)$, e.g., is $\kappa^1_{\rm sct}(T) = \kappa_{\rm sct}(T)$ is $\kappa{\rm r}_{\rm cdt}(T) = \kappa{\rm r}_{\rm inp}(T) + \kappa{\rm r}_{\rm sct}(T)$?

EXERCISE 7.15: Show that if $\kappa_{\rm ctd}(T) < \infty$, then 1.3 holds.

CHAPTER IV

PRIME MODELS

IV.0. Introduction

We give here an axiomatic way to construct prime models (Sections 1 and 3) and prove that some cases satisfy the axioms (Section 2) and then give more information (mainly characterization (Section 4)) under stability assumptions. We first examine here the most common case.

So let T be a countable theory. We call M prime $=(\mathbf{F}^t_{\aleph_0}$-prime) over A if $A \subseteq |M|$ and whenever $A \subseteq |N|$, there is an elementary mapping $f: M \to N$, $f \restriction A =$ the identity. So, in a sense, a prime model "contains" only what every model should contain. It is easy to see that the sequences algebraic over A should be in the prime model, and no other elements should be in it. But, as we are considering elementary mappings, it is natural to look at the complete types over A which sequences from M realize. Clearly if M is prime over A, $p \in S^m(A)$ is realized in M iff it is realized in every model N, $A \subseteq |N|$. Which are those types? Clearly they include all the isolated types (p is isolated if for some $\varphi(\bar{x}; \bar{a}) \in p$, $\varphi(\bar{x}; \bar{a}) \vdash p$), and we prove (5.3) that for countable A the converse holds too. On the other hand if $\bar{a} \in A$, $\vDash (\exists x)\varphi(x, \bar{a})$, there should be an element c realizing $\varphi(x, \bar{a})$ in M. So we prove (5.10) that a necessary condition for the existence of prime models over every A, is that for any A, $\bar{a} \in A$, and consistent $\varphi(x, \bar{a})$ there is an isolated q, $\varphi(x, \bar{a}) \in q \in S(A)$. On the other hand this is sufficient, for then we can define inductively c_i ($i < \alpha_0$) such that $\mathrm{tp}(c_i, A \cup \{c_j : j < i\})$ is isolated, and $A \cup \{c_i : i < \alpha_0\}$ is the universe of a model M (such a model is called primary and the sequence a construction). If $A \subseteq |N|$, we can easily define $f: M \to N$, by defining $f(c_i)$ inductively. Naturally after proving existence, we want to prove uniqueness. For the primary model it is not too difficult; but for the prime model it fails in general (see [Sh 79a]).

Here we come again to stability. If T is $\aleph_0$-stable, we can prove that over every A a prime model exists (by 2.15(4), 2.17, 3.12) and characterize the prime model by "local" conditions, and prove its uniqueness (M is prime over A iff $A \subseteq |M|$, M is atomic over A (i.e., for each $\bar{c} \in |M|$, $\text{tp}(\bar{c}, A)$ is isolated) and there is no uncountable set $\subseteq M$ indiscernible over A). For superstable T, we cannot prove the existence of prime models, but if one always exists we can characterize it similarly (see 4.18). For stable T we cannot prove existence nor (natural) characterization, but when prime models always exist (or even if over A there is a primary model) the prime model is unique (5.6).

Those uniqueness theorems reprove the uniqueness of the algebraic closure of a field (which is a trivial application as this was the classical example model theorists had in mind), but also prove the uniqueness of the differential closure of a differential field (for characteristic zero we have an $\aleph_0$-stable theory, for other characteristics we have a stable theory). However the algebraic closure has an additional property: minimality. We call M minimal over A if $A \subseteq |M|$ and $A \subseteq |N| \subseteq |M|$ implies $N = M$. In general the prime model need not be minimal (e.g., when T is the theory of infinite models, with the equality sign only). However for $\aleph_0$-stable T, a minimal model N over A is necessarily prime over A; and a prime model over A is minimal iff in M there is no infinite indiscernible set over A.

Now a natural generalization of "prime model" is "model prime among the λ-saturated or λ-compact models", with suitable generalizations of isolated types. However for stable T, the notion $\mathbf{F}^a_\lambda$-saturated (= every type almost over a set of cardinality $< \lambda$ is realized) is better than λ-saturated. More exactly: for $\lambda > \aleph_0$ we get the same saturation, but the definition of isolated satisfies more axioms, insuring existence when $\lambda \geq \kappa(T)$ and helping to prove uniqueness. A fourth example is the following construction, analogous to the construction of primary models: if T is stable and countable, for every A there is M, $A \subseteq |M|$, such that for every $\bar{c} \in |M|$, $\varphi \in \mathrm{L}$ there is $\psi(\bar{x}; \bar{b}) \in \text{tp}(\bar{c}, A)$, $\psi(\bar{x}; \bar{b}) \vdash \text{tp}_\varphi(\bar{c}, A)$ (this is helpful for two cardinal theorems for stable theories, see V, Section 6).

Trying to develop each of these cases separately will cause much repetition, so some general setting is needed. We suggest here one, letting $\mathbf{F}$ be the set of pairs (p, B) where $p \in S^m(A)$, $B \subseteq A$, p is "isolated" over B in some proper sense (the main example was $\mathbf{F}^t_{\aleph_0}$ where $\mathbf{F}^t_\lambda = \{(p, B)$: there is $q \subseteq p \restriction B$, $|q| + |B| < \lambda$, $q \vdash p\}$). So an $\mathbf{F}$-construction over A is $\langle A, \langle \bar{c}_i : i < \alpha \rangle, \langle B_i : i < \alpha \rangle\rangle$, where $(\text{tp}(\bar{c}_i, A \cup \{\bar{c}_j : j < i\}), B_i) \in \mathbf{F}$; and we define naturally $\mathbf{F}$-saturated, -primary, -atomic.

In Section 1 we list the axioms, and give some basic definitions and trivial lemmas. We do not claim this set to be definitive, it is enough for us if it is quite useful; and it covers more cases than it was originally tailored for. More exactly, though we have "p is isolated over B" in mind, what we get is applicable to "p is definable over B". So in Section 2 we prove that certain $\mathbf{F}$'s satisfy most axioms (sometimes under additional hypotheses). In addition to the four previous cases, we deal with $\mathbf{F}^{f}_{\lambda} = \{(p, B): p$ does not fork over $B\}$ for stable T used in 5.15 and VIII, 2.7. In general, we prove that existence axioms hold using various stability assumptions. We sum up our results in a table.

In Section 3, we investigate the conclusions we can draw from the axioms, mentioning each time exactly on which axioms we rely (this is important as some of the examples fail to satisfy some of the axioms). We shall mention here only the uniqueness of the $\mathbf{F}$-primary model and its $\mathbf{F}$-atomicity.

In Section 4 we concentrate on stable theories and prove characterization and uniqueness theorems. The way is to find some properties of prime models, proved via their being included in primary models; and show that any two models having these properties, are isomorphic, e.g., for $\mathbf{F}^{a}_{\lambda}$-primeness we prove it when $\operatorname{cf}\lambda \geq \kappa(T)$; partitioning it to two cases ($\operatorname{cf}\lambda > \aleph_0$, $\operatorname{cf}\lambda = \aleph_0$) covered by two theorems with quite different proofs. (For $\operatorname{cf}\lambda = \aleph_0$, $\kappa(T) = \aleph_0$, so we can use induction on ranks; for $\operatorname{cf}\lambda > \aleph_0$ we use a different argument which fails when $\operatorname{cf}\lambda = \aleph_0$.) An important technical lemma is that if $p \in S^m(A)$ is $\mathbf{F}^{t}_{\lambda}$-isolated T stable, p does not fork over $B \subseteq A$ then $p \restriction B$ is $\mathbf{F}^{t}_{\lambda}$-isolated too. The case $\lambda \geq \kappa(T) > \operatorname{cf}\lambda$ remains obscure (we do not know even if $\mathbf{F}^{t}_{\lambda}$-primary models are $\mathbf{F}^{t}_{\lambda}$-atomic). The last section, 5, deals with various theorems. The most important theorem is the uniqueness theorem. In the characterization theorems we demand mainly $\operatorname{cf}\lambda \geq \kappa(T)$, in 5.6 for regular λ this can be replaced by $\lambda^{+} \geq \kappa(T)$. We use $\mathbf{F}^{f}_{\lambda}$ to construct models with an absolute indiscernible set (if $\lambda \geq \mu$, T stable in μ, then there is $I \subseteq |M|$, $|I| = \|M\| = \lambda$, I absolutely indiscernible) and investigate the possibilities of omitting types (e.g., if $|T| = \lambda$ is regular, $p_i \in D(T)$ are not $\mathbf{F}^{t}_{\lambda}$-isolated, ($i < i(0) < 2^{\lambda}$) then T has a model omitting every p_i (see 5.17)).

IV.1. The set of axioms

We give here a list of axioms for the concept of primeness, and mention some obvious connections. We will not assume all the axioms hold all the time, but specify each time the axioms used in proving a certain claim.

F will denote a family of pairs (p, A) where $p \in S^m(A')$ $(m < \omega)$ for some A', $A \subseteq A'$. We assume that **F** is closed under changing the names of the variables and their order, that is, if $p \in S^m(B)$, $A \subseteq B$, and $q = \{\varphi(y_0, \ldots, y_{m-1}; \bar{a}) : \varphi(x_{\sigma(0)}, \ldots, x_{\sigma(m-1)}; \bar{a}) \in p\}$ (σ a permutation of $\{0, \ldots, m-1\}$) then $(p, A) \in \mathbf{F} \Leftrightarrow (q, A) \in \mathbf{F}$. We let $\lambda(\mathbf{F})$ $[\lambda_r(\mathbf{F})]$ be the first [regular] cardinal λ (if it exists), such that $(p, A) \in \mathbf{F} \Rightarrow |A| < \lambda$. For any λ let $\mathbf{F}_\lambda = \{(p, A) : |A| < \lambda, (p, A) \in \mathbf{F}\}$. We call a type p *F-isolated over A* if $(p, A) \in \mathbf{F}$; p is **F**-isolated if there is A such that $(p, A) \in \mathbf{F}$.

Let $p \in \mathbf{F}(B)$ mean $(p, B) \in \mathbf{F}$. In order to get an intuitive feeling for the following axioms think of $(p, B) \in \mathbf{F}$ as meaning that the type p is isolated (in the usual sense) over B.

The Axioms

Isomorphism. Ax(I): **F** is mapped onto itself by any automorphism of $\mathfrak{C}$, i.e., for any pair (p, A) and automorphism F of $\mathfrak{C}$, $(p, A) \in \mathbf{F} \Leftrightarrow (F(p), F(A)) \in \mathbf{F}$.

Trivially Isolated Types

Ax(II.1): If $\bar{a} \in B \subseteq A$, $|B| < \lambda(\mathbf{F})$, $p = \mathrm{tp}(\bar{a}, A)$, then $(p, B) \in \mathbf{F}$.

Ax(II.2): If $B \subseteq A$, $\bar{a}$ is definable over B, $|B| < \lambda(\mathbf{F})$, $p = \mathrm{tp}(\bar{a}, A)$, then $(p, B) \in \mathbf{F}$.

Ax(II.3): If $\bar{a}$ is algebraic over A, $p = \mathrm{tp}(\bar{a}, A)$, then $(p, B) \in \mathbf{F}$ for some finite $B \subseteq A$.

Ax(II.4): If $\bar{c} \in B \subseteq A$, $\varphi(\bar{x}; \bar{c}) \equiv p = \mathrm{tp}(\bar{a}, A)$, $|B| < \lambda(\mathbf{F})$, then $(p, B) \in \mathbf{F}$.

Monotonicity

Ax(III.1): If $B \subseteq A \subseteq \mathrm{Dom}\, p$, $(p, B) \in \mathbf{F}$, then $p \restriction A \in \mathbf{F}(B)$.

Ax(III.2): If $B \subseteq A \subseteq \mathrm{Dom}\, p$, $|A| < \lambda(\mathbf{F})$, $(p, B) \in \mathbf{F}$ then $(p, A) \in \mathbf{F}$.

Ax(IV): If $q = \mathrm{tp}(\bar{a}^\frown\bar{b}, A)$, $p = \mathrm{tp}(\bar{a}, A)$, $B \subseteq A$ and $(q, B) \in \mathbf{F}$ then $(p, B) \in \mathbf{F}$.

Notation. If $p = \mathrm{tp}_*(C, A)$, $B \subseteq A$, then we say $(p, B) \in \mathbf{F}$ when for every $\bar{c} \in C$, $\mathrm{tp}(\bar{c}, A) \in \mathbf{F}(B)$. We say $\mathrm{tp}_*(C, A)$ is **F**-isolated if $\mathrm{tp}(\bar{c}, A)$ is **F**-isolated for every $\bar{c} \in C$. Note that when Ax(IV) holds, $\mathrm{tp}_*(\bar{c}, A) \in \mathbf{F}(B)$ iff $\mathrm{tp}(\bar{c}, A) \in \mathbf{F}(B)$.

Symmetry and Transitivity

Ax(V.1): If $q = \mathrm{tp}(\bar{a}^\frown\bar{b}, A)$, $p = \mathrm{tp}(\bar{a}, A \cup \bar{b})$ and $B \subseteq A$, and $(q, B) \in \mathbf{F}$ then $(p, B \cup \bar{b}) \in \mathbf{F}$.

Ax(V.2): If $|C| < \lambda(\mathbf{F})$, $q = \mathrm{tp}_*(\bar{a} \cup C, A)$, $p = \mathrm{tp}(\bar{a}, A \cup C)$, $B \subseteq A$ and $(q, B) \in \mathbf{F}$ then $(p, B \cup C) \in \mathbf{F}$.

Ax(VI): If $B, C \subseteq A$, $\mathrm{tp}(\bar{b}, A \cup \bar{a}) \in \mathbf{F}(C)$ and $\mathrm{tp}(\bar{a}, A) \in \mathbf{F}(B)$ then $\mathrm{tp}(\bar{a}, A \cup \bar{b}) \in \mathbf{F}(B)$.

Ax(VII): If $B \subseteq A$, $\mathrm{tp}(\bar{a}, A \cup C) \in \mathbf{F}(B \cup C)$, and $\mathrm{tp}_*(C, A) \in \mathbf{F}(B)$, then $\mathrm{tp}_*(\bar{a} \cup C, A) \in \mathbf{F}(B)$.

Continuity

Ax(VIII): If A_i $(i < \delta)$ is increasing with i, $p \in S^m(\bigcup_{i<\delta} A_i)$ and $p \restriction A_i \in \mathbf{F}(B)$ for $i < \delta$, then $p \in \mathbf{F}(B)$.

Ax(IX): If A_i, B_i $(i < \delta)$ are increasing with i, $\delta < \mathrm{cf}\,\lambda(\mathbf{F})$, $p \in S^m(\bigcup_{i<\delta} A_i)$ and $p \restriction A_i \in \mathbf{F}(B_i)$, then $(p, \bigcup_{i<\delta} B_i) \in \mathbf{F}$.

Existence

Ax(X.1): If $\bar{a} \in A$, $\models (\exists x)\varphi(x; \bar{a})$, then for some pair $(p, B) \in \mathbf{F}$, $B \subseteq A$, $\varphi(x; \bar{a}) \in p \in S(A)$.

Ax(X.2): If $A_i \subseteq A_j$ for $i < j \leq \alpha$, $\bar{a} \in A_0$, $\models (\exists x)\varphi(x; \bar{a})$, then for some $p \in S(A_\alpha)$, $\varphi(x; \bar{a}) \in p$, and $p \restriction A_i$ is $\mathbf{F}$-isolated for every $i \leq \alpha$.

Ax(XI.1): If $(p, B) \in \mathbf{F}$, $p \in S^m(A)$, $A \subseteq C$, then there are $B' \subseteq C$, and $q \in S^m(C)$ such that $p \subseteq q$, $(q, B') \in \mathbf{F}$.

Ax(XI.2): If $A_i \subseteq A_j$ for $i < j \leq \alpha$ and $p \in S^m(A_0)$ is $\mathbf{F}$-isolated, then for some $q \in S^m(A_\alpha)$ $p \subseteq q$, and $q \restriction A_i$ is $\mathbf{F}$-isolated for every $i \leq \alpha$.

Generalized Transitivity

Ax(XII): If $C_1, C_2 \subseteq B \subseteq A, p \in S^m(A), (p \restriction B, C_1) \in \mathbf{F}$, and $(p, C_2) \in \mathbf{F}$, then $(p, C_1) \in \mathbf{F}$.

Remark. We make no use of Ax(XII) and use Ax(X.2) and (XI.2) rarely.

LEMMA 1.1: (1) Ax(II.4) *implies* Ax(II.3) *and* Ax(II.2); Ax(II.2) *implies* Ax(II.1). *If* $\lambda(\mathbf{F}) = \aleph_0$, Ax(X.1) *implies* Ax(II.3).

(2) Ax(V.2) *and* (IV) *imply* Ax(V.1).

(3) Ax(II.1) *and* Ax(VII) *imply* Ax(IV).

(4) *If* Ax(V.1), (IX) *hold, then* Ax(V.2) *holds for* $|C| < \mathrm{cf}\,\lambda(\mathbf{F})$ *so it holds when* $\lambda(\mathbf{F})$ *is regular.*

(5) *Suppose* Ax(X.1) *holds for finite A. Then* Ax(XI.1) *implies* Ax(X.1), *and* Ax(XI.2) *implies* Ax(X.2).

(6) Ax(X.2) *implies* Ax(X.1); *and* Ax(XI.2) *implies* Ax(XI.1).

(7) *If* Ax(IX) *holds, then* Ax(XI.1) *implies that* Ax(XI.2) *holds for* $\alpha < \mathrm{cf}\,\lambda(\mathbf{F})$.

Remark. Except for trivial cases Ax(VII) implies Ax(IV) (as we may take $C \subseteq B$ assuming Ax(II.1)).

Proof. (3) Apply Ax(VII) with $B, A, \bar{a}{}^\frown\bar{b}, \emptyset$ for $B, A, \bar{a}, C$.

DEFINITION 1.1: (1) A set A is called *$(\mathbf{F}, \mu)$-saturated* if for every $(p, B) \in \mathbf{F}$, $\mathrm{Dom}\, p \subseteq A$, $|\mathrm{Dom}\, p| < \mu$, p is realized by some sequence from A.

(2) $\mu(\mathbf{F})$ is the first cardinal μ such that for every $\lambda \geq \mu$ and A; A is $(\mathbf{F}, \mu)$-saturated iff A is $(\mathbf{F}, \lambda)$ saturated (if there is no such cardinal, $\mu(\mathbf{F}) = \infty$ of course).

(3) A is *$\mathbf{F}$-semi-saturated* if it is $(\mathbf{F}, |A|)$-saturated.

(4) A is *$\mathbf{F}$-saturated* if it is $(\mathbf{F}, |A|^+)$-saturated (or equivalently, $(\mathbf{F}, \mu(\mathbf{F}))$-saturated; or $(\mathbf{F}, \mu)$-saturated for every μ).

LEMMA 1.2: (1) *Assume* Ax(IV) *and* (V.1). *A is $(\mathbf{F}, \mu)$-saturated iff when $p \in S(A')$, $A' \subseteq A$, $|A'| < \mu$, $B \subseteq A'$, $(p, B) \in \mathbf{F}$, p is realized in A.*

(2) *Assume* Ax(VIII), (III.1). *$(p, B) \in \mathbf{F}$ iff for every finite $C \subseteq$* $\mathrm{Dom}\, p$, $p \restriction (B \cup C) \in \mathbf{F}(B)$.

(3) *When $\mu < \lambda \leq \infty$, $(\mathbf{F}, \lambda)$-saturation implies $(\mathbf{F}, \mu)$-saturation.*

(4) *When $\mu \geq \mu(\mathbf{F})$, $(\mathbf{F}, \mu)$-saturation is equivalent to $\mathbf{F}$-saturation.*

(5) *Every $\mathbf{F}$-saturated set is $\mathbf{F}$-semi-saturated.*

(6) *Assume Axiom* (X.1). *Then every $(\mathbf{F}, \aleph_0)$-saturated set is a model.*

Proof. Immediate. The difference between (1) and the definition is that here p is a one-type, and not an arbitrary m-type. This is where Ax(IV) comes in. (2) is proved by induction on $|\mathrm{Dom}\, p|$.

DEFINITION 1.2: (1) An *$\mathbf{F}$-construction* is a triple

$$\mathscr{A} = \langle A, \{\bar{a}_i : i < \alpha\}, \langle B_i : i < \alpha\rangle\rangle$$

such that $p_i(\mathscr{A}) = \mathrm{tp}(\bar{a}_i, \bigcup \{\bar{a}_j : j < i\} \cup A) \in \mathbf{F}(B_i)$.

(2) $U_i = U_i(\mathscr{A}) = \{j < i : (B_i \cap \bar{a}_j) - (A \cup \bigcup \{\bar{a}_\beta : \beta < j\}) \neq \emptyset\}$.

A set $U \subseteq \alpha$ is *closed* (i.e., $\mathscr{A}$-closed) if $(\forall j \in U)(U_j \subseteq U)$. $U_i^* = U_i^*(\mathscr{A})$ is the smallest closed set containing $U_i \cdot B_i^* = B_i \cup \{B_j \cup \bar{a}_j : j \in U_i^*\}$. A set B of elements is *closed* if for some closed $U \subseteq \alpha$, $\bigcup_{j \in U} \bar{a}_j \subseteq B \subseteq \bigcup_{j \in U} \bar{a}_j \cup A$, and $j \in U$ implies $B_j \subseteq B$.

(3) For $U \subseteq \{j : j < \alpha\}$ let $\mathscr{A}(U) = \bigcup \{\bar{a}_i : i \in U\} \cup A$ and Dom $\mathscr{A} = \mathscr{A}(\alpha)$.

(4) When it is clear what $\mathscr{A}$ is, we omit it.

We denote $\mathscr{A}(\emptyset)$ by A; we also sometimes ignore $\langle B_i : i < \alpha \rangle$ in defining $\mathscr{A}$.

(5) Instead of α, we can use any well ordered set I, and in this case al(I) (= after last) will be an imaginary element of I which comes after all its (real) elements. In this case $\mathscr{A}(s)$ will mean $\mathscr{A}(\{t : t < s, t \in I\})$; so Dom $\mathscr{A} = \mathscr{A}(\text{al}(I))$.

DEFINITION 1.3: C_0 is $\mathbf{F}$-constructible over A_0 if for some $\mathbf{F}$-construction $\mathscr{A}$, $\mathscr{A}(\emptyset) = A_0$, Dom $\mathscr{A} = C_0$.

DEFINITION 1.4: (1) If C is $\mathbf{F}$-constructible over A, and C is $\mathbf{F}$-saturated [$(\mathbf{F}, \mu)$-saturated] then we say C is $\mathbf{F}$-*primary* [$(\mathbf{F}, \mu)$-primary] over A.

(2) C is $\mathbf{F}$-*primitive* [$(\mathbf{F}, \mu)$-primitive] over A if $A \subseteq C$ and for every C', such that $A \subseteq C'$ and C' is $\mathbf{F}$-saturated [$(\mathbf{F}, \mu)$-saturated] there is an elementary mapping f, from C into C', where $f \restriction A$ is the identity.

(3) C is $\mathbf{F}$-*prime* over A if it is $\mathbf{F}$-primitive over A and $\mathbf{F}$-saturated, C is $(\mathbf{F}, \mu)$-prime over A if it is $(\mathbf{F}, \mu)$-primitive over A and $(\mathbf{F}, \mu)$-saturated.

(4) If in any of the above C is the universe of a model M, we say M is an $\mathbf{F}$-primary $[\cdots]$ model. The same holds for Definition 1.1.

LEMMA 1.3: *Assume* Ax(IV), (V.1).

(1) *If C is $\mathbf{F}$-constructible over A, then C has a $\mathbf{F}$-construction over A in which $l(\bar{a}_i) = 1$.*

(2) *If for $j < \alpha$, A_{j+1} is $\mathbf{F}$-constructible [primitive] over A_j, and for limit $\delta \leq \alpha$, $A_\delta = \bigcup_{i<\delta} A_i$, then A_α is $\mathbf{F}$-constructible [primitive] over A_0.*

(3) *If for $0 < j \leq \alpha$, A_j is $\mathbf{F}$-prime [primary] over $\bigcup_{i<j} A_i$, then A_α is $\mathbf{F}$-prime [primary] over A_0.*

(4) *A is $\mathbf{F}$-primitive over A, and $\mathbf{F}$-constructible over A. If A is $(\mathbf{F}, \mu)$-saturated, it is also $(\mathbf{F}, \mu)$-primary and $(\mathbf{F}, \mu)$-prime over A.*

Proof. Immediate.

LEMMA 1.4: (1) *If* $\lambda(\mathbf{F})$ *is regular, then in Definition 1.2* $\lambda(\mathbf{F}) > |B_i^*| + |U_i^*|$. *So if in addition* Ax(III.2) *holds; then* w.l.o.g. *we can assume* $U_i^* = U_i$, $B_i^* = B_i$.

(2) *Assume* Ax(III.1). *If* B *is* **F**-*constructible over* A, $C \subseteq B$, *then there is* $C' \subseteq B$, $C \subseteq C'$ *such that* $A \cup C'$ *is* **F**-*constructible over* A, *and* $|C'| < |C|^+ + \lambda_r(\mathbf{F})$.

DEFINITION 1.5: We say A is **F**-*atomic* over B if $B \subseteq A$, and for every $\bar{a} \in A$, $\mathrm{tp}(\bar{a}, B)$ is **F**-isolated.

QUESTION 1.1: Are there any more connections between the axioms?

DEFINITION 1.6: For a class K of models, we say "M is prime over A for K (or among the members of K) if $M \in K$, and $A \subseteq N \in K$ implies there is an elementary embedding of M into N, $f \restriction A$ = the identity; let κ-prime stand for "prime among the κ-compact models of T. Similarly for "primitive for K".

IV.2. Examples of F's

Here we present some examples of **F**'s, and show which axioms they satisfy. For $\mathbf{F} = \mathbf{F}^t_{\aleph_0}$ we get the usual concept of primeness.

DEFINITION 2.1: (1) Let p be an m-type, or even a type with infinitely many free variables.

(i) p is a $\mathbf{F}^t_\lambda$-type if $|p| < \lambda$.

(ii) p is a $\mathbf{F}^s_\lambda$-type if $|\mathrm{Dom}\, p| < \lambda$.

(iii) p is a $\mathbf{F}^a_\lambda$-type if p is almost over some B of cardinality $< \lambda$.

(2) For $x \in \{s, t\}$ $\mathbf{F}^x_\lambda = \{(p, B)$: for some A, m, q; $B \subseteq A$, $p \in S^m(A)$, $|B| < \lambda$, q an $\mathbf{F}^x_\lambda$-type over B, $q \subseteq p$, $q \vdash p\}$. $\mathbf{F}^a_\lambda$ is defined similarly, but with q almost over B.

(3) M is $\mathbf{F}^x_\lambda$-compact if every $\mathbf{F}^x_\lambda$-type over $|M|$ is realized in M.

Remark. For understanding $\mathbf{F}^a_\lambda$, let us for a moment work in $\mathfrak{C}^{eq}$, and define $\mathrm{card}^* A = \min\{|B|: B \subseteq \mathrm{acl}\, A = \mathrm{acl}\, B\}$. Then $\mathbf{F}^a_\lambda$ is defined as $\mathbf{F}^s_\lambda$ replacing $|A| < \lambda$ by $\mathrm{card}^* A < \lambda$.

DEFINITION 2.2: $\mathbf{F}^f_\lambda = \{(p, B)$: for some $A, m, p \in S^m(A)$, $B \subseteq A$, $|B| < \lambda$ and p does not fork over $B\}$.

DEFINITION 2.3: $\mathbf{F}^l_\lambda = \{(p, B)$: for every φ there is $q_\varphi \subseteq p \restriction B$, $|q_\varphi| < \lambda$, such that $q_\varphi \vdash p \restriction^+ \varphi =_{\text{def}} \{\varphi(\bar{x};\bar{a}): \varphi(\bar{x};\bar{a}) \in p\}$ and $|B| \le \lambda + |T|$; $\text{cf}\,\lambda > |T| \Rightarrow |B| < \lambda$.}

Note that $p \restriction {}^+\varphi$ differs from $p \restriction \varphi$ in that the latter also contains formulas of the form $\neg\varphi(\bar{x};\bar{a})$.

In the above definitions the letters t, s, a, f, and l stand for type, set, almost, forking, and local, respectively.

To be precise we should say "$\mathbf{F}^x_\lambda$ in $\mathfrak{C}$" or write $\mathbf{F}^x_\lambda[\mathfrak{C}]$ for all $x \in \{a, f, l, s, t\}$, that is the definitions depend on the model $\mathfrak{C}$. We use the square brackets to differentiate this from $\mathbf{F}^x_\lambda(|\mathfrak{C}|) = \{p: (p, |\mathfrak{C}|) \in \mathbf{F}^x_\lambda\}$. We use $\mathbf{F}^f_\lambda$ only when T is assumed to be stable.

We use the following lemma freely in dealing with $\mathbf{F}^a_\lambda$.

LEMMA 2.1: *Let $B \subseteq A$, $p = \text{tp}(\bar{c}, A)$. Then* (4) $\Rightarrow$ (1) $\Leftrightarrow$ (2) $\Leftrightarrow$ (3). *If T is stable all the conditions are equivalent.*

(1) $(p, B) \in \mathbf{F}^a_\infty$, *i.e., there is $q \subseteq p$, $q \vdash p$, q almost over B.*

(2) *For some type q almost over B, $q \vdash p$.*

(3) $\text{stp}(\bar{c}, B) \vdash p$.

(4) $\text{stp}(\bar{c}, B) \vdash \text{stp}(\bar{c}, A)$.

Proof. Trivially (1) $\Rightarrow$ (2). By the remark to Definition III, 2.1 (2) $\Rightarrow$ (3). Suppose T is stable, (3) holds and $\bar{c}'$ realizes $\text{stp}(\bar{c},B)$, by III, 2.6(1) $\text{stp}(\bar{c},B)$ does not fork over B, so by (3) (and III, 1.18) also p does not fork over B, but $\bar{c}'$ realizes p (by (3)); hence, by II, 2.6(1) $\text{stp}(\bar{c}',A)$ does not fork over B, so by III, 2.9 $\text{stp}(\bar{c},A) \equiv \text{stp}(\bar{c}',A)$. By the choice of $\bar{c}'$, (4) holds. Hence by III, 2.9 $\text{stp}(\bar{c},A) \equiv \text{stp}(\bar{c}',A)$; so (4) holds.

Let us assume (4) or (3) and we prove (1). Let $q = \{\varphi(\bar{x},\bar{a}): \varphi(\bar{x},\bar{a}) \in p$, $\varphi(\bar{x},\bar{a})$ is almost over $B\}$. It suffices to prove $q \vdash p$. For any $\theta(\bar{x};\bar{a}) \in p$, by (4) or (3) there is $E \in \text{FE}^m(B)$ such that $E(\bar{x},\bar{c}) \vdash \theta(\bar{x};\bar{a})$ (remember that by III, 2.2(1) $\text{FE}^m(B)$ is closed under conjunctions). Let $\psi(\bar{x}) = (\forall\bar{y})[E(\bar{x},\bar{y}) \to \theta(\bar{y};\bar{a})]$. Clearly the parameters of ψ are from A; also $E(\bar{x};\bar{z}) \vdash \psi(\bar{x}) \equiv \psi(\bar{z})$ hence ψ is almost over B. As $E(\bar{x},\bar{c}) \vdash \theta(\bar{x};\bar{a})$, $\vdash\psi[\bar{c}]$. As $\psi(\bar{x})$ is over A, $\psi(\bar{x}) \in p$ and as $\psi(\bar{x})$ is almost over B, $\psi(\bar{x}) \in q$. From ψ's definition it follows that $\psi(\bar{x}) \vdash \theta(\bar{x};\bar{a})$ (as $E(\bar{x},\bar{x})$) hence $q \vdash \theta(\bar{x};\bar{a})$. So $q \vdash p$ and we finish.

In the following lemma we indicate many trivial facts concerning the $\mathbf{F}^x_\lambda$, their relation to previous notions, and the connections among the various $\mathbf{F}^x_\lambda$.

LEMMA 2.2: (1) *A is $\mathbf{F}^t_{\aleph_0}$-saturated if it is the universe of a model; if* Ax(X.1) *holds, also the converse is true.*

(2) *We get equivalent definitions of $\mathbf{F}^t_{\aleph_0}$, $\mathbf{F}^l_{\aleph_0}$ if we assume q consists of one formula and in all cases (in Definition 2.1(2), 2.2 and 2.3) we can assume that q is closed under conjunctions (for $\mathbf{F}^a_\lambda$, by* III, 2.2(1)).

(3) *For $x = t, s, a, f$, $\lambda(\mathbf{F}^x_\lambda) = \lambda$, and $\lambda(\mathbf{F}^l_\lambda) = \lambda^+ + |T|^+$ when* $\mathrm{cf}\,\lambda \le |T|$, *and* $\lambda(\mathbf{F}^l_\lambda) = \lambda$ *otherwise.*

(4) $\mu(\mathbf{F}^t_\lambda) \le \lambda$, $\mu(\mathbf{F}^s_\lambda) \le \lambda$, $\mu(\mathbf{F}^a_\lambda) \le \lambda + \aleph_1$, $\mu(\mathbf{F}^f_\lambda) = \infty$, $\mu(\mathbf{F}^l_\lambda) \le \lambda^+ + |T|^+$ *and when* $\mathrm{cf}\,\lambda > |T|$, $\mu(\mathbf{F}^l_\lambda) \le \lambda$ *(for $\mathbf{F}^a_\lambda$ use the proof of* (7); *for $\mathbf{F}^f_\lambda$ use* III, 1.3, III, 1.4). (*See Exercise* 2.2.)

(5) *A is $\mathbf{F}^t_\lambda$-saturated if for some λ-compact M, $|M| = A$. For $x = s$, a, $[t]$, M is $\mathbf{F}^x_\lambda$-saturated iff [if] M is $\mathbf{F}^x_\lambda$-compact.*

(6) *A is $\mathbf{F}^s_\lambda$-saturated iff for some λ-saturated M, $|M| = A$.*

(7) *For $\lambda > \aleph_0$, A is $\mathbf{F}^a_\lambda$-saturated iff for some λ-saturated M, $|M| = A$.*

(8) *If $\lambda > |T|$, $\mathbf{F}^t_\lambda = \mathbf{F}^s_\lambda$, and if* $\mathrm{cf}\,\lambda > |T|$, $\mathbf{F}^l_\lambda = \mathbf{F}^t_\lambda$. *If $\lambda > |T|$, p is $\mathbf{F}^a_\lambda$-isolated iff it is $\mathbf{F}^t_\lambda$-isolated.*

(9) *If $\lambda \le \kappa$, $x = s, t, a, f, l$ then $\mathbf{F}^x_\lambda \subseteq \mathbf{F}^x_\kappa$.*

(10) $\mathbf{F}^l_\lambda \supseteq \mathbf{F}^t_\lambda \subseteq \mathbf{F}^s_\lambda \subseteq \mathbf{F}^a_\lambda \subseteq \mathbf{F}^f_\lambda$ *(the last inclusion for stable T, of course).*

(11) *For $x = t, s, a, f, l$, $(p, B) \in \mathbf{F}^x_\lambda$ in $\mathfrak{C}$ iff $(p^{\mathrm{eq}}, B) \in \mathbf{F}^x_\lambda$ in $\mathfrak{C}^{\mathrm{eq}}$ (where $A \subseteq |\mathfrak{C}|$, $p \in S^m(A, \mathfrak{C})$, $B \subseteq A$).*

(12) *For $x = t, s, a, f, l$, if $\mathscr{A}$ is an $\mathbf{F}^x_\lambda$-construction in $\mathfrak{C}$, then it is an $\mathbf{F}^x_\lambda$-construction in $\mathfrak{C}^{\mathrm{eq}}$. Hence, for $A, C \subseteq |\mathfrak{C}|$:*

(i) *If C is $\mathbf{F}^x_\lambda$-constructible (primitive) over A in $\mathfrak{C}$, then this holds also in $\mathfrak{C}^{\mathrm{eq}}$, also for* $\mathrm{acl}\, C$,

(ii) *For $x = s, a$, if C is $\mathbf{F}^x_\lambda$-saturated then in $\mathfrak{C}^{\mathrm{eq}}$* $\mathrm{acl}(C)$ *is $\mathbf{F}^x_\lambda$-saturated.*

(iii) *Similarly for $\mathbf{F}^x_\lambda$-primary (prime).*

(iv) *If $A \subseteq \mathfrak{C}$ and in $\mathfrak{C}^{\mathrm{eq}}$* $\mathrm{acl}(A)$ *is $\mathbf{F}^x_\lambda$-saturated and* $A = \mathrm{acl}\, A \cap \mathfrak{C}$, *then in $\mathfrak{C}$, A is $\mathbf{F}^x_\lambda$-saturated ($x = t, s, a$).*

(v) *If $B \subseteq A \subseteq \mathfrak{C}$, A is $\mathbf{F}^a_\lambda$-atomic over B in $\mathfrak{C}$ iff this holds in $\mathfrak{C}^{\mathrm{eq}}$, if this holds when we replace A and/or B by their algebraic closure.*

(vi) *Let $x = s, a$. If $B \subseteq A \subseteq \mathfrak{C}$, and in $\mathfrak{C}^{\mathrm{eq}}$ A is $\mathbf{F}^a_\lambda$-primitive over B, then in $\mathfrak{C}$ A is $\mathbf{F}^x_\lambda$-primitive over B.*

(vii) *If $B \subseteq M \subseteq \mathfrak{C}$, and in $\mathfrak{C}^{\mathrm{eq}}$ M^{eq} is $\mathbf{F}^x_\lambda$-prime over B, then in $\mathfrak{C}$ M is $\mathbf{F}^x_\lambda$-prime over B.*

(viii) *$A \subseteq \mathfrak{C}$ $A =$* $\mathrm{acl}\, A$ *in $\mathfrak{C}$. Then A is $\mathbf{F}^t_\lambda$-compact iff* $\mathrm{acl}\, A$ *(in $\mathfrak{C}^{\mathrm{eq}}$) is $\mathbf{F}^t_\lambda$-compact.*

(13) *If $p \in S^m(A)$, $p \in \mathbf{F}^a_\lambda(B)$, $p \restriction B$ is stationary then $p \in \mathbf{F}^s_\lambda(B)$ (for stable T).* (*By* III, 2.14.)

Proof. (7) One direction is trivial. Now assume M is λ-saturated, p is almost over $B \subseteq |M|$, $|B| < \lambda$. We must show that p is realized in M. Let $\bar{a}$ realize p. Define $\bar{a}_i \in |M|$ $(i \leq \omega)$ such that $\bar{a}_i$ realizes $\operatorname{tp}(\bar{a}, B \cup \bigcup_{j<i} \bar{a}_j)$. Then $\bar{a}_\omega$ realizes p. (The proof is like the proof for stable T, in III, 2.13(1). For $\varphi \in p$ let E be the equivalence relation over B that φ depends on. Now since E has finitely many classes there is i such that $\bar{a}_i$ and $\bar{a}$ are in the same class. Thus $\bar{a}_i$ satisfies $E(\bar{x}; \bar{a})$ and so does $\bar{a}_\omega$).

(12) Notice that all the **F**'s satisfy Ax(II.3), hence $\operatorname{acl}(A)$ is **F**-constructible over A.

Remark. Notice that although $\mathbf{F}^a_{\aleph_1}$-saturation and $\mathbf{F}^s_{\aleph_1}$-saturation are the same, not necessarily $\mathbf{F}^s_{\aleph_1} = \mathbf{F}^a_{\aleph_1}$ (see Exercise 2.2(4)).

Now we check the axioms one by one. In first reading you can omit $\mathbf{F}^l_\lambda$, $\mathbf{F}^a_\lambda$.

LEMMA 2.3: *Axiom* (I) *is satisfied by* $\mathbf{F}^x_\lambda$ *for* $x = t, s, a, f, l$.

Proof. Immediate.

LEMMA 2.4: (1) *Axioms* (II.j) $j = 1, 2, 3, 4$ *are satisfied by* $\mathbf{F}^x_\lambda$ *for* $x = s, t, a, f, l$.

(2) *Moreover, if* $\bar{a}$ *is algebraic over* B, $B \subseteq A$, $|B| < \lambda$ *then* $\operatorname{tp}(\bar{a}, A) \in \mathbf{F}^a_\lambda(B)$.

Proof. (1) Immediate.

(2) By 2.1.

LEMMA 2.5: *Axiom* (III.1) *is satisfied by* $\mathbf{F}^x_\lambda$, *for* $x = t, s, a, f, l$.

Proof. For $x = t, s, l$ this is immediate. For $x = a$ this follows from 2.1(3), and for $x = f$ this is immediate (see III, 1.1(5)).

LEMMA 2.6: *Axiom* (III.2) *is satisfied by* $\mathbf{F}^x_\lambda$, *for* $x = t, s, a, f, l$.

Proof. Immediate.

LEMMA 2.7: *Axiom* (IV) *is satisfied by* $\mathbf{F}^x_\lambda$ *for* $x = s, t, a, f, l$.

Proof. So let $p = p(\bar{x}) = \operatorname{tp}(\bar{a}, A)$, $q = q(\bar{x}, \bar{y}) = \operatorname{tp}(\bar{a}^\frown\bar{b}, A)$, $B \subseteq A$, and $(q, B) \in \mathbf{F}^x_\lambda$, and we should prove $(p, B) \in \mathbf{F}^x_\lambda$. For $x = f$ this holds by III, 1.1(6). If $x = t, s$, there is an $\mathbf{F}^x_\lambda$-type $r = r(\bar{x}, \bar{y})$ over B, $r \vdash q$. By 2.2(2) we can assume r is closed under conjunctions. So $r' = \{(\exists \bar{y})\varphi(\bar{x}, \bar{y}; \bar{c}): \varphi(\bar{x}, \bar{y}; \bar{c}) \in r\}$ is an $\mathbf{F}^x_\lambda$-type over B, $|r'| = |r|$ and $r' \vdash p$. Hence $(p, B) \in \mathbf{F}^x_\lambda$. The cases $x = a, l$ are treated similarly (for $x = a$ notice III, 2.2(1)).

LEMMA 2.8: (1) *Axiom* (V.1) *is satisfied by* $\mathbf{F}^x_\lambda$, *for* $x = s, t, a, f, l$.

(2) *Axiom* (V.2) *is satisfied by* $\mathbf{F}^x_\lambda$ *for* $x = s, a, f$; *and also for* $x = t$, *when* λ *is regular. If we assume* $\operatorname{cf} \lambda > |C|$ *it holds for* $x = t, s, a, f, l$. (*See Exercise* 2.3, 4.)

Proof. (1) Follows from (2), by 1.1(2) and 2.7.

(2) We assume $p = \operatorname{tp}(\bar{a}, A \cup C)$, $q = \operatorname{tp}_*(\bar{a} \cup C, A)$, $B \subseteq A$, $|C| < \lambda$ and $(q, B) \in \mathbf{F}^x_\lambda$. We must prove $(p, B \cup C) \in \mathbf{F}^x_\lambda$.

If $x = f$ this follows from III, 1.1(5), 4.14. If $x = s, t\,[x = a]$ then clearly there is a type r [almost] over B, such that $r \vdash q, r = \bigcup\{r_{\bar{c}}: \bar{c} \in \bar{a} \cup C\}$, where $r_{\bar{c}} \vdash \operatorname{tp}(\bar{c}, A)$ is an $\mathbf{F}^x_\lambda$-type.

Let

$$r' = \{\varphi(\bar{x}; \bar{c}, \bar{b}): \varphi(\bar{x}_a, \bar{x}_c; \bar{b}) \in r,\ \bar{b} \in B,\ \bar{c} \in C\}.$$

Now r' is a type [almost] over $B \cup C$, and $r' \vdash p$. If $x = s, a$ $|B \cup C| < \lambda$ hence r' is an $\mathbf{F}^x_\lambda$-type, so we finish. If $x = t$, $|C| < \operatorname{cf} \lambda$, then $|r'| = |r| < \lambda$ and we finish. So let $x = l$, $|C| < \operatorname{cf} \lambda$, and let $\varphi = \varphi(\bar{x}; \bar{z})$.

Assume $\bar{z} = \langle z_0, \ldots, z_{k-1}\rangle$, then there are formulas $\varphi_l = \varphi_l(\bar{x}; \bar{z}_{l_1}; \bar{z}_{l_2})$ for $l < n(=2^k)$ such that for all partitions $\{i_1, \ldots, i_s\} \cup \{j_1, \ldots, j_t\} = k$ there is $l < n$ such that $\varphi_l(\bar{x}; z_{i_1}, \ldots, z_{i_s}; z_{j_1}, \ldots, z_{j_t}) \equiv \varphi(\bar{x}; z_0, \ldots, z_{k-1})$.

By assumption for every $l < n$ and $\bar{c}_1 \in C$ there is a type $r_l(\bar{c}_1) \subseteq \operatorname{tp}(\bar{a}^\frown \bar{c}_1, B)$ such that $r_l(\bar{c}_1) \vdash \operatorname{tp}(\bar{a}^\frown \bar{c}_1, A) \restriction^+ \varphi_l$.

Define

$$r'_l(\bar{c}_1) = \{\theta(\bar{x}; \bar{c}_1, \bar{c}_2): \theta(\bar{x}_a, \bar{x}_{c_1}, \bar{c}_2) \in r_l(\bar{c}_1)\},$$

$$r = \bigcup \{r'_l(\bar{c}_1): \bar{c}_1 \in C,\ l < n\}.$$

It is easy to check that $|r| < \lambda$, $r \vdash p \restriction^+ \varphi$, so we finish the last case $x = l$.

LEMMA 2.9: *Axiom* (VI) *is satisfied by* $\mathbf{F}^x_\lambda$ *for* $x = t, s, f$, *if* λ *is regular or* $\operatorname{cf} \lambda > |T|$, *also by* $\mathbf{F}^l_\lambda$, *and if* T *is stable, by* $\mathbf{F}^a_\lambda$ *too.*

Proof. We should prove that if $B, C \subseteq A$, $(q_1, C) \in \mathbf{F}^x_\lambda$ where $q_1 = \operatorname{tp}(\bar{b}, A \cup \bar{a})$, and $(p, B) \in \mathbf{F}^x_\lambda$ where $p = \operatorname{tp}(\bar{a}, A)$ then $(p_1, B) \in \mathbf{F}^x_\lambda$, where $p_1 = \operatorname{tp}(\bar{a}, A \cup \bar{b})$. Let $q = q_1 \restriction A$.

For $x = f$ this is III, 4.13.

For $x = s, t$, we first claim that if $\bar{a}_1$ realizes p, $\bar{b}_1$ realizes q, then $\bar{a}_1{}^\frown\bar{b}_1$ realizes $\mathrm{tp}(\bar{a}{}^\frown\bar{b}, A)$. Because $\mathrm{tp}(\bar{a}_1, A) = \mathrm{tp}(\bar{a}, A)$ for some elementary mapping f, $f \restriction A =$ the identity and $f(\bar{a}_1) = \bar{a}$; so w.l.o.g. $\bar{a} = \bar{a}_1$. As $C \subseteq A$, $(q, C) \in \mathbf{F}^x_\lambda$, $\mathrm{tp}(\bar{b}, A) \vdash \mathrm{tp}(\bar{b}, A \cup \bar{a})$, so $\mathrm{tp}(\bar{a}{}^\frown\bar{b}, A) = \mathrm{tp}(\bar{a}{}^\frown\bar{b}_1, A)$, so the claim is proved. Hence if $\bar{a}_1$ realizes p necessarily $\mathrm{tp}(\bar{a}_1{}^\frown\bar{b}, A) = \mathrm{tp}(\bar{a}{}^\frown\bar{b}, A)$, so $\mathrm{tp}(\bar{a}_1, A \cup \bar{b}) = \mathrm{tp}(\bar{a}, A \cup \bar{b}) = p_1$. So $p \vdash p_1$. But for some $\mathbf{F}^x_\lambda$-type p' over B, $p' \subseteq p$, $p' \vdash p$, hence $p' \vdash p_1$ hence $(p_1, B) \in \mathbf{F}^x_\lambda$.

For $x = a$ the proof is similar. We first show that if $\bar{a}_1$, $\bar{b}_1$ realize $\mathrm{stp}(\bar{a}, B)$, $\mathrm{stp}(\bar{b}, C)$ respectively then $\mathrm{tp}(\bar{a}{}^\frown\bar{b}, A) = \mathrm{tp}(\bar{a}_1{}^\frown\bar{b}_1, A)$. Hence $\mathrm{stp}(\bar{a}, B) \vdash \mathrm{tp}(\bar{a}, A \cup \bar{b})$ so we finish by 2.1.

We are left with $x = l$. Let $\varphi(\bar{x}_a; \bar{z})$ be a formula. As in the proof of 2.8 we can find $n < \omega$ and, for $l < n$, $\varphi_l = \varphi_l(\bar{x}_a; \bar{x}_b, \bar{z}_l)$ such that for every $\bar{c} \in A \cup \bar{b}$, there are $l < n$ and $\bar{c}' \subseteq \bar{c}$, $\bar{c}' \in A$ such that $\vdash \varphi(\bar{x}_a; \bar{c}) \equiv \varphi_l(\bar{x}_a, \bar{b}, \bar{c}')$. As $(q_1, C) \in \mathbf{F}^l_\lambda$ there are types $q^l_1 \subseteq q_1 \restriction C$, $q^l_1 = q^l_1(\bar{x}_b)$, $q^l_1 \vdash q_1 \restriction^+ \psi_l$, $|q^l_1| < \lambda$, where $\psi_l = \psi_l(\bar{x}_b; \bar{x}_a, \bar{z}_l) = \varphi_l(\bar{x}_a; \bar{x}_b, \bar{z}_l)$. Let $q^l_1 = \{\theta_{l,i}(\bar{x}_b, \bar{c}^l_i) : i < |q^l_1|\}$ be closed under conjunctions (by 2.2(2)) where $\bar{c}^l_i \in C$. For every $l < n$, $i < |q^l_1|$ let

$$\theta^*_{i,l}(\bar{x}_a, \bar{z}_{i,l}, \bar{z}_l) = (\forall \bar{x}_b)[\theta_{i,l}(\bar{x}_b, \bar{z}_{i,l}) \to \varphi_l(\bar{x}_a, \bar{x}_b, \bar{z}_l)].$$

As $(p, B) \in \mathbf{F}^l_\lambda$, there are types $p_{i,l} = p_{i,l}(\bar{x}_a) \subseteq p \restriction B$ such that $p_{i,l} \vdash p \restriction^+ \theta^*_{i,l}$; and let $p^* = \bigcup_{i,l} p_{i,l}$. So as in the case $x = l$ we assume λ is regular or of $\lambda > |T|$, we have $|p^*| < \lambda$; and of course $p^* \subseteq p \restriction B$. Now it is easy to check $p^* \vdash p_1 \restriction^+ \varphi$.

THEOREM 2.10: *Axiom (VII) is satisfied by $\mathbf{F}^x_\lambda$ when $x = s$ or $x = a, f$, T is stable, or $x = t, l$, λ is regular.*

Proof. Let $B \subseteq A$, $p = \mathrm{tp}(\bar{a}, A \cup C)$, $q = \mathrm{tp}_*(C, A)$, $r = \mathrm{tp}_*(\bar{a} \cup C, A)$ and $(p, B \cup C) \in \mathbf{F}^x_\lambda$, $(q, B) \in \mathbf{F}^x_\lambda$. We must prove $(r_{a,c}, B) \in \mathbf{F}^x_\lambda$ for $\bar{c} \in C$ where $r_{a,c} = \mathrm{tp}(\bar{a}{}^\frown\bar{c}, A)$.

CLAIM 2.11: *For sets A, C, suppose for every $\bar{c} \in C$ $q'_c \vdash q_c = \mathrm{tp}(\bar{c}, A)$, and $p' \vdash p = \mathrm{tp}(\bar{a}, A \cup C)$, p' is over $A \cup C$, (p', q'_c are closed under conjunctions). Define*

$$r'_{a,c} = \{(\exists \bar{x}_d)[\psi(\bar{x}_a, \bar{x}_{\bar{c}}, \bar{x}_d; \bar{b}^1) \wedge \theta(\bar{x}_c, \bar{x}_d, \bar{b}^2)] : \\ \bar{b}_1 \in A,\ \bar{c}, \bar{d} \in C,\ \psi(\bar{x}_a; \bar{c}, \bar{d}, \bar{b}^1) \in p',\ \theta(\bar{x}_c, \bar{x}_d; \bar{b}^2) \in q'_{c^\frown d}\}.$$

Then $r'_{a,c} \vdash r_{a,c} = \mathrm{tp}(\bar{a}{}^\frown\bar{c}, A)$.

Proof of 2.11. Let $\varphi = \varphi(\bar{x}_a, \bar{x}_c, \bar{b}_0) \in r_{a,c}$, and we shall prove $r'_{b,c} \vdash \varphi$, and this will suffice. By the hypothesis $p' \vdash p$ and $\models \varphi[\bar{a}, \bar{c}, \bar{b}_0]$ hence $p' \vdash \varphi(\bar{x}_a, \bar{c}, \bar{b}_0) \in p$. Hence (as p' is closed under conjunctions) there exists $\psi = \psi(\bar{x}_a, \bar{c}, \bar{d}, \bar{b}_1) \in p'$ (we add dummy variables if necessary to make room for $\bar{c}$) such that $\bar{d} \in C$, $\bar{b}_1 \in A$ and $\psi(\bar{x}_a, \bar{c}, \bar{d}, \bar{b}_1) \vdash \varphi(\bar{x}_a, \bar{c}, \bar{b}_0)$. Hence $\models \varphi^*[\bar{c}, \bar{d}, \bar{b}_2]$ where $\varphi^*(\bar{x}_c, \bar{x}_d, \bar{b}_2) = (\forall \bar{x}_a)[\psi(\bar{x}_a, \bar{x}_c, \bar{x}_d, \bar{b}_1) \to \varphi(\bar{x}_a, \bar{x}_c, \bar{b}_0)]$.

Hence $\varphi^*(\bar{x}_c, \bar{x}_d; \bar{b}_2) \in q_{c\frown d}$, so $q'_{c\frown d} \vdash \varphi^*(\bar{x}_c, \bar{x}_d; \bar{b}_2)$ hence for some $\theta(\bar{x}_c, \bar{x}_d; \bar{b}_3) \in q'_{c\frown d}$, $\models (\forall \bar{x}_c)(\forall \bar{x}_d)[\theta(\bar{x}_c, \bar{x}_d; \bar{b}_3) \to \varphi^*(\bar{x}_c, \bar{x}_d, \bar{b}_2)]$.

It is easy to check that

$$(\exists \bar{x}_d)[\psi(\bar{x}_a, \bar{x}_c, \bar{x}_d, \bar{b}_1) \wedge \theta(\bar{x}_c, \bar{x}_d, \bar{b}_3)] \vdash \varphi(\bar{x}_a, \bar{x}_c, \bar{b}_0)$$

and that the antecedent of this implication belongs to $r'_{a,c}$. So we finish.

Continuation of the proof of 2.10. Clearly if p', q'_c are over $B \cup C$, B respectively (and w.l.o.g. $C \cap A \subseteq A$), then $r'_{\bar{a},\bar{c}}$ is over B; so the *case* $x = s$ is immediate. If $x = t$, then $|C| < \lambda$ as $(p, B \cup C) \in \mathbf{F}^t_\lambda$ and there are $p', q'_{\bar{c}}$ satisfying 2.11 such that $|p'|, |q'_{\bar{c}}| < \lambda$.

As λ is regular in that case it follows that $|r'_{a,c}| < \lambda$ so we finish.

The case $x = f$ follows by III, 4.15.

So now we deal with the *case* $x = l$; again we assume λ is regular. Let $\bar{c} \in C$, and $\varphi(\bar{x}_a, \bar{x}_c; \bar{z})$ be a formula. By the hypothesis there is $p' \subseteq p$ closed under conjunctions, $|p'| < \lambda$, p' is over $B \cup C$ and $p' \vdash p \restriction^+ \varphi$. Let $p' = \{\psi_i(\bar{x}_a, \bar{c}, \bar{d}(i), \bar{b}^0_i) : i < \alpha < \lambda\}$, and $\varphi^*_i(\bar{x}_c, \bar{x}_{d(i)}; \bar{b}^*, \bar{b}^0_i) = (\forall \bar{x}_a)[\psi_i(\bar{x}_a, \bar{x}_c, \bar{x}_{d(i)}, b^0_i) \to \varphi(\bar{x}_a, \bar{x}_c, \bar{b}^*)]$, for $\bar{d}(i) \in C$, $\bar{b}^0_i \in B$, $\bar{b}^* \in A$.

By the hypothesis $(q_{c\frown d(i)}, B) \in \mathbf{F}^l_\lambda$ so for some $q'_{c\frown d(i)} \subseteq q_{c\frown d(i)} \restriction B$, $|q'_{c\frown d(i)}| < \lambda$, $q'_{c\frown d(i)} \vdash q_{c\frown d(i)} \restriction^+ \varphi^*_i$. Let

$$q'_{c\frown d(i)} = \{\theta_{i,j}(\bar{x}_c, \bar{x}_{d(i)}; \bar{b}^1_{i,j}) : j < \alpha(i) < \lambda\}$$

(again assume it is closed under conjunctions), $\bar{b}^1_{i,j} \in B$.

Let

$$\theta^*_{i,j}(\bar{x}_a, \bar{x}_c; \bar{b}^2_{i,j}) = (\exists \bar{x}_{d(i)})[\psi_i(\bar{x}_a, \bar{x}_c, \bar{x}_{d(i)}, \bar{b}^0_i) \wedge \theta_{i,j}(\bar{x}_c, \bar{x}_{d(i)}; \bar{b}^1_{i,j})].$$

Let $r'_{a,c} = \{\theta^*_{i,j}(\bar{x}_a, \bar{x}_c; \bar{b}^2_{i,j}) : j < \alpha(i), i < \alpha\}$. It is easy to check $r'_{a,c} \subseteq r_{a,c}$ is over B and of cardinality $< \lambda$. It suffices to show $r'_{a,c} \vdash r_{a,c} \restriction^+ \varphi$. For let $\varphi(\bar{x}_a, \bar{x}_c; \bar{b}^*) \in r_{a,c}$, ($\bar{b}^* \in A$) so $\models \varphi[\bar{a}, \bar{c}, \bar{b}^*]$. Hence $\varphi(\bar{x}_a, \bar{c}, \bar{b}^*) \in p$, hence for some $i < \alpha$, $\psi_i(\bar{x}_a, \bar{c}, \bar{d}(i), \bar{b}^0_i) \vdash \varphi(\bar{x}_a, \bar{c}, \bar{b}^*)$ hence

$$\models (\forall \bar{x}_a)[\psi_i(\bar{x}_a, \bar{c}, \bar{d}(i), \bar{b}^0_i) \to \varphi(\bar{x}_a, \bar{c}, \bar{b}^*)].$$

So $\models \varphi^*_i[\bar{c}, \bar{d}(i), \bar{b}^*, \bar{b}^0_i]$, hence $\varphi^*_i(\bar{x}_c, \bar{x}_{d(i)}; \bar{b}^*, \bar{b}^0_i) \in q_{c\frown d(i)}$ hence for some $j < \alpha(i)$ $\theta_{i,j}(\bar{x}_c, \bar{x}_{d(i)}; \bar{b}^1_{i,j}) \vdash \varphi^*_i(\bar{x}_c, \bar{x}_{d(i)}; \bar{b}^*, \bar{b}^0_i)$. So by the definition of φ^*_i

$$\theta_{i,j}(\bar{x}_c, \bar{x}_{d(i)}; \bar{b}^1_{i,j}) \wedge \psi_i(\bar{x}_a, \bar{x}_c, \bar{x}_{d(i)}; \bar{b}^0_i) \vdash \varphi(\bar{x}_a, \bar{x}_c; \bar{b}^*)$$

hence

$$(\exists \bar{x}_{d(i)})[\psi_i(\bar{x}_a, \bar{x}_c, \bar{x}_{d(i)}, \bar{b}^0_i) \wedge \theta_{i,j}(\bar{x}_c, \bar{x}_{d(i)}, \bar{b}^1_{i,j})] \vdash \varphi(\bar{x}_a, \bar{x}_c; \bar{b}^*)]$$

hence

$$\theta^*_{i,j}(\bar{x}_a, \bar{x}_c; \bar{b}^2_{i,j}) \vdash \varphi(\bar{x}_a, \bar{x}_c; \bar{b}^*)$$

so we finish also the case $x = l$.

We are left with the case $x = a$. First we prove

LEMMA 2.12: *Let T be stable*

(1) *If $B \subseteq A$*, $\mathrm{tp}(\bar{a}, A) \in \mathbf{F}^a_\lambda(B)$, $\bar{c} \in \mathrm{acl}(B \cup \bar{a})$ *then* $\mathrm{tp}(\bar{a}^\frown\bar{c}, A) \in \mathbf{F}^a_\lambda(B)$.

(2) *Let $B \subseteq A$*, $\mathrm{tp}(\bar{a}, A) \in \mathbf{F}^a(B)$ *iff* $\mathrm{tp}(\bar{a}, \mathrm{acl}\, A) \in \mathbf{F}^a(\mathrm{acl}\, B)$.

Proof. (1) Let $\theta(\bar{x}; \bar{a}, \bar{b})$ be algebraic, $\bar{b} \in B$ and $\vdash\theta[\bar{c}, \bar{a}, \bar{b}]$. By 2.4(1) for $\mathbf{F}^t_{\aleph_0}$ we can assume $\theta(\bar{x}; \bar{a}, \bar{b}) \vdash \mathrm{tp}(\bar{c}, B \cup \bar{a})$. Let $\theta(|\mathfrak{C}|, \bar{a}, \bar{b}) = \{\bar{c}_0, \ldots, \bar{c}_n\}$, $\bar{c}_0 = \bar{c}$. By III, 2.6(3), (1) there are $\bar{a}', \bar{c}'_i$ such that $\mathrm{stp}(\bar{a}^\frown\bar{c}_0{}^\frown\cdots{}^\frown\bar{c}_n, B) \equiv \mathrm{stp}(\bar{a}'^\frown\bar{c}'_0{}^\frown\cdots{}^\frown\bar{c}'_n, B)$ and $\mathrm{stp}(\bar{a}'^\frown\bar{c}'_0{}^\frown\cdots{}^\frown\bar{c}'_n, A)$ does not fork over B. So $\mathrm{stp}(\bar{a}', A) \equiv \mathrm{stp}(\bar{a}, A)$, (by III, 2.6 and III, 2.9) hence we can assume $\bar{a}' = \bar{a}$. So $\langle \bar{c}'_0, \ldots, \bar{c}'_n \rangle$ is a permutation of $\langle \bar{c}_0, \ldots, \bar{c}_n \rangle$, hence we can assume $\bar{c}_i = \bar{c}'_i$. So $\mathrm{tp}(\bar{a}^\frown\bar{c}_i, A)$ does not fork over B.

By 2.1 it suffices to prove $\mathrm{stp}(\bar{a}^\frown\bar{c}_0, B) \vdash \mathrm{tp}(\bar{a}^\frown\bar{c}_0, A)$. If this is not true there is $\bar{a}^{*\frown}\bar{c}^*$, $\mathrm{stp}(\bar{a}^{*\frown}\bar{c}^*, B) \equiv \mathrm{stp}(\bar{a}^\frown\bar{c}_0, B)$, $\mathrm{tp}(\bar{a}^{*\frown}\bar{c}^*, A) \neq \mathrm{tp}(\bar{a}^\frown\bar{c}_0, A)$. So necessarily $\mathrm{tp}(\bar{a}^{*\frown}\bar{c}^*, A)$ forks over B. Again we can assume $\bar{a}^* = \bar{a}$, hence for some i $\bar{c}^* = \bar{c}_i$, contradiction.

(2) Immediate, by 2.1 part (4), and III, 6.3(4).

Conclusion of the proof of 2.10. We have to prove it for $x = a$. By 2.2(11), it suffices to prove it in $\mathfrak{C}^{\mathrm{eq}}$.

Let $C' = \mathrm{acl}(B \cup C)$, so by 2.12(1) $\mathrm{tp}_*(C', A) \in \mathbf{F}^a_\lambda(B)$ and by 2.12(2) $\mathrm{tp}(\bar{a}, A \cup C') \in \mathbf{F}^a(B \cup C') = \mathbf{F}^a(C')$. So $\mathrm{tp}(\bar{a}, C') \vdash \mathrm{tp}(\bar{a}, A \cup C')$, $\mathrm{stp}_*(C', B) \vdash \mathrm{tp}(C', A)$. Now use 2.11 with $p' = \mathrm{tp}_*(\bar{a}, B \cup C')$, q'_c a type almost over B, which is over A, and such that $q'_c \vdash \mathrm{tp}(\bar{c}, A)$ (q'_c exists by 2.1). This proves the assertion.

LEMMA 2.13: *Axiom* (VIII) *is satisfied by $\mathbf{F}^x_\lambda$ for $x = t, s, a, f$.*

Proof of 2.13. Let A_i $(i < \delta)$ be increasing, $p \in S^m(\bigcup_{i<\delta} A_i)$, $p_i = p \restriction A_i \in \mathbf{F}^x_\lambda(B)$. We must prove $(p, B) \in \mathbf{F}^x_\lambda$. For $x = f$, it follows by III, 1.1(7).

For $x = s, t$ let q_i be an $\mathbf{F}^x_\lambda$-type over B, $q_i \subseteq p_i$, $q_i \vdash p_i$.

If $x = s$, $p \upharpoonright B$ is an $\mathbf{F}^s_\lambda$-type, $p \upharpoonright B \vdash q_i \vdash p_i$ hence $p \upharpoonright B \vdash \bigcup_{i<\delta} p_i = p$.

If $x = t$, q_0 is an $\mathbf{F}^t_\lambda$-type, $q_0 \vdash p_0$; and clearly $q_i \subseteq p_0$, hence $p_0 \vdash q_i \vdash p_i$, hence $q_0 \vdash p_0 \vdash \bigcup_{i<\delta} p_i = p$.

For $x = a$, by 2.1, if $\bar{a}$ realizes p, $\mathrm{stp}(\bar{a}, B) \vdash p_i$ hence $\mathrm{stp}(\bar{a}, B) \vdash p$ hence $(p, B) \in \mathbf{F}^a_\lambda$.

LEMMA 2.14: *Axiom* (IX) *is satisfied by* $\mathbf{F}^x_\lambda$ *for* $x = t, s, a, f$; *for* l, *if* $\delta < \operatorname{cf} \lambda$, *its conclusion holds.*

Proof. Let $\delta < \operatorname{cf} \lambda$, A_i, B_i are increasing with i, $p \in S^m(\bigcup_{i<\delta} A_i)$, and $(p \upharpoonright A_i, B_i) \in \mathbf{F}^x_\lambda$. We must prove $(p, \bigcup_{i<\delta} B_i) \in \mathbf{F}^x_\lambda$.

For $x = f$, if p forks over $\bigcup_{i<\delta} B_i$, some finite $p' \subseteq p$ forks over $\bigcup_{i<\delta} B_i$, so for some $j < \delta$ p' is over A_j, so by III, 1.1(8) $p \upharpoonright A_j$ forks over B_j, contradiction.

For $x = s$, $t[a]$ there are $\mathbf{F}^x_\lambda$-types q_i [almost] over B_i, $q_i \subseteq p$, $q_i \vdash p \upharpoonright A_i$. Then $\bigcup_{i<\delta} q_i$ proves our conclusion. For $x = l$ the proof is similar.

DEFINITION 2.4: (1) $\lambda_1(T)$ is the first cardinal λ such that there are no A_i $(i < \lambda)$ and $p \in S^m(\bigcup_{i<\lambda} A_i)$ such that $i < j \Rightarrow A_i \subseteq A_j$, and $p \upharpoonright A_{i+1}$ splits over A_i (if there is such λ, and ∞ otherwise).

(2) $\lambda_2(T)$ is the first cardinal λ such that there are no A_η and $p_\eta \in S^m(A_\eta)$ for $\eta \in {}^{\lambda>}2$ such that

(i) $p_{\eta^\frown\langle 0\rangle} \neq p_{\eta^\frown\langle 1\rangle}$, $A_{\eta^\frown\langle 0\rangle} = A_{\eta^\frown\langle 1\rangle}$ for $\eta \in {}^{\lambda>}2$,

(ii) if η is an initial segment of ν then $p_\eta \subseteq p_\nu$.

(3) $\lambda_3(T)$ is the first cardinal λ such that there are no m-types p_η, $\eta \in {}^{\lambda>}2$ such that

(i) $p_{\eta^\frown\langle 0\rangle} \cup p_{\eta^\frown\langle 1\rangle}$ is inconsistent for $\eta \in {}^{\lambda>}2$,

(ii) if η is an initial segment of ν then $p_\eta \subseteq p_\nu$.

DEFINITION 2.5: (1) $\lambda^1(T)$ [$\lambda^2(T)$] is the first cardinal λ such that for every $B \subseteq A$, $p \in S^m(B)$ there is $C \subseteq A$, $|C| < \lambda$ and $q \in S^m(A)$, such that $p \subseteq q$, and q does not split over $B \cup C$ [and $q \upharpoonright (B \cup C) \vdash q$].

(2) $\lambda^3(T)$ is the first cardinal λ such that for every m-type p over A, there is an m-type q over A, $|q| < \lambda$, and $p \cup q \vdash p' \in S^m(A)$ for some p'.

THEOREM 2.15: (1) $\kappa(T) \leq \lambda_1(T) \leq \lambda_2(T) \leq \lambda_3(T)$.

(2) *T is unstable in every* $\mu < 2^{<\lambda_3(T)}$.

(3) *If T is stable then* $\lambda_3(T) \leq |T|^+$.

(4) $\lambda_3(T) = \aleph_0$ *iff* T *is totally transcendental* (*see Exercise* 2.23).
(5) $\lambda^1(T) \leq \lambda^2(T) \leq \lambda^3(T)$.

Remark. If $2^{|T|^+} > 2^{|T|}$, then (2) implies (3) trivially.

Proof. (1) If T is stable $\kappa(T) \leq \lambda_1(T)$ by III, 3.1, as strong splitting implies splitting. If T is unstable, it is an easy exercise. We have, in fact, proved that $\lambda_1(T) \leq \lambda_2(T)$ in I, 2.7. Now $\lambda_2(T) \leq \lambda_3(T)$ is trivial.

(2) Let $\mu < 2^{<\lambda_3(T)}$; for some $\lambda < \lambda_3(T)$, $\mu < 2^\lambda$; taking the first such λ, we have $2^{<\lambda} \leq \mu < 2^\lambda$. Hence there are p_η, $\eta \in {}^{\lambda>}2$, as in Definition 2.4(3). For every $\eta \in {}^{\lambda>}2$ choose finite B_η so that $(p_{\eta\frown\langle 0\rangle} \restriction B_\eta) \cup (p_{\eta\frown\langle 1\rangle} \restriction B_\eta)$ is inconsistent, and let $B = \bigcup_{\eta\in{}^{\lambda>}2} B_\eta$. For $\nu \in {}^\lambda 2$, $p_\nu = \bigcup_{\alpha<\lambda} p_{\nu\restriction\alpha}$ is consistent, so for some $q_\nu \in S^m(B)$, $p_\nu \cup q_\nu$ is consistent. Clearly $\nu \neq \eta \Rightarrow q_\nu \neq q_\eta$, hence $|S^m(B)| \geq 2^\lambda > \mu \geq 2^{<\lambda} \geq |B|$.

(3) Suppose T is stable, $\lambda_3(T) > |T|^+$, so for $\lambda = |T|^+$, there are p_η $(\eta \in {}^{\lambda>}2)$ as in Definition 2.4(3). Let $\{\Delta_i: i < |T|\}$ be a list of all finite sets of formulas in L. We define by induction on $i \leq |T|$ sequences $\eta_i \in {}^{\alpha(i)}2$, $\alpha(i) < |T|^+$ and $i < j \Rightarrow \alpha(i) < \alpha(j)$, $\eta_j \restriction \alpha(i) = \eta_i$. For $i = 0$, or i limit choose in any acceptable way. For $i + 1$ choose $\eta_{i+1} = \eta_i\frown\nu \in {}^{\lambda>}2$ so that $R^m(p_{\eta_{i+1}}, \Delta_i, 2)$ is minimal. For $\nu = \eta_{|T|}$, $p_{\nu\frown\langle 0\rangle}$, $p_{\nu\frown\langle 1\rangle}$ contradict our construction and II, 1.4(1).

(4) This is a restatement of II, 3.1 and 3.2.

(5) Trivial.

THEOREM 2.16: (1) $\lambda^l(T) \leq \lambda_l(T)$ *for* $l = 1, 2, 3$.

(2) *Suppose* $A_i \subseteq A_j$ *for* $i < j \leq \alpha$ *and* p *is an* m-*type over* A_0:

(i) *There is* $q \in S^m(A_\alpha)$, $p \subseteq q$ *and* $B \subseteq A_\alpha$, $|B| < \lambda_2(T)$ *such that* $p \cup q \restriction (B \cap A_i) \vdash q \restriction A_i$.

(ii) *There is* $q \in S^m(A_\alpha)$ $p \subseteq q$ *and* $r \subseteq q$, $|r| < \lambda_3(T)$ *such that* $p \cup r'_i \vdash q \restriction A_i$ *where* $r'_i = r \restriction A_i$.

(iii) *There are* $q \in S^m(A_\alpha)$, $p \subseteq q$ *and* $B \subseteq A_\alpha$, $|B| < \kappa(T)$, *and* $r \subseteq q \restriction B$, $|r| < \kappa(T)$, *such that* $p \cup (r \restriction A_i)$ *has no extension in* $S^m(A_i)$ *which forks over* $B \cap A_i$.

(iv) *There is* $q \in S^m(A_\alpha)$, $p \subseteq q$ *and* $B \subseteq A_\alpha$, $|B| < \lambda_1(T)$ *such that*: $p \cup q \restriction (B \cap A_i)$ *has no extension in* $S^m(A_i)$ *which splits over* $B \cap A_i$; *for each* i.

Proof. (1) For $l = 1$ it is easy, and for $l = 2, 3$ it follows from (2) for $\alpha = 0$.

(2) As the proofs are similar, we prove only (ii). Suppose the conclusion fails, and we get a contradiction. We define for every sequence η of ones and zeroes, $l(\eta) < \lambda_3(T)$ by induction on $l(\eta)$, an ordinal $\beta(\eta) \le \alpha$ and an m-type q_η over $A_{\beta(\eta)}$, such that $|q_\eta| < |l(\eta)|^+ + \aleph_0$ and $p \cup q_\eta$ is consistent, and for $i < \beta(\eta)$, $p \cup (q_\eta \restriction A_i)$ has a unique extension in $S^m(A_i)$, and $\nu \lessdot \eta \Rightarrow q_\nu \subseteq q_\eta$.

For the empty η, $q_\eta = \emptyset$, $\beta(\eta) = 0$ and when $l(\eta)$ is a limit ordinal δ let $q_\eta = \bigcup_{i<\delta} q_{\eta|i}$, $\beta(\eta) = \bigcup_{i<\delta} \beta(\eta \restriction i)$. If q_η, $\beta(\eta)$ are defined, let $\beta \le \alpha$ be the first ordinal such that $p \cup q_\eta$ has at least two extensions in $S^m(A_\beta)$. Such β exists, as we assume that the conclusion of (2)(ii) fails, and $\beta \ge \beta(\eta)$ by the induction hypothesis. So for some $\bar{a}_\eta \in A_\beta$ and φ_η, $p \cup q_\eta \cup \{\varphi_\eta(\bar{x}; \bar{a}_\eta)^{\mathbf{t}}\}$ is consistent for $\mathbf{t} \in \{0, 1\}$. Define $q_{\eta^\frown\langle l\rangle} = q_\eta \cup \{\varphi_\eta(\bar{x}; \bar{a}_\eta)^l\}$, $\beta(\eta^\frown\langle l\rangle) = \beta$. Clearly all the induction hypotheses are satisfied. So the q_η's, $\eta \in \bigcup \{{}^\alpha 2: \alpha < \lambda_3(T)\}$, contradict the definition of $\lambda_3(T)$.

LEMMA 2.17: (1) *If* $\lambda \ge \lambda^3(T)$, *then* $\mathbf{F}^t_\lambda$ *satisfies axioms* (X.1) *and* (XI.1), *and if* $\lambda \ge \lambda_3(T)$, *then* $\mathbf{F}^t_\lambda$ *satisfies also axioms* (X.2) *and* (XI.2).

(2) *If* $\lambda \ge \lambda^2(T)$, *then* $\mathbf{F}^s_\lambda$ *satisfies axioms* (X.1) *and* (XI.1) *and if* $\lambda \ge \lambda_2(T)$, *then* $\mathbf{F}^s_\lambda$ *satisfies also axioms* (X.1) *and* (XI.2).

Proof. Immediate, by 2.16(2)(i), (ii) (for $\lambda \ge \lambda_l(T)$) and Definition 2.5 (when $\lambda \ge \lambda^l(T)$).

LEMMA 2.18: (1) *Axioms* (X.1), (XI.1), (X.2) *and* (XI.2) *are satisfied by* $\mathbf{F}^t_\lambda$, $\mathbf{F}^s_\lambda$ *if* $\lambda > |T|$, T *stable or* T *is stable in some* $\mu < 2^\lambda$.

(2) *Axioms* (X.1), (XI.1), (X.2) *and* (XI.2) *are satisfied by* $\mathbf{F}^a_\lambda$ *if* $\lambda \ge \kappa(T)$.

(3) *Axioms* (X.1), (XI.1), (X.2) *and* (XI.2) *are satisfied by* $\mathbf{F}^f_\lambda$ *if* T *is stable.*

(4) Ax(X.1) *is satisfied by* $\mathbf{F}^l_\lambda$ *if* T *is stable,* $\lambda \ge |T|$ (*e.g., by every countable stable* T).

(5) *Axioms* (X.1), (XI.1), (X.2) *and* (XI.2) *are satisfied by* $\mathbf{F}^t_{\aleph_0}$ *if* T *is totally transcendental.*

Proof. Part (1) follows by 2.15(2)(3), 2.16(1) and 2.17(1), (2). Part (2) follows by 2.16(2)(iii); part (3) by III, 1.4. Part (5) follows by 2.15(4), 2.16 and 2.17(1). So we are left with (4).

Let p be an m-type over A, and $\{\varphi_i(\bar{x}; \bar{y}_i): i < |T|\}$ a list of the formulas in L. Define by induction on $i < |T|$ formulas $\psi_i(\bar{x}; \bar{a}_i)$ ($\bar{a}_i \in A$)

such that $p_i = p \cup \{\psi_j(\bar{x}; \bar{a}_j): j < i\}$ is consistent. If we have defined $\psi_i(\bar{x}, \bar{a}_i)$ for $i < \alpha < |T|$, clearly p_α is consistent also when α is a limit ordinal. Choose $\psi_\alpha(\bar{x}; \bar{a}_\alpha)$ so that $R^m(p_{\alpha+1}, \varphi_\alpha, 2)$ is minimal. It is easy to check that for some $q_\alpha \in S^m_{\varphi_\alpha}(A)$, $p_{\alpha+1} \vdash q_\alpha$. Hence $p_{|T|}$ has a unique extension in $S^m(A)$: $\bigcup_{\alpha<|T|} q_\alpha$. So Ax(X.1) (and a little more) holds.

LEMMA 2.19: *When* $x = t, s, l, f$, $\mathbf{F} = \mathbf{F}^x_\lambda$, $\mathbf{F}$ *satisfies* Ax(XII). *Also for* $x = a$, *T stable this holds.*

Proof. For $x = t, s$, for some $\mathbf{F}^x_\lambda$-type $q \subseteq p \restriction C_1$, $q \vdash p \restriction B$ (as $(p \restriction B, C_1) \in \mathbf{F}^x_\lambda$) hence $q \vdash p \restriction C_2$; but as $(p, C_2) \in \mathbf{F}^x_\lambda$, $p \restriction C_2 \vdash p$, hence $q \vdash p$, so $(p, C_2) \in \mathbf{F}^x_\lambda$. For $x = a$ the proof is similar using 2.1(4). For $x = f$ clearly p does not fork over B (as $C_2 \subseteq B$, $(p, C_2) \in \mathbf{F}^f_\lambda$) and $p \restriction B$ does not fork over C_1, hence by the transitivity of forking (III, 4.4) p does not fork over C_1. We are left with $x = l$, so for every φ, for some $q \subseteq p \restriction C_2$, $|q| < \lambda$, $q \vdash p \restriction^+ \varphi$, it is also clear that $p \restriction C_1 \vdash p \restriction C_2$, hence $p \restriction C_1 \vdash q$, so for every $\psi \in q$ there is a finite $r_\psi \subseteq p \restriction C_1$, $r_\psi \vdash \psi$. Now $r = \bigcup \{r_\psi: \psi \in q\} \subseteq p \restriction C_1$, and $r \vdash q \vdash p \restriction^+ \varphi$, so $(p, C_1) \in \mathbf{F}^l_\lambda$.

DEFINITION 2.6: $\mathbf{F}^p_\lambda = \{(p, A): p \in S^m(B)$, where $m < \omega$, $A \subseteq B$, $|A| < \lambda$ and p does not split over $A\}$.

LEMMA 2.20: (1) $\mathbf{F}^p_\lambda$ *satisfies Axioms* (I), (II.1, 2, 3, 4), (III.1, 2), (IV), (VII), (VIII) *and* (IX).

(2) $\lambda(\mathbf{F}^p_\lambda) = \lambda$, $\mu(\mathbf{F}^p_\lambda) = \infty$.

(3) *If* $\lambda \geq \lambda^1(T)$, *then* $\mathbf{F}^p_\lambda$ *satisfies axiom* (X.1), *and if* $\lambda \geq \lambda_1(T)$ *then* $\mathbf{F}^p_\lambda$ *satisfies axioms* (X.2), (XI.1), (XII.2) *and every* $p \in S^m(A)$ *is* $\mathbf{F}^p_\lambda$-*isolated.*

(4) *If T does not have the independence property then* $\lambda^1(T) \leq |T|^+$.

Proof. (1), (2) are immediate, and (4) is a restatement of III, 7.5. (3) is similar to 2.16.

DEFINITION 2.7: $\mathbf{F}^c_\lambda = \{(p, B)$: $p \in S^m(A)$ for some $m < \omega$, $B \subseteq A$, $|B| < \lambda$, and there is $q \subseteq p \restriction B$, $|q| < \lambda$ such that q has no extension in $S^m(A)$ which forks over $B\}$.

Remark. We can assume q is closed under conjunctions.

LEMMA 2.21: (1) $\mathbf{F}^c_\lambda$ *satisfies axioms* (I), (II.1, 2, 3, 4), (III.1, 2), (IV), (V.1) *and* (IX).

(2) *If* λ *is regular* $\mathbf{F}^c_\lambda$ *satisfies also* (V.2).

TABLE 1

	F							
Axiom	$\mathbf{F}^t_\lambda$	$\mathbf{F}^s_\lambda$	$\mathbf{F}^a_\lambda$	$\mathbf{F}^f_\lambda$	$\mathbf{F}^l_\lambda$	$\mathbf{F}^p_\lambda$	$\mathbf{F}^g_\lambda$	$\mathbf{F}^b_\lambda$
I	+	+	+	T stable	+	+	+	+
II.1	+	+	+	T stable	+	+	+	+
II.2	+	+	+	T stable	+	+	+	+
II.3	+	+	+	T stable	+	+	+	Skolem functions
II.4	+	+	+	T stable	+	+	+	Skolem functions
III.1	+	+	+	T stable	+	+	+	+
III.2	+	+	+	T stable	+	+	+	+
IV	+	+	+	T stable	+	+	+	+
V.1	+	+	+	T stable	+	+	+	+
V.2	λ regular	+	+	T stable	×	+	λ regular	×
VI	+	+	T stable	T stable	λ regular	×	×	×
VII	λ regular	+	T stable	T stable	λ regular	+	?	+
VIII	+	+	+	T stable	×	+	?	+
IX	+	+	+	T stable	×	+	+	+
X.1	$\lambda \geq \lambda^3(T)$	$\lambda \geq \lambda^2(T)$	$\lambda \geq \kappa(T)$	T stable	T stable $\lambda \geq \|T\|$	$\lambda \geq \lambda^1(T)$	$\lambda \geq \kappa(T)$	Skolem functions
X.2	$\lambda \geq \lambda_3(T)$	$\lambda \geq \lambda_2(T)$	$\lambda \geq \kappa(T)$	T stable	×	$\lambda \geq \lambda_1(T)$	$\lambda \geq \kappa(T)$	Skolem functions
XI.1	$\lambda \geq \lambda^3(T)$	$\lambda \geq \lambda^2(T)$	$\lambda \geq \kappa(T)$	T stable	×	×	?	+
XI.2	$\lambda \geq \lambda_3(T)$	$\lambda \geq \lambda_2(T)$	$\lambda \geq \kappa(T)$	T stable	×	×	?	+
XII	+	+	T stable	T stable	+	×	?	×

(3) *Let* $\kappa(T) \leqslant \lambda$, $\mathbf{F}^c_\lambda$ *satisfies* Ax(X.1) *and* (X.2). *Then* A *is* $\mathbf{F}^c_\lambda$*-saturated iff* A *is the universe of a* λ*-compact model.*

Proof. Easy.

We can sum up (most of) our results (see Table 1) which includes also results on **F**'s which will be defined later. In each place we write a plus if the axiom is true for that **F** and a cross if there is a counter-example. In the remaining cases we give a sufficient condition for the axiom to hold or just put a question mark. On $\mathbf{F}^b_\lambda$ see VII, 4.4, Exercise VII, 4.2. See also 4.4–6.

The exercises, usually, show that appropriate lemmas cannot be improved, and are ordered approximately in the same way. Whenever λ is singular $\lambda = \sum_{i<\kappa} \lambda_i$ where $\lambda_i, \kappa < \lambda$, $\kappa = \mathrm{cf}\,\lambda$.

EXERCISE 2.1: Show that in Lemma 2.1, for unstable T, not necessarily (3) $\Leftrightarrow$ (4). [Hint: Let M be a model of T_{ind} (see II, 4.8(2)). $R(x, y, z) = P(x) \wedge P(y) \wedge \neg P(z) \wedge [zEx \equiv zEy]$, $N = (|M|, P, R)$,

$T = \mathrm{Th}(N)$ (so $N \prec \mathfrak{C}$), and choose a, c such that $N \models P(c) \wedge \neg P(a)$. Then $B = \emptyset$, $A = \{a\}$, $\bar{c} = \langle c \rangle$ form a counterexample.]

Deduce in $\mathfrak{C}^{\mathrm{eq}}$ a counterexample to Ax(IV), $\mathbf{F}^a_{\aleph_0}$.

EXERCISE 2.2: (1) Show that for every λ for some T, $\mu(\mathbf{F}^t_\lambda) = \mu(\mathbf{F}^l_\lambda) = \aleph_0$.

(2) Show that if $\mathbf{F}^t_\lambda$ satisfies "every $\mathbf{F}^t_\lambda$-type over A has a $\mathbf{F}^t_\lambda$-isolated complete extension over A" then $\mu(\mathbf{F}^t_\lambda) \leq \lambda$.

(3) Show $\mu(\mathbf{F}^s_\lambda) = \lambda$, $\mu(\mathbf{F}^a_\lambda) \geqslant \lambda$ and for $\lambda > \aleph_0$, $\mu(\mathbf{F}^a_\lambda) = \lambda$.

(4) Show that for some stable T, $\mu(\mathbf{F}^a_{\aleph_0}) = \aleph_1$.

(5) For every κ, show that there is a $\mathbf{F}^t_\kappa$-saturated, not $\aleph_1$-compact model.

Remark. See Lemma 2.2(4). [Hint: (1) Take T_{ind} (see II, 4.8(2)) choose $A \subseteq |\mathfrak{C}|$, $|A \cap P(\mathfrak{C})| = |A - P(\mathfrak{C})| = \lambda$, and let $T = \mathrm{Th}(\mathfrak{C}, \ldots, a, \ldots)_{a \in A}$.

(2) (3) By 4.10 and VIII, 4.8 for each regular $\mu < \lambda$, T has a μ-saturated model which is not λ^+-saturated.

(4) Let $T = \mathrm{Th}({}^{\omega}2, \ldots, E_n, \ldots)$ when $\eta E_n \nu$ iff $\eta \upharpoonright n = \nu \upharpoonright n$.

(5) like (1).]

EXERCISE 2.3: For λ singular, Ax(V.2) may fail for $\mathbf{F}^l_\lambda$ even for stable T (by 2.2(8), 2.8(2), $\lambda \leq |T|$ of course). [Hint: Let Q, P_i $(i < \kappa)$ be disjoint one-place predicates, F_i a (partial) one-place function, from Q onto P_i. E^i_α $(\alpha < \lambda_i)$ are independent equivalence relations on P_i $(i < \kappa)$. For every $c_i \in P_i(\mathfrak{C})$, $\{F_i(x) = c_i \wedge Q(x) : i < \kappa\}$ is realized in $\mathfrak{C}$. Let $c_i \in P_i(\mathfrak{C})$, $a \in Q(\mathfrak{C})$, $F_i(a) \neq c_i$, $A = B = \emptyset$, $C = \{c_i : i < \kappa\}$, $\bar{a} = a$.]

EXERCISE 2.4: Show that for λ satisfying cf $\lambda \leq |T|$, Ax(V.2) fail for $\mathbf{F}^l_\lambda$ (when cf $\lambda > |T|$ it holds by 2.2(8), 2.8(2).) [Hint: Let I be indiscernible, $|I| = \lambda$, $a \in I$, $C = I - \{a\}$, and $B = A = \emptyset$. Remember $\lambda(\mathbf{F}^l_\lambda) = \lambda^+ + |T|^+$ by 2.2(3).]

QUESTION 2.5: Show that Ax(VI) may fail for $\mathbf{F}^l_\lambda$, when λ is singular cf $\lambda \leq |T|$.

EXERCISE 2.6: Show that for λ singular Ax(VII) may fail for $\mathbf{F}^t_\lambda$, (notice that by 2.2(8), $\lambda \leq |T|$). [Hint: As in Exercise 2.3 but $F_i(a) = c_i$, $B = \emptyset$, $A = \{a^i_\alpha : \alpha < \lambda_i, i < \kappa\}$, $a^i_\alpha \in P_i(\mathfrak{C})$ and $E^i_\alpha(c_i, a^i_\beta)$ iff $\alpha = \beta$.]

QUESTION 2.7: Show that for λ singular Ax(VII) may fail for $\mathbf{F}^l_\lambda$, with $|T| = \kappa$ $(= \mathrm{cf}\ \lambda)$.

EXERCISE 2.8: Show that Ax(VII) may fail for $\mathbf{F}^a_\lambda$ when T is unstable. [Hint: As in Exercise 2.1, and let $b \in R(\mathfrak{C})$, $\mathfrak{C} \vDash \neg E(b, c, a)$. Let $B = \emptyset$, $A = \{b\}$, $C = \{a\}$, $\bar{a} = \langle c \rangle$.]

EXERCISE 2.9: Show that Ax(IX) may fail for $\mathbf{F}^l_\lambda$, when $\mathrm{cf}\,\lambda \le |T|$, and we can choose $|T| = \mathrm{cf}\,\lambda$. [Hint: Let L($T$) consist of equality only; a_i ($i \le \lambda$) be distinct elements. $B_i = A_i = \{a_j : j < i\}$ for $i < \lambda$ and $p = \mathrm{tp}(a_\lambda, \bigcup_{i<\lambda} A_i)$.]

EXERCISE 2.10: Show that Ax(VIII) may fail for $\mathbf{F}^l_\lambda$ when $\mathrm{cf}\,\lambda \le |T|$, and we may choose $|T| = \mathrm{cf}\,\lambda$. [Hint: Let P_i ($i < \kappa$) be infinite pairwise disjoint one-place predicates $B_i \subseteq P_i(\mathfrak{C})$, $|B_i| = \lambda_i$, $\delta = \kappa$, $A_i = \bigcup_{j<i} B_j$, $a \notin \bigcup_i P_i(\mathfrak{C})$, $p = \mathrm{tp}(a, \bigcup_i B_i)$.]

EXERCISE 2.11: Show that Ax(X.1) (hence (X.2)) may fail for $\mathbf{F}^x_\lambda$ $x = t, s, a$ even for stable T, $\lambda = |T|$. [Hint: (1) $x = t$: T will have independent one-place predicates P_i ($i < \lambda$). $A = \emptyset$, $\varphi = (x = x)$ contradicts Ax(X.1).

(2) $x = s$: P_i ($i < \lambda$) and Q will be disjoint one-place predicates, $|P_i(\mathfrak{C})| = 2$, F_i ($i < \lambda$) one-place partial function from Q onto P_i, such that for every $c_i \in P_i(\mathfrak{C})$, $\{F_i(x) = c_i \wedge Q(x) : i < \lambda\}$ is realized. Then $A = \bigcup_{i<\lambda} P_i(\mathfrak{C})$, $\varphi = Q(x)$ contradicts Ax(X.1).

(3) $x = a$: the example of Exercise 2.3, with λ instead of κ, deleting the E^i_α's, $A = \{c_i : i < \lambda\}$, $\varphi = Q(x)$.]

EXERCISE 2.12: Show that Ax(XI.1) (hence Ax(XI.2) may fail for $\mathbf{F}^x_\lambda$, $x = t, s, a$, even for stable T, $|T| = \lambda$. [Hint: (1) $x = t$: Use T from Exercise 2.11(2), $B = A = \emptyset$, $C = \bigcup_{i<\lambda} P_i(\mathfrak{C})$, $p = \mathrm{tp}(a, A)$ for any $a \in Q(\mathfrak{C})$.

(2) $x = s$: as in 1.

(3) $x = a$: T as in Exercise 2.11(3), $B = A = \emptyset$, $C = \{c_i : i < \lambda\}$, $p = \mathrm{tp}(a, A)$ for any $a \in Q(\mathfrak{C})$.]

EXERCISE 2.13: Show that Ax(X.1) may fail for $\mathbf{F}^l_\lambda$ ($|T| = \aleph_0$, *or* T stable, but $|T| > \lambda$; in fact we can choose a superstable T). [Hint: (1) T_{ind}, (2) Let T be T^{λ^+} but the counterexample is for T^{eq} (for T^{λ^+} see Exercise II, 2.3, for "eq" see III, Section 6). Clearly $\kappa(T^{\mathrm{eq}}) = \lambda^{++}$, $|T^{\mathrm{eq}}| = \lambda^+$. Let I_α ($\alpha < \lambda^+$) be the set of one-to-one functions from α into λ, $I = \bigcup_\alpha I_\alpha$, and choose $a_s \in \mathfrak{C}$ for $s \in I$ such that $a_s E_i a_t$ iff

$s \restriction i = t \restriction i$; then $A = \{a_s/E_i: s \in I, i < \lambda^+\} \subseteq \mathfrak{C}^{eq}$, $\varphi = P_=(x)$ form a counterexample,

(3) Let λ be regular

$$T = \mathrm{Th}(M), \qquad M = ({}^\lambda\lambda \times \omega, \ldots, P_\eta, \ldots, \ldots, E_\nu, \ldots)_{\eta \in {}^{\lambda >}\lambda, \nu \in {}^\lambda\lambda}$$

where $P_\eta = \{\langle \rho, n\rangle \in |M|: \eta \lessdot \rho\}$ and $\langle \rho, n\rangle E_\nu \langle \eta, m\rangle$ iff for some α $\rho \restriction \alpha = \eta \restriction \alpha = \nu \restriction \alpha$, $\rho(\alpha) = \eta(\alpha) \neq \nu(\alpha)$ or $\rho = \eta$. The counterexample is for T^{eq} (assuming $M \prec \mathfrak{C}$): $A = \{\langle \rho, n\rangle/E_\nu: \nu \in {}^\lambda\lambda, \langle \rho, n\rangle \in |M|, \nu \neq \rho\}$, and $\varphi = P_=(x)$.]

EXERCISE 2.14: Show that Ax(X.2) may fail for $\mathbf{F}^l_\lambda$, T stable, $|T| = \lambda$. [Hint: Let $T = T^\lambda$, λ regular (see Exercise II, 2.3), choose $a_\eta \in \mathfrak{C}$ for $\eta \in {}^\lambda(\lambda + 1)$ such that $a_\eta E_i a_\nu$ iff $\eta \restriction i = \nu \restriction i$. The counterexample is for T^{eq}: $A_0 = \emptyset$, $A_1 = \{a_\eta/E_i: i < \lambda, \eta \in {}^\lambda\lambda\}$, $A_2 = A_1 \cup \{a_\eta: \eta \in {}^\lambda(\lambda + 1)$, $(\lambda + 1) \in \mathrm{Range}(\eta)\}$ and $\varphi = P_=(x)$.]

EXERCISE 2.15: Show that Ax(XI.1) (hence (XI.2)) may fail for $\mathbf{F}^l_\lambda$, T stable, $\lambda = |T|$. [Hint: Let P_i ($i < \lambda$) be pairwise disjoint, infinite one-place predicates $c \in |\mathfrak{C}| - \bigcup_i P_i(\mathfrak{C})$, $c_i \in P_i(\mathfrak{C})$, $A = \emptyset$, $C = \{c_i: i < \lambda\}$. So tp(c, A) is $\mathbf{F}^l_{\aleph_0}$-isolated, it has a unique extension in $S(C)$: tp(c, C), which is not $\mathbf{F}^l_\lambda$-isolated.]

EXERCISE 2.16: Show that Ax(XII) may fail for $\mathbf{F}^a_\lambda$ (T unstable). [Hint: Use T, a, b, c from Exercise 2.8: we let $C_1 = \emptyset$, $B = C_2 = \{a\}$, $A = \{a, c\}$, $p = \mathrm{tp}(b, A)$.]

EXERCISE 2.17: Show that (VI) may fail for $\mathbf{F}^p_\lambda$. [Hint: Use the theory of the rationals as an ordered set, $B = C = \emptyset$, $A = \{1\}$, $\bar{b} = \langle 3\rangle$, $\bar{a} = \langle 2\rangle$.]

EXERCISE 2.18: Show that Ax(XI.1) (and XI.2)) may fail for $\mathbf{F}^p_\lambda$. [Hint: Let T be as in Exercise 2.11, hint 2. We let $A = \emptyset$, $B = \emptyset$, $C = \bigcup P_i(\mathfrak{C})$, choose $a \in Q(\mathfrak{C})$ and let $p = \mathrm{tp}(a, A)$. (We use the notation of Ax(XI.1).)

EXERCISE 2.19: Show that Ax(XII) may fail for $\mathbf{F}^p_\lambda$. [Hint: Use the previous example. $C_1 = \emptyset$, $A = C_2 = \{F_0(a)\}$, $B = P_0(\mathfrak{C})$, $p = \mathrm{tp}(a, B)$.]

EXERCISE 2.20: Prove 2.12 for $\mathbf{F}^l_\lambda$ instead $\mathbf{F}^a_\lambda$.

EXERCISE 2.21: Assume Ax(III.1) and (VII). If Ax(X.1) [or Ax(X.2)] holds, then it holds for $\varphi = \varphi(\bar{x}; \bar{a})$ too.

EXERCISE 2.22: Show that $\lambda_3(T) = \aleph_0$ iff $\lambda^3(T) = \aleph_0$. [Hint: Suppose $\lambda_3(T) > \aleph_0$, then T is not totally transcendental (by 2.15(4)); let M be a $|T|^+$-saturated model of T. Then $A = |M|$, $q = \{\neg\varphi(x; \bar{a})$: $R(\varphi(\bar{x}; \bar{a}), \mathrm{L}, 2) < \infty\}$ exemplifies $\lambda^3(T) > \aleph_0$. Now $\lambda^3(T) > \aleph_0$ implies $\lambda_3(T) > \aleph_0$ by 2.16(2), so as $\lambda^3(T), \lambda_3(T) \geq \aleph_0$ we finish.]

QUESTION 2.23: Show that between $\kappa_r(T)$, $|T|$, $\lambda^l(T)\, \lambda_l(T)$ $(l = 1, 2, 3)$ there are no connections except those from 2.15(1), (5), 2.16(1) Exercise 2.22. [Hint: (1) T_1: Let E_i $(i < \lambda_1)$ be independent equivalence relations, each with two equivalence classes. Then $\kappa_r(T_1) = \aleph_0$, $|T_1| = \lambda_1$, $\lambda_1(T_1) = \lambda^1(T_1) = \lambda_2(T_1) = \lambda^2(T_1) = \aleph_0$. $\lambda^3(T_1) = \lambda_3(T_1) = \lambda_1^+$.

(2) T_2: Let $M = ({}^{\lambda(2)}3, \ldots, P_\eta, \ldots)$ where, for $\eta \in {}^{\lambda(2)>}2$ $P_\eta = \{\nu \in {}^{\lambda(2)}3 \colon \eta \lessdot \nu\}$, and $\lambda_2 = 2^{<\lambda(2)}$ and $T_2 = \mathrm{Th}(M)$. Then $\kappa_r(T_2) = \aleph_0$, $|T_2| = 2^{<\lambda(2)}$, $\lambda_1(T_2) = \lambda^1(T_2) = \lambda_2(T_2) = \lambda^2(T_2) = \aleph_0$, $\lambda^3(T_2) = \aleph_1$, $\lambda_3(T_2) = \lambda(2)^+$.

(3) T^3: Is the T from Exercise 2.11, hint (2). Clearly $\kappa(T) = \aleph_0$, $|T| = \lambda$, $\lambda_l(T) = \lambda^l(T) = \lambda^+$ for $l = 1, 2, 3$.]

QUESTION 2.24: What happens if we change the definition of $\lambda^3(T)$ to: $\lambda^3(T)$ is the first cardinal λ such that if Γ is a set of m-formulas, $p \subseteq \Gamma$ an m-type, then there is $q \subseteq \Gamma$, $|q| < \lambda$, $p \cup q$ is consistent but for every $\varphi \in \Gamma$ either $p \cup q \vdash \varphi$ or $p \cup q \vdash \neg\varphi$?

(i) Does the value of $\lambda^3(T)$ change?

(ii) Do we have nicer or better results with this definition?

QUESTION 2.25: What happens if in the definition of $\lambda^2(T)$ (Definition 2.5(1)) we replace $q \restriction (B \cup C) \vdash q$ by $p \cup q \restriction C \vdash q$?

(i) Does the value of $\lambda^2(T)$ change?

(ii) Do we have nicer results with this definition?

EXERCISE 2.26: Prove that T is transcendental iff $\lambda_2(T) = \aleph_0$ (see Definition II, 3.4).

EXERCISE 2.27: Suppose T is stable, M is $\mathbf{F}^a_{\kappa(T)}$-saturated, $p \in S^m(|M|)$. Prove that there is a $B \subseteq |M|$, $|B| < \kappa(T)$ such that p does not fork over B and $p \restriction B$ is stationary.

EXERCISE 2.28: Show that if $\mathbf{F} = \{(p, B): p$ is definable over $B\}$ then Ax(XI.1) holds for $\mathbf{F}$ if for some M, $B \subseteq |M| \subseteq A$.

EXERCISE 2.29: Let $\mathbf{F} = \mathbf{F}^t_\lambda$,

(1) If cf $\lambda \geq \kappa(T)$, $\mathbf{F}$ satisfies Ax(XI.1), then $\mathbf{F}$ satisfies Ax(XI.2).

(2) If $\lambda \geq \lambda^3(T) + \kappa(T)$ *and* cf $\lambda > \kappa(T)$ or $\lambda > \lambda^3(T)$, then $\mathbf{F}$ satisfies Ax(XI.2).

Remark. (1) For (X.2) see 1.1(5).

(2) For $x = s$ we can get a result using $\lambda^3(T)$, but with $\lambda^2(T)$ it is not clear. [Hint: Use 4.3(1).]

EXERCISE 2.30: Suppose $p = \text{tp}(\bar{a}, A)$ is $\mathbf{F}^a_\lambda$-isolated.

(1) If cf $\lambda > \aleph_0$, and $\mathbf{F}^s_\lambda$ satisfies Ax(XI.1), then p is $\mathbf{F}^s_\lambda$-isolated.

(2) If cf $\lambda = \aleph_0 < \lambda$ but $\lambda^2(T) < \lambda$ (or $\mathbf{F}^s_\mu$ satisfies Ax(XI.1) for arbitrarily large $\mu < \lambda$), then p is $\mathbf{F}^s_\lambda$-isolated. [Hint: see 3.2.]

EXERCISE 2.31: Show that Ax(VII) holds for $\mathbf{F}^t_\lambda$ when $|C| <$ cf λ.

EXERCISE 2.32: $\mathbf{F}^c_\lambda$ may fail to satisfy Ax(VI).

Proof. Let P_i $(i < \lambda)$, P, Q be one-place predicates, $P_i(\mathfrak{C})$ pairwise disjoint, $P_i(\mathfrak{C}) \subseteq P(\mathfrak{C})$, $P(\mathfrak{C}) \cap Q(\mathfrak{C}) = \emptyset$, F_i a function from Q onto P_i, and for every $c_i \in P_i(\mathfrak{C})$, $\{Q(x) \wedge F_i(x) = c_i: i < \lambda\}$ is realized. Let $B = C = A = \emptyset$, $b \in Q(\mathfrak{C})$, $a \in P(\mathfrak{C}) - \bigcup_{i<\lambda} P_i(\mathfrak{C})$. If we add $G_i: P_0 \to P_i$ one-to-one, onto, $Q(x) \to G_i F_0(x) = F_i(x)$ we get T superstable.

IV.3. General properties of F-primary models

We use the notation of Definition 1.2 and Section 1 in general. Before each theorem the axioms used are listed and also assumptions on $\lambda(\mathbf{F})$.

THEOREM 3.1: (1) [Ax(XI.1)] *Over every A, for any μ there is an $(\mathbf{F}, \mu)$-primary set (so by* 1.2(6), *if* Ax(X.1) *holds it is an $(\mathbf{F}, \mu)$-primary model).*

(2) [Ax(XI.1)] *If $\mu(\mathbf{F}) < \infty$, then over every A there is an $\mathbf{F}$-primary set.*

(3) [Ax(XI.1)] *Suppose $|B| < \mu \Rightarrow |S^m(B)| \leq \mu$ or at least $|B| < \mu \Rightarrow |\{p: p \in S^m(B), p$ is an $\mathbf{F}$-isolated type$\}| \leq \mu$.*

If μ is regular then over every A, $|A| \le \mu$ there is an $(\mathbf{F}, \mu)$-primary set of cardinality μ, and so it is **F**-*semi-saturated.*

(4) [Ax(III.1)] *In* (3), *instead of requiring μ to be regular we can assume T stable and;* (i) $\mathrm{cf}\,\mu \ge \lambda(\mathbf{F}) + \kappa(T)$ *or* (ii) $\mathrm{cf}\,\mu \ge \lambda(\mathbf{F})$, *and also $p \in \mathbf{F}(B)$ implies p does not fork over B.*

(5) [Ax(X.1)] *Over every A, there is an* **F**-*constructible model.*

Proof. (1) We define by induction on i, sequences $\bar{a}_i$, and sets B_i, such that $p_i = \mathrm{tp}(\bar{a}_i, A_i) \in \mathbf{F}(B_i)$ where $A_i = A \cup \bigcup \{\bar{a}_j : j < i\}$. Suppose we have defined $\bar{a}_i$ for $i < \alpha$, and A_α is not $(\mathbf{F}, \mu)$-saturated. Then there is a pair $(q, C) \in \mathbf{F}$, $C \subseteq \mathrm{Dom}\, q \subseteq A_\alpha$, $|\mathrm{Dom}\, q| < \mu$, such that q is not realized in A_α. Among these pairs choose one, (q_α, C_α) with minimal $j(q) = \min\{j : j \le \alpha,\ \mathrm{Dom}\, q \subseteq A_j\}$. By Ax(XI.1) we can find a pair $(p_\alpha, B_\alpha) \in \mathbf{F}$, $q \subseteq p_\alpha$, $p_\alpha \in S^m(A_\alpha)$, and choose a sequence $\bar{a}_\alpha$ realizing p_α.

If for some α A_α is $(\mathbf{F}, \mu)$-saturated then clearly we are through. We shall show this holds for some $\alpha < ((|A| + 2)^{\mu + |T|})^+$. The number of pairs $(q, C) \in \mathbf{F}$, $\mathrm{Dom}\, q \subseteq A_j$, $|\mathrm{Dom}\, q| < \mu$ is $\le |A_j|^{\mu + |T|}$. Hence $|\{\alpha : j(q_\alpha) \le j\}| \le |A_j|^{\mu + |T|}$, so for some $\beta_j < (|A_j|^{\mu + |T|})^+$, $(\forall \alpha)(\alpha \ge j + \beta_j \Rightarrow j(q_\alpha) > j)$. We can easily prove by induction on $j < ((|A| + 2)^{\mu + |T|})^+$ that $\beta_j < (|A| + 2)^{\mu + |T|})^+$. Hence there is $\beta^* < ((|A| + 2)^{\mu + |T|})^+$, $\mathrm{cf}\, \beta^* = \mu^+$, and $j < \beta^* \Rightarrow \beta_j < \beta^*$. Clearly A_{β^*} is $(\mathbf{F}, \mu)$-saturated, so we finish.

(2) Immediate, by (1) for $\mu = \mu(\mathbf{F})$.

(3), (4) We shall define $\bar{a}_i$ for $i < \mu^2$ (ordinal exponentiation) such that:

(a) $\mathscr{A} = \langle A, \{\bar{a}_i : i < \mu^2\} \rangle$ is an **F**-construction.

(b) For $i < \mu$ write $\mathscr{A}(\mu \cdot i)$ as $\bigcup_{j < \mu} C^i_j$ where $|C^i_j| < \mu$ and $\alpha < \beta \Rightarrow C^i_\alpha \subseteq C^i_\beta$. We then require that for each i each **F**-isolated type in $S^m(C^i_j)$ $(j < \mu)$ is realized by some sequence in $\mathscr{A}(\mu \cdot (i + 1))$.

If μ is regular clearly $\mathscr{A}(\mu^2)$ is $(\mathbf{F}, \mu)$-saturated as desired. So we have to prove (4) only. If $(p, C) \in \mathbf{F}$, $|\mathrm{Dom}\, p| < \mu$ there are q, $p \subseteq q \in S^m(\mathscr{A}(\mu^2))$ and C', such that $(q, C') \in \mathbf{F}$. If $\mathrm{cf}\, \mu \ge \lambda(\mathbf{F}) + \kappa(T)$, then for some $\alpha < \mu$, $j < \mu$, $C' \subseteq C^\alpha_j$ and q does not fork over C^α_j. Now we can find for each β, $\alpha < \beta < \mu$, $\bar{a}_\beta \in \mathscr{A}(\mu \cdot (\beta + 1))$ realizing $q \restriction \mathscr{A}(\mu\beta)$. As we can assume $\mu > \aleph_0$, $\{\bar{a}_\beta : \alpha + \omega \le \beta < \mu\}$ is an indiscernible set, whose average is q. Hence by III, 3.5 some $\bar{a}_\beta$ realizes p. The other case ((ii)) is handled similarly.

(5) Easy, like (1), using the Tarski–Vaught test.

THEOREM 3.2 [Ax(II.1), (III.1), (III.2) and (VII); $\lambda(\mathbf{F})$ regular]: (1) *If C is* **F**-*constructible over A then C is* **F**-*atomic over A.*

(2) *A is* **F**-*atomic over A.*

(3) *If* $\operatorname{tp}(\bar{b}, A)$ *is* **F**-*isolated then $A \cup \bar{b}$ is* **F**-*atomic over A.*

(4) *If $A_1 \subseteq A_2 \subseteq A_3$, A_3 is* **F**-*atomic over A_2, A_2* **F**-*atomic over A_1, then A_3 is* **F**-*atomic over A_1.*

(5) *If for $i < \delta$ A_i is* **F**-*atomic over A_0, $i < j < \delta \Rightarrow A_i \subseteq A_j$, then $\bigcup_{i<\delta} A_i$ is* **F**-*atomic over A_0.*

Remark. For (5) we need no axiom at all, and for (2) we need Ax(II.1) only.

Proof. (1) Let $\mathscr{A}$ be an **F**-construction of C over A. It suffices to prove by induction on α that $\mathscr{A}(\alpha)$ is **F**-atomic over $A = \mathscr{A}(0)$. For $\alpha = 0$ this follows by (2), and for α a limit ordinal, by (5). For $\alpha = \beta + 1$, $\mathscr{A}(\beta)$ is **F**-atomic over A by the induction hypothesis and $\mathscr{A}(\alpha)[=\mathscr{A}(\beta) \cup \bar{a}_\beta]$ is **F**-atomic over $\mathscr{A}(\beta)$ by the construction, and (3). Hence by part (4) $\mathscr{A}(\alpha)$ is **F**-atomic over A.

(2) Trivial.

(3) If $\bar{c} \in A \cup \bar{b}$, then $\bar{c} = \bar{a}^\frown \bar{b}_1$, where $\bar{b}_1 \subseteq \bar{b}$ and $\bar{a} \in A$. Let $\operatorname{tp}(\bar{b}, A) \in \mathbf{F}(B)$. So by Axiom (III.2) $\operatorname{tp}(\bar{b}, A) \in \mathbf{F}(B \cup \bar{a})$. By Axiom (II.1) $\operatorname{tp}_*(\bar{a}, A) \in \mathbf{F}(\bar{a} \cup B)$, hence by Axiom (VII) (as $A \cup \bar{a} = A$) $\operatorname{tp}_*(\bar{a}^\frown \bar{b}_1, A) \in \mathbf{F}(\bar{a} \cup B)$. So $\operatorname{tp}(\bar{c}, A) = \operatorname{tp}(\bar{a}^\frown \bar{b}_1, A) \in \mathbf{F}(\bar{a} \cup B)$.

(4) Let $\bar{c} \in A_3$, hence for some $B \subseteq A_2$, $\operatorname{tp}(\bar{c}, A_2) \in \mathbf{F}(B)$. For every $\bar{b} \in B$ $\operatorname{tp}(\bar{b}, A_1)$ is **F**-isolated, hence $\operatorname{tp}(\bar{b}, A_1) \in \mathbf{F}(C_{\bar{b}})$ for some $C_{\bar{b}} \subseteq A_1$. By Axiom (III.2) and the regularity of $\lambda(\mathbf{F})$ $\operatorname{tp}_*(B, A_1) \in \mathbf{F}(C)$ $(C = \bigcup_{\bar{b} \in B} C_{\bar{b}})$. By Ax(III.1), (III.2), $\operatorname{tp}(\bar{c}, A_1 \cup B) \in \mathbf{F}(C \cup B)$ hence by Ax(VII), $\operatorname{tp}_*(B \cup \bar{c}, A_1) \in \mathbf{F}(C)$, in particular $\operatorname{tp}(\bar{c}, A_1) \in \mathbf{F}(C)$, so we finish.

(5) Trivial.

THEOREM 3.3 [Axioms (III.1), (VI) and (VIII)]: *Suppose I is a well-ordered set, $\mathscr{A} = \langle A, \{\bar{a}_s: s \in I\}, \{B_s: s \in I\}\rangle$ an* **F**-*construction. If J is another well order, with the same universe, and for $s \in I$, $U_s(\mathscr{A}) \subseteq \{t: J \models t < s\}$ then $\mathscr{A}^* = \langle A, \{\bar{a}_s: s \in J\}, \{B_s: s \in J\}\rangle$ is an* **F**-*construction.*

Proof. It suffices to prove by induction on $s \in J$ (on the order J) that $(p_s(\mathscr{A}^*), B_s) \in \mathbf{F}$. Suppose we have proved it for every $s' < s$ and we shall prove it for s. For $t \in I$, let $C_t = B_s \cup [\mathscr{A}(t) \cap \mathscr{A}^*(s)]$. We prove by induction on $t \in I$ (in fact $t \in I \cup \{\text{al } I\}$) that $(p_s(\mathscr{A}^*) \restriction C_t, B_s) \in \mathbf{F}$.

Notice that C_t is increasing with t, and $C_{\mathrm{al}(I)} = \mathscr{A}^*(s)$ so $p_s(\mathscr{A}^*) \restriction C_{\mathrm{al}(I)} = p_s(\mathscr{A}^*)$, so this will suffice.

For the first t in I, and in fact when $I \models t \le s$, $A \cup B_s \subseteq C_t \subseteq \mathscr{A}(s)$. As by definition $(p_s(\mathscr{A}), B_s) \in \mathbf{F}$, this holds by Axiom (III.1). If t is limit, it follows by Axiom (VIII). So suppose t is the successor of $t(1)$ $I \models t > s$. If $J \models s \le t(1)$ then $C_t = C_{t(1)}$, so there is nothing to prove. Otherwise $C_t = C_{t(1)} \cup \bar{a}_{t(1)}$, and the result is an application of Axiom VI.

So we finish the induction on t, hence on s, hence finish the theorem.

CONCLUSION 3.4 [Ax(III.1), (VI) and (VIII)]: *If* $\mathscr{A} = \langle A, \{\bar{a}_i : i < \alpha\}\rangle$ *is an* **F**-*construction,* $B \subseteq \mathscr{A}(\alpha)$, $|B| \ge \lambda_r(\mathbf{F})$, *then for some well ordering* I *of* α, $\mathscr{A}^* = \langle A, \{\bar{a}_i : i \in I\}\rangle$ *is an* **F**-*construction, and* B *is included in* $\mathscr{A}^*(s_{|B|})$, $s_{|B|}$ *the* $|B|$*th element of* I.

Proof. Immediate by 3.3.

LEMMA 3.5 [Ax(II.1)]: *If* $\mathscr{A} = \langle A, \{\bar{a}_i : i < \alpha\}\rangle$ *is an* **F**-*construction,* $\beta \le \alpha$ *then* $\mathscr{A}' = \langle \mathscr{A}(\beta), \{\bar{a}_i : i < \alpha\}\rangle$ *is also an* **F**-*construction.*

Proof. Immediate.

LEMMA 3.6 [Ax(II.1), (III.1), (III.2), (V.2) and (VII); $\lambda(\mathbf{F})$ regular]: *If* $\mathscr{A} = \langle A, \{\bar{a}_i : i < \alpha\}\rangle$ *is an* **F**-*construction,* $C \subseteq \mathscr{A}(\alpha)$, $|C| < \lambda(\mathbf{F})$ *then* $\mathscr{A}' = \langle A \cup C, \{\bar{a}_i : i < \alpha\}\rangle$ *is also an* **F**-*construction.*

Proof. By 3.2 and 3.5, for every $i < \alpha$, and $\bar{c} \in \mathscr{A}(\alpha)$ $p^i_{\bar{c}} = \mathrm{tp}(\bar{c}, \mathscr{A}(i))$ is **F**-isolated, hence for some $B^i_{\bar{c}} \subseteq \mathscr{A}(i)$, $|B^i_{\bar{c}}| < \lambda(\mathbf{F})$, $(p^i_{\bar{c}}, B^i_{\bar{c}}) \in \mathbf{F}$. Let $B^i = \bigcup\{B^i_{\bar{c}} : \bar{c} \in C \cup \bar{a}_i\}$ so $|B^i| < \lambda(\mathbf{F})$ by the regularity of $\lambda(\mathbf{F})$. By Ax(III.2) $(p^i_{\bar{c}}, B^i) \in \mathbf{F}$ for every $\bar{c} \in C \cup \bar{a}_i$. So $\mathrm{tp}_*(C \cup \bar{a}_i, \mathscr{A}(i)) \in \mathbf{F}(B^i)$, hence by Ax(V.2) $\mathrm{tp}(\bar{a}_i, \mathscr{A}(i) \cup C) \in \mathbf{F}(B^i \cup C)$. So we finish.

THEOREM 3.7 [Ax(I), (II.1), (III.1), (III.2), (V.2) and (VII); $\lambda(\mathbf{F})$ regular]: (1) *If* B *is an* **F**-*primary set over* A *then* $(B, a)_{a \in A}$ *is* $\lambda(\mathbf{F})$-*homogeneous: i.e., if* $a_i, b_i \in B$ *for* $i < \alpha < \lambda(\mathbf{F})$ *and* $\mathrm{tp}_*(\langle a_i : i < \alpha\rangle, A) = \mathrm{tp}_*(\langle b_i : i < \alpha\rangle, A)$ *then for every* $a_\alpha \in B$ *there is* $b_\alpha \in B$ *such that* $\mathrm{tp}_*(\langle a_i : i \le \alpha\rangle, A) = \mathrm{tp}_*(\langle b_i : i \le \alpha\rangle, A)$.

(2) *Suppose* B_l *is* **F**-*primary over* A_l $(l = 1, 2)$, f *an elementary mapping,* $A_1 \subseteq \mathrm{Dom}\, f \subseteq B_1$, $A_2 \subseteq \mathrm{Range}\, f \subseteq B_2$, $|\mathrm{Dom}\, f - A_1| < \lambda(\mathbf{F})$, $|\mathrm{Range}\, f - A_2| < \lambda(\mathbf{F})$. *Then for every* $\bar{a} \in B_1$ *there is* f' *extending* f, *satisfying all the above-mentioned conditions and* $\bar{a} \in \mathrm{Dom}\, f'$.

(3) *In* (2) *for every* $C_l \subseteq B_l$, $|C_l| \leqslant \lambda(\mathbf{F})$ *we can find an elementary* f'; $\operatorname{Dom} f' \supseteq A_1 \cup C_1$, $\operatorname{Range} f' \supseteq A_2 \cup C_2$.

(4) ***Instead of "F-primary", we can in* (2), (3) *assume only* "(F, μ)-*primary" where*** $\mu \geq |A_1|^+ + \lambda(\mathbf{F})$.

(5) ***In* (1) *it suffices to assume B is* (F, μ)-*saturated***, $\mu \geq |A|^+ + \lambda(\mathbf{F})$, ***and every sequence from B realizes over A an F-isolated type (i.e., B is F-atomic over A).***

Remark. Concerning Theorem 3.7, note that in (2), (3) it suffices to assume B_l is $(\mathbf{F}, \mu)$-saturated, $(\mathbf{F}, \mu)$-atomic over A_l, $\mu \geq |A_1|^+ + \lambda(\mathbf{F})$. Also in (3), for any $C_l \subseteq B_l$ $(l = 1, 2)$ $|C_1| + |C_2| \leqslant \lambda(F)$ there is an elementary f', $\operatorname{Dom} f' \supseteq A_1 \cup C_1$, $\operatorname{Range} f' \supseteq A_2 \cup C_2$.

Proof. (1) Follows from (2).

(2) By 3.6 B_1 is **F**-primary over $\operatorname{Dom} f$, so by 3.2 $p = \operatorname{tp}(\bar{a}, \operatorname{Dom} f)$ is **F**-isolated, so for some $C \subseteq \operatorname{Dom} f$, $(p, C) \in \mathbf{F}$. Hence by Axiom (I), $(f(p), f(C)) \in \mathbf{F}$. As B_2 is **F**-saturated some $\bar{b} \in B_2$ realizes $f(p)$. Extend f to f' by defining $f'(\bar{a}) = \bar{b}$. Clearly f' satisfies our requirements.

(3) Prove by repeating (2) $\lambda(\mathbf{F})$ times.

(4), (5) The same proof, in fact, as of (1), (2).

THEOREM 3.8 [Ax(I), (II.1), (III.1), (III.2), (V.2), (VI), (VII) and (VIII); $\lambda(\mathbf{F})$ regular]: (1) *Suppose* B_l *is* **F**-*primary over* A_l $(l = 1, 2)$ *and* f *is an elementary mapping from* A_1 *onto* A_2 *then we can extend* f *to an elementary mapping from* B_1 *onto* B_2.

(2) *The assertion is true also for* B_l $(\mathbf{F}, \mu)$-*primary over* A_l, *when* $|A_l| < \mu$, $|B_l| \leq \mu$.

CONCLUSION 3.9 [The same assumptions as in 3.8]: *The* **F**-*primary set over A is unique up to isomorphism over A (i.e., up to isomorphisms which are the identity on A).*

Proof of 3.8. (1) Let $\mathscr{A}_l = \langle A_l, \{\bar{a}^l_i : i < \alpha_l\}, \{B^l_i : i < \alpha_l\}\rangle$ be an **F**-construction of B_l over A_l. We define by induction on $\beta < \alpha^*$ elementary mappings f_β where $\alpha^* = \max\{\alpha_1, \alpha_2\}$ such that:

(1) $f_0 = f$.

(2) For $\beta < \alpha$, f_α extends f_β, and for limit δ, $f_\delta = \bigcup_{\beta<\delta} f_\beta$.

(3) $\operatorname{Dom} f_\beta \subseteq B_1$, $\operatorname{Range} f_\beta \subseteq B_2$.

(4) $\operatorname{Dom} f_\beta$ is closed, i.e., for some V^1_β, $\operatorname{Dom} f_\beta = A \cup \bigcup \{\bar{a}^1_i : i \in V^1_\beta\}$ and $i \in V^1_\beta \Rightarrow B^1_i \subseteq \operatorname{Dom} f_\beta$.

(5) Range f_β is closed (the same as in (4) for V^2_β).

(6) $\bar{a}^1_\alpha \in \operatorname{Dom} f_{2\alpha+1}$, $\bar{a}^2_\alpha \in \operatorname{Range} f_{2\alpha+2}$ (when $\bar{a}^l_\alpha$ is defined).

For $\beta = 0$, β a limit ordinal, there is no problem. So suppose f_β is defined. By 3.3, 3.5

$$\mathscr{A}^\beta_1 = \langle \operatorname{Dom} f_\beta, \{\bar{a}^1_i : i < \alpha_1\}\rangle, \qquad \mathscr{A}^\beta_2 = \langle \operatorname{Range} f_\beta, \{\bar{a}^2_i : i < \alpha_2\}\rangle$$

are **F**-constructions. Now we define by induction on $\gamma < \lambda(\mathbf{F})$ elementary mappings f^γ_β such that:

(i) f^γ_β is increasing in γ, $f^0_\beta = f_\beta$, $f^\delta_\beta = \bigcup_{\gamma<\delta} f^\gamma_\beta$,

(ii) $\operatorname{Dom} f^\gamma_\beta \subseteq B_1$, $\operatorname{Range} f^\gamma_\beta \subseteq B_2$,

(iii) $|\operatorname{Dom} f^\gamma_\beta - \operatorname{Dom} f_\beta| < \lambda(\mathbf{F})$.

Let $\operatorname{Dom} f^\gamma_\beta = \operatorname{Dom} f_\beta \cup \bigcup \{\bar{a}^1_j : j \in V^1_{\beta,\gamma}\}$, $\operatorname{Range} f^\gamma_\beta = \operatorname{Range} f_\beta \cup \bigcup \{\bar{a}^2_j : j \in V^2_{\beta,\gamma}\}$. For $\gamma = 0$, γ a limit ordinal—the definition is by (i).

By 3.6, B_1, B_2 are **F**-primary over $\operatorname{Dom} f^\gamma_\beta$, $\operatorname{Range} f^\gamma_\beta$ respectively. Hence by 3.7(2) we can extend f^γ_β to $f^{\gamma+1}_\beta$ satisfying (i), (ii), (iii) and such that for γ even:

$$V^2_{\beta,\gamma+1} = V^2_{\beta,\gamma} \cup \bigcup \{U_j(\mathscr{A}_2) : j \in V^2_{\beta\gamma}\} \cup \{\alpha\}$$

($\{\alpha\}$ appears only when $\beta = 2\alpha + 1$, $\alpha < \alpha_2$). For γ odd, replace 2 by 1, and $\{\alpha\}$ appears only when $\beta = 2\alpha$, $\alpha < \alpha_1$.

Clearly $f_{\beta+1} = \bigcup_{\gamma<\lambda} f^\gamma_\beta$ ($\lambda = \lambda(\mathbf{F})$) satisfies our requirements. Now f_{α^*} is the required mapping.

(2) The proof is similar.

Remark. In fact the regularity of $\lambda(\mathbf{F})$ is needed for 3.3, 3.6 but not in this proof.

THEOREM 3.10 [Ax(I)]: (1) *If B is* **F**-*constructible over A, then B is* **F**-*primitive over A.*

(2) *If B is* **F**-*primary over A, then B is* **F**-*prime over A.*

Proof. (1) Let $\mathscr{A} = \langle A, \{\bar{a}_i : i < \alpha\}\rangle$ be an **F**-construction of B over A and let B^* be an **F**-saturated set, $A \subseteq B^*$. We should find an elementary mapping f from B into B^*, $f \restriction A =$ the identity. Define by induction on $\beta \le \alpha$ elementary mappings f_β from $\mathscr{A}(\beta)$ into B^* so that f_γ extends f_β for $\beta < \gamma$. Let f_0 be the identity over A, and for limit δ, $f_\delta = \bigcup_{\beta<\delta} f_\beta$. If f_β is defined, $(p_\beta, B_\beta) \in \mathbf{F}$ ($p_\beta = \operatorname{tp}(\bar{a}_\beta, \mathscr{A}(\beta))$) then by Axiom (I), $(f_\beta(p_\beta), f_\beta(B_\beta)) \in \mathbf{F}$, and $\operatorname{Dom}[f_\beta(p_\beta)] \subseteq B^*$, so $f_\beta(p_\beta)$ is realized by some $\bar{b}_\beta \in B^*$ (as B^* is **F**-saturated). Extend f_β to $f_{\beta+1}$ by defining $f_{\beta+1}(\bar{a}_\beta) = \bar{b}_\beta$. Clearly f_α is the required mapping.

(2) Immediate.

LEMMA 3.11: (1) *If $A \subseteq B \subseteq C$, C is* **F**-*primitive over A then B is* **F** *primitive over A.*

(2) *If $A \subseteq B \subseteq C$, C is* **F**-*prime over A, B* **F**-*saturated then B is* **F**-*prime over A.*

(3) *The same for* $(\mathbf{F}, \mu)$ *instead of* **F**.

Proof. (1) Let B' be an **F**-saturated set, $A \subseteq B'$. Then there is an elementary mapping f from C into B', $f \restriction A =$ the identity. Let $g = f \restriction B$, so g is an elementary mapping from B into B', $g \restriction A =$ the identity. Hence B is **F**-primitive over A.

(2) Immediate from (1) and the definition. (3) Proved similarly.

CONCLUSION 3.12: *Let $\mu(\mathbf{F}) < \infty$ then:*

(1) [Ax(I), (X.1) and (XI.1)] *Over every A there is an* **F**-*prime model.*

(2) [Ax(II.1), (III.1), (III.2), (VII) and (XI.1); $\lambda(\mathbf{F})$ regular] *In every* **F**-*prime set B over A, every sequence realizes over A an* **F**-*isolated type.*

(3) [Ax(I), (II.1), (III.1), (III.2), (V.2), (VII) and (XI.1); $\lambda(\mathbf{F})$ regular] *If B is* **F**-*prime over A, $C \subseteq B$, $|C| < \operatorname{cf} \lambda(\mathbf{F})$ then B is* **F**-*prime over $A \cup C$. The same holds for* **F**-*atomic.*

Proof. (1) Immediate, by 3.1(2), 3.10 and Exercise 3.1.

(2) By 3.1(2) there is an **F**-primary set B^* over A. So by the definition of **F**-primeness there is an elementary mapping f from B into B^*. $f \restriction A =$ the identity. By 3.2 for every $\bar{b} \in B^*$ $\operatorname{tp}(\bar{b}, A)$ is **F**-isolated. Hence for every $\bar{b} \in B$, $\operatorname{tp}(f(\bar{b}), A)$ is **F**-isolated. But as $f \restriction A =$ the identity, $\bar{b}, f(\bar{b})$ realize the same type over A, so we finish.

(3) By 3.1(2) over A there is an **F**-primary set B^*; and by the definition of **F**-primeness there is an elementary mapping f from B into B^*, $f \restriction A =$ the identity. By 3.6 B^* is **F**-primary over $A \cup f(C)$, and by 3.10 B^* is **F**-prime over $A \cup f(C)$. As $A \cup f(C) \subseteq f(B) \subseteq B^*$ by 3.11(2) $f(B)$ is **F**-prime over $A \cup f(C)$. (By Ax(I) the property of being **F**-saturated hence **F**-prime over a set is preserved by elementary mapping.) Hence B is **F**-prime over $A \cup C$.

The proof for **F**-atomic is similar.

LEMMA 3.13 [Ax(III.1), (VI) and (VIII)]: *Suppose A, C_i, $(i < \alpha)$ are given sets, and for every $\bar{c} \in C_i$, for some $B_{\bar{c}}^i \subseteq A$, $\operatorname{tp}(\bar{c}, A \cup \bigcup_{j<i} C_j) \in \mathbf{F}(B_{\bar{c}}^i)$. Then for every $i < \alpha$, $\bar{c} \in C_i$, $\operatorname{tp}(\bar{c}, A \cup \bigcup_{j \neq i} C_j) \in \mathbf{F}(B_{\bar{c}}^i)$.*

Proof. By 1.2(2) (and Ax(III.1) and (VIII)) it suffices to prove that for

$i < \alpha$, $\bar{c} \in C_i$, $\bar{d} \in \bigcup_{j \neq i} C_j$, $\operatorname{tp}(\bar{c}, A \cup \bar{d}) \in \mathbf{F}(B_{\bar{c}}^i)$. Let $\bar{d} = \bar{d}_0 \frown \cdots \frown \bar{d}_k$, where $\bar{d}_l \in C_{j(l)}$ $0 \le l \le k$, and $j(0) < \cdots < j(n) < i < j(n+1) < \cdots < j(k)$. We prove by induction on $l \le k+1$ that $\operatorname{tp}(\bar{c}, A \cup \bar{d}_0 \frown \cdots \frown \bar{d}_{l-1}) \in \mathbf{F}(B_{\bar{c}}^i)$. For $l \le n+1$ this holds by Axiom (III.1). If it is true for l, $l \le k$, it is true for $l+1$ by Axiom (VI). For $l = k+1$ we get what we want.

THEOREM 3.15 [Ax(I), (III.1), (III.2), (V.2), (VIII) and (XI.1); $\lambda(\mathbf{F})$ regular]: (1) *Suppose $A \subseteq |M|$, M is* **F**-*saturated, and* **F**-*atomic over A. Let $p_i \in S(A)$ for $i < \alpha$, and $A^* = A \cup \{c: c \in |M|$, c realizes a p_i for an $i < \alpha\}$. Then M is* **F**-*atomic over A^*.*

(2) *For $\mathbf{F} = \mathbf{F}_{\lambda}^{a}$, we can assume that p_i's are types almost over A, when T is stable.*

Proof. (1) Let $\bar{a} \in |M|$, by Ax(XI.1) as $p = \operatorname{tp}(\bar{a}, A)$ is **F**-isolated, p has an extension $q \in S^m(A^*)$ which is also **F**-isolated. So for some $C \subseteq A^*$, $(q, C) \in \mathbf{F}$ (so $|C| < \lambda(\mathbf{F})$) and some $\bar{b} \in |M|$ realizes q. Let $C = \{c_i: i < |C|\}$. By the hypothesis for every $\bar{c} \in C \cup \bar{b}$, $\operatorname{tp}(\bar{c}, A)$ is **F**-isolated, say $\operatorname{tp}(\bar{c}, A) \in \mathbf{F}(B_{\bar{c}})$. By Axiom (III.2) and the regularity of $\lambda(\mathbf{F})$, $\operatorname{tp}_*(C \cup \bar{b}, A) \in \mathbf{F}(B)$, where $B = \bigcup \{B_{\bar{c}}: \bar{c} \in C \cup \bar{b}\}$. Hence by Ax(V.2), for every $\alpha < |C|$,

$$p_\alpha = \operatorname{tp}(c_\alpha, A \cup \bar{b} \cup \{c_i: i < \alpha\}) \in \mathbf{F}(B \cup \bar{b} \cup \{c_i: i < \alpha\}).$$

Now we define, by induction on i, $c_i' \in |M|$ such that the mapping $f_\alpha (\alpha \le |C|)$, $f_\alpha(a) = a$ for $a \in A$, $f_\alpha(\bar{b}) = \bar{a}$ and $f_\alpha(c_i) = c_i'$, for $i < \alpha$, is elementary. It is trivial to define f_0 and f_δ (δ a limit ordinal). For $f_{\alpha+1}$ note that by Ax(I), $f_\alpha(p_\alpha)$ is **F**-isolated, (as $\alpha < \lambda(\mathbf{F})$) hence some $c_\alpha' \in |M|$ realizes it. Let $C' = \{c_i': i < |C|\}$; as $\operatorname{tp}(c_i', A) = \operatorname{tp}(c_i, A)$, $c_i \in A^*$ clearly $C' \subseteq A^*$, and $|C'| < \lambda(\mathbf{F})$. We shall prove $p^* = \operatorname{tp}(\bar{a}, A^*) \in \mathbf{F}(C')$, and thus finish. By Ax(VIII) it suffices to prove $\operatorname{tp}(\bar{a}, A \cup C' \cup \bar{c}) \in \mathbf{F}(C')$ for any $\bar{c} \in A^*$. As before, we can extend $f_{|C|}$ to an elementary mapping f', whose domain is $\subseteq |M|$, and whose range is $\bar{c} \cup \operatorname{Range} f_{|C|}$. By the choice of C, and Ax(III.1), $\operatorname{tp}(\bar{b}, \operatorname{Dom} f') \in \mathbf{F}(C)$, hence by Ax(I) $\operatorname{tp}(\bar{a}, \operatorname{Range} f') \in \mathbf{F}(C')$ so $\operatorname{tp}(\bar{a}, A \cup C' \cup \bar{c}) \in \mathbf{F}(C')$.

(2) Work in $\mathfrak{C}^{\mathrm{eq}}$. By 2.12(2) M^{eq} is **F**-saturated and also **F**-atomic over A, hence it is **F**-atomic over acl A (see 2.1). Now w.l.o.g. each p_i is over acl A; hence, by 3.14(1) M^{eq} is **F**-atomic over (acl A) $\cup A^*$, and by 2.1 M^{eq} is **F**-atomic over A^* so M is **F**-atomic over A^*.

LEMMA 3.16 [Ax(III.2) and (V.2); $\lambda(\mathbf{F})$ regular or [Ax(III.2), (V.1) and

(IX)]]: *If C is* **F***-atomic over A, and $|C - A| \le \lambda(\mathbf{F})$ [or $|C - A| \le \mathrm{cf}\,\lambda(\mathbf{F})$ when $\lambda(\mathbf{F})$ is singular], then C is* **F***-constructible over A.*

Proof. Let $C - A = \{c_i : i < \alpha = |C|\}$, and let $A_j = A \cup \{c_i : i < j\}$. It suffices to prove $\mathrm{tp}(c_i, A_i)$ is **F**-isolated. As $\lambda(\mathbf{F})$ is regular [or as $|C - A| \le \mathrm{cf}\,\lambda(\mathbf{F})$] and Ax(III.2) holds, $\mathrm{tp}_*(\{c_j : j \le i\}, A) \in \mathbf{F}(B)$ for some $B \subseteq A$, $|B| < \lambda(\mathbf{F})$. Hence, by Ax(V.2) [or by 1.1(4)] $\mathrm{tp}(c_i, A_i) \in \mathbf{F}(B \cup \{c_j : j < i\})$.

THEOREM 3.17: (1) [Ax(XI.2) or Ax(XI.1), $\alpha < \mathrm{cf}\,\lambda(\mathbf{F})$, and Ax(IX)] *If $A_i \subseteq A_j$ for $i < j \le \alpha$ then there are sets B_i ($i \le \alpha$) such that $A_i \subseteq B_i$, $i < j \Rightarrow B_i \subseteq B_j$, $B_i \cup A_\beta$ is* **F***-constructible over $A_\beta \cup \bigcup_{j<i} B_j$ when $i \le \beta \le \alpha$, and B_i is $(\mathbf{F}, \mu)$-saturated. (So $B_i \cup A_\beta$ is* **F***-constructible over A_β when $i \le \beta \le \alpha$.)*

(2) [Ax(X.2) or Ax(X.1), $\alpha < \mathrm{cf}\,\lambda(\mathbf{F})$, and Ax(IX)] *If $A_i \subseteq A_j$ for $i < j \le \alpha$, then there are models M_i, $A_i \subseteq |M_i|$, $i < j \Rightarrow |M_i| \subseteq |M_j|$, $|M_i| \cup A_\beta$ is* **F***-constructible over $A_\beta \cup \bigcup_{j<i} |M_j|$ when $i \le \beta \le \alpha$.*

Proof. Similar to 3.1 (remember 1.1(7)).

CLAIM 3.18: [Ax(I), (IV), (V.1) and (VII)] *If* Ax(XI.1) [or (XI.2)] *are satisfied for* 1*-types then it is satisfied.*

Proof. Easy.

THEOREM 3.19: (1) [Ax(I), (X.1), (XI.1)] *Let M be* **F***-saturated. If M is* **F***-minimal over A (see Definition* 4.4*), then in M there is no infinite indiscernible set over A.*

(2) *We can replace the three axioms by the assumption "over any A there is an* **F***-prime model".*

Proof. Just the proof of one implication in 4.21.

EXERCISE 3.1 [Ax(X.1)]: If A is **F**-primary over B, then A is an **F**-primary model over B.

EXERCISE 3.2: In 3.15(1) let p_i be an m_i-type over A, and $A^* = A \cup \bigcup_i I_i$, $I_i = \{\bar{c} \in M : \bar{c}$ realizes $p_i\}$. Show the conclusion still follows. Similarly 3.15(2).

EXERCISE 3.3: Show that 3.6 holds for any closed C (assuming the appropriate axioms).

QUESTION 3.4: Is there in any of the lemmas of Section 3 an unnecessary axiom?

EXERCISE 3.5 [Ax(I)]: Suppose B is $\mathbf{F}$-primitive over A, f an elementary mapping, $\operatorname{Dom} f = A$, and $\operatorname{Range} f \subseteq C$, C is $\mathbf{F}$-saturated, then we can extend f to an elementary mapping g, $\operatorname{Dom} g = B$, $\operatorname{Range} g \subseteq C$.

EXERCISE 3.6: Suppose λ is singular, and Ax(VII) holds when $|C| < \operatorname{cf} \lambda$, and Ax(II.1), (III.1) and (III.2). Then:

If $|C| < \operatorname{cf} \lambda$, C is $\mathbf{F}$-atomic over A, $\operatorname{tp}(\bar{a}, A \cup C)$ is $\mathbf{F}$-isolated, then $C \cup \bar{a}$ is $\mathbf{F}$-atomic over A.

IV.4. Prime models for stable theories

In this section, we assume T is stable, except in the starred theorems.

LEMMA 4.1*: *If $\varphi(\bar{x}; \bar{a})$ is almost over B, $B \subseteq A$, then there is a formula $\psi(\bar{x}; \bar{b})$ such that*

(i) *$\psi(\bar{x}; \bar{b})$ is over A and almost over B;*

(ii) *$\vdash \varphi(\bar{x}; \bar{a}) \rightarrow \psi(\bar{x}; \bar{b})$;*

(iii) *if $\bar{c} \in A$ then $\vdash \varphi(\bar{x}; \bar{a}) \rightarrow \theta(\bar{x}; \bar{c})$ iff $\vdash \psi(\bar{x}; \bar{b}) \rightarrow \theta(\bar{x}; \bar{c})$. (Note that if $\bar{a} \in A$ we can take $\psi = \varphi$.)*

Proof. Let $\bar{a}_0, \ldots, \bar{a}_n$ be a maximal set, such that $\bar{a}_0 = \bar{a}$, the $\varphi(\bar{x}; \bar{a}_l)$ ($l \leq n$) are not equivalent, and $\operatorname{tp}(\bar{a}_l, A) = \operatorname{tp}(\bar{a}, A)$ ($n < \aleph_0$ by III, 2.3(1), using the fact that $\varphi(\bar{x}; \bar{a})$ is also almost over A). Let $\psi'(\bar{x}; \bar{b}') = \bigvee_{l \leq n} \varphi(\bar{x}; \bar{a}_l)$. Clearly by III, 2.3(2), $\psi'(\bar{x}; \bar{b}') \equiv \psi(\bar{x}; \bar{b})$ for some ψ and $\bar{b} \in A$. It is easy to check (i), (ii), (iii).

LEMMA 4.2: *Suppose $\varphi(\bar{x}; \bar{a})$ does not fork over C, $C \subseteq A$. Then there is a formula $\psi(\bar{x}; \bar{b})$ such that;*

(1) *$\psi(\bar{x}; \bar{b})$ is over A, but does not fork over C.*

(2) *$\{\varphi(\bar{x}; \bar{a}), \neg\psi(\bar{x}; \bar{b})\}$ forks over A (hence over C).*

(3) *For every $\bar{c} \in A$ and θ, $\vdash \varphi(\bar{x}; \bar{a}) \rightarrow \theta(\bar{x}; \bar{c})$ implies $\vdash \psi(\bar{x}; \bar{b}) \rightarrow \theta(\bar{x}; \bar{c})$.*

(4) *If $C \subseteq B \subseteq A$, $\operatorname{tp}(\bar{a}, A)$ does not fork over B, then we can choose $\psi(\bar{x}; \bar{b})$ which is almost over B.*

Proof. Let B be as in (4); for (1), (2) and (3), $B = A$ satisfies that $\operatorname{tp}(\bar{a}, A)$ does not fork over B. By III, 1.12, there is an indiscernible set

$I = \{\bar{a}_n : n < \omega\}$ over A, based on B, $\bar{a} = \bar{a}_0$. Hence by III, 2.5, for some n, $\varphi_n(\bar{x}; \bar{a}^*) = \bigvee_{w \subseteq 2n-1, |w|=n} \bigwedge_{l \in w} \varphi(\bar{x}; \bar{a}_l)$ is almost over B. Clearly by III, 1.4, there is $q \in S^m(\bigcup I \cup A)$, $\varphi(\bar{x}; \bar{a}_0) \in q$, q does not fork over C; for every l, $\varphi(\bar{x}; \bar{a}_l) \in q$ (as q does not split strongly over C by III, 1.6(1)) hence $\varphi_n \in q$ so $\varphi_n(\bar{x}; \bar{a}^*)$ does not fork over C. Clearly for every $l > 0$, $\{\varphi(\bar{x}; \bar{a}), \neg\varphi(\bar{x}, \bar{a}_l)\}$ splits strongly over A, hence forks over A. Hence for every $w \subseteq 2n - 1$ $\{\varphi(\bar{x}; \bar{a}), \bigvee_{l \in w} \neg\varphi(\bar{x}; \bar{a}_l)\}$ forks over A (by III, 1.1(9)), hence $\{\varphi(\bar{x}; \bar{a}), \neg\varphi_n(\bar{x}; \bar{a}^*)\}$ forks over A. Also, if $\bar{c} \in A$, $\vdash \varphi(\bar{x}; \bar{a}) \rightarrow \theta(\bar{x}; \bar{c})$ then $\vdash \varphi(\bar{x}; \bar{a}_l) \rightarrow \theta(\bar{x}, \bar{c})$ hence $\vdash \varphi_n(\bar{x}; \bar{a}^*) \rightarrow \theta(\bar{x}; \bar{c})$.

Now apply 4.1 to $\varphi_n(\bar{x}; \bar{a}^*)$, B and A. We get $\psi(\bar{x}; \bar{b})$ satisfying (i), (ii) and (iii) from 4.1. It is easy to check that it satisfies the conditions (1)–(4) from 4.2.

THEOREM 4.3: *Suppose $p \in S^m(A)$, $C \subseteq B \subseteq A$, p does not fork over C.*

(1) *If p is $\mathbf{F}^t_\lambda$-isolated, then $p \upharpoonright B$ is $\mathbf{F}^t_\lambda$-isolated.*

(2) *If p is $\mathbf{F}^a_\lambda$-isolated, $\lambda \geq \kappa_r(T)$, then $p \upharpoonright B$ is $\mathbf{F}^a_\lambda$-isolated.*

(3) *If $p \upharpoonright C$ is stationary, $|C| < \lambda$, p is $\mathbf{F}^s_\lambda$-isolated, then $p \upharpoonright B$ is $\mathbf{F}^s_\lambda$-isolated*, provided that $\lambda \geq \kappa_r(T)$.

(4) *If p is $\mathbf{F}^c_\lambda$-isolated $\lambda \geqslant \kappa_r(T)$, then $p \upharpoonright B$ is $\mathbf{F}^c_\lambda$-isolated.*

(5) *If p is $\mathbf{F}^l_\lambda$-isolated, then so is $p \upharpoonright B$.*

Proof. (1) As p is $\mathbf{F}^t_\lambda$-isolated, there is $q \subseteq p$, $q \vdash p$, $|q| < \lambda$, and as in 2.2(2) assume q is closed under conjunctions, and let $q = \{\varphi_i(\bar{x}; \bar{a}_i) : i < \alpha < \lambda\}$. As p does not fork over C and $\varphi_i(\bar{x}; \bar{a}_i) \in p$, by Lemma 4.2 there is $\psi_i(\bar{x}; \bar{b}_i)$, $\bar{b}_i \in B$ satisfying (1)–(3) from 4.2 (B here corresponds to A there). As $\varphi_i(\bar{x}; \bar{a}_i) \in p$ and (by (2) from 4.2) $\{\varphi_i(\bar{x}; \bar{a}_i), \neg\psi_i(\bar{x}; \bar{b}_i)\}$ forks over C, but p does not fork over C, we have that $\psi_i(\bar{x}; \bar{b}_i) \in p$. So $r = \{\psi_i(\bar{x}; \bar{b}_i) : i < \alpha\} \subseteq p$, so $r \subseteq p \upharpoonright B$; and clearly $|r| < \lambda$. If $\theta(\bar{x}; \bar{c}) \in p \upharpoonright B$, then for some $i < \alpha$, $\vdash \varphi_i(\bar{x}; \bar{a}_i) \rightarrow \theta(\bar{x}; \bar{c})$, hence by (3) from 4.2, $\vdash \psi_i(\bar{x}; \bar{b}_i) \rightarrow \theta(\bar{x}; \bar{c})$. Hence $r \vdash p \upharpoonright B$.

(2) As p is $\mathbf{F}^a_\lambda$-isolated, for some $C_1 \subseteq A$, $|C_1| < \lambda$, there is a type $q \subseteq p$ almost over C_1, $q \vdash p$. For every $\bar{c} \in C_1$ there is $B_{\bar{c}} \subseteq B$, $|B_{\bar{c}}| < \kappa(T)$ such that $\mathrm{tp}(\bar{c}, B)$ does not fork over $B_{\bar{c}}$ (by III, 3.2). Let $B^* = \bigcup_{\bar{c} \in C_1} B_{\bar{c}} \cup C$; then as $\lambda \geqslant \kappa_r(T)$, $|B^*| < \lambda$. Now for every $\varphi = \varphi(\bar{x}; \bar{c}) \in q$ there is $\psi_\varphi(\bar{x}; \bar{b}_\varphi)$ satisfying 4.2(4) (A, B there correspond to B, B^* here). Let $r = \{\psi_\varphi : \varphi \in q\}$. Clearly r is almost over B^*, and as in part (1), $r \subseteq p$, $r \vdash p \upharpoonright B$. As $|B^*| < \lambda$ we finish the proof.

(3) By (2), for some $B^* \subseteq B$, $p \upharpoonright B \in \mathbf{F}^a_\lambda(B^*)$. As $|C| < \lambda$ we can assume $C \subseteq B^*$. By 2.2(13) $p \upharpoonright B \in \mathbf{F}^s_\lambda(B^*)$.

Proof. (4) So suppose $q = \{\varphi_i(\bar{x}; \bar{a}_i): i < \alpha\} \subseteq p$, $\alpha < \lambda$, $\bigcup_{i<\alpha} \bar{a}_i \subseteq C^* \subseteq A$, $|C^*| < \lambda$, and q has no extension in $S^m(A)$ forking over C^*. W.l.o.g. q is closed under finite conjunctions. We produce $\psi_i(\bar{x}; \bar{b}_i)$ ($\bar{b}_i \in B$) as in the proof of 4.3(1), $r = \{\psi_i(\bar{x}; \bar{b}_i): i < \alpha\} \subseteq p$, and w.l.o.g. p, $\mathrm{tp}_*(C^*, B)$ does not fork over $C^* \cap B$. Now suppose for some $\bar{a} \in B$ and θ, $r \cup \{\theta(\bar{x}, \bar{c})\}$ is consistent. For each i, not $\psi_i(\bar{x}, \bar{b}_i) \vdash \neg\theta(\bar{x}; \bar{a})$ hence $\varphi_i(\bar{x}; b_i) \vdash \neg\theta(\bar{x}; \bar{a})$ fail, so as q was closed under conjunction, $q_1 = q \cup \{\theta(\bar{x}; \bar{a})\}$ is consistent. By the choice of q, q_1 does not fork over C^*, so let $\bar{c}$ realize q_1, $\mathrm{tp}(\bar{c}, A)$ does not fork over C^*. As $\{\varphi_i(\bar{x}; \bar{a}_i), \neg\psi_i(\bar{x}; \bar{b}_i)\}$ fork over C^*, $\psi_i(\bar{x}, \bar{b}_i) \in \mathrm{tp}(\bar{c}, A)$, hence $r \cup \{\theta(\bar{x}; a\} \subseteq \mathrm{tp}(\bar{c}, A)$.

As $p' = \mathrm{tp}(\bar{c}, A)$ does not fork over C^*, also $\mathrm{tp}(\bar{c}, B \cup C^*)$ does not fork over C^*, and as in addition $\mathrm{tp}_*(C^*, B)$ does not fork over $B \cap C^*$; $\mathrm{tp}_*(C^* \cup \bar{c}, B)$ does not fork over $B \cap C^*$, hence $\mathrm{tp}(\bar{c}, B)$ does not fork over $B \cap C^*$. So also $r \cup \{\theta(\bar{x}, \bar{c})\} \subseteq \mathrm{tp}(\bar{c}, B)$ does not fork over $B \cap C^*$.

So $r \subseteq p \restriction B$, $C^* \cap B \subseteq B$, $|r|$, $|C^* \cap B| < \lambda$ and for every $\theta, \bar{a} \in B$, $r \cup \{\theta(\bar{x}; \bar{a})\}$ consistent implies $r \cup \{\theta(\bar{x}; \bar{a})\}$ does not fork over A; clearly $p \restriction B$ is $\mathbf{F}^c_\kappa$-isolated.

(5) Just like 4.3(2).

EXERCISE 4.1: Does a similar assertion hold for $\mathbf{F}^s_\lambda$ (i.e., an assertion like 4.3(1) or at least 4.3(2))? Prove that the answer is no.

Remark. The question is interesting mainly for $\lambda = \aleph_0$ since if $\lambda > \aleph_0$ then $\mathbf{F}^c_\lambda$-saturation is equivalent to $\mathbf{F}^s_\lambda$-saturation (by 2.2(6), (7)). Similarly for other concepts (primeness, etc.).

EXERCISE 4.2: Show that in 4.3(2) we cannot omit the assumption that $\lambda \geq \kappa_r(T)$.

DEFINITION 4.1: For $x = t, s, a$, $\lambda \geq \kappa$, let

$$\mathbf{F}^x_{\lambda,\kappa} = \{(p, B): \text{for some } A, m;\ p \in S^m(A),\ B \subseteq A,\ |B| < \kappa,\ p \text{ does not fork over } B \text{ and } p \text{ is an } \mathbf{F}^x_\lambda\text{-isolated type}\}.$$

Note that $(p, B) \in \mathbf{F}^x_{\lambda,\kappa}$ does not imply $(p, B) \in \mathbf{F}^x_\lambda$.

LEMMA 4.4: (1) *For* $x = t, s, a$, $\kappa \geq \kappa(T)$, $p \in S^m(A)$, p *is* $\mathbf{F}^x_\lambda$*-isolated iff* p *is* $\mathbf{F}^x_{\lambda,\kappa}$*-isolated.*

(2) *In the following concepts for* $x = t, s, a$; *it does not matter whether* $\mathbf{F} = \mathbf{F}^x_\lambda$, *or* $\mathbf{F} = \mathbf{F}^x_{\lambda,\kappa}$: $\mathbf{F}$*-construction* $\langle A, \langle a_i : i < \alpha\rangle\rangle$, $\mathbf{F}$*-saturated*, $\mathbf{F}$*-primary*, $\mathbf{F}$*-primitive*, $\mathbf{F}$*-prime and* $\mathbf{F}$*-atomic provided that* $\kappa \geq \kappa(T)$.

Proof. Immediate.

Remark. But for $\mathbf{F}$-construction $\langle A, \langle a_i : i < \alpha\rangle, \langle B_i : i < \alpha\rangle\rangle$ it matters.

THEOREM 4.5: *If* $x = t, a$, $\kappa = \kappa_r(T) < \infty$ *then* $\mathbf{F}^x_{\lambda,\kappa}$ *satisfies the following axioms;* (I), (II.l) [$l = 1, 2, 3, 4$], (III.1), (III.2), (IV), (V.1) *and* (V.2) [*when* $\mathrm{cf}\,\lambda \geq \kappa(T)$], (VI), (VII) [*when* $\mathrm{cf}\,\lambda \geq \kappa(T)$], Ax(VIII) *for* δ, $\mathrm{cf}\,\delta < \mathrm{cf}\,\lambda$ *and* (IX) [*when* $\mathrm{cf}\,\lambda \geq \kappa$] *and* (XII). *Also* $\lambda(\mathbf{F}^x_{\lambda,\kappa}) = \kappa$ *and* $\mu(\mathbf{F}^x_{\lambda,\kappa}) = \mu(\mathbf{F}^x_\lambda)$.

Proof. Most of them follow by combining the results on $\mathbf{F}^x_\lambda$, and $\mathbf{F}^f_\kappa$. Ax(III.1) follows from 4.3. For Ax(V.2) notice Lemma 2.8(2), as $|C| < \kappa \leq \mathrm{cf}\,\lambda$. For Ax(VII) note that $|C| < \kappa \leq \mathrm{cf}\,\lambda$ so there is a $B' \subseteq A$ such that $\mathrm{tp}_*(C, A) \in \mathbf{F}^x_\lambda(B')$ and the rest follows by Ax(III.2). For Ax(VI), use Ax(IV) and (VII) to show $\mathrm{tp}(\bar{a}^\frown \bar{b}, A)$ is $\mathbf{F}^x_\lambda$-isolated, and (V.1) to show $\mathrm{tp}(\bar{a}, A \cup \bar{b})$ is $\mathbf{F}^x_\lambda$-isolated.

THEOREM 4.6: *If* $x = t, a$, $\lambda \geq \kappa = \kappa_r(T)$, $\mathbf{F}^x_\lambda$ *satisfies* Ax(X.1) [Ax(XI.1)] *then also* $\mathbf{F}^x_{\lambda,\kappa}$ *satisfies* Ax(X.1) [Ax(XI.1)]. *The same holds for* Ax(X.2) *and* (XI.2).

Proof. Immediate.

CONCLUSION 4.7: *If* $\kappa(T) \leq \mathrm{cf}\,\lambda$, A *is* $\mathbf{F}^x_\lambda$*-constructible over* B, $x = t, a$, *then* A *is* $\mathbf{F}^x_\lambda$*-atomic over* B.

Proof. Immediate, by 4.4 and 4.5, applying 3.2(1) to $\mathbf{F}^x_{\lambda,\kappa}$ ($\kappa = \kappa_r(T)$).

CLAIM 4.8: *Let* $\mathbf{F}^x_\lambda$ *satisfy* Ax(X.1), (XI.1)*;* $x = t, a$, $\mathrm{cf}\,\lambda \geq \kappa(T)$.

(1) *If* M *is* $\mathbf{F}^x_\lambda$*-prime over* A, $C \subseteq |M|$, $|C| < \mathrm{cf}\,\lambda$ *then* M *is* $\mathbf{F}^x_\lambda$*-prime over* $A \cup C$.

(2) *If* $\mathscr{A} = \langle A, \{a_i : i < \alpha\}\rangle$ *is an* $\mathbf{F}^x_\lambda$*-construction,* $C \subseteq \mathscr{A}(\alpha)$, $|C| < \mathrm{cf}\,\lambda$, *then* $\mathscr{A} = \langle A \cup C, \{a_i : i < \alpha\}\rangle$ *is an* $\mathbf{F}^x_\lambda$*-construction.*

Proof. (1) Like 3.12(3), using (2) instead of 3.6; (2) like the proof of 3.6, using 4.7 instead of 3.2, and 2.8(2) instead of Ax(V.2).

THEOREM 4.9: *Suppose* $x = t, a$, $\mathbf{F} = \mathbf{F}^x_\lambda$ *satisfies* Ax(X.1) *and* (XI.1). M *is an* $\mathbf{F}$*-prime model over* A, *and* $I \subseteq |M|$ *is an infinite indiscernible set over* A. (*So w.l.o.g.* I *is maximal in* M.)

(1) *If* cf $\lambda \geq \kappa(T)$, *then* $\dim(I, A, M) \geq \lambda$ *except, possibly, when:*
(i) cf $\lambda = \aleph_0 < \lambda$, *and*
(ii) $\dim(I, A, M) = \aleph_0$, *and*
(iii) *for no* $B \subseteq A \cup I$ *and* $J \subseteq I$, $|J| = \aleph_0$, *is* $\mathrm{Av}(J, A \cup \bigcup J) \in \mathbf{F}(B)$, *and*
(iv) *if* $x = a$, *for no* $B \subseteq A$, $\mathrm{tp}_*(\bigcup I, A) \in \mathbf{F}(B)$.
(2) $\dim(I, A, M) \leq \lambda$ *when* cf $\lambda \geq \kappa_r(T)$.
(3) *If* cf $\lambda \geq \kappa(T) + \aleph_1$ *or* $\lambda = \aleph_0 = \kappa(T)$, *then* $\mathrm{Av}(I, |M|)$ *is* $\mathbf{F}^x_{\lambda^+}$*-isolated.*
(4) *If* cf $\lambda \geq \kappa(T) + \aleph_1$, *then* M *is* λ*-homogeneous over* A *(as in* 3.7(1)).
(5) dim $(I,A,M) \leqslant \lambda$ *when* λ *is regular,* $\lambda^+ \geqslant \kappa(T)$.

Proof. (1) Assume dim $(I,A,M) < \lambda$ and we shall get a contradiction. Clearly, $\lambda > \aleph_0$. We can assume $|I| = \dim(I,A,M)$. Choose distinct $\bar{a}_n \in I$ $(n < \omega)$. By 3.12(2) (using 4.7 instead of 3.2) and Section 2, there are $B_n \subseteq A$ such that $\mathrm{tp}(\bar{a}_0 \frown \cdots \frown \bar{a}_n, A) \in \mathbf{F}(B_n)$. So when cf$\lambda > \aleph_0$, $B = \bigcup_{n<\omega} B_n$ has cardinality $< \lambda$ hence, $\mathrm{tp}_*(\bigcup I, A) \in \mathbf{F}(B)$ (by the indiscernibility), so when $x = a$, we get a contradiction to (iv). Similarly, when $x = t$ let $J = \{\bar{a}_n; 0 < n < \omega\}$, so by Ax(V.1) $\mathrm{tp}(\bar{a}_0, A \cup \bigcup_{l=1}^n \bar{a}_l) \in \mathbf{F}(B \cup \bigcup_{l=1}^n \bar{a}_l)$, hence $\mathrm{tp}(\bar{a}_0, A \cup \bigcup J) = \mathrm{Av}(J, A \cup \bigcup J) \in \mathbf{F}(B)$, contradiction to (iii). So (iii) [(iv)] implies cf $\lambda = \aleph_0$ (i.e., (i)); but they follow from 4.10(1) below. So we can assume cf $\lambda = \aleph_0$ hence $\kappa(T) = \aleph_0$, and have to prove (ii); so suppose $|I| > \aleph_0$. Choose disjoint $J_0, J_1 \subseteq I$, $|J_0| < \aleph_0$, $|J_1| = \aleph_0$, such that $\mathrm{Av}(I, \bigcup I)$ does not fork over $\bigcup J_0$, $\mathrm{Av}(I, \bigcup J_0)$ is stationary. By 4.8 M is $\mathbf{F}$-prime over $A \cup \bigcup J_0$, so w.l.o.g. $J_0 = \emptyset$. Let N be $\mathbf{F}$-prime over $A \cup \bigcup J_1$, so there is an elementary embedding f of M into N, $f \restriction A$ = the identity. As $|I| > \aleph_0$ for some $\bar{a} \in I$ $f(\bar{a})$ realizes $\mathrm{Av}(J, A \cup \bigcup J_1)$, so by 4.7 $\mathrm{Av}(J, A \cup \bigcup J_1)$ is $\mathbf{F}$-isolated, hence by 4.10(1) below $\dim(I, A, M) \geq \lambda$.

CLAIM 4.10: (1) *Under the assumption of* 4.9(1) *if* $x = a$, $\mathrm{tp}_*(\bigcup I, A) \in \mathbf{F}(B)$ *or for some infinite* $J \subseteq I$, $\mathrm{Av}(J, A \cup \bigcup J)$ *is* $\mathbf{F}$*-isolated, then* $\mathrm{Av}(I, A \cup \bigcup I)$ *is* $\mathbf{F}^x_\mu$*-isolated where* $\mu = \lambda + |I|^+$*; so* $|I| \geq \lambda$.

Proof. For simplicity we concentrate on $x = a$, $\mathrm{tp}_*(\bigcup I, A) \in \mathbf{F}(B)$.

By the indiscernibility of I, $q = \mathrm{Av}(I, B \cup \bigcup I) \vdash \mathrm{Av}(I, A \cup \bigcup I)$. So it suffices to prove that $\mathrm{Av}(I, B \cup \bigcup I)$ is $\mathbf{F}^x_\mu$-isolated. By the indiscernibility of I and Ax(V.1), for every finite $J \subseteq I$ there is an

$\mathbf{F}$-type $q_J \subseteq q$ almost over $B \cup \bigcup J$, $q_J \vdash \mathrm{Av}(I, B \cup \bigcup J)$. Clearly $q = \bigcup\{q_J: J \subseteq I,\ J \text{ finite}\}$ is an $\mathbf{F}^x_\mu$-type almost over $B \cup \bigcup I$, $q \vdash \mathrm{Av}(I, B \cup \bigcup I)$, so we finish.

For 4.9(2) we prove:

CLAIM 4.10: (2) *If* $\bigcup J \subseteq B \subseteq C$, *$J$ an infinite indiscernible set,* $|J| \geq \lambda \geq \kappa_r(T)$, *$C$ is $\mathbf{F}$-constructible over B,* $\mathbf{F} = \mathbf{F}^x_\lambda$, $x = t, s, a$ *then* $\mathrm{Av}(J, B) \vdash \mathrm{Av}(J, C)$.

(3) *The claim holds for* $\mathbf{F}^c_\lambda, \lambda \geqslant \kappa_r(T)$, *too.*

Proof. Let $\mathscr{B} = \langle B, \{\bar{b}_i: i < \alpha\}, \{B_i: i < \alpha\}\rangle$ be an $\mathbf{F}$-construction of C over B. We prove by induction on i that $\mathrm{Av}(J, B) \vdash \mathrm{Av}(J, \mathscr{B}(i))$. For $i = 0$ it is trivial, for $i = \delta$ a limit ordinal it is immediate. For $i = j + 1$, $B_j \subseteq \mathscr{B}(j)$, $|B_j| < \lambda$, so by III, 3.5, for every $\bar{a} \in \mathscr{B}(j)$ there is $J_a \subseteq J$. $|J_a| < |B_j|^+ + \kappa_r(T) \leq \lambda \leq |J|$, such that $J - J_a$ is indiscernible over $\bar{a} \cup B_j \cup \bigcup J_a$. Let $\bar{c}$ realize $\mathrm{Av}(J, \mathscr{B}(i))$. In all cases there is an $\mathbf{F}$-type $p \subseteq p_j = \mathrm{tp}(\bar{b}_j, \mathscr{B}(j))$ [almost] over B_j, $p \vdash p_j$. Now $p \vdash \mathrm{tp}(\bar{b}_j, \mathscr{B}(j) \cup \bar{c})$ (for otherwise for some $\bar{a} \in \mathscr{B}(j) \models \varphi[\bar{b}_j; \bar{c}, \bar{a}]$ but $p \cup \{\neg\varphi(\bar{x}; \bar{c}, \bar{a})\}$ is consistent. So for any $\bar{c}' \in J - J_a$, $p \cup \{\neg\varphi(\bar{x}; \bar{c}', \bar{a})\}$ is consistent, and, of course $\models\varphi[\bar{b}_j; \bar{c}', \bar{a}]$. Contradiction to $p \vdash p_j$). By Ax(VI) for $\mathbf{F}^s_\infty$

$$\mathrm{Av}(J, \mathscr{B}(j)) = \mathrm{tp}(\bar{c}, \mathscr{B}(j)) \vdash \mathrm{tp}(\bar{c}, \mathscr{B}(j) \cup \bar{b}_j) = \mathrm{Av}(J, \mathscr{B}(i)).$$

By the induction hypothesis $\mathrm{Av}(J, B) \vdash \mathrm{Av}(J, \mathscr{B}(j))$.

Hence $\mathrm{Av}(J, B) \vdash \mathrm{Av}(J, \mathscr{B}(i))$. So we finish the induction, hence for $i = \alpha$ we get our result.

(3) Because an $\mathbf{F}^c_\lambda$-construction is an $\mathbf{F}^a_\lambda$-construction.

EXERCISE 4.3: In 4.10(2) for $x = t$ show that $\lambda \geq \kappa(T)$ is sufficient.

Continuation of 4.9. (2) Suppose not, so we can assume $|I| = \lambda^+$. By III, 4.17(4) for some $J \subseteq I$, $|J| < \min\{\kappa(T), \aleph_1\}$, $p = \mathrm{Av}(I, |M|)$ does not fork over $\bigcup J$ and $p \restriction (\bigcup J)$ is stationary. By 4.8 M is $\mathbf{F}$-prime over $A \cup \bigcup J$. So we can replace A by $A \cup \bigcup J$, I by $I - J$, retaining all our previous hypotheses and in addition $p \restriction A$ is stationary, p does not fork over A. Let $I = \{\bar{a}_i: i < \lambda^+\}$, M^* be an $\mathbf{F}$-primary model over $A \cup \bigcup I_1$ where $I_1 = \{\bar{a}_i: i < \lambda\}$. M^* exists by 3.1; and by 4.10, $\mathrm{Av}(I, A \cup \bigcup I_1) \vdash \mathrm{Av}(I, |M^*|)$. As $A \subseteq |M^*|$, M^* is $\mathbf{F}$-saturated, there is an elementary mapping $f: M \rightarrow M^*$, $f \restriction A =$ the identity. So $\{f(\bar{a}_i): i < \lambda^+\}$ is an indiscernible set based on $p \restriction A$, hence equivalent to I. As $\kappa_r(T) \leq \lambda$, for some i $f(\bar{a}_i)$ realizes $\mathrm{Av}(I, A \cup \bigcup I_1)$ contradiction to $\mathrm{Av}(I, A \cup \bigcup I_1) \vdash \mathrm{Av}(I, |M^*|)$.

(3) Here the hypotheses of (1) and of (2) hold. We can assume $\mathrm{Av}(I, |M|)$ does not fork over A, $\mathrm{Av}(I, A)$ is stationary, as in (2) [by (1) we can assume $\lambda \le |I|$, hence $|J| < \aleph_1 \le \mathrm{cf}\,\lambda \le \lambda$ when $\mathrm{cf}\,\lambda \ge \kappa(T) + \aleph_1$, and $|J| < \aleph_0 = |I|$ when $\lambda = \aleph_0 = \kappa(T)$]. We can have $I = \{\bar{a}_i : i < \lambda\}$ (by (1) and (2)), and define M^*, f as in (2). Let the image of M by f be N. So it suffices to prove $\mathrm{Av}(I, |N|)$ is $\mathbf{F}^{x}_{\lambda^+}$-isolated. By Claim 4.10 $\mathrm{Av}(I, A \cup \bigcup I)$ is $\mathbf{F}^{x}_{\lambda^+}$-isolated, but $\mathrm{Av}(I, A \cup \bigcup I) \vdash \mathrm{Av}(I, |M^*|)$ hence $\mathrm{Av}(I, |M^*|)$ is $\mathbf{F}^{x}_{\lambda^+}$-isolated. Now as $\mathrm{Av}(I, |M^*|)$ does not fork over $A \subseteq |N|$, by 4.3, $\mathrm{Av}(I, |N|)$ is $\mathbf{F}^{x}_{\lambda^+}$-isolated.

(4) Let f be an elementary mapping, $A \subseteq \mathrm{Dom}\, f \subseteq |M|$, $\mathrm{Range}\, f \subseteq |M|$, $f \restriction A$ is the identity and $|\mathrm{Dom}\, f - A| < \lambda$; and $a \in M$. We have to find an elementary mapping f_1 extending f, $\mathrm{Dom}\, f_1 = \mathrm{Dom}\, f \cup \{a\}$, $f_1(a) \in |M|$. Let M^1 be $\mathbf{F}^{x}_{\lambda}$-primary over $\mathrm{Dom}\, f$, so we can assume $|M^1| \subseteq |M|$. As M is $\mathbf{F}^{x}_{\lambda}$-saturated, by 3.10(2), Exercise 3.5 we can extend f to an elementary mapping f^1, $\mathrm{Dom}\, f^1 = |M^1|$, $\mathrm{Range}\, f^1 \subseteq |M|$. Choose $C \subseteq |M_1|$, $C < \kappa(T)$ such that $\mathrm{tp}(a, |M^1|)$ does not fork over C. We now define by induction on $\alpha < \omega + \omega$, $a_\alpha \in |M^1|$ realizing $\mathrm{tp}(a, A \cup C \cup \{a_i : i < \alpha\})$. (Remember M is $\mathbf{F}^{x}_{\lambda}$-atomic over A by 4.7, so the above mentioned type is $\mathbf{F}^{x}_{\lambda}$-isolated by 2.8(2), as $|C| + \aleph_0 < \kappa(T) + \aleph_1 \le \mathrm{cf}\,\lambda$, hence it is realized in M^1.) So let $I = \{a_\alpha ; \omega \le \alpha < \omega + \omega\}$; clearly I is indiscernible over $A \cup C$, $\mathrm{Av}(I, |M^1|) = \mathrm{tp}(a, |M^1|)$. By 4.9(1) $\lambda \le \dim(I, A, M)$, so let $J \supseteq I$ be indiscernible, $|J| = \lambda$. Now some $b \in J$ should realize $\mathrm{tp}(a, \mathrm{Dom}\, f)$; so we can define f_1 by: $f_1(a) = f^1(b)$, $f_1 \supseteq f$.

(5) Since $\mathbf{F}$ satisfies Ax(X.1), (XI.1), there is over A an $\mathbf{F}$-primary model N. Since M is $\mathbf{F}$-prime over A, w.l.o.g. $M \prec N$. Suppose $I \subseteq M$ is indiscernible over A, $|I| = \lambda^+$. As in the proof of 4.9(2), w.l.o.g. $\mathrm{Av}(I, A \cup I)$ does not fork over A, $\mathrm{Av}(I, A)$ is stationary. As $\lambda^+ \geqslant \kappa(T)$ clearly

(*) For every $\bar{c}$, all but $\leqslant \lambda$ of the members of I realize $\mathrm{Av}(I, A \cup \bar{c})$.

By 3.4, since $I \subseteq N$, there is an $\mathbf{F}$-construction over A, $\mathscr{A} = \langle A, \langle a_i : i < \lambda^+ \rangle\rangle$, such that $I \subseteq \mathscr{A}(\lambda^+)$. Choose $\alpha < \lambda^+$, $\mathrm{cf}\,\alpha = \lambda$, such that $|I \cap \mathscr{A}(\alpha)| = \lambda$ and for every $\beta < \alpha$ some $\bar{c} \in I \cap \mathscr{A}(\alpha)$ realizes $\mathrm{Av}(I, \mathscr{A}(\beta))$ (this is possible by (*)). Again by (*) some $\bar{d} \in I$ realizes $\mathrm{Av}(I, \mathscr{A}(\alpha))$ but by 3.2, $\mathscr{A}(\lambda^+)$ is $\mathbf{F}$-atomic over $\mathscr{A}(\alpha)$ (and $\bar{d} \in I \subseteq \mathscr{A}(\lambda^+)$) so $\mathrm{tp}(\bar{d}, \mathscr{A}(\alpha))$ is $\mathbf{F}$-isolated. So for some $D \subseteq \mathscr{A}(\alpha)$, $\mathrm{tp}(\bar{d}, \mathscr{A}(\alpha)) \in \mathbf{F}(D)$, so $|D| < \lambda$. Since $\mathrm{cf}\,\alpha = \lambda$ for some $\beta < \alpha, D \subseteq \mathscr{A}(\beta)$. But by the choice of α some $\bar{d}' \in I$ belong to $\mathscr{A}(\alpha)$ and realizes $\mathrm{Av}(I, \mathscr{A}(\beta))$, i.e. it realizes $\text{tp}(\bar{d}, \mathscr{A}(\beta))$. This contradicts $\text{tp}(\bar{d}, \mathscr{A}(\alpha)) \in \mathbf{F}(D), D \subseteq \mathscr{A}(\beta)$ (for $x = a$ remember $\text{Av}(I, A)$ is stationary, hence $\text{tp}(\bar{d}, \mathscr{A}(\beta))$ is stationary).

We now prove some preliminaries to the characterization theorems.

LEMMA 4.11: *Let $x = t, a$. Assume $\mathbf{F}^x_\lambda$ satisfies* Ax(X.1), (XI.1), $\lambda \geq \kappa(T)$ *and* cf $\lambda > \aleph_0$.

If M is $\mathbf{F}^x_\lambda$-compact and $\mathbf{F}^x_\lambda$-atomic over A, $I \subseteq |M|$ an infinite indiscernible set over A and $p = \text{Av}(I, |M|)$ is $\mathbf{F}^x_{\lambda^+}$-isolated then there is an indiscernible set J over A, equivalent to I, $J \subseteq |M|$, $|J| = \lambda$ and $\text{Av}(J, A \cup \bigcup J) \vdash \text{Av}(J, |M|)$.

Proof. By III, 4.17(4) there is a $C \subseteq |M|$, such that $p = \text{Av}(I, |M|)$ does not fork over C and $p \restriction C$ is stationary, and $|C| < \aleph_1$. By a hypothesis there is $B \subseteq |M|$, $|B| = \lambda$, $(p, B) \in \mathbf{F}^x_{\lambda^+}$. We can assume $C \subseteq B$ [by Ax(III.2)] and let $\bar{c}$ realize p. So $\text{tp}(\bar{c}, B) \vdash \text{stp}(\bar{c}, B)$ (by III, 2.14) hence $\text{tp}(\bar{c}, B) \vdash \text{tp}(\bar{c}, |M|) = p$ (see 2.1 for $x = a$).

Let $\{\bar{b}_i : i < \lambda\}$ be a list of all finite sequences from B, each sequence appearing λ times. Now define by induction on $i < \lambda$, $\bar{a}_i \in |M|$ such that $\bar{a}_i$ realizes $\text{Av}(I, A_i)$ where $A_i = C \cup A \cup \{\bar{a}_j : j < i\}$.

If $\bar{a}_i$ is defined for $i < \alpha$ then as in the proof of 4.9 above, there is an $\mathbf{F}$-type $q_\alpha \subseteq p$ over $C \cup A \cup \bigcup \{\bar{a}_j : j < \alpha\}$, $q_\alpha \vdash p \restriction A_\alpha$. If $q_\alpha \vdash p \restriction \bar{b}_\alpha$, $\bar{a}_\alpha$ will be any element of $|M|$ which realizes q_α (there is one, as M is $\mathbf{F}$-compact, q_α an $\mathbf{F}$-type). Otherwise there is a formula $\varphi_\alpha(\bar{x}; \bar{b}_\alpha)$ such that $q^{\mathbf{t}}_\alpha = q_\alpha \cup \{\varphi_\alpha(\bar{x}; \bar{b}_\alpha)^{\mathbf{t}}\}$ is consistent for $\mathbf{t} \in \{0, 1\}$. As $p \restriction C \subseteq q_\alpha$ $p \restriction C$ is stationary, for one such $\mathbf{t}$, $q^{\mathbf{t}}_\alpha$ forks over C. W.l.o.g. it occurs for $\mathbf{t} = 0$, and let $\bar{a}_\alpha \in |M|$ realize q^0_α. As $q_\alpha \vdash p \restriction A_\alpha$ and $p \restriction C$ is stationary, $J = \{a_i : i < \lambda\}$ is an indiscernible set over A equivalent to I (as both are based on $p \restriction C$) by III, 1.10.

As $\lambda \geq \kappa(T)$, by III, 2.4 for every $\bar{b}_i$ there is $J' \subseteq J$, $|J'| < \kappa(T) \leq |J|$ such that $J - J'$ is indiscernible over $\bar{b}_i \cup C \cup A \cup \bigcup J'$. Hence for every $\bar{b} \in B$ the set $\{i < \lambda : \bar{b} = \bar{b}_i$, there is $\varphi_i(\bar{x}; \bar{b}_i)$ as mentioned above$\}$ has cardinality $< \lambda$. Hence for some α, $\bar{b}_\alpha = \bar{b}$ and $q_\alpha \vdash p \restriction \bar{b}_\alpha$, hence $p \restriction (A \cup \bigcup J) \vdash p \restriction \bar{b}_\alpha$ ($q_\alpha \subseteq p \restriction (A \cup \bigcup J)$ of course). As this holds for any α, $p \restriction (A \cup \bigcup J) \vdash p \restriction B$, but $p \restriction B \vdash p$ by the choice of B. Hence $p \restriction (A \cup \bigcup J) \vdash p$ so we finish.

THEOREM 4.12: (1)* *Suppose* $\text{tp}(\bar{a}_i, A) \vdash \text{tp}(\bar{a}_i, A_i)$ where for $S \subseteq \mu$, $A_S = A \cup \bigcup \{\bar{a}_j : j \in S\}$ *(notice $i = \{j : j < i\}$). (The $\bar{a}_i$'s may be infinite, in fact).*

Assume
case (i): *for no* $i < \mu$ *does* $\text{tp}(\bar{b}, A) \vdash \text{tp}(\bar{b}, A \cup \bar{a}_i)$, *or*
case (ii): $\aleph_0 < \mu$ *and for no* $S \subseteq \mu$, $|S| < \mu$ *does* $\text{tp}(\bar{b}, A_S) \vdash \text{tp}(\bar{b}, A_\mu)$.
Then $\mu < \lambda^2(T)$ (*see Definition* 2.5).

(2)* *Suppose* $\mu \geq \lambda$ *and that in the hypothesis of* (1) *we add to:*
case (i): $\text{tp}(\bar{b}, A)$ *is* $\mathbf{F}^x_\lambda$*-isolated* ($x = t, s$),
case (ii): *for every finite* $S \subseteq \mu$, $\text{tp}(\bar{b}, A_S)$ *is* $\mathbf{F}^x_\lambda$*-isolated; and* (A) $\mu > \lambda$, *or* (B) $\mu = \lambda$ *is regular* ($> \aleph_0$).
Then $\mathbf{F}^x_\lambda$ ($x = t, s$) *does not satisfy* Ax(XI.1).

(3) *The same as* (1) *replacing* tp *by* stp *and the conclusion is* $\mu < \kappa(T)$.

(4) *If in* (3) $\mu \geq \lambda$ *and in the hypothesis we add to:*
case (i): $\text{tp}(\bar{b}, A)$ *is* $\mathbf{F}^a_\lambda$*-isolated,*
case (ii): *for every finite* $S \subseteq \mu$, $\text{tp}(\bar{b}, A_S)$ *is* F^a_λ*-isolated; and* (A) $\mu > \lambda$ *or* (B) $\mu = \lambda$ *is regular* $> \aleph_0$.
Then $\mathbf{F}^a_\lambda$ *does not satisfy* Ax(XI.1).

Proof. (1) Suppose case (i) holds but $\mu \geq \lambda^2(T)$. Then there is $q \in S^m(A_\mu)$, $p = \text{tp}(\bar{b}, A) \subseteq q$ and $\alpha < \mu$ such that $q \restriction A_\alpha \vdash q$ (this can be done by the definition of $\lambda^2(T)$ and by renaming the $\bar{a}_i$'s which is possible by 3.13 applied to $\mathbf{F}^s_\infty$). By renaming, we can assume $\bar{b}$ realizes q, without changing the hypothesis.

By Ax(VI) for $\mathbf{F}^s_\infty$, as $\text{tp}(\bar{b}, A) \nvdash \text{tp}(\bar{b}, A \cup \bar{a}_\alpha)$ also $\text{tp}(\bar{a}_\alpha, A) \nvdash \text{tp}(\bar{a}_\alpha, A \cup \bar{b})$. So there are $\bar{c} \in A$ and φ such that $\models \varphi[\bar{b}; \bar{a}_\alpha, \bar{c}]$ but $\text{tp}(\bar{a}_\alpha, A) \cup \{\neg\varphi(\bar{b}, \bar{x}, \bar{c})\}$ is consistent, and let $\bar{a}'_\alpha$ realize it. By assumption, there is an elementary mapping $f, f \restriction A_\alpha =$ the identity, $f(\bar{a}'_\alpha) = \bar{a}_\alpha$. We can extend it to an elementary mapping g whose domain includes $\bar{b}$. Then $g(\bar{b})$ realizes $q \restriction A_\alpha$ but not $q \restriction A_{\alpha+1}$, a contradiction.

So we are left with case (ii), and we reduce it to case (i) as follows: Let $S_0 = \{\alpha < \mu: \text{tp}(\bar{b}, A_\alpha) \nvdash \text{tp}(\bar{b}, A_{\alpha+1})\}$. Now w.l.o.g. $S_0 = \mu$. [We define by induction on $\alpha \leq \mu$ sets $S(\alpha) \subseteq \mu$, $S(\alpha)$ increasing and continuous, $|S(\alpha + 1) - S(\alpha)| < \aleph_0$. If $S(\alpha)$ is defined $\text{tp}(\bar{b}, A_{S(\alpha)}) \nvdash \text{tp}(\bar{b}, A_\mu)$ so for some $S(\alpha + 1)$, $S(\alpha) \subseteq S(\alpha + 1)$, $|S(\alpha + 1) - S(\alpha)| < \aleph_0$, $\alpha \in S(\alpha + 1)$, $\text{tp}(\bar{b}, A_{S(\alpha)}) \nvdash \text{tp}(\bar{b}, A_{S(\alpha+1)})$, and $S(\alpha + 1)$ is minimal under those conditions. Let $S(\alpha + 1) - S(\alpha) = \{\bar{a}_{j(\alpha,0)}, \ldots, \bar{a}_{j(\alpha,k(\alpha))}\}$, $\bar{a}^*_\alpha = \bar{a}_{j(\alpha,0)}{}^\frown\bar{a}_{j(\alpha,1)}{}^\frown\cdots{}^\frown\bar{a}_{j(\alpha,k(\alpha))}$; now the $\bar{a}^*_\alpha$'s satisfies all the hypotheses on the $\bar{a}_\alpha$'s, and for them $S_0 = \mu$.] So for every α there are $n(\alpha) < \omega$, $i(l, \alpha) < \alpha$ (for $l < n(\alpha)$), formulas φ_α, and $\bar{c}_\alpha \in A$ such that $\models \varphi_\alpha[\bar{b}, \bar{a}_\alpha; \bar{a}_{i(0,\alpha)}, \ldots; \bar{c}_\alpha]$ but $\text{tp}(\bar{b}, A_\alpha) \cup \{\neg\varphi_\alpha(\bar{x}; \bar{a}_\alpha, \bar{a}_{i(0,\alpha)}, \ldots, \bar{c}_\alpha)\}$ is consistent. If $\aleph_0 < \chi \leq \mu$, χ regular, by 1.3(1) of the Appendix there is a stationary $S_\chi \subseteq \chi$, such that $\alpha \in S_\chi \Rightarrow n(\alpha) = n$, $i(l, \alpha) = i(l)$ (for $l < n$). Then let $A' =$

$A \cup \bigcup_{l<n} \bar{a}_{i(l)}$ and $\bar{a}'_\alpha$ is $\bar{a}_\beta$, where β is the αth element in S_χ. Clearly our hypothesis holds for $\bar{a}'_\alpha$, A', $\bar{b}$ $(\alpha < \chi)$ hence by (i) $\chi < \lambda^2(T)$.

So we are left with the case $\mu = \lambda^2(T) > \aleph_0$ is singular. Let $\mu = \sum_{i<\mathrm{cf}\,\mu} \chi(i)$, $\mathrm{cf}\,\mu < \chi(i) < \mu$, $\chi(i)$ regular.

For every regular $\chi < \mu$, find S_χ, $n = n_\chi$, $i(l) = i_\chi(l)$ as above. We can assume that for $\zeta \neq \xi < \mathrm{cf}\,\mu$, $i_{\chi(\xi)}(l) \notin S_{\chi(\zeta)}$. We let $A' = A \cup \{\bar{a}_i$: for some $\zeta < \mathrm{cf}\,\mu$. $l < n_{\chi(\zeta)}$, $i = i_{\chi(\zeta)}(l)\}$, and $\{\bar{a}'_\alpha: \alpha < \mu\}$ be $\{\bar{a}_i: i \in S_{\chi(\zeta)}, \zeta < \mathrm{cf}\,\mu\}$.

(2) The same proof.

(3) The same proof as (1).

(4) The same proof as (2).

THEOREM 4.13: *Suppose* $\mathbf{F}^x_\lambda = \mathbf{F}$, $x = t, s, a$, $\mathbf{F}$ *satisfies* Ax(XI.1), *and*

(A) $\lambda > \aleph_0$ *is regular, or*

(B) $x = t, s$; $\mathrm{cf}\,\lambda \geq \lambda^2(T) + \aleph_1$, *or*

(C) $x = a$, $\mathrm{cf}\,\lambda \geq \kappa(T) + \aleph_1$.

If for every $i < \delta$, $\bar{c} \in C_i$, *for some* $B^i_{\bar{c}} \subseteq A$ $\mathrm{tp}(\bar{c}, A_i) \in \mathbf{F}^x_\infty(B^i_{\bar{c}})$, *where* $A_i = A \cup \bigcup_{j<i} C_j$, *and for* $i < \delta$, $\mathrm{tp}(\bar{b}, A_i)$ *is* $\mathbf{F}$*-isolated, then* $\mathrm{tp}(\bar{b}, A_\delta)$ *is* $\mathbf{F}$*-isolated.*

Proof. We prove it by induction on $\mathrm{cf}\,\delta$. If $\mathrm{cf}\,\delta < \mathrm{cf}\,\lambda$, this is immediate by Axiom (IX). Suppose the conclusion fails. Then we can easily define by induction on $i < \mathrm{cf}\,\delta$, $\alpha(i) < \delta$, and $\bar{a}_i \in C_{\alpha(i)}$ such that

(1) $j < i \Rightarrow \alpha(j) < \alpha(i)$,

(2) $\mathrm{tp}(\bar{b}, A_{\alpha(i)}) \nvdash \mathrm{tp}(\bar{b}, A_{\alpha(i)} \cup \bar{a}_i)$ for $x = t, s$,

(3) $\mathrm{stp}(\bar{b}, A_{\alpha(i)}) \nvdash \mathrm{stp}(\bar{b}, A_{\alpha(i)} \cup \bar{a}_i)$ for $x = a$.

W.l.o.g. $\delta = \mathrm{cf}\,\delta$ (by renaming), so we already know $\delta \geqslant \mathrm{cf}\,\lambda$. If λ is regular, $x = t, s$ we get by 4.12(2) case (ii) (for $\mu \stackrel{\text{def}}{=} \delta$) that Ax(XI.1) fails for $\mathbf{F}^x_\lambda$ contradicting an assumption. If λ is regular $x = a$, we get a similar contradiction by 4.12(2) case (ii). So we have proved case (A) of 4.13. In case (B) of 4.13 apply 4.12(1) case (ii) and in case (C) of 4.13 apply 4.12(3) case (ii).

Remark. In 4.13 for $x = t, s$ we need not assume T is stable.

DEFINITION 4.2: Let $\mathrm{cf}\,\lambda \geq \kappa(T) + \aleph_1$, $x = t, s, a$. Then M is called $\mathbf{F}^x_\lambda$-admissible over A if:

(1) M is $\mathbf{F}^x_\lambda$-saturated.

(2) $A \subseteq |M|$, and M is $\mathbf{F}^x_\lambda$-atomic over A.

(3) For any infinite indiscernible set I (of elements) over A, $I \subseteq |M|$, $\mathrm{Av}(I, |M|)$ is $\mathbf{F}^x_{\lambda^+}$-isolated.

Note. If there is no I as in (3) then of course (3) holds vacuously.

THE FIRST CHARACTERIZATION THEOREM 4.14: *Suppose* $\mathrm{cf}\,\lambda \geq \kappa(T) + \aleph_1$, $x = t, s, a$, $\mathbf{F} = \mathbf{F}^x_\lambda$, $\mathbf{F}$ *satisfies Axioms* (X.1) *and* (XI.1). *If λ is singular, $x = t$, assume $\mathbf{F}^x_\mu$ satisfies* Ax(XI.1) *for arbitrarily large regular $\mu < \lambda$. For $x = t$, $\lambda \geq \lambda^3(t)$ or at least F^t_λ-saturated implies λ-compact. Then:*

(1) *M is $\mathbf{F}$-prime over A iff M is $\mathbf{F}$-admissible over A.*

(2) *The $\mathbf{F}$-prime model over A is unique up to isomorphism over A.*

(3) *If M_l is $\mathbf{F}$-admissible over A_l, $l = 1, 2$; f an elementary mapping from A_1 onto A_2 then we extend f to an isomorphism from M_1 onto M_2.*

Remark. In the proof, if $x = a$, you should replace tp by stp in many cases.

Proof. (1) If M is $\mathbf{F}$-prime over A, then M is $\mathbf{F}$-saturated, by definition, M is $\mathbf{F}$-atomic over A by 3.12(2) (using 4.7 instead of 3.2), and if $I \subseteq |M|$ is an indiscernible set of elements over A then $\mathrm{Av}(I, |M|)$ is $\mathbf{F}^x_{\lambda^+}$-isolated by 4.9(3). Hence M is $\mathbf{F}$-admissible over A. As there is an $\mathbf{F}$-prime model over A by 3.12(1), the other direction follows by (3).

(2) Follows by (1), (3) and the existence of an $\mathbf{F}$-prime model over A by 3.12(1).

(3) Assume $x = t$ or $x = a$ (the result for $x = s$ follows by $x = a$ as $\lambda > \aleph_0$ (see Exercise 2.30, 2.2(6), (7))). Let $\lambda = \sum_{i<\chi} \lambda_i$ be such that if λ is regular, $\chi = \kappa_r(T) + \aleph_1$, $\lambda_i = \lambda$; and if λ is singular then λ_i is regular, $\chi = \mathrm{cf}\,\lambda$; and if $x = t$, λ singular then $\mathbf{F}^t_{\lambda(i)}$ satisfies Ax(XI.1); and $\lambda(i) = \lambda_i > |i|$, $\lambda_i \geq \sum_{j<i} \lambda_j + \kappa(T) + \aleph_1$. For every $\bar{c} \in |M_l|$ $l = 1, 2$, there is $\lambda_i = h(\bar{c})$ such that $\mathrm{tp}(\bar{c}, A_l)$ is $\mathbf{F}^x_{\lambda(i)}$-isolated (we can assume $|M_1| \cap |M_2| = \emptyset$). We shall define elementary mappings f_α ($\alpha < \chi$) such that

(i) $\beta < \alpha$ implies f_α extends f_β,

(ii) $A_1 \subseteq \mathrm{Dom}\, f_\alpha = A^1_\alpha \subseteq |M_1|$, $A_2 \subseteq \mathrm{Range}\, f_\alpha = A^2_\alpha \subseteq |M_2|$,

(iii) M_1, M_2 are $\mathbf{F}$-admissible over $\mathrm{Dom}\, f_\alpha$, $\mathrm{Range}\, f_\alpha$ respectively,

(iv) if $l = 1, 2$ $\bar{c} \in |M_l|$, then $\mathrm{tp}(\bar{c}, A^l_\alpha)$ is $\mathbf{F}^x_\mu$-isolated, where $\mu = \lambda_\alpha + h(\bar{c})$.

For the checking of $\mathbf{F}$-admissibility in (iii) notice that by (ii) conditions (1) and (3) from Definition 4.2, are immediate; hence only (2) (on being $\mathbf{F}$-atomic) needs proof, and it follows from (iv).

Case 1. $\alpha = 0$.

Let $f_0 = f$; trivially all conditions hold.

Case 2. $\alpha = \delta$, a limit ordinal.

Let $f_\delta = \bigcup_{\alpha<\delta} f_\alpha$. Then (i) and (ii) are immediate. For (iv), as $\delta < \mathrm{cf}\,\lambda_\delta$ this follows by Ax(IX).

Case 3. $\alpha = \delta + 2n$, δ is 0 or limit, f_α is defined; we shall define $f_{\alpha+1}$.

Choose by induction on i elements $a_i^1 = a_{\alpha,i}^1 \in |M_1| - A_\alpha^1$ such that for some $B_{\alpha,i}^1 \subseteq A_\alpha^1$, $\mathrm{tp}(a_{\alpha,i}^1, A_\alpha^1 \cup \{a_{\alpha,j}^1 : j < i\}) \in \mathbf{F}_\mu^x(B_{\alpha,i}^1)$ where $\mu = \lambda_\alpha$. The first i for which we cannot find such $a_{\alpha,i}^1$ is $i(\alpha)$. Now we define $a_{\alpha,i}^2 \in |M_2|$ which will realize $f_\alpha[\mathrm{tp}(a_{\alpha,i}^1, A_\alpha^1 \cup \{a_{\alpha,j}^1 : j < i\})]$ by induction on $i < i(\alpha)$ (this is possible as M_2 is $\mathbf{F}$-saturated). We extend f_α to $f_{\alpha+1}$ by letting $f_{\alpha+1}(a_{\alpha,i}^1) = a_{\alpha,i}^2$. Clearly $f_{\alpha+1}$ is an elementary mapping and (i) and (ii) are clear. As for (iv), if $\bar{c} \in |M_l|$, by the induction hypothesis $\mathrm{tp}(\bar{c}, A_\alpha^l)$ is $\mathbf{F}_\mu^x$-isolated, $\mu = \lambda_\alpha + h(\bar{c})$, and the $\lambda(j)$ are regular $> \aleph_0$. So (iv) follows by 4.13(A), by induction on $i(\alpha)$.

Case 4. $\alpha = \delta + 2n + 1$, ($\delta = 0$ or δ limit) f_α is defined and we shall define $f_{\alpha+1}$.

We define by induction on i infinite non-trivial indiscernible sets (of elements) over A_α^1, $J_{\alpha,i}^1$ such that:

(A) For some infinite $I_{\alpha,i}^1 \subseteq A_\alpha^1$, $I_{\alpha,i}^1 \cup J_{\alpha,i}^1$ is indiscernible.

(B) For any n, $\bar{a}_1, \ldots, \bar{a}_n \in J_{\alpha,i}^1$, for some $B_{\alpha,i}^n \subseteq A_\alpha^1$,

$$\mathrm{tp}\Big(\bar{a}_1{}^\frown\cdots{}^\frown\bar{a}_n, A_1 \cup \bigcup_{j<i} J_{\alpha,j}^1\Big) \in \mathbf{F}_{\lambda(\alpha)}^x(B_{\alpha,i}^n).$$

(C) $\mathrm{Av}(J_{\alpha,i}^1, A_\alpha^1 \cup \bigcup J_{\alpha,i}^1) \vdash \mathrm{Av}(J_{\alpha,i}^1, |M_1|)$.

Let $i(\alpha)$ be the first i for which there is no such $J_{\alpha,i}^1$. By 3.13 clearly $J_{\alpha,i}^1$ is indiscernible over $\bigcup\bigcup_{j\neq i} J_{\alpha,j}^1 \cup A$.

Now we prove by induction on i that when $\mu = \lambda_\alpha + h(\bar{c})$, then $\mathrm{tp}(\bar{c}, A \cup \bigcup\bigcup_{j<i} J_{\alpha,j}^1)$ is $\mathbf{F}_\mu^x$-isolated. For $i = 0$ this is by the induction hypothesis on α, for i limit, by 4.13, and i successor by 4.15 (see below) (as $\lambda_j \geq \kappa(T) + \aleph_1$, λ_j is regular). So first we prove

CLAIM 4.15: *Suppose* $\mathrm{cf}\,\lambda \geq \kappa(T)$, $A \cup \bigcup J \cup \bar{c}$ *is* $\mathbf{F}_\lambda^x$*-atomic over* A

$(x = t, s, a)$ and $\mathrm{Av}(J, A \cup \bigcup J) \vdash \mathrm{Av}(J, A \cup \bigcup J \cup \bar{c})$, *and J is indiscernible over A.*

(1) $\mathrm{tp}(\bar{c}, A \cup \bigcup J)$ *is $\mathbf{F}^x_\lambda$-isolated, provided that* cf $\lambda \geq \kappa(T)$.

(2) *If J is based on A,* $\mathrm{Av}(J, A)$ *is stationary,* $\mathrm{tp}(\bar{c}, A \cup \bigcup J)$ *does not fork over A, then* $\mathrm{tp}(\bar{c}, A) \vdash \mathrm{tp}(\bar{c}, A \cup \bigcup J)$*;* $\mathrm{stp}(\bar{c}, A) \vdash \mathrm{stp}(\bar{c}, A \cup \bigcup J)$.

CLAIM 4.16: *Suppose $B \subseteq A$,* $\mathrm{tp}(\bar{a}_l, A)$ *does not fork over B, and* $\mathrm{tp}(\bar{a}_1, B)$ *is stationary, $l = 1, 2$.*

(1) *If* $\mathrm{tp}(\bar{a}_1, A) \vdash \mathrm{tp}(\bar{a}_1, A \cup \bar{a}_2)$ *then* $\mathrm{tp}(\bar{a}_1, B) \vdash \mathrm{tp}(\bar{a}_1, B \cup \bar{a}_2)$.

(2) *If* $\mathrm{stp}(\bar{a}_1, A) \vdash \mathrm{stp}(\bar{a}_1, A \cup \bar{a}_2)$ *then* $\mathrm{stp}(\bar{a}_1, B) \vdash \mathrm{stp}(\bar{a}_1, B \cup \bar{a}_2)$.

Remark. As Ax(VI) is satisfied by $\mathbf{F}^a_\infty$, the assumption and conclusion of (1) and (2) are symmetric for $\bar{a}_1$, $\bar{a}_2$. See also Ex. 4.10.

Proof of 4.16. (1) Suppose that the conclusion fails, so there are $\bar{c} \in B$ and a formula φ, such that $\mathrm{tp}(\bar{a}_1, B) \cup \{\varphi(\bar{x}; \bar{a}_2, \bar{c})^t\}$ is consistent for $t \in \{0, 1\}$. So there are $\bar{a}^t_1$, $t \in \{0, 1\}$, such that $\mathrm{tp}(\bar{a}^t_1, B) = \mathrm{tp}(\bar{a}_1, B)$, $\models\varphi[\bar{a}^t_1, \bar{a}_2, \bar{c}]^t$. By III, 2.6(3) there are sequences $\bar{b}^0_1, \bar{b}^1_1, \bar{b}_2$ such that $\mathrm{tp}(\bar{b}^0_1{}^\frown\bar{b}^1_1{}^\frown\bar{b}_2, A)$ does not fork over B, and $\mathrm{stp}(\bar{b}^0_1{}^\frown\bar{b}^1_1{}^\frown\bar{b}_2, B) \equiv \mathrm{stp}(\bar{a}^0_1{}^\frown\bar{a}^1_1{}^\frown\bar{a}_2, B)$. Hence $\mathrm{stp}(\bar{b}_2, B) \equiv \mathrm{stp}(\bar{a}_2, B)$. But the last type is stationary over B, by III, 2.9, hence $\mathrm{tp}(\bar{b}_2, A) = \mathrm{tp}(\bar{a}_2, A)$. Also $\mathrm{tp}(\bar{b}^i_1, B) = \mathrm{tp}(\bar{a}^i_1, B) = \mathrm{tp}(\bar{a}_1, B)$ $(i = 0, 1)$ and the last type is stationary, by hypothesis. So $\mathrm{tp}(\bar{b}^i_1, A) = \mathrm{tp}(\bar{a}_1, A)$. But clearly $\models\varphi[\bar{b}^t_1, \bar{b}_2, \bar{c}]^t$ $(t \in \{0, 1\})$. So $\mathrm{tp}(\bar{a}_1, A) = \mathrm{tp}(\bar{b}^0_1, A) \nvdash \mathrm{tp}(\bar{b}^0_1, A \cup \bar{b}_2)$. As $\mathrm{tp}(\bar{a}_2, A) = \mathrm{tp}(\bar{b}_2, A)$, $\mathrm{tp}(\bar{a}_1, A) \nvdash \mathrm{tp}(\bar{a}_1, A \cup \bar{a}_2)$, so the assumption fails.

(2) If $\mathrm{stp}(\bar{a}_1, A) \vdash \mathrm{stp}(\bar{a}_1, A \cup \bar{a}_2)$ then $\mathrm{stp}(\bar{a}_1, A) \vdash \mathrm{tp}(\bar{a}_1, A \cup \bar{a}_2)$. Hence $\mathrm{tp}(\bar{a}_1, A \cup \bar{a}_2)$ does not fork over A, and even over B. As $\mathrm{tp}(\bar{a}_1, B)$ is stationary, by III, 2.14, $\mathrm{tp}(\bar{a}_1, A) \vdash \mathrm{stp}(\bar{a}_1, A \cup \bar{a}_2)$ hence $\mathrm{tp}(\bar{a}_1, A) \vdash \mathrm{tp}(\bar{a}_1, A \cup \bar{a}_2)$. So by (1), $\mathrm{tp}(\bar{a}_1, B) \vdash \mathrm{tp}(\bar{a}_1, B \cup \bar{a}_2)$. Again by III, 2.14, $\mathrm{tp}(\bar{a}_1, B) \vdash \mathrm{tp}(\bar{a}_1, B \cup \bar{a}_2) \vdash \mathrm{stp}(\bar{a}_1, B \cup \bar{a}_2)$, so we finish.

Proof of Claim 4.15. (1) Choose $I \subseteq J$, $|I| < \kappa(T)$ such that $\mathrm{tp}(\bar{c}, A \cup \bigcup J)$ does not fork over $A \cup \bigcup I$, $\mathrm{Av}(J, A \cup \bigcup J)$ does not fork over $A \cup \bigcup I$, and $\mathrm{Av}(J, A \cup \bigcup I)$ is stationary (by Theorem III, 3.2 and III, 4.17(4)). So $|I \cup \bar{c}| < \kappa_r(T)$ hence by Ax(V.2) for $\mathbf{F}^x_{\lambda,\kappa(T_r)}$, it suffices to prove (2).

(2) For simplicity, let $x = s, t$. Let $J = \{\bar{a}_i : i < \mu\}$, $\bar{a}_\mu$ realizes $\mathrm{Av}(J, A \cup \bigcup J \cup \bar{c})$, and $A_\alpha = A \cup \bigcup \{\bar{a}_i : i < \alpha\}$. By assumption $\mathrm{tp}(\bar{a}_\mu, A_\mu) \vdash \mathrm{tp}(\bar{a}_\mu, A_\mu \cup \bar{c})$ hence by Ax(VI) (for $\mathbf{F}^s_\infty$) $\mathrm{tp}(\bar{c}, A_\mu) \vdash$

$\mathrm{tp}(\bar{c}, A \cup \bar{a}_\mu)$. By Claim 4.16(1) for every $\alpha < \mu$, $\mathrm{tp}(\bar{c}, A_\alpha) \vdash \mathrm{tp}(\bar{c}, A_\mu \cup \bar{a}_\mu)$. As $\{\bar{a}_i: i \le \mu\}$ is indiscernible over $A \cup \bar{c}$ (by III, 2.10) $\mathrm{tp}(\bar{a}_\alpha, A_\alpha \cup \bar{c}) = \mathrm{tp}(\bar{a}_\mu, A_\alpha \cup \bar{c})$ hence $\mathrm{tp}(\bar{c}, A_\alpha) \vdash \mathrm{tp}(\bar{c}, A_\alpha \cup \bar{a}_\alpha)$. So by induction $\mathrm{tp}(\bar{c}, A) \vdash \mathrm{tp}(\bar{c}, A_\mu)$. (For $x = a$, use 4.16(2), and III, 2.14.)

Continuation of the proof of 4.14. As in Case 3, we can define $J^2_{\alpha,i} \subseteq |M_2|$ $[i < i(\alpha)]$ such that $J^2_{\alpha,i}$ is indiscernible over A^2_α, and for every distinct $a^l_1, \dots, a^l_n \in J^l_{\alpha,i}$ $a^2_1 {}^\frown \dots {}^\frown a^2_n$ realizes $f_\alpha[\mathrm{tp}(a^1_1 {}^\frown \dots {}^\frown a^1_n, B^n_{\alpha,i})]$ (if $x = a$ write stp instead of tp) and $\mathrm{Av}(J^2_{\alpha,i}, A^1_\alpha \cup \bigcup J^2_{\alpha,i}) \vdash \mathrm{Av}(J^2_{\alpha,i}, |M_2|)$. Such $J^l_{\alpha,i}$ exist by 4.11. So $|J^2_{\alpha,i}| = \lambda = |J^1_{\alpha,i}|$, so we can extend f_α to $f_{\alpha+1}$, with domain $A^1_\alpha \cup \bigcup \bigcup_i J^1_{\alpha,i}$, which maps $J^1_{\alpha,i}$ onto $J^2_{\alpha,i}$. By 3.13 $f_{\alpha+1}$ is elementary. We have proved that for $\bar{c} \in |M_1|$, $\mu = \lambda_\alpha + h(\bar{c})$, $\mathrm{tp}(\bar{c}, A \cup \bigcup \bigcup_i J^1_{\alpha,i})$ is $\mathbf{F}^x_\mu$-isolated. A similar claim can be proved on any $\bar{c} \in |M_2|$. So (iv) holds.

So we finish the definition. Notice that if we interchange M_1 with M_2, A_1 with A_2, $a^1_{\alpha,i}$ with $a^2_{\alpha,i}$, $J^1_{\alpha,i}$ with $J^2_{\alpha,i}$ f_α with f_α^{-1} all conditions still hold. Also $f_\chi = \bigcup_{\alpha<\chi} f_\alpha$ is an elementary mapping, $\mathrm{Dom} f_\chi \subseteq |M_1|$, $\mathrm{Range} f_\chi \subseteq |M_2|$, f_χ extends f. So by symmetry it suffices to prove $|M_1| = \mathrm{Dom} f_\chi$. Let $c \in |M_1| - \mathrm{Dom} f_\chi$, and we shall get a contradiction. Now we prove

(*) If $\alpha < \chi$, $p \in S(A^1_\alpha)$ is realized in M_1, then it is realized in A^1_χ.

As M_1 is $\mathbf{F}$-atomic over A^1_α, p is $\mathbf{F}$-isolated, so by Ax(XI.1) there are $q \in S(A^1_\chi)$, $p \subseteq q$ such that q is $\mathbf{F}$-isolated. As $\mathrm{cf}\,\chi \ge \kappa(T)$ q does not fork over A^1_β for some $\beta < \chi$. Clearly for some j, q is $\mathbf{F}^x_{\lambda(j)}$-isolated.

If λ is singular choose $\gamma = \lambda(i) < \lambda$, such that $\alpha, \beta < i$, $\lambda(j) < \gamma$. So by 4.3 $q' = q \restriction A^1_\gamma$ is $\mathbf{F}^x_{\lambda(j)}$-isolated, so for some $B \subseteq A^1_\gamma$, $(q', B) \in F^x_{\lambda(j)}$, and as $\lambda(i)$ is regular $> \lambda(j)$, $B \subseteq A^1_\xi$ for some $\xi < \gamma$. We can assume ξ is even so we could have chosen as $a^1_{\xi,i(\alpha)}$ an element of M_1 realizing q', except when q' is realized in A^1_ξ, so we get the conclusion of (*). If λ is regular the proof is similar but easier.

As $c \in |M_1| - \mathrm{Dom} f_\chi$, for some $\alpha(0) < \chi$, $\mathrm{tp}(c, A^1_\chi)$ does not fork over $A^1_{\alpha(0)}$. By (*) we can define by induction on $i < \omega + \omega + \omega$, $a_i \in A^1_{\alpha(i)}$, $\alpha(i) < \chi$ increasing, $\alpha(i+1) \ge \alpha(i) + \omega$ (remember $\mathrm{cf}\,\chi > \aleph_0$), so that a_i realizes $\mathrm{tp}(c, A^1_{\alpha(i)})$, $a_i \in A_{\alpha(i+1)}$. By III, 2.12 and III, 1.10, $I = \{a_i: \omega \le i < \omega + \omega\}$ is an indiscernible set, $\mathrm{tp}(c, A^1_{\alpha(\omega)})$ is stationary, and $\mathrm{Av}(I, A^1_\chi) = \mathrm{tp}(c, A^1_\chi)$. Let $\beta = \bigcup_{i<\omega+\omega} \alpha(i)$ (so β is a limit ordinal). By 4.11 there is an indiscernible set $J \subseteq |M|$ over $A^1_{\beta+1}$, based on

$\text{tp}(c, A^1_{\alpha(\omega)})$, $\text{Av}(J, A^1_{\beta+1} \cup \bigcup J) \vdash \text{Av}(J, |M_1|)$. ($J$ is non-trivial as $c \notin A^1_\chi$.) As J is not $J^1_{\beta+1,i(\beta+1)}$, for some distinct $b_1, \ldots, b_n \in J$ $\text{tp}(\bar{b}, A^1_{\beta+1}) \nvdash \text{tp}(\bar{b}, A^1_{\beta+2})$ where $\bar{b} = \langle b_1, \ldots, b_n \rangle$, w.l.o.g. n is minimal. As $A^1_{\beta+2} = A^1_{\beta+1} \cup \bigcup \bigcup_{j<i(\beta+1)} J^1_{\beta+1,j}$, there is a minimal $\xi < i(\beta+1)$ such that $\text{tp}(\bar{b}, B) \nvdash \text{tp}(\bar{a}, B \cup J^1_{\beta+1,\xi})$ $B = A^1_{\beta+1} \cup \bigcup \bigcup_{j<\xi} J^1_{\beta+1,j}$.

But $\langle a_{\omega+\omega+1}, \ldots, a_{\omega+\omega+n-1}, c \rangle$ realizes $\text{tp}(\bar{b}, B)$,

$$\text{tp}(\langle a_{\omega+\omega+1}, \ldots, a_{\omega+\omega+n-1} \rangle, B) \vdash \text{tp}(\langle a_{\omega+\omega+1}, \ldots, a_{\omega+\omega+n-1} \rangle, B \cup J^1_{\beta+1,\xi})$$

by the minimality of n, and

$$\text{tp}(c, B \cup \{a_{\omega+\omega+1}, \ldots, a_{\omega+\omega+n-1}\}) \vdash \text{tp}(c, B \cup J^1_{\beta+1,\xi} \cup \{a_{\omega+\omega+1}, \ldots, a_{\omega+\omega+n-1}\})$$

by Claim 4.15(2), a contradiction by Ax(VII) (for λ singular reduce it to some regular $\lambda' < \lambda$).

DEFINITION 4.3: M is **F**-*admissible over A ($x = t, s, a$, $\mathbf{F} = \mathbf{F}^x_\lambda$) if:

(1) M is **F**-saturated.

(2) $A \subseteq |M|$ and M is **F**-atomic over A.

(3) If $I \subseteq |M|$ is an infinite indiscernible set (of elements) over A, then $\dim(I, A, M) \le \lambda$.

(4) If $I = \{a_n : n < \omega\} \subseteq |M|$ is indiscernible over A, but $\text{tp}(a_0, A \cup \{a_n : 0 < n < \omega\})$ is not **F**-isolated then $\dim(I, A, M) \le \aleph_0$ (this can happen only when $\text{cf}\,\lambda = \aleph_0$, and is not needed for $\lambda = \aleph_0$). We demand (4) only when $\aleph_0 = \text{cf}\,\lambda < \lambda$.

CLAIM 4.17: *Let $\mathbf{F} = \mathbf{F}^x_\lambda$, $x = t, s, a$ and M be* **F**-**admissible* [**F**-*atomic*, **F**-*saturated*] *over A, and p_i ($i < \alpha$) types over A (if $x = a$ almost over A), and $A^* = A \cup \{\bar{a} : \bar{a} \in |M|$, $\bar{a}$ realizes some p_i, $i < \alpha\} \cup B$, where $B \subseteq |M|$, $|B| < \text{cf}\,\lambda$ and for some regular $\mu \leqslant \lambda$, $\text{tp}_*(B, A) \in \mathbf{F}^x_\mu(C)$, $|C| < \mu$. Then M is* **F**-**admissible* [**F**-*atomic*, **F**-*saturated*] *over A^*, provided at least one of the following conditions holds:*

(1) *λ is regular and* **F** *satisfies* Ax(X.1) *and* (XI.1).

(2) *λ is singular, for arbitrarily large regular $\mu \le \lambda$, $\mathbf{F}^x_\mu$ satisfies* Ax(X.1), (XI.1) *and $x = a \Rightarrow \text{cf}\,\lambda \ge \kappa(T)$, $x = s \Rightarrow \text{cf}\,\lambda \ge \lambda^2(T)$ and $x = t \Rightarrow \text{cf}\,\lambda \ge \lambda^3(T)$.*

(3) *λ is singular, for arbitrarily large regular $\mu \le \lambda$, $\mathbf{F}^x_\mu$ satisfies* Ax(X.1), (XI.1), *and $x = t$, and for some $\mu < \lambda$ each p_i is a $\mathbf{F}^t_\mu$-type (or an $\mathbf{F}^t_\mu$-isolated type).*

Proof. Clearly it suffices to prove the version with "**F**-atomic", and assuming $B = \emptyset$.

(1) Immediate, by 3.15.

(2) Similar proof, e.g., if $x = a$, and $\bar{c} \in |M|$, then for some $B_1 \subseteq A$, $|B_1| < \lambda$, $\operatorname{stp}(\bar{c}, B_1) \vdash \operatorname{stp}(\bar{c}, A)$. By 2.16(2), (iii), we can find $B_2 \subseteq A^*$, $|B_2| < \kappa(T)$ and $\bar{c}'$ such that $\bar{c}'$ realizes $\operatorname{stp}(\bar{c}, A)$, and $\operatorname{stp}(\bar{c}', B_1 \cup B_2) \vdash \operatorname{stp}(\bar{c}', A^*)$. As $|B_2| < \operatorname{cf} \lambda$ we can continue as in the proof of 3.15.

(3) Let $\bar{c} \in |M|$, and we must prove that $\operatorname{tp}(\bar{c}, A^*)$ is **F**-isolated. By the hypothesis, there is a regular $\mu < \lambda$ such that $\mathbf{F}^t_\mu$ satisfies Ax(X.1) and (XI.1), $\operatorname{tp}(\bar{c}, A)$, and p_i $(i < \alpha)$ are $\mathbf{F}^t_\mu$-isolated. So $\operatorname{tp}(\bar{c}, A)$ has a $\mathbf{F}^t_\mu$-isolated extension $q \in S^m(A^*)$ realized by $\bar{c}' \in |M|$. So for some $r \subseteq q$, $|r| < \mu$, $r \vdash q$, and let $r = \{\varphi_i(\bar{x}, \bar{a}_i) : i < \beta < \mu\}$. For each i, there are $\bar{a}^1_i, \bar{a}^2_i, p^i_0, \ldots, p^i_{n(i)}$ such that $\bar{a}_i = \bar{a}^1_i{}^\frown\bar{a}^2_i$, $\bar{a}^1_i \in A$, $\bar{a}^2_i \in A_i = \bigcup\{\bar{a} \in M : \bar{a}$ realized p^i_l, $l \leq n(i)\}$ and $p^i_l \in \{p_j : j < \alpha\}$. By II, 2.2(8) for every i there is $\bar{b}_i \in A_i$ and θ_i such that for every $\bar{a} \in A_i$, $\vDash \varphi_i[\bar{c}', \bar{a}^1_i, \bar{a}] \equiv \theta_i[\bar{a}, \bar{a}^1_i, \bar{b}_i]$. As M is $\mathbf{F}^t_\lambda$-saturated, p^i_l is $\mathbf{F}^t_\mu$-type, there are $\psi^i_l(\bar{x}^i_l, \bar{d}^i_l) \in p^i_l$, such that the above mentioned fact holds for every $\bar{a} \in A'_i = \bigcup\{\bar{a} \in M :$ $\bar{a}$ realizes ψ^i_l for some $l \leq n(i)\}$. So we can define by induction on $i < \beta$, $\bar{b}'_i \in A_i$ such that $\operatorname{tp}(\bar{b}'_i, A \cup \bar{c}' \cup \{\bar{b}_j : j < i\})$ is $\mathbf{F}^t_\mu$-isolated, and for every $\bar{a} \in A'_i$, $\vDash \varphi_i[\bar{c}'; \bar{a}^1_i, \bar{a}] \equiv \theta_i[\bar{a}; \bar{a}^1_i, \bar{b}'_i]$. So clearly we can assume r is over $A \cup \{\bar{b}'_i : i < \beta\}$ and the rest is as in 3.15.

THE SECOND CHARACTERIZATION THEOREM 4.18: *Suppose* $\kappa(T) = \aleph_0$, $\mathbf{F} = \mathbf{F}^x_\lambda$, $x = t, s, a$ *and* **F** *satisfies Axioms* (X.1) *and* (XI.1) *and when* λ *is singular, for arbitrarily large regular* $\mu < \lambda$, $\mathbf{F}^x_\mu$ *satisfies* Ax(X.1) *and* (XI.1). *Assume*

(i) $x = a$; *or*

(ii) $x = s, t$ *and* $\lambda > \aleph_0$; *or*

(iii) $x = s, t$, T *totally transcendental.*

Then

(1) M *is* **F**-*prime over* A *iff* M *is* **F**-*$*$admissible over* A.

(2) *The* **F**-*prime model over* A *is unique up to isomorphism over* A.

(3) *If* M_l *is* **F**-*$*$admissible over* A_l, $l = 1, 2$, *and* f *is an elementary mapping from* A_1 *onto* A_2 *then we can extend* f *to an isomorphism from* M_1 *onto* M_2.

Proof. As in 4.14, it suffices to prove (3). For simplicity restrict ourselves to the case $x = a$. Clearly it suffices to prove Claim 4.19(2) below, by applying it to $q_0 = \{x = x\}$, $\gamma = 1$.

CLAIM 4.19: *Suppose M_l, A_l ($l = 1, 2$) and f are as in 4.18(3).*

(1) *If $a_l^* \in |M_l|$, $\mathrm{tp}(a_2^*, A_2) = f[\mathrm{tp}(a_1^*, A_1)]$ then we can extend f to an elementary mapping from $A_1 \cup \{\bar{a} \in |M_1|$: a realizes $\mathrm{stp}(a_2^*, A_1)\}$ onto $A_2 \cup \{a \in |M_2|$: a realizes $\mathrm{stp}(a_1^*, A_2)\}$.*

(2) *If q_i ($i < \gamma$) are 1-types over A_1, then we can extend f to an elementary mapping from $A_1 \cup \{a \in |M_1|$: a realizes some $q_i\}$ onto $A_2 \cup \{a \in |M_2|$: a realizes $f(q_i)\}$.*

Proof of 4.19. We prove by induction on α that: 4.19(2) holds, when $D(q_i, \mathrm{L}, \infty) < \alpha$ and 4.19(1) holds, when $D[\mathrm{stp}(a_1^*, A_1), \mathrm{L}, \infty] \leq \alpha$.

Case I. $\alpha = 0$.

Then (2) is empty, and in case (1) by definition of degree $\mathrm{stp}(a_1^*, A_1)$ is algebraic, hence is realized by a^* only, so extend f by letting $f(a_1^*) = a_2^*$.

Case II. $\alpha > 0$, we have proved (1) and (2) for all $\beta < \alpha$, and we now prove (2) for α.

Choose $a_i^1 \in |M_1|$ ($i < \zeta$) such that for $i < j < \zeta$, $\mathrm{stp}(a_i^1, A_1) \neq \mathrm{stp}(a_j^1, A_1)$, a_i^1 realizes some q_α and ζ is maximal. Let $A_l^j = A_l \cup \{a : a \in |M_l|$, a realizes some $\mathrm{stp}(a_i^l, A_l)$, $i < j\}$ (so for $l = 2$ it is defined only when $(a_i^2 : i < j)$ is defined). We define by induction on j elements $a_i^2 \in |M_2|$ ($i < j$) and an elementary mapping f_j from A_1^j onto A_2^j, such that $f_0 = f$, $i < j$ implies f_j extends f_i and $f_{j+1}(a_j^1) = a_j^2$. So clearly f_ζ is the mapping we need. Notice that by Claim 4.17, M_l is $\mathbf{F}$-*admissible over A_l^j.

For $j = 0$, j limit ordinal, there is no problem, so let $j = i + 1$. Choose $a_i^2 \in |M_2|$, which realizes $f_i(\mathrm{tp}(a_i^1, A_1^i))$ (possible as M_1, M_2 are $\mathbf{F}$-*admissible over $\mathrm{Dom}\, f_i$, $\mathrm{Range}\, f_i$, respectively). By the induction hypothesis we can extend f_i to an elementary mapping f_{i+1} from A_1^{i+1} onto A_2^{i+1} (first, since M_1, M_2 is $\mathbf{F}$-admissible over Dom f_i, Range f_i, respectively, we can extend f_i to an elementary mapping f_i^1 with domain $A_1^i \cup \bar{a}_i^1$, $\bar{a}_i^2 \overset{\mathrm{def}}{=} f_i^1(\bar{a}_i^1) \in M_2$; second, by 4.7 and (2) for $D(\mathrm{stp}(\bar{a}_i^1, A_1^i), L, \infty)$ we can find f_{i+1}).

Case III. $\alpha > 0$. We have proved (1) for every $\beta < \alpha$, and (2) for $\beta \leq \alpha$, and we shall prove (1) for α.

Let $I_l \subseteq |M_l|$ be a maximal independent set over A_l whose elements realize $\operatorname{stp}(a_l^*, |M_l|)$. Let $I_l = \{a_j^l : j < \alpha_l\}$ and $a_0^l = a_l^*$ so if $\alpha_l \geq \omega$, I_l is indiscernible over A_l (by III, 1.10 and III, 2.9(1)). By Definition 4.3, 4.9 and 4.10, we can assume that

(A) $\omega_1 \leq \alpha_l$ so, for some regular $\mu \leq \lambda$, for every $i < \lambda$ $\operatorname{tp}(a_i^l, A_1 \cup \{a_j^l : j < i\})$ is $\mathbf{F}^a_{\mu+|i|^+}$-isolated hence w.l.o.g. $\alpha_l = \lambda$, or

(B) $\alpha_l \leq \omega$.

By $\mathbf{F}$-homogeneity of $(M_l, a)_{a\in A_l}$ it is easy to see that the same case occurs for $l = 1, 2$.

If (B) holds, define by induction elementary mappings f_n, $n < \omega$, such that:

(i) $f_0 = f$, f_{n+1} extends f_n, $M_1[M_2]$ is $\mathbf{F}$-admissible over $\operatorname{Dom} f_n$ [$\operatorname{Range} f_n$].

(ii) a_i^1, $i < n$ belong to the domain of f_{3n+1}.

(iii) a_i^2, $i < n$, belong to the range of f_{3n+2}.

(iv) $\operatorname{Dom} f_{3n+3} = \operatorname{Dom} f_{3n+2} \cup \{a \in |M_1| : a$ realizes $\operatorname{stp}(a_1^*, A_1)$, and $\operatorname{tp}(a, \operatorname{Dom} f_{3n+2})$ forks over $A_1\}$.

This is easily done. If f_{3n} is defined, $\operatorname{tp}(a_n^1, \operatorname{Dom} f_{3n})$ is $\mathbf{F}$-isolated (by Definition 4.3, part (2)), hence $f_{3n}[\operatorname{tp}(a_n^1, \operatorname{Dom} f_{3n})]$ is $\mathbf{F}$-isolated, so by Definition 4.3, part (2), it is realized in M_2, by some b_n^2, and let $f_{3n+1}(a_n^1) = b_n^2$, $\operatorname{Dom} f_{3n+1} = \operatorname{Dom} f_{3n} \cup \{a_n^1\}$. If $n = 0$ we can let $b_n^2 = a_0^2 = a_2^*$. The definition of f_{3n+2} is similar by Definition 4.3(2) and that of f_{3n+3} is by the induction hypothesis on $3n + 2$, by (1) [notice that if a realizes $\operatorname{stp}(a_1^*, A_1)$ then $\operatorname{tp}(a, \operatorname{Dom} f_1) \vdash \operatorname{stp}(a_1^*, A_1)$; and by III, 1.2(4), forking implies decreasing in degree]. Now $\bigcup_{n<\omega} f_n$ is the desired mapping (by the maximality of I_l).

So assume (A) holds, i.e., $\alpha_l = \lambda$. So I_l is indiscernible over A_l, and, using 4.17, for every $i < \operatorname{cf} \lambda$, M_l is $\mathbf{F}$-$*$admissible over $B_i^l = A_l \cup \{a_j^l : j < i\}$. Let $C_i^l = A_l \cup \{a \in M_l : a$ realizes $\operatorname{stp}(a_l^*, A_l)$, $\operatorname{tp}(a, B_i^l)$ forks over $A_l\}$. Clearly for every $i \leq \lambda$, we can extend f to f_i by letting $f_i(a_j^1) = a_j^2 (j < i)$, and if $i < \operatorname{cf} \lambda$ then extend it by 4.18(1) to an elementary mapping f_i from C_i^1 onto C_i^2.

If $\operatorname{tp}(a_\omega^1, C_\omega^1)$ forks over A_l, then by $\mathbf{F}$-homogeneity it is easy to prove that if a realizes $\operatorname{stp}(a_l^*, A_l)$, $a \in M_l$ then $\operatorname{stp}(a, C_\omega^l)$ forks over A_l. Hence we can prove (1) as in (B), (with $C_n^l \cup \{a \in |M_l| : \operatorname{tp}(a, C_n^l)$ forks over $A_l\}$ included in $\operatorname{Dom} f_{3n}$, $\operatorname{Range} f_{3n}$ for $l = 1, 2$ resp.

So assume $\operatorname{tp}(a_\omega^1, C_\omega^1)$ does not fork over A. Then, again, $\operatorname{tp}(a_i^l, C_i^l)$ does not fork over A_l for $i < \lambda$. Also if $\bar{c} \in C_i^l$ $(i < \lambda)$ then $\operatorname{stp}(\bar{c}, B_i^l) \vdash \operatorname{stp}(\bar{c}, B_{i+1}^l)$ (by the non-forking assumption, i.e. otherwise, since M_l is $\mathbf{F}$-saturated, we can find $\bar{c}' \in M_1$ realizing $\operatorname{stp}(\bar{c}, B_i^l)$, but not $\operatorname{stp}(\bar{c},$

B^l_{i+1}) but necessarily $\bar{c}' \in C^l_i$ and tp$(\bar{a}^l_i, C^l_i)$ fork over A_l, a contradiction) hence stp$(\bar{c}, B^l_i) \vdash$ stp$(\bar{c}, B^l_\lambda)$. In fact for every $\bar{c} \in |M_l|$, for some finite $J \subseteq I_l$, stp$(\bar{c}, A_l \cup J) \vdash$ stp$(\bar{c}, A_l \cup I)$. So M_l is **F**-atomic over B^l_λ, and we can easily show that M_l is **F**-*admissible over B^l_λ (notice we can reorder the a^l_i's). So by 4.18(2) (for α) we can extend f_l to the desired mapping.

Sketch of the proof for other cases. **Let $x = s$; by 2.2 $\mathbf{F}^s_\lambda$-primeness (saturation) is equivalent to $\mathbf{F}^a_\lambda$-primeness (saturation). By Exercise 4.8 $\mathbf{F}^s_\lambda$-*admissibility implies $\mathbf{F}^a_\lambda$-*admissibility. So as we have proved 4.18 for $x = a$; it suffices to prove that over any A there is an $\mathbf{F}^s_\lambda$-*admissible model, but this follows by Ax(XI.1) and Exercise 4.14 and 4.15. If $x = s, t$ T totally transcendental, use $R(p, \mathrm{L}, 2)$ instead of $D(p, \mathrm{L}, \infty)$, and the proof is similar.**

If $x = t$, $\lambda > \aleph_0$, in proving 4.19, case II, we first, as in the proof of 4.14, extend f to f', such that (M_1, Dom f'), (M_2, Range f') are F-*admissible, and Dom f' is the universe of a model; so each p_i is stationary.

If $x = s, t$ $\lambda >$ cf $\lambda = \aleph_0$, in proving 4.19, case III(B) let $\lambda = \sum_{n<\omega} \mu_n$, $\mu_n < \mu_{n+1}$, $\mathbf{F}^t_{\mu_n}$ satisfies Ax(X.1), (XI.1), μ_n regular, and tp$(\langle a^1_0, \ldots, a^1_n \rangle, A)$ is $\mathbf{F}^t_{\mu_n}$-isolated, and e.g.,

(iv)$'$ Dom f_{3n+3} = Dom $f_{3n+2} \cup \{a \in M_1$: a realizes stp(a^*_1, A_1) and tp$(a$, Dom $f_{3n+2})$ forks over A_l and is $F^t_{\mu_{3n+3}}$-isolated$\}$.

In cases II, III(A), the change is similar, and when cf $\lambda > \aleph_0$—is easier.

DEFINITION 4.4: *M* is F-minimal over *A*, if *M* is F-saturated, but there is no F-saturated model N, $A \subseteq |N| \subseteq |M|$, $|N| \neq |M|$.

LEMMA 4.20: *If over A there are an* F*-prime model and an* F*-minimal model, then, up to isomorphism over A, there is a unique* F*-prime model over A, and a unique* F*-minimal model over A and they are isomorphic over A.*

Proof. **Let M_0 be F-prime over A, N_0 F-minimal over A. By the definition of F-prime there is an elementary mapping $f: M_0 \to N_0$ $f \restriction A$ = the identity. By the F-minimality of N_0, $f(M_0) = N_0$, so they are isomorphic over A.**

Hence if M_1, M_2 are F-prime over A, they are isomorphic to N_0 over A, hence isomorphic over A. Similarly any two F-minimal models over A are isomorphic over A.

THEOREM 4.21: *Let* $\mathbf{F} = \mathbf{F}^x_\lambda$, $x = t, s, a$, $\mathrm{cf}\,\lambda \geq \kappa(T)$, $\mathbf{F}$ satisfies Axioms (X.1) and (XI.1), and: T is totally transcendental or $x = s \Rightarrow \lambda > \aleph_0$, $x = t \Rightarrow \mathrm{cf}\,\lambda > \aleph_0$. *Suppose M is* **F**-*saturated and* **F**-*atomic over A. Then M is* **F**-*minimal over A iff in M there is no infinite non-trivial indiscernible set over A.*

Proof. Suppose $I \subseteq |M|$ is an infinite non-trivial indiscernible set over A. We can assume I is a maximal one, and let $I = \{\bar{a}_i: i < \alpha\}$ ($\alpha \geq \omega$). Let N be any **F**-prime model over $A \cup \bigcup I$ (exists by 3.1). We can assume $N \prec M$, hence N omits $\mathrm{Av}(I, A \cup \bigcup I)$, by the maximality of I. Hence also any **F**-prime model over $B = A \cup \{a_i: 0 < i < \alpha\}$ omits $\mathrm{Av}(I, B)$; and let $N_1 \subseteq M$ be one such model.

So N_1 is **F**-saturated, $A \subseteq |N_1| \subseteq |M|$, but $a_0 \notin |N_1|$. We conclude M is not **F**-minimal.

Now assume M is not **F**-minimal, so there is an **F**-saturated N, $A \subseteq |N| \subseteq |M|$ $|N| \neq |M|$. Choose $c \in |M| - |N|$, and $C \subseteq |N|$, $|C| < \kappa(T)$ such that $\mathrm{tp}(c, |N|)$ does not fork over C. For simplicity let $x = a$, choose $c_n \in |N|$ which realizes $\mathrm{stp}(c, A \cup C \cup \{c_i: i < n\})$ (exists, as M is **F**-atomic of A, and by Ax(V.2) (or 2.8), $\mathrm{tp}(c, A \cup C \cup \{c_i: i < n\})$ is **F**-isolated). So clearly $\{c_n: n < \omega\} \subseteq |M|$ is an infinite non-trivial indiscernible set over A (see III, 1.10 and III, 2.9).

EXERCISE 4.4: Prove 4.18, when $x = s$, $\lambda = \aleph_0$, and T is transcendental (see Definition II, 3.4).

QUESTION 4.5: In the characterization theorems, can we weaken the assumption "**F** satisfies Ax(X.1) and (XI.1)" by "over A (A_l) there is an **F**-primary model"? (see 5.6).

EXERCISE 4.6: Let M be **F**-$**$admissible over A, $\mathbf{F} = \mathbf{F}^x_{\aleph_0}$, $x = s, t, a$ if M is **F**-saturated, **F**-atomic over A, and for every finite $B \subseteq |M|$, and $p \in S(A \cup B)$, $\dim(p, A \cup B, M)$ is $\leq \aleph_0$. Prove a characterization theorem parallel to 4.14, 4.18, for countable superstable T. (Instead of countable, you can assume $\lambda^*(p) \leq \aleph_0$ for every $\mathbf{F}^*_{\aleph_0}$-isolated complete type p; $\lambda^*(p)$ is defined in Definition III, 5.1). Generalize similarly 4.21.

QUESTION 4.7: In 4.12, try to replace $\kappa(T)$ by $\kappa_{\mathrm{ind}}(T)$, at least for stable T.

EXERCISE 4.8: Prove that if M is $\mathbf{F}^s_\lambda$-admissible over A, $\lambda > \aleph_0$ *then*

M is $\mathbf{F}^a_\lambda$-admissible over A; and similarly for -*admissible and -**admissible. [Hint: Use 2.2; note that $\mathbf{F}^s_\lambda$-isolation (atomicity) implies $\mathbf{F}^a_\lambda$-isolation as for Definition 4.3(4) note that $\operatorname{tp}(a_0, A \cup \{a_n: 0 < n < \omega\}) \vdash \operatorname{stp}(a_0, A \cup \{a_n: 0 < n < \omega\})$, so if $\operatorname{tp}(a_0, A \cup \{a_n: 0 < n < \omega\})$ is not $\mathbf{F}^s_\lambda$-isolated, it is not $\mathbf{F}^a_\lambda$-isolated.]

QUESTION 4.9: In the proof of 4.14, λ regular, can we choose $\chi = \kappa_r(T)$ for λ singular, can we choose $\chi = \operatorname{cf} \lambda$?

EXERCISE 4.10: Show 4.16(2) holds also when $\operatorname{tp}(\bar{a}_1, B)$ is not stationary.

PROBLEM 4.11: In the characterization theorem 4.14, can we weaken Definition 4.2(3) to "$\dim(I, A, M) \le \lambda$"?

EXERCISE 4.12: Show that 4.18 may fail for $x = s, t$, $\lambda = \aleph_0$. [Hint: $T = \operatorname{Th}(M)$, $M = ({}^{\omega}2 \times \omega, \ldots, E_n, \ldots)$, where $\langle \eta, k\rangle\ E_n \langle \nu, l\rangle$ iff $\eta \restriction n = \nu \restriction n$. This model contradicts 4.18(1).]

EXERCISE 4.13: Show that 4.14 may fail if $\lambda < \kappa(T)$. [Hint: Use T^κ (see Exercise II, 2.3).]

EXERCISE 4.14: Suppose $\kappa(T) \le \operatorname{cf} \lambda$, A is $\mathbf{F}^s_\lambda$-constructible over B *then* A is $\mathbf{F}^s_\lambda$-atomic over B, provided that

(A) Ax(XI.1) holds, but not $\operatorname{cf} \lambda = \aleph_0 < \lambda$ or

(B) λ is a limit cardinal, and Ax(XI.1) holds for $\mathbf{F}^s_\mu$ for arbitrarily large $\mu < \lambda$. [Hint: See 4.7; for (A) use Exercise 2.30. For (B) let $\lambda = \sum_{i<\operatorname{cf}\lambda} \lambda_i$, $\sum_{j<i} \lambda_j < \lambda_i < \lambda$, $\mathbf{F}^s_{\lambda_i}$ satisfies Ax(XI.1). Now we can define inductively $\mathbf{F}^s_{\lambda_i}$-saturated models M_i, $|M_i|$ $\mathbf{F}^s_{\lambda_i}$-constructible over $A \cup \bigcup_{j<i} M_j$, and let $M = \bigcup_{i<\operatorname{cf}\lambda} M_i$. By III, 3.11 M is $\mathbf{F}^s_\lambda$-saturated, by 1.3(2) M_i is $\mathbf{F}^s_{\lambda_i}$-constructible over A, hence $F^s_{(\lambda_i^+)}$-constructible over A, hence by 3.2(1) $|M_i|$ is $\mathbf{F}^s_{(\lambda_i^+)}$-atomic over A, hence M is $\mathbf{F}^s_\lambda$-atomic over A. Now as in 3.12(2) we can prove A is $\mathbf{F}^s_\lambda$-atomic over B.]

EXERCISE 4.15: Suppose (α) λ is regular and $\mathbf{F}^s_\lambda$ satisfies Ax(XI.1) or (β) λ is singular, and for arbitrarily large $\mu < \lambda$ $\mathbf{F}^s_\lambda$ satisfies Ax(XI.1). Prove that 4.9 holds for $x = s$.

EXERCISE 4.16: Use for 4.14, 4.18, etc. (i) or (ii) from Exercise 4.15, instead "for arbitrarily large regular $\mu \le \lambda$ $\mathbf{F}^x_\mu$ satisfies Ax(XI.1)".

IV.5. Various results

THEOREM 5.1: *Suppose* $|T| = \lambda$, $\mu \le \lambda$, μ *regular. Then* (1) $\Rightarrow$ (2) $\Rightarrow$ (3) $\Rightarrow$ (4) *where:*

(1) $|D(T)| < 2^{\mu}$.

(2) *There are no* $\bar{x}$ *and formulas* $\varphi_\eta(\bar{x})$ $(\eta \in {}^{\mu >}2)$ *(without parameters) such that for every* $\eta \in {}^{\mu}2$, $\{\varphi_{\eta|i}(\bar{x})^{\eta[i]}: i < \mu\}$ *is consistent with* r.

(3) *For every* m*-type* p *over* $\emptyset$, $|p| < \lambda$, *there is an* m*-type* q *over* $\emptyset$ *such that* $|q| < \mu$, *and* $p \cup q \vdash r$ *for some* $r \in D_m(T)$.

(4) (A) T *has a model* M *of cardinality* λ, *which is* $\mathbf{F}^t_\lambda$*-constructible (over* $\emptyset$*) hence* $\mathbf{F}^t_\lambda$*-primitive and* $\mathbf{F}^t_\lambda$*-atomic (over* $\emptyset$*). In fact there is such a model over any* A, $|A| < \lambda$.

(B) *If* $\lambda^{<\kappa} = \lambda$, *we can have* M κ*-compact (hence* $\mathbf{F}^t_\kappa$*-saturated).*

(C) *If* $\lambda = \lambda^{<\lambda}$, *or* T *has a* λ*-compact model of cardinality* λ, *then* M *is* $\mathbf{F}^t_\lambda$*-primary and therefore* $\mathbf{F}^t_\lambda$*-prime and* λ*-prime.*

Remark. By III, 5.14 we can assume $|T| \le |D(T)|$.

Proof. We leave it to the reader, as it is similar to previous proofs, except (4) (C), where for λ regular it is easy; for λ singular, see Ex. 5.17, 5.18.

Remark. Notice that every $\mathbf{F}^t_\lambda$-saturated model has cardinality $\geqslant \lambda$. A converse to 5.1 is:

THEOREM 5.2: *Suppose* $|T| = \lambda$, *and* T *has a* λ*-compact model of cardinality* λ *(e.g., when* $\lambda = \lambda^{<\lambda}$*), and* λ *is regular.*

(1) *Among the* λ*-compact models, there is a prime one iff condition* 5.1(3) *holds for* $\mu = \lambda$ *(see Exercise 5.16).*

(2) *If there is a* λ*-prime model then it is unique (up to isomorphism) and* λ*-homogeneous. See Definition* 1.6.

(3) *If there is a* λ*-prime model* M, *it is characterized by*

(A) M *is* $\mathbf{F}^t_\lambda$*-saturated,*

(B) M *is* $\mathbf{F}^t_\lambda$*-atomic,*

(C) $\|M\| = \lambda$.

Proof. (1) By 5.1, if 5.1(3) holds, T has a λ-prime model of cardinality λ. Now suppose M is a λ-prime model, but the m-type p is a counter-example to 5.1(3). As $|p| < \lambda$, M λ-compact, some $\bar{c} \in |M|$ realizes it, thus $p_0 = \operatorname{tp}(\bar{c}, \emptyset)$ is not $\mathbf{F}^t_\lambda$-isolated. The following theorem, 5.3, shows that there is a λ-compact model which omits p_0, thus M cannot be elementarily embedded into it, contradiction. The rest of the proof will follow Theorem 5.3.

THEOREM 5.3: *Suppose p_i is an m_i-type over A for $i < \alpha \leq \lambda$ and for no m_i-type q over A, $q \vdash p_i$, $|q| < \lambda$. Assume $|A| + |T| = \lambda$.*

(1) *T has a model M, ($A \subseteq |M|$) of cardinality λ omitting each p_i (i.e., no sequence from the model realizes it).*

(2) *If $\lambda = \lambda^{<\lambda}$, or T has a λ-compact model of power λ, $A = \emptyset$, then in (1) we can choose the model to be $\mathbf{F}^t_\lambda$-saturated, and even λ-compact.*

Proof of 5.3. (1) For simplicity let $A = \emptyset$. Let $\{\bar{y}_i: i < \lambda\}$ be a list of all finite sequences from $\{x_i: i < \lambda\}$, each sequence appearing λ times and let $\{\varphi_i(x, \bar{y}_{\alpha(i)}): i < \lambda\}$ be a list of the formulas of the form $\varphi(x, \bar{y}_j)$ ($\varphi \in L, j < \lambda$) such that $\bar{y}_i, \bar{y}_{\alpha(i)} \subseteq \{x_j: j < i\}$. We now define by induction on $i < \lambda$ a consistent set Γ_i of formulas (with no parameters) such that in Γ_i appear only the variables $\{x_j: j < 2i\}$, and $|\Gamma_i| < |i|^+ + \aleph_0$. Let $\Gamma_0 = \emptyset$, $\Gamma_\delta = \bigcup_{i<\delta} \Gamma_i$, clearly the induction assumptions hold. So suppose Γ_i is defined, and let

$$\Gamma_i^1 = \Gamma_i \cup \{[(\exists x)\varphi_i(x, \bar{y}_{\alpha(i)})] \rightarrow \varphi_i(x_{2i}, \bar{y}_{\alpha(i)})\} \cup \{x_{2i+1} \neq x_j: j \leq 2i\}.$$

Trivially Γ_i^1 is consistent, and the variables that appear in it are only from $\{x_i: i < 2(i + 1)\}$ and its cardinality is $< |i|^+ + \aleph_0$. Let Γ_i^2 be the closure of Γ_i^1 under conjunctions, and let

$$r_i = \{(\exists \bar{z})\varphi(\bar{y}_i, \bar{z}): \bar{z} \text{ disjoint to } \bar{y}_i, \text{ but } \subseteq \{x_j: j < 2(i + 1)\},$$
$$\text{and } \varphi(\bar{y}_i, \bar{z}) \in \Gamma_i^2\}.$$

So $|r_i| < \lambda$, r_i an $l(\bar{y}_i)$-type. Hence when $l(\bar{y}_i) = m_j$, $j < \alpha$, $r_i \nvdash p_j$ (changing the names of variables accordingly). Suppose $\bar{y}_i$ is the jth appearance of $\bar{y}_i$ in $\{\bar{y}_\beta: \beta < \lambda\}$: if $j \geq \alpha$ or $l(\bar{y}_i) \neq m_j$, let $\Gamma_{i+1} = \Gamma_i^2$; otherwise for some $\psi_i(\bar{y}_i) \in p_j$, $r_i \cup \{\neg\psi_i(\bar{y}_i)\}$ is consistent, so let $\Gamma_{i+1} = \Gamma_i^2 \cup \{\neg\psi_i(\bar{y}_i)\}$. Clearly the induction assumptions are satisfied, and $\Gamma_\lambda = \bigcup_{i<\lambda} p_i$. As Γ_λ is consistent there is an assignment $x_i \mapsto a_i$ realizing it. By the Tarski–Vaught test $\{a_i: i < \lambda\}$ is a universe of a model M, which by the construction has cardinality λ and omits the p_i's.

(2) Similar proof.

Remark. For countable T the condition on the p_i's is also necessary for (1). Similarly if $\lambda = \lambda^{<\lambda}$ it is necessary for (2).

Continuation of the proof of 5.2. (2) If there is an $\mathbf{F}^t_\lambda$-prime model M_1, by 5.2(1) condition 5.1(3) holds, hence by 5.1 there is an $\mathbf{F}^t_\lambda$-primary model M_2. As M_1 can be elementarily embedded into M_2, and by 3.2(1) M_2 is $\mathbf{F}^t_\lambda$-atomic, also M_1 is $\mathbf{F}^t_\lambda$-atomic. So by 3.16 M_1 is $\mathbf{F}^t_\lambda$-constructible (over $\emptyset$), hence M_1 is $\mathbf{F}^t_\lambda$-primary too. So by 3.9 M_1, M_2 are isomorphic. As we could have replaced M_1 by any $\mathbf{F}^t_\lambda$-prime model, the $\mathbf{F}^t_\lambda$-prime model is unique. Its homogeneity follows by 3.7.

(3) Follows from the proof of (2). (Note the remark before 5.2.)

CONCLUSION 5.4: *Let T be countable. T has an $\mathbf{F}^t_{\aleph_0}$-prime model iff for every formula $\varphi(\bar{x})$, $(\exists\bar{x})\varphi(\bar{x}) \in T$ there is a formula $\psi(\bar{x})$, $(\exists\bar{x})[\varphi(\bar{x}) \wedge \psi(\bar{x})] \in T$, and $\varphi(\bar{x}) \wedge \psi(\bar{x}) \vdash r \in D_m(T)$ ($l(\bar{x}) = m$) for some r.*

THEOREM 5.5: *Suppose $\lambda^+ = |T| \leq |D(T)|$, λ regular. Then* (1) $\Rightarrow$ (2) $\Rightarrow$ (3) *where:*

(1) $|D(T)| < 2^\lambda$.

(2) *For every $m < \omega$, and m-type p, $|p| < \lambda$, there is an m-type q, $|q| < \lambda$, $p \cup q$ is consistent, and $p \cup q \vdash r \in D_m(T)$ for some r.*

(3) *T has an $\mathbf{F}^t_\lambda$-atomic model.*

Remark. In (3) we cannot demand that the model has cardinality λ^+ even if $\lambda < |D(T)|$, see Exercise 5.13. If $\lambda = \lambda^{<\lambda}$ we can demand in (3) that the model is λ-compact.

Proof. (1) $\Rightarrow$ (2) similar to 2.16.

(2) $\Rightarrow$ (3) It is easy to see that it suffices to prove:

(*) If A is $\mathbf{F}^t_\lambda$-atomic, $|A| < \lambda^+$, $\bar{a}_0 \in A$, $\models(\exists x)\varphi(x, \bar{a}_0)$, (where $\varphi \in L(T)$), then there is b, such that $A \cup \{b\}$ is $\mathbf{F}^t_\lambda$-atomic and $\models\varphi[b, \bar{a}_0]$.

For $\bar{a} \in A$, let $q_{\bar{a}}$ be an $l(\bar{a})$-type over $\emptyset$, $q_{\bar{a}} \vdash \mathrm{tp}(\bar{a}, \emptyset)$, $|q_{\bar{a}}| < \lambda$, $q_{\bar{a}}$ closed under conjunctions. Let us prove (*). Let $\{\bar{a}_i: i < \alpha \leq \lambda\}$ be the set of all finite sequences from A. We now define by induction on i, a 1-type p_i over A, $|p_i| < \lambda$. Let $p_0 = \{\varphi(x; \bar{a})\}$, $p_\delta = \bigcup_{i<\delta} p_i$. If p_i is defined let p'_i be the closure of p_i under conjunctions and

$$q_i = \{(\exists\bar{z})[\varphi(x, \bar{y}, \bar{z}) \wedge \psi(\bar{y}, \bar{z})]: \varphi(x; \bar{a}_i, \bar{c}) \in p'_i,$$
$$\bar{c} \text{ disjoint to } \bar{a}_i, \psi(\bar{y}, \bar{z}) \in q_{\bar{a}_i{}^\frown\bar{c}},$$
$$\bar{c} \subseteq \mathrm{Dom}\, p_i\}.$$

Clearly $|q_i| \leq |p'_i| + \sum\{|q_{\bar{a}_i{}^\frown\bar{c}}|: \bar{c} \subseteq \mathrm{Dom}\, p_i\} < \lambda$ and q_i is consistent.

So by (2) there is $r_i = r_i(x, \bar{y})$, $q_i \subseteq r_i$, $|r_i| < \lambda$, r_i consistent, and $r_i \vdash r^i \in D_m(T)$ for some r^i, $m = l(\bar{a}_i) + 1$. Let

$$p_{i+1} = p_i \cup \{\psi(x, \bar{a}): \psi(x, \bar{y}) \in r_i\}.$$

It is easy to check p_{i+1} is consistent, so if b realizes p_α, we finish (*) hence the theorem.

THEOREM 5.6 (The Uniqueness Theorem): (1) *Let T be stable and λ regular, $\lambda^+ \geq \kappa(T)$. If T has an $\mathbf{F}^t_\lambda$-primary model over A then all the $\mathbf{F}^t_\lambda$-prime models over A are isomorphic over A.*

(2) *Let T be stable, $\mu^+ \geq \kappa(T)$. Let M be an $\mathbf{F}^x_\lambda$-primary model over A ($x = t, a$) for which there are μ relations R_i on $|M|$ such that $C \subseteq B$, $|B| \leq \mu$, $(B, R_i)_{i<\mu} \prec (|M|, R_i)_{i<\mu} \Rightarrow C \cup A$ is $\mathbf{F}^t_\lambda$-constructible over A. Then all the $\mathbf{F}^t_\lambda$-prime models over A are isomorphic over A.*

Remark. Notice we do not demand that our $\mathbf{F}$ satisfies Ax(XI.1) or (X.1). In [Sh 79a] a simpler alternative proof appears. Theorem 5.6 is a generalization of:

THEOREM 5.6′: *If T is stable and countable and T has a primary model over A, then all the prime models over A are isomorphic over A.*

QUESTION 5.1: What about $x = s$?

Proof. In (2) we can assume $\mu \leq \lambda$. Let μ be λ and x be t in (1), μ be μ and x be t or a in (2). In (1) let M be an $\mathbf{F}^t_\lambda$-primary model over A.

(1) Let N be $\mathbf{F}^x_\lambda$-prime over A; we shall prove it is isomorphic (over A) to M. We can assume, by the definition of $\mathbf{F}$-primeness, that $A \subseteq |N| \subseteq |M|$. As M is $\mathbf{F}^x_\lambda$-primary, let $\mathscr{A} = \langle A, \{a_i : i < \alpha\}, \{B^1_i : i < \alpha\}\rangle$ be a $\mathbf{F}^x_\lambda$-construction of $|M|$. We can assume $a_i \notin A \cup \{a_j : j < i\}$. Clearly $\mathrm{tp}(a_i, \mathscr{A}(i))$ does not fork over B^1_i (by III, 4.4). So $|B^1_i| \leq \mu$. For $B, C \subseteq |M|$, call the pair (B, C) good if

(i) $a_i \in B$ [$a_i \in C$] implies $B^1_i \subseteq C$[$B \subseteq B^1_i$].

(ii) for every $\bar{b} \in B$, $\mathrm{tp}(b, |N|)$ does not fork over $B \cap |N|$,

(iii) for every $\bar{b} \in B \cap C$, $\mathrm{tp}(\bar{b}, |N|)$ does not fork over $B \cap C \cap |N|$.

If a set $D \subseteq M$ satisfies (i) we call it closed (as we have a fixed construction $\mathscr{A}$).

Now we prove

CLAIM 5.7: *Suppose (B, C) is a good pair, $|B| > \mu$. Then we can find sets B_i, $i < |B|$ such that $B = \bigcup_i B_i$, $|B_i| \leq \mu$ and forever y $i < |B|$,*

$(B_{i+1}, C \cup \bigcup_{j\le i} B_j)$ is a good pair, and $B_\delta \subseteq \bigcup_{i<\delta} B_i$ for limit $\delta < |B|$.

Proof of 5.7. We prove it by induction on $|B| > \mu$. Suppose we have proved the assertion for every cardinality $< |B|$ (and $> \mu$). Let $B = \{b_i: i < |B|\}$, and define by induction on i, sets $B_i \subseteq B$, such that

(1) $j < i \Rightarrow b_j \in B_i$, and $B_j \subseteq B_i$; $B_\delta = \bigcup_{j<\delta} B_j$ for limit δ.

(2) $\mu \le |B_i| \le |i| + \mu$, B_i is closed.

(3) If $\bar{b} \in B_i$, then $\text{tp}(\bar{b}, B \cap |N|)$ does not fork over $B_i \cap |N|$.

(4) If $\bar{b} \in B_i \cap C$ then $\text{tp}(\bar{b}, B \cap C \cap |N|)$ does not fork over $B_i \cap C \cap |N|$.

(5) If $\bar{b} \in B_i$ then $\text{tp}(\bar{b}, B \cap C)$ does not fork over $B_i \cap C$; $\text{tp}(\bar{b}, (B \cap |N|) \cup (B \cap C))$ does not fork over $(B_i \cap |N|) \cup (B_i \cap C)$. As $\mu^+ \ge \kappa(T)$, $|B_j^1| \le \mu$ this can be done.

Let $C_i = C \cup B_i$, and we shall prove (B_{i+1}, C_i) is a good pair. As for (i), B_{i+1} is closed by (2), and C_i as a union of closed sets. For (ii), let $\bar{b} \in B_{i+1}$, then $\text{tp}(\bar{b}, |N|)$ does not fork over $B \cap |N|$, as (B, C) is good; $\text{tp}(\bar{b}, B \cap |N|)$ does not fork over $B_{i+1} \cap N$ by (3), hence by III, 4.4 $\text{tp}(\bar{b}, |N|)$ does not fork over $B_{i+1} \cap |N|$, so (ii) holds. So we are left with (iii). So let $\bar{b} \in A^*$ where $A^* = B_{i+1} \cap C_i = B_{i+1} \cap (C \cup B_i) = B_i \cup (B_{i+1} \cap C)$, and we should prove that $\text{tp}(\bar{b}, |N|)$ does not fork over $A^* \cap |N|$. Note that:

(a) For every $\bar{c} \in B_i$, $\text{tp}(\bar{c}, B \cap |N|)$ does not fork over $B_i \cap |N|$ [by (3)].

(b) For every $\bar{c} \in B_i$, $\text{tp}(\bar{c}, B \cap |N|)$ does not fork over $A^* \cap |N|$ by (a) and monotonicity because $B_i \subseteq A^*$].

(c) For every $\bar{a} \in B_{i+1} \cap C$, $\text{tp}(\bar{a}, |N|)$ does not fork over $B \cap C \cap |N|$ [by the goodness of (B, C) as $B_{i+1} \subseteq B$].

(d) For every $\bar{a} \in B_{i+1} \cap C$, $\text{tp}(\bar{a}, B \cap C \cap |N|)$ does not fork over $B_{i+1} \cap C \cap |N|$ [by (4)].

(e) For every $\bar{a} \in B_{i+1} \cap C$, $\text{tp}(\bar{a}, B \cap |N|)$ does not fork over $A^* \cap |N|$ [by (c) and (d), using the monotonicity and transitivity (III, 4.4) of forking, remembering $B_{i+1} \cap C \cap |N| \subseteq A^* \cap |N|$; $B \cap C \cap |N| \subseteq B \cap |N| \subseteq |N|$].

Now as $\bar{b} \in A^* = B_i \cup (B_{i+1} \cap C)$ let $\bar{b} = \bar{c}^\frown\bar{a}$, $\bar{c} \in B_i$, $\bar{a} \in B_{i+1} \cap C$.

(f) $\text{tp}_*[\bar{c} \cup (B_i \cap C), B \cap |N|]$ does not fork over $A^* \cap |N|$ [by (b)].

(g) $\text{tp}[\bar{c}, (B_i \cap |N|) \cup (B_i \cap C]$ does not fork over $(A^* \cap |N|) \cup (B_i \cap C)$ [by (f) using Ax(V.2) for $\mathbf{F}^f_\infty$].

(h) $\text{tp}_*[\bar{a} \cup (B_i \cap C), B \cap |N|]$ does not fork over $A^* \cap |N|$ [by (e)].

(i) $\operatorname{tp}[\bar{c}, (B \cap |N|) \cup (B \cap C)]$ does not fork over $(B_i \cap |N|) \cup (B_i \cap C)$ [by (5)].

(j) $\operatorname{tp}[\bar{c}, (B \cap |N|) \cup (B_i \cap C) \cup \bar{a}]$ does not fork over $(B \cap |N|) \cup (B_i \cap C)$ [by (i) and the monotonicity of forking].

(k) $\operatorname{tp}[\bar{c}, (B \cap |N|) \cup (B_i \cap C) \cup \bar{a}]$ does not fork over $(A^* \cap |N|) \cup (B_i \cap C)$, hence over $(A^* \cap |N|) \cup (B_i \cap C) \cup \bar{a}$ [by (g) and (j), using the transitivity of forking III, 4.4].

(l) $\operatorname{tp}_*[\bar{c} \cup \bar{a} \cup (B_i \cap C), B \cap |N|]$ does not fork over $A^* \cap |N|$ [by (h) and (k), using Axiom VII for $\mathbf{F}^f_\infty$].

(m) $\operatorname{tp}(\bar{b}, B \cap |N|)$ does not fork over $A^* \cap |N|$ [by (l), as $\bar{b} = \bar{c}^\frown\bar{a} \in \bar{c} \cup \bar{a} \cup (B_i \cap C)$].

(n) $\operatorname{tp}(\bar{b}, |N|)$ does not fork over $A^* \cap |N|$ [by (m) and (ii), using III, 4.4].

Thus (B_{i+1}, C_i) is good. If $|B| = \mu^+$, then $|B_i| = \mu$ and we finish. Otherwise we can assume $|B_i| > \mu$, and let $\alpha(0) = 0$, $\alpha(i+1) = \alpha(i) + |B_i|$ (ordinal addition), $\alpha(\delta) = \bigcup_{i<\delta} \alpha(i)$, (so $\alpha(i) < |B|$ for $i < |B|$) and use the induction hypothesis on (B_{i+1}, C_i) to define B'_j, $\alpha(i) \le j < \alpha(i+1)$, and the B'_j show the claim holds.

Continuation of the proof of 5.6. Now clearly $(|M|, \emptyset)$ is a good pair. Assume $\|M\| > \mu$, as $\|M\| \le \mu$ is a trivial case; so by 5.7 there are $B_i\, i < \|M\|$, such that $|B_i| \le \mu$, and letting $C_i = \bigcup_{j<i} B_j, C_{\|M\|} = |M|$, $B_\delta \subseteq C_\delta$ (for limit δ) and (B_{i+1}, C_i) is a good pair. By 1.3(2) it suffices to show that $|N| \cap [A \cup C_{i+1}] = |N| \cap [A \cup C_i \cup B_{i+1}]$ is $\mathbf{F}^x_\lambda$-constructible over $|N| \cap [A \cup C_i]$. Now

(p) $\operatorname{tp}_*(B_{i+1} \cap C_i, |N|)$ does not fork over $B_{i+1} \cap C_i \cap |N|$ [by the goodness of (B_{i+1}, C_i)].

(q) $\operatorname{tp}_*(B_{i+1} \cap C_i, |N| \cap [A \cap B_{i+1}])$ does not fork over $|N| \cap [A \cup (B_{i+1} \cap C_i)]$ (by (p) and monotonicity of forking].

(r) $\operatorname{tp}_*(|N| \cap B_{i+1}, [A \cup (B_{i+1} \cap C_i)])$ does not fork over $|N| \cap [A \cup (B_{i+1} \cap C_i)]$ [by (q) and Ax(VI) for $\mathbf{F}^f_\lambda$ with $A^* = B^* = C^* = |N| \cap [A \cup (B_{i+1} \cap C_i)]$, and any $\bar{a} \in |N| \cap B_{i+1}$, $\bar{b} \in (B_{i+1} \cap C_i)$].

(s) $\mathscr{A}$ is also an $\mathbf{F}^f_\infty$-construction and B_{i+1}, C_i are closed for this construction [by B^1_i's definition and (i)].

(t) $\operatorname{tp}_*(B_{i+1}, A \cup C_i)$ does not fork over $A \cup (B_{i+1} \cap C_i)$ [by (s) 3.3 and Exercise 3.5].

(u) $\operatorname{tp}_*(|N| \cap B_{i+1}, A \cup C_i)$ does not fork over $|N| \cap [A \cup (B_{i+1} \cap C_i)]$ [by (r), (t) and transitivity of non-forking III, 4.4].

(v) $\operatorname{tp}_*(|N| \cap B_{i+1}, A \cup C_i)$ does not fork over $|N| \cap (A \cup C_i)$ [by monotonicity of forking].

(w) By (i), B_{i+1}, C_i are closed in the $\mathbf{F}^x_\lambda$-construction $\mathscr{A}$, hence [by 3.3] B_{i+1} is $\mathbf{F}^x_\lambda$-constructible over $A \cup C_i$, hence (3.2) $\mathbf{F}^x_\lambda$-atomic over $A \cup C_i$ [remember we are in 5.6(1)].

(x) $|N| \cap B_{i+1}$ is $\mathbf{F}^x_\lambda$-atomic over $|N| \cap (A \cup C_i)$ [by (v), (w) and 4.3].

(y) $|N| \cap B_{i+1}$ is $\mathbf{F}^x_\lambda$-constructible over $|N| \cup A \cup C_i)$ [by (x) and 3.16, λ is regular].

So we finish the proof of 5.6(1).

(2) Left to the reader.

CONCLUSION 5.8: *If T is stable, $|T|$ is regular and over A there is an $\mathbf{F}^t_{|T|}$-primary model, then the $\mathbf{F}^t_{|T|}$-prime model over A is unique.*

THEOREM 5.9: *Let $S^x_\kappa = \{\lambda : |A| = \lambda$ imply over A there is a prime model among the $\mathbf{F}^x_\kappa$-compact ones$\}$.*

(1) *$\mu \geqslant \lambda = \lambda^{<\kappa} \geqslant |T|$, $\lambda \in S^t_\kappa$ implies $\mu \in S^t_\kappa$,*

(2) *$\lambda^t_{\aleph_0} = \min S^t_{\aleph_0} < \infty$ implies $\lambda^t_{\aleph_0} < \mu(|T|, 1)$ (see VII, Definition 5.1),*

(3) *For $x = a, s$, similar results hold with $2^{|T|}$ instead of $|T|$.*

Proof. (1), (2).

5.9A DEFINITION: (i) A type $p \in S(A)$ is $\mathbf{F}^x_\kappa$-inevitable if every $\mathbf{F}^x_\kappa$-compact model M, $A \subseteq |M|$, realizes it. For $F^x_\kappa = F^t_{\aleph_0}$, we omit it.

(ii) $Q^x_\kappa(\lambda)$ is the assertion that for every A of cardinality λ, and $\mathbf{F}^x_\lambda$-type q over A, there is an $\mathbf{F}^x_\lambda$-inevitable $p \in S(A)$, $q \subseteq p$.

5.9B CLAIM: (1) *If $\lambda = \lambda^{<\kappa} + |T|$, then $\lambda \in S^t_\kappa$ iff $Q^t_\kappa(\lambda)$.*

(2) *If $Q^t_\kappa(\mu)$ holds whenever $\lambda \leq \mu < \lambda_1 = \lambda_1^{<\kappa} + |T| + \lambda^+$, then $\lambda \in S^t_\kappa$.*

Proof. (2) Given A, $|A| = \lambda$, we define by induction on i elements a_i such that $\text{tp}(a_i, A_i)$ is inevitable where $A_i = A \cup \{a_j : j < i\}$, and for each i, and each 1-type q over A_i, $|q| < \kappa$, for some $j < \lambda_1$, q is realized by a_j. It is easy to see A_{λ_1} is the universe of a κ-compact model M, $A \subseteq |M|$, M is $\mathbf{F}^x_\kappa$-prime over A.

(1) The "if" part follows from (2) (for $\lambda_1 = \lambda^+$). For the "only if" part, let $|A| = \lambda$, q a 1-type over A, $|q| < \kappa$. Let M be a κ-prime model over A (exist by an assumption), and $a \in M$ realizes q, then $p = \text{tp}(a, A) \in S(A)$ extend q and is $\mathbf{F}^t_\kappa$-inevitable (because if $A \subseteq M'$, M' $\mathbf{F}^t_\kappa$-compact, then M can be elementarily embedded into M' over A, hence M' realizes p).

5.9C CLAIM: *The following are equivalent*

(i) *q is a 1-type over A, $|q| < \kappa$, q has no $\mathbf{F}^t_\kappa$-inevitable completion in $S(A)$,*

(ii) *there are $\mathbf{F}^t_\kappa$-compact models M, M_a ($a \in M$) such that: $A \subseteq M$, $A \subseteq M_a$, and for every $a \in M$, if a realizes q, then M_a omit* $\mathrm{tp}(a, A)$.

Proof. (i) $\Rightarrow$ (ii): Let M be any $\mathbf{F}^t_\kappa$-compact model, $A \subseteq M$. If $a \in M$, a realizes q, then by (i), $\mathrm{tp}(a, A)$ is not inevitable, so there is a $\mathbf{F}^t_\kappa$-compact model M_a, $A \subseteq M_a$, M_a omit $\mathrm{tp}(a, A)$. If $a \in M$ does not realize q, let $M_a = M$.

(ii) $\Rightarrow$ (i): If q has an $\mathbf{F}^t_\kappa$-inevitable extension $p \in S(A)$, then p is realized in M (by the definition of $\mathbf{F}^t_\kappa$-inevitable), so let a realize it, but then M_a omit p, contradiction.

5.9D COROLLARY: *Suppose $Q^t_\kappa(\lambda)$ fail, or even $\lambda \notin S^t_\kappa$, then for every $\mu = \mu^{<\kappa} + |T| \leq \lambda$, $Q^t_\kappa(\mu)$ fail and so $\mu \notin S^t_\kappa$.*

Proof. If $\lambda \notin S^t_\kappa$, then by Claim B, $Q^t_\kappa(\lambda')$ fail for some $\lambda' \geq \lambda$, so w.l.o.g. $Q^t_\kappa(\lambda)$ fail. So for some A, $|A| = \lambda$ and 1-type q over A, $|q| < \kappa$, q has no inevitable extension in $S(A)$. So by Claim C, there are M, M_a ($a \in A$) as mentioned there. By the downward Löwenheim–Skolem theorem to the appropriate model, we get that Claim C(ii) holds for some $A' \subseteq A$, $|A'| = \mu$, so Claim C(i) holds so $Q^t_\kappa(\mu)$ fail, hence by Claim B, $\mu \notin S^t_\kappa$.

5.9E COROLLARY: *There is a sentence $\psi \in L_{|T|^+,\omega}$ which has a model of power λ iff not $Q^t_{\aleph_0}(\lambda)$.*

Proof. Formalize Claim B(ii).

PROBLEM 5.2: What is the first cardinal λ, such that if $\mathbf{F}^t_{\aleph_0}$ satisfies Ax(X.1), $|T| = \aleph_0$, and over some A the $\mathbf{F}^t_{\aleph_0}$-prime model is not unique, then there is such A, $|A| < \lambda$ (similar problems exist for other $\mathbf{F}$'s, and for uncountable T).

THEOREM 5.10: *Let $\lambda > |T|$, $\lambda = \lambda^{<\lambda}$. Then the following conditions are equivalent.*

(1) *$\mathbf{F}^t_\lambda$ satisfies Axioms* (XI.1), (X.1) *and $\mathbf{F}^t_\lambda$-saturation implies λ-compactness.*

(2) *Over any A there is a λ-prime model.*

(3) *Over any A, $|A| \leqslant \lambda$, there is a λ-prime model.*

(4) *If* $|p| < \lambda$, *p an m-type over A*, $|A| \leq \lambda$, *p has an* $\mathbf{F}^t_\lambda$*-isolated extension in* $S^m(A)$.

(5) *The same as* (4) *for arbitrary A.*

Proof. (1) ⇒ (2) By 3.1(2), 3.10(2).

(2) ⇒ (3) Trivial.

(4) ⇒ (5) Suppose A is a counter-example to (5).

As $\lambda = \lambda^{<\lambda}$, λ is regular. Define by induction on $i \leq \lambda$, $A_i \subseteq A$, $|A_i| \leq \lambda$. Let $A_0 = \operatorname{Dom} p$, and $A_\delta = \bigcup_{i<\delta} A_i$. If A_i is defined, for every m-type q over A_i, $p \subseteq q$, $|q| < \lambda$, q has at least two extensions in $S^m(A)$, so for some $\varphi_q(\bar{x}; \bar{a}_q)$ $(\bar{a}_q \in A)$. $q \cup \{\varphi_q(\bar{x}; \bar{a}_q)^{\mathbf{t}}\}$ is consistent for $\mathbf{t} \in \{0, 1\}$. Let $A_i = A_{i+1} \cup \{\bar{a}_q : p \subseteq q, |q| < \lambda, q \text{ an } m\text{-type over } A_i\}$. Clearly $|A_{i+1}| \leq \lambda$ as $|T| < \lambda = \lambda^{<\lambda}$. Now p and A_λ provide a counter-example to (4).

(5) ⇒ (1) Clear.

(3) ⇒ (4) Let p be an m-type over A, $|A| \leq \lambda$, $|p| < \lambda$. By (3) there is an λ-prime model M over A. So some $\bar{a} \in |M|$ realizes p. If $q = \operatorname{tp}(\bar{a}, A)$ is $\mathbf{F}^t_\lambda$-isolated we finish. Otherwise by 5.3 there is an λ-compact model N, $A \subseteq |N|$, N omits q. But as M is λ-prime there is an elementary mapping f from M into N, $f \restriction A =$ the identity, so $f(\bar{a}) \in |N|$ realizes q, contradiction.

PROBLEM 5.3: Find a similar theorem when $\lambda < \lambda^{<\lambda}$.

PROBLEM 5.4: What can $\{\lambda : \mathbf{F}^x_\lambda \text{ satisfies Ax(XI.1)}\}$ be, $x = t, s, a$?

THEOREM 5.11: *If T is countable* $\mathbf{F}^l_{\aleph_0}$ *does not satisfy* Ax(X.1), *then* $\mathbf{F}^x_\mu$ *does not satisfy* Ax(X.1) *for* $x = t, s, a$ *and any* μ *and* $\mathbf{F}^x_\mu$ *does not satisfy* Ax(XI.1) *for* $x = s, a$ *and any* μ *or* $x = t$, $\mu > |T|$.

Proof. So there are A, $\bar{a} \in A$ and $\psi(x, \bar{a})$, φ such that $(\exists x)\psi(x; \bar{a})$, but there are no $p \in S(A)$, $q \subseteq p$, such that $|q| < \lambda$, $\psi(x; \bar{a}) \in p$, and $q \vdash p \restriction^{+} \varphi$. Let $P = A$, and extend $(\mathfrak{C}, P)$ to a $\bar{\kappa}$-saturated model $(\mathfrak{C}_1, p_1)$. We can assume $\mathfrak{C}_1 = \mathfrak{C}$, and let $P_1 = A_1$. As $(\mathfrak{C}, P_1)$, $(\mathfrak{C}, P)$ are elementarily equivalent, for every finite 1-type q, over A, such that $(\exists x)[\psi(x, \bar{a}) \wedge \bigwedge q]$ there is $\bar{b} \in A_1$, such that $\{\psi(\bar{x}; \bar{a})\} \cup q \cup \{\varphi(x, \bar{b})^{\mathbf{t}}\}$ is consistent for $\mathbf{t} \in \{0, 1\}$. By the μ-saturation of $(\mathfrak{C}_1, P_1)$, this holds for any q, $|\operatorname{Dom} q| < \mu$. The conclusions are immediate.

DEFINITION 5.1: The set $I \subseteq |M|$ is *absolutely indiscernible* in M, if every permutation of I is induced by an automorphism of M.

LEMMA 5.12: (1) *An absolutely indiscernible set I in M, is indiscernible.*

(2) *If M is strongly $|I|^+$-saturated, (see Definition I, 1.4(2)) $I \subseteq |M|$ I an indiscernible set, then I is an absolutely indiscernible set in M.*

Proof. Immediate.

Remark. So the real problem is to find M and I absolutely indiscernible in M, such that $|I| = \|M\|$.

THEOREM 5.13: *Let I be an indiscernible set, T stable in $|I|$. Then I is absolutely indiscernible in some M, $\|M\| = |I|$. Moreover, I is a maximal indiscernible set in M.*

Proof. Let $\kappa = \kappa_r(T)$; so as T is stable in $|I|$, $|I|^{<\kappa} = |I|$ (by III, 3.7) so $\kappa \le |I|$ and even $\kappa \le \mathrm{cf}|I|$. Let M be $\mathbf{F}^a_\kappa$-primary over $\bigcup I$. M exists by 3.1, 2.18(2); clearly $\|M\| = |I|$ (as T is stable in $|I|$). By 4.10 I is a maximal indiscernible set in M, and by 3.8(1) I is absolutely indiscernible in M.

THEOREM 5.14: *Suppose T is stable, $\lambda = \lambda^{<\kappa}$, and T is stable in some μ, $\kappa \le \mu \le \lambda$.*

(1) *Over any A, $|A| \le \lambda$ there is a $\mathbf{F}^f_\kappa$-constructible model M over A, $\|M\| = \lambda$ such that: if $B \subseteq |M|$, $|B| < \lambda$, $p \in S^m(A \cup B)$ does not fork over $C \subseteq A \cup B$, $|C| < \kappa$ then p is realized in M.*

(2) *The model M in (1) is unique up to isomorphism over A.*

Proof. (1) The proof is similar to the proof of 3.1(3), (4).

(2) The proof is similar to the proof of 3.9 (in fact, by 3.1, 3.8(2) resp. applied to $\mathrm{Th}(M, a)_{a \in A}$).

CONCLUSION 5.15: *If I is a (non-trivial) indiscernible set of sequences, and T is stable in some $\mu \le |I|$ then $|I|$ is an absolutely indiscernible set in a model M, $\|M\| = |I|$.*

Proof. By 5.14(1) and 5.14(2).

THEOREM 5.16: *Suppose $|T| = \lambda < \chi$, $\Delta \subseteq \mathrm{L} = \mathrm{L}(T)$, $|\Delta| = \lambda$, and one of the following conditions holds.*

(A) *λ is regular, $\chi = 2^\lambda$.*

(B) *There is a tree $\mathbf{I} \subseteq {}^{\lambda\ge}2$ (i.e., $\eta \lessdot \nu \in \mathbf{I} \to \eta \in \mathbf{I}$), so that $|\mathbf{I} \cap {}^{\alpha}2| < \lambda$ for $\alpha < \lambda$ but $|\mathbf{I} \cap {}^{\lambda}2| = \chi$ (for $\chi = 2^\lambda$, $\mathbf{I} \cap {}^{\lambda>}2$ is a Kurepa tree).*

(C) $\chi = \lambda^+$, λ *regular.*

(D) *There is a tree* $I \subseteq {}^{\lambda \geq}2$, $\chi = |I \cap {}^{\lambda}2| > |I \cap {}^{\lambda >}2|^+$.

Then (1) *there are models* M_i $(i < \chi)$ *of* T, *each of cardinality* λ, *such that:* $i(1) \neq i(2) < \chi$, $\bar{a}_l \in |M_{i(l)}|$, $\mathrm{tp}_\Delta(\bar{a}_1, \emptyset) = \mathrm{tp}_\Delta(\bar{a}_2, \emptyset)$ *implies that there are* $p_l \subseteq \mathrm{tp}(\bar{a}_l, \emptyset)$, $|p_l| < \lambda$, $p_l \vdash \mathrm{tp}_\Delta(\bar{a}_l, \emptyset)$ $(l = 1, 2)$.

(2) *If* $\lambda^{<\kappa} = \lambda$, *we can assume each* M_i *is* κ*-compact.*

Remark. Case B applies, e.g., for singular λ when $(\forall \mu < \lambda)(\mu^{<\mathrm{cf}\lambda} < \lambda)$ $\chi = \lambda^{\mathrm{cf}\lambda}$; so for strong limit λ, we get $\chi = 2^\lambda$, and for λ of cofinality $\aleph_0$, we get $\chi = \lambda^{\aleph_0}$.

We can conclude:

CONCLUSION 5.17: *If* $|T| = \lambda < \chi$, λ, χ *satisfy* (A) *or* (B) *or* (C) *or* (D) *from* 5.16 *each* $p_i \in D(T)(i < i_0 < \chi)$ *is not* $\mathbf{F}^t_\lambda$*-isolated then* T *has a model of cardinality* λ *which omits every* p_i.

Proof of 5.16. We can assume w.l.o.g. that Δ is closed under Boolean combinations. We leave (2) to the reader and concentrate on (1).

Case C. We define the models M_i $(i < \chi = \lambda^+)$ by induction on i. If we have defined M_j for $j < i$ let $\Phi_i = \{\mathrm{tp}_\Delta(\bar{a}, \emptyset) : \bar{a} \in |M_j|,\ j < i\}$ (so $|\Phi_i| \leq \lambda$) and it suffices to find a model M_i of T of cardinality λ, such that if $\mathrm{tp}_\Delta(\bar{a}, \emptyset) \in \Phi_i$, $\bar{a} \in |M_i|$ then for some $q \subseteq \mathrm{tp}_\Delta(\bar{a}, \emptyset)$, $|q| < \lambda$, $q \vdash \mathrm{tp}_\Delta(\bar{a}, \emptyset)$ and a little more. This follows by

CLAIM 5.18: *If* $|T| = \lambda$, p_i $(i < i_0 \leq \lambda)$ *types* (*in* $\mathrm{L}(T)$, *over* $\emptyset$) *then* T *has a model* M *of cardinality* λ, *such that:*

(i) $\bar{a} \in M$, $\bar{a}$ *realizes* p_i *implies the existence of* $q \subseteq \mathrm{tp}(\bar{a}, \emptyset)$, $|q| < \lambda$ *for which* $q \vdash p_i$.

(ii) *If* $\bar{a} \in {}^m|M|$ *and for some m-type* q (*over* $\emptyset$), $|q| < \lambda$, $q \vdash \mathrm{tp}_\Delta(\bar{a}, \emptyset)$ *then this holds for some* $q \subseteq \mathrm{tp}(\bar{a}, \emptyset)$.

Proof. The proof is like that of 5.3(2), but we also do an assignment like (iii) of the proof of Case A. If $\lambda = \aleph_0$ the set of types forbidden by (ii) of 5.18 is countable, 5.3 applies directly.

Case B. Let $\{\varphi_i(x, \bar{y}_i) : i < \lambda\}$ be a list of all formulas $\varphi(x, \bar{y}) \in \mathrm{L}$ $\bar{y} \in \{x_j : j < \lambda\}$ and assume $\bar{y}_i \in \{x_j : j < 2i\}$. We define by induction on $\alpha < \lambda$, for each $\eta \in I \cap {}^\alpha 2$ a consistent set Γ_η of formulas in the variables $\{x_i : i < 2\alpha\}$, for $\lambda = \aleph_0$, Γ_η is finite, and for $\lambda > \aleph_0$, $|\Gamma_\eta| \leq \aleph_0 +$

$|I \cap {}^{\alpha}2|$ (note that w.l.o.g. $|I \cap {}^{\alpha}2|$ is non-decreasing). For $\alpha = 0$ let $\Gamma_{\langle\rangle} = \emptyset$, and for limit α, $\eta \in {}^{\alpha}2$ let $\Gamma_\eta = \bigcup_{\beta<\alpha} \Gamma_{\eta|\beta}$. Suppose we have defined for α, and let us prove for $\alpha + 1$. Let $\{\langle\eta_1^i, \eta_2^i, \bar{z}_1^i, \bar{z}_2^i\rangle : i < i(0)\}$ be a list of all quadrupals $\langle\eta_1, \eta_2, \bar{z}_1, \bar{z}_2\rangle$ where $\eta_1, \eta_2 \in {}^{\alpha}2 \cap I$ are distinct, $\bar{z}_1, \bar{z}_2$ are finite sequences of variables from $\{x_j : j < 2\alpha\}$ of equal length. So for $\lambda = \aleph_0$, $i(0) < \aleph_0$, and for $\lambda > \aleph_0$, $i(0) < \aleph_0 + |I \cap {}^{\alpha}2|^+$. We now define by induction on $i \leq i(0)$, consistent set of formulas Γ_η^i ($\eta \in {}^{\alpha}2$) in the variables $\{x_j : j < 2\alpha\}$ such that $|\Gamma_\eta^i| < \aleph_0 + |I \cap {}^{\alpha}2|^+$, and $j < i$ implies $\Gamma_\eta^j \subseteq \Gamma_\eta^i$. Let $\Gamma_\eta^0 = \Gamma_\eta$, and for limit $\Gamma_\eta^i = \bigcup_{j<i} \Gamma_\eta^j$. If we have defined for i, and there is a formula $\varphi(\bar{x}) \in \Delta$ such that $\Gamma_{\eta_l}^i \cup \{\varphi(\bar{z}_l)^{\text{if }(l=1)}\}$ $l = 1, 2$, are consistent, choose $\Gamma_{\eta_1}^{i+1} = \Gamma_{\eta_1}^i \cup \{\varphi(\bar{x})\}, \Gamma_{\eta_2}^{i+1} = \Gamma_{\eta_2}^i \cup \{\neg\varphi(\bar{x})\}$. For all other η's, and if there is not such φ for $\eta = \eta_1, \eta_2$ too, $\Gamma_\eta^{i+1} = \Gamma_\eta^i$. Now let, for $\eta \in {}^{\alpha}2$, $l \in \{0, 1\}$

$$\Gamma_{\eta\frown\langle l\rangle} = \Gamma_\eta^{i(0)} \cup \{(\exists x)\varphi_\alpha(x, \bar{y}_\alpha) \to \varphi_\alpha(x_{2\alpha}, \bar{y}_\alpha)\} \cup \{x_j \neq x_{2\alpha+1} : j < 2\alpha\}.$$

For each $\eta \in {}^{\lambda}2 \cap I$, $\Gamma_\eta = \bigcup_{\beta<\lambda} \Gamma_{\eta|\beta}$ is consistent so some assignment $x_i \mapsto a_\eta^i$ satisfies it. By Tarski–Vaught test there is a model M_η with universe $\{a_\eta^i : i < \lambda\}$. It is easy to check that the models are as required.

Case A. As Case B was proved, we can assume $\lambda > \aleph_0$, and as Case C was proved we can assume $2^\lambda > \lambda^+$.

Let $\{\varphi_i(x, \bar{y}_i) : i < \lambda, i \text{ a successor}\}$ be a list of all formulas $\varphi(x, \bar{y}) \in \mathrm{L}$, $\bar{y} \in \{x_j : j < \lambda\}$ so that $\bar{y}_i \in \{x_j : j < 2i\}$. Let $\{\bar{z}_\alpha : \alpha < \lambda\}$ be a list of all $\bar{z} \in \{x_j : j < \lambda\}$, such that for each such $\bar{z}$, $\{\alpha : \bar{z}_\alpha = \bar{z}\}$ is a stationary subset of λ (possible by 1.3 of the Appendix). We assume $\bar{z}_\alpha \in \{x_j : j < 1 + \alpha\}$, and let $\mathrm{L} = \bigcup_{\alpha<\lambda} \mathrm{L}_\alpha$, $|\mathrm{L}_\alpha| < \lambda$, L_α ($\alpha < \lambda$) is increasing and continuous.

We shall define by induction on $\alpha < \lambda$, for each $\eta \in {}^{\alpha}2$ a consistent set of formulas Γ_η, in the variables $\{x_i : i < 2\alpha\}$ such that $|\Gamma_\eta| < \lambda$.

Subcase (i) ($\alpha = 0$). $\Gamma_{\langle\rangle} = \emptyset$.

Subcase (ii) (α limit). For each $\eta \in {}^{\alpha}2$, $\Gamma_\eta = \bigcup_{\beta<\alpha} \Gamma_{\eta|\beta}$.

Subcase (iii) ($\alpha = \beta + 1$, β successor). Let $\{\varphi^i(\bar{z}^i) : i < i(0)\}$ be a list of the formulas from L_β with $\bar{z}^i \in \{x_i : i < 2\alpha\}$. For each $\eta \in {}^{\beta}2$ we define by induction on $i < i(0)$ formulas $\theta_\eta^i(\bar{z}^i)$ such that $\Gamma_\eta \cup \{\theta_\eta^j(\bar{z}^j) : j \leq i\}$ is consistent, and $\theta_\eta^i \in \Delta$. If we have defined for $j < i$, and there is $\theta \in \Delta$ such that $\Gamma_\eta \cup \{\theta_\eta^j(\bar{z}^j) : j < i\} \cup \{\theta(\bar{z}^i)\}$ is consistent, and T, $\varphi^i(\bar{z}^i) \vdash \neg\theta(\bar{z}^i)$, choose such $\theta = \theta_\eta^i$, otherwise choose $\theta_\eta^i \in \Delta$ such that

$\Gamma_\eta \cup \{\theta^j_\eta(\bar{z}^j): j \leq i\}$ is consistent (note that consistency is preserved for limit i's).

Now let for $\eta \in {}^\beta 2$, Γ^1_η be $\Gamma_\eta \cup \{\theta^j_\eta(\bar{z}^j): j < i(0)\} \cup \{\neg(\exists x)\varphi_\beta(x, \bar{y}^\beta)\}$ if consistent and $\Gamma_\eta \cup \{\theta^j_\eta(\bar{z}^j): j < i(0)\} \cup \{\varphi_\beta(x_{2\beta}, \bar{y}^\beta)\}$ otherwise. For $l = 0, 1$ let $\Gamma_{\eta^\frown\langle l\rangle} = \Gamma^1_\eta \cup \{x_i \neq x_{2\beta+1}: i < 2\beta\}$.

Subcase (iv) ($\alpha = \beta + 1$, β limit: and for no $\gamma < \beta$, $\bar{z}_\gamma = \bar{z}_\beta$). For $\eta \in {}^\beta 2$ let $p_\eta = \{(\exists \bar{v})\varphi(\bar{z}_\beta, \bar{v}): \varphi$ is a finite conjunction of formulas of Γ_η, $\bar{z}_\beta \cap \bar{v} = \emptyset\}$.

If there is a type $q = q(\bar{z}_\beta)$, $|q| < \lambda$, such that $p_\eta \cup q$ is consistent, and for every $\theta(\bar{x}) \in \Delta$, $l(\bar{x}) = l(\bar{z}_\beta)$, for some $\mathbf{t} \in \{0, 1\}$ $p_\eta \cup q \vdash \theta(\bar{z}_\beta)^{\mathbf{t}}$, then choose such q_η; and if there is no such q, let $q_\eta = \emptyset$. Let for $l = 0, 1$

$$\Gamma_{\eta^\frown\langle l\rangle} = \Gamma_\eta \cup q_\eta.$$

Subcase (v) ($\alpha = \beta + 1$, β limit, and for some $\gamma < \beta$, $\bar{z}_\gamma = \bar{z}_\beta$). Let us define p_η ($\eta \in {}^\beta 2$) as in the previous case, and let $S_\beta = \{\eta \in {}^\beta 2$: there is $\theta \in \Delta$ such that $p_\eta \cup \{\theta(\bar{z}_\beta)^{\mathbf{t}}\}$ is consistent for each $\mathbf{t} \in \{0, 1\}\}$. By the previous case, as $|p_\eta| < \lambda$, for each $q = q(\bar{z}_\beta)$, $|q| < \lambda$, $\eta \in S_\beta$, if $p_\eta \cup q$ is consistent, then for some $\theta \in \Delta$, $p_\eta \cup q \cup \{\theta(\bar{z}_\beta)^{\mathbf{t}}\}$ are consistent ($\mathbf{t} = 0, 1$). So we can find $\theta^\rho_\eta \in \Delta$ such that for every $\nu \in {}^\beta 2$, $p_\eta \cup \{\theta^{\nu|\gamma}_\eta(\bar{z}_\beta)^{\nu[\gamma]}: \gamma < \beta\}$ is consistent. We can assume, that if $\eta_1, \eta_2 \in S_\beta$ and for every $\theta(\bar{x}) \in \Delta$, $p_{\eta_1} \vdash \theta(\bar{z}_\beta) \Leftrightarrow p_{\eta_2} \vdash \theta(\bar{z}_\beta)$, then for every $\nu \in {}^{\beta >}2$, $\theta^\nu_{\eta_1} = \theta^\nu_{\eta_2}$ (remember that we assume Δ is closed under Boolean combinations).

Now for every $\eta \in S$, $l = 0, 1$ let

$$\Gamma_{\eta^\frown\langle l\rangle} = \Gamma_\eta \cup \{\theta^{\eta|\gamma}_\eta(\bar{z}_\beta)^{\eta[\gamma]}; \gamma < \beta\}.$$

For $\eta \in {}^\beta 2 - S$, let $\Gamma_{\eta^\frown\langle l\rangle} = \Gamma_\eta$.

$\therefore$ Let for each $\eta \in {}^\lambda 2$, $\Gamma_\eta = \bigcup_{\alpha<\lambda} \Gamma_{\eta|\alpha}$, so clearly it is consistent and let $x_i \mapsto a^i_\eta$ be an assignment satisfying it. So by the Tarski–Vaught test (and Subcase (iii)) there is a model M_η of cardinality λ whose universe is $\{a^i_\eta: i < \lambda\}$. We now prove

(*) Suppose $\bar{z} = \langle x_{i(0)}, \ldots, x_{i(k)}\rangle$, $\eta_1 \neq \eta_2 \in {}^\lambda 2$, and let $\bar{a}_l = \langle \bar{a}^{i(0)}_{\eta_l}, \ldots, \bar{a}^{i(k)}_{\eta_l}\rangle$. If $\mathrm{tp}_\Delta(\bar{a}_1, \varphi) = \mathrm{tp}_\Delta(\bar{a}_2, \varphi)$ then for at least one l, there is $q \subseteq \mathrm{tp}(\bar{a}_l, \varphi)$ $|q| < \lambda$, such that $q \vdash \mathrm{tp}_\Delta(\bar{a}_l, \varphi)$.

For suppose (*) fails, choose α_0 such that $\eta_1 \restriction \alpha_0 \neq \eta_2 \restriction \alpha_0$, and for some $\gamma < \alpha_0$, $\bar{z}_\gamma = \bar{z}$. So if $\alpha_0 < \beta < \lambda$, β limit $\bar{z}_\beta = \bar{z}$, then $\eta_1|\beta$, $\eta_2|\beta \in S_\beta$.

Clearly there is a closed unbounded set $W \subseteq \lambda$, such that for $\beta \in W$, $l = 1, 2$, $\Gamma_{\eta_l|\beta}$ is $\subseteq L_\beta$ and, for it, it is complete (i.e., for each formula $\psi(\bar{z})$, $\bar{z} \in \{x_i: i < 2\beta\}$, $\psi \in L_\beta$, for some $t \in \{0, 1\}$, $\Gamma_{\eta_l|\alpha} \vdash \psi(\bar{z})^t$)). We can also assume that (for each $\beta \in W$) β is limit, $\beta > \alpha_0$.

As $\{\beta: \bar{z}_\beta = \bar{z}\}$ is a stationary subset of λ there is a $\beta \in W$, $\bar{z}_\beta = \bar{z}$. So $p_{\eta_l|\beta} \in S_\beta$ (as $\bar{a}_1, \bar{a}_2$ contradicts (*)) and for every $\theta \in \Delta$, $p_{\eta_1|\beta} \vdash \theta(\bar{z}_\beta) \Leftrightarrow p_{\eta_2|\beta} \vdash \theta(\bar{z}_\beta)$. [For otherwise we can suppose $p_{\eta_1|\beta} \vdash \theta(\bar{z}_\beta)$ but $p_{\eta_2|\beta} \nvdash \theta(\bar{z}_\beta)$, hence for some $\psi(\bar{z}_\beta) \in p_{\eta_1|\beta}$, $\psi(\bar{z}_\beta) \vdash \theta(\bar{z}_\beta)$ but $p_{\eta_2|\beta} \cup \{\neg\theta(\bar{z}_\beta)\}$ is consistent. For some γ, $\alpha_0 < \gamma < \beta$, $\psi \in L_\gamma$; so by subcase (iii) for some $\theta_1 \in \Delta$, $\psi(\bar{z}_\beta) \vdash \theta_1(\bar{z}_\beta)$, $\neg\theta_1(\bar{z}_\beta) \in p_{\eta_1|(\gamma+1)}$ hence $\neg\theta_1(\bar{z}_\beta) \in p_{\eta_2|\beta}$. But as $\psi(\bar{z}_\beta) \in p_{\eta_1|\beta}$, $\theta(\bar{z}_\beta) \in p_{\eta_1|\beta}$ contradiction (to $\mathrm{tp}_\Delta(a_1, \varphi) = \mathrm{tp}_\Delta(\bar{a}_2, \varphi)$)].

So, by Subcase (v) $\theta^\nu_{\eta_1} = \theta^\nu_{\eta_2}$ for every $\nu \in {}^{\beta>}2$, and $\theta^{\eta_l|\gamma}_{\eta_l}(\bar{z}_\beta)^{\eta_l[\gamma]} \in \Gamma_{\eta_l|(\beta+1)} \subseteq \Gamma_{\eta_l}$ where $\gamma = \min\{\gamma: \eta_1[\gamma] \neq \eta_2[\gamma]\}$, and we get a contradiction (to $\mathrm{tp}_\Delta(\bar{a}_1, \emptyset) = \mathrm{tp}_\Delta(\bar{a}_2, \emptyset)$). So we proved (*). Now for each $\eta \in {}^\lambda 2$ we define $g(\eta)$ as the set of $\nu \in {}^\lambda 2$ such that for some $\bar{a} \in M_\nu$, $\bar{b} \in M_\eta$, $\mathrm{tp}_\Delta(\bar{a}, \emptyset) = \mathrm{tp}_\Delta(\bar{b}, \emptyset)$, but for no $q \subseteq \mathrm{tp}(\bar{a}_\nu, \emptyset)$, $|q| < \lambda$ does $q \vdash \mathrm{tp}_\Delta(\bar{a}, \emptyset)$. Now,

(**) $\quad |g(\eta)| \leq \lambda$.

Otherwise there are distinct $\nu_\alpha \in g(\eta)$, $(\alpha < \lambda^+)$ and for each of them there are witnesses $\bar{a}_\alpha, \bar{b}_\alpha$, and let $\bar{a}_\alpha = \langle \bar{a}^{i(\alpha,0)}_{\nu_\alpha}, \ldots, a^{i(\alpha,k(\alpha))}_{\nu_\alpha} \rangle$. The number of possible $k(\alpha), i(\alpha, 0), \ldots, i(\alpha, k(\alpha)), \bar{b}_\alpha$ is λ, hence there are $\alpha \neq \beta$ for which they are equal. Clearly $\mathrm{tp}_\Delta(a_\alpha, \emptyset) = \mathrm{tp}_\Delta(a_\beta, \emptyset)$ and we get easily a contradiction to (*).

Now as we have assumed $2^\lambda > \lambda^+$ (by case (C)) by 2.8 of the Appendix there is $U \subseteq {}^\lambda 2$, $|U| = 2^\lambda$ such that $\eta \neq \nu \in U$ implies $\eta \notin g(\nu)$. It is easy to check that $\{M_\alpha: \alpha \in U\}$ is a family of models exemplifying our assertion.

Case D. Use the proof of Theorem 5.19, Case I (with I instead ${}^{\lambda>}2$, omitting condition (4) there). In the end use AP 2.8.

We can improve Theorem 5.16 (hence 5.18) by:

THEOREM 5.19: *Suppose* $|T| = \lambda$, $\chi = 2^\lambda$, $\Delta \subseteq \mathrm{L} = \mathrm{L}(T)$, $|\Delta| = \lambda$, *then*:

(1) *There are models* M_i $(i < \chi)$ *each of cardinality* λ, *such that:* $i(1) > i(2), \bar{a}_l \in M_{i(l)}, \mathrm{tp}_\Delta(\bar{a}_1, \emptyset) = \mathrm{tp}_\Delta(\bar{a}_2, \emptyset)$ *implies there is* $p_1 \subseteq \mathrm{tp}(\bar{a}_1, \emptyset)$, $|p_1| < \lambda$, $p_1 \vdash \mathrm{tp}_\Delta(\bar{a}_1, \emptyset)$. *If* λ *is regular or* $2^{<\lambda} < 2^\lambda$ *there is also* $p_2 \subseteq \mathrm{tp}_\Delta(\bar{a}_2, \emptyset)$, $|p_2| < \lambda$, $p_2 \vdash \mathrm{tp}(\bar{a}_2, \emptyset)$.

(2) *If* $\lambda^{<\kappa} = \lambda$, *we can assume each* M_i *is* κ*-compact.*

Remark. So if $|T| = \lambda$, $P_i \in S^m_\Delta(\emptyset)$ $(i < \alpha < 2^\lambda)$, then for some M, $\|M\| = \lambda$, and $\bar{a} \in M$, $p_i = \text{tp}_\Delta(\bar{a}, \emptyset)$ implies $q \vdash p_i$ for some $q \subseteq \text{tp}(\bar{a}, \emptyset)$, $|q| < \lambda$. We can have λ pairs (Δ, m)'s.

Proof. (1) By Theorem 5.16 we can assume w.l.o.g. λ is singular.

Case I: $2^{<\lambda} < 2^\lambda$. Clearly $\langle 2^\mu : \mu < \lambda \rangle$ is not eventually constant so we can choose λ_α $(\alpha < \kappa = \text{cf}\,\lambda)$, $\alpha < \beta \Rightarrow 2^{\lambda_\alpha} < 2^{\lambda_\beta}$, $\lambda = \sum_{\alpha<\kappa} \lambda_\alpha$. Let $\{\varphi_i(x; \bar{y}_i) : i < \lambda\}$ be a list of all formulas $\varphi(x; \bar{y}) \in \mathrm{L}$, $\bar{y} \in \{x_j : j < \lambda\}$, so that $\bar{y}_i \in \{x_j : j < 2i\}$.

We define by induction on $i < \lambda$, for each $\eta \in {}^i2$ a set Γ_η of formulas in the variables $\{x_j : j < 2i\}$, consistent with T, such that

(1) $|\Gamma_\eta| \le |l(\eta)| + \aleph_0$, and when $\delta = l(\eta)$ is limit, $\Gamma_\eta = \bigcup_{\alpha<\delta} \Gamma_{\eta\restriction\alpha}$,

(2) $\nu \lessdot \eta$ implies $\Gamma_\nu \subseteq \Gamma_\eta$,

(3) if $i + 1 < l(\eta)$, then $(\exists x)\, \varphi_i(x; \bar{y}_i) \to \varphi_i(x_{2i}; \bar{y}_i) \in \Gamma_\eta$ and

$$\{x_{2i+1} \ne x_j : j \le 2i\} \subseteq \Gamma_\eta,$$

(4) if $\lambda_{\alpha+1} \le i < \lambda_{\alpha+2}$, $\eta \in {}^i2$, $\nu \in {}^{(\lambda_\alpha)}2$, $\bar{z} \in \{x_j : j < \lambda_\alpha\}$, and $\Gamma_\nu \vdash r(\bar{z})$ $r(\bar{z}) \in S^{l(\bar{z})}_\Delta(\emptyset)$, but not $\Gamma_\eta \vdash r(\bar{z})$, then for some $\theta(z) \in r$, $\Gamma_\nu \in \neg\theta(\bar{z})$, for any $\theta \in {}^{\lambda>}2)$.

If such r exists, we denote it by $r_\nu(\bar{z})$.

(5) If $\eta \in {}^i2$, $\bar{z} \in \{x_j : j < 2i\}$, $l \in \{0, 1\}$, $r_{\eta^\frown\langle l\rangle}(\bar{z})$ does not exist, then for some $\theta \in \Delta$, $\theta(\bar{z})^t \in r_{\eta^\frown\langle t\rangle}(\bar{z})$.

(6) If $\eta \in {}^i2$, $\bar{z} \in \{x_j : j < i\}$, and for some $q(\bar{z})$, $r \in S^{l(\bar{z})}_\Delta(\emptyset)$, $|q(\bar{z})| \le |i|$, $\bar{q}(\bar{z}) \vdash r$, $\Gamma_\eta \cup q(\bar{z})$ is consistent, then $\Gamma_\eta \vdash r$.

Let for $\eta \in {}^\lambda2$, $x_i \mapsto a^i_\eta$ be an assignment satisfying Γ_η, and M_η a model with universe $\{a^i_\eta : i < \lambda\}$, and let $g(\eta) = \{\nu \in {}^\lambda2$: For some $\bar{a} \in M_\nu$, $\bar{b} \in M_\eta$, $\text{tp}_\Delta(a, \emptyset) = \text{tp}_\Delta(\bar{b}, \emptyset)$ but for no $q \subseteq \text{tp}(\bar{a}_\nu, \emptyset)$, $|q| < \lambda$, does $q \vdash \text{tp}_\Delta(a, \emptyset)\}$. Now $|g(\eta)| \le 2^{<\lambda}$ by (5) above. So there are $\eta_\xi \in {}^\lambda2$ $(\xi < 2^\lambda)$ such that $\xi < \zeta \Rightarrow \eta_\zeta \notin g(\eta_\xi)$. But by (4) above, $[\eta_\delta \in g(\eta_\xi)] \Leftrightarrow [\eta_\xi \in g(\eta_\delta)]$. Hence $\{M_{\eta_\xi} : \xi < 2^\lambda\}$ is as required.

Case II: $2^{<\lambda} = 2^\lambda$. Like Case I, omitting condition (4), and try to define inductively $\eta_\xi \in {}^\lambda2$ $(\xi < 2^\lambda)$, $\xi < \delta \Rightarrow \eta_\delta \notin g(\eta_\xi)$. We call $A \subseteq {}^\lambda2$ rare if for some $\gamma = \gamma(A) < \lambda$

$$(\forall \eta, \nu)\,(\eta \in A \wedge \nu \in A \wedge \eta \restriction \gamma = \nu \restriction \gamma \Rightarrow \eta = \nu),$$

and clearly $g(\eta)$ is the union of λ rare sets. It is enough to prove ${}^\lambda2$ is not the union of $< 2^\lambda$ rare sets; and by assumption for some $\mu < \lambda$, $2^\mu = 2^\lambda$, and w.l.o.g. $\mu < \lambda_0$. If $A_i \subseteq {}^\lambda2$ is rare, $i < i_0 < 2^\lambda$, let $\gamma_i = \gamma(A_i)$, and choose inductively $\eta_\alpha \in {}^{(\lambda_\alpha)}2$ $(\alpha < \text{cf}\,\lambda)$ increasing by $\lessdot$, such that $i < i_0$, $\gamma_i < \lambda_\alpha$, $\nu \in A_i$ implies $\nu \restriction \lambda_{\alpha+1} \ne \eta_{\alpha+1}$ (for α zero or limit there is no problem, for successor-by cardinality consideration). The limit of the η_α is $\eta \in {}^\lambda 2$, $\eta \notin \bigcup_{i<i_0} A_i$.

QUESTION 5.5: Can we assume in 5.15 only that T is stable $|T| \le |I|$?

EXERCISE 5.6: Let $A \subseteq |\mathfrak{C}|$. Define L($A$) as L $\cup$ A, i.e., we add the elements of A to L as individual constants,

$$T(A) = T \cup \{\varphi(\bar{a}) : \bar{a} \in A, \vDash \varphi[\bar{a}]\}.$$

(1) Investigate the connections between the fulfillment of the axioms by $\mathbf{F}^x_\lambda$ for T and $T(A)$.
(2) Use (1) to generalize the uniqueness Theorem 5.6.

EXERCISE 5.7: Let T be a theory in L, $|\mathrm{L}| = \lambda$, $\mathrm{L} = \bigcup_{i<\lambda} \mathrm{L}_i$, L_i increasing, p_i is an $m(i)$-type for $i < \alpha \le \lambda$, and for each i and q, $m(i)$-type q $\{\beta < \lambda : T \cup (q \restriction L_\beta) \nvdash p_i \restriction L_{\beta+1}\}$ has cardinality λ. Show that T has a model omitting each p_i. Check generalization to the situation in 5.16–5.19.

EXERCISE 5.8: Show the consistency with ZFC of the existence of a countable T, and types p_i ($i < \omega_1$) as in 5.3(1) such that no model of T omits every p_i. (Hint: See Hechler [He 73].)

EXERCISE 5.9: Assume MA, and generalize 5.3 for a countable T, and $< 2^{\aleph_0}$ types.

QUESTION 5.10: Can we in Exercise 5.8 demand the types are pairwise contradictory.

EXERCISE 5.11: Generalize 5.16 for λ Δ's at once.

EXERCISE 5.12: Show that if in 5.16 we assume $\Delta = \mathrm{L}$, then we can simplify the proof (e.g., omitting Subcase (iii)).

EXERCISE 5.13: (1) Give an example of a theory T, $|T| = \aleph_1 \le |D(T)|$ such that T has a $\mathbf{F}^t_{\aleph_0}$-atomic model of cardinality $\aleph_0$ but not $> \aleph_0$.

(2) In 5.5(3) we can assume the model has cardinality $\ge \lambda$ and is κ-compact, when $\kappa < \lambda$, $\lambda^{<\kappa} \le \lambda^+$. [Hint: Let $T = \mathrm{Th}(M)$, $M = ({}^{\omega >}2, \ldots, P_\eta, \ldots)_{\eta \in {}^\omega 2}$ where $P_\eta = \{\eta \restriction n : n < \omega\}$.]

DEFINITION 5.2: M is a full model if for every formula $\varphi(\bar{x})$ in $L(M)$ $|\{\bar{a} \in |M| : M \models \varphi[\bar{a}]\}| < \aleph_0$ or $= \|M\|$.

EXERCISE 5.14: **Let T be a countable complete theory. Give a necessary and sufficient condition for the existence of a full model of T cardinality $\aleph_1$ which omits the types p_i $i < \omega$. Generalize 5.17. [Hint: The condition is that for no $p_i(\bar{x})$ and $\varphi = \varphi(\bar{x}, y_0, \bar{z}_0, \ldots, y_k, \bar{z}_k)$ the following hold:**

(i) $T \vdash (\exists^{\geq n} y_0)(\exists \bar{z}_0)(\exists^{\geq n} y_1)(\exists \bar{z}_1), \ldots, (\exists^{\geq n} y_k)(\exists \bar{z}_k)(\exists \bar{x})\ \varphi$ for every $n < \omega$.

(ii) For every $\psi(\bar{x}) \in p_i(\bar{x})$, for some $n < \omega$

$$T \vdash \neg(\exists^{\geq n} y_0)(\exists \bar{z}_0)(\exists^{\geq n} y_1)(\exists \bar{z}_1) \cdots$$

$$(\exists^{\geq n} y_k)(\exists \bar{z}_k)(\exists \bar{x})[\neg\psi(\bar{x}) \wedge \varphi(\bar{x}, y_0, \bar{z}_0, \ldots).]$$

EXERCISE 5.15: **Show that if 5.1(3) holds, any $\mathbf{F}^t_\lambda$-atomic, $\mathbf{F}^t_\lambda$-**saturated model is λ-compact.

EXERCISE 5.16: **Concerning Theorem 5.2(1), show that for some T,** it has a λ-compact model, but 5.1(3) fails. [Hint: See Exercise 2.2, hint (1), with A the universe of a non λ-compact model.]

EXERCISE 5.17: **Suppose $|T| = \lambda$, singular but not strong limit, and T has a λ-compact model M_0 of cardinality λ. Choose $\mu < \lambda$, $2^\mu > \lambda$.**

(1) $\kappa(T) \leq \mathrm{cf}\,\lambda$.

(2) There are no formulas $\varphi_\eta(\bar{x}, \bar{a}_\eta)$ ($\eta \in {}^{\mu>}2$, $\bar{a}_\eta \in |M_0|$) such that for any $\eta \in {}^{\mu}2$ $\{\varphi_{\eta|\alpha}(\bar{x}, \bar{a}_\eta)^{\eta[\alpha]} : \alpha < \mu\}$ is consistent.

(3) If $A \subseteq |M_0|$, $m < \omega$, q an m-type over A, then for some m-type p over A, $|p| < \mu$ and $q \cup p$ has a unique extension in $S^m(A)$ (so a condition stronger than 5.1(3) holds).

(4) T has a λ-prime, $\mathbf{F}^t_\lambda$-atomic model $M \prec M_0$ (over $\emptyset$) of cardinality λ.

(5) If $\mathrm{cf}\,\lambda > \aleph_0$, any $\mathbf{F}^t_\lambda$-atomic (over $\emptyset$) λ-compact model is λ-homogeneous and the conclusion of 5.2(2), (3) holds.

(6) The λ-prime model of T is unique.

Hints: (1) See VIII, 4.7 for suitable reducts of M_0.

(2) As $\|M_0\| = \lambda$, M_0 is λ-compact.

(3) By (2) (as in the proof of 2.16(1).

(4) By (3) (M is $\mathbf{F}^t_\lambda$-constructible) and 4.7.

(5) By (1), (3) and as 4.9(4), any $\mathbf{F}^t_\lambda$-atomic over $\emptyset$, λ-compact model

is λ-homogeneous. The uniqueness and characterization of the λ-prime model, are trivial now.

(6) Proved as 4.18.

EXERCISE 5.18: **Suppose $|T| = \lambda$, λ singular and strong limit and T** has a λ-compact model of cardinality λ, M_0. Then

(1) $\kappa(T) \leqslant \operatorname{cf}\lambda$, and even $\lambda^{<\kappa(T)} = \lambda$,

(2) T is stable in λ,

(3) The statement Exercise 5.17(3) holds for $\mu = \lambda$,

(4) Exercise 5.17(4); (5), (6) holds.

[Hint: (1) See VIII, 4.7, for suitable reducts of M_0.

(2) If not, by III, 5.15, by (1), $|D(T)| > \lambda$, hence for some m, $|D_m(T)| > \lambda$. Let M be a model of T of cardinality λ, so $|M| = \bigcup_{i<\operatorname{cf}\lambda} A(i)$, $A(i)$ increasing $|A(i)| < \lambda$. For each i, define an equivalence relation E_i on the set of m-formulas: $\varphi_1(x)E_i\varphi_2(\bar{x})$ iff for any $\bar{a} \in A_i$, $M \models \varphi_1[\bar{a}] \equiv \varphi_2[\bar{a}]$. So E_i has $\leq 2^{|A(i)|}$ equivalence classes; so the number of $p \in D_m(T)$ such that $\varphi_1(\bar{x})E_i\varphi_2(\bar{x})$ implies $\varphi_1(\bar{x}) \in p \Leftrightarrow \varphi_2(\bar{x}) \in p$, is $\leq 2^{2^{|A(i)|}} < \lambda$ (as $|A(i)| < \lambda$, λ strong limit). Hence for some $p \in D_m(T)$, for every i, there are m-formulas $\varphi_1^i(\bar{x})$, $\varphi_2^i(\bar{x})$ such that $\neg\varphi_1^i(\bar{x}) \in p \Leftrightarrow \varphi_2^i(\bar{x}) \in p$ and $\varphi_1^i(\bar{x})E_i\varphi_2^i(\bar{x})$. Hence $\psi_i(\bar{x}) = \varphi_1^i(\bar{x}) \equiv$ $\neg\varphi_2^i(\bar{x}) \in p$ but no $\bar{a} \in A(i)$ satisfies $\psi_i(\bar{x})$. So M omits $\{\psi_i(\bar{x}) : i < \operatorname{cf}\lambda\}$ **$\subseteq p$, hence is not λ-compact.**

(3) We first prove:

(*) There are no formulas $\varphi_{\eta,\nu,\alpha} = \varphi_{\eta,\nu,\alpha}(\bar{x}, \bar{a}_{\eta,\nu,\alpha})$ over M_0 for $i < \operatorname{cf}\lambda$, $\eta \in \prod_{j<i}\lambda_j$, $\nu \in \prod_{j<i}\lambda_j$, $\alpha < \lambda_i$ (where $\lambda = \sum_{i<\operatorname{cf}\lambda}\lambda_i$), such that:

if

$$\eta, \nu \in \prod_{i<\operatorname{cf}\lambda} \lambda_i, \quad (\forall i < \operatorname{cf}\lambda)\, \eta(i) < \nu(i),$$

then

$$\{\varphi_{\eta|i,\nu|i,\eta(i)} \wedge \neg\varphi_{\eta|i,\nu|i,\nu(i)} : i < \operatorname{cf}\lambda\}$$

is consistent.

This is proved using Exercise 5.18(2), with similar technique. Next we prove:

(**) If $p = p(\bar{x})$ is an m-type over $A \subseteq |M_0|$, $|p| < \lambda$, there is a $q(\bar{x})$, $|q(\bar{x})| < \operatorname{cf}\lambda$, such that $p(\bar{x}) \cup q(\bar{x})$ is an m-type over A, and: $r \in S^m(A)$, $r \supseteq p(\bar{x}) \cup q(\bar{x})$ implies r is $\mathbf{F}_\lambda^t$-isolated.

If (**) fail we can construct a counterexample to (*), and using (**), Exercise 5.18(3) is easy.

(4) As in Exercise 5.17.]

EXERCISE 5.19: **Show that Theorem 5.2 holds also for singular λ, when cf $\lambda > \aleph_0$, but when cf $\lambda = \aleph_0$ some parts may fail. [Hint: See Exercises 5.17, 5.18.]**

EXERCISE 5.20: **Generalize Theorem 5.2(2) and (3) and the previous exercises to the case $\lambda < |T|$.**

CHAPTER V

MORE ON TYPES AND SATURATED MODELS

V.0. Introduction

In this chapter we shall deal with stable theories only; note that for the concepts we investigate here, there is no need to distinguish between parallel types, nor between equivalent indiscernible sets.

In Sections 1–5 we investigate the notion "dimensions", and also regular, orthogonal and minimal types for stable and superstable theories. In Sections 5 and 2 we use this to investigate theories with few quite saturated models. In Section 6 we deal with cardinality quantifiers and strong transfer theorems. At last, in Section 7, we deal with a generalization of "algebraic", to "of small cardinality by a cardinality quantifier", with appropriate rank; and deal again with ranks.

The "classical" example of those who dealt with categoricity was the algebraically closed fields. There what is the (transcendence) dimension is well understood. The natural first try (see Definition III, 4.5) is as follows: let $p \in S^m(A)$, and we define on $\{\bar{a} \in \mathfrak{C}: p = \mathrm{tp}(\bar{a}, A)\}$ a dependence relation: $\bar{a}$ depends on I if $\mathrm{tp}\,(\bar{a}, A \cup \bigcup I)$ forks over A. Unfortunately not all axioms of dependence relations which enable one to define dimension are satisfied (transitivity is lacking). When there are (in a given model or set) large independent sets we still get a true dimension (see III, 4.21(2)), but when, e.g., $A \subseteq M$, and every independent set ($\subseteq p(M)$) is finite, this does not help. So it is natural to try to deal with the types p for which transitivity is satisfied, and we shall call them regular (see Definition 1.2 and 1.9 for equivalent definitions, including the one above). For developing this we define when $p_l \in S^m(A)$ are weakly orthogonal ($p_1(\bar{x}) \cup p_2(\bar{y})$ is complete) and when p_1, p_2 are orthogonal (every stationarizations are orthogonal).

The main result is contained in 1.14 and 1.15: among the sequences realizing over A, a regular type, the axioms of dependence (needed to define dimension) are satisfied.

Naturally we want to have existence theorems for regular types. For stable T, we can prove that every non-algebraic type can be extended to a minimal type q (i.e., for every φ for some $\mathbf{t}$, $q \cup \{\varphi^{\mathbf{t}}\}$ is algebraic). Minimal types are regular and we can understand them better (see 1.17 and 1.18). But we want to know that there are really many regular types. For superstable T, $M_1 \prec M_2$, $M_1 \neq M$, M_l $\mathbf{F}^a_{\aleph_0}$-saturated; for some $a \in |M_1| - |M_2|$, $\operatorname{tp}(a, |M_1|)$ is regular; and for T totally transcendental, $\mathbf{F}^a_{\aleph_0}$-saturativity is not needed (see 3.5 and 3.19). So regular types are important mainly for superstable T. In this context it is interesting that if M is $\mathbf{F}^a_\kappa$-saturated, $\kappa \geq \kappa_r(T)$, $p_l = \operatorname{tp}(\bar{a}_l, |M|)$ is regular, then p_1 is realized in the $\mathbf{F}^a_\kappa$-prime model over $|M| \cup \bar{a}_2$ iff p_1 is not orthogonal to p_2.

From this we can prove non-orthogonality is an equivalence relation among stationary regular types, and it is a meaningful equivalence relation.

After this theorem it is natural to define the following orders among indiscernible sets: $I \leq_w J$ if $\dim(I, M) \leq \dim(J, M)$ (for every M) and $I \leq_s J$ if when $I \cup J \subseteq M_1 \prec M_2$ are $\mathbf{F}^a_\kappa$-saturated ($\kappa = \kappa_r(T)$) $\operatorname{Av}(I, |M_1|)$ is realized in M_2, then $\operatorname{Av}(J, |M_1|)$ is realized in M_2 (but this implies $\operatorname{Av}(I, |M_1|)$ is more complicated than $\operatorname{Av}(J, |M_1|)$, so maybe it would have been better to reverse the orders). Those (quasi-) orders are investigated in Section 2. Our intuition is that for "simple" theories (in an appropriate sense) the $\mathbf{F}^a_{\kappa_r(T)}$-saturated models are characterized (up to isomorphism) by some dimensions. So the simplest theories will be the unidimensional, (i.e., $I \leq_w J$ for any indiscernible sets I, J), and we would like some dicotomy, saying meaningful assertions on unidimensional and non unidimensional theories. So the climax of Section 2 is Theorem 2.10, which says:

(1) T is unidimensional, and every $\mathbf{F}^a_{\kappa_r(T)}$-saturated model is a $\mathbf{F}^a_{\kappa_r(T)}$-prime model over $\bigcup I_\lambda$ for some λ, where I_λ is an indiscernible set of cardinality λ and has a fixed type, and I_λ is based on $\emptyset$. Or

(2) T is not undimensional and for every $\mu \geq \kappa_r(T)$, T has a $\mathbf{F}^a_\mu$-saturated, not $\mathbf{F}^a_{\mu^+}$-saturated models of arbitrarily large cardinals, which are not even μ^+-universal.

In Section 3 we shall show that for those orders, for superstable T, above each element there are only finitely many elements up to equivalence by the natural equivalence.

Though we get true dimension for regular types, we still want to get somehow the true dimension for the other types; and this is the aim of Section 3. Maybe the following example will clarify it. Suppose M is an algebraically closed field, $\{a_1, a_2\}$ is transcendentally independent, and $\bar{a} = \langle a_1, a_1 \rangle$ and $p = \mathrm{tp}(\bar{a}, \emptyset)$. Now this type is not regular; but it is clearly a matter of "notation" essentially. We should define a weighted dimension; i.e., let $w(\bar{a}, A)$ be the transcendence dimension of $\mathrm{acl}(A \cup \bar{a})$ over $\mathrm{acl}(A)$, and the parallel to the dimension being true is

(*) for $I = \{\bar{b}_i : i < \alpha\}$,

$$w\Big(\bigcup I, A\Big) = \sum_{i<\alpha} w\Big(\bar{b}_i, A \cup \bigcup_{j<i} \bar{b}_j\Big).$$

(We can define naturally w for infinite sets or sequences.)

The first question is how to define the weight $w(\bar{a}, A)$. Our suggestion is as follows: Assume $A = |M|$, M $\mathbf{F}^a_{\kappa_r(T)}$-saturated, N is $\mathbf{F}^a_{\kappa_r(T)}$-prime over $|M| \cup \bar{a}$. If we can find a set $J \subseteq |N|$ independent over $|M|$, of sequences realizing over $|M|$ regular types such that N is $\mathbf{F}^a_{\kappa_r(T)}$-prime over $|M| \cup \bigcup J$ we define $w(\bar{a}, |M|) = |J|$ (and for any p, $w(p)$ is $w(\bar{a}, |M|)$ where $\mathrm{tp}(\bar{a}, |M|)$ is a stationarization of p, $w(\bar{a}, A) = w(\mathrm{tp}(\bar{a}, A))$). Now the main properties we succeed to prove are:

(A) If T is superstable, $w(\bar{a}, A)$ is always well defined and finite (see 3.9).

(B) If I is independent over A, for every $\bar{a}$ there is $J \subseteq I$, $|J| \le w(\bar{a}, A)$ such that $I - J$ is independent over $(A \cup \bar{a}, A)$ (see 3.15).

Note that previously we could get $|J| < \kappa(T)$; however we then get $I - J$ is independent over $(A \cup \bar{a} \cup \bigcup J, A)$, and this cannot be done here (even the vector space over a finite field can serve as a counterexample).

Now by (B) we can show that dimensions are "almost" true, i.e.:

(C) If $p \in S^m(A)$, $J \subseteq p(\mathfrak{C})$, $I_l \subseteq J$ is a maximal set independent over A $(l = 1, 2)$ then $|I_1| \le w(p)|I_2|$.

However, the weight defined in Section 3 does not satisfy (*); so we continue our search in Section 4. First let us contemplate an example. Suppose M is the disjoint union of two algebraically closed fields: F_1 and F_2, and $a_l \in F_l$. Now clearly $\mathrm{tp}(\langle a_1, a_2 \rangle, \emptyset)$ has two dimensions, and the natural way is to try to decompose it. So (already in 3) for a stationary regular type p, we define $\mathrm{low}_p(\bar{a}, A)$ just as $w(\bar{a}, A)$, however it is not $|J|$ but $|\{\bar{c} \in J : \mathrm{tp}(\bar{c}, |M|)$ is not orthogonal to $p\}$. Note that then $w(\bar{a}, A) = \sum_p \mathrm{low}_p(\bar{a}, A)$, where among any family $\{p : p$ regular, stationary not orthogonal to $p_0\}$ just one appears. We call

$q = \text{tp}(\bar{a}, A)$ unidimensional if $q \leq_w p \Rightarrow p \leq_w q$; for superstable T this means that for some p, $p \leq_w q$, p regular. We may now ask:

(a) For $\bar{a}$ and A, and stationary regular p not orthogonal to $\text{tp}(\bar{a}, A)$, can we find equivalence relations E over A, such that $\text{tp}(\bar{a}/E, A)$ is unidimensional not orthogonal to p?

(b) Does the parallel of (∗) to low_p hold? i.e., let p be stationary and regular: $I = \{\bar{b}_i : i < \alpha\}$, then

$$\text{low}_p(\bigcup I, A) = \sum_{i<\alpha} \text{low}_p\Big(\bar{b}_i, A \cup \bigcup_{j<i} \bar{b}_j\Big).$$

(b′) Does (b) hold at least when $\bigcup I \subseteq q(\mathfrak{C})$, $q \in S^m(A)$, q unidimensional not orthogonal to p?

Unfortunately the answer to all those questions is negative (see Exercise 4.11). However some weakening has positive answers.

The most important notion of Section 4 is a semi-regular type (see Definition 4.1). Its important properties are:

(i) If p is semi-regular, it is unidimensional, so all the regular types not orthogonal to it are pairwise not orthogonal.

(ii) (b′) holds when q is semi-regular.

(iii) ($\mathfrak{C}^{eq}$) If $\bar{a} \notin \text{acl}\, A = A$, $\text{tp}(\bar{a}, A)$ not orthogonal to a minimal type q, then for some equivalence relation E over A, $\text{tp}(\bar{a}/E, A)$ is semi-regular (semi-minimal, in fact) not orthogonal to q.

(iv) ($\mathfrak{C}^{eq}$) If T is superstable, $\bar{a} \notin A = \text{acl}\, A$, then for some equivalence relation E over A, $\text{tp}(\bar{a}/E, A)$ is semi-regular (hence not algebraic) (see 4.11).

(v) If M is $\mathbf{F}^a_\kappa$-saturated, $\kappa \geq |A|^+ + \kappa_r(T)$, $p = \text{tp}(\bar{a}, A)$ semi-regular, then for some indiscernible set $I \subseteq M$ over A, based on p, $\text{Av}(I, A \cup \bigcup I) \vdash \text{Av}(I, |M|)$ (see 4.22).

Note that (v) would not hold for a non-unidimensional type. However if we are ready to wave (i), (v), we can replace semi-regular by $\text{cl}^2_A\{p\}$-simple (p some regular (and stationary) m-type) and (ii), (iv) holds, the family of such types is closed under supertypes, and a strengthening of (iv) holds.

(iv′) ($\mathfrak{C}^{eq}$) If $\bar{a} \notin A = \text{acl}\, A$, $\text{tp}(\bar{a}, A)$ is not orthogonal to the stationary regular type p, then for some equivalence relation E over A, $\text{tp}(\bar{a}/E, A)$ is $\text{cl}^2_A\{p\}$-simple and $\text{low}_p(\bar{a}/E, A) > 0$.

Note that the family of regular types fail to satisfy (iv), and it should be easy to find families satisfying (iv) (or (iv′)) but not (ii). The distinct roles of regular, semi-regular and simple will be exemplified in the proof of IX, 2.3 and 2.4. (Simplicity is defined in fact for a regular family of types, see Definition 4.3.)

Sections 3 and 4 are relevant mainly to superstable theories (as for stable theories we do not have existence theorems for regular types). So in Section 5 we return to stable theories. In Section 2 we have dealt with unidimensional theories, and in Section 5 we continue this by dealing with theories with a bound number of dimensions called non-multi-dimensional (this means, e.g., that $\{\dim(I, M): I \subseteq M\}$ M $\mathbf{F}^a_{\kappa_r(T)}$-saturated, has a bound λ_T). The main result is again a dichotomy (see 5.8 and 5.9):

(α) If T is multidimensional, $\aleph_\alpha \geq \aleph_\beta \geq \kappa_r(T)$, T stable in $\aleph_\alpha$, then T has at least $2^{|\alpha-\beta+1|}$ non-isomorphism $\mathbf{F}^a_{\aleph_\beta}$-saturated models of cardinality $\aleph_\alpha$.

(β) If T is not multidimensional, for every $\mathbf{F}^a_{\kappa_r(T)}$-saturated model there is $N \prec M$, $\|N\| \leq 2^{|T|}$, N $\mathbf{F}^a_{\kappa_r(T)}$-saturated, and $J \subseteq |M|$ independent over $|N|$, such that M is $\mathbf{F}^a_{\kappa_r(T)}$-prime over $|N| \cup \bigcup J$.

By (β) we get the number of $\mathbf{F}^a_{\aleph_\beta}$-saturated model of T of cardinality $\aleph_\alpha$ quite accurately (when $\alpha - \beta$ is big enough, T stable in $\aleph_\alpha$-accurately). In IX, 2.3 we get accurate results for superstable T.

The main lemma in the proof is as follows. We ask for an indiscernible set I over A, what can be $\{\dim(J,M): A \subseteq |M|, M\,\mathbf{F}^a_{\kappa_r(T)}\text{-saturated}, \operatorname{stp}_*(J, A) \equiv \operatorname{stp}_*(I, A)\}$. So we choose I_α realizing

$$\operatorname{stp}_*(I, A) \quad \text{such that} \quad \operatorname{tp}_*\Bigl(I_\alpha, A \cup \bigcup_{j<i} I_j\Bigr)$$

does not fork over A. Maybe we can choose the dimensions $\dim(I_\alpha, M)$ at random. Otherwise we prove that for some $\alpha < \kappa_r(T)$, $\{a_i : i < \alpha\}$ is independent over M, a_i realizes $\operatorname{Av}(I_i, M)$ and N is $\mathbf{F}^a_{\kappa_r(T)}$-prime over $M \cup \{a_i : i < \alpha\}$ then N realizes $\operatorname{Av}(I_\alpha, M)$ (see 5.3).

In Section 6 we try to deal with cardinality quantifiers, and two-cardinals theorems. Think of a complete (stable) theory T_1 in a language $L(\exists^{<\chi})$, χ regular; and we assume that every formula is equivalent to a predicate (in T). So we can replace T_1 by its first order part, T, together with the following requirement on the model M: if $(\forall \bar{y})[(\exists^{<\chi} x)\varphi(x, \bar{y}) \equiv \psi(\bar{y})] \in T_1$, φ, ψ first order then $M \models \psi[\bar{a}] \Leftrightarrow |\varphi(M, \bar{a})| < \chi$ for each $\bar{a} \in |M|$.

We generalize this somewhat in order to include more cases (e.g., we want just $P(x)$ to have a fixed cardinality). So T will be a (complete, stable) first order theory, $K = \langle \mathbf{F}, W, \lambda, \mu \rangle$ where: $\mathbf{F}$ is as in Chapter IV; W is a set of triples $\langle \varphi(x, \bar{y}), \psi(\bar{y}), \chi \rangle$ which means that for good M, $M \models \psi[\bar{b}] \Rightarrow |\varphi(M, \bar{a})| < \chi$ (note we demand only $\Rightarrow$ and not $\Leftrightarrow$).

We do not require anything from W, but for any formula $\varphi(x, \bar{a})$ we define $C(\varphi(x, \bar{a}))$, which is the minimal χ (or ∞) such that our previous requirements imply $|\varphi(M, \bar{a})| < \chi$ (for good M; we just notice algebraic formulas are $< \chi$, and the union of less than χ sets each of cardinality $<\chi$ is $<\chi$; we use just the rules corresponding to this). We formulate six conditions C1–6 from which we prove various results. For a function h, $h(\chi) < \chi$, we define an h-good model, assuming C3 it says that for $\bar{a} \in |M|$, $h(C(\varphi(x, \bar{a}))) \leq |\varphi(M, \bar{a})| < C(\varphi(x, \bar{a}))$. The main result is 6.7 which says that: any good model can be extended to an $\mathbf{F}$-saturated h-good model. Note that the question whether $|\varphi(M, \bar{a})| < \chi$ in any good M, is not an elementary property of $\bar{a}$, but is equivalent to "$\bar{a}$ does not satisfy a type p_χ".

Now if $\mathbf{F} = \mathbf{F}^a_\kappa$, $\kappa \geq \kappa_r(T)$, $\mu = \kappa_r(T)$, $\lambda = (2^{|T|})^+$ and $\lambda \leq \chi$, T stable in χ for each χ appearing in W, *then* all the conditions are satisfied. But there are other cases which interest us, where this is not the case. Hence we suggest another set of conditions C*1–8, which maybe are less nice looking. Their consequences are somewhat weaker, but they are satisfied in more cases. In 6.12 we prove the existence of h-good models. Note that in treating C*1–8, we use more good sets and less good models. In 6.14 we find when the various conditions are satisfied, and we go on to "applications" to two-cardinality quantifiers, which should exemplify the flexibility of our treatment. Those assertions fail in general for unstable theories.

THEOREM: (A) *If T is (stable and) countable, $M \prec N$, $M \neq N$, $P(M) = P(N)$ there is a model N^* of any cardinality $> \|N\|$, such that $N \prec N^*$, $P(N) = P(N^*)$ (see 6.14, the countability is essential; it is generalized to $\mathbf{F}$-saturated models).*

(B) *If $|P_l(M)| = \chi_l$, $\chi_0 < \cdots < \chi_n$, and $|T| \leq \chi^0 \leq \cdots \leq \chi^n$ then T has a model N, $|P_l(N)| = \chi^l$ (see 6.15).*

(C) *Suppose $\chi_0 < \cdots < \chi_n$ are regular, for any $\varphi(x, \bar{y})$ and l for some $\psi^l_\varphi(\bar{y})$, for every $\bar{a} \in M$, $M \models \psi^l_\varphi[\bar{a}] \Leftrightarrow |\varphi(M, \bar{a})| < \chi_l$. Then for any χ^l, $|T| < \chi^0 < \cdots < \chi^n$ T has a model N, such that for any φ, l and $\bar{a} \in |M|$, $|\varphi(M, \bar{a})| < \chi^l \Leftrightarrow M \models \psi^l_\varphi[\bar{a}]$ (See 6.15).*

(D) *If T is superstable, $|P(M)| < \lambda = \|M\|$, λ regular for simplicity, then there is $N \prec M$, $\|N\| = \lambda$, $|P(M)| \leq 2^{|T|}$ (see 6.16, in fact we can get $|P(M)| \leq |T|$; for T stable this fails).*

In Section 7 we return to ranks. First, we note that in Section 6, the property "$|\varphi(M, \bar{a})| < \chi$ for each good M" is quite similar to "$\varphi(x, \bar{a})$

is algebraic". Part of Section 7 is devoted to this generalization, so we have K-minimal, and other concepts in the same vein, and a rank $D(p, L, K)$ where now $D(p, L, K) \geq \alpha + 1$, for finite p, means that for any χ, for some good M, p has $> \chi$ pairwise contradictory extensions q over M, $D(q, L, K) \geq \alpha$. Some of the theorems are meaningful for the usual notation. In 7.3 we show that when $\varphi \leq_K \psi$ (i.e., for good M, $|\varphi(M)| \leq |\psi(M)| + \chi_{\varphi,\chi}$), then we can partition φ to "small" parts, indexed by ψ, and when φ (or ψ) is (weakly) K-minimal, we get better results. So in 7.4 we can characterize when $\varphi \leq_K \psi$. In 7.5 (and 7.7), we deal with $\varphi(x, \bar{y})$ and q such that

(**) for any good M, $\{|\varphi(M, \bar{a})| : \bar{a}$ realizes q, $\varphi(x, \bar{a})$ not K-algebraic$\}$ has cardinality $< \lambda_0$, where $\lambda_0 < \lambda_K$.

We show that then we can replace λ_0 by some $k < \omega$, q by a finite subtype. When $\varphi(x, \bar{a})$ is weakly K-minimal for $\bar{a}$ realizing q then also "$|\varphi(M, \bar{a})| = |\varphi(M, \bar{b})|$ for any good M, or $\varphi(x, \bar{a})$, $\varphi(x, \bar{b})$ are K-algebraic" is a first-order property of $\bar{a}^\frown\bar{b}$ with finitely many equivalence classes. It follows that $|\{|\varphi(M, \bar{a})| : \varphi(x, \bar{a})$ not K-algebraic$\}| < \lambda_0 < \lambda_K$ for good M, implies that "$\varphi(x, \bar{a})$ is not K-algebraic" is a first order property of $\bar{a}$. We can get (still assuming (**)) also that "$\varphi(x, \bar{a})$ is weakly K-minimal" is first-order property of $\bar{a}$ for the trivial K; and if $\varphi(x, \bar{a})$ is minimal, T totally transcendental, we can get in fact a similar result (see 7.7(2)).

Later we try to prove theorems of the form: if $\varphi(\mathfrak{C})$ is divided, the parts indexed by $\psi(\mathfrak{C})$, each part has a bounded rank, then we can bound φ's rank using ψ's rank and the bound on the ranks of the parts. More accurately, suppose

$$\theta(x, y) \vdash \varphi(x) \wedge \psi(y), \quad \varphi(x) \vdash (\exists y)\theta(x, y).$$

By 7.8, if $\beta = R[\psi(x), L, \aleph_0]$, $R[\theta(x, b), L, \aleph_0] \leq \alpha$ for every $b \in \psi(\mathfrak{C})$, then $R[\varphi(x), L, \aleph_0] \leq \alpha(\beta + 1)$ or $\alpha = 0$, $R[\varphi(x), L, \aleph_0] \leq \beta$. For the rank $R(-, L, \infty)$ we get a similar result when $\beta < \omega$, and a better bound $(\alpha + \beta)$ when $\alpha, \beta < \omega$. It follows (see 7.10) that when $\varphi \leq \psi$,

$$R(\psi, L, \aleph_0) < \omega \Rightarrow R(\varphi, L, \aleph_0) < \omega,$$

$$R(\psi, L, \infty) < \omega \Rightarrow R(\varphi, L, \infty) < \omega.$$

At last we deal with another rank, L, defined for complete types, without taking finite subtypes (so we lose the finite characteristic). We let $L(p) \geq \alpha + 1$ if $L(q) \geq \alpha$ for some extension q of p which forks over Dom p, and let $L(\bar{a}, A) = L(\text{tp}(\bar{a}, A))$. This ranks satisfies more than the above: $L(\bar{a}^\frown \bar{b}, A) \leq L(\bar{a}, A \cup \bar{b}) \oplus L(\bar{b}, A)$ ($\oplus$—the "natural sum" of ordinals, see Definition 7.6). The rank has other natural properties, e.g., T is superstable iff $(\forall p)L(p) < \infty$, if $\text{tp}(\bar{a}, A)$ does not fork over $B \subseteq A$ then $L(\bar{a}, A) = L(\bar{b}, A)$; $L(p) \leq R(p, L, \infty)$.

In this chapter T will be stable.

V.1. Orthogonality, regularity and minimality of types

DEFINITION 1.1: (1) If $p(\bar{x}_1)$, $q(\bar{x}_2)$ are complete types over A, p an m-type, q an n-type, we call p *weakly orthogonal* to q iff $p(\bar{x}_1) \cup q(\bar{x}_2)$ is complete (over A).

(2) Let p_1 be complete or stationary and p_2 complete or stationary. Then p_1 is *orthogonal* to p_2, if for every A, $\text{Dom}\, p_1 \cup \text{Dom}\, p_2 \subseteq A$, A the universe of a $\mathbf{F}^a_{\kappa(T)}$-saturated model, and any stationarizations q_l of p_l, $l = 1, 2$ over A; q_1 is weakly orthogonal to q_2 (see Definition III, 4.2(2)).

(3) The infinite indiscernible sets I, J are orthogonal if $\text{Av}(I, \bigcup I)$, $\text{Av}(J, \bigcup J)$ are orthogonal (see Lemma 1.1(3)).

(4) The type p is orthogonal to the set A if p is orthogonal to every complete type over A.

LEMMA 1.1: (1) *If p is algebraic then p is orthogonal to any type q (when both are stationary or complete). If $p \in S^m(A)$, p is realized by just one sequence, then p is weakly orthogonal to any complete type over A.*

(2) *If $p_l = p_l(\bar{x}_l) = \text{tp}(\bar{a}_l, A)$ then p_1 is weakly orthogonal to p_2 iff* $\text{tp}(\bar{a}_2, A) \vdash \text{tp}(\bar{a}_2, A \cup \bar{a}_1)$ *iff* $\text{tp}(\bar{a}_1, A) \vdash \text{tp}(\bar{a}_1, A \cup \bar{a}_2)$.

(3) *Weak orthogonality and orthogonality are symmetric relations.*

(4) *If* $\text{tp}(\bar{a}_1{}^\frown \bar{a}_2, A)$ *is [weakly] orthogonal to* $\text{tp}(\bar{b}_1{}^\frown \bar{b}_2, A)$ *then* $\text{tp}(\bar{a}_1, A)$ *is [weakly] orthogonal to* $\text{tp}(\bar{b}_1, A)$ *(for all $l(\bar{a}_1)$, $l(\bar{a}_2)$, $l(\bar{b}_1)$, $l(\bar{b}_2)$).*

(5) *If p_l $(l = 1, 2)$ are complete types over A, and they are not orthogonal, then for some finite B there are stationarizations q_l of p_l over $A \cup B$, such that q_1, q_2 are not weakly orthogonal.*

(6) *When $p_l = \text{tp}(\bar{a}_l, A_l)$, p_1 is orthogonal to p_2 iff for any $\bar{a}'_l$ realizing p_l,* $\text{stp}(\bar{a}'_1, A_1)$ *is orthogonal to* $\text{stp}(\bar{a}'_2, A_2)$.

Proof. Clearly (2) is a restatement of Ax(VI) for $\mathbf{F}^s_\infty$ (see Lemma IV, 2.9). Then (3) follows; and (4) follows from Ax(IV) for $\mathbf{F}^s_\infty$. Now (1), (5) are trivial and (6) is easy.

THEOREM 1.2: (1) *Let $B \subseteq A$, and for $l = 1, 2$, p_l is a complete type over A, p_l does not fork over B. If $p_2 \restriction B$ is stationary, p_1, p_2 are weakly orthogonal, then $p_1 \restriction B$, $p_2 \restriction B$ are weakly orthogonal.*

(2) *If p_1, p_2 are complete types over A, and they are orthogonal, and one of them is stationary then they are weakly orthogonal.*

(3) *If M is $\mathbf{F}^a_{\kappa(T)}$-saturated, p_l are complete types over M then p_1, p_2 are orthogonal iff p_1, p_2 are weakly orthogonal.*

(4) *If p_l is parallel to q_l for $l = 1, 2$ (hence they are stationary) then p_1, p_2 are orthogonal iff q_1, q_2 are orthogonal.*

Proof. (1) This is a restatement of Claim IV, 4.16.

(2) Follows by the definition and (1).

(3) By III, 2.15(2) p_1, p_2 are stationary, so if they are orthogonal, they are weakly orthogonal by (2).

So suppose they are not orthogonal. Then there is a $\mathbf{F}^a_{\kappa(T)}$-saturated N, $M \subseteq N$, and stationarizations q_l of p_l over N such that q_1, q_2 are not weakly orthogonal. By III, 3.2 and III, 2.9 there is $C \subseteq M$, $|C| < \kappa(T)$, such that p_l does not fork over C, $p_l \restriction C$ is stationary, hence q_l does not fork over C, hence q_l is definable over C, by III, 4.8(2).

There is $\bar{b} \in N$ such that $q_1(\bar{x}) \cup q_2(\bar{y}) \cup \{\varphi(\bar{x}, \bar{y}; \bar{b})^{\mathbf{t}}\}$ is consistent, for $\mathbf{t} \in \{0, 1\}$. Choose $\bar{b}_0 \in M$, $\operatorname{tp}(\bar{b}_0, C) = \operatorname{tp}(\bar{b}, C)$ (since M is $\mathbf{F}^a_{\kappa(T)}$-saturated). By the $\mathbf{F}^a_{\kappa(T)}$-saturation of N and the definability of q_l over C, $q_1(\bar{x}) \cup q_2(\bar{y}) \cup \{\varphi(\bar{x}, \bar{y}; \bar{b}_0)^{\mathbf{t}}\}$ is consistent (for $\mathbf{t} \in \{0, 1\}$) hence $p_1(\bar{x}) \cup p_2(\bar{y}) \cup \{\varphi(\bar{x}, \bar{y}; \bar{b}_0)^{\mathbf{t}}\}$ is consistent, hence p_1, p_2 are not weakly orthogonal.

(4) Choose a $\mathbf{F}^a_{\kappa(T)}$-saturated model M, $\operatorname{Dom} p_l$, $\operatorname{Dom} q_l \subseteq M$ for $l = 1, 2$. Let r_l be the stationarization of p_l (and q_l) over M. By the symmetry it suffices to prove r_1, r_2 are orthogonal iff p_1, p_2 are orthogonal. The "only if" can be proved by (1), (3) and the "if" by the definition.

LEMMA 1.3: *The types p, q are orthogonal (in $\mathfrak{C}$) iff $p^{\mathrm{eq}}, q^{\mathrm{eq}}$ are orthogonal in $\mathfrak{C}^{\mathrm{eq}}$.*

Proof. Use 1.2(3) for $\mathfrak{C}^{\mathrm{eq}}$, remembering $|M^{\mathrm{eq}}| = \operatorname{dcl}|M|$ in $\mathfrak{C}^{\mathrm{eq}}$.

LEMMA 1.4: (1) *Suppose $\{\bar{a}_i : i < n\}$ is independent over A (see* III, Definition 4.4(1)). *Then* $\operatorname{tp}(\bar{b}, A)$ *is orthogonal to* $\operatorname{tp}(\bar{a}_i; A)$ *for $i < n$ iff* $\operatorname{tp}(\bar{b}, A)$ *is orthogonal to* $\operatorname{tp}(\bar{a}_0{}^\frown\bar{a}_1{}^\frown\cdots{}^\frown\bar{a}_{n-1}, A)$.

(2) *If $p_i = \operatorname{tp}(\bar{a}_i, A)$ are stationary and pairwise orthogonal ($i < \alpha$) then $\bigcup_{i<\alpha} p_i(\bar{x}_i)$ is complete, or equivalently* $\operatorname{tp}(\bar{a}_i, A) \vdash \operatorname{tp}(\bar{a}_i, A \cup \bigcup_{j<i} \bar{a}_j)$.

Proof. (1) If $\operatorname{tp}(\bar{b}, A)$ is orthogonal to $\operatorname{tp}(\bar{a}_0 \frown \cdots \frown \bar{a}_{n-1}, A)$ then by 1.1(4) $\operatorname{tp}(\bar{b}, A)$ is orthogonal to $\operatorname{tp}(\bar{a}_l, A)$. For the other direction, by 1.1(6), 1.2(4) we can assume $A = |M|$, M is $\mathbf{F}^a_{\kappa(T)}$-saturated.

Now assume $\operatorname{tp}(\bar{a}_l, A)$, $\operatorname{tp}(\bar{b}, A)$ are orthogonal for $l < n$, and prove by induction on $k < n$ that $\operatorname{tp}(\bar{b}, A) \vdash \operatorname{tp}(\bar{b}, A \cup \bigcup_{l<k} \bar{a}_l)$. For $k = 0$ it is immediate; for $k \geq 1$ it follows by 1.2(2), (4). So $\operatorname{tp}(\bar{b}, A)$ and $\operatorname{tp}(\bar{a}_0 \frown \cdots \frown \bar{a}_{n-1}, A)$ are weakly orthogonal, hence by 1.2(3) they are orthogonal.

(2) The proof is immediate by (1).

CONCLUSION 1.5: (1) *Let I be an indiscernible set over A, based on A, and $\operatorname{tp}(\bar{c}, A)$, $\operatorname{Av}(I, A)$ are orthogonal. Then $\operatorname{tp}(\bar{c}, A)$, $\operatorname{tp}_*(I, A)$ are orthogonal (i.e., for $\bar{a}_l \in I$, $\operatorname{tp}(\bar{c}, A)$, $\operatorname{tp}(\bar{a}_0 \frown \cdots \frown \bar{a}_n, A)$ are orthogonal for all $n < \omega$).*

(2) *If I, J are indiscernible sets over A, based on A, and $\operatorname{Av}(I, A)$ is orthogonal to $\operatorname{Av}(J, A)$ then $\operatorname{tp}_*(I, A)$ is orthogonal to $\operatorname{tp}_*(J, A)$.*

Remark. For a strong converse to 1.5(2) see 2.7.

Proof. (1) By 1.4(1).

(2) By a double use of 1.5(1).

LEMMA 1.6: (1) *If for $i < \alpha$, $\operatorname{tp}(\bar{b}, A)$ is orthogonal to*

$$\operatorname{tp}\left(\bar{a}_i, A \cup \bigcup_{j<i} \bar{a}_j\right),$$

and $\operatorname{tp}(\bar{b}, A)$ is stationary then $\operatorname{tp}(\bar{b}, A) \vdash \operatorname{tp}(\bar{b}, A \cup \bigcup_{i<\alpha} \bar{a}_i)$. So $\operatorname{tp}(\bar{b}, A)$ is orthogonal to $\operatorname{tp}(\bar{a}_0 \frown \cdots \frown \bar{a}_n, A)$ (use (1.2(3)).

(2) *We can replace in* (1) tp *by* stp *and omit the assumption that $\operatorname{tp}(\bar{b}, A)$ is stationary.*

Proof. (1) Immediate by induction on α, using 1.2(2). The second part by 1.2(3).

(2) Can be easily proved in a similar way, or through $\mathfrak{C}^{eq}$.

LEMMA 1.7: *If p_i, $(i < \alpha)$ are pairwise orthogonal, but q is not orthogonal to p_i, q is stationary then $\alpha < \kappa(T)$, and even $< \kappa_{\text{ind}}(T)$.*

Proof. By 1.2(4), 1.1(6) we can assume q, p_i are complete types over a $\mathbf{F}^a_{\kappa(T)}$-saturated model M. Then use IV, 4.12(3).

DEFINITION 1.2: (1) A non-algebraic type $p \in S^m(A)$ is regular if for every $B \supseteq A$ and r such that $p \subseteq r \in S^m(B)$; if r forks over A then p, r are orthogonal.

(2) A stationary type p is regular if its stationarization over Dom p is regular.

LEMMA 1.8: (1) *If p, q are parallel (and stationary) then p is regular iff q is regular.*

(2) *If* $\text{tp}(\bar{a}, A)$ *is regular, then* $\text{stp}(\bar{a}, A)$ *is regular.*

(3) *If* $\text{tp}(\bar{a}, A) = \text{tp}(\bar{b}, A)$, *then* $\text{stp}(\bar{a}, A)$ *is regular iff* $\text{stp}(\bar{b}, A)$ *is regular.*

(4) $\text{tp}(\bar{a}, A)$ *is regular in* $\mathfrak{C}$ *iff it is regular in* $\mathfrak{C}^{eq}$.

Proof. (1) By the definition, we can assume p, q are complete. Let $C =$ Dom $p \cup$ Dom q, and let r be the stationarization of p (and q) over C. By the symmetry it suffices to prove that r is regular iff p is regular.

If p is regular, and $r \subseteq r^1 \in S^m(A)$, r^1 forks over $C \subseteq A$, then r^1 forks over Dom p, hence is orthogonal to r.

So suppose p is not regular; then there are $p_1 \in S^m(B)$, Dom $p \subseteq B$, $p \subseteq p_1, p_1$ forks over Dom p but p_1 is not orthogonal to p. We now know p and r are stationary and parallel. By III, 2.6(3) there is an elementary mapping f of $\mathfrak{C}$ such that $f \restriction \text{Dom } p =$ the identity, and for every $\bar{b} \in B$ $\text{stp}(\bar{b}, \text{Dom } p) \equiv \text{stp}(f(\bar{b}), \text{Dom } p)$, and $\text{tp}(f(\bar{b}), C)$ does not fork over Dom p. Hence $p_2 = f(p_1)$ is not orthogonal to p, so p_2 is not orthogonal to r. Let p_3 be the stationarization of p_2 over $C \cup$ Dom p_2; $p_3 \restriction C$ does not fork over Dom p, by Ax(VII) for $\mathbf{F}^l_{\infty}$. But $p \subseteq p_2 \subseteq p_3$, hence, as p is parallel to r, $r \subseteq p_3$. Now p_3, being parallel to p_2, is not orthogonal to r. This shows r is not regular.

(2), (3) Immediate.

(4) By 1.3.

QUESTION 1.1: Does the converse of (2) hold?

THEOREM 1.9: *Let $p \in S^m(A)$. The following conditions are equivalent:*

(1) *p is regular.*

(2) *If $A \subseteq B$, $I \cup \{\bar{c}\}$ a set of sequences realizing p and for every $\bar{a} \in I$,* $\text{tp}(\bar{a}, B)$ *forks over A, and* $\text{tp}(\bar{c}, B \cup \bigcup I)$ *forks over A, then* $\text{tp}(\bar{c}, B)$ *forks over A.*

(3) *If $B = A \cup \bigcup J$, $J \cup \{\bar{a}, \bar{c}\}$ is a set of sequences realizing p, and* $\text{tp}(\bar{a}, B)$ *forks over A and* $\text{tp}(\bar{c}, B \cup \bar{a})$ *forks over A, then* $\text{tp}(\bar{c}, B)$ *forks over A.*

(4) *If $\kappa \geq \kappa_r(T)$, $\kappa > |A|$, $A \subseteq M$, M is $\mathbf{F}^a_\kappa$-saturated, $p \subseteq q = \mathrm{tp}(\bar{a}, M)$, q does not fork over A, and N is $\mathbf{F}^a_\kappa$-prime over $M \cup \bar{a}$; then for no $\bar{b} \in |N| - |M|$, $\bar{b}$ realizes p and $\mathrm{tp}(\bar{b}, M)$ forks over A. (If p is stationary, then for all $\bar{b} \in |N| - |M|$, $\bar{b}$ realizes $p \Rightarrow \bar{b}$ realizes q).*

(5) *The hypothesis and conclusion of (4) hold for some M, $\bar{a}$, N, κ.*

Remark. If p is stationary, we can in (4) omit the condition $\kappa > |A|$.

QUESTION 1.2: Can we in (4) replace $\mathbf{F}^a_\kappa$ by $\mathbf{F}^s_\kappa$? Can we replace $\kappa_r(T)$ by $\kappa(T)$?

Proof. (1) $\Rightarrow$ (2) Suppose p is regular, $A \subseteq B$, $\mathrm{tp}(\bar{a}, B)$ forks over A for every $\bar{a} \in I$, and $\mathrm{tp}(\bar{c}, B)$ does not fork over A, and we shall prove $\mathrm{tp}(\bar{c}, B \cup \bigcup I)$ does not fork over A (clearly this is sufficient). By III, 2.6(1) and III, 1.4 $\mathrm{stp}(\bar{c}, B)$ has an extension q, $q \supseteq r \in S^m(B \cup \bigcup I)$, which does not fork over A. Hence it suffices to prove $\mathrm{stp}(\bar{c}, B) \vdash \mathrm{stp}(\bar{c}, B \cup \bigcup I)$, so it suffices to prove this for finite I. Let $I = \{\bar{a}_i : i < n\}$; so as $\mathrm{tp}(\bar{a}_i, B \cup \{a_j : j < i\})$ forks over A and extends p, it is orthogonal to p, hence to $\mathrm{stp}(\bar{c}, B)$. Hence by 1.6(2), $\mathrm{stp}(\bar{c}, B) \vdash \mathrm{stp}(\bar{c}, B \cup \bigcup I)$.

(2) $\Rightarrow$ (3) Immediate, as (3) is a particular case of (2); taking $I = \{\bar{a}\}$.

Before proving (3) $\Rightarrow$ (4) we need:

CLAIM 1.10: (1) *If M is $\mathbf{F}^a_\kappa$-saturated, $\kappa \geq \kappa_r(T)$, $p = \mathrm{tp}(\bar{a}, |M|)$ is orthogonal to $r \in S^m(|M|)$, r is non-algebraic and N is $\mathbf{F}^a_\kappa$-prime over $|M| \cup \bar{a}$, then r has a unique extension in $S^m(|N|)$, hence is not realized in N.*

(2) *Moreover, for every $\bar{b} \in |N|$, $\bar{b} \notin |M|$, $\mathrm{tp}(\bar{b}, |M| \cup \bar{a})$ forks over $|M|$.*

Proof of 1.10. (1) As M is $\mathbf{F}^a_\kappa$-saturated, there is an infinite indiscernible set I, $|I| = \kappa$, $I \subseteq |M|$, $\mathrm{Av}(I, |M|) = r$. Hence by the hypothesis and IV, 4.10(2) (remembering $\kappa \geq \kappa_r(T)$) $\mathrm{Av}(I, |M|) = r \vdash \mathrm{Av}(I, |M| \cup \bar{a}) \vdash \mathrm{Av}(I, |N|)$.

(2) The same proof.

Continuation of the proof of 1.9. (3) $\Rightarrow$ (4) Suppose (4) fails, and let $r = \mathrm{tp}(\bar{b}, |M|)$, which is stationary; hence by the claim $\mathrm{tp}(\bar{b}, |M| \cup \bar{a})$ forks over $|M|$, hence $\mathrm{tp}(\bar{a}, |M| \cup \bar{b})$ forks over $|M|$ (by Ax(VI) for $\mathbf{F}^f_\infty$). Let $J = \{\bar{c} : \bar{c} \in |M|, \bar{c}$ realizes $p\}$, $B = A \cup \bigcup J$; we shall show that A, B, $\bar{a}$ (for $\bar{c}$) and $\bar{b}$ (for $\bar{a}$) form a counter-example to (3). Clearly

$\operatorname{tp}(\bar{a}, B)$ does not fork over A, and $\operatorname{tp}(\bar{a}, |M| \cup \bar{b})$ forks over A, and $\operatorname{tp}(\bar{b}, |M|)$ forks over A. We need only to replace M by B in the last two types. By the following claim, $\operatorname{tp}(\bar{a}^\frown\bar{b}, |M|)$ does not fork over B, thus by III, 4.14, the result follows.

CLAIM 1.11: *If $x = s, t, a$, M a $\mathbf{F}^x_\lambda$-compact model, p_i $(i < n)$ $\mathbf{F}^x_\lambda$-types over $|M|$, and $\bar{a}_i$ realizes p_i, and $B_i = \bigcup \{\bar{c}\colon \bar{c} \in M,\ \bar{c}$ realizes $p_i\}$, then $\operatorname{tp}(\bar{a}_0{}^\frown \cdots {}^\frown\bar{a}_n, \bigcup_{i<n} \operatorname{Dom} p_i \cup \bigcup_{i<n} B_i) \vdash \operatorname{tp}(\bar{a}_0{}^\frown \cdots {}^\frown\bar{a}_n, M)$.*

Proof. W.l.o.g. p_i is a 1-type (if it is an m type, let

$$p_i^l = \{(\exists x_0 \cdots x_{l-1} x_{l+1} \cdots x_{m-1}) \bigwedge q\colon q \text{ a finite subset of } p_i\}$$

and replace the p_i's by the p_i^l's and the $\bar{a}_i$'s by $\bar{a}_i[l]$). We can also assume $p_i = p$ (replace p_i by $p = \{\bigvee_{i<n} \varphi_i\colon \varphi_i \in p_i\}$, clearly an element realizes p iff it realizes some p_i).

Let $B = B_i$, $\bar{a}_i = \langle a_i \rangle$, $\bar{a} = \langle a_0, \ldots, a_{n-1}\rangle$. Suppose $\models\varphi[\bar{a}, \bar{c}]$, $\bar{c} \in M$; then by II, 2.2 and II, 2.13 for some $\bar{b} \in B$ and ψ, $\models\varphi[\bar{a}', \bar{c}] \Leftrightarrow \models\psi[\bar{a}', \bar{b}]$ for any $\bar{a}' \in B$. As $M \prec \mathfrak{C}$, M $\mathbf{F}^x_\kappa$-compact, $\models\psi[\bar{a}, \bar{b}]$.

Suppose $\operatorname{tp}(\bar{a}, B \cup \operatorname{Dom} p) \nvdash \varphi(\bar{x}, \bar{c})$, so some $\bar{a}^*$ realizes $\operatorname{tp}(\bar{a}, B \cup \operatorname{Dom} p)$ (hence $\bar{a}^*[l]$ realizes p) but $\models\neg\varphi(\bar{a}^*, \bar{c})$. So,

$$q = \{\varphi(x_l; \bar{b})\colon l < n,\ \varphi(x; \bar{b}) \in p\} \cup \{\psi(x_0, \ldots, x_{n-1}; \bar{b}), \neg\varphi(x_0, \ldots, x_{n-1}, \bar{c})\}$$

is realized by $\bar{a}^*$. Clearly q is a $\mathbf{F}^x_\lambda$-type, hence q is realized in M, by $\bar{a}^{**}$. So $\bar{a}^{**} \in B$, $\models\psi[\bar{a}^{**}, \bar{b}]$, $\models\neg\varphi[\bar{a}^{**}, \bar{c}]$ contradiction to the definition of ψ.

Continuation of the proof of 1.9. (4) $\Rightarrow$ (5) Trivial.

(5) $\Rightarrow$ (1) Suppose p is not regular (i.e., (1) fails), $\kappa \geq \kappa_r(T) + |A|^+$, $A \subseteq |M|$, M $\mathbf{F}^a_\kappa$-saturated, and $\operatorname{tp}(\bar{a}, |M|)$ is a stationarization of p and N a $\mathbf{F}^a_\kappa$-prime model over $|M| \cup \bar{a}$. As p is not regular there is B, $A \subseteq B$, and $q \in S^m(B)$, $p \subseteq q$, q forks over A, q is not orthogonal to p. We can assume that $|B - A| < \min\{\kappa(T), \aleph_1\}$, and q is stationary.

As M is $\mathbf{F}^a_\kappa$-saturated, we can assume $B \subseteq M$, and let $q_1 \in S^m(|M|)$ be a stationarization of q. So q_1, $\operatorname{tp}(\bar{a}, |M|)$ are not orthogonal, (by 1.2(4)). By 1.2(3) there is a $\bar{b}$ realizing q_1 such that $\operatorname{tp}(\bar{b}, |M| \cup \bar{a})$ forks over A (as q_1 is stationary, being over a model). So (by III, 1.1(5)), for some $\psi, \bar{c} \in M$, $\models\psi(\bar{b}, \bar{a}, \bar{c})$ and $\psi(\bar{x}, \bar{a}, \bar{c})$ forks over $|M|$. Clearly $q \cup \{\psi(\bar{x}, \bar{a}, \bar{c})\}$ is a consistent $\mathbf{F}^a_\kappa$-type over N, hence realized by some $\bar{b} \in N$. This $\bar{b}$ contradicts (5), so not (1) implies not (5). This completes the proof of Theorem 1.9.

THEOREM 1.12: *Suppose p, q are complete types over $|M|$, $\bar{b}$ realizes q, M is $\mathbf{F}^a_\kappa$-saturated, $\kappa \geq \kappa_r(T)$, N is $\mathbf{F}^a_\kappa$-prime over $|M| \cup \bar{b}$. If p, q are not orthogonal and p is regular, then p is realized in N.*

Proof. By 1.2(3), as p, q are not orthogonal, they are not weakly orthogonal. Hence p has an extension $p_1 \in S^m(|M| \cup \bar{b})$, which forks over $|M|$; let $\bar{a}$ realize p_1 ($\bar{a}$ is not necessarily in N). Choose $C \subseteq |M|$, $|C| < \kappa(T)$ such that $p = \text{tp}(\bar{a}, |M|)$ does not fork over C and $\text{tp}(\bar{a}, C)$ is stationary (by the saturation of M). Let $B = \bigcup \{\bar{c}: \bar{c} \in N, \bar{c} \text{ realizes } p \restriction C\}$. Then $\text{tp}(\bar{a}, C \cup (B \cap M))$ does not fork over C. If $\text{tp}(\bar{a}, C \cup B)$ does not fork over C then by 1.11 also $\text{tp}(\bar{a}, N)$ does not fork over C, contradiction as $\bar{b} \in N$. So $\text{tp}(\bar{a}, C \cup B)$ forks over C, hence over $C \cup (B \cap M)$ (by III, 4.4). So for some $\bar{d}_0, \ldots, \bar{d}_n \in N$ realizing p, and $\bar{c} \in C$, and φ, $\varphi(\bar{x}, \bar{d}_0, \ldots, \bar{d}_n, \bar{c}) \in \text{tp}(\bar{a}, N)$ forks over $C \cup (B \cap M)$. If for every i the type $\text{tp}(\bar{d}_i, C \cup (B \cap M) \cup \bigcup_{j<i} \bar{d}_j)$ fork over C then it would be orthogonal to p (as p is regular) hence by 1.6

$$\text{tp}(\bar{a}, C \cup (B \cap M)) \vdash \text{tp}\left(\bar{a}, C \cup (B \cap M) \cup \bigcup_{i \leq n} \bar{d}_i\right).$$

But the second type forks over $C \cup (B \cap M)$ contradiction. So the $\bar{d}_i$ for which the above type does not fork over C realizes p and is, of course, in N.

CONCLUSION 1.13: (1) *The relation of non-orthogonality among the stationary regular types, is an equivalence relation.*

(2) *If p, q are stationary, regular, and not orthogonal, then r is orthogonal to p iff r is orthogonal to q.*

Proof. (1) Reflexivity is trivial (using non-algebraity). Symmetry was proved in 1.1; and transitivity is immediate by 1.12 (going to parallel types).

(2) Also easy.

Remember that $\bar{a}$ depends on I over A if $\text{tp}(\bar{a}, A \cup \bigcup I)$ forks over A.

THEOREM 1.14: *For a set A the relation of dependence over A among sequences $\bar{c}$ for which* $\text{stp}(\bar{c}, A)$ *is regular satisfies the axioms of dependency relation* (Def. AP 3.3). *If p_0 is regular, on* $\{\bar{c}: \text{stp}(\bar{c}, A)$ *regular not orthogonal to p_0\} it is a nice dependence relation.*

Proof. (1) The exchange principle. It suffices to show that if $\{a_j: j < \alpha\}$ is not independent over A, then some $\bar{a}_j$ depends on $\{\bar{a}_i: i < j\}$ over A. This is III, 4.19(5).

(2) The finite character. That is, if $\bar{a}$ depends on I over A then $\bar{a}$ depends on some finite $J \subseteq I$ over A. Easy, see III, 1.1(5).

(3) Transitivity: Here we use regularity. Suppose $\bar{a}$ depends on I over A, each $\bar{b} \in I$ depends on J over A and I, J are independent over A. We can replace I by a finite subset, and then J by a finite subset, without changing our assumptions. We can work in $\mathfrak{C}^{eq}$, and replace A by acl A, (see 1.8(4) and III, 6.3(3)), so every complete type over A is stationary. Let $I_1 = \{\bar{b}: \bar{b} \in I, \operatorname{tp}(\bar{b}, A)$ is not orthogonal to $\operatorname{tp}(\bar{a}, A)\}$, $J_1 = \{\bar{c}: \bar{c} \in J, \operatorname{tp}(\bar{c}, A)$ is not orthogonal to $\operatorname{tp}(\bar{a}, A)\}$. Now $\bar{a}$ depends on I_1, for otherwise let $I - I_1 = \{\bar{b}_l: l < n\}$, and as $\operatorname{tp}(\bar{a}, A \cup \bigcup I_1)$ does not fork over A and is stationary 1.6 shows $\operatorname{tp}(\bar{a}, A \cup \bigcup I_1) \vdash \operatorname{tp}(\bar{a}, A \cup \bigcup I_1 \cup \bigcup \{\bar{b}_l: l < n\})$. Hence the latter does not fork over A, contradiction. Similarly, every $\bar{b} \in I_1$ depends on J_1 over A. Let $I_1 = \{\bar{b}^i: i < k\}$, and assume $\bar{a}$ does not depend on J_1 over A. As $\operatorname{tp}(\bar{b}^i, A \cup \bigcup J_1)$ forks over A, $\operatorname{tp}(\bar{b}^i, A \cup \bigcup J_1 \cup \bigcup \{\bar{b}^j: j < i\})$ forks over A, hence (by the regularity of $\operatorname{tp}(\bar{b}^i, A)$) is orthogonal to $\operatorname{tp}(\bar{b}^i, A)$, hence to $\operatorname{tp}(\bar{a}, A)$, hence to $\operatorname{tp}(\bar{a}, A \cup \bigcup J_1)$ (using 1.13, 1.8(1) and 1.2(4)). So by 1.6 $\operatorname{tp}(\bar{a}, A \cup \bigcup J_1) \vdash \operatorname{tp}(\bar{a}, A \cup \bigcup J_1 \cup \bigcup I_1)$, so the latter does not fork over A, so $\operatorname{tp}(\bar{a}, A \cup \bigcup I_1)$ does not fork over A, contradiction.

(4) Full transitivity (for $\{\bar{c}: \operatorname{stp}(\bar{c}, A)$ regular not orthogonal to $p_0\}$): Repeat the last part of the proof of (3).

CONCLUSION 1.15: (1) *If* $p \in S^m(A)$, $A \subseteq M$, p *regular* (*or every stationarization of* $\bar{a}$ *is regular*) *then* $\dim(p, M)$ *is true.*

(2) *If* $\bar{c} \in I \Rightarrow \operatorname{stp}(\bar{c}, A)$ *is regular*, J_1, J_2 *are maximal subsets of* I *independent over* A, *then* $|J_1| = |J_2|$.

Proof. By the previous theorem and AP 3.10.

LEMMA 1.16: *Let* $A \subseteq |M|, p \in S^m(A), I \subseteq |M|$ *a maximal independent set over* A *of sequences realizing* p *in* M, p *regular*, M *is* $\mathbf{F}^a_\kappa$*-saturated*, $|A|^+ + \kappa(T) \leq \kappa$, κ *regular.*

(1) *For every stationarization* $q = \operatorname{tp}(\bar{a}, |M|)$ *of* p *over* M, $\operatorname{stp}(\bar{a}, A \cup \bigcup I) \vdash q$, *and if* p *is stationary*, $q \restriction (A \cup \bigcup I) \vdash q$.

(2) *For every* $\bar{c} \in M$ *there is* $J \subseteq I$, $|J| < \kappa(T)$ *such that* $\operatorname{tp}(\bar{c}, A \cup \bigcup J) \vdash \operatorname{tp}(\bar{c}, A \cup \bigcup I)$.

(3) *Let* p *be regular*, M_i $(i \leq \alpha)$ $\mathbf{F}^a_\kappa$*-saturated models*, $i < j \Rightarrow M_i \prec M_j$; M_δ *is* $\mathbf{F}^a_\kappa$*-primary over* $\bigcup_{i<\delta} |M_i|$, $M_0 = M$, *and* $I_i \subseteq |M_{i+1}|$ *is a maximal independent set over* $|M_i|$ *of sequences realizing some stationarization of* p *over* $|M_i|$. *Then* $\bigcup_{i<\alpha} I_i \cup I$ *is a maximal subset of* M_α *independent over* A *of sequences realizing* p.

Proof. Easy, using 1.11 and 1.14.

DEFINITION 1.3: (1) An m-type p is minimal if p is not algebraic but for no $\varphi(\bar{x}, \bar{a})$ are $p \cup \{\varphi(\bar{x}, \bar{a})^t\}$, $t = 0, 1$, both non-algebraic.

(2) An infinite indiscernible set I is minimal if $\mathrm{Av}(I, \bigcup I)$ is minimal.

(3) A formula $\varphi = \varphi(\bar{x}, a)$ is weakly minimal if φ is not algebraic but it has a bounded $(<\bar{\kappa})$ number of non-algebraic pairwise contradictory extensions.

THEOREM 1.17: (1) *The m-type p is minimal iff there is a finite* Δ_0 *such that for every finite* $\Delta \supseteq \Delta_0$, $R^m(p, \Delta, \aleph_0) = 1$, $\mathrm{Mlt}(p, \Delta, \aleph_0) = 1$ *iff for every* φ, $R^m(p, \varphi, \aleph_0) \leq 1$, *and* $R^m(p, \varphi, \aleph_0) = 1 \Rightarrow \mathrm{Mlt}(p, \varphi, \aleph_0) = 1$, *and for some* ψ, $R^m(p, \psi, \aleph_0) = 1$. *If p is complete,* $\mathrm{Mlt}(p, \varphi, \aleph_0) = 1$ *for every* φ.

(2) *The m-type p is algebraic iff for every* Δ, $R^m(p, \Delta, \aleph_0) = 0$ *iff for every* φ, $R^m(p, \varphi, \aleph_0) = 0$ *iff* $R^m(p, \mathrm{L}, \lambda) = 0$ (*any* $\lambda \geq \aleph_0$) *iff* $D^m(p, \mathrm{L}, \infty) = 0$.

(3) *The formula* $\varphi(\bar{x}; \bar{a})$ *is weakly minimal iff* $D^m[\varphi(\bar{x}; \bar{a}), \mathrm{L}, \infty] = 1$. *If p is stationary,* $\varphi(\bar{x}, \bar{a}) \in p$ *is weakly minimal, p is not algebraic, then p is minimal.*

(4) *If* $R^m(p, \mathrm{L}, \aleph_0) = 1$, $\mathrm{Mlt}(p, \mathrm{L}, \aleph_0) = 1$, *then p is minimal; and the converse holds for finite p.*

Proof. (1) Suppose p is minimal, then let $\Delta_0 = \{\bar{x} = \bar{y}\}$. As p is not algebraic, there are distinct $\bar{a}_n$ $(n < \omega)$ which realizes it, hence $p \cup \{\bar{x} = \bar{a}_n\}$ is consistent and pairwise contradictory. So for every $\Delta \supseteq \Delta_0$, $R^m(p, \Delta, \aleph_0) \geq 1$. Now it is easy to prove (by (2)) that if Δ is finite, $R^m(p, \Delta, \aleph_0) > 1$ or $R^m(p, \Delta, \aleph_0) = 1$, $\mathrm{Mlt}(p, \Delta, \aleph_0) > 1$ then for some $\varphi(\bar{x}, \bar{y}) \in \Delta$, and $\bar{a}$, $p \cup \{\varphi(\bar{x}, \bar{a})^t\}$ is not algebraic for $t = 0, 1$, contradicting the minimality of p. This proves that the first condition implies the second and third. The other direction is easy.

(2) If p is algebraic, some finite $q \subseteq p$ is realized only by $n < \aleph_0$ sequences; hence q does not have $n + 1$ pairwise contradictory extensions, and no $kn + 1$, k-contradictory extensions. On the other hand if p is not algebraic, for every λ there are distinct $\bar{a}_i$ $(i < \lambda)$ which realizes p, so $p \cup \{\bar{x} = \bar{a}_i\}$ are consistent and the formulas $\bar{x} = \bar{a}_i$ are pairwise contradictory. Observing this, (2) is easy.

(3), (4) Easy too.

THEOREM 1.18: (1) *if p is a minimal m-type over A, then it has a unique non-algebraic extension* $q \in S^m(A)$.

(2) *If p is a minimal type, then there is a unique (up to equivalence) non-trivial infinite indiscernible set of sequences realizing p.*

(3) *If p is minimal, $p \subseteq q$, q forks over* Dom p *then q is algebraic.*

(4) *If $p \in S^m(A)$ is minimal, then p is stationary and regular.*

(5) *A complete type parallel to a minimal type is minimal.*

(6) *If* $\mathrm{tp}(\langle a_1, \ldots, a_n \rangle, A)$ *is minimal, then for some l $\bar{a}$ is algebraic over $A \cup \{a_l\}$ (where $\bar{a} = \langle a_1, \ldots, a_n \rangle$).*

(7) *If M is $\mathbf{F}^t_\lambda$-compact, $\lambda_3(T) \le \lambda$, p an m-type over M, then p has an extension q, $|q| < \lambda$, even $|q - p| < \lambda_3(T)$, q a type over M, q is minimal.*

Remark. In view of (1), (4) and (5), we shall call minimal types stationary, and extend the definition of parallel accordingly; (and also accompanying concepts, "I based on p").

Proof. (1) As p is not algebraic, we can find $> 2^{|A|+|T|}$ elements realizing it, so infinitely many of them realizes some $q \in S^m(A)$, which is the right extension. Clearly it is unique.

(2) As p is not algebraic, there is at least one such I. If I, J are two such sets, let Dom $p \cup I \cup J \subseteq M$; then $\mathrm{Av}(I, M)$, $\mathrm{Av}(J, M)$ are non-algebraic extensions of p in $S^*(M)$, hence they are equal, hence (by III, 4.17(1)), I, J are equivalent.

(3) Follows by 1.17(1) and III, 1.2(4) and 1.17(2).

(4) By III, 4.2, and 1.17(1) p is stationary. By 1.1 every type is orthogonal to an algebraic type hence p is regular.

(5) Easy, using III, 4.2.

(6) By III, 6.2(5) for some l, $a_l \notin \mathrm{acl}\, A$. If $a_t \notin \mathrm{acl}\,(A \cup \{a_l\})$, then $\mathrm{tp}(\bar{a}, A \cup \{a_l\})$ forks over A (as $a_l \notin \mathrm{acl}\, A$, $x_l = a_l$ forks over A) but it is not algebraic, contradicting (3).

(7) Easy, like IV, 2.16(2), remembering the finite character of being an algebraic type.

EXERCISE 1.3: In 1.16(1), we can get $q \restriction (A \cup \bigcup I) \vdash q$ when some $\bar{a}_1 \in I$ realizes $\mathrm{stp}(\bar{a}, A)$.

EXERCISE 1.4: Suppose $\mathrm{tp}(\bar{a}^\frown \bar{b}, A)$ is regular, $\bar{a} \notin \mathrm{acl}\, A$. Prove $\mathrm{tp}(\bar{a}, A)$ is regular. [Hint: Let $A \subseteq |M|$, M $\mathbf{F}^a_\lambda$-saturated, $\lambda > |T| + |A|$, and w.l.o.g. $\mathrm{tp}(\bar{a}^\frown \bar{b}, |M|)$ does not fork over A. Suppose $\mathrm{tp}(\bar{a}_1, |M|)$ extend $\mathrm{tp}(\bar{a}, A)$ and fork over A; and it suffices to prove that $\mathrm{tp}(\bar{a}, |M|)$, $\mathrm{tp}(\bar{a}_1, |M|)$ are weakly orthogonal. Choose $\bar{b}_1$ such that $\mathrm{tp}(\bar{a}^\frown \bar{b}, A) = \mathrm{tp}(\bar{a}_1{}^\frown \bar{b}_1, A)$. Clearly also $\mathrm{tp}(\bar{a}_1{}^\frown \bar{b}_1, |M|)$ forks over A. Hence

$\text{tp}(\bar{a}_1{}^\frown\bar{b}_1, |M|)$ is weakly orthogonal to $\text{tp}(\bar{a}{}^\frown\bar{b}, |M|)$. So by 1.1(4) $\text{tp}(\bar{a}_1, |M|)$, $\text{tp}(\bar{a}, |M|)$ are weakly orthogonal as desired.]

EXERCISE 1.5: Suppose M is $\mathbf{F}^a_\kappa$-saturated, $\kappa \geq \kappa_r(T)$, $\text{tp}(\bar{a}_l, |M|)$ are orthogonal and $\text{tp}(\bar{b}_l, |M| \cup \bar{a}_l)$ are $\mathbf{F}^a_\kappa$-isolated (for $l = 1, 2$). Prove $\text{tp}(\bar{a}_l{}^\frown\bar{b}_l, |M|)$ ($l = 1, 2$) are orthogonal; and generalize by replacing the sequences by sets. (Hint: Use 3.2.)

V.2. Dimensions and orders between indiscernible sets

DEFINITION 2.1: (1) Let I, J be infinite indiscernible sets. $I \leq_w J$ if for every $\kappa \geq \kappa_r(T)$ and $\mathbf{F}^a_\kappa$-saturated model M and $I_1, J_1 \subseteq M$ (infinite) indiscernible sets equivalent to I, J respectively, $\dim(I_1 M) \leq \dim(J_1, M)$ (w stands for "weakly").

(2) Let I, J be infinite indiscernible sets, $I \leq_s J$ when for every $\kappa \geq \kappa_r(T)$ and $\mathbf{F}^a_\kappa$-saturated model M and $I_1, J_1 \subseteq M$ equivalent to I, J respectively and $\mathbf{F}^a_\kappa$-saturated model N, $M \prec N$ if $\text{Av}(I_1, M)$ is realized in N then also $\text{Av}(J_1, M)$ is realized in N (s stands for "strongly").

(3) For stationary types p, q, $p \leq_w q$ ($p \leq_s q$) if the corresponding relations hold between indiscernible sets based on them.

LEMMA 2.1: (1) *For any infinite indiscernible set* I, $I \leq_s I$, $I \leq_w I$.

(2) $\leq_w$, $\leq_s$ *are transitive.*

(3) $I \leq_s J \Rightarrow I \leq_w J$.

(4) *If* I_1, I_2 *are equivalent and* J_1, J_2 *are equivalent then for* $x = s, w$, $I_1 \leq_x J_1 \Leftrightarrow I_1 \leq_x J_2$.

Proof. (1) Immediate.

(2) Suppose I_1, I_2, I_3 are infinite indiscernible sets, M a $\mathbf{F}^a_\kappa$-saturated model, $I'_1 \subseteq M$, $I'_3 \subseteq M$ and I'_1, I'_3 are equivalent to I_1, I_3, respectively (where $\kappa \geq \kappa_r(T)$), w.l.o.g. $I_1 = I'_1$, $I_3 = I'_3$.

We can find, for $l = 1, 3$ sets $A_l \subseteq \bigcup I_l$, such that I_l is based on A_l, $\text{Av}(I_l, A_l)$ is stationary and $|A_l| < \min\{\kappa(T), \aleph_1\}$ (see III, 4.17(4)). So we can find an elementary mapping f, $f \restriction (A_1 \cup A_3) =$ the identity, and f maps A_2 into M(B exists as M is $\mathbf{F}^a_\kappa$-saturated).

So we can find in M an infinite indiscernible set I'_2 based on $f(\text{Av}(I_2, A_2))$. Clearly, I'_2 is not necessarily equivalent to I_2. But for $x = w, s$, $I_2 \leq_x I_3$ iff $I'_2 \leq_x I_3$; and $I_1 \leq_x I_2$ iff $I_1 \leq_x I'_2$.

So if $I_1 \leq_w I_2 \leq_w I_3$, $\dim(I_1, M) \leq \dim(I'_2, M)$ and $\dim(I'_2, M) \leq \dim(I_3, M)$. As M, I'_1, I'_2 were arbitrary, it follows that $I_1 \leq_w I_3$.

Also if $I_1 \leq_s I_2 \leq_s I_3$, $\bar{a}_1$ realizes $\mathrm{Av}(I_1, M)$, N is $\mathbf{F}^a_\kappa$-saturated, $|M| \cup \bar{a}_1 \subseteq |N|$ then $\mathrm{Av}(I'_2, M)$ is realized by some $\bar{a}_2 \in N$ (as $I_1 \leq_s I'_2$), hence some $\bar{a}_3 \in N$ realizes $\mathrm{Av}(I_3, M)$ (as $I'_2 \leq_s I_3$). Again $I_1 \leq_s I_3$ follows.

(3) Suppose $I \nleq_w J$, so for some $\mathbf{F}^a_\kappa$-saturated model M, $\kappa \geq \kappa_r(T)$ and $I_1, J_1 \subseteq M$ equivalent to I, J respectively, but $\dim(I_1, M) > \dim(J_1, M) \geq \kappa$; w.l.o.g. I_1, J_1 are maximal indiscernible sets $\subseteq M$, $|J_1| = \dim(J_1, M)$, $|I_1| = \dim(I_1, M)$. Hence $|I_1| > |J_1| \geq \kappa$. So by III, 3.5 there is $I_2 \subseteq I_1$, $|I_2| = |J_1|$ such that $I_1 - I_2$ is indiscernible over $\bigcup I_2 \cup \bigcup J_1$. By IV, 3.1 there is a $\mathbf{F}^a_\kappa$-prime model N over $\bigcup I_2 \cup \bigcup J_1$, $N \subseteq M$. By IV, 4.10 N omits $p = \mathrm{Av}(J_1, \bigcup I_2 \cup \bigcup J_1)$. Choose $\bar{a} \in I_1 - I_2$, so $\bar{a}$ realizes p; but M omits (by the definition of J_1) $\mathrm{Av}\,(J_1, N)$. This situation shows $I \nleq_s J$.

(4) Trivial.

LEMMA 2.2: *Assume $\kappa \geq \kappa_r(T)$, M is $\mathbf{F}^a_\kappa$-saturated $I, J \subseteq M$ infinite indiscernible sets, and $\bar{a}$ realizes $\mathrm{Av}(I, M)$. Then $I \leq_s J$ iff in the $\mathbf{F}^a_\kappa$-prime model N over $M \cup \bar{a}$, $\mathrm{Av}(J, M)$ is realized.*

Proof. The "only if" part follows by the definition. So suppose $\mathrm{Av}(J, M)$ is realized by $\bar{b} \in N$, and we shall prove $I \leq_s J$.

So let $\kappa(1) \geq \kappa_r(T)$, M_1 is $\mathbf{F}^a_{\kappa(1)}$-saturated, $I_1, J_1 \subseteq M_1$ are equivalent to I, J respectively, $\bar{a}_1$ realizes $\mathrm{Av}(I_1, M_1)$, N_1 be a $\mathbf{F}^a_\lambda$-saturated model, $|M_1| \cup \bar{a} \subseteq N_1$. We must prove N_1 realizes $\mathrm{Av}(J_1, M)$. As there is a $\mathbf{F}^a_{\kappa(1)}$-prime model $N_1 \subseteq N$ over $|M_1| \cup \bar{a}_1$, we can assume N_1 is $\mathbf{F}^a_{\kappa(1)}$-prime over $|M_1| \cup \bar{a}_1$. Choose $\lambda \geq \kappa(1) + \kappa$, λ regular $\lambda > |T| + \|M\| + \|M_1\|$, $\lambda^{|T|} = \lambda$, and let M^* be a saturated model of power λ, $|M| \cup |M_1| \subseteq |M^*|$. We can assume that $\bar{a}_1 = \bar{a}$ realizes $\mathrm{Av}(I, M^*) = \mathrm{Av}(I_1, M^*)$. Let N^* be a $\mathbf{F}^a_\lambda$-prime model over $|M^*| \cup \bar{a}$.

By the symmetry in assumptions, it suffices to prove that $\mathrm{Av}(J, M)$ is realized in N iff $\mathrm{Av}(J, M^*)$ is realized in N^*.

Suppose $\mathrm{Av}(J, M)$ is realized in N by $\bar{c} \in N$. Let

$$\mathscr{A} = \langle |M| \cup \bar{a}, \{b_i : i < \alpha\}, \{B_i : i < \alpha\}\rangle$$

be a $\mathbf{F}^a_\kappa$-construction of N. We prove by induction on β that $\mathscr{A}^*_\beta = \langle |M^*| \cup \bar{a}, \{b_i : i < \beta\}\rangle$ is a $\mathbf{F}^a_\kappa$-construction, and $\mathrm{tp}_*[\bar{a} \cup \{b_i : i < \beta\}, |M^*|]$ does not fork over $|M|$. So $\mathscr{A}^*_\beta$ is a $\mathbf{F}^a_\kappa$-construction, hence a $\mathbf{F}^a_\lambda$-construction, hence $\mathrm{tp}(\bar{c}, |M^*| \cup \bar{a})$ is $\mathbf{F}^a_\lambda$-isolated, hence $\mathrm{tp}(\bar{c}, |M^*|)$ is realized in N^*, does not fork over M, and extends $\mathrm{Av}(J, M)$, so

$\text{tp}(\bar{c}, |M^*|) = \text{Av}(J, M^*)$ so this will finish one direction. We prove by induction on β; for $\beta = 0$ and β a limit ordinal it is trivial. For $\beta + 1$ we know $q = \text{stp}(b_\beta, B_\beta) \vdash \text{stp}(b_\beta, |M| \cup \bar{a} \cup \{b_i : i < \beta\})$. It suffices to show $q \vdash \text{tp}(b_\beta, |M^*| \cup \bar{a} \cup \{b_i : i < \beta\})$ (the non-forking of

$$\text{tp}_*(\bar{a} \cup \{b_i : i \le \beta\}, |M^*|)$$

over $|M| \cup \bar{a}$ will then follow).

So let $\bar{c}_1 \in \bar{a} \cup \{b_i : i < \beta\}$, $\bar{c}_2 \in |M^*|$, and $\varphi(x, \bar{y}, \bar{z})$ be a formula; and we shall prove $q \vdash \varphi(x, \bar{c}_1, \bar{c}_2)^{\mathbf{t}}$ for some $\mathbf{t}$. Choose $C \subseteq |M|$ such that $\text{tp}(B_\beta \cup \bar{c}_1 \cup \bar{a}, |M^*|)$ does not fork over C, $|C| < \kappa$ (C exists by the induction hypothesis, and as $\kappa_r(T) \le \kappa$). Choose $\bar{c}_2' \in |M|$ such that $\text{stp}(\bar{c}_2', C) \equiv \text{stp}(\bar{c}_2, C)$ (possible as M is $\mathbf{F}^a_\kappa$-saturated). By the definition of C, $q \vdash \varphi(x, \bar{c}_1, \bar{c}_2) \Leftrightarrow q \vdash \varphi(x, \bar{c}_1, \bar{c}_2')$. As for some $\mathbf{t}$, $\varphi(x, \bar{c}_1, \bar{c}_2')^{\mathbf{t}} \in \text{tp}(b_\beta, |M| \cup \bar{a} \cup \{b_i : i < \beta\})$, $q \vdash \varphi(x, \bar{c}_1, \bar{c}_2)^{\mathbf{t}}$.

Now we prove the other direction, i.e., suppose that $\bar{c} \in N^*$ realizes $\text{Av}(J, |M|^*)$. As λ is regular $\ge \kappa_r(T)$, $\text{tp}(\bar{c}, |M^*| \cup \bar{a})$ is $\mathbf{F}^a_\lambda$-isolated. So there are $C \subseteq B \subseteq |M^*|$, $|C| < \kappa(T)$, $|B| < \lambda$, $\text{tp}(\bar{c}, B \cup \bar{a}) \vdash \text{tp}(\bar{c}, |M^*| \cup \bar{a})$, $\text{tp}(\bar{a}\frown\bar{c}, |M^*|)$ does not fork over C, and $\text{tp}(\bar{a}\frown\bar{c}, C)$ is stationary. As M^* is a saturated model, there is an automorphism f of M, $f \restriction (I \cup J) =$ the identify, $f(C) \subseteq M$, (w.l.o.g. I, J are countable) and for $\kappa = \aleph_0$, we can demand just that $f(I)$, $f(J)$ are equivalent to I, J resp.; so we can assume $C \subseteq M$. Now claim III, 4.22(2) shows $\text{tp}(\bar{c}, C \cup \bar{a}) \vdash \text{tp}(\bar{c}, |M^*|)$ (letting $C \cup \bar{a} = A$, $B \cup \bar{a} = B$, $|M^*| \cup \bar{a} = C$).

THEOREM 2.3: *The following conditions on the infinite indiscernible sets I, J, and infinite cardinals $\mu > \lambda \ge \kappa_r(T)$ are equivalent:*

(1) $I \le_w J$.

(2) *If M is $\mathbf{F}^a_\lambda$-saturated, $I, J \subseteq M$, then* $\dim(I, M) \le \dim(J, M)$.

(3) *There is no $\mathbf{F}^a_\lambda$-saturated model M, with $I_1, J_1 \subseteq M$ equivalent to I, J respectively,* $\dim(I_1, M) = \mu$, $\dim(J_1, M) = \lambda$.

(4) *For every I_1, J_1 equivalent to I, J respectively,* $|I_1|, |J_1| \ge \kappa_r(T)$, *$I_1(J_1)$ indiscernible over $\bigcup J_1 (\bigcup I_1)$,* $\text{Av}(J_1, \bigcup J_1)$ *has a $\mathbf{F}^s_{\kappa(T)}$-isolated extension in* $S^m(\bigcup I_1 \cup \bigcup J_1)$.

Remark. So if there is a $\mathbf{F}^a_{\lambda_1}$-saturated model M_1 with $I_1, J_1 \subseteq M_1$ equivalent to I, J respectively such that $\dim(I_1, M_1) = \mu_1 > \dim(J_1, M_1) \geqslant \lambda_1 \geqslant \kappa_r(T)$ then for all $\mu_2 > \lambda_2 \geqslant \kappa_r(T)$ there is a $\mathbf{F}^a_{\lambda_2}$-saturated M_2 with $I_2, J_2 \subseteq M_2$ equivalent to I, J respectively such that $\dim(I_2, M_2) = \mu_2$, $\dim(J_2, M_2) = \lambda_2$.

Proof of Theorem 2.3: (1) $\Rightarrow$ (2) by the definition.

(1) $\Rightarrow$ (3) by the definition.

(2) or (3) $\Rightarrow$ (4) Suppose I_1, J_1 contradict (4). Without changing the hypothesis on I_1, J_1, we can assume $|I_1| = \mu$, $|J_1| = \lambda$ (by III, 4.22). Let N be $\mathbf{F}^a_\lambda$-primary over $\bigcup I_1 \cup \bigcup J_1$. By IV, 4.10 N omits $\mathrm{Av}(J_1, \bigcup I_1 \cup \bigcup J_1)$ and $\mathrm{Av}(I_1, \bigcup I_1 \cup \bigcup J_1)$. From the first omitting it follows that $\dim(J_1, N) \leq |I_1| + |J_1| = \mu$; but $\dim(J_1, N) \geq |J_1| = \mu$, so $\dim(J_1, N) = \mu$. (See III, 3.9). On the other hand $\dim(I_1, N) \geq |I_1| = \lambda$; suppose $\dim(I_1, N) > \lambda$, then there is $I^* \subseteq N$ equivalent to I_1, $|I^*| = \lambda^+$. As $\lambda \geq \kappa_r(T)$, some $\bar{c} \in I^*$ realizes $\mathrm{Av}(I_1, \bigcup I_1)$. Let $\mathscr{A} = \langle I_1 \cup J_1, \{a_i: i < \alpha_0\}\rangle$ be a $\mathbf{F}^a_\lambda$-construction of N. Going through $\mathbf{F}^a_{\lambda,\kappa_r(T)}$ (see IV, 4.5–4.7) we can assume $\bar{c} \in \{a_i: i < \alpha\}$, $\alpha < \kappa_r(T)$ (see IV, 1.4(2)) so there are $I_2 \subseteq I_1$, $J_2 \subseteq J_1$, $|I_2 \cup J_2| < \kappa_r(T)$ such that $\mathrm{tp}_*(\{a_i: i < \alpha\}, \bigcup I_1 \cup \bigcup J_1)$ does not fork over $\bigcup (I_2 \cup J_2)$. We can also assume $\mathrm{Av}(I_1, \bigcup I_1)$ does not fork over $\bigcup I_2$, $\mathrm{Av}(I_1, \bigcup I_2)$ is stationary, and similarly for J_1, J_2 (see III, 4.17(4)) and for every $\bar{c} \in \bigcup (I_1 \cup J_1)$, $\mathrm{tp}(\bar{c}, \bigcup (I_2 \cup J_2)$ is stationary.

Now we prove by induction on i that

$$p_i = \mathrm{tp}\big(a_i, \bigcup (I_2 \cup J_2) \cup \{a_j: j < i\}\big) \vdash \mathrm{tp}\big(a_i, \bigcup (I_1 \cup J_1) \cup \{a_j: j < i\}\big).$$

This holds by III, 4.22(2) as p_i is $\mathbf{F}^a_\lambda$-isolated. So clearly

$$\mathrm{tp}_*\big(\{a_i: i < \alpha\}, \bigcup (I_2 \cup J_2)\big) \vdash \mathrm{tp}_*\big(\{a_i: i < \alpha\}, \bigcup (I_1 \cup J_1)\big)$$

hence $\mathrm{tp}_*(\bar{c}, \bigcup (I_2 \cup J_2)) \vdash \mathrm{tp}(\bar{c}, \bigcup (I_1 \cup J_1))$. So we could choose $\alpha < \omega$, so we could choose I_2, J_2 such that $|I_2 \cup J_2| < \kappa(T)$, and remember $\mathrm{tp}(\bar{c}, I_2 \cup J_2) \vdash \mathrm{tp}(\bar{c}, I_1 \cup J_1)$. So $\mathrm{tp}(\bar{c}, I_1 \cup J_1)$ is $\mathbf{F}^s_{\kappa(T)}$-isolated, and it extends $\mathrm{Av}(I_1, \bigcup I_1)$, contradiction to the assumption on I_1, J_1.

So $\dim(I_1, N) = \lambda$, $\dim(J_1, N) = \mu$, and (3) fails. If we choose $\lambda > |I| + |J|$, we can assume $I, J \subseteq |M|$, hence getting a contradiction to (2). So we proved that (2), or (3) implies (4).

(4) $\Rightarrow$ (1) Suppose $\kappa \geq \kappa_r(T)$, M is $\mathbf{F}^a_\kappa$-saturated, $I_1, J_1 \subseteq M$ are equivalent to I, J respectively; we shall prove $\dim(I_1, M) \leq \dim(J_1, M)$. Suppose not, and w.l.o.g. I_1, J_1 are maximal indiscernible sets in M. By III, 3.9 $|I_1| = \dim(I_1, M) > \dim(J_1, M) = |J_1|$, and of course $|J_1| \geq \kappa_r(T)$. By III, 3.5 there is $I_2 \subseteq I_1$, $|I_2| \leq |J_1| + \kappa_r(T) = |J_1|$, such that $I_1 - I_2$ is indiscernible over J_1. By III, 4.17(4) there is $J_3 \subseteq J_1$, $|J_3| < \kappa_r(T)$, $\mathrm{Av}(J_1, \bigcup J_1)$ does not fork over $\bigcup J_3$,

$\mathrm{Av}(J_1, \bigcup J_3)$ is stationary; and similarly $I_3 \subseteq I_1 - I_2$, and we can assume $\mathrm{tp}_*(I_3, \bigcup J_1)$ does not fork over $\bigcup J_3$. Then $\mathrm{tp}_*(I_1 - I_3 - I_2, I_3 \cup J_1)$ does not fork over $I_3 \cup J_3$, hence $\mathrm{tp}_*(J_1 - J_3, (I_1 - I_2) \cup J_3)$ does not fork over $I_3 \cup J_3$. So $I_1' = I_1 - I_3 - I_2$, and $J_1' = J_1 - J_3$ satisfies the hypothesis of (4) and $|I_1'| = |I_1|$, $|J_1'| = |J_1|$.

Now we define by induction on $\beta < |J_1|^+$ sequences $\bar{c}_\beta \in |M|$ and sets $I_\beta^* \subseteq I_1'$ such that $J_\beta^* = J_1' \cup \{\bar{c}_i : i < \beta\}$ is indiscernible, $|I_\beta^*| \le |J_1|$ and $I_1' - I_\beta^*, J_\beta^*$ satisfy the hypothesis of (4). For $\beta = 0$, β limit this is clear. For $\beta + 1$, by (4) there is a $\mathbf{F}^s_{\kappa(T)}$-isolated type $q \in S^m(J_\beta^* \cup (I_1' - I_\beta^*))$ which extend $\mathrm{Av}(J_\beta^*, \bigcup J_\beta^*)$, so some $\bar{c}_\beta \in M$ realizes it, and for some $I_{\beta+1}^* \subseteq I_1$, $|I_{\beta+1}^*| \le |J_1|$, $I_\beta^* \subseteq I_{\beta+1}^*$ and $\mathrm{tp}(\bar{c}_\beta, (I_{\beta+1}^* - I_\beta^*) \cup J_\beta^*) \vdash \mathrm{tp}(\bar{c}_\beta, (I_1' - I_\beta^*) \cup J^*)$. Clearly the induction hypothesis are satisfied, and for $\beta = |J_1|^+$, J_β^* contradicts $\dim(J_1, M) = |J_1|$.

LEMMA 2.4: (1) *If p, q are stationary, not orthogonal, and q is regular then $p \leqslant_s q$ (hence $p \leqslant_w q$).*

(2) *If p, q are stationary and orthogonal then $p \not\leqslant_w q$ (hence $p \not\leqslant_s q$).*

Proof. (1) Immediate by 1.12.

(2) Immediate, by IV, 4.10, and 1.4, 1.10.

THEOREM 2.5: *Let I be a (non-trivial) indiscernible set, $|I| = \mu > \lambda \ge \kappa_r(T)$, M a $\mathbf{F}^a_\lambda$-prime model over I, $J \subseteq M$ a (non-trivial) infinite indiscernible set. Then* $\dim(J, M) = \lambda$ *iff* $\dim(J, M) \ne \mu$ *iff* $I \not\leqslant_w J$.

Proof. As M is $\mathbf{F}^a_\lambda$-saturated, $\dim(J, M) \ge \lambda$. If $J' \subseteq M$ is equivalent to J, $|J'| > \mu$, then for some $J'' \subseteq J'$, $|J''| \le \mu$, $J' - J''$ is indiscernible over $\bigcup I$, and its cardinality is, of course $> \mu$, contradiction to IV, 4.9. Hence $\lambda \le \dim(J, M) \le \mu$, $\dim(I, M) = \mu$.

If $I \le_w J$ then $\dim(J, M) \ge \dim(I, M) = \mu$ hence $\dim(J, M) = \mu$. If $\dim(J, M) = \mu$ clearly $\dim(J, M) \ne \lambda$. To complete the proof we assume $I \not\leqslant_w J$ and prove $\dim(J, M) = \lambda$. By 2.3(3) there is a $\mathbf{F}^a_\lambda$-saturated model N, and $I_1, J_1 \subseteq N$ equivalent to I, J respectively, $\dim(I_1, N) = \mu$, $\dim(J_1, N) = \lambda$. Choose $J_2 \subseteq J$, $|J_2| < \kappa(T)$, such that $\mathrm{Av}(J, \bigcup J_2)$ is stationary, $\mathrm{Av}(J, \bigcup J)$ does not fork over $\bigcup J_2$ (by III, 4.17(4)). By IV, 3.12(3) M is $\mathbf{F}^a_\lambda$-prime over $I \cup J_2$. W.l.o.g. we can find an elementary mapping f of $I \cup J_2$ into N, such that $f(I)$ is equivalent to I_1, and $f(J_2) \subseteq J_1$. By the primeness we can extend f to an elementary mapping f' of M into N. So $\lambda \le \dim(J, M) \le \dim[f'(J), f'(M)] \le \dim(J_1, N) = \lambda$. So $\dim(J, M) = \lambda$, and we finish.

THEOREM 2.6: *Suppose I is a non-trivial indiscernible set, $|I| \geq \mu > \lambda \geq \kappa_r(T)$, M is $\mathbf{F}^a_\lambda$-prime over I, and $J \subseteq M$ an infinite indiscernible set.*

Then I is orthogonal to J iff $\mathrm{Av}(J, M)$ *is $\mathbf{F}^a_{\lambda^+}$-isolated iff* $\mathrm{Av}(J, M)$ *is $\mathbf{F}^a_\chi$-isolated for some $\chi \leq \mu$.*

Proof. We can assume $|J| = \lambda$, $I_1 \subseteq I$, $|I_1| = \lambda$, $I - I_1$ is indiscernible over $I_1 \cup J$, $I_1 \cup J \subseteq N \subseteq M$, N is $\mathbf{F}^a_\lambda$-primary over $I_1 \cup J$, and is $\mathbf{F}^a_\lambda$-constructible over $I \cup J$, and M is $\mathbf{F}^a_\lambda$-constructible over $I \cup N$ (see IV, 3.17 and IV, 3.9). Clearly $I - I_1$ is indiscernible over N.

If I, J are orthogonal then by 1.4 $\mathrm{Av}(J, N) \vdash \mathrm{Av}(J, N \cup I)$ and by IV, 4.10 $\mathrm{Av}(J, N \cup I) \vdash \mathrm{Av}(J, M)$ and $\mathrm{Av}(J, J \cup I_1) \vdash \mathrm{Av}(J, N)$. As $|J| + |I_1| = \lambda$, it follows that $\mathrm{Av}(J, N)$ is $\mathbf{F}^a_{\lambda^+}$-isolated, hence $\mathrm{Av}(J, M)$ is $\mathbf{F}^a_{\lambda^+}$-isolated, hence it is $\mathbf{F}^a_\chi$-isolated for some $\chi < \mu$.

Now assume $\mathrm{Av}(J, M)$ is $\mathbf{F}^a_\chi$-isolated, $\chi \leq \mu$, and we prove I, J are orthogonal, and so we shall finish. Let $A \subseteq M$, $|A| < \chi \leq \mu$ be such that $\mathrm{Av}(J, A) \vdash \mathrm{Av}(J, M)$. By IV, 3.17 and IV, 3.9 there is a model N, $A \subseteq |N| \subseteq |M|$, N $\mathbf{F}^a_\lambda$-prime over I_1 for some $I_1 \subseteq I(|I_1| < \mu)$ and $\mathbf{F}^a_\lambda$-constructible over I, M is $\mathbf{F}^a_\lambda$-constructible over $I \cup N$, $|N \cap I| \geq \aleph_0$, $|N \cap J| \geq \aleph_0$. So $\mathrm{Av}(I, N)$ is realized in M (by any $\bar{c} \in I - I_1$) and $\mathrm{Av}(J, N) \vdash \mathrm{Av}(J, M)$. Hence $\mathrm{Av}(J, N)$, $\mathrm{Av}(I, N)$ are weakly orthogonal, hence by 1.2(3) they are orthogonal.

CONCLUSION 2.7: *If I, J are (non-trivial, infinite) indiscernible sets over A, I, J are not orthogonal, then* $\mathrm{tp}_*(J, A)$, $\mathrm{tp}_*(I, A)$ *are not weakly orthogonal.*

Remark. See 1.5(2) for a converse.

Proof. Suppose $\mathrm{tp}_*(I, A)$, $\mathrm{tp}_*(J, A)$ are weakly orthogonal. As I, J are infinite, we can assume $|I| > |J| > \kappa = \kappa_r(T) + |A|^+$, I indiscernible over $A \cup J$, and J indiscernible over $A \cup I$. By Ax(V.2) for $\mathbf{F}^s_\infty$, for any $\bar{c} \in J$ (letting $J' = J - \{\bar{c}\}$) $\mathrm{tp}(\bar{c}, A \cup J') \vdash \mathrm{tp}(\bar{c}, A \cup J' \cup I)$. So if N is $\mathbf{F}^a_\kappa$-primary model over $A \cup J' \cup I$, then $\mathrm{Av}(J', A \cup J') = \mathrm{tp}(\bar{c}, A \cup J') \vdash \mathrm{Av}(J', N)$. So $\mathrm{Av}(J', N)$ is $\mathbf{F}^a_\lambda$-isolated, ($\lambda = |J|$) hence by the previous theorem I, J are orthogonal.

CONCLUSION 2.8: *If $p \in S^m(A)$, $B \subseteq A$, p orthogonal to B or to* $\mathrm{tp}_*(I, A)$. *I an (infinite) indiscernible set over A,* $\mathrm{tp}_*(I, A)$ *does not fork over B, then p is orthogonal to* $\mathrm{Av}(I, \bigcup I)$.

Proof. We can assume $A = |M|$, M $\mathbf{F}^a_{\kappa_r(T)}$-saturated. Let J be an indiscernible set based on p. By 1.4 $\text{tp}_*(I, M)$, $\text{tp}_*(J, M)$ are weakly orthogonal, hence by the previous conclusion p, $\text{Av}(I, \bigcup I)$ are orthogonal.

THEOREM 2.9: *Let I be a (non-trivial) indiscernible set, $|I| = \mu > \lambda \geq \kappa_r(T)$, M a $\mathbf{F}^a_\lambda$-prime model over $\bigcup I$, $A \subseteq M$, $|A| < \lambda$, $p \in S^m(A)$, λ_1 the first cardinal $\geq \lambda$ in which T is stable, $\mu > \lambda_1$, and*

$$J = \{\bar{b} \in M : \bar{b} \text{ realizes } p\}.$$

Then $|J| \geq \mu$ if $|J| > \lambda_1$ iff p has an extension q such that $\text{Av}(I, \bigcup I) \leq_w q$.

Proof. If p has an extension q, $\text{Av}(I, \bigcup I) \leq_w q$; then let $q \in S^m(A \cup B)$ $|B| < \kappa(T)$, so we can assume $B \subseteq M$, (of course, q is stationary). So $\mu = \dim(I, M) \leq \dim(q, M) \leq |J|$. If $|J| \geq \mu$ then $|J| > \lambda_1$, and if $|J| > \lambda_1$ then by I, 2.8 there is an indiscernible $J_0 \subseteq J$, $|J_0| = \lambda_1^+$. So $\dim(J_0, M) > \lambda_1 \geq \lambda$, hence by 2.5 $I \leq_w J_0$ so letting $q = \text{Av}(J_0, A \cup \bigcup J_0)$, $\text{Av}(I, \bigcup I) \leq_w q$ and q extends p (of course J_0 is not trivial, q not algebraic).

Notation. (for Theorem 2.10). (1) Let $I_{\aleph_0}$ be a non-trivial indiscernible set, $|I_{\aleph_0}| = \aleph_0$. For every λ let $I_{\aleph_0} \subseteq I_\lambda$, I_λ indiscernible, $|I_\lambda| = \lambda$, and let $M(I_\lambda)$ be a $\mathbf{F}^a_{\kappa_r(T)}$- prime model over $\bigcup I_\lambda$.

(2) Let $\chi \geq \kappa_r(T)$, T stable in χ.

THEOREM 2.10: *The following statements on T are equivalent (remember T is stable).*

(1) *For every $\lambda \geq \kappa_r(T)$, T has maximally$-\mathbf{F}^a_\lambda$-saturated models of arbitrarily high cardinality.*

(2) *As* (1), *but the models are in addition λ-universal but not λ^+-universal.*

$(3)_{I_{\aleph_0}}$ *Not every $\mathbf{F}^a_{\kappa_r(T)}$-saturated model is isomorphic to $M(I_\lambda)$ for some λ.*

$(4)_\chi$ *Not every two $\mathbf{F}^a_{\kappa_r(T)}$-saturated models of T of cardinality χ^+ are isomorphic.*

(5) *For some (non-trivial infinite) indiscernible sets I, J, $I \not\leq_w J$.*

(6) *There are (non-trivial infinite) indiscernible sets I, J such that J is minimal and I is orthogonal to J.*

(7) *There are (non-trivial infinite) indiscernible sets I, J, $I \not\leq_w J$, J is based on $\emptyset$ such that*

(A) *I is minimal, or*

(B) *$J^* \leq_w J \leq_w J^*$ for some minimal J^*, and I, J are orthogonal.*

DEFINITION 2.2: If T fails to satisfy the above conditions, it is called *uni-dimensional*.

Proof of 2.10. Trivially (2) $\Rightarrow$ (1); (6) $\Rightarrow$ (5) (by 2.4(2)).

(5) $\Rightarrow$ (1) By 2.3.

(1) $\Rightarrow$ $(4)_\chi$ By III, 3.12 T has a saturated model of cardinality χ^+; and by (1) and as T is stable in χ^+, T has a maximally-$\mathbf{F}^a_{\kappa_r(T)}$-saturated model of cardinality χ^+. As $\lambda \geq \kappa_r(T)$ they are not isomorphic.

$(4)_\chi \Rightarrow (3)_{I\aleph_0}$ Clearly $\mu \leq |I_\mu| \leq \|M(I_\mu)\|$; and if $\mu \leq \chi$, then as T is stable in χ, $\|M(I_\mu)\| \leq \chi$. So $\|M(I_\mu)\| = \chi^+$ iff $\mu = \chi^+$; hence one of the $\mathbf{F}^a_{\kappa_r(T)}$-saturated models of T of cardinality χ^+, is not isomorphic to any $M(I_\mu)$.

$(3)_{I\aleph_0} \Rightarrow (6)$ Let J be a countable (non-trivial) minimal indiscernible set (exists by 1.18(7)). Let M be a $\mathbf{F}^a_{\kappa_r(T)}$-saturated model not isomorphic to any $M(I_\mu)$. We can assume $J \subseteq M$ and let $J_1 \subseteq M$ be a maximal indiscernible set in M extending J. By 1.16 and 1.18(4) for every $\bar{c} \in |M|$ tp($\bar{c}$, $\bigcup J_1$) is $\mathbf{F}^a_{\kappa_r(T)}$-isolated. If $I^* \subseteq |M|$ is a (non-trivial) infinite indiscernible set over $\bigcup J_1$, then we can assume that for some $J_2 \subseteq J_1$, $|J_2|, |J_1 - J_2| \geq \aleph_0$, $|J_2| < \kappa_r(T) + \aleph_1$, $\text{tp}_*(I^*, \bigcup J_2) \vdash \text{tp}_*(I^*, J_1)$ so by 2.7 I^*, J_1 are orthogonal, so (6) holds. So suppose there is no such I^*, then by IV, 4.14 (the characterization theorem) M is $\mathbf{F}^a_{\kappa_r(T)}$-prime over $\bigcup J_1$.

Let $\mu = |J_1|$, then in $M(I_\mu)$ we can find a maximal indiscernible set J_2, such that for distinct $\bar{a}^l_1, \ldots, \bar{a}^l_n \in J_l$, $l = 1, 2$, $\text{tp}(\bar{a}^1_1 \frown \cdots \frown \bar{a}^1_n) = \text{tp}(\bar{a}^2_1 \frown \cdots \frown \bar{a}^2_n)$. As before we can assume $M(I_\mu)$ is $\mathbf{F}^a_{\kappa_r(T)}$-prime over J_2. If $|J_2| = \mu$, there is an elementary mapping f from $\bigcup J_1$ onto $\bigcup J_2$. By IV, 4.14 we can extend it to an isomorphism from M onto $M_\mu(I)$, contradicting the choice of M. If $|J_2| \neq \mu$ then as dim $(J_2, M) = |J_2|$, by 2.5 $J_2 \not\leq_w I_\mu$, hence J_2, I_μ are orthogonal (by 1.12) so (6) holds.

(6) $\Rightarrow$ (7) Let I, J be as in (6), and J_1 be a non-trivial infinite indiscernible set, based on $\emptyset$ (exists by III, 1.12). If $J \not\leq_w J_1$ then taking J, J_1 for I, J shows that (7)(A) holds. So assume $J \leq_w J_1$. If J_1 is orthogonal to I then $I \not\leq_w J_1$ so taking I, J_1, J for I, J, J^* shows that (7)(B) holds. ($J_1 \leq_w J$ holds as they are not orthogonal, and J is minimal hence regular). But J_1 is orthogonal to I by the following claim:

CLAIM 2.11: *If $J \leq_w J_1$, I is orthogonal to J then I is orthogonal to J_1.*

Proof of 2.11. Let M be λ-saturated, $\|M\| = \lambda > \kappa_r(T)$ (by stability) and w.l.o.g. J_1, J, I be countable and $\subseteq M$. Define J^* so that $|J^*| = \lambda^+$,

$J^* \cup J$ is indiscernible and J^* is indiscernible over $|M|$; and let N be $\mathbf{F}^a_\lambda$-prime over $|M| \cup \bigcup J^*$. Clearly $\dim(J, N) \geq |J^*| = \lambda^+$, hence $\dim(J_1, N) \geq \lambda^+ > \lambda$, so some $\bar{c} \in N$ realizes $\mathrm{Av}(J_1, |M|)$. On the other hand, as I is orthogonal to J, $\mathrm{Av}(I, |M|) \vdash \mathrm{Av}(I, |M| \cup \bigcup J^*)$ and by IV, 4.10(2) $\mathrm{Av}(I, |M| \cup \bigcup J^*) \vdash \mathrm{Av}(I, |N|)$, hence $\mathrm{Av}(I, |M|) \vdash \mathrm{Av}(I, |M| \cup \bar{c})$, so by 1.2(3) J_1, I are orthogonal.

Continuation of the proof. (7) $\Rightarrow$ (2) Let I, J be as in (7) and if possible, as in (7)(A); and $\mu > \lambda \geqslant \kappa_r(T)$. We can assume J, I are indiscernible over $\bigcup I$, $\bigcup J$ resp. and $|I| = \mu$, $|J| = \lambda$. Let N be a $\mathbf{F}^a_\lambda$-primary model over $\bigcup I \cup \bigcup J$; so by 2.3 we know N is maximally $\mathbf{F}^a_\lambda$-saturated, and of course $\|M\| \geq |I| = \mu$. As M is $\mathbf{F}^a_\lambda$-saturated, it is λ-saturated, hence λ-universal.

Choose an indiscernible set J^1 over $\bigcup I$, $J \subseteq J^1$, $|J^1| = \lambda^+$, and choose $I^1 \subseteq I$, $|I^1| = \lambda^+$. Suppose f is an elementary mapping from $\bigcup I^1 \cup \bigcup J^1$ into N; we try to get a contradiction, and thus show that N is not λ^+-universal. Clearly $\dim(f(I^1), N)) \text{ is } > \lambda$ thus as in 2.5 $I \leq_w f(I^1)$; and similarly $I \leq_w f(J^1)$. If $J^* \leq_w J \leq_w J^*$, J^* minimal then for some minimal $J^{**} \subseteq |M|$, $f(J^1) \leq_w J^{**}$, hence $I \leq_w J^{**}$. As $\dim(J, M) = \lambda$, $\dim(J^{**}, M) = \mu$, clearly $J^{**} \not\leq_w J$, so we could have started with J, J^{**} instead of J, I, and then (7)(A) holds, so I is minimal.

If I is minimal, then $f(I^1)$ is minimal too, hence by 2.4, 1.18(4) $f(I^1) \leq_w I$, thus (as $I \leq_w f(J^1)$ and by 2.1(2)) $f(I^1) \leq_w f(J^1)$ hence $I^1 \leq_w J^1$ contradiction.

As we proved (6) $\Rightarrow$ (5) $\Rightarrow$ (1) $\Rightarrow$ $(4)_\chi$ $\Rightarrow$ $(3)_{I\aleph_0}$ $\Rightarrow$ (6) $\Rightarrow$ (7) $\Rightarrow$ (2) $\Rightarrow$ (1), we finish.

PROBLEM 2.1: Is there a uni-dimensional, stable, but not super-stable theory? (see Exercise 5.4).

QUESTION 2.2: In which places in this section can we replace $\kappa_r(T)$ by $\kappa(T)$? Or should we find a natural $\kappa^x(T)$, $\kappa(T) \leq \kappa^x(T) \leq \kappa_r(T)$ for those theorems?

PROBLEM 2.3: (1) For every stable T and $\lambda < \kappa_r(T)$ does T have a λ-saturated model which is not λ^+-universal?

(2) For every stable theory T with $|D(T)| > \aleph_0$ does T have a model which is not $\aleph_0$-universal, or at least not $|T|$-universal?

EXERCISE 2.4: (1) Prove (1) above when $\lambda \geq \kappa_r(T)$.

(2) Show that if $\kappa_r(T) < |T|$ then T has a model which is not $|T|$-universal.

EXERCISE 2.5: Suppose $p \leq_w q$, $r \leq_w q$, and show that p, r are not orthogonal, except when they are algebraic.

V.3. Weighted dimensions and superstability

THEOREM 3.1: *Suppose* $\{\bar{a}, \bar{b}\}$ *is independent over* A *and* $\operatorname{stp}(\bar{c}, A)$ *is regular. Then it is impossible that both* $\operatorname{tp}(\bar{a}, A \cup \bar{c})$, *and* $\operatorname{tp}(\bar{b}, A \cup \bar{c})$ *fork over* A.

Proof. W.l.o.g. we work in $\mathfrak{C}^{eq}$, $A = \operatorname{acl} A$. Let M be λ-saturated, $|A| + |T| < \lambda = \|M\|$, $A \cup \bar{a} \cup \bar{b} \cup \bar{c} \subseteq M$, and $J = \{\bar{d}: \bar{d} \in M, \bar{d}$ realizes $\operatorname{stp}(\bar{c}, A)\}$. By II, 2.2, 13 there is $J_1 \subseteq J$, $|J_1| \leqslant |T|$ such that $\operatorname{tp}(\bar{a}, A \cup \bigcup J)$ is $(A \cup \bigcup J_1)$-definable (see Definition II, 2.1(4)). Clearly if $\operatorname{tp}_*(J_1', A \cup \bar{a}) = \operatorname{tp}_*(J_1, A \cup \bar{a})$, $J_1' \subseteq J$ (with suitable indexing) then there is an automorphism f of M, $f \restriction A =$ the identity, $f(\bar{a}) = \bar{a}$, f maps J_1 onto J_1'; hence we can replace J_1 by J_1'. So we can assume that $\operatorname{tp}(J_1, A \cup \bar{a} \cup \bar{b})$ does not fork over $A \cup \bar{a}$. Hence $\operatorname{tp}(\bar{b}, A \cup \bar{a} \cup \bigcup J_1)$ does not fork over $A \cup \bar{a}$ (by Ax(VI) for $\mathbf{F}^f_\infty$) so it does not fork over A (by III, 4.4). Similarly we can define $J_2 \subseteq J$ so that $\operatorname{tp}(\bar{b}, A \cup \bigcup J)$ is $(A \cup \bigcup J_2)$-definable, and

$$\operatorname{tp}_*\left(J_2, A \cup \bar{a} \cup \bar{b} \cup \bigcup J_1\right)$$

does not fork over $A \cup \bar{b}$. Hence (by Ax(VII) for $\mathbf{F}^f_\infty$))

$$\operatorname{tp}_*\left(\bar{b} \cup J_2, A \cup \bar{a} \cup \bigcup J_1\right)$$

does not fork over A, so $\operatorname{tp}_*(J_2, A \cup J_1)$ does not fork over A, in particular $J_1 \cap J_2 = \emptyset$. Choose maximal $I_l \subseteq J_l$ independent over A. So $I_1 \cup I_2$ is independent over A, $I_1 \cap I_2 = \emptyset$.

We can find $\bar{c}' \in M$, $\bar{c}'$ realizes $\operatorname{stp}(\bar{c}, A)$, but

$$\operatorname{tp}\left(\bar{c}', A \cup \bigcup J_1 \cup \bigcup J_2 \cup \bar{a} \cup \bar{b}\right)$$

does not fork over A. So $\{\bar{a}, \bar{c}'\}$ is independent over A. But $\{\bar{a}, \bar{c}\}$ is not independent over A, hence $\operatorname{tp}(\bar{a}{}^\frown\bar{c}, A) \neq \operatorname{tp}(\bar{a}{}^\frown\bar{c}', A)$. As $\bar{c}, \bar{c}' \in J$, and $\operatorname{tp}(\bar{a}, A \cup \bigcup J)$ is $(A \cup \bigcup J_1)$-definable, it must be that $\operatorname{tp}(\bar{c}, A \cup J_1) \neq \operatorname{tp}(\bar{c}', A \cup J_1)$. As $\operatorname{stp}(\bar{c}, A) \equiv \operatorname{stp}(\bar{c}', A)$ is stationary $\operatorname{tp}(\bar{c}, A \cup J_1)$ forks over A. So $\bar{c}$ depends on I_1 over A (by 1.9). Similarly $\bar{c}$ depends on I_2 over A. Contradiction by 1.14.

LEMMA 3.2: *Suppose M is $\mathbf{F}^a_\lambda$-saturated, $\lambda \geq \kappa_r(T)$, $\mathrm{tp}_*(A_1, |M| \cup A_2)$ does not fork over $|M|$ and N_l is $\mathbf{F}^a_\lambda$-prime over $|M| \cup A_l$ ($l = 1, 2$). Then:*

(1) (i) $\mathrm{tp}_*(|N_1|, |N_2|)$ *does not fork over* $|M|$.

(ii) $\mathrm{stp}_*(|N_1|, |M| \cup A_1) \vdash \mathrm{tp}_*(|N_1|, |M| \cup A_1 \cup |N_2|)$.

(2) *If* $\mathrm{tp}(\bar{b}, |M| \cup A_1)$ *is* $\mathbf{F}^a_\lambda$*-isolated then:*

(i) $\mathrm{tp}(\bar{b}, |M| \cup A_2)$ *does not fork over* $|M|$.

(ii) $\mathrm{stp}(\bar{b}, |M| \cup A_1) \vdash \mathrm{tp}(\bar{b}, |M| \cup A_1 \cup A_2)$.

Proof. (1) We can in the lemma replace $|N_l|$ by sets B_l $\mathbf{F}^a_\lambda$-constructible over $|M| \cup A_l$, and then prove it by induction on the length of the constructions. So it suffices to prove (2).

(2) If this is true for every finite subset of A_2, it is true for A_2; so we can assume $A_2 = \bar{c}$. Similarly we can assume $|A_1| < \lambda$.

As $\kappa_r(T) \leq \lambda$, there is $B \subseteq |M|$, $|B| < \lambda$ such that $\mathrm{tp}_*(A_1 \cup \bar{c}, |M|)$ does not fork over B, $\mathrm{tp}_*(A_1 \cup \bar{c}, B)$ is stationary and $\mathrm{stp}(\bar{b}, B \cup A_1) \vdash \mathrm{stp}(\bar{b}, |M| \cup A_1)$. As $\mathrm{tp}(\bar{c}, |M|)$ is stationary by III, 2.15(2) and $\mathrm{tp}(\bar{c}, |M| \cup A_1)$ does not fork over $|M|$ (by assumption) $\mathrm{tp}(\bar{c}, |M| \cup A_1)$ is stationary. If (ii) fails, then there is $\bar{a} \in |M|$ such that $\mathrm{stp}(\bar{b}, B \cup A_1) \nvdash \mathrm{tp}(\bar{b}, B \cup A_1 \cup \bar{a} \cup \bar{c})$ and w.l.o.g. $\bar{a} \in B$. Now as M is $\mathbf{F}^a_\lambda$-saturated there is $\bar{c}' \in |M|$, $\mathrm{stp}(\bar{c}', B) \equiv \mathrm{stp}(\bar{c}, B)$, so

$$\mathrm{stp}(\bar{c}', B \cup A_1) \equiv \mathrm{stp}(\bar{c}, B \cup A_1)$$

(remember $\mathrm{tp}_*(A_1, |M| \cup \bar{c})$ does not fork over $|M|$, hence over B (by III, 4.4)). Hence $\mathrm{stp}(\bar{b}, B \cup A_1) \nvdash \mathrm{tp}(\bar{b}, B \cup A_1 \cup \bar{c}')$, contradicting $\mathrm{stp}(\bar{b}, B \cup A_1) \vdash \mathrm{stp}(\bar{b}, |M| \cup A_1)$.

Remark. Really 3.2(2)(ii) implies $\mathrm{tp}(\bar{b}, M \cup A_1) \vdash \mathrm{tp}(\bar{b}, M \cup A_1 \cup A_2)$, as $\mathrm{tp}_*(A_1 \cup \bar{b}, M)$, $\mathrm{tp}_*(A_2, M)$ are stationary.

CLAIM 3.3: *If M is $\mathbf{F}^a_\lambda$-saturated, $\lambda \geq \kappa_r(T)$ regular, $\mathrm{tp}(\bar{a}, |M|)$ regular, N is $\mathbf{F}^a_\lambda$-primary over $|M| \cup \bar{a}$, and $\bar{b} \in |N|$, $\bar{b} \notin |M|$ then N is $\mathbf{F}^a_\lambda$-primary over $|M| \cup \bar{b}$.*

Proof. Choose $B \subseteq |M|$, $|B| < \lambda$ so that $\mathrm{tp}(\bar{a}{}^\frown\bar{b}, M)$ does not fork over B, $\mathrm{tp}(\bar{a}{}^\frown\bar{b}, B)$ is stationary, and $\mathrm{stp}(\bar{b}, B \cup \bar{a}) \vdash \mathrm{stp}(\bar{b}, |M| \cup \bar{a})$ (such B exists by III, 3.2, III, 2.9(1) and IV, 3.2). By III, 6.14 $\mathrm{tp}(\bar{b}, B \cup \bar{a}) \vdash \mathrm{tp}(\bar{b}, |M| \cup \bar{a})$($B$, $|M|$, $\bar{a}$ correspond to B, C, A.) If $\mathrm{stp}(\bar{a}, B \cup \bar{b}) \nvdash \mathrm{tp}(\bar{a}, |M| \cup \bar{b})$ then some $\bar{a}' \in N$ realizes $\mathrm{stp}(\bar{a}, B \cup \bar{b})$, but $\mathrm{tp}(\bar{a}{}^\frown\bar{b}, |M|) \neq \mathrm{tp}(\bar{a}'{}^\frown\bar{b}, |M|)$. As $\mathrm{tp}(\bar{a}{}^\frown\bar{b}, B) = \mathrm{tp}(\bar{a}{}^\frown{}'\bar{b}, B)$, $\mathrm{tp}(\bar{b}, B \cup \bar{a}')$ forks over B, so $\bar{a}' \notin M$, hence by 1.9 $\mathrm{tp}(\bar{a}', |M|) = \mathrm{tp}(\bar{a}, |M|)$; $\mathrm{tp}(\bar{a} \frown \bar{b}, B) = \mathrm{tp}(\bar{a}' \frown \bar{b}, B)$, $\mathrm{tp}(\bar{a} \frown \bar{b}, M) \neq \mathrm{tp}(\bar{a}' \frown \bar{b}, M)$, $\mathrm{tp}(\bar{b}, B \cup \bar{a}) \vdash \mathrm{tp}(\bar{b}, M \cup \bar{a})$, a contradiction.

Hence $\mathrm{tp}(\bar{a}, |M| \cup \bar{b})$ is $\mathbf{F}^a_\lambda$-isolated, and the rest is easy by IV, 3.16 and IV, 3.6.

THEOREM 3.4: *Suppose $A \subseteq B$, f an elementary mapping whose domain is B, $f \restriction A$ is the identity, and $p \in S^m(B)$ is stationary, and* $\mathrm{stp}_*(B, A) \equiv \mathrm{stp}_*(f(B), A)$, $\mathrm{stp}_*(f(B), B)$ *does not fork over A. Then p is orthogonal to A iff p is orthogonal to $f(p)$.*

Proof. Define elementary mappings $f_i (i < |T|^+)$ so that $\mathrm{stp}_*(B, A) \equiv \mathrm{stp}_*(f_i(B), A)$, and $\mathrm{tp}_*(f_i(B), \bigcup_{j<i} f_j(B))$ does not fork over A, and we let f_0 = the identity over B, $f_1 = f$.

Suppose p is orthogonal to $f(p)$, so clearly the $f_i(p)$ $(i < |T|^+)$ are pairwise orthogonal. So if p is not orthogonal to A, for some $\bar{a}$, p is not orthogonal to $\mathrm{stp}(\bar{a}, A)$, hence $\mathrm{stp}(\bar{a}, A)$ is not orthogonal to every $f_i(p)$, contradicting 1.7. So we have proved: if p is orthogonal to $f(p)$ then p is orthogonal to A.

Now suppose p is orthogonal to A. W.l.o.g. there is $I \subseteq B$, $p = \mathrm{Av}(I, B)$, I indiscernible over A. So p is orthogonal to $\mathrm{tp}_*(f(I), A)$, hence by 2.8 p is orthogonal to $\mathrm{Av}(f(I), \bigcup f(I))$, which is parallel to $f(p) = \mathrm{Av}(f(I), f(B))$.

DEFINITION 3.1: Then pair (p, φ) is called regular, if $\varphi = \varphi(\bar{x}, \bar{c}) \in p$ and (clearly the definition does not depend on M)

(*) Let p be over $|M|$, M $\mathbf{F}^a_\kappa$-saturated ($\kappa = \kappa_r(T)$), $\mathrm{tp}(\bar{a}, |M|)$ a stationarization of p, N $\mathbf{F}^a_\kappa$-primary over $|M| \cup \bar{a}$. Then for every $\bar{b} \in |N|$, $\bar{b} \notin |M|$, such that $\bar{b}$ realizes φ; $\mathrm{tp}(\bar{b}, |M|)$ does not fork over $\bar{c}$.

Remark. Clearly if (p, φ) is regular, then p is regular. Also if φ is weakly minimal, $\varphi \in p$, p is minimal then (p, φ) is regular.

CLAIM 3.5: *Suppose M, N are $\mathbf{F}^a_{\kappa_r(T)}$-saturated, $|M| \subset |N|$, $\bar{a} \in |N|$, $\bar{a} \notin |M|$, $\alpha = R[\mathrm{tp}(\bar{a}, |M|), \mathrm{L}, \infty] < \infty$ and $\alpha = R[\varphi(\bar{x}, \bar{c}), \mathrm{L}, \infty]$ where $\varphi(\bar{x}, \bar{c}) \in \mathrm{tp}(\bar{a}, |M|)$. Suppose also* (i) *for every $\bar{b} \in |N|$, $\bar{b} \notin |M|$, $\alpha \leq R[\mathrm{tp}(\bar{b}, |M|), \mathrm{L}, \infty]$, or* (ii) *for every $\bar{b} \in \varphi(N, \bar{c})$, $\bar{b} \notin |M|$;*

$$\alpha \leq R[\mathrm{tp}(\bar{b}, |M|), \mathrm{L}, \infty].$$

Then p is regular and even $(\mathrm{tp}(\bar{a}, |M|), \varphi(\bar{x}, \bar{c}))$ *is regular.*

Proof. Choose $A \subseteq |M|$, $|A| < \kappa(T)$ so that $\operatorname{tp}(\bar{a}, |M|)$ does not fork over A, $\operatorname{tp}(\bar{a}, A)$ is stationary and $\bar{c} \in A$ and $\alpha = R[\operatorname{tp}(\bar{a}, A), \mathrm{L}, \infty]$ (A exists by III, 3.2, III, 2.9 and II, 1.2). If $q \in S^m(|M|)$ is not algebraic (where $m = l(\bar{a})$), $\operatorname{tp}(\bar{a}, A) \subseteq q$, and $q \neq \operatorname{tp}(\bar{a}, M)$ then q forks over A, hence by III, 1.2(4) $R(q, \mathrm{L}, \infty) < \alpha$, so N omits q by hypothesis. Thus we can conclude, by 1.9, that $\operatorname{tp}(\bar{a}, |M|)$ is regular. Similarly $(\operatorname{tp}(\bar{a}, |M|), \varphi)$ is regular.

Definition 3.2: The weight of p, $w(p)$ (p stationary or complete) is the minimal cardinal $\mu \geq 0$ (possibly finite) such that for some regular $\lambda \geq \kappa_r(T)$, some $\mathbf{F}^a_\lambda$-saturated model M, and a stationarization $\operatorname{tp}(\bar{a}, |M|)$ of p, there is an independent set $\{\bar{a}_i: i < \mu\}$ over $|M|$, $\bar{a}_i \in N$ where N is $\mathbf{F}^a_\lambda$-primary over $|M| \cup \bar{a}$, such that N is $\mathbf{F}^a_\lambda$-primary over $|M| \cup \{\bar{a}_i: i < \mu\}$ and $\operatorname{tp}(\bar{a}_i, |M|)$ is regular. If there are no such M, $\bar{a}, \bar{a}_i, N$ $w(p)$ is said to be ∞ or not defined. We let $w(\bar{a}, A) = w(\operatorname{tp}(\bar{a}, A))$.

LEMMA 3.6: (1) *If $w(p)$ is defined, (i.e., $<\infty$), it is $<\kappa(T)$ (and even $<\kappa_{\text{ind}}(T)$) and if $\bar{a}$, M, $\bar{a}_i$ are as in the definition,* $\operatorname{tp}(\bar{a}_i, |M| \cup \bar{a})$ *forks over* $|M|$.

(2) *If $\lambda \geq \kappa_r(T)$ is regular, M is $\mathbf{F}^a_\lambda$-saturated,* $\operatorname{tp}(\bar{a}, |M|)$ *does not fork over A, $A \subseteq |M|$, N is $\mathbf{F}^a_\lambda$-prime over $|M| \cup \bar{a}$, $\mu = w(\bar{a}, A) < \infty$ then we can find $\bar{a}_i$ ($i < \mu$) as in Definition* 3.2.

(3) *If M is $\mathbf{F}^a_\lambda$-saturated, $\lambda = \kappa_r(T)$,* $\operatorname{tp}(\bar{a}_i, M)$ *regular, $\{\bar{a}_i: i < \mu\}$ independent over M, and* $\operatorname{tp}(\bar{a}, |M| \cup \{\bar{a}_i: i < \mu\})$, $\operatorname{tp}_*(\{\bar{a}_i: i < \mu\}, |M| \cup \bar{a})$ *are $\mathbf{F}^a_\lambda$-isolated then $w(\bar{a}, |M|) = \mu$.*

Proof. (1) By III, 4.21 it suffices to prove the second part. If $\operatorname{tp}(\bar{a}_i, |M| \cup \bar{a})$ does not fork over M, N omits it by IV, 4.10(2).

(2) Left to the reader as an exercise.

(3) Easy.

LEMMA 3.7: *When T is superstable, $\lambda \geq \aleph_0$ regular, M is $\mathbf{F}^a_\lambda$-saturated, $\bar{a}$ a sequence then we can find $n < \omega$, $\bar{a}_l$ ($l < n$), M_l ($l \leq n$) such that $M_0 = M$, M_{l+1} is $\mathbf{F}^a_\lambda$-primary over $|M_l| \cup \bar{a}_l$,* $\operatorname{tp}(\bar{a}_l, |M_l|)$ *is regular; and M_n is $\mathbf{F}^a_\lambda$-primary over $|M| \cup \bar{a}$.*

Proof. We define by induction on $l < \omega$, $\bar{a}_l$ and M_l, M_l being $\mathbf{F}^a_\lambda$-saturated.

Let $M_0 = M$, and if M_l is defined let N_l be $\mathbf{F}^a_\lambda$-primary over $|M_l| \cup \bar{a}$. If $\bar{a} \in M_l$, we finish, otherwise among $\bar{a}_l \in |N_l| - |M_l|$ there is one with minimal $\alpha = R[\operatorname{tp}(\bar{a}_l, |M_l|), \mathrm{L}, \infty]$. Thus by 3.5 $\operatorname{tp}(\bar{a}_l, |M_l|)$

is regular. By 1.10(2) $\text{tp}(\bar{a}_l, |M_l| \cup \bar{a})$ forks over $|M_l|$, hence (by Ax(VI) for $\mathbf{F}^f_\infty$) $\text{tp}(\bar{a}, |M_l| \cup \bar{a}_l)$ forks over $|M_l|$. Let M_{l+1} be $\mathbf{F}^a_\lambda$-primary over $|M_l| \cup \bar{a}_l$, $\mathbf{F}^a_\lambda$-constructible over $|M_l| \cup \bar{a}$ (exists by IV, 3.17). So clearly $\text{tp}(\bar{a}, |M_{l+1}|)$ forks over $|M_l|$.

Clearly $M_0 \subseteq M_1 \subseteq M_2 \subseteq \cdots$

As T is superstable, $\kappa(T) = \aleph_0$ so for some $n < \omega$, M_{n+1} is not defined, so $\bar{a} \in M_n$. As M_{l+1} is $\mathbf{F}^a_\lambda$-constructible over $|M_l| \cup \bar{a}$, M_n is $\mathbf{F}^a_\lambda$-primary over $|M| \cup \bar{a}$ (by IV, 1.3(2)).

LEMMA 3.8: *Suppose that $\lambda \geq \kappa_r(T)$ is regular, M_i $(i \leq \alpha)$ is $\mathbf{F}^a_\lambda$-saturated, M_{i+1} is $\mathbf{F}^a_\lambda$-prime over $|M_i| \cup \bar{a}_i$ $(i < \alpha)$, M_δ is $\mathbf{F}^a_\lambda$-prime over $\bigcup_{i<\delta} M_i$ for δ $(\leq \alpha$ limit) and $\text{tp}(\bar{a}_i, |M_i|)$ is regular. Then for every $\bar{a} \in M_\alpha$, we can find a set $\{\bar{b}_i : i < \alpha\} \subseteq |M_\alpha|$ independent over $|M_0|$, $\text{tp}(\bar{b}_i, |M_0|)$ is regular, $\text{tp}[\bar{a}, |M_0| \cup \{\bar{b}_i : i < \beta\}]$ is $\mathbf{F}^a_\lambda$-isolated and $\beta \leq \alpha$. In fact $\beta < \kappa(T)$.*

Proof. We first prove that we can assume that for every i, $\text{tp}(\bar{a}_i, |M_i|)$ either is orthogonal to $|M_0|$ or does not fork over $|M_0|$. For each i, choose $B \subseteq |M_i|$, $|B| < \kappa(T)$, and $A \subseteq |M_0|$, $|A| < \kappa_r(T)$ such that: $\text{tp}(\bar{a}_i, |M_i|)$ does not fork over B, $\text{tp}_*(B \cup \bar{a}_i, |M_0|)$ does not fork over A, $\text{tp}(\bar{a}_i, B)$, $\text{tp}_*(B \cup \bar{a}_i, A)$ are stationary, and w.l.o.g. $A \subseteq B$: and if $\text{tp}(\bar{a}_i, |M_i|)$ is not orthogonal to $|M_0|$, it is not orthogonal to A. As M_0 is $\mathbf{F}^a_\lambda$-saturated, we can choose $B' \subseteq |M_0|$, $\bar{a}'_i \in |M_0|$ $\text{tp}_*(B \cup \bar{a}_i, A) = \text{tp}_*(B' \cup \bar{a}'_i, A)$. By 3.4 $\text{tp}(\bar{a}_i, B)$ is orthogonal to A iff it is orthogonal to $\text{tp}(\bar{a}'_i, B')$. So if $\text{tp}(\bar{a}_i, |M_i|)$ is not orthogonal to $|M_0|$, it is not orthogonal to A, hence not orthogonal to $\text{tp}(\bar{a}'_i, B')$. As $\text{tp}(\bar{a}_i, B)$ is regular, also $\text{tp}(\bar{a}'_i, B')$ is regular, hence for for some $\bar{a}^*_i \in M_{i+1}$, $\text{tp}(\bar{a}^*_i, |M_i|)$ extends $\text{tp}(\bar{a}'_i, B')$ and it does not fork over B' (by 1.12). By Claim 3.3 M_{i+1} is $\mathbf{F}^a_\lambda$-primary over $|M_i| \cup \bar{a}^*_i$, and clearly $\text{tp}(\bar{a}^*_i, |M_i|)$ does not fork over $|M_0|$. So we can replace $\bar{a}_i$ by $\bar{a}^*_i$. So we can assume that $\text{tp}(\bar{a}_i, |M_i|)$ is either orthogonal to $|M_0|$ or does not fork over $|M_0|$. So $|M_\alpha|$ is $|M_0| \cup \{\bar{c}_i : i < \beta\}$ where for each i one of the following cases occurs:

(i) $\text{tp}(\bar{c}_i, A_i)$ is $\mathbf{F}^a_\lambda$-isolated, where $A_i = |M_0| \cup \{\bar{c}_j : j < i\}$.

(ii) $\text{tp}(\bar{c}_i, A_i)$ is regular, orthogonal to $|M_0|$ and stationary; moreover, it does not fork over some $B_i \subseteq A_i$, $|B_i| < \kappa_r(T)$, where $\text{tp}(\bar{c}_i, B_i)$ is stationary.

(iii) $\text{tp}(\bar{c}_i, A_i)$ is regular and does not fork over $|M_0|$.

As in IV, 1.4(2), 4.4–7 we can assume that for some $\gamma < \kappa_r(T)$, $\gamma < \beta$ $\bar{a} \in A_\gamma$. By III, 3.2 and III, 2.9, we can choose $B \subseteq |M_0|$, $|B| < \kappa_r(T)$

such that $\text{tp}_*[\{\bar{c}_j : j < \gamma\}, |M_0|]$ does not fork over B, and $\text{tp}_*[\{\bar{c}_j : j < \gamma\}, B]$ is stationary and if case (ii) occurs for $\bar{c}_i$, $\text{tp}(\bar{c}_i, B \cup \{\bar{c}_j : j < i\})$ is stationary. Let $C = \bigcup \{\bar{c}_i : \text{tp}(\bar{c}_i, A_i)$ does not fork over $|M_0|$, and is regular$\}$. We now prove that for $i < \gamma$, $\text{tp}(\bar{c}_i, A_i \cup C)$ is $\mathbf{F}^a_\lambda$-isolated. If case (iii) holds, $\bar{c}_i \in C \subseteq A_i \cup C$, so this is trivial. If case (i) holds, then $\text{tp}_*[C - A_i, A_i]$ does not fork over $|M_0|$, so our conclusion follows by IV, 4.10(2). If case (ii) holds, as $\text{tp}(\bar{c}_i, A_i)$ is orthogonal to $|M_0|$, also $\text{tp}(\bar{c}_i, B \cup \{\bar{c}_j : j < i\})$ is orthogonal to B. As $\text{tp}_*[\{\bar{c}_j : j < i\}, |M_0|]$ does not fork over B, for every $\bar{b} \in |M_0|$ $\text{tp}(\bar{b}, B \cup \{\bar{c}_j : j < i\})$ does not fork over B, hence it is orthogonal to $\text{tp}(\bar{c}_i, B \cup \{\bar{c}_j : j < i\})$; but the latter is stationary, so they are weakly orthogonal by 1.2(2). Hence

$$\text{tp}(\bar{c}_i, B \cup \{\bar{c}_j : j < i\}) \vdash \text{tp}(\bar{c}_i, B \cup \{\bar{c}_j : j < i\} \cup \bar{b}).$$

As $\bar{b} \in |M_0|$ was arbitrary,

$$\text{tp}(\bar{c}_i, B \cup \{\bar{c}_j : j < i)\} \vdash \text{tp}(\bar{c}_i, |M_0| \cup \{\bar{c}_j : j < i\}) = \text{tp}(\bar{c}_i, A_i).$$

As $i < \gamma < \kappa_r(T)$, $|B| < \kappa_r(T) \leq \lambda$, $\text{tp}(\bar{c}_i, A_i)$ is $\mathbf{F}^s_\lambda$-isolated hence $\mathbf{F}^a_\lambda$-isolated; so as before $\text{tp}(\bar{c}_i, A_i \cup C)$ is $\mathbf{F}^a_\lambda$-isolated. So we finish the three cases hence A_γ is $\mathbf{F}^a_\lambda$-constructible over $|M_0| \cup C$, so by IV, 3.2 $\text{tp}(\bar{a}, |M_0| \cup C)$ is $\mathbf{F}^a_\lambda$-isolated so we finish.

THEOREM 3.9: (1) *If T is superstable, for every $\bar{a}$, A $w(\bar{a}, A)$ is well-defined and finite.*

(2) *Assuming the conditions of Lemma* 3.8, *$w(\bar{a}, |M_0|)$ exists and is* $\leq |\beta| \leq |\alpha|$.

Proof. (1) By 3.7 it suffices to prove part (2).

Before proceeding to (2) we make two observations:

Observation 3.9A: Suppose $|A| + |B| + |C| < \kappa$, $C \subseteq C^*$, $\mathbf{F} = \mathbf{F}^x_\kappa$ for $x = t, s, a$ and $\kappa \geq \kappa_r(T)$ is regular. If

(i) $\text{tp}_*(A, C^* \cup B)$ is $\mathbf{F}$-isolated over $C \cup B$,

(ii) $\text{tp}_*(B, C \cup A) \vdash \text{tp}_*(B, C^*)$,

(iii) $\text{tp}_*(B, C \cup A)$ is $\mathbf{F}$-isolated,

then $\text{tp}_*(B, C^* \cup A)$ is $\mathbf{F}$-isolated over $C \cup A$.

Proof of 3.9A. The idea appears in the proof of 3.3, so we leave it to the reader.

Observation 3.9B: Suppose $q \in S^m(|M|)$ is regular; I, J are sets of sequences realizing q, independent over $|M|$, $|I| + |J| < \kappa$, M is $\mathbf{F}^a_\kappa$-saturated where $\kappa \geq \kappa(T)$ is regular, and $A \subseteq |M|$

If for every $\bar{c} \in J$, $\operatorname{tp}(\bar{c}, |M| \cup \bigcup I)$ forks over A then there is $B \subseteq |M|$, $|B| < \kappa$ such that $\operatorname{tp}_*(\bigcup J, B \cup \bigcup I) \vdash \operatorname{tp}_*(\bigcup J, |M|)$.

Proof of 3.9B. Choose $B \subseteq |M|$, $|B| < \kappa$ such that $\operatorname{tp}_*(\bigcup (I \cup J), |M|)$ does not fork over B, and $q \restriction B$ is stationary, and let $I^* \subseteq |M|$ be a maximal independent set over B of sequences realizing $q \restriction B$. Clearly $\operatorname{tp}_*(\bigcup J, B \cup \bigcup I)$ is orthogonal to q, hence to $\operatorname{tp}_*(\bigcup I^*, B)$ hence

(i) $\operatorname{tp}_*(\bigcup J, B \cup \bigcup I) \vdash \operatorname{tp}_*(\bigcup J, B \cup \bigcup I^*)$.

Now for any $\bar{b}_i \in I^+ = \{\bar{b} \in |M|: \bar{b}$ realizes $q \restriction B\}$ $\operatorname{tp}(\bar{b}_0{}^\frown \cdots {}^\frown \bar{b}_n, B \cup \bigcup I^*)$ is orthogonal to q, hence

(ii) $\operatorname{tp}(\bigcup J, B \cup \bigcup I^*) \vdash \operatorname{tp}(\bigcup J, B \cup \bigcup I^+)$.

By claim 1.11

(iii) $\operatorname{tp}(\bigcup J, B \cup \bigcup I^+) \vdash \operatorname{tp}(\bigcup J, |M|)$.

Combining (i), (ii), and (iii) we are finished.

Proof of 3.9(2). We can easily add to the conclusion of 3.8 that (letting $M = M_0$) $\operatorname{tp}(\bar{b}_i, |M_0|)$, $\operatorname{tp}(\bar{b}_j, |M_0|)$ are orthogonal or equal. So let $I = \{\bar{b}_i: i < \beta\} = \bigcup_{j<\gamma} I_j$, the I_j's are pairwise disjoint, and for every $\bar{c} \in I_j$, $\operatorname{tp}(\bar{c}, |M|) = p_j$, and for $j(1) \neq j(2)$, $p_{j(1)}, p_{j(2)}$ are orthogonal. Let N be $\mathbf{F}^a_\kappa$-prime over $|M| \cup \{\bar{b}_i: i < \alpha\}$ such that $\bar{a} \in N$.

Now choose $N_1 \subseteq N$ $\mathbf{F}^a_\lambda$-prime over $|M| \cup \bar{a}$ and choose a maximal $J = \bigcup_{j<\gamma} J_j \subseteq N_1$, independent over $|M|$, such that each sequence in J_j realizes p_j. Let $N_2 \subseteq N_1$ be $\mathbf{F}^a_\lambda$-prime over $|M| \cup \bigcup J$.

By the maximality of J, no sequence in N_1 realizes over N_2 a complete type which is a stationarization of some p_j; hence by 1.12 $\operatorname{tp}(\bar{a}, |N_2|)$ is orthogonal to each p_j. By 1.14, there are $I^1_j \subseteq I_j$ such that for every $\bar{c} \in I^1_j$, $\operatorname{tp}(\bar{c}, |M| \cup J_j \cup (I_j - I^1_j))$ forks over $|M|$, hence is orthogonal to p_j, and $J_j \cup (I_j - I^1_j)$ is independent over M. As clearly $|J| \leq |I| < \kappa(T)$, by observation 3.9B there is $B \subseteq |M|$, $|B| < \lambda$ such that letting $I^*_j = J_j \cup (I_j - I^1_j)$, $I^* = \bigcup_{j<\gamma} I^*_j$:

(i) $\operatorname{tp}_*(\bigcup I_j, B \cup \bigcup I^*_j) \vdash \operatorname{tp}_*(\bigcup I_j, |M|)$.

As $\operatorname{tp}_*(\bigcup I_j, |M|)$ $(j < \gamma)$ are pairwise orthogonal

(ii) $\bigcup_{j<\gamma} \operatorname{tp}_*(\bigcup I_j, |M|) \vdash \operatorname{tp}_*(\bigcup I, |M|)$.

Hence

(iii) $\operatorname{tp}_*(\bigcup I, B \cup \bigcup I^*) \vdash \operatorname{tp}_*(\bigcup I, |M|)$.

As N is $\mathbf{F}^a_\lambda$-prime over $|M| \cup \bigcup I$, $\operatorname{tp}_*(\bigcup I^*, |M| \cup \bigcup I)$ is $\mathbf{F}^a_\lambda$-isolated (w.l.o.g. over $B \cup \bigcup I$), and $\operatorname{tp}_*(\bigcup I, B \cup \bigcup I^*)$ is $\mathbf{F}^a_\kappa$-isolated (as $|B \cup \bigcup I^*| < \lambda$); so by Observation 3.9A (with $B, M, \bigcup I, \bigcup I^*$

standing for C, C^*, B, A, respectively) $\mathrm{tp}_*(\bigcup I, |M| \cup \bigcup I^*)$ is $\mathbf{F}^a_\lambda$-isolated.

By IV, 3.6 as N is $\mathbf{F}^a_\lambda$-prime over $|M| \cup \bigcup I$, it is also $\mathbf{F}^a_\lambda$-prime over $|M| \cup \bigcup I \cup \bigcup I^*$, but as $\mathrm{tp}(\bigcup I, |M| \cup \bigcup I^*)$ is $\mathbf{F}^a_\lambda$-isolated, N is F^a_λ-prime over $|M| \cup \bigcup I^*$, hence $\mathrm{tp}(\bar{a}, |M| \cup \bigcup I^*)$ is $\mathbf{F}^a_\lambda$-isolated.

By 1.4(1) I^* is independent over $|M|$, so as N_2 is $\mathbf{F}^a_\lambda$-prime over $|M| \cup J$, by 3.2, $I^* - J$ is independent over $(|N_2|, |M|)$. As $\mathrm{tp}(\bar{a}, |N_2|)$ is orthogonal to each p_j, it is orthogonal to $\mathrm{tp}_*(\bigcup (I^* - J), |N_2|)$, hence $\mathrm{tp}_*(\bigcup (I^* - J), |N_2| \cup \bar{a})$ does not fork over $|N_2|$, hence over $|M|$, hence $\mathrm{tp}(\bar{a}, |M| \cup \bigcup I^*)$ does not fork over $|M| \cup \bigcup J$. So by IV, 4.3, as $\mathrm{tp}(\bar{a}, |M| \cup I^*)$ is $\mathbf{F}^a_\lambda$-isolated also $\mathrm{tp}(\bar{a}, |M| \cup \bigcup J)$ is $\mathbf{F}^a_\lambda$-isolated.

On the other hand, as $J \subseteq N_1$, N_1 $\mathbf{F}^a_\kappa$-prime over $|M| \cup \bar{a}$, also $\mathrm{tp}_*(J, |M| \cup \bar{a})$ is $\mathbf{F}^a_\lambda$-isolated, so by 3.6(3) $w(\bar{a}, |M|)$ is well-defined and equal to $|J| \le |I| = |\beta| \le |\alpha|$.

LEMMA 3.10: *Let p be stationary or complete.*

(1) *The type p is algebraic iff $w(p) = 0$.*

(2) *If p is regular then $w(p) = 1$.*

EXERCISE 3.1: Prove that the converse of (2) fails. (Hint: Exercise 4.11.)

Proof of 3.10. (1) Trivial by Definition 3.1.

(2) Immediate by 3.3 and Definition 3.1.

LEMMA 3.11: (1) *If $\{\bar{a}, \bar{b}\}$ is independent over A then $w(\bar{a}^\frown\bar{b}, A) = w(\bar{a}, A) + w(\bar{b}, A)$ (i.e., if the right side is defined, then so is the left side and they are equal.)*

(2) $w(\bar{a}, A) \le w(\bar{a}^\frown\bar{b}, A) \le w(\bar{a}, A) + w(\bar{b}, A \cup \bar{a})$.

Remark. Notice that if $\{\bar{a}, \bar{b}\}$ is independent over A then $w(\bar{b}, A) = w(\bar{b}, A \cup \bar{a})$.

Proof. (1) We can assume $A = |M|$, M is $\mathbf{F}^a_\lambda$-saturated, $\lambda = \kappa_r(T)$. Let N_1 be $\mathbf{F}^a_\lambda$-primary over $|M| \cup \bar{a}$ and also over $|M| \cup \{\bar{c}_i : i < w(\bar{a}, A)\}$, and N_2 be $\mathbf{F}^a_\lambda$-primary over $|M| \cup \bar{b}$ and also over

$$|M| \cup \{\bar{c}_i : w(\bar{a}, A) \le i < w(\bar{a}, A) + w(\bar{b}, A)\},$$

where $\mathrm{tp}(\bar{c}_i, M)$ is regular and $\{\bar{c}_i : i < w(\bar{a}, A)\}$, $\{\bar{c}_i : w(\bar{a}, A) \le i < w(\bar{a}, A) + w(\bar{b}, A)\}$ are independent sets over A. By 3.2 $\mathrm{tp}_*(|N_1|, |N_2|)$ does not fork over $|M|$, so $\{\bar{c}_i : i < w(\bar{a}, A) + w(\bar{b}, A)\}$ is independent over $|M|$. By 3.2 $|N_1| \cup |N_2|$ is $\mathbf{F}^a_\lambda$-constructible over $|M| \cup \bar{a} \cup \bar{b}$, so $\mathrm{tp}_*[\{c_i: i < w(\bar{a}, A) + w(\bar{b}, A)\}, |M| \cup \bar{a} \cup \bar{b}]$ is $\mathbf{F}^a_\lambda$-isolated. In a similar way $\mathrm{tp}(\bar{a}^\frown\bar{b}, |M| \cup \{\bar{c}_i: i < w(\bar{a}, A) + w(\bar{b}, A)\}$ is $\mathbf{F}^a_\lambda$-isolated. So we finish.

(2) Left to the reader.

LEMMA 3.12: (1) *If* $\{\bar{c}_i: i < \alpha\}$ *is independent over* A, (*and* $\bar{c}_i \notin \mathrm{acl}\, A$) $\mathrm{tp}(\bar{c}_i, A \cup \bar{a})$ *forks over* A (*for* $i < \alpha$) *then* $|\alpha| \le w(\bar{a}, A)$.

(2) *If* N *is* $\mathbf{F}^a_\lambda$*-prime over* $|M| \cup \bar{a}$, M *is* $\mathbf{F}^a_\lambda$*-saturated*, $\lambda = \kappa_r(T)$, $\{\bar{c}_i: i < \alpha\}$ *independent over* $|M|$, $\bar{c}_i \in |N| - |M|$ *then* $|\alpha| \le w(\bar{a}, A)$.

Proof. (1) We can clearly assume $w(\bar{a}, A) < \infty$, i.e., defined and that $A = |M|$, M is $\mathbf{F}^a_\lambda$-saturated, $\lambda = \kappa_r(T)$. Let N be $\mathbf{F}^a_\lambda$-primary over $|M| \cup \bar{a}$, and $\{\bar{a}_i: i < \mu = w(\bar{a}, A)\}$ be as in Definition 3.1. Now define by induction on $i < \mu$ sets $u_i \subseteq \alpha$ such that $\mathrm{tp}_*[\{\bar{c}_j: j \in u_i\}, |M| \cup \{\bar{a}_j: j < i\}]$ does not fork over $|M|$, and $|\alpha - u_i| \le |i|, j < i \Rightarrow u_i \subseteq u_j$. For $i = 0$, $u_i = \alpha$, and for limit δ, $u_\delta = \bigcap_{i<\delta} u_i$. If u_i is defined, by renaming let it be $\{j: j < |u_i|\}$. If for every $\zeta < |u_i|$, $\mathrm{tp}[\bar{c}_\zeta, |M| \cup \{\bar{c}_j: j < \zeta\} \cup \{\bar{a}_j: j \le i\}]$ does not fork over $|M| \cup \{\bar{a}_j: j < i\}$, then by Ax(VII) for $\mathbf{F}^f_\infty$ $\mathrm{tp}_*[\{\bar{c}_j: j < |u_i|\}, |M| \cup \{\bar{a}_j: j \le i\}]$ does not fork over $|M| \cup \{\bar{a}_j: j < i\}$, hence by III, 4.4 and the induction hypothesis it does not fork over $|M|$. So we can let $u_{i+1} = u_i$. So suppose $\mathrm{tp}[\bar{c}_\zeta, |M| \cup \{\bar{c}_j: j < \zeta\} \cup \{\bar{a}_j: j \le i\}]$ forks over $|M| \cup \{\bar{a}_j: j < i\}$, and this is the first such ζ. For every $\bar{b} \in \bigcup \{\bar{c}_j: \zeta < j < |u_i|\}$, $\{\bar{c}_\zeta, \bar{b}\}$ is independent over $|M| \cup \{\bar{c}_i: i < \zeta\} \cup \{\bar{a}_j: j < i\}$ (because $\{\bar{c}_j: j < \alpha\}$ is independent over $|M|$ and the induction hypothesis). Now $\mathrm{tp}(\bar{a}_i, |M| \cup \{\bar{c}_j: j < \zeta\} \cup \{\bar{a}_j: j < i\})$ is regular (it does not fork over $|M| \cup \{\bar{a}_j: j < i\}$ by the choice of ζ, and over $|M|$ as $\{\bar{a}_j: j < \mu\}$ is independent over $|M|$ using III, 4.4). By the choice of ζ, $\mathrm{tp}[\bar{a}_i, |M| \cup \{\bar{c}_j: j < \zeta\} \cup \{\bar{a}_j: j < i\} \cup \bar{c}_\zeta]$ forks over $|M| \cup \{\bar{c}_j: j < \zeta\} \cup \{\bar{a}_j: j < i\}$. Combining those three facts and remembering 3.1 we get that

$$\mathrm{tp}_*[\{\bar{c}_j: \zeta < j < |u_i|\}, |M| \cup \{\bar{c}_j: j < \zeta\} \cup \{\bar{a}_j: j \le i\}]$$

does not fork over $|M| \cup \{\bar{c}_j: j < \zeta\} \cup \{\bar{a}_j: j < i\}$, hence it does not fork over $|M|$. So clearly $\mathrm{tp}_*[\{\bar{c}_j: j < |u_i|, j \ne \zeta\}, |M| \cup \{\bar{a}_j: j \le i\}]$ does not fork over $|M|$. So let $u_{i+1} = u_i - \{\zeta\}$. For any $j \in u_\mu$ we get $\mathrm{tp}[\bar{c}_j, |M| \cup \{\bar{a}_j: j < \mu\}]$ does not fork over $|M|$. So by 3.2, $\mathrm{tp}(\bar{c}_j, |M| \cup \bar{a})$ does not fork over $|M|$, contradiction. So $u_\mu = \emptyset$ hence $|\alpha| \le \mu$.

(2) Follows by (1) and 1.10(2).

CONCLUSION 3.13: *Let I, A be given.*

(1) *If J_1, J_2 are maximal subsets of I independent over A, then* $|J_2| \le \sum \{w(\bar{a}, A): \bar{a} \in J_1\}$.

(2) *If $p \in S^m(A)$, J a maximal subset of $\{\bar{c}: \bar{c} \in I, \bar{c}$ realizes $p\}$ independent over A then* $\dim(p, A, I) \le |J| \le w(p) \dim(p, A, I)$.

(3) *If $p \in S^m(A)$, $w(p) = 1$ then* $\dim(p, A, I)$ *is true.*

Remark. See Definition III, 4.5 for $\dim(p, A, I)$.

Proof. (1) Clearly if for some $\bar{a} \in J_1$, $w(\bar{a}, A) = \infty$, the conclusion is trivial, so we can assume $w(\bar{a}, A) < \infty$. First assume J_1 is finite and let $J_1 = \{\bar{c}_l : l < n\}$. Let $\bar{c}^* = \bar{c}_0 {}^\frown \cdots {}^\frown \bar{c}_{n-1}$, so by 3.11(1) $k \overset{\text{def}}{=} w(\bar{c}^*, A) = \sum_{l<n} w(\bar{c}_l, A)$. So if $k < |J_2|$, by 3.12(1) for some $\bar{c} \in J_2$, $\operatorname{tp}(\bar{c}, A \cup \bar{c}^*)$ does not fork over A, hence $\{\bar{c}_i: i < n\} \cup \{\bar{c}\}$ is independent over A, contradiction to the maximality of J_1. Hence

$$|J_2| \le k \le \sum \{w(\bar{c}, A): \bar{c} \in J_1\}.$$

So we finish for finite J_1 and for the infinite J_1, if $\lambda \le |J_2|$ is regular, $|J_2| > |J_1|$, then for some finite $J_1' \subseteq J_1$, for λ many $\bar{c} \in J_2$, $\operatorname{tp}(\bar{c}, A \cup \bigcup J_1')$ forks over A, and we get a contradiction as above.

(2) is immediate from (1) and the definition of dim,

(3) follows from (2).

LEMMA 3.14: *If $|M| \subseteq |N|$, M, N are $\mathbf{F}^a_{\kappa_r(T)}$-saturated then there is a cardinal $\mu > 0$ such that for every maximal set $\{\bar{a}_i: i < \alpha\}$ independent over $|M|$, of sequences $\bar{a}_i \in N$, $w(\bar{a}_i, |M|) < \infty$, this equation holds; $\mu = \sum_{i<\alpha} w(\bar{a}_i, |M|)$.*

Proof. Through the regular types and is left to the reader.

THEOREM 3.15: (1) *If T is superstable, I independent over A, $\bar{a}$ a sequence then for some $J \subseteq I$, $|J| \le w(\bar{a}, A)$, $(I - J) \cup \{\bar{a}\}$ is independent over A.*

(2) *For superstable T,*

$$\dim(p, A \cup \bar{a}, I) \le \dim(p, A, I) \le \dim(p, A \cup \bar{a}, I) + w(\bar{a}, A).$$

Remark. Work for stable T too.

Proof. (1) We can assume w.l.o.g. that $A = |M|$, M is $\mathbf{F}^a_\lambda$-saturated $\lambda = \kappa_r(T)$. Let $I = \{\bar{a}^i : 0 < i < \alpha\}$, $\bar{a} = \bar{a}^0$; by 3.6(2) there is a family

$\{\bar{a}^i_k\colon k < w(\bar{a}^i, A)\}$ independent over A, $\mathrm{tp}(\bar{a}^i_k, A)$ regular, and $\mathrm{tp}[\bar{a}^i, |M| \cup \{\bar{a}^i_k\colon k < w(\bar{a}^i, A)\}]$, $\mathrm{tp}_*[\{\bar{a}^i_k\colon k < w(\bar{a}, A)\}, A \cup \bar{a}^i]$ are $\mathbf{F}^a_\lambda$-isolated. By 3.2 $\{\bar{a}^i_k\colon 0 < i < \alpha, k < w(\bar{a}^i, A)\}$ is independent over A. By 1.14 there is $u \subseteq \{i\colon 0 < i < \alpha\}$, $|u| \le w(\bar{a}^0, A)$, such that $\{a^i_k\colon i \notin u, k < w(\bar{a}^i, A)\}$ is independent over $(A \cup \{\bar{a}^i_k\colon i = 0 \text{ and } k < w(\bar{a}^i, A)\}, A)$. By 3.2 the results are immediate.

(2) Left to the reader.

Now we define upper and lower weights.

DEFINITION 3.3: Let p be stationary or complete. Define the upper weight p, $\mathrm{upw}(p)$, as the supremum of the cardinals μ such that there exists M, N, I satisfying the following:

(1) M is $\mathbf{F}^a_\kappa$-saturated, $\kappa = \kappa_r(T)$.

(2) p has a stationarization which is a complete type over $|M|$ realized in N by $\bar{a}$.

(3) N is a $\mathbf{F}^a_\kappa$-primary model over $|M| \cup \bar{a}$.

(4) I is a set of sequences realizing non-algebraic types over $|M|$ and I is independent over $|M|$.

(5) $|I| = \mu$.

Define $\mathrm{upw}(p) = \mathrm{upw}(\bar{a}, A) = \mathrm{upw}(\mathrm{tp}(\bar{a}, A))$.

DEFINITION 3.4: For p as above define the lower weight, low (p), as the maximal cardinal μ such that there are $M, N, I, \bar{a}$ satisfying:

(1) M is $\mathbf{F}^a_\kappa$-saturated, $\kappa = \kappa_r(T)$.

(2) $\mathrm{tp}(\bar{a}, |M|)$ is a stationarization of p.

(3) N is $\mathbf{F}^a_\kappa$-primary over $|M| \cup \bar{a}$.

(4) $I \subseteq |N|$ is a set of sequences realizing regular types over $|M|$, I independent over $|M|$, and is a maximal such set.

(5) $|I| = \mu$.

If in addition, q is regular and stationary define $\mathrm{low}_q(p)$ to be the maximal μ satisfying the above conditions plus $\bar{c} \in I \Rightarrow \mathrm{tp}(\bar{c}, |M|)$ is not orthogonal to q.

LEMMA 3.16: (1) *In Definitions* 3.3 *and* 3.4, *if the appropriate weight is* μ *then for any* M, N, I, *satisfying* (1)–(4), *also* $|I| = \mu$.

(2) $\mathrm{upw}(p) = \mathrm{upw}(r)$, $\mathrm{low}(p) = \mathrm{low}(r)$, $\mathrm{low}_q(p) = \mathrm{low}_q(r)$ *if* $p = \mathrm{tp}(\bar{a}, A)$, $r = \mathrm{stp}(\bar{a}, A)$, *or* p, q *are parallel.*

(3) *If* $w(p)$ *is defined, then* $\mathrm{upw}(p) = w(p) = \mathrm{low}(p)$ (*so this always holds for superstable* T). *Also* $\mathrm{low}(p) \le \mathrm{upw}(p) < \kappa(T)$.

(4) $\mathrm{low}_q(p) = 0$ *iff* p *is orthogonal to* q (q *regular, of course*).

(5) *If q, r are non-orthogonal, and regular then* $\mathrm{low}_q(p) = \mathrm{low}_r(p)$.

(6) *If $\{q_i: i < \alpha\}$ is a maximal family of stationary, regular, pairwise orthogonal types not orthogonal to p then* $\sum_i \mathrm{low}_{q_i}(p) = \mathrm{low}(p)$.

(7) $\mathrm{upw}(\bar{a}\frown\bar{b}, A) \geq \mathrm{upw}(\bar{a}, A) + \mathrm{upw}(\bar{b}, A \cup \bar{a})$ *when $\{\bar{a}, \bar{b}\}$ is independent over A.*

(8) $\mathrm{low}_q(\bar{a}\frown\bar{b}, A) \leq \mathrm{low}_q(\bar{a}, A) + \mathrm{low}_q(\bar{b}, A \cup \bar{a})$.

(9) $\mathrm{low}(\bar{a}\frown\bar{b}, A) \leq \mathrm{low}(\bar{a}, A) + \mathrm{low}(\bar{b}, A \cup \bar{a})$.

(10) *If $\{\bar{a}, \bar{b}\}$ is independent over A, then equality holds in* (8) *and* (9).

(11) $\mathrm{low}_q(\bar{a}, A) \leq \mathrm{low}_q(\bar{a}\frown\bar{b}, A)$ *and similarly for* low *and* upw.

Proof. We leave to the reader (1)–(7) and (10), (11).

(8) W.l.o.g. $A = |M|$, M is $\mathbf{F}^a_\kappa$-saturated, $\kappa = \kappa_r(T)$, q is over $|M|$. Let N_1 be $\mathbf{F}^a_\kappa$-primary over $|M| \cup \bar{a}$, let $\bar{b}'$ be such that $\mathrm{tp}(\bar{a}\frown\bar{b}, |M|) = \mathrm{tp}(\bar{a}\frown\bar{b}', |M|)$ and $\mathrm{tp}(\bar{b}', N_1)$ does not fork over $|M| \cup \bar{a}$. Let N_2 be $\mathbf{F}^a_\kappa$-primary over $|N_1| \cup \bar{b}'$; and let q_1, q_2 be the stationarization of q over $|M|$, $|N_1|$ respectively (so clearly $q \subseteq q_1 \subseteq q_2$). Let I_l be a maximal set of sequences from $|N_l|$ realizing q_l which is independent over $|M|$ when $l = 1$, and over $|N_1|$ when $l = 2$. Clearly $I_1 \cup I_2$ is a set of sequences from $|N_2|$ realizing q_1 which is independent over $|M|$. It is a maximal such set. For suppose $\bar{c} \in |N_2|$, $\bar{c}$ realizes q_2. By the maximality of I_2, $\mathrm{tp}(\bar{c}, |N_1| \cup I_2)$ forks over $|N_1|$ hence over $|M|$. By 1.11 letting $J = \{\bar{d}: \bar{d} \in N_1, \bar{d}$ realizes $q_1\}$ $\mathrm{tp}(\bar{c}, |M| \cup J \cup I_2)$ forks over $|M|$. But for every $\bar{d} \in J$, $\mathrm{tp}(\bar{d}, |M| \cup I_1)$ forks over $|M|$, hence by 1.9 $\mathrm{tp}(\bar{c}, |M| \cup I_1 \cup I_2)$ forks over $|M|$. If $\bar{c}$ realizes q_1 but not q_2 this holds trivially so $I_1 \cup I_2$ is indeed maximal. Now let $N^* \subseteq N_2$ be $\mathbf{F}^a_\kappa$-primary over $|M| \cup \bar{a} \cup \bar{b}'$; and I be a maximal set of sequences from N^* realizing q_1 which is independent over $|M|$. Then by 1.14 $|I| = \mathrm{low}_q(\bar{a}\frown\bar{b}, A)$, and by 1.14, $|I| \leq |I_1 \cup I_2| = |I_1| + |I_2|$; and this proves our assertion.

(9) Follows by (6) and (8).

CLAIM 3.16A: (1) *In Definition* 3.4 *there is a maximum.*

(2) *Assume M is $\mathbf{F}^a_\kappa$-saturated, $\kappa = \mathrm{cf}(\kappa) \geqslant \kappa_r(T)$, $|A| < \kappa$, p is regular (and over M) $\boldsymbol{I}$ is a maximal family satisfying:*

(i) *for each $\bar{a} \in \boldsymbol{I}$,* $\mathrm{tp}(\bar{a}, M)$ *is regular not orthogonal to p,*

(ii) *for each $\bar{a} \in \boldsymbol{I}$,* $\mathrm{tp}(\bar{a}, M \cup A)$ *fork over M,*

(iii) *$\boldsymbol{I}$ is independent over M.*

Then (a) $|\boldsymbol{I}| = \mathrm{low}_p(\mathrm{tp}_*(A, M))$,

(b) $\mathrm{tp}(\bigcup \boldsymbol{I}, M \cup A)$ *is $\mathbf{F}^a_\kappa$-isolated,*

(c) *if N is $\mathbf{F}^a_\kappa$-constructible over $M \cup A$ $\mathbf{F}^a_\kappa$-primary over $M \cup \bigcup \boldsymbol{I}$, then* $\mathrm{tp}_*(A, N)$ *is orthogonal to p.*

(3) *Suppose for* $\rho \in \{1, 2\}$, κ, M, p, A_ρ, I_ρ, N_ρ *are as in* (2) *and* $\{A_1, A_2\}$ *is independent over* M *and* N *is* $\mathbf{F}^a_\kappa$*-primary over* $N_1 \cup N_2$. *Then* κ, M, p, $A_1 \cup A_2$, $I_1 \cup I_2$, N *are as in* (2).

Remark. We can replace p by a regular family $\mathscr{P}$ (see Definition 4.5)

Proof. Straightforward.

QUESTION 3.2: Is the independence assumption necessary in 3.16(7)? When equality holds?

QUESTION 3.3: In Definition 3.3 is the supremum obtained?

QUESTION 3.4: In Question 3.3, prove the answer is positive when $\operatorname{upw}(\bar{a}, A) = \aleph_0$.

PROBLEM 3.5: Try to generalize the results of this section on superstable theories to stable T for which $\kappa_{\text{ind}}(T) = \aleph_0$.

DEFINITION 3.5: (A) Define $\operatorname{lgw}(p)$ just as low (p) but μ is minimal and

(4) for $\bar{c} \in I$, in the $\mathbf{F}^a_\kappa$-primary model over $|M| \cup \bar{c}$ there is no independent set with more than one element, and

(6) I is a maximal set satisfying (4), of sequences $\bar{c} \in |N|$, $\bar{c} \notin |M|$.

(B) Define $\operatorname{ugw}(p)$ just as $\operatorname{lgw}(p)$, omitting (6), and

(3) N is $\mathbf{F}^a_\kappa$-primary over $|M| \cup I$.

If there are no such M, N, I, $\operatorname{ugw}(p) = \infty$.

LEMMA 3.17: (1) *If* $\kappa_{\text{ind}}(T) = \aleph_0$, *then* $\operatorname{lgw}(p) < \aleph_0$.

(2) *We get the same* $|I|$ *for every* M, $\bar{a}$, N *in Definition* 3.5.

(3) *If* $\{\bar{a}, \bar{b}\}$ *is independent over* A *then:*

$$\operatorname{ugw}(\bar{a}^\frown\bar{b}, A) \le \operatorname{ugw}(\bar{a}, A) + \operatorname{ugw}(\bar{b}, A).$$

(4) *If* T *is superstable then* $w(p) = \operatorname{ugw}(p) = \operatorname{lgw}(p)$.

(5) $\operatorname{lgw}(p) \le \operatorname{ugw}(p)$.

Proof. We leave it as an exercise to the reader.

QUESTION 3.6: Try to adapt the theorems and definitions of Section 2 (e.g., $\le_w$, $\le_s$), Section 3 (e.g., $w(p)$) and also Section 1 to other kinds of $\mathbf{F}$, mainly $\mathbf{F}^t_{\aleph_0}$ (replace, if necessary "T superstable" by "T totally transcendental" or add it) and also $\mathbf{F}^l_{\aleph_0}$ (for countable T). (See below.)

THEOREM 3.18: *Suppose* $\varphi = \varphi(\bar{x}, \bar{c}) \in p \in S^m(A)$; *then the following conditions satisfy:* (1) $\Rightarrow$ (2) $\Leftrightarrow$ (3) $\Leftrightarrow$ (4) $\Leftrightarrow$ (5) *and if* $|A| + |T| \leq \aleph_0$ *then also* (6) $\Leftrightarrow$ (5); *and if* T *is totally transcendental then also* (7) $\Leftrightarrow$ (5).

(1) *The pair* (p, φ) *is not regular.*

(2) *There are* $\varphi_1(\bar{x}; \bar{c}_1)$, $\psi_i(\bar{x}; \bar{a}_i)$ (*for* $i < k$), $\theta_i(\bar{x}; \bar{y}_0, \ldots, \bar{y}_{n(i)}, \bar{a}_i)$ ($i < l \leq k$) *and finite* Δ_i ($i < k$), *such that, letting* $\alpha_i = R^m(p_1, \Delta_i, 2)$ ($i < k$) *for any stationarization* p_1 *of* p, *the following conditions hold:*

(i) $\varphi_1(\bar{x}; \bar{c}_1) \in p$, $\varphi_1(\bar{x}, \bar{c}_1) \vdash \varphi(\bar{x}; \bar{c})$.

(ii) $\psi_i(\bar{x}; \bar{a}_i) \vdash \varphi(\bar{x}; \bar{c})$ *for* $i < k$.

(iii) $\varphi_1(\bar{x}; \bar{c}_1) \vdash \bigvee_{i<l} (\exists \bar{y}_0, \ldots, \bar{y}_{n(i)})[\bigwedge_{j \leq n(i)} \psi_i(\bar{y}_j; \bar{a}_i)$
$\wedge\ \theta_i(\bar{x}; \bar{y}_0, \ldots, \bar{y}_{n(i)}, \bar{a}_i)] \vee \bigvee_{l \leq i < k} \psi_i(\bar{x}; \bar{a}_i)$.

(iv) $R^m[\psi_i(\bar{x}; \bar{a}_i), \Delta_i, 2] < \alpha_i$ *when* $i < k$.

(v) *For any* $i < l$, *and* $\bar{c}_0, \ldots, \bar{c}_{n(1)}$

$$R^m[\theta_i(\bar{x}; \bar{c}_0, \ldots, \bar{c}_{n(i)}, \bar{a}_i), \Delta_i, 2] < \alpha_i.$$

(3) *The same as* (2) *but* $n(i) = 0$, $\psi_i = \psi$, $\theta_i = \theta$, $\Delta_i = \Delta$ *for* $i < l$; *and for* $l \leq i < k$, $\Delta_i = \Delta^0$, $\psi_i = \psi$; *and* $\bar{a}_i$ *is a concatenation of sequences satisfying* $\varphi(\bar{x}; \bar{c}) \bigvee \bar{x} = \bar{c}$ *for* $i < k$.

(4) *If* $A \subseteq |M|$, $\mathrm{tp}(\bar{b}, |M|)$ *is a stationarization of* p, $|M| \cup \bar{b} \subseteq |N|$ *then for some* $\bar{b}' \in |N|$, $\bar{b}' \notin |M|$, $\bar{b}'$ *satisfies* $\varphi(\bar{x}; \bar{c})$ *but* $\mathrm{tp}(\bar{b}', |M|)$ *is not a stationarization of* p.

(5) *There are* $\kappa > |A| + |T|$, $A \subseteq |M|$, M *a* $\mathbf{F}^a_\kappa$*-saturated model,* N $\mathbf{F}^a_\kappa$*-primary over* $|M| \cup \bar{b}$, $\mathrm{tp}(\bar{b}, |M|)$ *is a stationarization of* p, *and* $\bar{b}' \in |N|$, $\bar{b}' \notin |M|$, $\bar{b}'$ *satisfies* $\varphi(\bar{x}; \bar{c})$, *but* $\mathrm{tp}(\bar{b}', |M|)$ *is not a stationarization of* p.

(6) *There are a countable* M *and* $\bar{b}$ *such that* $\mathrm{tp}(\bar{b}, |M|)$ *is a stationarization of* p, $A \subseteq |M|$, *and for every* N, *for which* $|M| \cup \bar{b} \subseteq |N|$, *some* $\bar{b}' \in |N|$, $\bar{b}' \notin |M|$ *satisfies* $\varphi(\bar{x}; \bar{c})$ *but* $\mathrm{tp}(\bar{b}', |M|)$ *is not a stationarization of* p.

(7) *like* (6), *for not necessarily countable* M.

Remark. By III, 4.2, for all stationarizations p_1 of p, $R^m(p_1, \Delta_i, 2)$ is the same number.

Proof. (2) $\Rightarrow$ (4) Let A, M, N, $\bar{b}$ be as in (4). By II, 2.12(2) the properties of $\bar{a}_i$ ($i < k$) are expressable by finitely many formulas, with parameters from $\bar{c}_1$. Hence we can assume $\bar{a}_i \in |M|$; $\bar{c}_1 \in |M|$ as $A \subseteq |M|$.

As $q = \mathrm{tp}(\bar{b}, |M|)$ is a stationarization of p, $R^m(q, \Delta_j, 2) = \alpha_j$ (for

$j < k$), hence by (2)(iv) $\neg\psi_i(\bar{x}; \bar{a}_i) \in q$ when $l \leq i < k$; and clearly by (2)(i) $\varphi_1(\bar{x}; \bar{c}_1) \in q$. So by (2)(iii) necessarily for some $i < l$,

$$N \vDash (\exists \bar{y}_0, \ldots, \bar{y}_{n(i)}) \left[\bigwedge_{j \leq n} \psi_i(\bar{y}_j; \bar{a}_i) \wedge \theta_i(\bar{b}; \bar{y}_0, \ldots, \bar{y}_{n(i)}, \bar{a}_i) \right].$$

Hence there are $\bar{d}_0, \ldots, \bar{d}_{n(i)} \in |N|$ such that $N \vDash \psi_i[\bar{d}_j; \bar{a}_i]$ (for $j \leq n(i)$) and $N \vDash \theta_i(\bar{b}; \bar{d}_0, \ldots, \bar{d}_{n(i)}; \bar{a}_i)$. By (2)(v)$R^m$ $[\theta_i(\bar{x}; \bar{d}_0, \ldots, \bar{d}_{n(i)}, \bar{a}_i), \Delta_i, 2] < \alpha_i$, hence $\theta_i(x; \bar{d}_0, \ldots, \bar{d}_{n(i)}, \bar{a}_i) \notin q$; but $q = \mathrm{tp}(\bar{b}, |M|)$ so for some $j \leq n(i)$ $\bar{d}_j \notin |M|$. But as $N \vDash \psi_i[\bar{d}_j; \bar{a}_i]$, by (2)(ii) $N \vDash \varphi[\bar{d}_j; \bar{c}]$. So $\bar{d}_j$ satisfies $\varphi(\bar{x}, \bar{c})$, and as $\psi_i(\bar{x}, \bar{a}_i) \in \mathrm{tp}(\bar{d}_j, |M|)$, clearly $R^m[\mathrm{tp}(\bar{d}_j, |M|), \Delta_i, 2] \leq R^m[\psi_i(\bar{x}; \bar{a}_i), \Delta_i, 2] < \alpha_i$ hence $\mathrm{tp}(\bar{d}_j, |M|)$ is not a stationarization of p. So $\bar{d}_j$ satisfies the requirements on $\bar{b}'$, hence (4) holds.

(4) $\Rightarrow$ (5) Trivial.

(5) $\Rightarrow$ (3) Let $\{p_i: i < \beta \leq 2^{|T|}\}$ be a list of the stationarizations of p in $S^m(|M|)$ and $p_0 = \mathrm{tp}(\bar{b}, |M|)$. Choose $\psi_0(\bar{x}; \bar{a}_0) \in \mathrm{tp}(\bar{b}', |M|)$ such that $\neg\psi_0(\bar{x}, \bar{a}_0) \in p_0$ and we can assume that $\bar{a}_0 \in \cup I \cup \bar{c}$ (by 1.11) where $I = \{\bar{d} \in |M|: \vDash\varphi[\bar{d}; \bar{c}]\}$, and $\psi_0(\bar{x}, \bar{a}_0) \vdash \varphi(\bar{x}; \bar{c})$. So by 1.11 $\mathrm{tp}(\bar{b}', A \cup \bigcup I \cup \bar{b})$ is $\mathbf{F}^a_\kappa$-isolated, hence it forks over $A \cup \bigcup I$ hence $\mathrm{tp}(\bar{b}, A \cup \bigcup I \cup \bar{b}')$ forks over $A \cup \bigcup I$. So we can find $\bar{a}^0 \in A \cup \bigcup I \subseteq |M|$ and θ such that $\vDash\theta[\bar{b}, \bar{b}', \bar{a}^0]$ and $\theta(\bar{x}; \bar{b}', \bar{a}^0)$ forks over $A \cup \bigcup I$.

Hence by III, 4.2 we can find a finite Δ such that

$$R^m[p_0 \cup \{\theta(\bar{x}; \bar{b}', \bar{a}^0)\}, \Delta, 2] < R^m(p_0, \Delta, 2).$$

As by 1.11 $\mathrm{tp}(\bar{b}, A \cup \bigcup I) \equiv p_0$, we can replace p_0 by $\mathrm{tp}(\bar{b}, A \cup \bigcup I)$. By replacing $\theta(\bar{x}; \bar{b}', \bar{a}^0)$ by $\theta(\bar{x}; \bar{b}', \bar{a}^0) \wedge \psi^1(\bar{x}, \bar{a}^1)$ where $\psi^1(\bar{x}; \bar{a}^1) \in \mathrm{tp}(\bar{b}, A \cup \bigcup I)$,

$$R^m[p_0 \cup \{\theta(\bar{x}; \bar{b}', \bar{a}^0)\}\Delta, 2] = R^m[\{\psi^1(\bar{x}; \bar{a}^1), \theta(\bar{x}; \bar{b}', \bar{a})\}, \Delta, 2]$$

we can assume that $R^m[\theta(\bar{x}; \bar{b}', \bar{a}^0), \Delta, 2] < R^m(p_0, \Delta, 2)$. Similarly by II, 2.12,3 we can assume that for every $\bar{b}'', \bar{a}'', R^m[\theta(\bar{x}; \bar{b}'', \bar{a}''), \Delta, 2] < R^m(p_0, \Delta, 2)$.

So clearly for every $\bar{b}^*$ realizing p_0 there is $\bar{b}^{**}$ satisfying $\psi_0(\bar{x}; \bar{a}_0)$ such that $\vDash\theta[\bar{b}^*; \bar{b}^{**}, \bar{a}^0]$. Clearly for every $i < \beta$ there are $\bar{a}_i, \bar{a}^i \in |M|$ such that $p_i, \psi_0(\bar{x}; \bar{a}_i), \theta(\bar{x}; \bar{y}, \bar{a}^i)$ satisfy the parallel assertion.

Hence for every $\bar{b}^*$ satisfying $\varphi(\bar{x}; \bar{c})$ $\bar{b}^*$ satisfies some formula ψ, $\neg\psi \in \bigcap_{i<\beta} p_i$. *or* for some $i < \alpha \vDash (\exists \bar{y})[\psi_0(\bar{y}; \bar{a}_i) \wedge \theta(\bar{b}^*; \bar{y}, \bar{a}^i)]$. By a compactness argument we get $i(n) < \alpha_0$ $(n < l)$ and $\neg\psi_n(\bar{x}; \bar{a}^1_n) \in \bigcap_{i<\beta} p_i$ $(l \leq n < k)$ such that:

$$\varphi(\bar{x}, \bar{c}) \vdash \bigvee_{l \leq n < k} \psi_n(\bar{x}, \bar{a}^1_n) \wedge \bigvee_{n<l} (\exists \bar{y})[\psi_0(\bar{y}, \bar{a}_{i(n)}) \wedge \theta(\bar{x}; \bar{y}, \bar{a}^{i(n)})].$$

Clearly for each n, $l \leq n < k$, $p \cup \{\psi_n(\bar{x}; \bar{a}_n^1)\}$ forks over A, hence there is $\varphi_1(\bar{x}; \bar{c}_1) \in p$ such that $\varphi_1(\bar{x}; c_1) \vdash \varphi(\bar{x}; \bar{c})$ and $\varphi_1(\bar{x}, \bar{c}_1) \wedge \psi_n(\bar{x}; \bar{a}_n^1)$ forks over A. By III, 4.2 there is a finite Δ^0 such that for every $\bar{b}''$ realizing p, and $n < l$ $R^m[\mathrm{stp}(\bar{b}'', A) \cup \{\varphi_1(\bar{x}, \bar{c}_1) \wedge \psi_n(\bar{x}; \bar{a}_n^1)\}, \Delta^0, 2] < R^m[\mathrm{stp}(\bar{b}'', A), \Delta^0, 2]$. By III, Definition 2.1 there is $E(\bar{x}, \bar{y}, \bar{a}^*) \in FE(A)$ such that for any $\bar{b}''$ realizing p $R^m[\mathrm{stp}(\bar{b}'', A), \Delta^0, 2] = R^m[E(\bar{x}, \bar{b}'', \bar{a}^*), \Delta^0, 2]$.

Now by notational changes we can assume $\bar{a}_j = \bar{a}^l$, and

$$\vDash (\forall \bar{x}, \bar{y}) \bigwedge_{l \leq k < n} [\psi_n(\bar{x}; a_n^1) \wedge \psi_n(\bar{y}; a_n^1) \rightarrow E(\bar{x}; \bar{y}, \bar{a}^*)].$$

Now (3) is clear (by 1.11 it is easy to get $a_l \subseteq \bigcup I \cup \bar{c}$).

(3) $\Rightarrow$ (2) Trivial.

(1) $\Rightarrow$ (5) Immediate by the definition of a regular pair.

(4) $\Rightarrow$ (6) (when $|A| + |T| \leq \aleph_0$) Immediate.

(6) $\Rightarrow$ (5) (when $|A| + |T| \leq \aleph_0$) For each $\psi = \psi(\bar{x}, \bar{a})$ such that $p \cup \{\psi(\bar{x}; \bar{a})\}$ forks over A, $\bar{a} \in |M|$ (e.g., is inconsistent) and $\vDash (\exists \bar{x})(\varphi \wedge \psi)$ let $r_\psi = \{\varphi(\bar{x}; \bar{c}), \psi(\bar{x}; \bar{a})\} \cup \{\bar{x} \neq \bar{d}: \bar{d} \in |M|\}$ and Γ be the set of such r_ψ's. So Γ is countable and there is no model N, $|M| \cup \bar{b} \subseteq |N|$ omitting all types in Γ. Hence by IV, 5.3(1) there is $r_\psi \in \Gamma$, $\bar{a}_1 \in |M|$ and a formula $\varphi_1(\bar{x}; \bar{a}_1, \bar{b})$ such that $\vDash (\exists \bar{x})\varphi_1(\bar{x}; \bar{a}_1, \bar{b})$ and $\varphi_1(\bar{x}; \bar{a}_1, \bar{b}) \vdash r_\psi$. So choose a $|T|^+$-saturated model M_1, $M \subseteq M_1$, such that $\mathrm{tp}(\bar{b}, |M_1|)$ does not fork over $|M|$.

As clearly $\varphi_1(\bar{x}; \bar{a}_1, \bar{b}) \vdash \bar{x} \neq \bar{d}$ for every $\bar{d} \in |M_1|$, in the $\mathbf{F}^a_{\kappa_r(T)}$-prime model over $|M_1| \cup \bar{b}$, some $\bar{b}' \notin |M_1|$ will satisfy $\varphi(\bar{x}, \bar{c})$ but $\mathrm{tp}(\bar{b}, |M_1|)$ will not be a stationarization of p. Hence (5) holds.

(4) $\Rightarrow$ (7), (7) $\Rightarrow$ (5) (for T totally transcendental) Left to the reader.

CONCLUSION 3.19: (1) *If T is totally-transcendental, $M \subseteq N$, $M \neq N$ then for some $b \in |N| - |M|$, $\mathrm{tp}(b, |M|)$ is regular.*

(2) *If $M \subseteq N$, $b \in |N| - |M|$, $\varphi(x, \bar{c}) \in \mathrm{tp}(b, |M|)$, $R[\varphi(x, \bar{c}), L, \aleph_0] = R[\mathrm{tp}(b, |M|), L, \aleph_0] < \infty$ and for no $b_1 \in |N| - |M|$ which satisfies $\varphi(x, \bar{c})$ is $R[\mathrm{tp}(b_1, |M|), L, \aleph_0] < R[\mathrm{tp}(b, |M|), L, \aleph_0]$ then $[\mathrm{tp}(b, |M|), \varphi(x, \bar{c})]$ is regular.*

Proof. (1) By (2).

(2) Use 3.18 for $A = |M|$, $p = \mathrm{tp}(b, |M|)$.

Clearly condition (4) fails, by our hypothesis, hence (1) fails, but not (1) is our conclusion.

EXERCISE 3.7: Suppose $\bar{a}_1$, $\bar{a}_2$ are sequences, M is κ-saturated.

(1) If $\mathrm{tp}(\bar{a}_1, |M| \cup \bar{a}_2)$ is $\mathbf{F}^s_\kappa$-isolated then $\mathrm{tp}(\bar{a}_1, |M| \cup \bar{a}_2)$ is $\mathbf{F}^l_{\aleph_0}$-isolated [i.e., for every φ there is $\psi_\varphi(\bar{x}; \bar{c}_\varphi) \in \mathrm{tp}(\bar{a}_1, |M| \cup \bar{a}_2)$ such that $\psi_\varphi(\bar{x}; c_\varphi) \vdash \mathrm{tp}_\varphi(\bar{a}_1, |M| \cup \bar{a}_2)$].

(2) The converse of (1) holds when $\kappa \geq \kappa(T)$.

(3) Suppose there are M, $\bar{a}_1$, $\bar{a}_2$, $\psi_\varphi(\bar{x}; \bar{c}_\varphi)$ as above, and $\mathrm{tp}(\bar{a}'_l, |N|)$ is parallel to $\mathrm{tp}(\bar{a}_l, |M|)$, N is $\mathbf{F}^a_{\kappa(T)}$-saturated.

Then we can find $\bar{a}''_1$, $\bar{c}'_\varphi$ such that $\mathrm{tp}(\bar{a}''_1, |N|) = \mathrm{tp}(\bar{a}'_1, |N|)$, and $\psi_\varphi(\bar{x}, \bar{c}'_\varphi) \in \mathrm{tp}(\bar{a}''_1, |N| \cup \bar{a}'_2)$, $\psi_\varphi(\bar{x}, \bar{c}'_\varphi) \vdash \mathrm{tp}_\varphi(\bar{a}''_1, |N| \cup \bar{a}'_2)$.

EXERCISE 3.8: (1) Suppose $I \leq_s J$; then

$$w\left(\mathrm{Av}\left(I, \bigcup I\right)\right) \geq w\left(\mathrm{Av}\left(J, \bigcup J\right)\right),$$

and for superstable T if they are finite and equal then $J \leq_s I$. (Hint: See 3.9(2), and its proof.)

(2) For superstable T $I \leq_s J$ iff for every regular q, $\mathrm{low}_q(\mathrm{Av}(I, \bigcup I)) \geq \mathrm{low}_q(\mathrm{Av}(J, \bigcup J))$, and $I \leq_w J$ iff for every regular q,

$$\mathrm{low}_q\left(\mathrm{Av}\left(I, \bigcup I\right)\right) < 1 \Rightarrow \mathrm{low}_q\left(\mathrm{Av}\left(J, \bigcup J\right)\right) < 1.$$

(3) The relation $I \approx J =_{\mathrm{def}} I \leq_s J \wedge J \leq_s I$ is an equivalence relation, and for superstable T, for each I, the number of $J/\approx$, $I \leq_s J$ is finite (in fact $\leq 2^{w(\mathrm{Av}(I, \cup I))}$). A similar result holds for $\leq_w$.

In a series of exercises, we now check whether Section 3 Exercise 3.8(2) generalizes to $\mathbf{F}^t_{\aleph_0}$-primeness.

EXERCISE 3.9: If M is κ-compact, $|M| \subseteq A$, B is $\mathbf{F}^t_\kappa$-constructible over A, $\bar{b} \in B$ then $\mathrm{tp}(\bar{b}, A)$ forks over $|M|$. (Compare with 1.10(2). What about F^s_κ?) [Hint: It suffices to prove that if $\mathrm{tp}(\bar{b}, A)$ does not fork over $|M|$, $\mathrm{tp}(c, A)$ is $\mathbf{F}^t_\kappa$-isolated, then $\mathrm{tp}(\bar{b}, A \cup \{c\})$ does not fork over $|M|$; for this it suffices to prove $\mathrm{tp}(c, A) \vdash \mathrm{tp}(c, A \cup \bar{b})$. Suppose not, then for some $p \subseteq \mathrm{tp}(c, A)$ $|p| < \kappa$, $p \vdash \mathrm{tp}(c, A)$, p is closed under conjunctions and for some θ and $\bar{a} \in A$, $p \cup \{\theta(x, \bar{b}, a)^{\mathbf{t}}\}$ is consistent for $\mathbf{t} = 0, 1$. Let $r = \{(\exists x)(\varphi \wedge \theta(x, \bar{y}, \bar{a})) \wedge (\exists x)(\varphi \wedge \neg\theta(x, \bar{y}, \bar{a}) : \varphi \in p\}$ so $r \subseteq \mathrm{tp}(\bar{b}, A)$, $|r| < \kappa$. As $\mathrm{tp}(\bar{b}, A)$ does not fork over $|M|$, it is finitely satisfiable in it. Hence by the definability lemma and the κ-compactness of M, some $\bar{b}' \in |M|$ realizes it. So $p \cup \{\theta(x, \bar{b}', \bar{a})^{\mathbf{t}}\}$ is consistent for $\mathbf{t} = 0, 1$, contradiction.]

DEFINITION 3.6: The pair (p, φ) ($\varphi \in p \in S^m(A)$ for some A) is strongly regular if it fails to satisfy conditions (2)–(5) of 3.18. We call the complete type p strongly regular if for some φ (p, φ) is strongly regular.

EXERCISE 3.10: Show that the pair (p, φ) is strongly regular, iff $\varphi \in p \in S^m(A)$ for some m, A, and every stationary and complete q containing φ is orthogonal to p or is a stationarization of p.

EXERCISE 3.11: Suppose T is totally transcendental $M \subseteq N$, $M \neq N$. Show that for some $c \in |N| - |M|$, $\text{tp}(c, |M|)$ is strongly regular.

EXERCISE 3.12: If $(\varphi(\bar{x}; \bar{c}), p)$ $(p \in S^m(A))$ is strongly regular, the dependence relation on $\varphi(\mathfrak{C}, \bar{c})$ "$\text{tp}(\bar{a}, A \cup \bar{a}_1 \cup \cdots \cup \bar{a}_n)$ does fork over A, or do not extend p" satisfies the axioms of a nice dependency relation (Def. AP 3.4). Prove a similar assertion for several (φ_i, p_i) $p_i \in S^{m(i)}(A)$ strongly regular, pairwise not orthogonal; together.

EXERCISE 3.13: Suppose N is $\mathbf{F}^t_{\aleph_0}$-primary over $|M| \cup \bar{a}$, $\text{tp}(\bar{a}, |M|)$ is strongly regular, $\bar{b} \in |N|$, $\bar{b} \notin |M|$. Then N is $\mathbf{F}^t_{\aleph_0}$-primary over $|M| \cup \bar{b}$. (Hint: See 3.3 and 3.18(7).)

EXERCISE 3.14: For totally transcendental T, show that in 3.7 we can assume $\text{tp}(\bar{a}_i, |M_i|)$ is strongly regular; and does not fork over $|M|$ or is orthogonal to $|M|$. We can replace $\mathbf{F}^a_\lambda$ by $\mathbf{F}^t_{\aleph_0}$. (Hint: Use Exercise 3.11 and Exercise 3.9. Or see XI, §3.)

EXERCISE 3.15: Suppose T is totally transcendental, $\text{tp}(\bar{a}, |M|)$ is strongly regular, and $\text{tp}(\bar{b}, |M| \cup \bar{a})$ is orthogonal to $|M|$. Show that not necessarily $\text{tp}(\bar{b}, |M| \cup \bar{a})$ is $\mathbf{F}^t_{\aleph_0}$-isolated.

Remark. It is $\mathbf{F}^a_{\aleph_0}$-isolated by the proof of 3.9. In the example one equivalence relations and $\aleph_0$ disjoint one place predicates suffice.

EXERCISE 3.16: Suppose $M \subseteq N$, $\bar{a} \in |N|$, $\bar{a} \notin |M|$, $p \in S^m(|M|)$ is strongly regular and not orthogonal to $\text{tp}(\bar{a}, |M|)$. Then p is realized in N. (Hint: Use Exercise 3.12. This is similar to 1.12 and see.)

EXERCISE 3.17: Show that in 3.15(1) we cannot demand $I - J$ is independent over $(A \cup \bigcup J \cup \bar{a}, A)$. (Hint: Use vector spaces over a finite field, I a linearly independent set, $A = \emptyset$, $J \subseteq I$ finite, $a = \sum_{x \in J} x$.)

EXERCISE 3.18: If $\bar{b} \in \text{acl}(A \cup \bar{a})$ then $\text{stp}(\bar{a}, A) \leq_s \text{stp}(\bar{b}, A)$ hence $w(\bar{b}, A) \leq w(\bar{a}, A)$. Check the parallel assertions for other weights.

EXERCISE 3.19: Suppose $\kappa \geq \kappa_r(T)$, M is $\mathbf{F}^a_\kappa$-saturated, N $\mathbf{F}^a_\kappa$-prime over $|M| \cup \bigcup I$, I a set of sequences realizing over $|M|$ regular types, and $J \subseteq |N|$ a maximal set independent over $|M|$, of sequences realizing over $|M|$ regular types. Then N is $\mathbf{F}^a_\kappa$-prime over $|M| \cup \bigcup J$. (Hint: See the proof of 3.9(2).)

V.4. Semi-regular and semi-minimal types

THEOREM 4.1: *Assume:*

(i) $\kappa \geq \kappa(T)$ *is regular.*

(ii) $M \subseteq N$ *are* $\mathbf{F}^a_\kappa$*-saturated,* $A \subseteq |M|$, $|A| < \kappa$.

(iii) p *is a stationary regular type.*

(iv) $\bar{c}_l \in N$ $(l < k < \aleph_0)$ *are such that* $p_l = \text{tp}(\bar{c}_l, |M|)$ *is regular and not orthogonal to* p.

(v) p_l *does not fork over* A $(l < k)$.

(vi) $p_l \restriction A$ *is stationary* $(l < k)$.

(vii) *Every* $\bar{b} \in |N| - |M|$ *which realizes* $p_l \restriction A$ *realizes* p_l $(l < k)$.

(viii) $I \subseteq |N|$ *is a maximal independent (over* $|M|$*) set of sequences which realize over* $|M|$ *regular types not orthogonal to* p.

(ix) $\bar{c}$ *is algebraic over* $|M| \cup \bigcup_{l<k} \bar{c}_l$.

Then $\text{tp}(\bar{c}, |M| \cup \bigcup I)$ *is* $\mathbf{F}^a_\kappa$*-isolated, and even* $\mathbf{F}^s_\kappa$*-isolated.*

DEFINITION 4.1: (1) A stationary type q is sem-regular if it is not algebraic and there are κ, M, N, A, k, $\bar{c}$, $\bar{c}_l$ as in Theorem 4.1 and $\text{tp}(\bar{c}, |M|)$ is parallel to q. Clearly we can assume there is an $n \leq k$ such that $\bar{c}_l$ depends on $\{\bar{c}_i : i < l\}$ over $|M|$ iff $l \geq n$; and $l < n \Rightarrow p_l = p_0$, and $I = \{\bar{c}_i : i < n\}$.

(2) If in addition the types $\text{tp}(\bar{c}_l, |M|)$ are minimal, q is called semi-minimal, and then we can assume $n = k$. Clearly every stationary regular (minimal) type is semi-regular (minimal).

Proof of Theorem 4.1. By III, 3.5 there is $J \subseteq I$, $|J| < \kappa(T)$ such that $\text{tp}(\bar{c}', |M| \cup \bigcup I)$ does not fork over $|M| \cup \bigcup J$, where $\bar{c}' = \bar{c}_0{}^\frown \cdots {}^\frown \bar{c}_{k-1}$; let $J = \{\bar{a}_i : i < \beta\}$. Choose $B \subseteq |M|$, $|B| < \kappa$, $A \subseteq B$ such that $\text{tp}_*(\bar{c} \cup \bigcup_{l<k} \bar{c}_l \cup \bigcup J, |M|)$ does not fork over B and its restriction to B is stationary (possible by III, 2.9(1) and III, 3.2), and $\bar{c}$ is algebraic over $B \cup \bar{c}'$.

By the maximality of I, for every $l < k$, $\text{tp}(\bar{c}_l, |M| \cup I \cup \bigcup_{i<l} \bar{c}_i)$ forks over $|M|$, hence $\text{tp}(\bar{c}_l, |M| \cup \bigcup J \cup \bigcup_{i<l} \bar{c}_i)$ forks over $|M|$, hence

it is orthogonal to p_l (as p_l is regular). As non-orthogonality is an equivalence relation among the stationary regular types (by 1.13(1)) it is orthogonal to p. So by 1.6(1) $\mathrm{tp}(\bar{c}_0{}^\frown\cdots{}^\frown\bar{c}_{k-1}, |M| \cup \bigcup J)$ is orthogonal to p, so $\mathrm{tp}(\bar{c}', M \cup \bigcup J)$ is orthogonal to p, hence by 1.4(1) to $\mathrm{tp}_*(I - J, |M| \cup \bigcup J)$. So by 1.2(1):

(1) $\mathrm{tp}(\bar{c}', |M| \cup \bigcup J) \vdash \mathrm{tp}(\bar{c}', |M| \cup \bigcup I)$. Let $I_0 = \{\bar{b}_i: i < \gamma\}$ be a maximal set of sequences in $|M|$, each realizing some $p_l \upharpoonright B$, which is independent over B. By the Definition of B, $\mathrm{tp}(\bar{c}', |M| \cup \bigcup J)$ does not fork over $B \cup \bigcup J$, I_0 is independent over $B \cup \bigcup J$ and $\mathrm{tp}(\bar{c}', B \cup \bigcup J)$ is also orthogonal to $p_l \upharpoonright B$. As the latter is stationary:

(2) $\mathrm{tp}(\bar{c}', B \cup \bigcup J) \vdash \mathrm{tp}(\bar{c}', B \cup \bigcup J \cup \bigcup I_0)$. Now suppose $\bar{d} \in C = \bigcup\{\bar{a}\colon \bar{a} \in |M|,\ \bar{a}$ realizes $p_l \upharpoonright B,\ l < k\}$ and we shall prove that $\mathrm{tp}(\bar{c}', B \cup \bigcup J \cup \bigcup I_0)$, $\mathrm{tp}(\bar{d}, B \cup \bigcup J \cup \bigcup I_0)$ are weakly orthogonal. Let $J_0 \subseteq I_0$, $|J_0| < \kappa(T)$ be such that $\mathrm{tp}(\bar{d}, B \cup \bigcup I_0)$ does not fork over $B \cup \bigcup J_0$, and we can assume $\bar{d} = \bar{d}^0{}^\frown,\ldots,{}^\frown\bar{d}_n$, $\bar{d}_l \in |M|$, $\bar{d}_l$ realizes $p_{i(l)} \upharpoonright B$. If the above mentioned types are not weakly orthogonal there are J_0', $J_0 \subseteq J_0' \subseteq I_0$, $|J_0'| < \kappa(T)$ and a sequence $\bar{d}' = \bar{d}_0'{}^\frown\cdots{}^\frown\bar{d}_n'$ in N such that

$$\mathrm{tp}\Big(\bar{d}', B \cup \bigcup J \cup \bigcup J_0'\Big) = \mathrm{tp}\Big(\bar{d}, B \cup \bigcup J \cup \bigcup J_0'\Big)$$

but $\mathrm{tp}(\bar{c}'{}^\frown\bar{d}, B \cup \bigcup J \cup \bigcup J_0') \neq \mathrm{tp}(\bar{c}'{}^\frown\bar{d}', B \cup \bigcup J \cup \bigcup J_0')$ (because $\kappa > |B| + |J| + |J_0|$ and N is $\mathbf{F}^a_\kappa$-saturated). By the maximality of I_0, $\mathrm{tp}(\bar{d}_l, B \cup \bigcup I_0)$ forks over B, hence $\mathrm{tp}(\bar{d}_l, B \cup \bigcup J_0) = \mathrm{tp}(\bar{d}_l', B \cup \bigcup J_0)$ forks over B, hence $\mathrm{tp}(\bar{d}_l', B \cup \bigcup J_0')$ forks over B, hence $\mathrm{tp}(\bar{d}_l', |M|) \neq p_{i(l)}$. By hypothesis this implies $\bar{d}_l' \in |M|$, hence $\bar{d}' \in |M|$. Now by the choice of $\bar{d}'$, for some $\bar{b} \in \cup J$, $\mathrm{tp}(\bar{b}{}^\frown\bar{c}', B \cup \bigcup I_0 \cup \bar{d} \cup \bar{d}')$ forks over $B \cup \bigcup J_0$, remember $\mathrm{tp}(\bar{b}{}^\frown\bar{c}', B)$ is stationary, hence $\mathrm{tp}(\bar{b}{}^\frown\bar{c}', |M|)$ forks over B. But by the choice of B, $\mathrm{tp}(\bar{b}{}^\frown\bar{c}', |M|)$ does not fork over B, contradiction. So for every $\bar{d} \in C$,

$$\mathrm{tp}\Big(\bar{c}', B \cup \bigcup I_0 \cup \bigcup J\Big) \vdash \mathrm{tp}\Big(\bar{c}', B \cup \bigcup I_0 \cup \bigcup J \cup \bar{d}\Big)$$

hence

(3) $\mathrm{tp}(\bar{c}', B \cup \bigcup I_0 \cup \bigcup J) \vdash \mathrm{tp}(\bar{c}', B \cup \bigcup J \cup C)$. Now let $C_1 = \bigcup\{\bar{a}\colon \bar{a} \in N,\ \bar{a}$ realizes $p_l \upharpoonright B,\ l < k\}$, so by 1.11 for every $\bar{b} \in |M|$, $\mathrm{tp}(\bar{b}, B \cup C) \vdash \mathrm{tp}(\bar{b}, B \cup C_1)$ hence (noting $\bar{c}' \cup J \subseteq C_1$)

$$\mathrm{tp}_*(\bar{c}' \cup J, B \cup C) \vdash \mathrm{tp}_*(\bar{c}' \cup J, |M|)$$

(as this holds for every $\bar{b}$). Thus

(4) $\mathrm{tp}(\bar{c}', B \cup \bigcup J \cup C) \vdash \mathrm{tp}(\bar{c}', |M| \cup \bigcup J)$.

Combining (2), (3), (4) and (1) we see that

$$\operatorname{tp}\left(\bar{c}', B \cup \bigcup J\right) \vdash \operatorname{tp}\left(\bar{c}', |M| \cup \bigcup I\right).$$

As $\bar{c}$ is algebraic over $B \cup \bar{c}'$, and $|B|$, $|J| < \kappa$, we finish.

LEMMA 4.2: *For every semi-regular type q, there is a stationary regular type p such that:*

(1) *p, q are not orthogonal.*

(2) *For any type r, r is orthogonal to p iff r is orthogonal to q.*

(3) *If I, J are infinite indiscernible sets based on q, p resp., then $I \leq_s J \leq_w I$.*

Proof. Easy. (For (3) use Theorem 4.1.)

CONCLUSION 4.3: *The relation of being non-orthogonal, is an equivalence relation among the semi-regular types.*

Proof. By the preceding lemma and the fact that non-orthogonality among the stationary regular types is an equivalence relation.

Notation. Hence if p is stationary and regular, q is semi-regular and they are not orthogonal, we shall define $\operatorname{low}_q(r)$ as $\operatorname{low}_p(r)$.

LEMMA 4.4: *If* $\operatorname{tp}(\bar{c}_l, A)$ *are minimal and pairwise not orthogonal, then* $\operatorname{tp}(\bar{c}_0 {}^\frown \cdots {}^\frown \bar{c}_{k-1}, A)$ *is semi-minimal.*

Proof. Easy (as here Condition (vii) in Theorem 4.1 holds trivially.)

CONCLUSION 4.5: *If p is semi-regular, then $w(p)$ is defined is finite and* $w(p) = \operatorname{low}_p(p)$.

Proof. We use the notation of 4.1, and assume in addition that N is $\mathbf{F}^{\mathrm{a}}_{\kappa}$-primary over $|M| \cup \bigcup J$. Then by 1.14, $|J| \leq k < \aleph_0$. Hence by 3.9 $w(\bar{c}, |M|) \leq k < \aleph_0$.

LEMMA 4.6: (1) *Let* $\bar{b} \in \operatorname{acl}(A \cup \bar{c})$, $\bar{b} \notin \operatorname{acl} A$. *If* $\operatorname{stp}(\bar{c}, A)$ *is semi-regular [semi-minimal] then* $\operatorname{stp}(\bar{b}, A)$ *is semi-regular [semi-minimal] too,* $\operatorname{stp}(\bar{b}, A)$ *is not orthogonal to* $\operatorname{stp}(\bar{c}, A)$*; and* $w(\bar{b}, A) \leq w(\bar{c}, A)$, *and* $\operatorname{low}_p(\bar{b}, A) \leq \operatorname{low}_p(\bar{c}, A)$ *for any p.*

(2) *If* $\{\bar{c}, \bar{b}\}$ *is independent over* A *and* $\mathrm{stp}(\bar{c}, A)$, $\mathrm{stp}(\bar{b}, A)$ *are semi-regular and not orthogonal then* $\mathrm{stp}(\bar{b}^\frown\bar{c}, A)$ *is semi-regular, not orthogonal to either of them, and* $w(\bar{b}^\frown\bar{c}, A) = w(\bar{c}, A) + w(\bar{b}, A)$. *Similarly for semi-minimal.*

Proof. W.l.o.g. we can assume, in both parts, that $A = |M|$, M is $\mathbf{F}^a_\kappa$-saturated, where $\kappa = \kappa_r(T)$ and then the proof is easy (using Exercise 4.8.).

THEOREM 4.7: *Suppose* $p = \mathrm{stp}(\bar{a}, A)$ *is semi-regular and* $A \subseteq B$. *If* $\mathrm{stp}(\bar{a}, B)$ *forks over* A *then* $\mathrm{low}_p(\bar{a}, B) < w(\bar{a}, A) = w(p)$.

Proof. W.l.o.g. we can assume $A = |M|$, M is $\mathbf{F}^a_\kappa$-saturated where $\kappa = \kappa_r(T)$. By 4.5, as p is semi-regular there are $\bar{c}_l$ $(l < k)$ and N such that: for no $l < k$, there is $\bar{c}'_l \in |N|$, $\bar{c}'_l \notin |M|$, realizing $\mathrm{tp}(\bar{c}_l, \mathrm{Cb}(\mathrm{tp}(\bar{c}_0{}^\frown\cdots{}^\frown\bar{c}_{k-1}, |M|))$ but not $\mathrm{tp}(\bar{c}_l, |M|)$; $\mathrm{tp}(\bar{c}_l, |M|)$ is regular, not orthogonal to p, $\{\bar{c}_l: l < k_0 = w(\bar{a}, |M|)\}$ is independent over $|M|$, N is $\mathbf{F}^a_\kappa$-primary over $|M| \cup \bar{a}$ and also over $|M| \cup \bigcup_{l<k_0} \bar{c}_l$; $\bar{c}_l \in |N|$ for $l < k$, and $\bar{a}$ is algebraic over $|M| \cup \bigcup_{l<k} \bar{c}_l$. We can assume that $\mathrm{tp}(\bar{c}_l, B \cup \bigcup_{i<l} \bar{c}_i)$ forks over $|M|$ iff $l \geq k_1$; so $0 \leq k_1 \leq k_0 \leq k$. Notice that for $l \geq k_1$, $\mathrm{tp}(\bar{c}_l, B \cup \bigcup_{i<l} \bar{c}_i)$ is orthogonal to p, as $\mathrm{tp}(\bar{c}_l, |M|)$ is regular, not orthogonal to p. As $\bar{a}$ is algebraic over $B \cup \bigcup_{l<k} \bar{c}_l$, clearly $\mathrm{low}_p(\bar{a}, A) = w(\bar{a}, A) = k_0$ and $\mathrm{low}_p(\bar{a}, B) \leq \mathrm{low}_p(\bar{c}_0{}^\frown\cdots{}^\frown\bar{c}_{k-1}, B)$; and by 3.16(8) it is easy to check that the latter is k_1. Hence it suffices to prove $k_1 < k_0$; so assume $k_1 = k_0$. By 4.1 for some $C \subseteq |M|$, $|C| < \kappa$, $\bar{a} \in \mathrm{acl}(C \cup \bigcup_{l<k} \bar{c}_l)$, $\mathrm{stp}(\bar{c}_0{}^\frown\cdots{}^\frown\bar{c}_{k-1}, C \cup \bigcup_{i<k_0} \bar{c}_i) \vdash \mathrm{tp}(\bar{c}_0{}^\frown\cdots{}^\frown\bar{c}_{k-1}, M \cup \bigcup_{i<k_0} \bar{c}_i)$. We can assume that $\mathrm{tp}(\bar{c}_0{}^\frown\cdots{}^\frown\bar{c}_{k-1}, |M|)$ does not fork over C. As $\mathrm{tp}(\bar{c}_0{}^\frown\cdots{}^\frown\bar{c}_{k_0-1}, B)$ does not fork over $|M|$, it does not fork over C. Thus clearly (by the $\mathbf{F}^a_\kappa$-saturativity of M) $\mathrm{stp}(\bar{c}_0{}^\frown\cdots{}^\frown\bar{c}_{k-1}, C \cup \bigcup_{i<k_0} \bar{c}_i) \vdash \mathrm{tp}(\bar{c}_0{}^\frown\cdots{}^\frown\bar{c}_{k-1}, B \cup \bigcup_{i<k_0} \bar{c}_i)$. So $\mathrm{tp}(\bar{c}_0{}^\frown\cdots{}^\frown\bar{c}_{k-1}, B)$ does not fork over M, thus $\mathrm{tp}(\bar{a}, B)$ does not fork over $|M|$, contradiction.

CONCLUSION 4.8: *A type* p *is regular iff it is semi-regular, and* $w(p) = 1$.

THEOREM 4.9: (1) *Suppose for every* $b \in B$, $\mathrm{stp}(b, A)$ *is semi-regular, not orthogonal to the stationary regular type* p. *Then there is a cardinal* $\mu = \mathrm{low}_p(B, A)$ *such that for every* $I = \{\bar{b}_i: i < \alpha\}$, $\bigcup I = B$, *the following holds:* $\mu = \sum_i \mathrm{low}_p(\bar{b}_i, A \cup \bigcup_{j<i} \bar{b}_j)$.

Remark. We define $\mathrm{low}_p(B, A)$ by the natural extension of Definition 3.4.

(2) *If p is regular and stationary* $\mathrm{tp}_*(B, A)$ *is orthogonal to p, then for every $\bar{a}$,* $\mathrm{low}_p(\bar{a}, A) \le \mathrm{low}_p(\bar{a}, B)$.

(3) *If for every $\bar{b} \in I$,* $\mathrm{stp}(\bar{b}, A)$ *is regular, not orthogonal to p, $I = \{\bar{b}_i: i < \alpha\}$ then* $\sum_{i<\alpha} \mathrm{low}_p(\bar{b}_i, A \cup \bigcup_{j<i} \bar{b}_j) = \max\{|J| : J \subseteq I,\ J$ *independent over A*$\}$.

Proof. (1) Choose a regular $\kappa \ge \kappa_r(T)$, let M be $\mathbf{F}^a_\kappa$-saturated, $\mathrm{tp}_*(B, |M|)$ does not fork over $A \subseteq |M|$, and let N be $\mathbf{F}^a_\kappa$-prime over $|M| \cup B$. W.l.o.g. $p \in S^m(|M|)$, and let $J \subseteq |N|$ be a maximal set of sequences realizing p, independent over $|M|$. So $|J| = \mathrm{low}_p(B, A)$ by the natural extension of Definition 3.4, and it clearly depends on $\mathrm{tp}_*(B, A)$ alone (and not on M, N, J). Let $\mu = \sum_{i<\alpha} \mathrm{low}_p(\bar{b}_i, A \cup \bigcup_{j<i} \bar{b}_j)$.

Now we prove $|J| \le \mu$ (this does not depend on the assumption "$\mathrm{stp}(b, A)$ is semi-regular for every $b \in B$"). We define by induction on $i \le \alpha$ models N_i: $N_0 = M$, N_i is $\mathbf{F}^a_\kappa$-prime over $\bigcup_{j<i} (|N_j| \cup \bar{b}_j)$, but we choose it such that $\mathrm{tp}_*(B, N_i)$ does not fork over $A \cup \bigcup_{j<i} \bar{b}_j$. Now let $J_i \subseteq |N_{i+1}|$ be a maximal set, independent over $|N_i|$, of sequences realizing the stationarization of p over N_i. Clearly $|J_i| = \mathrm{low}_p(\bar{b}_i, A \cup \bigcup_{j<i} \bar{b}_j)$, and $\bigcup_{j<\alpha} J_j \subseteq |N_\alpha|$ is a maximal set, independent over $|N_0| = |M|$, of sequences realizing p (see 1.16(3)). Clearly there is an elementary embedding f of N into N_α over M, hence $|J| \le \sum_{i<\alpha} |J_i| = \mu$.

Now let us prove $\mu \le |J|$, thus finishing. As for each $b \in B$, $\mathrm{tp}(b, |M|)$ is semi-regular, there is a finite set $J_b \subseteq N$ of elements realizing over $|M|$ regular (complete) types not orthogonal to p, such that $b \in \mathrm{acl}(|M| \cup \bigcup J_b)$. So $B \subseteq \mathrm{acl}(|M| \cup \bigcup J^*)$ where $J^* = \bigcup_{b \in B} J_b$. Now we define by induction on $i < \alpha$, $J^*_i \subseteq J_{b_i}$, such that J^*_i is a minimal set (by $\subseteq$), independent over $(|M| \cup \bigcup J^0_i, |M|)$ satisfying $\bar{b}_i \in \mathrm{acl}(|M| \cup \bigcup J^1_{i+1})$ where $J^1_j = \{\bar{c} \in J^*: \mathrm{tp}(\bar{c}, |M| \cup \bigcup J^0_j)$ fork over $|M|\}$, $J^0_j = \bigcup_{\xi<j} J^*_\xi \subseteq J^*$.

Now $|J| \ge |J^0_\alpha| = \sum_{i<\alpha} |J^*_i|$, so it suffices to prove

$$|J^*_i| \ge \mathrm{low}_p\Big(\bar{b}_i, |M| \cup \bigcup_{j<i} \bar{b}_j\Big).$$

But $|J^*_i| \ge \mathrm{low}_p(\bar{b}_i, |M| \cup \bigcup J^1_i) = \mathrm{low}_p(\bar{b}_i, |M| \cup \bigcup J^1_i \cup \bigcup_{j<i} \bar{b}_j) \ge \mathrm{low}_p(\bar{b}_i, |M| \cup \bigcup_{j<i} \bar{b}_j)$ (first relation as $\bar{b}_i \in \mathrm{acl}(|M| \cup \bigcup J^1_{i+1})$, (see 3) and 4.6(1)), second relation as $\bar{b}_j \in \mathrm{acl}(|M| \cup \bigcup J^1_i)$ for $j < i$, and the third relation by (2) below, as for each j, and $c \in J^1_{j+1} - J^1_j$, $\text{tp}(c, |M| \cup \bigcup J^1_j \cup \bigcup_{\xi \leq j} \bar{b}_\xi)$ forks over $|M|$, in fact over $|M| \cup \bigcup J^1_j \cup \bigcup_{\xi < j} \bar{b}_\xi$ (for $c \in J^*_i$ as $J^*_i \subseteq J_{b_i}$ (and we could choose such J_{b_i}) otherwise by 1.9, hence is orthogonal to p, and use 1.6.)

(2) We can assume $A = |M|$, M $\mathbf{F}^a_\kappa$-saturated, $\kappa = \kappa_r(T)$, and $B \subseteq N$, N $\mathbf{F}^a_\kappa$-prime over $M \cup B$ and $\text{tp}(\bar{a}, |N|)$ does not fork over B. Let N_1 be $\mathbf{F}^a_\kappa$-saturated, $\mathbf{F}^a_\kappa$-constructible over $|M| \cup \bar{a}$ and also over $|N| \cup \bar{a}$ (see IV, 3.17) and N_2 $\mathbf{F}^a_\kappa$-prime over $|N_1| \cup |N|$, hence also over $|N| \cup \bar{a}$. We can assume $p \in S^m(|M|)$ and $J \subseteq |N_1|$ is a maximal set, independent over $|M|$ of sequences realizing p. Clearly $|J| = \text{low}_p(\bar{a}, A)$. Now by 1.4 $\text{tp}_*(\bigcup J, |M|)$, $\text{tp}_*(B, |M|)$ are orthogonal and so by 3.2 $\text{tp}_*(\bigcup J, |M|)$, $\text{tp}_*(|N|, |M|)$ are orthogonal. Hence J is independent over $(|N|, |M|)$ hence $\text{low}_p(\bar{a}, B) = \text{low}_p(\bar{a}, N) \geq |J| = \text{low}_p(\bar{a}, A)$. Note also $\text{Av}(J, |M| \cup J) \vdash \text{Av}(J, |N| \cup J)$, and $\text{Av}(J, |M| \cup J) \vdash \text{Av}(J, |N_1|)$.

(3) Easy by 1.14.

LEMMA 4.10: *If* $p = \text{stp}(\bar{a}, A)$ *is semi-minimal*, $A \subseteq B$, *then* $\text{stp}(\bar{a}, B)$ *is either algebraic, or semi-minimal and not orthogonal to* p.

Proof. W.l.o.g. $A = |M|$, M is $\mathbf{F}^a_\kappa$-saturated, $\kappa = \kappa_r(T)$. So there are $\bar{c}_l$ $(l < k)$ such that $\text{tp}(\bar{c}_l, |M|)$ is minimal not orthogonal to p, and $\bar{a} \in \text{acl}(|M| \cup \bigcup_{l<k} \bar{c}_l)$. By renaming we can assume that $\text{tp}(\bar{c}_l, B \cup \bigcup_{i<l} \bar{c}_i)$ forks over A iff $l \geq k_0$. Hence $\bar{c}_0 \frown \cdots \frown \bar{c}_{k-1}$ is algebraic over $B \cup \bigcup_{l<k_0} \bar{c}_l$, thus also $\bar{a} \in \text{acl}(B \cup \bigcup_{l<k_0} \bar{c}_l)$. As $\{\bar{c}_l : l < k_0\}$ is independent over (B, A), $\text{tp}(\bar{c}_l, B)$ is minimal, so $\text{tp}(\bar{c}_0 \frown \cdots \frown \bar{c}_{k_0-1}, B)$ is semi-minimal, hence $\text{tp}(\bar{a}, B)$ is semi-minimal; except when $k_0 = 0$ and both are algebraic.

THEOREM 4.11 (*in* $\mathfrak{C}^{eq}$): (1) *If* $\bar{a} \notin \text{acl}\, A$, T *is superstable then there is an equivalence relation* E *which is almost over* A *such that* $\text{stp}(\bar{a}/E, A)$ *is semi-regular (so* $\bar{a}/E \notin \text{acl}\, A$).

(2) *If* $\text{tp}(\bar{a}, A)$ *is not orthogonal to some minimal type* q, *then for some equivalence relation* E *which is almost over* A, $\text{stp}(\bar{a}/E, A)$ *is semi-minimal, and not orthogonal to* q.

Remark. E is a formula $E(\bar{x}, \bar{y}, \bar{b})$, $l(\bar{x}) = l(\bar{y})$, $\bar{b} \in A$.

Proof. Choose a regular $\kappa > |A| + \kappa(T)$, and a $\mathbf{F}^a_\kappa$-saturated model M such that $A \subseteq M$, $\text{tp}(\bar{a}, |M|)$ does not fork over A, and in part (2) the minimal type q is over $|M|$. Let N be $\mathbf{F}^a_\kappa$-primary over $|M| \cup \bar{a}$. For part (1) choose $c \in |N| - |M|$ such that $R(p, \mathrm{L}, \infty)$ is minimal where $p = \mathrm{tp}(c, |M|)$ and for part (2) so that p is a minimal type not orthogonal to $\mathrm{stp}(\bar{a}, A)$. In both cases p is regular (by 3.5, 1.18(4)). Now choose $B \subseteq |M|, A \subseteq B, |B| < \kappa_r(T) + |A|^+$ such that $\mathrm{tp}(\bar{a}^\frown\langle c\rangle, |M|)$ does not fork over B, $\mathrm{tp}(\bar{a}^\frown\langle c\rangle, B)$ is stationary, $R[\mathrm{tp}(c, B), \mathrm{L}, \infty] = R[\mathrm{tp}(c, |M|), \mathrm{L}, \infty]$.

Now we can find elementary mappings $f_i (i < \kappa)$ such that $\mathrm{Dom}\, f_i = B$, $\mathrm{Range}(f_i) \subseteq |M|$ and $\mathrm{stp}_*(f_i(B), A) \equiv \mathrm{stp}_*(B, A)$ and $\mathrm{stp}_*[f_i(B), \bigcup_{j<i} f_j(B)]$ does not fork over A, and $f_0 =$ the identity. Let $B_i = f_i(B)$ and p_i be the stationarization of $f_i(p \upharpoonright B)$ over $|M|$, so $p_0 = p$. Clearly as $\mathrm{stp}(\bar{a}, A)$ is not orthogonal to $p \upharpoonright B$, it is not orthogonal to each p_i, hence by 3.4 the p_i's are pairwise not orthogonal. If $b \in |N| - |M|$ realizes $p_i \upharpoonright B_i$, it necessarily realizes p_i (in part (1) because of the minimality of the rank, in part (2) by the minimality of $p_i \upharpoonright B_i$). Clearly it suffices to prove:

CLAIM 4.12 (*in* $\mathfrak{C}^{\mathrm{eq}}$): *Let* $\bar{a}, c, A, M, N, f_i, B_i, p_i, \kappa$ *be as above (the minimality and superstability conditions are not necessary). Then there is an equivalence relation* E *almost over* A *so that* $\mathrm{stp}(\bar{a}/E, A)$ *is semi-regular and not orthogonal to* p*, and if the* p_i*'s are minimal, it is semi-minimal.*

Proof. By the choice of $B = B_0$, $\mathrm{tp}(c, B_0 \cup \bar{a})$ forks over B_0 so there is a $\bar{b} \in B_0$, such that $\vDash\varphi[c, \bar{a}, \bar{b}]$, and $\{\varphi(x, \bar{a}, \bar{b})\}$ forks over B. Let $C_i = \{d: d \in N, d \text{ realizes } p_i \upharpoonright B_i\}$; clearly $\mathrm{Cb}[\mathrm{stp}(\bar{a}, B \cup C_0)]$ is algebraic over $B \cup \bar{a}$ (see 1.11, Cb-canonical base, see III, 6) and there are θ and $\bar{c} \in \mathrm{Cb}[\mathrm{stp}(\bar{a}, B_0 \cup C_0)] \subseteq \mathrm{acl}(B_0 \cup C_0) \cap \mathrm{acl}(B_0 \cup \bar{a})$ such that for $d \in C_0$, $\vDash\varphi[d, \bar{a}, \bar{b}]$ iff $\vDash\theta[\bar{c}, d, \bar{b}]$. Let $\psi(\bar{x}, \bar{a}, \bar{b}_0)$ $(\bar{b}_0 \in B_0)$ be an algebraic formula such that $\vDash\psi[\bar{c}, \bar{a}, \bar{b}_0]$ and $\psi(\bar{x}, \bar{a}, \bar{b}_0) \vdash \mathrm{tp}(\bar{c}, B_0 \cup \bar{a})$. Now let $E(\bar{x}, \bar{y})$ say that for infinitely many $i < \kappa$, $(\forall \bar{z})[\psi(\bar{z}, \bar{x}, f_i(\bar{b}_0)) \equiv \psi(\bar{z}, \bar{y}, f_i(\bar{b}_0))]$. By III, 2.5 $E(\bar{x}, \bar{y})$ is almost over A. If $\psi(\bar{z}, \bar{a}, f_i(\bar{b}_0))$ is realized by $\bar{c}^l_i$ $(l < k < \aleph_0)$, then $\bar{a}/E$ is algebraic over $\bigcup_{i<\kappa}(B_i \cup \bigcup_{l<k} \bar{c}^l_i)$, because $E(\bar{x}, \bar{a})$ iff for infinitely many i's $(\forall \bar{z})[\psi(\bar{z}, \bar{x}, f_i(\bar{b}_1)) \equiv \bigvee_{l<k} \bar{z} = \bar{c}^l_i]$. Hence $\mathrm{tp}(\bar{a}/E, |M|)$ is semi-regular (and if the p_i's are minimal, semi-minimal) provided we show it is not algebraic (as clearly $\bar{c}^l_i \in |N|$ as it is algebraic over $B_i \cup C_i$). But for big enough n, $\bar{a}/E$ is algebraic over $\bigcup_{i<n}(B_i \cup \{\bar{c}^l_i: l < k\})$. Clearly $\mathrm{tp}_*(\{\bar{c}^l_i: l < k, i < n\}, |M|)$ does not fork over $\bigcup_{i<n} B_i$ (as $\bar{c}^l_i$ is algebraic over $B_i \cup \bar{a}$), and it is not algebraic by the choice of φ. But clearly $\mathrm{tp}(\{\bar{c}^l_i: l < k, i < n\}, |M| \cup \bar{a}/E\})$ forks over

$|M|$, thus $\bar{a}/E \notin M$. Clearly $\bar{a}/E \in \mathrm{dcl}[(\mathrm{acl}\, A) \cup \bar{a}]$ hence $\bar{a}/E \in \mathrm{acl}(A \cup \bar{a})$. Of course, when p is minimal, $\mathrm{stp}(\bar{a}/E, A)$ is semi-minimal. More fully, $\mathrm{tp}(\bar{a}/E, M)$ is semi-regular and $\bar{a}/E \subseteq \mathrm{dcl}\,(M \cup \bigcup_{i<\kappa, l<k} \bar{c}_i^l) \subseteq \mathrm{dcl}(M \cup \bigcup_{i<k} C_i$. Hence, $\mathrm{tp}(\bar{a}/E, A)$ is semi-regular as $\mathrm{tp}(\bar{a}/E, M)$ does not fork over A and $\mathrm{tp}(\bar{a}/E, M)$ is semi-regular.

CONCLUSION 4.13 ($\mathfrak{C}^{eq}$): *Suppose* $\mathbf{F}$ *satisfies Axioms* (X.1) *and* (XI.1).

(1) *If T is superstable, for every construction $\mathscr{A}$ there is a construction $\mathscr{A}'$ such that $\mathscr{A}$, $\mathscr{A}'$ have the same domain and range, and* $\mathrm{tp}(a_i', \mathscr{A}_i')$ *is algebraic or semi-regular.*

(2) *For unidimensional T, the same conclusion with* $\mathrm{tp}(a_i', A_i')$ *algebraic or semi-minimal.*

Proof. Immediate.

DEFINITION 4.2: The m-formula $\varphi(\bar{x}, \bar{a})$ is semi-weakly-minimal if for some weakly-minimal $\psi(\bar{y}, \bar{c})$, every $\bar{b} \in \varphi(\mathfrak{C}, \bar{a})$ belongs to $\mathrm{acl}(\bar{c} \cup \psi(\mathfrak{C}, \bar{c}))$.

THEOREM 4.14 (*in* $\mathfrak{C}^{eq}$): *Suppose* $\mathrm{tp}(\bar{a}, A)$ *is not orthogonal to some minimal type q to which some weakly-minimal formula φ belongs. Then for some equivalence relation E which is over A, $\bar{a}/E$ satisfies some semi-weakly-minimal formula ψ which is over A, but* $\bar{a}/E \notin \mathrm{acl}\, A$. *Also every stationary type which contains ψ is not orthogonal to some stationary type to which φ belongs.*

Proof. The proof is the same as that of 4.11(2) with the following additions. Using the notation of 4.11(2), let $\varphi(x; \bar{b}_0^*) \in \mathrm{tp}(c, B_0)$ be weakly minimal, and let $\varphi_1(\bar{x})$ "say" w.l.o.g. that for infinitely many $i < \kappa$,

$$(\exists^{<k} z)\psi(z; \bar{x}, f_i(\bar{b}_0)) \wedge (\forall z)[\psi(z; \bar{x}, f_i(b_0)) \rightarrow \varphi_0(z, f_i(\bar{b}_0^*))].$$

Let $\varphi_2(y)$ "say" that $y = \bar{x}/E$ for some $\bar{x}$ satisfying $\varphi_1(\bar{x})$. Then $\varphi_2(y)$ is semi-weakly minimal, but it and E are only almost over A. This can be corrected (see III, 2.4). For the last phrase, see 4.6.

THEOREM 4.15: *Suppose p, q are complete types over A, and are stationary. If p, q are semi-minimal and not orthogonal then there are $l_0 < \omega$, and $\bar{a}_l$ ($l < l_0$) realizing p such that every sequence realizing p is algebraic over* $\bigcup_{l<l_0} \bar{a}_l \cup A \cup \bigcup \{\bar{b}\colon \bar{b}$ *realizes* $q\}$.

Proof. Let $\lambda > |A| + |T|$ be regular, and M a $\mathbf{F}^a_\lambda$-saturated model, $A \subseteq |M|$, and $r \in S^m(|M|)$ a minimal type not orthogonal to p (nor q). Let $I = \{\bar{a} \in \mathfrak{C}: \bar{a} \text{ realizes } p\}$, $J = \{\bar{b} \in \mathfrak{C}: \bar{b} \text{ realizes } q\}$. Let p_1, q_1 be stationarizations of p, q (resp.) over $|M|$. Clearly for every $\bar{a}$ realizing p_1 there are $\bar{c}_l(l \leqslant n)$ such that $r_l = \text{tp}(\bar{c}_l, M)$ is minimal, not orthogonal to r, $\bar{a} \in \text{acl}(|M| \cup \bigcup_{l \leqslant n} \bar{c}_l)$. Also for any $\bar{c}$ realizing r_l there is an $\bar{a}$ realizing p_1 such that $\bar{c} \in \text{acl}(|M| \cup \bar{a})$. Similar assertions holds for q_1; so for every $\bar{a}$ realizing p_1 there are $\bar{b}_0, \ldots, \bar{b}_l$ realizing q_1 such that $\bar{a} \in \text{acl}(|M| \cup \bigcup_{i \leq l} \bar{b}_i)$; hence for some $\bar{c} \in |M|$ we can have $\bar{a} \in \text{acl}(\bar{c} \cup \bigcup_{i<l} \bar{b}_i)$. Clearly there are $C_1 \subseteq (\bigcup I) \cap |M|$, $C_2 \subseteq (\bigcup J) \cap |M|$ where $|C_1| + |C_2| \leq |T|$ such that $\text{tp}(\bar{c}, (A \cup \bigcup I \cup \bigcup J))$ is definable over $C_1 \cup C_2$ (as M is $\mathbf{F}^a_\lambda$-saturated, $|A| + |T| \leqslant \lambda$). By 1.11 for each $\bar{a}$ realizing p_1 there are $\bar{b}_0, \ldots, \bar{b}_l$ realizing q_1 such that $\bar{a} \in \text{acl}(\bigcup_{i \leq l} \bar{b} \cup C_1 \cup C_2 \cup A)$. We can replace C_1, C_2 by finite subsets without changing the last conclusion, by a compactness argument.

So let $C_1 = \bigcup_{i<l(0)} \bar{a}^0_i$, $C_2 = \bigcup_{i<l(1)} \bar{b}^0_i$, and choose $\bar{a}^\alpha_i$, $\bar{b}^\alpha_i$ so that

$$\text{stp}_*(\{\bar{a}^0_i: i < l(0)\} \cup \{\bar{b}^0_i: i < l(1)\}, A) \\ \equiv \text{stp}_*(\{\bar{a}^\alpha_i: i < l(0)\} \cup \{\bar{b}^\alpha_i: i < l(1)\}, A),$$

$$\text{tp}_*\Big(\{\bar{a}^\alpha_i: i < l(0)\} \cup \{\bar{b}^\alpha_i: i < l(1)\}, A \cup \bigcup_{\beta<\alpha} \Big(\bigcup_i \bar{a}^\beta_i \cup \bigcup_i \bar{b}^\beta_i\Big)\Big)$$

does not fork over A. Then for every $\bar{a}$ realizing p there is $j < \kappa(T)$, such that $\text{tp}(\bar{a}, A \cup \bigcup_i \bar{a}^j_i \cup \bigcup_i \bar{b}^j_i)$ does not fork over A. As p is stationary for every $\bar{a}'$ realizing p_1, $\bar{a}$ and $\bar{a}'$ realize the same type over $A \cup \bigcup_i \bar{a}^j_i \cup \bigcup_i \bar{b}^j_i$; hence for every $\bar{a}$ realizing p there are $\bar{b}_0, \ldots, \bar{b}_l$ realizing q such that $\bar{a} \in \text{acl}[(A \cup \bigcup_{i,j} \bar{a}^j_i) \cup (\bigcup_{i,j} \bar{b}^j_i \cup \bigcup_i \bar{b}_i)]$; so we finish.

QUESTION 4.1: Try to replace in 4.9 semi-regular by a weaker concept, and investigate this concept. (See Definition 4.2 and 4.4 for a suggestion).

EXERCISE 4.2: Generalize 4.9 to semi-weakly-minimal formulas.

QUESTION 4.3: It is true that in 4.11 we cannot replace semi-regular (semi-minimal) by regular (minimal) even if we replace $\bar{a}/E$ by any $b \in \text{acl}(A \cup \bar{a}) - \text{acl}\, A$? What if T is $\aleph_0$-categorical?

EXERCISE 4.4 ($\mathfrak{C}^{eq}$): Suppose $p \in S(A)$ is minimal, I and J are sets of elements realizing p, which are independent over A, but $I \cup J$ is not independent over A. Show $\mathrm{acl}(A \cup I)$, $\mathrm{acl}(A \cup J)$ not necessarily have an element in common, which is not in $\mathrm{acl}\, A$ [e.g. T = theory of the complex field].

QUESTION 4.5: Generalize 4.15 to semi-regular types.

EXERCISE 4.6: Prove that if $\varphi(\bar{x}, \bar{a})$ is semi-weakly minimal, then $D^m[\varphi(\bar{x}; \bar{a}), \mathrm{L}, \infty] < \omega$. [Hint: See Section 7.]

QUESTION 4.7: Prove 4.11 does not hold for stable T in general; and try to find a substitute.

Remark. Let T be defined by the following axioms:

(i) P_η ($\eta \in {}^{\omega >}2$) are pairwise disjoint,

(ii) F^l_η ($l = 0, 1$) are partial one-place functions, from P_η onto $P_{\eta \frown \langle l \rangle}$,

(iii) G_η is a partial two-place function,

$$(\exists z)[G_\eta(x, y) = z] \equiv P_{\eta \frown \langle 0 \rangle}(x) \wedge P_{\eta \frown \langle 1 \rangle}(y),$$

$$P_\eta(x) \rightarrow G_\eta(F_{\eta \frown \langle 0 \rangle}(x), F_{\eta \frown \langle 1 \rangle}(x)) = x,$$

T exemplifies the non-existence of regular types.

EXERCISE 4.8: Suppose $M_1 \subseteq M_2 \subseteq M_3$, $p \in S^m(A)$, $A \subseteq M_1$, and for $l = 1, 2$, every $\bar{b} \in |M_{l+1}|$, $\bar{b} \notin |M_l|$ which realizes p, realizes a stationarization of p over M_l. Show that every $\bar{b} \in |M_3|$ which realizes p, realizes a stationarization of p over $|M_1|$.

EXERCISE 4.9: Show that 4.4 does not hold for regularity, even for $\aleph_0$-stable T. [Hint: Let $M = (|M|, P, Q, E, F, +)$, $|M|$ is the disjoint union of P and Q which are infinite, E an equivalence relation on Q, $+$ is a two place function from P to P, $(P, +)$ is an abelian group, each element of exponent two and F is a two-place (partial) function, $F(a, b)$ is defined iff aEb, $a, b \in Q$, and $F(a, b) + F(b, c) = F(a, c)$. Choose $a_1, a_2 \in Q(\mathfrak{C}) - |M|$, $\models \neg E(a_1, c) \wedge E(a_1, a_2) \wedge F(a_1, a_2) \neq c$ for every $c \in M$. Then $\mathrm{tp}(a_l, |M|)$ is regular but $\mathrm{tp}(\langle a_1, a_2 \rangle, |M|)$ is not semi-regular.]

DEFINITION 4.3: (1) A formula $\varphi(\bar{x}, \bar{a})$ is (strongly) semi-minimal if for every $\bar{c}$ satisfying it, $\operatorname{stp}(\bar{c}, \bar{a})$ is semi-minimal not orthogonal to some (fixed) minimal type containing a minimal formula, or is algebraic.

(2) A formula $\varphi(x, \bar{a})$ is almost minimal iff it is equivalent to a disjunction of minimal formulas.

EXERCISE 4.10: (1) $\varphi(\bar{x}, \bar{a})$ is semi-minimal iff there is an almost minimal $\psi(\bar{x}, \bar{b})$ and $\bar{c}$ such that $\varphi(\mathfrak{C}, \bar{a}) \subseteq \operatorname{acl}(\bar{c} \cup \psi(\mathfrak{C}, \bar{b}))$.

(2) The formula $\varphi(\bar{x}; \bar{a})$ is almost-minimal iff $R^m(\varphi, L, \aleph_0) = 1$.

(3) In 4.14, if in the hypothesis we replace weakly minimal by minimal, then in the conclusion we get a semi-minimal formula. For $A = \operatorname{acl} A$, even strongly semi-minimal.

(4) The following families of formulas are closed under disjunction: almost minimal, semi-minimal, weakly minimal and semi-weakly minimal.

EXERCISE 4.11: Let V be an infinite vector space over the field with two-elements. Let us define for each n, $M(=M_n)$:

$$|M| = V \cup V \times V \times \cdots \times V \ (n \text{ times}),$$
$$P = V \ (\text{a one place predicate}),$$
$$E_l = \{\langle (a_1, \ldots, a_l, a_{l+1}, \ldots, a_n), (a_1, \ldots, a_l, a'_{l+1}, \ldots, a'_n)\rangle : a_1, \ldots, a_n, a'_{l+1}, \ldots, a'_n \in V\},$$
$$R = \{\langle a, b, c\rangle : a, b, c \in V, a + b = c\},$$
$$R_l = \{\langle c, (a_1, \ldots, a_l, a_{l+1}, \ldots, a_n), (a_1, \ldots, a_l, a_{l+1} + c, a'_{l+2}, \ldots, a'_n\rangle : c, a_1, \ldots, a_n, a'_{l+2}, \ldots, a'_n \in V\},$$

so

$$M = (|M|, P, E_l, R, R_l)_{l<n}, \ T = \operatorname{Th}(M).$$

Choose N $\aleph_1$-saturated.

(A) Prove T is totally transcendental and unidimensional, and categorical in every λ.

(B) Show that here it is possible that (when $n \geq 3$) $w(\bar{b}, A) = w(\bar{a}^\frown\bar{b}, A) = w(\bar{a}, A \cup \bar{b}) = 1$.

(C) Show that 4.15 cannot be substantially improved.

DEFINITION 4.4: (1) Let $\mathscr{P}$ denote a subset of $\mathscr{P}^* = \{\operatorname{stp}(\bar{a}, A) : A, \bar{a}\}$. $\mathscr{P}$ is called regular if:

(i) Whenever $A \subseteq B$, $\operatorname{stp}(\bar{a}, A) \in \mathscr{P}$, $\operatorname{stp}(\bar{a}, B)$ fork over A, then $\operatorname{stp}(\bar{a}, B)$ is orthogonal to $\mathscr{P}$ (see 2).

(ii) If $A \subseteq B$, $\operatorname{stp}(\bar{a}, B)$ does not fork over A, $\operatorname{stp}(\bar{a}, A) \in \mathscr{P}$ then $\operatorname{stp}(\bar{a}, B) \in \mathscr{P}$.

(2) A type q is orthogonal to $\mathscr{P}$ if it is orthogonal to every $p \in \mathscr{P}$. If $\mathscr{P} = \{p\}$, we write p instead of $\{p\}$.

(3) An m-type q is strongly orthogonal to $\mathscr{P}$ if $q \subseteq q_1$, q_1 complete implies q_1 orthogonal to $\mathscr{P}$.

(4) For a regular set $\mathscr{P}$ $\mathrm{low}_{\mathscr{P}}(q) = \sum_{p \in \mathscr{P}(1)} \mathrm{low}_p(q)$, where $\mathscr{P}(1) \subseteq \mathscr{P}$ is a maximal subset of pairwise orthogonal types (it is well defined) (see 4.16(13)). Similarly $\mathrm{low}_{\mathscr{P}}(\bar{a}, A)$ etc.

(5) For a set of types $\mathscr{P}$ let

$\mathrm{cl}^0(\mathscr{P}) = \{p: p = \mathrm{stp}(\bar{a}, B)$ is a stationarization of $q \in \mathscr{P}, q \subseteq p\} \cap \mathscr{P}^*$,

$\mathrm{cl}^1(\mathscr{P}) = \{p: p$ parallel to some $q \in \mathscr{P}\} \cap \mathscr{P}^*$,

$\mathrm{cl}^2_A(\mathscr{P}) = \{F(p): p \in \mathscr{P}, F$ an automorphism of $\mathfrak{C}$ and $F^{\mathrm{eq}} \restriction \mathrm{acl}\, A =$ the identity$\} \cap \mathscr{P}^*$,

$\mathrm{cl}^3(\mathscr{P}) = \{p: p$ regular, stationary not orthogonal to $\mathscr{P}\} \cap \mathscr{P}^*$.

Now $\mathrm{cl}^{i,j}(\mathscr{P}) = \mathrm{cl}^i\, \mathrm{cl}^j(\mathscr{P})$ and we let $\mathrm{cl}^2(\mathscr{P}) = \mathrm{cl}^2_{\emptyset}(\mathscr{P})$.

DEFINITION 4.5: Let $\mathscr{P}$ be a regular family.

(1) An m-type q is $\mathscr{P}$-simple when for some A, every $\bar{a}$ realizing q is algebraic over

$A \cup \bigcup\{\bar{b}: \mathrm{stp}(\bar{b}, A)$ is $\mathscr{P}$-regular (see (4) below)$\} \cup \{\bar{b}: \mathrm{stp}(\bar{b}, A)$ **is strongly orthogonal to $\mathscr{P}\}$.**

(2) A stationary m-type q is strongly $\mathscr{P}$-simple if there are $\mathbf{F}^a_\kappa$-saturated models $M, N (\kappa = \kappa_r(T))$, $M \prec N$, and $\bar{a}, \bar{c}_l \in N (l < n)$ such that

(i) $\mathrm{tp}(\bar{a}, |M|)$ is a stationarization of q,

(ii) $\mathrm{tp}(\bar{c}_l, |M|)$ is $\mathscr{P}$-regular (see (4) below),

(iii) $\bar{a} \in \mathrm{acl}(|M| \cup \bigcup_{l<n} \bar{c}_l)$.

(3) An m-type q is $\mathscr{P}$-semi-regular, if (2) holds with the addition

(iv) if $\bar{c} \in |N|$ realizes $\mathrm{tp}(\bar{c}_l, \mathrm{Cb}(\mathrm{tp}(\bar{c}_l, |M|))$, $\bar{c} \notin |M|$, then $\bar{c}$ realizes $\mathrm{tp}(\bar{c}_l, |M|)$.

(4) An m-type q is $\mathscr{P}$-regular if every extension of q which forks over $\mathrm{Dom}\, q$ is strongly orthogonal to $\mathscr{P}$, but q is not strongly orthogonal to $\mathscr{P}$.

(5) We say $\bar{a}$ is $\mathscr{P}$-simple over A if $\mathrm{stp}(\bar{a}, A)$ is $\mathscr{P}$-simple. Similarily for $\mathscr{P}$-semi-regular, strongly $\mathscr{P}$-simple and $\mathscr{P}$-regular.

LEMMA 4.16: (1) ***If p is a regular stationary type then: q is semi-regular not orthogonal to p iff q is $\mathrm{cl}^3\{p\}$-semi-regular, not algebraic (so $\mathrm{cl}^3\{p\}$ is regular).***

(2) *If q is $\mathscr{P}$-semi-regular, then it is strongly $\mathscr{P}$-simple, and if $\mathscr{P} = \mathrm{cl}^2(\mathscr{P})$ any complete strongly $\mathscr{P}$-simple q, is $\mathscr{P}$-simple. Also if q is*

$\mathscr{P}$-regular, it is $\mathscr{P}$-simple and if all stationarizations of a complete p are in $\mathscr{P}$, p is $\mathscr{P}$-regular. Any stationary $\mathscr{P}$-regular type is $\mathscr{P}$-semi-regular ($\mathscr{P}$ is a regular family, of course).

(3) ***If $p = \mathrm{tp}(\bar{a}, A)$ is regular, $p \subseteq q$, q forks over A, $p \in \mathscr{P}$, $\mathscr{P}$ regular,*** *then q is strongly orthogonal to $\mathscr{P}$. Also p is $\mathscr{P}$-regular.*

(4) ***If $r \vdash q$, q is $\mathscr{P}$-simple, $\mathscr{P}$ is regular then r is $\mathscr{P}$-simple. So if $A \subseteq B$, $\bar{a}$ $\mathscr{P}$-simple over A then $\bar{a}$ is $\mathscr{P}$-simple over B. If $r \vdash q$, q is $\mathscr{P}$-regular then r*** *is $\mathscr{P}$-regular or is strongly orthogonal to $\mathscr{P}$. If $\bar{a}$ is $\mathscr{P}$-simple, then $w_{\mathscr{P}}(\bar{a}, A) < \omega$.*

(5) ***If q is strongly orthogonal to $\mathscr{P}$, $r \vdash q$ then r is strongly orthogonal to $\mathscr{P}$.***

(6) ***If $\bar{a}$ is $\mathscr{P}$-semi-regular over A then $\mathrm{low}_{\mathscr{P}}(\bar{a}, A) = w(\bar{a}, A)$.***

(7) ***If q is $\mathscr{P}$-regular, q over A, $q \subseteq q_1$, q_1 forks over A then q_1 is strongly orthogonal to $\mathscr{P}$.***

(8) ***Suppose q, r are parallel, then q is strongly $\mathscr{P}$-simple iff r is strongly $\mathscr{P}$-simple, and q is $\mathscr{P}$-semi-regular iff r is $\mathscr{P}$-semi-regular.***

(9) ***If $\mathscr{P}$ satisfies (i) of Definition 4.4(1) then $\mathrm{cl}^0(\mathscr{P})$ is regular. If*** *$\mathscr{P} \subseteq \mathscr{P}_1$ then: $\mathscr{P}_1$ is regular, implies $\mathrm{cl}^0(\mathscr{P})$ is regular.*

(10) *If p is minimal, $\mathscr{P} = \mathrm{cl}^3\{p\}$, q complete and stationary then: q is semi-minimal not orthogonal to p iff q is strongly $\mathscr{P}$-simple not algebraic.*

(11) *If p, q are parallel and complete ($\mathscr{P} = \mathrm{cl}^2(\mathscr{P})$, $\mathscr{P}$ regular), then p is $\mathscr{P}$-simple iff q is $\mathscr{P}$-simple, and p is $\mathscr{P}$-regular iff q is $\mathscr{P}$-regular.*

(12) *Suppose $\mathscr{P}_1 = \mathrm{cl}^2(\mathscr{P}_1) \subseteq \mathscr{P}_2$ are regular families: if q is $\mathscr{P}_1$-regular and complete [$\mathscr{P}_1$-semi-regular] [strongly $\mathscr{P}_1$-simple], then q is $\mathscr{P}_2$-regular [$\mathscr{P}_2$-semi-regular] [strongly $\mathscr{P}_2$-simple].*

Proof. Easy. In (2), note that even a stationary q may have a stationarization in $\mathscr{P}$ but not be in $\mathscr{P}$ (instead $\mathscr{P} = \mathrm{cl}^2(\mathscr{P})$ also $\mathscr{P} = \mathrm{cl}^1(\mathscr{P})$ suffice). In (4), use (13) below.

(13) *Let T be superstable. Every complete stationary $\mathscr{P}$-regular type p is regular.*

Remark. Part (13) sounds, at first, a truism, but an extension may be orthogonal to $\mathscr{P}$ while it is not orthogonal to the type.

Proof of 4.16(13). W.l.o.g. $p \in S^m(M)$, M is $\mathbf{F}^a_{\aleph_0}$-saturated, $\bar{c} \in M$, r is over M and $r \in \mathscr{P}$ is regular, not orthogonal to p and $\bar{d}$ realizes p. If p has weight > 1, there are $n < \omega, n > 1$, and $\bar{d}_i (i < n)$ such that $\mathrm{tp}(\bar{d}_i, M)$ is regular, $\{d_i : i < n\}$ is independent over M and $\mathrm{tp}(\bar{d}_0 {}^\frown \bar{d}_1 {}^\frown \cdots \bar{d}_{n-1}, M \cup \bar{d})$, $\mathrm{tp}(\bar{d}, M \cup \bigcup_{i<n} \bar{d}_i)$ are $\mathbf{F}^a_{\aleph_0}$-isolated (by V,

3.9(2) and see Def. 3.2). Since we have assumed p is not orthogonal to r, w.l.o.g. (by 3.9(1)'s proof) $\operatorname{tp}(\bar{d}_0, M)$ is not orthogonal to r. Now $p_1 = \operatorname{tp}(\bar{d}, M \cup \bar{d}_1)$ forks over M, whereas $\operatorname{tp}(\bar{d}_0, M \cup \bar{d}_1)$ does not fork over M, so $\operatorname{tp}(\bar{d}_0, M \cup \bar{d} \cup \bar{d}_1)$ forks over M, and hence should be orthogonal to r. But $\operatorname{tp}(\bar{d}_0, M \cup \bar{d} \cup \bar{d}_1)$ forks over M, whereas $\operatorname{tp}(\bar{d}_0, M \cup \bar{d}_1)$ does not fork over M so $\operatorname{tp}(\bar{d}_0, M \cup \bar{d} \cup \bar{d}_1)$ forks over $M \cup \bar{d}_1$: hence, $\operatorname{tp}(\bar{d}, M \cup \bar{d}_0 \cup \bar{d}_1)$ forks over $M \cup \bar{d}_1$. So $\operatorname{stp}(\bar{d}, M \cup \bar{d}_1)$, $\operatorname{stp}(\bar{d}_0, M \cup \bar{d}_1)$ are not orthogonal. But the latter is regular and parallel to $\operatorname{tp}(\bar{d}_0, M)$ which is not orthogonal to r, so $\operatorname{stp}(\bar{d}, M \cup \bar{d}_1)$ is not orthogonal to r, contradicting the $\mathscr{P}$-regularity of p.

So p should have weight 1. If it is not regular it has an extension q, forking over M, not orthogonal to a regular type r' iff r' is not orthogonal to r (as $\omega_0(p) = 1, p$ not orthogonal to r). We conclude that q is not orthogonal to r, but that q forks over M and hence over $\bar{c}$, a contradiction.

So p is regular, so we have proved 4.16(13).

LEMMA 4.17: (1) ***If*** $\bar{a} \in \operatorname{acl}(A \cup \bar{b})$, $\bar{b}$ ***is*** $\mathscr{P}$***-simple over*** A ***then*** $\bar{a}$ ***is*** $\mathscr{P}$***-simple over*** A, ***and*** $\operatorname{low}_{\mathscr{P}}(\bar{a}, A) \le \operatorname{low}_{\mathscr{P}}(\bar{b}, A)$. ***Similarly for*** $\mathscr{P}$***-semi-regular,*** *$\mathscr{P}$-regular and (except $\bar{a} \in \operatorname{acl} A$) strongly $\mathscr{P}$-simple.*

(2) ***If*** q ***is*** $\mathscr{P}$***-simple,*** q ***is over*** A, $\bar{a}$ ***realizes*** q, ***then*** $\operatorname{low}_{\mathscr{P}}(\bar{a}, A)$ ***is finite.***

(3) $\mathscr{P} = \{p: p$ ***a minimal type***$\}$ ***is a regular family. Only algebraic types are strongly orthogonal to*** $\mathscr{P}$.

(4) $\operatorname{tp}(\bar{a}{}^\frown\bar{b}, A)$ ***is*** $\mathscr{P}$***-simple iff*** $\operatorname{tp}(\bar{a}, A)$, $\operatorname{tp}(\bar{b}, A)$ ***are*** $\mathscr{P}$***-simple.***

(5) *For any ordinal α, $\mathscr{P}^*_\alpha$ is a regular family, where $\mathscr{P}^*_\alpha \stackrel{\text{def}}{=} \{p{:}\, p$ a regular type, $R(p, L, \alpha) = \alpha$, and p is orthogonal to every stationary q for which $R(q, L, \infty) < \infty\} \cap \mathscr{P}^*$.*

(6) *$\mathscr{P}^*_\alpha = \operatorname{cl}^0(\mathscr{P}^*_\alpha) = \operatorname{cl}^1(\mathscr{P}^*_\alpha) = \operatorname{cl}^2(\mathscr{P}^*_\alpha)$, and any formula $\psi(x, \bar{c})$ with $R(\psi(x, \bar{c}), L, \infty) = \alpha$ is $\mathscr{P}^*_\alpha$-regular or is strongly orthogonal to $\mathscr{P}^*_\alpha$. Also, every regular $p \in \mathscr{P}^*_\alpha$ contains (if it is closed under finite conjunctions) a $\mathscr{P}^*_\alpha$-regular formula.*

(7) *For superstable T, for any stationary regular r, for a unique α the type r is not orthogonal to $\mathscr{P}^*_\alpha$, so some $\mathscr{P}^*_\alpha$-regular formula $\varphi(x, \bar{c})$ is not strongly orthogonal to r.*

Proof. Easy.

LEMMA 4.17(8): ($\mathfrak{C}^{\text{eq}}$) *Suppose r is regular and* $\operatorname{stp}(\bar{a}, A)$ *is not orthogonal to r, r not orthogonal to $\mathscr{P}^*_\alpha$. Then*

(i) *There is a formula* $E = E(\bar{x}, \bar{y})$ *an equivalence relation over* A *such that* $\mathrm{tp}(a/E, A)$ *is* $\mathscr{P}^*_\alpha$*-simple but not orthogonal to* r.

(ii) *Moreover,* $\mathrm{tp}(a/E, A)$ *contains a formula* $\theta(y, \bar{b})$ *that is* $\mathscr{P}^*_\alpha$*-simple.*

(iii) *Also, for every* $\bar{b}'$ *realizing* $\mathrm{tp}(\bar{b}, \phi), \theta(y, \bar{b}')$ *is* $\mathscr{P}^*_\alpha$*-simple.*

4.17 A remark: Instead of $\mathscr{P}^*_\alpha$ we can use any regular $\mathscr{P} = \mathrm{cl}^{2,1}_A(\mathscr{P})$ if $A = \mathrm{acl} A$ and any $p \in \mathscr{P}$ has a $\mathscr{P}$-regular formula in it, e.g. any $\mathrm{cl}^{2,1}(\mathscr{P})$, $\mathscr{P} \subseteq \mathscr{P}^*_\alpha$, is O.K. and $\mathrm{cl}^3(p)$, p stationary regular.

Proof. Choose $\varphi(x, \bar{c})$ as in Lemma 4.17(7). Clearly, $\{\varphi(x, \bar{c})\}$ has a stationary completion q which is regular, not orthogonal to r, and w.l.o.g. $q(x, \bar{c}) \in S(\bar{c})$ (as we can increase $\bar{c}$). So $q(x, \bar{c})$, $\mathrm{stp}(\bar{a}, A)$ are not orthogonal, hence w.l.o.g. their stationarization over $A \cup \bar{c}$ are not weakly orthogonal. Let $\{\bar{c}_n : n < \omega\}$ be an indiscernible set based on $A, \bar{c}_0 = \bar{c}$. So each $q_n = q(x, \bar{c}_n)$ is regular, not orthogonal, to A (as q_0 is). Hence, by 3.4 the q_n's are pairwise, not orthogonal; hence, by 1.13(1) (as q_0, r are not orthogonal) each q_n is not orthogonal to r. Also for each n, $\mathrm{stp}(\bar{a}, A)$ has non-orthogonal stationarizations over $A \cup \bar{c}_n$.

Now repeating the proof of 4.11 we get E, an equivalence relation over acl A as required in (i).

So let $E_1, \ldots, E_n$ be the images of E under the automorphisms over A. As $\mathscr{P}^*_\alpha$ is preserved by the automorphism, for every $\bar{d}$ and l the type $\mathrm{stp}(\bar{d}/E_l, A)$ is $\mathscr{P}^*_\alpha$-simple. Let $E^*(\bar{x}, \bar{y}) \stackrel{\mathrm{def}}{=} \bigwedge_{l=1}^{k} E_l(\bar{x}, \bar{y})$, then $\bar{a}/E \in \mathrm{dcl}(A \cup \bigcup_l \bar{a}/E_l)$ so $\mathrm{tp}(a/E, A)$ is $\mathscr{P}^*_\alpha$-simple.

Now $\mathrm{tp}(\bar{a}/E, A)$ contains a formula $\theta(y, \bar{b})$ such that for some $n, \bar{c}_l (l < n)$ realizing $\mathrm{tp}(\bar{c}, A)$

$$\theta(\mathfrak{C}^{\mathrm{eq}}, \bar{b}) \subseteq \mathrm{acl}\left[A \cup \bigcup_l \bar{c}_l \cup \bigcup_l \varphi(\mathfrak{C}^{\mathrm{eq}}, \bar{c}_l)\right].$$

As each $\varphi(\bar{x}, \bar{c}_l)$ is $\mathscr{P}^*_\alpha$-regular (by Lemma 4.17(6)) clearly $\theta(y, \bar{b})$ is $\mathscr{P}^*_\alpha$-simple. So (ii) holds. Also for every automorphism F of $\mathfrak{C}^{\mathrm{eq}}, F(\bar{c}_l)$ realizes $\mathrm{tp}(\bar{c}, \emptyset)$, hence each $\varphi(x, F(\bar{c}_l))$ is $\mathscr{P}^*_\alpha$-regular, hence $\theta(x, F(\bar{b}))$ is $\mathscr{P}^*_\alpha$-simple. This proves (iii).

THEOREM 4.18: ***Suppose*** $\mathscr{P}$ ***is regular,*** $\mathscr{P} = \mathrm{cl}^2_A(\mathscr{P})$ ***and*** $I = \{\bar{b}_i : i < \alpha\}$. ***Then*** $\mathrm{low}_{\mathscr{P}}(\bigcup I, A) = \sum_{i<\alpha} \mathrm{low}_{\mathscr{P}}(\bar{b}_i, A \cup \bigcup_{j<i} \bar{b}_j\}$ ***if every*** $b \in \bigcup_{i<\alpha} \bar{b}_i$ ***is*** $\mathscr{P}$***-simple over*** A.

Proof. **Like 4.9.**

Remark. This really generalizes 4.9(1).

LEMMA 4.19: (1) ***If*** $A \subseteq B$, $\mathscr{P}$ ***regular***, $\mathscr{P} = \mathrm{cl}^2_A(\mathscr{P})$, $\bar{a}$ $\mathscr{P}$***-simple over*** A ***then*** $\mathrm{low}_{\mathscr{P}}(\bar{a}, B) \leq \mathrm{low}_{\mathscr{P}}(\bar{a}, A)$.

(2) *If in addition* $\bar{a}$ *is* $\mathscr{P}$*-semi-regular over* A, $\mathrm{tp}(\bar{a}, B)$ *fork over* A *then* $\mathrm{low}_{\mathscr{P}}(\bar{a}, B) < \mathrm{low}_{\mathscr{P}}(\bar{a}, A)$.

Proof. We prove (2) like 4.7, and (1) like 4.9(2).

THEOREM 4.20 (*in* $\mathfrak{C}^{eq}$): ***Suppose*** p ***is stationary and regular, and*** $\mathrm{stp}(\bar{a}, A)$ ***is not orthogonal to*** p***. Then for some equivalence relation*** E *almost over* A, $\mathrm{stp}(\bar{a}/E, A)$ *is strongly* cl^3 $\{p\}$*-simple*, $\mathrm{low}_p(\bar{a}/E, A) > 0$. *It is also* $\mathrm{cl}^3\{p\}$*-simple.*

Proof. **Like 4.11.**

Similarly

CLAIM 4.21: ***Suppose*** $I \subseteq \varphi(\mathfrak{C}, \bar{b})$, I ***an infinite indiscernible set,*** $\mathrm{tp}_*(I^1, A) = \mathrm{tp}_*(I, A)$ ***implies*** I^1, I ***are not orthogonal. Then there are*** n, $\bar{b}_l (l < n)$ $\bar{a} \in A$, ***and*** ψ ***such that:***

(A) $\mathrm{tp}(\bar{b}_l, A) = \mathrm{tp}(\bar{b}, A)$.

(B) ***For some*** $\bar{c}$, $\psi(\mathfrak{C}, \bar{a}) \subseteq \mathrm{acl}(\bar{c} \cup \bigcup_{l<n} \varphi(\mathfrak{C}, \bar{b}_l))$, ***in fact there is a set*** $\{\bar{b}^\alpha_l{}^\frown \bar{c}^\alpha_l : l < n, \alpha < \kappa\}$ ***(any*** κ***) independent over*** A $\mathrm{stp}(\bar{b}^\alpha_l, A) \equiv \mathrm{stp}(\bar{b}^0_l, A)$ ***such that for every*** $\alpha(l) < \kappa$, $\psi(\mathfrak{C}, \bar{a}) \subseteq \mathrm{acl}(\bigcup_{l<n} \bar{c}^{\alpha(l)}_l \bigcup_{l<n} \psi(\mathfrak{C}, \bar{b}^{\alpha(l)}_l))$.

(C) ***There is an (infinite) indiscernible set*** $J \subseteq \psi(\mathfrak{C}, \bar{a})$ ***based on*** A ***such*** *that* $\mathrm{tp}_*(I^1, A) = \mathrm{tp}_*(I, A)$ *implies* I^1, J *are not orthogonal.*

(D) ***If*** $J_1 \subseteq \psi(\mathfrak{C}, \bar{a})$ ***is an infinite indiscernible set, then for some*** l, ***and*** $I_1 \subseteq \varphi(\mathfrak{C}, \bar{b}_l)$, J_1, I_1 ***are not orthogonal.***

(E) ***If*** $\mathrm{Av}(I, \bigcup I)$ ***is regular, then*** $\mathrm{Av}(J, \bigcup J)$ ***is semi-regular.***

(F) ***If*** $\varphi(x; \bar{a})$ ***is*** $\mathscr{P}$***-regular,*** $\psi(x; \bar{a})$ ***is*** $\mathscr{P}$***-simple.***

LEMMA 4.22: ***Suppose*** $\kappa = |A|^+ + \kappa_r(T)$, M ***is*** $\mathbf{F}^a_\kappa$***-saturated,*** $p = \mathrm{stp}(\bar{a}, A)$ ***is semi-regular, then there is an indiscernible set*** $I \subseteq |M|$ ***based on*** p, $\mathrm{Av}(I, A \cup \bigcup I) \vdash \mathrm{Av}(I, M)$.

Proof. First choose a maximal set I of sequences from M realizing p, **which is independent over** A. **Let** B **be a maximal subset of** M **satisfying**

(i) $A \cup \bigcup I \subseteq B$,

(ii) B **is** $\mathbf{F}^a_\kappa$**-atomic over** $A \cup \bigcup I$

(exists as $A \cup \bigcup I$ **satisfies those conditions, and they are preserved by** unions of increasing chains). By IV, 3.2, for no $b \in |M| - B$ is

$\mathrm{tp}(b, B)$ $\mathbf{F}^a_\kappa$-isolated, hence $B = |N|$, N $\mathbf{F}^a_\kappa$-saturated. We can find a complete regular type q over $|N|$ not orthogonal to p, and let $J_1 = \{\bar{c}_i : i < \alpha\} \subseteq |M|$ be a maximal set of sequences realizing q independent over $|N|$. Let p_1 be the stationarization of p over $|N|$, clearly M omits p_1, hence $\alpha < \aleph_0$, and even $\alpha < w(p)$.

Choose $B \subseteq N$, $|B| < \kappa(T)$ such that q does not fork over B, $q \restriction B$ is stationary, $A \subseteq B$, $\mathrm{tp}(B, A \cup \bigcup I)$ does not fork over $A \cup (B \cap I)$, and $p_1 \restriction B, q \restriction B$ are not weakly orthogonal. Let $J \subseteq |N|$ be a maximal set of sequences realizing $q \restriction B$ and independent over B. Now $J \cup J_1$ is a maximal set of sequences from $|M|$ realizing $q \restriction B$ and independent over B (1.16(3)). Next choose $I_0 \subseteq I$, $J_0 \subseteq J$, $|I_0| = |J_0| = \kappa$ such that $I - I_0$ is independent over $(A \cup I_0 \cup B \cup J_0, A)$ and $J - J_0$ is independent over $(A \cup I_0 \cup B \cup J_0, B)$. Let $N^* \subseteq M$ be $\mathbf{F}^a_\kappa$-prime over $A \cup I_0 \cup J_0 \cup J_1$. By IV, 4.11 there is a set $I_2 \subseteq |N^*|$ of sequences realizing p, independent over A, $\mathrm{Av}(I_2, A \cup \bigcup I_2) \vdash \mathrm{Av}(I_2, |N^*|)$ (the hypothesis of IV, 4.11 holds as N^* is $\mathbf{F}^a_\kappa$-prime over $\emptyset$, by IV, 4.9(3)).

Now $I - I_0$ is independent also over $(|N^*|, A)$, and by IV, 4.15 N^* is $\mathbf{F}^a_\kappa$-atomic over $A \cup \bigcup I_2$, hence $N^* \cup \bigcup I$ is $\mathbf{F}^a_\kappa$-atomic over $A \cup \bigcup I_3$, where $I_3 = (I - I_0) \cup I_2$. Clearly for every $\bar{c} \in J$, $\mathrm{tp}(\bar{c}, B \cup \bigcup I)$ fork over B hence $\mathrm{tp}(\bar{c}, |N^*| \cup I_3)$ fork over B, hence is orthogonal to p so also $|N^*| \cup \bigcup I \cup \bar{c}$ is $\mathbf{F}^a_\kappa$-atomic over $A \cup \bigcup I_3$. Repeating (and noting $J_0 \subseteq |N^*|$) we get that $|N^*| \cup \bigcup I_3 \cup \bigcup (J \cup J_0)$ is $\mathbf{F}^a_\kappa$-atomic over $A \cup I_3$. So we can find a maximal set $B_1 \subseteq |M|$ $\mathbf{F}^a_\kappa$-atomic over $A \cup I_3$, $|N^*| \cup \bigcup (I_3 \cup J \cup J_0) \subseteq B_1$ and again $B_1 = |N_1|$, N_1 $\mathbf{F}^a_\kappa$-saturated, and let p', q' be the stationarizations of p, q, resp., over $|N_1|$. But $J \cup J_0 \subseteq |N_1|$ was a maximal set of sequences in M realizing $q \restriction B$ independent over B, hence M omits q', hence p' has a unique extension over $|M|$. As by 4.10(2) $\mathrm{Av}(I_3, A \cup I_3) \vdash \mathrm{Av}(I_3, |N_1|)$ we finish.

EXERCISE 4.12: Show that in 4.18 and 4.19(1) the demand $\mathscr{P} = \mathrm{cl}^2_A(\mathscr{P})$ is necessary. [Hint: Let T be a theory with only one equivalence relation, with infinitely many equivalence classes, each of them infinite, and choose distinct $a_l \in \mathfrak{C}$, $\mathfrak{C} \models a_1 E a_2 \wedge \neg a_1 E a_3$.

For 4.18 let $\mathscr{P} = \mathrm{cl}^3\{\mathrm{tp}(a_3, a_1), \mathrm{tp}(a_2, a_1)\}$, $A = \emptyset$, $\alpha = 2$, $\bar{b}_0 = \langle a_1 \rangle$, $\bar{b}_1 = \langle a_2 \rangle$: and for 4.19(1) $\mathscr{P} = \mathrm{cl}^3\{\mathrm{tp}(a_2, a_1)\}$, $A = \emptyset$, $B = \{a_1\}$, $\bar{a} = \langle a_2 \rangle$.]

EXERCISE 4.13: Show the if in Definition 4.3(1) (i) we demand only "$\mathrm{stp}(\bar{a}, B)$ is orthogonal to every $p \in \mathscr{P}$ which does not fork over A," then 4.18 may fail. (Hint: Use the example from Exercise 4.12.)

EXERCISE 4.14: Look through Sections 1–4 and generalize the relevant theorems to the notions from Definition 4.5 and 4.4 (in addition to Theorems 4.16–4.22).

EXERCISE 4.15: Suppose q, A are as in Definition 4.4(1) A, $\operatorname{Dom} q \subseteq B \subseteq C$, $\bar{a}$ realizes q. Then $\operatorname{low}_{\mathscr{P}}(\bar{a}, C) \leq \operatorname{low}_{\mathscr{P}}(\bar{a}, B)$.

V.5. Multi-dimensional theories

DEFINITION 5.1: Let $\{J_i: i \in S\} \leq_w I$ *if* for every $\mathbf{F}^a_\kappa$-saturated model (where $\kappa = \kappa_r(T) + |S|^+$), such that $I, J_i \subseteq |M|$.

$$\min_{i \in S} \dim(J_i, M) \leq \dim(I, M)$$

LEMMA 5.1: *The following conditions on I, J_i ($i \in S$) are equivalent (letting $\kappa = \kappa_r(T) + |S|^+$):*

(1) $\{J_i: i \in S\} \leq_w I$.

(2) *If M is $\mathbf{F}^a_\kappa$-saturated, $J^* = \{\bar{a}^i_j: i \in S, j < \kappa\}$ is independent over $|M|$, $I, J_i \subseteq M$ and $\bar{a}^i_j$ realizes* $\operatorname{Av}(J_i, |M|)$ *and N is $\mathbf{F}^a_\kappa$-prime over $|M| \cup \bigcup J^*$, then* $\operatorname{Av}(I, |M|)$ *is realized in N.*

(3) *There are a $\mathbf{F}^a_{\kappa(T)}$-saturated model M, $I' \subseteq M$ equivalent to I, and $S' \subseteq S$, $|S'| < \kappa(T)$ and $J'_i \subseteq M (i \in S')$ equivalent to J_i, and $J^* = \{\bar{a}^i_j: i \in S', j < \kappa(T)\}$ independent over $|M|$, a^i_j realizes* $\operatorname{Av}(J'_i, |M|)$ *such that in the $\mathbf{F}^a_{\kappa(T)}$-prime model over $|M| \cup \bigcup J^*$,* $\operatorname{Av}(I', |M|)$ *is realized.*

(4) *There are J'_i equivalent to J_i ($i \in S$) and I' equivalent to I such that*

(i) $|I'|, |J'_i| = \kappa_r(T)$,

(ii) *I' is indiscernible over $\bigcup_{i \in S} J'_i$ and J'_i is indiscernible over $I' \cup \bigcup \{J'_i: i \neq j \in S\}$,*

(iii) $\operatorname{Av}(I', I')$ *has a $\mathbf{F}^a_{\kappa(T)}$-isolated complete extension over* $I' \cup \bigcup \{J_i: i \in S\}$.

(5) *For some $\lambda \geq \mu \geq \kappa$, there is no $\mathbf{F}^a_\kappa$-saturated M, $I, J_i \subseteq M$ such that* $\dim(I, M) = \mu$, $\dim(J_i, M) = \lambda$ ($i \in S$).

Proof. As the ideas essentially appear in Section 2, we leave it to the reader as an exercise.

CONCLUSION 5.2: (1) $\{J_i: i \in S\} \leq_w I$ *iff for some $|S'| \subseteq S$, $|S'| < \kappa(T)$,* $\{J_i: i \in S'\} \leq_w I$.

(2) *The holding of $\{J_i: i \in S\} \leq_w I$ depend on I, J_i only up to equivalence.*

Proof. Immediate by 5.1.

THEOREM 5.3: *Suppose* I, J_i^α *are indiscernible over* A $(i < \kappa(T), \alpha < \kappa(T))$ *and* $|I|, |J_i^\alpha| \geq \kappa_r(T)$, $|A| < \kappa_r(T)$ *and* $\mathrm{tp}_*(J_i^\alpha, A \cup \bigcup (I \cup \bigcup \{J_j^\beta : \beta < \alpha$ *or* $\beta = \alpha, j < i\})$ *does not fork over* A, *and* $\mathrm{stp}_*(J_i^\alpha, A) \equiv \mathrm{stp}_*(J_i^\beta, A)$. *Let* $J^* = \bigcup_{i,\alpha} J_i^\alpha$, $B = A \cup \bigcup (I \cup J^*)$. *Assume* $\{J_i^\alpha : \alpha, i < \kappa(T)\} \leq_w I$.

(1) *If* $J = \{\bar{a}_i^\alpha : \alpha, i < \kappa(T)\}$ *is independent over* B, $\bar{a}_i^\alpha$ *realizes* $\mathrm{Av}(J_i^\alpha, B)$, *then* $\mathrm{Av}(I, A \cup I)$ *has an* $\mathbf{F}^a_{\kappa(T)}$-*isolated complete extension* q *over* $B \cup \bigcup J$.

(2) *If in addition* M *is* $\mathbf{F}^a_{\kappa_r(T)}$-*saturated*, $B \subseteq |M|$, $\{\bar{a}_i^\alpha : \alpha, i < \kappa(T)\}$ *is independent over* $(|M|, B)$, *then* $q \vdash \mathrm{Av}(I, |M|)$.

Proof. (1) W.l.o.g. we assume $|I| = |J_i^\alpha| = \lambda = (2^{|T|})^+$, and let $p = \mathrm{Av}(I, B)$. Clearly there are $J' \subseteq J$, $|J'| < \kappa(T)$ and q_1, $p \subseteq q_1 \in S^m(B \cup \bigcup J')$, such that q_1 has a unique extension $q_2 \in S^m(B \cup \bigcup J)$ and let $\bar{b}$ realize q_1. By notational changes we can assume that $J' = \{\bar{a}_i^\alpha : \alpha, i < \alpha_0 < \kappa(T)\}$, and that we can choose $C \subseteq B$, $|C| < \kappa(T)$ $C \subseteq B_0 = A \cup \bigcup (J' \cup I \cup \bigcup \{J_i^\alpha : \alpha, i < \alpha_0\})$ such that q_1 does not fork over C. Clearly also q_2 does not fork over C. Choose $J_j^+ = \{\bar{b}_{i,j}^\alpha : \alpha_0 \leq \alpha < \kappa(T), i < \kappa(T)\}$ such that $J^+ = \bigcup \{J_j^+ : j < \lambda^+\}$ is independent over $B \cup \bigcup J$ and $\bar{b}_{i,j}^\alpha$ realizes $\mathrm{Av}(J_i^\alpha, B \cup \bigcup J)$. We shall show that $q_1 \vdash \mathrm{tp}(\bar{b}, B \cup \bigcup (J \cup J^+))$, for this it suffices to prove that for every $n < \omega$, $q_1 \vdash q^n = \mathrm{tp}(\bar{b}, B \cup (J \cup \bigcup_{j<n} J_j^+))$, and q^n does not fork over C and this we shall prove by induction on n. For $n = 0$ there is nothing to prove. For $n + 1$ choose an elementary mapping f such that f is the identity over B_0, and for $i, \alpha, \alpha_0 \leq \alpha < \kappa(T)$, f maps J_i^α onto $J_i^\alpha \cup \{a_{i,j}^\alpha : j < n\}$ and $f(\bar{a}_i^\alpha) = \bar{a}_{i,n}^\alpha$. As $q_1 \vdash q_2$ clearly $f(q_1) \vdash f(q_2)$, so if $f(q_1) = q^n, f(q_2) = q^{n+1}$ we finish. But this holds as q_1 does not fork over C, $C \subseteq B_0$, $f \restriction B_0$ is the identity, clearly $f(q_1) \supseteq q^1$ does not fork over C, hence does not split strongly over C, but also q_n satisfies this and $q_1 \subseteq q^n$; hence $q^n = f(q_1)$; similarly $f(q_2) = q^{n+1}$.

We can conclude that $q_1 \vdash \mathrm{tp}(\bar{b}, B \cup (J \cup J^+))$. Let M be $\mathbf{F}^a_\lambda$-prime over B and $\mathbf{F}^a_\lambda$-constructible over $B \cup J^+$; and N be $\mathbf{F}^a_\lambda$-prime over $|M| \cup J^+$. As for $\alpha \geq \alpha_0$, $\dim(J_i^\alpha, N) \geq \lambda^+$, $\dim(I, N) \geq \lambda^+$ [clearly there is an elementary mapping f, $f \restriction A \cup \bigcup I$ is the identity, and f maps J_i^α onto $J_i^{\alpha_0+\alpha}$ $(\alpha, i < \kappa(T))$ hence $\{J_i^\alpha : \alpha, i < \kappa(T)\} \leq_w I$ implies $\{J_i^\alpha : \alpha_0 \leq \alpha < \kappa(T), i < \kappa(T)\} \leq_w I$]. So there is $I^+ \subseteq |N|$ indiscernible over $|M| \cup \bar{b}$, $|I^+| = \lambda^+$, $I \cup I^+$ is indiscernible over A.

By IV, 4.10(2) $\mathrm{tp}(\bar{b}, B \cup \bigcup J^+) \vdash \mathrm{tp}(\bar{b}, |N|)$; hence

$$q_1 \vdash \mathrm{tp}(\bar{b}, \bigcup (I \cup I^+) \cup A).$$

We can assume that if $\bar{c} \in I$ is not disjoint to $C - A$ then $\bar{c} \in C$. By the proof of IV, 4.3 we can deduce that $\operatorname{stp}(\bar{b}, A \cup C) \vdash \operatorname{tp}(\bar{b}, A \cup \bigcup (I_1 \cup I^+))$, $(I_1 = \{\bar{c} \in I : \bar{c} \in C\})$ hence $\operatorname{stp}(\bar{b}, A \cup C) \vdash \operatorname{tp}(\bar{b}, A \cup \bigcup I)$.

(2) The proof follows easily now.

DEFINITION 5.2: The (infinite) indiscernible set I is called multi-dimensional over A if whenever I_j $(j \leq \kappa(T))$ are such that $\operatorname{tp}_*(I_j, A \cup \bigcup_{i<j} I_i)$ does not fork over A and $\operatorname{stp}_*(I_j, A) \equiv \operatorname{stp}_*(I, A)$ then $\{I_j : j < \kappa(T)\} \not\geq_w I_{\kappa(T)}$.

DEFINITION 5.3: The theory T is multi-dimensional if some I is multi-dimensional over some A.

LEMMA 5.4: *If I is orthogonal to A then I is multi-dimensional over A.*

Proof. Immediate by 1.5 and 2.7.

THEOREM 5.5: *Suppose T is not multi-dimensional, and M is $\mathbf{F}^a_{\kappa(T)}$-saturated, and $\lambda_0 \leq 2^{|T|}$ is the first cardinal in which T is stable. Then there is $N \subseteq M$, $\|N\| \leq \lambda_0$ and a set $J \subseteq M$ independent over M such that M is a $\mathbf{F}^a_{\kappa(T)}$-prime over $|N| \cup \bigcup J$.*

We first prove a claim:

CLAIM 5.6: *Suppose λ_0 is as in 5.5, $N_0 \subseteq M_0 \subseteq M$, and they are $\mathbf{F}^a_{\kappa(T)}$-saturated and $\|N_0\| \leq \lambda_0$, and for every $\bar{c} \in M$, $\operatorname{tp}(\bar{c}, |M_0|)$ forks over N_0 or $\bar{c} \in |N_0|$. There are N_1, $N_0 \subseteq N_1 \subseteq M_0$, $\|N_1\| \leq \lambda_0$ and $J \subseteq M$ independent over $(|M_0|, |N_1|)$ such that if $\bar{c} \in |M|$, $\bar{c} \notin |M_0|$, $\bar{c}$ realizes $\operatorname{Av}(I^*, |M_0|)$, I^* any countable set indiscernible over N_0, I^* not multi-dimensional over $|N_0|$ and $\operatorname{Cb}(\operatorname{Av}(I^*, \bigcup I^*)) \subseteq |M_0|$, then $\operatorname{tp}(\bar{c}, M_0 \cup \bigcup J)$ fork over $|M_0|$ or $\operatorname{tp}_*(I^*, |N_1|)$ forks over $|N_0|$. Also N_1 is $\mathbf{F}^a_{\kappa(T)}$-saturated.*

Proof of 5.6. Let $\{\langle I_\alpha, J_\alpha \rangle : \alpha < \alpha_0\}$ be a maximal family such that:

(i) I_α is countable, and it is indiscernible over $|N_0|$, and I_α is not multi-dimensional over $|N_0|$ and $\operatorname{Cb}(\operatorname{Av}(I_\alpha, \bigcup I_\alpha)) \subseteq |M_0|$.

(ii) $J_\alpha \subseteq M$ is a maximal set independent over $(|M_0| \cup \bigcup_{\beta<\alpha} J_\beta, |M_0|)$ of sequences realizing $\operatorname{Av}(I_\alpha, |M_0|)$, and J_α is not empty.

(iii) $\operatorname{tp}_*(I_\alpha, \bigcup_{\beta<\alpha} I_\beta \cup |N_0|)$ does not fork over $|N_0|$. The number of possible $\operatorname{tp}_*(I_\alpha, |N_0|)$ is $\leq \lambda_0$. [If $\lambda_0^{\aleph_0} = \lambda_0$ by the λ_0-stability of T. If $\lambda_0^{\aleph_0} > \lambda_0$ then necessarily $\kappa(T) = \aleph_0$, so choose $I'_\alpha \subseteq I_\alpha$, I'_α finite so that $\operatorname{Av}(I_\alpha, \bigcup I_\alpha)$ does not fork over I'_α, and let $I'_\alpha = \{\bar{a}^\alpha_0, \ldots, \bar{a}^\alpha_{n(\alpha)-1}\}$, $\bar{a}^\alpha_{n(\alpha)} \in I_\alpha - I'_\alpha$; now $\operatorname{tp}(\bar{a}_0 {}^\frown \cdots {}^\frown \bar{a}_{n(\alpha)}, |N_0|)$ determine $\operatorname{tp}_*(I_\alpha, |N_0|)$.]

Similarly, for each i there is $C_i \subseteq |N_0|$, $|C_i| < \kappa_r(T)$ and $I_i^* \subseteq |N_0|$ such that $\text{tp}_*(I_i, |N_0|)$ does not fork over C_i and $\text{stp}_*(I_i^*, C_i) \equiv \text{stp}_*(I_i, C_i)$.

So if $\alpha_0 \geq \lambda_0^+$ then w.l.o.g. for every $\alpha < \lambda_0^+$, $\text{tp}_*(I_\alpha, |N_0|) = \text{tp}_*(I_0, |N_0|)$. As I_0 is not multi-dimensional over $|N_0|$, we get a contradiction by 5.3. (As $\{I_j : j < \kappa(T)\} \leq_w I_0^*$, $\text{Av}(I_0^*, |M_0|)$ is realized in M.) So $\alpha_0 < \lambda_0^+$ and we choose N_1, $N_1 \subseteq M$, $\|N_1\| \leq \lambda_0$,

$$|N_0| \cup \bigcup_{i<\alpha_0} \text{Cb}(\text{Av}(I_\alpha, \bigcup I_\alpha)) \subseteq |N_1|$$

and our conclusion follows.

Proof of 5.5. We define by induction on $i \leq \kappa_r(T)$, N_i, M_i, J_i so that:

(i) $N_i \subseteq M$, $\|N_i\| \leq \lambda_0$ and N_i is $\mathbf{F}^a_{\kappa_r(T)}$-saturated.

(ii) $J_i \subseteq M$, J_i is a maximal set independent over $(|N_i| \cup \bigcup_{j<i} |M_j|, |N_i|)$.

(iii) M_i is $\mathbf{F}^a_{\kappa(T)}$-prime over $\bigcup J_i \cup \bigcup_{j<i} |M_j|$, and $M_i \subseteq M$.

(iv) $N_{i+1} \subseteq M_i$, $N_\delta \supseteq \bigcup_{i<\delta} N_i$ for limit δ.

(v) For every $\bar{c} \in |M|$, $\bar{c} \notin |M_i|$, and countable indiscernible set I, $\text{Cb}(\text{Av}(I, \bigcup I)) \subseteq |M_i|$, $\text{Av}(I, |M_i|) = \text{tp}(\bar{c}, |M_i|)$, at least one of the following holds:

(A) $\text{tp}(\bar{c}, |M_i| \cup \bigcup J_i)$ forks over $|M_i|$.

(B) $\text{tp}_*(I, |N_{i+1}|)$ forks over $|N_i|$.

We choose N_0 to satisfy (i), and M_i by (iii). J_i and N_{i+1} are defined by Claim 5.6, and N_δ by (iv).

If $M_{\kappa_r(T)} \neq M$ choose $\bar{c} \in M$, $\bar{c} \notin M_{\kappa_r(T)}$ and a countable indiscernible set I over $N_{\kappa_r(T)}$, $\text{Cb}(\text{Av}(I, \bigcup I)) \subseteq M_{\kappa_r(T)}$, such that $\text{Av}(I, M_{\kappa_r(T)}) = \text{tp}(\bar{c}, M_{\kappa_r(T)})$. By (v) we get a contradiction (when $\kappa_r(T) = \aleph_0$, we have to work a little more).

Remark. In 5.6 we can assume only M is $\mathbf{F}^a_{\kappa(T)}$-saturated.

THEOREM 5.8: *For every non-multidimensional T, there is a cardinal $\lambda \leq (2^{|T|})^+$ such that:*

(1) *λ is the first cardinal for which there are no $I_\alpha (\alpha < \lambda)$ such that $\{I_\beta : \alpha < \beta < \lambda\} \not\leq I_\alpha$.*

(2) *Let $\kappa_r(T) \leq \aleph_\alpha < \aleph_\beta$, T stable in $\aleph_\beta$, and μ is the number of non-isomorphic $\mathbf{F}^a_{\aleph_\alpha}$-saturated models of T of cardinality $\aleph_\beta$. Then*

$$\min\left\{\sum_{\chi<\lambda} |\beta - \alpha + 1|^\chi, 2^{|\beta-\alpha|}\right\} \leq \mu \leq \sum_{\chi<\lambda} |\beta - \alpha + 1|^\chi + 2^{\lambda_0}$$

(λ_0-the first cardinality in which T is stable).

Proof. Left to the reader as an exercise.

PROBLEM 5.1: Make 5.8(2) more precise. In particular try to prove that λ is always a successor. (See IX, Section 2.)

THEOREM 5.9: *If T is multi-dimensional, T stable in $\aleph_\beta$, $\kappa_r(T) \leq \aleph_\alpha$, $\alpha < \beta$ then T has $\geq 2^{|\beta-\alpha|}$ non-isomorphic $\mathbf{F}^a_{\aleph_\alpha}$-saturated models of cardinality $\aleph_\beta$ provided that $\beta < \aleph_\alpha$.*

Proof. Let I be (an infinite indiscernible set over A which is) multi-dimensional over A; and we can assume $|A| < \kappa_r(T)$, I countable. Choose $I^\mu_\xi(\aleph_\alpha \leq \mu \leq \aleph_\beta, \xi < \aleph_\beta)$, $|I^\mu_\xi| = \mu$, such that $\text{stp}_*(I^\mu_\xi, A) \equiv \text{stp}_*(I, A)$, and $\text{tp}_*(I^\mu_\xi, A \cup \bigcup_{\lambda<\mu} I^\lambda_\zeta \cup \bigcup_{\zeta<\xi} I^\mu_\zeta)$ does not fork over A. For any set $S \subseteq S_0 = \{\lambda: \aleph_\alpha \leq \lambda < \aleph_\beta, \lambda \text{ regular or } \lambda = \aleph_\alpha\}$, let $\mu = \min(S \cup \{\aleph_\beta\})$ and M_s be a $\mathbf{F}^a_\mu$-prime model over

$$A \cup \bigcup \{I^\lambda_\xi: \lambda \in S \cup \{\aleph_\beta\}, \xi < \aleph_\beta\}.$$

Clearly M_S is a $\mathbf{F}^a_{\aleph_\alpha}$-saturated model (of T) of cardinality $\aleph_\beta$, and it is not hard to prove that $S(M_S) \cap S_0 = S$ where $S(M) = \{\dim(J, M)$: J an infinite indiscernible set, $J \subseteq M\}$.

So we finish.

PROBLEM 5.2: In Definition 5.2, can we replace "over some A" by "over $\emptyset$".

PROBLEM 5.3: Suppose A, I, J^α_i are as in Theorem 5.3. Is it necessarily true that for some $\alpha, \{J^\alpha_i : i < \kappa(T)\} \leqslant_s I$.

EXERCISE 5.4: Construct a two-dimensional (i.e., $\lambda = 3$ in Theorem 5.8) countable, stable but not superstable theory.

EXERCISE 5.5: In 5.3 replace $\kappa_r(T)$ by any $\kappa \geq \kappa_r(T)$ (so $|A| < \kappa$ now).

EXERCISE 5.6: Suppose $A \subseteq B$, $\text{tp}(\bar{a}, A)$ has no extension over B which fork over A, and $\text{tp}_*(B, A)$ is stationary. Prove $\text{tp}(\bar{a}, A) \vdash \text{tp}(\bar{a}, B)$.

QUESTION 5.7: Can we in 5.3 get $\mathbf{F}^s_{\kappa(T)}$-isolation?

EXERCISE 5.8: Suppose I is indiscernible over A. Prove that for some $B \subseteq A, |B| < \kappa_r(T)$, $\mathrm{tp}_*(\bigcup I, A)$ does not fork over B: provided that I is based on A.

V.6. Cardinality-quantifiers and two-cardinal theorems

We first define a general context, and then get conclusions on the logics with cardinality quantifiers and two-cardinal theorems.

Let $K = (\mathbf{F}, W, \lambda, \mu)$ be a quadruple (we write $\lambda = \lambda_K$, $\mu = \mu_K$, etc.) satisfying the following conditions:

Condition 1. $\mathbf{F}$ is as in IV, Section 1, satisfying Axioms (I), (III.1), (IV), (VI), (VIII) and (X1).

Condition 2. W is a set of triples $\langle \varphi(x, \bar{y}), \psi(\bar{y}), \chi \rangle$ where χ is a regular cardinal $\geq \lambda_K$ and $|A| < \chi$ implies that for any $\mathbf{F}$-maximal $\mathbf{F}$-constructible model M over A, $\|M\| < \chi$, and if M is $\mathbf{F}$-maximal, $A \subseteq M, |A| < \chi$, then there is an $\mathbf{F}$-maximal $N, A \subseteq N \subseteq M$, $\|N\| < \chi$.

Remark. A model M is called $\mathbf{F}$-maximal, if for no $\bar{c} \notin M$, is $\mathrm{tp}(\bar{c}, |M|)$ $\mathbf{F}$-isolated; similarly for a set.

Notation. Let Car K be the set of cardinals appearing in W and ∞. We say $\vDash$"$(\exists^{<\chi} x)\varphi(x, \bar{a})$" (or $\vDash_K$, or $\vDash_W$ instead $\vDash$) if for some triple $\langle \varphi(x, \bar{y}), \psi(\bar{z}), \chi \rangle \in W$, $\vDash \psi[\bar{a}]$. Clearly this is an elementary property of $\bar{a}$. Define inductively $C_n(\varphi(\bar{x}; \bar{a})) = \min\{\kappa\colon |\varphi(\mathfrak{C}, \bar{a})| = \kappa < \aleph_0$ or there are $\psi(y; \bar{b})$, $\theta(\bar{x}; y; \bar{c})$ such that: $\theta(\bar{x}; y; \bar{c}) \vdash \varphi(\bar{x}; \bar{a}) \wedge \psi(y; \bar{b})$, $\varphi(\bar{x}, \bar{a}) \vdash (\exists y)\theta(\bar{x}; y; \bar{c})$, $\vDash$"$(\exists^{<\chi} y)\psi(y, \bar{b})$" for some $\chi \leq \kappa$ and for each b_1: $n > 0$, $C_{n-1}(\theta(\bar{x}; b_1, \bar{c})) \leq \kappa$ or $\theta(\bar{x}; b_1, \bar{c})$ is algebraic$\}$. Let $C(\varphi(\bar{x}; \bar{u})) = \min_n C_n(\varphi(\bar{x}\colon \bar{a})), C(p) = \min\{C(\varphi(\bar{x}, \bar{a}))\colon p \vdash \varphi(\bar{x}; \bar{a})\}, C(\bar{a}, A) = C(\mathrm{tp}(\bar{a}, A))$ and let for an (infinite) indiscernible set (of sequences) I, $H(I) = \sup\{\kappa\colon$ if J is an infinite indiscernible set and $\varphi(x, \bar{a}) \in \mathrm{Av}(J, \bar{a})$, $\vDash$"$(\exists^{<\chi} x)\varphi(x, \bar{a})$", $\chi < \kappa$ then I, J are orthogonal$\}$. For a complete or stationary type p, let $H(p)$ be $H(I)$, for any indiscernible set based on some stationarization of p, let $H(\bar{a}, A) = H(\mathrm{tp}(\bar{a}, A))$. Let $C^*(\varphi(x, \bar{a})) = \min\{\kappa\colon$ there is a $\mathbf{F}$-maximal model M, $\bar{a} \in M$, such that for every p, $\varphi(x, \bar{a}) \in p \in S(|M|)$ implies $H(p) \leq \kappa\}$ (clearly $C^*(\varphi(x, \bar{a})) \leq C(\varphi(x, \bar{a}))$. Let $B \prec^{\lambda} A$ (or $B \prec^{\lambda}_{K} A$) means that for any $\bar{a} \in A$, $H(\bar{a}, B) \geq \lambda$ (B, A can be replaced by models.) The sup in the definition of $H(I)$ is taken on $\kappa \in \mathrm{Car}\ K$; similarly elsewhere.

Condition 3. For any formula $\varphi(x, \bar{a})$ $C^*(\varphi(x, \bar{a})) = C(\varphi(x, \bar{a}))$, i.e., for any $\mathbf{F}$-maximal model M, $\bar{a} \in |M|$, there is a type r such that $\varphi(x, \bar{a}) \in r \in S(|M|)$ and $H(r) = C(\varphi(x, \bar{a}))$.

Condition 4. $\lambda \geq \mu^+ + |T|^+ + \lambda(\mathbf{F})$.

Condition 5. If M is $\mathbf{F}$-maximal $|M| \subseteq B$, N is a $\mathbf{F}$-constructible over B, $c \in |N| - |M|$ then $\operatorname{tp}(c, B)$ forks over $|M|$.

Condition 6. If M_i $(i < \mu)$ is increasing, and each M_i is $\mathbf{F}$-maximal, then $\bigcup_{i<\mu} M_i$ is maximal.

In each theorem $(Cn_1, n_2, \dots)$ will denote the conditions on which the proof of the theorem depends:

CLAIM 6.1: (1) *For formulas $\psi(x, \bar{a})$, $\varphi(x, \bar{c})$ the following conditions are equivalent:*

(i) *If $\bar{a}, \bar{c} \in |M|$ then $|\psi(M, \bar{a})| < |\varphi(M, \bar{c})|^+ + \aleph_0$ (this we denote by $\psi(x,\bar{a}) \leq \varphi(x, \bar{c})$).*

(ii) *There are a model M, and $k, l < \omega$, and finite Δ such that in $\psi(M, \bar{a})$ there is no (Δ, l)-indiscernible set over $\varphi(M, \bar{c})$ of cardinality k. ($\bar{a}, \bar{c} \in |M|$, of course.)*

(iii) *In $\psi(\mathfrak{C}, \bar{a})$ there is no infinite set indiscernible over $\varphi(\mathfrak{C}, \bar{c})$.*

(iv) *There are no models M, N, $\bar{a} \cup \bar{c} \subseteq |M| \subseteq |N|$, such that $\varphi(M, \bar{c}) = \varphi(N, \bar{c})$, $\psi(M, \bar{a}) \neq \psi(N, \bar{a})$.*

(v) *There are M, $\bar{a} \cup \bar{c} \subseteq M$ and finite k, Δ such that there are no $\bar{b}_i \in \psi(M; \bar{a})$ $(i < k)$ for which $\operatorname{tp}_\Delta(\bar{b}_i, \bigcup_{j<i} \bar{b}_j \cup \varphi(M, \bar{c}))$ increase with i and $\bar{b}_i \neq \bar{b}_{i+1}$ and $(\bar{x} = \bar{y}) \in \Delta$.*

(2) *For every cardinal $\kappa \geq \aleph_0$ and formula $\varphi = \varphi(\bar{x}, \bar{y})$ there is a set p^κ_φ such that for every $\bar{b}$, $C(\varphi(\bar{x}; \bar{b})) \leq \kappa$ iff $\bar{b}$ satisfies some $\psi(\bar{y}) \in p^\kappa_\varphi$.*

(3) *In* (2) $p^\kappa_\varphi = \bigcup_{n<\omega} p^{\kappa,n}_\varphi$, *where*

(i) $C_n(\varphi(\bar{x}; \bar{b})) \leq \kappa$ *iff $\bar{b}$ satisfies some $\psi(\bar{y}) \in p^{\kappa,n}_\varphi$,*

(ii) *$p^{\kappa,n}_\varphi$ is the set of the following formulas: for each $\psi(\bar{x}; \bar{y}_1)$, $\theta(\bar{x}; y; \bar{z})$ and finite $q \subseteq p^{\kappa,n-1}_{\theta(\bar{x};y;\bar{z})}$, and $k < \omega$ and $\langle \psi(x; \bar{y}_1), \psi_1(\bar{y}_1), \chi \rangle \in W$, $\chi \leq \kappa$ the formula saying: there are $\bar{y}_1$, $\bar{z}$ such that $(\forall \bar{x}\bar{y})(\theta(\bar{x}, y; \bar{z}) \to \varphi(\bar{x}; \bar{y}) \wedge \psi(y; \bar{y}_1))$ and $(\forall \bar{x})(\varphi(\bar{x}; \bar{y}) \to (\exists y)\theta(\bar{x}; y; \bar{z}))$ and $\psi_1(\bar{y}_1)$ and $(\forall y)(\bigvee q(y, \bar{z}) \vee (\exists^{\leq k}\bar{x})\theta(\bar{x}, y, \bar{z}))$ (for $n = 0$, $q = \emptyset$).*

(4) *p^κ_φ is closed under disjunctions (more exactly: finite disjunction is equivalent to another formula) and contains all formulas $(\exists^{<k}\bar{x})\varphi(\bar{x}, \bar{y})$. Clearly $C(\bigvee^n_{i=1} \varphi_i(\bar{x}, \bar{a}_i)) = \max_i C(\varphi_i(\bar{x}; \bar{a}_i))$. $C(\bigwedge \varphi_i) \leq \min_i C(\varphi_i)$.*

(5) *If* $\theta(\bar{x}; \bar{y}, \bar{c}) \vdash \varphi(\bar{x}; \bar{a}) \wedge \psi(\bar{y}; \bar{b})$, $\varphi(\bar{x}; \bar{a}) \vdash (\exists \bar{y})\theta(\bar{x}, \bar{y}; \bar{c})$, $C(\psi(\bar{y}; \bar{b})) \leq \kappa$ *and for every* $\bar{d}$ $C(\theta(\bar{x}, \bar{d}, \bar{c})) \leq \kappa$, *then* $C(\varphi(\bar{x}; \bar{a})) \leq \kappa$.

(6) *If* $p \in S^m(B)$, $A \subseteq B$, *p does not fork over A then* $C(p) = C(p \upharpoonright A)$ *and* $H(p) = H(p \upharpoonright A)$.

(7) *For every p, for some finite* $q \subseteq p$, $C(p) = C(\bigwedge q)$.

Proof. (1) Left as an exercise to the reader. (Hint: prove $\neg$(ii) $\Rightarrow$ $\neg$(iii) $\Rightarrow$ $\neg$(i) $\Rightarrow$ $\neg$(iv) $\Rightarrow$ $\neg$(ii) and, using II, 2.17 (ii) $\Leftrightarrow$ (v).)

(2) Follows by (3).

(3) Easy, by induction and compactness.

(4) Trivial.

(5) By induction on the first n for which $C_n((\psi(\bar{y}, \bar{b})) \leq \kappa$ and for every $\bar{d}$ $C_n(\theta(\bar{x}, \bar{d}, \bar{c})) \leq \kappa$ (this suffices by (2) and compactness).

(6) Left to the reader.

(7) Left to the reader.

LEMMA 6.2. *For any stationary type* $p \in S^m(A)$ *and formula* $\psi(\bar{x}; \bar{a})$ *the following conditions are equivalent:*

(i) *For every model M, if* $A \cup \bar{a} \subseteq |M|$ *then* $\dim(p, M) < |\psi(M; \bar{a})|^+ + \aleph_0$.

(ii) *There is an equivalence relation* $E = E(\bar{x}, \bar{y}, \bar{d})$, $\bar{d} \in A \cup \bar{a}$ *such that:*

(α) *if* $A \cup \bar{a} \subseteq B$, $\mathrm{tp}(\bar{b}, B)$ *is any stationarization of p, then* $\bar{b}/E \notin \mathrm{acl}\, B$ (*in* $\mathfrak{C}^{eq}$, *of course*),

(β) *for any* $\bar{b}$, $\bar{b}/E \in \mathrm{dcl}\, [\psi(\mathfrak{C}; \bar{a}) \cup \bar{d}]$ (*and by a fixed formula*).

(iii) *There is a model M*, $A \cup \bar{a} \subseteq |M|$, *M is* $\mathbf{F}^a_\kappa$-*saturated*, $\kappa \geq \kappa_r(T)$, *and* $\mathrm{tp}(\bar{b}, |M|)$ *is a stationarization of p, and N is* $\mathbf{F}^a_\kappa$-*prime over* $|M| \cup \bar{b}$, *and* $\psi(M, \bar{a}) \neq \psi(N, \bar{a})$.

(iv) *If I is an indiscernible set based on p, and is indiscernible over* $\bar{a}$, *then I is not indiscernible over* $\bar{a} \cup \psi(\mathfrak{C}, \bar{a})$.

(v) *not every* $J \subseteq \psi(\mathfrak{C}, \bar{a})$ *is orthogonal to p.*

(vi) *There is an equivalence relation* $E = E(\bar{x}, \bar{y}, \bar{d})$, $\bar{d} \in A$, *such that:*

(α) $\bar{b}/E \notin \mathrm{acl}\, A$, *where* $\bar{b}$ *realizes p,*

(β) $C((\exists \bar{y})x = \bar{y}/E) \leq C(\psi(\bar{x}, \bar{a}))$,

(γ) *some stationary type q*, $\psi(\bar{x}, \bar{a}) \in q$, *is not orthogonal to* $\mathrm{tp}(\bar{b}/E, A)$.

Proof. (ii) $\Rightarrow$ (i). Let I be an indiscernible set based on p, $I \subseteq |M|$, $|I| = \dim(p, M)$. By (ii), if $I = \{\bar{b}_i : i < |I|\}$ then the $\bar{b}_i/E$'s ($i < |I|$) are pairwise distinct, hence $|\{\bar{b}/E : \bar{b} \in |M|\}| \geq |I|$. By (ii)($\beta$) $|\{\bar{b}/E : \bar{b} \in |M|\}| < \aleph_0 + |\psi(M, \bar{a})|^+$; so (i) follows.

$\neg$(iv) $\Rightarrow$ $\neg$(i) If (iv) fails, let M be $\mathbf{F}^a_\kappa$-saturated (choose $\kappa =$

$(|A|^+ + |T|)^+)$, $I \cup \{\bar{a}\} \subseteq |M|$. So in the $\mathbf{F}^a_\kappa$-primary model over $\bigcup I \cup \bar{a} \cup \psi(M, \bar{a})$, no non-algebraic q, $\psi(\bar{x}, \bar{a}) \in q \in S(\bar{a} \cup \psi(M, \bar{a}))$, is realized. Choose a set J, indiscernible, over $\bar{a} \cup \psi(M, \bar{a})$, $I \subseteq J$, $|J| > |\psi(M, \bar{a})|$, and then the $\mathbf{F}^a_\kappa$-prime model over $\bigcup J \cup \bar{a} \cup \psi(M, \bar{a})$ contradicts (i) (mapping A to M).

$\neg$(iii) $\Rightarrow$ $\neg$(iv) Let M, $\bar{b}$, N be as in (iii), except that $\psi(M, \bar{a}) = \psi(N, \bar{a})$. Clearly $\operatorname{tp}(\bar{b}, A \cup \bar{a} \cup \psi(N, \bar{a}))$ is $(A \cup \bar{a})$-definable, hence we choose $\bar{b}_n \in M$, such that $\bar{b}$, $\bar{b}_n$ realizes the same strong type over $A \cup \bar{a} \cup \bigcup_{l<n} \bar{b}_l$, then they realize the same type over $A \cup \bar{a} \cup \bigcup_{l<n} \bar{b}_l \cup \psi(N, \bar{a})$. So $\{\bar{b}_n : n < \omega\}$ proves $\neg$(iv) (we assume w.l.o.g. $|A| < \kappa(T)$).

(iii) $\Rightarrow$ (ii) Choose $\bar{c} \in \psi(N, \bar{a}) - \psi(M, \bar{a})$ so $\operatorname{tp}(\bar{c}, |M| \cup \bar{b})$ forks over $|M|$ hence $\operatorname{tp}(\bar{c}, A \cup \bar{a} \cup \psi(M, \bar{a}) \cup \bar{b})$ forks over $A \cup \bar{a} \cup \psi(M, \bar{a})$, and let $\varphi(\bar{x}, \bar{b}, \bar{d}_1, \bar{d}_2)$ forks over $|M|$, where $\bar{d}_2 \in A \cup \bar{a}$, $\bar{d}_1 \in \psi(M, \bar{a})$ and $\models \varphi[\bar{c}, \bar{b}, \bar{d}_1, d_2]$. Let $\bar{d}_1 = \bar{d}^0_1 \frown \cdots \frown \bar{d}^k_1$, $\models \psi[\bar{d}^l_1, \bar{a}]$,

$$E(\bar{x}, \bar{y}, \bar{d}_2, \bar{a}) = (\forall \bar{z})(\forall \bar{z}_0, \ldots, \bar{z}_k)\Big[\psi(\bar{z}, \bar{a}) \wedge \bigwedge_{l \le k} \psi(\bar{z}_l, \bar{a}) \rightarrow \varphi(\bar{z}, \bar{x}, \bar{z}_0, \ldots, \bar{z}_k, \bar{d}_2) \equiv \varphi(\bar{z}, \bar{y}, \bar{z}_0, \ldots, \bar{z}_k, \bar{d}_2)\Big]$$

We leave the checking to the reader.

(v) $\Leftrightarrow$ (iii) Immediate.

(vi) $\Rightarrow$ (v) Immediate (we do not use (vi) (β)).

(ii) $\Rightarrow$ (vi) We work in $\mathfrak{C}^{\mathrm{eq}}$. Let $\bar{d} = \bar{d}_1 \frown \bar{a}$, $\bar{d}_1 \in A$, $I = \{\bar{a}_n : n < \omega\}$ an indiscernible set over A, based on A, $\bar{a}_0 = \bar{a}$. Clearly $C(\psi(x, a_n)) = C(\psi(x, \bar{a}))$ so for every n $C(\bigvee_{l<n} \psi(x; \bar{a}_l)) = C(\psi(x; \bar{a}))$; and for every n, if $E_n = E(\bar{x}, \bar{y}, \bar{d}_1, \bar{a}_n)$, then for every $\bar{b}$, $\bar{b}/E_n \in \operatorname{dcl}[\bar{a}_n \cup \psi(\mathfrak{C}, \bar{a}_n)]$. For each n let $E^n = \bigvee_{|w|=n, w \subseteq 2n-1} \bigwedge_{l \in w} E(\bar{x}, \bar{y}, \bar{d}_1, \bar{a}_l)$; so for every $\bar{b}$, $\bar{b}/E^n \in \operatorname{dcl}[\bigcup_{l<2n-1} (\bar{a}_l \cup \psi(\mathfrak{C}, \bar{a}_l))]$ hence $C((\exists \bar{y})(x = \bar{y}/E^n)) \le C(\psi(x, \bar{a}))$ (by 6.2). By III, 2.5 for some n, E^n is almost over A. By a cosmetic treatment we get an E^* which is over A and $C((\exists \bar{y})(x = \bar{y}/E)) \le C(\psi(x, \bar{a}))$.

Remark 6.2*A*. In (vi) (γ) we can have q regular and if $\operatorname{tp}(\bar{a}, A)$ is stationary we can have $\operatorname{tp}(\bar{b}/E, A)$ semi-regular (see 4.1).

LEMMA 6.3: (C2) (1) *The following conditions on M, A, where $|M| \subseteq A$, are equivalent:*

(i) $M \prec^\lambda A$,

(ii) *for no $\bar{a} \in A$, $\bar{b} \in |M|$, φ, ψ does $C(\varphi(x, \bar{b})) < \lambda$, $\models (\exists x)\psi(x, \bar{a}, \bar{b})$ and $\psi(x, \bar{a}, \bar{b}) \vdash \{\varphi(x, \bar{b}) \wedge x \ne d : d \in |M|\}$,*

(iii) *for no* $\bar{a} \in A$, $\bar{b} \in |M|$, φ, ψ *and* $\chi < \lambda$ *does* $\models$"$(\exists^{<\chi})\varphi(x, \bar{b})$", $\models (\exists x)\psi(x, \bar{a}, \bar{b})$ *and* $\psi(x, \bar{a}, \bar{b}) \vdash \{\varphi(x, \bar{b}) \wedge x \neq d : d \in |M|\}$.

(2) *Working in* $\mathfrak{C}^{eq}$, *if* $H(\bar{a}, A) < \lambda$, *then for some formulas* $E = E(\bar{x}, \bar{y}; \bar{b})$, $\varphi(x; \bar{b})$, *E is an equivalence relation*, $\bar{b} \in A$, $\mathrm{tp}(\bar{a}, A) \vdash \varphi(\bar{x}/E; \bar{b})$, $C(\varphi(z; \bar{b})) < \lambda$ *and* $\bar{a}/E \notin \mathrm{acl}\, A$.

(3) $H(I) = \sup\{\kappa$: *if J is indiscernible*, $\varphi(\bar{x}, \bar{a}) \in \mathrm{Av}(J, \bar{a})$, $C(\varphi(\bar{x}; \bar{a}))^{+} < \kappa$ *then I, J are orthogonal*$\}$.

Proof. (3) Let κ be the supremum mentioned in (3). Trivially $\kappa \leq H(I)$, so suppose $\varphi(\bar{x}; \bar{a}) \in \mathrm{Av}(J, \bar{a})$, $C(\varphi(\bar{x}; \bar{a}) < H(I))$. By 6.2(vi) and the definition of C, the rest is easy.

(2) Trivial by 6.2(vi).

(1) By (3), (i) $\Rightarrow$ (ii), and trivially (ii) $\Rightarrow$ (iii).

$\neg$(i) $\Rightarrow$ $\neg$(ii) So let $\bar{b} \in A$, $H(\bar{b}, |M|) < \lambda$, so by the definition there is $\psi(x, \bar{a})$, such that $C(\psi(x, \bar{a})) < \lambda$, and some $J \subseteq \psi(\mathfrak{C}, \bar{a})$ is not orthogonal to any I based on $\mathrm{tp}(\bar{b}, |M|)$. Using 6.2(ii) to see what is the requirement on $\bar{a}$, we see that we can assume $\bar{a} \in |M|$.

By the proof of 6.2(ii), for some φ and $\bar{c} \in |M|$, $\psi(x, \bar{a}, \bar{c}) \vdash \varphi(x, \bar{a})$ is consistent, but not realized in $|M|$.

$\neg$(ii) $\Rightarrow$ $\neg$(iii) So let $\varphi(x; \bar{b})$, $\psi(x, \bar{a}, \bar{b})$ contradicts (ii). As $C(\varphi(x; \bar{a})) < \lambda$, for some n $C_n(\varphi(x, \bar{a})) < \lambda$, and we can easily prove by induction on n. that the existence of such φ, ψ contradicts (iii).

COROLLARY 6.4: (C2) *If* $M \subseteq N$, *then* $M \prec^{\lambda} N$ *iff* $\bar{a} \in M$, $C(\varphi(x, \bar{a})) < \lambda$ *implies* $\varphi(N, \bar{a}) \subseteq |M|$, *iff* $\bar{a} \in M$, $\models$"$(\exists^{<\chi} x)\varphi(x, \bar{a})$", $\chi < \lambda$ *implies* $\varphi(N, \bar{a}) \subseteq |M|$.

LEMMA 6.5: (1) [C1, 2, 5]. *If* $M \prec^{\lambda} A$, *M is* $\mathbf{F}$-*maximal, N is* $\mathbf{F}$-*constructible over A, then* $M \prec^{\lambda} N$.

(2) [C2] *If* A_i $(i < \alpha)$ *is increasing*, $B \prec^{\lambda} A_i$ *then* $B \prec^{\lambda} \bigcup_{i<\alpha} A_i$.

(3) [C1, 2, 5] *If* $M_1 \prec^{\lambda} M_2 \prec^{\lambda} M_3$ *then* $M_1 \prec^{\lambda} M_3$; *and if* $i < j < \delta \Rightarrow M_i \prec^{\lambda} M_j$, *$M$ is* $\mathbf{F}$-*contructible over* $\bigcup_{i<\delta} M_i$ *then* $i < \delta \Rightarrow M_i \prec^{\lambda} M$.

(4) [C2] *The property* $H(p) \geq \lambda$ *is preserved by automorphisms of* $\mathfrak{C}$ *and by replacing p by a parallel type.*

(5) [C2] *If for each* $i < \alpha$, $H(\bar{a}_i, B \cup \bigcup_{j<i} \bar{a}_j) \geq \lambda$, *then* $B \prec^{\lambda} B \cup \bigcup_{i<\alpha} \bar{a}_i$.

(6) [C2] *If* $A \prec^{\lambda} B \prec^{\lambda} C$ *then* $A \prec^{\lambda} C$, *and if* $i < j < \delta \Rightarrow A_i \prec^{\lambda} A_j$ *then* $A_i \prec^{\lambda} \bigcup_{j<\delta} A_j$.

(7) [C1, 2, 5] *If* $\mathrm{tp}_*(N_1, N_2)$ *does not fork over* $|M| = |N_1| \cap |N_2|$ *and* M, N_2 *are* $\mathbf{F}$-*maximal*, $M \prec^{\lambda} N_1$, *and M' is* $\mathbf{F}$-*constructible over* $|N_1| \cup |N_2|$ *then* $N_2 \prec^{\lambda} M'$.

(8) *Condition* (6) *holds for* $\mu = \kappa_r(T)$ *if condition* 5 *holds, so we can wave condition* 6, *and replace condition* 4 *by* $\lambda \geq |T|^+ + \kappa_r(T)^+ + \lambda(\mathbf{F})$.

(9) [C1, 4] *Any regular* $\mu' \geq \lambda_K$ *satisfies condition* 6.

(10) *The condition on* χ *in* C2 *is equivalent to: if* B *is* **F**-*constructible over* A *and* $|A| < \chi$, *then* $|B| < \chi$.

(11) [C1] *Any* **F**-*maximal set is* (*the universe of*) *a model.*

Proof. (1) If the conclusion fails, then there are $\bar{a} \in |M|$, φ and $c \in |N| - |M|$ such that $\models$"$(\exists^{<\chi} x)\varphi(x, \bar{a})$", $\models\varphi[c, \bar{a}]$, $\chi < \lambda$ (by 6.4). By the definition of $\prec^\lambda$, $\mathrm{tp}_*(A, |M|)$, $\mathrm{tp}(c, |M|)$ are orthogonal, hence weakly orthogonal contradicting (C5).

(2) Trivial, by the definition.

(3) Immediate: first part by 6.4, second part by 6.5(1) and 6.5(2).

(4) Trivial, by the definition.

(5) Define inductively M_i such that: M_i is $|T|^+$-compact, $B \subseteq |M_0|$, $\bar{a}_i \in |M_{i+1}|$, $\mathrm{tp}_*(M_i, B \cup \bigcup_{j<\alpha} \bar{a}_j)$ does not fork over $B \cup \bigcup_{j<i} \bar{a}_j$ and M_{i+1} is $F^a_{|T|^+}$-constructable over $|M_i| \cup \bar{a}_i$, M_δ is $F^a_{|T|^+}$-constructible over $\bigcup_{i<\delta} M_i$. By 6.5(4) $H(\bar{a}_i, |M_i|) \geq \lambda$, hence $M_i \prec^\lambda M_i \cup \bar{a}_i$, so as 6.5(1) $M_i \prec^\lambda M_{i+1}$, so by 6.5(3) $M_0 \prec^\lambda M_\alpha$ so $H(\bar{a}_{i(0)} {}^\frown \cdots {}^\frown \bar{a}_{i(n)}, |M_0|) \geq \lambda$ for every $n < \omega$, $i(0), \ldots, i(n) < \alpha$. Hence we finish.

(6) Left to the reader.

(7) Left to the reader.

(8)–(11) Immediate.

DEFINITION 6.1: A model M will be called good (or K-good) if $\bar{a} \in |M|$, $\models$"$(\exists^{<\lambda} x)\varphi(x, \bar{a})$" implies $|\varphi(M, \bar{a})| < \lambda$; and it is called h-good if in addition for $\bar{a} \in |M|$ $h[C^*(\varphi(x, a))] \leq |\varphi(M, \bar{a})|$. (If condition 3 holds this is equivalent to $h[C(\varphi(x, \bar{a}))] \leq |\varphi(M, \bar{a})|$.)

LEMMA 6.6: [C2] *M is good iff* $\bar{a} \in M$ *implies* $|\varphi(M, \bar{a})| < C(\varphi(x, \bar{a}))$.
Proof. See 6.4.

THEOREM 6.7: [C1, 2, 4, 5, 6] *If M is* **F**-*maximal and good, h a function from* Car K *to cardinals,* $h(\lambda) < \lambda$, *then there is a model* N, $M \prec N$, N *is* h-*good*, **F**-*maximal*.

We first prove:

LEMMA 6.8: (1) [C1, 2, 4, 5, 6] *Suppose M is good,* $|A| < \lambda_K$, N *is* **F**-*constructible over* $|M| \cup A$. *Then N is good.*

(2) [C1, 2, 4, 5, 6] *Suppose* M_i $(i < \delta)$ *is good and* **F**-*maximal*, $i < j < \delta \Rightarrow M_i \prec M_j$; cf $\delta < \lambda_K$ *and* M^* *is* **F**-*constructible over* $\bigcup_{i<\delta} |M_i|$. *Then* M^* *is good, and if* $\bar{a} \in |M_i|$, $\varphi(M_j, \bar{a}) = \varphi(M_i, \bar{a})$ *for every* $i \le j < \delta$ *then* $\varphi(M^*, \bar{a}) = \varphi(M_i, \bar{a})$.

(3) [C1, 2, 4, 5, 6] *In* (1) *suppose in addition that* $\psi(x, \bar{d}) \in \mathrm{tp}(a, |M|)$, $A = \{a\}$, *and* $\bar{b} \in |M|$, $\varphi(N, \bar{b}) \ne \varphi(M, \bar{b})$, *and let* $c_0 \in \varphi(N, \bar{b}) - \varphi(M, \bar{b})$. *Then there is an equivalence relation* $E = E(x, y, \bar{b}_1)$ $(\bar{b}_1 \in \varphi(M, \bar{b})) \cup \bar{b}$ *such that* $|\{c/E: c \in \varphi(M, \bar{b})\}| \le |\psi(M, \bar{d})|$ *and* $c_0/E \notin M^{\mathrm{eq}}$.

Proof. (1) Suppose N is not good, so that is $\varphi_0(x, \bar{a})(\bar{a} \in |N|)$ such that $\vDash$"$(\exists^{<\chi} x)\varphi_0(x, \bar{a})$" but $|\varphi_0(N, \bar{a})| \ge \chi$. Choose $N^* \subseteq N$ such that $\|N^*\| < \chi$ (see C1), $A \cup \bar{a} \subseteq |N^*|$, $\mathrm{tp}(|N^*|, |M|)$ does not fork over $|M^*|$ where $|M^*| = |M| \cap |N^*|$, and $M^* \prec^{\chi^+} M$ (see 6.4) and N^* is **F**-constructible over $|M| \cup A$, $|N^*| \cup A$, and N is **F**-constructible over $|M| \cup |N^*|$ (see IV, 3.3) and N^* is **F**-maximal (use C6). By 6.5(7) $N^* \prec^{\chi^+} N$, but $|\varphi_0(N^*, \bar{a}_0)| < \chi \leqslant |\varphi_0(N, \bar{a})|$, contradiction.

(2) A similar proof.

(3) By (C5) $\mathrm{tp}(c_0, |M| \cup a)$ forks over $|M|$, so let $\vDash \theta[c_0, a, \bar{b}_1]$, $\theta(x, a, \bar{b}_1)$ forks over $|M|$, and $\theta(x, y, \bar{b}_1) \vdash \varphi(x, \bar{b})$ where $\bar{b}_1 \in M$, $\bar{d} \subseteq \bar{b}_1$. So let $E(x, y, \bar{b}_1) = (\forall z)[\psi(z, \bar{d}) \rightarrow \theta(x, z, \bar{b}_1) \equiv \theta(y, z, \bar{b}_1)]$. Then clearly $c_0/E \notin M^{\mathrm{eq}}$, and $|\{c/E: c \in \varphi(M, \bar{a})\}| \leqslant |\psi(M, \bar{d})|$. Using the definability Lemma II, 2.12, we can get $\bar{b}_1 \in \varphi(M, \bar{b}) \cup \bar{b}$.

Proof of 6.7. We define inductively models M_i $(i \le \alpha_0)$ and elements a_i such that $M_0 = M$, M_i is **F**-maximal, M_{i+1} is **F**-constructible over $M_i \cup \{a_i\}$ and M_δ (δ limit $\le \alpha_0$) is **F**-constructible over $\bigcup_{i<\delta} M_i$ and for each i there is $\psi_i(x, \bar{d}_i) \in \mathrm{tp}(a_i, |M_i|)$, $|\psi_i(M_i, \bar{d}_i)| < h[C^*(\psi_i(x, \bar{d}_i))]$, and $C^*(\psi_i(x, \bar{d}_i)) \le H(a_i, |M_i|)$.

Let us prove by induction that M_i is good. For $i = 0$ it is known, for $i + 1$ it follows by 6.8(1). For $i = \delta$ limit, let $M_\delta^1 = \bigcup_{i<\delta} M_i$. If cf $\delta < \lambda_K$, as M_δ is **F**-constructible over $|M_\delta^1|$, the assertion follows by 6.8(2). So suppose cf $\delta \ge \lambda_K$, hence by 6.5(9) M_δ^1 is **F**-maximal hence $M_\delta = M_\delta^1$.

So if $\vDash$"$(\exists^{<\chi} x)\varphi(x, \bar{b})$", $\bar{b} \in |M_\delta^1|$, then $|\varphi(M_i, \bar{b})| < \chi$ for each $i < \delta$, by the induction hypothesis hence cf $\delta = \chi$. As by (C2) χ is regular, and $\chi \ge \lambda_K > |T|$ [(C4)] there is $\alpha < \chi$ such that for each equivalence relation $E = E(x, y, \bar{d})$ $\bar{d} \in \varphi(M_\alpha, \bar{b}) \cup \bar{b}$ either $|\{c/E: c \in \varphi(M_\delta, \bar{b})\}| < \chi$ and for each $c \in \varphi(M_\delta, \bar{b})$ there is $c' \in \varphi(M_\alpha, \bar{b})$, $\vDash E[c, c', \bar{d}]$ or $|\{c/E: c \in \varphi(M_\delta, \bar{b})\}| = \chi$, $|\{c/E: c \in \varphi(M_\alpha, \bar{b})\}| \ge h(\chi)$. So by 6.8(2), (3) $\varphi(M_i, \bar{b})$ is constant for $\alpha \leqslant i \leqslant \delta$. In both cases we get that M_δ^1 is good. Let α_0

be the first α for which we cannot define $M_{\alpha+1}$ (clearly if $\lambda \geq \|M\|$, and $\lambda \geq \chi$ for every $\chi \in \text{Car } K$, $\chi < \infty$, T stable in χ then $\alpha_0 < \lambda^+$). Clearly M_{α_0} is as desired.

Definition 6.2: If h is a function into $\{\chi: \chi$ regular, and $|A| < \chi$, N $\mathbf{F}$-constructible over A implies $\|N\| < \chi\}$; let $K^h = \langle \mathbf{F}, W^h, h(\lambda), \mu \rangle$, where $W^h = \{\langle \varphi, \psi, h(\chi)\rangle: \langle \varphi, \psi, \chi\rangle \in W\}$.

LEMMA 6.9: *If h is increasing $h(\lambda_K) \geq |T|^+ + \mu^+ + \lambda(\mathbf{F})$ and as above; then K^h satisfies Cn if K satisfies Cn for $n = 1, 6$ and $h[C_K(\varphi(x, \bar{a}))] = C_{K^h}(\varphi(x, \bar{a}))$ and similarly for $C^*(\varphi(x, \bar{a}))$.*

Proof. Trivial.

CONCLUSION 6.10: [C1–6] *If h is as in Definition 6.2, $h(\lambda) = h_1(\lambda)^+$, $h(\infty) < \infty$, then there is a $\mathbf{F}$-maximal model M such that for every $\bar{a} \in |M|$, $h_1[C(\varphi(x, \bar{a}))] = |\varphi(M, \bar{a})|$.*

Proof. Apply 6.7 to K^h.

Unfortunately not all the conditions necessary for 6.10 are satisfied in all interesting cases. E.g., $\mathbf{F}^a_\kappa$ satisfies C1–6 for $\kappa \geq \kappa_r(T)$, but for $\mathbf{F}^c_\kappa$ we have difficulties in C1, 3, 5, and for $\mathbf{F}^l_{|T|}$ we have difficulties in C1, 3. Hence we suggest here an alternative set of conditions, denoted by C^*i (if it is the same as the previous condition, we do not redefine it). Then we prove a parallel to 6.7 and 6.8 this time ensuring only the existence of a suitable model. We work in $\mathfrak{C}^{eq}$.

Condition 1.* $\mathbf{F}$ is as in IV, Section 1, satisfying Ax(III.1) and Ax(X.1).

Condition 2.* (a) W is a set of triples $\langle \varphi(x, \bar{y}), \psi(\bar{y}), \chi\rangle$.

(b) If χ appears in W, $|A| < \chi$, B is $\mathbf{F}$-constructible over A then $|B| < \chi$; and $\chi > |T| + \lambda$, $\chi \geqslant \lambda_K$ and χ is regular also.

We define $C(\varphi(x, \bar{a}))$, $H(I)$, $C(p)$, $<^\lambda$ as before; and

Condition 6.* At least one of the following holds:

(a) If $A_i \subseteq B$ ($i < \mu$) is increasing and each A_i is $\mathbf{F}$-maximal in B, then $\bigcup_{i<\mu} A_i$ is $\mathbf{F}$-maximal in B. (A is $\mathbf{F}$-maximal in B if $\bar{c} \in B$, $\bar{c} \notin A$ implies $\text{tp}(\bar{c}, A)$ is not $\mathbf{F}$-isolated.)

(b) If A_i ($i \leq \chi \in \text{Card } K$) is increasing and continuous then for some $i < \chi$, A_i is $\mathbf{F}$-maximal in A_χ, provided that $|A_i| < \chi$.

Note that we weaken C1, change C6, and add:

*Condition** 7. If $p \in S^m(A)$, $\bar{a} \in A$, $A = \operatorname{dcl} A$, $\varphi(\bar{x}, \bar{a}) \in p$, $C = A \cap \operatorname{dcl}(\bar{a} \cup \varphi(\mathfrak{C}^{eq}, \bar{a}))$ and p is **F**-isolated *then* $p \restriction C$ is **F**-isolated.

*Condition** 8. If $c \in \operatorname{dcl}(A \cup \bar{a}) - \operatorname{dcl} A$, $\operatorname{tp}(\bar{a}, A)$ is **F**-isolated then $\operatorname{tp}(c, A)$ is **F**-isolated.

DEFINITION 6.3: $A = \operatorname{dcl} A$ is called good if for every φ and $\bar{a} \in A$, $|\varphi(A, \bar{a})| < C(\varphi(\bar{x}, \bar{a})$ (remember we are working in $\mathfrak{C}^{eq}$).

LEMMA 6.11: (1) (C*2, 6a) *If* $A = \{a_i: i < \chi\}$, χ *appears in* K, *then there is a closed unbounded* $S \subseteq \chi$ *such that* $\delta \in S$, $\operatorname{cf} \delta = \mu$, *implies* $\{a_j: j < \delta\}$ *is* **F**-*maximal in* A; *so* C*6(b) *holds*.

(2) *We can replace in* 6.3, 6.4, 6.5 *and* 6.9 (C2) *by* (C*2a), (C_n) *by* (C*n) (*for* $n \neq 2$) (*and get true lemmas*).

(3) (C*2) *If* A *is good*, $|B| < \lambda_K$, *then* $\operatorname{dcl}(A \cup B)$ *is good.*

(4) (C*2) *If* A_i $(i < \delta)$ *is increasing, each* A_i *is good and* $\operatorname{cf} \delta < \lambda_K$ *then* $\bigcup_{i<\delta} A_i$ *is good.*

(5) (C*1, 2, 6, 7, 8) *If* A *is good*, B *is* **F**-*constructible over* A *then* B *is good.*

Proof. (1) For every $\xi < \chi$ we define inductively on $\alpha \leq \mu$ $i_\alpha < \chi$, $A^\alpha \subseteq A$, $|A^\alpha| < \chi$. We let $i_0 = \xi$ and if A^α is define, $i_{\alpha+1}$ is minimal $i > i_\alpha$ such that $A^\alpha \subseteq \{a_j: j < i_{\alpha+1}\}$, and for limit α $i_\alpha = \bigcup_{\beta<\alpha} i_\beta$. Now A^α is a maximal set $\subseteq A$ which is **F**-constructible over $\{a_j: j < i_\alpha\}$, so by C*2 $|A^\alpha| < \chi$. Clearly i_μ satisfies the demand on δ; so $i_\mu \in S_1 = \{\delta < \chi: \operatorname{cf} \delta = \mu, \{a_i: i < \delta\}$ is **F**-maximal in $A\}$. Hence S_1 is an unbounded subset of χ. Clearly the closure of S_1 prove the lemma.

(2) Easy.

(3) Suppose not, then for some $\bar{c} \in C = \operatorname{dcl}(A \cup B)$, λ and φ, $C(\varphi(x, \bar{c})) \leq \lambda \leq |\varphi(C, \bar{c})|$, so w.l.o.g. we can assume $|A| = \lambda$, and we find $A_1 \subseteq A$, $|A_1| < \lambda$ such that $\bar{a} \in A_1$, $C(\psi(x, \bar{a})) \leq \lambda$ implies $\psi(A, \bar{a}) \subseteq A_1$. We can assume that $\operatorname{tp}_*(B, A)$ does not fork over A_1. By Exercise 6.7 we get $A_1 \prec^\lambda A$, hence clearly $C_1 \prec^\lambda C$ where $C_1 = \operatorname{dcl}(A_1 \cup B)$. So $\varphi(C, \bar{c}) \subseteq \varphi(C_1, \bar{c})$ but $|\varphi(C_1, \bar{c})| \leq |C_1| + |T| < \lambda$, contradiction.

(4) Trivial.

(5) Let $\mathscr{A} = \langle A, \{a_i: i < \alpha\}\rangle$ be a **F**-construction of B over A, s.t. $A \cup \{a_j: j < i\} \neq \operatorname{dcl}(A \cup \{a_j: j < i\})$ implies $a_j \in \operatorname{dcl}(A \cup \{a_j: j < i\})$, and $\operatorname{dcl}(B)$ is not good, and α is minimal under those conditions, so by 6.11(3) α is limit. Let $B^0 = \operatorname{dcl}(B)$.

Let $\bar{b} \in B$, χ and φ be such that $C(\varphi(x, \bar{b})) = \chi \leqslant |\varphi(B^0, \bar{b})|$, so by the minimality of α, $|\varphi(B^0, \bar{b})| = \chi$, and for every $\beta < \alpha$, $|\varphi(B_\beta, \bar{b})| < \chi$, where $B_\beta = \operatorname{dcl} \mathscr{A}_\beta$. Let $C = \operatorname{dcl}(\bar{b} \cup \varphi(\mathfrak{C}^{eq}, \bar{b}))$, $C_\beta = B_\beta \cap C$; again by the minimality of α $\beta < \alpha$ implies $|C_\beta| < \chi$, and clearly for limit δ $C_\delta = \bigcup_{\beta<\alpha} C_\beta$. Hence by 6.11(1) (or C*6(b)), as cf $\alpha = \chi$, for some $\delta(0) < \alpha$, $C_{\delta(0)}$ is **F**-maximal relative to C_α. Let $\gamma < \alpha$ be the last ordinal such that $C_\gamma = C_{\delta(0)}$ (there is one, as for limit δ, $C_\delta = \bigcup_{\beta<\delta} C_\beta$ and $|C_\alpha| > |C_{\delta(0)}|$) and let $c \in C_{\gamma+1} - C_\gamma$. By C*8 tp($c$, B_γ) is **F**-isolated, hence by C*7 (after a cosmetic treatment of φ) tp(c, C_γ) is **F**-isolated; so tp(c, $C_{\delta(0)}$) is **F**-isolated, $c \in C_\chi - C_{\delta(0)}$, contradiction to the choice of $\delta(0)$.

THEOREM 6.12: *The parallels of* 6.8(2), 6.8(3) *and* 6.7 *(even if we replace M by a good set) holds, hence also of* 6.10.

Proof. Easy, by checking the proofs.

THEOREM 6.13: *Let* $K = (\mathbf{F}, W, \lambda, \mu)$, *K satisfies* C2, *and* $\lambda > |T|$.

(1) *If* $\mathbf{F} = \mathbf{F}^a_\kappa$, $\kappa_r(T) \leq \kappa < \lambda$, $\mu = \kappa_r(T) < \lambda$ *then*

(a) C1–6 *are satisfied, and also* C*1–8, C*6(a),

(b) *in* C2, *the condition on* χ *is equivalent to:* χ *is regular, and T is stable in arbitrarily large* $\chi_1 < \chi$.

(c) **F**-*maximal is equivalent to* **F**-*saturated.*

(2) *If* $\mathbf{F} = \mathbf{F}^s_\kappa$, $\lambda_2(T) + \kappa_r(T) \leq \kappa < \lambda$, $\mu = \kappa_r(T) < \lambda$, *then* (a), (b), (c) *of* (1) *holds, except possibly* (C*7).

(3) *If* $\mathbf{F} = \mathbf{F}^t_\kappa$, $\lambda_3(T) + \kappa_r(T) \leq \kappa < \lambda$, $\mu = \kappa_r(T) < \lambda$, *then*

(a) C1–6 *holds, and* C*1–8 *holds,*

(b) *in* C2 *the condition on* χ *is equivalent to:* $\chi > |T|$ *is regular and for every* $\chi_1 < \chi$, $\chi_1^{<\kappa(T)} < \chi$, *and* $\lambda^{\lambda'} < \chi$ *when* $\lambda' < \lambda_3(T)$,

(c) **F**-*maximal is equivalent to* **F**-*saturated.*

(4) *If* $\mathbf{F} = \mathbf{F}^l_{|T|}$, $\lambda > |T|$, $\mu = |T|$, *and* C2 *is satisfied, then*

(a) C*1, 2, 4, 5, 7, 8 *are satisfied* (*and* C2, 4, 5) *and when* $|T|$ *is regular,* (*or* $\lambda > |T|^+$, *and we change* μ *to* $|T|^+$) C6, C*6 *too is satisfied,*

(b) *in* C2(b) *the condition on* χ *is equivalent to:* $\chi \geq \lambda$ *is regular,* min Car *K satisfies the condition as stated and for every* $\chi_1 < \chi$, $\chi_1^{<|T|} < \chi$ *or* $\chi_1^{<\kappa(T)} < \chi$,

(c) **F**-*maximal is equivalent to* $|T|$-*compact.*

Proof. (1) Parts (c) and (b) are quite immediate, so let us prove (a) remembering C1 $\Rightarrow$ C*1, C2 $\Rightarrow$ C*2, C3 = C*3, C4 = C*4, C5 = C*5, and C*6(a) $\Rightarrow$ C6, C*6(a) $\wedge$ C*2 $\Rightarrow$ C*6(b).

Condition 1. Immediate (see Table 1 in IV, Section 2).

Condition 2. Easy by (b).

Condition 3. Let M be **F**-maximal, $\bar{a} \in |M|$, $\models (\exists x)\varphi(x, \bar{a})$. We can clearly find $A \subseteq |M|$, $\bar{a} \in A$, $|A| < \kappa(T)$, and $r \in S^1(A)$ such that

(α) $C(r) = C(\varphi(x, \bar{a}))$, i.e., $r \vdash \psi(x, \bar{b})$ implies $C(\varphi(x, \bar{a})) \leq C(\psi(x, \bar{b}))$,

(β) no extension of r which is a type over $|M|$, satisfies (α) and forks over A.

Choose any r_1, $r \subseteq r_1 \in S^1(|M|)$, which satisfies ($\alpha$). Suppose $H(r_1) < C(\varphi(x, \bar{a}))$ and let c realize r_1, so for some $E = E(x, y, \bar{b})$ ($\bar{b} \in |M|$) $c/E \notin |M^{\text{eq}}|$ and for some $\psi(x, \bar{b}')$ (in M^{eq}), $C(\psi(x, \bar{b}')) < C(\varphi(x, \bar{a}))$, $\models \psi[c/E, \bar{b}']$. Choose $c' \in |M|$ such that $\text{tp}(c, A \cup \bar{b} \cup \bar{b}') = \text{tp}(c', A \cup \bar{b} \cup \bar{b}')$ and then $r \cup \{E(x, c', \bar{b})\}$ will contradict (α) or (β).

Condition 4. Immediate.

Condition 5. See IV, 4.10(2).

Condition 6*(a). Easy, by IV, 4.3(2).

Condition 7*. Let $(p, B) \in \mathbf{F}$, $|B| < \kappa$, and choose $B_1 \subseteq C$, $|B_1| < \kappa$ such that $\text{tp}_*(B, C)$ does not fork over B_1. By the definability Lemma II, 2.12 and IV, 4.3(2), it is clear that $(p \restriction C, B_1) \in \mathbf{F}$.

Condition 8*. If $\varphi(x, \bar{a}, \bar{b})$ ($\bar{b} \in A$) define c, and $r \vdash \text{tp}(\bar{a}, A)$, r closed under conjunctions then $r' \vdash \text{tp}(c, A)$ where $r' = \{(\exists \bar{y})(\psi(\bar{y}, \bar{b}') \wedge \varphi(x, \bar{y}, \bar{b})): \psi(\bar{y}; \bar{b}') \in r\}$ so the result is clear.

(2) The proof is like the proof of (1) only in the proof of C1, Ax(X,1) we need $\lambda^2(T) \leq \kappa$; and in the proof of C3 we need $\lambda_2(T) \leq \kappa$ (remember that by IV.2.16(1) $\lambda^2(T) \leq \lambda_2(T)$). Those are the only places we use $\lambda_2(T) \leq \kappa$.

(3) The proof is like the proof of (1), only in the proof of C1, Ax(X.1) we need $\lambda^3(T) \leq \kappa$ (remember $\lambda^3(T) \leq \lambda_3(T)$ by IV, 2.16(1) in the proof of C3 we need $\lambda_3(T) \leq \kappa$.

We use $\lambda_3(T) \leq \kappa$ in no other place.

(4) Parts (b), (c) are easy.

Condition 1*. Immediate (see Table 1 in IV, Section 2; for C1 Ax(VIII) is missing and Ax(VI) for singular κ).

Condition 2. Holds by an assumption.

Condition 4. Immediate.

Condition 5.* By IV, 3.2, N is $\mathbf{F}$-atomic over B, hence $\operatorname{tp}(c, B)$ is $\mathbf{F}$-isolated. By IV, 4.3(5), $\operatorname{tp}(c, M)$ is $\mathbf{F}$-isolated, contradicting M being $\mathbf{F}$-maximal.

Note that IV, 4.3(5), implies C*6a holds when $\mu > |T|$.

Condition 6.* Part (b) holds, so suppose A_i $(i \leq \chi)$ is as mentioned there, and for each i there is $\bar{c}_i \in A_\chi$, $\bar{c}_i \notin A_i$; and $\operatorname{tp}(\bar{c}_i, A_i)$ is $\mathbf{F}$-isolated. Then for each i there is $p_i \subseteq \operatorname{tp}(\bar{c}_i, A_i)$, $|p_i| < |T|$ such that $p_i \vdash \{\bar{x} \neq \bar{c} : \bar{c} \in {}^{m(i)}A_i\}$ where $m(i) = l(\bar{c}_i)$. If $\operatorname{cf} i = |T|$ there is $h(i) < i$ such that p_i is over $A_{h(i)}$, hence by 1.3 of the appendix there are m, j_0 so that $S_0 = \{i < \chi : \operatorname{cf} i = |T|, h(i) = j_0, m(i) = m\}$ is stationary. By C2 there is a $\mathbf{F}$-maximal M, $A_{j_0} \subseteq |M|$, $\|M\| < \chi$.

As $\chi \geqslant \lambda_K > |T|$, χ and $|T|$ are regular, there is $L' \subseteq L(T)$, $|L'| < |T|$, and a stationary $S_1 \subseteq S_0$ such that p_i is a L'-type for each $i \in S_1$. By part b of 6.13(4) it is easy to check that $|A| < \chi$ implies $|S^m_{L'}(A)| < \chi$, hence by Exercise I, 2.5 there is an unbounded $S_2 \subseteq S_1$ such that $\{\bar{c}_i : i \in S_2\}$ is L'-indiscernible over A_{j0}, an easy contradiction.

Condition 7.* Easy by the definability Lemma II, 2.12.

Condition 8.* As in 6.13(1).

THEOREM 6.13(5). *If* $\lambda > \kappa \geqslant \kappa_r(T)$, $\mathbf{F} = \mathbf{F}^c_\kappa$, $\mu = \kappa_r(T)$, *then*

(a) C*1, 2a, 4, 6, 7, 8 *hold and, in fact,* C*6a *holds.*

(b) *In* C2 *for regular* $\chi > 2^{|T|}$, *the condition on* χ *is equivalent to* $(\forall \chi_1 < \chi)\ (\chi_1^{<\kappa|T|} < \chi)$, *and for regular* $\chi \leqslant 2^{|T|}$ *the condition on* χ *is satisfied if*

$$(\forall \chi_1 < \chi)(\chi_1^{<\kappa} < \chi),\ |T| < \chi.$$

(c) $\mathbf{F}$ *maximal is equivalent to* κ-*compact.*

(d) *Though* C5 *is missing, the parallel of* 6.7 *holds.*

Proof. (5) (a), (b), (c) No new point (we use IV, 4.3(4)).

(d) As in the proof of 6.7 we define by induction on $i \leqslant \alpha_0$ $M_i, \bar{a}_i$ such that M_i is increasing, $M_0 = M$, M_{i+1} is $\mathbf{F}$-constructible over $M_i \cup \bar{a}_i$, for i limit M_i is $\mathbf{F}$-constructible over $\bigcup_{j<i} M_j$. By the correct choice of the $\bar{a}_i$'s we can ensure M_{α_0} is h-good provided we have proved each M_i is good. We prove this by induction on i, so by

Lemma 6.11 the only problematic case is $i = \delta, \delta$ limit, cf $\delta \geqslant \lambda_K$. Clearly, $\bigcup_{j<\delta} M_j$, is κ-compact (by 6.13(5)(c)), hence $M_\delta = \bigcup_{j<\delta} M_j$. So suppose $\bar{a} \in M_\delta \models$ "$(\exists^{<\chi} x)\varphi(x, \bar{a})$" and $|\varphi(M_\delta, \bar{a})| = \chi$ and we shall get a contradiction. Clearly, for some $j_0, \bar{a} \in M_{j_0}$, and for any $\beta, j_0 \leqslant \beta < \delta$ let $B_\beta = M_\beta \cap \mathrm{dcl}(\bar{a} \cup \varphi(\mathfrak{C}, \bar{a}))$. So $|B_\beta| < \chi$. By C*2 over any B_β there are $< \chi \mathbf{F}^c_\kappa$-isolated types. By 6.13(5) (b) there is $\beta < \delta$, cf $\beta = \kappa$ (or cf $\beta = \mu$ if κ is singular) such that:

for any $\bar{c} \in B_\delta$ and $r \subseteq \mathrm{tp}(\bar{c}, B_\gamma)$, $|r| < \kappa, \gamma < \beta$, finite Δ and $C \subseteq B_\beta$, $|c| < \lambda + \kappa_r(T)$, there is $\bar{c}' \in B_\beta$ realizing r such that

$$R^m(\mathrm{stp}(\bar{c}, B_\beta), \Delta, 2) = R^m[\mathrm{stp}(\bar{c}', C^*), \Delta, 2]$$

for some $C^* \subseteq A, C \subseteq C^*$.

Continue as in the proof of 6.11(5) using IX, 1.1, IX, 1.1A (if κ is singular use IV, 4.3(4)).

CONCLUSION 6.14: *Suppose $M \subseteq N$, $M \neq N$, $P(M) = P(N)$ (P—a one place predicate) and $P(x)$ is not algebraic.*

(1) *If T is countable, for every $\lambda \geq \|N\|$ there is a model N_1, $N \subseteq N_1$, $\|N_1\| = \lambda$, $P(N_1) = P(M)$.*

(2) *For every $\lambda \geq \mu \geq |T|$, there is a model N_1, $\|N_1\| = \lambda$, $|P(N_1)| = \mu$.*

(3) *If M, or N, is $|T|$-compact, then for every $\lambda \geq \|N\|$ there is N_1, $N \subseteq N_1$, $P(N_1) = P(M)$ and $\|N_1\| \geq \lambda$.*

(4) *If M or N is $\mathbf{F}$-maximal, and $\mathbf{F}$ satisfies* C5, 1 *then for every $\lambda \geq \|N\|$ there is N_1, $N \subseteq N_1$, $P(N_1) = P(M)$ and $\|N_1\| \geq \lambda$.*

Remark. Note that C5 deals with $\mathbf{F}$ only so it is meaningful to say "$\mathbf{F}$ satisfies C5".

Proof. (1) Let $K = \langle \mathbf{F}^l_{\aleph_0}, W^{\aleph_1}_P, \aleph_1, \aleph_0 \rangle$, $W^\chi_P = \{\langle P(x), (\forall y)(y = y), \chi \rangle\}$. Choose $c \in |N| - |M|$, $A \subseteq |M|$, $|A| < \kappa(T)$ such that $\mathrm{tp}(c, |M|)$ does not fork over A; now add the elements of A as individual constants of $\mathfrak{C}$, and get $\mathfrak{C}_1$. In $\mathfrak{C}_1$ for every model M_1, $A \subseteq |M_1|$ hence there is $r \in S(|M_1|)$ parallel to $\mathrm{tp}(c, |M|)$ so $H(r) = \infty$ hence $C^*(x = x) = \infty$. So apply 6.13(4), and 6.12.

(2) For countable T it follows easily by (1), for uncountable T apply 6.7 for $K = \langle \mathbf{F}^a_{\kappa_r(T)}, W_P, |T|^+ + \kappa_r(T)^+, \kappa_r(T) \rangle$ and get the result when T is stable in λ and μ. For the general case apply Exercise VII, 5.1 to the case just mentioned.

(3) Apply 6.14(4) to $\mathbf{F} = \mathbf{F}^l_{|T|}$ (by 6.13 the hypothesis of 6.14(4) holds).

(4) Define W_P^χ as in (1) and let $K = \langle \mathbf{F}^l, W_P^\chi, \chi, \chi \rangle$ where $\chi = (2^{|T|})^+$. Choose $c \in |M| - |N|$ and define c_i, N_i ($i \leq \lambda$) inductively: $N_0 = N$, N_δ is $\mathbf{F}$-constructible and $\mathbf{F}$-maximal model over $\bigcup_{j<\delta} N_j$; $\mathrm{tp}(c_i, N_i)$ is a stationarization of $\mathrm{tp}(c, |M|)$, and N_{i+1} is $\mathbf{F}$-constructible and $\mathbf{F}$-maximal over $N_i \cup \{c\}$. We should prove by induction on i that $N \prec^\chi N_i$ and $N \prec^\chi |N_i| \cup \{c_i\}$ which is easy by 6.5.

CONCLUSION 6.15: *Suppose $\lambda(i)$ ($i < \alpha$) is an increasing sequence of regular cardinals, M a model and for every $\varphi(x, \bar{y})$ and $i < \alpha$ there is ψ_φ^i such that for every $\bar{a} \in |M|$ $M \models \psi_\varphi^i[\bar{a}]$ iff $|\varphi(M, \bar{a})| \geq \lambda(i)$ (and of course $T = \mathrm{Th}(M)$ is stable). Then for any non-decreasing sequence $\mu(i)$ ($i < \alpha$) of cardinals $\geq |T|$, T has a model N; such that for any $a \in |N|$ and $\varphi(x, y)$ $|\varphi(N, \bar{a})| = \min\{\mu(i) : N \models \psi_\varphi^i[\bar{a}]\}$.*

Proof. Left to the reader (use VII, Section 5). Note that w.l.o.g. $|\alpha| \leqslant |T|$.

THEOREM 6.16: *Suppose T is superstable, $\|M\| = \lambda$, $|P(M)| = \mu$, $\lambda > \mu$. Suppose also that*

(i) $\lambda^* \geq \mu^* \geq 2^{|T|}$,

(ii) $\lambda \geq \lambda^*$, $\mu \geq \mu^*$,

(iii) $\lambda > \lambda^*$ *or* $\lambda = \lambda^*$, $\mathrm{cf}\,\lambda > \mu$ *or* $\lambda = \lambda^*$, $\mathrm{cf}\,\lambda \leq \mu^*$.

Then M has an (elementary) submodel N, $\|N\| = \lambda^*$, $|P(N)| = \mu^*$.

Remark. See 6.17.

Proof. Suppose first that $\lambda > \lambda^*$ or $\lambda = \lambda^* = \mathrm{cf}\,\lambda$, so by Exercise I, 2.6 there is $I \subseteq |M|$, $|I| = \lambda^*$, I indiscernible over $P(M)$. Hence I is orthogonal to any $J \subseteq P(\mathfrak{C})$, so if N^1 is the $\mathbf{F}_{\aleph_0}^c$-prime model over I then $|P(N_1)| \leq 2^{|T|}$. Now over I there is a $\mathbf{F}_{\aleph_0}^c$-constructible model $N_2 \subseteq M$, so there is a $\mathbf{F}_{\aleph_0}^c$-prime model N_1 over I, $N_2 \subseteq N_1$. Hence $|P(N_2)| \leq |P(N_1)| \leqslant 2^{|T|}$, and $\|N_2\| \geqslant |I| = \lambda^*$. If we choose $A \subset P(M)$, $|A| = \mu^*$, and N as $\mathbf{F}_{\aleph_0}^c$-constructible over $A \cup I$, $N \subseteq M$, then $\|N\| = \lambda^*$, $|P(N)| = \mu^*$.

Now suppose $\lambda = \lambda^*$, then let $\lambda = \sum_{i<\mathrm{cf}\,\lambda} \lambda_i$, ($\lambda_i < \lambda$), and choose inductively $J_i \subseteq |M|$, $|J_i| = \lambda_i^+$, J_i indiscernible over $P(M) \cup \bigcup_{j<i} J_j$. Choose $I_i \subseteq J_i$, $|I_i| = \aleph_0$, and choose $J_i' \subseteq J_i$, $|J_i'| = \lambda_i^+$, J_i' indiscernible over $P(M) \cup \bigcup_{j<i} J_j \cup \bigcup_{j<\mathrm{cf}\,\lambda} I_j$. It is easy to check that J_i' is indiscernible over $P(M) \cup \bigcup_{j\neq i} J_j'$; and if $\mathrm{cf}\,\lambda \leq \mu^*$, we can choose N as any submodel of M, which is $\mathbf{F}_{\aleph_0}^c$-constructible over $A \cup \bigcup_{i<\mathrm{cf}\,\lambda} J_i'$, where A is any subset of $P(M)$ of cardinality μ^* (we leave the checking that N satisfies our conclusion to the reader).

We are left with the case $\mathrm{cf}\,\lambda > |P(M)|$; and then we can choose $J''_i \subseteq J_i$, $|J''_i| < \aleph_0$ and J_i is based on $\bigcup J''_i$ and $\mathrm{Av}(J_i, \bigcup J''_i)$ is stationary we can assume that $|J''_i|$ does not depend on i, and that for some finite $A \subseteq M$, $tp_*(J''_i, P(M) \cup A \cup \bigcup_{j<i} J''_j)$ does not fork over A and $\mathrm{stp}(J''_i, A)$ is constant. Then we proceed as in the previous case.

THEOREM 6.17: *In* 6.16 *we can replace* (i) $\lambda^* \geq \mu^* \geq 2^{|T|}$ *by* (i′) $\lambda^* \geq \mu^* \geq |T|$.

Proof. Let $\mathbf{F} = \mathbf{F}^c_{\aleph_0}$, $W = \{\langle P(x), (\forall y)(y = y), (\mu^*)^+\rangle\}$,

$$K = \langle \mathbf{F}, W, |T|^+, \aleph_0\rangle,$$

so the conclusion of 6.13(5) holds and let us work in $\mathfrak{C}^{eq}$. Now we can find $I \subseteq M$, $|I| = \lambda^*$ and $\bar{c} \in M$ such that I is independent over $(P(M) \cup \bar{c}, \bar{c})$ (for $\lambda > \lambda^*$ or λ regular use a variant of III, 4.23, in the other cases work analogously to the proof of 6.16 and replace $\bar{c}$ by a set).

Choose $M_0 \prec M$, $\|M_0\| = \mu$, $\bar{c} \cup P(M) \subseteq |M_0|$, and w.l.o.g. I is independent over $(|M_0|, \bar{c})$, so it is easy to show $a \in I \Rightarrow H(a, \bar{c}) = \infty$.

Choose $M_1 \prec M_0$, $\bar{c} \in M_1$, $\|M_1\| = \mu^* = \|P(M_1)\|$, and I is independent over $(|M_1| \cup P(M) \cup \bar{c}, \bar{c})$. Let $A = \mathrm{dcl}(|M_1| \cup \bigcup I)$, so $A \subseteq M$, $A = \mathrm{dcl}(A)$, we shall prove that A is good, and this suffice (take $N \subseteq M$ $\mathbf{F}$-constructible over A, so by 6.11(5) N is good, so $|P(N)| \leq \mu^*$, by $P(N) \supseteq P(A) \supseteq P(M_1)$, so $|P(N)| \geq |P(M_1| \geq \mu^*$, hence $|P(N)| = \mu^*$, and $A \subseteq N \Rightarrow \|N\| \geq |A| = \lambda^*$, and clearly $\|N\| \leq |A| + |T| = \lambda^*$).

Suppose A is not good, so for some $\bar{a} \in |M_1| \cup \bigcup I$, $C(\varphi(x, \bar{a})) \leq \mu^*$ but $|\varphi(A, \bar{a})| > \mu^*$ (see III, 6.3(1)), and choose a finite $I_0 \subseteq I$, $\bar{a} \in |M_1| \cup \bigcup I_0$. As $|\varphi(A, \bar{a})| > \mu^* \geq |(|M_1| \cup \bigcup I_0)| + |T|$, there is $c \in \varphi(A, \bar{a})$, $c \in \mathrm{dcl}(|M_1| \cup \bigcup I) - \mathrm{dcl}(|M_1| \cup \bigcup I_0)$. By changing if necessary I_0, we can assume $c \in \mathrm{dcl}(|M_1| \cup I_0 \cup \{\bar{b}\}) - \mathrm{dcl}(|M_1| \cup \bigcup I_0)$ for some $\bar{b} \in I$. So clearly $\mathrm{tp}(\bar{b}, |M_1| \cup \bigcup I_0)$, $\mathrm{tp}(c, |M_1| \cup \bigcup I_0)$ are not orthogonal (as they are not weakly orthogonal and the first, not forking over M_1, is stationary). But the first, being a stationarization of $\mathrm{stp}(\bar{b}, |M_1|) \subseteq \mathrm{stp}(\bar{b}, M_0)$ satisfies $H(\bar{b}, |M_1| \cup \bigcup I_0) = \infty$, whereas by the choice of c,

$$H(c, |M_1| \cup \bigcup I_0) \leq C(\varphi(x, \bar{a}) \leq (\mu^*)^+$$

contradiction. Note that instead of 6.13(5) we can use IX, 1.2, 3.

Remark 6.17 A. The superstability assumption in 6.16 is necessary, counterexamples can be built as in IX, Exercise 1.2.

QUESTION 6.1: **Try to eliminate the stability assumption from the general presentation.**

EXERCISE 6.2: **Generalize the theorems in this section to the case which deals with dimensions of indiscernible sets instead of cardinalities of sets $\varphi(M, \bar{a})$. That is suppose not W is given, but a function C, whose domain is a family of indiscernible sets, and $C(I)$ is a infinite cardinality as in Condition 2. Assume that if $C(I), C(J)$ are defined and distinct then I, J are orthogonal. Let $H(I)$ be the supremum of λ such that $H(J) < \lambda$ implies I, J, are orthogonal; and define $H(p)$, $H(\bar{a}, A)$ and $B \not<^{\lambda} A$ as before; and $C^*(I)$ parallelly. M is good if for every $I \subseteq |M|$, $\mathrm{Av}(I, |M|)$ is $\mathbf{F}^a_{C(I)}$-isolated when $C(I)$ is defined, and h-good if $\dim(I, M) \geq h(C^*(I))$ in such case.**

Generalize also Exercises 6.10–6.20.

QUESTION **6.3: Try to weaken the demands we mention or get** naturally in Exercise 6.2. Alternatively, weaken the orthogonality demand to $\{J : C(I) < C(J)\} \not\leq_w I$ (see §5).

EXERCISE 6.4: **Prove that in 6.16 assumption (iii) cannot be deleted.**

EXERCISE 6.5: **Adapt 6.7 to deal also with $|\{a \in M : a \text{ realizes } p\}|$ for "small" types p over $|M|$ (.e.g., $\mathbf{F} = \mathbf{F}^c_{\lambda}$, λ regular $\geq \kappa(T)$, $|p| < \lambda$).**

EXERCISE 6.6: **There is a countable $\aleph_0$-stable theory T such that**

(i) T has a model M, $\|M\| > |P^M| \geq \aleph_0$.

(ii) T has a model M, such that for no N $M \subset N$, $M \neq N$, $P(M) = P(N)$.

[Hint: In $L(T)$ there are only an equivalence relation E and a one place **predicate P.]**

EXERCISE 6.7: **Prove 6.2 for not necessarily stationary p, when** $\bar{a} \in A$.

EXERCISE 6.8: Prove that in 6.14(1) the countability of T is necessary even when it is superstable.

[Hint: $N = (|N|, R^N, P^N, P^N_\eta, Q^N, E^N, Q^N_\eta, F^N_\nu,)_{\eta \in {}^{\omega>}2, \nu \in {}^{\omega}2}$, $T = Th(N)$ where:

$$P^N = {}^{\omega>}2,$$

$$Q^N = \bigcup_{n<\beta\leqslant\omega} {}^n2 \times \{\{n,\beta\}\},$$

$$R^N = \omega + 1,$$

$$|N| = P^N \cup Q^N \cup R^N,$$

$$P^N_\eta = \{v : v \lessdot \eta \quad \text{or } \eta \lessdot \nu \in {}^{\omega>}2\},$$

E is a three-place relation $\subseteq R^N \times R^N \times Q^N$,

$$E^N = \{\langle n, \beta, \langle \rho, \{n,\beta\}\rangle\rangle : n < \beta \leqslant \omega, \rho \in {}^n2, l < n, k < n\},$$

$$Q^N_\eta = \{\langle \rho, \{n,\beta\}\rangle : n < \beta \leqslant \omega, \eta \lessdot \rho \in {}^n2\},$$

F^N_ν is a partial one-place function,

$F^N_\nu(\langle \rho, \{n,\beta\}\rangle) = \rho \restriction k$ when k is maximal such that $\rho \restriction k \lessdot \nu$.

Using the automorphism of saturated models we can see that T is superstable. Let M be the submodel of N with universe

$$P^N \cup \omega \cup \bigcup_{n<\beta<\omega} {}^n2 \times \{\{n,\beta\}\}.$$

Then M, N is a counterexample.]

Remark. On categoricity for logics with generalized quantifiers, see the end of VI, §6.

V.7. Ranks revisited

We use the convention of Section 6, and $h = h^+_K$, be a function as in Definition 6.2, $K = K^h$. For simplicity we deal with $\varphi(x; \bar{y})$ rather than $\varphi(\bar{x}; \bar{y})$.

Definition 7.1: We call $\varphi(x; \bar{a})$ K-algebraic if $C(\varphi(x; \bar{a})) < \infty$; we call $\varphi(x; \bar{a})$ K-minimal if it is not K-algebraic but it has no two contradictory extensions which are not K-algebraic; we call $\varphi(x; \bar{a})$ weakly K-minimal if it is not K-algebraic but whenever $\psi(x, \bar{b})$ forks over $\bar{a}$, $\varphi(x; \bar{a}) \wedge \psi(x; \bar{b})$ is K-algebraic; and we say $\varphi(x; \alpha) \leq_K \psi(x; \bar{b})$ if for some χ for every K-good model $M \supseteq \bar{a} \cup \bar{b}$, $|\varphi(M; \bar{a})| \leq |\psi(M; \bar{b})| + \chi$.

CLAIM 7.1: (1) *The only information from K needed to Definition* 7.1 *is*

$$W_K^a = \{\langle\varphi(x;\bar{y}), \psi(\bar{y})\rangle: \text{for some } \chi\ \langle\varphi(x;\bar{y}), \psi(\bar{y}), \chi\rangle \in W_K\}$$

(*provided that K satisfies* C1, 2, 4, 5, 6) *for* $\leq_K$ *being as we want. Alternatively we can define* $\varphi(x;\bar{a}) \leq_K \psi(x;\bar{b})$ *by: in* $(\mathfrak{C}, \bar{b})$ *if we add to* W_K $\langle\psi(x;\bar{y}), \bar{y} = \bar{b}, \aleph_0\rangle$, *and get* K_1, $\varphi(x,\bar{a})$ *is* K_1*-algebraic.*)

(2) *For every set* $W^a = \{\langle\varphi_i(x;\bar{y}_i), \psi_i(\bar{y}_i)\rangle: i < \alpha\}$ *we can find a K satisfying conditions* 1–6, *such that* $W_K^a = W^a$. *So henceforth we assume K satisfies* C1–6, $\lambda_k > (2^{2^{|T|}})^+$.

(3) *In particular, for the empty* W^a, *in* (2) *we get the trivial K, which we denote by* $K_\emptyset$. *So* $K_\emptyset$*-algebraic,* $K_\emptyset$*-minimal, weakly* $K_\emptyset$*-minimal,* $\varphi(x;\bar{a}) \leq_{K_\emptyset} \psi(x;\bar{b})$, *means algebraic, minimal, weakly minimal,* $\varphi(x;\bar{a}) \leq \psi(x;\bar{b})$ *respectively; and for* $K_\emptyset$, $p^K_{\varphi(x,\bar{y})} = \{(\exists^{\leq k}x)\varphi(x;\bar{y}): k < \aleph_0\}$ (*see* 6.1(2)).

(4) *If* $\varphi(x;\bar{a})$ *is weakly K-minimal, the dependence relation on* $\varphi(\mathfrak{C},\bar{a})$ "*b depends on* $\{b_1, b_2, \ldots, b_n\}$ *iff* $H(b, \{b_1, \ldots, b_n) \cup \bar{a}) < \infty$ (*or equivalently,* $C(b, \bar{a} \cup \{b_1, \ldots, b_n\}) < \infty$) *satisfies the axioms of dependence relation mentioned in* 1.14 *with transitivity also for non-independent sets.*

(5) *If* $\varphi(x;\bar{a})$ *is weakly K-minimal and* $\models\varphi[c;\bar{a}]$ *and* $H(c,\bar{a}) = \infty$, *then* $\text{stp}(c;\bar{a})$ *is regular. The family* $\mathscr{P} = \{p: p \text{ a } K\text{-minimal type}\}$ *is regular and* $\varphi(x,\bar{a})$ *is* $\mathscr{P}$*-regular* (*see Definition* 4.5).

(6) *For every* $\varphi = \varphi(x;\bar{y})$: $\varphi(x;\bar{a})$ *is not K-algebraic iff* $\bar{a}$ *realizes* $\{\neg\varphi; p \in p_\varphi\}$ *where* $p^K_\varphi = p_\varphi =_{\text{def}} \{\varphi: (\exists\kappa)\varphi \in p^\kappa_\varphi\}$ see 6.1(2)).

(7) *The formula* $\varphi(x;\bar{a})$ *is not weakly K-minimal nor K-algebraic iff there are formulas* $\psi(x;\bar{b}_i)$ $(i < |T|^+)$ *such that*:

(i) $\psi(x;\bar{b}_i) \vdash \varphi(x;\bar{a})$,

(ii) $\psi(x;\bar{b}_i)$ *is not K-algebraic,*

(iii) *for* $i < j < |T|^+$, $\psi(x;\bar{b}_i) \wedge \psi(x;\bar{b}_j)$ *is K-algebraic.*

(8) $\varphi(x;\bar{a}) \leqslant_K \psi(x;\bar{b})$ *iff for any* $\mathbf{F}^a_{\kappa_r(T)}$*-saturated model* $M \prec^\infty_K N$, *if* $\bar{a} \cup b \subseteq |M|$, $\psi(N, b) \subseteq |M|$ *then* $\varphi(N,\bar{a}) \subseteq |M|$.

Proof. Easy, hence we leave it to the reader. (For (7) use II, 3.12.)

DEFINITION 7.2: (1) We define $D^m(p, \Delta, K)$ as Definition II, 3.2, but $D^m(p, \Delta, K) \geq \alpha + 1$ iff for every λ and finite $q \subseteq p$, there are m-contradictory $\psi_i(\bar{x};\bar{a}_i)$ for $i < \lambda$ such that:

(i) $D^m[q \cup \{\psi(x;\bar{a}_i)\}, L, K] \geq \alpha$,

(i) $I = \{\bar{a}_i: i < \lambda\}$ is an indiscernible set,

(iii) $H(I) = \infty$,

(iv) $\psi_i(\bar{x};\bar{y}) \in \Delta$.

Equivalently, there are a K-good M and pairwise explicitly contradictory r_i ($i < \lambda$) such that q, r_i are over $|M|$ and $D^m(q \cup r_i, \mathrm{L}, K) \geq \alpha$. We define $\mathrm{Mlt}^l(p, L, K)$ similarly.

(2) For a stationary type p, let $w_K(p) = \sum \{\mathrm{low}_q(p)\colon q$ a regular stationary complete type, $H(q) = \infty\}$ and $w_K(\bar{a}, A) = w_K(\mathrm{stp}(\bar{a}, A))$.

(3) We call $\varphi(x; \bar{a})$ K-weakly-minimal if $D[\varphi(x; \bar{a}), \mathrm{L}, K] = 1$. The following claim sum up the trivial facts concerning Definition 7.2.

CLAIM 7.2: (1) *The two versions of Definition* 7.2(1) *are really equivalent and the parallels of* II, 3.4 *and* II, 3.5 *holds. If M is κ-saturated, $\kappa(T) \leq \kappa$, $A \subseteq |M|$, $|A| < \kappa$, p is over A not κ-algebraic, then for some $q \in S^m(M)$ extending p, $H(q) = \infty$, $D(q, L, K) = D(p, L, K)$.*

(2) *If T is superstable, $w_K(p)$ is always defined, finite, and $\leq w(p)$; and it satisfies the parallels of* 3.10, 3.11, 3.13 *and* 3.15 *when "I independent over A" is replaced by "I K-independent over A" which means I is independent over A, and* $\mathrm{tp}(\bar{c}, A)$ *is not K-algebraic for any $\bar{c} \in I$.*

(3) *If K is trivial, $w_K(p) = w(p)$, and $D(p, L, K) = D(p, L, \infty)$.*

(4) *If $\varphi(x; \bar{a})$ is K-weakly-minimal, $\models\varphi[b; \bar{a}]$, then $w_K(b, \bar{a}) = 1$. Always*

$$w_K(\bar{b}, \bar{a}) \leq D[\mathrm{tp}(\bar{b}, \bar{a}), L, K].$$

(5) *Any weakly K-minimal formula is K-weakly-minimal, but not conversely. However for trivial K, a formula is K-weakly minimal iff it is weakly K-minimal iff it is weakly minimal.*

(6) *Suppose*

(i) $\theta(x, \bar{y}; \bar{c}) \vdash \varphi(\bar{x}; \bar{a}) \wedge \psi(\bar{y}; \bar{b})$,

(ii) $\varphi(x; \bar{a}) \vdash (\exists \bar{y})\theta(x, \bar{y}; \bar{c})$,

(iii) *for every $\bar{d}$, $\theta(x, \bar{d}, \bar{c})$ is K-algebraic.*

Then $D[\varphi(x; \bar{a}), L, K] \leq D[\psi(\bar{x}; \bar{a}), L, K]$.

Proof. Easy, so we prove only (6).

(6) Now we prove by induction on α that (for every suitable φ, ψ, θ, $\bar{c}$)

$$(*) \qquad D^m[\varphi(x; \bar{a}), L, K] \geq \alpha \text{ implies } D^m[\psi(\bar{y}; \bar{b}), L, K] \geq \alpha.$$

For $\alpha = 0$ or α limit there is no problem, so suppose $\alpha = \beta + 1$, and $(*)$ holds for β.

Let μ be any cardinality, and we can assume it is regular and bigger than 2^{2^χ}, $\chi = \chi_K$ (the supremum of all cardinalities χ appearing in W_K). So there are explicitly contradictory types p_i ($i \leq \mu$) over a K-good model M, such that $\varphi(x;\bar{a}) \in p_i$, and $D(p_i, L, K) \geq \beta$. We can assume p_i is closed under conjunctions, and let

$$q_i(\bar{y}) = \{(\exists x)[\theta(x, \bar{y}; \bar{c}) \wedge \varphi_1(x; \bar{a}_1)] : \varphi_1(x; \bar{a}_1) \in p_i\}$$

(so clearly $q_i(\bar{y}) \vdash \psi(\bar{y}; \bar{b})$). It suffices to prove:

(a) $D^m(q_i, L, K) \geq \beta$.

(b) The union of any μ q_i's is contradictory,

because then by 7.2(1) we can find $A \subseteq |M|$ and q_i^1, $q_i \subseteq q_i^1 \in S^m(A)$, $D(q_i^1, L, K) = D(q_i, L, K)$, hence $|\{q_i^1 : i \leq \mu\}| = \mu$. As this holds for every suitable μ, this implies $D^m(\psi(\bar{y}, \bar{b}), L, K) \geqslant \beta + 1 = \alpha$ by the parallel to Th. II, 3.11, Ex. II, 3.27.

Proof of (a). It suffices to prove that every finite subtype of q_i has degree $\geq \beta$, so as p_i is closed under conjunctions, it suffices to prove that for every $\varphi_1(x; \bar{a}_1) \in p_i$ $D^m[(\exists x)[\theta(x; \bar{y}; \bar{c}) \wedge \varphi_1(x; \bar{a}_1)], L, K] \geq \beta$. We can assume $\varphi_1(x; \bar{a}_1) \vdash \varphi(x; \bar{a})$. Now $\varphi_1(x; \bar{a}_1)$, $(\exists x)[\theta(x; \bar{y}; \bar{c}) \wedge \varphi_1(x; \bar{a}_1)]$, $\theta(x, \bar{y}; \bar{c}) \wedge \varphi_1(x; \bar{a}_1)$ satisfies the hypothesis (for $\varphi(x; \bar{a})$, $\psi(\bar{y}; \bar{b})$, $\theta(\bar{y}; \bar{c})$ resp.). Hence by the induction hypothesis

$$D^m[(\exists x)[\theta(x, \bar{y}; \bar{c}) \wedge \varphi_1(x; \bar{a}_1)], L, K] \geq D^m[\varphi_1(x; \bar{a}_1), L, K]$$
$$\geq D^m(p_i, L, K) \geq \beta.$$

Proof of (b). If not let $\bar{d}^*$ realizes μ q_i's, no by notational change we can assume it $\in M$ (6.8(1)) and satisfies all of them. So for each i $p_i \cup \{\theta(x, d^*; \bar{c})\}$ is finitely satisfiable in M; hence it defines a filter D_i on $\theta(M, \bar{d}^*, \bar{c})$, generated by $\{a \in \theta(M, \bar{d}^*; \bar{c}) : \models \varphi_1[a, \bar{a}_1]\}$ for $\varphi_1(x, \bar{a}_1) \in p_i$. Let $|\theta(M, \bar{d}^*, \bar{c})| = \chi$, so $2^{2^\chi} < \mu$, so for some $i \neq j$ $D_i = D_j$, but then $p_i \cup p_j$ is finitely satisfiable, contradicting their choice.

LEMMA 7.3: (1) *Suppose* $\varphi(x; \bar{a}) \leq_K \psi(x; \bar{b})$, *then there is a formula* $\theta(x, y; \bar{c})$ *such that:*

(i) $\theta(x, y; \bar{c}) \vdash \varphi(x; \bar{a}) \wedge \psi(y; \bar{b})$,

(ii) $\varphi(x; \bar{a}) \vdash (\exists y)\theta(x, y; \bar{c})$,

(iii) *for each* $a \in \varphi(\mathfrak{C}, \bar{a})$, *except* $\leqslant |T|$ *of them, there is* b *such that* $\models \theta[a, b, \bar{c}]$ *and* $\theta(x, b, \bar{c})$ *forks over* $\bar{c}$.

(2) *If in addition* $\varphi(x; \bar{a})$ *is weakly* K*-minimal, then for some* $\theta_1(y, \bar{z}) \in p_{\theta(x;y,\bar{z})}$, $\models (\forall y)\theta_1(y, \bar{c})$. (see 7.1(6)).

(3) *If in* (1), *in addition* $\psi(y, \bar{b})$ *is weakly* K*-minimal, then for some* $\theta_2(y, \bar{z}) \in p_{\theta_0(y;x,\bar{z})}$, $\models (\forall y)\theta_2(y, \bar{c})$, $\theta_0(y; x, \bar{z}) = \theta(x, y, \bar{z})$.

(4) *There is a set* $p(\bar{x}; \bar{y})$ *such that* $\varphi(x; \bar{a})$, $\psi(x; \bar{b})$ *satisfies the con-*

clusion of (1)(i), (ii) *iff* $\bar{a}^\frown\bar{b}$ *satisfies at least one formula of* $p(\bar{x}, \bar{y})$. *Similarly for* (2), *and for* (3), *and for their combination.*

Proof. There is no new point (see 7.1(8) and the proof of 6.2).

DEFINITION 7.3: An n-witness (or just a witness) for $\varphi(x; \bar{a}) \le_K \psi(x; \bar{b})$ is a sequence $\bar{\theta} = \langle \theta_l(x; \bar{y}_l, \bar{z}_l, \bar{c}) : l \le n \rangle$ such that

(i) $\bar{y}_l = \langle y_0, \ldots, y_{l-1} \rangle$, $\bar{z}_l = \langle z_0, \ldots, z_{l-1} \rangle$, $\bar{c} = \bar{a}^\frown\bar{b}$,

(ii) $\vdash \theta_0(x, \bar{y}_0, \bar{z}_0, \bar{c}) \equiv \varphi(x; \bar{a})$,

(iii) $\theta_{l+1}(x, \bar{y}_{l+1}, \bar{z}_{l+1}, \bar{c}) \vdash \theta_l(x, \bar{y}_l, \bar{z}_l, \bar{c}) \wedge \psi(y_l, \bar{a}) \wedge \theta_l(z_l, \bar{y}_l, \bar{z}_l, \bar{c})$,

(iv) $\theta_l(x, \bar{y}_l, \bar{z}_l, \bar{c}) \vdash (\exists y_l)(\forall z_l)[\theta_l(z_l, \bar{y}_l, \bar{z}_l, \bar{c}) \to \theta_{l+1}(x, \bar{y}_{l+1}, \bar{z}_{l+1}, \bar{c})]$,

(v) for some $\theta^*(\bar{y}_n, \bar{z}_n, \bar{z}) \in p^K_{\theta_n(x, \bar{y}_n, \bar{z}_n, \bar{z})}$, $\models (\forall \bar{y}_n, \bar{z}_n)\theta^*(\bar{y}_n, \bar{z}_n, \bar{c})$.

LEMMA 7.4: (1) *If* $\bar{\theta} = \langle \theta_l : l \le n \rangle$ *is an* n*-witness for* $\varphi(x; \bar{a}) \le_K \psi(x; \bar{b})$, *then for any* a, b, $\langle \theta_l : l \le n - 1 \rangle$ *is a witness for* $\varphi(x, \bar{a}) \wedge \theta_1(x, b, a, \bar{c}) \le_K \psi(x; \bar{b})$, *where*

$$\theta'_l = \theta_{l+1}(x; b, y_0, y_1, \ldots, y_{l-1}; a, z_0, z_1, \ldots, z_{l-1}; \bar{c})$$

(*so* y_l *take the role of* y_{l+1}).

(2) $\varphi(x; \bar{a}) \le_K \psi(x; \bar{b})$ *iff there is a witness for it.*

Proof. (1) Trivial.

(2) Easy, by 6.1(v) and the definability lemma.

LEMMA 7.5: (1) *Suppose* $q(\bar{y})$ *is an* m*-type,* $\varphi(x; \bar{y})$ *a formula and for every* $\bar{a}$ *realizing* q, $\varphi(x; \bar{a})$ *is not* K*-algebraic; and for some* λ_0, *for every* h_K*-good* M

$$\{|\varphi(M, \bar{a})| : \bar{a} \in M \text{ realizes } q\},$$

has cardinality $< \lambda_0 < \lambda_K$. *Then there are* $\psi(\bar{y}) \in q$ (*assuming* q *is closed under conjunctions*) *and* $k < \aleph_0$ *such that for every good* M, $\{|\varphi(M, \bar{a})| : M \models \psi[\bar{a}]\}$ *has cardinality* $\le k$, *and* $\models \psi[\bar{a}]$ *implies* $\varphi(x; \bar{a})$ *is not* K*-algebraic.*

(2) *If in* (1), *whenever* $\bar{a}$ *realizes* q, $\varphi(x; \bar{a})$ *is weakly* K*-minimal then we can assume:*

(i) $E(\bar{x}, \bar{y})$ *is an equivalence relation on* $\psi(\mathfrak{C})$, $E(\bar{x}, \bar{y}) \vdash \psi(\bar{x})$,

(ii) E *has finitely many equivalence classes, and in each of them there is a sequence* $\bar{a}$ *realizing* q,

(iii) *if* $\models \psi(\bar{a}) \wedge \psi(\bar{b})$, *then* $\models E(\bar{a}, \bar{b})$ *iff in any* h_K*-good* $M \supseteq \bar{a} \cup \bar{b}$, $|\varphi(M, \bar{a})| = |\varphi(M, \bar{b})|$,

(iv) *there are* $\theta(x, y, \bar{z})$, θ_1, θ_2 *such that if* $E(\bar{a}, \bar{b})$ *then for some* $\bar{c}$, 7.3(1)(i), (ii), (2), (3) *holds* (*with* $\psi = \varphi$),

(v) $\models \psi[\bar{a}]$ *implies* $\varphi(x; \bar{a})$ *is* K*-weakly-minimal,*

(vi) *if K is trivial*, $R[\varphi(x;\bar a), L, \lambda] = 1$ *for every $\bar a$ realizing q, then* $\models \psi[\bar a']$ *implies* $R[\varphi(x, \bar a'), \mathrm{L}, \lambda] = 1$.

Proof. (1) Let for every α, Γ_α be the union of the following sets of formulas:

(a) $\{\varphi(\bar z_i^n, \bar y_i): n < \omega, i < \alpha\}$,

(b) the set of formulas saying: $\{z_i^n: n < \omega\}$ is a non-trivial indiscernible set over $\bigcup_{j<i} \varphi(\mathfrak{C}, \bar y_j) \cup \bigcup_{j<i} \bar y_j$,

(c) the set of formulas saying $H(\{z_i^n: n < \omega\}) = \infty$, (i.e., for each formula $\psi = \psi(x; \bar y)$ and $\theta(\bar y) \in p_\psi$ and $i < \alpha$ let $n(0)$ be as n in II, 2.20 and we have the formula

$$(\forall \bar y)\Bigl[\theta(\bar y) \to \bigvee_{\substack{w \subseteq 2n(0)+1 \\ |w| > n(0)}} \bigwedge_{new} \neg\psi(z_i^n, \bar y)\Bigr],$$

(d) $q(\bar y_i)$.

Clearly if M is an h_K-good model, $\bar a_i$ $(i < n)$ realizes q, and $|\varphi(M, \bar a_0)| < |\varphi(M, \bar a_1)| < \cdots < |\varphi(M, \bar a_{n-1})|$, then Γ_n is consistent. It is also clear that if each Γ_n is consistent then each Γ_α is consistent. Lastly, it is clear that if Γ_λ is consistent, $\lambda < \lambda_K$, we can find a model M and $\bar a_i$, I_i $(i < \lambda)$ such that $\bar a_i$ realizes q, $I_i \subseteq \varphi(M, \bar a_i)$ is indiscernible over $\bigcup_{j<i} \varphi(M, \bar a_j) \cup \bigcup_{j<i} \bar a_j$, $H(I_i) = \infty$. Then we can contradict the hypothesis by $\lambda = \lambda_0$, and we find ψ easily.

(2) Let us define $\bar a \sim \bar b$ if for every h_K-good $M \supseteq \bar a \cup \bar b$, $|\varphi(M, \bar a)| = |\varphi(M, \bar a)|$. First we prove $\{\bar a/\sim : \bar a \text{ realizes } q\}$ is bounded, in fact $< \lambda^0 = (2^{2^{|T|}})^+ < \lambda_K$, otherwise there are $\bar a_i$ $(i < (2^{|T|})^+)$ realizing q, pairwise non $\sim$-equivalent. By I, 2.8 we can assume $\{\bar a_i: i < (2^{2^{|T|}})^+\}$ is an indiscernible set. As not $\bar a_0 \sim \bar a_1$, and $\mathrm{tp}(\bar a_0{}^\frown \bar a_1, \emptyset) = \mathrm{tp}(\bar a_1{}^\frown \bar a_0, \emptyset)$ for no $i \neq j$ does $\varphi(x, \bar a_i) \leq_K \varphi(x, \bar a_j)$, hence there is an indiscernible set $I_i^j \subseteq \varphi(\mathfrak{C}, \bar a_j)$, $H(I_i^j) = \infty$, I_i^j indiscernible over $a_j \cup \bar a_j \cup \varphi(\mathfrak{C}, a_i)$. By Erdös–Rado theorem (2.5 of the Appendix) (as the number of $J \subseteq \varphi(\mathfrak{C}, \bar a_i)$, $H(J) = \infty$, up to equivalence, is $\leq 2^{|T|}$) we can assume that for $\alpha < \beta < \gamma < |T|^+$, $I_\alpha^\gamma\, I_\beta^\gamma$ are equivalent. Using again the symmetry, for any i there is an indiscernible $I_i \subseteq \varphi(\mathfrak{C}, \bar a_i)$, $H(I_i) = \infty$, I_i indiscernible over $\bigcup_{j \neq i} \bar a_j \cup \bigcup_{j \neq i} \varphi(\mathfrak{C}, \bar a_j)$ (let $I_i = I_0^i$ for $i > 0$), and by this we get a contradiction to the hypothesis by 7.5(1). Let $\bar a_i$ $(i < \alpha < \lambda^0)$ be a maximal family of pairwise non-$\sim$-equivalent sequences realizing it. Let $r = r(\bar x_1, \bar x_2)$ be such that $\varphi(x, \bar b_1)$, $\varphi(x; \bar b_2)$ satisfies the conclusions of 7.3 (1) (i), (ii), (2), (3) iff $\bar b_1{}^\frown \bar b_2$ satisfies at least one formula of r. Let

$$r_1(\bar x_1, \bar x_2) = \{\psi_1(\bar x_1; \bar x_2) \wedge \psi_2(\bar x_2, \bar x_1): \psi_1(\bar x_1, \bar x_2), \psi_2(\bar x_1, \bar x_2) \in r\};$$

so if $\bar{b}_1, \bar{b}_2$ realizes q, $\bar{b}_1 \sim \bar{b}_2$ then $\bar{b}_1{}^\frown\bar{b}_2$ satisfies at least one formula of r_1. Hence $q(\bar{x}) \cup \{\neg\psi(\bar{x}, \bar{a}_i): i < \alpha, \psi(\bar{x}_1, \bar{x}_2) \in r_1\}$ is contradictory, so there is a finite, contradictory subset. Clearly it involves only finitely many $\bar{a}_i$'s so $\alpha < \omega$ and let $k = \alpha$, and we can replace q by some $\psi(\bar{x}) \in q$ (or $\psi = \bigwedge q_1$, $q_1 \subseteq q$, q_1 finite). Also only finitely many formulas $\psi_n^*(\bar{x}_1, \bar{x}_2) \in r_1(\bar{x}_1, \bar{x}_2)$ are involved. Let

$$E(\bar{x}_1, \bar{x}_2) = \psi(\bar{x}_1) \wedge \psi(\bar{x}_2) \wedge (\exists\bar{y})\left[\bigvee_n \psi_n^*(\bar{x}_1, \bar{y}) \wedge \bigvee_n \psi_n^*(\bar{x}_2, \bar{y})\right].$$

Now if $\vDash E[\bar{b}_1, \bar{b}_2]$, then for some $n(1), n(2), \bar{a}$, $\vDash \psi_{n(1)}^*[\bar{b}_1, \bar{a}] \wedge \psi_{n(2)}[\bar{b}_2, \bar{a}]$, hence:

(i) $\varphi(x, \bar{b}_i)$, $\varphi(x, \bar{a})$ are both K-algebraic, or for every h_K-good $M \supseteq \bar{b}_1 \cup \bar{a}$, $|\varphi(M, \bar{b}_1)| = |\varphi(M, \bar{a})|$,

(ii) similarly for $\varphi(x, \bar{b}_2)$, $\varphi(x, \bar{a})$.

Hence a similar conclusion holds for $\varphi(x, \bar{b}_1)$, $\varphi(x, \bar{b}_2)$.

Now if $\vDash\psi[\bar{b}_1]$ there are $\bar{a}_i$ and n such that $\vDash\psi_n^*[\bar{b}_1, \bar{a}_i]$, hence $\varphi(x, \bar{b}_1)$ is not K-algebraic and for every h_K-good M, $|\varphi(M, \bar{b}_1)| = |\varphi(M, \bar{a}_i)|$. Hence if $\psi_n^*[\bar{b}_1, \bar{a}_i] \wedge \psi_l^*[\bar{b}_2, \bar{a}_j]$, $i \neq j$ then $\vDash\neg E[\bar{b}_1, \bar{b}_2]$.

So clearly E is an equivalent relation on $\psi(\mathfrak{C})$, and $\bar{a}_0, \ldots, \bar{a}_{k-1}$ are representatives, and $\vDash E[\bar{b}_1, \bar{b}_2]$ iff $\vDash\psi[\bar{b}_1] \wedge \psi[\bar{b}_2]$ and for every h_K-good $M \supseteq \bar{b}_1 \cup \bar{b}_2$, $|\varphi(M; \bar{b}_1)| = |\varphi(M; \bar{b}_2)|$. So 7.5(i), (ii), (iii) holds and (iv) is proved similarly. For (v) use 7.2(6) and for (vi) Exercise II, 3.32.

LEMMA 7.6: (1) *Suppose that for every $\bar{a}$ realizing q $\varphi(x; \bar{a})$ is weakly minimal, and $R[\varphi(x, \bar{a}), L, \lambda] = 1$, $\mathrm{Mlt}[\varphi(x, \bar{a}), L, \lambda] = \mu$ and $|\{|\varphi(M, \bar{a})| : \bar{a}$ realizes $q, \bar{a} \in M\}| \leq \lambda_0$ for some λ_0. Then there are ψ, E as in 7.5(2) (for $K = K_\aleph$, $\lambda_K > \lambda_0$) and k such that in addition:*

$$\vDash\psi[\bar{a}] \text{ implies } R[\varphi(x, \bar{a}), L, \lambda] = 1, \mathrm{Mlt}[\varphi(x, \bar{a}), L, \lambda] \leq \mu k.$$

(2) *Suppose in addition that*

(i) *for $\bar{a}$ realizing q, $\varphi(x; \bar{a})$ is minimal,*

(ii) (α) $\vDash\psi[\bar{a}]$ *and* $\varphi_1(x, \bar{a}_1) \vdash \varphi(x; \bar{a})$ *implies* $\{|\varphi_1(M, \bar{a}')| : \bar{a}' \in M$ *realizes* $\mathrm{tp}(\bar{a}_1, \emptyset)\}$ *is bounded or*

(β) *T does not have the f.c.p.*

Then there are formulas $\psi_l(\bar{y}_1)$, $\varphi_1(x, \bar{y}_1)$ ($l < n$) such that:

(a) $\vDash\psi_l[\bar{b}]$ *implies $\varphi_1(x, \bar{b})$ is minimal,*

(b) *the formula*

$$(\exists\bar{y}_0, \ldots, \bar{y}_{n-1})\left[\bigwedge_l \psi_l(\bar{y}_l) \wedge (\forall x)(\varphi(x, \bar{y}) \equiv \bigvee_l \varphi_1(x, \bar{y}_l))\right]$$

is consistent with $\psi(\bar{y})$.

Proof. Left to the reader.

THEOREM 7.7: (1) *Suppose for some $\lambda_0 < \lambda_K$ for every h_K-good model M $\{|\varphi(M, \bar{a})| : \bar{a} \in M, \varphi(x, \bar{a})$ not K-algebraic$\}$ has cardinality $< \lambda_0$. Then for some formula ψ^*_φ, for every $\bar{a}$, $\varphi(x, \bar{a})$ is K-algebraic iff $\vDash \psi^*_\varphi[\bar{a}]$.*

(2) *Suppose the hypothesis of 1 holds for every φ. Suppose $\varphi(x, \bar{a}^*)$ is K-minimal, T totally transcendental. Then in addition to the conclusion of 7.5(2) (for $q = \emptyset$)) there are $\varphi_1(x; \bar{y}_1)$, E_1, ψ_1, also satisfying the conclusions of 7.5(2) and*

(v′) *$\vDash \psi_1[\bar{b}]$ implies $\varphi_1(x, \bar{b})$ is K-minimal,*

(vii) *if $\vDash \psi_1[\bar{b}]$ then for some $\bar{a}$, $\vDash \psi[\bar{a}]$ and $\varphi(x; \bar{a}) \leq_K \varphi_1(x; \bar{b}) \leq_K \varphi(x; \bar{a})$; and conversely for every such $\bar{a}$ there is such a $\bar{b}$.*

Proof. (1) Let $\varphi(x, \bar{y})$ be a formula, and q be a type such that $\varphi(x; \bar{a})$ is not algebraic iff $\bar{a}$ realizes q (see 7.1(6)). Now apply 7.5(1), so we get $\psi(\bar{y}) = \bigwedge q_1$, $q_1 \subseteq q$ finite, and ψ satisfies our demands.

(2) Let ψ be as in 7.5(2). We now define by induction on $n < \omega$, formulas $\varphi_\eta(x, \bar{y})$ for $\eta \in {}^{n>}2$, such that, letting $\varphi^\eta(x; \bar{y}^\eta) = \varphi^\eta = \varphi(x; \bar{y}) \wedge \bigwedge_{l<l(\eta)} \varphi_{\eta|l}(x; \bar{y}_{\eta|l})^{\eta[l]}$ for $\eta \in {}^{n\leq}2$:

$$(*) \qquad \vDash (\exists \bar{y})(\exists \cdots \bar{y}_\eta \cdots)_{\eta \in {}^{n>}2}\Big[E(\bar{y}, \bar{a}^*) \wedge \bigwedge_{\eta \in {}^{n}\geq 2} \psi^*_{\varphi^\eta}(\bar{y}^\eta)\Big]$$

which by (1) is equivalent to:

(*′) There are $\bar{a}$, $\bar{a}_\eta$ $(\eta \in {}^{n>}2)$ such that $\vDash E(\bar{a}, \bar{a}^*)$ and for each $\eta \in {}^{n}2$ $\varphi(x, \bar{a}) \wedge \bigwedge_{l<l(\eta)} \varphi_{\eta|l}(x, \bar{a}_{\eta|l})^{\eta[l]}$ is not K-algebraic.

If we can define them for every n, we get a contradiction to the total transcendency of T. So for some n φ_η is defined for $\eta \in {}^{n>}2$ but we cannot choose for $\eta \in {}^{n}2$. Choose a maximal set $S \subseteq {}^{n}2$, such that

(**′) there are $\bar{a}$, $\bar{a}_\eta$ $(\eta \in {}^{n>}2 \cup S)$ such that $E(\bar{a}, \bar{a}^*)$ and for every $\eta \in {}^{n}2 \cup \{\eta^\frown\langle i\rangle : \eta \in S; i \in \{0, 1\}\}$,

$$\varphi(x; \bar{a}) \wedge \bigwedge_{l<l(\eta)} \varphi_{\eta|l}(x, \bar{a}_{\eta|l})^{\eta[l]}$$

is not K-algebraic.

S exists as ${}^{n}2$ is finite by the choice of n, $S \neq {}^{n}2$, so we can choose $\nu \in {}^{n}2 - S$, and let $\theta^*(\bar{y}^\nu, \bar{y}^*)$ be $(\exists \cdots \bar{y}_\eta \cdots)_{\eta \in J}[E(\bar{y}, \bar{y}^*) \wedge \bigwedge_{\eta \in J} \psi^*_{\varphi^\eta}(\bar{y}^\eta)]$, where $J = {}^{n<}2 \cup S - \{\nu \restriction l; l < n\}$.

Now let $\psi_1(\bar{y}^\nu) = (\exists \bar{y}^*)[\theta^*(\bar{y}^\nu, \bar{y}^*) \wedge \psi(\bar{y}^*)]$, $\varphi_1 = \varphi^\nu$; we leave the rest to the reader.

THEOREM 7.8: *Suppose that*

(i) $\theta(x, y, \bar{c}) \vdash \varphi(x, \bar{a}) \wedge \psi(y, \bar{b})$,

(ii) $\varphi(x, \bar{a}) \vdash (\exists y)\theta(x, y, \bar{c})$,

(iii) $\beta = R[\psi(x, \bar{b}), \mathrm{L}, \aleph_0]$,

(iv) *for every* b, $R[\theta(x, b, \bar{c}), \mathrm{L}, \aleph_0] \leq \alpha$.

Then $R[\varphi(x, \bar{a}), \mathrm{L}, \aleph_0] \leq \alpha(\beta + 1)$ *or* $\alpha = 0$, $R[\varphi(x, \bar{a}), \mathrm{L}, \aleph_0] \leq \beta$.

Proof. We can assume $\alpha > 0$ as for $\alpha = 0$ the conclusion follows by Exercise II, 3.32 (or 7.2(6) for $K_\varnothing$). We prove the assertion by induction on β for all $\varphi(x; \bar{a})$, $\psi(y, \bar{b})$, $\theta(x, y; \bar{c})$ and α.

For $\beta = 0$ $\psi(y; \bar{b})$ is algebraic (by Exercise II, 1.3), so let $\psi(\mathfrak{C}, \bar{b}) = \{b_0, \ldots, b_n\}$, hence $\varphi(x; \bar{a}) \vdash \bigvee_{l \leq n} \theta(x, b_l, \bar{c})$, hence by II, 1.7

$$R[\varphi(x, \bar{a}), \mathrm{L}, \aleph_0] \leq \max_{l \leq n} R[\theta(x, b_l; \bar{c}), \mathrm{L}, \aleph_0] \leq \alpha \leq \alpha(0 + 1).$$

So suppose $\beta > 0$, and we have proved for every $\beta' < \beta$. We can assume $\mathrm{Mlt}[\psi(y; \bar{b}), L, \aleph_0] = 1$ (otherwise we can find $n < \aleph_0$ and $\psi_l(y; \bar{b}_l)$ ($l < n$) such that $\vdash \psi(y; \bar{b}) \equiv \bigvee_{l<n} \psi_l(y; \bar{b}_l)$, $R[\psi_l(y; \bar{b}_l), L, \aleph_0] = \beta$, $\mathrm{Mlt}[\psi_l(y; \bar{b}_l), L] = 1$, we use the assertion for $\varphi(x; \bar{a}) \wedge (\exists y)[\theta(x, y; \bar{c}) \wedge \psi_l(y; \bar{b}_l)]$, $\psi_l(y; \bar{b}_l)$, $\theta(x, y, \bar{c}) \wedge \psi_l(y; \bar{b}_l)$ instead $\varphi(x; \bar{a})$, $\psi(y; \bar{b})$, $\theta(x, y; \bar{c})$ resp., and use II, 1.7:

$$R[\varphi(x, \bar{a}), L, \aleph_0] = \max_{l<n} R[\varphi(x; \bar{a}) \wedge (\exists y)(\theta(x, y; \bar{c}) \wedge \psi_l(y, \bar{b}_l)), L, \aleph_0]$$

By Exercise II, 3.21 it is easy to prove:

CLAIM: $R^m[\varphi(x; \bar{a}), L, \aleph_0] \geq \gamma$ *iff there are formulas* $\varphi_\eta(x; \bar{a}_\eta)$ *for* $\eta \in ds(\omega\gamma)$ (see Def. II, 3.3) *such that*

(i) $\varphi_{\langle\rangle}(x; \bar{a}_{\langle\rangle}) = \varphi(x; \bar{a})$,

(ii) *for* $\nu \lessdot \eta$, $\varphi_\eta(x; \bar{a}_\eta) \vdash \varphi_\nu(x; \bar{a}_\nu)$,

(iii) *if* $\eta, \nu \in ds(\omega\gamma)$ *are incomparable by* $\lessdot$ *then* $\varphi_\eta(x; \bar{a}_\eta), \varphi_\nu(\bar{x}; \bar{a}_\psi)$ *are contradictory*,

(iv) *for* $\eta \in ds(\omega\gamma)$, $\models (\exists x)\varphi_\eta(x; \bar{a}_\eta)$.

Suppose $R[\varphi(x, \bar{a}), L, \aleph_0] > \alpha(\beta + 1) = \alpha\beta + \alpha$, and we shall get a contradiction. Let $\gamma = \alpha\beta + \alpha + 1$, and choose $\varphi_\eta(x, \bar{a}_\eta)$ for $\eta \in ds(\omega\gamma)$ as in the claim, and let

$$\psi_\eta(y, \bar{b}_\eta) = \psi(y, \bar{b}) \wedge (\exists x)(\theta(x, y, \bar{c}) \wedge \varphi_\eta(x, \bar{a}_\eta)).$$

Notice that $R[\varphi_\eta(x, \bar{a}_\eta), L, \aleph_0] \geq \gamma_\eta$ where $l(\eta) > 0$, $\omega\gamma_\eta \leq \eta(l(\eta) - 1) < \omega\gamma_\eta + \omega$, or $l(\eta) = 0$, $\gamma_\eta = \gamma = \alpha\beta + \alpha + 1$. Hence $\gamma_\eta > \alpha\beta$ implies $R[\psi_\eta(x, \bar{b}_\eta), L, \aleph_0] = \beta$ (if it is $\beta' < \beta$, by the induction hypothesis $\gamma_\eta \leq R[\varphi_\eta(x, \bar{a}_\eta), L, \aleph_0] \leq \alpha(\beta' + 1) \leq \alpha\beta$ contradiction). Now as $\mathrm{Mlt}[\psi(x, \bar{b}), L, \aleph_0] = 1$, there is a unique $q \in S(\bar{a} \cup \bar{b} \cup \bigcup \{\bar{a}_\eta : \eta \in ds(\omega\gamma)\})$

such that $\psi(x, \bar{b}) \in q$, $R(q, L, \aleph_0) = \beta$, and let b realize q. So if $\eta \in ds(\omega\gamma)$, $\gamma_\eta > \alpha\beta$ then $\psi_\eta(x, \bar{b}_\eta) \in q$ hence $\vDash \psi_\eta[b, \bar{b}_\eta]$, therefore $\varphi_\eta(x, \bar{a}_\eta) \wedge \theta(x, b, \bar{c})$ is consistent. If $\gamma_\eta > \alpha\beta$ implies $\varphi_\eta(x, \bar{a}_\eta) \wedge \theta(x, b, \bar{c})$ is not algebraic, then by the "if" part of the claim,

$$R(\varphi_{\langle\rangle}(x, \bar{a}_{\langle\rangle}) \wedge \theta(x, b, \bar{c}), L, \aleph_0) \geq \alpha + 1$$

contradiction. Hence there are $\nu \in ds(\omega\gamma)$, $k < \omega$; $\gamma_\nu > \alpha\beta$, $\vDash (\exists^{\leq k} x)[\varphi_\nu(x, \bar{a}_\nu) \wedge \theta(x, b, \bar{c})]$; hence $\psi^*(x, \bar{c}^*) = (\exists^{\leq k} y)[\varphi_\nu(y, \bar{a}_\nu) \wedge \theta(y, x, \bar{c})] \in q$.

By Exercise II, 3.32 and monotonicity,

$$R[\varphi_\nu(x, \bar{a}_\nu) \wedge (\exists z)[\theta(x, z, \bar{c}) \wedge \psi^*(z, \bar{c}^*)], L, \aleph_0]$$
$$\leq R[\psi^*(x, \bar{c}^*), \mathrm{L}, \aleph_0] \leq R[\psi(x, \bar{b}), L, \aleph_0] \leq \beta \leq \alpha\beta.$$

As $\beta' =_{\mathrm{def}} R[\psi(x, \bar{b}) \wedge \neg\psi^*(x, \bar{c}^*), L, \aleph_0] < \beta$, by the induction hypothesis

$$R[\varphi_\nu(x; \bar{a}_\nu) \wedge (\exists z)[\theta(x, z, \bar{c}) \wedge \neg\psi^*(z, \bar{c}^*)], L, \aleph_0] \leq \alpha(\beta' + 1) \leq \alpha\beta.$$

So by II, 1.7, $R[\varphi_\nu(x, \bar{a}_\nu), L, \aleph_0] \leq \alpha\beta < \gamma_\nu$, a contradiction.

LEMMA 7.9: *Let $\varphi(x; \bar{a})$, $\psi(x; \bar{b})$, $\theta(x, y, \bar{c})$ be as in 7.3(1) and assume $x = y \in \Delta$.*

(1) *If $D[\psi(x, \bar{b}), L, K] = \beta$, and for every b $D[\theta(x, b, \bar{c}), \Delta, K] \leqslant n$ then $D[\varphi(x, \bar{a}), \Delta, K] \leq n(\beta + 1)$ except when $n = 0$, and then*

$$D[\varphi(x, \bar{a}), \Delta, K] \leq \beta.$$

(2) *If in (1) $\beta = k < \omega$, then $D[\varphi(x; \bar{a}), \Delta, K] \leq n + k$.*

Proof. (1) Using 7.2(1), (6) we can assume $n > 0$, $\beta > 0$, and prove by induction on β; suppose we have proved for every $\beta' < \beta$, but $D[\varphi(x; \bar{a}), \Delta, K] > n(\beta + 1) = n\beta + n$, and we shall get a contradiction.

So we can find an h_K-good $\mathbf{F}^a_{\kappa_r(T)}$-saturated model M, $\bar{a} \cup \bar{b} \cup \bar{c} \subseteq |M|$, $m_l < \omega$, $\varphi_l \in \mathrm{cl}_1 \Delta$, $\bar{a}_\eta \in |M|$ for $\eta \in {}^l\chi$, $0 < l \leq n$, $\chi = \chi_K^+$, such that:

(i) $\varphi_{l+1}(x, \bar{a}_{\eta\frown\langle i\rangle})$ $(i < \chi)$ are m_l-contradictory for $l < n$, $\eta \in {}^l\chi$.

(ii) $D[\varphi^*(x, \bar{a}^\eta), \Delta, K] > n\beta$ where $\eta \in {}^n\chi$,

$$\varphi^*(x, \bar{a}^\eta) = \varphi(x, \bar{a}) \wedge \bigwedge_{0<l\leq n} \varphi_l(x, \bar{a}_{\eta|l}),$$

By 7.2(1) we can find a model N, $M \prec^\infty N$, and $\bar{a}_\eta \in N$, $(\forall\eta \in {}^n\chi)$ $D[\mathrm{tp}(a_\eta, |M|), \Delta, K] > n\beta$, $\vDash \varphi^*[a_\eta, \bar{a}^\eta]$, and $b_\eta \in |N|$, $\vDash \theta[a_\eta, b_\eta, \bar{c}]$. If for some η, $D[\mathrm{tp}(b_\eta, |M|), L, K] < \beta$, we get a contradiction by the induction hypothesis. Hence the number of possible $\operatorname{tp}(b_\eta, |M|)$ is $<\chi$, and χ is regular, hence by 2.6 of the Appendix we can assume that for every $\eta \in {}^n\chi$, $q = \operatorname{tp}(b_\eta, |M|)$; hence we can assume $b_\eta = b$. If for each $\eta \in {}^n\chi$, $\varphi^*(x, \bar{a}^\eta) \wedge \theta(x, b, \bar{c})$ is not K-algebraic, then $D[\varphi(x, \bar{a}) \wedge \theta(x, b, \bar{c}), \Delta, K] > n$ contradiction, hence for some η it is K-algebraic, so there is θ^*, $\vDash \theta^*[b, \bar{a}^\eta, \bar{c}]$ and $\vDash \theta^*[b', \bar{a}', \bar{c}']$ implies $\varphi^*(x; \bar{a}') \wedge \theta(x, b', \bar{c}')$ is K-algebraic. Let

$$\varphi^+(x; \bar{a}^+) = \varphi^*(x, \bar{a}^\eta) \wedge (\exists y)[\theta(x, y, \bar{c}) \wedge \theta^*(y, \bar{a}^\eta, \bar{c})],$$
$$\psi^+(x; \bar{b}^+) = \psi(x, \bar{b}) \wedge \theta^*(x, \bar{a}^\eta, \bar{c}),$$
$$\theta^+(x, y, \bar{c}^*) = \theta(x, y, \bar{c}) \wedge \varphi^+(x, \bar{a}^+) \wedge \psi^+(y, \bar{b}^+).$$

Clearly $\varphi^+(x, \bar{a}^+) \in \operatorname{tp}(a_\eta, |M|)$ hence $D[\varphi^+(x, \bar{a}^+), \Delta, K] > n\beta$, and clearly $D[\psi^+(x; \bar{b}^+), L, K] \leq D[\psi(x, \bar{b}), L, K] = \beta$, and for every b' $D[\theta^+(x, b', \bar{c}^+), \Delta, K] = 0$; and thus our hypotheses are satisfied; contradiction.

(2) It is easy to prove that if $D[\varphi(x, \bar{a}), \Delta, K] \geq l < \omega$, then for some finite $\Delta' \subseteq \Delta$, $D[\varphi(x, \bar{a}), \Delta', K] \geq l$. Hence we can assume Δ is finite; so for every formula $\varphi_1(x, \bar{y}_1)$ and $l < \omega$ there is a type $q^l(\varphi_1)$ such that $D[\varphi_1(x, \bar{a}_1), \Delta, K] \geq l$ iff $\bar{a}_1$ realizes $q^l(\varphi_1)$. We now prove the assertion by induction on k; and the cases $m = 0$ or $k = 0$ are easy.

So suppose $D[\varphi(x, \bar{a}), \Delta, K] > n + k$, hence there are $m < \omega$, $\varphi_1 \in \operatorname{cl}_1 \Delta$ and $\bar{a}_i$ $(i < \chi = \chi_K^+)$ and an h_K-good model M, $\bar{a} \cup \bar{b} \cup \bar{c}$, $\bigcup_i \bar{a}_i \subseteq |M|$, such that:

(a) $D[\varphi(x, \bar{a}) \wedge \varphi_1(x, \bar{a}_i), \Delta, K] \geq n + k$,

(b) the $\varphi_1(x, \bar{a}_i)$ are m-contradictory.

We can now continue as in (1) (for $n = 1$). The only difference is that instead of asserting "for some η, $\varphi^*(x, \bar{a}^\eta) \wedge \theta(x, b, \bar{c})$ is K-algebraic" we assert "for some i, $R(\varphi^0(x, b, \bar{a}^i, \bar{c}), \Delta, K) < n$," where

$$\varphi^0(x, b, \bar{a}^i, \bar{c}) = \varphi(x, \bar{a}) \wedge \varphi_1(x, \bar{a}_i) \wedge \theta(x, b, \bar{c}),$$

and we replace "$\vDash \theta^*[b', \bar{a}^\eta, \bar{c}]$ implies $\varphi^*(x, \bar{a}') \wedge \theta(x, b', \bar{c}']$ is K-algebraic" by "$\vDash \theta^*[b', \bar{a}', \bar{c}']$ implies $D[\varphi^0(x, b', \bar{a}', \bar{c}'), \Delta, K] < n$" (so $\neg\theta^* \in q^n(\varphi^0)$).

CONCLUSION 7.10: (1) *If $\varphi(x, \bar{a}) \leq_K \psi(x, \bar{b})$, $\psi(\bar{x}, \bar{b})$ is K-weakly-minimal or even $D[\psi(x; \bar{b}), L, K] = k < \omega$, then $D[\varphi(x; \bar{a}), L, K]$ is finite too. If there is an n-witness for $\varphi(x, \bar{a}) \leq_K \psi(x, \bar{b})$, then $D[\varphi(x; \bar{a}), L, K] \leqslant kn$.*

(2) *Suppose*

(i) $\psi(x, \bar{b})$ *is K-weakly-minimal,*

(ii) $\varphi(x, \bar{a}) \leq_K \psi(x, \bar{b})$,

(iii) *for every* $\theta(x, \bar{y})$ *there is* $\theta^*(\bar{y}, \bar{b})$ *such that* $\psi(x, \bar{b}) \wedge \theta(x, \bar{c})$ *is K-algebraic iff* $\models\theta^*[\bar{c}, \bar{b}]$.

Then there is a formula $\varphi^*(\bar{x}, \bar{b})$ *such that* $\models\varphi^*[\bar{a}, \bar{b}]$ *and* $\models\varphi^*[\bar{a}', \bar{b}]$ *implies* $D[\varphi(x, \bar{a}'), L, K] = D[\varphi(x, \bar{a}), L, K]$.

(3) *If for some* ψ^*, $\models\psi^*[\bar{b}^*]$ *implies* $\psi(x; \bar{b}^*) \leq \varphi(x; \bar{b}^*) \leq \psi(x; \bar{b}^*)$ *then we can omit* $\bar{b}$ *from* φ^*.

Proof. Left to the reader.

LEMMA 7.11: (1) *Suppose*

(i) $\varphi(x, \bar{a}) \leq \psi(x, \bar{b})$, $\varphi(x, \bar{a})$ *not algebraic,*

(ii) $R[\psi(x, \bar{b}), L, \infty] = R[\psi(x, \bar{a}), L, \lambda] = 1$.

Then $R[\varphi(x, \bar{a}), L, \infty] = R[\varphi(x, \bar{a}), L, \lambda]$.

(2) *If* $\varphi(x, \bar{a}) \leqslant \psi(x, \bar{b})$, *then* $R(\varphi(x, \bar{a}), L, \aleph_0) < [R(\psi(x, \bar{b}), L, \aleph_0)]^{\omega}$.

Proof. (1) By induction on the length of the witness, refining the proof of 7.10 to deals with multiplicity.

(2) By 7.8 and witnesses.

DEFINITION 7.5: (1) For every complete type $p \in S^m(A)$ we define a rank $L(p)$ as an ordinal or ∞ by:

$L(p) \geq 0$ if p is a type

$L(p) \geq \delta$ if $L(p) \geq \alpha$ for each $\alpha < \delta$,

$L(p) \geq \alpha + 1$ if there is p_1, $p \subseteq p_1$, p_1 forks over A and $L(p_1) \geq \alpha$.

(2) $L(\bar{a}, A)$ is $L(\mathrm{tp}(\bar{a}, A))$.

DEFINITION 7.6: We define $\alpha \oplus \beta$ (the natural sum) by double induction on α and β:

$$\alpha \oplus 0 = \alpha, \qquad 0 \oplus \beta = \beta,$$

$$\alpha \oplus \beta = \sup\{\alpha' \oplus \beta' + 1 : \alpha' \leqslant \alpha, \beta' < \beta \text{ or } \alpha' < \alpha, \beta' \leqslant \beta\}.$$

EXERCISE 7.1: Show that $\oplus$ is commutative and associative.

LEMMA 7.12: *Here the types will be complete.*

(1) $p \subseteq q$ *implies* $L(p) \geq L(q)$; *and if for no* $p \supseteq p_0$, $L(p) = \alpha$, *then* $p \supseteq p_0$, $L(p) > \alpha$ *implies* $L(p) = \infty$; $L(p) = 0$ *iff* p *is algebraic.*

(2) *If* $p \in S^m(A)$ *does not fork over* $B \subseteq A$ *then* $L(p) = L(p \upharpoonright B)$.
(3) $L(p) \leq R(p, L, \infty)$.
(4) *T is superstable iff for every* p, $L(p) < \infty$.
(5) *If* $R[\varphi(x, \bar{a}), L, \infty] \geq n$, *then for some* $\bar{b}$, $\models\varphi[\bar{b}; \bar{a}]$ *and* $L(\bar{b}; \bar{a}) \geq n$.
(6) $L(\bar{a}^\frown\bar{b}, A) \leq L(\bar{a}, A \cup \bar{b}) \oplus L(\bar{b}, A)$.

Proof. (1) Trivial.

(2) By (1) it suffices to prove by induction on α that $L(p \upharpoonright B) \geq \alpha$ implies $L(p) \geq \alpha$; and for $\alpha = 0$, $\alpha = \delta$ this is trivial so let $\alpha = \beta + 1$. As $L(p \upharpoonright B) \geq \alpha$ there are $\bar{c}$, C, $B \subseteq C$, $\bar{c}$ realizes $p \upharpoonright B$ and $L(\bar{c}, C) \geq \beta$, $\mathrm{tp}(\bar{c}, C)$ forks over B. Let $\bar{a}$ realizes p; so we can assume first that $\mathrm{stp}(\bar{a}, B) \equiv \mathrm{stp}(\bar{c}, B)$ and then also $\mathrm{tp}_*(C \cup \bar{c}, A)$ does not fork over B (by III, 2.6(3)). Then by III, 4.14 $\mathrm{tp}(\bar{c}, A \cup C)$ does not fork over $C = B \cup C$ hence by the induction hypothesis, $L(\bar{c}, A \cup C) \geq \beta$. Now $\mathrm{tp}(\bar{a}, A) \subseteq \mathrm{tp}(\bar{c}, A \cup C)$ and clearly $\mathrm{tp}(\bar{c}, A \cup C)$ forks over B; hence it forks over A (by III, 4.4, as $\mathrm{tp}(\bar{c}, A) = \mathrm{tp}(\bar{a}, A)$ does not fork over B). So by definition $L(\bar{a}, A) > L(\bar{c}, A \cup C) \geq \beta$, so $L(\bar{a}, A) \geq \alpha$.

(3) See III, 1.2(4).

(4) The only if follows by (3) (and II, 3.14). The if part is left to the reader.

(5) We prove by induction on n: for $n = 0$ it is trivial, for $n + 1$ we can find, n-contradictory formulas $\varphi^*(x, \bar{a}_i)$ $(i < (2^{|T|})^+)$ such that $R[\varphi(\bar{x}. \bar{a}) \wedge \varphi^*(\bar{x}, \bar{a}_i), L, \infty] \geq n$. So for some i $\varphi^*(\bar{x}, \bar{a}_i)$ forks over $\bar{a}$. By the induction hypothesis for some $\bar{b}$ $L(\bar{b}, \bar{a} \cup \bar{a}_i) \geq n$, $\models\varphi[\bar{b}, \bar{a}] \wedge \varphi^*[\bar{b}, \bar{a}_i]$. Clearly $\mathrm{tp}(\bar{b}, \bar{a} \cup \bar{a}_i)$ forks over $\bar{a}$, hence $L(\bar{b}, \bar{a}) \geq n + 1$.

(6) It suffices to prove by induction on γ that $L(\bar{a}^\frown\bar{b}, A) = \gamma$ implies $L(\bar{a}, A \cup \bar{b}) \oplus L(\bar{b}, A) \geq \gamma$. For $\gamma = 0$, $\gamma = \delta$ there is no problem; so let $\gamma = \beta + 1$. As $L(\bar{a}^\frown\bar{b}, A) > \beta$, there are B, $\bar{a}_1^*$, $\bar{b}_1^*$ such that $A \subseteq B$, $\mathrm{tp}(\bar{a}_1{}^\frown\bar{b}_1, B)$ extend $\mathrm{tp}(\bar{a}^\frown\bar{b}, A)$ and forks over A, and $L(\bar{a}_1{}^\frown\bar{b}_1, B) \geqslant \gamma$. By III, 4.15 $\mathrm{tp}(\bar{a}_1, B \cup \bar{b}_1)$ forks over A or $\mathrm{tp}(\bar{b}_1, B)$ forks over A. As clearly $L(\bar{a}_1, A \cup \bar{b}_1) = L(\bar{a}, A \cup \bar{b})$; this implies $L(\bar{a}_1, B \cup b_1) < L(\bar{a}_1, A \cup \bar{b}_1) = L(\bar{a}_1, A \cup \bar{b})$ or $L(\bar{b}, B) < L(\bar{b}_1, A)$ or $L(\bar{a}, A \cup \bar{b}) = \infty$ or $L(\bar{b}, A) = \infty$. Hence in all cases

$$L(\bar{a}_1, B \cup \bar{b}_1) \oplus L(\bar{b}_1, B) < L(\bar{a}, A \cup \bar{b}) \oplus L(\bar{b}, A).$$

But by the induction hypothesis

$$\beta \leq L(\bar{a}_1, B \cup \bar{b}_1) \oplus L(\bar{b}_1, B);$$

so

$$\gamma = \beta + 1 \leq L(\bar{a}, A \cup \bar{b}) \oplus L(\bar{b}, A).$$

DEFINITION 7.7: Let $\mathrm{acl}_K(A) = \{b : \mathrm{tp}(b, A) \text{ is } K\text{-algebraic}\}$.

THEOREM 7.14 ($\mathfrak{C}^{\mathrm{eq}}$): (1) *Suppose K satisfies* C1-6, $\mu = \kappa_r(T)$, Card $K = \{(2^{|T|})^+\}$, *T superstable non-multi-dimensional and $\varphi(x) = (x = x)$ is not K-algebraic. Then there is a formula $\theta(x)$ (with no parameters), types $q_i \in S(\mathrm{acl}\,\emptyset)$, $\bar{c}_i \subseteq q_i(\mathfrak{C})$, and types r_i, $q_i \subseteq r_i \in S(\bar{c}_i)$ such that*

(i) *If $M \prec_K N$, M, N are $F^a_{\kappa r(T)}$-saturated, then $\theta(M) = \theta(N)$ iff $q_i(M) = q_i(N)$ for every i.*

(ii) *r_i is K-minimal, q_i is $\mathrm{cl}^3\{r_i\}$-simple and $q_i(\mathfrak{C}^{\mathrm{eq}}) \subseteq \mathrm{acl}_K(r_i(\mathfrak{C}^{\mathrm{eq}}) \cup \bar{c}_i)$.*

(2) *Suppose in addition T is totally transcendental, then we can assume there are K_a-minimal $\varphi_i(x, \bar{c}_i) \in r_i$, and $\psi_i(x, \bar{c}_i) \in q_i$, $\bar{a}_i \in \mathrm{acl}\,\emptyset$, such that $\varphi_i(x, \bar{c}_i) \vdash \psi_i(x, \bar{a}_i)$, $\psi_i(\mathfrak{C}^{\mathrm{eq}}, \bar{a}_i) \subseteq \mathrm{acl}_K(\varphi_i(\mathfrak{C}^{\mathrm{eq}}, \bar{c}_k) \cup \bar{a}_i)$.*

Proof. We hereby prove Theorem 7.14(1) by a series of claims. Of course T is stable, but the use of stability is mentioned explicitly. We shall assume K satisfies C1–6, $\mu = \kappa_{r(T)}$, Card $K = \{(2^{|T|})^+\}$, and works in $\mathfrak{C}^{\mathrm{eq}}$.

FACT A: *$B = \mathrm{acl}_K A$ implies $B = \mathrm{acl}\,B = \mathrm{acl}_K B$ and $\bar{b} \in B$ if $\mathrm{tp}(\bar{b}, A)$ is K-algebraic.*

FACT B: *If $\bar{b} \in \mathrm{acl}_K(A \cup \bar{a}) - \mathrm{acl}_K A$, then $\mathrm{tp}(\bar{a}, A \cup \bar{b})$ forks over A.*

Proof. Suppose $\mathrm{tp}(\bar{a}, A \cup \bar{b})$ does not fork over A. Let $C_1 = \mathrm{acl}(A \cup \bar{a}) \cap \mathrm{acl}_K A$, $C_2 = \mathrm{acl}(A \cup \bar{b}) \cap \mathrm{acl}_K A$. As we are working in $\mathfrak{C}^{\mathrm{eq}}$, $\mathrm{tp}(\bar{a}, \mathrm{acl}_K A)$ does not fork over C_1, and similarly $\mathrm{tp}(\bar{b}, \mathrm{acl}_K A)$ does not fork over C_2. As we assume $\mathrm{tp}(\bar{a}, A \cup \bar{b})$ does not fork over A, also $\mathrm{tp}_*(\mathrm{acl}(A \cup \bar{a}), \mathrm{acl}(A \cup \bar{b}))$ does not fork over A, hence $\mathrm{tp}_*(\bar{a} \cup C_1, A \cup \bar{b} \cup C_2)$ does not fork over A, hence $\mathrm{tp}(\bar{a}, A \cup \bar{b} \cup C_1 \cup C_2)$ does not fork over $A \cup C_1$.

By 6.3(2) $A \cup C_1 \prec A \cup C_1 \cup \bar{a}$, and $H(\bar{a}, A \cup C_1) = \infty$. Similarly, $H(\bar{b}, A \cup C_2) = \infty$. Let M be h_K-good, $\{\bar{a}_i : i < \chi\}$ be an indiscernible set over $A \cup C_1 \cup C_2$ based on $\mathrm{stp}(\bar{a}, A \cup C_1)$ and $\{\bar{b}_i ; i < \chi\}$ be an indiscernible set over $A \cup C_1 \cup C_2$ based on $\mathrm{stp}(\bar{b}, A \cup C_2)$, where χ is large enough, w.l.o.g. $A \cup \bar{a} \cup \bar{b} \cup C_1 \cup C_2 \cup \{\bar{a}_i, \bar{b}_i : i < \chi\} \subseteq M$. Then $M \cap \mathrm{acl}_K(A)$ has power $< \chi$, hence for some $i < \chi$ $\mathrm{tp}(\bar{a}_i, \mathrm{acl}_K(A))$ does not fork over $A \cup C_1$. Now $M \cap \mathrm{acl}_K(A \cup \bar{a}_i)$ has power $< \chi$, hence for some j, $\mathrm{tp}(\bar{b}_j, M \cap \mathrm{acl}_K(A \cup \bar{a}_i))$ does not fork over $A \cup C_2$. Hence $\bar{b}_j \notin \mathrm{acl}_K(A \cup \bar{a}_i)$, but clearly $\mathrm{stp}(\bar{a}_i {}^\frown \bar{b}_j, A \cup C_1 \cup C_2) \equiv \mathrm{stp}(\bar{a} {}^\frown \bar{b}, A \cup C_1 \cup C_2)$, hence $\mathrm{tp}(\bar{a}_i {}^\frown \bar{b}_j, A) = \mathrm{tp}(\bar{a} {}^\frown \bar{b}, A)$, contradicting an assumption of Fact B.

FACT C: *If T is superstable there is a weakly K-minimal $\varphi(x, \bar{a})$.*

FACT D: *If T is not multi-dimensional then:*

(i) *for any $\varphi(x, \bar{y})$ and M, $|\{|\varphi(M, \bar{a})| : \bar{a} \in M\}| < \aleph_0$,*

(ii) *if $p(x, \bar{a}) \in S(\bar{a})$, $\text{stp}(\bar{b}, \emptyset) \equiv \text{stp}(\bar{a}, \emptyset)$, then $p(x, \bar{a})$, $p(x, \bar{b})$ are not orthogonal.*

FACT E: *Suppose $|\{|\varphi(M, \bar{b}) : \bar{b} \in M\}| < \aleph_0$ for any M, $\varphi(x, \bar{a})$ is weakly K-minimal, then there are $n < \omega$, $\bar{a}_l$ realizing $\text{tp}(\bar{a}, \emptyset)$ $(l < n)$ and non K-algebraic formula $\theta(x)$ (with no parameters) $\theta(\mathfrak{C}^{eq}) \subseteq \text{acl}(\varphi^*(\mathfrak{C}^{eq}, \bar{a}^*) \cup \bar{a}^*)$ where $\varphi^*(x, \bar{a}^*) = \bigvee_{l<n} \varphi(x, \bar{a}_l)$ is weakly K-minimal too.*

Proof. Like 4.11.

FACT F: *If $\varphi(x, \bar{a})$ is weakly K-minimal, $\theta(x)$ not K-algebraic, T superstable, $\theta(\mathfrak{C}^{eq}) \subseteq \text{acl}_K(\varphi(\mathfrak{C}^{eq}, \bar{a}) \cup \bar{a})$, then there are complete types p_i $(i < i_0)$, $\theta(x) \in p_i$, p_i K-minimal such that: if $M \prec_K^{\infty} N$ are $F^a_{\aleph_0}$-saturated, $\bigcup_i \text{Dom}\, p_i \cup \bar{a} \subseteq |M|$, then $\theta(M) \neq \theta(N)$ iff some $c \in |N| - |M|$ realizes some p_i.*

Proof of Fact F. Let $\{p_i : i < i_0\}$ be a maximal list of K-minimal types to which $\theta(x)$ belongs and are pairwise orthogonal. The "if" part is clear, so let us prove the "only if". Let $c \in \theta(N) - \theta(M)$, be with minimal $R(\text{tp}(c, |M|), \text{L}, \infty)$, so $\text{tp}(c, |M|)$ is regular. By a hypothesis there are $b_1, \ldots, b_n \in \varphi(\mathfrak{C}^{eq}, \bar{a})$ such that $c \in \text{acl}_K(\{b_1, \ldots, b_n\} \cup |M|)$, and w.l.o.g. we choose $n, c, b_1, \ldots, b_n$ such that n is minimal. As $c \in N - M$, $M \prec_K N$ necessarily $n \geq 1$. If $n = 1$, then clearly $\text{tp}(c, |M|)$ is K-minimal, hence not orthogonal to some p_i $(i < i_0)$, so some $c' \in |N| - |M|$ realizes p_i, so we finish.

Now we suppose $n > 1$, and we shall get a contradiction; let $\bar{b} = \langle b_1, \ldots, b_{n-1} \rangle$, and choose finite $A \subseteq |M|$ such that $\bar{a} \subseteq A$, $c \in \text{acl}_K(A \cup \bar{b} \cup \{b_n\})$, $\text{tp}(\bar{b} \frown \langle c, b_n \rangle, |M|)$ does not fork over A and $\text{tp}(\bar{b} \frown \langle c, b_n \rangle, A)$ is stationary. As n is minimal c, $b_n \notin \text{acl}_K(|M| \cup \bar{b})$, so (as $\varphi(x, \bar{a})$ is weakly K-minimal) $\{b_n, \bar{b}\}$ is independent over $(|M|, A)$. Choose $\bar{b}' \in |M|$, $\bar{b}'$ realizing $\text{stp}(\bar{b}, A)$, then

$$\text{tp}(\bar{b}, A \cup \{b_n\}) = \text{tp}(\bar{b}', A \cup \{b_n\}),$$

and we can find $c' \in N$, $\text{tp}(\bar{b}' \frown \langle c', b_n \rangle, A) = \text{tp}(\bar{b} \frown \langle c, b_n \rangle, A)$. So now $c' \in \text{acl}_K(A \cup \bar{b}' \cup \{b_n\}) - \text{acl}_K(A \cup \bar{b}')$, hence by Fact B,

$$\text{tp}(b_n, A \cup \bar{b}' \cup \{c'\})$$

fork over $A \cup \bar{b}'$, hence $\text{tp}(b_n, |M| \cup \{c'\})$ fork over $A \cup \bar{b}'$, but $\text{tp}(b_n, |M|)$ does not fork over A hence over $A \cup \bar{b}'$, so $\text{tp}(b_n, |M| \cup \{c'\})$ fork over $|M|$, so $c' \notin |M|$. So c', b_n contradict the minimality of n.

FACT G: *Suppose* $p = p(x, \bar{a})$ *is* K*-minimal, and* $\text{stp}(\bar{a}, \emptyset) \equiv \text{stp}(\bar{a}', \emptyset)$ *implies* $p(x, \bar{a})$, $p(x, \bar{a}')$ *are not orthogonal. Then there is* $q \in S(\text{acl}\,\emptyset)$, $\bar{c} \subseteq \text{acl}(\emptyset) \cup q(\mathfrak{C}^{\text{eq}})$, $q \subseteq r \in S(\bar{c})$ *such that:* r *is* K*-minimal,* q *is* $\text{cl}^3_\beta\{p\}$*-simple, and* $q(\mathfrak{C}^{\text{eq}}) \subseteq \text{acl}_K(r(\mathfrak{C}^{\text{eq}}) \cup \bar{c})$*; and there are* $n < \omega$, $\bar{a}_l$ $(l < n)$ *realizing* $\text{stp}(\bar{a}, \emptyset)$ *such that*

$$q(\mathfrak{C}^{\text{eq}}) \subseteq \text{acl}\left[\bigcup_{l<n} (p(\mathfrak{C}^{\text{eq}}, \bar{a}_l) \cup \bar{a}_l)\right].$$

So q is $\text{cl}^3\{r\}$-simple. (We assume T is superstable.)

The proof is like 4.11, except r. As q is not K-algebraic, there is a K-minimal $r \in S(\bar{c})$, $\bar{c} \subseteq q(\mathfrak{C}^{\text{eq}})$ extending q (we can have $\bar{c} \subseteq q(\mathfrak{C}^{\text{eq}})$ by the Definability lemma). Necessarily $r, p(x, \bar{a})$ are not orthogonal (as $q(M) \neq q(N)$, $M \subseteq N$ are $\mathbf{F}^a_{\aleph_0}$-saturated, $\bar{a}_l \subseteq M$ implies $p(N, \bar{a}) = p(M, \bar{a})$). As they are K-minimal necessarily for some finite B, $(\bar{a} \cup \bar{c} \subseteq B)\, p(\mathfrak{C}^{\text{eq}}, \bar{a}) \subseteq \text{acl}_K(B \cup r(\mathfrak{C}^{\text{eq}})$, hence $q(\mathfrak{C}^{\text{eq}}) \subseteq \text{acl}_K(B \cup r(\mathfrak{C}^{\text{eq}})$. Let $\bar{c}' \subseteq q(\mathfrak{C}^{\text{eq}})$ be such that $\text{tp}_*(B, q(\mathfrak{C}^{\text{eq}}))$ does not fork over $\bar{c}'$, $\bar{c} \subseteq \bar{c}'$. So $\text{tp}_*(B, \text{acl}(q(\mathfrak{C}^{\text{eq}}))$ is definable over $\text{acl}(\bar{c}')$, hence $q(\mathfrak{C}^{\text{eq}}) \subseteq \text{acl}_K(\bar{c}' \cup r(\mathfrak{C}^{\text{eq}}))$.

Remark. (1) We could, by choosing the p_l looking at $\varphi(x, \bar{a}_l)$, have that $\theta \in q_l$.

(2) The proof of 7.14(2) is similar, remembering 7.7(2).

THEOREM 7.14(3): *We can add to* 7.14(1) *and* 7.14(2).

(iii) $q_i(\mathfrak{C}^{\text{eq}}) \subseteq \text{dcl}\, \theta(\mathfrak{C})$.

(iv) *For any model* M, *and* $\bar{c}^1_i$, $\bar{c}^2_i \in M$ *realizing* $\text{tp}(\bar{c}_i, \text{acl}\emptyset)$ *letting* $\mathscr{P}_i = \text{cl}^3\{r_i\}$, *we have*: $\text{low}_{\mathscr{P}}(q_i(M), \text{acl}\emptyset) = \text{low}_{\mathscr{P}}(\bar{c}^1_i, \text{acl}\emptyset) + \text{low}\mathscr{P}(r_i(M, \bar{c}^1_i), \bar{c}^1_i \cup \text{acl}\emptyset)$.

(v) *For* $i \neq j$, q_j *is strongly orthogonal to* $\mathscr{P}_i$, *so* q_j *is* $\mathscr{P}_i$*-simple and for* $\bar{d} \in q_j(\mathfrak{C})$, $\text{low}_{\mathscr{P}_i}(\bar{d}) = 0$.

CHAPTER VI

SATURATION OF ULTRAPRODUCTS

VI. 0. Introduction

The ultraproduct construction is one of the most important methods in model theory, and unlike the primary model, the ultraproduct does not depend on the specific theories we deal with.

The ultraproduct $\prod_{i\in I} M_i/D$, for D an ultrafilter, M_i L-models, is the product when we identify any two elements of $\prod_{i\in I} M_i$ (i.e., functions from I) which are equal for "almost" every i. The important properties of the ultraproduct are:

(1) Łoš's Theorem: a (first-order) sentence is satisfied by the ultraproduct iff it is satisfied by almost every M_i.

(2) The L_1-reduct of the ultraproduct is the ultraproduct of the L_1-reducts.

It is also important to know:

(3) The ultraproduct is λ-compact if the ultrafilter is λ-good and $\aleph_1$-incomplete (thus we can construct λ-saturated models).

(4) Two L-models are elementarily equivalent iff they have isomorphic ultrapowers. So (4) gives an "algebraic" characterization of elementary equivalence. It is also important that by ultraproducts, we can get an "algebraic" proof of the compactness theorem.

The ultraproduct is not central in this book (and outside this chapter, we shall use it only once or twice), and the first three sections of this chapter have nothing to do with stability. However we have developed here a classification of theories using stability and associated notions, different from the one Keisler [K67] gives (using his ordering $\triangleleft$). We show that our classification is helpful in the analysis of Keisler's order, thus giving more evidence of the naturalness of the classification. We also deal with the categoricity of pseudo-elementary classes, and saturation of ultraproducts.

In Section 1 we give the basic properties of ultraproducts and reduced products and regular filters.

In Section 2 we deal mainly with λ-good filters. This property is defined combinatorically, but its main property is: an ultrafilter D is λ-good iff for every family of λ-saturated L-models M_i, $i \in I$, $\prod_{i\in I} M_i/D$ is λ-compact iff for every λ-saturated atomic Boolean algebra M, M^I/D is λ-saturated. If we require D to be $\aleph_1$-incomplete as well, we can omit the condition that the models are λ-saturated. The Boolean algebra comes quite naturally, but we can replace them by dense order. This is done as follows: suppose M^I/D is λ^+-compact for every dense order M. For any model N, we expand it by encoding the set of finite sequences of formulas (with parameters) which are consistent, and some natural relations and functions and get N_1. Note they form naturally a tree and by it we can "speak" on sequences of formula in N_1^I/D which are not necessarily "standard". We look in N_1^I/D, and any 1-type p in N^I/D, $|p| < \lambda$, and define by induction on $i \le |p|$ elements a_i of the tree, increasing with i, and "consistent" with p (in the inner sense we demand this for every finite subset of p separately and this is expressible): and letting $p_i = \{\varphi_i : i < |p|\}$, we want that φ_i will "appear" in the "sequence" a_i, the element "realizing" $a_{|p|}$ is as required. The translation of the tree to the dense (linear) order is by intervals.

Now from this we can deduce for singular λ that λ-goodness implies λ^+-goodness.

The importance of constructing ultrafilters is clear, and this is the subject of Section 3. Some of those problems are connected to problems on large cardinals, and consistency results and are outside our scope (e.g., $\aleph_1$-complete ultrafilters, non-regular ultrafilters).

It is easy to construct regular ultrafilters as they speak on the existence of a family of sets, and this is done in 1.3(4). Good ultrafilters posed a more difficult problem as their definition says: for every function into D there is a function into D.... Keisler [K 64], constructs a good ultrafilter on λ when $\lambda^+ = 2^\lambda$ in λ^+ steps; in each step we have a uniform filter generated by λ subsets of λ. When $2^\lambda > \lambda^+$, this seems to fail, but Kunen [Ku 72] suggests an alternative proof: he takes a family of 2^λ independent functions from λ onto λ; and construct the ultrafilter in 2^λ steps, after the ith step, we have to delete $\le \aleph_0 + |i|$ of them only. Now it is natural to ask where there is a (regular) μ-good ultrafilter over λ which is not μ^+-good. But $\mu \le \lambda^+$, and by 2.10 μ is regular. For μ a successor we use the product of ultrafilters, whose most important property is $M^{I\times J}/D_1 \times D_2 = (M^I/D_1)^J/D_1$, i.e., if we take ultrapower twice, it is like taking it once for the product. By this we get λ^+-good, not λ^{++}-good ultrafilters over $\mu \ge \lambda$.

We use independent functions to construct ultrafilters with various properties; e.g.,

(1) For $\mu \geq \lambda$, λ regular, over μ there is a regular ultrafilter which is λ-good but not λ^+-good.

(2) If $2^\mu \geq \lambda = \lambda^{\aleph_0}$, then there is a regular ultrafilter D over μ and $n_t < \omega$ $(t \in I)$ such that $\prod_{t \in I} n_t/D = \lambda$ (the conditions $\lambda = \lambda^{\aleph_0} \leq 2^\mu$ are necessary (for an infinite λ), see Exercise 2.10).

(3) If $2^\mu \geq \lambda$, $\mu \geq \kappa$; $\lambda > \kappa$ regular, then there is a regular ultrafilter D over μ such that $\lambda = \mathrm{lcf}(\kappa, D) =_{\mathrm{def}} \min\{\chi$: in $\{a \in \kappa^\mu/D: \kappa^\mu/D \models \alpha < a$ for every $\alpha < \kappa\}$ there is a set unbounded from below of cardinality $\chi\}$.

We prove also that T_{ind} is essentially "simpler" than the theory of linear order.

In the exercises, we indicate why any two elementarily equivalent models have isomorphic ultraproducts, and other results.

In Section 4 we define Keisler's order and prove some theorems on it. We could change the definition slightly, e.g., by omitting the requirement that D is an ultrafilter over λ, and still get the same results. We also show that some ultraproducts are not λ-compact.

Section 5 is the heavy section, utilizing the results from II on Δ-n-indiscernible sets and their dimensions, to determine how saturated are ultraproducts. We find which elementary classes (of countable type) contain categorical pseudo-elementary classes. E.g., we prove

THEOREM 0.1: *If M is a model of T, T countable and without the f.c.p., D an $\aleph_1$-incomplete ultrafilter over I, then M^I/D is $\aleph_0^I/D$-saturated.*

THEOREM 0.2: *If T is countable, superstable and without the f.c.p., then there is a theory T_1, $T \subseteq T_1$, $|T_1| = 2^{\aleph_0}$, such that any uncountable model of T_1 has a saturated* $\mathrm{L}(T)$*-reduct.*

Other theorems (5.1, VIII, 2.1) show this result is the best possible.

Summing up our results we get the following picture for Keisler's order on countable theories.

Let

$$K_{\min} = \{T\colon T \text{ countable, without the f.c.p.}\},$$

$$K_{\mathrm{scp}} = \{T\colon T \notin K_{\min} \text{ is countable and stable with the f.c.p.}\},$$

$$K_{\mathrm{ind}} = \{T\colon T \text{ countable, unstable } \kappa_{\mathrm{cdt}}(T) < \infty\},$$

$$K_{\mathrm{cdt}} = \{T\colon T \text{ countable, } \kappa_{\mathrm{cdt}}(T) = \infty, \ T \text{ without the strict order property}\},$$

$$K_{\max} = \{T\colon T \text{ countable and with the strict order property}\}.$$

THEOREM 0.3: (1) *If T_1, T_2 are both $K_{\min}[K_{\mathrm{scp}}][K_{\max}]$ then they are $\otimes$-equivalent, i.e., $T_1 \otimes T_2 \otimes T_1$.*

(2) *If* $T_1 \in K_{\min}$, $T_2 \notin K_{\min}$ *then not* $T_2 \otimes T_1$. *If* $T_1 \in K_{\mathrm{scp}}$, T_2 *unstable then not* $T_2 \otimes T_1$. *If* $T_2 \in K_{\mathrm{ctd}}$ *or* $T_2 \in K_{\max}$, *it is consistent with ZFC that not* $T_2 \otimes T_{\mathrm{ind}}$ (*we need* $MA + 2^{\aleph_0} > \aleph_1$).

(3) *If* $T_1 \in K_{\min}$, $T_2 \in K_{\mathrm{scp}}$, $T_3 \in K_{\mathrm{ind}} \cup K_{\mathrm{cdt}}$, $T_4 \in K_{\max}$, *then* $T_1 \otimes T_2 \otimes T_3 \otimes T_4$.

In Section 6 we find how compact are ultralimits.

PROBLEM 0.1: It would be very desirable to prove that (1A) $T_1, T_2 \in K_{\mathrm{ind}}$ implies T_1, T_2 are $\otimes$-equivalent, (1B) $T_1, T_2 \in K_{\mathrm{cdt}}$ (or we should ask also whether $\kappa_{\mathrm{inp}}(T) = \infty$, $\kappa_{\mathrm{sct}}(T) = \infty$) implies T_1, T_2 are $\otimes$-equivalent. This will complete the model-theoretic share of investigating Keisler's order for countable theories. For this it seems reasonable to try to find for $T \in K_{\mathrm{ind}}$ a theory parallel to II, Section 2 for stable theories.

PROBLEM 0.2: It would be desirable to replace in 0.3(2) the "consistent with" by "provable form". This is a problem in constructing ultrafilters. Another problem on ultrafilters is 5.1.

PROBLEM 0.3: On Keisler's order for uncountable T see Shelah [Sh 72]. We conjecture that for every T there are a countable T_1 and simple T_2 (Definition 6.4) such that $T' \otimes T \Leftrightarrow [T' \otimes T_1$ and $T' \otimes T_2]$.

PROBLEM 0.4: We conjecture: if T is not $\otimes$-minimal then T is not $\otimes_{|T|}$-minimal, and if M^I/D is $(2^{|I|})^+$-compact, M is λ-compact, then M^I/D is λ-compact (see 5.7).

In this chapter D will always denote a filter (or ultrafilter) over I (or sometimes J) (For the definition and some theorems see Section 1 of the Appendix.)

VI.1. Reduced products and regular filters

DEFINITION 1.1: Let D be a filter over I and for each $i \in I$ let M_i be an L-model.

(1) We define an equivalence relation $\approx_D$ (or $\approx$ for short) over $\prod_{i\in I} |M_i|$.

$$f \approx g \quad \text{iff} \quad \{i \in I : f(i) = g(i)\} \in D.$$

(2) We define the L-model $\prod_{i\in I} M_i/D$ (the reduced product of the M_i's):

(A) Its universe is $\{f/\approx : f \in \prod_{i\in I} |M_i|\}$.

(B) If R is an n-place predicate symbol in L then

$$R^M = \{\langle f_1/\approx, \ldots, f_n/\approx\rangle : \{i \in I : \langle f_1(i), \ldots, f_n(i)\rangle \in R^{M_i}\} \in D\}.$$

(C) If F is an n-place function symbol in L then $F^M(f_1, \ldots, f_n) = f$ where for every $i \in I, f(i) = F^{M_i}(f_1(i), \ldots, f_n(i))$ and $F^M(f_1/\approx, \ldots, f_n/\approx) = F^M(f_1, \ldots, f_n)/\approx$.

(3) If D is an ultrafilter, the reduced product is called an ultraproduct. If $M_i = N$ for every $i \in I$, then we write N^I/D for $\prod_{i\in I} M_i/D$ and call it a power instead of a product.

Remark. As D is a filter, $\approx$ really is an equivalence relation and $f_l \approx g_l$ for $1 \leq l \leq n$ implies $\{i: \langle f_1(i), \ldots, f_n(i)\rangle \in R^{M_i}\} \in D \Leftrightarrow \{i: \langle g_1(i), \ldots, g_n(i)\rangle \in R^{M_i}\} \in D$ and $F^M(f_1, \ldots, f_n) \approx F^M(g_1, \ldots, g_n)$ so $\prod_{i\in I} M_i/D$ is a well-defined L-model. Notice that for a term τ, $\tau(f_1, \ldots, f_n)(i) = \tau(f_1(i), \ldots, f_n(i))$.

Notation. (1) If M_i is an L-model for $i \in I$, $P \notin$ L, $N = \prod_{i\in I} M_i/D$ and (M_i, P_i) is an L $\cup \{P\}$-model for every $i \in I$, then in an abuse of notation $(N, P^N) = \prod_{i\in I} (M_i, P_i)/D$.

(2) We do not strictly distinguish between $a \in |\prod_{i\in I} M_i/D|$ and a representative of a in $\prod_{i\in I} |M_i|$. We write $\langle a[i]: i \in I\rangle$ for the representative of a. Also (in this chapter) if $\bar{a} = \langle a_0, \ldots, a_{n-1}\rangle \in |\prod_{i\in I} M_i/D|$, then $\bar{a}[i] = \langle a_0[i], \ldots, a_n[i]\rangle$.

(3) We identify a and $\langle a: i \in I\rangle$ and so regard M^I/D as an extension of M. The embedding $a \mapsto \langle a: i \in I\rangle$ is called natural.

Remark. Note that if $L_2 \subseteq L_1$, M_i^1 is an L_1-model and M_i^2 is the L_2-reduct of M_i^1 (for $i \in I$), then $\prod_{i\in I} M_i^2/D$ is the L_2-reduct of $\prod_{i\in I} M_i^1/D$.

THEOREM 1.1 (Łoš's Theorem): *Let D be an ultrafilter over I.*

(1) *M^I/D is elementarily equivalent to M.*

(2) *If M_i is an L-model (for each $i \in I$) and $\varphi(\bar{x}) \in$ L, $\bar{a} \in |\prod_{i\in I} M_i/D|$ then*

$$\prod_{i\in I} M_i/D \vDash \varphi[\bar{a}] \quad \textit{iff} \quad \{i \in I: M_i \vDash \varphi[\bar{a}[i]]\} \in D.$$

(3) *The natural embedding of M into M^I/D (defined by $a \mapsto \langle a: i \in I\rangle$) is elementary, so we can say $M \prec M^I/D$.*

Proof. (1) Follows by (2), for sentences.

(2) Easy by induction on formulas (only for negation does the proof fail for filters).

(3) Follows by (2).

DEFINITION 1.2: (1) A basic Horn formula is a formula of the form $\bigwedge_{i<n} \varphi_i \to \varphi$ where φ_i, φ are atomic, $n < \omega$, or $\neg\varphi$.

(2) The family of Horn formulas is the closure of the family of basic Horn formulas under conjunction, existential quantification and universal quantification (but not disjunction or negation).

Remark. In Definition 2.1(1) we can take $n = 0$, so every atomic formula is a basic Horn formula. Also, every negation of an atomic formula is a basic Horn formula.

THEOREM 1.2: *If* $\varphi(\bar{x})$ *is a Horn formula,* $M = \prod_{i \in I} M_i/D$, $\bar{a} \in |M|$ *and*

$$\{i \in I: M_i \models \varphi[\bar{a}[i]]\} \in D,$$

then $M \models \varphi[\bar{a}]$.

Proof. By induction on φ.

DEFINITION 1.3: (1) The family $\{X_i: i < \lambda\}$ of subsets of I is regular if for $w \subseteq \lambda$,

$$\bigcap_{i \in w} X_i \neq \emptyset \Leftrightarrow |w| < \aleph_0.$$

(2) The above-mentioned family regularizes D (which is a filter over I) if it is regular and $X_i \in D$ for $i < \lambda$.

(3) The filter D over I is λ-regular if some $\{X_i: i < \lambda\}$ regularizes it.

(4) D is regular if it is $|I|$-regular.

(5) D is λ-incomplete if there are $X_i \in D(\ i < \alpha < \lambda)$, $\bigcap_{i<\alpha} X_i = \emptyset$.

(6) D is λ-complete if for $X_i \in D$ $(i < \alpha < \lambda)$, $\bigcap_{i<\alpha} X_i \in D$.

LEMMA 1.3: (1) *If* D *is* λ*-regular,* $\mu \leq \lambda$, *then* D *is* μ*-regular.*

(2) *The filter* D *is* $\aleph_1$*-incomplete iff* D *is* $\aleph_0$*-regular.*

(3) D *is not* $|I|^+$*-regular.*

(4) *Over every infinite* I *there is a* $|I|$*-regular filter, hence a regular ultrafilter.*

Proof. (1) Immediate.

(2) Immediate.

(3) If $\{X_i: i < |I|^+\} \subseteq D$ is regular, each X_i, being a member of D, is non-empty, so choose $j_i \in X_i$. Clearly for some j^*, $|\{i < |I|^+ : j_i = j^*\}| = |I|^+$ but $\bigcap \{X_i: j_i = j^*\} \supseteq \{j^*\} \neq \emptyset$, contradiction.

(4) Clearly it suffices to find a set J, $|J| = |I|$ over which there is a regular filter. So let: $J = S_{\aleph_0}(I) = \{w: w \subseteq I, |w| < \aleph_0\}$, for $w \in J$,

$X_w = \{v: v \in J, w \subseteq v\} \subseteq J$. The filter D generated by $\{X_w: w \in J\}$ (see Definition 1.1 and Theorem 1.1(1), (2) of the Appendix) is regularized by $\{X_{\{i\}}: i \in I\}$. The second phrase follows by 1.1(4) of the Appendix.

Proof of the Compactness Theorem (I, 1.1): From 1.1 and 1.3 we get a proof of the Compactness Theorem. Let T be a set of sentences, such that every finite $t \subseteq T$ has a model M_t. Let $J = S_{\aleph_0}(T)$ and let D be the filter over J defined in the proof of 1.3(4), and D^* an ultrafilter over J, extending D (exists by 1.1(4) of the Appendix). Let $M = \prod_{w \in J} M_w/D^*$, then for every $\psi \in T$, $\{w \in J: M_w \models \psi\} \supseteq \{w \in J: \psi \in w\} = X_{\{\psi\}} \in D \subseteq D^*$. Hence, by Łoš's Theorem, $M \models \psi$, so M is a model of T.

DEFINITION 1.4: A type is atomic if all its formulas are atomic.

THEOREM 1.4: (1) *Let D be a filter over I. Then the following conditions are equivalent.*

(A) *D is λ-regular.*

(B) *For every* L *and every family M_i ($i \in I$) of* L*-models, every atomic type p over $\emptyset$ in each M_i, $|p| \leq \lambda$, is realized in $\prod_{i \in I} M_i/D$.*

(C) *For every M every atomic type p over $|M|$ in M^I/D, $|p| \leq \lambda$, is realized in M^I/D (notice: p is in M iff it is in M^I/D).*

(D) *Condition* (C) *holds for $M = (S_{\aleph_0}(\lambda), \subseteq)$.*

(2) *If in* (1) *D is an ultrafilter we can replace "atomic type" by "type" in the conditions.*

Proof. (1) (A) $\Rightarrow$ (B) Let $N = \prod_{i \in I} M_i/D$, $\{X_\alpha: \alpha < \lambda\}$ regularize D, $p = \{\varphi_\alpha(\bar{x}): \alpha < \lambda\}$ be the type. For any finite $w \subseteq \lambda$ let $\bar{a}_w[i] \in |M_i|$ realize $\{\varphi_\alpha(\bar{x}): \alpha \in w\}$. Now as $\{X_\alpha: \alpha < \lambda\}$ is regular, for every $i \in I$, $w(i) = \{\alpha < \lambda: i \in X_\alpha\}$ is finite. Define $\bar{a} \in |N|$ by $\bar{a}[i] = \bar{a}_{w(i)}[i]$. Hence for $\alpha < \lambda$ $\{i \in I: M_i \models \varphi_\alpha[\bar{a}[i]]\} \supseteq \{i \in I: \alpha \in w(i)\} = \{i \in I: i \in X_\alpha\} = X_\alpha \in D$. Hence $N \models \varphi_\alpha[\bar{a}]$, so $\bar{a}$ realizes p.

(B) $\Rightarrow$ (C) Use (B) for $M_i = (M, a)_{a \in |M|}$, for every $i \in I$. (Clearly p is a type in M.)

(C) $\Rightarrow$ (D) Holds by their definition.

(D) $\Rightarrow$ (A) Let $N = M^I/D$ and let $p = \{\{\alpha\} \subseteq x: \alpha < \lambda\}$, so clearly p is finitely satisfiable in M, (hence in N). Suppose $a \in |N|$ realizes p, and let for $\alpha < \lambda$

$$X_\alpha = \{i \in I: M \models \{\alpha\} \subseteq a[i]\}.$$

Then $X_\alpha \in D$ as $N \models \{\alpha\} \subseteq a$, and for $w \subseteq \lambda$, $\bigcap_{\alpha \in w} X_\alpha = \{i \in I : w \subseteq a[i]\}$, so clearly it is empty iff w is infinite.

(2) The same proof using Łoš's Theorem.

DEFINITION 1.5: M is (λ, Δ, m)-compact if every (Δ, m)-type p over $|M|$, in M, $|p| < \lambda$, is realized in M. For Δ the set of atomic formulas of $L(M)$, we write at. Similarly the sets of quantifier free formulas, conjunctions of atomic formulas, and formulas of quantifier depth $\leq n$ are denoted by qf, cnat, and qd_n respectively; if $m = 1$ we omit it.

DEFINITION 1.6: (1) For models M_0, M_1 and an ordinal α, ($L(M_0) \cap L(M_1) \supseteq L$) we define a game $G^\alpha_L(M_0, M_1)$ between the players I and II as follows:

At the βth move player I chooses $l \in \{0, 1\}$ and $a^l_\beta \in |M_l|$, and then player II chooses $a^{1-l}_\beta \in |M_{1-l}|$.

The play ends after α moves and then player II wins if for every atomic formula $\varphi(x_0, \ldots, x_n) \in L$ and ordinals $\beta(0), \ldots, \beta(n) < \alpha$, $M_0 \models \varphi[a^0_{\beta(0)}, \ldots, a^0_{\beta(n)}] \Leftrightarrow M_1 \models \varphi[a^1_{\beta(0)}, \ldots, a^1_{\beta(n)}]$, and player I wins otherwise.

(2) A strategy (of a player) is a sequence of functions f_β ($\beta < \alpha$) which "tells" him what to do (f_β for the βth move) depending only on the previous choices in the play. A *winning strategy* is a strategy such that in any play in which the player chooses according to it, he wins. A player *wins in the game* if he has a winning strategy.

(3) We omit L if $L = L(M_0) = L(M_1)$.

LEMMA 1.5: (1) *In the game $G^\alpha_L(M_0, M_1)$ not both players win.*

(2) *The relation "player* II *wins in $G^\alpha_L(M, N)$" is an equivalence relation among models.*

(3) *If player* II [I] *wins in $G^\alpha_L(M_0, M_1)$*, $L(1) \subseteq L$, $\beta \leqq \alpha$ [$\alpha \leq \beta$, $L \subseteq L(1)$] *then he wins in $G^\beta_{L(1)}(M_0, M_1)$.*

(4) *If M_0, M_1 are isomorphic, then player* II *wins in $G^\alpha(M_0, M_1)$.*

(5) *If M_0, M_1 are elementarily equivalent, and λ-saturated, then player* II *wins in $G^\lambda(M_0, M_1)$.*

Proof. (1) Suppose both players win, then in some play, both use their winning strategy, but one of them must lose this play, contradiction.

(2) Combine the strategies to prove transitivity. Symmetry is trivial, and reflexivity follows by (4).

(3) Trivial.

(4) If $g: |M_0| \to |M_1|$ is the isomorphism then player II chooses a_β^{1-l} so that $g(a_\beta^0) = a_\beta^1$.

(5) Player II has to choose a_β^{1-l} such that $\mathrm{tp}_*(\{a_\alpha^0: \alpha \leq \beta\}, \emptyset, M_0) = \mathrm{tp}_*(\{a_\alpha^1: \alpha \leq \beta\}, \emptyset, M_1)$.

THEOREM 1.6: *Suppose player II wins in* $G_L^\alpha(M_0, M_1)$, $L = L(M_0) = L(M_1)$.

(1) *If* $\alpha \geq \lambda = \|M_0\| + \|M_1\|$ *then* M_0, M_1 *are isomorphic.*

(2) *If* $\alpha \geq \lambda + 1$, $\lambda = \|M_0\|$ *then* M_0, M_1 *are isomorphic.*

(3) *If* $\alpha \geq \lambda + 1$, *then* M_0 *is* λ^+*-saturated iff* M_1 *is* λ^+*-saturated.*

(4) *If* $\alpha \geq \lambda + 1$ *then* M_0 *is* (λ, Δ, m)*-compact iff* M_1 *is* (λ, Δ, m)*-compact.*

Proof. Immediate.

THEOREM 1.7: *The* L^**-models* M_0, M_1 *are elementarily equivalent iff for every finite* $L \subseteq L^*$, $n < \omega$, *player* II *wins in* $G_L^n(M_0, M_1)$.

Proof. Let us define by induction on n, the sets $\Gamma_{n,m}^a(L)$, $\Gamma_{n,m}^c(L)$ of formulas $\varphi(x_0, \ldots, x_{m-1}) \in L$.

(i) $\Gamma_{0,m}^a(L)$ is the set of atomic formulas $\varphi(x_0, \ldots, x_{m-1}) \in L$.

(ii) If $\Gamma_{n,m}^a(L)$ is defined, $\Gamma_{n,m}^c(L)$ is the set of formulas of the form $\bigwedge \{\varphi(\bar{x})^{\text{if } (\varphi \in \Gamma)}: \varphi(\bar{x}) \in \Gamma_{n,m}^a(L)\}$ for $\Gamma \subseteq \Gamma_{n,m}^a(L)$.

(iii) If $\Gamma_{n,m+1}^c(L)$ is defined

$$\Gamma_{n+1,m}^a(L) = \{(\exists x_m)\varphi(x_0, \ldots, x_m): \varphi(x_0, \ldots, x_m) \in \Gamma_{n,m+1}^c(L)\}.$$

Clearly any sentence in L is equivalent to a disjunction of some of the sentences in $\Gamma_{n,0}^c$ for every sufficiently large n (in fact for its quantifier depth), and this holds for any formula.

If M_0, M_1 are elementarily equivalent, $L \subseteq L^*$ finite, $n < \omega$, a winning strategy for player II in $G_L^n(M_0, M_1)$ is to preserve the satisfaction of

(∗) $M_0 \models \varphi[a_0^0, \ldots, a_{k-1}^0]$ iff $M_1 \models \varphi[a_0^1, \ldots, a_{k-1}^1]$ for every $\varphi(x_0, \ldots, x_{k-1}) \in \Gamma_{n-k,k}^c(L)$.

For $k = 0$ this holds as M_0, M_1 are elementarily equivalent; and by the definition of $\Gamma_{n-k,k}^c(L)$ it is easy to play so that (∗) is preserved.

For the other direction, it suffices to prove that if $a_0^0, \ldots, a_{k-1}^0$, $a_0^1, \ldots, a_{k-1}^1$ have been chosen in a play in which player II uses a winning strategy for $G_L^n(M_0, M_1)$ then (∗) holds (because then for any

$\psi \in \bigcup \Gamma^c_{n,0}(L)$, $L \subseteq L^*$, L finite, $M_0 \models \psi \Leftrightarrow M_1 \models \psi$, and as any sentence ψ of L^* is a sentence of some finite sublanguage, M_0, M_1 are elementarily equivalent). The proof is easy, by induction on $n - k$.

THEOREM 1.8: *If M^0_t, M^1_t are elementarily equivalent* L*-models for $t \in I$, $|L| \le \lambda$, D a λ-regular filter over I, $\alpha < \lambda^+$, then player* II *wins in $G^\alpha(M^0, M^1)$, where $M^0 = \prod_{t\in I} M^0_t/D$, $M^1 = \prod_{t\in I} M^1_t/D$. (So M^0, M^1 are elementarily equivalent.)*

Proof. The number of formulas $\varphi(x_{\beta(0)}, \ldots, x_{\beta(n)})$, $\varphi \in L$, $\beta(l) < \alpha$, is $\le \lambda$, so let $\{\varphi_j(x_{\beta(0,j)}, \ldots, x_{\beta(n(j),j)}): j < \lambda\}$ be an enumeration of the set of these formulas (possibly with repetitions). Let $\{X_j: j < \lambda\} \subseteq D$ regularize D, and for $t \in I$, let $w(t) = \{j: t \in X_j\}$, (which is finite), let

$$\{x_{\gamma(0,t)}, \ldots, x_{\gamma(n(t),t)}\} = \bigcup_{j\in w(t)} \{x_{\beta(0,j)}, \ldots\} \quad \text{where } \gamma(0, t) < \cdots < \gamma(n(t), t).$$

Let L_t be the minimal sublanguage of L such that $\varphi_j \in L_t$ for $j \in w(t)$. So by the previous theorem, player II wins in $G^{n(t)+1}_{L_t}(M^0_t, M^1_t)$ for every t. Now we shall describe the winning strategy of player II in $G^\alpha_L(M^0, M^1)$: when he has to choose $a^{1-l}_\beta \in |M^{1-l}|$ he chooses each $a^{1-l}_\beta[t]$ separately. If $\beta \notin \{\gamma(k, t): k \le n(t)\}$, he chooses $a^{1-l}_\beta[t]$ arbitrarily. If $\beta = \gamma(k, t)$ he imagines he is playing $G^{n(t)+1}_{L_t}(M^0_t, M^1_t)$, that $a^0_{\gamma(0,t)}[t]$, $a^1_{\gamma(0,t)}[t], \ldots,$ $a^1_{\gamma(k-1,t)}[t]$, $a^0_{\gamma(k-1,t)}[t]$ and $a^l_{\gamma(k,t)}[t]$ have been chosen, and he chooses $a^{1-l}_{\gamma(k,t)}[t]$ by his winning strategy in $G^{n(t)+1}_{L_t}(M^0_t, M^1_t)$. It is easy to check that this is a winning strategy for $G^\alpha_L(M^0, M^1)$.

CONCLUSION 1.9: *If M, N are elementarily equivalent, D is a λ-regular filter over I then M^I/D is (λ, Δ, m)-compact iff N^I/D is (λ, Δ, m)-compact.*

LEMMA 1.10: *If for each $t \in I$ player* II *wins in $G^\alpha(M^0_t, M^1_t)$, D a filter over* I, *then player* II *wins in $G^\alpha(\prod_{t\in I} M^0_t/D, \prod_{t\in I} M^1_t/D)$.*

Proof. In view of the proofs of 1.8, 1.9, we leave it to the reader.

CONCLUSION 1.11: *If M_t, N_t are elementarily equivalent and λ-saturated, D a filter, then $M = \prod_{t\in I} M_t/D$ is (λ, Δ, m)-compact iff $N = \prod_{t\in I} N_t/D$ is (λ, Δ, m)-compact.*

Proof. By 1.5(5) player II wins in $G^\lambda(M_t, N_t)$. By 1.10 player II wins in $G^\lambda(M, N)$. Hence, if $\lambda > \aleph_0$, by 1.6(4) M is (λ, Δ, m)-compact iff N is (λ, Δ, m)-compact. If $\lambda = \aleph_0$, every model is (λ, Δ, m)-compact; so we finish.

EXERCISE 1.1: Show that if M_i^0, $i \in I$, are L_0-models and M_i^1, $i \in I$, are L_1-models such that $\|M_i^0\| = \|M_i^1\|$ for all $i \in I$ and D is a filter over I than $\|\prod_{t\in I} M_i^0/D\| = \|\prod_{t\in I} M_i^1/D\|$.

Notation. If A_i, $i \in I$ is a family of non-empty sets and D is a filter over I we define $\prod_{t\in I} A_i/D = \{f/\approx : f \in \prod_{t\in I} A_i\}$ where $\approx$ is the equivalence relation from Definition 1.1. If A_t is a cardinal λ_t for every $i \in I$, we write $\prod_{t\in I} \lambda_t/D$ for $|\prod_{t\in I} \lambda_t/D|$ and $\prod_{t\in I} \lambda_t$ for $|\prod_{t\in I} \lambda_t|$. So $\|\prod_{t\in I} M_t/D\| = \prod_{t\in I} \|M_t\|/D$.

If D is a filter over I, $J \subseteq I$ then $D \restriction J = \{X \cap J : X \in D\}$ so when $I - J \notin D$, $D \restriction J$ is not trivial.

EXERCISE 1.2: (1) If D is a filter over I, $J \in D$ then $\prod_{t\in I} M_t/D \cong \prod_{t\in J} M_t/D \restriction J$ (the isomorphism maps $\langle a[t] : t \in I\rangle/D$ to $\langle a[t] : t \in J\rangle/D$); so $\prod_{t\in I} \lambda_t/D = \prod_{t\in J} \lambda_t/D \restriction J$.

(2) If $J \subseteq I$, $I - J \notin D$, D a μ-regular filter over I then $D \restriction J$ is μ-regular.

Notation. If $\{t\colon \lambda_t = 0\} \neq \emptyset \bmod D$, let $\prod_{t\in I} \lambda_t/D = 0$, and if $J = \{t\colon \lambda_t \neq 0\} \in D$ let $\prod_{t\in I} \lambda_t/D$ be $\prod_{t\in J} \lambda_t/D \restriction J$.

EXERCISE 1.3: (1) Show that if $(M, P) = \prod_{t\in I} (M_t, P_t)/D$ then $|P| = \prod_{t\in I} |P_t|/D$.

(2) Show that $\prod_{t\in I} \lambda_t/D \leqslant \prod_{t\in I} \lambda_t$; and $\{t\colon \lambda_t \leqslant \mu_t\} \in D$ implies $\prod_{t\in I} \lambda_t/D \leq \prod_{t\in I} \mu_t/D$.

(3) Show that if $\prod_{t\in I} \lambda_t/D > 0$ then $\prod_{t\in I} \lambda_t/D \geqslant \prod_{t\in I} \lambda_t/D_1 \restriction J$, for any filter $D_1 \supseteq D$, $J \subseteq I$.

THEOREM 1.12: (1) *If D is a μ-regular filter over I, λ_t $(t \in I)$ (infinite) cardinals, and $\lambda = \prod_{t\in I} \lambda_t/D > 0$, $\chi = \min\{\chi\colon \{t \in I\colon \lambda_t \leq \chi\} \in D\}$ then $\lambda \geq \chi^\mu$.*

(2) *If in* (1) $\mu = |I|$ *then* $\lambda = \chi^\mu$.

(3) *Moreover there are natural numbers n_t such that $\prod_{t\in I} n_t/D \geq 2^\mu$, and if D is regular $\prod_{t\in I} n_t/D = 2^{|I|}$.*

Proof. (1) By Exercise 1.3(3) we can assume D is an ultrafilter (by extending the original filter to an ultrafilter containing $\{t\colon \lambda_t > \chi_1\}$ for each $\chi_1 < \chi$ (by 1.1(2), (3) of the Appendix).

Case I: $\chi > \mu$. By Exercise 1.2(1) we can assume that for every t, $\mu \leq \lambda_t \leq \chi$. Now we want to define models M_t $(t \in I)$ so that $\|M_t\| = \lambda_t$

and χ^μ pairwise contradictory types will be realized in $\prod_{t\in I} M_t/D$. Let L consist of the one-place predicates P^i_α ($i < \chi$, $\alpha < \mu$). We define M_t such that for each α, for $i < \lambda_t$ the $P^i_\alpha(M_t)$ are pairwise disjoint, for $\lambda_t \le i < \chi$ $P^i_\alpha(M_t) = |M_t|$ and for every finite $w \subseteq \mu$ and function $f: w \to \chi M_t \models (\exists x)[\bigwedge_{\alpha\in w} P^{f(\alpha)}_\alpha(x)]$. It is quite easy to construct such a model:

$$|M_t| = \{c^w_f: w \subseteq \mu, |w| < \aleph_0, f: w \to \lambda_t \text{ a function}\},$$

$$P^i_\alpha(M_t) = \{c^w_f \in |M_t|: \alpha \in w, f(\alpha) = i \text{ or } i \ge \lambda_t\}.$$

By the definition of χ, for every $i < j < \chi$, $\alpha < \mu$

$$\{t \in I: M_t \models \neg(\exists x)(P^i_\alpha(x) \wedge P^j_\alpha(x))\} \supseteq \{t \in I: |i|, |j| < \lambda_t\} \in D,$$

hence $M = \prod_{t\in I} M_t/D \models \neg(\exists x)[P^i_\alpha(x) \wedge P^j_\alpha(x)]$. Let for each $\eta \in {}^\mu\chi$, $p_\eta = \{P^{\eta[\alpha]}_\alpha(x): \alpha < \mu\}$, so the p_η's are pairwise contradictory; so it suffices to prove they are realized in M. But clearly each p_η is finitely satisfiable in each M_t, so by 1.4 it is realized in M.

Case II: $\chi \le \mu$. By Exercise 1.3(2) it follows from 1.12(3).

(2) $\lambda \ge \chi^\mu$ by (1) and $\lambda \le \chi^\mu$ as we can assume (again by Exercise 1.2(1)) $\lambda_t \le \chi$ for every t, so by Exercise 1.3(2), $\prod_{t\in I} \lambda_t/D \le \prod_{t\in I} \lambda_t \le \chi^{|I|} = \chi^\mu$.

(3) We use a method similar to the one used in (1). Let L consist of the one place predicates P^l_α ($l = 0, 1, \alpha < \mu$), $\{X_j: j < \mu\}$ regularize D, and $w(t) = \{j: t \in X_j\}$ which is finite. We define M_t such that $\|M_t\| = 2^{|w(t)|}$, and denote this number by n_t. We define it such that for $j \in w(t)$, P^0_j, P^1_j are complementary and for any finite $w \subseteq \mu$

$$M_t \models (\exists x)\Bigl(\bigwedge_{\alpha\in w} P^0_\alpha(x) \wedge \bigwedge_{\alpha\in w(t)-w} P^1_\alpha(x)\Bigr).$$

For this let

$$|M_t| = \{c_w: w \subseteq w(t)\},$$

$P^l_\alpha(M_t) = \{c_w: l = 0 \text{ and } \alpha \in w \text{ or } l = 1 \text{ and } \alpha \in w(t) - w \text{ or } \alpha \notin w(t)\}$. Let $M = \prod_{t\in I} M_t/D$, then $P^0_\alpha(M)$, $P^1_\alpha(M)$ are a partition of $|M|$, and for every $w \subseteq \mu$, $p_w = \{P^l_\alpha(x): \alpha < \mu \text{ and } \alpha \in w \Leftrightarrow l = 0\}$ is realized in M, so $\|M\| \ge 2^\mu$, but $\|M\| = \prod_{t\in I} \|M_t\|/D = \prod_{t\in I} n_t/D$. When D is regular we get equality as in (2).

CONCLUSION 1.13: *If D is a regular filter over I, then $\lambda^I/D = \lambda^{|I|}$.*

EXERCISE 1.4: Suppose the filter D is μ-regular for every $\mu < \lambda$, and λ is singular. Prove that D is λ-regular. [Hint: If $\{X_i^\mu : i < \mu\}$ regularizes D for each $\mu < \lambda$, $\kappa = \text{cf}\,\lambda < \lambda$,

$$\lambda = \sum_{\alpha<\kappa} \mu(\alpha),\ \mu(\alpha) < \lambda, \quad \text{then} \quad \{X_i^{\mu(\alpha)} \cap X_\alpha^\kappa : \alpha < \kappa, i < \mu(\alpha)\}$$

proves our assertion.]

EXERCISE 1.5: Suppose λ is a regular cardinal, prove there is a filter D which is μ-regular for each $\mu < \lambda$, but is not λ-regular. [Hint: D will be generated by $\{X_i^\mu : i < \mu < \lambda\}$, for each μ, $\{X_i^\mu : i < \mu\}$ is regular, but for every $\alpha \le \omega$, and distinct $\mu_n < \lambda$ $(n < \alpha)$, and finite sets $w_n \subseteq \{i : i < \mu_n\}$ $\bigcap_{n\le\alpha} \bigcap_{i\in w_n} X_i^{\mu_n} \ne \emptyset$. Use 1.4 of the Appendix to prove this is possible.]

QUESTION 1.6: In Exercise 1.5, if $\kappa \ge \lambda^{\aleph_0}$ clearly we can find such a D over κ. Can we always find such a D over λ? (For λ a successor it is trivial.)

EXERCISE 1.7: Prove that in 1.4(1) (and (2)) to the four equivalent conditions we can add:

(5) If Γ is a set of $\le\lambda$ atomic formulas (in $\le\lambda$ variables) which is finitely satisfiable in each M_t, then Γ is satisfiable in $\prod_{t\in I} M_t/D$.

(6) In (5) instead of atomic formulas we can allow Horn formulas.

EXERCISE 1.8: Prove that for any ultrafilter D and cardinal λ, D is λ-incomplete iff D is not λ-complete.

EXERCISE 1.9: If $M = (A, R_1, \ldots)$, $M^I/D = N = (A', R_1', \ldots)$ and $A \subseteq |M_1|$, then $N_1' = (M_1, A, R_1, \ldots)^I/D = (M_1^I/D, A', R_1', \ldots)$. (It is convenient to use this for $M_1 = (H(\kappa), \in, M)$, where $H(\kappa)$ is the family of sets hereditarily of power $<\kappa$, $M \in H(\kappa)$).

VI.2. Good filters and compactness of reduced products

DEFINITION 2.1: (1) A filter D (over I) is λ-*good* if for every $\mu < \lambda$, every $f : S_{\aleph_0}(\mu) \to D$ which is monotonic [i.e., $w \subseteq u \Rightarrow f(u) \subseteq f(w)$] has a refinement $g : S_{\aleph_0}(\mu) \to D$ [i.e., $g(w) \subseteq f(w)$] which is multiplicative [i.e., $g(w \cup u) = g(w) \cap g(u)$].

(2) D is good if it is $|I|^+$-good.

Remark. (1) Clearly we can add "g is monotonic", because multiplicity implies monotonicity.

(2) We can delete the assumption "f is monotonic" because for any $f: S_{\aleph_0}(\mu) \to D$, let f^* be defined by $f^*(w) = \bigcap_{u \subseteq w} f(u)$. As $|\{u: u \subseteq w\}| = 2^{|w|} < \aleph_0$, and D, being a filter, is closed under finite intersections, $f^*(w) \in D$, and any multiplicative refinement of f^*, is a multiplicative refinement of f; and f^* is monotonic.

CLAIM 2.1: *Every filter is* $\aleph_1$-good.

Proof. Let $f: S_{\aleph_0}(\omega) \to D$. Define $g(w) = \bigcap \{f(u): \max u \le \max w\}$. As for each $w \in S_{\aleph_0}(\omega)$ $\max w < \omega$, and the number of $u \subseteq \{n: n \le \max w\}$ is finite, clearly $g(w) \in D$. The function g is multiplicative as $\max(v \cup w) = \max\{\max v, \max w\}$.

THEOREM 2.2: *Let D be a filter over I, λ a cardinal, then the following conditions are equivalent:*

(1) *D is λ-good.*

(2) *For every family of λ-compact models M_t, $t \in I$, and $m < \omega$, $\prod_{t \in I} M_t/D$ is (λ, at, m)-compact.*

(3) *For every $\mu < \lambda$ and λ-saturated elementary extension M of $M_\mu = (S_{\aleph_0}(\mu), \subseteq, P)$ $(P(w) \Leftrightarrow w \ne \emptyset)$, M^I/D is (μ, at)-compact.*

Proof. (1) $\Rightarrow$ (2) Let $N = \prod_{t \in I} M_t/D$, p be an (at, m)-type over $|N|$ in N, $|p| < \lambda$, $p = \{\varphi_\alpha(\bar{x}; \bar{a}_\alpha): \alpha < \mu < \lambda\}$ (φ_α atomic). Let for $w \in S_{\aleph_0}(\mu)$ $f(w) = \{t \in I: M_t \models (\exists \bar{x}) \bigwedge_{\alpha \in w} \varphi_\alpha(\bar{x}; \bar{a}_\alpha[t])\}$. As p is finitely satisfiable in N, some $\bar{c}_w \in |N|$ realizes $\{\varphi_\alpha(\bar{x}; \bar{a}_\alpha): \alpha \in w\}$, so $X^\alpha_{\bar{c}_w} = \{t: M_t \models \varphi_\alpha[\bar{c}_w[t], \bar{a}_\alpha[t]]\} \in D$, (for $\alpha \in w$) hence $\bigcap_{\alpha \in w} X^\alpha_{\bar{c}_w} \in D$, but clearly it is a subset of $f(w)$, hence $f(w) \in D$.

As D is λ-good, there is a multiplicative function $g: S_{\aleph_0}(\mu) \to D$ refining f. Let for $t \in I$, $w(t) = \{\alpha < \mu: t \in g(\{\alpha\})\}$. Now, for every finite $w \subseteq w(t)$, $\alpha \in w \Rightarrow t \in g(\{\alpha\})$, hence by the multiplicativity of g, $t \in g(w)$; and as g refines f, $t \in f(w)$. Hence $M_t \models (\exists \bar{x}) \bigwedge_{\alpha \in w} \varphi_\alpha(\bar{x}, \bar{a}_\alpha[t])$, and as this holds for any finite $w \subseteq w(t)$, $p_t = \{\varphi_\alpha(\bar{x}; \bar{a}_\alpha[t]): \alpha \in w(t)\}$ is finitely satisfiable in M_t. As M_t is λ-compact, there is $\bar{c}[t] \in |M_t|$ which realizes p_t, and let $\bar{c} = \langle \dots, \bar{c}[t], \dots \rangle_{t \in I}/D$. Let us prove that $\bar{c}$ realizes p; for every α

$$\{t: M_t \models \varphi_\alpha[\bar{c}[t], \bar{a}_\alpha[t]]\} \supseteq \{t: \alpha \in w(t)\} = \{t: t \in g(\{\alpha\})\} = g(\{\alpha\}) \in D$$

hence $N \models \varphi_\alpha[\bar{c}, \bar{a}_\alpha]$.

(2) $\Rightarrow$ (3) Trivial.

(3) $\Rightarrow$ (1) Let $\mu < \lambda$ and $f : S_{\aleph_0}(\mu) \to D$ be monotonic. For each $t \in I$ we define $a_\alpha[t] \in |M|$ for $\alpha < \mu$ such that:

(*) for every finite $w \in S_{\aleph_0}(\mu)$,

$$M \vDash (\exists x)\Big(\bigwedge_{\alpha \in w} x \subseteq a_\alpha[t] \wedge P(x)\Big) \Leftrightarrow t \in f(w).$$

[If we want to define $a_\alpha[t]$ for finitely many α's only, we can easily choose them in M_μ. By the λ-saturation of M we can choose all $a_\alpha[t]$, $\alpha < \mu$].

So let $a_\alpha = \langle \ldots, a_\alpha[t], \ldots \rangle_{t \in I}/D$ and $p = \{x \subseteq a_\alpha : \alpha < \mu\} \cup \{P(x)\}$.

If $q \subseteq p$ is finite, then for some finite $w \subseteq \mu$ $q \subseteq \{x \subseteq a_\alpha : \alpha \in w\} \cup \{P(x)\}$, and, by (*) $\{t : M \vDash (\exists x)[\bigwedge_{\alpha \in w} x \subseteq a_\alpha[t] \wedge P(x)]\} = \{t : t \in f(w)\} = f(w) \in D$.

Hence $M \vDash (\exists x) \bigwedge q$, so p is finitely satisfiable in $N = M^I/D$ and by (3) it is realized in it, so let $c \in |N|$ realize p.

Define $g(w) = \{t \in I : \bigwedge_{\alpha \in w} c[t] \subseteq a_\alpha[t] \wedge P(c[t])\}$, and it is easy to check that $g : S_{\aleph_0}(\mu) \to D$ is multiplicative and refines f.

THEOREM 2.3: *Let D be a filter over I, λ a cardinal $> \aleph_0$ then the following conditions are equivalent:*

(1) *D is λ-good and $\aleph_1$-incomplete.*

(2) *For every family of* L*-models M_t $(t \in I)$, $\prod_{t \in I} M_t/D$ is* (λ, at, m)*-compact.*

(3) $N = M_\mu^I/D$ *is* (μ^+, at)*-compact for every $\mu < \lambda$ (M_μ—as defined in* 2.2(3)).

Proof. (2) $\Rightarrow$ (3) Trivial.

(3) $\Rightarrow$ (1) The set $\{\{\alpha\} \subseteq x : \alpha < \mu\}$ is a set of formulas over $|M_\mu|$, and it is finitely satisfiable in M_μ, hence some $c \in |N|$ realizes it. Let for $\alpha < \mu$

$$X_\alpha = \{t \in I : M_\mu \vDash \{\alpha\} \subseteq c[t]\}$$

so clearly $X_\alpha \in D$, and if $w \subseteq \mu$ is infinite, $t \in \bigcap_{\alpha \in w} X_\alpha$, then $M_\mu \vDash \{\alpha\} \subseteq c[t]$ for each $\alpha \in w$, hence $w \subseteq c[t] \in |M_\mu|$ contradiction. So $\{X_\alpha : \alpha < \mu\}$ regularizes D; hence, for any λ-saturated M, $M_\mu \prec M$, by 1.8 player II wins in $G^{\mu+1}(M_\mu^I/D, M^I/D)$; and, by 1.6(4), as M_μ^I/D is (μ, at)-compact also M^I/D is (μ, at)-compact. So by 2.2 D is λ-good. As D is $\aleph_0$-regular, it is $\aleph_1$-incomplete, by 1.3(2).

(1) $\Rightarrow$ (2) As in (3) $\Rightarrow$ (1) it suffices to prove that D is μ-regular for every $\mu < \lambda$.

CLAIM 2.4: *If D is μ^+-good and $\aleph_1$-incomplete then D is μ-regular.*

Proof. As D is $\aleph_1$-incomplete, there are $X_n \in D$ for $n < \omega$, $\bigcap_{n<\omega} X_n = \emptyset$. Let $f: S_{\aleph_0}(\mu) \to D$ be defined by $f(w) = X_{|w|}$, and let $g: S_{\aleph_0}(\mu) \to D$ be multiplicative refinement of f. Then for $\alpha < \mu$, $g(\{\alpha\}) \in D$, and for any infinite $w \subseteq \mu$, if $t \in \bigcap_{\alpha \in w} g(\{\alpha\})$, then for each $n < \omega$, choose $w(n) \subseteq w$, $|w(n)| = n$, and then $t \in \bigcap_{\alpha \in w(n)} g(\{\alpha\}) = g(w(n)) \subseteq X_n$, hence $t \in \bigcap_{n<\omega} X_n = \emptyset$, so necessarily $\bigcap_{\alpha \in w} g(\{\alpha\}) = \emptyset$, hence $\{g(\{\alpha\}): \alpha < \mu\}$ regularizes D.

THEOREM 2.5: *Let D be a λ-good, $\aleph_1$-incomplete filter over I, D_1 a filter over I, $D \subseteq D_1$. Let $|\mathrm{L}| < \lambda$, $\mathrm{L} = \mathrm{L}(M_t^0) = \mathrm{L}(M_t^1)$ for $t \in I$ and $M^l = \prod_{t \in I} M_t^l / D_1$. If M_t^0, M_t^1 are elementarily equivalent for every $t \in I$, then player* II *wins in $G^\lambda(M^0, M^1)$.*

Remark. Compare with 1.8, 1.10. We assume more and prove more. See Exercise 2.3.

Proof. We use the notation of Definition 1.6. Players II's strategy is to play so that for every formula $\varphi(x_1, \ldots, x_n) \in \mathrm{L}$, and $\alpha_1 > \cdots > \alpha_n$ $\{t \in I: M_t^0 \models \varphi(a_{\alpha_1}^0[t], \ldots, a_{\alpha_n}^0[t])\} = \{t \in I: M_t^1 \models \varphi(a_{\alpha_1}^1[t], \ldots, a_{\alpha_n}^1[t])\} \bmod D$ and therefore $\bmod D_1$ since $D \subseteq D_1$ (where $X = Y \bmod D$ means $I - (X - Y) - (Y - X) \in D$).

Suppose a_i^0, a_i^1 have been chosen for $i < \alpha < \lambda$, and player I has chosen $l \in \{0, 1\}$, $a_\alpha^l \in |M^l|$, and player II has to choose a_α^{1-l}. By the symmetry we can assume $l = 0$. Let Φ be the set of formulas $\varphi = \varphi(x, x_{i_1}, \ldots, x_{i_n})$, $\varphi \in \mathrm{L}$, $i_1, \ldots, i_n < \alpha$. Let $\Phi = \{\varphi_\beta: \beta < \mu = |\Phi|\}$; clearly $\mu < \lambda$, hence by 2.4 D is μ-regular, so some family $\{X_\beta: \beta < \mu\}$ regularizes D. Let

$$h(t) = \{\beta < \mu: M_t^0 \models \varphi_\beta(a_\alpha^0[t], a_{i(1,\beta)}^0[t], \ldots]\}.$$

For $w \in S_{\aleph_0}(\mu)$ we define

$$f(w) = \{t \in I: \beta \in w \Rightarrow t \in X_\beta, \text{ and } M_t^1 \models (\exists x) \bigwedge_{\beta \in w \cap h(t)} \varphi_\beta(x; a_{i(1,\beta))}^1[t], \ldots)\}.$$

By the induction assumption

$$\left\{t \in I: M_t^0 \models (\exists x) \bigwedge_{\beta \in w \cap h(t)} \varphi_\beta(x; a_{i(1,\beta)}^0[t], \ldots)\right\}$$

$$= \left\{t \in I: M_t^1 \models (\exists x) \bigwedge_{\beta \in w \cap h(t)} \varphi_\beta(x; a_{i(1,\beta)}^1[t], \ldots)\right\} \bmod D$$

hence it is easy to check that $f(w) \in D$. So some $g : S_{\aleph_0}(\mu) \to D$ is multiplicative and refines f.

Notice that $\{\beta : t \in g(\{\beta\})\} \subseteq \{\beta : t \in X_\beta\}$ which is finite. So for each $t \in I$, define $a^1_\alpha[t] \in |M^1|$ so that it realizes $\{\varphi_\beta(x; a^1_{i(1,\beta)}[t], \ldots) : \beta \in g(t)\}$ and $a^1_\alpha = \langle \ldots, a^1_\alpha[t], \ldots \rangle_{t \in I}/D$. For the induction hypothesis, for each φ we get only an inclusion; by applying it also for $\neg\varphi$ we get equality $\operatorname{mod} D$.

THEOREM 2.6: *The following conditions on the filter D over I, and the cardinal $\lambda > \aleph_0$ are equivalent:*

(1) *D is λ-good.*

(2) *For every λ-saturated model M of T_{ord} [the theory of dense linear order $\leq$ with no finite or last element], M^I/D is (λ, at)-compact (where $M = \langle |M|, \leq \rangle$).*

(3) *For every λ-saturated model M of T_{ord}, and for every set $A \subseteq |M^I/D|$ which is linearly ordered by $\leq$, every atomic 1-type p over A in M^I/D, $|p| < \lambda$, is realized in M^I/D.*

Proof. The implication (1) ⇒ (2) follows by 2.2 and (2) ⇒ (3) is trivial; so assume (3), and we shall prove (1).

CLAIM 2.7: *Assume* (3) *from* 2.6, *and let $N = \langle |N|, \leq \rangle$ be a λ-saturated rooted tree (i.e., $\leq$ is a partial order, and for every $a \in |N|$, $\{b \in |N| : b \leq a\}$ is linearly ordered, and $0 \leq a$ for every $a \in |N|$); such that each $a \in |N|$ has $\|N\| = \|N\|^{<\lambda}$ immediate successors, and for every $a, b \in |N|$ there is is a maximal c, $c \leq a$ and $c \leq b$. Then*

(1) *If $N^I/D \models c_i \leq c_j$ for $i < j < \alpha$, where $\alpha < \lambda$, then for some c, $N^I/D \models c_i \leq c$ for $i < \alpha$.*

(2) *If $N^I/D \models c^1_{i(1)} \leq c^1_{j(1)} \wedge c^1_{j(1)} \leq c^2_{j(2)} \wedge c^2_{j(2)} \leq c^2_{i(2)}$ for $i(1) < j(1) < \alpha$, $i(2) < j(2) < \beta$, where $\alpha, \beta < \lambda$, then for some c, $N^I/D \models c^1_i \leq c \leq c^2_j$ for $i < \alpha, j < \beta$.*

Proof. We first note that if A is a linearly ordered subset of $|M^I/D|$, M a model of T_{ord}, then if every (at, 1)-type over A in M^I/D is realized in M^I/D, then for every m, $1 \leq m < \omega$, every (cnat, m)-type over A in M^I/D is realized in M^I/D.

Now let us choose for each $a \in |N|$; a λ-saturated linear order $<_a$ on $S_a = \{b \in |N| : b$ an immediate successor of $a\}$ (this is possible because $|S_a| = \|N\| = \|N\|^{<\lambda}$). Define on $A^* = |N| \times \{0, 1\}$ an order $\leq^*$: $\langle a, l \rangle \leq^* \langle b, k \rangle$ iff one of the following conditions holds:

(1) $a = b$, $l \leq k$,

(2) $a < b, l = 0$,

(3) $b < a, k = 1$,

(4) let c be the maximal element for which $c \leq a, c \leq b; c < a, c < b$, and $a' <_c b'$ where $a', b' \in S_c$, $a' \leq a$, $b' \leq b$.

It is easy to check that $N_1 = (A^*, \leq^*)$ is a dense linear order, with first and last elements and it is λ-saturated. Hence it follows easily by our hypothesis that if $A \subseteq |N_1^I/D|$ is linearly ordered by $\leq^*$, then every atomic $(n+1)$-type p over A in N_1^I/D, $|p| < \lambda$, $n < \omega$ is realized in N_1^I/D.

Now define a function F.

(1) For $a \in |N|$, $F(a) = \langle F_0(a), F_1(a)\rangle$ where $F_0(a) = \langle a, 0\rangle$, $F_1(a) = \langle a, 1\rangle$ so $F_0(a), F_1(a) \in |N_1|$ and for $a \in |N^I/D|$,

$$F(a) = \langle\langle \ldots F_0(a[t]) \ldots\rangle_{t\in I}/D, \langle \ldots F_1(a[t]) \ldots\rangle_{t\in I}/D\rangle \in |N_1^I/D|.$$

(2) For atomic formulas $\varphi = \varphi(x; y)$, define

$$F(\varphi) = F(\varphi)(\bar{x}; \bar{y}), \qquad \bar{x} = \langle x_0, x_1\rangle, \qquad \bar{y} = \langle y_0, y_1\rangle,$$

$$F(x = y) = [x_0 = y_0 \wedge x_1 = y_1],$$

$$F(x \leq y) = [x_0 \leq y_0 \wedge y_0 \leq y_1 \wedge y_1 \leq x_1],$$

$$F(y \leq x) = [y_0 \leq x_0 \wedge x_0 \leq x_1 \wedge x_1 \leq y_1].$$

(3) For an atomic formula $\varphi(x; a)$ define

$$F(\varphi(x; a)) = F(\varphi)(x_0, x_1; F(a)).$$

(4) For a set p of atomic formulas $\varphi(x; a)$ let

$$F(p) = \{F(\varphi(x; a)) : \varphi(x; a) \in p\}.$$

Now it is easy to check that for every such type p in N, p is realized in N iff $F(p)$ is realized in N_1. Hence by 2.8 below for each atomic type p in N^I/D, p is realized in N^I/D iff $F(p)$ is realized in N_1^I/D. For each p as in 2.9(A) or (B), $F(p)$ is an atomic 2-type over some linearly ordered subset of $|N_1^I/D|$ in N_1^I/D, so $F(p)$ is realized in N_1^I/D, hence p is realized in N^I/D, hence we get our conclusion.

CLAIM 2.8: *Let M, N be models (possibly with distinct languages) and let F be a function, such that*

(1) $$\operatorname{Dom} F \subseteq \{\varphi(\bar{x}^1; \bar{a}) : \varphi \in L(M), \varphi \in \text{at}, \bar{a} \in |M|\},$$

$$\operatorname{Range} F \subseteq \{\varphi(\bar{x}^2; \bar{b}) : \varphi \in L(N), \varphi \in \text{cnat}, \bar{b} \in |N|\}$$

(2) $$F(\varphi(\bar{a}^1; \bar{a})) = F_1(\varphi)(\bar{x}^2, F_2(\varphi, \bar{a})),$$

i.e.,

$$F(\varphi_l(\bar{x}^1, \bar{a}_l)) = \psi_l(\bar{x}^2, \bar{b}_l)\ (l = 1, 2) \quad \text{implies} \quad \varphi_1 = \varphi_2 \Rightarrow \psi_1 = \psi_2.$$

(3) *For every* $p \subseteq \operatorname{Dom} F$, *$p$ is realized in M iff*

$$F(p) = \{F[\varphi(\bar{x}^1, \bar{a})]: \varphi(\bar{x}^1; \bar{a}) \in p\}$$

is realized in N.

Then if $p = \{\varphi_i(\bar{x}^1; \bar{a}_i): i < \alpha\}$ *is a set of atomic formulas in* M^I/D, *and for every* $i < \alpha$ $\{t \in I: \varphi_i(\bar{x}^1; \bar{a}_i[t]) \in \operatorname{Dom} F\} \in D$ *then p is realized in* M^I/D *iff* $F(p) = \{F[\varphi_i(\bar{x}^1; \bar{a}_i)]: i < \alpha\}$ *is realized in* N^I/D.

Proof. Suppose $\bar{c}$ realizes $p(\bar{c} \in |M^I/D|)$; for every $t \in I$ let $w(t) = \{i < \alpha: M \models \varphi_i(\bar{c}[t]; \bar{a}_i[t])$ and $\varphi_i(\bar{x}^1; \bar{a}_i[t]) \in \operatorname{Dom} F\} \in D$ and define $\bar{c}^1[t] \in |N|$ so that it realizes

$$\{F[\varphi_1(\bar{x}^1; \bar{a}_i[t])]: i \in w(t)\}.$$

It is easy to check that $\bar{c}^1$ realizes $F(p)$. The other direction is easy too.

Continuation of the proof of 2.6. Let M be any λ-saturated model, $|L(M)| \le \|M\|$, and we shall prove that M^I/D is (λ, at)-compact. This is sufficient by 2.2. Let

$$\Gamma = \{\langle \varphi_0(x, \bar{a}_0), \ldots, \varphi_{m-1}(x, \bar{a}_{m-1})\rangle: m < \omega, \bar{a}_i \in |M|, \varphi_i \in L(M),$$
$$\varphi_i \text{ atomic, and } M \models (\exists x) \bigwedge_{i<m} \varphi_i(x, \bar{a}_i)\}.$$

Clearly $|\Gamma| = \|M\|$, so there is a one-to-one function g from $|M|$ onto Γ.

Let us define some new relations and functions over $|M|$ (assuming that they $\notin L(M)$).

(0) c^* will be $g^{-1}(\langle\ \rangle)$.

(1) $\le$: $a \le b \Leftrightarrow g(a)$ is an initial segment of $g(b)$.

(2) F^1: $F^1(a)$ is an element of M realizing $g(a)$.

(3) P_φ (for each conjunction $\varphi(x; \bar{y})$ of atomic formulas): $P_\varphi(b, \bar{a}) \Leftrightarrow$ each conjunct of $\varphi(x; \bar{a})$ appears in $g(b)$.

(4) Q_φ (for each conjunction $\varphi(x; \bar{y})$ of atomic formulas): $Q_\varphi(b, \bar{a}) \Leftrightarrow$ $\varphi(x; \bar{a})$ is consistent with $g(b)$, i.e., $g(b)^\frown\langle\psi_0(x; \bar{a}^0), \ldots, \psi_{n-1}(x; \bar{a}^{n-1}\rangle \in \Gamma$ where $\varphi(x; \bar{a}) = \bigwedge_{i<n} \psi_i(x; \bar{a}^i)$.

(5) F^2_φ (for each $\varphi(x; \bar{y})$ as in (4): $F^2_\varphi(b, \bar{a}) = c$ if $Q_\varphi(b, \bar{a})$ and $g(c) = g(b)^\frown\langle\psi_0(x; \bar{a}^0), \ldots, \psi_{n-1}(x; \bar{a}^{n-1})\rangle$ or $\neg Q_\varphi(b, \bar{a})$ and $c = b$.

(6) F^3_φ (for each $\varphi(x;\bar{y})$ as in (4)): $F^3_\varphi(b,\bar{a}) = c$ if

$$g(b) = \langle \varphi_0(x;\bar{a}_0),\ldots,\varphi_{m-1}(x;\bar{a}_{m-1})\rangle,$$

$$g(c) = \langle \varphi_0(x;\bar{a}_0),\ldots,\varphi_{l-1}(x;\bar{a}_{l-1})\rangle$$

$$l = \max\left(k: M \models (\exists x)\left[\bigwedge_{i<k} \varphi_i(x;\bar{a}_i) \wedge \varphi(x;\bar{a})\right] \text{ or } k = 0\right)$$

(notice that in (6) if $M \models (\exists x)\varphi(x;\bar{a})$, then $M \models (\exists x)[\bigwedge_{i<0} \varphi_i(x;\bar{a}_i) \wedge \varphi(x;\bar{a})]$).

Let us define $N = (M, c^*, \leq, F^1, P_\varphi, Q_\varphi, F^2_\varphi, F^3_\varphi)_\varphi$, and let N_1 be a λ-saturated elementary extension of N, $\|N_1\|^\lambda = \|N_1\|$, and M_1 be the L(M)-reduct of N_1. Clearly M_1 is λ-saturated, hence by 1.11 M^I/D is (λ, at)-compact iff $M_2 = M_1^I/D$ is (λ, at)-compact, so it suffices to prove that M_1^I/D is (λ, at)-compact. Clearly M_2 is the L(M)-reduct of $N_2 = N_1^I/D$. It is easy to check that $(|N_1|, \leq^{N_1})$ satisfies the requirement of claim 2.7, hence also the conclusion of that claim.

Let $p = \{\varphi_i(x;\bar{a}_i): i < \alpha_0 < \lambda\}$ be a cnat-type in M_2 which is w.l.o.g. closed under conjunctions, and we will prove that it is realized in M_2. We define by induction on $i \leq \alpha_0$ elements $c_i \in |N_2|$ such that:

(A) $j < i \Rightarrow N_2 \models c_j \leq c_i$.

(B) For every $\beta < \alpha_0$, $N_2 \models Q_{\varphi_\beta}(c_i, \bar{a}_\beta)$.

(C) If $i = j + 1$, $N_2 \models P_{\varphi_j}(c_i, \bar{a}_j)$.

Case I: $i = 0$. Let $c_0 = c^*$ (the root of the tree). Conditions (A), (C) are vacuous, as for condition (B), for every β $N \models (\forall\bar{y})(\forall z)[\varphi_\beta(z,\bar{y}) \to Q_{\varphi_\beta}(c^*,\bar{y})]$ and this is a Horn sentence, so N_2 also satisfies it. As p is a type in M_2, $N_2 \models (\exists x)\varphi_\beta(x,\bar{a}_\beta)$ hence $N_2 \models Q_{\varphi_\beta}(c^*,\bar{a}_\beta)$.

Case II: $i + 1$. Let $c_{i+1} = F^2_{\varphi_i}(c_i,\bar{a}_i)$. For condition (A) notice that

(i) $N \models (\forall xyz)(x \leq y \wedge y \leq z \to x \leq z)$,

(ii) $N \models (\forall x\bar{y})(x \leq F^2_{\varphi_i}(x,\bar{y}))$,

and both are Horn sentences, hence N_2 satisfies them.

By (ii) $N_2 \models c_i \leq c_{i+1}$, and by (i) and the induction hypothesis $j < i \Rightarrow N_2 \models c_j \leq c_{i+1}$.

As for Condition (B), for each $\beta < \alpha_0$ there is $\gamma = \gamma(\beta) < \alpha_0$ such that $\varphi_\gamma(x;\bar{y}) = \varphi_\gamma(x;\bar{y}_1,\bar{y}_2) = \varphi_\beta(x;\bar{y}_1) \wedge \varphi_i(x,\bar{y}_2)$ and $\bar{a}_\gamma = \bar{a}_\beta{}^\frown\bar{a}_i$, (remember p is closed under conjunctions). Clearly

$$N \models (\forall z, z_1, \bar{y}_1, \bar{y}_2)[Q_{\varphi_\gamma}(z;\bar{y}_1,\bar{y}_2) \wedge z_1 = F^2_{\varphi_i}(z,\bar{y}_2) \to Q_{\varphi_\beta}(z_1,\bar{y}_1)]$$

and as this is a Horn sentence N_2 satisfies it. As $N_2 \models Q_{\varphi_\gamma}(c_i;\bar{a},\bar{a}_i) \wedge c_{i+1} = F^2_{\varphi_i}(c_i,\bar{a}_i)$ clearly $N_2 \models Q_{\varphi_\beta}[c_{i+1},\bar{a}_\beta]$.

As for Condition (C), $N \models (\forall \bar{y}, z_1, z)[Q_{\varphi_i}(z, \bar{y}) \wedge z_1 = F^2_{\varphi_i}(z, \bar{y}) \rightarrow P_{\varphi_i}(z_1, \bar{y})]$ so by condition (B) clearly $N_2 \models P_{\varphi_i}(c_{i+1}, \bar{a}_i)$.

Case III: $i = \delta$ is a limit ordinal. By Claim 2.7(1) and Condition (A) for $j < \delta$, clearly there is $d_0 \in |N_2|$ such that $j < \delta \Rightarrow N_2 \models c_j \leq d_0$. So d_0 satisfies Condition (A) on c_δ, but not necessarily Condition (B). We now define by induction of $j \leq \alpha_0$ elements $d_j \in |N_2|$ such that

(*) $\quad j_1 < j_2 \leq \alpha_0, \gamma < \delta \Rightarrow N_2 \models c_\gamma \leq d_{j_2} \wedge d_{j_2} \leq d_{j_1}$.

We have already defined d_0; for limit j, the existence of d_j follows by part (2) of Claim 2.7 (using (*) and Condition (A)). For $j + 1$ let $d_{j+1} = F^3_{\varphi_j}(d_j, \bar{a}_j)$, and we shall prove that (*) holds.

Notice that

$$N \models (\forall z)(\forall \bar{y})(F^3_{\varphi_j}(z, \bar{y}) \leq z)$$

and this is a Horn sentence, hence N_2 satisfies it, hence $N_2 \models d_{j+1} \leq d_j$, hence (as $\leq^{N_2}$ is transitive) $j_1 < j + 1 \Rightarrow N_2 \models d_{j+1} \leq d_{j_1}$. On the other hand,

$$N \models (\forall \bar{y})(\forall z_1, z_2)[z_1 \leq z_2 \wedge Q_{\varphi_j}(z_1, \bar{y}) \rightarrow z_1 \leq F^3_{\varphi_j}(z_2, \bar{y})]$$

and this is a Horn sentence, hence N_2 satisfies it.

As for $\gamma < \delta$ $N_2 \models c_\gamma \leq d_j$ (by the induction hypothesis on j) and $N_2 \models Q_{\varphi_j}(c_\gamma, \bar{a}_j)$ (by the induction hypothesis on $i = \delta$) clearly $N_2 \models c_\gamma \leq F^3_{\varphi_j}(d_j, \bar{a}_j)$, but $F^3_{\varphi_j}(d_j, \bar{a}_j) = d_{j+1}$. So we prove (*) for $j + 1$.

So we finish defining $d_j, j \leq \alpha_0$, and let $c_\delta = d_{\alpha_0}$. Condition (A) holds by (*), Condition (C) is vacuous and Condition (B) holds because for each $\beta < \alpha_0$,

$$N \models (\forall \bar{y})(\forall z_1, z)[Q_{\varphi_\beta}(z, \bar{y}) \wedge z_1 \leq z \rightarrow Q_{\varphi_\beta}(z_1, \bar{y})]$$

and this is a Horn sentence, hence N_2 satisfies it, and $N_2 \models d_{\alpha_0} \leq d_{\beta+1}$. Now $N_2 \models Q_{\varphi_\beta}(d_{\beta+1}, \bar{a}_\beta)$ because $N \models (\forall x, \bar{y}, z)[\varphi_\beta(x, \bar{y}) \rightarrow Q_{\varphi_\beta}(F^3_{\varphi_\beta}(z, \bar{y}), \bar{y})]$; this is a Horn sentence, and as p is a type in M_2, for some $b \in |N_2|$ $N_2 \models \varphi_\beta[b, \bar{a}_\beta]$ hence $N_2 \models Q_{\varphi_\beta}(F^3_{\varphi_\beta}(d_j, \bar{a}_\beta), a_\beta)$.

So we finish the definition of c_i, $i \leq \alpha_0$. We shall prove that $F^1(c_{\alpha_0})$ realizes p and thus finish the proof.

Now for each $\beta < \alpha_0$ $N \models (\forall \bar{y}, z_1, z_2)[P_{\varphi_\beta}(z_1, \bar{y}) \wedge z_1 \leq z_2 \rightarrow P_{\varphi_\beta}(z_2, \bar{y})]$ and $N \models (\forall z)(\forall \bar{y})[P_{\varphi_\beta}(z, \bar{y}) \rightarrow \varphi_\beta(F^1(z), \bar{y})]$ and they are Horn sentences hence N_2 satisfies it, and $N_2 \models P_{\varphi_\beta}(c_{\beta+1}, \bar{a}_\beta) \wedge c_{\beta+1} \leq c_{\alpha_0}$ by Condition (C). Hence $N_2 \models \varphi_\beta[F^1(c_{\alpha_0}), \bar{a}_\beta]$ for every $\beta < \alpha_0$.

THEOREM 2.9: *Let $\lambda > \aleph_0$, D a filter over I. Then the following conditions are equivalent.*

(1) *D is λ-good and $\aleph_1$-incomplete.*

(2) *For every model M of T_{ord}, M^I/D is* (λ, at)*-compact.*

(3) *N^I_μ/D is* (λ, at)*-compact, where $N_\mu = \langle {}^{\omega >}\mu, < \rangle$, $<$-lexicographic order, for each $\mu < \lambda$.*

Proof. (1) $\Rightarrow$ (2), (1) $\Rightarrow$ (3) Hold by 2.3.

(2) $\Rightarrow$ (1) By 2.6 it suffices to prove that D is $\aleph_1$-incomplete. As $(Q, \leq)^I/D$ is (λ, at)-compact (Q-the rationals) for some $c \in |\langle Q, \leq\rangle^I/D|$, $n \leq c$ for each $n < \omega$, and let

$$X_n = \{t \in I : n < c[t]\}. \text{ Clearly } X_n \in D, \bigcap_n X_n = \emptyset.$$

(3) $\Rightarrow$ (1) The proof is similar to that of 2.6, hence we leave it to the reader.

CONCLUSION 2.10: *If λ is a singular cardinal and the filter D is λ-good, then D is λ^+-good.*

Proof. We use 2.6(3), so let M be a λ^+-saturated model of T_{ord}, $A \subseteq |N| = |M^I/D|$ be linearly ordered by $\leq$, and p be an atomic 1-type over A in N, $|p| \leq \lambda$, and we shall prove that p is realized in N, thus finishing. If for some a, $(x = a) \in p$, a realizes p (because p is finitely satisfiable in N). So let $p = \{a \leq x : a \in A_1\} \cup \{x \leq a : a \in A_2\}$.

As $A_1 \subseteq A$, A_1 is linearly ordered by $\leq$, so it has a cofinal sequence $\{a_i : i < \mu_1\}$, μ_1 a regular cardinal (so $(\forall a \in A_1)(\exists i < \mu_1)(a \leq a_i)$ and $a_i \in A_1$). As μ_1 is regular, $\mu_1 < \lambda$. Similarly there are $a^i \in A_2$ $(i < \mu_2)$ such that $(\forall a \in A_2)(\exists i < \mu_2)(a^i \leq a)$, μ_2 regular. As M satisfies the Horn sentence saying $\leq$ is transitive, an element realizes p if it realizes

$$p_1 = \{a_i \leq x : i < \mu_1\} \cup \{x \leq a^i : i < \mu_2\}.$$

As $|p_1| < \lambda$, and D is λ-good, by 2.2 N realizes p_1, so we finish.

Remark. See Exercise 2.7.

Up to now we have dealt with filters and atomic compactness. By the following lemma, we do not need to deal separately with ultrafilters and compactness.

LEMMA 2.11: *Let D be an ultrafilter over I, $\lambda > \aleph_0$ a cardinal.*

(1) *For every M, M^I/D is* (λ, at)*-compact iff for every M, M^I/D is λ-compact.*

(2) *For every* L, *and* L-*models* M_t, $\prod_{t\in I} M_t/D$ *is* (λ, at)-*compact iff for every* L *and* L-*models* M_t, $\prod_{t\in I} M_t/D$ *is* λ-*compact.*

(3) *As* (1) [2] *but for* λ-*saturated* $M[M_t]$.

Proof. In all parts the if part is trivial. For the other direction, for each model M define M^* as follows:

$$L(M^*) = \{R_{\varphi(\bar{x})}: \varphi(\bar{x}) \in L(M)\},$$

($R_{\varphi(\bar{x})}$ has $l(\bar{x})$ places)

$$|M^*| = |M|, \qquad R^{M^*}_{\varphi(\bar{x})} = \{\bar{a} \in |M|: M \models \varphi[\bar{a}]\}.$$

By Łoš's Theorem it is easy to check that $(\prod_{t\in I} M_t/D)^* = \prod_{t\in I} M_t^*/D$ hence M^I/D is λ-compact iff M^{*I}/D is (λ, at)-compact and $\prod_{t\in I} M_t/D$ is λ-compact iff $\prod_{t\in I} M_t^*/D$ is (λ, at)-compact, and M is λ-saturated iff M^* is λ-saturated. Hence the "only if" part follows easily.

THEOREM 2.12: *Suppose* $2^\lambda = \lambda^+$, $|I| = \lambda$, D *a good,* $\aleph_1$-*incomplete, ultrafilter over* I; M, N *are elementarily equivalent models of cardinality* $\leq \lambda^+$, *and* $L(M) = L(N)$ *has power* $\leq \lambda$. *Then* M^I/D, N^I/D *are isomorphic (and saturated).*

Proof. By I, 1.11 every two elementarily equivalent λ^+-saturated models of power λ^+ are isomorphic. By Łoš's Theorem M^I/D, N^I/D are elementarily equivalent. So, by the symmetry, it suffices to prove that M^I/D has cardinality $\leq \lambda^+$ and is λ^+-saturated. Now

$$\|M^I/D\| \leq \|M\|^{|I|} \leq (\lambda^+)^\lambda \leq 2^{\lambda\cdot\lambda} = \lambda^+$$

and by 2.3 and 2.11 M^I/D is λ^+-compact. As $|L(M)| \leq \lambda$, M^I/D is λ^+-saturated, so we finish.

THEOREM 2.13: *If* D *is a good ultrafilter over* I, $\prod_{t\in I} n_t/D \geq \aleph_0$, *then* $\prod_{t\in I} n_t/D = 2^{|I|}$.

Proof. If D is $\aleph_1$-complete, clearly for some n $\{t: n_t = n\} \in D$ hence $\prod_{t\in I} n_t/D = n$; hence D is $\aleph_1$-incomplete (by Exercise 1.8) so by 2.4 D is regular.

For each n let M_n be the following model: $M_n = \langle n + 2^n, P_n, R_n\rangle$, where P_n, R_n are one-place and two place relations respectively, $P_n = \{0, \ldots, n-1\}$ and R_n is such that for every $w \subseteq P_n$ for some $i, n \leqslant i < n+2^n$, and for every $k < n, k \in w \Leftrightarrow R_n(k, i)$. Let for $t \in I$, $m(t) = [\log_2 n_t] - 1$, so $\|M_{m(t)}\| \leq n_t$, but for every n $\{t: \|M_{m(t)}\| \leq n\} \subseteq$

$\{t: n_t \leq 2^{n+1}\} \notin D$. Hence $M = \langle A, P, R\rangle = \prod_{t\in I} M_{m(t)}/D$ is an infinite model and is $|I|^+$-saturated. So clearly $|P| > |I|$ (otherwise $\{P(x) \wedge x \neq a : a \in P\}$ is omitted by M) and $|A| \geqslant 2^{|I|}$ (for let $C \subseteq P$, $|C| = |I|$, then for every $B \subseteq C$, $p_B = \{R(a;x)^{\text{if}(a\in B)} : a \in C\}$ is realized by some $b_B \in A$, so $|A| \geqslant 2^{|C|} = 2^{|I|}$. We can conclude

$$\prod_{t\in I} n_t/D \geq \prod_{t\in I} \|M_{m(t)}\|/D = \prod_{t\in I} \|M_{m(t)}/D\| = |A| = 2^{|I|}$$

and the other inequality is Exercise 1.3(2).

EXERCISE 2.1: Show that to the list of equivalent conditions in 2.3 we can add:

(4) If p is an atomic 1-type in $M = \prod_{t\in I} M_t/D$ of cardinality $< \lambda$, then p is realized in M.

EXERCISE 2.2: Show that in 2.5 instead of assuming M_t^0, M_t^1 are elementarily equivalent for every $t \in I$ it suffices to assume $\prod_{t\in I} M_t^0/D$ and $\prod_{t\in I} M_t^1/D$ are elementarily equivalent.

EXERCISE 2.3: Show that in 2.5 we can omit "D is $\aleph_1$-incomplete" but must then demand that for each $t \in I$, M_t^0, M_t^1 are λ-saturated and elementarily equivalent.

EXERCISE 2.4: Show that to the list of equivalent conditions in 2.9 we can add.

(4) (i) N_μ^I/D is (λ, at)-compact for some μ.

(ii) D is μ-regular for each $\mu < \lambda$.

EXERCISE 2.5: Let E be a family of subsets of I closed under finite intersection, and let E generate the filter D. Every monotonic $f : S_{\aleph_0}(\mu) \to E$ has a refinement $g : S_{\aleph_0}(\mu) \to E$ which is multiplicative *iff* D is μ^+-good.

EXERCISE 2.6: (1) Every λ-complete filter is λ^+-good. (Hint: See 2.1.)

(2) If D is μ-complete for each $\mu < \lambda$, λ singular, then D is λ^+-complete.

EXERCISE 2.7: If λ is regular $\leq |I|$, then over I there is a non-λ^+-complete filter D which is λ-complete. Moreover D is μ-good iff $\mu \leq \lambda^+$ (Hint: D is generated by $\{X_i: i < \lambda\}$, X_i decreasing, $\bigcap_i X_i = \emptyset$.)

EXERCISE 2.8: Suppose the filter D over I is μ^+-incomplete and λ^+-good. Show there are $X_i \in D$ $(i < \lambda)$ such that the intersection of any μ X_i's is empty.

EXERCISE 2.9: In Definition 2.1(1), when $\lambda = \mu_0^+$, it suffices to take $\mu = \mu_0$.

EXERCISE 2.10: Suppose D is an $\aleph_1$-incomplete filter over I, and $\lambda = \prod_{t\in I} n_t/D \geq \aleph_0$. Prove that $\lambda^{\aleph_0} = \lambda$. [Hint: Let $M = (\omega, <, +, \times, F)$. where $F(x) = [\sqrt[4]{x}]$, $N = M^I/D$, and $n^* = \langle \ldots, n_t, \ldots \rangle/D$. and for $a \in |N|$, $|a| = |\{b \in |N|: N \models a < b\}|$. Then $|n^*| = \lambda$, $|a| \geq \aleph_0 \Rightarrow |a| = |F(a)|$, and let $n_0^* = F(n^*)$, $n_{k+1}^* = F(n_k^*)$, and for every sequence $\bar{\mathbf{b}} = \langle b_n; n < \omega \rangle$, $b_k < n_k^*$, let

$p_{\bar{b}} = \{n_0^* b_0 + n_1^* b_1 + \cdots + n_k^* b_k < x < n_{b_0}^* b_0 + \cdots + n_k^* b_k + n_k^* : k < \omega\}$

Clearly $p_{\bar{b}}$ is realized by some $a_{\bar{b}}$, thus providing $\prod_k |n_k^*| = \lambda^{\aleph_0}$ distinct elements of $\{b: b < n^*\}$.]

EXERCISE 2.11: In Theorem 2.12 instead "D good" we can assume only "there is a good $\aleph_1$-incomplete filter $D_1 \subseteq D$."

EXERCISE 2.12: Show that no $\aleph_1$-incomplete filter over I is $|I|^{++}$-good.

VI.3. Constructing ultrafilters

We shall show that over every cardinality there is a good ultrafilter. We define the product of ultrafilters, and find how regular and good the product is. Then we assume $2^{\aleph_0} > \aleph_1$ and MA (Martin's axiom) to prove the existence of an ultrafilter D over ω, such that for every $\aleph_2$-saturated model M of T_{ind}, M^ω/D is $\aleph_2$-saturated.

THEOREM 3.1: *Over any λ there is a good $\aleph_1$-incomplete ultrafilter.*

Proof. We first give a definition and prove some claims.

Meanwhile let D denote a filter over λ, and $\mathscr{G}$ denote a family of functions from λ onto λ.

DEFINITION 3.1: $\mathscr{G}$ is called independent mod D *if* for every $n < \omega$ and distinct $g_0, \ldots, g_{n-1} \in \mathscr{G}$ and every $j_0, \ldots, j_{n-1} < \lambda$

$$\{\alpha < \lambda: g_0(\alpha) = j_0, \ldots, g_{n-1}(\alpha) = j_{n-1}\} \neq \emptyset \bmod D$$

(and D is non-trivial).

CLAIM 3.2: *Let D_0 be the filter over λ generated by $\{\lambda\}$. Then there is a family $\mathcal{G}$ of cardinality 2^λ which is independent* mod D_0.

Proof. This is a restatement of 1.5 of the Appendix for a particular case.

CLAIM 3.3: *Let $\mathcal{G}$ be independent* mod D, *and $S \subseteq \lambda$. Then for some finite $\mathcal{G}' \subseteq \mathcal{G}$, $\mathcal{G} - \mathcal{G}'$ is independent* mod D_1 *or* mod D_2, *where $D_1[D_2]$ is the filter generated by $D \cup \{S\}[D \cup \{\lambda - S\}]$.*

Proof. If $\mathcal{G}$ is not independent mod D_1, then for some $n < \omega$ there are distinct $g_0, \ldots, g_{n-1} \in \mathcal{G}$ and there are $j_0, \ldots, j_{n-1} < \lambda$ such that: $W_1 = \{\alpha < \lambda : g_0(\alpha) = j_0, \ldots, g_{n-1}(\alpha) = j_{n-1}\} = \emptyset \bmod D_1$, hence $W_1 \subseteq \lambda - S \bmod D$ (notice that $n = 0$ means D_1 is trivial). Let $\mathcal{G}' = \mathcal{G}$-$\{g_0, \ldots, g_{n-1}\}$, and assume $\mathcal{G}'$ is not independent mod D_2, so there are $m < \omega$ and $j^0, \ldots, j^{m-1} < \lambda$ and distinct $g^0, \ldots, g^{m-1} \in \mathcal{G}'$ such that $W_2 = \{\alpha < \lambda : g^0(\alpha) = j^0, \ldots, g^{m-1}(\alpha) = j^{m-1}\} = \emptyset \bmod D_2$, hence $W_2 \subseteq S \bmod D$. So $W_1 \cap W_2 = \emptyset \bmod D$, contradicting the independence of $\mathcal{G}$ mod D.

CLAIM 3.4: *Suppose $\mathcal{G}$ is independent* mod D, *$g \in \mathcal{G}$, $\mathcal{G}' = \mathcal{G}$-$\{g\}$, and $f : S_{\aleph_0}(\lambda) \to D$ is monotonic. Then there is a filter D', $D \subseteq D'$, and a multiplicative function $f' : S_{\aleph_0}(\lambda) \to D'$ refining f such that $\mathcal{G}'$ is independent* mod D'.

Proof. Let $\{w_\alpha : \alpha < \lambda\}$ be an enumeration of $S_{\aleph_0}(\lambda)$, let $f'(w) = \{\beta : g(\beta) = \alpha, \beta \in f(w_\alpha), w \subseteq w_\alpha\}$ and D' the filter generated by $D \cup \{f'(\{i\}) : i < \lambda\}$. Clearly D' is non-trivial and f' is multiplicative, into D', and a refinement of f.

Let us prove $\mathcal{G}'$ is independent mod D', so let $j_0, \ldots, j_{n-1} < \lambda$, and $g_0, \ldots, g_{n-1} \in \mathcal{G}'$, the g_i's being distinct and

$$W = \{\beta : g_0(\beta) = j_0, \ldots, g_{n-1}(\beta) = j_{n-1}\}.$$

We must prove $W \neq \emptyset \bmod D'$. For this it suffices to prove that if $w \in S_{\aleph_0}(\lambda)$, $W \cap f'(w) \neq \emptyset \bmod D$. Let $w = w_\alpha$, so

$$W \cap f'(w) \supseteq \{\beta : g(\beta) = \alpha, g_0(\beta) = j_0, \ldots, g_{n-1}(\beta) = j_{n-1}\} \cap f(w)$$

so we finish, as $\mathcal{G}$ is independent mod D and $f(w) \in D$.

Proof of 3.1. Let $\{S_\alpha : \alpha < 2^\lambda, \alpha > 0 \text{ even}\}$ be an enumeration of $\{S : S \subseteq \lambda\}$, and $\{f_\alpha : \alpha < 2^\lambda, \alpha \text{ odd}\}$ an enumeration of the functions $f : S_{\aleph_0}(\lambda) \to \{S : S \subseteq \lambda\}$, each appearing 2^λ times. Let D_0 be the filter over λ generated by $\{\lambda\}$, and $\mathscr{G}_0$ be independent mod D_0 and of cardinality 2^λ.

Now we define by induction on α D_α, $\mathscr{G}_\alpha$ such that:

(i) $\mathscr{G}_\alpha$ is independent mod D_α,

(ii) $|\mathscr{G} - \mathscr{G}_\alpha| \le \aleph_0 + |\alpha|$,

(iii) for $\beta < \alpha$, $D_\beta \subseteq D_\alpha$, $\mathscr{G}_\beta \supseteq \mathscr{G}_\alpha$,

(iv) if α is a limit ordinal $D_\alpha = \bigcup_{\beta<\alpha} D_\beta$, $\mathscr{G}_\alpha = \bigcap_{\beta<\alpha} \mathscr{G}_\beta$,

(v) if α is even, $\alpha > 0$, $S_\alpha \in D_{\alpha+1}$ or $\lambda - S_\alpha \in D_{\alpha+1}$,

(vi) if α is odd, $f_\alpha : S_{\aleph_0}(\lambda) \to D_\alpha$, f_α monotonic then there is a multiplicative refinement f'_α of f_α, $f'_\alpha : S_{\aleph_0}(\lambda) \to D_{\alpha+1}$.

The induction is easy: for $\alpha = 0$ we have defined, for α limit see (iv) (using iii), for $\alpha + 1$, α even, $\alpha > 0$ use Claim 3.3 and for $\alpha + 1$, α odd use Claim 3.4, and for $\alpha + 1$, $\alpha = 0$ let $D_1 = \{\{\alpha : n < g(\alpha) < \omega\} : n < \omega\}$, $\mathscr{G}_1 = \mathscr{G}_0\text{-}\{g\}$, where $g \in \mathscr{G}_0$.

Now D_{2^λ} is an ultrafilter by (v). If $f : S_{\aleph_0}(\lambda) \to D_{2^\lambda}$, then for some $\alpha < 2^\lambda$, $f : S_{\aleph_0}(\lambda) \to D_\alpha$, and for some β, $\alpha < \beta < 2^\lambda$, $f_\beta = f$, β is odd, so (when f is monotonic) there is a multiplicative refinement f' of f, $f' : S_{\aleph_0}(\lambda) \to D_{\beta+1} \subseteq D_{2^\lambda}$. So D_{2^λ} is a λ^+-good ultrafilter over λ. As $D_1 \subseteq D_{2^\lambda}$, D_{2^λ} is $\aleph_1$-incomplete.

CONCLUSION 3.5: *If M, N are elementarily equivalent models of T, of cardinality $\le \lambda^+$, $|T| \le \lambda$, $2^\lambda = \lambda^+$, then for some ultrafilter D over λ, $M^\lambda/D \cong N^\lambda/D$.*

Proof. Immediate by 2.12 and 3.1.

DEFINITION 3.2: If D_l is a filter over I_l $(l = 1, 2)$ then $D_1 \times D_2$ is the family of sets $S \subseteq I_1 \approx I_2$ such that:

$$\{i \in I_2 : \{j \in I_1 : \langle j, i\rangle \in S\} \in D_1\} \in D_2.$$

LEMMA 3.6: (1) *$D_1 \times D_2$ is a filter over $I_1 \times I_2$.*

(2) *If D_l is an ultrafilter over I_l $(l = 1, 2)$ then $D_1 \times D_2$ is an ultrafilter over $I_1 \times I_2$.*

(3) $\prod_{\langle i,j\rangle \in I_1 \times I_2} M_{i,j}/D_1 \times D_2 \cong \prod_{j \in I_2} (\prod_{i \in I_1} M_{i,j}/D_1)/D_2$.

(4) $M^{I_1 \times I_2}/D_1 \times D_2 \cong (M^{I_1}/D_1)^{I_2}/D_2$.

Proof. Immediate.

LEMMA 3.7: (1) *If D_1 or D_2 is λ-regular then $D_1 \times D_2$ is λ-regular.*

(2) *$D_1 \times D_2$ is λ-good iff D_2 is λ-good; provided that D_2 is $\aleph_1$-incomplete.*

Proof. (1) Immediate by 1.4(1)(A), (C) and 3.6(4).

(2) Immediate by 2.3 and 3.6(4).

If D_2 is λ-good, for any λ-saturated model M, $M^{I_1 \times I_2}/D_1 \times D_2 \cong (M^{I_1}/D_1)^{I_2}/D_2$ (by 3.6) is (λ, at)-compact (by 2.3). Hence by 2.3 $D_1 \times D_2$ is λ-good. If $D_1 \times D_2$ is λ-good we get the result similarly.

CONCLUSION 3.8: *For every $\lambda \leq \mu$, there is a regular ultrafilter D over μ which is λ^+-good but not λ^{++}-good. If $\lambda^+ < 2^\lambda$, D is not λ^{++}-good but $\prod_{i<\mu} n_i/D \geq \aleph_0$ implies $\prod_{i<\mu} n_i/D \geq 2^\lambda > \lambda^+$.*

Proof. Let D_1 be a regular ultrafilter over μ, and D_2 a λ^+-good $\aleph_1$-incomplete ultrafilter over λ (exists by 3.1) D_2 is not λ^{++}-good by 2.4 and 1.3(3). So $D_1 \times D_2$ is an ultrafilter over $\mu \times \lambda$ which is λ^+-good but not λ^{++}-good. As $|\lambda \times \mu| = \mu$, we finish. The second part follows by 2.13.

LEMMA 3.9 (MA): *There is an ultrafilter D over ω such that: If $|P^{M_n}| \leq \aleph_0$, and $M = \prod_{n<\omega} M_n/D$ and p is an m-type in M of cardinality $< 2^{\aleph_0}$ and $P(x_0) \wedge \cdots \wedge P(x_{m-1}) \in p$, then p is realized in M.*

Remark. On Martin's Axiom the reader can consult, e.g., Jech [Je 74]. It is consistent with ZFC + $2^{\aleph_0} = \aleph_\alpha$ for any regular $\aleph_\alpha$, if ZFC is consistent and it implies $\lambda < 2^{\aleph_0} \rightarrow 2^\lambda = 2^{\aleph_0}$. As we use it only rarely, we do not elaborate.

Proof. It is sufficient to prove the lemma just for models M_n such that $\|M_n\| < 2^{\aleph_0}$, $|L(M_n)| < 2^{\aleph_0}$.

So let $\{S_\alpha: \alpha < 2^{\aleph_0}, \alpha \text{ odd}\}$ be an enumeration of $\{S: S \subseteq \omega\}$ and $\{\langle p_\alpha, \langle M_n^\alpha: n < \omega\rangle\rangle: \alpha < 2^{\aleph_0}, \alpha \text{ even}\}$ be an enumeration of all pairs $\langle p, \langle M_n: n < \omega\rangle\rangle$ where $L = L(M_n)$ has cardinality $< 2^{\aleph_0}$, $\|M_n\| < 2^{\aleph_0}$, and p is a set of cardinality $< 2^{\aleph_0}$ of $\varphi(\bar{x}, \bar{a})$ ($l(\bar{x}) = m$, $\bar{a} \in \prod_{n<\omega} M_n$), and $|P^{M_n}| \leqslant \aleph_0$; each pair appearing $2^{\aleph_0}$ times.

We define by induction on $\alpha < 2^{\aleph_0}$, families E_α of subsets of ω, such that

(1) E_α generates a non-trivial filter $[E_\alpha]$ over ω which is regular,

(2) $\beta < \alpha \rightarrow E_\beta \subseteq E_\alpha$ and for limit α, $E_\alpha = \bigcup_{\beta<\alpha} E_\beta$ and $|E_\alpha| < 2^{\aleph_0}$,

(3) for α odd, $S_\alpha \in E_{\alpha+1}$ or $\omega - S_\alpha \in E_{\alpha+1}$,

(4) for α even if for every $\varphi_1(\bar{x}, \bar{a}_1), \ldots, \varphi_n(\bar{x}, \bar{a}_n) \in p_\alpha$,

$$\{i < \omega: M_i^\alpha \models (\exists \bar{x})(P(x_0) \wedge \cdots \wedge P(x_{m-1}) \wedge \varphi_1(\bar{x}, \bar{a}_1[i]) \wedge \cdots \wedge \varphi_n(\bar{x}, \bar{a}_n[i]))\} \in [E_\alpha]$$

then for some $\bar{b} \in \prod_{i\in\omega} M_i^\alpha$, $\bar{b} = \langle b_0, \ldots, b_{m-1}\rangle$, for every $\varphi_1(\bar{x}; \bar{a}_1) \in p_\alpha$

$$\{i < \omega: M_i^\alpha \models P(b_0[i]) \wedge \cdots \wedge P(b_{m-1}[i]) \wedge \varphi_1(\bar{b}[i], \bar{a}_1[i])\} \in [E_{\alpha+1}].$$

Clearly if we succeed then $[E_{2^{\aleph_0}}]$ is the ultrafilter we want: and there is no problem for $\alpha = 0$ or α limit, or α odd (by 1.1 of the Appendix). So assume α is even and p_α, M_i^α $(i < \omega)$ satisfy the hypothesis of (4) (the other cases are trivial).

Let us define a set V of "forcing conditions" a forcing condition is a finite set v of equations $x_l[i] = a$ where $l < m$, $i < \omega$, $a \in P^{M_i}$; such that $x_l[i] = a \in v$ and $x_l[i] = a' \in v$ implies $a = a'$. V is a partially ordered by inclusion, and it is countable, hence there are no $\aleph_1$ pairwise incompatible conditions. For each $\varphi(\bar{x}; \bar{a}) \in p'_\alpha = \{\bigwedge q : q \subseteq p_\alpha, q \text{ finite}\}$ and finite intersection S of members of E_α let

$$V(S, \varphi(\bar{x}; \bar{a}))$$
$$= \{v \in V: \text{for some } i \in S \text{ and } b_0, \ldots, b_{m-1} \in P(M_i^\alpha)$$
$$M_i^\alpha \models \varphi[b_0, \ldots, b_{m-1}, \bar{a}[i]], \{x_0[i] = b_0, \ldots, x_{m-1}[i] = b_{m-1}\} \subseteq v\}.$$

Clearly each $V(S, \varphi(\bar{x}; \bar{a}))$ is dense in V (by our hypothesis from (4)) and their number is $< 2^{\aleph_0}$. Hence by MA there is a generic $V^* \subseteq V$, i.e., every two members of V^* have a common upper bound, and $V^* \cap V(S, \varphi(\bar{x}; \bar{a})) \neq \emptyset$ for every $\varphi(\bar{x}; \bar{a}) \in p'_\alpha$, S a finite intersection of members of E_α. Let $b_l \in \prod_{i<\omega} M_i^\alpha$, $b_l[i] = a$ iff $x_l[i] = a \in \bigcup\{v: v \in V^*\}$ $(l < m)$. $\bar{b} = \langle b_0, \ldots, b_{m-1}\rangle$ is well defined as V^* is generic. Let

$$E_{\alpha+1} = E_\alpha \bigcup \{\{i < \omega: M_i^\alpha \models \varphi(\bar{b}[i], \bar{a}[i])\}: \varphi(\bar{x}; \bar{a}) \in p_\alpha\}.$$

$[E_{\alpha+1}]$ is a proper filter as V^* intersects each $V(S, \varphi(\bar{x}, \bar{a}))$ and the conclusion of (4) clearly holds.

THEOREM 3.10 (MA): (1) *There is a regular ultrafilter D over ω, such that if M is a λ-saturated model of T_{ind} (see II, 4.8), $\lambda \leq 2^{\aleph_0}$, then M^ω/D is λ-saturated too.*

(2) *If $\lambda < 2^{\aleph_0}$, there is a regular ultrafilter D over λ which is not $\aleph_2$-good such that for any model M of T_{ind}, M^λ/D is λ^+-saturated.*

Proof. (1) Let D be the ultrafilter from 3.9. If p is any 1-type over M^{ω}/D, $|p| = \mu < \lambda$, let $p = \{\varphi_{\alpha}(x; \bar{a}_{\alpha}): \alpha < \mu\}$, let $A_i = \bigcup \{\bar{a}_{\alpha}[i]: \alpha < \mu\}$ by 1.5 of the Appendix, and the elimination of quantifiers of T_{ind}, and the μ^+-saturation of M, there is $B_i \subseteq |M|$, $|B_i| = \aleph_0$, such that any non-algebraic finite type over A_i is realized by an element from B_i. Let us expand M to $M_i = (M, P^{M_i})$, $P^{M_i} = B_i$ and apply 3.9.

(2) Let D_1 be a good $\aleph_1$-incomplete ultrafilter over λ (exists by 3.1)) so D_1 is regular by 2.4, and D_2 the ultrafilter from (1), then $D = D_1 \times D_2$ satisfies our demands ($D_1 \times D_2$ is regular by 3.7(1) and $M^{\lambda \times \omega}/D = (M^{\lambda}/D_1)^{\omega}/D_2$, M^{λ}/D_1 is λ^+-saturated by 2.3 and 2.11 as D_1 is λ^+-good and $\aleph_1$-incomplete. So by (1) $(M^{\lambda}/D_1)^{\omega}/D_2$ is λ^+-saturated too).

Let $\mathscr{G}$ denote a family of functions $g: \lambda \to \lambda$, g onto λ, D a filter over λ. Exercises 3.2–3.4 are from [Sh 71c].

DEFINITION 3.3: $(\mathscr{G}_1, \mathscr{G}_2, D)$ is κ-independent if whenever $j_{\zeta} < \lambda$ $(\zeta < \zeta_0 < \kappa)$ $g^l \in \mathscr{G}_2$ $(l < n < \omega)$ and $f_{\zeta}, f^l \in \mathscr{G}_1$ are distinct $(\zeta < \zeta_0, l < n)$ then $\{\alpha < \lambda: f_{\zeta}(\alpha) = j_{\zeta}$ for $\zeta < \zeta_0, f^l(\alpha) = g^l(\alpha)$ for $l < n\} \neq \emptyset \bmod D$.

EXERCISE 3.1: There is a $\mathscr{G}_1$ of cardinality 2^{λ} such that $(\mathscr{G}_1, \emptyset, D_0)$ is κ-independent, $D_0 = \{\lambda\}$, provided that $\lambda = \lambda^{<\kappa}$.

EXERCISE 3.2: If $(\mathscr{G}_1, \emptyset, D)$ is κ-independent, $D = [E]$ (the filter generated by E), $\mathscr{G}_2$ is a family of functions from λ into α, $\alpha < \kappa$, then there is $\mathscr{G}'_1 \subseteq \mathscr{G}_1$, $|\mathscr{G}_1 - \mathscr{G}'_1| \leq |E| + |\mathscr{G}_2| + \kappa$, such that $(\mathscr{G}'_1, \mathscr{G}_2, D)$ is κ-independent.

EXERCISE 3.3: If $(\mathscr{G}_1, \emptyset, D)$ is κ-independent, $S \subseteq \lambda$ then for some $\mathscr{G}'_1 \subseteq \mathscr{G}_1$ $|\mathscr{G}_1 - \mathscr{G}'_1| < \kappa$ and $(\mathscr{G}_1, \emptyset, D')$ is κ-independent, where $D' = [D \cup \{S\}]$ or $D' = [D \cup \{\lambda - S\}]$.

EXERCISE 3.4: Assume $(\mathscr{G}_1, \emptyset, D)$ is κ-independent, $D = [E]$. Assume also that M_i $(i < \lambda)$ is an L-model, $|P^{M_i}| \leq \chi < \kappa$, $a_{\beta,m} \in \prod_{i<\lambda} M_i$ for $\beta < \beta_0 < 2^{\lambda}$, $1 \leq m \leq n(\beta)$. Assume $\varphi_{\beta} \in L$ and $\{\varphi_{\beta}(x, y_{\beta,1}, \ldots, y_{\beta,n(\beta)}): \beta < \beta_0\}$ is closed under conjunctions, and for every $\beta < \beta_0$

$$\{i < \lambda: M_i \models (\exists x)(P(x) \wedge \varphi_{\beta}(x, a_{\beta,1}[i], \ldots, a_{\beta,n(\beta)}[i]))\} \in D.$$

Then there are $a \in \prod_{i<\lambda} |M_i|$, $\mathscr{G}_1' \subseteq \mathscr{G}_1$, $D' \supseteq D$ such that $|\mathscr{G}_1 - \mathscr{G}_1'| \leq \kappa + |E| + |\beta_0|$, $D' = [E']$ and $|E'| \leq |E| + |\beta_0|$; and for every $\beta < \beta_0$,

$$\{i < \lambda: M_i \models P(a[i]) \wedge \varphi_\beta(a[i], a_{\beta,1}[i], \ldots, a_{\beta,n(\beta)}[i])\} \in D'.$$

(Hint: W.l.o.g. $P^{M_i} \subseteq \chi$.)

EXERCISE 3.5: (1) There is an ultrafilter D over λ such that, if $\lambda = \lambda^{<\kappa}$, then:

(i) If M, N are elementarily equivalent and of cardinality $<\kappa$, then $M^\lambda/D \cong N^\lambda/D$.

(ii) If $\|M\| < \kappa$, $2^\mu \leq 2^\lambda$ then M^λ/D is μ^+-saturated.

(iii) If $|P^{M_i}| \leq \kappa_0 < \kappa$, $2^\mu \leq 2^\lambda$, p a 1-type in $\prod_{i<\lambda} M_i/D$, $P(x_0) \in p$, $|p| \leq \mu$ then p is realized in $\prod_{i<\lambda} M_i/D$.

(iv) If $2^\mu \leq 2^\lambda$, $2^\chi = \lambda$, M a model of T_{ind} of cardinality $\leq \lambda$, then M^λ/D is μ^+-saturated.

(v) If $\|M_i^l\| \leq \chi < \kappa$, M^1, M^2 are elementarily equivalent where $M^l = \prod_{i<\lambda} M_i^l/D$ then M^1, M^2 are isomorphic ($l = 1, 2$) provided that $|L(M^1)| \leq \lambda$.

Remark. Notice the following difference between the proof of Exercise 3.5 sketched by the preceeding exercises, and the proof of 3.1. In the proof of 3.1 the functions in $\mathscr{G}$ are used as partitions of λ, whereas in the proof of Exercise 3.5(1) they are used as elements in the ultraproducts.

EXERCISE 3.5: (2) In Exercise 3.5(1) (and Exercise 3.1–4) instead of starting with the filter $D_0 = \{\lambda\}$, we can start with any κ-complete filter D_1 over λ provided the conclusion of Exercise 3.1 holds. (In Exercises 3.2 and 3.4 $[D_1 \cup E]$ replace $[E]$, and in Exercise 3.5 $D_1 \subseteq D$.] Of course, we can start also with $[D_1 \cup E]$, provided that $|E| < 2^\lambda$, and the conclusion of Exercise 3.1 holds. If there are $A_i \subseteq \lambda$, $A_i \neq \text{mod } D_1$, $A_i \cap A_j = \emptyset$ (for $i < j < \lambda$) then the conclusion of Exercise 3.1 holds.

EXERCISE 3.6: Let f be a 2-place function from $S_{\aleph_0}(\mu)$ to subset of I, let D be a μ^+-good filter over I, and X_i ($i < \mu$) be subsets of I. Suppose

(1) For any $s, t \in S_{\aleph_0}(\mu)$,

$$\bigcap_{i\in s} X_i \cap \bigcap_{i\in t} (I - X_i) \subseteq f(s, t) \text{ mod } D.$$

(2) If $s, t \in S_{\aleph_0}(\mu)$, $s \cap t \neq \emptyset$ then $f(s, t) = \emptyset$.

(3) $f(0, 0) = I$.

(4) If $s, t, u \in S_{\aleph_0}(\mu)$, $s \cap t = \emptyset$, $s \cup t \subseteq u$, then

$$f(s, t) = \bigcup \{f(s_1, t_1): s \subseteq s_1 \subseteq u, t \subseteq t_1 \subseteq u, s_1 \cup t_1 = u\}.$$

Then there are subsets Y^i ($i < \mu$) of I such that:

(A) $Y_i = X_i$ mod D for each $i < \mu$.

(B) For every $s, t \in S_{\aleph_0}(\mu)$,

$$\bigcap_{i \in s} Y_i \cap \bigcap_{i \in t} (I - Y_i) \subseteq f(s, t).$$

On Exercises 3.6 and 3.7 see [Sh 72c].

EXERCISE 3.7: Prove that the following conditions on the filter D over I and the cardinality $\lambda > \aleph_0$ are equivalent.

(1) For every set of L-models M_t ($t \in I$) $\prod_{t \in I} M_t/D$ is λ-compact.

(2) D is $\aleph_1$-incomplete, λ-good and $B(D)$ is λ-saturated, where $B(D)$ is the Boolean algebra of the subsets of I mod D. (Hint: Use the Feferman–Vaught theorem; see e.g. [CK 73] and Exercise 3.6.)

EXERCISE 3.8: In Exercise 3.7 we can replace λ-compact and λ-saturated by (λ, Δ)-compact for Δ the set of Σ_n (or Π_n) formulas (A Σ_n-formula is a formula of the form $\psi(\bar{y}) = (\exists \bar{x}_1)(\forall \bar{x}_2) \ldots \varphi(\bar{x}_1, \ldots, \bar{x}_n, \bar{y})$, φ quantifier free; a Π_n-formula is the negation of a Σ_n-formula).

EXERCISE 3.9: Prove that $B(D)$ is isomorphic to M_0^I/D where M_0 is the Boolean algebra with two elements.

EXERCISE 3.10: Show that in general, in 2.3(3) we cannot replace "for any $\mu < \lambda$" by "for some $\mu < \lambda$". (Hint: Use Exercise 3.5.)

EXERCISE 3.11: (1) Show that for any λ-good filter D, $B(D)$ is (λ, at)-compact (See Exercise 3.12).

(2) Suppose M_0 is a (λ^+, at)-compact Boolean algebra, of cardinality λ^+, and $2^\lambda = \lambda^+$. Prove that there is an $\aleph_1$-incomplete good filter D over λ such that $B(D)$ is isomorphic to M_0.

(3) We can suppose M_0 is a compact Boolean algebra of cardinality 2^λ, and $(\forall \mu < 2^\lambda)(2^\mu \leq 2^\lambda)$, and get the same conclusion.

Remark. In (3) the existence of M_0 implies $(\forall \mu < 2^\lambda)(2^\mu \leq 2^\lambda)$ by VIII, 4.7. Similarly in (2), it implies $2^\lambda = \lambda^+$.

[Proof: (2), (3). Similar to the proof of 3.1.

In the αth step we have a filter $D_\alpha = [E_\alpha]$, $|E_\alpha| \leq \lambda + |\alpha|$, a subalgebra N_α of M_0, $\|N_\alpha\| \leq |\alpha| + \aleph_0$, an embedding H_α of N_α to the

Boolean algebra of the subsets of I such that $0 \neq a \in |M_0|$ implies $H_\alpha(a) \neq \emptyset \bmod D_\alpha$, and a family $\mathscr{G}_\alpha$ of functions from λ to λ, such that for every distinct $g_l \in \mathscr{G}_\alpha$ and $j_l < \lambda$ $(l < n < \omega)$ and $a \in |N_\alpha|$, $a \neq 0$ $\{i < \lambda : g_0(i) = j_0, \ldots, g_{n-1}(i) = j_{n-1}\} \cap H(a) \neq \emptyset \bmod D_\alpha$ and $|\mathscr{G} - \mathscr{G}_\alpha| \leq \lambda + |\alpha|$. (Clearly N_α, H_α, E_α increase with α, and $\mathscr{G}_\alpha$ decreases with it, each of them continuously. In the end $D = D_{2^\lambda}$, $H = H_{2^\lambda}$.)

We have steps of three kinds:

Kind I: We want to include b in the domain of H. Choose $g \in \mathscr{G}_\alpha$, and let $N_{\alpha+1}$ be generated by N_α and b, $H_{\alpha+1}(b) = \{i : g(i) = 0\}$, $\mathscr{G}_{\alpha+1} = \mathscr{G}_\alpha - \{g\}$ and

$$E_{\alpha+1} = E_\alpha \cup \{I - H_{\alpha+1}(b_1) \cap H_{\alpha+1}(a) : a \in N_\alpha, a \cap b_1 = 0$$

and b_1 is b or its complement$\}$.

Kind II: We want to refine the monotonic function $f : S_{\aleph_0}(\lambda) \to D_\alpha$. We do it exactly as in 3.1.

Kind III: We want to decide the fate of $S \subseteq \lambda$ (that is, we want that for some $a \in N_{\alpha+1}$, $S = H_{\alpha+1}(a) \bmod D_{\alpha+1}$).

Let here h denote a function: $h = \{(g^i, j^i) : i < i_0\}$ where $i_0 < \lambda^+ + |\alpha|^+$ and $g^i \in \mathscr{G}$ are distinct. For such h let D^h be the filter generated by E^h,

$$E^h = E_\alpha \cup \{\{g^i(\alpha) = j^i : \alpha < \lambda\} : i < i_0\}$$

and

$$\mathscr{G}^h = \mathscr{G}_\alpha - \{g^i : i < i_0\}.$$

Case (i): For some $a \in N_\alpha$ and h, $H_\alpha(a) = S \bmod D^h$. We then define: $N_{\alpha+1} = N_\alpha$, $H_{\alpha+1} = H_\alpha$, $\mathscr{G}_{\alpha+1} = \mathscr{G}^h$, $E_{\alpha+1} = E^h$ (and of course $D_{\alpha+1} = [E_{\alpha+1}]$). It is easy to check our demands hold.

Case (ii): Not (i), but there are $a, b \in N_\alpha$, h such that $H_\alpha(a) \subseteq S \subseteq H_\alpha(b) \bmod D^h$, and $b - a$ is an atom of N_α (i.e., $(\forall c \in N_\alpha)(b - a \subseteq c \vee (b - a) \cap c = 0)$). We then define $N_{\alpha+1} = N_\alpha$, $\mathscr{G}_{\alpha+1} = \mathscr{G}^h$, $E_{\alpha+1} = E^h \cup \{\lambda - (S - H_\alpha(a))\}$. The only way in which our demands may fail is that for some h^1, $h \subseteq h^1$, and $c \in N_\alpha$, $c \neq 0$ such that $H_\alpha(c) \subseteq S - H_\alpha(a) \bmod D^{h^1}$. But then $H_\alpha(c) \cap H_\alpha(a) = \emptyset \bmod D^{h^1}$, hence $c \cap a = 0$; but also $H_\alpha(c) \subseteq H_\alpha(b) \bmod D^{h^1}$, hence $c \subseteq b$; so as $b - a$ is an atom of N_α, $c = b - a$. Now it follows that $H_\alpha(b) = H_\alpha(a) \cup H_\alpha(c) = S \bmod D^{h^1}$, so case (i) holds, contradiction.

Case (iii): Not (ii) or (ii) but for some $a \in N_\alpha$ and h, $H(a) \subseteq S \bmod D^h$, and for every $b \in N_\alpha$, $h^1 : h \subseteq h^1$, $H_\alpha(b) \subseteq S \bmod D^{h^1}$ implies $b \subseteq a$ (in N_α, of course). We then choose $N_{\alpha+1}$, $\mathscr{G}_{\alpha+1}$, $E_{\alpha+1}$ as in case (ii). The only way in which our demands may fail is that for some $c \in N_\alpha$, $c \neq 0$ and h^1, $h \subseteq h^1$, the following holds:

$$H(c) \subseteq S - H_\alpha(a) \bmod D^{h^1},$$

but then again $c \cap a = 0$, and $H(c \cup a) \subseteq S \bmod D^{h^1}$, so $c \cup a \subseteq a$, hence $c = 0$, contradiction.

Case (iv): Not (i), (ii) or (iii), but for some $b \in N_\alpha$, h: $S \subseteq H(b) \bmod D^h$, and $h \subseteq h^1$, $c \in N_\alpha$, $S \subseteq H(c) \bmod D^{h^1}$ implies $b \subseteq c$.

The proof is similar to that of case (iii); this time $E_{\alpha+1} = E^h \cup \{\lambda - (H(b) - S)\}$.

Case (v): None of the previous ones. Choose h such that if $c \in N_\alpha$, $c \neq 0$, and $S_1 \in \{S, \lambda - S\}$, $h \subseteq h^1$, and $H_\alpha(c) \cap S_1 = \emptyset \bmod D^{h^1}$ then $H_\alpha(c) \cap S_1 = \emptyset \bmod D^h$. Now let $p = \{x \subseteq b: b \in N_\alpha, S \subseteq H_\alpha(b) \bmod D^h\} \cup \{a \subseteq x: a \in N_\alpha, H_\alpha(a) \subseteq S \bmod D^h\}$. It is easy to check p is a type in N_α, and as $|p| \leq \|N_\alpha\| + \aleph_0 \leq \lambda$, p is atomic, and N is (λ^+, at)-compact, some $c \in M_0$ realizes p. Define $N_{\alpha+1}$ as the subalgebra of M_0 generated by $|N_\alpha| \cup \{c\}$, $H_{\alpha+1}$ extend H_α, $H_{\alpha+1}(c) = S$, and $E_{\alpha+1} = E^h$. We leave the reader the checking.]

EXERCISE 3.12: Show that the Boolean algebra M_0 is (λ, at)-compact iff for every $\alpha, \beta < \lambda$, $a_i, b_i \in M_0$ satisfying $M_0 \models a_{i(1)} \leq a_{i(2)} \leq b_{j(2)} \leq a_{j(1)}$ for $i(1) < i(2) < \alpha$, $j(1) < j(2) < \beta$ there is $c \in M_0$, $M_0 \models a_i \leq c \leq b_j$ for $i < \alpha$, $j < \beta$.

EXERCISE 3.13: Suppose M_0 is a Boolean algebra of cardinality 2^λ. Then for some regular, good filter D over λ, M_0 is elementarily embeddable into $B(D)$, and $B(D)$ is μ-saturated, where $\mu = \min\{\mu: 2^\mu > 2^\lambda\}$.

PROBLEM 3.14: For any λ, characterize the possible $B(D)$ for D a filter over I.

EXERCISE 3.15: Prove that for any n and λ, there is a filter D over λ such that $B(D)$ is (λ^+, Σ_n)-compact but not (λ^+, Π_n)-compact, and vice versa. (Hint for Exercises 3.13 and 3.15: The proof is like Exercise 3.11, but N_α $(\alpha < 2^\lambda)$ is not predetermined.)

EXERCISE 3.16: Define a λ-good filter in a Boolean algebra just as in Definition 2.1. Prove the parallel of Exercise 3.6. Show that if $\mathbf{B}_l$ are Boolean algebras, D_l a filter in $\mathbf{B}_l$, $h: \mathbf{B}_1 \to \mathbf{B}_2$ a homomorphism onto $h^{-1}(1) = D_1$, and D_2, D_1 are λ-good, then the filter $h^{-1}(D_2)$ is λ-good. [Hint: Let $g: S_{\aleph_0}(\mu) \to h^{-1}(D_2)$, $\mu < \aleph_0$; so $hg: S_{\aleph_0}(\mu) \to D_2$, so it has a multiplicative refinement $g_1 : S_{\aleph_0}(\mu) \to D_2$ and choose $X_i \in \mathbf{B}_1$, $h(X_i) = g_1(\{i\})$. Hence $\bigcap_{i \in s} X_i \subseteq g(s) \bmod D_1$. Define $g^1 : S_{\aleph_0}(\mu) \to D_1$ by

$$g^1(s) = \bigcup \left\{1 - \left(\bigcap_{i \in t} X_i - g(t)\right) : t \subseteq s\right\}, \qquad g^2 : S_{\aleph_0}(\mu) \to D_1$$

a multiplicative refinement of g^1, and $g^*: S_{\aleph_0}(\mu) \to h^{-1}(D_2)$ defined by $g^*(s) = \bigcap_{i \in s} X_i \cap g^2(s)$ is as regarded.]

EXERCISE 3.17: Suppose $\mu \leq \chi = \chi^{<\kappa} < 2^\mu$, and show there is a regular, κ-good filter over μ generated by a family of $\leq \chi$ sets. (Hint: Use 3.1, and show that any regular, good ultrafilter over μ has such subfilters, using Exercise 2.5.)

EXERCISE 3.18: Suppose E is a family of subsets of λ, $|E| \leq \lambda$ closed under finite intersections, $\emptyset \notin E$, and D is a λ-regular filter over I.

(1) There is a function $G : [E] \to D$ such that:

(i) for $A, B \in [E]$, $G(A \cap B) = G(A) \cap G(B)$, $A \subseteq B \Leftrightarrow G(A) \subseteq G(B)$,

(ii) if $A_i \in [E]$ $(i < \delta)$,

$$\bigcap_{i<\delta} A_i = \emptyset \Leftrightarrow \bigcap_{i>\delta} G(A_i) = \emptyset.$$

(2) $[E]$ is μ-good [μ-regular] iff $[\{G(A)\colon A \in E\}]$ is μ-good [μ-regular] (for each μ). (Hint: (2) See Exercise 2.5.)

EXERCISE 3.19: If D is λ-regular, $\lambda^{<\kappa} = \lambda$, then there is a κ-good, λ-regular filter $D_1 \subseteq D$.

DEFINITION 3.4: D is a uniform filter if all members have the same cardinality (it is $|I|$ when D is a filter over I).

EXERCISE 3.20: (1) A ultrafilter D over λ is uniform iff $D^0_\lambda \subseteq D$ where $D^0_\lambda = \{A \subseteq \lambda\colon |\lambda - A| < \lambda\}$ which is a uniform filter.

(2) If D is a uniform ultrafilter D over λ, $\mathrm{cf}[(\lambda, <)^\lambda/D] > \lambda$.

(3) If D is a regular ultrafilter over λ, δ_i limit, then

$$\mathrm{cf}\left[\prod_{i<\lambda} (\delta_i, <)/D\right] > \lambda.$$

EXERCISE 3.21: If for some ultrafilter D over λ, $D_0 \subseteq D$, D_0 a filter and $\mathrm{cf}[(\kappa, <)^\lambda/D] = \mu$, then there are functions $f_i : \lambda \to \kappa$, $(i < \mu)$ such that for no $g : \lambda \to \kappa$ for every $i < \mu$ $(\kappa, <)^\lambda/D_0 \vDash f_i < g$.

EXERCISE 3.22: If D_0 is a filter over λ and there are function $f_i : \lambda \to \kappa$ ($i < \mu$) such that for every $g : \lambda \to \kappa$ for some $i < \mu$ $(\kappa, <)^\lambda/D_0 \models g < f_i$, then for every ultrafilter D over λ, extending D_0 $\mathrm{cf}[(\kappa, <)^\lambda/D] \le \mu$. If also $i < \alpha < \mu$ implies $(\kappa, <)^\lambda/D_0 \models f_i < f_\alpha$, then $\mathrm{cf}[(\kappa, <)^\lambda/D] = \mu$.

Remark. The family of f_i's as in Exercise 3.22 are called a (D_0, μ)-scale for κ. On independence results concerning scales see, e.g. [He 74]. The existence of an $2^{\aleph_0}$-scale is weaker than MA.

The following exercises are an improvement of Theorem 3.10.

EXERCISE 3.23: Suppose D is λ-regular $M = \prod_{t \in I} M_t/D$, $A \subseteq |M|$, $|A| \le \lambda$ then there are finite $P_t \subseteq |M_t|$ such that $A \subseteq P^M$ where $(M, P^M) = \prod_{t \in I} (M_t, P_t)/D$.

EXERCISE 3.24: Suppose D an ultrafilter over I, $\|M_t\| \le \lambda$, $A \subseteq |M|$, $M = \prod_{t \in I} M_t/D$.

(1) If $|A| < \mathrm{cf}[(\lambda, <)^I/D]$, then for some $P_t \subseteq |M_t|$, $|P_t| < \lambda$, and $A \subseteq P^M$ where $(M, P^M) = \prod_{t \in I} (M_t, P_t)/D$.

(2) If $\lambda = \aleph_{\alpha+n}$, $|A| < \mathrm{cf}[(\aleph_{\alpha+l}, <)^I/D]$ for $0 \le l \le n$, then for some $P_t \subseteq |M_t|$, $|P_t| < \aleph_\alpha$ and $A \subseteq P^M$, where

$$(M, P^M) = \prod_{t \in I} (M_t, P_t)/D.$$

Remark. Now a combination of Exercise 3.5 and 3.24 gives results like 3.10 (for the existence of such filters the existence of scales is sufficient, see Exercise 3.22).

EXERCISE 3.25: Suppose $\lambda = 2^\mu < 2^{\mu^+}$, and some μ^+-complete filter D_1 over λ satisfies Exercise 3.1 and a (D_1, λ^{++})-scale for μ^+ exists. Then

(1) over λ there is a regular and good ultrafilter D, such that for every λ^{++}-saturated model of T_{ind}, M^λ/D is λ^{++}-saturated;

(2) over λ^+ there is a regular not good ultrafilter D, such that for every model M of T_{ind}, M^{λ^+}/D is λ^{++}-saturated.

EXERCISE 3.26: (1) If $\lambda = \aleph_0$ and *MA* holds, M, N are elementarily equivalent $|\mathrm{L}(M)| \le \lambda$, $\|M\| + \|N\| \le \lambda$ then for some regular ultrafilter D over λ, $M^\lambda/D \cong N^\lambda/D$.

(2) If, e.g., $\|M\| + \|N\| + |\mathrm{L}(M)| \le \mu^+$, M, N elementarily equivalent and the assumption of Exercise 3.25 holds, then for some regular ultrafilter D over λ $M^\lambda/D = N^\lambda/D$ provided that $\lambda^{++} = 2^\mu$.

CONJECTURE 3.26: (1) In Exercise 3.26(1) we can eliminate MA.

(2) It is consistent with ZFC, that for $\lambda = \aleph_0$ for every regular ultrafilter D over λ^+, $\mathrm{cf}(\lambda^+, <)^{\lambda^+}/D = \lambda^{++}$, $\mathrm{cf}(\lambda, <)^{\lambda^+}/D > \lambda^{++}$ (thus Exercise 3.5(1) is essentially best possible). We may replace λ, λ^+ by λ^+, 2^λ.

DEFINITION 3.5: For an ultrafilter D (over I) and regular cardinal κ; $\mathrm{lcf}(\kappa, D)$ is the smallest cardinality λ such that: there is a subset of

$$\{a \in \kappa^I/D: \kappa^I/D \models \alpha < a \text{ for each } \alpha < \kappa\}$$

which is unbounded from below, and has cardinality λ.

Remarks. (1) We do not distinguish between κ and the model $(\kappa, <)$ (till the end of this section).

(2) Clearly $\mathrm{lcf}(\kappa, D)$ is an infinite regular cardinal or 1.

(3) Till the end of this section, an ultrafilter is over I if not mentioned otherwise.

CLAIM 3.11: (1) *If* $\kappa > |I|$ *is regular,* $\mathrm{lcf}(\kappa, D) = 1$*; in fact* $\mathrm{lcf}(\kappa, D) = 1$ *iff D is* cf *κ-descendingly complete (see Definition* 6.1).

(2) κ^I/D *is not* $(\kappa + \mathrm{lcf}(\kappa, D))^+$*-compact.*

(3) *If D is a λ-good $\aleph_1$-incomplete ultrafilter and* $\kappa < \lambda$*, then* $\mathrm{lcf}(\kappa, D) \geq \lambda$.

Proof. Immediate.

THEOREM 3.12: *For every* $\mu = \mu^{\aleph_0} \leq 2^\lambda$, *and regular* κ, $\aleph_0 < \kappa \leq \mu$ *there is a regular ultrafilter D over λ such that* $\mathrm{lcf}(\aleph_0, D) = \kappa$ *and*

$$\mu = \min\left\{\prod_{i<\lambda} n_i/D: n_i < \omega, \prod_{i<\lambda} n_i/D \geq \aleph_0\right\}.$$

THEOREM 3.13: (1) *Suppose* $\aleph_0 = \lambda_0 < \lambda_1 < \cdots < \lambda_n = \lambda^+$ *each* λ_l *is regular and* $\lambda_{l+1} \leq \mu_l \leq 2^\lambda$, μ_l *regular (for* $l < n$*). Then for some regular λ_1-good ultrafilter D over λ,* $\mathrm{lcf}(\kappa, D) = \mu_l$ *whenever* $\lambda_l \leq \kappa < \lambda_{l+1}$. *In fact D is not* λ_1^+*-good.*

(2) *So in particular for every λ and regular* $\mu \leq \lambda$, *there is a regular ultrafilter over λ which is μ-good but not μ^+-good.*

THEOREM 3.14: *Suppose in the previous theorem that* $\lambda_1 = \aleph_1$ *and* $k < \omega$, $\chi_l = \chi_l^{\aleph_0}$ $(l \leq k)$ *are given* $\chi_0 < \chi_1 < \cdots < \chi_k = 2^\lambda$ *and* χ_l^i ($i < 2$, $l < k$ or $i = 0$, $l = k$) *are regular cardinals* $> \aleph_0$, $\chi_l^0 \leq \chi_l$, $\chi_l^1 \leq \chi_l$. *Let*

$$J_l = \{a \in \omega^\lambda/D: |a| = \chi_l\}$$

where

$$|a| = \prod_{i<\lambda} a[i]/D = |\{b \in \omega^\lambda/D : \omega^\lambda/D \vDash b < a\}|,$$

then we can assume in addition, that

(i) *for each* $a \in \omega^\lambda/D$, $|a|$ *is finite or* $\in \{\chi_l : l \leq k\}$,

(ii) *for* $l < k$ $\mathrm{cf}(J_l) = \chi_l^1$ (J_l *is order as a subset*),

(iii) *for* $l \leqslant k$ $\mathrm{cf}(J_l^*) = \chi_l^0$ (J_l^* *the inverse of* J_l).

DEFINITION 3.6: (1) In Definitions 3.1 and 3.3 we allow the functions in $\mathscr{G}$, $\mathscr{G}_1$ resp., not be necessarily onto λ, and then demand $j_i \in \mathrm{Range}(g_i)$, $j_\xi \in \mathrm{Range}(f_\xi)$ resp.

(2) For a family $\mathscr{G}$ with domain λ let (FI for finite intersection): $\mathrm{FI}(\mathscr{G}) = \{h\colon h$ a function, $\mathrm{Dom}\, h \subseteq \mathscr{G}$, $|\mathrm{Dom}\, h| < \aleph_0$ and $h(f) \in \mathrm{Range}(f)\}$.

For $h \in \mathrm{FI}(\mathscr{G})$ let $A_h = \{i < \lambda\colon f \in \mathrm{Dom}\, h$ implies $f(i) = h(f)\}$. $\mathrm{FI}_s(\mathscr{G}) = \{A_h : h \in \mathrm{FI}(\mathscr{G})\}$.

CLAIM 3.15: (1) *All relevant lemmas remain true under the new Definition* 3.6: 3.2, 3.3, 3.4, *Exercises* 3.1,2,3 and 4. *E.g.*, 3.2 *becomes if* $|J_i| \leq \lambda$ *for* $i < 2^\lambda$, *then there is a family* $\{f_i : i < 2^\lambda\}$, f_i *from* λ *onto* J_i, *independent* mod$\{\lambda\}$.

(2) *If* $\mathscr{G}$ *is independent* mod D_α *for* $\alpha < \delta$, D_α $(\alpha < \delta)$ *an increasing sequence of filters over* I *then* $\mathscr{G}$ *is independent* mod $\bigcup_{\alpha<\delta} D_\alpha$.

(3) *If* D *is a filter over* λ, $\mathscr{G}$ *independent* mod D, *then there is a maximal filter* $D^* \supseteq D$, *modulo which* $\mathscr{G}$ *is independent* (*i.e., there is no filter* $D^1 \neq D^*$, *and* $D^1 \supseteq D^*$ *such that* $\mathscr{G}$ *is independent* mod D^1).

(4) $\mathscr{G}$ *is independent* mod D *iff* $A_h \neq \emptyset$ mod D *for every* $h \in \mathrm{FI}(\mathscr{G})$; *and* $h_1 \subseteq h_2$ *iff* $A_{h_2} \subseteq A_{h_1}$. (*We discard the case* $|\mathrm{Range}\, f| = 1$.)

DEFINITION 3.7: (1) In a Boolean algebra, a partition is a maximal set of pairwise disjoint non-zero elements. An element is based on a partition W, if $b \in W$ *implies* $b \subseteq a$ or $b \cap a = 0$; the element is supported by a set W of elements if it is based on some partition $W_1 \subseteq W$ (notice if W is a partition, a is based on W iff W supports a). A set W is dense if for every non-zero a there is in W $b \subseteq a$.

(2) If D is a filter over I, instead of speaking on $B(D)$, $A/D, \ldots$, we speak on $\{A : A \subseteq I\}$, $A, \ldots$ (and then everything is mod D.)

(3) $\mathrm{CC}(B)$, for a Boolean algebra B, is the minimal regular cardinal $\lambda \geq \aleph_0$ such that every partition of B has cardinality $< \lambda$.

CLAIM 3.16: *In a Boolean algebra*

(1) *a is based on W iff* $1 - a$ *is based on W,*

(2) *if* a_l *is based on* W_l $(l = 1, 2)$ *then* $a_1 \cap a_2$, $a_1 \cup a_2$, $a_1 - a_2$ *are based on* $\{c \cap d: c \in W_1, d \in W_2, c \cap d \neq 0\}$ *which is a partition,*

(3) *for any set W of elements closed under finite intersections such that every* $a \in W$ *belong to a partition* $\subseteq W$, *the set of elements it supports is a sub-algebra, and include all elements of W,*

(4) *if a, b are supported by W and* $c \in W \Rightarrow [c \subseteq a \Leftrightarrow c \subseteq b]$ *then* $a = b$.

CLAIM 3.17: *Suppose D is a maximal filter over I modulo which* $\mathscr{G}$ *is independent.*

(1) $\mathrm{FI}_s(\mathscr{G})$ *is dense* mod D; *hence each element is supported by it.*

(2) *For each* $f \in G$, $\{f^{-1}(j): j \in \mathrm{Range}\, f\}$ *is a partition* mod D.

(3) *If* $h_n \in \mathrm{FI}(\mathscr{G})$, $\mathrm{Dom}\, h_n (n < \omega)$ *are pairwise disjoint,* $A \subseteq I$, *and* $A \cap A_{h_n} = \emptyset \bmod D$ *then* $A = \emptyset \bmod D$.

(4) *Let* $\mathscr{G}$ *be the disjoint union of* $\mathscr{G}_1$, $\mathscr{G}_2$, $A \subseteq I$ *is supported by* $\mathrm{FI}_s(\mathscr{G}_1)$, $h \in \mathrm{FI}(\mathscr{G})$, $h = h^1 \cup h^2$, $h^l \in \mathrm{FI}(\mathscr{G}_l)$. *If* $A_h \subseteq A \bmod D$ *then* $A_{h^1} \subseteq A \bmod D$.

(5) $\mathrm{CC}(B(D)) = \aleph_0$ *iff for only finitely many* $f \in G$ *is* $|\mathrm{Range}(f)| > 1$, *and for no one* $|\mathrm{Range}(f)| = \aleph_0$. *Otherwise* $\mathrm{CC}(D)$ *is the first regular* $\lambda > \aleph_0$ *such that* $f \in \mathscr{G} \Rightarrow |\mathrm{Range}(f)| < \lambda$. *Moreover, if* $\mu \geq \lambda$ *is regular,* $A_i \neq \emptyset \bmod D$ *for* $i < \mu$, *then there is* $S \subseteq \mu$, $|S| = \mu$ *such that for* $n < \omega$, *and distinct* $i(l) \in S$, $\bigcap_{l<n} A_{i(l)} \neq \emptyset \bmod D$.

(6) $|B(D)| \leq |\mathscr{G}|^{<\kappa} + 2^{<\kappa}$ *when* $\kappa = \mathrm{CC}(B(D))$.

Proof. (1) Otherwise for some $A \subseteq I$, $A \neq \emptyset \bmod D$ and $A_h \nsubseteq A \bmod D$ for each $h \in \mathrm{FI}(\mathscr{G})$ hence $A_h \cap (I - A) \neq \emptyset \bmod D$. Hence $\mathscr{G}$ is independent $\mathrm{mod}[D \cup \{I - A\}]$, contradicting the maximality of D (as $A \neq \emptyset \bmod D$).

(2) Clearly the $f^{-1}(j)$'s are pairwise disjoint mod D. Suppose $A \cap f^{-1}(j) = \emptyset \bmod D$ for each $j \in \mathrm{Range}(f)$, but $A \neq \emptyset \bmod D$. By (1) for some $h \in \mathrm{FI}(G)$ $A_h \subseteq A \bmod D$, and clearly for some h_1, $h \subseteq h_1 \in \mathrm{FI}(G)$, $f \in \mathrm{Dom}\, h_1$. So also $A_{h_1} \subseteq A \bmod D$; but let $h_1(f) = j_0$, so $A \cap A_{h_1} \subseteq A \cap f^{-1}(j_0) = \emptyset \bmod D$, contradiction.

(3) Suppose $A \neq \emptyset \bmod D$ hence by (1) for some $h \in \mathrm{FI}(\mathscr{G})$, $A_h \subseteq A \bmod D$; for all but finitely many n's, $\mathrm{Dom}\, h \cap \mathrm{Dom}\, h_n = \emptyset$, so $h \cup h_n \in \mathrm{FI}(\mathscr{G})$. Now $A_{h \cup h_n} \neq \emptyset \bmod D$ and $A_{h \cup h_n} \subseteq A \bmod D$, and $A \cap A_{h \cup h_n} \subseteq A \cap A_{h_n} = \emptyset \bmod D$ contradiction.

(4) Let A be based on the partition $\{A_{h_\zeta}: \zeta < \zeta(0)\}$, $h_\zeta \in \mathrm{FI}(\mathscr{G}_1)$. Now, $A_h \subseteq A \bmod D$, hence $A \cap A_{h_\zeta} = \emptyset \bmod D \Rightarrow A_{h_\zeta} \cap A_h = \emptyset \bmod D$

hence (as $\mathscr{G}_1 \cup \mathscr{G}_2$ is independent mod D) $A \cap A_{h_\zeta} = \emptyset \bmod D \Rightarrow A_{h^1} \cap A_{h_\zeta} = \emptyset \bmod D$ hence (otherwise $A_{h^1} - A$ contradicts the maximality of the partition) $A_{h^1} \subseteq A \bmod D$.

(5) Notice that if $f_n \in G$ are distinct, $|\text{Range} f_n| > 1$, $j_n^0 \neq j_n^1 \in \text{Range} f_n$, and h_n is defined by $h_n(f_l) = j_l^0$ for $l < n$ and $h_n(f_n) = j_n^1$, then $h_n \in \text{FI}(\mathscr{G})$ and A_{h_n}/D $(n < \omega)$ are $\aleph_0$ pairwise disjoint non-zero elements of $B(D)$. By 3.17(2) $f \in \mathscr{G} \Rightarrow |\text{Range} f| < \text{CC}(B(D))$, hence the first phrase follows, and the second follows from the third.

So let $\mu \geq \lambda$ be regular, $A_i \neq \emptyset \bmod D$ for $i < \mu$. By 3.17(1) there are $h_i \in \text{FI}(\mathscr{G})$ such that $A_{h_i} \subseteq A_i \bmod D$. By its definition $\lambda > \aleph_0$, hence $\mu > \aleph_0$ and we assume μ is regular; and Dom h_i is finite. So by 1.4 of the Appendix for some $n < \omega$ and $S_1 \subseteq \mu$, $|S_1| = \mu$, for every $i \neq j \in S_1$ Dom $h_i \cap$ Dom $h_j = \{f_0, \ldots, f_{n-1}\}$. As $|\text{Range} f_l| < \lambda \leq \mu$, there are $S \subseteq S_1$ and $j_l \in \text{Range} f_l$ $(l < n)$ such that $|S| = \mu$ and for every $i \in S$, $h_i(f_0) = j_0, h_i(f_1) = j_1, \ldots, h_i(f_{n-1}) = j_{n-1}$. Clearly S is as required.

(6) By 3.16(4) each partition W supports $\leq 2^{|W|}$ elements. By 3.17(1) every element of $B(D)$ is supported by $\{A_h/D: h \in \text{FI}(\mathscr{G})\}$. By 3.17(5) $|\text{FI}(\mathscr{G})| \leq |\mathscr{G}| + \kappa$, and every partition of $B(D)$ has cardinality $< \kappa$. Collecting those facts the result $|B(D)| \leq |\mathscr{G}|^{<\kappa} + 2^{<\kappa}$ becomes obvious.

LEMMA 3.18: *Suppose D a maximal filter modulo which $\mathscr{G}^* \cup \mathscr{G}$ is independent, $\mathscr{G}^*$, $\mathscr{G}$ are disjoint; $f \in \mathscr{G}^* \cup \mathscr{G}$ implies* $|\text{Range}(f)| <$ cf α: cf $\alpha > \aleph_0$, $\mathscr{G} = \bigcup_{\beta<\alpha} \mathscr{G}_\beta$, $\mathscr{G}_\beta$ *increasing and let* $\mathscr{G}^\beta = \mathscr{G} - \mathscr{G}_\beta$. *Suppose also that the filters $D_i (i < \alpha)$ satisfy $D_i \subseteq D_j$ for $i < j < \alpha$ and:*

(i) *D_i is generated by D and sets supported* (mod D) *by*

$$\text{FI}_s(\mathscr{G}^* \cup \mathscr{G}_i).$$

(ii) $\mathscr{G}^* \cup \mathscr{G}^\beta$ *is independent* mod D_β.

(iii) *D_i is maximal with respect to* (i) *and* (ii).

Then

(1) $D^* = \bigcup_{i<\alpha} D_i$ *is a maximal filter modulo which $\mathscr{G}^*$ is independent.*

(2) *If $\mathscr{G}^*$ is empty, D^* is an ultrafilter and* (ii) *is satisfied whenever D_i is non-trivial and satisfies* (i).

(3) *If D_i' satisfies* (i) *and* (ii) *we can extend it to a filter satisfying* (i), (ii) *and* (iii).

(4) *If $f \in \mathscr{G}^\beta$ then $\langle f^{-1}(t)/D_\beta : t \in \text{Range}(f)\rangle$ is a partition in $B(D_\beta)$.*

Proof. Part (3) is trivial (as the family of D's satisfying (i) and (ii) is closed under ascending chains), and part (2) first statement follows by (1). The second statement of (2) holds easily.

(1) As $\mathscr{G}^*$ is independent mod D_i for each i, clearly it is independent mod D^*. So we have to prove only the maximality. So let $A \subseteq I$, $A \neq \emptyset$ mod D^*; now for some $\mathscr{G}^+ \subseteq \mathscr{G}$, $|\mathscr{G}^+| < \mathrm{CC}(B(D))$, and A is supported by $\mathrm{FI}_s(\mathscr{G}^+ \cup \mathscr{G}^*)$ mod D; so for some $\gamma < \alpha$ $\mathscr{G}^+ \subseteq \mathscr{G}_\gamma$ (as $\mathrm{cf}\,\alpha > |\mathrm{Range}\,(f)|$ for $f \in \mathscr{G}^* \cup \mathscr{G}$, and $\mathrm{cf}\,\alpha > \aleph_0$, by 3.17(5) $\mathrm{cf}\,\alpha \geq \mathrm{CC}(B(D))$ hence $\mathrm{cf}\,\alpha > |\mathscr{G}^+|$). Now $[D_\gamma \cup \{I - A\}]$ is a filter properly extending D_γ, and satisfies (i), hence by (iii) should fail to satisfy (ii). That is for some $h \in \mathrm{FI}(\mathscr{G}^* \cup \mathscr{G}^\gamma)$ $A_h \subseteq A$ mod D_γ, hence (by (ii) and 3.16(2)), for some $B \in D_\gamma$, B/D is supported by $\mathrm{FI}_s(\mathscr{G}^* \cup \mathscr{G}_\gamma)$ and $A_h \cap B \subseteq A$ mod D (by 3.16(2)) so $A_h \subseteq A \cup (\lambda - B)$ mod D. Let $h = h^1 \cup h^2$, $h^1 \in \mathrm{FI}(\mathscr{G}^*)$, $h^2 \in \mathrm{FI}(\mathscr{G}^\gamma)$. By 3.17 (4) (where D, $A \cup (\lambda - B)$, $\mathscr{G}^* \cup \mathscr{G}_\gamma$, $\mathscr{G}^\gamma$ stands for $D, A, \mathscr{G}, \mathscr{G}_1$). $A_{h^1} \subseteq A \cup (\lambda - B)$ mod D hence $A_{h^1} \subseteq A$ mod D_γ. As this holds for every $A \neq \emptyset$ mod D^*, D^* is a maximal filter modulo which $\mathscr{G}^*$ is independent.

(4) Suppose $f \in \mathscr{G}_\beta$ is a counterexample. Then for some $A \subseteq I$, $A \neq \emptyset$ mod D_β and for every $t \in \mathrm{Range}(f)$ $A \cap f^{-1}(t) = \emptyset$ mod D_β so for some B_t supported by $FI_s(\mathscr{G}^* \cup \mathscr{G}_\beta)$, $A \cap f^{-1}(t) \subseteq B_t$ mod D and $I - B_t \in D_\beta$. By 3.17(2) there is a partition $\langle A_{h_i}/D : i < \xi\rangle$ of $B(D)$ on which A is based; by a hypothesis $\xi < \mathrm{cf}\,\alpha$ and w.l.o.g. for some $\zeta \leqslant \xi$, for $i < \zeta$, $A_{h_i} \subseteq A$ mod D, and for $i \geqslant \zeta (i < \xi)$ $A_{h_i} \cap A = \emptyset$ mod D. So for $i < \zeta$, $A_{h_i} \cap f^{-1}(t) \subseteq B_t$ mod D. Let $h_{i,0} = h_i \restriction (\mathscr{G}^* \cup \mathscr{G}_\beta)$, so as B_t is supported by $FI_s(\mathscr{G}^* \cup \mathscr{G}_\beta)$ and $f \in \mathscr{G}^\beta = \mathscr{G} - \mathscr{G}_\beta$, by 3.17(4) $A_{h_{i,0}} \subseteq B_t$ mod D. Define $A^* = \bigcup_{i<\zeta} A_{h_{i,0}}$, now:

Firstly, $A \subseteq A^*$ mod D [as for $i < \zeta$ $A_{h_i} \subseteq A_{h_{i,0}} \subseteq A^*$ so $A - A^*$ is disjoint to $A_{h_i} (i < \zeta)$ and for $i \geqslant \zeta$ (but $\leqslant \xi$) $(A - A^*) \cap A_{h_i} \subseteq A \cap A_{h_i} = \emptyset$ mod D, so $A - A^* = \emptyset$ mod D would contradict the choice of $\langle A_{h_i}; i < \xi\rangle$].

Secondly, A^* is supported by $FI_s(\mathscr{G}^* \cup \mathscr{G}_\beta)$ [for this it suffices to prove that A^*/D is the union of $A_{h_{i,0}}/D (i < \zeta)$ in $B(D)$, i.e. that if $h \in FI(\mathscr{G}^* \cup \mathscr{G})$, $A_h \cap A_{h_{i,0}} = \emptyset$ mod D for every $i < \zeta$, then $A_h \cap A^* = \emptyset$ mod D; but $A_h \cap A_{h_{i,0}} = \emptyset$ mod D iff $A_h \cap A_{h_{i,0}} = \emptyset$ iff $h \cup h_{i,0}$ is not a function, so this is obvious. More formally use a maximal antichain $\subseteq \{A_h : h \in FI(\mathscr{G}^* \cup \mathscr{G}_\beta)$, h extends some $h_{i,0}$ or $h \cup h_{i,0}$ is not a function for every $i < \zeta\}$].

Thirdly, for each $t \in \mathrm{Range}\,(f)$ $A^* \subseteq B_t$ mod D [otherwise for some $h \in FI(\mathscr{G}^* \cup \mathscr{G})$ $A_h \subseteq (A^* - B_t)$ mod D, then w.l.o.g. h extends some $h_{i,0}$, $i < \zeta$ (by the proof of "secondly") and we get contradiction to

$A_{h_{t,0}} \subseteq B_t \bmod D$]. But $I - B_t \in D_\beta$, hence $I - A^* \in D_\beta$, hence $I - A \in D_\beta$, a contradiction.

Remark. We have shown that $B(D)$ is complete (this holds generally when for some $\mathscr{G}$, D is maximal such that $\mathscr{G}$ is independent mod D).

CLAIM 3.19: (1) *Let $\mathscr{G}$ be in dependent* mod D, *and* $\langle f^{-1}(t)/D: t \in \text{Range } f\rangle$ *is a partition of* $B(D)$ *for every* $f \in \mathscr{G}$ [*which holds if* D *is maximal over which* $\mathscr{G}$ *is independent but also if* D, $\mathscr{G}$ *are* D_β, $\mathscr{G}_\beta$ *from* 3.18 (*see* 3.18(4))]. *If* $g: I \to \kappa$, *and* $\kappa^I/D \models \alpha < g/D$ *for every* $\alpha < \kappa$, *and* $f \in \mathscr{G}$, *then* $\kappa^I/D \models f/D < g/D$ (*hence for every ultrafilter* $D^* \supseteq D$, $\kappa^I/D^* \models f/D^* < g/D^*$).

(2) ***Suppose*** $\text{CC}(B(D)) \leq \kappa$, κ, μ ***regular,*** $f_i: I \to \kappa$ ***for*** $i < \mu$; ***and*** $\kappa^I/D \models \alpha < f_i/D < f_j/D$, ***for every*** $\alpha < \kappa$, $j < i < \mu$; ***but for any*** $g: I \to \kappa$ ***if for every*** $\alpha < \kappa$ $\kappa^I/D \models \alpha < g/D$ ***then for some*** i, $\kappa^I/D \models g/D > f_i/D$. ***Then for every ultrafilter*** $D^* \supseteq D$, $\text{lcf}(\kappa, D) = \mu$ ***(and this is exemplified by the*** f_i/D^****'s).***

(3) ***Part one holds for*** $[D \cup \{A_h\}]$ ***when*** $h \in \text{FI}(G)$, $f \notin \text{Dom } h$ ***(hence for*** $[D \cup \{A\}]$ ***when*** $A \neq \emptyset \bmod D$).

Proof. (1) Suppose not, then $A = \{t \in I: f(t) \geq g(t)\} \neq \emptyset \bmod D$, hence by a hypothesis for some $j_0 < \kappa A \cap f^{-1}(j_0) \neq \emptyset \bmod D$, hence

$$\{t: g(t) \leq j_0\} \supseteq A \cap f^{-1}(j_0) \neq \emptyset \bmod D,$$

contradicting $\kappa^I/D \models j_0 < g/D$.

(2) Otherwise there is $g/D^*(g: I \to \kappa)$ contradicting it. Let $A_\alpha = \{t: \alpha > g(t)\}$ and define inductively $\alpha(i) < \kappa (i < \kappa)$ as follows: $\alpha(0) = 0$, $\alpha(\delta) = \bigcup_{i<\delta} \alpha(i)$, $\alpha(i+1)$ the first $\alpha > \alpha(i)$ such that $A_\alpha - A_{\alpha(i)} \neq \emptyset \bmod D$. If $\alpha(i)$ is defined for every $i < \kappa$, $\{(A_{\alpha(i+1)} - A_{\alpha(i)})/D: i < \kappa\}$ contradicts $\text{CC}(B(D)) \leq \kappa$, hence for some i, $\alpha(i)$ is defined but not $\alpha(i+1)$. Define $g_1 \in I \to \kappa$, by

$$g_1(t) = \begin{cases} f_0(t) & t \in A_{\alpha(i)} \\ g(t) & t \in I - A_{\alpha(i)} \end{cases}$$

then g_1 contradicts the hypothesis.

(3) Easy.

CLAIM 3.20: ***Suppose*** $\mathscr{G}$ ***is a non-empty family of functions onto*** ω, D ***a maximal filter modulo which*** $\mathscr{G}$ ***is independent.***

(1) *If* $f \in \mathscr{G}$, *then* $\prod_{t \in I} (1 + f(t))/D \leq |\mathscr{G}|^{\aleph_0}$ *hence for any ultrafilter* $D^* \supseteq D$, $\prod_{t \in I} f(t)/D^* = |\{a: \omega^I/D \models a < f/D\}| \leq |\mathscr{G}|^{\aleph_0}$.

(2) *If* $f \in \mathscr{G}$, $g_l : I \to \omega$, $\omega^I/D \models g_l/D \leq f/D$ $(l = 0, 1)$ *and for every* n $g_0^{-1}(n) = g_1^{-1}(n)$ mod D *then* $\omega^I/D \models g_0/D = g_1/D$.

Proof. (1) The phase after "hence" is trivial, so suppose $\prod_{t \in I} (1 + f(t))/D > \mu =_{\text{def}} |\mathscr{G}|^{\aleph_0}$, so there are $g_i : I \to \omega$, $g_i(t) \leq f(t)$ such that $\{t: g_i(t) \neq g_j(t)\} \neq \emptyset$ mod D for $i < j < \mu^+$. The number of possible sequences $\langle g_i^{-1}(n)/D: n < \omega \rangle$ is $\leq \mu$ (it is $\leq |B(D)|^{\aleph_0}$; but by 3.17(5) $CC(B(D)) = \aleph_1$, hence by 3.17(6) $|B(D)| \leq \mu^{\aleph_0}$, but $\mu^{\aleph_0} = \mu$). So w.l.o.g. $g_i^{-1}(n)/D = g_j^{-1}(n/D)$ for $i < j < \mu^+$ but we get a contradiction by part (2).

(2) Suppose not, so $A = \{t \in I: g_0(t) \neq g_1(t)\} \neq \emptyset$ mod D. By 3.17(2) there is $n < \omega$ such that $A \cap f^{-1}(n) \neq \emptyset$ mod D. Let for $l, k \leq n$ $A_{k,l} = \{t \in A: f(t) = n,\ g_0(t) = k,\ g_1(t) = l\}$ so clearly $A \cap f^{-1}(n) = \bigcup_{k,l \leq n} A_{k,l}$, hence for some k, l $A_{k,l} \neq \emptyset$ mod D. As $A_{k,l} \subseteq A$, clearly $k \neq l$; it is also clear that $A_{k,l} \subseteq g_0^{-1}(k)$ and $A_{k,l} \subseteq g_1^{-1}(l)$; hence $g_0^{-1}(k) \cap g_1^{-1}(l) \neq \emptyset$ mod D, but $g_0^{-1}(k)$ is disjoint to $g_0^{-1}(l)$ and $g_0^{-1}(l) = g_1^{-1}(l)$ mod D, contradiction.

CLAIM 3.21: *Suppose* $\mathscr{G}^*$, $\mathscr{G}$, $\mathscr{G}_\beta$, $\mathscr{G}^\beta$, α, D_i, D^* *are as in Claim* 3.18. *Suppose furthermore that* $f \in \mathscr{G}^* \Rightarrow |\text{Range}(f)| < \kappa$; $\kappa > \aleph_0$ *or* $\mathscr{G}^* = \emptyset$, $\kappa = \aleph_0$; κ *is regular;* $f_\beta \in \mathscr{G}_{\beta+1} - \mathscr{G}_\beta$ *and for every* $\xi < \kappa$, $\{t: \xi < f_\beta(t) \in \kappa\} \in D_{\beta+1}$.

Then for every ultrafilter $D^1 \supseteq D^*$, $\text{lcf}(\kappa, D^1) = \text{cf}\,\alpha$ *and this is exemplified by* f_β/D^1 $(\beta < \alpha)$ *and* $\kappa^I/D^1 \models f_\beta/D^1 < f_\gamma/D^1$ *for* $\gamma < \beta < \alpha$.

Proof. By 3.19(1) $\kappa^I/D^* \models f_\beta/D^* < f_\gamma/D^*$ for $\gamma < \beta < \alpha$, and by 3.19(2) (for $\kappa > \aleph_0$) and 3.18(2) (for $\kappa = \aleph_0$) it suffices to prove

(*) If $g : I \to \kappa$, and for every $\xi < \kappa$, $\kappa^I/D^* \models \xi < g/D^*$ then for some $\beta < \alpha$ $\kappa^I/D^* \models g/D^* > f_\beta/D^*$.

Suppose g falsifies (*) then for any $\xi < \kappa$, $B_\xi = \{t: \xi < g(t)\} \in D^*$ hence for some $\alpha(\xi) < \alpha$, $B_\xi \in D_{\alpha(\xi)}$. As $\text{cf}\,\alpha > \kappa$, $\alpha(*) = \sup_{\xi < \kappa} \alpha(\xi) < \alpha$, so $B_\xi \in D_{\alpha(*)}$ for every ξ, so $\{t: \xi < g(t)\} \in D_{\alpha(*)}$ for every ξ, hence by 3.19(1) $\{t: f_{\alpha(*)}(t) \leqslant g(t)\} = I \bmod D_{\alpha(*)}$, hence $\{t: f_{\alpha(*)}(t) \leqslant g(t)\} = I$ mod D^*, contradiction.

CLAIM 3.22: *If* $|J_i| \leq \lambda$ *for* $j < j_0 \leq 2^\lambda$ *then there are functions* f_i *from* λ *onto* J_i, *and a regular* λ^+*-good filter* D, *such that* $\mathscr{G}$ *is independent* mod D, *and* D *is a maximal such filter.*

Proof. By 3.15(1) and 3.2 there are functions $f_j (j < j_0)$ from λ onto J_j and g_i from λ onto λ (for $i < 2^\lambda$) such that $\mathscr{G}^* \cup \mathscr{G}$ is independent mod$\{\lambda\}$, where $\mathscr{G}^* = \{f_j: j < j_0\}$, $\mathscr{G} = \{g_i: i < 2^\lambda\}$. Let D_0 be a maximal filter modulo which $\mathscr{G}^* \cup \mathscr{G}$ is independent $\{H_i: 0 < i < 2^\lambda\}$ a list of all functions from $S_{\aleph_0}(\lambda)$ into $\{S: S \subseteq \lambda\}$, each appearing 2^λ times. We define D_i as in 3.18 (letting $\mathscr{G}_i = \{f_j: j < i\}$). We use f_0 to define a regular D_1. If Range $(H_i) \subseteq D_i$, we work as in Claim 3.3, using f_i to get $D' \supseteq D$, and extend it to D_{i+1} by 3.18(3), then $\bigcup_i D_i$ is the required filter.

CLAIM 3.23: *Let D be a maximal filter modulo which $\mathscr{G}$ is independent, $\mu = \mathrm{CC}(B(D))$, for infinitely many $f \in \mathscr{G}$ $|\mathrm{Range}\, f| > 1$ and $D^* \supseteq D$ an ultrafilter. Then D^* is not μ^+-good.*

Proof. Choose distinct $f_n \in \mathscr{G}$ and then $j_n \in$ Range f_n such that $f_n^{-1}(j_n) \notin D^*$ (possible as $|\mathrm{Range}\ f_n| > 1$). Let $A_n = \bigcap_{l<n}(\lambda - f_n^{-1}(j_n)) \in D^*$ and define $g: S_{\aleph_0}(\mu) \to D^*$ by $g(s) = A_{|s|}$. We suppose $g_1 : S_{\aleph_0}(\mu) \to D^*$ refine g and is multiplicative, and get a contradiction.

As $g_1(\{\alpha\}) \in D^*$ (for $\alpha < \mu$), $g_1(\{\alpha\}) \neq \emptyset$ mod D. Hence for some $h_\alpha \in \mathrm{FI}(\mathscr{G})$ $A_{h_\alpha} \subseteq g_1(\{\alpha\})$ mod D. By 3.17(5) $\mu > \aleph_0$, and it is regular by definition; so for some $S \subseteq \mu$, $|S| = \mu$ and for some k Dom $h_\alpha \cap \{f_n: n < \omega\} \subseteq \{f_n: n < k\}$ for every $\alpha \in S$. By 3.17(5) for some $S_1 \subseteq S$, $|S_1| = \mu$, for any $n < \omega$, and distinct $\alpha(l) \in S_1 (l < n)$ $\bigcap_{l<n} A_{h_{\alpha(l)}} \neq \emptyset$ mod D. Choose $s = \{\alpha(l): l \leq k\}$, $\alpha(l) \in S_1$ distinct so $g(s) \subseteq I - f_k^{-1}(j_k)$, hence $g_1(s) \subseteq I - f_k^{-1}(j_k)$.

Now $g_1(s) = \bigcap_{l \leq k} g_1(\{\alpha(l)\}) \supseteq \bigcap_{l \leq k} A_{h_{\alpha(l)}} = A_h$, where $h = \bigcup_{l \leq k} h_{\alpha(l)}$ (by the choice of the $\alpha(l)$'s and S_1, $h \in \mathrm{FI}(\mathscr{G})$); but $f_k \notin$ Dom $h_{\alpha(l)}$ (by the choice of S) hence $f_k \notin$ Dom h, so extend h to $h_1 \in \mathrm{FI}(\mathscr{G})$ by defining $h_1(f_k) = j_k$. Clearly we get $A_{h_1} \subseteq A_h \subseteq g_1(s)$; but we had proved $g_1(s) \subseteq I - f_k^{-1}(j_k)$, hence $g_1(s)$ is disjoint to $A_{h_1} \subseteq f_k(j_k)$, contradiction.

CLAIM 3.24: *Suppose D is an $\aleph_1$-incomplete maximal filter over I modulo which $\mathscr{G} \cup \mathscr{G}^*$ is independent, $\mathscr{G} \cup \mathscr{G}^*$ a family of functions from I onto ω, $\mathscr{G}^* = \{f_i: i < \delta\}$*, of $\delta > \aleph_0$. *Define D_1 as the filter generated by*

(A) *the members of D,*

(B) *$A^i_g = \{t: g(t) < f_i(t)\}$ when $g: I \to \omega$, and for every $n < \omega$, $g^{-1}(n)$ is supported by $\mathrm{FI}_s(\mathscr{G} \cup \{f_j: j < i\})$, and for any $h \in \mathrm{FI}(\mathscr{G} \cup \{f_j: j < i\})$ for some n $A_h \cap \{t: g(t) \leq n\} \neq \emptyset$* mod D. *Then*

(1) *$\mathscr{G}$ is independent* mod D_1.

(2) *For any ultrafilter $D^* \supseteq D_1$.*

(i) *$\omega^I/D^* \vDash f_j/D^* < f_i/D^*$ for $j < i < \delta$,*

(ii) *for any* $a \in \omega^I/D^*$, *for some* $i < \delta$, $\omega^I/D^* \models a < f_i/D^*$ *or for some* $g : I \to \omega$, $\omega^I/D^* \models g/D^* = a$ *and* $\omega^I/D \models n < g/D$ *for every* $n < \omega$.

Proof. (1) Note that by this we prove D_1 is non-trivial (take h the empty function). So suppose g_l, $i(l)$ $(l < n)$ as in (B), $h \in \mathrm{FI}(\mathscr{G})$, and we have to prove

$$A_h \cap \bigcap_{l<n} A_{g_l}^{i(l)} \neq \emptyset \bmod D.$$

Let $\{j(k): k < m\} = \{i(l): l < n\}$, $j(k) < j(k+1)$ (note an ordinal may appear several times in the list $i(l)$ $(l < n)$). We now define by induction on $k \le m$ $h_k \in \mathrm{FI}(\mathscr{G} \cup \mathscr{G}^*)$, such that (let $j(m) = \delta$):

(i) $\mathrm{Dom}\, h_k \subseteq \mathscr{G} \cup \{f_j: j < j(k)\}$,

(ii) $h_0 = h$; h_k $(k \le m)$ is increasing,

(iii) if $i(l) \le j(k)$ then for some unique n_l,

$$A_{h_k} \subseteq g_l^{-1}(n_l) \bmod D,$$

(iv) $h_{k+1}(f_{j(k)})$ is the minimal natural number n, $n > n_l$ whenever $i(l) = j(k)$, $l < n$.

For doing this it suffices to prove that: if $h^1 \in \mathrm{FI}(\mathscr{G} \cup \{f_j: j < j^*\})$, $i(l) = j^*$, then for some h^2, n^*, $h^1 \subseteq h^2$, $h^2 \in \mathrm{FI}(\mathscr{G} \cup \{f_j: j < j^*\})$ and $A_{h^2} \subseteq g_l^{-1}(n^*) \bmod D$. But as $i(l)$, g_l satisfy (B) for some n^1, $A_{h^1} \cap \{t \in I:\ g_l(t) \le n^1\} \neq \emptyset \bmod D$, hence for some n^*, $A_{h^1} \cap g_l^{-1}(n^*) \neq \emptyset \bmod D$. Now A_{h^1}, $g_l^{-1}(n^*)$ are supported by $\mathrm{FI}_s(\mathscr{G} \cup \{f_j: j < j^*\})$, hence (by 3.16(2)) also $A_{h^1} \cap g_l^{-1}(n^*)$ is, hence there is $h^2 \in \mathrm{FI}(\mathscr{G} \cup \{f_j: j < j^*\})$, $A_{h^2} \subseteq A_{h^1} \cap g_l^{-1}(n^*) \bmod D$. So we can define the h_k's.

Now it is trivial to check that $A_{h_m} \subseteq A_h \cap \bigcap_{l<n} A_{g_l}^{i(l)}$ hence $A_h \cap \bigcap A_{g_l}^{i(l)} \neq \emptyset \bmod D$.

(2) (i) Clearly $f_i : I \to \omega$ for every n $f_i^{-1}(n)$ is supported by $\mathrm{FI}_s(\{f_i\})$, hence by $\mathrm{FI}_s(\mathscr{G} \cup \{f_\alpha: \alpha < j\})$ and by 3.17(2) $A \neq \emptyset \bmod D$ implies that for some n $A \cap f_i^{-1}(n) \neq \emptyset \bmod D$. Hence $A_{f_j}^i$ is one of the generators of D_1, so $\{t \in I: f_j(t) < f_i(t)\} = A_{f_j}^i \in D^1 \subseteq D^*$.

(ii) Suppose $a = f/D^*$, and let $\{A_\alpha: \alpha < \alpha_0\}$ be a maximal family of pairwise disjoint subsets of I mod D, such that $A_\alpha \neq \emptyset \bmod D$, but for every $n < \omega$, $\{t \in I: n \ge f(t)\} \cap A_\alpha = \emptyset \bmod D$. As $\mathrm{CC}(B(D)) = \aleph_1$ (by 3.17(5)) $\alpha_0 < \omega_1$, so w.l.o.g. $\alpha_0 \le \omega$. If $\alpha_0 < \omega$, let $A = \bigcup_{l<\alpha_0} A_l$; and if $\alpha_0 = \omega$, let $A = \bigcup_{l<\omega} (A_l - \{t: f(t) \le l\})$. In any case for each $n < \omega$, $l < \alpha_0$ $A \cap f^{-1}(n) = \emptyset \bmod D$, and $A_l \subseteq A \bmod D$, hence $\{f^{-1}(n)/D: n < \omega\} \cup \{A/D\}$ is a partition of $B(D)$. Clearly there is $\beta < \delta$ such that the $f^{-1}(n)$'s $(n < \omega)$ and A are supported by $\mathrm{FI}_s(\mathscr{G} \cup \{f_j: j < \beta\})$.

Case (i). $A \notin D^*$. Define

$$g : I \to \omega, \qquad g(t) = \begin{cases} f(t) & t \notin A, \\ 0 & t \in A. \end{cases}$$

Clearly A_g^β is one of the generators of D_1 (by 3.16(2)), hence $\omega^I/D^* \models a = f/D^* = g/D^* < f_\beta/D^*$.

Case (ii). $A \in D^*$. As D is $\aleph_1$-incomplete, then are $B_n \in D$, $\bigcap_{n<\omega} B_n = \emptyset$; w.l.o.g. $B_{n+1} \subseteq B_n$. Define

$$g : I \to \omega, \qquad g(t) = \begin{cases} f(t) & t \in A, \\ n & t \notin A, \quad t \in B_n - B_{n+1}. \end{cases}$$

Then $\omega^I/D^* \models a = g/D^*$, and $\omega^I/D \models n < g/D$ for every $n < \omega$.

Proof of Theorem 3.12. Let D, $\mathscr{G}$ be such that $\mathscr{G}$ is a family of μ functions from λ onto ω, independent mod D, D a maximal filter over λ modulo which $\mathscr{G}$ is independent, and D is regular (by 3.22). Let $\mathscr{G} = \{f_i : i < \mu\kappa\}$ ($\mu\kappa$-ordinal product) $\mathscr{G}_\beta = \{f_i: i < \beta\}$. We define D_i ($i < \mu\kappa$) as in 3.18 (for $\mathscr{G}^* = \emptyset$) such that $A_i^n = \{t : n \leqslant f_i(t)\} \in D_{i+1}$ for each $n < \omega$. We have to prove only that $D' = [D_i \cup \{A_i^n: n < \omega\}]$ satisfies (i) and (ii). D' is not trivial as $A_i^{n+1} \subseteq A_i^n$ and $A_i^n \supseteq f_i^{-1}(n) \neq \emptyset$ mod D, and as A_i^n is based on $\{f_i^{-1}(n): n < \omega\} \subseteq \mathrm{FI}_s(\mathscr{G}_{i+1})$, (i) holds. Hence by 3.18(2) (ii) holds for D' and by 3.18(3) we can define D_{i+1} properly. By 3.18(2) $D^* = \bigcup_i D_i$ is an ultrafilter.

By 3.21 $\mathrm{lcf}(\omega, D^*) = \mathrm{cf}(\mu\kappa) = \kappa$, and $f_\beta/D^*(\beta < \mu\kappa)$ exemplify this.

Now if $\omega^\lambda/D^* \models n < g/D$ for every $n < \omega$, let $i(*)$ be such that $\omega^\lambda/D^* \models f_{i(*)}/D^* < g/D^*$, so

$$\prod_{i<\lambda} g(i)/D^* = |\{f/D^*: \omega^\lambda/D^* \models f/D^* < g/D^*\}| \geq |\{f_i/D^*: i(*) < i < \mu\kappa\}| = \mu$$

(for $i < j$ $\omega^\lambda/D \models f_j/D^* < f_i/D^*$, hence $f_j/D^* \neq f_i/D^*$). On the other hand $\prod_{\alpha<\lambda} f_i(\alpha)/D^* \leq \mu^{\aleph_0} = \mu$ by 3.20(1).

Proof of Theorem 3.13. Let D be a maximal filter modulo which $\mathscr{G}$ is independent, D regular (see 3.22) such that $\mathscr{G} = \bigcup_{l<n} \mathscr{G}_l$, $\mathscr{G}_l = \{f_\kappa^i: \lambda_l \leq \kappa < \lambda_{l+1}, \kappa \text{ regular}, i < 2^\lambda \mu_l\}$ (ordinal multiplication) where f_κ^i is onto κ.

We define by downward induction, filters D_l ($l \leq n$) such that

(i) $D_n = D$, $D_{l+1} \subseteq D_l$,

(ii) D_l is a maximal filter modulo which $\bigcup_{k<l} \mathscr{G}_k$ is independent,

(iii) for any ultrafilter D^* extending D_l, $\lambda_l \leq \kappa < \lambda_{l+1}$, κ regular $\mathrm{lcf}(\kappa, D^*) = \mu_l$.

If D_{l+1} is defined, we define $D_{l,i}$ ($i < 2^{\lambda}\mu_l$) as in 3.18 (with $\bigcup_{k<l} \mathscr{G}_k$, $\{f_\kappa^i: \lambda_l \leq \kappa < \lambda_{l+1}, \kappa \text{ regular}, i < \beta\}$, $2^{\lambda}\,\mu_l$ for $\mathscr{G}^*$, $\mathscr{G}_\beta$, α resp.) such that $\{t \in I: \xi \leq f_\kappa^i(t)\} \in D_{l,i+1}$ for each $\xi < \kappa$. Let $D_l = \bigcup_i D_{l,i}$, so (i) and (ii) holds by 3.18, and by 3.21 (iii) holds. For $l = 0$, for even i's we do as above, and for odd i's we immitate the proof of 3.22 (and 3.1) to get a λ_1-good ultrafilter D_0. Now D_0 is not λ_1^+-good by 3.23 and 3.17(5).

Proof of Theorem 3.14. Combine the previous proofs, but use also Claim 3.24.

THEOREM 3.25: *For every regular $\lambda \leq |I|$, there is a regular ultrafilter D over I such that*

(i) *D is λ-good but not λ^+-good,*

(ii) $\mathrm{lcf}(\kappa, D) = \lambda$ *for every* $\kappa < \lambda$,

(iii) *every function $f: S_{\aleph_0}(|I|) \to D$, $|\mathrm{Range}\,(f)| < \lambda$, can be refined to a multiplicative one.*

Proof. Let $\mathscr{G} = \bigcup_{\alpha<\lambda} \mathscr{G}_\alpha$, every function $f \in \mathscr{G}_\alpha$ is from I onto α, $\mathscr{G}$ is independent mod D^0 and $|\mathscr{G}_\alpha| = 2^{\lambda}$.

We now define by induction on α filters D_α such that

(a) $\alpha < \beta$ implies $D_\alpha \subseteq D_\beta$,

(b) $D^0 \subseteq D_0$,

(c) D_α is a maximal filter modulo which $\bigcup_{\alpha \leq \beta < \lambda} \mathscr{G}_\beta$ is independent,

(d) for each $\kappa \leq |I|$ and $\alpha < \lambda$ for some $f \in \mathscr{G}_\alpha : \kappa^I/D_{\alpha+1} \models i < f/D_{\alpha+1}$ for every $i < \kappa$.

(e) D_α is $|\alpha|^+$-good

We leave the details to the readers.

EXERCISE 3.27: Generalize Claims 3.15–3.24, when we replace $\mathrm{FI}_s(\mathscr{G})$ be any family S of subsets I, "$\mathscr{G}$ independent mod D" by "$S \cap D^c = \emptyset$" (i.e., D^c disjoint to S, $D^c = \{A \subseteq I: \lambda - A \in D\}$) [Hint: (1) 3.17(5), (6), 3.19(1), 3.20 (assume $\mathrm{CC}(B(D))$) 3.21, 3.22, 3.23 and 3.24 (see 3.18) are left to the reader.

(2) 3.15(2), (3), (4), 3.17(1) and 3.19(2) are trivial.

(3) 3.17(2) becomes: if $S^1 \subseteq S$, $A \neq B \in S^1 \Rightarrow \lambda - A \cap B \in D$; $A \ni S - S^1 \Rightarrow (\exists B \in S^1)(\exists C \in S)(C \subseteq A \cap B)$, then S^1 is a partition.

(4) 3.17(3) becomes: if $S^1 \subseteq S$, and $(\forall A \in S)\,((\exists B, C \in S)\,(C \subseteq A \cap B)$ and $(\forall B \in S^1)\,(A \cap B = \emptyset \bmod D)$ then $A = \emptyset \bmod D$.

(5) 3.17(4) becomes: if D is a maximal filter disjoint to $\{\{B \cap A_h: B \in S, h \in FI(\mathscr{G})\}$, B_1, B_2 supported by S, $h \in \mathrm{FI}(\mathscr{G})$ and $B_1 \cap A_h \subseteq B_2 \bmod D$ then $B_1 \subseteq B_2 \bmod D$.

(6) In 3.18 replace $\mathscr{S}_1$, $\mathrm{FI}_s(\mathscr{S}_1 \cup \mathscr{S}^*)$)} by S_1, $\{B \cap A_h: B \in S_1, h \in \mathrm{FI}(\mathscr{S}^*)\}$ ($\mathscr{S}^* \subseteq \mathscr{S}$) resp., and assume $\aleph_0 + \mathrm{CC}(B(\mathrm{D})) < \mathrm{cf}\,\alpha$.]

EXERCISE 3.28: (1) Let $\mathbf{B}$ be a Boolean algebra, $\mathrm{CC}(\mathbf{B}) \le \lambda$ then $\mathbf{B}$ is (λ, at)-compact iff $\mathbf{B}$ is complete.

(2) if $\lambda = \mathrm{CC}(\mathbf{B}(D))$, D a λ-good filter, then $\mathbf{B}(D)$ is complete.

EXERCISE 3.29: Suppose $\mathbf{B}$ is a complete Boolean algebra, $\mathrm{CC}(\mathbf{B}) \le \lambda$, $\|\mathbf{B}\| \le 2^\lambda$. Then for some regular filter over λ, $\mathbf{B} \cong B(D)$. [Hint: If $\|\mathbf{B}\| \le \lambda$, this is easy by Exercise 3.27 (on 3.22) and Exercise 3.28. Otherwise let f_i be from λ onto λ, D a maximal filter modulo which $\{f_i: i < 2^\lambda\}$ is independent; let $\mathbf{B}_0$ be the subalgebra of $B(D)$ generated by $f_i^{-1}(j)/D$, h_0, a homomorphism from $\mathbf{B}_0$ onto $\mathbf{B}$, which we extend step by step, and get $h^*: B(D) \to B$, and let $D^* = \{A: h(A/D) = 1).]$

EXERCISE 3.30: Suppose D_i ($i < \lambda$) are filters over I, D a filter over λ, and let $D^* = \{A: \{i < \lambda: A \in D_i\} \in D\}$.

(1) D^* is a filter; and if $\mu \ge \mathrm{CC}(B(D_i))$, $\mu^* \to (\mu)^2_\lambda$ (e.g., $\mu^* = (2^{<\mu})^+$, $\mu \ge \lambda$, see 2.5 of the Appendix) then $\mu^* \ge \mathrm{CC}(\mathrm{B}(D^*))$. If D, D_i are ultrafilters, then D^* is an ultrafilter too (note D is an ultrafilter iff $\mathrm{CC}(B(D)) = 1$, so we can take $\mu^* = 1$).

(2) Suppose $\lambda < \kappa, \mu$ and κ, μ are regular, $f_\alpha: I \to \kappa$ for $\alpha < \mu$ and for some $A \in D$, for every $i \in A$ the following holds: $\kappa^I/D_i \models \gamma < f_\alpha/D_i < f_\beta/D_i$ for every $\gamma < \kappa$, $\beta < \alpha < \mu$ but for any $g: I \to \kappa$: if for every $\gamma < \kappa\, \kappa^I/D_i \models \gamma < g/D_i$, then for some α, $\kappa^I/D \models g/D_i > f_\alpha/D_i$. Then for any ultrafilter $D^1 \supseteq D^*$, $\mathrm{lcf}(\kappa, D) = \mu$.

(3) If $\mu_i = \prod_{t \in I} n_t/D_i$ then $\prod_{i<\lambda} \mu_i \ge \prod_{t \in I} n_t/D$.

(4) Suppose each D_i is a maximal filter modulo which $\mathscr{S}$ is independent, $\mu \ge \mathrm{CC}(B(D_i))$ (see 3.17(5)), χ regular $> 2^\lambda$ and $(\forall \chi_1 < \chi)$ $(\chi_1^{<\mu} < \chi)$. Then $\chi \ge \mathrm{CC}(B(D))$. [Hint: (1) If $A_\alpha \ne \emptyset \bmod D(\alpha < \chi)$ let $f(\alpha, \beta) = \min\{i: A_\alpha, A_\beta \ne \emptyset \bmod D_i, A_\alpha \cap A_\beta = \emptyset \bmod D_i\}$ and see Definition 2.1 of the Appendix.

(2) Use 3.19(2).

(4) Suppose $A_\alpha \ne \emptyset \bmod D(\alpha < \chi)$ are pairwise disjoint mod D. As $\chi > 2^\lambda$, w.l.o.g. $S = \{i: A_\alpha \ne \emptyset \bmod D_i\}$ does not depend on α. By 2.8 of the Appendix and the proof of 3.17(5) w.l.o.g. for each $i \in S$, and distinct $\alpha(l)$ ($l < n$), $\bigcap_{l<n} A_{\alpha(l)} \ne \emptyset \bmod D_i$. Hence $A_0 \cap A_1 \ne \emptyset \bmod D$, contradiction. The proof gives more than required.]

EXERCISE 3.31: (1) Suppose for each $\alpha < \lambda$, κ_α, μ_α are regular cardinals $>\lambda$, $\kappa_\alpha < \mu_\alpha$, and $\kappa_\alpha \neq \kappa_\beta$ for $\alpha \neq \beta$, $|I| \geq \sum_{\alpha \leq \lambda} \kappa_\alpha$, $\mu_\alpha \leq 2^{|I|}$. Then for some regular ultrafilter D over I, $\mathrm{lcf}(\kappa_\alpha, D) = \mu_\alpha$. (Hint: Use Exercise 3.30 and the proof of 3.13.)

(2) In (1) if $\mathscr{G}_\alpha(\mathscr{G})$ is a family of $2^{|I|}$ functions from I onto $\kappa_\alpha(\lambda)$, $\bigcup_\alpha \mathscr{G}_\alpha \cup \mathscr{G}$ is independent mod D_0, then there is a filter $D \supseteq D_0$, $\mathscr{G}$ is independent mod D, such that for every ultrafilter $D^0 \supseteq D$ and $\alpha < \lambda$, $\mathrm{lcf}(\kappa_\alpha, D^0) = \mu_\alpha$. [Hint: (2) By the proof of 3.13, for every finite $w \subseteq \lambda$ there is a maximal filter D_w modulo which $\mathscr{G}$ is independent, and for $\alpha \in w\, f^\alpha_\xi/D$ ($\xi < \mu_\alpha$) exemplifies $\mathrm{lcf}(\kappa_\alpha, D^1) = \mu_\alpha$ for any ultrafilter $D^1 \supseteq D_w$, and then use Exercise 3.30(2).]

EXERCISE 3.32 (G.C.H.): Suppose $\kappa < \mu_\kappa \leq 2^{|I|}$, μ_κ regular for each $\kappa \in S$; $\kappa \in S$ implies $\kappa \leq |I|$ is regular, and for inaccessible λ $\{\kappa \in S: \kappa < \lambda\}$ is bounded below λ. Then for some regular ultrafilter D over I, $\mathrm{lcf}(\kappa, D) = \mu_\kappa$ for each $\kappa \in S$. [Hint: Let $\{f^\alpha_\kappa: \alpha < 2^{|I|}, \kappa \in S\}$ is independent mod D_0, D_0 regular, and Range $(f^\alpha_\kappa) = \kappa$. For regular χ, $S_1 \subseteq \{\kappa \in S: \chi \leq \kappa\}$, let

$(*)^\chi_{S_1}$ there is a filter $D = D(S_1, \chi) \supseteq D_0$ such that
(i) $\{f^\alpha_\kappa: \alpha < 2^{|I|}, \kappa < \chi\}$ is independent mod D,
(ii) for every ultrafilter $D^* \supseteq D$, $\kappa \in S_1$, $\mathrm{lcf}(\kappa, D^*) = \mu_\kappa$ and this is exemplified by $f^\alpha_\kappa/D^*(\alpha < \mu_\alpha)$.

We prove it by induction on the order type of S_1 (for χ a successor of a singular, use χ^+ and then add χ^+, if necessary).]

DEFINITION 3.8: For a model M and a complete Boolean algebra $\mathbb{B}$ and ultrafilter D of it, we define the Boolean ultrapower $N = M^{(\mathbb{B})}/D$ as follows (the sup exists by the completeness):

(i) On the set $\{f$: Dom f is a partition of $\mathbb{B}$, Range $f \subseteq M\}$ we define an equivalence relation $\approx : f_1 \approx f_2$ iff $\sup\{a_1 \cap a_2: a_l \in \text{Range} f_l\ (l = 1, 2)\ f_1(a_1) = f_2(a_2)\} \in D$. By 3.16(2) $\approx$ is an equivalence relation, and for every $f_1, \ldots, f_n$ there are $f^1_l \approx f_l$ which have the same domain.

Now $|N|$ will be the set of equivalence classes of $\approx$ and

$$R^N = \Big\{\langle f_1/\approx, \ldots, f_n/\approx\rangle: \sup\Big\{\bigcap_l a_l: a_l \in \text{Dom} f_l, \langle f_1(a_1), \ldots, f_n(a_n)\rangle \in R\Big\} \in D\Big\}$$

(similarly for F^N).

EXERCISE 3.34: Prove that Boolean ultrapower is a particular case of limit-ultrapower from Definition 4.2, hence the parallel of Exercise 4.10 holds.

EXERCISE 3.35: Let D be a filter over I, $\mathbb{B} = B(D)$, $\aleph_1 = \mathrm{CC}(\mathbb{B})$, $D_1 \supseteq D$ an ultrafilter, and D^* an ultrafilter of $\mathbb{B}$, $D^* = \{A/D: A \in D_1\}$. We define a function $H: \omega^{(\mathbb{B})}/D^* \to \omega^I/D_1$ as follows: if $f/\approx\ \in \omega^{(\mathbb{B})}/D^*$, let $\mathrm{Dom} f = \{a_n: n < \alpha_0 \leq \omega\}$ (as $\mathrm{CC}(\mathbb{B}) = \aleph_1$), let $a_n = A_n/D$ and w.l.o.g. $I = \bigcup_{n<\omega} A_n$, $A_n \cap A_m = \emptyset$ for $n \neq m$ (for let $A'_0 = I - \bigcup_{n>0} A_n$, $A'_{n+1} = A_n - \bigcup_{l\leq n} A'_l$) and let $H(f) = g/D_1$ where $g(t) = n \leftrightarrow t \in A_n$. Prove

(1) H is well-defined as is an elementary embedding.

(2) The range of H is an initial segment ω^I/D_1.

Remark. By this, the problem on "what can be $\{\prod_{t\in I} n_t/D: n_t < \omega\}$" can be reduced to a problem of Boolean ultrapowers.

EXERCISE 3.36: In 3.19(2) replace the $\alpha < \kappa$ by f^j $(j < \kappa)$.

EXERCISE 3.37: Suppose $\mathscr{G}$ is a family of functions from I into the ordinals, and D is a maximal filter over I modulo which $\mathscr{G}$ is independent. Let $\mu = \mathrm{CC}(B(D))$, and suppose D is λ-good. Then for any ultrafilter $D_1 \supseteq D$, and regular $\kappa, \chi < \lambda$ satisfying $\kappa, \chi \geq \mu + |\mathscr{G}|^+$ and: $\kappa > 2^\chi$ or $\chi > 2^\kappa$, or $\kappa = \chi$ are weakly compact ω^I/D_1 has no (κ, χ)-Dedekind cut (see VII, Definition 1.10(B)).

EXERCISE 3.38: Use Exercise 3.37, Theorem 3.25 and Exercise 3.30 to investigate the possible $\{(\kappa, \chi): \omega^I/D$ has a (κ, χ)-Dedekind cut$\}$. Phrase the open problems.

QUESTION 3.39: Investigate the problems from Exercise 3.38.

VI.4. Keisler's order

DEFINITION 4.1: (1) For every λ Keisler's $\trianglelefteq_\lambda$-order on theories is defined as follows: $T_1 \trianglelefteq_\lambda T_2$ if when M_l is a model of T_l $(l = 1, 2)$ and D a regular ultrafilter over λ, the λ^+-compactness of M_2^λ/D implies the λ^+-compactness of M_1^λ/D.

(2) Keisler's order $\trianglelefteq$ on theories is defined as follows: $T_1 \trianglelefteq T_2$ if for every λ $T_1 \trianglelefteq_\lambda T_2$.

(3) T_1 and T_2 are $\trianglelefteq_\lambda$-equivalent ($\trianglelefteq$-equivalent) if $T_1 \trianglelefteq_\lambda T_2 \trianglelefteq_\lambda T_1$ ($T_1 \trianglelefteq T_2 \trianglelefteq T_1$).

LEMMA 4.1: (1) *The relations $\trianglelefteq_\lambda$, $\trianglelefteq$ are transitive and reflexive.*

(2) *$\trianglelefteq_\lambda$- and $\trianglelefteq$-equivalence are equivalence relations.*

Proof. (1) Easy to check the transitivity by the definition. The reflexivity follows by 1.9.

(2) Follows from (1).

Remark. So naturally, $\trianglelefteq$ ($\trianglelefteq_\lambda$) is an order on its equivalence classes.

LEMMA 4.2: (1) *T is $\trianglelefteq_\lambda$-maximal (i.e., $T_1 \trianglelefteq_\lambda T$ for every T_1) iff for every model M of T and regular ultrafilter D over λ,*

$$M^\lambda/D \text{ is } \lambda^+\text{-compact} \Leftrightarrow D \text{ is } \lambda^+\text{-good}.$$

(2) *T is $\trianglelefteq_\lambda$-minimal (i.e., $T \trianglelefteq_\lambda T_1$ for every T_1) iff for every model M of T and regular ultrafilter D over λ, M^λ/D is λ^+-compact.*

(3) *T is $\trianglelefteq$-maximal [-minimal] iff it is $\trianglelefteq_\lambda$-maximal [-minimal] for every λ.*

Proof. (1) Let M_1 be a model of T_1, M a model of T, and D a regular ultrafilter over λ. If T satisfies the condition mentioned in (1), and M^λ/D is λ^+-compact, then D is λ^+-good, hence by 2.3 and 2.11 M_1^λ/D is λ^+-compact.

As this holds for every M_1, D; $T_1 \trianglelefteq_\lambda T$, so T is $\trianglelefteq_\lambda$-maximal.

Suppose now T is $\trianglelefteq_\lambda$-maximal, D a regular ultrafilter over λ and M a model of T and assume M^λ/D is λ^+-compact. By the $\trianglelefteq_\lambda$-maximality of T, $\mathrm{Th}(M_\lambda) \trianglelefteq_\lambda T$ [M_λ from 2.2(3)] hence M_λ^λ/D is λ^+-compact. By 2.3 this implies that D is λ^+-good.

(2) If T satisfies the condition, clearly it is $\trianglelefteq_\lambda$-minimal. If T is $\trianglelefteq_\lambda$-minimal, then $T \trianglelefteq_\lambda T_=$ [the theory of $(\lambda, =)$]. But for every regular ultrafilter D over λ, $\|(\lambda, =)^\lambda/D\| = \lambda^\lambda/D = 2^\lambda > \lambda$, hence $(\lambda, =)^\lambda/D$ is λ^+-compact. So clearly T satisfies the condition.

(3) Immediate.

THEOREM 4.3: *Any theory with the strict-order property is $\trianglelefteq$-maximal.*

Proof. Just like 2.6 and then use 2.11 and 4.2. (Alternatively use Exercise 4.5.)

THEOREM 4.4: *Let M be a model of an unstable theory T, D an ultrafilter over I and suppose that there are finite cardinals m_i, $i \in I$, such that $\aleph_0 \leq \prod_{i \in I} m_i/D < 2^\lambda$. Then $N = M^I/D$ is not λ^+-compact.*

Proof. By II, 4.7 T has the strict order property or the independence property.

Case I. T has the strict order property. So there is a formula $\varphi(\bar{x}, \bar{y})$ which is a partial order, and not reflexive, and for every n there are $\bar{a}_0^n, \ldots, \bar{a}_{n-1}^n \in M$ such that $\models \varphi[\bar{a}_l^n, a_{l+1}^n]$. Let $P_i = \{\bar{a}_l^{m_i}: l < m_i\}$, $(N, P) = \prod_{i \in I}(M, P_i)$; so by assumption $|P| < 2^\lambda$. Notice that each (M, P_i), and hence (N, P), satisfies the sentence $(\forall \bar{x})[(\exists \bar{y})[P(\bar{y}) \wedge (\forall \bar{z})(P(\bar{z}) \rightarrow \varphi(\bar{x}, \bar{z}) \equiv \varphi(\bar{y}, \bar{z}))] \vee (\forall \bar{y})(P(\bar{y}) \rightarrow \neg\varphi(\bar{x}, \bar{y}))]$.

Suppose N is λ^+-compact, then we can define $\bar{a}_\eta, \bar{b}_\eta \in P$ $\eta \in {}^{\lambda \geq}2$ by induction on $l(\eta)$ such that:

(i) $N \models \varphi[\bar{a}_\eta, \bar{b}_\eta]$ and for every n

$$(N, P) \models (\exists \bar{x}_0, \ldots, \bar{x}_n)\left(\bigwedge_{l<n} \varphi(\bar{x}_l, \bar{x}_{l+1})\right) \wedge \bar{a}_\eta = \bar{x}_0 \wedge \bar{b}_\eta = \bar{x}_n \wedge \bigwedge_{l<n} P(\bar{x}_l)),$$

(ii) $\eta \lessdot \nu$ implies $N \models \varphi[\bar{a}_\eta, \bar{a}_\nu] \wedge \varphi[\bar{b}_\nu, \bar{b}_\eta] \wedge \varphi[\bar{b}_{\eta \frown \langle 0 \rangle}, \bar{a}_{\eta \frown \langle 1 \rangle}]$.

Clearly the $\bar{a}_\eta$'s are pairwise distinct and $\in P$, and their number is $\leq |P| < 2^\lambda$, but also $\geq |{}^\lambda 2| = 2^\lambda$, contradiction.

Case II. T has the independence property. So some $\varphi(x; \bar{y})$ has the independence property. Hence we can find sequences $\bar{a}_l^k \in |M|$ $(l < \kappa)$ so that $\models (\exists x)[\bigwedge_{l<k} \varphi(x; \bar{a}_l^k)^{\text{if}(l \in w)}]$ for each $w \subseteq k$, so let $b_w^k \in |M|$, be such that

$$\models \bigwedge_{l<k} \varphi(b_w^k; \bar{a}_l^k)^{\text{if}(l \in w)}.$$

Let $\mu = \min\{\prod_{i \in I} n_i/D: n_i < \omega \text{ for } i \in I \text{ and } \prod_{i \in I} n_i/D \geq \aleph_0\} < 2^\lambda$, $\mu = \prod_{i \in I} n_i/D$.

For every $i \in I$ define $k(i) = [\log_2 n_i]$ if $n_i \neq 1$, $k(i) = 1$ if $n_i = 1$. Let $P_i = \{\bar{a}_l^{k(i)}: l < k(i)\}$, $Q_i = \{b_w^{n(i)}: w \subseteq k(i)\}$, so clearly $n_i/2 - 1 \leq |Q_i| \leq n_i$. Let $(N, P, Q) = \prod_{i \in I}(M, P_i, Q_i)$, so clearly $|Q| = \prod_{i \in I} |Q_i|/D \leq \mu$, but clearly $|P| \leq |Q|$ and for every $l < \omega$

$$\{i \in I: |P_i| \geq l\} \supseteq \{i \in I: n_i \geq 2^l\} \in D$$

(as $\prod n_i/D \geq \aleph_0$). Hence $|P|$ is infinite; hence by μ's definition $|Q| = |P| = \mu$. Let $\lambda_1 = \min(\lambda, \mu)$ and choose $\mathbf{I} \subseteq P$, $|\mathbf{I}| = \lambda_1$. By the definition of the P_i's, clearly for every $\mathbf{J} \subseteq \mathbf{I}$, $p_{\mathbf{J}} = \{\varphi(x; \bar{a})^{\text{if}(\bar{a} \in \mathbf{J})}: \bar{a} \in \mathbf{I}\}$ is a type in N. Now by the definition of the Q_i, if p_J is realized in N, it is realized by some element of Q. Hence the number of types p_J, which are realized in N is $\leq |Q| = \mu$ (because $\mathbf{J}_1 \neq \mathbf{J}_2$ implies no element realizes both p_{J_1} and p_{J_2}). On the other hand the number of such types is $|\{\mathbf{J}: \mathbf{J} \subseteq \mathbf{I}\}| = 2^{|\mathbf{I}|} = 2^{\lambda_1}$. Clearly $2^{\mu} > \mu$, and by hypothesis and definition of μ, $2^{\lambda} > \mu$; hence $2^{\lambda_1} > \mu$. So for some $\mathbf{J} \subseteq \mathbf{I}$ N omits p_J, and as $|p_J| = \lambda_1 \leq \lambda$, N is not λ^+-compact.

EXERCISE 4.1: Prove that if $M = \prod_{i \in I} M_i/D$ is a model of an unstable theory, D an ultrafilter over I, $\aleph_0 \leq \prod_{i \in I} m_i/D < 2^{\lambda}$, then M is not λ^+-compact.

THEOREM 4.5: *Let T be stable and with the finite cover property. Let M be a model of T and D an ultrafilter over I; if $\aleph_0 \leq \mu = \prod_{i \in I} m_i/D$ then M^I/D is not μ^+-compact.*

Proof. Let $\varphi(x, y; \bar{z})$ be as in II, 4.4, i.e., for every $\bar{c} \in |M|$ $\varphi(x, y; \bar{c})$ is an equivalence relation and for $n < \omega$, $\varphi(x, y; \bar{c}_n)$ has $\geq n$ but $< \aleph_0$ equivalence classes. Let $\mathrm{Neq}(\bar{c}, M)$ be the number of equivalence classes of $\varphi(x, y; \bar{c})$ in M. Clearly for every family of models M_i $i \in I$ in which $\varphi(x, y; \bar{z})$ is always an equivalence relation

$$\mathrm{Neq}\left(\bar{c}, \prod_{i \in I} M_i/D\right) = \prod_{i \in I} \mathrm{Neq}(\bar{c}[i], M_i)/D.$$

Choose $\bar{c}[i]$ as $\bar{c}_{n[i]}$ where $n[i] = \max\{l: \mathrm{Neq}(\bar{c}_l, M) \leq m_i\}$ if such exists, $n[i] = 1$ otherwise. Clearly $\aleph_0 \leq \mathrm{Neq}(\bar{c}, M^I/D) \leq \prod_{i \in I} m_i/D$, so we finish.

EXERCISE 4.2: Prove the parallel to Exercise 4.1 for Theorem 4.5.

LEMMA 4.6: *If T is $\otimes_{\lambda}$-minimal, $\mu \leq \lambda$, then T is $\otimes_{\mu}$-minimal.*

Proof. Assume T is not $\otimes_{\mu}$-minimal. By 4.2(2) there is a regular ultrafilter on μ, and a model M of T such that M^{μ}/D is not μ^+-compact. Let D_1 be a regular ultrafilter on λ, $D_2 = D_1 \times D$, $I = \lambda \times \mu$, so D_2 is a regular ultrafilter on I, $|I| = \lambda$, and $M^I/D_2 \cong (M^{\lambda}/D_1)^{\mu}/D$. By 1.1 $M \equiv M^{\lambda}/D$, so by 1.9 and 2.11 M^{μ}/D is μ^+-compact iff $(M^{\lambda}/D_1)^{\mu}/D$ is μ^+-compact so M^I/D_2 is not μ^+-compact, hence not λ^+-compact. So T is not $\otimes_{\lambda}$-minimal. Contradiction.

THEOREM 4.7: *If* $\kappa r_{\rm cdt}(T) = \infty$, *$D$ is a regular ultrafilter over I, M is an $|I|^{++}$-saturated model of T, then M^I/D is not $|I|^{++}$-compact (see Definition* III, 7.2).

Proof. If T has the strict order property the theorem follows immediately from 4.3, 4.2, 2.4 and 1.3. Next assume $\kappa_{\rm inp}(T) = \infty$; w.l.o.g. we can assume (see III, Definition 7.3, III, Exercise 7.4 and 1.11) $\|M\| > \mu = \beth_{\lambda^+}$ where $|I| = \lambda$ and there is a formula $\varphi(\bar{x}; \bar{y})$ and $\bar{a}^\beta_\alpha \in |M|, \beta < \lambda, \alpha < \mu, l(\bar{a}^\beta_\alpha) = l(\bar{y})$ such that for every β $\{\varphi(\bar{x}; \bar{a}^\beta_\alpha): \alpha < \mu\}$ is 2-contradictory, and for every $\eta \in {}^\lambda\mu$ $\{\varphi(\bar{x}; \bar{a}^\beta_{\eta[\beta]}): \beta < \lambda\}$ is consistent, realized by $\bar{c}_\eta$. Let $P = \{\bar{a}^\beta_\alpha: \beta < \lambda, \alpha < \mu\}$, $Q = \{\bar{c}_\eta; \eta \in {}^\lambda\mu\}$. We define equivalence relations $E(\bar{u}, \bar{y})$, $l(\bar{x}) = l(\bar{y}) = l(\bar{a}^\beta_\alpha)$ as follows $E(\bar{a}^{\beta_1}_{\alpha_1}, \bar{a}^{\beta_2}_{\alpha_2})$ iff $\beta_1 = \beta_2$; $\neg E(\bar{c}, \bar{a}^\beta_\alpha)$ if $\bar{c} \notin P$, $E(\bar{c}, \bar{d})$ if $\bar{c}, \bar{d} \notin P$. E has λ equivalence classes in $P(M)$ and $\geq \lambda^+$ equivalence classes in $P(M^I/D)$ each one of cardinality $\geq \mu$. Let J_i, $i < \lambda^+$ be distinct equivalence classes of E, then for every sequence $\langle \bar{a}_i: \bar{a}_i \in J_i, i < \lambda^+\rangle$, $\{\varphi(\bar{x}; \bar{a}_i): i < \lambda^+\}$ is consistent, and if M^I/D is λ^{++}-compact it is realized in $Q(M^I/D)$. $\mu < \mu^+ \leq \prod_{i<\lambda^+} |J_i| \leq |Q(M^I/D| = |Q(M)|^I/D \leq \mu^\lambda = \mu$, contradiction. So we are left with the case $\kappa_{\rm inp}(T) < \infty$, $\kappa_{\rm cdt}(T) = \infty$, hence by III, 7.11 and 1.11 we can assume there are a formula $\varphi(\bar{x}; \bar{y})$ and $\bar{a}_\eta \in |M|$ $(\eta \in {}^{\omega >}\mu) l(\bar{a}_\eta) = l(\bar{y})$ such that if $\eta, \nu \in {}^{\omega >}\mu$ are $\triangleleft$-incompatible then $M \models \neg(\exists x)[\varphi(\bar{x}; \bar{a}_\eta) \wedge \varphi(\bar{x}; \bar{a}_\nu)]$ and for each $\eta \in {}^\omega\mu$, $\bar{c}_\eta \in |M|$ realizes $\{\varphi(\bar{x}; \bar{a}_{\eta|n}): n < \omega\}$. Let $P = \{\bar{a}_\eta: \eta \in {}^{\omega >}\mu\}$, $Q = \{\bar{c}_\eta: \eta \in {}^\omega\mu\}$, $E = \{\langle \bar{a}_\eta, \bar{a}_\nu\rangle$: for some $n < \omega$, $\eta, \nu \in {}^n\mu\}$, and $< = \{\langle \bar{a}_\eta, \bar{a}_\nu\rangle: l(\eta) < l(\nu); \eta, \nu \in {}^{\omega >}\mu\}$, $E_1 = \{\langle \bar{a}_{\eta\frown\langle i\rangle}, a_{\eta\frown\langle j\rangle}\rangle: i, j < \mu, \eta \in {}^{\omega >}\mu\}$. Let $N_1 = (N, P^N, Q^N, E^N, E^N_1, <^N) = M^I_1/D$, $M_1 = (M, P, Q, E, E_1, <)$: so as D is regular there are $\bar{a}^\alpha \in P^N$, $N_1 \models \bar{a}^\alpha < \bar{a}^\beta$ for $\alpha < \beta < \lambda^+$. Now for every $\eta \in {}^{\lambda^+ >}\mu$ we define $\bar{a}^\eta \in P^N$ such that $N_1 \models E(\bar{a}^\eta, \bar{a}^{l(\eta)})$ and the $\bar{a}^\eta$'s are distinct, and $N_1 \models (\exists \bar{x})(\varphi(\bar{x}; \bar{a}^\eta) \wedge \varphi(\bar{x}, \bar{a}^{\eta|\alpha})$, $N_1 \models E_1(\bar{a}^{\eta\frown\langle i\rangle}, \bar{a}^{\eta\frown\langle j\rangle})$. We do it by induction on $l(\eta)$. For $l(\eta) = 0$, $l(\eta) = \alpha + 1$ there are no problems. For $l(\eta) = \delta$ limit let $p_\eta = \{\varphi(\bar{x}; \bar{a}^{\eta|\alpha}): \alpha < \delta\}$; clearly it is consistent, so if it is not realized, M^I/D is not even λ^+-compact, so we finish. So assume $\bar{c}^\eta \in |M|$ realizes p_η, but

$$M_1 \models (\forall \bar{z})(\exists \bar{x})[Q(\bar{x}) \wedge (\forall \bar{y})(P(\bar{y}) \wedge \varphi(\bar{z}; \bar{y}) \rightarrow \varphi(\bar{x}; \bar{y})]$$

(by the requirements on the $\varphi(\bar{x}; \bar{a}_\eta)$'s) so we can assume $\bar{c}^\eta \in Q^N$, and the rest is easy. Similarly for every $\eta \in {}^{\lambda^+}\mu$ there is $\bar{c}^\eta \in Q^N$ realizing p_η, but $|Q^N| \leq |Q^M|^I/D \leq \mu^\lambda = \mu$; the p_η's are pairwise explicitly contradictory and their number is $\mu^{\lambda^+} > \mu$; contradiction.

EXERCISE 4.3: (1) Show that in 4.7 it suffices to assume D is uniform.

(2) Show that by 4.7, there is a model M_0 of T, $\|M_0\| = |I|^+$, such that $M_0 \prec M$ implies M^I/D is not $|I|^{++}$-compact.

DEFINITION 4.2: Let M be a model, D a filter over I; for $a \in |M^I/D|$, $\text{eq}(a) = \{(s, t) \in I \times I: a[s] = a[t]\}$ and for G a filter over $I \times I$, $M^I_D | G$ is the submodel of M^I/D whose set of elements is $\{a \in |M^I/D|: \text{eq}(a) \in G\}$. We always assume $\{\langle t, t\rangle : t \in I\} \in G$.

DEFINITION 4.3: $T_1 \otimes^* T_2$ if for every set I, $\aleph_1$-incomplete ultrafilter D over I, filter G over $I \times I$, cardinal λ, and $(\lambda^+ + |I|^+)$-saturated models M_1, M_2 of T_1, T_2 respectively: if $M^I_{2D}|G$ is λ^+-compact then $M^I_{1D}|G$ is λ^+-compact.

DEFINITION 4.4: Let $\Phi = \{\varphi_\gamma(\bar{x}; \bar{z}^\gamma): \gamma < \alpha\}$, $\Phi_2 = \{\psi_\gamma(\bar{y}; \bar{z}_\gamma): \gamma < \beta\}$ be indexed sets of formulas (possibly with repetitions) from $L(T_1)$ and $L(T_2)$ respectively; $l(\bar{x}) = m_1$, $l(\bar{y}) = m_2$. $(\Phi_1, m_1) \leq (\Phi_2, m_2)$ if there is $\eta \in {}^\alpha\beta$ such that for every model M_1 of T_1 and $\bar{a}^\gamma \in |M_1|$, $\gamma < \alpha$, there are a model M_2 of T_2 and $\bar{b}^\gamma \in |M_2|$, $\gamma < \beta$ such that: for every $w \subseteq \alpha$ $\{\varphi(\bar{x}; \bar{a}^\gamma): \gamma \in w\}$ is consistent with M_1 iff $\{\psi_{\eta[\gamma]}(\bar{y}; \bar{b}^{\eta[\gamma]}): \gamma \in w\}$ is consistent with M_2. We write $(\Phi_1, m_1) \leq (\Phi_2, m_2)$ by η.

Remarks. (1) Clearly by the compactness theorem $(\Phi_1, m_1) \leq (\Phi_2, m_2)$ by η iff for every finite $\Phi \subseteq \Phi_1$, $(\Phi, m_1) \leqslant (\Phi_2, m_2)$ by $\eta \restriction w_\Phi$ where $w_\Phi = \{\gamma < \alpha: \varphi_\gamma(\bar{x}; \bar{z}^\gamma) \in \Phi\}$.

(2) In Definition 4.4 we can take M_1, M_2 as fixed λ-universal models (for large enough λ).

(3) We write $\Phi \subseteq L$ (in this section only) to mean Φ is an indexed set of formulas, possibly with repetitions, from L.

EXERCISE 4.4: (1) If $(\Phi_1, m_1) \leq (\Phi_2, m_2)$, $\Phi^1 \subseteq \Phi_1$, $\Phi_2 \subseteq \Phi^2$ then $(\Phi^1, m_1) \leq (\Phi^2, m_2)$.

(2) If $\Phi^1[\Phi^2]$ is the closure of Φ_1 $[\Phi_2]$ under conjunction and disjunction, then $(\Phi_1, m_1) \leq (\Phi_2, m_2)$ implies $(\Phi^1, m_1) \leq (\Phi^2, m_2)$ (why do we not say "negation"?)

(3) If $(\Phi_1, m_1) \leq (\Phi_2, m_2)$ and $(\Phi_2, m_2) \leq (\Phi_3, m_3)$ then $(\Phi_1, m_1) \leq (\Phi_3, m_3)$.

EXERCISE 4.5: If for every $\Phi_1 \subseteq L(T_1)$, $|\Phi_1| = \lambda$, there is $\Phi_2 \subseteq L(T_2)$ and $m_2 < \omega$ such that $(\Phi_1, 1) \leq (\Phi_2, m_2)$ then $T_1 \otimes_\lambda T_2$.

DEFINITION 4.5: Let Φ_1, Φ_2, m_1, m_2 be as in Definition 4.4 $(\Phi_1, m_1) \leq^* (\Phi_2, m_2)$ if there is $\eta \in {}^\alpha\beta$ such that for every model M_1 of T_1 and $\bar{a}^\gamma_n \in |M_1|$, $\gamma < \beta$, $n < \omega$, there are a model M_2 of T_2 and $\bar{b}^\gamma_n \in |M_2|$, $\gamma < \beta$, $n < \omega$ such that: for every $w \subseteq \alpha \times \omega$ $\{\varphi_\gamma(\bar{x}; \bar{a}^\gamma_n): \langle\gamma, n\rangle \in w\}$ is consistent with M_1 iff $\{\psi_{\eta[\gamma]}(\bar{y}; \bar{b}^{\eta[\gamma]}_n): \langle\gamma, n\rangle \in w\}$ is consistent with M_2. We write $(\Phi_1, m_1) \leq^* (\Phi_2, m_2)$ by η.

EXERCISE 4.6: (1) In Definition 4.5, we can replace ω by any $\alpha \geq \omega$.

(2) $(\Phi_1, m_1) \leq^* (\Phi_2, m_2)$ by η implies $(\Phi_1, m_1) \leq (\Phi_2, m_2)$ by η.

(3) $(\Phi_1, m_1) \leq^* (\Phi_2, m_2)$ implies $(\Phi_1, m_1) \leq (\Phi_2, m_2)$.

(4) If Φ_l, Φ^l contain the same formulas (with a different number of repetitions) then

$$(\Phi_1, m_1) \leq^* (\Phi_2, m_2) \Leftrightarrow (\Phi^1, m_1) \leq^* (\Phi^2, m_2).$$

(5) If $\Phi^1 \subseteq \Phi_1$, $\Phi_2 \subseteq \Phi^2$ then $(\Phi_1, m_1) \leq^* (\Phi_2, m_2)$ implies $(\Phi^1, m_1) \leq^* (\Phi^2, m_2)$.

(6) $(\Phi_1, m_1) \leq^* (\Phi_1, m_1)$ (by the identity map).

EXERCISE 4.7: The following statements about T_1, T_2 are equivalent.

(1) For every $\Phi_1 \subseteq L(T_1)$ there are $\Phi_2 \subseteq L(T_2)$ and $m_2 < \omega$ such that $(\Phi_1, 1) \leq^* (\Phi_2, m_2)$.

(2) For every $\Phi_1 \subseteq L(T_1)$, $|\Phi_1| \leq |T_1| + |T_2|^+$ there are $\Phi_2 \subseteq L(T_2)$ and m_2 such that $(\Phi_1, 1) \leq (\Phi_2, m_2)$.

(3) For every $\Phi_1 \subseteq L(T_1)$ there are $\Phi_2 \subseteq L(T_2)$ and m_2 such that $(\Phi_1, 1) \leq (\Phi_2, m_2)$.

(4) For every $\Phi_1 \subseteq L(T_1)$, $|\Phi_1| \leq |T_1|$ there are $\Phi_2 \subseteq L(T_2)$ and m_2 such that $(\Phi_1, 1) \leq^* (\Phi_2, m_2)$.

(5) Let Φ_0 be the set of formulas $\varphi(x; \bar{y}) \in L(T_1)$ (clearly $|\Phi_0| = |T_1|$). There are $\Phi_2 \subseteq L(T_2)$, m_2 such that $(\Phi_0, 1) \leq^* (\Phi_2, m_2)$.

EXERCISE 4.8: (1) If Φ_0 is the set of all formulas in $L(T_1)$, and for some $\Phi_2 \subseteq L(T_2)$, $m_2 < \omega$, $(\Phi_0, 1) \leq^* (\Phi_2, m_2)$ then $T_1 \otimes^* T_2$.

(2) In fact it suffices to demand that there are Φ_i, $i < i_0$, such that: if M_1 is a non-λ^+-compact model of T_1, then there is a type p over M_1, $p = \{\varphi_\alpha(x, \bar{a}^\alpha): \alpha < \alpha_0 < \lambda^+\}$, such that for some $i < i_0$ every $\varphi_\alpha(x, \bar{y}^\alpha) \in \Phi_i$; and there are $\Phi_{2,i} \subseteq L(T_2)$, $m_{2,i} < \omega$ such that $(\Phi_i, 1) \leq^* (\Phi_{2,i}, m_{2,i})$.

EXERCISE 4.9: Suppose M_1 is a λ-universal model of T, M_2 is a λ^+-compact model of T, and D an ultrafilter over λ. If M_1^I/D is λ^+-compact then M_2^I/D is λ^+-compact. If D is uniform it suffices to demand M_1 is $(<\lambda)$-universal, M_2 is λ-compact.

EXERCISE 4.10: For M, I, D, G as in Definition 4.2, D an ultrafilter; $M_D^I|G$ is elementarily equivalent to M, moreover $a \mapsto \langle \ldots, a, \ldots \rangle_{t \in I}/D$ is an elementary embedding of M into $M_D^I|G$. Generalize 1.1.

DEFINITION 4.6: A complete theory T is simple if

(1) $L(T)$ contains only one-place predicates, the equality sign $=$, and one other two-place predicate E;

(2) for every model M of T, E^M is an equivalence relation over $|M|$;

(3) there is a model M of T such that for every $a \in |M|$, $[a]_M$ is infinite where

$$[a]_M = \{b \in M : M \models bEa, \text{ and for every predicate } P(x) \text{ of } L(T) \quad M \models P(a) \equiv P(b)\};$$

(4) there is a model M of T such that for every $a \in |M|$, there are infinitely many $b \in |M|$ from different E-equivalence classes which realize the same type over $\emptyset$.

EXERCISE 4.11: Let T be a simple theory. Show that:

(1) If M is a model of T, $a \in |M|$, then any permutation of $[a]_M$ is an automorphism of M.

(2) Every formula of $L(T)$ is equivalent to a Boolean combination of formulas of the following forms:

(i) $x = y$,

(ii) xEy,

(iii) $P(x)$,

(iv) $(\exists y)[xEy \wedge \bigwedge_{i<n} P_i(y) \wedge \bigwedge_{j<m} \neg P^j(y)]$.

(3) T is stable in every $\lambda \geq 2^{|T|}$, so T is superstable.

EXERCISE 4.12: Suppose M is a non-λ^+-compact model of a simple theory T. Show that M omits a type p, which is of one of the following forms:

(i) $p = \{xEa\} \cup \{P_\zeta(x)^{\eta[\zeta]} : \zeta < \zeta_0 \leq \min(\lambda, |T|)\} \cup \{x \neq c_\xi : \xi < \xi_0 \leq \lambda\}$,

(ii) $p = \{P_\zeta(x)^{\eta[\zeta]} : \zeta < \zeta_0 \leq \min(\lambda, |T|)\} \cup p_0 \cup \{\neg xEc_\xi : \xi < \xi_0 \leq \lambda\}$, where p_0 consists of formulas of the fourth form from the previous exercise and of negations of such formulas, and η is a sequence of zeroes and ones.

EXERCISE 4.13: If M is a λ-compact model of a simple theory T, and $N = M_D^I|G$ (see Definition 4.2) is $|T|^+$-compact, show that N is λ-compact. (In fact it is $\lambda_D^I|G$-compact.)

EXERCISE 4.14: Show that:

(1) A simple countable theory is $\otimes$-minimal.

(2) If M is a model of simple theory T, D a $|T|^+$-good ultrafilter on μ, then M^μ/D is $\aleph_0^\mu/D$-compact. Hence if D is μ-regular, M^μ/D is 2^μ-compact.

EXERCISE 4.15: Show that for every theory T_1 and cardinal λ there is a simple theory T_2 such that $T_1 \otimes_\lambda T_2 \otimes_\lambda T_1$. If $|T_1| \leq \lambda$ then also $|T_2| \leq \lambda$. Moreover if D is a λ-regular ultrafilter over μ, M_1 a model of T_1, M_2 a model of T_2 then M_1^μ/D is λ^+-compact iff M_2^μ/D is λ^+-compact. (Hint: See Exercise 4.5.)

EXERCISE 4.16: Show that for every set $\{T_\xi: \xi < \xi_0\}$ of theories there is a least upper bound for each of the orderings $\otimes$, $\otimes_\lambda$. Its cardinality is $\leq \sum_\xi |T_\xi|$.

EXERCISE 4.17: (1) Show that for every λ there is a simple theory T_λ, $|T_\lambda| = \lambda$ such that T_λ is $\otimes_\lambda$-maximal. Hence if $\lambda < \mu$, $T_\lambda \otimes T_\mu$ but not $T_\mu \otimes T_\lambda$. So there is an (uncountable) theory which is not $\otimes$-minimal nor $\otimes$-maximal.

(2) Show that if there is a countable theory which is not $\otimes$-minimal nor $\otimes$-maximal (see 5.9) then there are $\otimes$-incomparable theories.

EXERCISE 4.18: Prove case I of the proof of 4.4 by 4.3.

EXERCISE 4.19: (1) Suppose $\kappa_{\rm inp}(T) > |I|^+$, M an $|I|^{++}$-saturated model of T, and D a regular ultrafilter over I. Prove that M^I/D is not $|I|^{++}$-compact (see Theorem 4.7).

(2) Prove the parallel of Exercise 4.3.

QUESTION 4.20: Can we in Exercise 4.19 replace $\kappa_{\rm inp}(T)$ by $\kappa_{\rm cdt}(T)$?

EXERCISE 4.21: Show that $T_1 \otimes_{\aleph_0} T_2$ for any complete theories T_1, T_2.

THEOREM 4.8: *For any model M of an unstable theory T and $\aleph_1$-incomplete ultrafilter D over I, $\kappa = \mathrm{lcf}(\omega, D)$, M^I/D is not κ^+-compact.*

Proof. Choose an unstable formula $\varphi(x; \bar{y})$ and sequences $\bar{a}_l^n$ ($l < n < \omega$) such that $\{\varphi(x, \bar{a}_l^n)^{\mathrm{if}(l > k)}: l < n\}$ is consistent for every $k < n < \omega$. Choose a distinct $b_n \in |M|$ and let: $P^M = \{b_n: n < \omega\}$,

$$<^M = \{\langle b_k, b_n\rangle: k, < n < \omega\}$$

and F_l^M ($l < l(\bar{y})$) two place functions such that $\bar{a}_k^n = \langle F_0(b_k, b_n), F_1(b_k, k_n), \ldots\rangle$. Let

$$N_1 = (N, P^N, <^N, F_0^N, \ldots) = (M, P^M <^M, F_0^M, \ldots)^I/D,$$

and $c_i \in P^N$ ($i < \kappa$) be such that $N_1 \models b_n < c_i < c_j$ for $n < \omega, i < j < \kappa$ and for no $c \in P^N$, $N \models b_n < c < c_i$ for every $n < \omega$, $i < \kappa$. Let

$$p = \{\neg\varphi(x, F_0(b_n, c_0), \ldots): n < \omega\} \cup \{\varphi(x; F_0(c_i, c_0), \ldots): 0 < i < \kappa\}.$$

Clearly p is finitely satisfiable, but N omit it (if a realizes it, then the formula

$$P(y) \wedge (\forall z)[P(z) \wedge z < y \rightarrow \varphi(a, F_0(y, c_0)F_1(y, c_0), \ldots)]$$

define a bounded subset of P with no last element).

VI.5. Saturation of ultrapowers and categoricity of pseudo-elementary classes

THEOREM 5.1: *Let T be a countable theory, M_i a model of T for every $i \in I$, and D an $\aleph_1$-incomplete ultrafilter over I. Let $N = \prod_{i \in I} M_i/D$.*

(1) *If T does not have the* f.c.p., *$\lambda = \aleph_0^I/D$, then N is λ-saturated.*

(2) *If T is stable and has the* f.c.p., *then N is λ-saturated, but not λ^+-saturated, where $\lambda = \min\{\prod_{i \in I} n_i/D: \prod_{i \in I} n_i/D \geq \aleph_0\}$.*

(3) *If T does not have the* f.c.p., *each M_i is μ-saturated, and $\lambda = \mu^I/D$, then N is λ-saturated.*

(4) *For every finite $\Delta \subseteq \mathrm{L}(T)$ let $\lambda_i(\Delta) = \min\{|p|: p$ is a Δ-1-type over $|M_i|$ in M_i which is omitted by $M_i\}$ and*

$$\lambda^* = \min\left\{\prod_{i \in I} \lambda_i(\Delta)/D: \Delta \subseteq \mathrm{L}(T), |\Delta| < \aleph_0\right\}.$$

Let λ be an (infinite) cardinal such that $\lambda = \prod_{i \in I} \lambda^i/D$ for some λ^i, $i \in I$, where for every finite $\Delta \subseteq L(T)$, $\{i: \lambda^i \leq \lambda_i(\Delta)\} \in D$. Then if T does not have the f.c.p., *N is λ-saturated, but not $(\lambda^*)^+$-saturated.*

Remarks. (1) Clearly the results, except (4), are the best possible. For example in (1), if we choose the M_i as countable models, then $\|N\| = \aleph_0^I/D = \lambda$, so N is not λ^+-saturated. On (4) see Exercise 5.10.

(2) Instead of demanding T countable, we can require D to be $|T|^+$-good. See Exercise 5.2.

Proof. Notice that as T is countable, for every model M of T and cardinality $\kappa > \aleph_0$, M is κ-compact iff M is κ-saturated.

Now in part (2), N is not λ^+-saturated by Theorem 4.5. Similarly we can prove in part (4) that N is not $(\lambda^*)^+$-saturated. So it remains to prove that in all the parts N is λ-saturated.

N is $\aleph_1$-saturated by 2.1, 2.3 and 2.11. By III, 3.3, 3.9 and 3.10, as T is countable and stable, it suffices to prove: if $\{c_i: i < \omega\} \subseteq |N|$ is an indiscernible set, then it can be extended in N to an indiscernible set of cardinality λ. For every $i \in I$ let us choose a family S_i of subsets of $|M_i|$ such that:

(i) $|S_i| = \|M_i\|$,

(ii) every finite subset of $|M_i|$ belongs to S_i,

(iii) for every finite $\Delta \subseteq L(T)$, $n < \omega$, if $w \in S_i$ is a Δ-n-indiscernible set, $0 \leq \mu \leq \|M_i\|$, and there is a Δ-n-indiscernible set w', $w \subseteq w' \subseteq |M_i|$, $|w'| = \mu$, then there is $w'' \in S_i$, $|w''| = \mu$, $w \subseteq w'' \subseteq |M_i|$, and w'' is a Δ-n-indiscernible set. Let $|M_i| = \{a^i_j: j < \|M_i\|\}$, $S_i = \{w^i_j: j < \|M_i\|\}$. Let us define the relation $\in^i$ on $|M_i|$: $\in^i = \{\langle a^i_j, a^i_\alpha\rangle: a^i_j \in w^i_\alpha\}$. We shall write $x \in y$ instead of $\in(x, y)$. In the language $L = L(T) \cup \{\in\}$, clearly, for every finite $\Delta \subseteq L(T)$, $n < \omega$, there is a formula $\varphi_{\Delta,n}(x)$ meaning $\{y: y \in x\}$ is a Δ-n-indiscernible set.

Now for every $i \in I$ we define P^i according to the part of the theorem we want to prove;

in (1): $P^i = \{a^i_\alpha: |w^i_\alpha| \geq \aleph_0\}$,

in (2): $P^i = \{a^i_\alpha: \alpha < \|M_i\|\} = |M_i|$;

in (3): $P^i = \{a^i_\alpha: |w^i_\alpha| \geq \mu\}$;

in (4): $P^i = \{a^i_\alpha: |w^i_\alpha| \geq \lambda^i\}$;

where the λ^i are defined so that $|\prod_{i \in I} \lambda^i/D| = \lambda$, and for every finite $\Delta \subseteq L(T)$ $\{i: \lambda^i \leq \lambda_i(\Delta)\} \in D$.

Now the following hold:

(*) For every finite $\Delta \subseteq L(T)$, $n < \omega$, there is $m = m(\Delta, n) < \omega$, such that the set of i's for which the following holds belongs to D:

(**) For every Δ-n-indiscernible set w^i_α, $|w^i_\alpha| \geq m$, there is a Δ-n-indiscernible set w^i_j, $w^i_\alpha \subseteq w^i_j$ and $a^i_j \in P^i$.

Let us prove it. In part (2) it is trivial. In the other parts T does not have the f.c.p. so in part (1) it follows from II, 4.6(3) and in parts (3) and (4) from II, 4.6(2). Notice that except in part (4) (**) holds for every i.

Now clearly (**) is equivalent to a first-order sentence in $L' = L \cup \{\in\} \cup \{P\}$. Let $N' = (N, \in^N, P^N) = \prod_{i \in I} (M_i, \in^i, P^i)/D$. N' is $\aleph_1$-saturated (by 2.1, 2.3 and 2.11). By (*), clearly the sentences corresponding to (**) (for all finite Δ, n, m) are satisfied by N'. Remember it suffices to prove that $\{c_i: i < \omega\}$ can be extended in N to an indiscernible set of cardinality λ. As $\{c_i: i < \omega\}$ is an indiscernible set, it is a Δ-n-indiscernible set for every Δ, n. Hence every finite subset of $p = \{c_i \in x: i < \omega\} \cup \{\varphi_{\Delta,n}(x): \Delta \subseteq L(T), |\Delta| < \aleph_0, n < \omega\} \cup \{P(x)\}$ is satisfied in N', hence p is satisfied in N' say by b. As $N' \vDash \varphi_{\Delta,n}(b)$ for every Δ, n, clearly $w = \{a \in |N|: N' \vDash a \in b\}$ is an indiscernible set, and of course $\{c_i: i < \omega\} \subseteq w$. As $N' \vDash P[b]$, and $|w| \geq |\{c_i: i < \omega\}| = \aleph_0$, clearly $|w| \geq \lambda$ (the check for each part is easy). So we have proved the theorem.

It would be more satisfactory if in 5.1(4) $\lambda = \lambda^*$. (This is possible if $\forall i \in I, M_i = M$). For this it suffices to prove

CONJECTURE 5.1: Let Let $\langle J, <\rangle = \langle \mu, <\rangle^I/D$. ($<$-the natural order on ordinals.) For $a \in |J|$, let $|a| = |\{b \in |J|: b < a\}|$. Suppose $a_n \in |J|$ for $n < \omega$, $|a_n| = |a_0|$. Then there is $a \in |J|$, $a \leq a_n$ for every $n < \omega$, and $|a| = |a_0|$ (when D is an $\aleph_1$-incomplete ultrafilter). See Exercise 5.11 and 5.12.

THEOREM 5.2: *Let M be a λ-compact model of T, D an ultrafilter over I, $|T| \leq |I|$, $N = M^I/D$. If N is $(2^{|I|})^+$-saturated then N is λ^I/D-saturated.*

Remarks. (1) For countable T this theorem follows from 4.4, 4.5 and 5.1(3).

(2) Here the proof works also for D an $\aleph_1$-complete ultrafilter.

(3) See Exercise 5.13.

Proof. As N in $(2^{|I|})^+$-saturated, by 4.4 and 4.5 T is stable and without the f.c.p., and so clearly every infinite indiscernible set can be extended to one of cardinality $\geq (2^{|I|})^+$. By III, 3.10 (remembering that by III, 3.3, $\kappa(T) \leq |T|^+ \leq |I|^+$), it suffices to prove that:

If $\mathbf{I}$ is an indiscernible set in N, $|\mathbf{I}| \geq (2^{|I|})^+$, then there is an indiscernible set $\mathbf{J}$, $|\mathbf{I} \cap \mathbf{J}| \geq \aleph_0$, $|\mathbf{J}| \geq |\lambda^I/D|$. Let $\{a_\beta: \beta < (2^{|I|})^+\} \subseteq \mathbf{I}$. The following statement will be proved later.

(*) There is an infinite $w \subseteq (2^{|I|})^+$ such that for every $i \in I$, $\{a_\beta[i]: \beta \in w\}$ is an indiscernible set in M.

We can assume $\lambda > |T|$, as otherwise the conclusion of the theorem is trivial. For every $i \in I$ let P^i be a maximal indiscernible set such that $\{a_\beta[i]: \beta \in w\} \subseteq P \subseteq |M|$. As M is λ-compact, $\lambda > |T|$, clearly $|P^i| \geq \lambda$. Let $(N, P) = \prod_{t \in I} (M, P^t)/D$. Clearly $|P| = \prod_{t \in I} |P_t| \geq \lambda^I/D$. Now for every finite $\Delta \subseteq L(T)$, $n < \omega$, the statement "P is a Δ-n-indiscernible set" is elementary, hence P is an indiscernible set. So $\{a_\beta: \beta \in w\} \subseteq P \subseteq |N|$, hence $|P \cap \mathbf{I}| \geq |\{a_\beta: \beta \in w\}| \geq \aleph_0$. So P satisfies the requirement for $\mathbf{J}$. So we need only prove (*).

As T is stable, by II, 2.13, $|B| \leq 2^{|I|}$ implies $|S(B)| \leq (2^{|I|})^{|T|} = 2^{|I|}$. It is also clear that for $B_t \subseteq |M|$, $|B_t| \leq 2^{|I|}$, for every $t \in I$; $|\prod_{t \in I} S(B_t)| = \prod_{t \in I} |S(B_t)| \leq (2^{|I|})^{|I|} = 2^{|I|}$.

Define for $j \leq |I|^+$, sets $w_j \subseteq (2^{|I|})^+$ by induction:

(i) $w_0 = \emptyset$, $w_\delta = \bigcup_{j<\delta} w_j$ for a limit ordinal δ, and $w_i \subseteq w_j$ for $i < j \leq |I|^+$.

(ii) Let w_α be defined. Then for every $\beta < (2^{|I|})^+$ there is a unique $\gamma \in w_{\alpha+1}$ such that: for every $i \in I$, $a_\beta[i]$, $a_\gamma[i]$ realize the same type in M over $\{a_j[i]: j \in w_\alpha\}$. Clearly for every j, $|w_j| \leq 2^{|I|}$. Choose $\alpha_0 < (2^{|I|})^+$, $\alpha_0 \notin w_{|I|^+}$. For every $\alpha < |I|^+$, let $j(\alpha)$ be the ordinal such that for every $i \in I$, $a_{\alpha_0}[i]$, $a_{j(\alpha)}[i]$ realize the same type in M over $\{a_j[i]: j \in w_\alpha\}$ and $j(\alpha) \in w_{\alpha+1}$. Clearly for every i, $\alpha \leq \zeta \leq \xi < |I|^+$, $a_{j(\zeta)}[i]$, $a_{j(\xi)}[i]$ realize the same type in M over $\{a_{j(\gamma)}[i]: \gamma < \alpha\}$. By II, 2.18 for every i, there is $\gamma(i) < |I|^+$ such that $\{a_{j(\alpha)}[i]: \gamma(i) \leq \alpha < |I|^+\}$ is an indiscernible set. Let $\gamma_0 = \sup_{i \in I} \gamma(i)$, $w = \{j(\alpha): \gamma_0 \leq \alpha < |I|^+\}$. Clearly this is the w required in (*).

Remark. We can in fact find such w of cardinality $(2^{|I|})^+$. See Exercise I, 2.9.

DEFINITION 5.1: A model M is complete if for every $n < \omega$, every n-ary relation on $|M|$ and every n-ary function from $|M|$ to $|M|$ is the interpretation of some predicate symbol or function symbol, respectively, from $L(M)$.

Remark. Clearly for every set A there is a complete model M, $|M| = A$, $|L(M)| = \aleph_0 + 2^{|A|}$.

DEFINITION 5.2: $PC(T_1, T)$ is the class of $L(T)$-reducts of models of T_1, of cardinality $\geq |T_1|$, where T is complete, $T \subseteq T_1$ and T_1 has infinite models.

The restriction "of cardinality $\geq |T_1|$" is technical, and without it the class is denoted by $PC_*(T_1, T)$.

THEOREM 5.3: *If T is countable, superstable, and does not have the* f.c.p., *then there is T_1, $T \subseteq T_1$, $|T_1| = 2^{\aleph_0}$ such that* $PC(T_1, T)$ *is categorical in every cardinality $\geq 2^{\aleph_0}$.*

Proof. Let M be a countable model of T. We expand M to a complete model M_1, $|L(M_1)| = 2^{\aleph_0}$. Let L_1 be the language of M_1 and T_1 the theory of M_1 (i.e., the set of sentences from L_1 that M_1 satisfies). Clearly T_1 contains its Skolem functions (see Definition VII, 1.1).

Let N_1 be any uncountable model of T_1, and let N be the reduct of N_1 to $L(T)$. It suffices to prove that N is saturated (as by I, 1.11 every two saturated models of the same complete theory which are of the same cardinality are isomorphic). So let p be any 1-type in N, $|p| < \|N\|$, and it suffices to prove that p is realized in N.

Let p_1 be any extension of p to a complete type over $|N|$, and let $\varphi(x, \bar{a}) \in p_1$ be such that $R[\varphi(x; \bar{a}), L, \infty] = R(p_1, L, \infty)$. Let $|M| = \{a_i: i < \omega\}$, and let c_i, $i < \omega$ be individual constants in L_1 such that $c_i^{M_1} = a_i$. Clearly there is $a^0 \in |N_1|$, $a^0 \neq c_i^{N_1}$ for $i < \omega$. Define $A = \{F^{N_1}[\bar{a}, a^0]: F \text{ is a function symbol in } L_1\}$. Clearly the submodel N_1^* of N_1, $|N_1^*| = A$, is an elementary submodel of N_1 (by the definition of T_1 and the Tarski–Vaught test I, 1.2 or VII, 1.2). Let N^* be the reduct of N_1^* to $L(T)$. Clearly N^* is an elementary submodel of N.

We now show

(*) N_1^* is $\aleph_1$-compact hence N^* is $\aleph_1$-saturated.

So let q be a countable type in N_1^* and we must prove it is realized in N_1^*. Let $q = \{\varphi_i(x; a_0^i, \ldots, a_{n_i}^i): i < \omega\}$.

As every $a_j^i \in A$, for some $F_{i,j} \in L_1$, $a_j^i = F_{i,j}[\bar{a}, a^0]$. So by substituting we get $q = \{\psi_i(\bar{x}; \bar{a}, a^0): i < \omega\}$. Remembering that $|M| = \{a_i: i < \omega\}$,

$c_i^{M_1} = a_i$, M_1 is complete; it is clear that there is a function symbol G in L_1 such that for every a_n, $\bar{b}$, b^0 from $|M|$, $G^{M_1}(a_n, \bar{b}, b^0)$ realizes $\{\psi_i(x; \bar{b}, b^0); i < m\}$ for the maximal possible $m \leq n$. Clearly for every n

$$M_1 \models (\forall \bar{z})(\forall y)\left[\left(\bigwedge_{i<n} y \neq c_i \wedge (\exists x) \bigwedge_{i<n} \psi_i(x, \bar{z})\right) \rightarrow \bigwedge_{i<n} \psi_i(G(y, \bar{z}), \bar{z})\right].$$

As $a^0 \neq c_i^{N^*}$ for $i < \omega$, clearly $G^{N^*}(a^0, \bar{a}, a^0)$ realizes q. So we have proved (*).

As N^* is $\aleph_1$-saturated, by III, 1.2 and III, 2.15, we can find $B \subseteq |N^*|$, $|B| = \aleph_0$, $\bar{a} \in B$, such that $p_1 \restriction B$ is stationary, and we can define $b_i \in |N^*|$ for $i < \omega$, such that b_i realizes $p_1 \restriction (B \cup \{b_j: j < i\})$. By III, 1.10 $\mathrm{Av}(\{b_i: i < \omega\}, |N|) = p_1$, so if $\mathbf{I}$ is an indiscernible set in N, $b_i \in \mathbf{I}$ for $i < \omega$, then $\theta(x, \bar{c}) \in p_1$, implies $\{b \in \mathbf{I}: N \models \neg\theta[b, \bar{c}]\}$ is finite. So clearly it suffices to prove that $\{b_i: i < \omega\}$ can be extended in N (not N^*) to an indiscernible set of cardinality $\|N\|$ (because then all but $\leq |p| + \aleph_0$ elements of the set will realize p).

Let S be a family of subsets of $|M|$ such that:

(1) $|S| = \aleph_0$.

(2) Every finite subset of $|M|$ belongs to S.

(3) If $\mathbf{I}$ is a finite Δ-n-indiscernible subset of $|M|$, (Δ a finite subset of L), and $\mathbf{I}$ can be extended to an infinite Δ-n-indiscernible set in M, then there is such an extension which belongs to S.

Let $S = \{\mathbf{I}_i: i < \omega\}$, and noting $|M| = \{a_i: i < \omega\}$, let $\in^{M_1} = \{\langle a_i, a_j\rangle: a_i \in \mathbf{I}_j\}$, $P^{M_1} = \{a_j: |\mathbf{I}_j| = \aleph_0\}$, where $\in$, P belongs to L_1, and let $F \in L_1$ be such that for every $a_j \in P^{M_1}$, $F^{M_1}(x, a_j)$ is a function from $\mathbf{I}_j$ onto $|M|$; we write $x \in y$ instead of $\in(x, y)$. Clearly for every finite $\Delta \subseteq L(T)$, $n < \omega$, there is a formula $\varphi_{\Delta,n}(x)$ in L_1 saying that $\{y: y \in x\}$ is a Δ-n-indiscernible set. Let $q = \{\varphi_{\Delta,n}(x): \Delta \subseteq L(T), n < \omega, |\Delta| < \aleph_0\} \cup \{b_i \in x: i < \omega\} \cup \{P(x)\}$. It suffices to prove that q is consistent with N_1^*. Since then, as N_1^* is $\aleph_1$-compact, q is realized by some element $b \in N_1^*$. Hence $\mathbf{I} = \{c \in N_1: N_1 \models c \in b\}$ is an indiscernible set (as $N_1^* \models \varphi_{\Delta,n}[b]$, N_1^* is an elementary submodel of N_1). Clearly $b_i \in \mathbf{I}$ for $i < \omega$. Also $|\mathbf{I}| = \|N\|$ as $N_1 \models P[b]$ (using $F^{N_1}(x, b)$).

Now in order to prove that q is consistent with N_1^* it suffices to prove that every finite subset of it is consistent. By I, 2.3(3) instead of a finite number of $\varphi_{\Delta,n}(x)$ we need take only one. So it suffices to prove the consistency of $q' = \{P(x), \varphi_{\Delta,n}(x)\} \cup \{b_i \in x: i < m < \omega\}$. By II, 4.6(3) for every finite Δ, n there is $r = r(\Delta, n) < \omega$, such that: if $m \geq r$, $\{b_0, \ldots, b_m\}$ is a Δ-n-indiscernible set in M, then there is an infinite Δ-n-indiscernible set in M which extends $\{b_0, \ldots, b_m\}$. So for $r \geq r(\Delta, n)$

$$M_1 \models (\forall x)(\forall y_0 \cdots y_r)\Big[\Big(\bigwedge_{i<j} y_i \neq y_j \wedge \varphi_{\Delta,n}(x) \wedge \bigwedge_{i\le r} y_i \in x\Big) \\ \rightarrow (\exists y)\Big(\varphi_{\Delta,n}(y) \wedge P(y) \wedge \bigwedge_{i \le r} y_i \in y\Big)\Big].$$

This clearly implies the consistency of q', as $\{b_i: i < \omega\}$ is an indiscernible set (in $L(T)$) and for every $c_1, \ldots, c_n \in N_1$ there is $c \in N_1$ such that $N_1 \models (\forall x)(x \in c \equiv \bigwedge_{i=1}^{n} x = c_i)$.

Remarks. (1) T_1 has no models in any uncountable cardinal $< 2^{\aleph_0}$ and is categorical in $\aleph_0$.

(2) By 5.1(2), VIII, 1.7 and VIII, 2.1 the theorem is the best possible. The following theorems have similar proofs so we omit them.

THEOREM 5.4: (1) *If T is countable, without the* f.c.p., *and stable in $\aleph_0$ (i.e., totally transcendental) then there is T_1, $T \subseteq T_1$, $|T_1| = \aleph_0$, such that* PC(T_1, T) *is categorical in every $\lambda \geq \aleph_0$, and every model in it is saturated.*

(2) *If T has the* f.c.p., *is countable, and stable in $\aleph_0$, $\lambda \leq 2^{\aleph_0}$ then there is T_1, $T \subseteq T_1$, $|T_1| = \lambda$ such that* PC(T_1, T) *is categorical in λ and every model in it of cardinality λ is saturated.*

THEOREM 5.5: *If T is countable and superstable, then there is T_1, $T \subseteq T_1$, $|T_1| = 2^{\aleph_0}$ such that* PC(T_1, T) *is categorical in $2^{\aleph_0}$, and every model in it of cardinality $2^{\aleph_0}$ is saturated.*

Remark: We use the following fact: if M_1 is a complete model which expands $\langle \omega, < \rangle$, N_1 is an uncountable model of the theory of M_1, $a \in |N_1|$, $|\{b \in |N_1|: b < a\}| \geq \aleph_0$, then $|\{b \in |N_1|: b < a\}| \geq 2^{\aleph_0}$. We leave it as an exercise.

THEOREM 5.6: *Let M be a model of a countable and superstable theory T, M_1 a complete expansion of M, $N \in$* PC(Th(M_1), T). *Then N is $2^{\aleph_0}$-saturated.*

THEOREM 5.7: *If T is not $\circledS_\lambda$-minimal, then it is not $\circledS_\mu$-minimal for every $\mu \geq \min\{2^{|T|}, \lambda\}$.*

Proof. If $\mu \geq \lambda$ the conclusion follows by 4.6 so we can assume $\lambda > \mu \geq 2^{|T|}$; and by 4.6 again it suffices to prove the theorem for the case $\mu = 2^{|T|}$. So let $\lambda > \mu = 2^{|T|}$, T is $\circledS_\mu$-minimal but not $\circledS_\lambda$-minimal.

As T is not $\otimes_\lambda$-minimal, by 4.2(2) there is a regular ultrafilter D over λ such that for every model N of T, N^λ/D is not λ^+-compact. Let M_0 be a λ^+-saturated model of T, $M \prec M_0$, $\|M\| = |T|$.

Suppose first M^λ/D is not $|T|^+$-compact. Then there is $A \subseteq |M^\lambda/D|$, $|A| \leq |T|$, such that M^λ/D omits a type over A. W.l.o.g. (by D's regularity) there is an equivalence relation, eq, over λ, eq $\subseteq \lambda \times \lambda$, such that for every $a \in A$, eq $\subseteq \{\langle i, j\rangle \in \lambda \times \lambda : a[i] = a[j]\}$ and eq has $|T|$ equivalence classes. Let N be the submodel of M^λ/D defined by $|N| = \{a \in |M^I/D| :$ eq $\subseteq \{\langle i, j\rangle : a[i] = a[j]\}\}$. Then N is also not $|T|^+$-compact and clearly for some ultrafilter D_1 over $|T|$ N is isomorphic to $M^{|T|}/D_1$, so T is not $\otimes_{|T|}$-minimal hence not $\otimes_\mu$-minimal (as we can make D_1 regular).

Assume now M^λ/D is $|T|^+$-saturated. By 1.8, III, 3.10 there is an indiscernible set $\mathbf{I} = \{a_n : n < \omega\}$ in M^λ/D, $\dim(\mathbf{I}, M_0^\lambda/D) \leqslant \lambda$. Without loss of generality there is an equivalence relation eq over λ with $\leq |T|$ equivalence classes such that eq $\subseteq \{\langle i, j\rangle : a_n[i] = a_n[j]\}$ for every $n < \omega$. Clearly $\{a \in |M_0^\lambda/D| :$ eq $\subseteq \{\langle i, j\rangle : a[i] = a[j]\}\}$ is the universe of an elementary submodel N of M_0^λ/D and $\mathbf{I} \subseteq |N|$. It is also clear that for some ultrafilter D_1 which we can assume is regular, over $|T|$, N and $M^{|T|}/D_1$ are isomorphic. As M_0 is λ^+-saturated, $\lambda > 2^{|T|}$ it suffices to prove N is not $(2^{|T|})^+$-saturated. If it were, it would be λ^+-saturated by 5.2, hence $\lambda \geqslant \dim(\mathbf{I}, M_0^\lambda/D) \geqslant \dim(\mathbf{I}, N) \geqslant \lambda^+$. Contradiction.

Now we shall try to deduce some results on $\otimes$.

THEOREM 5.8: *Let T be countable.*

(1) *T is $\otimes$-minimal iff T does not have the* f.c.p.

(2) *For $\lambda \geq 2^{\aleph_0}$ T is $\otimes_\lambda$-minimal iff T does not have the* f.c.p.

(3) *If $\aleph_0 < \lambda < 2^{\aleph_0}$, T is $\otimes_\lambda$-minimal iff T is stable.*

Proof. (1) By 4.2(3) and Exercise 4.21, this follows from (2) and (3).

(2) We use the criterion in 4.2(2); i.e., T is $\otimes_\lambda$-minimal if for any model M of T and regular ultrafilter D over λ, M^λ/D is λ^+-compact. Suppose T does not have the f.c.p., M is a model of T and D is a regular ultrafilter over λ. By 5.1(1) M^λ/D is $\aleph_0^\lambda/D$-compact and by 1.12 $\aleph_0^\lambda/D = \aleph_0^\lambda \geq \lambda^+$, hence M^λ/D is λ^+-compact.

Suppose now T does have the f.c.p.; by 3.12 there is a regular ultrafilter D over λ and $n_i < \omega$ $(i < \lambda)$ such that $\prod_{i<\lambda} n_i/D = 2^{\aleph_0}$ (as

$2^{\aleph_0} \le \lambda$) and by 4.5, for any model M of T, M^λ/D is not $(2^{\aleph_0})^+$-compact hence not λ^+-compact.

(3) Similarly. For M a model of T, T stable; if D is a regular ultrafilter over λ, by 5.1(2) M^λ/D is μ-compact, where $\mu = \min\{\prod_{i<\lambda} n_i/D: \prod_{i<\lambda} n_i/D \ge \aleph_0\}$. But by 1.3(1), (2) D is $\aleph_1$-incomplete, hence by Exercise 2.10 $\mu^{\aleph_0} = \mu$, hence $\mu \ge 2^{\aleph_0}$, hence $\mu > \lambda$, hence M^λ/D is λ^+-compact.

Now let T be unstable, by 3.12 there is a regular ultrafilter D over λ such that $\mathrm{lcf}(\aleph_0, D) = \aleph_1$, hence by 4.8 M^λ/D is not $\aleph_2$-compact, but $\aleph_1 \le \lambda$, so it is not λ^+-compact.

THEOREM 5.9: *Among countable theories*

(1) *the theories without the* f.c.p. *form an equivalence class,*

(2) *the stable theories with the* f.c.p. *form an equivalence class,*

(3) *the theories with the strict order property are all equivalent,*

(4) *if T_1, T_2 and T_3 are as in* (1), (2) *and* (3) *respectively, and T satisfies none of them, then $T_1 \otimes T_2 \otimes T_{\mathrm{ind}} \otimes T \otimes T_3$, but not $T_2 \otimes T_1$, and not $T_{\mathrm{ind}} \otimes T_2$. If, e.g., $MA + 2^{\aleph_0} > \aleph_1$ holds then not $T_3 \otimes T_{\mathrm{ind}}$; moreover $\kappa_{\mathrm{odt}}(T) = \infty$ implies not $T \otimes T_{\mathrm{ind}}$.*

Remarks. (1) The theorem holds for $\otimes_\lambda$ instead of $\otimes$ for any $\lambda \ge 2^{\aleph_0}$; and for $\aleph_0 < \lambda < 2^{\aleph_0}$, the only difference is that all the stable theories become equivalent.

(2) For (4) see also Exercise 3.25.

Proof. Let T_1, T_2, T_3 and T be as in (4), with respective models M_1, M_2, M_3 and M and M_{ind} a model of T_{ind}.

(1) This is the previous theorem.

(2) M_2^λ/D is λ^+-compact iff $\prod_{i<\lambda} n_i/D \ge \aleph_0 \Rightarrow \prod_{i<\lambda} n_i/D > \lambda$ (by 5.1(2)) hence they are all equivalent.

(3) This is 4.3.

(4) $T_1 \otimes T_2$ is clear from the proof of 5.8 and 5.9(2), $T_2 \otimes T_{\mathrm{ind}}$ is clear by the proof of 5.8(2) and 4.8 (as $\prod_{i<\lambda} n_i/D \ge \aleph_0 \Rightarrow \prod_{i<\lambda} n_i/D \ge \mathrm{lcf}(\aleph_0, D)$). Now $T_{\mathrm{ind}} \otimes T$ can be proved by Exercise 4.5, and $T \otimes T_3$ follows by 4.3. Now $T_2 \otimes T_1$ fails by 5.8, $T_{\mathrm{ind}} \otimes T_2$ fails, as by 3.12, for every λ there is a regular ultrafilter D over λ, $\prod_{i<\lambda} n_i/D \ge \aleph_0 \Rightarrow \prod_{i<\lambda} n_i/D = 2^\lambda$, $\mathrm{lcf}(\aleph_0, D) = \aleph_1$, so M_2^λ/D is λ^+-compact by 5.1(2), but $M_{\mathrm{ind}}^\lambda/D$ is not $\aleph_2$-compact by 4.8. Now assume $MA + 2^{\aleph_0} > \aleph_1$. For not $T_3 \otimes T_{\mathrm{ind}}$ use 3.10(2) (by 4.3 and 4.2(1) M_3^λ/D is λ^+-compact iff D is λ^+-good). For not $T \otimes T_{\mathrm{ind}}$ when $\kappa_{\mathrm{odt}}(t) = \infty$, note

that in 3.10(2) $D = D_1 \times D_2$ so $M^\lambda/D = (M^\lambda/D_1)^\omega/D_2$ is not $\aleph_2$-compact by 4.7.

EXERCISE 5.2: Show that if we replace "T is countable" by "D is $|T|^+$-good" in the hypothesis of Theorem 5.1, the conclusions still hold.

EXERCISE 5.3: Let M be a model of a countable and superstable theory T, D an $\aleph_1$-incomplete ultrafilter over I, G a filter over $I \times I$, such that $\|M^I_D|G\| > \aleph_0$, $M^I_D|G \neq M$. Show that if T does not have the f.c.p. and M is λ-compact, then $M^I_D|G$ is $\lambda^I_D|G$-compact. (See Definition 4.2.)

EXERCISE 5.4: Suppose $E(\bar{x}, \bar{y}; \bar{z})$ is as in II, 4.4, M a countable model of a stable theory T which has the f.c.p., D an $\aleph_1$-incomplete ultrafilter over I and G a filter over $I \times I$. $n_{\bar{c}} = |\{\bar{x}/E(\bar{x}, \bar{y}; \bar{c}): \bar{x} \in |M|\}|$, $P = \{n_{\bar{c}}: \bar{c} \in |M|\}$. Show that there is $\bar{c} \in |M^I_D|G|$ $n_{\bar{c}} = \lambda$, iff there is $s \in |(\omega + 1, <, P)^I_D|G|$, such that $P(s)$ and

$$n_{\bar{c}} = |\{t: (\omega + 1, <)^I_D|G \models t < s\}| = \lambda.$$

EXERCISE 5.5: Let M be a model of a stable theory T which has the f.c.p. D an $\aleph_1$-incomplete ultrafilter over I, G a filter over $I \times I$, and let $\Delta \subseteq \mathrm{L}(T)$ be finite. Let p be a Δ-1-type over $N = M^I_D|G$ which is omitted by N; but every $q \subseteq p$, $|q| < |p|$ is realized in N; and $|p|$ is regular. Show there is $s \in (\|M\| + 1, <)^I_D|G$ such that $|p| = |\{t: (\|M\| + 1, <)^I_D|G \models t < s\}|$.

EXERCISE 5.6: Let M be a model, M_1 a complete expansion of M, $T = \mathrm{Th}(M)$, $T_1 = \mathrm{Th}(M_1)$. Then $\{M^I_D|G: D$ an ultrafilter over I, G a filter over $I \times I\}$ is equal to $\mathrm{PC}_*(T_1, T)$.

EXERCISE 5.7: Let T be countable and complete. Then: for some countable T_1, $T \subseteq T_1$ and $\mathrm{PC}(T_1, T)$ is categorical in $\aleph_0$ iff $|D(T)| = \aleph_0$. (Hint: Like 5.3.)

EXERCISE 5.8: Use Exercise 1.9 to get a better (?) presentation of the proofs of 5.1 and 5.3 (see next exercise).

EXERCISE 5.9: Let $L \subseteq L_1$, $P \in L_1$ a one place predicate, and L_1-model M_1. Let M_1^P be the submodel of the L-reduct of M_1 with universe $P(M_1)$. Then for every theory $T_1 \subseteq L_1$, $\{M_1^P;\ M_1 \models T_1,\ M_1$ an L_1-model$\}$ is $PC(T_2, T^2)$ for some T_2 (you can use VII, 1.1 and VII, 1.2).

EXERCISE 5.10: In Theorem 5.1(4) let $\mu = \sum_{t\in I} \|M_t\|$, and $\lambda(\Delta) \in \mu^I/D$ be $\langle \dots, \lambda_t(\Delta), \dots \rangle_{t\in I}/D$ and let $\lambda^{**} = |\{a \in \mu^I/D\colon$ for every finite $\Delta \subseteq L,\ \mu^I/D \models a < \lambda(\Delta)\}|$. Prove that N is λ^{**}-saturated. [Hint: Use Exercise 1.9 and 5.8. There is (in N_1') an order of type "ω" of the formulas of L (including non-standard ones) and for every "set" $\mathbf{I} \subseteq N$ there is maximal $n = n_{\mathbf{I}}$ such that $\mathbf{I}$ is φ_k-indiscernible for every $k < n_{\mathbf{I}}'$. It suffices to prove that each such $\mathbf{I}$ can be extended to an indescernible set of cardinality λ^{**}. We define by induction on $\alpha \in N_1'$, $\bar{a}_\alpha \in N$ such that $n_{\mathbf{I}_\alpha}$ is maximal where $\mathbf{I}_\alpha = \mathbf{I} \cup \{\bar{a}_\beta \mid \beta < \alpha\}$. Then $\mathbf{I} \cup \{a_\alpha\colon N_1' \models a < \lambda(\Delta)$ for each finite $\Delta\}$ is as required.]

EXERCISE 5.11: The regular ultrafilter D over I satisfies the conclusion of conjecture 5.1 when $(J<) \models \omega \le a_n$ for each n iff $\mathrm{lcf}(\kappa, D) > \aleph_0$, $\kappa \le |I|$ where $\kappa = \mathrm{lcf}(\aleph_0, D)$.

Remark: The conditions $(J, <) \models \omega \le a_n$ are satisfied in the case we are interested in. [Hint: Let $a_n = \langle \dots, \alpha_t^n, \dots \rangle_{t\in I}/D$, and w.l.o.g. α_t^n is a cardinal (see Exercise 1.1) and let $\chi_n = \min\{\chi\colon \{t\colon \chi < \alpha_t^n\} \notin D\}$; clearly $\aleph_0 \le \chi_{n+1} \le \chi_n$, so w.l.o.g. $\chi_n = \chi_0$; and by 1.12 $|a_n| = \chi^{|I|}$. If $\{t\colon \alpha_t^n = \chi\} \in D$ for infinitely many n's, $a = \langle \dots, \chi, \dots \rangle/D$ prove the assertion. So w.l.o.g. $\alpha_t^n < \chi$ for each n, t. Necessarily χ is a limit cardinal, $\mathrm{cf}\,\chi \le |I|$. Let $b_j = \langle \dots, k_t^j, \dots \rangle_{t\in I}/D$ $(j < \kappa)$ exemplify $\mathrm{lcf}(\aleph_0, D) = \kappa$; and $c_j = \langle \dots, \alpha_t^{(k_t^j)}, \dots \rangle_{t\in I}/D$; then $(J, <) \models c_j \le a_n \wedge c_i < c_j$ for $i < j < \kappa$ and let $\{d \in J\colon (J, <) \models d \le a_n$ for each $n\} = \bigcup_{j<\kappa} \{d \in J\colon (J, <) \models d < c_j\}$. So $\mathrm{cf}\,\chi = \kappa$. If a_n, D fails the conjecture $|c_j| < \chi^{|I|}$ for each j hence for some $\lambda < \chi$, $(J, <) \models c_j < \lambda$, and it is easy to show $\mathrm{lcf}(\kappa) = \aleph_0$; and the other direction should be easy.]

EXERCISE 5.12: Show that for good $\aleph_1$-incomplete ultrafilter D conjecture 5.1 holds. [Hint: If $(J, <) \models a_n < \omega$ for some n, by 2.3 there is c, $(J, <) \models n < c < a_n$ for each n, and use 2.13. Otherwise by 2.3. $\kappa \le |I| \Rightarrow \mathrm{lcf}(\kappa, D) > |I|$ and use Exercise 5.11.]

EXERCISE 5.13: Suppose that for some n_t, $\mu = \prod_{t\in I} n_t/D > \aleph_0$ and $N = M^I/D$ is $(|T| + \mu)^+$-saturated, and M is λ-compact. Then N is

λ^I/D saturated. [Hint: As in 5.2, let $\lambda > |T|$, T without the f.c.p., $\mathbf{I} \subseteq N$ indiscernible, $|\mathbf{I}| = (|T| + \mu)^+$. There are finite $P_t \subseteq |M|$ such that $|P^N| \leq \mu$, $|\mathbf{I} \cap P^N| \geq \aleph_0$, where $(N, P^N) = \prod_{t \in I} (M, P_t)/D$. There is $\bar{a} \in \mathbf{I}$ such that $\mathrm{stp}(\bar{a}, \cup P^N) \subseteq \mathrm{Av}(\mathbf{I}, N)$, and $Q_t = \mathbf{I}_t \subseteq |M|$ an indiscernible set over $\cup P_t$ based on $\mathrm{stp}(\bar{a}[t], \bigcup P_t)$, and $(M, P, Q) = \prod_{t \in I} (M, P_t, Q_t)/D$. Then Q is indiscernible, $|Q| = \lambda^I/D$, $\mathrm{Av}(Q, N) = \mathrm{Av}(\mathbf{I}, N)$.]

VI.6. Saturation of ultralimits

For every model M and ultrafilter D over I the natural elementary embedding of M into M^I/D, is defined by $a \mapsto \langle a : i \in I \rangle/D$. Hence we can look at M^I/D as an elementary extension of M; and so we get an elementary chain of models by repeatedly taking ultrapowers and taking unions at limit stages. For simplicity, all the ultrapowers will be with the same ultrafilter, which will be assumed $\aleph_1$-incomplete.

DEFINITION 6.1: Let M be a model, D an $\aleph_1$-incomplete ultrafilter over I. We define the ultralimit, $\mathrm{UL}(M, D, \alpha)$ by induction on α, so that for $\beta < \alpha$, $\mathrm{UL}(M, D, \beta)$ is an elementary submodel of $\mathrm{UL}(M, D, \alpha)$.

(1) $\mathrm{UL}(M, D, 0) = M$.

(2) For α a limit ordinal $\mathrm{UL}(M, D, \alpha) = \bigcup_{\beta < \alpha} \mathrm{UL}(M, D, \beta)$.

(3) For $\alpha = \beta + 1$, $\mathrm{UL}(M, D, \alpha)$ will be isomorphic to $\mathrm{UL}(M, D, \beta)^I/D$, and for each $a \in |\mathrm{UL}(M, D, \beta)|$, the isomorphism F_β takes $\langle a : i \in I \rangle/D \in |\mathrm{UL}(M, D, \beta)^I/D|$ to a. So $\mathrm{UL}(M, D, \beta) \prec \mathrm{UL}(M, D, \alpha)$.

Notation. As M and D are fixed for most of this section we let $M_\alpha = \mathrm{UL}(M, D, \alpha)$ and F_β be the isomorphism mentioned in (3) of the above definition. Clearly we can assume that for every α, β, $\mathrm{UL}(M, D, \alpha + \beta) = \mathrm{UL}(M_\alpha, D, \beta)$. We write T for $\mathrm{Th}(M)$. We shall try to find how compact the ultralimits are, for various properties of the ordinal, the ultrafilter, and the theory of the model. As $M_{\alpha+1}$ is isomorphic to M_α^I/D, we shall restrict ourselves to M_δ for limit ordinals δ.

THEOREM 6.1: *If* $\mathrm{cf}\,\delta \geq \mu$, *and for every* $\lambda < \mu$, D *is* λ*-regular, then* M_δ *is* μ*-compact.*

Proof. Let p be a type in M_δ of cardinality $< \mu$. Then clearly p is a type in M_β for some $\beta < \delta$. As D is $|p|$-regular, p is realized in $M_{\beta+1}$ by 1.4, hence p is realized in M_δ. So every type in M_δ of cardinality $<\mu$ is realized in M_δ; hence M_δ is μ-compact.

THEOREM 6.2: *If T is unstable, $\mu = \text{cf}\,\delta$, then M_δ is not μ^+-compact.*

Proof. M_1 is $\aleph_1$-compact by 2.1, 2.3 and 2.11 (remember D is $\aleph_1$-incomplete). As T is unstable, by II, 2.13 and II, 2.2; there is a formula $\varphi(x; \bar{y})$ and sequences $\bar{a}_0, \bar{a}_1, \ldots \bar{a}_n, \ldots$ from M_1 (all of the length of $\bar{y}$) such that: for every $m < \omega$, $\{\varphi(x; \bar{a}_n)^{\text{if}(n \geq m)}: n < \omega\}$ is consistent with M_1. As $\text{cf}\,\delta = \mu$, let $\delta = \bigcup_{j<\mu} \alpha(j)$, where $j < i < \mu$ implies $1 < \alpha(j) < \alpha(i) < \delta$. We shall now define by induction on j sequences $\bar{a}^j$ such that:

(1) $\bar{a}^j \in |M_{\alpha(j)+1}|$, $\bar{a}^j \notin |M_{\alpha(j)}|$.

(2) $q_j = \{\neg\varphi(x; \bar{a}_n): n < \omega\} \cup \{\varphi(x; \bar{a}^j)\}$ is not realized by any element of $M_{\alpha(j)}$.

(3) For every $m < \omega, p_j^m = \{\varphi(x; \bar{a}_n)^{\text{if}(n \geq m)}: n < \omega\} \cup \{\varphi(x; \bar{a}^i): i \leq j\}$ is consistent (with $M_{\alpha(j)+1}$).

If we succeed in defining the $\bar{a}^j$'s then clearly by (3) $p = \{\neg\varphi(x; \bar{a}_n): n < \omega\} \cup \{\varphi(x; \bar{a}^i): i < \mu\}$ is consistent (with M_δ), because every finite subset of p is a subtype of p_j^m for some m, j. But if p is realized in M_δ, then it is realized in M_β for some $\beta < \delta$, and there is $j < \text{cf}\,\delta$, $\beta < \alpha(j) < \delta$. Hence p is realized in $M_{\alpha(j)}$, in contradiction to (2). Hence p is a consistent type in M_δ, which M_δ omits, and $|p| = \aleph_0 + \mu < \mu^+$. So M_δ is not μ^+-compact.

It remains only to define $\bar{a}^j$, assuming $\bar{a}^i$ has been defined for $i < j$. As D is $\aleph_1$-incomplete there are $I_n \in D$, $I_{n+1} \subseteq I_n$, $I_0 = I$, $\bigcap_{n<\omega} I_n = \emptyset$. Let us define $\bar{b}^j \in |M^I_{\alpha(j)}/D|$: if $i \in I_n - I_{n+1}$, then $\bar{b}^j[i] = \bar{a}_n$, $\bar{b}^j = \langle \bar{b}^j[i]: i \in I\rangle/D$, and $\bar{a}^j = F_{\alpha(j)}(\bar{b}^j)$.

Let us check conditions (1), (2) and (3) are satisfied. Clearly $\bar{a}^j \in |M_{\alpha(j)+1}|$. Now for any $n < \omega$, $\{i \in I: \bar{b}^j[i] = \bar{a}_n\} = I_n - I_{n+1} \notin D$ so $\bar{a}^j \notin |M_{\alpha(j)}|$. So (1) is satisfied.

For proving (2), suppose, for some $j < \mu$, $c_j \in |M_{\alpha(j)}|$ realizes q_j. Then $\{i \in I: M_{\alpha(j)} \models \varphi[F^{-1}_{\alpha(j)}(c_j)[i], \bar{b}^j[i]]\} \in D$, that is, $\{i \in I: M_{\alpha(j)} \models \varphi[c_j, \bar{b}^j[i]]\} \in D$. Hence for some $n < \omega$ $M_{\alpha(j)} \models \varphi[c_j, \bar{a}_n]$, so c_j does not realize q_j, contradiction, hence

(2) holds; and

(3) has a similar proof.

DEFINITION 6.2: $\mu(D)$ is the least infinite cardinal μ such that D is μ-descendingly complete, that is, $\mu(D)$ is the least infinite cardinality μ

such that $I_\alpha \in D$, $\alpha < \mu$, and $\alpha < \beta \Rightarrow I_\beta \subseteq I_\alpha$ implies $\bigcap_{\alpha<\mu} I_\alpha \neq \emptyset$ (equivalently $\bigcap_{\alpha<\mu} I_\alpha \in D$).

Remark. If D is κ-regular then $\kappa < \mu(D)$; also $\mu(D) \leq |I|^+$. Note also that $\mu(D)$ is a regular cardinal.

THEOREM 6.3: *If $\mu \leq \mu(D)$, $\mu \leq \operatorname{cf} \delta$ then M_δ is μ-compact.*

Proof. Let p be a type in M_δ, $|p| < \mu$, and we shall prove that p is realized in M_δ, thus proving the theorem.

As $|p| < \mu \leq \operatorname{cf} \delta$, p is a type in M_α for some $\alpha < \delta$. Let $|p| = \aleph_\beta$. We shall prove by induction on $\gamma \leq \beta$ that:

(*) Every subtype of p of cardinality $\leq \aleph_\gamma$ is realized in $M_{\alpha+\gamma+1}$.

As $\beta \leq \aleph_\beta = |p| < \mu \leq \operatorname{cf} \delta$, $\alpha + \beta + 1 < \delta$, hence by proving (*) we shall prove that p is realized in M_δ.

Suppose we have proved (*) for every $\gamma_1 < \gamma$. Hence every subtype of p of cardinality $< \aleph_\gamma$ is realized in $M_{\alpha+\gamma}$ (remember every model is $\aleph_0$-compact, hence every finite subtype of p is realized in M_α). Clearly we can assume w.l.o.g. that p is a 1-type. So let $q = \{\varphi_j(x; \bar{a}_j): j < \aleph_\gamma\}$ be any subtype of p of cardinality $\aleph_\gamma$; we shall prove q is realized in $M_{\alpha+\gamma+1}$. By the induction hypothesis, for every $j < \aleph_\gamma$, there is $c_j \in |M_{\alpha+\gamma}|$ which realizes $\{\varphi_i(x; \bar{a}_i): i < j\}$. As $\aleph_\gamma \leq |p| < \mu \leq \mu(D)$ there is a decreasing sequence I_j, $j < \aleph_\gamma$, $I_j \in D$, $\bigcap_{j<\aleph_\gamma} I_j = \emptyset$, with $I_0 = I$. Let us define $c \in |(M_{\alpha+\gamma})^I/D|$: if $i \in \bigcap_{\beta<j} I_\beta - I_j$ then $c[i] = c_j$ (clearly c is well-defined). Now $F_{\alpha+\gamma}(c) \in |M_{\alpha+\gamma+1}|$ realizes q, as for every $j < \aleph_\gamma$, $\{i \in I: M_{\alpha+\gamma} \models \varphi[c[i], \bar{a}_j]\} \supseteq \bigcup_{\zeta>j} (\bigcap_{\beta<\zeta} I_\beta - I_\zeta) = I_{j+1} \in D$. So q is realized in $M_{\alpha+\gamma+1}$; so p is realized in M_δ.

DEFINITION 6.3: A model N strongly omits a type p (in it) if no subtype of p of cardinality $|p|$ is realized in N.

LEMMA 6.4: (1) *If M strongly omits p, $|p| = \mu(D)$, then M_1 also strongly omits p.*

(2) *if M_α strongly omits p, $|p| = \mu(D)$, and $\alpha < \beta$, then M_β also strongly omits p.*

(3) *In* (1) *and* (2) *instead of $|p| = \mu(D)$, its suffices to assume that there are no $I_\beta \in D$ for $\beta < |p|$, $\beta < \gamma \Rightarrow I_\gamma \subseteq I_\beta$, $\bigcap_{\gamma<|p|} I_\gamma = \emptyset$; and $|p|$ is regular.*

Proof. We shall prove that (1) as (2) and (3) have similar proofs. Suppose (1) fails, so $c_1 \in |M_1|$ realizes $q \subseteq p$, $|q| = |p|$. Let $c_1 = F_0(c)$, $q = \{\varphi_\xi(x; \bar{a}_\xi): \xi < |q|\}$. So clearly for every $\xi < |q| = |p|$ $\{i \in I: M \vDash \varphi_\xi[c[i]; \bar{a}_\xi]\} \in D$. It is also clear that for every $i \in I$, $q(i) = \{\varphi_\xi(x; \bar{a}_\xi): M \vDash \varphi_\xi[c[i], \bar{a}_\xi]\}$ is a subtype of q, hence of p, which is realized in M; hence $|q(i)| < |p|$. As $|p| = \mu(D)$ is regular, for every $i \in I$ there is a bound $\xi(i) < |p|$ to $\{\xi: M \vDash \varphi_\xi[c[i]; \bar{a}_\xi]\}$. Let for $\zeta < |p|$, $I_\zeta = \{i \in I: \xi(i) \geq \zeta\}$. Clearly I_ζ, $\zeta < |p|$ is a decreasing sequence, and by the definition of $\xi(i)$, $\bigcap_{\zeta < |p|} I_\zeta = \emptyset$. In addition each $I_\zeta \in D$ as $I_\zeta = \{i \in I: \xi(i) \geq \zeta\} \supseteq \{i \in I: M \vDash \varphi_\zeta[c[i]; \bar{a}_\zeta]\} \in D$. So we get a contradiction to the definition of $\mu(D)$.

THEOREM 6.5: *If T is unstable, $\delta \geq \mu(D)$, then M_δ is not $\mu(D)^+$-compact. (Moreover there is a type in $M_{\mu(D)}$ of cardinality $\mu(D)$ which M_δ strongly omits.)*

Proof. As it is similar to the proofs of 6.2 and 6.7 we omit it.

CONCLUSION 6.6: *If T is unstable, $\mu = \min(\mathrm{cf}\,\delta, \mu(D))$, then M_δ is maximally μ-compact. (i.e., μ-compact but not μ^+-compact).*

Proof. Immediate by 6.2, 6.3 and 6.5.

THEOREM 6.7: *If $\kappa(T) > \mu = \min(\mathrm{cf}\,\delta, \mu(D))$, T is stable, then M_δ is maximally μ-compact.*

Proof. By 6.3, M_δ is μ-compact so we need to prove only that M_δ is not μ^+-compact. By hypothesis T satisfies $\mu < \kappa(T)$, so there are A_ξ, $\xi \leq \mu$, $p \in S(A_\mu)$, $\varphi_\xi(x, \bar{y}_\xi)$, and $\bar{a}_{\xi,n}$, $(\xi < \mu, n < \omega)$; such that: $\xi < \zeta \Rightarrow A_\xi \subseteq A_\zeta$; for every $\xi < \mu$, $\{\bar{a}_{\xi,n}: n < \omega\}$ is an indiscernible set over A_ξ, $\bar{a}_{\xi,0}, \bar{a}_{\xi,1} \in A_{\xi+1}$ and $\neg\varphi_\xi(x; \bar{a}_{\xi,0})$, $\varphi(x; \bar{a}_{\xi,1}) \in p$. Clearly it suffices to prove the theorem for the case $L = L(T)$ is the minimal language containing all the formulas $\varphi_\xi(x; \bar{y}_\xi)$; so $|L| \leq \mu$. Choose $\alpha(\xi) < \delta$ for $\xi < \mathrm{cf}\,\delta$ such that $\delta = \bigcup_{\xi < \mathrm{cf}\,\delta} \alpha(\xi)$.

Now we define by induction on $\xi \leq \mu$ an increasing sequence of $(\mathfrak{C}, M_\delta)$-elementary mappings H_ξ and ordinals $\beta(\xi) \leq \delta$ such that:

(1) The domain of H_ξ is $\bigcup_{\zeta < \xi} \mathrm{Range}\,(\bar{a}_{\zeta,0}{}^\frown\bar{a}_{\zeta,1})$.

(2) The range of H_ξ is included in $|M_{\beta(\xi)+1}|$, $\beta(\xi) \geq \alpha(\xi)$.

(3) If $\xi < \zeta < \mu$ then $\beta(\xi) < \beta(\zeta) < \delta$, and for every $c \in |M_{\beta(\xi)}|$, $M_\delta \vDash \varphi_\zeta[c, H_{\zeta+1}(\bar{a}_{\zeta,0})] \equiv \varphi_\zeta[c, H_{\zeta+1}(\bar{a}_{\zeta,1})]$.

Let H_0 be the empty function, $\beta(0) = \alpha(0)$. For ξ a limit ordinal, $H_\xi = \bigcup_{\zeta<\xi} H_\zeta$, $\beta(\xi) = \max(\alpha(\xi), \bigcup_{\zeta<\xi}\beta(\zeta))$ if $\xi < \mu$, and $\beta(\mu) = \bigcup_{\zeta<\mu}\beta(\zeta)$. [If $\xi < \mu$, $\beta(\xi) < \delta$ because $\mu \leq \text{cf}\,\delta$.]

Suppose H_ξ, $\beta(\xi)$ are defined, $\xi < \mu$, and we shall define $H_{\xi+1}$, $\beta(\xi + 1)$. We first show:

(∗) There is $\beta < \delta$ such that we can extend H_ξ to a $(\mathfrak{C}, M_\beta)$-elementary mapping H^* from $\text{Dom}\, H_\xi \cup \bigcup \{\bar{a}_{\xi,n}: n < \omega\}$ into M_β.

If $\mu = \aleph_0$, this is true, as for every N, N^I/D is $\aleph_1$-compact, so $\beta = \beta(\xi) + 1$ will suffice. So assume $\mu > \aleph_0$. We now define, by induction on n, an increasing sequence of $(\mathfrak{C}, M_\delta)$-elementary mappings H^n from $\text{Dom}\, H_\xi \cup \bigcup \{\bar{a}_{\xi,m}: m < n\}$ into M_δ. If we have defined H^n, and cannot define H^{n+1}, we can deduce M_δ is not μ^+-compact [as it omits $\{\varphi(\bar{x}; H^n(\bar{c})): \varphi \in L, \bar{c} \in \text{Dom}\, H^n, \vDash \varphi[\bar{a}_{\xi,n}, \bar{c}]\}$] and so the conclusion of the theorem holds. So we can assume H^n is defined for every n and let $H^* = \bigcup_{n<\omega} H^n$. Clearly H^* is a $(\mathfrak{C}, M_\delta)$-elementary mapping with the appropriate domain into M_δ, Range $H_\xi \subseteq M_{\beta(\xi)}$, $\beta(\xi) < \delta$, cf $\delta \geq \mu$ and H^* is into M_β for some $\beta < \delta$. So we have proved (∗). But we cannot define $H_{\xi+1} = H^*$, as (3) may fail.

Define $\beta(\xi + 1) = \max(\beta, \alpha(\xi))$. As D is $\aleph_1$-incomplete there are $I_n \in D$, $n < \omega$, $I_0 = I$, $m < n \Rightarrow I_n \subseteq I_m$, $\bigcap_{n<\omega} I_n = \emptyset$. Define $H_{\xi+1}(\bar{a}_{\xi,0}{}^\frown\bar{a}_{\xi,1}) \in |M_{\beta(\xi+1)+1}|$ as $F_{\beta(\xi+1)}(\bar{c})$ where $\bar{c} \in M_{\beta(\xi)+1}{}^I/D$ is defined as follows: if $i \in I_n - I_{n+1}$, $\bar{c}[i] = H^*(\bar{a}_{\xi,n}{}^\frown\bar{a}_{\xi,n+1})$. It is easy to verify that $H_{\xi+1}$, $\beta(\xi + 1)$ satisfy the induction conditions (by II, 2.20).

Now $p = \{\varphi_\xi(x, H_{\xi+1}(\bar{a}_{\xi,0})) \equiv \neg\varphi_\xi(x, H_{\xi+1}(\bar{a}_{\xi,1})): \xi < \mu\}$ is a consistent type in $M_{\beta(\mu)}$, and it is strongly omitted by $M_{\beta(\mu)}$. As $M_\delta = \text{UL}(M_{\beta(\mu)}, D, \delta - \beta(\mu))$, by 6.4 M_δ strongly omits p, and this means M_δ is not μ^+-compact.

It is natural to conjecture that if $\kappa(T) \leq \min(\mu(D), \text{cf}\,\delta)$, and α, $\beta < \delta \Rightarrow \alpha + \beta < \delta$, then M is $\text{UL}(\aleph_0, D, \delta)$-saturated [$\text{UL}(\aleph_0, D, \delta)$ is the cardinality of $\text{UL}(M, D, \delta)$ for every countable M] which would generalize 5.1(1). But this is not true. T may be superstable ($\kappa(T) = \aleph_0$) or even simple (see Definition 6.4) while M or M_1 strongly omits a type of cardinality $\mu(D)$. However:

THEOREM 6.8: *Suppose* $\kappa(T) \leq \min(\mu(D), \text{cf}\,\delta)$, *$D$ is a $|T|$-regular ultrafilter;* $\alpha, \beta < \delta \Rightarrow \alpha + \beta < \delta$. *Then M_δ is λ-saturated, where* $\lambda = \text{UL}(\aleph_0, D, \delta)$.

Remarks. (1) For every δ_1 there are δ_2, δ such that $\delta_1 = \delta_2 + \delta$; $\alpha, \beta < \delta \Rightarrow \alpha + \beta < \delta$ and $\mathrm{UL}(M, D, \delta_1) = \mathrm{UL}(M_{\delta_2}, D, \delta)$. So the restriction on δ is natural.

(2) Clearly $\lambda > |T|$, so it suffices to prove M_δ is λ-compact.

Proof. Let p be a 1-type in M_δ, $|p| < \lambda$. It suffices to prove p is realized in M_δ. Let q be any extension of p in $S(|M_\delta|)$.

Notice that if $|B| < \kappa(T) \leq \mathrm{cf}\,\delta$, $B \subseteq |M_\delta|$, then for some $\alpha < \delta$, $B \subseteq |M_\alpha|$. Hence by III, 3.2 there is $\alpha < \delta$ and $B \subseteq |M_\alpha|$ such that q does not fork over B, and by III, 2.15 and III 4.18, $q \restriction |M_\alpha|$ is stationary. So by III, 4.2 there is a set $B \subseteq |M_\alpha|$, $|B| \leq |T|$ such that for every φ $R(q \restriction \varphi, \varphi, 2) = R[(q \restriction B) \restriction \varphi, \varphi, 2]$. Now we can define a_n for $n < \omega$ such that:

(1) a_n realizes $q \restriction (B \cup \{a_m : m < n\})$,

(2) if $\delta > \omega$, $a_n \in |M_{\alpha+n+1}|$.

This is possible since D is $|T|$-regular, by 1.4(2). Hence $\{a_n : a < \omega\}$ is an indiscernible set and $q = \mathrm{Av}(\{a_n : n < \omega\}, M_\delta)$. Suppose for a moment $\delta > \omega$. Let $P = \{a_n : n < \omega\} \subseteq |M_{\alpha+\omega}|$ (as $\alpha < \delta$, $\omega < \delta$; $\alpha + \omega < \delta$). Let $(M_\delta, P^\delta) = \mathrm{UL}((M_{\alpha+\omega}, P), D, \delta)$ (remember $\delta = \alpha + \omega + \delta$). Clearly P^δ extends P and is an indiscernible set. So $\varphi(x, \bar{b}) \in p$ implies $\varphi(x, \bar{b}) \in q$ implies $\{a : a \in P^\delta, \vDash \neg\varphi(a, \bar{b})\}$ is finite. So all except $|p| \cdot \aleph_0 < \lambda$ members of P^δ realize p. As $|P^\delta| = \mathrm{UL}(\aleph_0, D, \delta) = \lambda$, the theorem follows. So we remain only with the case $\delta = \omega$. But then we can define the a_n's simultaneously in $M_{\alpha+1}$ and the proof goes in the same way.

The following exercises will deal with the problem of categoricity in logics with generalized quantifiers.

Assumption: Let T be a complete theory in the logic $L(Q^1, \ldots, Q^n)$ (where if φ is a formula so is $(Q^l x \varphi)$). Let $P_1, \ldots, P_m$ be unary predicates in L, $\forall x P_m(x) \in T$ and let for any cardinals $\mu_1 < \cdots < \mu_m, \lambda_1, \ldots, \lambda_m$, $\boldsymbol{K}(\mu_1, \ldots, \lambda_1, \cdots) = \{M : M$ an L-model, that satisfies every $\psi \in T$ when $(Q^l x)\theta$

is interpreted as "there are at least λ_l x's such that $\theta(x)$, and $|P_l^M| = \mu_l\}$."

Suppose further that $\mu_l^{\aleph_0} = \mu_l$, each λ_l is regular and $(\forall \chi < \lambda_l)\chi^{\aleph_0} < \lambda_l$.

Suppose (*)$\boldsymbol{K}(\mu_1, \ldots, \lambda_1, \ldots)$ has a unique model up to isomorphism.

EXERCISE 6.1: We can assume w.l.o.g. that for any formula $\varphi(\bar{x})$ in $L(Q^1, \ldots, Q^n)$, for some predicate R, $(\forall \bar{x})[\varphi(\bar{x}) \equiv R(\bar{x})] \in T$. Let T' be the set of first order sentences of T.

EXERCISE 6.2: Prove T' is superstable. [Hint: It is well known (see [CK 73]) that $\boldsymbol{K}(\mu_1, \ldots, \lambda_1, \ldots)$ is closed under ultraproducts for ultrafilters over ω. Let M belong to $K(\mu_1, \ldots, \lambda_1, \ldots)$ (exists by a hypothesis), and let D be a non-principal ultrafilter over ω. So $M_1 \stackrel{\text{def}}{=} M^\omega/D$ and easily also $M_\omega \stackrel{\text{def}}{=} \mathrm{UL}(M, D, \omega)$ belong to the class $\boldsymbol{K}(\mu_1, \ldots, \lambda_1, \ldots)$, hence by a hypothesis they are isomorphic. Now by 2.1, 2.3, 2.11(1), M_1 is $\aleph_1$-compact. By 6.5 if T is not superstable M_ω is not $\aleph_1$-compact. We conclude T is superstable.] We now define:

$$W = \{\langle \varphi(x, \bar{y}), \psi(\bar{y}), \lambda^l \rangle : \varphi, \psi \text{ first-order formulas, } (\forall \bar{y})(\psi(\bar{y})$$
$$\equiv \neg (Q^l x)\varphi(x, \bar{y})) \text{ belong to } T'\} \cup \{\langle P^l(x), x = x, \mu_l^+ \rangle : l = 1, m\}.$$

EXERCISE 6.3: Suppose T is stable (but not necessarily (*)). Prove there is $M \in K(\mu_1, \ldots, \lambda_1, \ldots)$ which is the L-reduct of M_1, $|L(M_1)| = |L|$, M_1 the Skolem Hull of $\bigcup_{l=1}^{m+n} I_l, I_l$ indiscernible $\bigcup_{k<l} I_k$. [Hint: See VII, §2,5 and use V, 6.10.]

Now we apply the results of V, §6 to analyze the $(n+m)$-tuples of cardinals for which we get categoricity.

Example. Let $N = ({}^\omega 2, +^N, P_n^N)_{n<\omega}$ where: $P_n^N = \{\eta \in {}^\omega 2 : \eta(n) = 0\}$, $+^N$ is a two-place operation, $\eta +^N \nu = \rho$ iff for every n $\eta(n) + \nu(n) - \rho(n)$ is even. Let $M = (|M|, P^M, Q^M, E^M, +^N, G^N, H^N, F^M, P_0^N, \ldots, P_n^N, \ldots)_{n<\omega}$ where:

$P^m = N$,
$Q^M = N \times N \times N$,
$E^M = \{\langle \langle a_1, b_1, c_1 \rangle, \langle a_2, b_2, c_2 \rangle \rangle : a_1, b_1, c_1, a_2, b_2, c_2 \in N, a_1 = a_2, b_1 = b_2\}$,
$+^M$ is a two-place function from P^M to P^M, $+^M = +^N$,
H^N, G^N are one-place functions from Q^M to P^M, $H^N(\langle a, b, c \rangle) = b$, $G^N(\langle a, b, c \rangle) = a$,
F^M is a partial two-place function,
$F^M(\langle a, b, c_1 \rangle, \langle a, b, c_2 \rangle) = c_1 +^N c_2$,
$P_n^M = \{\langle a, b, c \rangle \in N \times N \times N : c \in P_n^N\}$.

CHAPTER VII

CONSTRUCTION OF MODELS

VII.0. Introduction

The character of this chapter is different from that of II–V as extending the theory does not change much; hence it is natural to use theories with Skolem functions, that is, when we want to construct models of T, we choose T_1, $T \subseteq T_1$, $|T_1| = |T|$, and T_1 has Skolem functions; we construct models of T_1 and go back to their L(T)-reducts. The importance of T_1 having Skolem functions is that: (1) we can find such T_1 for every T; (2) every model of T can be expanded to a model of T_1; (3) if M^1 is a model of T_1, every submodel of M^1 is an elementary submodel. (This is done in 1.1–1.4).

Saturation is a very useful device, but unfortunately if $\lambda < \lambda^{<\kappa}$, T does not necessarily have a κ-saturated model of cardinality λ. We have constructed in I, 1.7 a saturated model by an elementary chain M_i of models of T of cardinality λ such that in M_{i+1} we realize as many types over M_i as we can (remembering the cardinality restriction). We generalize this process (in Definition 1.5) and prove (in 1.7) that for many χ's we get some weakening of saturation, e.g., of the form: if p is an m-type over M, $|p| = \lambda$, and every $q \subseteq p, |q| < \mu$ is realized in M then some $r \subseteq p$, $|r| = \chi$ is realized in M. Notice that when $\lambda^{<\kappa} = \lambda$ we get a κ-compact model (when the length of the chain has cofinality $\geq \kappa$, of course); if $(\forall \mu < \aleph_\alpha)(\mu^{<\kappa} < \aleph_\alpha)$, $\aleph_\alpha^{<\kappa} > \aleph_\alpha$, and $\lambda = \aleph_{\alpha+\beta}$, our results become weaker as β increases, and we get none when $\beta \geq \gamma = \aleph_\gamma > \aleph_\alpha$. However for the theory of dense linear order we can get something (see Exercise 1.7) and a generalization of this is utilized in VIII, 3.2. (There are also other applications of Section 1 to VIII, Section 2, and 3, but we do not need them in the main presentation.)

From one point of view a saturated model realizes many types, but from a deeper point of view it realizes few. Let $TP(\bar{a}, M)$ be the set of types realized in M over $\bar{\mathbf{a}}$, $TP_\kappa(M) = \{TP(\bar{\mathbf{a}}, M): \bar{\mathbf{a}} = \langle a_i: i < \kappa\rangle, a_i \in |M|\}$. Then clearly for a λ-saturated model M of a countable theory ($\kappa < \lambda$), $TP_\kappa(M)$ has cardinality $\leq 2^{\kappa+|T|}$, but for a not necessarily λ-saturated model, it is possible that $|TP_\kappa(M)| = 2^{2^\kappa}$. We deal with these problems in 1.10–1.12 and prove, e.g.,

THEOREM 0.1: *If* $\mu < \aleph_\alpha \Rightarrow \mu^\kappa < \aleph_\alpha$, $\beta < \kappa^+$, T *countable*, $\|M\| \leq \aleph_{\alpha+\beta}$ *then M has an elementary extension N of cardinality* $\aleph_{\alpha+\beta}$ *for which* $|TP_\kappa(N)| = 2^\kappa$.

In Sections 2 and 3 we deal with Ehrenfeucht–Mostowski models, generalizations in two directions, and various applications; which are preparatory for VIII.

For a theory T_1 with Skolem functions, every set A is a set of generators of a model, and it is natural to look for "nice", "uniform", such sets. The classical example is a free algebra in which every permutation of the set of (the free) generators can be extended to an automorphism of the model. But for any extension of the theory of dense linear order we cannot have such sets. However, by Ramsey's theorem and compactness we can find models of T_1 generated by indiscernible sequences, so the order of the generators is important. Clearly the elementary type Φ of the sequence and the order type I determine the model up to isomorphism, so we denote it by $EM^1(I, \Phi)$, and its L(T)-reduct by $EM(I, \Phi)$.

Its importance is that we can easily find orders I with specific properties, and often the model inherits them; e.g., stability (of the model) and some of the generalizations of saturativity from Section 1. An application is

THEOREM 0.2: *For stable T with the* f.c.p., $|T_1| = \aleph_0$, $I(\aleph_\alpha, T_1, T) \geq 2^{|\alpha|}$.

Unfortunately this is true only for cardinalities $> |T_1|$, so when we try to construct models of cardinality $|T_1|$ (e.g., in VIII) we have to generalize this construction. We first show we can assume w.l.o.g. that in L(T_1) there are only countably many non-logical symbols which are not individual constants, then we form a model N from individual constants, and define $EM^1(I, N^1)$ for any order I, and elementary extension N^1 of N. Now we can determine properties of $EM^1(I, N^1)$ by those of I and N^1.

At first sight, we cannot expect anything better than an indiscernible sequence of generators. However, we sometimes know of a set of elements satisfying some conditions, and we want to find such a set as homogeneous as possible, which still satisfies those conditions. For proving this we always need the proper generalization of Ramsey's Theorem. In particular if T is not superstable there are $\varphi_l \in L(T)$, and $\bar{a}_\eta$ (for $\eta \in {}^{\omega\geq}\omega$) such that for $\eta \in {}^{\omega}\omega$, $\nu \in {}^{\omega>}\omega$, $\varphi_{l(\nu)}[\bar{a}_\eta, \bar{a}_\nu]$ iff $\nu \triangleleft \eta$. For dealing with such situations we define the indiscernibility of $\{\bar{a}_s \colon s \in I\}$ for a model I: the type of $\bar{a}_{s(0)} {}^\frown \ldots, {}^\frown \bar{a}_{s(n-1)}$ depends only on the atomic type of $\langle s(0), \ldots, s(n-1)\rangle$ in I. The general treatment appears in Section 2 and the specific cases in Section 3. We deal there with theories with the f.c.p., and indiscernible sets indexed by trees. An important case, which does not completely fit with the general case, is uncountable $D(T)$ for countable T_1. What we get is $\{\bar{a}_\eta \colon \eta \in {}^{\omega}2\}$ so that the $\bar{a}_\eta$'s realize distinct types from $D(T)$, and for every $\varphi(\bar{x}_0, \ldots, \bar{x}_{m-1}) \in L(T_1)$ for every $n \geq n(\varphi)$, if $\eta_l \restriction n (l < m)$ are distinct, the satisfaction of $\varphi(\bar{a}_{\eta_0}, \ldots)$ depends only on $\langle \eta_l \restriction n \colon l < m\rangle$ and α. For uncountable T's see a generalization VIII, 1.10.

The reader can relax while reading Section 4—almost no background is needed. Our main aim is

THEOREM 0.3: *If* $\lambda \geq 2^{\mu}$, $\mu \geq 2^{\aleph_0}$, T_1 *countable (for simplicity), then* T_1 *has a* μ*-universal model stable in* μ, *of cardinality* λ.

If I is a well-ordered set of cardinality λ, $M = EM^1(I, \Phi)$ is stable in every cardinality $\geq |T_1|$ (see 2.9). So if D is a good ultrafilter over κ, M^κ/D is κ^+-saturated, hence κ^+-universal, but it still is stable in every cardinality $\lambda = \lambda^\kappa$. So assuming G.C.H., we can prove the theorem for a successor μ. However, without it, we need finer methods.

Remember that (by III, 4.10) a type p does not fork over $|M|$ iff p is finitely satisfiable in $|M|$ (though p is not necessarily over $|M|$). For not necessarily stable theories the second condition is still meaningful, and we can prove for it the extension property, and that it is satisfied by any type over $|M|$. So if M_i $(i < \lambda)$ is a list of models of T_1, containing, up to isomorphism, every one of cardinality $\leq \mu$, we can find $M'_i \cong M_i$ such that for each $\bar{a} \in |M'_i|$, $\mathrm{tp}(\bar{a}, \bigcup_{j<i} M'_j)$ is finitely satisfiable in M'_0. Let $\|M_0\| = |T_1|$, if T_1 has Skolem functions $\bigcup_i M'_i$ generates a model of T_1. For $\mu \geqslant 2^{2^{|T_1|}}$ the model is μ-stable; but we work more and get it for $\mu \geqslant 2^{|T_1|}$. By such a construction we prove our theorem. Notice that by such a construction we can get a sequence of indiscernibles.

In the last section we deal with Morley numbers, i.e., Hanf numbers for omitting types.

PROBLEM 0.1: It will be interesting to find new kinds of indiscernibility. This may require proving new partition theorems or trying to find an application of a known partition theorem, e.g., Laver's generalization of Galvin's theorem $\eta \to [\eta]^2_3$, (i.e., $\eta \to [\eta]^n_{h(m)}$).

VII.1. Skolem functions and generalizations of saturativity

DEFINITION 1.1: (1) L_0 has Skolem functions in T_1 (= T_1 has Skolem functions for L_0) if for every formula $\varphi(x; \bar{y}) \in L_0$ there is in T_1 a sentence of the form

$$(\forall \bar{y})[(\exists x)\varphi(x; \bar{y}) \to \varphi(F(\bar{y}); \bar{y})]$$

(F a function symbol.) Such a sentence is called a Skolem sentence of $\varphi = \varphi(x; \bar{y})$ and F is called a Skolem function of $\varphi(x; \bar{y})$.

(2) T_1 has Skolem functions if it has Skolem functions for its language $L_1 = L(T_1)$.

THEOREM 1.1: (1) *For every language* L *there is a language* $L^{(1)} \supseteq L$ *and a theory* $T^{(1)} = T^{(1)}(L)$ *in* $L^{(1)}$ *such that:*

(A) $T^{(1)}$ *has Skolem functions for* L.

(B) *Every* L*-model can be expanded to an* $L^{(1)}$*-model of* $T^{(1)}$.

(C) $|L^{(1)}| = |T^{(1)}| = |L|$.

(2) *For every language* L *there is a language* $L^{SK} \supseteq L$ *and theory* $T^{SK} = T^{SK}(L)$ *in* L^{SK} *such that:*

(A) T^{SK} *has Skolem functions.*

(B) *Every* L*-model can be expanded to an* L^{SK}*-model of* T^{SK}.

(C) $|T^{SK}| = |L^{SK}| = |L|$.

Proof. (1) For every formula $\varphi = \varphi(x; \bar{y})$ in L we add a new function symbol F^L_φ (which $\notin$ L), and we let $L^{(1)} = L \cup \{F^L_\varphi : \varphi \in L\}$; $T^{(1)} = \{(\forall \bar{y})[(\exists x)\varphi(x; \bar{y}) \to \varphi(F^L_\varphi(\bar{y}); \bar{y})] : \varphi(x; \bar{y}) \in L\}$.

Condition (A) is satisfied by the definition of $T^{(1)}$. If M is an L-model we expand it to an $L^{(1)}$-model as follows: we define $F^L_\varphi(\bar{a})$ as any element b of M satisfying $M \models \varphi[b, \bar{a}]$ if $M \models (\exists x)\varphi(x, \bar{a})$; and we define $F^L_\varphi(\bar{a})$ as an arbitrary element of M otherwise. Clearly we succeed in expanding M to an $L^{(1)}$-model of $T^{(1)}$, so condition (B) holds. Condition (C) is immediate.

(2) We define, by induction on n, L^n and T^n: $L^0 = L$, $T^0 = \emptyset$; $T^{n+1} = T^{(1)}(L^n)$, $L^{n+1} = (L^n)^{(1)}$. We let $L^{SK} = \bigcup_{n<\omega} L^n$, $T^{SK} = \bigcup_{n<\omega} T^n$. Clearly L^n is an increasing sequence hence any formula φ in L^{SK} is in L^n for some n, hence in T^{n+1} we can find a Skolem sentence of φ, so in T^{SK} there is a Skolem sentence of φ; hence condition (A) holds. As for condition (B), if M is an L-model we define by induction on n, the L^n-model M^n: $M^0 = M$, M^{n+1} is an expansion of M^n to an L^{n+1}-model of T^{n+1} (by (1) this is possible). In the limit we get an L^{SK}-model of $T^{SK} = \bigcup_n T^n$. Condition (C) is immediate.

THEOREM 1.2: (1) *Suppose M is a model of T_1, N a submodel of M, (M, N are L_1-models) $L_1 = L(T_1)$ and $L_0 \subseteq L_1$, and T_1 has Skolem functions for L_0. Then the L_0-reduct of N is an elementary submodel of the L_0-reduct of M.*

(2) *Suppose M is a model of T_1, T_1 has Skolem functions, and N is a submodel of M. Then N is an elementary submodel of M.*

Proof. (1) By the Tarski–Vaught test I, 1.2 it suffices to prove that if $\bar{a} \in |N|$ $\varphi(x, \bar{y}) \in L_0$ and $M \models (\exists x)\varphi(x, \bar{a})$ then for some $b \in |N|$ $M \models \varphi[b, \bar{a}]$. But by assumption there is in L_1 a Skolem function F of φ (in T_1), hence $M \models (\exists x)\varphi(x; \bar{a})$ implies $M \models \varphi[F(\bar{a}), \bar{a}]$. As N is a submodel of M, $F(\bar{a}) \in |N|$, so we finish.

(2) A particular case of (1) where $L_0 = L_1$.

CONCLUSION 1.3: *Suppose* Th(M) *has Skolem functions, $A \subseteq |M|$. Then the Skolem closure (=Skolem Hull) of A, $\{\tau(\bar{a})\colon \bar{a} \in A, \tau$ a term$\}$ =* dcl A = acl A *is (the universe of) an elementary submodel N of M. So for every $\bar{b} = \langle b_0, \ldots, b_n\rangle \in |N|$ there are $\bar{a} \in A$, and terms $\tau_0(\bar{x}), \ldots, \tau_n(\bar{x})$ so that $b_i = \tau_i(\bar{a})$. We write it in short $\bar{b} = \bar{\tau}(\bar{a})$, $\bar{\tau} = \langle \tau_0, \ldots, \tau_n\rangle$. If A is (totally) ordered, we can assume $\bar{a}$ is an increasing sequence. In any case we can assume there are no repetitions in $\bar{a}$.*

THEOREM 1.4: *Suppose $|M| = \{a_i\colon i < \lambda\}$,* $L = L(M)$, $\lambda > |L|$, *λ is regular. Then the set C of $\alpha < \lambda$ which satisfy the following, is a closed unbounded subset of λ:*

$A_\alpha = \{a_i\colon i < \alpha\}$ is (the universe of) an elementary submodel of M.

Proof. By I, 1.3(2) C is closed. Let us prove it is unbounded. By 1.1(2) we can assume w.l.o.g. that T = Th(M) has Skolem functions. So

$\alpha \in C$ iff A_α is closed under the functions of M. Clearly $|\operatorname{dcl} A_\alpha| < \lambda$ hence for some $\beta = \beta(\alpha)$, $\operatorname{dcl} A_\alpha \subseteq A_\beta$. So for any $\alpha_0 < \lambda$, if $\alpha_{n+1} = \beta(\alpha_n)$, $\alpha = \bigcup_n \alpha_n$, then $\operatorname{dcl} A_\alpha = A_\alpha$, so we finish.

DEFINITION 1.2: (1) $S_\lambda(A) = \{B: B \subseteq A, |B| < \lambda\}$.

(2) $S_\lambda^*(A) = \{B: B \subseteq A, |B| \geq \lambda\}$.

(3) $S^\lambda(A) = \{B: B \subseteq A, |A - B| < \lambda\}$.

(4) $S_*^\lambda(A) = \{B: B \subseteq A, |A - B| \geq \lambda\}$.

DEFINITION 1.3: (1) H is a set function if it is defined on sequences $\{t_i: i < \lambda\}$ (t_i any entities),

$$H(\{t_i: i < \lambda\}) \subseteq \{S: S \subseteq \{t_i: i < \lambda\}\}$$

and

$$H(\{t_i: i < \lambda\}) = \{\{t_i: i \in S\}: S \in H(\lambda)\}.$$

(2) We sometimes use as set of indices sets other than λ, mainly orders. H is a pure set function if any permutation of λ maps $H(\lambda)$ onto itself. $H_1 \leq H_2$ if $H_1(I) \subseteq H_2(I)$ whenever $H_1(I)$ is defined.

Remark. Clearly S_λ, S_λ^*, S^λ, S_*^λ are pure set functions.

DEFINITION 1.4: Suppose H_1, H_2 are set functions, M a model, λ a cardinality. Then

(1) M is (λ, H_1, H_2)-saturated when for every set $A \subseteq |M|$ of cardinality $\leq \lambda$, ordered in some way in order type λ, and $p \in S^m(A)$; if for every $B \in H_1(A)$, $p \restriction B$ is realizes in M, then for some $B \in H_2(A)$ $p \restriction B$ is realized in M.

(2) M is (λ, H_1, H_2)-compact when for every sequence $\Gamma = \{\varphi_i(\bar{x}; \bar{a}_i): i < \lambda\}$ ($\bar{a}_i \in |M|$, $\varphi_i \in L(M)$) if every $p \in H_1(\Gamma)$ is realized in M, then some $p \in H_2(\Gamma)$ is realized in M.

LEMMA 1.5: *M is λ-compact iff it is $(\mu, S_{\aleph_0}, S^1)$-compact for every $\mu < \lambda$. M is λ-homogeneous iff it is $(\mu, S_{\aleph_0}, S^1)$-saturated and $\aleph_0$-homogeneous, for every $\mu < \lambda$.*

(2) *If $H_1 \leq H^1$, $H_2 \leq H^2$, M is (λ, H_1, H_2)-saturated [-compact] then M is (λ, H^1, H^2)-saturated [-compact].*

(3) *(λ, H_1, H_2)-saturation and compactness are preserved by definitional expansion.* (*See* III 5.14.)

(4) *(λ, H_1, H_2)-compactness is preserved by adding individual constants and by taking reducts.*

Proof. Immediate.

DEFINITION 1.5: Let M be a model of T, ($\mathrm{L} = \mathrm{L}(T) = \mathrm{L}(M)$) $\|M\| = \lambda \geq \kappa + |\mathrm{L}|$, κ is regular. Then M is called (λ, κ)-saturated if the following holds: $\langle R(\bar{\kappa}), \in\rangle$ is a model of ZFC, consisting of the sets of hereditary power $< \bar{\kappa}$. Some $N \in R(\bar{\kappa})$ is a $(2^{\lambda})^{+}$-saturated model of T and as a set N is of hereditary power $\leq \|N\|$, $\mathfrak{B}_{\alpha} = \mathfrak{B}_{\lambda,\alpha}$ ($\alpha \leq \kappa$) is an increasing and continuous sequence of elementary submodels of $\mathfrak{B} = \langle R(\bar{\kappa}), \in, N, \lambda\rangle$ (N, λ serve as individual constants) such that:

(1) $\mathfrak{B}_{\alpha}$ has cardinality λ.

(2) If $a \in |\mathfrak{B}_{\alpha}|$, and a is a set (in $R(\bar{\kappa})$) of cardinality $\leq \lambda$ then $b \in a \Rightarrow b \in |\mathfrak{B}_{\alpha}|$.

(3) $\langle |\mathfrak{B}_{\beta}|: \beta \leq \alpha\rangle \in \mathfrak{B}_{\alpha+1}$ (hence, e.g., there is an ordinal in $\mathfrak{B}_{\alpha+1}$ bigger than any ordinal $\in \mathfrak{B}_{\alpha}$, hence the set of ordinals $\in \mathfrak{B}_{\kappa}$ has cofinality κ).

(4) Let N_{α} be N as interpreted in $\mathfrak{B}_{\alpha}$. Then M is isomorphic to N_{κ}. (Notice $N_{\alpha} \in \mathfrak{B}_{\alpha+1}$ by (3), and that N_{α} is an elementary chain as $\mathfrak{B}_{\alpha}$ is, and $\mathrm{L} \in \mathfrak{B}_{\alpha}$.)

Remark. It almost always suffices to assume N is λ- or at most λ^{+}-saturated. We can replace "$\langle R(\bar{\kappa}), \in\rangle$ is a model of ZFC" by weaker conditions to ensure the existence of such $\bar{\kappa}$'s, but our theorems will not be changed.

THEOREM 1.6: (1) *If $\lambda \geq \kappa + |T|$, κ regular, M a model of T of cardinality $\leq \lambda$, then M has an elementary extension N which is (λ, κ)-saturated.*

(2) *If $\mathrm{L}_0 \subseteq \mathrm{L}$, the saturation of Definition* 1.5 *holds, and* $\mathrm{L}_0 \in \mathfrak{B}_{\kappa}$ *then the reduct of $M \cong N_{\kappa}$ to L_0 is also (λ, κ)-saturated.*

(3) *Let M be (λ, κ)-saturated. If $2^{|\mathrm{L}|} \leq \lambda$, M is $\aleph_0$-saturated. If $\lambda^{\chi} = \lambda$, $\kappa > \chi$, M is χ^{+}-compact and χ^{+}-homogeneous. If $2^{|\mathrm{L}|} \leq \lambda = \lambda^{\chi}$, M is χ^{+}-saturated. M is $\aleph_0$-homogeneous.*

Proof. (1) and (2) are immediate. For (3) see the proof of Theorem 1.7(1) below.

THEOREM 1.7: *Suppose M is a (λ, κ)-saturated model, μ a cardinal $\leq \lambda$. If $\lambda^{\mu} > \lambda > 2^{\mu}$ let $\aleph_{\alpha} = \min\{\chi: \chi^{\mu} \geq \lambda\}$ (hence $2^{\mu} < \aleph_{\alpha}$, $\mathrm{cf}(\aleph_{\alpha}) \leq \mu$; $\beta < \alpha \Rightarrow \aleph_{\beta}^{\mu} < \aleph_{\alpha}$). Let $\lambda = \aleph_{\gamma}$. Then:*

(1) *If $\lambda = \lambda^{<\mu}$, $\kappa \geq \mu$ then M is μ-compact and μ-homogeneous.*

(2) *If $2^{\mu} < \lambda < \lambda^{\mu}$, $\gamma - \alpha < \mathrm{cf}\,\mu$, $\mathrm{cf}\,\aleph_{\alpha} \neq \mathrm{cf}\,\mu$, and $\kappa \neq \mathrm{cf}\,\mu$ then M is $(\mu, S_{\aleph_0}, S_{\mu}^{*})$-compact.*

(3) *If $2^{\mu} < \lambda < \lambda^{\mu}$, $\gamma - \alpha < \mathrm{cf}\,\mu$, $\mathrm{cf}\,\aleph_{\alpha} \neq \mathrm{cf}\,\mu$, $\mu < \kappa$, then M is (μ, S_{μ}, S^{1})-compact.*

(4) *If* $2^\mu < \lambda < \lambda^\mu$; *and for any* β, $\alpha < \beta \leq \gamma \Rightarrow \beta - \alpha < \aleph_\beta$; $\mu < \chi \leq \lambda < \aleph_{\alpha+\chi}$, $\chi \neq \kappa$ *then* M *is* $(\chi, S_{\aleph_0}, S^*_\mu)$*-compact.*

(5) (i) *If* $2^\mu < \lambda < \lambda^\mu$; *and for every* β, $\alpha < \beta \leq \gamma \Rightarrow \beta - \alpha < \aleph_\beta$ *and* $\mu < \chi < \kappa \leq \lambda < \aleph_{\alpha+\chi}$, *then* M *is* (χ, S^χ_*, S^1)*-compact.*

(ii) *If* $\lambda = \lambda^{\aleph_0}$ *and* $\chi < \kappa \leq \lambda$, *then* M *is* (χ, S^χ_*, S^1)*-compact.*

(6) *The statements* (1)–(5) *hold with saturation instead of compactness; but for* (1) *we should assume* $2^{|L|} \leq \lambda$, *in addition.*

Remark. In parts (4) and (5)(i) we can replace "$\beta - \alpha$" by "β".

Proof. W.l.o.g. we shall use the notation of Definition 1.5 and assume $M = N_\kappa$.

(1) Suppose p is a sequence of m-formulas over M of length $\mu_1 < \mu$ (i.e., $p = \{\varphi_i(\bar{x}; \bar{a}_i): i < \mu_1\}$, $\varphi_i \in L$, $\bar{a}_i \in M$, $\bar{x} = \langle x_0, \ldots, x_{m-1}\rangle$) which is finitely satisfiable in N. As κ is regular and $>\mu_1$, p is over N_α for some $\alpha < \kappa$. Now L, $N_\alpha \in \mathfrak{B}_{\alpha+1}$, $|L| \leq \|N_\alpha\| = \lambda$, hence the set Φ of m-formulas over N_α belongs to $\mathfrak{B}_{\alpha+1}$, hence $S_\mu(\Phi) \in \mathfrak{B}_{\alpha+1}$, but $|S_\mu(\Phi)| = \lambda^{<\mu} = \lambda$, so $p \in S_\mu(\Phi) \Rightarrow p \in \mathfrak{B}_{\alpha+1}$ (by (2) in the definition). So p is realized in $N_{\alpha+1}$ (as $\mathfrak{B}_{\alpha+1}$ is an elementary submodel of $\langle R(\kappa), \in, N, \lambda\rangle$ and N is $(2^\lambda)^+$-saturated; we use such reasoning without saying). Hence p is realized in $M = N_\kappa$.

The μ-homogeneity is proved similarly.

CLAIM 1.8: *Suppose* $u \in |\mathfrak{B}_{\lambda,\beta}|$, $u \subseteq |\mathfrak{B}_{\lambda,\beta}|$, $S \subseteq u$, $|S| < \aleph_\zeta \leq |u|$.

(1) *If* $|u| < \aleph_{\zeta+\omega}$, *then for some* $v \in \mathfrak{B}_{\lambda,\beta}$ $S \subseteq v$, $|v| \leq \aleph_\zeta$ *and if* $\mathrm{cf}\, \aleph_\zeta > |S|$ *even* $|v| < \aleph_\zeta$.

(2) *If* $|u| \leq \aleph_{\zeta+j}$, $\mathrm{cf}\,|S| > j$, *then for some* $v \in \mathfrak{B}_{\lambda,\beta}$ $|S \cap v| = |S|$, $|v| \leq \aleph_\zeta$ *and if* $\mathrm{cf}\, \aleph_\zeta \neq \mathrm{cf}\,|S|$ *even* $|v| < \aleph_\zeta$.

(3) *If* $\mathrm{cf}\,\aleph_\zeta > |S|$; *and* $\chi < \aleph_\zeta \Rightarrow \chi^{|S|} \leqslant \lambda$, *then in* (1) $\aleph_{\zeta+n}^{|S|} \leqslant \lambda + \aleph_{\zeta+n}$ *so* $S \in \mathfrak{B}_{\lambda,\beta}$; *and if* $\mathrm{cf}\, \aleph_\zeta \neq \mathrm{cf}\,|S|$, $\chi < \aleph_\zeta \Rightarrow \chi^{|S|} \leqslant \lambda$, *then in* (2) *we can assume* $v \subseteq S$.

Proof of 1.8. (1) Let $|u| = \aleph_{\zeta+n}$, and we prove it by induction on n. Clearly $\aleph_{\zeta+n} \leq \lambda$, and there is in $\mathfrak{B}_{\lambda,\beta}$ a one-to-one function g from $\aleph_{\zeta+n}$ onto u. For $\xi < \aleph_{\zeta+n}$ let $u_\xi = \{g(i): i < \xi\}$, so clearly $u_\xi \in \mathfrak{B}_{\lambda,\beta}$ and if $|S| < \mathrm{cf}\, \aleph_{\zeta+n}$, then for some ξ, $S \subseteq u_\xi$; as $|u_\xi| < \aleph_{\zeta+n}$ the conclusion follows by the induction hypothesis. If $|S| \geq \mathrm{cf}\, \aleph_{\zeta+n}$, $n = 0$, $u = v$ is sufficient.

(2) Let $|u| = \aleph_{\zeta+\xi}$, and we prove it by induction on ξ.

If $\xi = 0$ or ξ is a successor the proof is as in (1). If ξ is a limit ordinal cf $|u| <$ cf $|S|$; hence defining u_i, $i < \aleph_{\zeta+\xi}$ as in (1), there are $\xi(i)$, $i < i_0 <$ cf $|S|$ so that $u = \bigcup_{i<i_0} u_{\xi(i)}$, hence for some i, $|S \cap u_{\xi(i)}| = |S|$, and of course $|u_{\xi(i)}| < |u|$, so using the induction hypothesis on $u_{\xi(i)}$, $S \cap u_{\xi(i)}$ we get the desired conclusion.

(3) Immediate, by Definition 1.5, condition (2).

Continuation of the proof of 1.7. (2) Let p be a set of m-formulas over N_κ, $|p| = \mu$, p finitely satisfiable. We should find some $p' \subseteq p$, $|p'| = \mu$ which is realized in N_κ; so we can replace p by any subset of cardinality μ. As cf $\mu \neq \kappa$ for some $\beta < \kappa$, the set of formulas in p which are over N_β, p_1, has cardinality μ. So p_1 is a subset of Φ = the set of m formulas over N_β, which $\in \mathfrak{B}_{\lambda,\beta+1}$. By Claim 1.8(2), (3) (taking $p_1 = S$, $\Phi = u$, $\zeta = \alpha$) and the assumptions, some $p_2 \subseteq p_1$, $|p_2| = \mu$, is a member of $\mathfrak{B}_{\lambda,\beta+1}$, hence realized in $N_{\beta+1}$, hence in N_κ.

(3) Let p be a set of m-formulas over N_κ of cardinality μ such that every $p' \in S_\mu(p)$ is realized in N_κ. We should prove p is realized in N. Let $p = \{\varphi_i(\bar{x}, \bar{a}_i) : i < \mu\}$ and choose $\bar{b}_j \in N_\kappa$ which realizes $p_j = \{\varphi_i(\bar{x}, \bar{a}_i) : i < j\}$.

As $\kappa > \mu$ there is β, $\alpha \leq \beta < \kappa$ such that $\bar{a}_i$, $\bar{b}_j \in N_\beta$. Using 1.8(2), (3) with $\{\bar{b}_j : j < \mu\} = S$, the set of finite sequences from $N_\beta = u$ and $\alpha = \zeta$; some $S' \subseteq \{\bar{b}_j : j < \mu\}$, $|S'| = \mu$, belongs to $\mathfrak{B}_\beta$. As $\mathfrak{B}$ satisfies "there is a sequence $\bar{b} \in N$ which satisfies any formula $\varphi(\bar{x}, \bar{a})$ ($\bar{a} \in N_\beta$) which is satisfied by $\bar{b}_i$ for every big enough $i \in S$" also $\mathfrak{B}_{\beta+1}$ satisfies it, so this $\bar{b}$ necessarily realizes p (in N_κ). So we have proved (3).

Now we need

CLAIM 1.9: *Let* $\mathfrak{B}_\zeta$ *be as in Definition* 1.5, $\aleph_\alpha = \min\{\aleph_\alpha : \aleph_\alpha^\mu > \aleph_\gamma\}$, $2^\mu < \aleph_\gamma < \aleph_\gamma^\mu$; $\alpha < \beta \leq \gamma \Rightarrow \beta < \aleph_\beta$; $\mu < \chi \leq \aleph_\gamma < \aleph_{\alpha+\chi}$. *If* $u \in |\mathfrak{B}_\zeta|$, $S \subseteq u \subseteq |\mathfrak{B}_\zeta|$, $|u| < \aleph_{\alpha+\chi}$, $|S| = \chi$ *then there is* $S' \subseteq S$, $|S'| = \mu$, $S' \in |\mathfrak{B}_\zeta|$.

Proof of 1.9. We prove it by induction on $|u|$ and for fixed u by induction on χ. Always there is in $\mathfrak{B}_\zeta$ a one-to-one function g from $|u|$ onto u, and we let $u_\xi = \{g(i) : i < \xi\}$ for $\xi < |u|$.

Case I: $|u| \leq \aleph_\alpha$. Then, cf $\aleph_\alpha \leq \mu < \chi$, for some $\xi < \aleph_\alpha$ $|u_\xi \cap S| \geq \mu$, hence (by Definition 1.5(2)) $S' = u_\xi \cap S \in \mathfrak{B}_\zeta$, so we get the conclusion.

Case II: $|u| > \aleph_\alpha$, χ a limit cardinal. Then for some $\chi(1) < \chi$, $\mu < \chi(1)$ and $|u| < \aleph_{\alpha+\chi(1)}$, so choose $S_1 \subseteq S$, $|S_1| = \chi(1)$ and use the induction hypothesis on u, S_1.

Case III: $|u| > \aleph_\alpha$, χ is a successor, $|u| > \chi$, $|u|$ singular. As $\aleph_\alpha < |u| < \aleph_{\alpha+\chi}$, χ regular, necessarily $|u|$ has confinality $<\chi$, hence for some ξ, $|u_\xi \cap S| = \chi$ so use the induction hypothesis on u_ξ, $u_\xi \cap S$.

Case IV: $|u| > \aleph_\alpha$, χ is a successor, $|u| > \chi$, $|u|$ regular. Then for some ξ, $u_\xi \supseteq S$, so use the induction hypothesis on u_ξ, S.

Case V: $|u| > \aleph_\alpha$, χ is a successor, $|u| = \chi$. Let $|u| = \lambda_1^+$, then for some ξ, $|u_\xi| = |u_\xi \cap S| = \lambda_1$, so use the induction hypothesis on u_ξ, $u_\xi \cap S$.

So we finish the proof of 1.9.

Continuation of the proof of 1.7. (4) Let p be an m-type over N_κ of cardinality χ (which is finitely satisfiable). If $\text{cf}\,\chi \neq \kappa$ then there are $\beta < \kappa$, $p^* \subseteq p$, $|p^*| = \chi$, p^* is over N_β; and then we proceed as in the proof of 1.7(2) using Claim 1.9 instead of 1.8. If $\text{cf}\,\chi = \kappa$, $\chi \neq \kappa$, χ is singular (so a limit cardinal) hence there is $\chi(1) < \chi$ such that $\mu < \chi(1) \leq \lambda < \aleph_{\alpha+\chi(1)}$, $\chi(1)$ is regular and $\neq \kappa$. So replace p by $p^* \subseteq p$, $|p^*| = \chi(1)$ and proceed as before.

(5) (i) Let p be a set of m-formulas over N of cardinality χ; such that every $q \in S_*^\chi(p)$ is realized in N_κ. Let $p = \bigcup_{i<\chi} p_i$, p_i has cardinality χ, the p_i's are pairwise disjoint. Let $\bar{b}_i \in |N_\kappa|$ realize $p - p_i$, and as $\kappa > \chi$ for some $\beta < \kappa$, $\bar{b}_i \in |N_\beta|$ and p is over N_β. The rest is like the proof of 1.7(3) using Claim 1.9 instead of 1.8.

(ii) Now the proof is trivial.

(6) The proofs for saturativity are similar.

DEFINITION 1.6: (1) Let $\bar{\mathbf{a}}$ be an infinite sequence from M, $\bar{\mathbf{a}} = \langle \ldots, a_i, \ldots \rangle_i$, $TP(\bar{\mathbf{a}}, M) = \{\text{tp}_*(\bar{\mathbf{a}}^\frown \bar{b}, \emptyset, M) \colon \bar{b} \in |M| \ (\bar{b} \text{ a finite sequence})\}$.

(2) $TP_\lambda(M) = \{TP(\bar{\mathbf{a}}, M) \colon \bar{\mathbf{a}}$ is a sequence from M of length $\lambda\}$.

DEFINITION 1.7: For $X = S, C$

$TPX^\kappa(\bar{\mathbf{a}}, M) = \{\Gamma \colon \Gamma$ is a set of formulas $\varphi(\bar{x}; x_{i(1)}, \ldots)$ such that for some $\bar{b} \in |M|$, $M \models \varphi[\bar{b}; a_{i(1)}, \ldots]$ for every $\varphi \in \Gamma$; and $X = C \Rightarrow |\Gamma| \leq \kappa$, $X = S \Rightarrow$ only $\leq \kappa$ variables appear in $\Gamma\}$

$TPX_\lambda^\kappa(M) = \{TPX^\kappa(\bar{\mathbf{a}}, M) \colon \bar{\mathbf{a}}$ a sequence of elements of M of length $\lambda\}$.

Remark. Clearly from $TPS_\lambda^\lambda(M)$ we can compute $TP_\lambda(M)$ and vice versa.

LEMMA 1.10: *Let* $\mathrm{L}(M) = \mathrm{L}(N) = \mathrm{L}$, *and add new variables* x_i, $i < \lambda$, *to* L.

(1) $2^\lambda \le |TP_\lambda(M)| \le \min[2^{2^{\lambda+|L|}}, \|M\|^\lambda]$.

(2) $TPX^\kappa_\lambda(M)$ *is a monotonic function in* κ, λ; *but if* $\kappa \geqslant \lambda$, $[X = C \Rightarrow \kappa \geqslant |\mathrm{L}|]$, *it is a constant function in* κ; *and if* $\kappa \geqslant |\mathrm{L}|: |TPC^\kappa_\lambda(M)| = |TPS^\kappa_\lambda(M)|$ *(in fact, one is computable from the other); and* $|TPS^\lambda_\lambda(M)| = |TP_\lambda(M)|$ *(in fact, one is computable from the other). Also* $|TPC^\kappa_\lambda(M)| \le |TPS^\kappa_\lambda(M)|$.

(3) $2^\lambda \le |TPC^\kappa_\lambda(M)| \le \min[2^{(\lambda+|\mathrm{L}|)^\kappa}, \|M\|^\lambda]$.

(4) $2^\lambda \le |TPS^\kappa_\lambda(M)| \le \min[2^{(\lambda^\kappa+2^{|\mathrm{L}|})}, \|M\|^\lambda]$.

(5) *If* M *is* κ^+*-compact,* $|TPC^\kappa_\lambda(M)| \le 2^{\lambda+|\mathrm{L}|}$, *and if* N *is elementarily equivalent to* M *and is* κ^+*-compact,* $TPC^\kappa_\lambda(M) = TPC^\kappa_\lambda(N)$.

(6) *If* M *is* κ^+*-homogeneous (e.g.,* κ^+*-saturated) then* $|TPS^\kappa_\lambda(M)| \le 2^{\lambda+|\mathrm{L}|}$.

(7) *If* M *is stable in* λ *(i.e.,* $m < \omega$, $A \subseteq |M|$, $|A| \le \lambda$ *implies* $|S^m(A, M)| \le \lambda$*) then* $|TP_\lambda(M)| \le 2^{\lambda+|\mathrm{L}|}$.

Proof. Immediate.

The next theorem will be used in Ch. VIII in computing the number of models for an unsuperstable theory.

THEOREM 1.11: *Let* $2^\mu < \lambda$; $\kappa < \lambda \Rightarrow \kappa^\mu < \lambda$; M *an* L*-model,* $\|M\| \le \lambda$, $|\mathrm{L}| \le \mu$, *then there are* $\chi \le \mu$, *a language* $\mathrm{L}_1 = \mathrm{L} \cup \{Q_i : i < \chi \le \mu\}$ *(*Q_i*—one place predicates) and an* L_1*-model* M_1 *such that*

(1) M_1 *is* μ^+*-homogeneous; moreover if* $\mathrm{L}_0 \subseteq \mathrm{L}_1$, $Q_i \in \mathrm{L}_0$, *then the* L_0*-reduct of* M_1 *is* μ^+*-homogeneous.*

(2) *The reduct of* M_1 *to* L *is an elementary extension of* M.

(3) M_1 *has cardinality* λ.

(4) $|M_1| = \bigcup_{i<\chi} Q_i(M_1)$, and $i < j \Rightarrow Q_i(M_1) \subseteq Q_j(M_1)$.

Proof. W.l.o.g. $\|M\| = \lambda$. If $\lambda^\mu = \lambda$, take $\chi = 1$ and extend M to a μ^+-saturated model M_1 of cardinality λ. So assume $\lambda^\mu > \lambda$, and let $\chi = \mathrm{cf}\,\lambda \le \mu$. Choose an increasing continuous sequence M^i, $i \le \chi$ so that $\mathrm{L}(M^i) = \mathrm{L}^i = \mathrm{L} \cup \{Q_j : j < i\}$, $M^i \prec (M, Q_j)_{j<i}$ where $Q_j = |M^j|$, and $|M^\chi| = |M|$, $\|M^i\| < \lambda$ for $i < \chi$. Expand M to an L_1-model M_2 by defining $Q_i(M_2) = |M^i|$. Let D be a good $\aleph_1$-incomplete ultrafilter over μ, (exists by VI, 3.1) $M_3 = M_2^\mu/D$, and let M_1 be the submodel of M_3 with universe $\bigcup_{i<\chi} Q_i(M_3)$. By VI, 2.3 M_3 is μ^+-saturated. As $Q_i(M_2)$ is (the universe of) an L^i-elementary submodel of M_2, $Q_i(M_3)$ is (the universe of) an L^i-elementary submodel of M_3, hence M_1 is an L_1-elementary submodel of M_3 hence an L_1-elementary extension of M_2. As $|Q_i(M_2)| < \lambda$ also $|Q_i(M_3)| = |Q_i(M_2)|^{\mu}/D < \lambda$, hence $\|M_1\| = \lambda$. Now if $A \subseteq |M_1|$, $|A| \leq \mu$, $p \in S^m(A)$, then p is realized in M_1 iff it is realized in M_3 in some $Q_i(M_1)$ iff it is finitely satisfiable in some $Q_i(M_1)$ (by the μ^+-saturativity of M_3); and this depends on the type of A and on p only. Hence M_1 is μ^+-homogeneous.

Remark. If the other conditions are satisfied but $\mu < |L| \leq \lambda$ we can get M_1 satisfying (2), such that for any $L_0 \subseteq L_1$, $|L_0| \leq \mu$, $Q_i \in L_0$ the reduct of M_1 to L_0 is μ^+-homogeneous.

THEOREM 1.12: *Suppose M is (λ, κ)-saturated, $(2^{\chi+|L|})^+ < \kappa$, $\aleph_0 < \aleph_\alpha \leq \lambda = \aleph_\gamma$, $\gamma - \alpha < \chi^+ < \aleph_\alpha$; and for $\beta < \alpha$, $\aleph_\beta^\mu < \aleph_\alpha$. Then $2^\chi \leq |TPC_\chi^\mu(M)| \leq |TPS_\chi^\mu(M)| \leq 2^{\chi+|L|}$.*

(By 1.10(2), we can assume $\mu \leq \chi$).

Remark. If $\lambda \geq 2^{\chi+|L|}$, we can find suitable κ, or $\lambda^{\chi+|L|} = \lambda$ and the conclusion holds by 1.10(6) as a (λ, λ)-saturated model is χ^+-saturated by 1.6(3). But we cannot always satisfy "$\gamma - \alpha < \chi^+$".

Proof. We use the notation of Definition 1.5. Note that as $2^{|L|} < \kappa \leq \lambda$, M is $\aleph_0$-saturated by 1.6(3). By 1.10(2), (3) it suffices to prove $|TPS_\chi^\mu(M)| \leq 2^{\chi+|L|}$. Suppose not; then there are sequences $\bar{a}_i$ from M of length χ for $i < (2^{\chi+|L|})^+$ such that for $i \neq j$ $TPS^\mu(\bar{a}_i, M) \neq TPS^\mu(\bar{a}_j, M)$. As $(2^{\chi+|L|})^+ < \kappa$, and $TPS^\mu(\bar{a}_i, M)$ has cardinality $\leq 2^{\chi+|L|} < \kappa$, for some $\zeta < \kappa$, $\bar{a}_i$ is from N_ζ and $TPS^\mu(\bar{a}_i, M) = TPS^\mu(\bar{a}_i, N_\zeta)$. W.l.o.g. $\aleph_\alpha = \min\{\aleph_\beta\colon \aleph_\beta^\mu \geq \lambda\}$; so we can assume cf $\aleph_\alpha \leq \mu$ (otherwise $\aleph_\alpha^\mu = \aleph_\alpha$ hence $\lambda = \aleph_\alpha$, hence M is μ^+-saturated, so our conclusion follows from 1.10(6).), and $(2^{\chi+|L|})^+ < \aleph_\alpha$. Let D be a regular ultrafilter over $\mu(1) = \mu + |L|$ (exists by VI, 1.3), which belongs to $\mathfrak{B}_0$; and its power is $2^{\mu+|L|} \leq 2^{\chi+|L|} \leq \kappa \leq \lambda$ so it is $\subseteq \mathfrak{B}_0$. As $\mathfrak{B}_{\zeta+1} \models$ "N_ζ has cardinality λ", in $\mathfrak{B}_{\zeta+1}$ there is an elementary embedding of $N_\zeta^{\mu(1)}/D$ (as interpreted in $\mathfrak{B}_{\zeta+1}$) into $N_{\zeta+1} = N^{\mathfrak{B}_{\zeta+1}}$, which is an inverse to the natural embedding of N_ζ into $N_\zeta^{\mu(1)}/D$.

Now define $TPS^\mu(\bar{a}, B, M)$ as we defined $TPS^\mu(\bar{a}, M)$, but restricting ourselves to $\bar{b} \in B$, and let $TPS_\chi^\mu(C, B, M) = \{TPS^\mu(\bar{a}, B, M)\colon \bar{a}$ a sequence of length χ of elements from $C\}$. Let us prove by induction on $|B|$

(*) If $C, B \in \mathfrak{B}_\xi$, $C, B \subseteq N_\xi$ then there is $B' \in \mathfrak{B}_\xi$, $B \subseteq B' \subseteq N_\xi$, $|B'| < |B|^+ + \aleph_\alpha$ and $TPS^\mu_\chi(C, B', N_\xi)$ has cardinality $\leq 2^{\chi+|L|}$. Moreover there is a formula $\theta \in L(\mathfrak{B})$, with parameters from $\mathfrak{B}_{\xi+1}$ such that for suitable $B, C, \mathfrak{B}_\xi$ $(\exists! x)\theta(x, B, C)$, and $\mathfrak{B}_\xi \vDash \theta[B', B, C]$.

Proof of (*). For the definition of θ use the construction below and a choice function (in $\mathfrak{B}_\xi$) for the family of sets of hereditary cardinality $\leq \|N\|$ (we assumed N is in it).

Case I: $|B| < \aleph_\alpha$. B' will be the image of $B^{\mu(1)}/D$ (ultrapower computed in $\mathfrak{B}_\xi$) by an elementary embedding f of $N_*^{\mu(1)}/D$ (ultrapower in $\mathfrak{B}_\xi$) into N_ζ where $B \cup C \subseteq |N_*|$ $N_* \prec N$, $N_* \in \mathfrak{B}_\xi$, N_* is $\aleph_0$-saturated, $\mathfrak{B}_\xi \vDash$ "$\|N_*\| = \lambda$"; where f is an inverse to the natural embedding of N_* into $N_*^{\mu(1)}/D$. As $|B| < \aleph_\alpha$, all functions from $\mu(1)$ into B with range of cardinality $\leq \mu$ are included, as $|B|^\mu < \aleph_\alpha$, $2^{\mu(1)} \leq \lambda$. As D was regular, clearly any m-type over $A \subseteq C$, $|A| \subseteq \mu$, is realized in B' iff it was finitely satisfiable in B. It is easy to check that B' satisfies our demands.

Case II: $|B| \geq \aleph_\alpha$, $|B|$ is regular. So $|B| > (2^{\chi+|L|})^+$. In $\mathfrak{B}_\xi$ there is a well ordering of B, so let $B = \{b_i: i < |B|\}$, $B_j = \{b_i: i < j\}$, and let $B_j^* = (B_j \cup \bigcup_{i<j} B_i^*)'$; and $B' = \bigcup_{i<|B|} B_i^*$ satisfies our demands; otherwise, as cf $|B| > (2^{\chi+|L|})^+$, some B_j^* fail our demands. (Notice that $|B_j \cup \bigcup_{i<j} B_i^*| < |B|$, so B_j^* is well defined by the induction hypothesis.) [The primes are from (*).]

Case III: $|B| \geq \aleph_\alpha$, $|B|$ is singular. Define B' as in Case II: by assumptions cf $|B| \leq \chi$, so let $\alpha(i)$, $i < \delta =$ cf $|B|$, be an increasing unbounded sequence of ordinals $< |B|$, then $B' = \bigcup_{i<\delta} B^*_{\alpha(i)}$ and $B^*_{\alpha(i)} \subseteq B^*_{\alpha(j)}$ for $i < j$. Hence $|TP(C, B', N_\xi)| \leq \prod_{i<\delta} |TP(C, B^*_{\alpha(i)}, N_\xi)| \leq (2^{\chi+|L|})^\chi = 2^{\chi+|L|}$. So we have proved (*).

Let us finish the proof of 1.12. Choose $B = C = |N_\zeta|$, $\xi = \zeta + 1$, then, we get a contradiction by B' and the properties of the $\bar{a}_i$'s.

DEFINITION 1.8: H is a bi-set function if $H(\{a_i: i < \lambda\})$ is a family of pairs of subsets of $\{a_i: i < \lambda\}$, and $H(\{a_i: i < \lambda\}) = \{(A_1, A_2)$: for some $(w_1, w_2) \in H(\lambda)$, $A_l = \{a_i: i \in w_l\}\}$. We can use index sets other than λ. A set function H is used also as the bi-set function $\hat{H}$, $\hat{H}(\{a_i: i < \lambda\}) = \{(A, \emptyset): A \in H(\{a_i: i < \lambda\})\}$.

DEFINITION 1.9: M is (λ, H_1, H_2)-compact, provided that for any set $\Gamma = \{\varphi_i(\bar{x}, \bar{a}_i): i < \lambda\}$ of m-formulas over M, if for every $(\Gamma_1, \Gamma_2) \in H_1(\Gamma)$ some sequence in M realizes (Γ_1, Γ_2) (i.e., realizes Γ_1, and fails to satisfy any formula in Γ_2), then some sequence from M realizes some pair from $H_2(\Gamma)$.

DEFINITION 1.10: For an ordered set I:

(A) $\mathrm{DC}(I) = \{I_a: a \in I\}$, $I_a = (\{b \in I: b < a\}, \{b \in I: b > a\})$.

(B) The pair (A, B) is a (λ, μ)-Dedekind cut of I if $A < B$ (i.e., $a \in A$, $b \in B \Rightarrow a < b$) and $A \cup B = I$, λ is the cofinality of A, μ is the lower cofinality of B (i.e., the cofinality of B with inverse order).

(C) If W is a family of pairs of regular cardinals

$$\mathrm{DC}_W(I) = \{(A, B): (A, B) \text{ is a } (\lambda, \mu)\text{-Dedekind cut of } I,\ (\lambda, \mu) \in W\},$$
$$\mathrm{DC}^*_W(I) = \{(A, B): A, B \subseteq I,\ A < B, \text{ and for no } c \in I\ A < c < B \text{ and } (\mathrm{cf}(A, <), \mathrm{cf}(B, >)) \in W\}.$$

DEFINITION 1.11: (1) A (λ, κ)-family is a family of sets of cardinality λ such that the intersection of any two distinct members has cardinality $< \kappa$.

(2) $\mathrm{AD}(\chi, \lambda, \mu, \kappa)$ holds if there is a (μ, κ)-family of χ subsets of λ; where $\chi > \lambda \geq \mu \geq \kappa$.

Remark. Baumgartner proved the consistency of $\mathrm{AD}(2^{\aleph_0}, \aleph_1, \aleph_1, \aleph_0) + 2^{\aleph_0} > \aleph_1$.

EXERCISE 1.1: (1) If $\mathrm{AD}(\chi, \lambda, \mu, \kappa)$ holds, $\chi' \leq \chi$, $\lambda' \geq \lambda$, $\mu' \leq \mu$, $\kappa' \geq \kappa$, and $\chi' > \lambda' \geq \mu' \geq \kappa'$ then $\mathrm{AD}(\chi', \lambda', \mu', \kappa')$ holds.

(2) If $\mathrm{AD}(\chi, \lambda, \mu, \kappa)$ holds then $\lambda^\kappa \geq \chi$; if $\chi = 2^\lambda$ then $\lambda^\kappa = 2^\lambda$.

(3) Prove that $\mathrm{AD}(\lambda^+, \lambda, \mu, \mu)$ holds when cf $\lambda = \mu$.

(4) Use (3) to show 1.8 cannot be improved in this case.

CONJECTURE 1.2: Claims 1.8 and 1.9 cannot be improved (at least in ZFC).

EXERCISE 1.3: Find the order between the set functions S_μ, S^*_μ, S^μ, S^μ_* (also different μ's).

QUESTION 1.4: Investigate the logical connection between the various notions of saturativity (implications, and independence).

QUESTION 1.5: Investigate the various kinds of compactness for dense linear order.

QUESTION 1.6: Let $I = \sum_{s \in J} I_s$. Investigate the connection between the compactness (and saturativity) (with various notions) of I; J, I_s.

EXERCISE 1.7: For every order J, $|J| \leq \kappa$ there is an order $I \supseteq J$, $|I| = \kappa$ such that $DC_{\langle\langle\lambda,\mu\rangle\rangle}(I) = \emptyset$ if $\lambda, \mu \neq \aleph_0$, $\lambda \neq \mu$.

QUESTION 1.8: If we use the construction of Definition 1.5 but without assuming that N is saturated, does the big model still bequeath its properties (such as $(\mu, S_{\aleph_0}, S_\mu^*)$-compactness, -saturation, etc.) to its submodels.

EXERCISE 1.9: Show that in 1.6(3), 1.7(1) and 1.7(6) we can replace "χ-homogeneous" by "strongly χ-homogeneous", if $\kappa \leq \chi$ or N is strongly χ-saturated.

VII.2. Generalized Ehrenfeucht–Mostowski models

Here we shall deal with the pseudo-elementary class $PC(T_1, T)$ where

DEFINITION 2.1: (1) $PC(T_1, T)$ is the class of L-reducts of models of T_1, of cardinality $\geq |T_1|$, $L = L(T)$, $L_1 = L(T_1)$, T is as usual complete, $T \subseteq T_1$, T_1 has infinite models.

(2) T_2 is a conservative extension of T_1 if any L_1-model of T_1 of cardinality $\geq |T_1|$ can be expanded to a model of T_2.

Clearly if T_2 is a conservative extension of T_1 and $|T_2| = |T_1|$, then $PC(T_2, T) = PC(T_1, T)$; hence we can replace T_1 by T_2 without loss of generality. Also if we want to prove the existence of some models in $PC(T_1, T)$ we can replace T_1 by any extension of the same cardinality. Hence if T_1 has an extension of power $|T_1|$ which has some additional properties we can assume T_1 has them. We list below some such assumptions on T_1. Note that if T_{i+1} is a conservative extension of T_i, $T_\delta = \bigcup_{i<\delta} T_i$, $|T_i| = |T_0|$, then T_α is a conservative extension of T_0.

Assumption I: T_1 has Skolem functions. (See Theorem 1.1(2).)

Assumption II: T_1 is complete.

Assumption III: There is a countable $L_1^c \subseteq L_1$ such that for every formula $\varphi(\bar{x})$ and term $\tau(\bar{x})$ in L_1, there are a formula $R(\bar{x}, y)$ and function symbol $F(\bar{x}, z)$ from L_1^c and individual constants c_1, c^1 in L_1 such that

(1) $(\forall \bar{x})[\varphi(\bar{x}) \equiv R(\bar{x}, c_1)] \in T_1$,

(2) $(\forall \bar{x})[\tau(\bar{x}) = F(\bar{x}, c^1)] \in T_1$.

For this it suffices to prove:

CLAIM 2.1: *Some conservative extension T_2 of T_1, $|T_2| = |T_1|$ satisfies Assumption* III.

Proof. It suffices to define by induction on n theories T^n, and countable languages $L_n^* \subseteq L_n = L(T^n)$ such that $T^0 = T_1$, T^{n+1} is a conservative extension of T^n, $|T^n| = |L_n| = |T_1|$ and for every $\varphi(\bar{x})$, $\tau(\bar{x})$ in L_n there are $R(\bar{x}, y)$, $F(\bar{x}, y)$ in L_{n+1}^* and individual constants c_1, c^1 in L_{n+1} such that (1) and (2) in assumption III holds for T^{n+1}. (Then $T_2 = \bigcup_{n<\omega} T^n$, $L_1^c = \bigcup L_n^*$ satisfies the claim.) If T^n is defined let L_{n+1}^* consist of $(m + 1)$-place predicates $R_n^m(\bar{x}, y)$ and function symbols $F_n^m(\bar{x}, y)$ for $m < \omega$, and

$$T^{n+1} = T^n \cup \{(\forall \bar{x})[\varphi(\bar{x}) \equiv R_n^m(\bar{x}, c_\varphi^n)]: \varphi(\bar{x}) \in L_n, l(\bar{x}) = m\}$$
$$\cup \{(\forall \bar{x})[\tau(\bar{x}) = F_n^m(\bar{x}, c_n^\tau)]: \tau(\bar{x}) \in L_n, l(\bar{x}) = m\}.$$

Clearly all the demands are satisfied, so we have proved the claim.

I denotes an index-model, i.e., a model whose elements serve as indexes; usually I is just an order.

DEFINITION 2.2: If $J \subseteq I$, $\bar{s} \in I$, then the atomic type of $\bar{s}$ over J (in I) is

$$\text{atp}(\bar{s}, J, I) = \{\varphi(\bar{x}; \bar{t}): l(\bar{x}) = l(\bar{s}), \bar{x} = \langle x_0, \ldots \rangle, \bar{t} \in J, I \models \varphi[\bar{s}; \bar{t}] \text{ and } \varphi(\bar{x}; \bar{y}) \text{ is an atomic, or negation of atomic, formula in } L(I)\}$$

We omit I when its identity is clear. Clearly $\text{atp}(\bar{s}, J, I) \equiv \text{tp}_\Delta(\bar{s}, J, I)$ where $\Delta =$ the set of quantifier free formulas.

DEFINITION 2.3: If $\bar{s}, \bar{t} \in I$, $J \subseteq I$, $\bar{s} \sim \bar{t} \pmod J$ (in I) if $\text{atp}(\bar{s}, J) = \text{atp}(\bar{t}, J)$.

In both definitions, if J is empty we omit it.

EXERCISE 2.1: Check the meanings for I an order.

Notation. If $\{\bar{b}_s: s \in I\}$ is an indexed set, $\bar{s} \in I$, $\bar{s} = \langle s(0), s(1), \ldots \rangle$ then $\bar{b}_{\bar{s}} = \bar{b}_{s(0)} {}^\frown \bar{b}_{s(1)} {}^\frown \cdots$.

DEFINITION 2.4: (1) The indexed set $\{\bar{b}_s: s \in I\}$ ($\bar{b}_s \in \mathfrak{C}$) is called Δ-n-indiscernible over A:

(A) If $s \sim t$ then $l(\bar{b}_s) = l(\bar{b}_t)$.

(B) If $\bar{s} \sim \bar{t}$, $l(\bar{s}) = l(\bar{t}) = n$, $\varphi(\bar{x}; \bar{y}) \in \Delta$, $\bar{a} \in A$ then $\mathfrak{C} \models \varphi[\bar{b}_{\bar{s}}; \bar{a}] \equiv \varphi[\bar{b}_{\bar{t}}; \bar{a}]$.

(2) If $s \neq t$ implies $\bar{b}_s \neq \bar{b}_t$ we call our indexed set non-trivial (we use (2) rarely—mainly in estimations of cardinalities).

(3) We adopt the same conventions for shortening as in Definitions I, 2.3 and I, 2.4. (Note that this definition and Definitions I, 2.3 and I, 2.4 are consistent.)

Assumption IV: When I is an index-model we assume for every $\bar{s} \in I$ there is a quantifier-free formula $\theta(\bar{x})$ in $\mathrm{L}(I)$ such that for $\bar{t} \in I$, $\bar{s} \sim \bar{t}$ iff $I \models \theta[\bar{t}]$.

DEFINITION 2.5: $K(I)$ is the class of $\mathrm{L}(I)$-index models I_1 such that for any $\bar{s} \in I_1$ there is $\bar{t} \in I$ for which $\mathrm{atp}(\bar{s}, \emptyset, I_1) = \mathrm{atp}(\bar{t}, \emptyset, I)$. So for every I there is $I_1 \in K(I)$, $|I_1| \leq |\mathrm{L}(I)|$, such that $K(I_1) = K(I)$ (without assumption IV, we would have only $|I_1| \leq 2^{|\mathrm{L}(I)|}$). The class of orders is $K_{\mathrm{ord}} = K(\langle \omega, < \rangle) = K(\omega)$.

LEMMA 2.2: *If $I_1 \in K(I)$, $\{\bar{b}_s : s \in I\}$ is an indiscernible indexed set in a model M of T_1, then there is an indiscernible indexed set $\{\bar{a}_s: s \in I_1\}$ in a model M' of T_1, such that if $\bar{s} \in I$, $\bar{t} \in I_1$,* $\mathrm{atp}(\bar{s}, \emptyset, I) = \mathrm{atp}(\bar{t}, \emptyset, I_1)$ *then* $\mathrm{tp}(\bar{b}_{\bar{s}}, \emptyset, M) = \mathrm{tp}(\bar{a}_{\bar{t}}, \emptyset, M')$. *As T_1 has Skolem functions we can assume* $|M'| = \mathrm{cl}\{\bar{a}_s: s \in I_1\}$ *and then M' is uniquely determined (up to isomorphism preserving the $\bar{a}_s$).*

Proof. Immediate, by the compactness theorem and the definition of $K(I)$.

DEFINITION 2.6: If $\{\bar{b}_s: s \in I\}$ is non-trivial and indiscernible in M, then the Φ of $\{\bar{b}_s: s \in I\}$ which we define as $\{\varphi(\bar{x}_{\bar{s}}): M \models \varphi[\bar{b}_{\bar{s}}]\}$ and the index-model $I_1 \in K(I)$ determine M' so we denote M' by $N^1 = EM^1(I_1, \Phi)$, and its reduct to L by $N = EM(I_1, \Phi)$. Such Φ is called proper for (I, T_1). We call $\{\bar{a}_s: s \in I_1\}$ the skeleton of N^1 (and N) (of course, really we cannot reconstruct the skeleton from the model, but we behave as if we can).

LEMMA 2.3: *There is Φ proper for (ω, T_1).*

Proof. It suffices to prove in some model of T_1 there is a non-trivial indiscernible sequence $\{a_i: i < \omega\}$. Let M_1 be a model of T_1, $\bar{b}_i$ distinct sequences from M_1. It suffices to prove

$$\Gamma = T_1 \cup \{\varphi(\bar{x}_{\bar{s}}) \equiv \varphi(\bar{x}_{\bar{t}}): \varphi(\bar{x}) \in \mathrm{L}_1, \bar{s} \sim \bar{t}; \bar{s}, \bar{t} \in \omega\}$$
$$\cup \{\bar{x}_s \neq \bar{x}_t : s \neq t < \omega\}$$

is consistent, for this it suffices to prove the consistency of a finite subset $\Gamma' \subseteq T_1 \cup \{\varphi_l(\bar{x}_{\bar{s}(l)}) \equiv \varphi_l(\bar{x}_{\bar{t}(l)}): l < n\} \cup \{\bar{x}_s \neq \bar{x}_t: s \neq t < \omega\}$ where $\bar{s}(l) \sim \bar{t}(l)$. Let $\Delta = \{\varphi_l: l < n\}$, $m = \max_{k<n} l(\bar{s}(k))$, and let $\{\bar{b}_{t(k)}: k < \omega\}$ be a Δ-$\leq m$-indiscernible sequence (exists by I, 2.4 and I, 2.3(3)) and interpret $\bar{x}_k$ by $\bar{b}_{t(k)}$; so in M_1, Γ' is satisfied.

Assumption V: If T is unstable, for some formula $\bar{x} < \bar{y}$ in L and some $\bar{b}_i, i < \omega$, $(l(\bar{b}_i) = l(\bar{x}) = l(\bar{y})), \bar{b}_n < \bar{b}_k$ iff $n < k$. Then $(\bar{x} < \bar{y}) \in \mathrm{L}_1^c$. If T has the f.c.p. then $T \cap \mathrm{L}_1^c$ has the f.c.p.

LEMMA 2.4: *If T is unstable, there is Φ proper for (ω, T_1) which is ordered by $<$, i.e., $(\bar{x}_n < \bar{x}_m) \in \Phi$ iff $n < m$.*

Proof. As of 2.3, only to Γ we add $\{(\bar{x}_n < \bar{x}_m)^{\text{if}(n<m)}: n < m < \omega\}$, and use the $\bar{b}_n$'s from assumption V.

Similarly we can prove

LEMMA 2.5: *If Φ_1 is proper for (ω, T_1) and T_1 is complete, $T_2 \supseteq T_1$, then there is $\Phi_2 \supseteq \Phi_1$ which is proper for (ω, T_2).*

For Φ proper for (ω, T_1), we can investigate properties of $EM(I, \Phi)$ $(I \in K(\omega))$, but for cardinals $\leq |T_1|$ we cannot get much. Hence we first introduce a generalization.

DEFINITION 2.7: If Φ is proper for (I, T_1) let $N^1 = EM^1(I, \Phi)$, N' be the submodel of N^1 with universe the set of interpretations of terms (with no free variables). (As T_1 has Skolem functions, N' is an elementary submodel of N^1.) Let N_Φ be N' expanded by adding the relations:

$$R_{\theta,\varphi,\bar{\tau}} = \{\bar{c} \in N': N^1 \vDash \varphi[\bar{\tau}(\bar{a}_{\bar{s}}, \bar{c})], \text{ when } \bar{s} \in I, I \vDash \theta[\bar{s}]\}$$

for every atomic formula $\varphi(\bar{z}) \in \mathrm{L}_1^c$, term $\bar{\tau}(\bar{x}, \bar{y}) \in \mathrm{L}_1^c(l(\bar{z}) = l(\bar{\tau}))$ and quantifier free formula $\theta(\bar{x}) \in \mathrm{L}(I)$ satisfying assumption IV (we choose one $\theta(\bar{x})$ for each $\mathrm{atp}(\bar{s}, \emptyset, I)$, $\bar{s} \in I$). Let $T_\Phi = \mathrm{Th}(N_\Phi)$, $\mathrm{L}_\Phi = \mathrm{L}(N_\Phi)$, $\mathrm{L}_\Phi^c = \mathrm{L}_1^c \cup \{R_{\theta,\varphi,\bar{\tau}}: \theta, \varphi, \bar{\tau}$ as mentioned above$\}$, so $|\mathrm{L}_\Phi^c| \leq \aleph_0 + |\mathrm{L}(I)|$ and T_Φ satisfies assumption III when $\mathrm{L}(I)$ is countable.

We behave as if T_Φ determines Φ, and say, T_Φ, or a model of T_Φ has a property, if Φ has it.

LEMMA 2.6: *Suppose N is a model of T_Φ (hence an elementary extension of N_Φ) where Φ is proper for (I_0, T_1), and $I \in K(I_0)$.*

Then there is a unique L_1*-model* M^1, $|M^1| = \mathrm{cl}(|N| \cup \{\bar{a}_s : s \in I\})$ *such that*

(1) *For every* $R_{\theta,\varphi,\bar{\tau}} \in \mathrm{L}_\Phi - \mathrm{L}_1$, *and* $\bar{c} \in |N|$, $\bar{s} \in I$

$$(*) \qquad \text{when } I \vDash \theta[\bar{s}], \quad N \vDash R_{\theta,\varphi,\bar{\tau}}[\bar{c}] \Leftrightarrow M^1 \vDash \varphi[\bar{\tau}(\bar{a}_{\bar{s}}, \bar{c})].$$

We denote this model by $EM^1(I, N)$, *and its* L*-reduct by* $M = EM(I, N)$. *We call* $\{\bar{a}_s : s \in I\}$ *the skeleton and* N *the basis of* M^1 *(and of* M*).*

(2) M^1 *is a model of* T_1, *and an elementary extension of the* L_1*-reduct of* N. *The indexed set* $\{\bar{a}_s : s \in I\}$ *is indiscernible over* $|N|$ *in* M^1.

(3) *For every* $\bar{c} \in |M^1|$ *there are* $\bar{s} \in |I|$, $\bar{b} \in |N|$ and $\bar{\tau} \in L_1^c$ *(we consider them as determined by* $\bar{c}$*) such that* $\bar{c} = \bar{\tau}(\bar{a}_{\bar{s}}, \bar{b})$. *Also* $(*)$ *holds for any* $\bar{\tau}$, $\varphi \in \mathrm{L}_1$ *for suitable formulas* $R_{\theta,\varphi,\bar{\tau}}$ *(using suitable individual constants). The elementary type of* $\bar{c}$ *is determined by* $\mathrm{atp}(\bar{s}, \emptyset, I)$, $\mathrm{tp}(\bar{b}, \emptyset, N)$. *Also* $\|M^1\| = \|N\| + \|I\|$.

(4) *If in I only finitely many atomic types of sequences of length m are realized for each m: then for every formula* $\varphi(\bar{\tau}(\bar{a}_{\bar{s}}, \bar{y}))$ $(\varphi, \bar{\tau} \in \mathrm{L}_1$, $\bar{s}$ *varied over* I, $\bar{y}$ *over* N) *there is a Boolean combination* $\psi(\bar{s}, \bar{y})$ *of formulas* $\theta_1(\bar{s}) \in \mathrm{L}(I)$, $\theta_2(\bar{y}) \in \mathrm{L}_\Phi$ *such that* $M^1 \vDash \varphi(\bar{\tau}(\bar{a}_{\bar{t}}, \bar{c}))$ *iff* $\psi(\bar{t}, \bar{c})$. [Being formalistic ψ is not a well defined formula, but its meaning is clear.]

Proof. There is no problem in the proof.

Remark. If M_1 satisfies (1) it automatically satisfies (2), (3) and (4).

DEFINITION 2.8: I is λ-atomically-stable if $J \subseteq I$, $|J| \le \lambda$ implies $|S^m_\Delta(J, I)| \le \lambda$ where $\Delta = \Delta_{\mathrm{qf}}(\mathrm{L}(I))$ is the set of quantifier free formulas of $\mathrm{L}(I)$. (Note that if $\|I\| \le \lambda$, I is λ-atomically stable.)

LEMMA 2.7: (1) *If N is proper for (I, T_1), N stable in λ, and I λ-atomically-stable then* $M^1 = EM^1(I, N)$ *is stable in* λ.

(2) *If N is proper for (ω, T_1), $\|N\| \le \lambda$ and I is a well ordering then* $EM^1(I, N)$ *is stable in* λ.

Proof. (1) Let $A \subseteq |M^1|$, $|A| \leq \lambda$, then there are $J \subseteq I$, $|J| \leq \lambda$ and $B \subseteq N$, $|B| \leq \lambda$, such that for every $a \in A$ there are $\bar{s}(a) \in J$, $\bar{b}_a \in B$ and $\tau_a \in L_1^c$ such that $a = \tau_a(\bar{a}_{\bar{s}(a)}, \bar{b}_a)$. Noticing that if $\bar{a}^l = \bar{\tau}(\bar{a}_{\bar{s}(l)}, \bar{b}_l)$ $\bar{s}(l) \in I$, $\bar{b}_l \in |N|$ for $l = 1, 2$, and $\bar{s}(1) \sim \bar{s}(2) \bmod J$, $\mathrm{tp}(\bar{b}_1, B, N) = \mathrm{tp}(\bar{b}_2, B, N)$, then $\mathrm{tp}(\bar{a}^1, A, M^1) = \mathrm{tp}(\bar{a}^2, A, M^1)$; the conclusion is clear.

(2) Immediate from (1) and part (5) of the following claim.

CLAIM 2.8: *Let $I \in K(\omega)$, $J \subseteq I$.*

(1) $s_1 \sim s_2 \bmod J$ *iff for any* $t \in J$, $t < s_1 \Leftrightarrow t < s_2$, $t = s_1 \Leftrightarrow t = s_1$ *and* $t > s_1 \Leftrightarrow t > s_2$.

(2) $\langle s_0, \ldots, s_m\rangle \sim \langle t_0, \ldots, t_m\rangle \bmod J$ *iff* $s_i \sim t_i \bmod J$ *for* $i = 0, \ldots, m$ *and* $s_i < s_j \Leftrightarrow t_i < t_j$ *for* $0 \leq i, j \leq m$.

(3) *The number of atomic types of sequences of length m from I is finite (even* $< (2m)!$).

(4) *I is λ-atomically stable if* $|S_\Delta^1(J, I)| \leq |J|$ *for* $J \subseteq I$, $|J| \leq \lambda$, *where* $\Delta = \Delta_{\mathrm{qf}}(\mathrm{L}(I))$.

(5) *If I is well ordered, I is λ-atomically stable for each* $\lambda \geq \aleph_0$.

(6) *If* $\lambda \geq 2^{\aleph_0}$, *I is λ-atomically-stable iff I is λ-stable.*

(7) *If* $I = \sum_{s \in J} I_s$; *J*, I_s *are λ-atomically-stable orders then I is λ-atomically stable.*

Proof. (1), (2) and (3) are immediate. (6) will not be used hence we leave it as an exercise to the reader. (Hint: use the Feferman–Vaught theorem, see, e.g., [CK 73].)

(4) Suppose I, J are a counterexample.

So $|J| = \lambda$, $J \subseteq I$, $p_i = \mathrm{atp}(\bar{s}_i, J, I)$ $l(\bar{s}_i) = m$, for $i < \lambda^+$, $p_i \neq p_j$ for $i \neq j$. As we can replace $\{p_i: i < \lambda^+\}$ by any subfamily of same cardinality we can assume w.l.o.g.

(A) $\mathrm{atp}(\bar{s}_i, \emptyset, I)$ is constant.

(B) Letting $\bar{s}_i = \langle s_i^0, \ldots, s_i^{m-1}\rangle$, then for each k, $0 \leq k < m$, $\mathrm{atp}(s_i^k, J, I)$ depends only on k (for this use m times the assumption $|\{\mathrm{atp}(s, J, I): s \in I\}| \leq \lambda$).

So we get a contradiction by (2).

(5) Using (4) we prove $|S_\Delta^1(J, I)| \leq \lambda$ when $|J| \leq \lambda$.

Define a function h on I: $h(s) = \min\{t: t \in J, s < t\}$ and if for no $t \in J$ $t > s$, let $h(s) = \infty$ (h is well defined as I is well ordered). So the range of h has cardinality $|J| + 1$, and if $s, t \notin J$, $h(s) = h(t)$ then by (1) $\mathrm{atp}(s, J, I) = \mathrm{atp}(t, J, I)$. Hence we finish.

(7) by (4) it is immediate.

CONCLUSION 2.9: (1) *If* $\lambda \geq |T_1|$ *then there is* $M \in \mathrm{PC}(T_1, T)$ *of cardinality* λ, *which is stable in every* $\mu \geq |T_1|$; *so if* T *is unstable in* $|T_1|$, M *is not* $|T_1|^+$*-universal.*

(2) *If* $\mu \geq 2^\lambda$, $\lambda \geq |T_1|$ *then there is* $M \in \mathrm{PC}(T_1, T)$ *of cardinality* μ, *which is* λ^+*-universal,* χ*-stable for* $\chi \geq 2^\lambda$, *hence when* T *is unstable* M *is not* $(2^\lambda)^+$*-universal.*

Remark. In (2) if $2^\lambda = \lambda^+$ the result on universality is sharp, but in general not. If $\mu^\lambda = \mu$ we can assume M is λ^+-saturated χ-stable for $\chi = \chi^\lambda$ (using ultrapowers). Sharp results, using finer methods, appear in Section 4.

Proof. (1) Let N be proper for (ω, T_1), and be of cardinality $|T_1|$. Then $EM(\lambda, N)$ is of cardinality λ by 2.6(3); it is stable in $\mu \geq |T_1| = \|N\|$ by 2.7(2).

If T is unstable in $|T_1|$, there is a model M of T which is unstable in $|T_1|$, of cardinality $|T_1|^+$, hence M cannot be elementarily embedded in $EM(\lambda, N)$.

(2) Use a λ^+-saturated model N of cardinality 2^λ which is proper for (ω, T_1). Take $EM(\mu, N)$ as the required model and use Theorem II, 2.13(2).

LEMMA 2.10: *If* $M^1 = EM^1(I, N)$, N *proper for* (I, T_1)*; then* $|TPX^\mu_\chi(M^1)| \leq |TPX^\mu_\chi(N)| + |ATPX^\mu_\chi(I)|$ *where* $X = S$, C *and where* ATP *means that in definition* 1.6 *we replace* tp *by* atp.

Proof. Similar to that of 2.7.

LEMMA 2.11: *Suppose* I *is a dense and* λ^+*-saturated order and* N *is* λ^+*-compact and proper for* (ω, T_1). *Then* $M^1 = EM^1(I, N)$ *is* $(\lambda, S_{\aleph_1}, S^1)$*-compact.*

Proof. Let p be an m-type over M^1 of cardinality λ, and every $q \in S_{\aleph_1}(p)$ is realized. W.l.o.g. $\lambda \geq |L_1|$.

(*) There is $\tau = \tau(\bar{y}, \bar{z}) \in L_1^c$ such that any $q \in S_{\aleph_1}(p)$ is realized by some $\tau(\bar{a}_{\bar{s}}, \bar{b})$, $\bar{s} \in I$, $\bar{b} \in N$.

Otherwise for each $\tau \in L_1^c$ there is a counterexample q_τ, so $q = \bigcup_\tau q_\tau$ is a countable subtype of p, and is not realized, contradiction. So for every $q \in S_{\aleph_1}(p)$ there are $\bar{s}(q) \in I$, $\bar{b}(q) \in N$ such that $\tau(\bar{a}_{\bar{s}(q)}, \bar{b}(q))$ realizes q. Let D be an ultrafilter over $S_{\aleph_0}(p)$, such that $\{q \in S_{\aleph_0}(p): \varphi \in q\} \in D$ for any $\varphi \in p$. Let $J \subseteq I$, $|J| \leq \lambda$, $B \subseteq |N|$, $|B| \leq \lambda$ be such that p is a type over $\mathrm{dcl}(B \cup \{\bar{a}_s: \bar{s} \in J\})$. We can find, by the λ^+-saturation of I and N, $\bar{t} \in I$ and $\bar{c} \in |N|$ such that:

(1) For every $\varphi(\bar{x}, \bar{b})$, $\bar{b} \in B$, $\varphi \in \mathrm{L}(N)$,

$$N \vDash \varphi[\bar{c}, \bar{b}] \Leftrightarrow \{q \in S_{\aleph_0}(p): N \vDash \varphi[\bar{b}(q), \bar{b}]\} \in D.$$

(2) For every $\psi \in \mathrm{L}(I)$, $\bar{s} \in J$,

$$I \vDash \psi[\bar{t}, \bar{s}] \Leftrightarrow \{q \in S_{\aleph_0}(p): I \vDash \psi[\bar{s}(q), \bar{s}]\} \in D.$$

It is easy to check $\bar{\tau}(\bar{a}_{\bar{t}}, \bar{c})$ realizes p.

QUESTION 2.2: Investigate the connection between (λ, H_1, H_2)-compactness (or saturativity) of N, I and of $EM(I, N)$; in particular when the H's are S_μ, S^*_μ, S^μ and S^μ_*.

QUESTION 2.3: (A) Investigate the connection between non-(λ, H_1, H_2)-compactness (or saturativity) of N, I and of $EM(I, N)$ when $N \vDash T_\Phi$, Φ is ordered (see Lemma 2.4).

(B) Show, for unstable T, the existence of models which satisfy simultaneously various compactness and non-compactness conditions (also for saturativity; we mean of course for the various (λ, H_1, H_2)-compactness notions).

EXERCISE 2.4: Suppose $\lambda > \mu > |T_1|$ are regular, $M = EM(I, \Phi)$ I a dense order, Φ proper for (ω, T_1). Show M is $(\lambda + \mu^*, DC, DC_{\langle\lambda,\mu\rangle})$-compact, when I has no (i) (λ, μ) or (μ, λ) Dedekind cut, (ii) $(1, \lambda)$ or $(\lambda, 1)$ or $(0, \lambda)$ or $(\lambda, 0)$ *and* $(1, \mu)$ or $(\mu, 1)$ or $(0, \mu)$ or $(\mu, 0)$ Dedekind cut.

EXERCISE 2.5: Show that if T_i $i < \delta$ is an increasing sequence of complete theories, Φ_i proper for (I, T_i) $\Phi_i \subseteq \Phi_j$ for $i < j < \delta$, then $\bigcup_{i<\delta} \Phi_i$ is proper for $(I, \bigcup_{i<\delta} T_i)$.

DEFINITION 2.9: I has the extension property if: When T_1 is complete, Φ_1 proper for (I, T_1), $T_2 \supseteq T_1$, then there is $\Phi_2 \supseteq \Phi_1$ proper for (I, T_1).

EXERCISE 2.6: Show that if Φ_1 is proper for (I, T_1), I has the extension property, then there are $T_2 \supseteq T_1$, $|T_2| = |T_1|$, $\Phi_2 \supseteq \Phi_1$ proper for I, T_2, such that for any $J \in K(I)$, $EM(J, \Phi_2)$ is $\aleph_0$-homogeneous.

EXERCISE 2.7: Suppose $T_1 = \mathrm{Th}(M_1)$, $M_1 = (\omega, <, R_1, \dots)$, T_1 has Skolem functions, and that we get Φ, proper for (ω, T_1) as in 2.3 for $b_n = n$. Then for any $I \in K(\omega)$, $N = EM(I, \Phi)$, $s \in I$

$$\begin{aligned}|\{t \in I : t < s\}| &\le |\{b \in |N| : N \models b < a_s\}| \\ &\le |\{t \in I : t < s\}| + \min\{2^{\aleph_0}, |T_1|\}.\end{aligned}$$

Moreover for each κ we can find such a T_1, $|T_1| = \kappa$, such that the second inequality becomes an equality. (Hint: See the proof of 3.1.)

EXERCISE 2.8: Let w denote here a finite set, and say M is $(\lambda, H_1, H_2)^*$-compact if for any indexed set $\{\varphi_w : w \subseteq \lambda\}$ of formulas over M if for every $S \in H_1(\lambda)$, $p_S = \{\varphi_w : w \subseteq S\}$ is realized in M then for some $S \in H_2(\lambda)$, p_S is realized.

(1) If $H_1(\lambda)$ is closed under countable unions, N is λ^+-compact, I is $(\lambda, H_1, H_2)^*$-compact (or vice versa) and N is proper for (ω, T_1), then $EM(I, N)$ is $(\lambda, H_1, H_2)^*$-compact.

(2) Generalize from $I \in K(\omega)$ to $I \in K(I_0)$.

(3) Generalize to saturativity and Definition 1.8.

(4) Check the implications among these notions.

VII.3. On the f.c.p., uniform trees, and $|D(T)| > |T| = \aleph_0$

We continue the conventions of Section 2, with various additional assumptions on T.

THEOREM 3.1: *Suppose T has the* f.c.p. (*finite cover property*). *If Φ_1 is proper for (ω, T_1) then there are $T_2 \supseteq T_1$, $|T_2| = |T_1|$ and Φ_2 and $\varphi(x; \bar{y}) \in \mathrm{L}$, $\theta(\bar{y}; \bar{z}) \in \mathrm{L}_2$ such that:*

(1) *$\Phi_2 \supseteq \Phi_1$ is proper for (ω, T_2).*

(2) *If $N^1 = EM^1(I, \Phi_2)$, $s \in I$, then $\theta(N^1; \bar{a}_s)$ has cardinality*

$$|\{t \in I : t < s\}| + \min\{|T_1|, 2^{\aleph_0}\}.$$

(3) *$\{\varphi(x, \bar{c}) : \bar{c} \in N^1, N^1 \models \theta[\bar{c}, \bar{a}_s]\}$ is not realized in N^1 but any proper subset of it is realized in N^1.*

Proof. Let $M^1 = EM^1(\omega, \Phi_1)$. By assumption for some $\varphi(x, \bar{y}) \in \mathrm{L}$ and infinite set $W \subseteq \omega$ for any $n \in W$ there are $\bar{b}_{n,i}$ $(i < n)$ such that

$$M^1 \models \bigwedge_{i<n} (\exists x)\Big[\bigwedge_{\substack{j<n \\ j \ne i}} \varphi(x, \bar{b}_{n,j})\Big] \wedge \neg(\exists x)\Big[\bigwedge_{i<n} \varphi(x, \bar{b}_{n,i})\Big].$$

We expand M^1 by: $P = \{\bar{a}_n : n \in W\}$, sequence of functions $\bar{F}$, $\bar{F}(\bar{a}_n, \bar{a}_i) = \bar{b}_{n,i}$ and $Q = \{\bar{b}_{n,i} : i < n, n \in W\}$ and $R = \{\bar{b}_{n,i}{}^\frown\bar{a}_n : i < n, n \in W\}$, and by adding Skolem and other functions so that we get a model M^2, such that $T^2 = \mathrm{Th}(M^2)$ satisfies all the assumptions from Section 2. Now for any finite $\Delta \subseteq \mathrm{L}_2 = \mathrm{L}(T^2)$, by Ramsey's theorem (see I, 2.4) we can find an infinite $J_\Delta \subseteq \{\bar{a}_n : n \in W\}$ which is Δ-indiscernible. So we can find a maximal consistent set of formulas Φ_2 in L_2 in the variables $\bar{x}_n$, $n < \omega$, such that for any formula $\psi(\bar{x}_0, \ldots, \bar{x}_k) \in \mathrm{L}_2$ if for some finite $\Delta_0 \subseteq \mathrm{L}_2$ for all finite Δ, $\Delta_0 \subseteq \Delta \subseteq \mathrm{L}_2$, and all $\bar{a}_{n(0)}, \ldots, \bar{a}_{n(k)} \in J_\Delta(n(0) < n(1) < \cdots)$, $M^2 \models \psi[\bar{a}_{n(0)}, \ldots]$ then $\psi(\bar{x}_0, \ldots) \in \Phi_2$. So clearly $\Phi_2 \supseteq \Phi_1$, and Φ_2 is proper for (ω, T_2).

Let $\theta(\bar{x}; \bar{y}) = R(\bar{y}; \bar{x})$, then (3) is easy to check so only (2) remains. Suppose s is a counterexample. Using $\bar{F}$ we see $|\theta(|N^1|; \bar{a}_s)| \geq |\{\bar{F}(\bar{a}_s, \bar{a}_t) : t < s\}| = |\{t : t < s\}|$.

By defining M^2 properly we can ensure that $|\theta(|N^1|; \bar{a}_s)| \geq \min\{|T_1|, 2^{\aleph_0}\}$. So we prove one inequality. Now suppose $|\theta(N^1; \bar{a}_s)| > \lambda_0 = |\{t : t < s\}| + \min\{|T_1|, 2^{\aleph_0}\}$. So there are distinct $\tau_i(\bar{a}_{\bar{s}(i)}) \in \theta(|N^1|; \bar{a}_s)$ for $i < \lambda_0^+$. As we can replace them by any subset of cardinality λ_0^+, we can assume all the $\bar{s}(i)$ are similar over $\{t : t < s\}$. Also, if $\lambda_0 \geq |T_1|$ we can assume $\tau_i = \tau$ for any i. If $\lambda_0 < |T_1|$ necessarily $\lambda_0 \geq 2^{\aleph_0}$. Define an equivalence relation among the τ_i: $\tau_i \sim \tau_j$ if for any $\bar{t} \in W$, $\tau_i(\bar{a}_{\bar{t}}) = \tau_j(\bar{a}_{\bar{t}})$, or $\tau_i(\bar{a}_{\bar{t}}), \tau_j(\bar{a}_{\bar{t}}) \notin Q$ in the model M^2. Clearly the number of equivalence classes is $\leq 2^{\aleph_0}$, so we can assume all the τ_i's are equivalent. As $\tau_i(\bar{a}_{\bar{s}(i)}) \in Q(N^1)$, $\tau_i(\bar{a}_{\bar{s}(i)}) = \tau_0(\bar{a}_{\bar{s}(i)})$; we can assume $\bar{s}(i) = \bar{t}{}^\frown\langle s\rangle{}^\frown\bar{t}(i)$, $\bar{t}[l] < s$, $\bar{t}(i)[j] > s$. Now if $R[\bar{y}, \tau_0(\bar{x}, \bar{y}, \bar{z}_1)] \rightarrow [\tau_0(\bar{x}, \bar{y}, \bar{z}_1) = \tau_0(\bar{x}, \bar{y}, \bar{z}_2)] \in \Delta$ and $\bar{n}$, m, $\bar{k}$, $\bar{l} \in \{i : \bar{a}_i \in J_\Delta\}$, $(l(\bar{n}) = l(\bar{t})$, $l(\bar{k}) = l(\bar{l}) = l(\bar{t}(i))$, and $\bar{n}[i] < m < \bar{k}[j], \bar{l}[j]$ *then* necessarily

$$R[\bar{a}_m, \tau_0(\bar{a}_{\bar{n}}, \bar{a}_m, \bar{a}_{\bar{k}})] \rightarrow [\tau_0(\bar{a}_{\bar{n}}, \bar{a}_m, \bar{a}_{\bar{k}}) = \tau_0(\bar{a}_{\bar{n}}, \bar{a}_m, \bar{a}_{\bar{l}})]$$

(otherwise $n \geq |\{\bar{c} : R(\bar{a}_m, \bar{c})\}| \geq |\{\tau_0(\bar{a}_{\bar{n}}, \bar{a}_m, \bar{a}_{\bar{k}}) : \text{suitable } \bar{k}\}| \geq \aleph_0$). So all the $\tau_0(\bar{a}_{\bar{s}(i)})$ are equal, contradiction.

THEOREM 3.2: *Suppose T is stable, $M^1 = EM^1(I, N)$, N proper for (ω, T_1).*

(1) *Let $\varphi(\bar{x}, \bar{y}) \in \mathrm{L}$, $\bar{\sigma}$, $\bar{\tau} \in \mathrm{L}_1$. If $\bar{t}_1$, $\bar{t}_2$, $\bar{s}_1$, $\bar{s}_2$ are increasing sequences from I, $l(\bar{s}_1) = l(\bar{t}_1) = m_1$, $l(\bar{s}_2) = l(\bar{t}_2) = m_2$ and $\bar{s}_1[i] = \bar{s}_2[j] \Leftrightarrow \bar{t}_1[i] = \bar{t}_2[j]$, and $\bar{b}, \bar{c} \in N$ then $M^1 \models \varphi[\bar{\tau}(\bar{a}_{\bar{s}_1}, \bar{c}), \bar{\sigma}(\bar{a}_{\bar{s}_2}, \bar{b})] \equiv \varphi[\bar{\tau}(\bar{a}_{\bar{t}_1}, \bar{c}), \bar{\sigma}(\bar{a}_{\bar{t}_2}, \bar{b})]$*

(*Remark: Instead of $\bar{s}_1$, $\bar{s}_2$, we could have taken $\bar{s}_1, \ldots, \bar{s}_k$.*)

(2) *Suppose* $\lambda > \|N\|$ *is regular, and every interval of* I *has cardinality* $\neq \lambda$ *(including the intervals with an endpoint* $\pm\infty$*). Then* M*, the* L*-reduct of* M^1*, is* (λ, S^1_*, S^1)*-compact; and moreover* $(\lambda, S_\lambda, S^1)$*-compact. (The same holds for saturation.)*

(3) *In* (2), *instead of* λ *regular, it suffices to assume* I *has no interval of cardinality* cf λ*; or for some* $\lambda_0 < \lambda$*, any interval of cardinality* $\geq \lambda_0$ *has cardinality* $> \lambda$*. We can replace "*$\lambda > \|N\|$*" by "*N λ^+*-compact".*

Proof. (1) Here we can assume I is dense. Suppose we have a counter-example. Then, by assumptions we can find $\bar{s}^0, \ldots, \bar{s}^k$ such that:

(i) $\bar{s}^0 = \bar{s}_2$, and $\bar{s}_1{}^\frown\bar{s}^k \sim \bar{t}_1{}^\frown\bar{t}_2$.

(ii) $\bar{s}^l[i] = \bar{s}_1[j] \Rightarrow \bar{t}_2[i] = \bar{t}_1[j]$ $(0 \leqslant l \leqslant k,\ 0 \leqslant i < m_1,\ 0 \leqslant j < m_2)$.

(iii) For each l there are $i(l), j(l)$ such that for $n \neq i(l)$ $\bar{s}^l[n] = \bar{s}^{l+1}[n]$ and $\bar{s}_1[j(l)] < \bar{s}^l[i(l)]$ iff not $\bar{s}_1[j(l)] < \bar{s}^{l+1}[i(l)]$, but for $j \neq j(l)$, $\bar{s}_1[j] < s^l[i(l)]$ iff $\bar{s}_1[j] < \bar{s}^{l+1}[i(l)]$.

By (i)

$$M^1 \models \varphi[\bar{\tau}(\bar{a}_{\bar{s}_1}, \bar{c}), \bar{\sigma}(\bar{a}_{\bar{s}_2}, \bar{b})] \Leftrightarrow M^1 \models \varphi[\bar{\tau}(\bar{a}_{\bar{s}_1}, \bar{c}), \bar{\sigma}(\bar{a}_{\bar{s}^0}, \bar{b})],$$

$$M^1 \models \varphi[\bar{\tau}(\bar{a}_{\bar{t}_1}, \bar{c}), \bar{\sigma}(\bar{a}_{\bar{t}_2}, \bar{b})] \Leftrightarrow M^1 \models \varphi[\bar{\tau}(\bar{a}_{\bar{s}_1}, \bar{c}), \bar{\sigma}(\bar{a}_{\bar{s}^k}, \bar{b})].$$

But as we are dealing with a counter example

$$M^1 \models \varphi(\bar{\tau}(\bar{a}_{\bar{s}_1}, \bar{c}), \bar{\sigma}(\bar{a}_{\bar{s}_2}, \bar{b})) \Leftrightarrow M^1 \models \neg\varphi(\bar{\tau}(\bar{a}_{\bar{t}_1}, \bar{c}), \bar{\sigma}(\bar{a}_{\bar{t}_2}, \bar{b})).$$

So for some $l \leq k$

$$M^1 \models \varphi(\bar{\tau}(\bar{a}_{\bar{s}_1}, \bar{c}), \bar{\sigma}(\bar{a}_{\bar{s}^l}, \bar{b})) \equiv \neg\varphi(\bar{\tau}(\bar{a}_{\bar{s}_1}, \bar{c}), \bar{\sigma}(\bar{a}_{\bar{s}^{l+1}}, \bar{b})).$$

Let $U_n \in I$, $U_n < U_{n+1}$, and U_n is between $\bar{s}^l[i(l)]$ and $\bar{s}^{l+1}[i(l)]$. Let $\bar{V}^1_n, \bar{V}^2_n \in I$ for $n < \omega$ be defined such that: they are of length $l(\bar{s}_1), l(\bar{s}_2)$ if $j \neq j(l)$, $\bar{V}^1_n[j] = \bar{s}_1[j]$, if $j \neq i(l)$, $\bar{V}^2_n[j] = \bar{s}^l[j]$ and $\bar{V}^1_n[j(l)] = U_n$, $\bar{V}^2_n[i(l)] = U_n$. It is easy to check that for $n \neq k$ $M^1 \models \varphi[\bar{d}^1_n, \bar{d}^2_k] \equiv \neg\varphi[\bar{d}^1_k, \bar{d}^2_n]$ where $\bar{d}^1_n = \bar{\tau}(\bar{a}_{\bar{V}^1_n}, \bar{c})$, $\bar{d}^2_n = \bar{\sigma}(\bar{a}_{\bar{V}^2_n}, \bar{b})$. Hence for $\psi = \varphi$ or $\psi = \neg\varphi$ $M^1 \models \psi[\bar{d}^1_n, \bar{d}^2_k] \wedge \bar{d}^1_n \neq \bar{d}^1_k$ iff $n < k$, so T has the order property contradiction by II, 2.2 and II, 2.13.

(2) Suppose every $q \in S_\lambda(\Gamma)$, $\Gamma = \{\varphi_i(\bar{x}, \bar{b}_i): i < \lambda\}$ is realized in M^1. So let $\bar{c}_j \in |M^1|$, $M^1 \models \varphi_i[\bar{c}_j, \bar{b}_i]$ for $i < j$. So let $\bar{c}_i = \bar{\tau}_i(\bar{a}_{\bar{s}_i}, \bar{c}^i)$ $(\bar{c}^i \in |N|$, $\bar{\tau}_i \in \mathrm{L}^c_1)$. Of course, each $\bar{s}_i$ is increasing.

As $\lambda > \|N\|$ by renaming we can assume that for $i < \lambda$ $\bar{\tau}_i = \bar{\tau}$, $\bar{c}^i = \bar{c}$, (as we can replace $\bar{c}_i$ by $\bar{c}_j$ when $j > i$). In the same way, we can assume that for each l, *either* all $\bar{s}_i[l]$, $i < \lambda$ are equal, *or* for every i, j $\bar{s}_i[l] = \bar{s}_j[l] \Rightarrow i = j$.

Let $l(0) < \cdots < l(k)$ be those for which the first case occurs, and again we can assume that for $n, m \neq l(0), \ldots, l(k)$ $\bar{s}_i[n] = \bar{s}_j[m]$ iff $i = j$, $n = m$. So $l(n) + 1 < l(n+1)$ implies $\{s \in I: s_0[l(n)] < s < s_0[l(n+1)]\}$ has cardinality $\geq \lambda$ hence $> \lambda$ (for $-1 \leq n \leq k+1$, where we stipulate $l(-1) = -1$, $l(k+1) =$ the length of $\bar{s}_0$). Hence we can define $\bar{s} \in I$, $l(\bar{s}) = l(\bar{s}_0)$, $\bar{s}[l(n)] = \bar{s}_0[l(n)]$, and if $n \neq l(0), \ldots, l(k)$, then $\bar{s}[n]$ does not appear in $\bar{t}_i$, where $\bar{b}_i = \tau^i(\bar{a}_{\bar{t}_i}, \bar{b}^i)$. Then by part (1) $\tau(\bar{a}_{\bar{s}}, \bar{c})$ realizes p.

(3) Proof similar to (2).

CONCLUSION 3.3: *Suppose T is stable, with the* f.c.p. $\lambda \geq |T_1|$, *and S a set of cardinals $\lambda \in S$. Then there is $M \in \mathrm{PC}(T_1, T)$ such that:*

(1) *If* $\min\{|T_1|, 2^{\aleph_0}\} \leq \mu \leq \lambda$, $\mu \in S$, *then for some* $\varphi(x, \bar{y}) \in \mathrm{L}$, *$M$ omits a φ-type p of cardinality μ, such that any proper subtype of p is realized in M.*

(2) *If* $\min\{2^{\aleph_0}, |T_1|\} \leq \mu < \lambda$, $\mu \notin S$, *μ is regular or* $\mu = \min\{2^{\aleph_0}, |T_1|\}$, *then M is $(\lambda, S_\lambda, S^1)$-compact.*

Proof. When $|T_1| < 2^{\aleph_0}$ this is immediate by 3.1 and 3.2, as there is an order I, $|I| = \lambda$ such that for every μ, $|T_1| \leq \mu \leq \lambda$: $\mu \in S \Rightarrow$ for some $s \in I$ $|\{t \in I: t < s\}| = \mu$, and $\mu \notin S \Rightarrow I$ has no interval of cardinality μ. When $|T_1| \geq 2^{\aleph_0}$ we can use Exercise VI, 5.4, 5, Exercise 2.9 and take an elementary submodel, or 3.2(3), last phrase.

CONCLUSION 3.4: *Suppose T is stable with the* f.c.p.

If $\aleph_\alpha \geq \aleph_\beta + |T_1|$, $\aleph_\beta = \min\{|T_1|, 2^{\aleph_0}\}$, *then there are in* $\mathrm{PC}(T_1, T)$ *at least $2^{|\alpha - \beta|}$ non-isomorphic models of cardinality $\aleph_\alpha$.*

Proof. If $\alpha - \beta \geq \omega$, this is immediate by 3.3. When $\alpha - \beta < \omega$ we look at $\{\mu$: for some finite $\Delta \subseteq \mathrm{L}$, $n < \omega$, there is a finite Δ-n-indiscernible set $I \subseteq M$ such that $\dim(I, \Delta, n, \mu) = \mu\}$. For suitable T_1 this is not difficult, and we leave it to the reader. (Compare with VI, 3.9.)

DEFINITION 3.1: For an order J, I is called a (J, β)-tree if I is a model with universe $\subseteq J^{\leq \beta} = \{\eta$: η a sequence of elements of J of length $\leq \beta\}$, such that $\eta \in |I| \Rightarrow \eta \restriction \alpha \in |I|$; and with the relations $P_\alpha = \{\eta \in |I|: l(\eta) = \alpha\}$ for $\alpha \leq \beta$, the lexicographic order $<$, and order $\lessdot$ of being an initial segment, and the function h, $h(\eta, \nu) =$ the longest common initial segment of η, ν. I is the full (J, β)-tree when $|I| = J^{\leq \beta}$. We denote I by $J^{\leq \beta}$. Note that up to isomorphism the class of (J, β)-trees for all J's is $K(\omega^{\leq \beta})$. If for some J, I is a (J, β)-tree, then I is a β-tree.

Remark. If $\bar{s}, \bar{t} \in I$; $\bar{s} \sim \bar{t}$ iff $l(\bar{s}[i]) = l(\bar{t}[i])$ and for any i, j; $\alpha_{i,j} =_{\text{def}} \max\{\alpha: \bar{s}[i] \restriction \alpha = \bar{s}[j] \restriction \alpha\} = \max\{\alpha: \bar{t}[i] \restriction \alpha = \bar{t}[j] \restriction \alpha\}$. (In another notation, $l(h(\bar{s}[i], \bar{s}[j])) = l(h(\bar{t}[i], \bar{t}[j]))$ and $\bar{s}[i] < \bar{s}[j] \Leftrightarrow \bar{t}[i] < \bar{t}[j]$).

DEFINITION 3.2: (A) T has a uniform β-tree of the form $\langle\langle\varphi_\alpha, m_\alpha\rangle$: $\alpha < \beta$, α successor$\rangle$ ($\varphi_\alpha \in$ L, $m_\alpha \leq \omega$) if there are $\bar{a}_\eta$ for $\eta \in \lambda^{\leq\beta}$ such that:

(1) $\bar{a}_\eta$ realizes $p_\eta = \{\varphi_\alpha(\bar{x}, \bar{a}_{\eta\restriction\alpha})$: $\alpha < \beta$, α successor$\}$ for $\eta \in {}^\beta\lambda$.

(2) For any $\alpha, \eta \in \lambda^\alpha, \nu \in \lambda^\beta, \bar{a}_\nu$ realizes $< m_\alpha + 1$ of the formulas $\varphi_{\alpha+1}(\bar{x}, \bar{a}_{\eta\frown\langle i\rangle})$, $i < \lambda$. (We can assume that if $\delta < \beta$ is not a successor $\eta \in \lambda^\delta$, then $\bar{a}_\eta$ is empty.)

(B) The tree (and the form) are called *strong* if for any set $w \subseteq \lambda$, $|w| \geq m_\alpha$,

$$\mathfrak{C} \models \neg(\exists \bar{x})\Big[\bigwedge_{i \in w} \varphi_{\alpha+1}(\bar{x}, \bar{a}_{\eta\frown\langle i\rangle})\Big].$$

LEMMA 3.5: (1) *Definition* 3.2 *holds for some* λ *iff for every* λ.

(2) *If* $\beta < \kappa(T)$ *then* T *has a uniform* β-*tree*, $m_\alpha = 1$ *for all* α.

(3) *If* T *is stable*, $\beta < \kappa(T)$ *then* T *has a strong uniform* β-*tree*, $m_\alpha = 1$, *for all* α.

(4) *If* T *has the strict order property then* T *has a strong uniform* β-*tree, with* $m_\alpha = 1$ *for all* α.

(5) *If* $D(x = x, \text{L}, \infty) = \infty$ *then* T *has a strong uniform* $\aleph_0$-*tree*, $m_\alpha = 1$, *for all* α.

(6) *If* $\kappa < \kappa r_{\text{cdt}}(T)$, κ *is regular, then* T *has a strong uniform* κ-*tree*, $m_\alpha = 1$.

Proof. Immediate, see III, 3.1 for (3), II, 3.9 for (5) and III, 7.10 for (6). Clearly, (1) and (4) are immediate.

For (2), note that if on $J = \lambda^{\leq\beta} \times \{0, 1\}$ we define an ordering $<$: $\langle\eta, i\rangle < \langle\nu, j\rangle$ if $\eta \lneqq \nu$, $i = 0$ or $\eta = \nu$, $i = 0$, $j = 1$, *or* $\rho \lessdot \eta, \nu$ where $\rho = h(\eta, \nu) \neq \eta, \nu$ and $\eta < \nu$ (in the lexicographic order). Then assuming T is unstable (as (3) has been proved) by the order property, there are $\varphi \in$ L, $\bar{a}_s$ ($s \in J$) such that $\models\varphi[\bar{a}_s, \bar{a}_t]$ iff $s < t$. Then let

$$\bar{a} = \bar{a}_{\langle\eta,0\rangle}{}^\frown\bar{a}_{\langle\eta,1\rangle}$$

and

$$\varphi(\bar{x}_1{}^\frown\bar{x}_2, \bar{a}_\eta) = [\varphi(\bar{x}_1, \bar{a}_{\langle\eta,0\rangle}) \equiv \neg\varphi(\bar{x}_1, \bar{a}_{\langle\eta,1\rangle})].$$

THEOREM 3.6: (1) *If* T *has a uniform* β-*tree of the form* $\langle\langle\varphi_\alpha, m_\alpha\rangle$: $\alpha < \beta$, α *a successor*$\rangle$, *then it has such a tree which is indiscernible (when* *we view the β-tree as an index set). In fact, $J^{\leq\beta}$ has the extension property (see Definition 2.9).*

(2) *By the assumption of* (1), *there is a Φ proper for β-trees and T_1, such that if $M = EM(\lambda^{\leq\beta}, \Phi)$, then $\{\bar{a}_\eta: \eta \in \lambda^{\leq\beta}\}$ is a uniform β-tree of the form $\langle\langle\varphi_\alpha, m_\alpha\rangle: \alpha < \beta: \alpha$ successor$\rangle$.*

(3) *If we start in* (1) *and* (2) *with a strong tree, we get one.*

(4) *By replacing $\varphi_\alpha(\bar{x}; \bar{y})$ by $\varphi'_\alpha(\bar{x}; \bar{y}_1, \bar{y}_2) = \varphi_\alpha(\bar{x}; \bar{y}_1) \wedge \neg\varphi_\alpha(\bar{x}; \bar{y}_2)$, we can assume in* (2) *that if $M = EM(I, \Phi)$, $\eta \in I \cap \lambda^\alpha$, no sequence in M satisfies infinitely many formulas $\varphi_{\alpha+1}(\bar{x}, \bar{a}_{\eta^\frown\langle i\rangle})$, $\eta^\frown\langle i\rangle \in I$. That is, we get a strong uniform tree of the form $\langle\langle\varphi_\alpha, \aleph_0\rangle: \alpha < \beta$, α a successor$\rangle$.*

Proof. (1) Suppose $\{\bar{a}_\eta: \eta \in \lambda^{\leq\beta}\}$ is a uniform β-tree; and choose $\lambda \geq \beth_{\alpha+\omega}$, $\beth_\alpha \geq |T|$. By adding dummy variables to the φ_α's we can assume the a_η's are distinct.

It suffices to prove the consistency of

$$\begin{aligned}\Gamma = T &\cup \{\varphi(\bar{y}_{\bar{s}}) \equiv \varphi(\bar{y}_{\bar{t}}): \bar{s} \sim \bar{t};\ \bar{s}, \bar{t} \in \lambda^{\leq\beta};\ \varphi \in \mathrm{L}\}\\ &\cup \{\varphi_\alpha(\bar{y}_\eta, \bar{y}_\nu): \alpha \text{ a successor},\ l(\nu) = \alpha,\ l(\eta) = \beta,\ \nu \lessdot \eta\}\\ &\cup \{\neg\varphi_\alpha(\bar{y}_\eta, \bar{y}_\nu): \alpha = \gamma + 1,\ l(\nu) = \alpha,\ l(\eta) = \beta,\\ &\qquad \nu \restriction \gamma = \eta \restriction \gamma,\ \nu(\gamma) \neq \eta(\gamma)\}\\ &\cup \{\bar{y}_\eta \neq \bar{y}_\nu : \eta \neq \nu,\ l(\bar{y}_\eta) = l(\bar{y}_\nu)\}.\end{aligned}$$

So it suffices to prove the consistency of all finite $\Gamma' \subseteq \Gamma$. In Γ' the set of α which are $l(\eta)$ where $\bar{y}_\eta$ appears in Γ' is finite. So by renaming we can assume all such α's are $< n_0$. (We rename in a corresponding way the $\bar{a}_\eta$'s.) By the combinatorial theorem 2.6 of the Appendix we can find an assignment showing the consistency of Γ'.

(2), (3) Follow easily. For (4) notice that for the Φ from 2, for no $\bar{b} \in M$ $\aleph_0 \leq |\{i: \varphi_\alpha(\bar{b}, \bar{a}_{\eta^\frown\langle i\rangle}) \equiv \neg\varphi_\alpha(\bar{b}; \bar{a}_{\eta^\frown\langle i+1\rangle})\}|$.

THEOREM 3.7: *Suppose $|T| = \aleph_0$ and let $\varphi_\eta(\bar{x}) \in \mathrm{L}(T)$ for $\eta \in 2^{<\omega}$ be such that*

(α) $\eta \lessdot \nu \Rightarrow \vDash(\forall\bar{x})[\varphi_\nu(\bar{x}) \rightarrow \varphi_\eta(\bar{x})]$,

(β) $\vDash\neg(\exists x)[\varphi_{\eta^\frown\langle 0\rangle}(\bar{x}) \wedge \varphi_{\eta^\frown\langle 1\rangle}(\bar{x})]$,

(γ) $\vDash(\exists x)\varphi_\eta(\bar{x})$.

Then T has a model M and $\bar{a}_\eta \in M$ for $\eta \in 2^\omega$, and a function $h: 2^{<\omega} \rightarrow 2^{<\omega}$ such that:

(1) *$\eta \lessdot \rho$ implies $h(\eta) \lessdot h(\rho)$, and $h(\eta^\frown\langle 0\rangle)$, $h(\eta^\frown\langle 1\rangle)$ are $\lessdot$-incomparable, so for $\eta \in 2^\omega$ we can define $h(\eta) \in 2^\omega$ as the unique $\nu \in 2^\omega$, such that $h(\eta \restriction n) \lessdot \nu$ for $n < \omega$.*

(2) $\bar{a}_\eta$ *realizes* $p_\eta^0 = p_{h(\eta)}^1$, $p_\nu^1 = \{\varphi_{\nu\restriction n}(\bar{x}): n < \omega\}$.

(3) *Call* $\bar{\eta}, \bar{\nu} \in 2^\omega$ *similar over* m, $\bar{\eta} \sim \bar{\nu}(\mathrm{mod}\, m)$ *if*

(i) $\bar{\eta}[l] \restriction m \neq \bar{\eta}[k] \restriction m$ *for* $k \neq l$,

(ii) $\bar{\eta}[l] \restriction m = \bar{\nu}[l] \restriction m$ *for any* l.

Then for $\varphi(\bar{y}) \in \mathrm{L}$ *there is* m_φ *such that if* $\bar{\eta}, \bar{\nu}$ *are similar over* $m \geq m_\varphi$, *then* $M \models \varphi[\bar{a}_{\bar{\eta}}] \equiv \varphi[\bar{a}_{\bar{\nu}}]$.

(4) *If* $\bar{\tau} \in \mathrm{L}$, $\bar{\tau}(\bar{a}_{\bar{\eta}})$ *realizes* p_ν^0 *then* $\nu = \bar{\eta}[l]$ *for some* l.

Moreover there is $m_{\bar{\tau}}$ *such that if* $\bar{\eta} \sim \bar{\eta}(\mathrm{mod}\, m)$, $m > m_{\bar{\tau}}$ *and* $\nu \in 2^m$, $\nu \neq \bar{\eta}[0] \restriction m, \ldots,$ *then* $M \models \neg\varphi_{h(\nu)}[\bar{\tau}(\bar{a}_{\bar{\eta}})]$.

(5) *If* Γ_n *are finite sets of formulas in* L, $\Gamma_n \subseteq \Gamma_{n+1}$, $\bigcup_{n<\omega} \Gamma_n$ *is the set of formulas of* L, *then in* (3), *if* $\varphi(\bar{y}) \in \Gamma_n$ *we can choose* $m_\varphi = n$.

Proof. Choose a model M_0 of T, and $\bar{c}_\eta \in |M_0|$, $M_0 \models \varphi_{\bar{\eta}}[\bar{c}_\eta]$ for $\eta \in {}^{\omega>}2$. We use 2.4 of the Appendix for the reasonable f_m's (for each $\varphi(\bar{x}_1, \ldots, \bar{x}_n)$ there is an m such that $f(\bar{\eta}) = 0$ iff $f(\bar{\eta}) \neq 1$ iff $M \models \varphi[\bar{c}_{\bar{\eta}}]$). Now letting $M = M_0^\omega/D$, $\bar{a}_\eta = \langle \ldots, \bar{c}_{h^1(\eta\restriction n)}, \ldots \rangle/D$, D be any non-principal ultrafilter over ω, clearly (1), (2) and (3) are satisfied. Now it is trivial to get (5) by renaming, and (4) is quite easy.

EXERCISE 3.1: We can assume in 3.7, that there are distinct $b_n \in M$ such that

(i) $\{b_n: n < \omega\}$ is indiscernible over $\bigcup \{\bar{a}_\eta; \eta \in {}^\omega 2\}$,

(ii) $(M, b_0, b_1, \ldots)$ satisfies (1)–(5) of 3.7.

[Hint: Let $N \prec M$, $b_n \in |M|$, $\{b_n: n < \omega\}$ indiscernible over N. Apply 3.7 to $T' = \mathrm{Th}(M, P, b_0, \ldots)$ (for $P = N$) and $\varphi_n(x)$ such that $\models P[a_n]$ (more exactly, we have to repeat the proof).]

EXERCISE 3.2: Show that in 3.2(2), we can replace "$\|N\| < \lambda$" by "N is $(\lambda, S_\lambda, S^1)$-compact."

EXERCISE 3.3: Suppose $\psi_n \in \mathrm{L}(T)$, T countable, p an m-type over $\emptyset$, N a model of T, omitting p, and $\{\{\psi_n(\bar{x}): N \models \psi_n[\bar{b}]\}: \bar{b} \in |N|\}$ has cardinality $2^{\aleph_0}$ and $2^{\aleph_0}$ is real valued measurable (see [So 71]).

Show that T has a model M omitting p, and there are $\varphi_\eta \in \mathrm{L}(\eta \in {}^{\omega>}2)$ which are Boolean combinations of the ψ_n's, and $\bar{a}_\eta \in |M|$ as in 3.7. [Hint: Let D be a normal measure over $2^{\aleph_0}$, and $\bar{b}_i \in |N|$, $(i < 2^{\aleph_0})$ such that $p_i = \{\psi_n(\bar{x}): N \models \psi_n[\bar{b}_i]\}$ are distinct; and w.l.o.g. T has Skolem functions. Now we define inductively $S_\alpha \in D$ $(\alpha < \omega_1)$ such that $S_\delta = \bigcap_{i<\delta} S_i$, $S_{i+1} \subseteq S_i$, $S_0 = 2^{\aleph_0}$, and: if $i(0), \ldots, i(n-1) \in S_{\alpha+1}$ are distinct, and $\varphi \in \mathrm{L}$, $N \models \varphi[\bar{b}_{i(0)}, \ldots, \bar{b}_{i(n-1)}]$, then for some $j \in S_\alpha - S_{\alpha+1}$

$N \models \varphi[\bar{b}_j, \bar{b}_{i(1)}, \ldots, \bar{b}_{i(n-1)}]$. Now we shall define inductively $\varphi_n(\ldots, x_\eta, \ldots)_{\eta \in 2^n}$ such that

(i) for every α for some $i(\eta) \in S_\alpha$, $N \models \varphi_n[\ldots, \bar{b}_{i(\eta)}, \ldots]$;

(ii) If $\nu(\eta) \in \{\eta^\frown\langle 0\rangle, \eta^\frown\langle 1\rangle\}$ for $\eta \in 2^n$, then

$$\varphi_{n+1}(\ldots, x_\rho, \ldots)_{\rho \in 2^{n+1}} \vdash \varphi_n(\ldots, x_{\nu(\eta)}, \ldots)_{\eta \in 2^n};$$

(iii) When $\eta \lessdot \nu(\eta) \in 2^\omega$ for $\eta \in 2^n$, $\varphi_n(\ldots, x_\eta, \ldots)$ is an approximation to the desired properties of $\cdots{}^\frown\bar{a}_{\nu(\eta)}{}^\frown\cdots$ (i.e., we allow only finitely many terms, and note here we should decide why a term does not realize p). For further help see [Sh 75].]

VII.4. Semi-definability

DEFINITION 4.1: (1) If D is an ultrafilter over $\mathbf{I}$ ($\mathbf{I} \subseteq |\mathfrak{C}|^m$) then

$$\mathrm{Av}(D, A) = \{\varphi(\bar{x}; \bar{a}) : \bar{a} \in A, \{\bar{b} \in \mathbf{I} : \models \varphi[\bar{b}; \bar{a}]\} \in D\}$$

(Clearly it $\in S^m(A)$, it is consistent as D is closed under intersection).

(2) An m-type p is semi-definable over $\mathbf{I}$ if there is an ultrafilter D over $\mathbf{I}$ such that $p \subseteq \mathrm{Av}(D, \mathrm{Dom}\, p)$.

(3) An m-type p is semi-definable over A if it is semi-definable over A^m.

(4) $\mathrm{tp}_*(C, A)$ is semi-definable over $B \subseteq A$, if for any $\bar{c} \in C$, $\mathrm{tp}(\bar{c}, A)$ is semi-definable over $B \subseteq A$. (This naturally extends to not-necessarily complete types with infinitely many variables.)

(5) $p \in S^m(A)$ ($B \subseteq A$) is stationary over B, if p is semi-definable over B, and it has no two contradictory extensions which are semi-definable over B. (This is not necessarily consistent with the previous definition of stationary Definition III, 1.7, but our meaning will always be clear.)

Remark. Notice that if T is stable $\mathbf{I}$ an indiscernible set, D a non-principal ultrafilter over $\mathbf{I}$ then $\mathrm{Av}(D, A) = \mathrm{Av}(\mathbf{I}, A)$.

LEMMA 4.1: (1) *A type p (possibly with infinitely many variables) is semi-definable over $\mathbf{I}(A)$ iff every finite subtype of p is realized in $\mathbf{I}(A)$.*

(2) *Every type over a model M, is semi-definable over $|M|$.*

(3) *If p is a type over A, and is semi-definable over B, then there is a complete type over A extending p which is semi-definable over B (the same holds for $\mathbf{I}$ instead of B).*

Remark. Part (1), implies, by III, 4.10(1) that if T is stable, $B = |M|$, then p is semi-definable over B iff p does not fork over B.

Proof. (1) Let p be an m-type semi-definable over $\mathbf{I}$. If $p \subseteq \mathrm{Av}(D, \mathrm{Dom}\, p)$ for some ultrafilter D, and $q \subseteq p$ is finite, then

$$\{\bar{c} \in \mathbf{I}: \bar{c} \text{ realizes } q\} = \bigcap_{\varphi(\bar{x};\bar{a}) \in q} \{\bar{c} \in \mathbf{I}: \models\varphi[\bar{c}; \bar{a}]\} \in D$$

(as D is closed under intersection). So $\{\bar{c} \in \mathbf{I}: \bar{c} \text{ realizes } q\} \neq \emptyset$ so q is realized in $\mathbf{I}$.

Suppose every finite $q \subseteq p$ is realized in $\mathbf{I}$. For each $\varphi \in p$ let $\mathbf{J}_\varphi = \{\bar{c} \in \mathbf{I}: \bar{c} \text{ satisfies } \varphi\}$. By assumption the intersection of any finitely many $\mathbf{J}_\varphi$ is non-empty, hence there is an ultrafilter D over $\mathbf{I}$ such that $\varphi \in p \Rightarrow \mathbf{J}_\varphi \in D$, by 1.1(2) and (3) of the Appendix. Clearly $\varphi \in p \Rightarrow \varphi \in \mathrm{Av}(D, \mathrm{Dom}\, p)$.

For types with infinitely many variables the proof is similar.

(2) Every type over a model is finitely satisfiable in M, hence the result follows from (1).

(3) It suffices to note the following; and use (1):

(i) If p_i, $i < \delta$, is an increasing sequence of types, which are finitely satisfiable in B, then $\bigcup_{i<\delta} p_i$ is finitely satisfiable in B. (The proof is immediate.)

(ii) If p is finitely satisfiable in B, $\varphi(\bar{x}; \bar{a})$ a formula, then $p_1 = p \cup \{\varphi(\bar{x}; \bar{a})\}$ or $p_2 = p \cup \{\neg\varphi(\bar{x}; \bar{a})\}$ is finitely satisfiable in B (or both are).

Because otherwise there are finite $q_1 \subseteq p_1$, $q_2 \subseteq p_2$ which are not satisfiable in B. So $q = (q_1 \cup q_2) \cap p \subseteq p$ and is finite, hence satisfied by some $\bar{b} \in B$. If $\models\varphi[\bar{b}; \bar{a}]$ then $\bar{b}$ realizes q_1, and otherwise it realizes q_2, contradiction.

LEMMA 4.2: (1) *If* $\bar{c}$ *is a subsequence of* $\bar{b}$, $\mathrm{tp}(\bar{b}, A)$ *is semi-definable over* B *then* $\mathrm{tp}(\bar{c}, A)$ *is semi-definable over* B.

(2) *If* $p \vdash q$ *(e.g.,* $q \subseteq p$*) and* p *is semi-definable over* B*, then* q *is semi-definable over* B.

(3) *If* $B' \subseteq B$, p *is semi-definable over* B'*, then* p *is semi-definable over* B.

Proof. Immediate. Note this lemma shows the consistency of the definitions of semi-definability.

LEMMA 4.3: (1) *If p is semi-definable over B, then it does not split over B.*

(2) *If p is a complete type over A, semi-definable over B, $B \subseteq A$, and every $q \in S^m(B)$, $m < \omega$, is realized by some $\bar{c} \in A$, then p is stationary over B.*

Proof. (1) Suppose $p \subseteq \mathrm{Av}(D, A)$, and $\mathrm{tp}(\bar{b}, B) = \mathrm{tp}(\bar{c}, B)$, $\bar{b}, \bar{c} \in A$. If $\varphi(\bar{x}; \bar{b}) \in p$, then $S = \{\bar{a} \in B: \vDash \varphi[\bar{a}; \bar{b}]\} \in D$. But for any $\bar{a} \in B \vDash \varphi[\bar{a}; \bar{b}] \equiv \varphi[\bar{a}; \bar{c}]$, hence $S = \{\bar{a} \in B: \vDash \varphi[\bar{a}, \bar{c}]\}$, hence $\varphi(\bar{x}; \bar{c}) \in \mathrm{Av}(D, A)$ hence $\neg\varphi(\bar{x}; \bar{c}) \notin p$, so p does not split over B.

(2) Suppose p_1, p_2 are contradictory extensions of p which are semi-definable over B. By 4.1(3) we can assume they are complete types over some $C \supseteq A$, so there is $\bar{a} \in C$ $\varphi(\bar{x}, \bar{a}) \in p_1$, $\neg\varphi(\bar{x}, \bar{a}) \in p_2$. Choose $\bar{a}_1 \in A$, $\mathrm{tp}(\bar{a}_1, B) = \mathrm{tp}(\bar{a}, B)$ and w.l.o.g. $\varphi(\bar{x}; \bar{a}_1) \in p$. So p_2 splits over B, contradiction.

DEFINITION 4.2: $\mathbf{F}^{b}_{\lambda} = \{(p, B): p \in S^m(A), B \subseteq A, |B| < \lambda, p$ is semi-definable over $\mathrm{dcl}\, B\}$.

LEMMA 4.4: (1) *$\mathbf{F}^{b}_{\lambda}$ satisfies the following axioms (see* IV, *Section* 1): Ax (I), (II, 1), (II, 2), (III, 1), (III, 2), IV, (VII), (VIII), (IX), (XI, 1) *and* (XI, 2).

(2) $\lambda(\mathbf{F}^{b}_{\lambda}) = \lambda$; $\mu(\mathbf{F}^{b}_{\lambda}) = \infty$.

(3) *If T has Skolem functions, then also* Ax(X.1) *and* (X.2) *hold. Moreover if $|\mathrm{Dom}\, p| < \lambda$, p is over $B \subseteq A$, $|B| < \lambda$, then there is a complete type q over A extending p such that $(q, B) \in \mathbf{F}^{b}_{\lambda}$. Also* Ax(II.3) *and* (II.4) *hold.*

Proof. (1) For Ax(II.1) notice that if $\bar{a} \in B \subseteq A$, $l(\bar{a}) = m$; then $D = \{\mathbf{J} \subseteq B^m: \bar{a} \in \mathbf{J}\}$ is an ultrafilter over B^m and $\mathrm{Av}(D, A) = \mathrm{tp}(\bar{a}, A)$. Ax(IV) is Lemma 4.2(1). For Ax(VII), it suffices to prove the following: suppose $\mathrm{dcl}\, B = B$, $\mathrm{dcl}\, C = C$, $B \subseteq C$; and $B \subseteq A$, $\mathrm{tp}(\bar{a}, A \cup C)$ is semi-definable over $B \cup C$, $\mathrm{tp}_*(C, A)$ is semi-definable over B, $\bar{d} \in A$ and $\varphi(\bar{x}; \bar{d})$ is realized by some $\bar{c} \in \bar{a} \cup C$. We should prove $\varphi(\bar{x}; \bar{d})$ is realized by some $\bar{c}' \in B$. We can assume $\bar{c} = \bar{a}{}^\frown\bar{c}_1$, $\bar{c}_1 \in C$, so $\vDash\varphi[\bar{a}, \bar{c}_1; \bar{d}]$, hence for some $\bar{a}_1 \in C \vDash \varphi[\bar{a}_1, \bar{c}_1; \bar{d}]$. As $\bar{a}_1, \bar{c}_1 \in C$, for some $\bar{a}'_1, \bar{c}'_1 \in B$, $\vDash \varphi[\bar{a}'_1, \bar{c}'_1; \bar{d}]$ and we finish.

The other axioms are trivial.

(2) Immediate.

(3) If T has Skolem functions, for any B, $\mathrm{cl}\, B$ is the universe of a model. So by 4.1(2) every type p over $\mathrm{cl}\, B$ is semi-definable over $\mathrm{cl}\, B$. So 4.1(3) implies our conclusion. The proof of Ax(II.3) and (II.4) is easy.

LEMMA 4.5: *Let $P = |N|$, and (N^*, P) be an elementary submodel of $(\mathfrak{C}, P)$ such that $|N| \subseteq |N^*|$. Then for any sets A, B there are elementary mappings f, g which are the identity over $|N|$, $|N^*|$ resp. such that if $A' = f(A)$, $B' = g(B)$ then:*

(1) $\operatorname{tp}_*(B', |N^*| \cup A')$ *is semi-definable over* $|N^*|$,
(2) $\operatorname{tp}_*(A', |N^*| \cup B')$ *is semi-definable over* $|N|$.

Proof. By 4.1(2) and (3) we can extend $\operatorname{tp}_*(A, |N|)$ to a complete type over $|N^*|$ which is semi-definable over $|N|$. Hence we can define an elementary mapping f, which is the identity over $|N|$, such that $\operatorname{tp}_*(A', |N^*|)$ is semi-definable over $|N|$, where $A' = f(A)$. Now let

$$\Gamma = \operatorname{tp}_*(B, |N^*|) \cup \Phi \cup \Psi, \text{ where}$$

$$\Phi = \{\neg\varphi(\bar{x}_{\bar{b}}; \bar{a}, \bar{c}): \bar{b} \in B, \bar{a} \in A', \bar{c} \in |N^*| \ (\varphi \in \mathrm{L}) \text{ and } \varphi(\bar{x}_{\bar{b}}; \bar{a}, \bar{c}) \text{ is not realized in } N^*\},$$

$$\Psi = \{\neg\psi(\bar{x}_{\bar{b}}; \bar{a}, \bar{c}): \bar{b} \in B, \bar{a} \in A', \bar{c} \in |N^*| \ (\psi \in \mathrm{L}) \text{ and } \psi(\bar{b}, \bar{y}, \bar{c}) \text{ is not realized in } N\}.$$

Clearly it suffices to prove Γ is consistent (if $x_b \mapsto b'$ is an assignment satisfying Γ, let g map c to c for $c \in |N^*|$ and b to b' for $b \in B$. Now g is elementary as $\operatorname{tp}_*(B, |N^*|) \subseteq \Gamma$; it satisfies (1) as $\Phi \subseteq \Gamma$ and (2) as $\Psi \subseteq \Gamma$).

Let Γ' be a finite subset of Γ. As $\operatorname{tp}_*(B, |N^*|)$, Φ, Ψ are closed under conjunctions, we can assume, by adding dummy variables,

$$\Gamma' = \{\theta(\bar{x}_{\bar{b}}, \bar{c}), \neg\varphi(\bar{x}_{\bar{b}}, \bar{a}, \bar{c}), \neg\psi(\bar{x}_{\bar{b}}, \bar{a}, \bar{c})\}$$

where $\bar{a} \in A'$, $\bar{b} \in B$, $\bar{c} \in |N^*|$, $\models\theta[\bar{b}, \bar{c}]$, $\varphi(\bar{x}_{\bar{b}}, \bar{a}, \bar{c})$ is not realized in N^*, $\psi(\bar{b}, \bar{y}, \bar{c})$ is not realized in N. So, as $P = |N|$,

$$(\mathfrak{C}, P) \models (\forall \bar{y})\Big[\bigwedge_i P(\bar{y}[i]) \to \neg\psi(\bar{b}, \bar{y}, \bar{c})\Big] \wedge \theta(\bar{b}, \bar{c}).$$

As (N^*, P) is an elementary submodel of $(\mathfrak{C}, P)$ and $\bar{c} \in |N^*|$, there is $\bar{b}' \in |N^*|$

$$(N^*, P) \models (\forall \bar{y})\Big[\bigwedge P(\bar{y}[i]) \to \neg\psi(\bar{b}', \bar{y}, \bar{c})\Big] \wedge \theta(\bar{b}', \bar{c}).$$

By the second conjunct, $\bar{b}'$ satisfies $\theta(\bar{x}_{\bar{b}}, \bar{c})$; by the first one for every $\bar{a}' \in |N|$, $\models\neg\psi[\bar{b}', \bar{a}', \bar{c}]$, but $\operatorname{tp}(\bar{a}, |N^*|)$ is semi-definable over $|N|$ hence by 4.1(1) finitely satisfiable in N, hence $\models\neg\psi[\bar{b}', \bar{a}, \bar{c}]$. So in order to realize Γ', $\bar{b}'$ has to satisfy $\neg\varphi(\bar{x}_{\bar{b}}, \bar{a}, \bar{c})$, but as $\bar{b}' \in |N^*|$ this follows from Φ's definition.

THEOREM 4.6: *For every* $\mu \geq 2^{\lambda_1}$, $\lambda_1 \geq 2^{|T_1|}$, $\mathrm{PC}(T_1, T)$ *has a* λ_1*-universal model M of cardinality* μ *which is stable in* λ_1. (*Hence if T is* λ_1*-unstable, M is* λ_1*-universal but not* λ_1^+*-universal.*)

Remark. The construction here will serve additional theorems.

Proof. We shall describe here a construction for a cardinal $\lambda > |T_1|$, $2^{<\lambda} \leq \mu$ such that for $\lambda = \lambda_1^+$ we get the desired model. Of course, we can assume T_1 has Skolem functions.

Let N be a model of T_1 of cardinality $|T_1|$, $P = |N|$, and (N^*, P) an elementary submodel of $(\mathfrak{C}_1, P)$ of cardinality $|T_1|$. Let M^1, M^2 be μ-saturated models of T_1 extending N. By Lemma 4.5 (and 4.2) we can assume $\mathrm{tp}_*(|M^1|, |M^2|)$, $\mathrm{tp}_*(|M^2|, |M^1|)$ are semi-definable over $|N^*|$, $|N|$ resp.; hence do not split over them (by 4.3(1)). By 4.3(2) $\mathrm{tp}_*(|M^2|, |M^1|)$ is stationary over N. Note that if $\bar{a}, \bar{b} \in |M^2|$, $\mathrm{tp}(\bar{a}, |N^*|) = \mathrm{tp}(\bar{b}, |N^*|)$ then $\mathrm{tp}(\bar{a}, |M^1|) = \mathrm{tp}(\bar{b}, |M^1|)$ (as for any $\bar{c} \in |M^1|$ and $\varphi(\bar{x}, \bar{y})$, $\vDash \varphi(\bar{a}, \bar{c}) \equiv \varphi(\bar{b}, \bar{c})$ as $\mathrm{tp}(\bar{c}, |M^2|)$ does not split over $|N^*|$); hence the number of $p \in S^m(M^1)$, $m < \omega$, realized in M^2 is $\leqslant |\bigcup_m S^m(N^*)| \leqslant 2^{|T_1|}$. We can interchange the roles of M^1 and M^2 (with N instead of N^*).

Let $\{M_i : i < \mu\}$ be a list of models of T_1 of cardinality $< \lambda$, such that up to isomorphism, each model of T_1 of cardinality $< \lambda$ appears in it. As M^2 is λ-saturated, we can assume all the M_i are elementary submodels of M^2. We define by induction on i models M_i^* such that:

(i) $\mathrm{tp}_*(|M_i^*|, |M^1|) = \mathrm{tp}_*(|M_i|, |M^1|)$; more exactly there is an elementary mapping f_i, f_i is the identity over M^1, f_i maps M_i onto M_i^*.

(ii) The type of M_i^* over $M^1 \cup \bigcup_{j<i} M_j^*$ is semi-definable over $|N|$.

Let M be the Skolem closure of $\bigcup_{i<\mu} M_i^*$. Clearly M has cardinality $\leq \mu$ (as $2^{<\lambda} \leq \mu$) and M is χ-universal for every $\chi < \lambda$. Suppose $\chi^+ \geq \lambda$ so $\|M_i\| \leq \chi$ for each i, and we shall show M is χ-stable. Let $A \subseteq |M|$, $|A| \leq \chi$, but $S^m(A, M)$ has cardinality $> \chi$. W.l.o.g. A is the Skolem closure of $\{M_i^* : i \in w\}$ where $|w| \leq \chi$. Suppose $\bar{a}_i \in |M|$, $i < \chi^+$ realize different types over A, so by the definition of M $\bar{a}_j = \tau_j(\bar{b}_{j,1}, \ldots, \bar{b}_{j,n})$ where $n = n_j$, $\bar{b}_{j,l} \in |M^*_{\beta(j,l)}|$. As we can replace $\{\bar{a}_i : i < \chi^+\}$ by any subset of the same cardinality we can assume $n_i = n$, $\tau_i = \tau$. Let $\bar{b}_{j,l} = f_{\beta(j,l)}(\bar{c}_{j,l})$ $\bar{c}_{j,l} \in M_{\beta(j,l)}$ hence $\bar{c}_{j,l} \in M^2$. As the number of complete types sequences from M^2 realize over M^1 is $\leq 2^{|T_1|}$, we can assume, if $\chi \geq 2^{|T_1|}$, that $\mathrm{tp}(\bar{c}_{j,l}, |M^1|) = p_l$. As the order μ is atomically stable in χ (see 2.8(5)) we can assume the atomic type of $\langle \beta(j, 1), \ldots, \beta(j, n) \rangle$ over w is constant, and $\beta(j, l) \in w \Rightarrow b_{j,l} = b_l$. So it suffices to show that under all those conditions all the $\bar{a}_i$ realize the same type over A. For this it suffices to show:

(*) If $i(1) < \cdots < i(m) < \mu, j(1) < \cdots < j(m) < \mu, \bar{a}_l \in |M_{i(l)}|$,

$$\bar{b}_l \in |M_{j(l)}|, \text{tp}(\bar{a}_l, |M^1|) = \text{tp}(\bar{b}_l, |M^1|)$$

$$\bar{a}'_l = f_{i(l)}(\bar{a}_l), \quad \bar{b}'_l = f_{j(l)}(b_l),$$

then $\text{tp}(\bar{a}'_1 \frown \cdots \frown \bar{a}'_m, |M^1|) = \text{tp}(\bar{b}'_1 \frown \cdots \frown \bar{b}'_m, |M^1|)$.

Proof of (*). We prove it by induction on m. For $m = 1$, $\text{tp}(\bar{a}'_1, |M^1|) = \text{tp}(\bar{a}_1, |M^1|) = \text{tp}(\bar{b}_1, |M^1|) = \text{tp}(\bar{b}'_1, |M^1|)$. Suppose we have proven for m, and we shall prove for $m + 1$. By the symmetry we can assume $i(m + 1) \leq j(m + 1)$. Now $\text{tp}(\bar{a}'_{m+1}, |M^1|) = \text{tp}(\bar{b}'_{m+1}, |M^1|)$ is stationary and $\text{tp}(\bar{a}'_{m+1}, |M^1| \cup \bigcup \{|M^*_j|: j < i(m + 1)\})$ $\text{tp}(\bar{b}'_{m+1}, |M^1| \cup \bigcup \{|M_j|: j < j(m + 1)\})$ are semi-definable over $|N|$, hence the former is a subtype of the latter, and the latter does not split over $|N|$, so by the induction hypothesis we get (*) and finish the proof.

(By checking that when $\lambda = \lambda_1^+$, $\lambda_1 \geq 2^{|T_1|}$, we get the result easily.)

Note that w.l.o.g. $M_i \not\subseteq N$, hence $\|M\| = \mu$.

EXERCISE 4.1: Let T be the theory of the rational order, $|M| = (0, 1)$, $a = 1$, $b = 2$. Show there is no a', such that: $\text{tp}(a', M \cup b)$ extends $\text{tp}(a, M)$ and is semi-definable over M and $\text{tp}(b, M \cup a')$ is semi-definable over M.

EXERCISE 4.2: Show that $\mathbf{F}^{\mathrm{b}}_\lambda$ cannot (in general) satisfy more axioms than stated in Lemma 4.4. (Hint: Ax(II.3, 4): Choose $A = B = \emptyset$, $a \in \text{acl}\, A - \text{dcl}\, A$.)

Ax(V): Let $M = EM^1(\omega_1, \Phi)$ where we get the theory T_1 and Φ as in Exercise 2.7. So $M \models a < a_\delta$ implies: $M \models a < a_i$ for some $i < \delta$, and $a \in \text{acl}\{a_j: j < \delta\}$ ($\{a_j: j < \omega_1\}$ is the skeleton). Let $B = \text{acl}\{a_i: i < \omega$ or $i > \omega + \omega\}$, $A = \text{acl}\{a_i: i \leq \omega$ or $i > \omega + \omega\}$, $\bar{a} = \langle a_{\omega+\omega}\rangle$, $\bar{b} = \langle a_{\omega+7}\rangle$.

Ax(IV): By the previous example, for $A = B = C = \text{acl}\{a_i: i < \omega\}$ $\bar{b} = \langle a_\omega\rangle$, $\bar{a} = \langle a_{\omega+1}\rangle$.

Ax(X.1), (X.2): Choose $A = \text{dcl}\, A = \emptyset$ $(A_0 = \text{dcl}\, A_0)$, $\varphi = x = x$.

Ax(XII): $C_1 = \text{acl}\{a_i: i < \omega\}$ $C_2 = \text{acl}\{a_i: \omega \cdot 2 < i < \omega \cdot 3\}$ $B = C_1 \cup C_2$, $A = \text{acl}(B \cup \{a_\omega\})$, $p = \text{tp}(a_{\omega+1}, A)$ (in the notation above).]

VII.5. Hanf numbers of omitting types

κ need not be infinite

DEFINITION 5.1: (1) We define $\mu(\lambda, \kappa)$ as the first cardinal μ such that: if $|L_0| \le \lambda$, T_0 a theory in L_0, Γ a set of $(<\aleph_0)$-types in L_0, $|\Gamma| \le \kappa$; and for every $\chi < \mu$ there is a model in $EC(T_0, \Gamma)$ of cardinality $\ge \chi$, where $EC(T_0, \Gamma)$ is the class of L_0-models of T_0 omitting each $p \in \Gamma$, then there are in $EC(T_0, \Gamma)$ models of arbitrarily large cardinality.

(2) We define $\delta(\lambda, \kappa)$ as the first ordinal α such that: if $|L_0| \le \lambda$, $< \in L_0$, T_0 a theory in L_0, Γ a set of $(<\aleph_0)$-types in L_0, $|\Gamma| \leqslant \kappa$ and there is a model M in $EC(T_0, \Gamma)$ such that $(|M|, <^M)$ has order type $\ge \alpha$, then there is $M \in EC(T_0, \Gamma)$ which is not well-ordered by $<^M$.

Remark: If no such $\mu(\alpha)$ exists we let $\mu(\lambda, \kappa) = \infty$ $(\delta(\lambda, \kappa) = \infty)$. (3)$\mu(\lambda) = \mu(\lambda, 1)$, $\delta(\lambda) = \delta(\lambda, 1)$.

Remark. The difference between the two definitions ("for every $\chi < \mu$" but not "for every $\delta < \delta(\lambda, \kappa)$") is inessential, see 5.2.

LEMMA 5.1: (1) *If in* $EC(T_0, \Gamma)$ *there is a model of cardinality* λ, $|T_0| \le \mu \le \lambda$, *then in* $EC(T_0, \Gamma)$ *there is a model of cardinality* μ.

(2) *If* $\lambda \le \lambda_1, \kappa \le \kappa_1$, *then* $\mu(\lambda, \kappa) \le \mu(\lambda_1, \kappa_1)$, *and* $\delta(\lambda, \kappa) \le \delta(\lambda_1, \kappa_1)$.

(3) $\delta(\lambda, 1) \ge \lambda^+$.

(4) $\delta(\lambda, 1) = \delta(\lambda, \lambda)$.

(5) $\delta(\lambda, 0) = \omega$.

Proof. (1) Because $N \prec M$, $M \in EC(T_0, \Gamma)$ implies $N \in EC(T_0, \Gamma)$ using the downward Löwenheim–Skolem theorem I, 1.4.

(2) Immediate.

(3) For $\alpha < \lambda^+$, let M_α be $\langle \alpha, <, \ldots, i, \ldots \rangle_{i<\alpha}$, $T_\alpha = \mathrm{Th}(M_\alpha)$, $p = \{x \ne i : i < \alpha\}$. Then every model in $EC(T_\alpha; \{p_\alpha\})$ is well ordered and has order type α as it is isomorphic to M_α. So $\delta(\lambda, 1) > \alpha$; and as this holds for every $\alpha < \lambda^+$, $\delta(\lambda, 1) \ge \lambda^+$.

(4) By (2) $\delta(\lambda, 1) \le \delta(\lambda, \lambda)$. To prove the converse, assume $|L_0| \le \lambda$, T_0 a theory in L_0, Γ a set of $(<\aleph_0)$-types in L_0, $|\Gamma| \le \lambda$, $M \in EC(T_0, \Gamma)$ has order type $\delta(\lambda, 1)$ and we shall prove that there is a non-well-ordered model in $EC(T_0, \Gamma)$.

Let $\Gamma = \{p_\alpha : \alpha < \alpha_0 \le \lambda\}$, $p_\alpha = \{\varphi_i^\alpha(\bar{x}_\alpha) : i < i_\alpha \le \lambda\}$, p_α an m_α-type. Let $c_i^\alpha (\alpha < \alpha_0, i < i_\alpha)$ be new individual constants, Q, P_α $(\alpha < \alpha_0)$

new one place predicates, and for $\alpha < \alpha_0$, F_α a new m_α-place function-symbol. Let $L_1 = L_0 \cup \{Q\} \cup \{P_\alpha, F_\alpha: \alpha < \alpha_0\} \cup \{c_i^\alpha: i < i_\alpha, \alpha < \alpha_0\}$,

$$T_1 = T_0 \cup \{Q(c_i^\alpha), P_\alpha(c_i^\alpha): i < i_\alpha, \alpha < \alpha_0\} \cup \{(\forall \bar{x}_\alpha)P_\alpha(F_\alpha(\bar{x}_\alpha)): \alpha < \alpha_0\}$$
$$\cup \{(\forall \bar{x}_\alpha)[F_\alpha(\bar{x}_\alpha) = c_i^\alpha \rightarrow \neg\varphi_i^\alpha(\bar{x}_\alpha)]: i < i_\alpha, \alpha < \alpha_0\}$$
$$\cup \{(\forall x)(P_\alpha(x) \rightarrow Q(x)): \alpha < \alpha_0\}$$
$$\cup \{(\forall x)[P_\alpha(x) \rightarrow \neg P_\beta(x)]: \alpha \neq \beta < \alpha_0\},$$
$$p = \{Q(x) \wedge x \neq c_i^\alpha: i < i_\alpha, \alpha < \alpha_0\}.$$

Clearly every model in $\mathrm{EC}(T_0, \Gamma)$ of cardinality $\geq \lambda$ can be expanded to a model of $\mathrm{EC}(T_1, \{p\})$, and the L_0-reduct of every model of $\mathrm{EC}(T_1, \{p\})$ belongs to $\mathrm{EC}(T_0, \Gamma)$. So we can expand M to $M' \in \mathrm{EC}(T_1, \{p\})$; and then as T_1, p are in L_1, $|L_1| \leq \lambda$, and M' has order type $\delta(\lambda, 1)$, there is $N' \equiv M'$, $N' \in \mathrm{EC}(T_1, \{p\})$ which is not well-ordered. The L_0-reduct of N', N, is the desired model.

(5) If in (3) we take $\alpha < \omega$, then every model in $\mathrm{EC}(T_\alpha, \emptyset)$ is isomorphic to M_α, hence $\delta(\lambda, 0) > \alpha$; so $\delta(\lambda, 0) \geq \omega$. By the compactness theorem if for every $n < \omega$ T_0 has a model M_n of order tyoe $\geq n$, then $T_0 \cup \{c_{n+1} < c_n: n < \omega\}$ has a model, so T_0 has a non-well-ordered model.

LEMMA 5.2: *Suppose P, $< \in L_0$(P—a one place predicate, $<$ a two place predicate) $|L_0| \leq \lambda$, T_0 a theory in L_0, Γ a set of $\leq \kappa$ ($< \aleph_0$)-types. If for every $\alpha < \delta = \delta(\lambda, \kappa)$ there is in $\mathrm{EC}(T_0, \Gamma)$ a model M_α such that $(P^{M_\alpha}, <^{M_\alpha})$ has order-type $\geq \alpha$ then in $\mathrm{EC}(T_0, \Gamma)$ there is a model M in which $<^M$ does not well-order P^M.*

Proof. For $\kappa = 0$ this follows by the proof of 5.1(5), so let $\kappa > 0$. Let F be the function such that $F(\alpha) = M_\alpha$ for $\alpha < \delta(\lambda, \kappa)$, w.l.o.g. the $|M_\alpha|$ are pairwise disjoint and F_1 be the function such that for each $\alpha < \delta(\lambda, \kappa)$ $F_1(\alpha, x)$ is an isomorphism from $(\alpha, <)$ into $(P^{M_\alpha}, <^{M_\alpha})$. Let $\mathfrak{B}$ be an elementary submodel of $\langle R(\bar{\kappa}), \in, F, F_1, \delta, T_0, \ldots, \varphi, \ldots \rangle_{\varphi \in L_0}$ of cardinality $|\delta|$, $\{i: i < \delta\} = \delta \subseteq |\mathfrak{B}|$, and let G be a one-to-one function from δ onto $|\mathfrak{B}|$, and $<^*$ an order on $|\mathfrak{B}|$ such that G is an isomorphism from $\langle \delta, < \rangle$ onto $\langle |\mathfrak{B}|, <^* \rangle$. Let $\mathfrak{B}_1 = (\mathfrak{B}, <^*, G)$ and $T_1 = \mathrm{Th}(\mathfrak{B}_1)$. For $\varphi = \varphi(\bar{x}) \in L_0$ let $\varphi^1 = \varphi^1(y, \bar{x})$ be an $L(\mathfrak{B}_1)$-formula saying that y is $M_\alpha = F(\alpha)$ for some $\alpha < \delta$ and $\bar{x}$ is from $|M_\alpha|$ and $\varphi(\bar{x})$ is satisfied in M_α. For $p \in \Gamma$, $p = p(\bar{x})$ let $p^1 = \{\varphi^1: \varphi \in p\}$, and $\Gamma^1 = \{p^1: p \in \Gamma\} \cup \{\{x \in L_0 \wedge x \neq \varphi: \varphi \in L_0\}\}$. So in $\mathrm{EC}(T_1, \Gamma^1)$ there is a model $(\mathfrak{B}_1)$ such that $<^*$ orders it in order-type

$\delta = \delta(\lambda, \kappa)$, $|L_1| \le \lambda$, $|\Gamma^1| \le \kappa$ (or if $\kappa < \lambda$, $|\Gamma^1| \le \lambda$, so by 5.1(4) $\delta(\lambda, |\Gamma_1|) = \delta(\lambda, \kappa)$) hence there is $\mathfrak{B}^* \in \mathrm{EC}(T_1, \Gamma^1)$ which is not well-ordered by $<^*$. Hence $\mathrm{ord}(\mathfrak{B}^*) = \{a \in |\mathfrak{B}^*|: \mathfrak{B}^* \vDash$ "a is an ordinal $< \delta$"$\}$ is not well ordered by $<$ (use G). Choose $a \in \mathrm{ord}\, B^*$ so that $\{b \in \mathrm{ord}\, \mathfrak{B}^*: b < a\}$ is not well ordered. Then $F(a)$ is, in fact, the required model.

THEOREM 5.3: *Suppose* $|L_0| \le \lambda$, T_0 *a theory in* L_0, Γ *a set of* $\le \kappa$ $(< \aleph_0)$-*types in* L_0 *and* P *a one-place predicate in* L_0. *If for every* $\alpha < \delta(\lambda, \kappa)$ *there is* $M_\alpha \in \mathrm{EC}(T_0, \Gamma)$, $\|M_\alpha\| \ge \beth_\alpha(|P^{M_\alpha}|)$ *then for every* $\mu \ge |L_0|$, *there is* $M \in \mathrm{EC}(T_0, \Gamma)$, $\|M\| = \mu$, $|P^M| = |L_0|$.

Remark. Notice that we can replace $\beth_\alpha(|P^{M_\alpha}|)$ by $\beth_\alpha(|P^{M_\alpha}| + \lambda)$ for $\kappa > 0$, because $\beth_\lambda(0) \ge \lambda$, so for $\alpha \ge \lambda + \omega$ they are equal.

Proof. We proceed as in the proof of 5.2. Let F be a function from $\delta = \delta(\lambda, \kappa)$, $F(\alpha) = M_\alpha$ and let $\mathfrak{B}_1 = \langle R(\bar{\kappa}), \in, F, \delta, T_0, \ldots, \varphi, \ldots \rangle_{\varphi \in L_0}$.

So again we can find $\mathfrak{B}$ elementarily equivalent to $\mathfrak{B}_1$ such that in $\mathfrak{B}$ there are "ordinals" $\alpha_n < \delta$, $\alpha_{n+1} < \alpha_n$ and for each $\alpha < \delta$, $F^{\mathfrak{B}}(\alpha)$ is a model of T_0 omitting each $p \in \Gamma$. Now w.l.o.g. we can assume that T_0 has Skolem functions and $\mathfrak{B} \vDash$ "$\alpha_{n+1} + n + 1 < \alpha_n$". Let $<$ be, in $\mathfrak{B}$, a well ordering of $F(\alpha_0)$. Now define by induction on n, elements X_n of $\mathfrak{B}$ such that $\mathfrak{B} \vDash$ "X_n is an increasing sequence of elements from the universe of $F(\alpha_0)$, it has cardinality $\beth_{\alpha_n}(|P(F(\alpha_0))|)$, and is n-indiscernible over $P(F(\alpha_0))$" and $\mathfrak{B} \vDash$ "X_{n+1} is a subset of X_n".

This is possible by the Erdös-Rado Theorem (2.5 of the Appendix) (more exactly—as $\mathfrak{B}$ satisfies a first-order sentence saying it).

Now we define Φ proper for (ω, T_0) so that if $\varphi(x_1, \ldots, x_n) \in \Phi$, then if $\mathfrak{B} \vDash$ "$\bigwedge_{i=1}^{n} a_i \in X_n \wedge \bigwedge_{i=1}^{n-1} a_i < a_{i+1}$" then $\mathfrak{B} \vDash$ "$\varphi(a_1, \ldots, a_n)$ is satisfied by $F(\alpha_0)$". Clearly there is such Φ, and $EM(\lambda, \Phi)$ is the required model. (It does not realize types omitted by $F(\alpha_0)$ because of Lemma 2.2; hence, it omits the set of types Γ.)

THEOREM 5.4: $\mu(\lambda, \kappa) = \beth_{\delta(\lambda,\kappa)}$ for $\kappa > 0$; and $\mu(\lambda, 0) = \aleph_0$, $\delta(\lambda, 0) = \omega$.

Proof. For $\kappa = 0$ this is by 5.1(5) and the compactness theorem. So let $\kappa > 0$. By the last theorem it is easy to see that $\mu(\lambda, \kappa) \le \beth_{\delta(\lambda,\kappa)}$. For the other direction assume that every model in $\mathrm{EC}(T_0, \Gamma)$ is well ordered by $<$; T_0, Γ are in L_0, $|L_0| = \lambda$, $|\Gamma| \le \kappa$. Now let Q be a new one place predicate, for every ψ let ψ^Q be ψ relativized to Q, $T_0^Q = \{\psi^Q: \psi \in T_0\}$, $p^Q = \{\psi^Q: \psi \in p\}$, $\Gamma^Q = \{p^Q: p \in \Gamma\}$.

Let $\in$ be a new two place relation, P a new one place relation, and F a new one place function symbol, and $c_i, i < \aleph_0$ new constant symbols. Let

$$T_1 = T_0^Q \cup \{(\forall x)[(\exists y) y \in x \equiv \neg P(x)],$$
$$(\forall x, y)[\neg P(x) \wedge \neg P(y) \wedge (\forall z)(z \in x \equiv z \in y) \rightarrow x = y],$$
$$(\forall x)(Q(F(x)), (\forall x, y)(x \in y \rightarrow F(x) < F(y))\} \cup \{P(c_i): i < \aleph_0\},$$
$$\Gamma_1 = \Gamma^Q \cup \{\{P(x) \wedge x \neq c_i: i < \aleph_0\}\}.$$

Clearly if $M \in \mathrm{EC}(T_1, \Gamma_1)$ then its submodel with universe Q^M belongs to $\mathrm{EC}(T_0, \Gamma)$ and if the order type of the latter is α, then $\|M\| \leq \beth_\alpha$. Conversely, if some $N \in \mathrm{EC}(T_0, \Gamma)$ has order type $\geq \alpha + 1$, then some $M \in \mathrm{EC}(T_1, \Gamma_1)$ has cardinality $\beth_\alpha + |T_1|$.

THEOREM 5.5: *Let* $\kappa > 0$.

(1) $\delta(\lambda, \kappa)$ *is a limit ordinal of cofinality* $> \lambda$.

(2) $\lambda^+ \leq \delta(\lambda, \kappa) \leq (2^\lambda)^+$.

(3) *If* $\mathrm{cf}\,\lambda > \aleph_0$ *then* $\delta(\lambda, 1) > \lambda^+$.

(4) *If* $\mathrm{cf}\,\lambda = \aleph_0$, $\mu = \sum_{\chi<\lambda} 2^\chi$ *then* $\delta(\lambda, 1) \leq \mu^+$.

(5) *If* $\lambda = \aleph_0$ *or* λ *is a strong limit cardinal of cofinality* $\aleph_0$, *then* $\delta(\lambda, 1) = \lambda^+$.

(6) $\delta(\lambda, 2^\lambda) = (2^\lambda)^+$.

(7) $\delta(\lambda, 1) < (2^\lambda)^+$ *(and* $\delta(\lambda, \kappa) < (2^\lambda)^+$ *when* $2^\kappa \leqslant 2^\lambda$*)*.

(8) *For a strong limit cardinal of cofinality* $> \aleph_0$, $2^\lambda < \delta(\lambda)$*; and generally* of $\lambda > \aleph_0$, $(\forall \mu < \lambda)(\mu^{\mathrm{cf}\,\lambda} < \lambda)$ *implies* $\delta(\lambda, 1) > \lambda^{\mathrm{cf}\,\lambda}$.

Proof. (1) Easy, left as an exercise to the reader.

(2) We have already proved that $\delta(\lambda, \kappa) \geq \lambda^+$ (in 5.1(3)). For the other inequality let $|L_0| \leq \lambda$, T_0 be in L_0, Γ a set of types in L_0, $|\Gamma| \leq \kappa$ and suppose $M \in \mathrm{EC}(T_0, \Gamma)$ has order type $\geq (2^\lambda)^+$ (ordered by $<$). We can assume $\mathrm{Th}(M)$ has Skolem functions, and let $a_i \in |M|$, $i < (2^\lambda)^+$, be such that for $i < j$, $M \models a_i < a_j$. Now we define by induction on $n < \omega$, sets $S_n \subseteq (2^\lambda)^+$, $|S_n| = (2^\lambda)^+$, and for each $i \in S_n$, ordinals $\alpha_i(n, 0), \ldots, \alpha_i(n, n-1)$ for which $i < \alpha_i(n, n-1) < \cdots < \alpha_i(n, 0)$ such that: if $i \in S_n$, $j \in S_m$, $n \leq m$ then $\mathrm{tp}(\langle a_{\alpha_i(n,0)}, \ldots, a_{\alpha_i(n,n-1)}\rangle, \emptyset, M) = \mathrm{tp}(\langle a_{\alpha_j(m,0)}, \ldots, a_{\alpha_j(m,n-1)}\rangle, \emptyset, M)$.

For $n = 0$ let $S_0 = (2^\lambda)^+$, (we have no $\alpha_i(0, l)$ to define). If we have defined for n; for each $j < (2^\lambda)^+$ we choose an i, $j < i \in S_n$ and let $\alpha_j(n+1, l) = \alpha_i(n, l)$ for $0 \leq l < n$ and $\alpha_j(n+1, n) = i$. The number of possible types $p_j^{n+1} = \text{tp}(\langle a_{\alpha_j(n+1,0)}, \ldots, a_{\alpha_j(n+1,n)}\rangle, \emptyset, M)$ is $\leq 2^{|L_0|} \leq 2^\lambda < (2^\lambda)^+$ whereas the number of j's is $(2^\lambda)^+$. So there is a type p^{n+1} such that $S_{n+1} = \{j < (2^\lambda)^+ : p_{j_i}^{n+1} = p^{n+1}\}$ has cardinality $(2^\lambda)^+$.

Now by the compactness theorem and our construction there is a model N_1 of $\text{Th}(M)$, and in it elements a^n, $n < \omega$, such that $\text{tp}(\langle a^0, \ldots, a^n\rangle, \emptyset, N_1) = p^{n+1}$. As $\text{Th}(M)$ has Skolem functions, we can assume that $|N_1|$ is the closure of $\{a^n : n < \omega\}$. So for every $\bar{b} \in |N_1|$ there are n and τ so that $N_1 \models \bar{b} = \bar{\tau}(a^0, \ldots, a^n)$ so if $i \in S_{n+1}$, $\bar{b}' = \bar{\tau}(a_{\alpha_i(n,0)}, \ldots)$ (in M) then $\text{tp}(\bar{b}, \emptyset, N_1) = \text{tp}(\bar{b}', \emptyset, M)$ so N_1 omits every $p \in \Gamma$ so $N_1 \in \text{EC}(T_0, \Gamma)$.

As for each $i \in S_n$, $\alpha_i(n, 0) > \alpha_i(n, 1) > \cdots$ clearly $N_1 \models a_n > a_{n+1}$, so N_1 is not well ordered. All this proves that $\delta(\lambda, \kappa) \leq (2^\lambda)^+$.

(3) Let $\mathfrak{B}_\lambda = \langle R(\bar{\kappa}), \in, <, \lambda^+, \lambda, i\rangle_{i<\lambda}$ ($<$—the order between the ordinals $\leq \lambda^+$) and let $p = \{x < \lambda \wedge x \neq i : i < \lambda\}$, and $\Gamma = \{p\}$ and $T = \text{Th}(\mathfrak{B}_\lambda)$. Suppose $\mathfrak{B} \in \text{EC}(T, \Gamma)$, and we shall prove that $\{x : \mathfrak{B} \models x < \lambda^+\}$ is well ordered, thus by 5.2 proving that $\delta(\lambda, 1) > \lambda^+$. For suppose $\mathfrak{B} \models x_{n+1} < x_n < \lambda^+$. Clearly in $\mathfrak{B}$ there is an element f such that $\mathfrak{B} \models$ "f is a one-to-one mapping from $\{x : x < x_0\}$ onto $\{x : x < \lambda\}$".

Let $\mathfrak{B} \models$ "$f(x_n) = a_n$". As cf $\lambda > \aleph_0$, $\mathfrak{B}$ omits p, and $\{a_n : n < \omega\}$ is countable it is bounded by some $a^* < \lambda$. As $\mathfrak{B} \models$ "the set of $x < \lambda^+$ such that $f(x) < a^*$ has cardinality $< \lambda$" there is g in $\mathfrak{B}$ such that $\mathfrak{B} \models$ "g is an order-preserving function from $\{x < x_0 : f(x) < a^*\}$ into λ" so $\mathfrak{B} \models$ "$g(x_{n+1}) < g(x_n) < \lambda$" contradiction to the omitting of p by $\mathfrak{B}$ (as λ is well-ordered).

(4) Let $M \in \text{EC}(T_0, \Gamma)$ have order type $\geq \mu^+$, and $\Gamma = \{p\}$, $p = \{\varphi_i(\bar{x}) : i < i_0 \leq \lambda\}$, $l(\bar{x}) = m$. Let $\bar{\tau}_\alpha^n(x_0, \ldots, x_{n-1})$, $\alpha < \lambda$ be a list of all terms from L_0 where $l(\bar{\tau}_\alpha^n) = m$. Let for $\bar{a} \in {}^m|M|$, $G(\bar{a}) = \min\{i : M \models \neg\varphi_i(\bar{a})\}$. For $\lambda > \aleph_0$ as cf $\lambda = \aleph_0$ let $\lambda = \sum_n \lambda_n$, $\lambda_n < \lambda_{n+1} < \lambda$, and let $H(\bar{a}) = \min\{n : G(\bar{a}) < \lambda_n\}$.

The rest of the proof is just like that of (2), only μ^+ replaces $(2^\lambda)^+$, and instead of $\text{tp}(\langle a_{\alpha_i(n,0)}, \ldots, a_{\alpha_i(n,n-1)}\rangle, \emptyset, M)$ we use when $\lambda = \aleph_0$

$$\{\langle l, \bar{\tau}_\alpha^l, G(\bar{\tau}_\alpha^l(a_{\alpha_i(n,0)}, \ldots, a_{\alpha_i(n,l-1)}))\rangle : l \leq n, \alpha \leq n\}$$

and when $\lambda > \aleph_0$,

$$\{\langle l, \bar{\tau}_\alpha^l, H(\bar{\tau}_\alpha^l(a_{\alpha_i(n,0)}, \ldots, a_{\alpha_i(n,l-1)}))\rangle : l \leq n, \alpha \leq \lambda_n\}$$

$$\cup \{\langle l, \bar{\tau}_\alpha^l, G(\bar{\tau}_\alpha^l(a_{\alpha_i(n,0)}, \ldots, a_{\alpha_i(n,l-1)}))\rangle : l \leq n, \alpha \leq \lambda_n$$

$$\text{and } H(\bar{\tau}_\alpha^l(a_{\alpha_i(n,0)}, \ldots)) \leqslant n\}.$$

At the end we choose N_1 as a model of

$$\begin{aligned}\mathrm{Th}(M) &\cup \{a^{n+1} < a^n : n < \omega\}\\ &\cup \{\neg\varphi_\beta(\bar{\tau}^n_\alpha(a^0, \ldots, a^{n-1})) : \beta = G(\bar{\tau}^n_\alpha(a_{\alpha_i(n,0)}, \ldots))\\ &\qquad \text{for every } i \in S_n,\ n \text{ big enough}\}.\end{aligned}$$

(5) It is immediate by parts (2) and (4).

(6) By (2) it suffices to prove $\delta(\lambda, 2^\lambda) \geq (2^\lambda)^+$, or that for every $\alpha < (2^\lambda)^+$, $\delta(\lambda, 2^\lambda) \geq \alpha$. So let $\alpha < (2^\lambda)^+$ and choose for each $\beta < \alpha$ a subset S_β of λ so that $\beta \neq \gamma \Rightarrow S_\beta \neq S_\gamma$. Let for every $\beta < \gamma < \alpha$ $p_{\beta,\gamma} = \{\neg(x < y)\} \cup \{P_i(x)^{\mathrm{if}(i \in S_\beta)} : i < \lambda\} \cup \{P_i(y)^{\mathrm{if}(i \in S_\gamma)} : i < \lambda\}$ and for every $S \subseteq \lambda$ $p_S = \{P_i(x)^{\mathrm{if}(i \in S)} : i < \lambda\}$. Let $\mathrm{L}_0 = \{P_i : i < \lambda\} \cup \{<, =\}$, T_0 just "says" $<$ is a linear order,

$$\begin{aligned}\Gamma = \{p_{\beta,\gamma} : \beta < \gamma < \alpha\} &\cup \{p_S : S \subseteq \lambda, S \neq S_\beta \text{ for each } \beta < \alpha\}\\ &\cup \{x \neq y \wedge P_i(x) \equiv P_i(y) : i < \lambda\}.\end{aligned}$$

Clearly there is $M_0 \in \mathrm{EC}(T_0, \Gamma)$ of order type α ($|M| = \alpha$, $\beta \in P_i$ iff $i \in S_\beta$, $<$ is the usual order) and every model in $\mathrm{EC}(T_0, \Gamma)$ is a submodel of M_0 (up to isomorphism) hence is well ordered.

(7) Immediate by cardinality considerations (there are essentially only 2^λ pairs $(T_0, \{p\})$ and similarly for $\delta(\lambda, \kappa)$).

(8) Let $\kappa = \mathrm{cf}\,\lambda$, and D be any $\aleph_1$-complete filter over κ. For any function f from κ to λ (in fact, to the ordinals) we define its D-rank $R_D(f)$: $R_D(f) \geq 0$ always, $R_D(f) \geq \delta$ if for every $\alpha < \delta$, $R_D(f) \geq \alpha$, and $R_D(f) \geq \alpha + 1$ if there is $g : \kappa \to \lambda$. $g/D < f/D$ (i.e., $\{\alpha < \kappa : g(\alpha) < f(\alpha)\} \in D$) and $R_D(g) \geq \alpha$. Now $R_D(f) = \alpha$ if $R_D(f) \geq \alpha$ but not $R_D(f) \geq \alpha + 1$; as D is $\aleph_1$-complete, there is no descending sequence f_n/D ($n < \omega$) so $R_D(f)$ is always an ordinal. Let

$$\alpha(\lambda) = \sup\{R_D(f) : f : \kappa \to \lambda\}.$$

Let us define now a model M': its universe is the disjoint union $P^M \cup Q^M \cup Q_1^M$ where ignoring trivialities:

$P^M = \alpha(\lambda)$,
$Q^M = \{f : f \text{ a function from } \kappa \text{ to } \lambda\}$,
$Q_1^M = \{A : A \subseteq \kappa\}$,
$<^M$ = the order on ordinals,
F = a partial function such that $F(f, i) = f(i)$ for $i < \kappa$, $f \in {}^\kappa\lambda$,
i ($i \leq \lambda$) = an individual constant,
A ($A \subseteq \kappa$) = an individual constant,

$\in^M = \{(i, A): i < \kappa, A \subseteq \kappa, i \in A\}$,
$Q_2^M = D$,
$G =$ a partial function such that $G(f) = R_D(f) \in P^M$.

Let $T = \mathrm{Th}(M)$ and $p = \{x < \lambda \wedge x \neq i: i < \lambda\}$ and let N be a model of T omitting p; so by renaming we can assume the interpretation of i $(i \leq \lambda)$ is i itself and any member of Q^N is a function from κ to λ, and $F(f, i) = f(i)$, and

$$Q_1^N = Q_1^M, \quad \in^N = \{(i, A): i < \kappa, A \subseteq \kappa, A \in Q_1^N, i, \in A\},$$

so $Q_2^N = D$. Now we prove P^M is well ordered. If $c_n > c_{n+1}$ $(n < \omega)$ is a counterexample, then there is $f_0 \in Q^M$, $G(f) = c_0$. (As $(\forall x \in P)(\exists y \in Q)(G(y) = x) \in T$, as $P^M = \alpha(\lambda)$.) Similarly, by the rank's definition, we can define $f_n \in Q^M$, such that $G(f_n) = c_n$, $\{i < \kappa: f_n(i) > f_{n+1}(i)\} \in Q_2^N = D$; and we get a contradiction. So in every model $N \in EC(T, \{p\})$, P^N is well-ordered by $<^N$. By 5.2 and T's definition this implies $\alpha(\lambda) < \delta(\lambda)$.

Up to now we have not used any assumption on λ, κ except $\kappa \leq \lambda$, cf $\kappa > \aleph_0$, $2^\kappa < \lambda$; and if D is easily defined (e.g., the filter of closed unbounded subsets of κ, or D_κ^{Ub} the filter of co-bounded subsets of κ) the last restriction is not necessary.

Now we prove $\alpha(\lambda) \geq \lambda^{\mathrm{cf}\,\lambda}$, e.g., for $D = D_\kappa^{\mathrm{Ub}}$ thus finishing. First note that there are $\lambda^{\mathrm{cf}\,\lambda}$ functions $f_i : \kappa \to \lambda$ such that $(\forall i \neq j)(\exists \alpha < \kappa)(\forall \beta)(\alpha < \beta < \kappa \to f_i(\beta) \neq f_j(\beta))$ (let $\lambda = \sum_{i<\kappa} \lambda_i$, $\lambda_i < \lambda$, and let for $i < \kappa$ h_α be a one-to-one function from $\prod_{j<\alpha} \lambda_j$ into λ (exists as $(\forall \mu < \lambda)(\forall \chi < \mathrm{cf}\,\lambda)(\mu^\chi < \lambda)$). Let $\{f_i^1; i < \lambda^{\mathrm{cf}\,\lambda}\}$ be a list of all functions from κ to λ, and define $f_i(\alpha) = h_\alpha(\langle f_i^1(\gamma): \gamma < \alpha\rangle)$. The f_i's are as necessary. If $\alpha(\lambda) < \lambda^\kappa$, there are $(2^\kappa)^+$ f_i's with the same D-rank, say $\{f_i: i < (2^\kappa)^+\}$. For any $i < (2^\kappa)^+$, cf $i = \kappa$, and $\alpha < \kappa^+$, we define by induction on $n < \omega$, $\xi_n(i, \alpha)$ such that

(i) $\xi_n(i, \alpha) < \xi_{n+1}(i, \alpha) < i$,
(ii) for $\xi = \xi_n(i, \alpha)$, $f_\xi(\alpha) > f_i(\alpha)$,
(iii) for any $m < n$, letting $\zeta = \xi_m(i, \alpha)$, $\xi = \xi_n(i, \alpha)$

$$f_\zeta(\alpha) > f_i(\alpha) \Leftrightarrow f_\zeta(\alpha) > f_\xi(\alpha).$$

By (iii) and (ii), $f_{\xi_n(i,\alpha)}(\alpha)(n < \omega)$ is strictly decreasing hence for some $n = n(i, \alpha)$ $\xi_n(i, \alpha)$ is not defined. Let $S_\alpha^i = \{\xi_n(i, \alpha): n < n(i, \alpha)\}$, $S^i = \bigcup_{\alpha<\kappa} S_\alpha^i$, so $|S^i| \leq \kappa$, $S^i \subseteq i$. As cf $i = \kappa^+$ some $h(i) < i$ bounds it, so by 1.3 of the Appendix, for some γ $\{i < (2^\kappa)^+: h(i) = \gamma\}$ is stationary and also for some $S: W = \{i < (2^\kappa)^+ : h(i) = \gamma, S^i = S\}$ is stationary. We can find $j < i$, $j, i \in W$ such that for every $\xi \in S$, $\beta < \kappa$,

$$f_\xi(\beta) > f_i(\beta) \Leftrightarrow f_\xi(\beta) > f_j(\beta)$$

If for some α, $f_j(\alpha) > f_i(\alpha)$, j will satisfy the conditions on $\xi_{n(i,\alpha)}(i, \alpha)$, contradiction. So for every β $f_j(\beta) \le f_i(\beta)$; but $|\{\beta: f_j(\beta) = f_i(\beta)\}| < \kappa$ hence $\{\beta < \kappa: f_j(\beta) < f_i(\beta)\} \in D(= D_\kappa^{\text{ub}})$. But this implies $f_j/D < f_i/D$ hence $R_D(f_j) < R_D(f_i)$, contradiction.

Remark. Barwise and Kunen [BK 71] have shown that it is consistent with ZFC that:

(1) $\lambda^+ < 2^\lambda$, $\delta(\lambda, 1) < \lambda^{++}$,
(2) $\lambda^+ < 2^\lambda < \delta(\lambda, 1)$.

EXERCISE 5.1: Suppose $|L_0| \le \lambda$, T_0 a theory in L_0, Γ a set of $(<\aleph_0)$-types in L_0, $|\Gamma| \le \kappa$. Show that if for all $\alpha < \delta(\lambda, \kappa)$ there is a model M_α in $EC(T_0, \Gamma)$ such that $|P^{M_\alpha}| \ge \beth_\alpha$ and $\|M_\alpha\| \ge \beth(|P^{M_\alpha}|, \alpha)$ then for all $\lambda \ge \mu \ge |T_0|$ there is a model M in $EC(T_0, \Gamma)$ such that $\|M\| = \lambda$, $|P^M| = \mu$. If we omit "$|P^{M_\alpha}| \ge \beth_\alpha$" we can still get $|P^M| \le |T|$.

EXERCISE 5.2: Show that if in the definition of $\mu(\aleph_0, 1)$ we require T_0 to be a complete theory and $p(\Gamma = \{p\})$ to be a complete type the value of $\mu(\aleph_0, 1)$ is unaltered.

CHAPTER VIII

THE NUMBER OF NON-ISOMORPHIC MODELS IN PSEUDO-ELEMENTARY CLASSES

VIII.0. Introduction

The essence of this chapter is the construction of many non-isomorphic models in a pseudo-elementary class $\mathrm{PC}(T_1, T)$ and we shall use methods developed in VII; the proofs have a combinatorial flavor.

At the beginning of Section 1 we make two important observations (Lemmas 1.3 and 1.4).

(1) If after adding κ individual constants to T, there are μ many $(> \lambda^\kappa)$ non-isomorphic models of cardinality λ in the pseudo-elementary class, then the adding does not change this number.

(2) If we have $\lambda \geq |T_1|$ types in $D(T)$ which are independent (i.e., for any subset of them, there is a model realizing them but not the others) then $I(\lambda, T_1, T) = 2^\lambda$ (see Definition 1.1).

Now we can show that some properties of T imply the existence of large families of independent types, sometimes over a set A (1.5) and then by (1) and (2) we prove that $I(\lambda, T_1, T)$ is big (1.7). A more difficult theorem is 1.8 (the main case is $2^\lambda = 2^{\aleph_0}$, $|D(T)| = \aleph_0$).

THEOREM 0.1: *If T is countable, not $\aleph_0$-stable, $|T_1| = \aleph_0$, $\aleph_0 < \lambda \leq 2^{\aleph_0}$, then $I(\lambda, T_1, T) = 2^\lambda$.*

If instead of $|T_1| = \aleph_0$ we demand $|T_1| < 2^{\aleph_0}$ we still get a result when $|T_1| \leq \lambda < 2^{\aleph_0}$:

(1) *If λ satisfies the combinatorial condition* (*) ("*not* $\mathrm{AD}(2^{\aleph_0}, \lambda, \lambda, \aleph_0)$"), *then $I(\lambda, T_1, T) \geqslant 2^{\aleph_0}$ (see* 1.9).

(2) *If Martin Axiom (see* [MS 70]) *holds, $I(\lambda, T_1, T) \geqslant 2^\lambda$ (see Exercise* 2.6).

More interesting is the fact that we can generalize the theorem to higher cardinals, but the hypothesis is quite strong:

THEOREM 0.2: *Suppose*

(i) *T has λ independent formulas,*

(ii) $T \subseteq T_1$, $\lambda = |T_1| = |T|$,

(iii) *There is a λ-Kurepa tree with χ branches (e.g., λ strong limit, or* $\operatorname{cf} \lambda = \aleph_0$, $\chi = \lambda^{\aleph_0}$),

Then for $\mu \geq \lambda$, $I(\mu, T_1, T) \geq \min\{2^\chi, 2^\mu\}$.

It would be nicer to get such a theorem assuming only $|D(T)| = \chi > \lambda = |T_1|$.

The main aim of Sections 2 and 3 is

THEOREM 0.3: *Any unsuperstable T, has 2^λ non-isomorphic models of cardinality λ, for any $\lambda \geq |T| + \aleph_1$.*

We prove it also for pseudo-elementary classes in many cases (e.g., countable T, T_1). The proof is by cases.

The easiest case is:

THEOREM 0.4: *If T is unsuperstable, λ is regular and $> |T_1|$, then in* $\mathrm{PC}(T_1, T)$ *there is a family of 2^λ models of cardinality λ, no one elementarily embeddable in another.*

In the proof note that if $M = \bigcup_{i<\lambda} M_i^l$ $(l = 0, 1)$, M_i^l $(i < \lambda)$ inincreasing and continuous, λ regular $\|M\| = \lambda > \|M_i^l\|$ then $\{i: M_i^0 = M_i^1\}$ is a closed unbounded subset of λ. Hence if P is a property of pairs of models then $\{i < \lambda: (M, M_i^0)$ satisfies $P\}$ is determined, mod $D(\lambda)$ by the isomorphism type of M; and in fact we can make P to depend on more information. We use the indiscernible tree of sequences we have constructed in VII, Section 3. Choose for each $\delta < \lambda$, $\operatorname{cf} \delta = \omega$, an increasing sequence of ordinals of length ω converging to it, η_δ, and for $S \subseteq \{\delta < \lambda: \operatorname{cf} \delta = \omega\}$ let $I_S = {}^{\omega >}\lambda \cup \{\eta_\delta: \delta \in S\}$. Now from the isomorphism type of $EM(I_S)$ we can reconstruct $S/D(\lambda)$.

The other cases have each one its specific trick, and the proofs are somewhat complicated.

If $\mu < \lambda \leq \mu^{\aleph_0}$, $2^\mu < 2^\lambda$, the method of Section 1, on independent types, applies. If λ is singular, but no previous case applies we choose a proper $\mu < \lambda \leq 2^\mu$, and make a construction similar to the first one, but λ-fold. If λ is a singular strong limit cardinal, we use games. For an elementary class, T stable, $\lambda = |T|$, use $\mathbf{F}^t_{\aleph_0}$-primary models over indiscernible trees, and repeat the previous cases (all this in Section 2).

Then for $\lambda = |T_1|$, T unstable we note that skeletons of Ehrenfeucht–Mostowski models are $\aleph_0$-skeleton like (Definition 3.1 or below), and if in an insomorphism of one such model onto another, the intersection of one skeleton with the image of another is big, the orders of the skeletons are not contradictory (Definition 3.2). Proving the existence of large families of pairwise contradictory orders, we finish except when λ satisfies a strong set theoretic condition, which implies it is $< 2^{\aleph_0}$ or is high in the sequence of $\aleph_\alpha$'s (this is 3.2, case I). Note that $M^1 = EM^1(I)$ satisfies:

(A) $\langle \bar{a}_s : s \in I \rangle$ is $\aleph_0$-skeleton like, i.e., for every $\bar{b} \in |M^1|$ there is a finite $J \subseteq I$, such that if $s, t \in I - J$, $(\forall v \in J)\ (s < v \equiv t < v)$ then $\bar{b}^\frown \bar{a}_s$, $\bar{b}^\frown \bar{a}_t$ realizes the same type.

(B) Hence if $I = I_1 + I_2$, then for every $\bar{b} \in M_1$, for some $t_l \in I_l$, $\bar{b}^\frown \bar{a}_s$ realizes the same type for all $s \in I$ satisfying $t_1 < s < t_2$.

The case left is T unstable, $|T_1| = \lambda$, AD$(2^\lambda, \lambda, \lambda, \aleph_0)$; we can assume T is countable. For understanding the proof let us concentrate on showing $I(\lambda, T_1, T) > 1$ for λ regular $> \aleph_1$. Let N be a $(\lambda, \aleph_1)$-saturated model (see Definition VII, 1.5), $N = \bigcup_{i<\omega_1} N_i$: and $M = EM(I)$, $I = I_1 + I_2, I_1 \cong \lambda$, $I_2 \cong \omega^*$, and let $\langle \bar{a}_s : s \in I \rangle$ be the skeleton of M (order by $<$, of course) and $f : M \to N$ the isomorphism, and $\bar{b}_s = f(\bar{a}_s)$, so $\langle \bar{b}_s : s \in I \rangle$ is an $\aleph_0$-skeleton like sequence in N. We want to show for some $\bar{b}$, $\models \bar{b} < \bar{b}_s$ for $s \in I_2$, and $\models \bar{b}_s < \bar{b}$ for λ elements $s \in I_1$. As I_2 is countable, $|I_1| = \lambda$, $\mathrm{cf}\,\lambda > \aleph_1$, for some $\zeta < \omega_1$, J: $\bigcup_{s \in I_2} \bar{b}_s \subseteq N_\zeta$, $\bigcup_{s \in J} \bar{b}_s \subseteq N_\zeta$, $J \subseteq I_1$, $|J| = \lambda$. In $\mathfrak{B}_{\zeta+1}$ there is a sequence $\langle A_\alpha : \alpha < \lambda \rangle$, increasing, continuous, $|A_\alpha| < \lambda$, $\bigcup_{\alpha<\lambda} A_\alpha = |N_\zeta|$. We can find $\xi < \lambda$, such that $\bigcup_{s \in I_2} \bar{b}_s \subseteq A_\xi$. As $\langle \bar{b}_s : s \in I \rangle$ is $\aleph_0$-skeleton like all except $< |A_\xi|^+ + \aleph_0$ members of $\{\bar{b}_s : s \in J\}$ realizes the same type over A_ξ, let $\bar{b}_t$ be one of them and $J_1 = \{s \in J : \mathrm{tp}(\bar{a}_s, A_\xi) = \mathrm{tp}(\bar{a}_t, A_\xi)\}$. Now it suffices to prove some consistent type extending $\{\bar{x} < \bar{b}_s : s \in I_2\} \cup \{\bar{b}_s < \bar{x} : s \in J_1\}$ belong to $\mathfrak{B}_{\zeta+1}$. But we can define such a type using N_ζ, A_ξ, and $\bar{b}_t$:

$$p = \{\bar{x} < \bar{b} : \bar{b} \in A_\xi, N_\zeta \models \bar{b}_t < \bar{b}\} \cup \{\bar{b} < \bar{x} : \bar{b} \in |N_\zeta|, \text{ and for every } \bar{b}' \in A_\xi, N_\zeta \models \bar{b}_t < \bar{b}' \text{ implies } N_\zeta \models \bar{b} < \bar{b}'\}$$

(every finite subset of p is realized by all $\bar{a}_s$, $s \in I_2$ except finitely many). In the actual proof we have to take care for getting 2^λ non-isomorphic models, and making them κ-compact, for suitable κ.

In Section 4 the reader can relax; here we mainly use previous results to prove theorems e.g., on categoricity, we prove:

Theorem. If $\mathrm{PC}(T_1, T)$ is categorical in $\lambda > |T_1|$, then T is superstable, without the f.c.p. and stable in $|T_1|$.

We also give a partial solution to the problem: does the categoricity of $\mathrm{PC}(T_1, T)$ in λ imply its categoricity in μ? Show its equivalence to a two-cardinal theorem with omitting a type. We get similar results when we replace "categorical in λ" by "each model in $\mathrm{PC}(T_1, T)$ of cardinality λ is homogeneous".

Then we use the construction of VII, Section 4 to prove results on universality, i.e.:

THEOREM 0.5: *If T is unsuperstable, λ is regular $> 2^{|T_1|}$ then in* $\mathrm{PC}(T_1, T)$ *there are arbitrarily large $(<\lambda)$-universal not λ-universal models.*

On the other hand we shall show that if $\mathrm{cf}\,\lambda = \aleph_0$ the conclusion does not hold.

We also characterize the class of cardinalities in which a theory has saturated models, by using various previous results.

THEOREM 0.6: *T has a saturated model of cardinality λ iff $\lambda = \lambda^{<\lambda} + |D(T)|$ or T is λ-stable.*

PROBLEM 0.1: Though we get a complete result for elementary classes, this is not true for pseudo-elementary classes, when we look for κ-compact models $\kappa < \kappa(T)$, and when we look for 2^λ pairwise non-elementary embeddable models; it will be interesting to complete them.

PROBLEM 0.2: It will be desirable to simplify and unify the proofs. A possible way is in the proof of 2.1 to try to replace the filter of closed unbounded subsets of λ, by a filter on $S_\kappa(\lambda)$ or on the family of increasing sequences of members of $S_\kappa(\lambda)$, of length δ (see Kueker [Kk 77] and Shelah [Sh 75a]).

PROBLEM 0.3: In many cases we have parallel proofs; in 2.7 an $\mathbf{F}^t_{\aleph_0}$-constructible set, replaces the Skolem Hull. Also clearly different proofs use different sets of assumptions on the construction. Maybe an axiomatization is in order.

PROBLEM 0.4: For uncountable T_1, we know little on $I(\lambda, T_1, T)$ when $\mu = |D(T)| > |T_1|$. It is natural to conjecture that for $\lambda > |T_1|$, $I(\lambda, T_1, T) \geqslant \min\{2^\lambda, 2^\mu\}$, but it is reasonable to try also an independence proof.

VIII.1. Independence of types

DEFINITION 1.1: (1) $I(\lambda, T_1, T)$ is the number of non-isomorphic models in $\mathrm{PC}(T_1, T)$ of cardinality λ.

(2) $IE(\lambda, T_1, T)$ is the maximal μ such that there is a family of $M_i \in \mathrm{PC}(T_1, T)$, $i < \mu$, $\|M_i\| = \lambda$, such that there is no elementary embedding of M_i into M_j for $i \neq j$.

(3) If we omit T_1 this means $T_1 = T$.

Remark. The "maximal" in (2) is inaccurate, as there may be none, and we may get, e.g., $\aleph_n < IE(\lambda, T_1, T) < \aleph_\omega$ for any n.

Remember $D(T) = \bigcup_m S^m(\emptyset)$.

LEMMA 1.1: (1) $2^\lambda \geq I(\lambda, T_1, T) \geq IE(\lambda, T_1, T) \geq 1$ *for* $\lambda \geq |T_1|$.

(2) *If* $T \subseteq T' \subseteq T_1' \subseteq T_1$, *then* $I(\lambda, T_1, T) \leq I(\lambda, T_1', T')$ *and* $IE(\lambda, T_1, T) \leq IE(\lambda, T_1', T')$.

Proof. Remember T_1 is consistent, and any model of T_1 is isomorphic to a model with universe $\{i: i < \lambda\}$, and there are only

$$\prod_{R \in L_1} 2^{(\lambda^{m(R)})} \leq (2^\lambda)^\lambda = 2^\lambda,$$

such L_1-models where $m(R)$ is the number of places in R (we consider here m-place functions as $(m+1)$-place relations). The rest is even more trivial.

Remark. We will be mainly interested in I and not in IE.

THEOREM 1.2: (1) *If* $|D(T)| > |T_1|$, $\lambda \geq |T_1|$, *then* $I(\lambda, T_1, T) \geq |D(T)|$.

(2) *If* $|D(T)| > |T_1|^+$, $\lambda \geq |T_1|$, *then* $IE(\lambda, T_1, T) \geq |D(T)|$.

Proof. By VII, 2.9(1), for every $p \in D(T)$ there is a model $M_p \in \mathrm{PC}(T_1, T)$ of cardinality λ which realizes p and is stable in $|T_1|$. (Replace T_1 by $T_1 \cup \{\varphi(\bar{c}): \varphi \in p\}$, $\bar{c}$ a sequence of individual constants.) Now $M_p \cong M_q$ induces an equivalence relation over $D(T)$. Each equivalence class has cardinality $\leq |T_1|$ as $|\{q \in D(T): M_q \cong M_p\}| \leq |\{q \in D(T): q \text{ is realized in } M_p\}| \leq |T_1|$ (by the $|T_1|$-stability of M_p). Hence $|D(T)|$ is at most $|T_1|$ plus the number of equivalence classes (by cardinal arithmetic). As $|D(T)| > |T_1|$, $|D(T)|$ is equal to the number of equivalence classes. So there are $|D(T)|$ non-isomorphic models M_p, so $I(\lambda, T_1, T) \geq |D(T)|$. Let $g(p) = \{q\colon q \in D(T), M_q$ has an elementary embedding into $M_p\}$. So $g(p) \subseteq D(T)$ and again $|g(p)| \leq |T_1|$; hence, when proving (2), by theorem 2.8 of the Appendix there is $S \subseteq D(T)$, $|S| = |D(T)|$, such that $p \neq q \in S$ implies $p \notin g(q)$, hence $IE(\lambda, T_1, T) \geqslant |\{M_p : p \in S\}| = |D(T)|$.

PROBLEM 1.1: Does (2) hold when $|D(T)| = |T_1|^+$?

LEMMA 1.3: *If* $\lambda \geq |T_1|$, $\mu = I(\lambda, T_1 \cup T(A), T(A)) > \lambda^{|A|}$ *then* $I(\lambda, T_1, T) \geq \mu$; *where* $T(A) = T \cup \{\varphi(\bar{a})\colon \bar{a} \in A, \vDash\varphi[\bar{a}]\}$ (*the* $a \in A$ *serve also as individual constants*).

Proof. Every model M_1 of T_1 of cardinality λ may be expanded to a model of $T_1 \cup T(A)$ in $\leqslant \lambda^{|A|}$ forms, and we can get each model of $T_1 \cup T(A)$ in this way, from just one M_1 (up to isomorphism) hence

$$\mu = I(\lambda, T_1 \cup T(A), T(A)) \leq I(\lambda, T_1, T) + \lambda^{|A|}.$$

But $\mu > \lambda^{|A|}$, so our result follows.

Remark. Notice that we cannot use 1.2 for 1.3 as the estimation in 1.2 is too weak. Hence the following concept is interesting:

DEFINITION 1.2: (1) If S is a family of types over A (in L), $T \subseteq T_1$, then S or (S, A) is called (T_1, T)-free (or free in (T_1, T)) if for every $S' \subseteq S$ there are models M_1 of arbitrarily large cardinality, $A \subseteq M_1$ (of course $M_1 \restriction L \prec \mathfrak{C}$) such that $p \in S$ is realized in M_1 iff $p \in S'$ (hence in every $\lambda \geq |T_1| + |S'|$ there is such a model).

(2) S is free if it is (T_1, T)-free for every T_1.

(3) The pair (T_1, T) has (μ, λ)-freedom if there are A, $|A| = \lambda$, S, $|S| = \mu$ satisfying (1). If this holds for any T_1 with the same A, S we omit T_1.

LEMMA 1.4: (1) *If* (T_1, T) *has* (μ, χ)-*freedom*, $\mu \geq \lambda \geq |T_1|$, $2^\lambda > \lambda^\chi$, *then* $I(\lambda, T_1, T) = 2^\lambda$.

(2) *If* (T_1, T) *has* (μ, χ)-*freedom*, $\lambda \geq \mu$, $2^\mu > (\mu + |T_1|)^\chi$, *then* $I(\lambda, T_1, T) \geq 2^\mu$.

(3) *If* $(S, \emptyset)$ *is* (T_1, T)-*free*, $\lambda \geq |T_1|$, *then* $IE(\lambda, T_1, T) \geq \min\{2^{|S|}, 2^\lambda\}$.

Proof. For (1), we estimate $I(\lambda, T_1 \cup T(A), T(A))$, from below, (where (S_0, A) is (T_1, T)-free, $|S_0| = \mu$, $|A| = \chi$) by looking at $\{p \in S_0\colon M$ realizes $p\}$, and then use 1.3. Part (3) is immediate as by 1.5(2) of the

Appendix there are $2^{|S|}$ subsets of S, no one is a subset of the other. We are left with (2); by assumption for every $S \subseteq S_0$ there is a model $M^1(S)$ of T_1 of arbitrarily large cardinality, $A \subseteq M^1(S)$, each $p \in S$ is realized in $M^1(S)$ by $a_p = a(p, S)$, and no $p \in S_0 - S$ is realized in it. By VII, 5.3 proof there is a model $M^2(S)$ of $T_2 = T_1 \cup T(A \cup \{a_p : p \in S\})$ omitting each $p \in S_0 - S$, which is $EM(\lambda, \Phi)$ Φ proper for (ω, T_2). Let $M^3(S)$ be its L-reduct. By VII, 2.10 and 2.8(5) $TP_\chi(M^3(S))$ has cardinality $\leq (\mu + |T_1|)^\chi$ which is $< 2^\mu$. As $TP_\chi(M^3(S))$ depends on $M^3(S)$ only up to isomorphism and the number of possible S's is 2^μ, we finish.

THEOREM 1.5: (1) *T has (λ, μ)-freedom if T is unstable and* $\lambda <$ Ded μ (*i.e., there is an order of power λ, with a dense subset of power μ*).

(2) *T has (λ, μ)-freedom if $\kappa < \kappa(T)$ and there is a (λ, κ)-tree I, $|I \cap \lambda^{<\kappa}| \leq \mu < |I \cap \lambda^\kappa| = \lambda$ (e.g., if $\aleph_0 < \kappa(T)$, $\lambda \leq \mu^{\aleph_0}$).*

(3) *T has (λ, μ)-freedom if T has the independence property and $\mu \leq \lambda \leq 2^\mu$.*

(4) *(T_1, T) has $(2^{\aleph_0}, \aleph_0)$-freedom if $|T_1| = \aleph_0$, T $\aleph_0$-unstable. It has $(2^{\aleph_0}, \emptyset)$-freedom when $|D(T)| > \aleph_0$.*

Proof. (1) Let $J \subseteq I$ be orders, $|J| = \mu$, $|I| = \lambda$, J dense in I. Let $J \subseteq J_1 \subseteq I$, and we can assume J, I are dense orders (by extending), and let I_1 be an order of type χ. By VII, 2.4 there are $(\bar{x} < \bar{y}) \in$ L, and Φ proper for (ω, T_1) such that in $M = EM(J_1 + I_1, \Phi)$, $M \models \bar{a}_s < \bar{a}_t$ iff $s < t$. Clearly $\|M\| \geq \chi$ which was arbitrary. Let $A = \{\bar{a}_s : s \in J\}$, $S = \{p_s : s \in I - J\}$, where $p_s = \{(\bar{x} < \bar{a}_t)^{\text{if}(s<t)} : t \in J\}$. In M p_s is realized if $s \in J_1$ (by $\bar{a}_s$); the converse is also true, for if $\bar{a} = \tau(\bar{a}_{\bar{t}}) \in M$, as $s \notin J_1$ and J is dense, there are $s(l) \in J$, $s(l) \sim s(\text{mod } \bar{t})$ for $l = 1, 2$, and $s(1) < s < s(2)$ hence $(\bar{a} < \bar{a}_{s(1)}) \equiv (\bar{a} < \bar{a}_{s(2)})$ so $\bar{a}$ does not realize p_s. We only have to choose A independently of χ, this can be done by VII, 2.5 or by:

CLAIM 1.6: *If for every $T_1 \supseteq T$ and χ there are a pair (S, A), $S \subseteq \bigcup_m S^m(A)$, $|S| = \lambda$, $|A| = \mu$ and for every $S' \subseteq S$ a model M, $\|M\| \geq \chi$, M realizes $p \in S$ iff $p \in S'$ then T has (λ, μ)-freedom.*

Proof. We should prove that we can choose (S, A) independently of T_1. Up to isomorphism (of $\mathfrak{C}$) there are $\leq 2^\lambda$ such pairs, so if each such pair (S, A) fails for $T_1(S, A)$, $S'(S, A)$ and $\chi(S, A)$ then (as T is complete) we can find $T_1 \supseteq T$, such that up to change of names of predicates and function symbols not in T, it extends each $T_1(S, A)$ (such T_1, is consistent as $\mathfrak{C}$ can be expanded to models of $T_1(S, A)$ for any S, A by

I, 1.12). For T_1 we get some suitable (S, A) for $\chi = \sum \{\chi(S, A)$: possible $(S, A)\}$; contradiction.

Continuation of the proof of 1.5. (2) Let I be the (λ, κ)-tree mentioned there, and $\chi \geq \lambda$. T has a strongly uniform κ-tree of the form $\langle\langle\varphi_\alpha, m_\alpha\rangle$: α successor, $\alpha < \kappa\rangle$ and let Φ be as in VII, 3.6(1) and (2). Then let

$$A = \{\bar{a}_\eta : \eta \in I \cap \lambda^{<\kappa}\},$$

$$S = \{p_\eta : \eta \in I \cap \lambda^\kappa\},\ p_\eta = \{\varphi_\alpha(x, \bar{a}_{\eta\restriction\alpha}) : \alpha < \kappa \text{ successor}\}.$$

If $S' \subseteq S$ let $I(S') = \chi^{<\kappa} \cup \{\eta : \eta \in I \cap \lambda^\kappa, p_\eta \in S'\}$ and $M = EM(I(S'), \Phi)$. Clearly $\|M\| \geq \chi$, $A \subseteq |M|$, and if $p_\eta \in S'$, $\bar{a}_\eta \in |M|$ realizes p_η. If $\nu \in \lambda^\kappa$ $\tau(\bar{a}_{\bar{\eta}})$ realizes p_ν, $p_\nu \notin S'$ there is $\alpha < \kappa$ such that $\bar{\eta}[l] \restriction \alpha \neq \nu \restriction \alpha$ (or $\bar{\eta}[l]$ has length $<\alpha$). As $M \vDash \varphi_\beta(\tau(\bar{a}_{\bar{\eta}}), \bar{a}_{\nu\restriction\beta})$ $(\beta = \alpha + 1)$ for every $i < \chi$, letting $\rho = (\nu \restriction \alpha)^\frown\langle i\rangle$ $M \vDash \varphi_\beta(\tau(\bar{a}_{\bar{\eta}}), \bar{a}_\rho)$ (by the indiscernibility of the indexed set $\{\bar{a}_\eta : \eta \in I(S')\}$). But $\tau(\bar{a}_{\bar{\eta}})$ can satisfy such formulas only for finitely many i's, contradiction.

(3) Let $\chi \geq \mu$ be arbitrary, $T_1 \supseteq T$, with Skolem functions, of course. As T has the independence property there are $\varphi(\bar{x}; \bar{y}) \in$ L, and $\bar{a}_n$ such that for every $w \subseteq \omega$ $\{\varphi(\bar{x}; \bar{a}_n)^{\text{if}(n \in w)} : n < \omega\}$ is consistent. So we can assume $\{\bar{a}_n : n < \omega\}$ is an indiscernible sequence. So as in VII, 2.4 there is Φ proper for (ω, T_1) such that in $M = EM(\chi; \Phi)$ for $w \subseteq \chi$, $\{\varphi(\bar{x}; \bar{a}_s)^{\text{if}(s \in w)} : s \in \chi\}$ is consistent. Let $A = \{\bar{a}_s : s \in \mu\}$ and $\{w(i) : i < 2^\mu\}$ be a family of subsets of μ, such that for all distinct $i_0, i_1, \ldots, i_n$ $w(i_0) - \bigcup_{l=1}^n w(i_l)$ is infinite (exists by 1.5(2) of the Appendix); and $p_i = \{[\varphi(\bar{x}; \bar{a}_{2\alpha}) \equiv \neg\varphi(\bar{x}; \bar{a}_{2\alpha+1})]^{\text{if}\alpha \in w(i)} : \alpha < \mu\}$ and $S = \{p_i : i < 2^\mu\}$. There are Skolem functions $\bar{F}_n = \bar{F}_n(\bar{y}_0, \bar{y}_1, \ldots, \bar{y}_{2n-1})$ of $(\exists\bar{x}) \bigwedge_{l<n} [\varphi(\bar{x}; \bar{y}_{2l}) \equiv \neg\varphi(\bar{x}; \bar{y}_{2l+1})]$. So $\bar{F}_n(\bar{a}_{\alpha(1)}, \bar{a}_{\alpha(1)+1}, \bar{a}_{\alpha(2)}, \bar{a}_{\alpha(2)+1}, \ldots)$ realizes

$$\{[\varphi(\bar{x}, \bar{a}_{\alpha(l)}) \equiv \neg\varphi(\bar{x}, \bar{a}_{\alpha(l)+1})] : l < n\}$$

(when, e.g., $\alpha(l) \neq \alpha(k) + 1$ for $l \neq k$.) Let for $i < 2^\mu$, $U \subseteq \mu$, $|U| < \aleph_0$ $\bar{b}_i(U) = \bar{F}_n(\ldots, \bar{a}_{2\alpha}, \bar{a}_{2\alpha+1}, \ldots)_{\alpha \in w(i) \cap U}$ where $n = |w(i) \cap U|$. So for $\alpha \in U$; $\alpha \in w_i$ iff $\varphi[\bar{b}_i(U), \bar{a}_{2\alpha}] \equiv \neg\varphi[\bar{b}_i(U), \bar{a}_{2\alpha+1}]$. Let D be an ultrafilter over $W = S_{\aleph_0}(\mu)$, such that for any finite $w_0 \subseteq \mu$, $\{w \in W : w_0 \subseteq w\} \in D$; and let $M^* = M^W/D$, making the natural identification of a and $\langle \ldots, a, \ldots\rangle/D$; and for $i < 2^\mu$ let $\bar{b}_i = \langle \ldots, \bar{b}_i(U), \ldots\rangle_{U \in W}/D$. For $S' \subseteq S$ let $M(S')$ be the Skolem closure of $|M| \cup \{\bar{b}_i : p_i \in S'\}$ in M^*. Clearly $A \subseteq M(S')$, and each $p_i \in S'$ is realized in $M(S)$. Suppose $p_i \notin S'$, but $\bar{b} = \tau(\bar{a}_{\bar{s}}, \bar{b}_{j(0)}, \ldots, \bar{b}_{j(n)})$ realizes p_i, $\bar{s} \in \chi$. By assumption on the w_i's, there is $\alpha \notin \bigcup_{l \leq n} w_{j(l)}$, $2\alpha \notin \bar{s}$, $2\alpha + 1 \notin \bar{s}$, $\alpha \in w_i$.

We can assume that for every $U \in W$, $\bar{b}[U] = \tau(\bar{a}_{\bar{s}}, \bar{b}_{j(0)}[U], \ldots)$ so if $\alpha \in U \in W$ $\bar{b}[U] = \bar{\tau}(\bar{a}_{\bar{l}})$ where $\bar{l} \subseteq \{2\beta, 2\beta+1: \beta \in w_{j(l)}, l \leqslant n\} \cup \bar{s}$, hence 2α, $2\alpha+1 \notin \bar{l}$, hence $M \models \varphi(\bar{\tau}(\bar{a}_{\bar{l}}), \bar{a}_{2\alpha}) \equiv \varphi(\bar{\tau}(\bar{a}_{\bar{l}}), \bar{a}_{2\alpha+1})$ so $\{U \in W: \models \varphi(\bar{b}[U], \bar{a}_{2\alpha}) \equiv \varphi(\bar{b}[U], \bar{a}_{2\alpha+1})\} \supseteq \{U: \alpha \in U\} \in D$ hence $M^* \models \varphi(\bar{b}, \bar{a}_{2\alpha}) \equiv \varphi(\bar{b}, \bar{a}_{2\alpha+1})$, so $\bar{b}$ does not realize p_i, contradiction.

(4) Suppose $T_1 \supseteq T$ has Skolem functions, $|T_1| = \aleph_0$. If $|D(T)| > \aleph_0$ by II, 3.15 there are formulas $\varphi_\eta(\bar{x}) \in \mathrm{L}$ for $\eta \in 2^{<\omega}$ such that $(\exists \bar{x})\varphi_\eta(\bar{x}) \in T$; $\nu \lessdot \eta$ implies $(\forall \bar{x})(\varphi_\eta(\bar{x}) \rightarrow \varphi_\nu(\bar{x})) \in T$ and $\varphi_{\nu^\frown\langle 0\rangle}(\bar{x})$, $\varphi_{\nu^\frown\langle 1\rangle}(\bar{x})$ are contradictory. We use theorem VII, 3.7, Exercise 3.1 on T_1, φ_η (and its notation), we let $S = \{p^0_\eta: \eta \in 2^\omega\}$. Now $(S, \emptyset)$ is (T_1, T)-free, as for any χ and $S' \subseteq S$ let $(M^1, \ldots, b_n, \ldots)_{n<\omega}$ satisfies the conclusion of VII, 3.7 where (in M_1) $\{b_i: i < \omega\}$ is an indiscernible sequence over $\{a_\eta: \eta \in 2^\omega\}$, and M^2 an elementary extension of M^1 in which $\{b_i: i < \chi\}$ is an indiscernible sequence over $\{a_\eta: \eta \in 2^\omega\}$. Then let $M(S')$ be the Skolem closure of $\{\bar{a}_\eta: p^0_\eta \in S'\} \cup \{b_i: i < \chi\}$. Clearly each $p^0_\eta \in S'$ is realized by $\bar{a}_\eta \in M(S')$, and by VII, 3.7(4) no $p^0_\eta \notin S'$ is realized in $M(S')$. If T is unstable in $\aleph_0$ we choose A, $|S(A)| > |A| = \aleph_0$, and proceed as before with $T(A)$.

Clearly 1.5 and 1.4 have many conclusions, e.g.

CONCLUSION 1.7: (1) *If T is not superstable, $\lambda^{\aleph_0} \geq \mu \geq \lambda + |T_1|$, $2^\mu > 2^\lambda$ then $I(\mu, T_1, T) = 2^\mu$.*

(2) *If $T_1 \supseteq T$, T_1 countable, T $\aleph_0$-unstable then $\lambda \geq 2^{\aleph_0}$ implies $I(\lambda, T_1, T) \geq 2^{2^{\aleph_0}}$, and $2^\lambda > 2^{\aleph_0} \geq \lambda$ implies $I(\lambda, T_1, T) = 2^\lambda$.*

THEOREM 1.8: *Suppose $T_1 \supseteq T$ are countable, T $\aleph_0$-unstable $2^{\aleph_0} \geq \lambda > \aleph_0$, then $I(\lambda, T_1, T) = 2^\lambda$.*

Proof. We assume T_1 has Skolem functions. If $2^\lambda > 2^{\aleph_0}$ or $|D(T)| > \aleph_0$ the conclusion follows from 1.7(2), 1.5 and 1.4(3); if T is not superstable by the next sections (Theorem 2.1). So assume T superstable, $\aleph_0$-unstable, $2^\lambda = 2^{\aleph_0}$ and $|D(T)| = \aleph_0$. The first two assumptions imply, essentially by III, 5.1 that there are equivalence relations $E_n(\bar{x}; \bar{y}) \in \mathrm{L}$ with finitely many equivalence classes (in $\mathfrak{C}$) E_{n+1} refining E_n and $\bar{c}_\eta$ $\eta \in 2^{<\omega}$ such that $\eta \lessdot \nu$, $\eta \in 2^n$ implies $E_n(\bar{c}_\eta, \bar{c}_\nu)$ but $\eta \neq \nu \in 2^n$ implies $\neg E_n(\bar{c}_\eta, \bar{c}_\nu)$. Let $\varphi_\eta(\bar{x}) = E_n(\bar{x}; \bar{c}_\eta)$ where $\bar{c}_\eta$ will be individual constants of T_1.

Using VII, 3.7 (to T_1), and renaming, there is a model M of T_1 and $\bar{a}_\eta$, $\eta \in 2^\omega$ satisfying (1)–(5) from that theorem, and by renaming we get h is the identity. We can also assume (for (*2) by taking a subtree of $2^{<\omega}$ and renaming)

(*) For every $\tau(\bar{x}_1, \ldots, \bar{x}_m) \in L_1$ there is n_τ such that for every $n \geq n_\tau$:

(1) There are in L_1 individual constants c_i^n $i < n^i$ representing the E_n-equivalence classes such that

$$E_n(\tau(\bar{x}_1, \ldots, \bar{x}_n), c_i^n) \in \Gamma_n$$

(see VII, 3.7(5) for the Γ_n's).

(2) If $\bar{\eta}$ is a sequence of length m of distinct members of 2^n, and the same holds for $\bar{\nu}$, and there are k, $n \leq k < \omega$, and $\bar{\eta}_1$, $\bar{\nu}_1$ such that $l(\bar{\eta}_1) = l(\bar{\nu}_1) = m$, $\bar{\eta}[l] \lessdot \bar{\eta}_1[l] \in 2^\omega$, $\bar{\nu}[l] \lessdot \bar{\nu}_1[l] \in 2^\omega$, and $\bar{\eta}_1[l_1] = \bar{\nu}_1[l_2] \Leftrightarrow \bar{\eta}[l_1] = \bar{\nu}[l_2]$ and $\neg E_k(\tau(\bar{a}_{\bar{\eta}_1}), \tau(\bar{a}_{\bar{\nu}_1}))$, then this holds for $k = n$, and any such $\bar{\eta}_1$, $\bar{\nu}_1$.

Remark. We can first take care of (2) then of (1). Note that $M \models E_n(\bar{a}_\eta, \bar{a}_\nu)$ $(\eta, \nu \in 2^\omega)$ iff $\eta \upharpoonright n = \nu \upharpoonright n$.

Now for any $S \subseteq 2^\omega$ let $M_1(S)$ be the Skolem closure of $\{\bar{a}_\eta \colon \eta \in S\}$ and $M(S)$ its L-reduct. Suppose $f : M(S) \to M(S^*)$ is an elementary embedding; $S, S^* \subseteq 2^\omega$ uncountable. For any $\nu \in S$ there are $\tau = \tau_\nu$, $n = n_\nu$ and $\bar{\eta} = \bar{\eta}_\nu$ of length n such that $M_1(S^*) \models f(\bar{a}_\nu) = \tau_\nu(\bar{a}_{\bar{\eta}})$. We can assume $\bar{\eta}[l] \neq \bar{\eta}[k]$ for $l \neq k$.

As S is uncountable there is an uncountable $S_1 \subseteq S$ and τ, n_0, k_0, $\bar{\eta}_0$ such that:

(**) For any $\nu \in S_1$, $\tau_\nu = \tau$, $n_\nu = n_0$, and $\bar{\eta}_\nu[l] \upharpoonright k_0 = \bar{\eta}_0[l] \upharpoonright k_0$ and $\bar{\eta}_0[l_1] \upharpoonright k_0 \neq \bar{\eta}_0[l_2] \upharpoonright k_0$ for $l_1 \neq l_2$. We can assume also that n_τ (from (*)) is $\leq k_0$ and that for $\eta \neq \nu \in S_1$, $h(\eta, \nu) > k_0$, where we let $h(\eta, \nu) = \max\{n : \eta \upharpoonright n = \nu \upharpoonright n\}$, so $\eta(n) \neq \nu(n)$ when $n = h(\eta, \nu)$.

Suppose $\rho \neq \nu \in S_1$ $h(\rho, \nu) = n$, so $M(S) \models E_n(\bar{a}_\nu, \bar{a}_\rho) \wedge \neg E_{n+1}(\bar{a}_\nu, \bar{a}_\rho)$. If for every l $h(\bar{\eta}_\nu[l], \bar{\eta}_\rho[l]) < n$ then by (*2) $\neg E_{n+1}[f(\bar{a}_\nu), f(\bar{a}_\rho)]$ implies $\neg E_n[f(\bar{a}_\nu), f(\bar{a}_\rho)]$. Suppose for every l $h(\bar{\eta}_\nu[l], \bar{\eta}_\rho[l]) \neq n$, so $\bar{\eta}_\nu[l] \upharpoonright n = \bar{\eta}_\rho[l] \upharpoonright n$ implies $\bar{\eta}_\nu[l] \upharpoonright (n+1) = \bar{\eta}_\rho[l] \upharpoonright (n+1)$.

Now define a sequence $\bar{\eta}$: if $\bar{\eta}_\nu[l] \upharpoonright n \neq \bar{\eta}_\rho[l] \upharpoonright n$ *then* $\bar{\eta}[l] = \bar{\eta}_\rho[l]$, and otherwise $\bar{\eta}[l] = \bar{\eta}_\nu[l]$. Let $\bar{b} = \tau(\bar{a}_{\bar{\eta}})$; notice that by the supposition above $\bar{\eta}[l] \upharpoonright (n + 1) = \bar{\eta}_\rho[l] \upharpoonright (n + 1)$, hence by (*1) $M(S^*) \models E_{n+1}(\bar{b}, f(\bar{a}_\rho))$, hence $M(S^*) \models \neg E_{n+1}(\bar{b}, f(\bar{a}_\nu))$. Now $\bar{b} = \tau(\bar{a}_{\bar{\eta}})$, $f(\bar{a}_\nu) = \tau(\bar{a}_{\bar{\eta}_\nu})$ and for any l $\bar{\eta}[l] = \bar{\eta}_\nu[l]$ or $h(\bar{\eta}[l], \bar{\eta}_\nu[l]) < n$ (by $\bar{\eta}$'s definition) and $k_0 < n$, $\bar\eta[l_1] \upharpoonright k_0 \neq \bar\eta[l_2] \upharpoonright k_0$, $\bar\eta_\nu[l_1] \upharpoonright k_0 \neq \bar\eta_\nu[l_2] \upharpoonright k_0$ for $l_1 \neq l_2$ (by (**)). So by (*2), $\neg E_{n+1}(\bar b, f(\bar a_\nu))$ implies $\neg E_n(\bar b, f(\bar a_\nu))$. But $\vDash E_n[\bar b, f(\bar a_\rho)] \wedge E_n[f(\bar a_\rho), f(\bar a_\nu)]$ contradiction.

So we proved that for any $\nu \neq \rho \in S_1$, for some l $h(\bar\eta_\nu[l], \bar\eta_\rho[l]) = h(\nu, \rho)$.

Choose infinite subsets w_i, $i < 2^{\aleph_0}$, of ω, such that for $i \neq j$ $|w_i \cap w_j| < \aleph_0$, and let $S^i \subseteq \{\eta \in 2^\omega : \eta[l] = 0$ when $l \notin w_i\}$, be a set of cardinality λ; and $M_i = M(S^i)$. Clearly $\|M_i\| = \lambda$. Suppose $i \neq j$ and there is an elementary embedding $f: M_i \to M_j$. By what we have proved till now there is an uncountable $S_1^i \subseteq S^i$ such that for $\eta \neq \nu \in S_1^i$ there are $\eta' \neq \nu' \in S^j$ such that $h(\eta, \nu) = h(\eta', \nu')$. Now $w = \{h(\eta, \nu): \eta \neq \nu \in S_1^i\}$ is infinite (as S_1^i is) and w is a subset of w_i (by the definition of S^i) and is a subset of $\{h(\eta, \nu): \eta \neq \nu \in S^j\}$ (by the above) which is a subset of w_j (by the definition of w_j). Hence $|w_i \cap w_j| \geq |w| = \aleph_0$, contradiction. So

$$I(\lambda, T_1, T) \geq IE(\lambda, T_1, T) = 2^{\aleph_0}.$$

But we have restricted ourselves to the case $2^{\aleph_0} = 2^\lambda$ in the beginning, so we finish.

QUESTION 1.2: Is there a parallel for 1.8 when $2^{\aleph_0} \geq \lambda > |T_1| > \aleph_0$? (Add if necessary $|D(T)| > |T_1|$).

THEOREM 1.9: *Suppose*

(*) $2^{\aleph_0} > \lambda > \aleph_0$, *and there is no family of $2^{\aleph_0}$ subsets of λ, each of power λ, the intersection of any two of which is finite. (This is just not* AD$(2^{\aleph_0}, \lambda, \lambda, \aleph_0)$.)

If $T_1 \supseteq T$, T superstable but not totally transcendental $\lambda \geq |T_1|$, then $I(\lambda, T_1, T) \geq 2^{\aleph_0}$.

Remarks: (1) Baumgartner [Ba 76] proved the consistency of "$\lambda = \aleph_1$ satisfies (*)" and of "$\lambda = \aleph_1$ does not satisfy (*)" with ZFC.

(2) The assumption on superstability is needed only for proving the existence of the E_n.

Proof. Let $\mu \leq 2^{\aleph_0}$ be regular. As in 1.8 we can find equivalence relations $E_n(\bar x, \bar y)$ $(n < \omega)$ and $\bar a_\eta$ $\eta \in 2^\omega$ such that $M_1 \vDash E_n(\bar a_\eta, \bar a_\nu)$ iff $\eta \upharpoonright n = \nu \upharpoonright n$; in some model M_1 of T_1. Assume for simplicity, $l(\bar a_\eta) = 1$ so $\bar a_\eta = a_\eta$.

Let, for $S \subseteq 2^\omega$, $M(S)$ be the Skolem closure of $\{a_\eta : \eta \in S\}$. As in 1.8 choose $w_i \subseteq \omega$ for $i < 2^{\aleph_0}$, w_i infinite but $|w_i \cap w_j| < \aleph_0$ for $i \neq j$, and let $S_i \subseteq \{\eta \in 2^\omega : \eta[l] = 0 \text{ for } l \notin w_i\}$, of cardinality λ, $M_i^1 = M(S_i)$. So $\|M_i^1\| = \lambda$, M_i^1 a model of T_1, and let M_i be the L-reduct of M_i^1.

Suppose $I(\lambda, T_1, T) < 2^{\aleph_0}$; as $\mu \leq 2^{\aleph_0}$ is regular, for some i $W = \{j : M_j \cong M_i\}$ has cardinality $\geq \mu$, and let $f_j : M_j \to M_i$ be the isomorphism (we use only its being an elementary embedding). Let, for $j \in W$, $A_j = \{f_j(a_\eta) : \eta \in S_j\}$. Clearly $|A_j| = \lambda$, $A_j \subseteq |M_i|$ where $\|M_i\| = \lambda$; the number of the A_j's is μ, and for $\alpha \neq \beta$ $A_\alpha \cap A_\beta$ is finite (otherwise this contradicts $|w_\alpha \cap w_\beta| < \aleph_0$). If $2^{\aleph_0}$ is regular choose $\mu = 2^{\aleph_0}$ and we get a contradiction. Otherwise combine our conclusion on all μ and by 3.1 of the Appendix we get a contradiction to (*).

THEOREM 1.10: *Suppose $T \subseteq T_1$, $|T_1| = \lambda$, and there is a family $\{\varphi_i(\bar{x}) : i < \lambda\}$ of λ independent formulas of* L(T) *(i.e., for every finite $w \subseteq \lambda$ and $h : w \to \{0, 1\}$, $T \vdash (\exists \bar{x}) \bigwedge_{i \in w} \varphi_i(\bar{x})^{h(i)}$). Suppose J is a subtree of $2^{\leq \lambda}$ (closed under initial segments) such that, letting $J_\alpha = J \cap 2^\alpha$, for $\alpha < \lambda$, $|J_\alpha| < \lambda$, and $\chi = |J_\lambda|$ (for $\chi > \lambda$, the interesting case, this is a Kurepa tree). Then*

(1) *For every $\mu \geqslant \lambda$, $IE(\mu, T_1, T) \geqslant \min\{2^\mu, 2^\chi\}$. Moreover if $(\exists \alpha)(\alpha < \lambda \leqslant 2^{|\alpha|} \wedge |\alpha|^{<\kappa(0)} = |\alpha|)$, there are, for each $\mu = \mu^{<\kappa(0)} \geqslant \lambda$, $\min\{2^\mu, 2^\chi\}$ L(T)-models, the reducts of $\kappa(0)$-compact models of T_1, of cardinality μ, no one elementarily embeddable in another.*

(2) *There are $M_1 \models T_1$, $\bar{y}_\eta \in M_1 (\eta \in J_\lambda)$, $z_i \in M_1$ $(i < \lambda)$ such that*

(A) *For each formula $\varphi(\bar{x}_1, \ldots, \bar{x}_{n(1)}, x^1, \ldots, x^{n(2)}) \in$ L(T_1) there is $\alpha_\varphi < \lambda$ such that if $\nu_1, \ldots, \nu_{n(1)}, \eta_1, \ldots, \eta_{n(1)} \in J_\lambda$, $\beta \geq \alpha_\varphi$ and $\nu_l \upharpoonright \beta = \eta_l \upharpoonright \beta$ but for $l \neq k$, $\nu_l \upharpoonright \beta \neq \nu_k \upharpoonright \beta$, then*

$$M_1 \models \varphi[\bar{y}_{\nu_1}, \ldots, \bar{y}_{\nu_{n(1)}}, z_1, \ldots, z_{n(2)}] \equiv \varphi[\bar{y}_{\eta_1}, \ldots, \bar{y}_{\eta_{n(1)}}, z_1, \ldots, z_{n(2)}].$$

(B) *$\{z_i : i < \mu\}$ is an indiscernible sequence over $\{\bar{y}_\eta : \eta \in J_\lambda\}$.*

(C) *For any $\nu \in J_\alpha$, $\alpha < \lambda$, there is $\varphi_\nu(\bar{x}) \in \{\varphi_i(\bar{x}) : i < \lambda\}$ such that $\nu \lessdot \eta \in J_\lambda$ implies $M_1 \models \varphi_\nu[\bar{y}_\eta]^{\eta(\alpha)}$.*

(D) *For every term $\tau(\bar{x}, \ldots, \bar{x}_n, x^1, \ldots, x^m)$ there is a $\alpha_\tau < \lambda$, such that if $\eta_1, \ldots, \eta_{n+1} \in J_\lambda$, $\beta \geq \alpha_\tau$ and $\eta_l \upharpoonright \beta$ $(l = 1, n+1)$ are pairwise distinct, then $\tau(\bar{y}_{\eta_1}, \ldots, \bar{y}_{\eta_n}, z_1, \ldots, z_m)$ does not realize $p_{\eta_{n+1}}$, where*

$$p_\eta = \{\varphi_i(\bar{x})^{\mathbf{t}} : M_1 \models \varphi_i[\bar{y}_\eta]^{\mathbf{t}}, i < \lambda, \mathbf{t} \in \{0, 1\}\}.$$

(E) *For each $\kappa < \lambda$ such that $2^\kappa = \lambda$ choose a good ultrafilter D_κ over κ, and assume $(\forall \alpha < \lambda)(|\alpha|^{<\kappa(0)} < \lambda)$ (and $\kappa(0)$ is regular). We can assume that if $S \subseteq J_\lambda$, $|S| < \kappa(0)$, M_S^κ is the Skolem Hull of $\{\bar{y}_\eta : \eta \in S\} \cup \{z_i : i < \kappa(0)\}$, $\nu \in J_\lambda - S$ then p_ν is not realized in $M_S^{\kappa(0)}$, where M_S^κ $(\kappa \leqslant \kappa(0))$ is increasing and continuous, $M_S^0 = M_S$, $M_S^{\kappa^+}$ is isomorphic to $(M_S^\kappa)^\kappa / D_\kappa$ over M_S^κ.*

Remark. **When does such a tree exist? If λ is strong limit (i.e., $\beth_\delta$) $J = 2^{\leq\lambda}$ will serve, so $\chi = 2^\lambda$. Clearly $J \subseteq 2^{\leq\lambda}$ is not really ncessary, it suffices $J \subseteq \alpha^{\leq\beta}$, $\mathrm{cf}\,\beta = \mathrm{cf}\,\lambda$, $\gamma < \beta \to |J_\gamma| < \lambda$, $\chi = |J_\beta|$. So if $\mathrm{cf}\,\lambda = \aleph_0$, $J = \prod_n \lambda_n$, where $\lambda_n < \lambda = \sum_{n<\omega} \lambda_n$ is sufficient, so $\chi = \lambda^{\aleph_0} > \lambda$.**

Instead assume J is a λ-Kurepa tree (i.e., $\alpha < \lambda \Rightarrow |J_\alpha| < \lambda$), it suffices to assume: there are $J'_\alpha \subseteq J_\alpha (\alpha < \lambda)$, $|\bigcup_{\alpha<\beta} J'_\alpha| < \lambda$ for $\beta < \lambda$, and for every distinct $\eta_1, \ldots, \eta_n \in J_\alpha$, $\{\alpha < \lambda : \eta_1 \restriction \alpha, \ldots, \eta_n \restriction \alpha$ are in $J'_\alpha\}$ is unbounded (for part (E) we should replace $n < \omega$ by $\kappa < \kappa(0)$). For $\chi = 2^\lambda$ this follows from $\diamondsuit_\lambda$ (λ regular), and if $\lambda = \lambda^{<\lambda}$ it holds with $\chi = \lambda^+$.

Proof. (1) Follows by taking Skolem hulls of $\{\bar{y}_\eta : \eta \in J\} \cup \{z_i : i < \mu\}$ for $J \subseteq J_\lambda$, using (2) (b) (the $\kappa(0)$-compact case by (2) (E)), using 1.4.

(2) *Case* I: There is κ, $\kappa < \lambda \leqslant 2^\kappa$.

Let $S = \{w(i) : i < \lambda\}$ be an independent family of subsets of μ (see 1.5 of the Appendix). For $j < \kappa$ let $\bar{a}_j$ realize $\{\varphi_i(\bar{x})^{\mathrm{if}(j \in w(i))} : i < \lambda\}$, let $\mathbf{I} = \{\bar{a}_j : j < \kappa\}$, and let M_0 be a model of $T_1, \bar{a}_j \in |M_0|$, $\|M_0\| = \mu$ and $\{z_i : i < \mu\}$ be an indiscernible sequence over $\bigcup_{j<\kappa} \bar{a}_j$ (in M_0). We define by induction on $\alpha < \lambda$, for each $\eta \in J \cap 2^\alpha$ a filter D_η over κ, and a set **$S_\eta \subseteq S$ independent mod D_η, (see Definition VI, 3.1) such that $|S - S_\eta|$** $\leqslant \kappa + |l(\eta)|$; and $\nu \lessdot \eta$ implies $D_\nu \subseteq D_\eta$, $S_\eta \subseteq S_\nu$. For $\eta \in J_\lambda$ let D_η **be a completion of $\bigcup_{\alpha<\lambda} D_{\eta \restriction \alpha}$ to an ultrafilter. Let M_1 be a $(\chi + \lambda)^+$-saturated model of T_1, $M_0 \prec M_1$, and for each $\eta \in J_\lambda$, $\bar{y}_\eta$ will realize $\mathrm{Av}(D'_\eta, B_\eta)$ where $B_\eta = \bigcup \{\bar{y}_\nu : \nu < \eta$ (in lexicographic order) $\nu \in J_\lambda\}$ $\cup \bigcup_{j<\kappa} \bar{a}_j \cup \{z_i : i < \mu\}$, $D'_\eta = \{\{\bar{a}_j : j \in W\} : W \in D_\eta\}$. It should be clear that the only nontrivial point is (and parts (C) (D) (E) are left to the reader):**

CLAIM 1.11: *Let λ, κ, S, $\mathbf{I}$, M_0, be as above, and let $D_1, \ldots, D_n$ be filters over κ, S_l independent* mod D_l, *$S_l \subset S$, and $\bar{c} \in |M_0|$, $\varphi \in \mathrm{L}(T_1)$. Then we* ***can find filters D_l^1, and families S_l^1 such that***

(A) $D_l \subseteq D_l^1$;

(B) $S_l^1 \subseteq S_l$, $|S_l - S_l^1| \leqslant \kappa$; *and for* $l = 1, n, S_l^1$ *is independent* mod D_l^1;

(C) *There is* $\mathbf{t} \in \{0, 1\}$, *such that* $\{\langle i(1), \ldots, i(n)\rangle : i(1), \ldots, i(n) < \kappa$, $M_1 \models \varphi[\bar{a}_{i(1)}, \ldots, \bar{a}_{i(n)}; \bar{c}]^{\mathbf{t}}\} \in D_1^1 \times D_2^1 \times \cdots \times D_n^1$.

Proof. We prove by induction on n.

For $n = 1$ this follows by VI, 3.3. For $m = n + 1$, we define by induction on $\alpha \leqslant \kappa, D_{l,\alpha}, S_{l,\alpha}$ for $1 \leqslant l \leqslant n$ such that

(1) $D_l \subseteq D_{l,\beta} \subseteq D_{l,\alpha}$ for $\beta < \alpha$.

(2) $S_{l,\alpha} \subseteq S_{l,\beta} \subseteq S_l, |S_l - S_{l,\alpha}| \leqslant \kappa$ for $\beta < \alpha$.

(3) $S_{l,\alpha}$ is independent $D_{l,\alpha}$.

(4) There is $\mathfrak{t}(\alpha) \in \{0, 1\}$ such that

$$\{\langle i(1), \ldots, i(n)\rangle : i(1), \ldots, < \kappa, M_1 \models \varphi[\bar{a}_{i(1)}, \ldots, a_{i(n)}, \bar{a}_\alpha, \bar{c}]^{\mathfrak{t}(\alpha)}\} \in D_{1,\alpha} \times \cdots \times D_{n,\alpha}.$$

We can do it by the induction hypothesis on n. Let $D_l^1 = D_{l,\mu}$ $(1 \leq l \leq n)$, and let D_m^1, S_m^1 be such that $D_m \subseteq D_m^1, S_m^1 \subseteq S_m, |S_m - S_m^1| \leqslant \kappa, S_m^1$ is independent $\operatorname{mod} D_m^1$, and for some $\mathbf{t} \in \{0, 1\}$, $\{\alpha < \kappa: \mathbf{t}(\alpha) = \mathbf{t}\} \in D_m^1$. Clearly this proves the claim, hence Claim I.

Case II: $\lambda > \aleph_0$, not I; so λ is strong limit. Here in the induction step-α, we have a set Γ_α $(\alpha < \lambda)$ of formulas of L_1 in the variables $\bar{y}_\eta$ $(\eta \in J_\alpha)$ $z_i (i < \omega)$ and set $U_\alpha \subseteq \lambda$, $|U_\alpha| \leq |\alpha| + \aleph_0$, a model M_1^α of T_1, and sequences $\bar{b}_\eta^V$ $(\eta \in J_\alpha, V \subseteq \lambda - U_\alpha)$ and $c_i \in M_1^\alpha$ $(i < \lambda)$ such that:

(i) for every $\varphi(\bar{y}_{\eta_1}, \ldots, \bar{y}_{\eta_n}, z_1, \ldots, z_m) \in \Gamma_\alpha$, and $V(1), \ldots, V(n) \subseteq \lambda - U_\alpha$, $M_1^\alpha \models \varphi[\bar{b}_{\eta_1}^{V(1)}, \ldots, \bar{b}_{\eta_n}^{V(n)}, c_1, \ldots, c_m]$; and $M_1^\alpha \models \varphi_i(\bar{b}_\eta)^{\text{if}\,(i \in V)}$ for $i \in \lambda - U_\alpha$, $V \subseteq \lambda - U_\alpha$,

(ii) $\{c_i : i < \omega\}$ is indiscernible over $\{\bar{b}_\eta^V : \eta \in J_\alpha, V \subseteq \lambda - U\}$.

The connection between Γ_α and Γ_β $(\alpha < \beta)$ is that $\varphi(\bar{y}_{\eta_1}, \ldots, \bar{y}_{\eta_n}, z_1, \ldots, z_m) \in \Gamma_\alpha$ the η_l distinct, $\eta_l \lessdot \nu_l \in J_\beta$ implies $\varphi(\bar{y}_{\nu_1}, \ldots, \bar{y}_{\nu_n}, z_1, \ldots, z_m) \in \Gamma_\beta$. Again the point is the assertion parallel to claim. More exactly we want to add $\varphi(\bar{x}_{\eta_1}, \ldots, \bar{x}_{\eta_n}, z_1, \ldots, z_m)$ or its negation, extend U_α by $\leq \chi = |\alpha| + \aleph_0$ indices, and preserve (i) and (ii). Let N be the model $(H((2^\lambda)^+), \in)$ expanded by individual constant for M_1^α (assuming w.l.o.g. $\|M_1^\alpha\| \leq 2^\lambda$), $\{\langle\eta, \bar{y}_\eta\rangle : \eta \in J_\alpha\}$, $\{\langle\eta, V, \bar{b}_\eta^V\rangle : \eta \in J_\lambda, V \subseteq \lambda - U_\alpha\}$, $\{z_i : i < \lambda\}$, Γ. Our desire can be expressed by an $(L(N))_{\chi^+,\chi^+}$-sentence, hence if it fails, it fails in some $N_1 \prec N$, $\|N_1\| = 2^\chi$, $\|N_1 \cap \lambda\| = 2^\chi$; and $b \subseteq |N_1| \wedge |b| \leq \chi \Rightarrow b \in N_1$. But then we can apply the claim, and find the truth value of φ, and the subset of $\lambda - U_\alpha$ of cardinality $\leq \chi$ we want, contradiction.

Case III: $\lambda = \aleph_0$. This follows by 1.5.

THEOREM 1.11: *Suppose* $(*)$ *from* 1.9, $\lambda \geq |T_1|$, T *stable not superstable. Then* $I(\lambda, T_1, T) \geqslant 2^{\aleph_0}$.

Remark. We can omit "T stable" by 3.4 and replace $2^{\aleph_0}$ by 2^λ by 1.7(1), and replace "not superstable" by "not totally transcendental" by 1.9.

Proof. Let $M = EM(\lambda^{\le\omega}, \Phi)$ where Φ is as in 1.5(2) and w.l.o.g. we can assume that if $A = \bigcup\{\bar{a}_\eta : \eta \in \lambda^{<\omega}\}$, $A_\eta = \bigcup_{n<\omega} \bar{a}_{\eta \restriction n}$ dim(p_η, A_η, M') is λ, where $M' = EM(\lambda^{<\omega} \cup \{\eta\}, \Phi)$, $p_\eta = \text{stp}(\bar{a}_\eta, A_\eta)$ for $\eta \in \lambda^\omega$, and p_η is stationary, and let $I_\eta \subseteq M$ be an indiscernible set based over p_η. We can assume also that $L(T) = \bigcup_{n<\omega} \Delta_n$ (Δ_n finite and increasing) and $\text{tp}_{\Delta_n}(\bar{a}_\eta, EM(\lambda^{<\omega}, \Phi)) = \text{tp}_{\Delta_n}(\bar{a}_\nu, EM(\lambda^{<\omega}, \Phi))$ iff $\eta \restriction n = \nu \restriction n$.

Choose $w_i \subseteq \omega$ $(i < 2^{\aleph_0})$ $i \ne j \Rightarrow |w_i \cap w_j| < \aleph_0$, and let S_i be a subset of $\{\eta \in 2^\omega : n \notin w_i \rightarrow \eta[n] = 0\}$ of cardinality λ, and let $M_i = EM(\lambda^{<\omega} \cup S_i, \Phi)$. Suppose $\lambda < \mu \le 2^{\aleph_0}$, μ regular and

$$U = \{j < 2^{\aleph_0} : M_j \cong M_{j(0)}\}$$

has cardinality $\ge \mu$, and let $f_j : M_j \rightarrow M_{j(0)}$ be the isomorphism. Let $M_{j(0)} = \bigcup_{\alpha<\lambda} B^\alpha$, B^α increasing, $\|B^\alpha\| < \lambda$. For each $j \in V$ there is $\alpha(j) < \lambda$ such that f_j maps A_η $(\eta \in 2^{<\omega})$ into $B^{\alpha(j)}$, provided that cf $\lambda > \aleph_0$, hence for some β, $U' = \{j \in U : \alpha(j) = \beta)\}$ has cardinality $\ge \mu$. For each $j \in U'$ and $\eta \in S_j$ choose $\bar{b}^j_\eta \in f_j(I_\eta)$ such that $\text{tp}(\bar{b}^j_\eta, |B^\beta|)$ does not fork over $f_j(A_\eta)$. So for some $j(1) \ne j(2) \in U'$

$$\{\bar{b}^{j(1)}_\eta : \eta \in S_{j(1)}\} \cap \{\bar{b}^{j(2)}_\eta : \eta \in S_{j(2)}\}$$

is infinite. But clearly for $j \in U'$, $\text{tp}_{\Delta_n}(\bar{b}^j_\eta, B^\beta) = \text{tp}(\bar{b}^j_\nu, B^\beta)$ iff $\eta \restriction n = \nu \restriction n$, hence we get a contradiction as in 1.8 and 1.9. So assume cf $\lambda = \aleph_0$, $\lambda = \sum_{n<\omega} \lambda_n$, $\lambda_n < \lambda$. For each n, j there is $\alpha(j, n) < \lambda$ such that

$$|\{\eta \in S_j : |f_j(I_\eta) \cap M^{\alpha(j,n)}| \ge \aleph_0\}| \ge \lambda_n.$$

We can also find $\beta(n)$ such that $U^n = \{j \in U : \alpha(j, n) = \beta(n)\}$ has cardinality $\ge \mu$. So as before we get AD$(\mu, \lambda, \lambda_n, \aleph_0)$, and as this holds for each n, μ it is easy to prove AD$(2^{\aleph_0}, \lambda, \lambda, \aleph_0)$, i.e., not (*) of 1.9.

QUESTION 1.3: Can we in 1.4 deduce something about IE?

EXERCISE 1.4: Prove in 1.9 and 1.11 that if $2^{\aleph_0}$ is a singular cardinal then $IE(\lambda, T_1, T) \ge 2^{\aleph_0}$ and if $2^{\aleph_0}$ is regular, $\chi < 2^{\aleph_0}$, and in (*) there is no such family of cardinality χ then $IE(\lambda, T_1, T) \ge 2^{\aleph_0}$.

QUESTION 1.5: Can we prove in 1.8 that $IE(\lambda, T_1, T) = 2^\lambda$? (The remaining case is: T superstable, $\aleph_0$-unstable, $2^{\aleph_1} > 2^{\aleph_0}$, and $|D(T)| = \aleph_0$.)

QUESTION 1.6: Try to improve 1.7(2) to results on $IE(\mu, T_1, T)$.

EXERCISE 1.7: Suppose T has (λ, μ)-freedom by one of the cases 1.5 $\lambda \geqslant |T_1|$, $\lambda = 2^\mu = \lambda^\alpha$. Then $IE(\lambda, T_1, T) = 2^\lambda$.

Remark. Instead 1.5 it may suffice to assume that in Definition 1.1, there is a model $M_1 \supseteq A$, and $\bar{a}_p \in M$ realizes p, $\{b_i : i < \omega\}$ is indiscernible over $A \cup \{a_p : p \in S\}$, and for each p, p is not realized in the Skolem Hull of $A \cup \{\bar{a}_q : q \neq p\} \cup \{b_i : i < \omega\}$.

EXERCISE 1.8: Rewrite the proof of 1.8 where you prove only that for any $\nu \neq \rho \in S_1$ for some $lh(\nu, \rho) \leq h(\bar{\eta}_\nu[l]),\ \bar{\eta}_\rho[l]) \leq h(\nu, \rho) + 100$, assume (∗1) holds for even n's, (∗2) for odd n. Is it simpler?

EXERCISE 1.9: Show that in Definition 1.1, S and A can be chosen independently of the cardinality.

VIII.2. Unsuperstable theories

Our main theorem here is

THEOREM 2.1: *$I(\lambda, T_1, T) = 2^\lambda$ provided that the following condition holds:*

(∗) *T is unsuperstable, $\lambda \geq |T_1| + \aleph_1$ and at least one of the following cases occurs:* (i) $\lambda > |T_1|$; (ii) $\lambda^{\aleph_0} = \lambda$; (iii) $\lambda = \sum_{n<\omega} \lambda_n, \lambda_n < \lambda, \lambda_n^{\aleph_0} = \lambda_n$.

Proof. By the subsequent theorems. If λ is regular by 2.2 for $\mu = \lambda$. Also if there is $\mu < \lambda$, $2^\mu = 2^\lambda$, 2.2 implies our conclusion. If there is $\mu < \lambda \leq \mu^{\aleph_0}$, $2^\mu < 2^\lambda$ the result follows from 1.7(1). If none of the previous cases occur and there is $\mu < \lambda$, $2^\mu \geq \lambda$, 2.3 gives our result. If $\mu < \lambda \Rightarrow 2^\mu < \lambda$, but λ is singular, there is always χ, cf $\lambda \leq \chi < \lambda$, $\chi = \aleph_0$ or χ a strong limit cardinal of cofinality $\aleph_0$. Noticing that $\lambda^{\text{cf}\,\lambda} = 2^\lambda$, Theorem 2.6(1) gives our result. We covered all possibilities, thus prove the theorem.

When (∗) (from 2.1) holds there is a model N_0 of cardinality $|T_1|$, proper for $(\omega^{\leq\omega}, T_1)$, such that the skeleton of $EM^1(\omega^{\leq\omega}, N_0)$ is a uniform $\aleph_0$-tree of the form $\langle\langle\varphi_n, 1\rangle : n < \omega\rangle$, $\varphi_n \in \mathrm{L}$ (by VII, 3.6(2) and 3.5(2)); let $\varphi_n = \varphi_n(x; \bar{y}_n) = \varphi_n(\bar{x}; \bar{y})$. In Case (ii) we can assume

N_0 is $\aleph_1$-compact; (by I, 1.7, and as we can replace N_0 by any elementarily equivalent model of the same cardinality). In Case (iii), by VII, 1.11 we can assume there are $P_n \subseteq |N_0|$, $(n < \omega)$ $P_n \subseteq P_{n+1}$, $|N_0| = \bigcup_{n<\omega} P_n$ such that $N_1 = (N_0, P_0, P_1, \ldots)$ satisfies: if a countable type over N_1 is finitely satisfiable in P_n, then it is realized in P_n (i.e., by a finite sequence from P_n). In Cases (i) and (ii) let $P_n = |N_0|$, $N_1 = (N_0, P_0, \ldots)$. In all cases let L_2 be a countable sublanguage of $L(N_1)$, such that $P_n \in L_2$, and the formulas $R_{\theta,\varphi,\bar{\tau}} \in L_2$ when $\theta \in L(\omega^{\leq\omega})$, $\varphi \in \{\varphi_n : n < \omega\}$, $\bar{\tau} \in L_1^c$. We use those conventions in this section, except in 2.7.

THEOREM 2.2: *Suppose T is not superstable.*

(1) *If $\lambda \geq \mu > |T_1|$, μ regular then $IE(\lambda, T_1, T) \geq 2^\mu$.*

(2) *If* (*) (ii) or (iii) *holds $\aleph_0 < \mu \leq \lambda$, μ regular, then $I(\lambda, T_1, T) \geq 2^\mu$.*

Proof. For every ordinal $\delta < \mu$ cf $\delta = \aleph_0$ choose a (strictly) increasing sequence of ordinals $\eta_\delta = \eta(\delta)$ whose limit is δ. For every $w \subseteq \mu$ let I_w be the $\aleph_0$-tree $\mu^{<\omega} \cup \{\eta_\delta : \delta \in w\} \cup \{\langle\alpha\rangle : \mu + |T_1| \leq \alpha < \lambda\}$ ($\mu + |T_1|$-cardinal addition, so in (ii) and (iii) the third part disappears). Let $M_w^1 = EM^1(I_w, N_0)$ and M_w be the L-reduct of M_w^1 (we use the notation of 1.1). By 1.3 of the Appendix there are pairwise disjoint stationary subsets of μ, $u_i \subseteq \{\delta < \mu : \text{cf}\,\delta = \aleph_0\}$ $(i < \mu)$, and by 1.5 of the Appendix there are $S(\alpha) \subseteq \mu$, $(\alpha < 2^\mu)$ such that $S(\alpha) \subseteq S(\beta) \Rightarrow \alpha = \beta$. Let $w(\alpha) = \bigcup_{i \in S(\alpha)} u_i$. Our family of models is $\{M_{w(\alpha)} : \alpha < 2^\mu\}$. To prove that it exemplifies our conclusion, it suffices to prove: supposing $w, u \subseteq \{\delta < \mu : \text{cf}\,\delta = \aleph_0\}$, $w - u$ $u - w$ are stationary, and $f : M_w \to M_u$ is an isomorphism for (2), and for (1) an elementary embedding, we get a contradiction ("$u - w$ stationary" is not needed for (1) and can be weakened to "$\lambda - w$ stationary" for (2)).

Clearly for every $\eta \in I_w$ there are $\bar{\tau}_\eta \in L_1^c$, $\bar{\nu}(\eta) = \bar{\nu}_\eta \in \mu^{\leq\omega} \cap I_u$ and $\bar{s}(\eta) = \bar{s}_\eta \in \{\langle\alpha\rangle : \mu + |T_1| \leq \alpha < \lambda\}$, $\bar{c}_\eta \in N_0$ such that

$$f(\bar{a}_\eta) = \bar{\tau}_\eta(\bar{a}_{\bar{\nu}(\eta)}, \bar{a}_{\bar{s}(\eta)}, \bar{c}_\eta).$$

Now we can define by induction on $i < \mu$, $A_i \subseteq |N_0|$, $\alpha(i) < \mu$ and $X_i \subseteq \{\langle\beta\rangle : \mu + |T_1| \leq \beta < \lambda\}$ such that:

(A) X_i, A_i are increasing (by $\subseteq$) and continuous sequence in i.

(B) $|X_i| < \mu$, $|A_i| < \mu$, and if $|T_1| < \mu$, $A_i = |N_0|$.

(C) $\alpha(i)$ is a (strictly) increasing and continuous sequence, $\alpha(i) \geq i$.

(D) If $\eta \in I_w \cap \bigcup_{\beta<i} \beta^{\leq\omega}$, then $\bar{s}_\eta[l] \in X_i$, $\bar{c}_\eta \in A_i$, and $\bar{\nu}_\eta[l] \in \bigcup_{\beta<\alpha(i)} \beta^{\leq\omega}$.

(E) If $\eta \in \alpha(i+1)^{<\omega}$ and $\beta < \mu$ but $\beta \geq \alpha(i+1)$, then *there* are infinitely many $\gamma < \alpha(i+1)$ such that:

(α) $\mu > |T_1| \Rightarrow \bar{c}_{\eta^\frown\langle\beta\rangle} = \bar{c}_{\eta^\frown\langle\gamma\rangle}$.

(β) $\bar{s}_{\eta^\frown\langle\beta\rangle} \sim \bar{s}_{\eta^\frown\langle\gamma\rangle} \bmod X_i$ (in I_u, so, in fact, in the ordered set $\{\langle\alpha\rangle : \mu + |T_1| \leq \alpha < \lambda\}$).

(γ) $\bar{\nu}_{\eta^\frown\langle\beta\rangle} \sim \bar{\nu}_{\eta^\frown\langle\gamma\rangle} \bmod \alpha(i)^{\leq\omega}$.

(δ) $\bar{\tau}_{\eta^\frown\langle\beta\rangle} = \bar{\tau}_{\eta^\frown\langle\gamma\rangle}$

It is easy to define $\alpha(i)$ inductively; for $i+1$ notice that the demand is only that in the Skolem closure of $A_{i+1} \cup \{\bar{a}_\eta : \eta \in I_w \cap \bigcup_{j<\alpha(i+1)} j^{\leq\omega}\} \cup X_{i+1}$ there will be some $<\mu$ elements, and as $\mu > \aleph_0$ is regular, clearly such $\alpha(i+1)$ exists; for E) notice that I_u is χ-atomically stable for any χ.

Now clearly $S = \{i < \mu : j < i \Rightarrow \alpha(j+\omega) < i\}$ is a closed unbounded subset of μ, hence $(w-u) \cap S$ is stationary.

Proof for (1). ($\mu > |T_1|$). Choose $\delta \in (w-u) \cap S$.

As $\delta \notin u$, $\eta_\delta \notin I_u$ hence if $\nu_l = \bar{\nu}_{\eta_\delta}[l] \in \mu^\omega$, then (as it is increasing) there is $\alpha_l < \delta$ such that $\nu_l[n] < \alpha_l \Leftrightarrow \nu_l[n] < \delta \Leftrightarrow n < n_l$ where $n_l \leq \omega$. If $\nu_l = \bar{\nu}_{\eta_\delta}[l] \in \mu^{<\omega}$ there is $\alpha_l < \delta$ such that $\nu_l[n] < \alpha_l \Leftrightarrow \nu_l[n] < \delta$. Let $\alpha^* = g(\delta)$ be $<\delta$, $\geq \alpha_l$ and such that $x_l = \min\{x \in X_\delta : x \geq \bar{s}_{\eta_\delta}[l]\} \in X_{\alpha^*}$ or $x_l = \infty$ (exists as $X_\delta = \bigcup_{i<\delta} X_i$). By the definition of S there are i, n; $\alpha^* < \alpha(i) < \delta$, $\eta_\delta[m] < \alpha(i)$ for $m < n$ but $\eta_\delta[n] > \alpha(i+1)$. Let $\rho = \eta_\delta \restriction n$, $\beta = \eta_\delta[n]$, then by (E) there are, for $m < \omega$, distinct $\gamma(m)$, $\alpha^* < \alpha(i) \leq \gamma(m) < \alpha(i+1) < \beta$, such that $\bar{s}_{\rho^\frown\langle\gamma(m)\rangle} \sim \bar{s}_{\rho^\frown\langle\beta\rangle} \bmod X_i$, $\bar{c}_{\rho^\frown\langle\gamma(m)\rangle} = \bar{c}_{\rho^\frown\langle\beta\rangle}$ and $\bar{\nu}_{\rho^\frown\langle\gamma(m)\rangle} \sim \bar{\nu}_{\rho^\frown\langle\beta\rangle} \bmod \alpha(i)^{\leq\omega}$ and $\bar{\tau}_{\eta^\frown\langle\gamma(m)\rangle} = \bar{\tau}_{\eta^\frown\langle\beta\rangle}$; hence $\bar{\nu}_{\rho^\frown\langle\gamma(m)\rangle} \sim \bar{\nu}_{\rho^\frown\langle\beta\rangle} \bmod \bar{\nu}_{\eta_\delta}$ (by the definition of α^*). Hence, by the indiscernibility of the skeleton $M_u \models \varphi_{n+1}[f(\bar{a}_{\eta_\delta}), f(\bar{a}_{\rho^\frown\langle\gamma(m)\rangle})]$ for $m < \omega$, hence $M_w \models \varphi_{n+1}[\bar{a}_{\eta_\delta}; \bar{a}_{\rho^\frown\langle\gamma(m)\rangle}]$ (as f is an elementary embedding); contradicting the definition of a uniform $\aleph_0$-tree of the form $\langle\langle\varphi_n, 1\rangle : n < \omega\rangle$. So we proved (1) as no sequence in the model realizes $\varphi_{n+1}(\bar{x}; \bar{a}_{\rho^\frown\langle\gamma\rangle})$ for infinitely many γ's (see VII, 3.6(4)).

Proof of (2). As $\mu \leq |T_1|$ clearly $X_i = \emptyset$. We define g as in the proof of (1). As $(w-u) \cap S$ is a stationary subset of μ and for each δ in it $g(\delta) < \delta$, by 1.3 of the Appendix for some $\alpha_* < \mu$, $w_0 = \{\delta : \delta \in (w-u) \cap S, g(\delta) = \alpha_*\}$ is stationary. As the number of $\eta \in (\alpha_*)^{\leq\omega} \cap I_u$ is $<\mu$; and if we partition a stationary set to $<\mu$ parts, at least one is stationary (see 1.2 of the Appendix), there is a stationary $w_1 \subseteq w_0$ such that for all $\delta \in w_1 : \bar{\tau}_{\eta(\delta)} = \bar{\tau}$ is constant (as $\bar{\tau}_{\eta(\delta)} \in \mathrm{L}_1^c$, L_1^c is countable); $(\bar{\nu}_{\eta(\delta)}[l]) \restriction n$ is constant or always $\notin (\alpha_*)^{\leq\omega}$; the similarity type of $\bar{\nu}_{\eta(\delta)}$ is constant;

and there is a constant $n(\delta) = n(*)$ such that $\bar{c}_{\eta(\delta)} \in P_{n(\delta)}(N_1)$. Clearly we can find a closed unbounded $S_1 \subseteq S$ such that

(1) if $\alpha \in S_1$, $\eta \in \alpha^{<\omega}$ and there is $\delta \in w_1$ such that $\eta \lessdot \eta_\delta$ then there is such $\delta < \alpha$.

(2) if $\alpha \in S_1$, $\eta \in \alpha^{<\omega}$ and $\{\delta \in w_1 : \eta \lessdot \eta_\delta\}$ is stationary then there are $\nu \in \alpha^{<\omega}$, with arbitrarily large $(<\alpha)$ last element such that $\{\delta < \mu : \eta^\frown\nu \lessdot \eta_\delta\}$ is stationary (the existence of such ν's follows from 1.3 of the Appendix).

(3) If $\delta \in S_1 \cap w_1, \eta \in \alpha^{<\omega}, \eta \lessdot \eta_\delta$, then $\{\delta \in w_1 : \eta \lessdot \eta_\delta\}$ is stationary.

So for each limit point δ^* of S_1, cf $\delta^* = \aleph_0$, we can easily find a strictly increasing η of length ω with limit δ^*, and $\delta(n) < \delta^*$, $\delta(n) \in w_1$, and $\delta(n) < \beta(n) < \delta(n+1)$ where $\beta(n) \in S_1$; such that $\eta \restriction n = \eta_{\delta(n)} \restriction n$ (define $\delta(n), \beta(n)$ simultaneously by induction on $n \leq \omega$, using the properties of S_1). Choose also $\delta > \delta^*, \delta \in w_1$. Then we can check that $\tau(\bar{a}_{\bar{\nu}(\eta(\delta))}, \bar{c}_{\eta(\delta(n))})$ satisfies $\{\varphi_m(\bar{x}; f(\bar{a}_{\eta\restriction m})) : m < n\}$ (as $\bar{\nu}_{\eta\restriction m}{}^\frown\bar{\nu}_{\eta(\delta)}, \bar{\nu}_{\eta\restriction m}{}^\frown\bar{\nu}_{\eta(\delta(m))}$ are similar, by α_*'s definition). Now choose $\bar{c} \in P_{n(*)}(N_1)$ which satisfies every formula of L_2 all but finitely many $\bar{c}_{\eta(\delta(n))}$ satisfy ($\bar{c}$ exists by N_1's definition). So clearly $\tau(\bar{a}_{\bar{\nu}(\eta(\delta))}, \bar{c})$ realizes $\{\varphi_m(\bar{x}, f(\bar{a}_{\eta\restriction m})) : m < \omega\}$. As in (2) f is an isomorphism, in M_w $\{\varphi_m(\bar{x}; \bar{a}_{\eta\restriction m}) : m < \omega\}$ is realized. Hence $\delta^* \in w$. As δ^* was arbitrary $S_1 \subseteq w$. So we finish the proof of (2) too.

THEOREM 2.3: *Suppose* $(*)$ *from* 2.1, *and that* $\mu \leqq \lambda$ *is regular;* $\lambda > |T_1| \Rightarrow \mu > |T_1|$; $2^\mu \geq \lambda$, $\mu > \aleph_0$; *and* $\chi < \mu \Rightarrow \chi^{\aleph_0} < \mu$. *Then* $I(\lambda, T_1, T) \geq 2^\lambda$.

Proof. For $\bar{w}$ a sequence of subsets of $\{\delta < \mu : \text{cf } \delta = \aleph_0\}$ of length λ, let $I_{\bar{w}}$ be the $\aleph_0$-tree $\{\eta \in \lambda^{<\omega} : \eta[0] < \lambda, (\forall n > 0)(\eta[n] < \mu)\} \cup \{\eta : \eta \in \lambda^\omega;$ $n > 0 \Rightarrow \eta[n] < \eta[n+1] < \mu$; η's limit is δ and $\delta \in \bar{w}[\eta[0]]\}$.

Let $M^1_{\bar{w}} = EM^1(I_{\bar{w}}, N_0)$ and $M_{\bar{w}}$ its L-reduct. Choose for each $\delta < \mu$, cf $\delta = \aleph_0$ a (strictly) increasing sequence of ordinals of length ω, η_δ, whose limit is δ, and $\eta_\delta[0] = 0$. For a sequence η define η^* such that $\eta = \langle\eta[0]\rangle^\frown\eta^*$. Let u_i, $i < \mu$, be pairwise disjoint stationary subsets of $\{\delta < \mu : \text{cf } \delta = \aleph_0\}$, and $\{S(\alpha) : \alpha < 2^\mu\}$ be a family of subsets of μ, no one of which includes an intersection of finitely many others. (See 1.3 and 1.5 of the Appendix for existence). Let $\{u^i : i < 2^\lambda\}$ be a family of subsets of λ, no one a subset of the other. Let $\bar{w}_i$ be a sequence of length λ whose range is $\{\bigcup_{i \in S(\alpha)} u_i : \alpha < \lambda, \alpha \in u^i\}$. Our family is $\{M_{\bar{w}(i)} : i < 2^\lambda\}$.

So it suffices to prove that if f is an isomorphism from $M_{\bar{w}}$ onto $M_{\bar{v}}$ then for every $i < \lambda$ there are $n < \omega$; $j_1, \ldots, j_n < \lambda$ and closed unbounded $S \subseteq \mu$ such that $\bar{w}[i] \supseteq \bar{w}[j_1] \cap \cdots \cap \bar{w}[j_n] \cap S$. For simplicity let $i = 0$, $\mu > |T_1|$ and let $f(\bar{a}_\eta) = \bar{\tau}_\eta(\bar{a}_{\bar{\nu}(\eta)}, \bar{c}_\eta)$ for $\eta \in I_{\bar{w}}$, (so $\bar{\nu}_\eta = \bar{\nu}(\eta) \in I_{\bar{v}}$, $\bar{c}_\eta \in N_0$). Now define by induction on $i < \mu$, ordinals $\alpha(i)$ and sets X_i such that:

(A) X_i is an increasing and continuous (by $\subseteq$) sequence of subsets of λ of cardinality $< \mu$.

(B) $\alpha(i)$ is a (strictly) increasing and continuous sequence of ordinals $< \mu$.

(C) If $\eta[0] = 0$, $\eta \in I_{\bar{w}} \cap \bigcup_{\beta < i} \beta^{\leq \omega}$ then $(\bar{\nu}_\eta[l])[0] \in X_i$, and $(\bar{\nu}_\eta[l])^* \in \bigcup_{\beta < \alpha(i)} \beta^{\leq \omega}$ (this is possible as $\chi < \mu \Rightarrow \chi^{\aleph_0} < \mu$). We know (see 1.2 and 3 of the Appendix) that if $w \subseteq \mu$ is stationary then if g is a function from w to a set of cardinality $< \mu$ then for some t, $\{\alpha \in w : g(\alpha) = t\}$ is stationary; and if g is a function from w, $g(\alpha) \in X_\alpha$; X_α is an increasing continuous sequence of sets of cardinality $< \mu$ then for some t, $\{\alpha \in w : g(\alpha) = t\}$ is stationary. By successive use of this we can find a stationary $w_1 \subseteq \bar{w}[0]$ and $\alpha(*) < \mu$ such that:

(α) For all $\delta \in w_1$ $\bar{\tau}_{\eta(\delta)} = \bar{\tau}$, $\bar{c}_{\eta(\delta)} = \bar{c}$.

(β) For all $\delta \in w_1$, $\bar{\nu}_{\eta(\delta)}$ has the same similarity type.

(γ) For all $\delta \in w_1$ for each l, $\beta^l_\delta = (\bar{\nu}_{\eta(\delta)}[l])[0]$ is constant and $\in X_{\alpha(*)}$ or always it $\notin X_\delta$, but $\min\{\beta \in X_\delta : \beta > \beta^l_j\}$ is fixed (maybe as ∞). Let $\{\beta^l_\delta : l\} \cap X_{\alpha(*)}$ be $\{j_1, \ldots, j_n\}$.

(δ) Let $\nu^\delta_l = (\bar{\nu}_{\eta(\delta)}[l])^*$. Then for some n_l, $\nu^\delta_l \restriction n_l$ is constant and $\in \alpha(*)^{\leq \omega}$, and $\nu^\delta_{l}[n_l]$, if defined, is $\geq \delta$.

Now, as in the proof of 2.2(2) we can find a closed unbounded $S \subseteq \mu$, such that for every $\delta \in S \cap \bar{v}[j_1] \cap \cdots \cap \bar{v}[j_n]$ for some $\eta \in \mu^\omega$, $\eta[0] = 0$, η strictly increasing with limit δ, $\{\varphi_n(\bar{x}, f(\bar{a}_{\eta \restriction n})) : n < \omega\}$ is realized in $M_{\bar{v}}$, hence $\delta \in w[0]$, what we want to prove (if $\mu \leq |T_1|$ the changes are like the proof of 2.2(2) and we can have $\|N_0\| = |T_1|^{\aleph_0}$).

DEFINITION 2.1: For models M, N cardinal χ and ordinal α we define a game $GE^\alpha_\chi(M, N)$ $[GI^\alpha_\chi(M, N)]$ between the players I and II as follows: in the βth move player I chooses $i_\beta < \chi$ and $a^\beta_i \in M$ $i < i_\beta$, and then player II chooses $b^\beta_i \in N$, for $i < i_\beta$. The play ends after α moves, and then player II wins if $\mathrm{tp}_*(\{b^\beta_i : i < i_\beta, \beta < \alpha\}, \emptyset, N) = \mathrm{tp}_*(\{a^\beta_i : i < i_\beta, \beta < \alpha\}, \emptyset, M)$,

$$[TPS^\chi(\{b^\beta_i : i < i_\beta, \beta < \alpha\}, \emptyset, N) = TPS^\chi(\{a^\beta_i : i < i_\beta, \beta < \alpha\}, \emptyset, M)]$$

and player I wins otherwise. A strategy (of a player) is a sequence of functions f_β $(\beta < \alpha)$ which "tells" him what to do (f_β for the βth move) depending only on the previous choices in the play. A winning strategy is a strategy such that in any play in which the player "behaves" according to it, he wins. A player wins in the game if he has a winning strategy.

Remark. Sometimes we denote that player I chooses sequences $\bar{a}_i^\beta$ instead of elements.

LEMMA 2.4: (1) *If there is an elementary embedding f of M into N; then player* II *wins in $GE_\chi^\alpha(M, N)$.*

(2) *If M, N are isomorphic, player* II *wins in $GI_\chi^\alpha(M, N)$.*

(3) *In the games from Definition* 2.1, *at most one player wins.*

(4) *If player* II *wins in $GX_\chi^\beta(M,N)$ he wins in $GX_\chi^\alpha(M, N)$ where $\beta \geq \alpha$, $X \in \{I, E\}$.*

(5) *If $X \in \{E, I\}$, player* II *wins in $GX_\chi^\alpha(M_l, M_{l+1})$ $l = 1, 2$ then player* II *wins in $GX_\chi^\alpha(M_1, M_3)$.*

Proof. (1) In the βth move player II chooses $b_i^\beta = f(a_i^\beta)$. Also the other proofs are immediate.

To clarify the proof of 2.6, we prove first:

LEMMA 2.5: *Suppose $\lambda = \lambda^{\aleph_0} \geq |T_1|$, M_i are models of T of cardinality $\leq \lambda$, for $i < \lambda$. Suppose T is unsuperstable. Then there is $N \in \mathrm{PC}(T_1, T)$ such that player* I *wins in $GI_{\aleph_0}^\omega(N, M_i)$ for each $i < \lambda$.*

Proof. Let g be a one-to-one function from $\{\langle i, \bar{a}\rangle : i < \lambda, \bar{a} \in |M_i|\}$ onto λ. Let $g(g_1(\alpha), g_2(\alpha)) = \alpha$ for $\alpha < \lambda$, and

$$I = \lambda^{<\omega} \cup \{\eta \in \lambda^\omega : g_1(\eta[n]) = i \text{ for every } n, \text{ and } g_2(\eta[n]) \in |M_i| \text{ and } \{\varphi_n(x; g_2(\eta[n])) : n < \omega\} \text{ is (well defined and) omitted by } M_i\}$$

and let $N = EM(I, N_0)$ (N_0-proper for $(\omega^{\leq\omega}, T_1)$ and of cardinality $|T_1|$). Let us show how player I wins in $\mathrm{GI}_{\aleph_0}^\omega(N, M_i)$: for $\beta = 0$ player I chooses the sequence $\bar{a}_\eta$ $\eta = \eta_0 = \langle\ \rangle$; for $\beta = n + 1$, if in the mth move player II chooses $\bar{b}^m$ for $m \leq n$, player I now chooses $\bar{a}_\eta$, $\eta = \eta_{n+1} = \langle g(i, \bar{b}^0), \ldots, g(i, \bar{b}^n)\rangle$. If in such play player II wins, necessarily N realizes $p_\nu = \{\varphi_n(x; \bar{a}_{\eta_n}) : n < \omega\}$ iff M_i realizes $\{\varphi_n(x; \bar{b}^n) : n < \omega\}$, (letting $\nu \in \lambda^\omega$, $\eta_n \lessdot \nu$); then N realizes p_ν iff $\nu \in I$ iff M_i omits $\{\varphi_n(x; g_2(\nu[n])) : n < \omega\} = \{\varphi_n(x; \bar{b}^n) : n < \omega\}$, contradiction.

THEOREM 2.6: *Suppose* $\lambda \geq \mu = \chi^{\aleph_0} = 2^{\chi}$, $(\forall \kappa < \lambda)(\kappa^{\chi} \leq \lambda)$ *and* $\lambda \geq |T_1|$. *Then:*

(1) $I(\lambda, T_1, T) \geq \lambda^{\mu}$.

(2) *There are models* $M_h \in \mathrm{PC}(T_1, T)$ *for* $h : \mu \to \lambda$ *such that if for some* i $h_1(i) < h_2(i)$ *then in* $GI^{\omega}_{\chi}(M_{h_2}, M_{h_1})$ *player* I *wins.*

Proof. Clearly (2) implies (1). By the assumptions on λ we can assume N_0, N_1 are as mentioned in 2.1(ii) or (iii) except that maybe $\|N_0\| > |T_1|$ (but $\|N_0\| \leq \lambda$). We can assume $\chi = \sum_{n<\omega} \chi_n$, $\chi_n^{\aleph_0} = \chi_n$ or $\chi = \aleph_0$ and let $\chi_n = n$. We define by induction on $\alpha < \lambda$ pairwise disjoint well-ordered sets J^i_α $(i < \mu)$, $|J^i_\alpha| = |\alpha|^{\chi} + \mu \leq \lambda$, and for $\alpha > 0$ the universe of J^i_α is the set of quadruples $s = \langle \alpha, \zeta, u, \gamma \rangle$ where $\zeta < \chi$ and u is a subset of $(J^i_\gamma)^{\leqslant\omega}$ of cardinality $< \chi$ and $\gamma < \alpha$, and let $g_1(s) = \alpha$, $g_2(s) = \zeta$, $g_3(s) = u$, $g_4(s) = \gamma$. For $\alpha = 0$, we let $J^i_\alpha = \{\langle 0, \zeta, i, -1 \rangle : \zeta < \chi\}$ and, to simplify notation, let distinct J^i_α's have distinct empty subsets. The well ordering is arbitrary. For $\eta \in (J^i_\alpha)^{\leqslant\omega}$ let $\mathrm{His}(\eta)$ be the smallest set S such that $\eta \in S$, and $\nu \in S, n < l(\nu), \rho \in g_3(\nu[n]) \Rightarrow \rho \in S$. Let $\mathrm{His}^*(\eta) = \{\nu[n] : \nu \in S, n < \omega\}$. Let $DP(\eta)$ be the order type of $T(\eta) = \{\beta : \mathrm{His}^*(\eta) \cap J^i_\beta \neq \emptyset$ for some i or $g_4(s) = \beta$ for $s \in \mathrm{His}^*(\eta)\}$. Clearly $\mathrm{His}^*(\eta)$, $\mathrm{His}(\eta)$ have cardinality $\leqslant \chi$ hence $DP(\eta) < \chi^+$.

We shall define sets $S^i_\gamma \subseteq \chi^{\omega}$ for $i < \mu, \gamma < \chi^+$. Now let, for $h : \mu \to \lambda$, $J_h = \sum_{i<\mu} J^i_{h(i)}$, and $I_h = \bigcup_{i<\mu} (J^i_{h(i)})^{<\omega} \cup \{\eta :$ for some $i < \mu, \eta \in (J^i_{h(i)})^{\omega}$, and $g_2(\eta) \in S^i_\gamma$ where $\gamma = DP(\eta)\}$, where we define $g_2(\eta)$ by $g_2(\eta)[l] = g_2(\eta[l])$; and let $M_h = EM(I_h, N_0)$. Suppose $h_1(i) < h_2(i) = \alpha_0$ and we describe the winning strategy of player I in $GI^{\omega}_{\chi}(M_{h_2}, M_{h_1})$.

In the nth move player I chooses $\{\bar{a}_\eta : \eta \in u_n\}$ which is defined by induction as follows: let player II choose in the mth move $\{\bar{b}_\eta : \eta \in u_m\}$; and let $\bar{b}_\eta = \tau_\eta(\bar{a}_{\bar{\nu}_\eta}, \bar{c}_\eta)$, $\bar{\nu}_\eta \in I_{h_1}$, $\bar{c}_\eta \in N_0$, and let

$$v_m = (J^i_{h_1(i)})^{\leq\omega} \cap \{\bar{\nu}_\eta[l] : l < l(\bar{\nu}_\eta), \eta \in u_m\}.$$

Let $u_0 = \{\langle\langle \alpha_0, \zeta, \emptyset, h_1(i) \rangle\rangle : \zeta < \chi_0\}$, and

$$u_{n+1} = \{\eta \frown \langle\langle \alpha_0, \zeta, v_n, h_1(i) \rangle\rangle : \zeta < \chi_{n+1}, \eta \in u_n\}.$$

Clearly the strategy is well defined and the nth move of the player I depends only on what player II has done in the previous moves. Let the history H of this play be $\bigcup \{\mathrm{His}(\bar{\nu}_\eta[l]) : \eta \in \bigcup_{n<\omega} u_n, l < l(\bar{\nu}_\eta)\}$; (Clearly $|H| \leq \chi$) $H_* = \bigcup \{\mathrm{His}^*(\bar{\nu}_\eta[l]) : \eta \in \bigcup_{n<\omega} u_n, l < l(\bar{\nu}_\eta)\}$. So it suffices to prove we can define the S^i_γ's so that those strategies are winning strategies for player I.

Suppose we have another such play, (where player I plays by the strategy we mention) with $h'_1, \ldots, \bar{c}'_\eta, \ldots, H'$, instead of $h_1, \ldots, \bar{c}_\eta, \ldots, H$. We call the two plays isomorphic if there are functions $f: H \to H'$ $f_\gamma(\gamma < \mu)$ with domain $\{\alpha: J^\gamma_\alpha \cap H_* \neq \emptyset$ or $\alpha = g_4(s)$ for $s \in H_*\}$, $f_*: H_* \to H'_*, g^n: u_n \to u'_n$ such that:

(1) f, f_*, g_n are one-to-one and onto H', H'_*, u'_n, resp.,

(2) If $s \in J^\gamma_j, f(s) \in J^\gamma_{f_\gamma(j)}$ and f_γ is increasing.

(3) If $s = \langle \beta, \zeta, u, \alpha \rangle \in J^\gamma_\beta$ then $f(s) = \langle f_\gamma(\beta), \zeta, f_*(u), f_\gamma(\alpha) \rangle$ where $f_*(u) = \{f_*(t): t \in u\}$.

(4) $(f(\eta))[n] = f_*(\eta[n]), f(\alpha_0) = \alpha'_0, i = i'$.

(5) $g^0(\langle\langle \alpha_0, \zeta, \emptyset, h_1(i)\rangle\rangle) = \langle\langle \alpha'_0, \zeta, \emptyset, h'_1(i')\rangle\rangle$, and

$$g^{n+1}(\eta^\frown\langle\langle \alpha_0, \zeta, v_n, h_1(i)\rangle\rangle) = g^n(\eta)^\frown\langle\langle \alpha'_0, \zeta, v'_1, h'_n(i')\rangle\rangle.$$

(6) If $\eta \in u_n, f_*(\bar{v}_\eta[l]) = \bar{v}'_\rho[l]$ where $\rho = g^n(\eta)$.

(7) If $\eta(l) \in u_{m(l)}$ $l \leq n$ then; denoting $\rho(l) = g_{m(l)}(\eta(l))$; $\bar{c}_{\eta(0)}{}^\frown \cdots {}^\frown \bar{c}_{\eta(n)}, \bar{c}'_{\rho(0)}{}^\frown \cdots {}^\frown \bar{c}'_{\rho(n)}$ realize the same type in the L_2-reduct of N_1.

(8) $f_* \upharpoonright J^\gamma_j: J^\gamma_j \to J^\gamma_{f_\gamma(j)}$ is order preserving; moreover if $s_1, s_2 \in J^\gamma_j \cap \operatorname{Dom} f_*$, then

$$\min\{\aleph_0, |\{t \in J^\gamma_j: s_1 \leq t < s_2\}|\} = \min\{\aleph_0, |\{t \in J^\gamma_{f_\gamma(j)}: f_*(s_1) \leq t < f_*(s_2)\}|\}.$$

Clearly, by the definition of the I_h's, the winner in two isomorphic games is the same (as the L_2-reduct of N_1 is $\aleph_1$-homogeneous). It is also clear that the number of isomorphism types of such plays is $\leq 2^\chi = \mu$. Let $\{G^j: j < \mu\}$ be a set of representatives. We define by induction on j, a set Γ_j of non-contradictory "requirements" $\eta \in S^\gamma_\xi, \eta \notin S^\gamma_\xi$, such that $|\Gamma_j - \Gamma_0| \leq |j| + \chi$ and Γ_{j+1} "ensures" the victory of player I in the play G^j.

Let Γ_0 be $\{\eta \in S^\gamma_\xi: \eta \in \chi^\omega, \eta$ is eventually zero$\}$. If Γ_j is defined, let us use for G^j the notation in the definition of the strategy of player I. Choose $\eta \in \prod_{n<\omega} \chi_n$ such that η does not "appear" in Γ_j (there is one by cardinality considerations). Let η_0 be defined by $\eta_0 \upharpoonright n \in u_n$, and $\langle \ldots, g_2(\eta_0[n]), \ldots \rangle = \eta$, and let $p = \{\varphi_n(x; \bar{a}_{\eta_0 \upharpoonright n}): n < \omega\}$, and $q = \{\varphi_n(x, \bar{b}_{\eta_0 \upharpoonright n}): n < \omega\}$, and $\gamma = DP(\eta_0)$, $i = g_3(\eta[0])$.

Choose, if possible, S^β_ξ's so that they satisfy the requirement of Γ_j and so that q is well defined and realized, in M_{h_1} say by $\tau(\bar{a}_{\bar{v}}, \bar{c})$; note $\eta_0 \upharpoonright n \in I_{h_1}$. Then define, when $\gamma = DP(\eta_0)$ $\Gamma_{j+1} = \Gamma_j \cup \{\eta \notin S^\gamma_i\} \cup \{\rho \in S^\beta_\xi$: for some $m < l(\bar{v}), l(\bar{v}[m]) = \omega, \rho = g_2(\bar{v}[m]), \beta = DP(\bar{v}[m])$ and $\bar{v}[m] \in \bigcup_{\alpha<\lambda} (J^\xi_\alpha)^{<\omega}\}$. There is no contradiction as we can assume, in the third term of the union above, that $\xi = i \Rightarrow \beta < \gamma$ because if $l(\bar{v}[m])$ is ω, $g_2(\bar{v}[m]) \in S^\beta_i$ then $(\bar{v}[m]) \upharpoonright l \in \bigcup_{k<\omega} v_k$ for every $l < \omega$ [otherwise we can make $\bar{v}[m]$ eventually zero but still $\tau(\bar{a}_{\bar{v}}, \bar{c})$ will realize q] hence by the strategy definitions, $\mathrm{His}^*(\eta) \supseteq \mathrm{His}^*(\bar{\nu}[m])$, $\alpha_0 = \max T(\eta) \notin \mathrm{His}^*(\bar{\nu}[m])$, so $\beta < \gamma$.

So Γ_{j+1} ensures q is realized in M_{h_1}, but p is not realized in M_{h_1} hence player I wins. But if there are no such S_ξ^β's, let $\Gamma_{j+1} = \Gamma_j \cup \{\eta \in S_\xi^j\}$. For limit δ, $\Gamma_\delta = \bigcup_{j<\delta} \Gamma_j$; so Γ_μ shows us how to define the S's properly:

$$S_\xi^j = \{\eta : \{\eta \in S_\xi^j\} \in \Gamma_\mu\}.$$

THEOREM 2.7: *Suppose T is stable but not superstable, $\lambda \geq |T| + \aleph_1$, $\lambda > \mu > \aleph_0$, μ regular. Then $I(\lambda, T) \geq 2^\mu$.*

Proof. The proof is similar to that of 2.2.

By III, Exercise 4.13 we can find $\bar{a}_\eta \in \mathfrak{C}$ for $\eta \in \mu^{\leq\omega}$ and $\varphi_n \in L$ such that:

(1) For $\eta \in \mu^\omega$, $\nu \in \mu^n$, $(n > 0)$ $\models \varphi_n[\bar{a}_\eta; \bar{a}_\nu]$ iff $\nu \lessdot \eta$.

(2) For $\eta \in \mu^n$, the $\varphi_{n+1}(\bar{x}; \bar{a}_{\eta\frown\langle i\rangle})$ $i < \mu$ are pairwise contradictory.

(3) For $\eta \in \mu^{\leq\omega}$, $tp(\bar{a}_\eta, A_\eta^1)$ does not fork over A_η^2 where $A_\eta^2 = \bigcup \{\bar{a}_{\eta\restriction n} : n < l(\eta)\}$, $A_\eta^1 = \bigcup \{\bar{a}_\nu : \nu \neq \eta$ but not $\eta \lessdot \nu\}$.

Let, for $\delta < \mu$, $\mathrm{cf}\,\delta = \omega, \eta_\delta$ be an increasing sequence of length ω of ordinals with limit δ, and for $w \subseteq \{\delta < \mu : \mathrm{cf}\,\delta = \aleph_0\}$ let M_w be a $\mathbf{F}^f_{\aleph_0}$-constructible model over $A_w = C_w^\lambda$ of cardinality λ, where $C_w^\alpha = \{\bar{a}_\eta : \eta \in \alpha^{<\omega}$ or $\eta = \eta_\delta$, $\delta \in w$, $\delta < \alpha\}$. Let the construction-sequence of M_w be $\{a_i^w : i < i(0) < \lambda^+\}$. (See table in IV, 2 and IV, 3.1). Let $A_w^\alpha = \{a_i^w : i < \alpha\}$, and let $B_\alpha^w \subseteq A_w \cup A_w^\alpha$ be a finite set over which $\mathrm{tp}(a_\alpha^w, A_w \cup A_w^\alpha)$ does not fork. So if $(\forall \alpha < \delta)(B_\alpha^w \subseteq C_w^\delta \cup A_w^\alpha)$, $\delta < \mu$, $\mathrm{cf}\,\delta = \omega$, $\delta \notin w$, $\eta_l \in \mu^{\leq\omega}$, $\eta_l \restriction n_l \in \delta^{\leq\omega}$ but $n_l = \omega$ or $\eta_l \restriction (n_l + 1) \notin \delta^{<\omega}$ for $l < m$; then $C = \bigcup \{\bar{a}_\nu : \nu \lessdot \eta_l \restriction n_l, l < m$, or $\nu = \eta_l$, $n_l = \omega\}$ is finite, and $\mathrm{tp}(\bar{a}_{\bar{\eta}}, C_w^\delta)$ does not fork over C, where $\bar{\eta} = \langle \eta_0, \ldots, \eta_{m-1} \rangle$, hence, by III, 4.13, and as $B_\alpha^w \subseteq C_w^\delta \cup A_w^\delta$, also $\mathrm{tp}(\bar{a}_{\bar{\eta}}, C_w^\delta \cup A_w^\delta)$ does not fork over C. Now by IV, 3.2 if $\bar{b} \in A_w \cup A_w^\mu$ there is finite $B \subseteq A_w \cup A_w^\delta$ such that $\mathrm{tp}(\bar{b}, A_w \cup A_w^\delta)$ does not fork over B, and there is $\bar{\eta} \in \mu^{\leq\omega}$ such that $B \cap A_w \subseteq \bar{a}_{\bar{\eta}}$, and there is finite C as above; so $\mathrm{tp}_*(\bar{b} \cup B, C_w^\delta \cup A_w^\delta)$, hence $\mathrm{tp}(\bar{b}, C_w^\delta \cup A_w^\delta)$ does not fork over $C \cup (B \cap (C_w^\delta \cup A_w^\delta))$. So for every $\bar{b} \in A_w \cup A_w^\mu$ $\mathrm{tp}(\bar{b}, C_w^\delta \cup A_w^\delta)$ is $\mathbf{F}^f_{\aleph_0}$-isolated.

As in 2.2 it suffices to prove that if $f : M_w \to M_u$ is an elementary embedding, then for some closed unbounded $S \subseteq \mu$, $w \cap S \subseteq u \cap S$. By renaming we can assume f maps $A_w \cup A_w^\mu$ onto $A_u \cup A_u^\mu$ (see IV, 3.3), hence $S_1 = \{\delta : \delta < \mu$, f maps $C_w^\delta \cup A_w^\delta$ onto $C_u^\delta \cup A_u^\delta\}$, $S = \{\delta \in S_1 :$

δ a limit point of $S_1\}$ are closed and unbounded. If $\delta \in w \cap S$ but $\delta \notin u$, then $\bar{b} = f(\bar{a}_{\eta(\delta)})$ will give us a contradiction, as $\operatorname{tp}(\bar{b}, C^\delta_w \cup A^\delta_w)$ splits strongly over any $C^\alpha_w \cup A^\alpha_w$ for any $\alpha < \delta$ (because $\{\bar{a}_{\eta^\frown\langle i\rangle}: \delta \le i < \mu\}$ is indiscernible over $C^\delta_w \cup A^\delta_w$, also $\{f(\bar{a}_{\eta^\frown\langle i\rangle}): \delta \le i < \mu\}$ is indiscernible over $C^\delta_u \cup A^\delta_u$ for $\delta \in S$).

EXERCISE 2.1: Suppose $\lambda = |T_1| = \aleph_{\alpha+\beta}$, $(\forall\kappa < \aleph_\alpha)(\kappa^{\aleph_0} < \aleph_\alpha)$ $\lambda \ge \mu = \chi^{\aleph_0} = 2^\chi$; $\beta < \chi^+$ or $2^{|\beta|} \le 2^\chi$; T not superstable, then $I(\lambda, T_1, T) \ge 2^\mu$.

PROBLEM 2.2: Uniformize the proofs (maybe use the filters from Kueker [Kk 77] and Shelah [S 75] Section 3).

CONJECTURE 2.3: $IE(\lambda, T_1, T) = 2^\lambda$ for T unsuperstable, $\lambda > |T_1|$.

EXERCISE 2.4: Show that for λ regular $> |T|$, T not superstable the partial order $(P(\lambda), \subseteq)$ can be embedded into $\{M: M \models T, \|M\| = \lambda\}$ quasi-ordered by elementary embeddability.

EXERCISE 2.5: For each cardinal $\lambda = \lambda^{\aleph_0}$, show there is no Boolean algebra M_0 of cardinality λ, such that any other Boolean algebra M of cardinality $\le \lambda$ has an embedding in M_0, preserving countable intersections. (Hint: See 2.5, Th. 12, Grossberg and Shelah [GSh 83].)

EXERCISE 2.6: Generalize VII, 3.7 and 1.7 to the case $|T_1| < 2^{\aleph_0}$, assuming MA (Martin Axiom).

VIII.3. Saturated models and the case $\lambda = |T_1|$

DEFINITION 3.1: The indexed-set $\langle \bar{a}_s: s \in I\rangle$ (I an index-model) is called κ-skeleton-like in (a model) M when:

(1) It is indiscernible.

(2) For any $\bar{c} \in M$ there is $J \subseteq |I|$, $|J| < \kappa$ such that if $\bar{s}, \bar{t} \in I$, $\bar{s} \sim \bar{t} \bmod J$, then for any $\varphi \in L(M)$ $M \models \varphi[\bar{a}_{\bar{s}}, \bar{c}] \equiv \varphi[\bar{a}_{\bar{t}}, \bar{c}]$.

DEFINITION 3.2: The orders I_1, I_2 are called κ-contradictory if there is no model M, $m < \omega$, an antisymmetric formula in $L(M)$ $[\bar{x} < \bar{y}](l(\bar{x}) = l(\bar{y}) = m)$, an order J with lower cofinality $\ge \kappa$ and sequences $\bar{a}_s \in M$ for $s \in I_1 \cup I_2 \cup J$ (we assume for notational simplicity I_1, I_2, J are disjoint) such that:

(1) $\langle \bar{a}_s : s \in I_1 + J\rangle$, $\langle \bar{a}_s : s \in I_2 + J\rangle$ are κ-skeleton like in M.

(2) For $l = 1, 2$, $s, t \in I_l + J$, $M \models [\bar{a}_s < \bar{a}_t]^{\mathrm{if}(s \le t)}$.

LEMMA 3.1: (1) *Suppose M_0, M are models of T_1, κ is regular $|M| = |M_0| \cup \{a_i : i < \alpha\}$, $\bar{a}_s \in |M_0|$ for $s \in I$ and $\langle \bar{a}_s : s \in I\rangle$ is κ-skeleton like in M_0. If the type each a_i realizes over $|M_0| \cup \{a_j : j < i\}$ is $\mathbf{F}^b_\kappa$-isolated, or even $\mathbf{F}^p_\kappa$-isolated, then $\langle \bar{a}_s : s \in I\rangle$ is κ-skeleton like in M.*

(2) *The skeleton of $EM^1(I, \Phi)$, or of $EM^1(I, N)$ are κ-skeleton like, for every $\kappa \ge \aleph_0$.*

(3) *If $\langle \bar{a}_s : s \in I\rangle$ is the skeleton of $M = EM^1(I, N)$, M_1 the submodel of M^λ/D (D ultrafilter over λ) whose universe is $\{h/D$: for some $J \subseteq I$, $|J| < \kappa$, $\{h(i) : i < \lambda\} \subseteq \mathrm{dcl}\{\bar{a}_s : s \in J\}\}$ (clearly it exists), then $\langle \bar{a}_s : s \in I\rangle$ is κ-skeleton like in M_1 and $M_1 \prec M^\lambda/D$(dcl—in L_1).*

Proof. (1) Prove for $\bar{c} \in |M_0| \cup \{a_i : i < \beta\}$ by induction on β.

(2) and (3) immediate.

THEOREM 3.2: *If $\lambda \ge |T_1| + \kappa^+$, $\lambda^{<\kappa} = \lambda$, T unstable, κ regular then there are $M_i \in \mathrm{PC}(T_1, T)$ for $i < 2^\lambda$ such that M_i has cardinality λ and*

(1) *For $i \ne j$, M_i, M_j are not isomorphic.*

(2) *M_i is κ-compact, and κ-homogeneous and if $|D(T)| \le \lambda$, M_i is κ-saturated.*

(3) *M_i is the L-reduct of M^1_i, M^1_i is κ-compact, κ-homogeneous; and if $|D(T_1)| \le \lambda$, also κ-saturated.*

DEFINITION 3.3: $(*)^\kappa_\lambda$ means that there is a family of 2^λ subsets of λ, each of cardinality λ, the intersection of any two is $< \kappa$.

Remark. This is just not AD$(2^\lambda, \lambda, \lambda, \kappa)$ (see Definition VII, 1.11). A related property appears in 1.9. On independence results concerning it see Baumgartner [Ba 76].

Proof. Let Φ be proper for (ω, T_1), such that the antisymmetric formula $[\bar{x} < \bar{y}] \in L$ orders the skeleton of $EM(\omega, \Phi)$. We assume the assumptions from VII, Section 2.

Case I: not $(*)^\kappa_\lambda$. By 3.3 of the Appendix there are 2^λ pairwise κ-contradictory orders I_α, $\alpha < 2^\lambda$; each of cardinality λ. Let, for $\alpha < 2^\lambda$, $i \le \lambda$ I^i_α be an order isomorphic to the converse of I_α, and $s^i_\alpha \in I_\alpha$. Let, for $\alpha < 2^\lambda$, $J_\alpha = \sum_{i \le \lambda} I^i_\alpha$. Let $N^1_\alpha = EM^1(J_\alpha, \Phi)$, so $\|N^1_\alpha\| = \lambda$ (T_1 has

Skolem functions). Clearly $\langle \bar{a}_s : s \in J_\alpha \rangle$ is the skeleton of N^1_α. Let $N^1_\alpha \prec M^1_\alpha$, M^1_α satisfy (2) and (3) from the theorem, but such that $\langle \bar{a}_s : s \in J_\alpha \rangle$ is still κ-skeleton like (this can be done by 3.1 and VII, 4.4), and $\lambda = \|M^1_\alpha\|$ (by cardinality considerations this is possible). Suppose the number of non-isomorphic M_α's is $< 2^\lambda$ and $\mu \leq 2^\lambda$ is regular, where M_α is the L-reduct of M^1_α. Then for some M $S = \{\alpha : M_\alpha \cong M\}$ has cardinality $\geq \mu$ and let $f_\alpha : M_\alpha \to M$ be such an isomorphism. Let $\bar{b}^\alpha_s = f_\alpha(\bar{a}_s)$ for $s \in J_\alpha, \alpha \in S$. So $\langle \bar{b}^\alpha_s : s \in J_\alpha \rangle$ is κ-skeleton like in M. Let $W_\alpha = \{\bar{b}^\alpha_s : s = s^i_\alpha, i < \lambda\}$. Suppose $\alpha \neq \beta$; $\alpha, \beta \in S$, $|W_\alpha \cap W_\beta| \geq \kappa$; then, as λ is well ordered, there are for $\xi < \kappa$ $i(\xi) < \lambda$, $j(\xi) < \lambda$ such that $\bar{b}^\alpha_{s(\xi)} = \bar{b}^\beta_{t(\xi)}$ where $s(\xi) = s^{i(\xi)}_\alpha$, $t(\xi) = s^{j(\xi)}_\beta$, and $i(\xi), j(\xi)$ are strictly increasing.

Let $\delta(1) = \sup\{i(\xi) : \xi < \kappa\} \leq \lambda$, $\delta(2) = \sup\{j(\xi) : \xi < \kappa\} \leq \lambda$. Now clearly

$$\langle \bar{b}^\alpha_s : s = s(\xi),\ \xi < \kappa \text{ or } s \in I^{\delta(1)}_\alpha \rangle,$$

$$\langle \bar{b}^\beta_s : s = t(\xi),\ \xi < \kappa \text{ or } s \in I^{\delta(2)}_\alpha \rangle$$

are κ-skeleton like in M. Looking at Definition 3.2 this clearly proves I_α, I_β are not κ-contradictory; contradiction.

So $\alpha \neq \beta \in S \Rightarrow |W_\alpha \cap W_\beta| < \kappa$; also $|W_\alpha| = \lambda$, and all those W_α are subsets of $|M|^m$ for suitable m, which has cardinality λ. If 2^λ is regular, choose $\mu = 2^\lambda$, so we have proved $(*)^\kappa_\lambda$, hence finished, otherwise use 3.1 of the Appendix to get the same contradiction.

Case II: $\lambda = \mu^+$, λ and μ are regular. Let I_0 be the set $\{\langle i, j, \gamma \rangle : i < \lambda, j < \lambda, \gamma < \mu\}$ ordered by: $\langle i_1, j_1, \gamma_1 \rangle < \langle i_2, j_2, \gamma_2 \rangle$ iff $i_1 < i_2$ *or* $i_1 = i_2$, $j_1 > j_2$, *or* $i_1 = i_2, j_1 = j_2, \gamma_1 < \gamma_2$. We identify $\langle i, j, 0 \rangle$ with $\langle i, j \rangle$. For every $I \subseteq I_0$ let $M_0(I) = EM^1(I, \Phi)$, $M_1(I) = M_0(I)^\lambda / D$ where D is a regular, good ultrafilter over λ (exists by VI, 3.1). Let $d_0(\langle i, j, \gamma \rangle) = \{i, j, \gamma\}$, and for $J \subseteq I_0$, $d_0(J) = \bigcup_{s \in J} d_0(s)$. For every $I \subseteq I_0$, $h/D \in M_1(I)$ let $d_1(h) = \{s \in I$: for some $i < \lambda$, $h(i) = \tau(\bar{a}_{\bar{t}})$, $\bar{t}$ minimal, and $s = \bar{t}[l]$ for some $l\}$, (note w.l.o.g. $\bar{t}$ is uniquely chosen). Now for $I \subseteq I_0$, let $M_2(I)$ be the submodel of $M_1(I)$ whose universe is $\{h/D : |d_1(h)| < \mu\}$. By 3.1(3) $M_2(I) \prec M_1(I)$ and $\langle \bar{a}_s : s \in I \rangle$ is μ-skeleton-like in $M_2(I)$. It is also clear that $M_2(I)$ is κ-saturated because if $J \subseteq I$, $|J| < \mu$ then $M_2(J) = M_1(J) \prec M_2(I)$ is even μ-saturated (by VI, 2.3 and VI, 2.11). In the same way we can prove that for such J, if p is an m-type over $M_2(I)$ finitely satisfiable in $M_2(J)$, $|p| \leqslant \lambda$ then p is realized in $M_2(J)$.

Let $\mathfrak{B} = \langle R(\bar{\kappa}), \in \rangle$, and define $\mathfrak{B}_\alpha$ for $\alpha \leq \mu$ such that:

(1) $\|\mathfrak{B}_\alpha\| = \lambda$, $\mathfrak{B}_\alpha \prec \mathfrak{B}$.

(2) If $a \in |\mathfrak{B}_\alpha|$ and a has cardinality $\leq \lambda$ (in $R(\bar{\kappa})$), then $b \in a \Rightarrow b \in |\mathfrak{B}_\alpha|$.

(3) $\langle \mathfrak{B}_\beta : \beta \leq \alpha \rangle \in |\mathfrak{B}_{\alpha+1}|$, $|\mathfrak{B}_\delta| = \bigcup_{i<\delta} |\mathfrak{B}_i|$.

(4) L, I_0, D, $M_0(I_0)$, $J \mapsto M_1(J)$, $\langle \bar{a}_s : s \in I_0 \rangle$ all belong to $\mathfrak{B}_0$; and $I \in \mathfrak{B}_1$.

(5) In fact, $\mathfrak{B}_\alpha$ depends on I so when confusion may arise we write $\mathfrak{B}_\alpha = \mathfrak{B}_\alpha(I)$, but $\mathfrak{B}_0(I) = \mathfrak{B}_0(I_0)$.

Let for $\alpha \leq \mu$, $M_2^\alpha(I)$ be $M_2(I)$ as interpreted in $\mathfrak{B}_\alpha$. Clearly $M_2^\alpha(I) \prec M_2(I)$ and $M_2^\delta(I) = \bigcup_{i<\delta} M_2^i(I)$.

Let $N_2^\alpha(I)$ be the L-reduct of $M_2^\alpha(I)$. Notice that when $\mathrm{cf}\,\delta \geq \kappa$, $N_2^\delta(I)$ satisfies conditions (2) and (3) from the theorem.

For $w \subseteq \lambda$ let $I(w) = \{\langle i, j, \gamma \rangle \in I_0 : i \notin w \Rightarrow j < \mu\}$. Suppose $w_1, w_2 \subseteq \{\delta : \delta < \lambda, \mathrm{cf}\,\delta = \mu\}$, $w = w_1 - w_2$ is stationary and there is an isomorphism f from N_1^μ onto N_2^μ where $N_l^\alpha = N_2^\alpha(I(w_l))$ (for $l = 1, 2$). We shall get a contradiction, and this is sufficient to prove the theorem in this case (as in the proof of 2.2).

Let $\bar{b}_s = f(\bar{a}_s)$. We can find a stationary $w' \subseteq w$, and $\zeta < \mu$ such that for $\alpha \in w'$, $\bar{b}_{\langle \alpha, 0 \rangle} \in N_2^\zeta$ and $U_\alpha = \{i < \lambda : \bar{b}_{\langle \alpha, i \rangle} \in N_2^\zeta\}$ is a stationary subset of λ. Let, for $a \in M_1(I_0)$, $d_2(a)$ be a set $J \subseteq I$ of minimal cardinality such that $d_1(h) \subseteq J$ where $a = h/D$ (we can assume the function d_2 belong to $\mathfrak{B}_0$). Also, for $\bar{a} \in M_1(I_0)$, $d_2(\bar{a}) = \bigcup_l d_2(\bar{a}[l])$. So for each $\bar{a} \in M_2(I_0)$, $|d_2(\bar{a})| < \mu$. Let $d_3(\bar{a}) = \{\alpha < \lambda$: for some β, $\langle \alpha, \beta \rangle \in d_2(\bar{a})$ or $\langle \beta, \alpha \rangle \in d_2(\bar{a})\}$. So clearly, for $\bar{a} \in M_2(I_0)$, $d_3(\bar{a})$ is a subset of λ of cardinality $< \mu$. Hence by a double use of 1.8 of the Appendix, there are w'', U'_α, ξ_0 and V_α, V such that:

(1) $V \subseteq \lambda$, $|V| < \mu$; $w'' \subseteq w'$, $U'_\alpha \subseteq U_\alpha$; and w'', U'_α are stationary, for $\alpha \in w''$ and $V_\alpha \subseteq \lambda$, $|V_\alpha| < \mu$.

(2) Let $\alpha, \beta \in w''$, $\alpha < \beta$.

(i) $d_3(\bar{b}_{\langle \alpha, 0 \rangle}) \cap d_3(\bar{b}_{\langle \beta, 0 \rangle}) \subseteq V$, $V_\alpha \cap V_\beta \subseteq V$, and $V \subseteq V_\alpha$, $V_\alpha \cap \mu \subseteq V$,

(ii) if $i \in d_3(\bar{b}_{\langle \alpha, 0 \rangle})$, $i \notin V$ then $i \geq \alpha$,

(iii) if $i \in d_3(\bar{b}_{\langle \alpha, 0 \rangle})$, then $i < \beta$.

(3) Let $\alpha \in w''$, $i, j \in U'_\alpha$, $i < j$.

(i) $d_3(\bar{b}_{\langle \alpha, i \rangle}) \cap d_3(\bar{b}_{\langle \alpha, j \rangle}) \subseteq V_\alpha$,

(ii) if $\xi \in d_3(\bar{b}_{\langle \alpha, i \rangle})$, $\xi \notin V_\alpha$, then $\xi \geq i$,

(iii) if $\xi \in d_3(\bar{b}_{\langle \alpha, i \rangle})$, then $i, \xi < j$.

(4) If $i \in U'_\alpha$, then $i > \sup V$, and $i > \alpha + \alpha$, and $\alpha > \mu$.

(5) There are sets $B_\alpha \in \mathfrak{B}_{\zeta+1}$ $(\alpha < \lambda)$, $B_\alpha \subseteq N_2^\zeta$, $|B_\alpha| < \lambda$; $\beta \in w'$ $\beta < \alpha \Rightarrow \bar{b}_{\langle \beta, 0 \rangle} \in B_\alpha$, and $B_\delta = \bigcup_{\alpha<\delta} B_\alpha$ (this only because $\mathfrak{B}_\zeta \in \mathfrak{B}_{\zeta+1}$).

(6) $\alpha \in w''$, $c \in B_\alpha \Rightarrow d_3(c) \subseteq \alpha$.

(7) $\xi_0 > \sup(V \cap \mu)$, $\xi_0 < \mu$.

Choose $\delta \in w''$ which is an accumulation point of w'', (so bigger than ξ_0, and than μ). So $\delta \in w_1$, $\delta \notin w_2$, $\mathrm{cf}(\delta) = \mu$ and $\langle \delta, i \rangle \in I(w_1) \Leftrightarrow i < \lambda$ but $\langle \delta, i \rangle \in I(w_2) \Leftrightarrow i < \mu$. Let $J_0 = \{\langle i, j, \gamma \rangle : \langle i, j, \gamma \rangle \in I_0; i, j, \gamma \in V\}$. Define $J = J_0 \cup \{\langle \delta, \xi_0 + n \rangle : 0 < n < \omega\} \cup \{\langle \alpha, \delta, \xi_0 + n \rangle : n < \omega, \alpha \in V, \alpha \in w_2\}$. Clearly $|J| < \mu$, $J \subseteq I(w_2)$.

Let (for $\alpha < \delta$)

$$A = \{a: a \in N_2^\xi, d_2(a) \subseteq \{\langle \alpha, i, \gamma \rangle : \alpha = \delta \Rightarrow i < \xi_0, \gamma < \xi_0\}\},$$

$$A_\alpha = \{a \in A : d_3(a) \subseteq \alpha \cup (\lambda - \delta)\},$$

$$p = \{\varphi(\bar{x}; \bar{c}) : \bar{c} \in A, \varphi \in \mathrm{L}, \text{ and for all sufficiently large } \alpha < \delta, \; N_2^\mu \models \varphi[\bar{b}_{\langle \alpha, 0 \rangle}; \bar{c}]\},$$

$$p_\alpha = p \restriction A_\alpha.$$

As f is an isomorphism, $\langle \bar{b}_s : s \in I(w_1) \rangle$ is μ-skeleton-like in N_2^μ, so p is a complete type over A. We now show that p is realized in $M_2(I(w_2))$, and even in $M_2(J)$. As $|J| < \mu$, $|p| \le \lambda$, it suffices to prove p is finitely satisfiable in $M_2(J)$; so as $p = \bigcup_{\alpha < \delta} p_\alpha$, p_α increasing it suffices to prove any finite $q \subseteq p_\alpha$ is realized in it. By the choice of δ, p there is $\beta \in w''$, $\alpha < \beta < \delta$, $\beta > \mu$ such that $\bar{b}_{\langle \beta, 0 \rangle}$ realizes q and let it be $\langle h_0/D, \ldots, h_{n-1}/D \rangle$ where $d_1(h_l) \subseteq d_2(\bar{b}_{\langle \beta, 0 \rangle})$ for $l < n$. Now define $h_0', \ldots, h_{n-1}'$ such that:

If $h(i) = \tau(\bar{a}_{\bar{s}})$ (minimal $\bar{s}$) then $\bar{h}'(i) = \tau(\bar{a}_{\bar{t}})$ and $\bar{t}$ satisfies the following (here $\bar{h}(i) = \langle h_l(i) : l < n \rangle$):

(α) $\bar{t} \sim \bar{s}$,

(β) $\bar{t} \in J$,

(γ) if $\bar{s}[m] \in J_0$, then $\bar{t}[m] = \bar{s}[m]$,

(δ) if $\bar{s}[m] = \langle i, j, \gamma \rangle$, $i > \beta$, then $\bar{t}[m] \in \{\langle \delta, \xi_0 + k \rangle : 0 < k < \omega\}$,

(ε) if $\bar{s}[m] = \langle i, j, \gamma \rangle$, $i < \beta$ (so $i \in V$), $\bar{s}[m] \notin J_0$ (so $\beta \le j < \delta$) then $\bar{t}[m] \in \{\langle i, \delta, \xi_0 + k \rangle : k < \omega\}$.

Clearly this can be done, and $\bar{s} \sim \bar{t} \bmod \{\langle i, j, \gamma \rangle : i = \delta \Rightarrow j < \xi_0$, and $i < \delta \Rightarrow i \le \alpha\}$ hence if $\bar{b} = \langle h_0'/D, \ldots, h_{n-1}'/D \rangle$ then $\bar{b}^\alpha$ realizes q, and by the above $\bar{b} \in M_2(J)$.

So by previous remarks there is $\bar{b} \in M_2(I(w_2))$ realizing p. As $\langle \bar{b}_s : s \in I(w_1) \rangle$ is μ-skeleton-like in N_2^μ, $|B_\delta| < \lambda$, there is $\xi(1) < \lambda$ such that all $\bar{b}_{\langle \delta, i \rangle}$ ($\xi(1) \le i < \lambda$) realize the same type (in N_2^μ) over B_δ. Let

$$q = \{\bar{c} < \bar{x} : \bar{c} \in B_\delta, N_2^\xi \models \bar{c} < \bar{b}_{\langle \delta, \xi(1) \rangle}\}$$

$$\cup \{\bar{x} < \bar{c} : \bar{c} \in A, \mathrm{tp}(\bar{c}, B_\delta, N_2^\xi) = \mathrm{tp}(\bar{b}_{\langle \delta, \xi(1) \rangle}, B_\delta, N_2^\xi)\}.$$

It is easy to check that $q \subseteq p$ (by (6)) so $\bar{b}$ realizes q, and $q \in \mathfrak{B}_{\zeta+1}$. So, as

$$\mathfrak{B} \models \text{“in } M_2(I(w_2)) \text{ there is an element realizing } q\text{”}$$

also $\mathfrak{B}_{\zeta+1}$ satisfies this, so some $\bar{b}' \in N_2^{\zeta+1}$ realizes q, so $\langle \bar{b}_s: s \in I(w_1)\rangle$ is not μ-skeleton-like (as U'_δ is an unbounded subset of λ, and $i \in U'_\delta \Rightarrow \bar{b}_{\langle\delta,i\rangle} \in A$). Contradiction.

Case III: $\lambda = \lambda_1$ is a limit cardinal, or $\lambda = \lambda_1^+$, λ_1 singular; and for some $\mu < \lambda_1$, $2^\mu = 2^\lambda$. We can choose regular $\mu < \lambda$, $\mu > \kappa$, $\mu^+ \neq \text{cf}\,\lambda_1$, $2^\mu = 2^\lambda$. The proof is similar to the previous case. W.l.o.g. $\lambda = |EM(\varphi, \Phi)|$.

Define I_0, $M_0(I)$, $M_1(I)$ as in Case II. Let, for $J \subseteq I_0$ $pr_1(J) = \{\alpha$: for some β, $\langle\alpha, \beta\rangle \in J\}$ and $M_2(I)$ will be the submodel of $M_1(I)$ with universe $\{h/D$: $pr_1[d_1(h)]$ has cardinality $<\mu$, $d_1(h)$ has cardinality $\leq\mu\}$. $\mathfrak{B}_\alpha$, $M_2^\alpha(I)$, are defined as in case II. For any function $g: \mu^+ \to \lambda$, such that each $g(i)$ is a regular cardinal $>\mu^+$, $<\lambda_1$, let

$$I(g) = \{\langle i,j\rangle \in I_0 : j < g(i), i < \mu^+\}.$$

Now we suppose g_1, g_2 are such functions, and for every $\chi < \lambda_1$ $\{i < \mu^+$: $\text{cf}\, i = \mu$, $g_1(i) > g_2(i) > \chi\}$ is stationary, and f is an isomorphism from N_1^μ onto N_2^μ, and get a contradiction, and this is sufficient [where $N_l^\alpha = N_2^\alpha(I(g_l))$].

Easily we can find $\zeta < \mu$, $B \in \mathfrak{B}_\zeta$, $\mu^+ < |B| < \lambda_1$ and

$$S = \{\alpha < \mu^+ : \bar{b}_{\langle\alpha,0\rangle} \in B, \text{cf}\,\alpha = \mu\}$$

is stationary, and let $S = \{\beta_\alpha: \alpha < \mu^+\}$ where β_α is increasing (so $\beta_\alpha \geq \alpha$). Let $w' = \{\delta < \mu^+ : \text{cf}\,\delta = \mu, g_1(\delta) > g_2(\delta) > |B|\}$ (so w' is stationary) and let w'', V be such that (exists by 1.8 of the Appendix):

(1) w'' is stationary, $w'' \subseteq w'$.

(2) $V \subseteq \mu^+$, $|V| < \mu$.

(3) If $\alpha \in w''$, $i \in pr_1[d_2(\bar{b}_{\langle\beta_\alpha,0\rangle})]$, $i \notin V$, then $i \geq \alpha$.

(4) If $\alpha \in w''$, $i \in V$, then $i < \alpha$.

(5) Let $B_\alpha = B \cap \{c: c \in N_2^\mu, pr_1[d_2(c)] \subseteq \alpha\}$ so $B_\alpha \in \mathfrak{B}_{\zeta+1}$, B_α is increasing and continuous.

(6) If $\delta \in w''$, $\alpha < \delta$, then $\bar{b}_{\langle\beta_\alpha,0\rangle} \in B_\zeta$.

Choose $\delta \in w''$, δ an accumulation point of w'' (so $\text{cf}(\delta) = \mu$, $g_1(\delta) > g_2(\delta) > |B|$). Let $J_0 = \{\langle i,j\rangle: \langle i,j\rangle \in I_0$, $i \in V$; and for some $\alpha \in w''$, $\alpha < \delta$, $\langle i,j\rangle \in d_2(\bar{b}_{\langle\beta_\alpha,0\rangle})\}$. Now there are $\xi_0 < g_2(\delta)$ and $U \subseteq g_1(\delta)$, $|U| = g_1(\delta)$ such that for $i \in U$, $d_2(\bar{b}_{\langle\delta,i\rangle}) \cap \{\langle\delta,j\rangle: \xi_0 \leq j < g_2(\delta)\} = \emptyset$.

Let $J = J_0 \cup \{\langle\delta, \xi_0 + n\rangle: n < \omega\}$. Clearly $|J| \leq \mu$, $|pr_1(J)| < \mu$ (as $\mu > \aleph_0$). Let

$$A = \{a: a \in N_2^\mu, d_2(a) \cap \{\langle\delta, i\rangle: \xi_0 \leq i < g_2(\delta)\} = \emptyset\}.$$

A_α, p, p_α are defined as in Case II, and so is the existence of $\bar{b} \in M_2(I(g_2))$ which realizes p (only part (ε) falls). The rest is like Case II—we prove q is realized in N_2^μ and so get the contradiction.

So it suffices to prove that Cases I–III, exhaust all possibilities.

Suppose not $(*)_\lambda^\kappa$, and it suffices to prove that for some regular μ, $\mu^+ \leq \lambda$, $2^\mu = 2^\lambda$. By 3.2(1) of the Appendix $\lambda^\kappa = 2^\lambda$, so if $\lambda \leq 2^\kappa$, $2^\kappa = 2^\lambda$ so $\mu = \kappa$ will suffice. If $\lambda > 2^\kappa$, by 3.2(2) of the Appendix for some $\chi, \kappa^+ \leqslant \chi^+ < \lambda, \chi^\kappa \geqslant \lambda$ so $\chi^\kappa = 2^\lambda$, and $2^\chi = 2^\lambda$. So we finish.

COROLLARY 3.3: *If $\lambda \geq |T_1| + \aleph_1$, T unstable, then $I(\lambda, T_1, T) = 2^\lambda$.*

Proof. Take $\kappa = \aleph_0$ in 3.1.

COROLLARY 3.4: *If $\lambda \geq |T_1| + \aleph_1$, T not superstable, then $I(\lambda, T) = 2^\lambda$.*

Proof. By 3.3, 2.1, 2.7 and 1.7(1).

THEOREM 3.5: *Suppose $\lambda \geq \mu \geq |T_1| + \kappa^+$, $\lambda^{<\kappa} = \lambda$, $\kappa < \kappa(T)$, κ, μ are regular; and $\chi < \mu \Rightarrow \chi^{<\kappa} < \mu$. Suppose also that $\mu > |T_1|$ or $\chi < \mu \Rightarrow \chi^\kappa \leq \lambda$. Then there are models M_i^1, $i < 2^\mu$ of T_1, with L-reducts M_i, of cardinality λ such that:*

(1) *The M_i's are pairwise non-isomorphic.*

(2) *If $\mu > |T_1|$, then no M_i is elementarily embeddable in M_j for $i \neq j$.*

(3) *M_i is κ-compact, κ-homogeneous, and, if $|D(T)| < \mu$, κ-saturated. Similarly for M_i^1.*

Proof. For simplicity we assume $\mu > |T_1|$. Using VII, 3.5(3) and 3.6 we can find Φ proper for $(\omega^{\leq\kappa}, T_1)$ such that the skeleton of $EM(\omega^{\leq\kappa}, \Phi)$ has the form $\langle\langle\varphi_\alpha, 1\rangle: \alpha < \kappa$ a successor$\rangle$. For $\delta < \mu$, cf $\delta = \kappa$ choose $\eta_\delta \in \mu^\kappa$ increasing with limit δ, and $I_w = \mu^{\leq\kappa} \cup \{\eta_\delta: \delta \in w\}$, and let $N_w^1 = EM^1(I_w, \Phi)$, M_i^w a $\mathbf{F}_\kappa^b$-primary model over N_w^1 with constructing sequence $\{a_i^w: i < \lambda\}$, satisfying (3). (For proving (2) when $\lambda \leq 2^{|T_1|}$ we should assume that if the type of a_i^w over $A_i^w = \{\bar{a}_\eta: \eta \in I_w\} \cup \{a_j^w; j < i\}$ is $\subseteq \mathrm{Av}(D_i^w, \mathrm{dcl}\, B_i^w)$, $|B_i^w| < \kappa$, then $|\{D_i^w: B_i^w = B\}| < \lambda$.)

Let M_w be the L-reduct of M_w^1, $f: M_w \to M_u$ an isomorphism (for 2) elementary embedding). For (2) let $\mu = \lambda$ and the proof is as in 2.2. For (1) there is $V \subseteq \lambda$, $|V| = \mu$ such that f maps $\{\bar{a}_\eta: \eta \in I_w\} \cup \{a_i^w: i \in V\}$ onto $\{\bar{a}_\eta: \eta \in I_u\} \cup \{a_i^u: i \in V\}$.

Let $V = \{i_\alpha: \alpha < \mu\}$. Then there is a closed unbounded $S \subseteq \mu$ such that for $\alpha \in Sf$ maps $\{\bar{a}_\eta: \eta \in I_w, \eta \in \bigcup_{\beta<\kappa} \beta^{\leq\kappa}\} \cup \{a_i^w: i = i_\beta \in V, \beta < \alpha\}$ onto $\{\bar{a}_\eta: \eta \in I_u, \eta \in \bigcup_{\beta<\kappa} \beta^{\leq\kappa}\} \cup \{a_i^u: i = i_\beta \in V, \beta < \alpha\}$. The rest is like 2.7.

THEOREM 3.6: *Suppose* $\lambda \geq \mu + |T| + \kappa^+$; $\lambda^{<\kappa} = \lambda$, $\kappa < \kappa(T)$; κ, μ *are regular,* $\chi < \mu \Rightarrow \chi^{<\kappa} < \mu$. *Then* T *has* 2^μ *non-isomorphic models of cardinality* λ, *which are* κ*-homogeneous,* κ*-compact, and if* $|D(T)| \leq \lambda$ *also* κ*-saturated.*

Proof. Similar to 3.5 and 2.7.

CONJECTURE 3.1: (1) If $\lambda = \lambda^{<\kappa} \geq |T_1| + \kappa^+$, $\kappa < \kappa(T)$, κ is regular, then there are 2^λ non-isomorphic models $M \in \text{PC}(T_1, T)$ of cardinality λ satisfying (2) and (3) from Theorem 3.2.

(2) Moreover, if $\lambda > |T_1|$, no one of them is elementarily embedded in any other.

EXERCISE 3.2: Prove conjecture 3.1(1) for λ strong limit cardinal $> \beth_\kappa$. (Hint: look at 2.6.)

EXERCISE 3.3: Prove conjecture 3.1(1) when there is μ, $\mu^{<\kappa} = \mu < \lambda$, $2^\mu = 2^\lambda$.

EXERCISE 3.4: Prove conjecture 3.1(1) when $\kappa < \kappa_{\text{ind}}(T)$ and λ is regular, or $\mu < \lambda \leq \mu^\kappa$, $2^\mu < 2^\lambda$.

PROBLEM 3.6: Axiomatize the proofs in Sections 2 and 3 (to include also 1.7(1)).

Remark. Clearly in 2.5, 6 and 1.4, 7 we have less than in, e.g., 2.2. Of course, it will be desirable to use fewer axioms in each proof.

VIII.4. Categoricity, saturation and homogeneity up to a cardinality

THEOREM 4.1: *Suppose* $\text{PC}(T_1, T)$ *is categorical in* λ.

(1) *If* $\lambda \geq |T_1| + \aleph_1$, *then* T *is stable and if* $\lambda > 2^{\aleph_0}$ *has not the* f.c.p.

(2) *If* $\lambda > |T_1|$ *then* T *is superstable without the* f.c.p., *and stable in every* $\mu \geq |T_1|$.

Proof. (1) T is stable by 3.3; hence has not the f.c.p. by VII, 3.4.

(2) T is superstable by 2.1. By VII, 2.7 there is $M \in \mathrm{PC}(T_1, T)$ of cardinality λ, stable in $|T_1|$. If T is not stable in $|T_1|$, there is $M \in \mathrm{PC}(T_1, T)$ unstable in $|T_1|$, $\|M\| = \lambda$, contradicting the categoricity. So T should be stable in $|T_1|$, hence by III, 0.1 stable in every $\mu \geq |T_1|$. By VII, 3.4, T is without the f.c.p.

THEOREM 4.2: (1) *If T is not superstable, then for every $\lambda > |T_1|$ there is a non-$|T_1|^+$-model-homogeneous model in* $\mathrm{PC}(T_1, T)$ *of cardinality λ (so it is not $|T_1|^+$-homogeneous, nor homogeneous). See below.*

(2) *If T is not superstable, $|T_1| < \mu \leq \lambda$, μ regular, $(\forall \chi < \mu)(\chi^{\aleph_0} \leq \lambda)$, then there are models M_i^1 of T_1 of cardinality λ, for $i < 2^\mu$ such that*

(i) *for $i \neq j$, the* L*-reduct of M_i^1 cannot be elementarily embedded into the* L*-reduct of M_j^1,*

(ii) *the set of isomorphism types of elementary submodels of M_i^1 of cardinality $<\mu$ does not depend on i.*

Remark. M is λ-model-homogeneous if $M_1 \prec M$, $\|M_1\| < \lambda, f: M_1 \cong M_2$, $M_1 \prec N_1 \prec M$, $\|N_1\| < \lambda$, then for some f' extending f, and N_2, $M_2 \prec N_2 \prec N, f': N_1 \cong N_2$.

Proof. (1) We use the terminology of 2.2. Define $\alpha(i)$ for $i < |T_1|^+$ as follows: $\alpha(0) = 0$, $\alpha(i+1) = \alpha(i) + i$, $\alpha(\delta) = \bigcup_{i<\delta} \alpha(i)$. We say $\nu = \beta + \eta$ if $\nu[i] = \beta + \eta[i]$ for any $i < l(\nu) = l(\eta)$. Let

$$I_0 = \lambda^{<\omega} \cup \{\alpha(i_{n-1}) + \dots + \alpha(i_0) + \eta_\delta : \delta < \alpha(i_0) < \dots < \alpha(i_{n-1}) < |T_1|^+\},$$

$$I_1 = |T_1|^{<\omega} \cup \{\eta_\delta : \delta < |T_1|^+\},$$

$$M^l = EM^1(I^l, N_0).$$

As $\alpha(i)$ is an increasing and continuous function $\{\delta < |T_1|^+ :$ there is $\eta \in I^0, \eta \in \delta^\omega, \sup_n \eta[n] = \delta\}$ is not stationary (each $\alpha(i)$ does not belong to it).

As in 2.2 this implies that $M^1 \restriction \mathrm{L}$ cannot be elementarily embedded into $M^0 \restriction \mathrm{L}$. Let, for $\gamma < |T_1|^+$, $I_\gamma^1 = I^1 \cap \gamma^{\leq\omega}$, $M_\gamma^1 = EM^1(I_\gamma^1, N_0)$. Clearly every elementary submodel of M^1 of cardinality $\leq |T_1|$ is an elementary submodel of some M_γ^1, but M_γ^1 can be elementarily embedded into M^0, by the embedding induced by $\bar{a}_\eta \mapsto \bar{a}_{\alpha(\gamma+1)+\eta}$. Easily, those two facts imply $M^0 \restriction \mathrm{L}$ is not $|T_1|^+$-model homogeneous.

(2) For $w \subseteq \mu$ let $I_w = \lambda^{<\omega} \cup \{\eta_\delta : \delta \in w\} \cup \{\mu + \alpha(i) + \eta : i < \mu, \eta \in \beta^\omega$ for some $\beta < \alpha(i)\}$. The rest is like 2.2 and 4.2(1).

THEOREM 4.3: *The following assertions on the cardinals,* μ, λ, χ, *where* $\lambda > \chi$, $\mu > \chi$, *are equivalent:*

(1) *If* $|L| \leq \chi$, $L(T) \subseteq L$, p *a* 1*-type in* L, P *a one place predicate in* L, *and* T *has a model* M *omitting* p, $\|M\| = \lambda > |P(M)|$, *then* T *has a model* N *omitting* p, $\|N\| = \mu > |P(N)|$.

(2) *If* $T \subseteq T_1, |T_1| \leqslant \chi$, $\mathrm{PC}(T_1, T)$ *is categorical in* μ, *then* $\mathrm{PC}(T_1, T)$ *is categorical in* λ.

(3) *If* $T \subseteq T_1$, $|T_1| \leq \chi$, *and every model in* $\mathrm{PC}(T_1, T)$ *of cardinality* μ *is homogeneous, then every model in* $\mathrm{PC}(T_1, T)$ *of cardinality* λ *is homogeneous.*

(4) *If* $T \subseteq T_1$, $|T_1| \leq \chi$, p *a* 1*-type,* $|p| \leq \chi$ *and every model of* T_1 *of cardinality* μ *omitting* p *has homogeneous* L*-reduct then every model of* T_2 *omitting* p *of cardinality* μ *has homogeneous* L*-reduct.*

CONCLUSION 4.4: *If* $\lambda = \beth_\delta$, $(\forall \alpha < \delta)(\alpha + \delta(\chi) \leq \delta)$, λ, $\mu > \chi$ *then* (1)–(4) *from* 4.2 *holds.*

Proof of 4.4: By VII, 5.3, (1) holds in this case.

Proof of 4.3. (1) $\Rightarrow$ (4) Suppose M^1 is a model of T_1, omitting p, of cardinality λ, with a non-homogeneous L-reduct M. So there are $\kappa < \lambda$, $a_i \in M$ $(i \leq \kappa)$ $b_i \in M$ $(i < \kappa)$ such that

$$\mathrm{tp}_*(\langle b_i : i < \kappa \rangle, M) = \mathrm{tp}_*(\langle a_i : i < \kappa \rangle, M)$$

but for no $b \in M$

$$\mathrm{tp}_*(\langle b_i : i < \kappa \rangle ^\frown \langle b \rangle, M) = \mathrm{tp}_*(\langle a_i : i \leq \kappa \rangle, M).$$

Let $M^* = (M, P, F, c)$ where $P = \{a_i : i < \kappa\}$ (one place relation), $F(a_i) = b_i$ for $i < \kappa$ and $F(a) = c$ for $a \notin P$, $c = a_\kappa$. So by (1) there is a model N^* elementarily equivalent to M^*, $\|N^*\| = \mu > |P(N^*)|$ and omitting the types p and

$$q = \left\{ (\forall y_1, \ldots, y_n) \left[\bigwedge_i P(y_i) \rightarrow \varphi(c, y_1, \ldots, y_n) \equiv \varphi(x, F(y_1), \ldots, F(y_n)) \right] : \varphi \in \mathrm{L} \right\}$$

(because omitting p and q is equivalent to omitting $r = \{\varphi(x) \vee \psi(x) : \varphi(x) \in p, \psi(x) \in q\}$). Clearly (by q) N is not homogeneous. So we prove (4).

(4) $\Rightarrow$ (3) Immediate.

(1) ⇒ (2) Suppose $PC(T_1, T)$ is categorical in μ. By 4.1 T is stable in every $\lambda \geq |T_1|$, hence in μ, hence by III, 3.12 T has a saturated model in μ, so by I, 1.13 $PC(T_1, T)$ has a saturated model in μ, hence every model in $PC(T_1, T)$ of cardinality μ is saturated. Suppose M^1 is a model of T_1 of cardinality λ and its L-reduct M is not saturated. So there are $A \subseteq |M|$, and $p \in S(A, M)$; p is omitted by M, $|A| < \lambda$. For every $\varphi(x; \bar{y}) \in L$ let $R_\varphi = \{\bar{a}: \varphi(x; \bar{a}) \in p\}$, and $P = A$; let $M^* = (M, P, \ldots, R_\varphi, \ldots)_{\varphi \in L}$. By (1) there is N^* elementarily equivalent to M^*; $|N^*| = \mu > |P(N^*)|$ and it omits

$$q = \{(\forall \bar{y})[R_\varphi(\bar{y}) \rightarrow \varphi(x, \bar{y})]: \varphi \in L\}.$$

Clearly $N^* \restriction L \in PC(T_1, T)$ and it is not saturated as it omits $p' = \{\varphi(x, \bar{a}): N^* \vDash R_\varphi[\bar{a}], \varphi \in L, \bar{a} \in |N^*|\}$ (p' is consistent as N^*, M^* are elementarily equivalent). Contradiction. So every $M \in PC(T_1, T)$ of power λ is saturated. Hence (by I, 1.11) $PC(T_1, T)$ is categorical in λ.

not (1) ⇒ not (2), not (3). Suppose L, T, p, P form a counterexample to (1) and M is a model of T omitting p, $\|M\| = \lambda > |P(M)|$, and let $p = \{\varphi_i(x): i < \chi\}$. Let F be a one-to-one function from $|M| \times |M|$ onto $M - P(M)$, with converses F_1, F_2 [i.e., for $a, b \in |M|$, $c \in |M| - P(M)$, $F_1(F(a, b)) = a$, $F_2(F(a, b)) = b$, $F(F_1(c), F_2(c)) = c$]. Let $Q_i = \{a \in |M|: a \notin P(M)$ and $M \vDash \varphi_j[F_1(a)]$ for $j < i$, but $M \vDash \neg\varphi_i[F_1(a)]\}$.

Let $M_1^* = (M, F, F_1, F_2, \ldots, Q_i, \ldots)_{i<\chi}$ and $L_1^* = L(M^*)$, $T_1^* = Th(M^*)$, $L^* = \{Q_i: i < \chi\}$, $T^* = T_1^* \cap L^*$. Clearly each Q_i has cardinality λ or is empty, and $|M| - \bigcup_{i<\chi} Q_i = P(M)$. As $|P(M_1^*)| < \lambda = |M_1^*|$, the L-reduct of M_1^* is not saturated nor homogeneous. (If $P(M) = \{b_i: i < \kappa\}$, let $a_i = b_{i+1}$, $a_\kappa = b_0$, and then we cannot find a suitable b_κ.)

Suppose now N_1^* is a model of T_1^* of cardinality μ. Being a model of T_1^*, each $Q_i(N_1^*)$ is empty or has cardinality $\|N_1^*\|$ (use F, F_1, F_2). By assumption N_1^* realizes p or $|P(N_1^*)| = \lambda$. In each case $A = N_1^* - \bigcup_{i<\chi} Q_i(N_1^*)$ has cardinality λ (in the first case, if a realizes p, as $\{F(a, c): c \in N_1^*\} \subseteq A$; and the second case as $P(N_1^*) \subseteq A$). As T_1^* has elimination of quantifiers, it is easy to check that the L-reduct of N_1^* is saturated, hence homogeneous. So T_1^*, T^* provides the counterexample for (2) and (3).

As we prove (1) ⇒ (4) ⇒ (3) ⇒ 1, (1) ⇒ (2) ⇒ (1) we finish.

THEOREM 4.5: *There is a model M_1 of T_1 of cardinality μ which is $(<\lambda)$-universal, but its* L*-reduct is not λ-universal, if:*

(1) *T unsuperstable, $2^{|T_1|} < \lambda$, $2^{<\lambda} \leqslant \mu$; and λ is regular or $\lambda^{\aleph_0} = \lambda$; or*

(2) *$T = T_1$ is stable but not superstable, $|T| \leqslant \lambda$, $2^{<\lambda} \leqslant \mu$; and λ is regular or $\lambda^{\aleph_0} = \lambda$.*

Proof. Our model M_1 will be the model constructed in VII, 4.6.

Clearly it is $(<\lambda)$-universal. We should show only that its L-reduct, M_0 is not λ-universal.

(1) If λ is regular, the proof is similar to that of 2.2. If $\lambda^{\aleph_0} = \lambda$, choose Φ, φ_n as in the proof of 2.2, and let $N_0 = EM(\lambda^{\leq\omega}, \Phi)$, so N is a model of T of cardinality λ, hence it suffices to show N_0 is not elementarily embedded into M_0. Suppose f is such an embedding, and $\bar{b}_\eta = f(\bar{a}_\eta)$ for $\eta \in \lambda^{\leq\omega}$. We can find finite $w_\eta \subseteq \mu$ such that $\nu \lessdot \eta \Rightarrow w_\nu \subseteq w_\eta$, and $\bar{b}_\eta = \tau_\eta(\bar{c}_{\eta,0}, \ldots, \bar{c}_{\eta,k(\eta)})$, $\bar{c}_{\eta,l} \in M^*_{i(\eta,l)}$, $w_\eta \supseteq \{i(\eta, l) : l < k(\eta)\}$. We now define by induction $\eta_n = \eta(n) \in \lambda^n$, such that $\eta_n \lessdot \eta_{n+1}$ and

$\{\eta \in \lambda^{n+1}: \tau_\eta = \tau_{\eta(n+1)}, k(\eta) = k(\eta_{n+1});$

$\langle i(\eta, 0), \ldots\rangle \sim \langle i(\eta_{n+1}, 0), \ldots\rangle \bmod w_{\eta(n)};$

$\bar{c}_{\eta,l}, \bar{c}_{\eta(n+1),l}$ realize the same type over M^1, moreover if

$\bar{a} = \cup \{\bar{c}_{\eta|i,j}: i < n, j \leq k(\eta \upharpoonright i), i(\eta \upharpoonright i, j) = \eta[n]\}$,

then $\bar{a}^\frown\bar{c}_{\eta,l}, \bar{a}^\frown\bar{c}_{\eta(n+1),l}$ realizes the same type over $|M^1|\}$

has cardinality $\geq \aleph_0$. (Of course, $l_1 \neq l_2 \Rightarrow i(\eta, l_1) \neq i(\eta, l_2)$).

Let $\eta \in \lambda^\omega$, $\eta_n \lessdot \eta$ for every n. Then $\{\varphi_n(\bar{x}, \bar{b}_{\eta\upharpoonright n}): n < \omega\}$ is realized by b_η, but cannot be realized in M_0 by the definition of η, contradiction. (2) A similar proof.

LEMMA 4.6: (1) *There are complete countable theories $T_1 \supseteq T$, T unstable, such that: if $M \in \mathrm{PC}(T_1, T)$ is $(<\lambda)$-universal, λ strong limit, $\mathrm{cf}\,\lambda = \aleph_0$ then M is λ-universal.*

(2) *There are complete countable theories $T_1 \supseteq T$, T stable but not superstable, such that: if $M \in \mathrm{PC}(T_1, T)$ is $(<\lambda)$-universal, λ strong limit, $\mathrm{cf}\,\lambda = \aleph_0$ then M is λ-universal.*

Proof. (1) Let T be the theory of the rational order, and T_1 says that any two intervals are isomorphic. We leave the proof to the reader.

(2) Define the relation E_n over ω^ω: $\eta E_n \nu \Leftrightarrow \eta \upharpoonright n = \nu \upharpoonright n$. Let $M = (\omega^\omega, E_0, E_1, \ldots, E_n, \ldots)$. Let $F_n(\nu, \rho) = \langle \nu[0], \ldots, \nu[n-1), \rho[n], \rho[n+1], \ldots\rangle$, and $M_1 = (M, F_0, F_1, \ldots)$. Let $T = \mathrm{Th}(M)$, $T_1 = \mathrm{Th}(M_1)$. We leave the proof to the reader.

THEOREM 4.7: *The following conditions on* λ, T *are equivalent:*

(1) $\lambda = \lambda^{<\lambda} + |D(T)|$ *or* T *is stable in* λ.

(2) T *has a saturated model of cardinality* λ.

(3) *If* $T_1 \supseteq T$, $|T_1| \leq \lambda$ *then there is a saturated* $M \in \mathrm{PC}(T_1, T)$ *of cardinality* λ.

Proof. (1) $\Rightarrow$ (2) If $\lambda = \lambda^{<\lambda} + |D(T)|$, the proof of I, 1.7 shows that (2) holds. If T is stable in λ III, 3.12 shows that (2) holds.

(3) $\Rightarrow$ (2) This is immediate.

(2) $\Rightarrow$ (3) This is I, 1.13 (in its proof we use (2) $\Rightarrow$ (1) only, so no vicious circle arises).

(2) $\Rightarrow$ (1) Let M be the saturated model of T of cardinality λ. So clearly $\lambda = \|M\| \geq |D(T)|$, and if $|A| < \lambda$, $|S(A)| \leq \lambda$ (as M is universal w.l.o.g. $A \subseteq M$, now M is saturated). Now we prove by cases.

Case I: T unstable, $\lambda < 2^{<\lambda}$. By II, 2.2(5) there are φ, and $\bar{a}_\eta$, $\eta \in {}^{\lambda>}2$ such that for each $\eta \in {}^{\lambda}2$ $\{\varphi(x, \bar{a}_{\eta\restriction\alpha})^{\eta[\alpha]}: \alpha < \lambda\}$ is consistent. Now we define $\bar{b}_\eta \in M$ by induction on $l(\eta)$, such that if $\eta_0 \lessdot \eta_1 \lessdot \cdots \lessdot \eta_n$ then $\bar{a}_{\eta(0)} {}^\frown \bar{a}_{\eta(1)} {}^\frown \cdots {}^\frown \bar{a}_{\eta(n)}$, and $\bar{b}_{\eta(0)} {}^\frown \bar{b}_{\eta(1)} {}^\frown \cdots {}^\frown \bar{b}_{\eta(n)}$ realize the same type. (This is possible as the only requirement on $\bar{b}_\eta$ is that it realizes a certain type over $\bigcup \{\bar{b}_\nu: \nu \lessdot \eta\}$. The type is consistent by the induction hypothesis, and is realized as M is λ-saturated). So for every $\eta \in {}^{\lambda>}2$, some $c_\eta \in M$ realizes $p_\eta = \{\varphi(x, \bar{a}_{\eta\restriction\alpha})^{\eta[\alpha]}: \alpha < l(\eta)\}$. If $l(\eta) = l(\nu)$ then $c_\eta \neq c_\nu$ [for if $\alpha = \min\{\alpha: \eta[\alpha] \neq \nu[\alpha]\}$, $\rho = \eta \restriction \alpha = \nu \restriction \alpha$, then $\varphi(c_\eta, \bar{a}_\rho) \equiv \neg\varphi(c_\nu, \bar{a}_\rho)$]. So for every $\alpha < \lambda$, $\lambda = \|M\| \geqslant |\{c_\eta : \eta \in 2^\alpha\} = |2^\alpha|$, hence $\lambda \geq 2^{<\lambda}$, contradiction.

Case II: T unstable $\lambda \geq 2^{<\lambda}$, λ regular. It is easy to check $\lambda = \lambda^{<\lambda}$, so we finish.

Case III: T unstable, $\lambda \geq 2^{<\lambda}$, λ singular. It is easy to check that $\mu < \lambda \Rightarrow 2^\mu < \lambda$ (otherwise $2^{\mu + \mathrm{cf}\lambda} > \lambda, \mu + \mathrm{cf}\lambda < \lambda$). So it suffices to prove (as a saturated model is universal by I, 1.9(3)).

LEMMA 4.8: *Suppose* $\lambda \geq |T_1|$, λ *strong limit cardinal and* cf $\lambda < \kappa(T)$. *Then there is* $M \in \mathrm{PC}(T_1, T)$, $\|M\| = \lambda$, *such that no elementary extension* N *of* M *of cardinality* λ *is* $(\mathrm{cf}\,\lambda)^+$*-saturated.*

Proof of Lemma 4.8. Let $\kappa = \mathrm{cf}\,\lambda$. We can define suitable M and $\bar{a}_\eta \in M$ for $\eta \in {}^{\kappa>}\lambda$, such that for every $\eta \in {}^{\kappa}\lambda$, $q_\eta = \{\varphi(x, \bar{a}_\nu)^{\mathrm{if}(\eta \lessdot \nu)}: \nu \in {}^{\kappa>}\lambda\}$ is consistent (by III, 7.6, 7 and the definition of the independence "property"). Suppose $M \prec N$, $\|N\| = \lambda$, so let $|N| = \bigcup_{i<\kappa} A_i$, $|A_i| < \lambda$, hence $|S(A_i)| \leq 2^{|A_i|} < \lambda$. Now define by induction η_α, ν_α for $\alpha \leq \kappa$ such that $\alpha < \beta \Rightarrow \eta_\alpha \lessdot \eta_\beta$, $l(\nu_\alpha) = \alpha$, $l(\eta_\alpha) = \alpha$, $\eta_\alpha \lessdot \nu_{\alpha+1}$, and $\text{tp}(\bar{a}_{\eta(\alpha+1)}, A_\alpha) = \text{tp}(\bar{a}_{\nu(\alpha+1)}, A_\alpha)$, hence no element of A_α realizes $\varphi(x, \bar{a}_{\eta(\alpha+1)}) \wedge \neg\varphi(x, \bar{a}_{\nu(\alpha+1)})$. Clearly $q = \{\varphi(x, \bar{a}_{\eta(\alpha+1)}), \neg\varphi(x, \bar{a}_{\nu(\alpha+1)}):$ $\alpha < \kappa\}$ is $\subseteq q_{\eta(\kappa)}$, hence consistent, but is not realized in N as $|N| = \bigcup_{\alpha<\kappa} A_\alpha$. So N is not κ^+-saturated.

Case IV: T stable, $\lambda^{<\kappa(T)} = \lambda$. So $\lambda = \lambda^{<\kappa(T)} + |D(T)|$. If T is stable in λ we finish, otherwise, by III, 5.15 it follows that $\lambda < 2^{\aleph_0}$, and for some countable A, $|S(A)| \geq 2^{\aleph_0}$, contradiction.

Case V: T stable, $\lambda^{<\kappa(T)} > \lambda$. It suffices to prove the following lemma.

LEMMA 4.9: *Suppose $\lambda \geq |T_1|$, $\lambda^\kappa > \lambda$ and $\kappa < \kappa(T) < \infty$ or T has the strict order property or $\kappa < \kappa r_{\text{cdt}}(T)$. Then there is $M \in \text{PC}(T_1, T)$, $\|M\| = \lambda$ such that no elementary extension N of M of cardinality λ is κ^+-saturated.*

Remark. In fact if $\kappa < \kappa(T) < \infty$ or T has the strict order property, then $\kappa < \kappa r_{\text{cdt}}(T)$.

Proof of Lemma 4.9. We can assume $\kappa = \min\{\mu: \lambda^\mu > \lambda\}$, hence $\lambda = \lambda^{<\kappa}$. So there is $M \in \text{PC}(T_1, T)$, $\varphi_\alpha \in \text{L}$, and $\bar{a}_\eta \in M$, for $\eta \in {}^{<\kappa}\lambda$ such that:

(1) $\{\varphi_\alpha(\bar{x}; \bar{a}_\eta)^{\text{if}(\eta \lessdot \nu)}: \alpha < \kappa, \eta \in {}^\alpha\lambda, \alpha \text{ successor}\}$ is consistent, for every $\nu \in {}^\kappa\lambda$.

(2) If $\eta \in {}^\alpha\lambda$, $i < j < \lambda$ $\{\varphi_{\alpha+1}(\bar{x}; \bar{a}_{\eta\frown\langle\zeta\rangle}): \zeta \in \{i, j\}\}$ is inconsistent. (This is by III, 7.7, III, 7.6(1) and (5).)

Suppose $M \prec N$, $\|N\| = \lambda$, and for $\eta \in {}^\kappa\lambda$ let $p_\eta = \{\varphi_\alpha(\bar{x}, \bar{a}_{\eta\restriction\alpha}):$ $\alpha < l(\eta)$, α a successor$\}$.

If N is κ^+-saturated, each p_η is realized in N; but the p_η's are pairwise contradictory. So $\|N\| = \lambda^\kappa$ contradiction. So we finish the proof of Theorem 4.6.

THEOREM 4.10: *Suppose $\lambda \geq |T_1| + \aleph_1$, and every model $M \in$ $\text{PC}(T_1, T)$ of cardinality λ is $\aleph_1$-homogeneous. Then T is superstable.*

Proof. Our assumption implies that every $M \in \text{PC}(T_1, T)$ of cardinality

$\geq \lambda$ is $\aleph_1$-homogeneous [Because if $\langle a_i: i \leq \omega\rangle$, $\langle b_i: i < \omega\rangle$ is a counterexample, i.e., $\mathrm{tp}_*(\langle a_i: i < \omega\rangle) = \mathrm{tp}_*(\langle b_i: i < \omega\rangle)$ but for no $b_\omega \in M$ $\mathrm{tp}_*(\langle a_i: i \leq \omega\rangle) = \mathrm{tp}_*(\langle b_i: i \leq \omega\rangle)$ and M is the L-reduct of M_1, M_1 a model of T_1, then by any elementary submodel M_1' of M_1 of cardinality λ such that $a_i, b_i \in |M_1'|$, we arrive at a contradiction]. But there is a strong limit cardinal $\mu > \lambda$ of cofinality $\aleph_0$; so by 4.8 T is stable, and by 4.9 superstable (remember I, 1.9(4)).

CONJECTURE 4.1: If $\mathrm{PC}(T_1, T)$ is categorical in λ, $\lambda \geq |T_1| + \aleph_1$, then T is superstable (but see [Sh 80]).

PROBLEM 4.2: Can we improve 4.2(2) (i.e., weaken the conditions on μ)?

PROBLEM 4.3: Close the gap between 4.5 and 4.6. That is characterize the cardinals λ, μ, χ such that: if $|T_1| \leq \chi$, $T \subseteq T_1$, T unstable [stable but not superstable] [has the independence property], then there is a model M_1 of T_1 of cardinality μ which is $(<\lambda)$-universal but its L(T)*-reduct is not* λ*-universal.*

PROBLEM 4.4: Characterize the T, λ, κ such that:

(*) For every model M of T of cardinality λ there is N, $M \prec N$, $\|N\| = \lambda$, N κ-saturated (note 4.7, 4.8, I, 1.7 and Exercise 4.5) (but see [Sh 80a]).

EXERCISE 4.5: Suppose $\mu^\kappa = \mu$, $\mu \leq \lambda \leq 2^\mu$, $T = T_{\mathrm{ind}}$ (see II, 4.8). Then T, λ satisfies (*) from Problem 4.4. [Hint: See the solution in [Sh 81].]

PROBLEM 4.6: (1) Characterize the cardinals μ, λ, κ, χ such that for every unstable T, $T \subseteq T_1$, $|T_1| \leq \chi$, there is a model $M \in \mathrm{PC}(T_1, T)$, $\|M\| = \mu$, M is $(<\lambda)$-universal but not λ-universal, M is κ-saturated but not κ^+-saturated.

(2) Replace "unstable" by $\kappa < \kappa(T) < \infty$, or by "has the independence property".

EXERCISE 4.7: Solve those cases of 4.6 which can be solved by inessential changes in the proofs here. E.g., assume G.C.H., μ, λ, κ are successor cardinals.

CONJECTURE 4.9: Remove the exception in 3.4.

CONJECTURE 4.10: In 3.5 waive the condition $(\forall \chi < \lambda)(\chi^{<\kappa} < \lambda)$.

CHAPTER IX

CATEGORICITY AND THE NUMBER OF MODELS IN ELEMENTARY CLASSES

IX.0. Introduction

In this chapter we return to elementary classes so in some respects it is a natural continuation of Chapter V.

Part of Section 1 is devoted to categoricity theorems, etc., which are, in a sense, a summation of previous results with some additions. We characterize the countable theories categorical in $\aleph_0$ (1.6). We prove that for countable T, T is categorical in some $\lambda > \aleph_0$ iff T is categorical in every $\lambda > \aleph_0$ iff T is totally transcendental and for $\bar{a} \in |M|$ $\varphi(M, \bar{a}) \geq \aleph_0 \Rightarrow |\varphi(M, \bar{a})| = \|M\|$ (see 1.8 and 1.4).

For not necessarily countable T:

for some A, $T(A)$ is categorical in some $\lambda > T(A)$

iff for some A, $|A| \leq 2^{|T|}$, $T(A)$ is categorical in every $\lambda > |T(A)|$,

iff every $2^{|T|}$-universal model of T is saturated,

iff T is superstable and undimensional.

If those conditions fail, for every $\lambda \geq 2^{\mu}$, $\mu \geq 2^{|T|}$ T has a μ-universal not μ^+-universal model of cardinality λ (see 1.14).

We prove also that T is categorical in some $\lambda > |T|$ iff T is categorical in every $\lambda > |T|$, and that if T is categorical in $|T| > \aleph_0$ it is a definitional extension of some $T' \subseteq T$, $|T'| < |T|$. Note that the property "T is categorical in $|T|^+$" is not absolute. The properties "T categorical in $\aleph_1$," for countable T, and "$T(A)$ is categorical in $|T(A)|^+$ for some A" are absolutes by the proofs of the previous theorems.

We also prove that if every model of T of cardinality λ is $|T|^+$-universal, T is $|T|^+$-categorical; (see 1.15) and if every model of T of cardinality $\lambda_0 > |T|$ is homogeneous, then: every model of T of cardinality $> \lambda(T)$ is homogeneous, and T has a non-homogeneous model in λ when $|T| + \aleph_1 \leq \lambda \leq \lambda(T)$ (see 1.16).

The method of those proofs is as follows: by Chapter VIII we get the results on unstable and unsuperstable theories, and by V, Section 2 on stable non-unidimensional T. So we concentrate on unidimensional theores, and by V, Section 7 we find weakly minimal formulas $\varphi(x, \bar{a})$. By the unidimensionality for no $M \prec N$, $M \neq N$ is $\bar{a} \in |M|$, $\varphi(N, \bar{a}) = \varphi(M, \bar{a})$, so M is "almost prime" over $\bar{a} \cup \varphi(M, \bar{a})$, whereas it is easy to handle $\varphi(M, \bar{a})$ by the weak minimality. For using this we need theorems assuring, for a suitable A, the existence of M, $A \subseteq |M|$, $\varphi(M, \bar{a}) \subseteq A$, see 1.1, Exercise 1.10(2).

In Section 1 we deal with two more problems. We prove that if $|T| < \lambda(T)$ $(= \sup\{|S(A)|: |A| \le |T|\})$ then for every $\mu \ge |T|$, $I(\mu, T) \ge \min\{2^{\mu}, 2^{\lambda(T)}\}$. In VIII, 1.7(2), 8 we have proved this even for $\mathrm{PC}(T_1, T)$ but we demand T_1 is countable; the most difficult case is $2^{\mu} = 2^{\aleph_0}$, $|D(T)| \le |T|$.

The second result is that for superstable T, and if, e.g., $\|M\| > |T|$, $\delta < \|M\|^+$, then M is the union of a strictly increasing elementary chain of length δ (see 1.3). (For stable T this may fail, see Exercise 1.2.)

In Section 2 we try to compute $I(\lambda, T)$ in some cases. We try to count the models by their dimension (or dimensions).

We prove that for T totally transcendental, if $|T| = \aleph_\alpha \le \aleph_\beta$, T not categorical in $|T|^+$, then $I(\aleph_\beta, T) \ge |\beta + 1|$. $I(\aleph_\beta, T) \ge |\beta + 1| + \aleph_0$ except, possibly, when T is countable, categorical in $\aleph_0$ (the idea is that T is not unidimensional by 1.8, so we have a freedom to define one dimension and for appropriate T, also the finite dimensions are possible). We also prove that for superstable countable T, not categorical in $\aleph_0$, $I(\aleph_0, T) \ge \aleph_0$ (the idea is that if $p \in S^m(\emptyset)$ is omitted by the prime model of T, there is a model M_n, in which $\{\bar{a}_l: l < n\} \subseteq p(M_n)$ which are prime over $I_n = \{\bar{a}_l: l < n\}$, when I_n is included in an infinite indiscernible set based on p, and $\dim(p, M_n)$ is n.

In V, Section 5 we deal with multidimensionality. We prove multidimensional theories have many models (in $\aleph_\alpha > |T|$ at least $2^{|\alpha|}$). For non-multidimensional T we prove they have quite few $\mathbf{F}^a_{\kappa_r(T)}$-saturated models. For superstable T, we get here a sharper result, of course for non-multidimensional T. We also get a better structure for the models. Every $\mathbf{F}^a_{\aleph_0}$-saturated models is (in $\mathfrak{C}^{eq}$) $\mathbf{F}^a_{\aleph_0}$-prime over $\bigcup_i J_i$, each J_i based on $\emptyset$, $\mathrm{Av}(J_i, \bigcup J_i)$ semi-regular, and the $\mathrm{Av}(J_i, \bigcup J_i)$ are pairwise orthogonal (we see somewhat more by the inductive definition of the J_i's).

In 2.4 we get a similar theorem for T totally transcendental, for all models, or for the $\aleph_\beta$-compact models.

IX.1. Superstable theories and categoricity

THEOREM 1.1: *Suppose $p = \mathrm{tp}(\bar{a}_0, A)$ does not fork over $B \subseteq A$, p is stationary, $|B| < \kappa(T) < \infty$. If A' is $\mathbf{F}^c_\lambda$-constructible over A then p is not realized in A', moreover p has a unique extension in $S^m(A')$ provided that:*

$(*)^\lambda$ *For every type $r \subseteq p$, $|r| < \lambda$ and $C \subseteq A$, $|C| < \lambda + \kappa_r(T)$ and finite Δ, there is $\bar{a} \in A$ which realizes r such that*

$$R^m(p, \Delta, 2) = R^m[\mathrm{stp}(\bar{a}, C^*), \Delta, 2]$$

for some C^*, $B \cup C \subseteq C^* \subseteq A$.

FACT 1.1A: *If $p = \mathrm{stp}(\bar{a}_0, A)$ satisfies $(*)^\lambda$ for some B, λ or even is just finitely satisfiable in A, then p is stationary* [*because for every $E \in FE^m(A)$, $E = E(\bar{x}, \bar{y}, \bar{b})$, choose by induction on $l < \omega, \bar{a}^l \in A$ realizing $\{\neg E(\bar{x}, \bar{a}^m, \bar{b}) : m < l, \neg E(\bar{a}_0, \bar{a}^m, \bar{b})\}$, so for some l $E(\bar{a}_0, \bar{a}^l, \bar{b})$, hence $p \vdash E(\bar{x}, \bar{a}^0, \bar{b})$*].

So we can omit "p stationary" from the assumptions of 1.1.

Proof. This is proved by induction on the length of the construction; for limit length, or length zero it is immediate. For the successor step, note that if p has a unique extension $p' \in S^m(A')$ then necessarily p' does not fork over B. So clearly it suffices to prove the following two claims. For simplicity we assume $p \in S^m(A)$. Note that as p is stationary, $R^m(p, \Delta, 2) = R^m(\mathrm{stp}(\bar{a}_0, A), \Delta, 2)$ for every Δ.

CLAIM 1.2: *Suppose $A \subseteq A^1$, $p \subseteq p^1 \in S^m(A^1)$, p^1 does not fork over A (and the conditions of* 1.1 *holds). Then also B, A^1, p^1 satisfy $(*)^\lambda$.*

CLAIM 1.3: *Suppose B, A, p are as in* 1.1 *and $\mathrm{tp}(\bar{b}, A)$ is $\mathbf{F}^c_\lambda$-isolated. Then p has a unique extension in $S^m(A \cup \bar{b})$.*

Proof of 1.2. Let $r^1 \subseteq p^1$, $|r^1| < \lambda$, r^1 is over C^1, $C^1 \subseteq A^1$ $|C^1| < \lambda + \kappa_r(T)$, and Δ be finite, and let $r^1 = \{\varphi_i(\bar{x}; \bar{a}_i): i < \alpha < \lambda\}$. By IV, 4.2 (with B, A here for C, A there) for each i there is a formula $\psi_i(\bar{x}; \bar{b}_i) \in p$ such that for every $\bar{c} \in A$ and θ $\varphi_i(\bar{x}; \bar{a}) \vdash \theta(\bar{x}; \bar{c})$ implies $\psi_i(\bar{x}; \bar{b}_i) \vdash \theta(\bar{x}, \bar{c})$. So for any $\bar{c} \in A$ $\models \psi_i[\bar{c}; \bar{b}_i] \to \varphi_i[\bar{c}; \bar{a}_i]$ (use $\theta = (\bar{x} \neq \bar{y})$). Choose C, $B \subseteq C \subseteq A$, $|C| < \lambda + \kappa_r(T)$ such that $\mathrm{tp}_*(C^1, A)$ does not fork over C, and $\bar{b}_i \in C$, and $R^m(p, \Delta, 2) = R^m(p \restriction C, \Delta, 2)$. By the hypothesis, $(*)^\lambda$ holds, and clearly $r = \{\psi_i(\bar{x}; \bar{b}_i): i < \alpha\} \subseteq p$ and C satisfy the assumptions of $(*)^\lambda$. Hence there is $\bar{a} \in A \subseteq A^1$ realizing r such that for some C^*, $C \subseteq C^* \subseteq A$.

$$R^m(p, \Delta, 2) = R^m[\mathrm{stp}(\bar{a}, C^*), \Delta, 2].$$

By the choice of C, $\mathrm{tp}(C^1, C^* \cup \bar{a})$ does not fork over C, hence over C^*, hence $\mathrm{tp}(\bar{a}, C^* \cup C^1)$ does not fork over C^*. So by III, 4.2,

$$R^m[\mathrm{stp}(\bar{a}, C^*), \Delta, 2] = R^m[\mathrm{stp}(\bar{a}, C^* \cup C^1), \Delta, 2].$$

We can conclude that (again by III, 4.2),

$$R^m(p_1, \Delta, 2) = R^m(p, \Delta, 2) = R^m[\mathrm{stp}(\bar{a}, C^* \cup C^1), \Delta, 2]$$

and as $\bar{a} \in A$ realizes r, by the choice of the $\psi_i(\bar{x}; \bar{b}_i)$'s it realizes r^1, so we finish.

Proof of 1.3. Suppose not, then some $\bar{a}$ realizes p but $\mathrm{tp}(\bar{a}, A \cup \bar{b})$ forks over A (remember p is stationary). So (by III, 4.2) for some finite Δ and formula $\varphi(\bar{x}, \bar{y}, \bar{c})$, $\models \varphi[\bar{a}, \bar{b}, \bar{c}]$, $\bar{c} \in A$ and

$$n^* = R^m(p, \Delta, 2) > R^m[\varphi(\bar{x}; \bar{b}, \bar{c}), \Delta, 2].$$

As $\mathrm{tp}(\bar{b}, A)$ is $\mathbf{F}^c_\lambda$-isolated there are $q \subseteq \mathrm{tp}(\bar{b}, A)$ $|q| < \lambda$ and $C \subseteq A$, $|C| < \lambda$ such that q is over C, and no type over A extending q forks over C. We can assume q is closed under finite conjunctions, and that for any $\bar{b}', \bar{c}'$, $R^m[\varphi(x; \bar{b}', \bar{c}'), \Delta, 2] < n^*$ (by II, 2.9). Let $r = \{(\exists \bar{y})[\varphi(\bar{x}, \bar{y}, \bar{c}) \wedge \psi(\bar{y}, \bar{d})]: \psi(\bar{y}, \bar{d}) \in q\}$, so $|r| < \lambda$, r is over C, and clearly $r \subseteq p$. Hence by $(*)^\lambda$ there are $\bar{a}^*, C^*; C \subseteq C^* \subseteq A$, $\bar{a}^*$ realizes r, $\bar{a}^* \in A$ and $n^* = R^m[\mathrm{stp}(\bar{a}^*, C^*), \Delta, 2]$. Clearly $q^1 = q \cup \{\varphi(\bar{a}^*, \bar{x}, \bar{c})\}$ is consistent and let $\bar{b}^*$ realize it; as $\models \varphi[\bar{a}^*, \bar{b}^*, \bar{c}]$, and $R^m[\varphi(\bar{x}, \bar{b}^*, \bar{c}), \Delta, 2] < n^* = R^m[\mathrm{stp}(\bar{a}^*, C^*), \Delta, 2]$ it follows that $\mathrm{tp}(\bar{a}^*, C^* \cup \bar{b}^*)$ forks over C^*, hence $\mathrm{tp}(\bar{b}^*, C^* \cup \bar{a}^*)$, fork over C^*, so also over C, contradiction.

THEOREM 1.4: *Suppose M is κ-compact,* cf $\delta \geq \kappa \geq \kappa(T)$, *and at least one of the following conditions*

(i) *there is $\mu = \mu^{<\kappa}$ such that $|\delta| + |T| \leq \mu < \|M\|$,*

(ii) *there is an indiscernible set $I \subseteq M$, $|\delta| \leq |I|$,*

(iii) *there is $C \subseteq |M|$, and $I \subseteq |M|$ independent over C, and $|T|^{<\kappa} + |C|^{<\kappa} < |I|$, $|\delta| \leq |I|$.*

Then we can find a strictly increasing elementary chain of κ-compact models M_i, $i < \delta$, such that $M = \bigcup_{i<\delta} M_i$.

Remark. Theorem 1.4, Example 1.4, can be adopted to prove the non-existence of Jonsson models.

First we prove Claim 1.5.

CLAIM 1.5: (1) *Suppose* A_i $(i < \delta)$ *is increasing*, $\bigcup_{i<\delta} A_i \subseteq A$, $\mathrm{cf}\, \delta \geq \kappa(T)$.

Then we can find B_i $(i < \delta)$ *such that* B_i *is increasing* $A = \bigcup_{i<\delta} B_i$ *and* $\mathrm{tp}_*(B_i, \bigcup_{j<\delta} A_j)$ *does not fork over* A_i.

(2) *Moreover, if* A *is the universe of a* κ*-compact model, then the* B_i*'s are too, provided that for each* i

(∗) *For every type* r *over any* $C \subseteq A_i$, $|r| < \kappa$, $|C| < \kappa$, *for any* $\bar{a} \in \bigcup_{j<\delta} A_j$ *realizing* r *and any finite* Δ, *there is* $\bar{a}_1 \in A_i$ *realizing* r *such that*

$$R[\mathrm{stp}(\bar{a}, C), \Delta, 2] = R[\mathrm{stp}(\bar{a}_1, C), \Delta, 2].$$

Proof of 1.5. (1) Let us define B_i inductively: B_i is maximal subset of A such that $\bigcup_{j<i} B_j \,\dot{\cup}\, A_i \subseteq B_i$, and $\mathrm{tp}_*(B_i, \bigcup_{j<\delta} A_j)$ does not fork over A_i. So we have to prove only that $\bigcup_{i<\delta} B_i = A$. Suppose $c \in A - \bigcup_{i<\delta} B_i$. For every i, as $c \notin B_i$, necessarily $B_i \cup \{c\}$ does not satisfy our requirement, so $\mathrm{tp}_*(B_i \cup \{c\}, \bigcup_{j<\delta} A_j)$ forks over A_i, hence $\mathrm{tp}(c, B_i \cup \bigcup_{j<\delta} A_j)$ forks over $B_i \cup A_i = B_i$, hence for some $j < \delta$, $\mathrm{tp}(c, B_i \cup A_j)$ forks over B_i, hence $\mathrm{tp}(c, B_j)$ forks over B_i. As $\kappa(T) \leq \mathrm{cf}\, \delta$, we get a contradiction.

(2) We suppose A is the universe of κ-compact model, and prove that B_i is too. So let $r \in S^m(B_i)$ be an $\mathbf{F}^c_\lambda$-isolated type, which is not realized in B_i; and let $\bar{c} \in A$ realize it. For every $\bar{d} \in \bigcup_{j<\delta} A_j$, A_i and $\mathrm{stp}(\bar{d}, A_i)$ satisfy the requirement $(*)^\kappa$ from 1.1 on A, p resp. (for any suitable B). By 1.2 we can replace A_i by B_i, and by 1.3 $\mathrm{tp}(\bar{d}, B_i) \vdash \mathrm{tp}(\bar{d}, B_i \cup \bar{c})$, hence $\mathrm{tp}(\bar{d}, B_i \cup \bar{c})$ does not fork over B_i, hence $\mathrm{tp}(\bar{c}, \bigcup_{j<\delta} A_j)$ does not fork over B_i, hence $B_i \cup \bar{c}$ contradict the maximality of B_i.

Proof of 1.4. We prove Case (i), and leave (ii) and (iii) to the reader.

We can easily find a strictly increasing sequence M_i $(i < \delta)$ such that $\|M_i\| \leq \mu$, M_i is κ-compact, $M_i \subseteq M$, and for every m-type r over any $C \subseteq |M_i|$, $|r| < \kappa$, $|C| < \kappa$, and any finite Δ, for any $\bar{a} \in |M|$ realizing r there is $\bar{a}_1 \in |M_i|$ realizing r such that

$$R[\mathrm{stp}(\bar{a}, C), \Delta, 2] = R[\mathrm{stp}(\bar{a}_1, C), \Delta, 2].$$

Now apply 1.5(2).

EXERCISE 1.1: (1) Show that in 1.1(*)$^\lambda$ we can replace "and finite Δ" by "and finite Δ_1 there is a finite $\Delta \supseteq \Delta_1$, and" (and 1.1, 1.2 and 1.3 holds).

(2) Show that also we can replace "$R^m(p,\Delta,2) = R^m[\text{stp}(\bar{a},C^*),\Delta,2]$" by $R^m(p,\Delta,\aleph_0) = R^m[\text{tp}(\bar{a},C^*),\Delta,\aleph_0]$" and remember $R(\text{tp}(\bar{b},\bar{c}),\Delta,\aleph_0) = R(\text{stp}(\bar{b},\bar{c}),\Delta,\aleph_0)$.

(3) Show similar assertions on 1.4.

EXERCISE 1.2: Find a countable stable theory T such that T has a model of power λ, which is not the union of a strictly increasing elementary chain of length μ, μ regular when at least one of the following holds:

(i) there is such countable T' (not necessarily stable),

(ii) $\lambda^\mu = \lambda$, or

(iii) cf $\lambda \neq \mu < \lambda$, and $\chi < \lambda \Rightarrow \chi^\mu < \lambda$.

[Hint: Let $M = (\lambda \cup {}^\omega\lambda, P^M, Q^M, \ldots, F_n, \ldots)_{n<\omega}$ where $P^M = \lambda$, $Q^M = {}^\omega\lambda$, $F_n(\eta) = \eta[n]$; $F_n(\alpha) = \alpha$. $T = \text{Th}(M)$ does not depend on λ and have elimination of quantifiers. So let, e.g., $\lambda = \lambda^\mu$ and construct a model N, $|P(N)| = \|N\| = \lambda$, such that for every strictly increasing sequence $\langle A_i : i < \mu\rangle$, $A_i \subseteq P(N)$, $|A_i| \leq \mu$ there is some $a \in P^N$, such that if $c \in Q^N$, $F_0(c) = a$, $F_1(c) \in A_i - \bigcup_{j<i} A_j$, then $\{F_n(c): n < \omega\} \cap (\bigcup_{j<\mu} A_j - \bigcup_{j\leq i} A_j) \neq \emptyset$. This suffices for the relativized version: if $N = \bigcup_{i<M} N_i$, N_i increasing, then $\bigvee_{i<M} P^N \subseteq A_i$. To get the full version we have to use a more complicated M: it is the disjoint union of P_n^M, $P_0^M = \lambda$, $P_{n+1}^M = {}^\omega(P_n^M)$, with all the projection functions.]

PROBLEM 1.3: Characterize, or at least investigate the possible $\{\langle\lambda,\mu\rangle: \lambda \geq \mu, \mu \text{ regular}, \lambda \geq |T|$, and every model of T of cardinality λ is the union of a strictly increasing elementary chain of length $\mu\}$.

THEOREM 1.6: *The following are equivalent for a countable T:*

(1) *T is categorical in $\aleph_0$.*

(2) *For each m, $D_m(T)$ is finite.*

(3) *Every $p \in D(T)$ is $\mathbf{F}^t_{\aleph_0}$-isolated.*

(4) *Every model of T of cardinality $\aleph_0$ is atomic (i.e., $\mathbf{F}^t_{\aleph_0}$-atomic).*

(5) *Every model of T of cardinality $\aleph_0$ is saturated.*

Proof. (1) $\Rightarrow$ (3) If $p \in D(T)$ is not $\mathbf{F}^t_{\aleph_0}$-isolated T has a countable model which realizes it, and T has a countable model which omits it (by IV, 5.3).

(3) $\Rightarrow$ (2) Let $\Gamma_m = \{\varphi(\bar{x}): \bar{x} = \langle x_0, \ldots, x_{m-1}\rangle$ and $\varphi(\bar{x}) \vdash r$ for some $r \in D_m(T)\}$. So by (3) $\{\neg\varphi(\bar{x}): \varphi(\bar{x}) \in \Gamma_m\}$ is inconsistent, so there is a

finite subset Γ'_m of Γ_m which is inconsistent; clearly $|D_m(T)| \le |\Gamma'_m| < \aleph_0$.

(2) ⇒ (3) For every m, $p \ne q \in D_m$ choose $\varphi_{p,q}(\bar{x}) \in p$, such that $\neg\varphi_{p,q}(\bar{x}) \in q$. Clearly $p \in D_m(T)$ is isolated by $\bigwedge \{\varphi_{p,q}(\bar{x}) : q \ne p, q \in D_m(T)\}$ (which is a formula by the finiteness of $D_m(T)$).

(3) ⇒ (4) By definition.

(4) ⇒ (1) A particular case of IV, 5.2(2).

(3) ⇒ (5), (5) ⇒ (1) Trivial.

THEOREM 1.8: *The following conditions on a totally transcendental T are equivalent:*

(1) *T is categorical in at least one $\lambda > |T|$.*

(2) *T is categorical in every $\lambda > |T|$.*

(3) *Every model of T of cardinality $> |T|$ is saturated.*

(4) *T is undimensional.*

(5) *If $M \prec N$, $M \ne N$, $\bar{a} \in M$, $\varphi(x, \bar{a})$ is not algebraic, then $\varphi(M, \bar{a}) \ne \varphi(N, \bar{a})$.*

(6) *There is no $\varphi(x, \bar{y})$ such that for every $\lambda > \mu \ge |T|$ there is a μ-saturated model M, $\|M\| = \lambda$ and $\bar{a} \in M$, $|\varphi(M, \bar{a})| = \mu$.*

(7) *Working in $\mathfrak{C}^{eq}$, there is a strongly semi-minimal formula $\varphi(x)$, such that for every M, M is prime over $\varphi(M)$, φ almost over $\emptyset$.*

Proof. (1) ⇒ (4) Immediate as V, 2.10(1), Definition V, 2.2 says not (4) implies not (1).

(4) ⇒ (6) Immediate.

(6) ⇒ (5) By V, 6.14.

(5) ⇒ (7) Let $\psi(x, \bar{a})$ be a minimal formula, and M be a saturated model $\bar{a} \in M$, and let $p = \text{tp}(a, M)$ not fork over $\emptyset$, and not algebraic. In the prime model N over $|M| \cup \{a\}$ there is $c \in \psi(N, \bar{a}) - |M|$ (by (5)) hence p, $\text{tp}(c, M)$ are not orthogonal, hence by V, Exercise 4.10(3) there is E such that $\text{stp}(a/E, \emptyset)$ is semi-minimal, (working in $\mathfrak{C}^{eq}$) and there is in it a strongly semi-minimal $\varphi(x)$, and there is a $\bar{c} \in M$ such that $\psi(M, \bar{a})$ is included in the algebraic closure of $\bar{c} \cup \varphi(\mathfrak{C}^{eq})$. There is no infinite indiscernible set $I \subseteq |M|$ over $\psi(M, \bar{a})$ (otherwise the prime model over $|M| \cup b$, b realizing $\text{Av}(I, |N|)$) contradicts (5)) hence in M^{eq} there is no infinite indiscernible set over $\varphi(M^{eq})$. Let M be a model.

Let N^{eq} be $\mathbf{F}^t_{\aleph_0}$-prime over $\varphi(M^{eq})$, $N^{eq} \prec M^{eq}$, hence by (5) $N^{eq} = M^{eq}$, hence for every $\bar{c} \in M^{eq}$ $\text{tp}(\bar{c}, \varphi(M^{eq}))$ is $\mathbf{F}^t_{\aleph_0}$-isolated and as in M there is no infinite indiscernible set over $\varphi(M^{eq})$; by the characterization of prime models (IV, 4.18) M^{eq} is prime over $\varphi(M^{eq})$.

(7) $\Rightarrow$ (3) For each M, $\|M\| > |T|$ clearly:

(*) If p is a type of cardinality $< \mu = \|M\|$, finitely satisfiable in $\varphi(M^{eq})$, then p is realized.

Let N^{eq} be $\mathbf{F}^t_\mu$-prime over $\varphi(M^{eq})$; then clearly $B = \varphi(N^{eq}) = \varphi(M^{eq})$, and we know by (7) that N^{eq} and M^{eq} are prime over B, hence isomorphic over B, hence M^{eq} too is saturated, hence M is saturated.

(3) $\Rightarrow$ (2) By I, 1.11.

(2) $\Rightarrow$ (1) Trivial.

CONCLUSION 1.9: *If* $|T| < 2^{\aleph_0}$, *T categorical in at least one* $\lambda > |T|$ *then T is categorical in every* $\lambda > |T|$.

Proof. By VIII, 4.1 T is stable in $|T|$ hence by Definition II, 3.1 and II, 3.2 T is totally transcendental, hence the previous theorem applies.

LEMMA 1.10: *Suppose T is stable and unidimensional.*

(1) *For any finite* Δ, *model* M, *and* (Δ, m)-*type* p *over* $|M|$, $|p| < \|M\|$, *p is realized in M.*

(2) *The conclusion in* (1) *implies that T does not have the* f.c.p. *and if* $\bar{a} \in M \precneqq N$, $\varphi(x, \bar{a})$ *is not algebraic then* $\varphi(M, \bar{a}) \neq \varphi(N, \bar{a})$; *and for every* φ, *for some* k_φ $|\varphi(M, \bar{a})| \geq k_\varphi \Rightarrow |\varphi(M, \bar{a})| = \|M\|$.

Proof. (1) Suppose p, a (Δ, m)-type over M, is a counterexample. We can assume $p \in S^m_\Delta(A)$, $A \subseteq M$, $|A| < \|M\|$, and for each $\varphi(\bar{x}; \bar{y}) \in \Delta$ let $\psi_\varphi(\bar{y}; \bar{a}_\varphi)$ define $p \restriction \varphi$.

For every set Δ_1 we define a theory $T_\alpha(\Delta_1)$, consisting of the following

(i) T.

(ii) $\{a_\beta: \beta < \alpha\}$ is a non-trivial Δ_1-indiscernible set over Q.

(iii) $\{\varphi(\bar{x}; b_0, b_1, \ldots): \varphi \in \Delta, b_0, \ldots \in Q, \vDash \psi_\varphi(b_0, \ldots; \bar{a}_\varphi)\}$ is consistent; $\bar{a}_\varphi \subseteq Q$. ($a_\beta, \bar{a}_\varphi$ serve as additional individual constants, Q as a new one-place predicate.)

(iv) The type described in (iii) is omitted.

For every finite Δ_1 $T_\omega(\Delta_1)$ is consistent: we expand M to a model of $T_\alpha(\Delta_1)$ by interpreting Q as A, $\bar{a}_\varphi$ as itself; and for a_β—by II, 2.19—there is in M an infinite set indiscernible over A. As $T_\alpha(\mathrm{L}) = \bigcup \{T_\alpha(\Delta_1): \Delta_1 \subseteq \mathrm{L}$ finite$\}$ there is a model M^1 of $T_{|T|^+}(\mathrm{L})$, and we can assume M^1 is $|T|^+$-saturated $\|M^1\| = 2^{|T|}$. So we can find a $|T|^+$-saturated model M of T, $A \subseteq M$, $q \in S^m_\Delta(A)$ is omitted by M, and $I = \{a_\alpha: \alpha < |T|^+\}$ is indiscernible over A, and $\|M\| = 2^{|T|}$. Hence the $\mathbf{F}^s_{|T|^+}$-prime model over $A \cup I$ omits q. Let J be such that $J \cup I$ is indiscernible over A, $|J| = (2^{|T|})^+$; so clearly also the $\mathbf{F}^s_{|T|^+}$-prime model over $A \cup J$ omits q. We find that T has a $|T|^+$-saturated model which is not saturated, and has cardinality $(2^{|T|})^+$ (as $|A| \leqslant \|M\| \leqslant 2^{|T|}$); hence T is not unidimensional.

(2) The first phrase follows by VI, 4.4, 5, the second by V, 6.14. For the third phrase note that if there are $\bar{a}_n \in M_n$, $n \leqslant |\varphi(M_n, \bar{a}_n)| < \aleph_0$, w.l.o.g. $\|M_n\| > 2^{\aleph_0}$ and by an ultraproduct get a contradiction to (1).

CONCLUSION 1.11: *Suppose T is superstable and unidimensional. Then:*

(1) $R^m(\bar{x} = \bar{x}, \mathrm{L}, \infty) < \omega$.

(2) *For every $\varphi(\bar{x}; \bar{y})$ and n there is a formula $\psi(\bar{y})$ such that $R[\varphi(\bar{x}; \bar{a}), \mathrm{L}, \infty] = n$ iff $\vDash \psi[\bar{a}]$.*

(3) *For every model M of T there are φ, and $\bar{a} \in A$ such that $\varphi(x, \bar{a})$ is weakly minimal.*

(4) *If T is totally transcendental, in* (1) *and* (2) *we can replace ∞ by $\aleph_0$, and in* (3) *"weakly minimal" by "minimal".*

Proof. By 1.10, V, 7.6, V, 7.10, and compactness.

LEMMA 1.12: (1) *Suppose T is superstable and unidimensional, $\theta(x, \bar{a})$ is a formula, $\bar{a} \in A$ and*

$(*)_\kappa$ *If p is a 1-type over A, $p \vdash \theta(x, \bar{a})$, $|p| < \kappa$, then some $c \in A$ realizes p (for $\kappa = \aleph_0$ we can assume $p = \{\varphi(x; \bar{b})\}$, $\bar{b} \in A$).*

Then for some κ-compact $\mathbf{F}^c_\kappa$-constructible model M over A, $\theta(M, \bar{a}) \subseteq A$

(2) *If we assume instead*

$(*)^a_\kappa$ *if p is an $\mathbf{F}^a_\kappa$-type over A, $p \vdash \theta(x; \bar{a})$ then some $c \in A$ realizes p,*

then for every $\mathbf{F}^c_\kappa$-prime model M over A, $\theta(M, \bar{a}) \subseteq A$.

Proof. (1) It suffices to prove that if $\mathrm{tp}(b, A)$ is $\mathbf{F}^c_\kappa$-isolated, for some $q \subseteq \mathrm{tp}(b, A)$, $|q| < \lambda$ and for no extension $q \in S(A)$ of q is $R(q_1, \mathrm{L}, \infty) < R(q, \mathrm{L}, \infty)$, then $A \cup b$ satisfies $(*)_\kappa$ and $b \notin \theta(\mathfrak{C}, \bar{a})$ (i.e., $\theta(\mathfrak{C}, \bar{a}) \cap A = \theta(\mathfrak{C}, \bar{a}) \cap (A \cup b)$).

This is because we can find a κ-compact M, $|M| = A \cup \{a_i : i < \alpha\}$ such that $\mathrm{tp}(a_i, A_i)$ is as above, where $A_i = A \cup \{a_j : j < i\}$. Now we prove by induction on i that A_i satisfies $(*)_\kappa$ and $\theta(\mathfrak{C}, \bar{a}) \cap A_i \subseteq A$. The induction step is by the above mentioned assertion. For limit stages notice that by the definability Lemma II, 2.12 A satisfies $(*)_\kappa$ iff A satisfies $(*)_{\aleph_0}$ and $\bar{a} \cup (A \cap \theta(\mathfrak{C}, \bar{a}))$ satisfies $(*)_\kappa$. Hence the property "$\theta(\mathfrak{C}, \bar{a}) \cap A \subseteq A_0 \subseteq A$ and A satisfies $(*)_\kappa$" is preserved by union of increasing chains of A's.

So let $B \subseteq A$, $|B| < \kappa$, $q \subseteq p \restriction B$, $|q| < \kappa$ where $p = \mathrm{tp}(b, A)$ be such that q has no extension q_1 in $S(A)$ for which $R(q_1, \mathrm{L}, \infty) < R(q, \mathrm{L}, \infty)$. We can assume $b \notin A$, hence q cannot be realized in A, hence $b \notin \theta(\mathfrak{C}, \bar{a})$. Now suppose $A \cup \{b\}$ does not satisfy $(*)_\kappa$, so there is a type r over $A \cup \{b\}$, $|r| < \kappa$, $r \vdash \theta(x, \bar{a})$, but r is not realized in A. Let c realize r, then $\mathrm{tp}(c, A \cup b)$ forks over A (otherwise by IV, 4.2 r is realized by some $c' \in A$), hence $\mathrm{tp}(b, A \cup c)$ forks over A, hence by III, 1.2(4) $n =_{\mathrm{def}} R[\mathrm{tp}(b, A \cup c), \mathrm{L}, \infty] < R[\mathrm{tp}(b, A), \mathrm{L}, \infty] \leq R(q, \mathrm{L}, \infty)$. So we can find $\bar{a}_1 \in A$ and φ such that $\models \varphi[b, c, \bar{a}_1]$ and $R[\varphi(x, c, \bar{a}_1), \mathrm{L}, \infty] = n$. Let $\varphi^*(y, \bar{z})$ be such that $R[\varphi(x, c', \bar{a}_1'), \mathrm{L}, \infty] = n$ iff $\models \varphi^*[c', \bar{a}_1']$ (by 1.11). We can assume q is closed under conjunctions and let

$$r^* = \{(\exists y)(\varphi(y, x, \bar{a}_1) \wedge \varphi^*(y, \bar{a}^*) \wedge \theta(x, \bar{a})) \colon \varphi^*(y, \bar{a}^*) \in q\}.$$

So $r^* \vdash \theta(x, \bar{a})$, $|r^*| < \kappa$, r^* is consistent (c realizes it: choose b as y), hence some $c^* \in A$ realizes it. So $q \cup \{\varphi(x, c^*, \bar{a}_1)\}$ is consistent, $R[\varphi(x, c^*, \bar{a}_1), \mathrm{L}, \infty] = n$, and this contradicts the choice q.

(2) A similar proof, using IV, 4.2.

CONCLUSION 1.13: *If T is superstable and unidimensional $\bar{a} \in M$, $\theta(x, \bar{a})$ is not algebraic and $|M|$, $\theta(x, \bar{a})$ satisfy $(*)_\kappa$ then M is κ-compact $[(*)^a_\kappa$ then M is $\mathbf{F}^a_\kappa$-saturated$]$.*

Proof. Otherwise let N be an $\mathbf{F}^c_\kappa$-constructible, over M, κ-compact, such that $\theta(N, \bar{a}) \subseteq \theta(M, \bar{a})$ (exists by the proof of 1.12). So $N \neq M$, and we get a contradiction to 1.11.

CONCLUSION 1.14: *The following properties of T are equivalent:*

(1) *T is superstable and unidimensional.*

(2) *For some λ_0, every λ_0-universal model of T is saturated.*

(3) *As (2) for $\lambda_0 = 2^{|T|}$.*

(4) *For some set A, $|A| \leq \lambda(T) \leq 2^{|T|}$ ($\lambda(T)$ is the first cardinal in which T is stable) $T(A)$ is categorical in every $\lambda > \lambda(T)$.*

(5) *For some λ, μ, $\lambda \geq 2^\mu$, $\mu \geq 2^{|T|}$, T has no μ-universal, non-μ^+-universal model of cardinality λ (for superstable T, $\lambda > \mu \geq \lambda(T)$ is enough).*

Proof. Trivially (3) $\Rightarrow$ (2) $\Rightarrow$ (5), and by VIII, 4.7, VIII, 4.1 (4) $\Rightarrow$ (3). Assuming (5), by VIII, 4.5 T is stable and even superstable, and by V, 2.10 (as λ_0-saturated implies λ_0-universal) T is unidimensional; so (5) $\Rightarrow$ (1).

So assume (1) and we shall prove (4). Let $A = |M|$, where M is a saturated model of T of cardinality $\lambda(T)$ (exists by VIII, 4.7). It suffices to prove that any $N \supseteq M$ is saturated. By 1.11(3) for some $\bar{a} \in M$, and θ, $\theta(x, \bar{a})$ is weakly minimal. Now if r is an $\mathbf{F}^a_{\lambda(T)}$-type over $|N|$, which is omitted, and $r \vdash \theta(x, \bar{a})$ choose $r_1 \in S(|N|)$, $r \subseteq r_1$; so r_1 is not algebraic, and $\theta(x, \bar{a}) \in r_1$, hence r_1 does not fork over $\bar{a}$ hence (using IV, 4.2) r is realized in M, contradiction. So N satisfies $(*)^a_{\lambda(T)}$ of 1.12; hence by 1.13 N is $\mathbf{F}^a_{\lambda(T)}$-saturated; hence (as T is unidimensional, and stable in every $\lambda \geqslant \lambda(T)$) N is saturated.

THEOREM 1.15: *Suppose T is categorical in some $\lambda > |T|$ or every model of T of cardinality λ (for some $\lambda > |T|$) is $|T|^+$-universal. Then T is categorical in every $\mu > |T|$, and every model of T of cardinality $> |T|$ is saturated.*

Proof. If T is categorical in $\lambda > |T|$ every model of T of cardinality λ is necessarily $|T|^+$-universal, and this implies T is stable in $|T|$ (by VII, 2.9(1)) and superstable (by VIII, 4.5) and unidimensional (by V, 2.10). Now it suffices to prove that every model M of cardinality $> |T|$ be saturated. So suppose M is a counterexample, and by 1.11 there are $\bar{a} \in M$ and θ such that $\theta(x, \bar{a})$ is weakly minimal. As M is not saturated, it is not $|T|^+$-saturated (by V, 2.10) by 1.13 there is $q \in S(B)$, $\bar{a} \in B \subseteq M$, $\theta(x, \bar{a}) \in q$, $|B| \leq |T|$, M omits q. As $|\theta(M, \bar{a})| = \|M\| > |T|$, there is a set $I \subseteq \theta(M, \bar{a})$, indiscernible over B, $|I| > |T|$. Let J be any indiscernible set over $|M|$ such that $I \cup J$ is indiscernible, $|J| \geq \lambda$ and let $A = \operatorname{acl}(|M| \cup J)$. If we prove A omits q, and we prove the existence of N; $A \subseteq N$, $\theta(N, \bar{a}) \subseteq A$, then we have a non-$|T|^+$-saturated model of cardinality $\geq \lambda$, hence there is such a model of cardinality λ. As T is superstable and stable in $|T|$, hence in λ, and by VIII, 4.7, T also has a saturated model of cardinality M_0 of power $|T|^+$. If $M_0 \prec N$ we can see that every type r over N, $|r| \leqslant |T|$, $\theta(x, \bar{a}) \in r$, is realized in N (either it is algebraic or it is realized in M_0), contradicting 1.13.

By 1.12 in order to prove the existence of N, it suffices to prove only the following:

(i) $A = \operatorname{acl}(|M| \cup J)$ satisfies $(*)_{\aleph_0}$ of 1.12. For suppose $\bar{c} \in A$, $\varphi(x, \bar{c}) \vdash \theta(x, \bar{a})$: if $\varphi(x, \bar{c})$ forks over $\bar{a}$, it is algebraic, hence $\varphi(\mathfrak{C}, \bar{c}) \subseteq A$ (as $\operatorname{acl}(\operatorname{acl} B) = \operatorname{acl} B$ by III, 6.2(4)). If $\varphi(x, \bar{c})$ does not fork over $\bar{a}$, it is realized in M, by III, 4.10.

(ii) No $c \in A$ realizes q. Otherwise there are $a_1, \ldots, a_n \in J$, $\bar{b} \in M$, and φ such that $\varphi(x, a_1, \ldots, a_n, \bar{b}) \vdash \operatorname{tp}(c, |M| \cup J)$ (by IV, 2.4(1)).

Clearly there is a finite $I_1 \subseteq I$ such that $J \cup (I - I_1)$ is indiscernible over $B \cup I_1 \cup \bar{b}$. Choose distinct $a'_1, \ldots, a'_n \in I - I_1$. Then $\varphi(x, a'_1, \ldots, a'_n, \bar{b}) \vdash \mathrm{tp}(c, B)$, hence some $c' \in M$ realizes q, contradiction.

QUESTION 1.4: Suppose T is superstable, not $|T|^+$-categorical. Does T have a non-$\aleph_1$-universal model in every cardinality?

THEOREM 1.16: *Suppose $\lambda > |T|$ and every model of T of cardinality λ is homogeneous, or at least $|T|^+$-homogeneous. Then (when $\lambda(T)$ is the first cardinality in which T is stable):*

(i) *every model of T of cardinality $> \lambda(T)$ is homogeneous,*

(ii) *if $|T| + \aleph_1 \leq \lambda \leq \lambda(T)$, then T has a non $\aleph_1$-homogeneous model of cardinality λ.*

Proof. By VIII, 4.10 T is superstable. If T is not unidimensional, T has a model M, $\|M\| = \lambda$, and a maximal indiscernible set $I \subseteq M$, $|I| = \aleph_0$; so M is not $\aleph_1$-homogeneous.

Now suppose $\|M\| > \lambda(T)$, M is not homogeneous, so it is not μ^+-homogeneous, for some $\mu < \|M\|$, $\mu \geq \lambda(T)$. Choose a weakly minimal $\theta(x, \bar{a})$, $\bar{a} \in M$, and also $N \prec M$ such that $\|N\| = \mu$, and if $c \in \theta(M, \bar{a}) - |N|$, then $\mathrm{tp}(c, |N|)$ is realized by $\geq \mu^+$ elements of M (possible as each such $\mathrm{tp}(c, |N|)$ is minimal), and there are $b^l_j \in N$ ($l = 0, j \leq \mu$ or $l = 1, j < \mu$). $\mathrm{tp}_*(\langle b^0_j : j < \mu \rangle, \emptyset) = \mathrm{tp}_*(\langle b^1_j : j < \mu \rangle, \emptyset)$ but for no $b^1_j \in M$ $\mathrm{tp}_*(\langle b^0_j : j \leq \mu \rangle, \emptyset) = \mathrm{tp}_*(\langle b^1_j : j \leq \mu \rangle, \emptyset)$.

Now choose M_1, $N \prec M_1 \prec M$, $\|M_1\| = \mu^+$, and for every $c \in \theta(M, \bar{a}) - |N|$, $\mathrm{tp}(c, |N|)$ is realized by $\geq \mu^+$ elements of M_1. Now choose $\lambda_1 > \mu^+$, $\lambda_1 = \beth_\delta$, δ is divisible by $(2^{|T|})^+ \times \omega$; and let M_2 be a saturated model of T, $\|M_2\| = \lambda_1$, $M_1 \prec M_2$.

Let $A = |M_1| \cup \{c \in M_2 : \mathrm{tp}(c, |N|)$ is realized by some $c \in \theta(M_1; \bar{a}) - |N|\}$ and it is easy to check that A satisfies $(*)_{\aleph_0}$ from 1.12(1) hence there is a model M_3, $A \subseteq M_3$, $\theta(M_3, \bar{a}) \subseteq A$, so $\|M_3\| = \lambda_1$ hence by VIII, 4.4 M_3 is homogeneous. So there is M_4, $M_1 \prec M_4 \prec M_3$, $\|M_4\| = \mu^+$, M_4 is homogeneous (possible as T is stable in μ^+). Clearly each $c \in \theta(M_4, \bar{a}) - |N|$ realize over $|N|$ a type realized by μ^+ elements of M_1.

Now it is easy to find an elementary mapping F, Dom $F = |N| \cup \theta(M_1, \bar{a})$, Rng $F = |N| \cup \theta(M_4, \bar{a})$, $F \restriction |N| =$ the identity. Now it is not hard to extend F to an elementary embedding F_1 of M_1 into M_4.

By 1.10 Rng $F_1 = |M_4|$, hence M_1 is homogeneous too; contradiction. So we have proved (i). For (ii) note:

LEMMA 1.17: *If* $|T| + \aleph_1 \le \lambda \le \lambda(T)$ (= *the first cardinal in which* T *is stable*), $\aleph_0 = \kappa(T) \leqslant \kappa \leqslant \lambda$, *then* T *has a model* M_λ *of cardinality* λ, *in which there is an indiscernible set of cardinality* κ, *but no more.*

Proof. Let I be indiscernible, $|I| = \kappa$, and let M be $\mathbf{F}^a_{\kappa(T)}$-primary over I, so clearly $\|M\| = \lambda(T)$ and let $N \prec M$ be such that $I \subseteq N$, $\|N\| = \lambda$. N is as required (see IV, 4.9).

QUESTION 1.5: Can we in 1.16 replace "$|T|^+$-homogeneous" by "$\aleph_1$-homogeneous"?

EXERCISE 1.6: Show that any counterexample to Question 1.5 is superstable and unidimensional.

PROBLEM 1.7: Can we in 1.16 replace homogeneous by model-homogeneous?

CONCLUSION 1.18: *If* T *is countable, and every model of* T *of cardinality* $\aleph_1$ *is homogeneous, then* T *is categorical in* $\aleph_1$.

Remark. We can replace $\aleph_1$ by any λ, $\aleph_0 < \lambda \le 2^{\aleph_0}$.

Proof. By 1.16 $\lambda(T) < \aleph_1$, so T is $\aleph_0$-stable. By 1.16 T is also unidimensional, so by 1.7 T is categorical in $\aleph_1$.

THEOREM 1.19: *If* T *is categorical in* $|T| > \aleph_0$, *then* T *is a definitional extension of some* $T' \subseteq T$, $|T| < |T|$ (*see* III, 5.14).

Proof. By VIII, 3.4 T is superstable, so by 1.17 necessarily $\lambda(T) < |T|$ but $|D(T)| \le \lambda(T)$; so we can apply III, 5.14.

PROBLEM 1.8: Suppose $|T| > \aleph_0$, $|T| \le |D(T)|$; prove T has a non-universal model of cardinality $|T|$.

EXERCISE 1.9: Suppose $\operatorname{cf} \lambda = \aleph_0$, and $(\forall \mu < \lambda)(2^\mu < \lambda)$ (e.g., $\beth_\omega$), hence $2^\lambda = \lambda^{\aleph_0}$.

(A) If $|S^m(A)| > |T| + |A| \ge \lambda$ then $|S^m(A)| \ge 2^\lambda$ (instead λ strong limit it suffices to assume that for

$$A' \subseteq A, \quad L' \subseteq L(T), \quad |A'| + |L'| < \lambda, \quad |S^m_{L'}(A')| \le |T| + |A|$$

and get $|S^m(A')| \ge \lambda^{\aleph_0}$). [Hint: Note that if $S \subseteq S^m(A)$, $|S| > |T| +$

$|A|$, $\mu < \lambda$, and we define $\varphi_\eta(\bar{x}, \bar{a}_\eta)$ by induction on $l(\eta)$ for $\eta \in {}^{\mu>}2$, such that, when

$$\eta \in W = \{\nu: |\{p \in S: (\forall \alpha < l(\nu))(\varphi_{\nu|\alpha}(\bar{x}, \bar{a}_{\nu|\alpha}) \in p\}| > |T| + |A|\},$$

then $\eta^\frown\langle 0\rangle, \eta^\frown\langle 1\rangle \in W$; and then $|W \cap {}^{\mu}2| \geq \mu$ and even $\eta \in W \cap {}^{\mu>}2$ implies $|W \cap \{\nu \in {}^{\mu}2: \eta \lessdot \nu\}| \geq \mu$.]

(B) Suppose $\lambda < |T| < 2^\lambda$, $|D(T)| = |T|$ then for every $\mathrm{L}' \subseteq \mathrm{L}(T)$, $|D(T \cap \mathrm{L}')| \leq |\mathrm{L}'| + \lambda$.

(C) If T is as in (B), $|T|$ regular, then there is an m-type p over $\emptyset$, $|p| < \lambda$, and p has a unique extension q in $D_m(T)$ which is not $\mathbf{F}^t_\lambda$-isolated. (Hint: Otherwise, act like in (A), where "$= |T|$" replaces "$> |T| + |A|$".)

(D) Prove that if $\lambda < |T| < 2^\lambda$, $|T|$ regular, then T has a model of cardinality $|T|$ which omits some type (from $D(T)$) hence is not universal (except when $|D(T)| < |T|$ hence T is a definitional extension of some $T' \subseteq T$, $|T'| < |T|$).

DEFINITION 1.1: Let $TV^x_\kappa(\theta, A)$ where $\theta = \theta(x, \bar{a})$ is a formula over A, means that any $\mathbf{F}^x_\kappa$-type p over A, $\theta \in p$, is realized by some $c \in A$.

EXERCISE 1.10: (1) $TV^t_\kappa(\theta, A)$ iff $TV^t_{\aleph_0}(\theta, A)$ and $TV^t_\kappa(\theta, \theta(A, \bar{a}) \cup a)$ for stable T (use II, 2.12).

(2) If A_i is increasing ($i < \alpha$) $\theta(A_i, \bar{a}) = \theta(A_0, \bar{a})$ and $TV^t_\kappa(\theta, A_i)$, then $TV^t_\kappa(\theta, \bigcup_{i<\kappa} A_i)$ provided T is stable.

(3) Suppose $TV^t_\kappa(\theta, A)$ $\kappa \geq \kappa(T)$, and N is an $\mathbf{F}^c_\kappa$-constructible model over A. Then $\theta(M, \bar{a}) \subseteq A$ when one of the following holds:

(α) for every finite Δ, $C \subseteq A$, $|C| < \kappa$, a type p over C, $|p| < \kappa$, $\theta \in p$, and $c \in \theta(\mathfrak{C}, \bar{a})$ realizing p, there is $c' \in A$ realizing p such that

$$R[\mathrm{stp}(c, C), \Delta, 2] = R[\mathrm{stp}(c', C), \Delta, 2],$$

(β) every equivalence relation $\varphi(x, y, \bar{c})$, over $\theta(\mathfrak{C}, \bar{a})$, $\bar{c} \in A$, has $< \aleph_0$ or $> |T|$ equivalence class in $p(A)$ whenever p is a 1-type over A, $|p| < \kappa$, $\theta \in p$ (use Exercise III, 4.11).

(4) If T is unidimensional and stable, $TV^t_{\aleph_0}(\theta, A)$, then every equivalence relation $\varphi(x, y, \bar{c})$ over $\theta(\mathfrak{C}, \bar{a})$, $\bar{c} \in A$, has $< \aleph_0$ or $|\theta(A, \bar{a})|$ equivalence classes in $\theta(A, \bar{a})$.

(5) If T is unidimensional (and stable), $\kappa \geq \kappa(T)$, $TV^t_\kappa(\theta, A)$, $|\theta(A, \bar{a})| > |T|$ then for any $\mathbf{F}^c_\kappa$-constructible model N over A, $\theta(N, \bar{a}) \subseteq A$.

EXERCISE 1.11: (1) Use Exercise 1.10(3)(β) to get another version of V, 6.14 and V, 6.16.

(2) Show some condition (as (α), (β)) is needed in Exercise 1.10(3). [Hint: Let $M = (|M|, P^M, P^M_\eta, Q^M, Q^M_\eta, F^M_\nu, c^M_\eta)_{\eta \in {}^{\omega>}2, \nu \in {}^{\omega}2}$, $T = \mathrm{Th}(M)$ where:

$$P^M = {}^{\omega>}2 \cup ({}^{\omega}2 \times \omega),$$

$$Q^M = {}^{\omega}2 \times \omega \times \omega,$$

$$|M| = P^M \cup Q^M,$$

$$P^M_\eta = \{\nu : \nu \lessdot \eta \text{ or } \eta \lessdot \nu \text{ or } \eta \lessdot \nu \in {}^{\omega>}2\} \cup \{\langle \rho, n\rangle : \eta \lessdot \rho \in {}^{\omega}2, n < \omega\},$$

$$Q^M_\eta = \{\langle \rho, n, m\rangle : \eta \lessdot \rho \in {}^{\omega}2, n < \omega, m < \omega\},$$

F^M_ν is a partial one-place function: its domain is Q^M,

$$F^M_\nu(\langle \rho, n, m\rangle) = \begin{cases} \rho \restriction k & \rho \restriction k = v \restriction k, \\ & \rho \restriction k \neq v(k), \\ \langle \rho, n\rangle & \rho = \nu, \end{cases}$$

$$c^M_\eta = \eta \text{ (which is in } P^M).$$

Building an automorphism of saturated models we can prove that T is superstable. Also for some model N of T, $\|N\| > |P^M|$ (remember F_ν is not one-to-one even on (for any n) the set $\{\nu\} \times \{n\} \times \omega$ which we can "blow up"). Let $A = {}^{\omega>}2, \theta = P(x)$, now for every $a \in Q^{\mathfrak{C}}$, for some $\nu \in {}^{\omega}2, F_\eta(a) \in P^{\mathfrak{C}} - A$. Lastly $TV_{\aleph_0}(\theta, A)$ holds [note that it suffices to prove this for the reduct to every finite sublanguage, where we have enough automorphisms].]

EXERCISE 1.12: **Through the following series of assertions we reprove 1.18.**

(A) If every model of T of cardinality λ is homogeneous $|T| < 2^{\aleph_0}$, $\lambda > |T|$; then for every p, CR(p, 2) $< \infty$ (see Exercise II, 3.8) (use $EM(\lambda)$).

(B) If for every p, CR(p, 2) $< \infty$ then there is an infinite set of indiscernibles, and for every A, there is an $\mathbf{F}^s_{\aleph_0}$-atomic model M over A.

(C) Use (B) to show that under the assumptions of (A), T has a non-$\aleph_1$-homogeneous model M of T, which is $\aleph_0$-saturated, hence $\|M\| \geq |D(T)|$.

(D) Show that under the hypothesis of (A), there is $T_1 \supseteq T$, $|T_1| = |D(T)|$ such that PC(T_1, T) is categorical in λ.

EXERCISE 1.13: (1) If T is countable and $\aleph_0$-categorical prove that every model of T is $\mathbf{F}^a_{\aleph_0}$-saturated (see III, 5.16).

(2) If T is totally transcendental, every $\aleph_0$-saturated model is $\mathbf{F}^a_{\aleph_0}$-saturated.

EXERCISE 1.14: Find a totally transcendental T such that for some model M of T, $\|M\| < |T| = |D(T)|$. [Hint: Let I be a tree, $\{B_i : i < \lambda\}$ be a set of branches of I, let $\mathrm{L} = \{P_i : i < \lambda\}$, P_i a one place predicate, and $T = \mathrm{Th}(M)$, where $|M| = I$, $P_i(M) = B_i$.]

EXERCISE 1.15: Suppose $\lambda = |T| = \lambda^{<\lambda}$ is regular, then the following conditions are equivalent.

(1) All $\mathbf{F}^t_\lambda$-compact models of T of power $|T|$ are isomorphic.

(2) Every $p \in D(T)$ is $\mathbf{F}^t_\lambda$-isolated.

(3) Every $\mathbf{F}^t_\lambda$-compact model of T is λ-saturated.

THEOREM 1.20: *Suppose T is superstable and $\lambda > |T|$ is the first cardinality in which T is stable. Then for every $\mu \geq |T| + \aleph_1$, $I(\mu, T) \geq \min\{2^\mu, 2^\lambda\}$.*

Proof. Let χ be any regular cardinal $|T| < \chi \leq \lambda$, and choose a formula $\varphi(x, \bar{a})$ (existing by III, 5.16) such that:

(i) For some A, $\bar{a} \in A$, $|A| \leq |T|$, and $|\{p \in S(A): \varphi(x, \bar{a}) \in p\}| \geq \chi$.

(ii) $\alpha^* = R(\varphi(x, \bar{a}), \mathrm{L}, \infty)$ is minimal (for φ's satisfying (i)). Choose a model M_0, $\|M_0\| = |T|$, $\bar{a} \in |M_0|$, such that if B, Δ, n, p are finite, $B \subseteq |M_0|$, p a finite type over B, and for some c realizing p, $n = R[\mathrm{stp}(c, B), \Delta, 2]$ then there is such $c \in M_0$. Choose also I indiscernible over $|M_0|$, $|I| = \aleph_0$. Note that by the regularity of χ and the minimality of α^*:

(*) If $\bar{a} \in A$, $|A| < \chi$, then $\{p \in S(A): \varphi(x, \bar{a}) \in p, R(p, \mathrm{L}, \infty) < \alpha^*\}$ has cardinality $<\chi$.

Now there are two possibilities.

Possibility A. Some $q^* \in S(\bar{a})$ has $\geq \chi$ extensions in $S(|M_0|)$ and $\varphi(x, \bar{a}) \in q^*$.

Possibility B. Every $q \in S(\bar{a})$ has $<\chi$ extensions in $S(|M_0|)$ if $\varphi(x, \bar{a}) \in q$.

When Possibility A occurs, there is a set A^*, $|A^*| = \lambda(q^*)$ (see Definition III, 5.1, Lemma III, 5.1, 3) such that every extension of q^* in $S(A^*)$ which does not fork over $\bar{a}$ is stationary; and $|\{p \in S(M_0) : q^* \subseteq p, p$ does not fork over $\bar{a}\}| = 2^{\lambda(q^*)}$. We can assume $A^* \subseteq M_0$ and that for any $E(x, y, \bar{a}) \in E(A)$ and c realizing q^*, there is an indiscernible set $I \subseteq A^*$, based on some $q_1 \supseteq q^* \cup \{E(x, c, \bar{a})\}$. If Possibility B occurs let $A^* = \bar{a}$. Now we define by induction on $\alpha < \chi$, p_α and a_α realizing p_α such that (α) $\varphi(x, \bar{a}) \in p_\alpha \in S(A^*)$, $R(p_\alpha, \mathrm{L}, \infty) = \alpha^*$, and if Possibility A holds $q^* \subseteq p_\alpha$, (β) no extension of p_α in $S(|M_0| \cup I \cup \{\bar{a}_i: i < \alpha\})$ forks over $\bar{a}$. Now p_α exists as the number of p's satisfying (α) is $\geq \chi$ (check each possibility) whereas by (*)

$$|\{p \in S(A^*): p \text{ satisfies } (\alpha), \text{ but not } (\beta)\}| < \chi$$

because $|M_0| \cup I \cup \{\bar{a}_i: i < \alpha\}| < \chi$ so we can choose an appropriate p_α and also a_α. Now w.l.o.g. for every α, no extension of p_α in $S(|M_0| \cup I \cup \{a_i : i < \chi, i \neq \alpha\})$ forks over A^* [if Possibility A holds, as p_α is stationary $p_\alpha = \mathrm{tp}(\bar{a}_\alpha, A^*) \vdash \mathrm{tp}(\bar{a}_\alpha, |M_0| \cup I \cup \{a_i : i < \alpha\})$; if Possibility B holds, we choose by induction on $\alpha, \gamma_\alpha < \chi$ and $\bar{a}_{\alpha,\gamma} (\gamma < \gamma_\alpha)$ such that (i) holds, and we replace: (ii) by (ii)′ no extension of p_α in $S(|M_0| \cup I \cup \{\bar{a}_{\beta,\gamma} : \beta < \alpha, \gamma < \gamma_\beta)$ fork over A^*, and (iii) $\bar{a}_{\alpha,\gamma}$ realizes p_α, $\{\bar{a}_{\alpha,\gamma} : \gamma < \gamma_\alpha\}$ is independent over $|M_0| \cup I \cup \{\bar{a}_{\beta,\gamma} : \beta < \alpha, \gamma < \gamma_\beta\}$, and every $p \in S^1(|M_0| \cup I \cup \{\bar{a}_{\beta,\gamma} : \beta < \alpha, \gamma < \gamma_\beta\})$ extending p_α is realized by some $\bar{a}_{\alpha,\gamma}$, this is possible and $\bar{a}_\alpha \overset{\text{def}}{=} a_{\alpha,0}$ are as required—if $q \in S(|M_0| \cup I \cup \{a_{\beta,0} : \beta < \chi, \beta \neq \alpha\}$ is an extension of p_α forking over A^*, for some $\gamma < \gamma_\alpha$, $q \vdash \mathrm{stp}(a_{\alpha,\gamma}, A^*)$ and by IV, 3.13 applied to $\mathbf{F}^a_\infty$ we get a contradiction]. Clearly, I is indiscernible over $|M_0| \cup \{a_i : i < \chi\}$.

So let $\mu \geq |T| + \aleph_1$, and I_μ be an indiscernible set over $|M_0| \cup \{a_i: i < \chi\}$, $I \subseteq I_\mu$, $|I_\mu| = \mu$. Clearly for every α, no extension of p_α in $S(|M_0| \cup I_\mu \cup \{a_i: i < \chi, i \neq \alpha\})$ forks over $\bar{a}$. For every $S \subseteq \chi$ let M^μ_S be any $\mathbf{F}^c_{\aleph_0}$-constructible model over $|M_0| \cup I_\mu \cup \{a_i: i \in S\}$, so by 1.1 M^μ_S realizes p_α iff $\alpha \in S$ (for $\alpha \notin S$ if Possibility A holds, each p_α is stationary, if Possibility B holds, show that for each c realizing p_α, $\mathrm{stp}(c, A^*)$ is omitted). Clearly $|M^\mu_S| = \mu + |S|$, thus we have proved that (T, T) has $(\chi, |A^*|)$-freedom. All this depends on the parameter χ which we suppresss up to now.

Let $\kappa = \min\{\mu, \lambda\}$; and we split the proof to cases.

Case (i). There is a regular $\chi \leq \kappa$, $2^\chi = 2^\kappa$. Apply the above construction for $\chi + |T|^+$. If Possibility B holds, for some $k < \aleph_0$, (T, T) has (χ, χ)-freedom hence by VIII, 1.4 $I(\mu, T) > 2^\chi = 2^\kappa = \min\{2^\mu, 2^\kappa\}$. So, suppose Possibility A holds. If $2^\chi > 2^{|A^*|}$ again by VIII, 1.4 $I(\mu, T) \geq 2^\chi = \min\{2^\mu, 2^\kappa\}$, so assume $2^\chi = 2^{|A^*|}$. Now $|A^*| = \lambda(q^*)$ so if $\lambda(q^*) > \aleph_0$, by III, 5.13's proof $|D(T)| \geq 2^{\lambda(q^*)} = 2^\chi = \min\{2^\lambda, 2^\mu\} > |T|$ hence by VIII, 1.2 $I(\mu, T) \geq |D(T)| \geq 2^{\lambda(q^*)} = 2^{|A^*|} = 2^\chi = \min\{2^\lambda, 2^\mu\}$. So we can assume $\lambda(q^*) \leq \aleph_0$ (hence $\lambda(q^*) = \aleph_0$), so $2^\chi = 2^{\aleph_0}$, $|A^*| = \aleph_0$, but $2^\kappa = 2^\chi$ hence $|T| \leq \kappa < 2^{\aleph_0}$. We can also assume $|D(T)| < 2^\chi$, hence by II, 3.2. $|D(T)| \leq |T|$. By II, 3.15 as $\lambda > |T|$, in fact $\lambda = 2^{\aleph_0}$, hence $\kappa = \mu < 2^{\aleph_0}$. We shall deal with this in Case (iii).

Case (ii). Not Case (i). So κ is singular and $\chi < \kappa \Rightarrow 2^\chi < 2^\kappa$; let $\kappa = \sum_{i < \text{cf}\kappa} \chi(i)$, cf $\kappa < \chi(i) < \kappa$, $\chi(i)$ regular, and increasing. For each i there are $\alpha^*_{\chi(i)}$, $M_{\chi(i)}$, $\varphi_{\chi(i)}(\chi; \bar{a}_{\chi(i)})$ as in the construction above and choose M_0, I, $\|M_0\| = \text{cf}\,\kappa + |T|$, $\bigcup_{i<\text{cf}\kappa} M_{\chi(i)} \subseteq M_0$, as before. Now we define inductively on $\alpha < \kappa$ elements a_α, such that if $\sum_{j<i} \chi(j) \leq \alpha < \chi(i)$ then $R^m(p_\alpha, \text{L}, \infty) = \alpha^*_{\chi(i)} \vDash \varphi_{\chi(i)}[a_\alpha, \bar{a}_{\chi(i)}]$ and $p_\alpha = \text{tp}(a_\alpha, |M_0|)$ has a unique extension in $S(|M_0| \cup I \cup \{a_i: i < \alpha\})$. Now let I_μ be an indiscernible set over $|M_0| \cup \{a_i: i < \kappa\}$, $I \subseteq I_\mu$, $|I_\mu| = \mu$, and for each $S \subseteq \kappa$ let M^μ_S be any $\mathbf{F}^c_{\aleph_0}$-constructible model over $|M_0| \cup I_\mu \cup \{a_\beta: \beta \in S\}$. As before M^μ_S realizes p_α iff $\alpha \in S$ (by 1.1) hence (T, T) has $(\kappa, \text{cf}\,\kappa + |T|)$-freedom as $2^{\text{cf}\kappa+|T|} < 2^\kappa$ (provided that $|T| < \kappa$) by VIII, 1.4 $I(\mu, T) \geqslant 2^\kappa = \min\{2^\lambda, 2^\mu\}$. Suppose $|T| = \kappa$. We can assume $|D(T)| \geq \lambda$ (otherwise we return to Case (iii)), and as $\lambda > |T|$, we can work with $\chi = |T|^+$, so if $2^{|A^*|} < 2^{|T|}$ we can finish as in Case (i). Otherwise, $D(T) \geq 2^{\lambda^*(q)} \geq 2^{|A^*|} = 2^{|T|}$, so by VIII, 1.2 we finish.

Case (iii). $|D(T)| \leq |T| \leq \mu < 2^{\aleph_0}$, $2^\mu = 2^{\aleph_0}$. Here we choose $\chi = |T|^+$, and $\varphi(x; \bar{a})$ as before; by VIII, 1.3 we can assume $\bar{a}$ is empty, so $\varphi = \varphi(x)$. By the choice of φ, if $\psi(x, \bar{b}) \vdash \varphi(x)$, $\psi(x; \bar{b})$ forks over $\emptyset$ (or even $R[\psi(x; \bar{b}), \text{L}, \infty] < \alpha^*$) then for every A, $|\{p \in S(A): \psi(x; \bar{b}) \in p\}| \leq |T| + |A|$, hence by II, 3.2 $R[\psi(x; \bar{b}), \text{L}, \aleph_0] < \infty$. Choose q, $\varphi(x) \in q \in D_1(T)$, such that for some A, $|A| = |T|$ and $|\{p \in S(A): q \subseteq p\}| = 2^{\aleph_0}$; by III, Section 5 as $\lambda^*(q) \leq \aleph_0$, hence $\lambda^*(q) = \aleph_0$, there are $E_n \in E(\emptyset)$ such that E_{n+1} refines E_n, and for each c realizing q, $q \cup \{E_n(x, c): n < \omega\}$ is stationary. Choose b_l ($l < \omega$) realizing q, such that for every b realizing q, and n, for some $l > n$, $\vDash E_n[b_l, b]$, and $\{b_l : l < \omega\}$ is independent over $\emptyset$. Now we define by induction on l, a natural number $n(l)$ and elements a_η ($\eta \in {}^l2$) realizing q such that

(i) if $\eta \neq \nu \in {}^l2$, then $\vDash \neg E_{n(l)}(a_\eta, a_\nu)$,

(ii) if $\nu \lessdot \eta$, $\eta \in {}^l2$, $\nu \in {}^k2$, then $\vDash E_{n(k)}(a_\eta, a_\nu)$,

(iii) $\{b_i: i < \omega\} \cup \{a_\eta: \eta \in {}^{l \geq} 2\}$ is independent over $\emptyset$,
(iv) if for each $\eta \in {}^l 2$, $\vDash E_{n(l)}(a_\eta, a'_\eta)$ and
$\{b_i: i < l\} \cup \{a_\nu: \nu \in {}^{l>} 2\} \cup \{a'_\eta: \eta \in {}^l 2\}$
is independent over $\emptyset$, then

$$\mathrm{tp}(\langle \ldots, a_\eta, \ldots \rangle_{\eta \in {}^l 2}, \{b_i: i < l\} \cup \{a_\nu: \nu \in {}^{l>} 2\})$$
$$= \mathrm{tp}(\langle \ldots, a'_\eta, \ldots \rangle_{\eta \in {}^l 2}, \{b_i: i < l\} \cup \{a_\nu: \nu \in {}^{l>} 2\})$$

(for (iv) use $|D(T)| \leq |T| < 2^{\aleph_0}$).

Now for each $\eta \in {}^\omega 2$ choose a_η such that $\{b_l: l < \omega\} \cup \{a_\eta: \eta \in {}^{\omega \geq} 2\}$ is independent over $\emptyset$ and a_η realizes $q \cup \{E_{n(l)}(x, a_{\eta | l}): l < \omega\}$.

For each $S \subseteq {}^\omega 2$, $|S| = \mu$, let $M_1(S)$ be an $\mathbf{F}^a_{\aleph_0}$-prime model over $A_S = \{b_l: l < \omega\} \cup \{a_\eta: \eta \in S\}$, and let $B_S = \{c \in M_1(S);\ c$ realizes q, and $\mathrm{tp}(c, \bar{c})$ is realized in $M_1(S)$ by $\leq |T|$ elements for some $\bar{c} \in A_S\}$ so $A_S \subseteq B_S$. Now let $M(S)$ be a $\mathbf{F}^c_{\aleph_0}$-constructible model over B_S. The proof is now like that of VIII, 1.8, when we observe the following, whose proof we leave to the reader.

Let c depend on C (here always $C \cup \{c\} \subseteq q(\mathfrak{C})$) if for some $\bar{c} \in C$, in the $\mathbf{F}^a_{\aleph_0}$-prime model over $\bar{c}$, $\mathrm{tp}(c, \bar{c})$ is realized $\leq |T|$ times.

Fact 1: If $\mathrm{tp}(c, C)$ forks over $\emptyset$, then c depends on C.

Fact 2: This dependence relation is transitive and of finite character.

Fact 3: Any $\mathbf{F}^a_{\aleph_0}$-type over B_S which forks over $\emptyset$ is realized by some $c \in B_S$, hence by 1.1 $|\{c \in M(S): c \text{ realizes } q\}| = B_S$.

Fact 4: For any $\bar{c}$, $\{\mathrm{stp}(c, \emptyset): c \text{ realizes } q, c \text{ depends on } \bar{c}\}$ is countable.

IX.2. On the lower parts of the spectrum

Remark. We have used here, for $p \in S^m(B)$, $B \subseteq A$, "a stationarization of p over A" as "a complete type extending A not forking over B".

THEOREM 2.1: (1) *If T is superstable, countable and not categorical is $\aleph_0$, then $I(\aleph_0, T) \geq \aleph_0$.*

(2) *If $|T| = \aleph_\alpha$ (and of course $|T| \leq |D(T)|$), then $I(\aleph_\alpha, T) \geq |\alpha + 1|$; and if T is totally transcendental except when $\alpha = 0$, T categorical in $\aleph_0$, then $I(\aleph_\alpha, T) \geq |\alpha + 1| + \aleph_0$.*

(3) *If T is totally transcendental, $|T| = \aleph_\alpha$ (and of course $|T| \leq |D(T)|$), $\beta \geq \alpha$ and T is not categorical in $|T|^+$, then $I(\aleph_\beta, T) \geq |\beta + 1|$ and, except when $\alpha = 0$, T categorical in $\aleph_0$, $I(\aleph_\beta, T) \geq |\beta + 1| + \aleph_0$.*

Remark. On $|T| \leq |D(T)|$ see III, 5.18.

Proof. (1) If $|D(T)| > \aleph_0$, the conclusion follows by VIII, 1.2. So assume $|D(T)| \leq \aleph_0$, hence by IV, 5.1 T has a $\mathbf{F}^t_{\aleph_0}$-prime model M_0. As T is not $\aleph_0$-categorical, there is a $p \in D_m(T)$ which is not isolated, hence M_0 omits p; and let $p_0 \in S^m(M_0)$ be a stationarization of p. Define inductively $\bar{a}_n$ $(n < \omega)$ such that $p_n = \text{tp}(\bar{a}_n, |M_0| \cup \bigcup_{l<n} \bar{a}_l)$ is a stationarization of p_0 (hence $\{\bar{a}_n : n < \omega\}$ is indiscernible over $|M_0|$); and let N_n be an $\mathbf{F}^t_{\aleph_0}$-prime model over $\bigcup_{l<n} \bar{a}_l$ (exists by IV, 5.1(4)). Now we shall prove that N_n omits

$$r = \bigcap \left\{ r_1 : r_1 \text{ a stationarization of } p \text{ over } \bigcup_{l<n} \bar{a}_l \right\}.$$

Otherwise there is $\varphi(\bar{x}, \bar{a}_0, \ldots, \bar{a}_{n-1}) \vdash r$, $\varphi(\bar{x}, \bar{a}_0, \ldots)$ does not fork over $\emptyset$. So $\varphi(\bar{x}, \bar{a}_0, \ldots)$ does not fork over $|M_0|$ hence some $\bar{c} \in M_0$ realize it (by III, 4.10(1)). But M_0 omits p, hence for some $\psi(\bar{x}) \in p$, $\models \neg\psi[\bar{c}]$ hence $\models (\exists \bar{x})(\varphi(\bar{x}, \bar{a}_0, \ldots, \bar{a}_{n-1}) \wedge \neg\psi[\bar{x}])$, contradiction. So in N_n, $I_n = \{\bar{a}_0, \ldots, \bar{a}_{n-1}\}$ is a maximal set independent over $\emptyset$ of sequences realizing p.

Hence by V, 3.13(2),

$$\dim(p, \emptyset, N_n) \leq |I_n| = n \leq w(p) \dim(p, \emptyset, N_n)$$

and by V, 3.9(1) $w(p)$ is finite.

We can conclude that $\{n : N_n \cong N_{n_0}\}$ is finite; hence $I(\aleph_0, T) \geq \aleph_0$.

(2) The assertion $I(\aleph_\alpha, T) \geq |\alpha + 1|$, follows by VIII, 3.4 if T is not superstable, by 1.20 if T is superstable, not stable in $|T| = \aleph_\alpha$ except when $\alpha = 0$, and then the assertion is trivial. Now if T is superstable, T stable in $|T|$ (see 1.17, IV, 4.9) for every $\beta \leq \alpha$ T has a maximally $\aleph_\beta$-saturated model of cardinality $\aleph_\alpha$, hence $I(\aleph_\alpha, T) \geq |\alpha + 1|$. So now we assume T is totally transcendental, and $|T| > \aleph_0$ *or* T not categorical in $\aleph_0$, and have to prove, in fact, $I(\aleph_\alpha, T) \geq \aleph_0$. As $|T| \leq |D(T)$ and by 1.6, for some m $D_m(T)$ is infinite, hence as in 1.6 some $p \in D_m(T)$ is not isolated, hence M_0, the prime model of T, omits p; and M_0 exists (by IV, 3.1, 10 and IV, 2.18). We can proceed as in the proof of (1) but maybe $\|N_n\| < |T|$. If for some $\bar{b}$, the prime model over $\bar{b}$ has cardinality $|T|$, we define $\bar{a}_n$ such that $\text{tp}(\bar{a}_n, |M_0| \cup \bar{b} \cup \bigcup_{l<n} \bar{a}_l)$ is a stationarization of p_0, let N_n be a $\mathbf{F}^t_{\aleph_0}$-prime model over $\bar{b} \cup \bigcup_{l<n} \bar{a}_l$; then $I_n = \{a_l : l < n\}$, is, in N_n, a maximal set independent over $(\bar{b}, \emptyset)$ of sequences realizing p, hence by V, 3.13(2) and V, 3.15(2),

$$\dim(p, \emptyset, N_n) \leq |I_n| = n \leq w(p) \dim(p, \emptyset, N_n) + w(\bar{b}, \emptyset)$$

hence as in (1) $I(\aleph_\alpha, T) \geq \aleph_0$.

So suppose for every $\bar{b}$, the $\mathbf{F}^t_{\aleph_0}$-prime model over $\bar{b}$, has cardinality $< |T|$; we can also assume, w.l.o.g., $0 < \alpha < \omega$ (by (1) and the first part of (2)). For each non-algebraic $q \in D(T)$, let I_q be an indiscernible set based on q of cardinality $\aleph_\alpha$, $I_q = \{\bar{a}^i_q : i < \aleph_\alpha\}$. Let M_q be $\mathbf{F}^t_{\aleph_0}$-prime over I_q, so clearly $|\{r \in D(T) : M_q \text{ realizes } r\}| = |\{r \in D(T)$: for some n, r is realized in the $\mathbf{F}^t_{\aleph_0}$-prime model over $\bigcup_{i<n} \bar{a}^i_q\}| < \aleph_\alpha$. Also the number of algebraic $q \in D(T)$, is $\leqslant \|M_0\| < \aleph_\alpha$. As $\aleph_\alpha \leqslant |D(T)|$, clearly there are $\geqslant \aleph_\alpha$ non-isomorphic M_q's, so we finish.

(3) By 1.8, T is not unidimensional, hence by V, 2.10 for every $\gamma \leq \beta$, T has a maximally $\aleph_\gamma$-saturated model of cardinality $\aleph_\beta$, hence $I(\aleph_\beta, T) \geq |\beta + 1|$. So as in (2) it suffices to prove $I(\aleph_\beta, T) \geq \aleph_0$.

As in (2) let M_0 be a $\mathbf{F}^t_{\aleph_0}$-prime model of T, $p \in D_m(T)$ omitted by M_0. We can assume p is stationary. (By III, 2.9, for some $E(x, y) \in \mathrm{FE}(\emptyset)$, $p \cup \{E(x, a)\}$ is stationary for any a realizing p; we add the names of the E-equivalence classes as new relations. As each model can be so expanded in finitely many ways the proof is sufficient.) As in (2) we choose appropriate $\bar{b}$, choose as before a_n, $N_n (n < \omega)$, and then find N^1_n, $\bar{b} \cup \bigcup_{l<n} \bar{a}_l \subseteq N^1_n$, $\|N^1_n\| = \aleph_\beta$, such that no $\bar{a} \in N^1_n$ realizes a stationarization of p over $\bar{b} \cup \bigcup_{l<n} \bar{a}_l$, or at least $\dim(p, \emptyset, N^1_n) < \aleph_0$ and the rest is as in (2). So what remains are the choices of $\bar{b}$ and N^1_n.

***Case* A. For some $\bar{b}$, in the $\mathbf{F}^t_{\aleph_0}$-prime model over $\bar{b}$, $M_{\bar{b}}$ there is an infinite indiscernible set I.**

Clearly we can find an infinite indiscernible set $I_n \subseteq |N_n|$, and by III, 3.5 we can assume it is indiscernible over $\bar{b} \cup \bigcup_{l<n} \bar{a}_l$. Let J_n be an indiscernible set over $\bar{b} \cup \bigcup_{l<n} \bar{a}_l$, $I_n \subseteq J_n$, $|J_n| = \aleph_\beta$, and let N^1_n be $\mathbf{F}^t_{\aleph_0}$-prime over $J_n \cup \bar{b} \cup \bigcup_{l<n} \bar{a}_l$; clearly $\|N^1_n\| = \aleph_\beta$, and every type over $\bar{b} \cup \bigcup_{l<n} \bar{a}_l$ realized in N^1_n, is realized in N_n; so N^1_n is as required.

***Case* B. There is a non-algebraic type orthogonal to p. Let q be such a complete type, $q \in S^m(\bar{b})$, such that q is stationary; choose $\{c_i : i < \aleph_\beta\}$ be independent over $(\bar{b} \cup \bigcup_{l<\omega} \bar{a}_l, \bar{b})$, c_i realizing q, and let N^1_n be $\mathbf{F}^t_{\aleph_0}$-prime over $\bar{b} \cup \bigcup_{l<n} \bar{a}_l \cup \{c_i : i < \aleph_\beta\}$. If r_j is any stationarization of p over $\bar{b} \cup \bigcup_{l<n} \bar{a}_l \cup \{c_i : i < j\}$, it is weakly orthogonal to the stationarization of q over the same set (by V, 2.1(2)). So r_0 has a unique extension in $S^m(\bar{b} \cup \bigcup_{l<n} \bar{a}_l \cup \{c_i : i < \aleph_\beta\})$ so $r_{\aleph_\beta}$ does not fork over $\emptyset$, hence as in (1) it is omitted by N^1_n. As clearly $\|N^1_n\| = \aleph_\beta$ we finish.**

***Case* C. Not Case A nor Case B. By 1.8 T is not unidimensional, so by V, 2.10, there are a minimal type r, and a type q which are orthogonal.**

As $\kappa(T) = \aleph_0$, we can assume that r and q are complete types over some $\bar{b}$, and they are stationary; and the stationarization of p over $\bar{b}$ is not weakly orthogonal to r nor to q (by not Case B) and as T is totally transcendental, some $\varphi(x, \bar{b}) \in r$ is a minimal formula. Let J be an indiscernible set over $|N_n|$ based on q, and let N_n^1 be $\mathbf{F}^t_{\aleph_0}$-prime over $|N_n| \cup J$. Clearly the stationarization of r over N_n has a unique extension over $|N_n| \cup J$, hence is omitted by N_n^1, hence $\varphi(N_n^1, \bar{b}) = \varphi(N_n, \bar{b})$. If $\dim(p, N_n^1) = \aleph_0$, then $\dim(p, \bar{b}, N_n^1) = \aleph_0$, so there is $\{\bar{a}^n: n < \omega\}$ exemplifying it. As $\text{tp}(\bar{a}^n, \bar{b})$, r are not weakly orthogonal, there are $c_n \in \text{acl}(\bar{a}^n \cup b) - \text{acl}(\bar{b})$, $c_n \in \varphi(N_n^1, \bar{b})$; and clearly $\{c_n: n < \omega\}$ is independent over $\bar{b}$, hence indiscernible, contradiction to not Case A.

EXERCISE 2.1: In 2.1(1) replace "superstable" by "stable and $\kappa_{\text{ind}}(T) = \aleph_0$" and show $N_n \cong N_m$ iff $n = m$.

THEOREM 2.2: *Suppose T is countable and categorical in $\aleph_1$.*

(1) *If T is not categorical in $\aleph_0$, then $I(\aleph_0, T) = \aleph_0$.*

(2) *Every model M of T has minimal and prime proper extension N (minimal means for no N', $M \precneqq N' \precneqq N$; prime means that for each N', $M \prec N'$ implies N can be elementary embedded into N' over M). Note N is necessarily unique.*

(3) *Suppose T is not categorical in $\aleph_0$. There are countable models M_α $(\alpha \leq \omega)$ of T such that:*

(i) *Every countable model of T is isomorphic to one of them.*

(ii) *They are pairwise non-isomorphic, M_ω is saturated and each M_n is homogeneous.*

(iii) *M_{n+1} is the prime minimal proper extension of M_n, and $M_\omega = \bigcup_{n<\omega} M_n$, and M_0 is the $\mathbf{F}^t_{\aleph_0}$-prime model of T.*

(iv) *M_α can be elementary embedded into M_β iff $\alpha \leq \beta$ (for $\alpha, \beta \leq \omega$).*

Proof. (1) By VIII, 4.1 T is totally transcendental, hence by 2.1 $I(\aleph_0, T) \geq \aleph_0$; the other direction follows by (3)(i).

(2) Let M be any model of T, and let $\theta(x, \bar{a})$ be a minimal formula, $\bar{a} \in M$ (at least one exists by 1.11(4)). Let $b \in \theta(\mathfrak{C}, \bar{a}) - |M|$, and N be $\mathbf{F}^t_{\aleph_0}$-prime over $M \cup b$.

If $M \precneqq N'$ then by 1.8(5) $\theta(M, \bar{a}) \neq \theta(N', \bar{a})$, hence there is $c \in \theta(N', \bar{a}) - \theta(M, \bar{a})$. Let F be the following mapping: $F \restriction |M| =$ the identity, $F(b) = c$. Clearly F is elementary, and we can extend it to an elementary mapping from N into N' as N is $\mathbf{F}^t_{\aleph_0}$-prime over $|M| \cup b$. So we have proved the primeness.

Now suppose $M \precneqq N' \precneqq N$; so again choose $c \in \theta(N', \bar{a}) - |M|$; hence $\operatorname{tp}(c, |M| \cup b)$ forks over $|M|$, hence $\operatorname{tp}(b, |M| \cup c)$ forks over $|M|$, hence $b \in \operatorname{acl}(|M| \cup c)$ hence $\theta(N, \bar{a}) \subseteq \operatorname{acl}(|M| \cup b) \subseteq \operatorname{acl}(|M| \cup c) \subseteq |N'|$; hence by 1.8(5) $N = N'$; so we prove the minimality. The uniqueness follows as in IV, 4.20.

(3) Define M_α by induction: let M_0 be the $\mathbf{F}^t_{\aleph_0}$-prime model of T, and $\theta(x, \bar{a})$, $\bar{a} \in M_0$, be a minimal formula. If M_n is defined let $b_n \in \theta(\mathfrak{C}, \bar{a}) - M_n$, and M_{n+1} be $\mathbf{F}^t_{\aleph_0}$-prime over $|M_n| \cup b_n$, and let $M_\omega = \bigcup_{n<\omega} M_n$.

By 1.8(5), $n = \dim(\operatorname{tp}(b_0, \bar{a}), M_0) < \aleph_0$ (M_0 is not saturated, as T is not $\aleph_0$-categorical) so w.l.o.g. $n = 0$. Clearly M_n is the $\mathbf{F}^t_{\aleph_0}$-prime model over $\bar{a} \cup \{b_l : l < n\}$.

If M is a countable model of T, we can assume $M_0 \subseteq M$, and let $I \subseteq \theta(M, \bar{a}) - \operatorname{acl} \bar{a}$ be a maximal set independent over $\bar{a}$; and clearly $M \cong M_\alpha$ where $\alpha = |I|$, so (i) holds.

Clearly (iii) follows by (2).

Now let us prove (ii); by (iii) if $M_n \cong M_k$, $n < k$ then for every l $M_{n+l} \cong M_{k+l}$ (proof—by induction on l), hence every $M_{n(1)}$ is isomorphic to some M_l, $l \leq k$, contradiction to what we proved from (1) by (i). Hence M_n $(n < \omega)$ are pairwise non-isomorphic. By 1.8(5) it is easy to see that if $\theta(x, \bar{a})$ is minimal, $I \subseteq \theta(\mathfrak{C}, \bar{a})$—$\operatorname{acl} \bar{a}$ is independent over $\bar{a}$, $|I| = \aleph_0$ and $\bar{a} \cup I \subseteq |M|$ then M is $\aleph_0$-saturated. Hence M_ω is saturated, and any countable model into which M_ω can be elementarily embedded is saturated. So from (ii) we have to prove only the homogeneity of M_n. Assume $\bar{a}_l \in M_n$, $\operatorname{tp}(\bar{a}_1, \emptyset, |M_n|) = \operatorname{tp}(\bar{a}_2, \emptyset, |M_n|)$ and we shall prove that M_n has an automorphism F, $F(\bar{a}_1) = \bar{a}_2$. We define by induction on k $M^l_k \prec M_n$: M^l_0 is an $\mathbf{F}^t_{\aleph_0}$-prime model over $\bar{a}_l$, $M^l_0 \prec M_l$.

If M^l_k is defined it is not saturated (as M_n is not) and, by (2) if $M^l_k \neq M_n$ there is a prime minimal proper extension M^l_{k+1} of it, $M^l_k \precneqq M^l_{k+1} \prec M_n$. If for some l, M^l_k is defined for every k again $\bigcup_{k<\omega} M^l_k$ is saturated, hence M_n is saturated, contradiction. So let M^l_k be defined just for $k \leq k(l)$; clearly there is an isomorphism F_0 from M^1_0 onto M^2_0, $F_0(\bar{a}_1) = \bar{a}_2$, and we can extend it inductively to the isomorphism F_m from M^1_m onto M^2_m. If $k(1) = k(2)$ we finish. Now let $M^1_0 \cong M_{n(1)}$ hence $M^l_0 \cong M_{n(1)}$ hence (by (iii)) $M^l_m \cong M_{n(1)+m}$, so $k(1) \neq k(2)$ implies $M_{n(1)+k(1)} \cong M_{n(1)+k(2)}$, contradicting the part of (ii) we have proved.

So only part (iv) remains, for $\alpha = \omega$ we have already proved it. Now suppose $M_\alpha \cong N \prec M_\beta$ defines inductively N_l: $N_0 = N$. $N_{l+1} \prec M_\beta$ is a prime minimal proper extension of N_l. If N_l is defined for every l,

M_β is saturated, so $\beta = \omega$; if $N_{l(0)} = M_\beta$, $M_\beta \cong N_{l(0)} \cong M_{\alpha+l(0)}$ hence $\alpha + l(0) \le \beta$.

EXERCISE 2.2: Extend 2.2(2) to any $|T|^+$-categorical theory when $\|M\| > |T|$ or M is $\mathbf{F}^a_{\aleph_0}$-saturated or M extends such a model.

EXERCISE 2.3: Suppose T is totally transcendental $|T| = \aleph_\alpha \ge \aleph_\beta$, and let M be the $\mathbf{F}^t_{\aleph_\beta}$-prime model of T. Then (1) $\Leftrightarrow$ (2) $\Rightarrow$ (3), (1) $\Rightarrow$ (4), cf $\aleph_\beta > \aleph_0$ implies (4) $\Rightarrow$ (1) and when $\aleph_\beta$ is regular they are equivalent.

(1) M is not $\aleph_0$-saturated.
(2) Some $p \in D(T)$ is not $\mathbf{F}^t_{\aleph_\beta}$-isolated.
(3) For some m $|D_m(T)| \ge \aleph_\beta$.
(4) M is not $\aleph_\beta$-saturated.

DEFINITION 2.1: Let $I(\lambda, \mathbf{F}, T)$ be the number of non-isomorphic $\mathbf{F}$-saturated models of T of cardinality λ.

When $\mathbf{F} = \mathbf{F}^x_\kappa$, let $I^x(\lambda, \kappa, T) = I(\lambda, \mathbf{F}^x_\kappa, T)$.

MAIN THEOREM 2.3: *Let T be superstable not multidimensional (see V, 5.8) then there is a cardinal $\mu = \mathrm{ND}(T)$ (=the number of dimensions of T) and a group $G = G(T)$ of permutations of μ, such that:*

(1) $1 \le \mu \le 2^{|T|}$ *and even* $\mu \le \lambda(T)$, *and when T is countable,* $\mu \le \aleph_0$ *or* $\mu = 2^{\aleph_0}$.

(2) $\mu = 1$ *iff T is unidimensional.*

(3) *If* $\aleph_\alpha > \lambda(T)$, *or* $\aleph_\alpha = \lambda(T)$, $\alpha - \beta + \mu \ge \omega$, $I^a(\aleph_\alpha, \aleph_\beta, T) = |\alpha - \beta + 1|^\mu/G$ *(see below).*

(4) *If* $\aleph_\alpha = \lambda(T)$, $0 \le \alpha - \beta < \omega$, $\mu < \aleph_0$, *then* $I^a(\aleph_\alpha, \aleph_\beta, T) = \sum_{\gamma=\beta}^{\alpha} |\gamma - \beta + 1|^\mu/G$.

(5) *If* $\aleph_\alpha < \lambda(T)$, *then* $I^a(\aleph_\alpha, \aleph_\beta, T) = 0$,

where $\chi^\mu/G = \{f/\sim_G : f : \mu \to \chi, 0 \in \mathrm{Range}\, f\}$; *for G a group of permutations of μ; χ, μ possibly finite and $\sim_G$ is the equivalence relation,* $f_1 \sim_G f_2$ *iff for some* $\sigma \in G$, $f_1 = f_2\sigma$.

THEOREM 2.4: *Suppose T is totally transcendental,* $\mu = \mathrm{ND}(T)$. *There are functions* $h_1, h_2 : \mu \to \{\aleph_\beta : \aleph_0 \le \aleph_\beta \le |T|^+\}$ *such that* $h_1(i) = h_2(i)$ *is* $\aleph_0$ *or a successor except when* $\mathrm{cf}\,[h_1(i)] = \aleph_0$, $h_2(i) = h_1(i)^+$ *and* h_1, h_2 *preserved by the permutations of G, satisfying:*

(1) *The following cardinals are equal or both finite, when,* $\aleph_\alpha \ge |T| + \aleph_\beta$, $\mu > 1$*:*

(i) $I^t(\aleph_\alpha, \aleph_\beta, T)$.

(ii) $\prod_{i<\mu} \kappa_i$ *where* $\kappa_i = \begin{cases} |\alpha| + \aleph_0 & \aleph_\beta < h_1(i), \\ |\alpha| + 1 & \aleph_\beta = h_1(i) < h_2(i), \\ |\alpha - \beta + 1| & \aleph_\beta \geq h_2(i). \end{cases}$

(2) *If in* (1) *they are finite, every model is* $\mathbf{F}^a_{\aleph_0}$*-saturated as* $h_1(i) \leqslant \aleph_\beta$ *for every* i. *If for no* $i, \aleph_\beta = h_1(i) < h_2(i)$, *then every* $\aleph_\beta$*-compact model of* T *is* $\mathbf{F}^a_{\aleph_\beta}$*-saturated, hence* $I^t(\aleph_\alpha, \aleph_\beta, T) = I^a(\aleph_\alpha, \aleph_\beta, T)$ *and* 2.3 *applies. In the remaining case necessarily* $\alpha < \omega$, *and if* $\aleph_\alpha > |T|$, *easily*

$$I^t(\aleph_\alpha, \aleph_\beta, T) = \prod_{i<\mu} \kappa_i/G(T) - \prod_{i<\mu} (\kappa_i - 1)/G(T).$$

If $\mu = 1$, $\aleph_\alpha > |T|$ *then* $I(\aleph_\alpha, \aleph_\beta, T) = 1$; *if* $\mu = 1$, $\aleph_\alpha = |T| = |D(T)|$, *then* $I^t(\aleph_\alpha, \aleph_\beta, T) = \kappa_0$.

(3) *If* $\alpha \geq \beta + \omega$, *or* $\alpha > \beta$, $\mu \geq \aleph_0$, $\aleph_\alpha \geq |T|$, *then* $I^t(\aleph_\alpha, \aleph_\beta, T) = |\alpha - \beta + 1|^\mu$.

Proof. The proof is broken to several stages, which we may use later. In fact, they contain information not summed up in the theorem.

For 2.3 let $x = a$ and for 2.4 $x = t$, but we concentrate on the first.

Stage A. By V, 1.13(1) the relation of non-orthogonality among regular types is an equivalence relation. Let $\{p_i: i < \mu\}$ be a set of representatives and $\mathrm{ND}(T) = \mu$, and let I_i be based on p_i. Every automorphism F of $\mathfrak{C}$ induced a permutation H_F of μ: $H_F(p_i) = p_j$ iff $F(p_i)$, p_j are not orthogonal. The set of permutations H_F of μ is a group we call $G = G(T)$. As T is not multi-dimensional clearly $\mu \leq 2^{|T|}$.

Now in order to get the precise results we shall find "nice" p_i's.

Stage B. For any $\varphi(x, \bar{y})$ there is k_φ such that for any model M, $|\{|\varphi(M, \bar{a})| : \bar{a} \in M\}| \leqslant k_\varphi$ (by V, 7.5 for the trivial K). By VII, 3.4, V, 5.8, 9. T does not have the f.c.p.; and for any suitable K (see V, Section 7) and weakly K-minimal $\varphi(x; \bar{a})$ the conclusion of V, 7.5 (2) holds.

The proof is easy by the non-multidimensionality of T.

Stage C. We shall define by induction on α formulas $\varphi_\alpha(x; \bar{y}_\alpha)$, $\psi_\alpha(\bar{y}_\alpha)$, E_α and a natural number $n(\alpha)$ for $\alpha < \alpha_0 < \mu^+$. Let

$$W_\alpha = \{\langle \varphi_\gamma(x, \bar{y}_\gamma), \psi_\gamma(\bar{y}_\gamma)\rangle : \gamma < \alpha\}$$

so by V, 7.1(2) we can find a K_α satisfying C1–6 from V, Section 6, $W^a_K = W_\alpha$, so K_α-minimality, etc., are defined. As T is superstable, either every formula is K_α-algebraic, or there is a non-K_α-algebraic formula $\varphi(x; \bar{a})$ with minimal $R(\varphi(x; \bar{a}), \mathrm{L}, \infty)$, hence $\varphi(x; \bar{a})$ is weakly

K_α-minimal. If T is totally transcendental we can assume $\varphi(x;\bar{a})$ is K_α-minimal. By V, 7.7 (possibly after changing φ) there are formulas $\psi(\bar{y})$, E such that:

(i) E is an equivalence relation on $\psi(\mathfrak{C})$ with $n(\alpha) < \aleph_0$ equivalence classes.

(ii) $\models\psi[\bar{a}']$ implies $\varphi(x;\bar{a}')$ is K_α-weakly-minimal and, for totally transcendental T, $\varphi(x;\bar{a}')$ is even K_α-minimal. Also there is $\bar{a}''$, $\models E[a', \bar{a}'']$, $\operatorname{tp}(\bar{a}'', \emptyset) = \operatorname{tp}(\bar{a}, \emptyset)$.

(iii) If $\models\psi[\bar{a}_1]$, then $\varphi(x;\bar{a}_1) \le_{K_\alpha} \varphi(x;\bar{a}_2)$ iff $\models E[\bar{a}_1, \bar{a}_2]$.

Let $\varphi_\alpha, \psi_\alpha, E_\alpha, n(\alpha)$ be φ, ψ, E, n resp. Now α_0 exists as clearly $\alpha_0 \le \mu^+$, because if $I_\alpha \subseteq \varphi_\alpha(\mathfrak{C}, \bar{a}_\alpha)$, $\models\psi_\alpha[\bar{a}_\alpha]$ $H_{K_\alpha}(I_\alpha) = \infty$, then the I_α's are pairwise orthogonal.

Clearly each $\langle\varphi, \psi\rangle$ appear at most once, so $\alpha_0 < |T|^+$.

Stage D. We define by induction on α, $\bar{a}_\alpha = \bar{a}_{\alpha,0}{}^\frown\cdots{}^\frown\bar{a}_{\alpha,n(\alpha)-1}$ $\varphi^\alpha(x;\bar{a}_\alpha) = \bigvee_{l<n(\alpha)} \varphi_\alpha(x, \bar{a}_{\alpha,l})$ such that

(i) $\varphi_\alpha(x, \bar{a}_{\alpha,l})$ is weakly K_α-minimal when $x = a$,

(ii) $\models \bigwedge_{l<n(\alpha)} \psi[\bar{a}_{\alpha,l}] \wedge \bigwedge_{l<k<n(\alpha)} \neg E_\alpha(\bar{a}_{\alpha,l}, \bar{a}_{\alpha,k})$,

(iii) $\operatorname{tp}(\bar{a}_\alpha, A_\alpha)$ is $\mathbf{F}^x_{\aleph_0}$-isolated where $A_\alpha = \bigcup\{\bar{a}_\beta : \beta < \alpha\}$.

Clearly this is possible by IV, 2.18, (ii) of Stage C and (i) above. Now for each α we choose a maximal family of pairwise orthogonal, regular, infinite indiscernible sets I^α_γ $(\gamma < \gamma(\alpha))$ such that

(iv) $I^\alpha_\gamma \subseteq \varphi^\alpha(\mathfrak{C};\bar{a}_\alpha)$, I^α_γ based on $\bar{a}_\alpha$,

(v) $H_{K_\alpha}(I^\alpha_\gamma) = \infty$, so I^α_γ is indiscernible over $\bigcup_{\beta<\alpha} \varphi_\beta(\mathfrak{C}, \bar{a}_\beta) \cup A_{\alpha+1}$,

(vi) if each $\varphi_\alpha(x, \bar{a}_{\alpha,l})$ is K_α-minimal then $\gamma(\alpha) = n_\alpha$ and $I^\alpha_l \subseteq \varphi_\alpha(\mathfrak{C}, \bar{a}_{\alpha,l})$.

Clearly the I^α_γ's are pairwise orthogonal, and regular.

Let us work in $\mathfrak{C}^{\mathrm{eq}}$, and let $A^* = \operatorname{acl}(A_{\alpha_0})$ and $p^\alpha_\gamma = \operatorname{Av}(I^\alpha_\gamma, A^*)$.

Stage E. Let N be any $\mathbf{F}^x_{\aleph_0}$-saturated model of T. By the inductive choice of the $\bar{a}_\alpha$'s we can assume $A^* \subseteq |N^{\mathrm{eq}}|$, and for every α, γ, let $J^\alpha_\gamma \subseteq N$ be a maximal indiscernible set over A^* based on p^α_γ.

Let $M \subseteq N$ be $\mathbf{F}^x_{\aleph_0}$-constructible over $A^* \cup \bigcup\{J^\alpha_\gamma : \gamma < \gamma(\alpha), \alpha < \alpha^0\}$, and we shall prove $N = M$. We prove by induction on α that $\varphi^\alpha(N;\bar{a}_\alpha) \subseteq M$. If α is the first for which this fails choose $c \in \varphi^\alpha(N, \bar{a}_\alpha) - |M|$. Clearly $M \prec_{K_\alpha} N$ (by V, 6.4) hence $H_{K_\alpha}(c, |M|) = \infty$, hence $\operatorname{tp}(c, |M|)$ is not orthogonal to some p^α_γ, hence if $x = a$ by V, 1.12 $\operatorname{Av}(J^\alpha_\gamma, |M|)$ is realized in N, contradicting the maximality of I^α_γ. If $x = t$, T is totally transcendental so for some $l < \gamma(\alpha) = n(\alpha)$, $c \in \varphi_\alpha(\mathfrak{C}, \bar{a}_{\alpha,l})$ and c realizes $\operatorname{Av}(I^\alpha_l, |M|)$, as $\varphi_\alpha(x, \bar{a}_{\alpha,l})$ is K_α-minimal we get again a contradiction.

Stage F. By Stage E, the function $h^{A^*}_M$ defined by $h^{A^*}_M(\alpha, \gamma) = |J^\alpha_\gamma| = \dim(p^\alpha_\gamma, M)$ determines M up to isomorphism (remember the p^α_γ are stationary (by III, 6.9) and pairwise orthogonal). Moreover no regular type is orthogonal to all p_γ^α (otherwise let M be $|T|^+$-saturated, tp$(c, |M_1|)$ such a type, M_1 an $\mathbf{F}^a_{\aleph_0}$-prime over $|M| \cup \{c\}$; then necessarily not all J_γ^α are maximal in M_1). So necessarily by Stage A, $\mu = \Sigma\{|\gamma(\alpha)|: \alpha < \alpha_0\}$, and the p_γ^α are a set of representatives.

We can conclude $I^x(\aleph_\alpha, \aleph_\beta, T) \leq |\alpha - \beta|^\mu + \aleph_0^\mu$.

Stage G. The two first assertions of 2.3(1) are immediate. As for the third as $\alpha_0 < |T|^+ = \aleph_1$ it suffices to prove for each α, $|\gamma(\alpha)| \leq \aleph_0$ or $|\gamma(\alpha) = 2^{\aleph_0}$. Choose N countable, $\bar{a}_\alpha \in N$, and let $S = \{p \in S(N): \varphi^\alpha(x, \bar{a}_\alpha) \in p, p \text{ is not } K_\alpha\text{-algebraic}\}$; we know each $p \in S$ is regular, hence non-orthogonality is an equivalence relation on S, and γ_α is the number of equivalence classes. Now S is a closed set (in the natural topology on $S(N)$) and non-orthogonality is a Borel equivalence relation on S, hence by a theorem of Silver (see e.g. [Sh 84]) $|\gamma_\alpha| \leqslant \aleph_0$ or $|\gamma_\alpha| = 2^{\aleph_0}$.

Stage H. Suppose $\mu < \aleph_0$, then we can find $\bar{a}^* \in A^*$ such that each p_γ^α does not fork over $\bar{a}^*$ and $p_\gamma^\alpha \restriction \bar{a}^*$ is stationary, and let us rename the $p_\gamma^\alpha \restriction \bar{a}^*$ as $\{p_i: i < \mu\}$, and we allow ourselves to write $p_i = p_i(x, \bar{a}^*)$. Now for every sequence $\langle \lambda_i: i < \mu \rangle$, $\lambda_i \geq \aleph_0$ there is a unique, up to isomorphism, model $M = M(\bar{a}^*, \langle \lambda_i: i < \mu \rangle)$ such that $\dim(p_i, M) = \lambda_i$. But we have here an arbitrary choice of $\bar{a}^*$. However it is easy to check that $M(\bar{a}^*, \langle \lambda_i: i < \mu \rangle) \cong M(\bar{a}^*, \langle \lambda^i: i < \mu \rangle)$ iff there is $\bar{b}^*$, tp$(\bar{a}^*, \emptyset) = $ tp$(\bar{b}^*, \emptyset)$ and a permutation σ of μ such that $\lambda_i = \lambda^{\sigma(i)}$ and $p_i(x, \bar{a}^*)$ is not orthogonal to $p_{\sigma(i)}(x, \bar{b}^*)$ (because tp$(\bar{b}^*, \bar{a}^*)$ is always $\mathbf{F}^a_{\aleph_0}$-isolated) and III, 3.5. (The proof here does not work apparently for $\mu \geq \aleph_0$ as it is not clear why B^* is $\mathbf{F}^a_{\aleph_0}$-constructible over A^*).

It is also easy to check that

$$\|M(\bar{a}^*, \langle \lambda_i: i < \mu \rangle)\| = \sum_{i<\mu} \lambda_i + \lambda(T).$$

This clearly proves 2.3(3), (4) and (5).

To make the proof work for $\mu \geq \aleph_0$ (and $\alpha - \beta$ small) we use a better canonization. In stage I assume $x = a$.

Stage I. We repeat stage C more carefully, working in $\mathfrak{C}^{eq}$. We define by induction on α, formulas $\theta_\alpha(x)$ (with no parameters) ordinals $\gamma(\alpha)$, types $q_\gamma^\alpha \in S(\operatorname{acl}\emptyset)$, $q_\gamma^\alpha \subseteq r_\gamma^\alpha \in S(\bar{c}_\gamma^\alpha)$, $\bar{c}_\gamma^\alpha \subseteq q_\gamma^\alpha(\mathfrak{C}^{eq})$, such that (letting $W_\alpha = \{\langle \theta_\beta(x), \ \rangle: \beta < \alpha\}$ and K_α is defined by V, 7.1(2) to satisfy C, 1–6 from V, Section 6, $W^\alpha_{K_0} = W_\alpha$) using V, 7.14:

(i) If $M <_{K_\alpha} N$, M, N are $F^a_{\operatorname{kr}(T)}$-saturated, then $\theta_\alpha(M) = \theta_\alpha(N)$ iff $q_\gamma^\alpha(M) = q_\gamma^\alpha(N)$ for every $\gamma < \gamma(\alpha)$,

(ii) r^α_γ is K_α-minimal, q^α_i is $\mathscr{P}^\alpha_\gamma$-simple and $q^\alpha_i(\mathfrak{C}^{eq}) \subseteq \mathrm{acl}_{K_i}(r^\alpha_i(\mathfrak{C}^{eq}) \cup \bar{c}_i)$, where $\mathscr{P}^\alpha_\gamma = \mathrm{cl}^3\{r^\alpha_i\}$ and r^α_i, r^α_j are orthogonal if $i \neq j$. Hence

(iii) For $A \subseteq q^\alpha_i(\mathfrak{C})$, $i \neq j$, we have $\mathrm{low}_{\mathscr{P}^\alpha_j}(A \cdot \mathrm{acl}\emptyset) = 0$.

For a model M let $\lambda^\alpha_\gamma(M) = \mathrm{low}_{\mathscr{P}^\alpha_\gamma}(q_i(M), \emptyset)$, which is equal to $\dim(r^\alpha_\gamma, \bar{c}^\alpha_\gamma, M)$ (as $q^\alpha_\gamma(M) \subseteq \mathrm{acl}_{K_\alpha}(r^\alpha_a(M) \cup \bar{c})$ using V, 4.18 or V7.14(3)).

Now we shall prove that for $\mathbf{F}^a_{\aleph_0}$-saturated M_1, M_2, they are isomorphic over acl $\emptyset$ iff $\lambda^\alpha_\gamma(M_1) = \lambda^\alpha_\gamma(M_2)$ for every $\alpha < \alpha_0, \gamma < \gamma(\alpha)$. The only if part is clear. For the if part we define by induction on $\alpha \leq \alpha_0$ an elementary mapping F_α, Dom $F_\alpha = \bigcup_{\beta<\alpha} \theta_\beta(M_1) \cup \mathrm{acl}\ \emptyset$ onto $\bigcup_{\beta<\alpha} \theta_\beta(M_2) \cup \mathrm{acl}\ \emptyset$, $F_\alpha \restriction \mathrm{acl}\ \emptyset =$ is the identity, increasing with α. For $\alpha = 0$, α limit, there is no problem. For $\alpha = \beta + 1$, choose by induction on $\gamma < \gamma(\alpha)$, $\bar{c}^*_\gamma \in q^\alpha_\gamma(M_1) \cup \mathrm{acl}\ \emptyset$, realizing $\mathrm{tp}(\bar{c}^\alpha_\gamma, \mathrm{acl}\ \emptyset)$, such that $\mathrm{tp}(\bar{c}^*_\gamma, \mathrm{Dom}\, F_\beta \cup \bigcup\{\bar{c}^*_i : i < \gamma\})$ is $\mathbf{F}^a_{\aleph_0}$-isolated, extend F_β to an (M_1, M_2)-elementary mapping F'_β,

$$\mathrm{Dom}\, F'_\beta = \mathrm{Dom}\, F_\beta \cup \bigcup \{\bar{c}^*_\gamma : \gamma < \gamma(\alpha)\}.$$

Using $\lambda^\alpha_\gamma(M_1) = \lambda^\alpha_\gamma(M_2)$, r^α_γ K-minimal, $q^\alpha_\gamma(M_1) \subseteq \mathrm{acl}_{K_\alpha}(r^\alpha_\gamma(M_1) \cup \bar{c}^*_\gamma)$ and the parallel fact for M_2, and (iii) we can extend F'_β to F''_β, an (M_1, M_2)-elementary mapping extending F'_β, Dom $F''_\beta =$ Dom $F_\beta \cup \bigcup_\gamma q^\alpha_\gamma(M^*_1)$, Range $F''_\beta =$ Range $F_\beta \cup \bigcup_\gamma q^\alpha_\gamma(M_2)$; you may wonder what to do when qr^α_γ has small dimension, then e.g. use V7.14(3) or Example X, 5.3A (or [Sh 88a, 1.4]). Let $N_1 \prec M_1$ ($N_2 \prec M_2$) be $\mathbf{F}^a_{\aleph_0}$-prime over Dom F''_β (Range F''_β), F'''_β an isomorphism from N_1 onto N_2 extending F''_β, and $F_\alpha = F'''_\beta \restriction (\bigcup_{i\leq\beta} \theta_i(N_1))$ is onto $\bigcup_{i\leq\beta} \theta_i(N_2)$. So clearly F_α is as required (as $\theta_\beta(N_l) = \theta_\beta(M_l)$).

It is also clear that for each sequence $\bar{\lambda} = \langle \lambda^\alpha_\gamma : \gamma < \gamma(\alpha), \alpha < \alpha_0 \rangle$ of infinite cardinals, there is an $\mathbf{F}^a_{\aleph_0}$-saturated model $M = M(\bar{\lambda})$, $\lambda^\alpha_\gamma(M) = \lambda^\alpha_\gamma$, and $\|M\| = \lambda(T) + \sum_{\alpha,\gamma} \lambda^\alpha_\gamma$ and M is $\mathbf{F}^a_{\aleph_\beta}$-saturated iff $\aleph_\beta \leq \lambda^\alpha_\gamma$ for any $\alpha < \alpha_0$, $\gamma < \gamma(\alpha)$. The only remaining question is when $M(\bar{\lambda}) \cong M(\bar{\mu})$. Notice that any automorphism of $\mathfrak{C}^{eq}$ preserve the $\theta_\alpha(x)$'s, but may take q^α_γ to another type. If $F: M(\bar{\lambda}) \cong M(\bar{\mu})$ and w.l.o.g. $\bar{c}^\alpha_\gamma \in M(\bar{\lambda})$, then $f(r^\alpha_\gamma)$ should be not orthogonal to some $r^\alpha_{f_\alpha(\gamma)}$, $\lambda^\alpha_\gamma = \mu^\alpha_{f_\alpha(\gamma)}$, and $F \restriction \mathrm{acl}\ \emptyset$ determine $f_\alpha(\gamma)$. The rest is easy.

Stage J: Now we return to the case $x = t$, T totally transcendental, and we repeat stage I, using now V, 7.14(2), using $\mathbf{F}^t_{\aleph_0}$ instead $\mathbf{F}^a_{\aleph_0}$, with no special problems.

The only point that remains is the possible values of λ^α_γ. Let $I^{\alpha,0}_\gamma$ be an indiscernible set on $\bar{a}_\alpha$ based on p^α_γ, $I^\alpha_\gamma = \{a^\alpha_{\gamma,\zeta} : \zeta \leq \omega\}$. Let $h_1(\alpha, \gamma)$ be the first cardinal κ such that for every n $\mathrm{tp}(a^\alpha_{\gamma,n}, \bar{a}_\alpha \cup \{a^n_{\gamma,l} : l < n\})$ is

$\mathbf{F}^t_\kappa$-isolated and $h_2(\alpha, \gamma)$ is the first κ such that $\mathrm{tp}(a^\alpha_{\gamma,\omega}, \bar{a}_\alpha \cup \{a^\alpha_{\gamma,l}: l < \omega\})$ is $\mathbf{F}^t_\kappa$-isolated. For (1) we do not divide by $\sim G$, as we can easily prove that the result is the same: as T is totally transcendental, each $\gamma(\alpha)$ is finite; so the equivalence relation on $\mu: i \sim j$ iff for some automorphism F of $\mathfrak{C}$, $F(I_i), I_j$ are not orthogonal (see Stage A) has only finite equivalence classes, and $i \sim j \Rightarrow h(i) = h(j)$, and if we take the product $(\prod_{i<\mu})$ on a maximal set of non-$\sim$-equivalent ordinals we get the same result. Also, the fact that the model should have power $\aleph_\alpha$ and not less does not influence the computation.

EXERCISE 2.4: In 2.4 replace "totally transcendental" by "$\lambda_3(T) \le \aleph_\beta$, T superstable". Also replace it by $\lambda_2(T) = \aleph_0$, and investigate $I^s(\aleph_\alpha, \aleph_0, T)$.

EXERCISE 2.5: (1) For $\beta > 0$, $I^a(\aleph_\alpha, \aleph_\beta, T) = I^s(\aleph_\alpha, \aleph_\beta, T)$.

(2) For countable $\aleph_0$-categorical T,

$$I^a(\aleph_\alpha, \aleph_\beta, T) = I^t(\aleph_\alpha, \aleph_\beta, T).$$

EXERCISE 2.6: (1) In 2.3, prove that for countable T, $|\alpha - \beta + 1|^\mu/G = |\alpha - \beta + 1|^\mu$ except, possibly, when μ and $\alpha - \beta$ are finite; and $|\gamma(\alpha)| \ge \aleph_0 \Rightarrow |\gamma(\alpha)| = 2^{\aleph_0}$.

(2) Show that for every χ, finite μ, and a group G of permutation of G, for some T, $|T| = \chi$, $\mathrm{ND}(T) = \mu$, $G(T) = G$.

(3) In (2) we can also have arbitrary h_1, h_2 (when they satisfy the conditions mentioned in 2.4 and $h_1 = h_2$).

(4) Show that if there is a Boolean algebra B, with $\le \chi$ elements and μ ultrafilters (this occurs, e.g., whenever $\mu \le \chi$ or $\mu = 2^\kappa$, or $\mu = \chi^\kappa$ for some $\kappa \le \chi$), then for some T, $\mathrm{ND}(T) = \mu$, $|T| = \chi$.

(5) In 2.4(2) complete the case $\aleph_\alpha = |T| < \aleph_\omega, \mu, \kappa_i < \aleph_0$ (look at the proof of 2.1(3)).

CHAPTER X

CLASSIFICATION FOR $\mathbf{F}^a_{\aleph_0}$-SATURATED MODELS

X.0. Introduction

We introduce here two dividing lines: having the dop, and being deep. They are meaningful for stable theories, but at present important for superstable theories. The dop (dimensional order property) says that the relation: "there is an indiscernible set $\boldsymbol{I}$ realizing the type $p = p(\langle x_0, x_1, \ldots \rangle, \bar{a}_i, \bar{a}_j)$ with $\dim(\boldsymbol{I}, M) > |T|$" (where $\bar{a}_i \in M$) can define an arbitrary two-place relation on $\{\bar{a}_i : i < \alpha\}$. This is very similar to the order property from Chapter II, only the order is not defined by a first-order formula (hence, does not imply the complexity of pseudo-elementary classes) but it implies T has 2^λ pairwise non-isomorphic $\mathbf{F}^a_{\aleph_0}$-saturated models of power $\lambda > \lambda(T)$. The positive property equivalent to "not dop" is that for $\mathbf{F}^a_{\kappa_r(T)}$-saturated $M_0 \prec M_1, M_2$ such that $\{M_1, M_2\}$ is independent over M_0, the $\mathbf{F}^a_{\kappa_r(T)}$-prime model is $\mathbf{F}^a_{\kappa_r(T)}$-minimal. This is developed in Section 2. Its importance is revealed in Section 3 where we prove the decomposition lemma for superstable T without the dop: any $\mathbf{F}^a_{\aleph_0}$-saturated model is $\mathbf{F}^a_{\aleph_0}$-prime, $\mathbf{F}^a_{\aleph_0}$-saturated over $\bigcup_{\eta \in I} N_\eta$, where $\langle N_\eta : \eta \in I \rangle$ is a non-forking tree of $\mathbf{F}^a_{\aleph_0}$-saturated models, $I \subseteq {}^{\omega >}\alpha$, for some α, is closed under initial segments (non-forking trees means $\operatorname{tp}_*(N_{\eta\frown\langle\alpha\rangle}, \bigcup \{N_\nu : \eta\frown\langle\alpha\rangle \not\leq \nu\})$ does not fork over N_η).

Why do we assume T is superstable? Because we use regular types to show that the tree "exhausts" the model.

Note that the class of $\mathbf{F}^a_{\aleph_0}$-saturated models is very nice from our point of view. The saturation demand is weak, local (e.g. preserved by increasing unions), it is still easy to prove for unsuperstable the existence of many non-isomorphic models, and superstability implies the existence of an $\mathbf{F}^a_{\aleph_0}$-primary model over any A.

The existence of such a decomposition is a structure theorem, e.g. it bounds the number of pairwise non-elementarily embeddable $\mathbf{F}^a_{\aleph_0}$-

saturated models. Thus, the question naturally arises as to whether I may contain an infinite branch. If this is the case, T is called deep, if not, T is called shallow. Shallow (superstable without the dop) T has quite a few $\mathbf{F}^a_{\aleph_0}$-saturated non-isomorphic models, whereas deep T has 2^λ $\mathbf{F}^a_{\aleph_0}$-saturated non-isomorphic models of power $\lambda \geqslant \lambda(T)+\aleph_1$. In Section 4 we investigate the depth (which measures how far are we from the deep case), which really is the depth of a type, and get the upper bound. In Section 5 we prove that any deep T has many models. For simplifying the proof, in Section 7 we introduce "trivial regular types" (when the dependency relation is trivial: $\bar{a}$ depends on $\boldsymbol{I}$ iff it depends on some $\bar{a} \in \boldsymbol{I}$). In 7.2 and 7.3 we investigate them: if T does not have dop, $\mathrm{Dp}(p) > 0$, p regular, then p is trivial, and we have better canonization theorems (e.g. if $\mathrm{stp}(\bar{a}, A)$ is not orthogonal to a trivial p, then for some E almost over A, $\mathrm{stp}(\bar{a}/E, A)$ is $\mathrm{cl}^3(p)$-simple with weight 1).

In Section 6 we compute exactly the number of non-isomorphic $\mathbf{F}^a_{\aleph_0}$-saturated models of T of power λ, when T is superstable, without dop, shallow of depth $\geqslant \omega$. Note that for countable $\aleph_0$-categorical T every model is $\mathbf{F}^a_{\aleph_0}$-saturated, so we can (essentially) compute $I(\lambda, T)$. However, whereas usually the depth can be any countable ordinal, here it is finite. This is as if T is superstable and $\aleph_0$-categorical, it is $\aleph_0$-stable (see III, 5.17). Hence by Cherlin, Harrington and Lachlan [CHL 86] $R(x=x, L, \aleph_0) < \omega$, so by 4.8 $\mathrm{Dp}(T) < \omega$.

X.1. Preliminaries

In what follows, $\boldsymbol{I}, \boldsymbol{J}$ are sets of sequences from $\mathfrak{C}$, I, J are sets of indexes (ordered sets, or trees or sets of sequences of ordinals, usually).

(Hyp) In this section T will be stable.

CLAIM 1.1: *If $A \subseteq B_l$ $(l=0,1)$, $\mathrm{tp}(B_1, B_0)$ does not fork over A, $p \in S^m(B_0)$ is orthogonal to A, then it is orthogonal to B_1.*

Proof. W.l.o.g. we work in $\mathfrak{C}^{\mathrm{eq}}$ and w.l.o.g. $B_1 = \mathrm{acl}\, B_1$, $A = \mathrm{acl}\, A$. Choose any $q \in S(B_1)$. Construct an infinite indiscernible set $\boldsymbol{I}$ of elements realizing q such that $\mathrm{tp}_*(\boldsymbol{I}, B_0 \cup B_1)$ does not fork over B_1 and $\mathrm{Av}(\boldsymbol{I}, \boldsymbol{I})$ is a stationarization of q. Similarly, choose a set $\boldsymbol{J}$ of indiscernibles over B_0 realizing p. Since $\text{tp}_*(\boldsymbol{I}, B_0 \cup B_1)$ does not fork over B_1, and $\text{tp}(B_1, B_0)$ does not fork over A, by III, 0.1 $\text{tp}_*(B_1 \cup \boldsymbol{I}, B_0)$ does not fork over A, hence $\text{tp}_*(\boldsymbol{I}, B_0)$ does not fork over A. As $p \in S^m(B_0)$ is orthogonal to A by V, 1.5(1) $\text{tp}_*(\boldsymbol{J}, B_0)$ is orthogonal to A. Hence $\text{tp}_*(\boldsymbol{J}, B_0)$, $\text{tp}_*(\boldsymbol{I}, B_0)$ are orthogonal, hence by V, 1.2(1) (since we work in $\mathfrak{C}^{\text{eq}}$) weakly orthogonal, hence by V, 2.7 $\text{Av}(\boldsymbol{I}, \cup \boldsymbol{I})$, $\text{Av}(\boldsymbol{J}, \cup \boldsymbol{J})$ are orthogonal, so p, q are orthogonal.

DEFINITION 1.2: For $A \subseteq B \subseteq C$ we say $B <_A C$ iff for every $\bar{c} \in C$, $\text{tp}(\bar{c}, B)$ is orthogonal to A.

LEMMA 1.3: (1) *Let $N \subseteq M \subseteq A$, $M <_N A$, M and N are $\mathbf{F}^a_\kappa$-saturated, $\kappa \geqq \kappa_r(T)$ and M' is $\mathbf{F}^a_\kappa$-prime over A, then $M <_N M'$.*

(2) *If $B <_A C$, $A \subseteq A_0$, $B \subseteq B_0$, $A_0 \subseteq B_0 \subseteq C_0$, $C_0 = C \cup B_0$, $\text{tp}_*(C, A_0)$ does not fork over A, and $\text{tp}_*(C, B_0)$ does not fork over B, then $B_0 <_{A_0} C_0$.*

Proof. (1) By V, 3.2 and V, 1.2(3).

(2) By 1.1 it is easy.

CLAIM 1.4: *Let $\kappa \geqq \kappa_r(T)$. If N is $\mathbf{F}^a_\kappa$-saturated, $p \in S^m(B)$ is regular, stationary, not orthogonal to N, then p is not orthogonal to some regular $q \in S^m(N)$.*

Proof. W.l.o.g. $p \in S^m(M)$, $N \subseteq M$ and M is $\mathbf{F}^a_\kappa$-saturated. Let $C \subseteq M$, $|C| < \kappa$, p does not fork over C, $p \upharpoonright C$ stationary. Let $A \subseteq N$, $|A| < \kappa$, $\text{tp}_*(C, N)$ does not fork over A, $\text{tp}_*(C, A)$ stationary.

Choose by induction on $i < \omega$ elementary mappings f_i, $\text{Dom} f_i = A \cup C$, $f_i \upharpoonright A = \text{id}$, $\text{stp}_*(f_i(C), A) \equiv \text{stp}_*(C, A)$, and $f_0 = \text{id}$, $\text{Rang}(f_1) \subseteq N$, and for $i < \omega$, $i \geqq 2$, $tp_*(f_i(C), M \cup \bigcup_{j<i} f_j(C))$ does not fork over A.

Let $p_i = f_i(p \upharpoonright (A \cup C))$.

Clearly for $i < j$, p_i is orthogonal to p_j iff p_0 is orthogonal to p_1 (by indiscernibility).

Extend the domain of f_2 to $N \cup C$ by fixing $N - C$. By V, 3.4, p_0, p_2 are not orthogonal ($p, f, f(p)$, A, B there, stand for p_0, f_2, p_2, N, $N \cup C$ here). Hence p_0, p_1 are not orthogonal. Now p is parallel to p_0, not orthogonal to p_1, parallel to the stationarization q of p_1 in $S^m(N)$, q regular. So we finish.

DEFINITION 1.5: Let $A \subseteq B$, $p \in S^m(B)$, then we say that p is almost orthogonal to A *if* for every $\bar{c}$, s.t. $\text{tp}(\bar{c}, B)$ does not fork over A and every $\bar{b}$ realizing p, $\text{tp}(\bar{b}, B \cup \bar{c})$ does not fork over B.

CLAIM 1.6: (1) *Let $A \subseteq B$, then* $\text{tp}(\bar{b}, B)$ *is almost orthogonal to A* iff $\text{tp}(\bar{b}, \text{acl}\, B)$ *is almost orthogonal to A if* $\text{tp}(\bar{b}, B)$ *is orthogonal to A.*

(2) *If $A \subseteq B$, $p = \text{tp}(\bar{b}, B)$, then p is almost orthogonal to A* iff *for any $\bar{c}$,* $\text{tp}(\bar{c}, B)$ *does not fork over A implies* $\text{stp}(\bar{b}, B) \vdash \text{stp}(\bar{b}, B \cup \bar{c})$ iff *for any C,* $\text{tp}_*(C, B)$ *does not fork over A implies* $\text{stp}(\bar{a}, B) \vdash \text{stp}(\bar{a}, B \cup C)$.

(3) *Let $A \subseteq B$, $p = \text{tp}(\bar{b}, B)$, then p is almost orthogonal to A* iff *for every $\bar{a}$,* $\text{tp}(\bar{a}, B)$ *does not fork over A implies* $\text{tp}(\bar{a}, B \cup \bar{b})$ *does not fork over A.*

(4) *For every $\bar{a}$,* $\text{tp}(\bar{a}, B)$ *does not fork over A implies* $\text{tp}(\bar{b}, B) \vdash \text{tp}(\bar{b}, B \cup \bar{a})$ iff $\text{tp}(\bar{b}, B)$ *is almost orthogonal to A and in* $\mathfrak{C}^{\text{eq}}$, $\text{tp}(\bar{b}, B) \vdash \text{tp}(\bar{b}, B \cup \text{acl}\, A)$.

(5) *If $A = |M| \subseteq B$, M is $\mathbf{F}^a_\lambda$-saturated, $p \in S^m(B)$ is $\mathbf{F}^a_\lambda$-isolated, $\lambda \geqq \kappa_r(T)$, then p is almost orthogonal to A.*

(6) *If $A = |M| \subseteq B$, M is $\mathbf{F}^t_\lambda$-compact (i.e. λ-compact), $p \in S^m(B)$ is $\mathbf{F}^t_\lambda$-isolated, then p is almost orthogonal to A.*

CLAIM 1.7: *If N is $\mathbf{F}^a_\lambda$-prime over $\emptyset$, $\lambda \geqq \kappa_r(T)$, $|A| < \lambda$, $\text{cf}\, \lambda \geqq \kappa(T)$ and M is $\mathbf{F}^a_\lambda$-prime over $N \cup A$, then M is $\mathbf{F}^a_\lambda$-prime over $\emptyset$.*

Proof. We concentrate on the case $\lambda = \aleph_0$, so $A = \bar{a}$. W.l.o.g. M is $\mathbf{F}^a_\lambda$*-constructible over* $N \cup A$. By IV, 4.9(3), IV, 4.10(2), IV, 4.18 we can find $\boldsymbol{I} \subseteq N$, $\boldsymbol{I}$ an infinite indiscernible set, $\text{Av}(\boldsymbol{I}, N) = \text{tp}(\bar{a}, N)$, and $\text{Av}(\boldsymbol{I}, \cup \boldsymbol{I}) \vdash \text{Av}(\boldsymbol{I}, N)$. Now $\|\boldsymbol{I}\| = \lambda$ as N is $\mathbf{F}^a_\lambda$-prime over $\emptyset$ (see IV, 4.9(2)). By IV, 4.18, N is $\mathbf{F}^a_\lambda$-prime over $\bigcup \boldsymbol{I}$, and by IV, 4.10(2) $\mathbf{F}^a_\lambda$-constructible over $\boldsymbol{I} \cup \{\bar{a}\}$. Hence M is $\mathbf{F}^a_\lambda$*-constructible over* $\boldsymbol{I} \cup \{\bar{a}\}$, but as $|\boldsymbol{I} \cup \{\bar{a}\}| = \lambda$, $\boldsymbol{I} \cup \{\bar{a}\}$ is $\mathbf{F}^a_\lambda$-constructible over $\emptyset$. Hence M is $\mathbf{F}^a_\lambda$-constructible over $\emptyset$, hence M is $\mathbf{F}^a_\lambda$-prime over $\emptyset$.

DEFINITION 1.8: (1) $I(\lambda, \mathbf{F}, T)$ is the number of $\mathbf{F}$-saturated models of T of power λ, up to isomorphism. If $\mathbf{F} = \mathbf{F}^x_\kappa$ we write $I^x_\kappa(\lambda, T)$.

(2) $IE(\lambda, \mathbf{F}, T)$ is the maximal number of pairwise non-elementarily embeddable $\mathbf{F}$-saturated models of T of power λ. If a maximum is not obtained, and the supremum is a limit cardinal μ, we write the value as μ^-. If $\mathbf{F} = \mathbf{F}^x_\kappa$ we write $IE^x_\kappa(\lambda, T)$. We omit λ if we do not restrict the cardinality (so the value may be ∞ or ∞^-).

X.2. The dimensional order property

(Hyp) In this section T will be stable and κ be $\kappa_r(T)$.

Remember that T is unstable iff it has the order property, and unstable theories are complicated in some respects, e.g. they have many non-isomorphic models. However, a stable T may have an order hidden in it. For example, consider for $\lambda > \aleph_0$, $A \subseteq \lambda^2$, the theory T of the model (B, F_1, F_2) where

$$B = \lambda \cup \{\langle \alpha, \beta, \gamma \rangle : \alpha, \beta, \gamma < \lambda, \text{ and } \langle \alpha, \beta \rangle \in A \Rightarrow \gamma < \omega\},$$

$$F_1(\alpha) = \alpha, \quad F_1(\langle \alpha, \beta, \gamma \rangle) = \alpha, \quad F_2(\alpha) = \alpha, \quad F_2(\langle \alpha, \beta, \gamma \rangle) = \beta.$$

Clearly T is not only stable, but even $\aleph_0$-stable, and $\aleph_0$-categorical; however, by cardinality quantifiers we can define an order (if A is an order).

We shall consider here a property, which clearly means there is a hidden order property. From later sections we can see that for T superstable and $\mathbf{F}^a_{\aleph_0}$-saturated models it is the only one. Note that here the order and independence properties coincide.

Definition 2.1: T has the dimensional order property (dop for short) *if* there are models M_l ($l = 0, 1, 2$), each $\mathbf{F}^a_\kappa$-saturated, $M_0 \subseteq M_1$, M_2, and $\{M_1, M_2\}$ is independent over M_0, *and* the $\mathbf{F}^a_\kappa$-prime model over $M_1 \cup M_2$ is not $\mathbf{F}^a_\kappa$-minimal over $M_1 \cup M_2$.

In this section we first develop a number of equivalent forms of the dimensional order property. These are summarized in Lemma 2.4. Condition 2.4(d)$_{\aleph_0}$ is the form of the dimensional order property used in Theorem 2.5 to show that if a superstable T has dop, then $I^a_{\aleph_0}(T, \lambda) = 2^\lambda$ (when T is stable in λ).

LEMMA 2.2: *Let $M_0 \prec M_1, M_2$, each M_l $\mathbf{F}^a_\lambda$-saturated, $\lambda \geqq \mu \geqq \kappa$, $\{M_1, M_2\}$ independent over M_0, M $\mathbf{F}^a_\mu$-atomic over $M_1 \cup M_2$ and M is $\mathbf{F}^a_\mu$-saturated (and $M_1 \cup M_2 \subseteq M$). Then the following conditions are equivalent:*

(a) *M is not $\mathbf{F}^a_\mu$-minimal over $M_1 \cup M_2$.*
(b) *There is an infinite indiscernible $\boldsymbol{I} \subseteq M$ over $M_1 \cup M_2$.*
(c) *There is $p \in S^m(M)$ orthogonal to M_1 and to M_2, p not algebraic.*
(d) *There is an infinite $\boldsymbol{I} \subseteq M$ indiscernible over $M_1 \cup M_2$ such that*

Av$(\boldsymbol{I},M)$ *is orthogonal to* M_1 *and* M_2. *Notes by* 1.6(5), $\mathrm{tp}_*(\boldsymbol{I},M_1\cup M_2)$ *is almost orthogonal to* M_1 *and to* M_2.

Proof. The equivalence of (a) and (b) is the content of IV, 4.21, and (d) $\Rightarrow$ (b), (c) are trivial; we now prove (b) $\Rightarrow$ (d), (c) $\Rightarrow$ (d). The "notes" in (d) can be proved by 1.6(5).

(b) $\Rightarrow$ (d). We can assume $|\boldsymbol{I}|=\aleph_0$, and suppose $\boldsymbol{I}\subseteq M$ is indiscernible over $M_1\cup M_2$ but not orthogonal to M_1 (by symmetry). So by definition, $\mathrm{Av}(\boldsymbol{I},\boldsymbol{I})$ is not orthogonal to some $r\in S^m(M_1)$.

W.l.o.g. $\boldsymbol{I}=\boldsymbol{J}\cup\{\bar a_n:n<\omega\}$, $|\boldsymbol{J}|<\kappa$, $\mathrm{Av}(\boldsymbol{I},\boldsymbol{I})$ does not fork over $\boldsymbol{J}$, $\mathrm{Av}(\boldsymbol{I},\boldsymbol{J})$ is stationary, and let $\{\bar b_n:n<\omega\}$ be an independent set over $(M_1\cup M_2\cup\boldsymbol{I},M_1)$ of sequences realizing r (hence indiscernible over $M_1\cup M_2\cup\boldsymbol{I}$). By V, 2.7 for some k, $\mathrm{tp}(\bar b_0{}^\frown\dots{}^\frown\bar b_k,M_1\cup M_2\cup\boldsymbol{J})$, $\mathrm{tp}(\bar a_0{}^\frown\dots{}^\frown\bar a_k,M_1\cup M_2\cup\boldsymbol{J})$ are not weakly orthogonal.

As M is $\mathbf{F}^a_\mu$-atomic over $M_1\cup M_2$, for some $B_1\subseteq M_1\cup M_2$, $|B_1|<\mu$, $\mathrm{tp}_*(\boldsymbol{I}',B_1)\vdash\mathrm{tp}_*(\boldsymbol{I}',M_1\cup M_2)$ where $\boldsymbol{I}'=\boldsymbol{J}\cup\{\bar a_l:l\leq k\}$ (for μ singular use 2.3(1)). Hence for some $C\subseteq M_1\cup M_2$, $|C|<\aleph_0$, $\mathrm{tp}(\bar b_0{}^\frown\dots{}^\frown\bar b_k,B_1\cup C\cup\boldsymbol{J})$ and $\mathrm{tp}(\bar a_0{}^\frown\dots{}^\frown\bar a_k,B_1\cup C\cup\boldsymbol{J})$ are not weakly orthogonal. Now $\mathrm{tp}(\bar b_0{}^\frown\dots{}^\frown\bar b_k,M_1\cup M_2\cup\boldsymbol{J})$ does not fork over M_1 (by the choice of the $\bar b$'s) and M_1 is $\mathbf{F}^a_\mu$-saturated and $|B_1\cup C\cup\boldsymbol{J}|<\mu+\aleph_0+\kappa=\mu$, so some $\bar c_0{}^\frown\dots{}^\frown\bar c_k\in M_1$ realizes $\mathrm{tp}(\bar b_0{}^\frown\dots{}^\frown\bar b_k,B_1\cup C\cup\boldsymbol{J})$ (see III, 0.1). So $\mathrm{tp}(\bar a_0{}^\frown\dots{}^\frown\bar a_k,B_1\cup C\cup\boldsymbol{J})$, $\mathrm{tp}(\bar c_0{}^\frown\dots{}^\frown\bar c_k,B_1\cup C\cup\boldsymbol{J})$ are not weakly orthogonal, hence

$$\mathrm{tp}(\bar a_0{}^\frown\dots{}^\frown\bar a_k,B_1\cup C\cup\boldsymbol{J})\not\vdash\mathrm{tp}(\bar a_0{}^\frown\dots{}^\frown\bar a_k,B_1\cup C\cup\boldsymbol{J}\cup\bar c_0{}^\frown\dots{}^\frown\bar c_k),$$

hence (by Ax V2 for F^s_∞) $\mathrm{tp}_*(\boldsymbol{I}',B_1\cup C)\not\vdash\mathrm{tp}_*(\boldsymbol{I}',B_1\cup C\cup(\bar c_0{}^\frown\dots{}^\frown\bar c_k))$, hence by monotonicity $\mathrm{tp}_*(\boldsymbol{I}',B_1)\not\vdash\mathrm{tp}(\boldsymbol{I}',M_1\cup M_2)$ contradicting the choice of B_1.

(c) $\Rightarrow$ (d). So let $r\in S^m(M)$ be (not algebraic) and orthogonal to M_1,M_2, It suffices to prove that $r\restriction(B\cup M_1\cup M_2)$ is realized in M for every $B\subseteq M$, $|B|<\kappa$. [This is because then we can choose $B_0\subseteq M$, $|B_0|<\kappa$, r does not fork over B, $r\restriction B_0$ stationary, and $\bar b_n\in M$ realizing $r\restriction(B_0\cup\{\bar b_1:l<n\}\cup M_1\cup M_2)$, and then $\boldsymbol{I}=\{\bar b_n:n<\omega\}$ is as required – indiscernible by III, 1.10(1).]

We can, of course, increase B as long as $|B|<\kappa$. So w.l.o.g. r does not fork over B, $r\restriction B$ is stationary, for $l=0,1,2$, $\mathrm{tp}_*(B,M_l)$ does not fork over $B\cap M_l$, and let $\bar c$ realize r, $B_l=B\cap M_l$ and $\mathrm{tp}_*(B,M_1\cup M_2)$ does not fork over $B_1\cup B_2$.

As $|B|<\kappa$, M is $\mathbf{F}^a_\kappa$-saturated, it suffices to prove

$$\mathrm{stp}(\bar c,B)\vdash r\restriction(B\cup M_1\cup M_2).$$

For this let $\bar{b}_l \in M_l$ $(l = 1,2)$ and it suffices to prove

$$\mathrm{stp}(\bar{c},B) \vdash r \restriction (B \cup \bar{b}_1 \cup \bar{b}_2).$$

We can find $C \subseteq M_0$, $|C| < \kappa$ such that $\mathrm{tp}(\bar{b}_l, M_0 \cup B_l)$ does not fork over $B_l \cup C$ for $l = 1,2$.

Now $\mathrm{tp}_*(B,M_0)$ does not fork over $B \cap M_0 = B_0$ and $C \subseteq M_0$ so $\mathrm{tp}_*(B, B_0 \cup C)$ does not fork over B_0. By symmetry, $\mathrm{tp}_*(C, B_0 \cup B) = \mathrm{tp}_*(C,B)$ does not fork over B_0. Extend $\mathrm{tp}_*(C,B)$ to a type q over M which does not fork over B_0. Then $q \restriction M_1$ is orthogonal to r and parallel to $\mathrm{stp}_*(C,B)$ so by V, 1.2(4), $\mathrm{stp}_*(C,B)$ is orthogonal to r. Hence,

(1) $\mathrm{stp}_*(\bar{c},B) \vdash \mathrm{stp}_*(\bar{c}, B \cup C)$.

Now $\mathrm{tp}(\bar{b}_1 \cup B_1, M_2)$ does not fork over M_0 [as it is $\subseteq \mathrm{tp}_*(M_1,M_2)$] hence $\mathrm{stp}_*(\bar{b}_1, M_2 \cup B_1)$ does not fork over $M_0 \cup B_1$. Also $\mathrm{tp}(\bar{b}_1, M_0 \cup B_1)$ does not fork over $B_1 \cup C$ by the choice of C, hence by III, 0.1(2), $\mathrm{tp}(\bar{b}_1, M_2 \cup B)$ does not fork over $B_1 \cup C$. Now $B_1 \cup C \subseteq M_1$, $B_1 \cup C \subseteq B \cup C \subseteq M_2 \cup B$, hence $\mathrm{stp}_*(\bar{b}_1, C \cup B)$ is parallel to some complete type over M_1, hence is orthogonal to r, hence to $\mathrm{stp}_*(\bar{c}, C \cup B)$. So

(2) $\mathrm{stp}_*(\bar{c}, C \cup B) \vdash \mathrm{stp}_*(\bar{c}, C \cup B \cup \bar{b}_1)$.

Now $\mathrm{tp}_*(\bar{b}_2 \cup B_2, M_1)$ does not fork over M_0 [as it is $\subseteq \mathrm{tp}_*(M_2,M_1)$], hence $\mathrm{tp}(\bar{b}_2, M_1 \cup B_2)$ does not fork over $M_0 \cup B_2$.

Also $\mathrm{tp}(\bar{b}_2, M_0 \cup B_2)$ does not fork over $B_2 \cup C$ (by C's choice), hence $\mathrm{tp}(\bar{b}_2, M_1 \cup B_2)$ does not fork over $B_2 \cup C$. As $B_2 \cup C \subseteq M_2$, and $B_2 \cup C \subseteq B \cup C \cup \bar{b}_1 \subseteq M_1 \cup B$, clearly $\mathrm{stp}(\bar{b}_2, C \cup B \cup \bar{b}_1)$ is parallel to some complete type over M_2 hence is orthogonal to r, hence to $\mathrm{stp}(\bar{c}, C \cup B \cup \bar{b}_1)$. So

(3) $\mathrm{stp}(\bar{c}, C \cup B \cup \bar{b}_1) \vdash \mathrm{stp}(\bar{c}, C \cup B \cup \bar{b}_1 \cup \bar{b}_2)$.

By (1), (2), (3) clearly

(4) $\mathrm{stp}(\bar{c},B) \vdash \mathrm{stp}(\bar{c}, B \cup \bar{b}_1 \cup \bar{b}_2)$

which, as mentioned above, is sufficient.

CLAIM 2.3: *Suppose $\lambda \geq \kappa$, M_l $(l = 0,1,2)$ are $\mathbf{F}^a_\lambda$-saturated, $M_0 \prec M_1$, M_2, and $\{M_1, M_2\}$ is independent over M_0. Then*

(1) *Every $\mathbf{F}^a_\lambda$-isolated $p \in S^m(M_1 \cup M_2)$ is $\mathbf{F}^a_\kappa$-isolated.*

(2) *A model M is $\mathbf{F}^a_\lambda$-atomic over $M_1 \cup M_2$ iff it is $\mathbf{F}^a_\kappa$-atomic over $M_1 \cup M_2$. Hence if $\kappa \leq \mu$, $\chi \leq \lambda$, and an $\mathbf{F}^a_\mu$-prime model M over $M_1 \cup M_2$ is $\mathbf{F}^a_\mu$-minimal, then M is $\mathbf{F}^a_\chi$-prime over $M_1 \cup M_2$ (and $\mathbf{F}^a_\chi$-minimal).*

Proof. (1) Suppose $B \subseteq M_1 \cup M_2$, $|B| < \lambda$, $\bar{c}$ realizes p and $\mathrm{stp}(\bar{c},B) \vdash p = \mathrm{tp}(\bar{c}, M_1 \cup M_2)$. We can assume that $\mathrm{tp}_*(B,M_0)$ does not fork over $B \cap M_0$.

Let $C \subseteq B, |C| < \kappa$ be such that p does not fork over C, $\mathrm{tp}_*(C, M_0)$ does not fork over $C \cap M_0$. Now by III, 4.22 it suffices to prove:

(*) for any $\bar{a} \in M_1 \cup M_2$, there is $\bar{a}' \in M_1 \cup M_2$ such that $\mathrm{stp}(\bar{a}', B)$ is a stationarization of $\mathrm{stp}(\bar{a}, C)$.

We can find $A \subseteq M_0$ such that $\mathrm{tp}_*(\bar{a} \cup C, M_0)$ does not fork over A, its restriction to A is stationary, $|A| < \kappa$, and let $\bar{a} = \bar{a}_1 {}^\frown \bar{a}_2, \bar{a}_l \in M_l$ $(l = 1, 2)$. Let $A = \{a_i : i < i(0)\}$, and we can find $a'_i \in M_0$, such that $\mathrm{stp}_*(\langle a'_i : i < i(0)\rangle, B \cap M_0)$ extends $\mathrm{stp}_*(\langle a_i : i < i(0)\rangle, C \cap M_0)$ and does not fork over $C \cap M_0$, hence over C. Now find $\bar{a}'_l \in M_l$ such that $\mathrm{stp}_*(\langle a_i : i < i(0)\rangle {}^\frown \bar{a}'_l, B \cup \bar{a}'_{3-l})$ extends $\mathrm{stp}_*(\langle a_i : i < i(0)\rangle {}^\frown \bar{a}_l, C)$ and does not fork over C. It is easy to check that $\bar{a}' = \bar{a}'_1 {}^\frown \bar{a}'_2$ is as required.

(2) trivial from (1).

LEMMA 2.4: *The following properties of T are equivalent for $\lambda \geqq \chi \geqq \kappa$:*

(a) *T has the dop (= dimensional order property).*

$(\mathrm{b})_{\lambda,\chi}$ *There are $\mathbf{F}^a_\lambda$-saturated models $M_0, M_1, M_2, M_0 \prec M_1, M_2$, $\{M_1, M_2\}$ independent over M_0, such that the $\mathbf{F}^a_\chi$-prime model M_3 over $M_1 \cup M_2$ is not $\mathbf{F}^a_\chi$-minimal.*

$(\mathrm{c})_{\lambda,\chi}$ *There are $\mathbf{F}^a_\lambda$-saturated models $M_0, M_1, M_2, M_0 \prec M_1, M_2$, $\{M_1, M_2\}$ independent over M_0, and there is an $\mathbf{F}^a_\chi$-atomic non-$\mathbf{F}^a_\chi$-minimal model M_3 over $M_1 \cup M_2$.*

$(\mathrm{d})_\lambda$ *There are sets A_0, A_1, A_2 such that $A_0 \subseteq A_1, A_2, |A_l| < \lambda, \{A_1, A_2\}$ is independent over A_0; and there is an infinite $\boldsymbol{I}$ indiscernible over $A_1 \cup A_2$, orthogonal to A_1 and to A_2, and $\mathrm{tp}_*(\boldsymbol{I}, A_1 \cup A_2)$ is almost orthogonal to A_1 and to A_2. Moreover, if $\bar{a}_l$ $(l = 0, 1, 2)$ are such that $\{A_1, A_2, \bar{a}_0\}$ is independent over A_0, and $\mathrm{tp}(\bar{a}_l, A_0 \cup A_1 \cup A_2 \cup \bar{a}_0 \cup \bar{a}_{3-l})$ does not fork over $A_l \cup \bar{a}_0$ for $l = 1, 2$, then*

$$\mathrm{stp}_*(\boldsymbol{I}, A_1 \cup A_2) \vdash \mathrm{tp}_*(\boldsymbol{I}, A_1 \cup \bar{a}_1 \cup A_2 \cup \bar{a}_2).$$

We can replace $\bar{a}_l$ by B_l.

Remark. Note that (a) does not depend on the cardinals λ, χ.

Proof. As (a) is $(\mathrm{b})_{\kappa,\kappa}$ it suffices to prove:

(i) $(\mathrm{c})_{\lambda,\chi} \Rightarrow (\mathrm{d})_\kappa$; (ii) $(\mathrm{d})_\kappa \Rightarrow (\mathrm{d})_\chi$; (iii) $(\mathrm{d})_\chi \Rightarrow (\mathrm{b})_{\lambda,\chi}$; (iv) $(\mathrm{b})_{\lambda,\chi} \Rightarrow (\mathrm{c})_{\lambda,\chi}$.

(i) $(\mathrm{c})_{\lambda,\chi} \Rightarrow (\mathrm{d})_\kappa$. So let M_0, M_1, M_2, M_3 exemplify $(\mathrm{c})_{\lambda,\chi}$. Now by 2.2(d) there is an infinite $\boldsymbol{I} \subseteq M_3$ indiscernible over $M_1 \cup M_2$ such that $\mathrm{Av}(\boldsymbol{I}, M_3)$ is orthogonal to M_1 and to M_2. First assume $\kappa > \aleph_0$, and w.l.o.g. $|\boldsymbol{I}| = \aleph_0$.

By 2.3(1), $\operatorname{tp}_*(\boldsymbol{I}, M_1 \cup M_2)$ is $\mathbf{F}^a_\kappa$-isolated, and so for some $A \subseteq M_1 \cup M_2$, $|A| < \kappa$, with $\operatorname{stp}_*(\boldsymbol{I}, A) \vdash \operatorname{stp}_*(\boldsymbol{I}, M_1 \cup M_2)$. We can also assume that $\operatorname{tp}_*(A, M_0)$ does not fork over $A \cap M_0$ and $\operatorname{tp}_*(A, A \cap M_0)$ is stationary. Let $A_l = A \cap M_l$. It is easy to check that $\{A_1, A_2\}$ is independent over A_0 (as $\{M_1, M_2\}$ is independent over M_0, and use III, 0.1). Clearly $\operatorname{tp}_*(\boldsymbol{I}, A_1 \cup A_2)$ is almost orthogonal to A_1 and to A_2, and even the stronger assertion in 2.4(d) holds.

If $\kappa = \aleph_0$, T is superstable and so there is a finite $\boldsymbol{J} \subseteq \boldsymbol{I}$ such that $\operatorname{Av}(\boldsymbol{I}, \boldsymbol{I})$ does not fork over $\bigcup \boldsymbol{J}$ and $\operatorname{Av}(\boldsymbol{I}, \boldsymbol{J})$ is stationary, and continue as before with $\boldsymbol{J}$ instead of $\boldsymbol{I}$, noticing that $\operatorname{tp}_*(\boldsymbol{I}, A_1 \cup A_2 \cup \boldsymbol{J})$ does not fork over $\bigcup \boldsymbol{J}$, is stationary, and is orthogonal to M_1 and to M_2.

(ii) $(\mathrm{d})_\kappa \Rightarrow (\mathrm{d})_\chi$. As $\kappa \leqq \chi$, if A_l $(l = 1,2,3)$, $\boldsymbol{I}$ exemplify $(\mathrm{d})_\kappa$, then they exemplify $(\mathrm{d})_\chi$.

(iii) $(\mathrm{d})_\chi \Rightarrow (\mathrm{b})_{\lambda,\chi}$. Let $A_0, A_1, A_2, \boldsymbol{I}$ exemplify $(\mathrm{d})_\chi$.

Let M_0 be an $\mathbf{F}^a_\lambda$-saturated model of $T, A_0 \subseteq M_0$. By using automorphism of $\mathfrak{C}$ we can assume $\operatorname{tp}_*(A_l, M_0 \cup A_{3-l})$ does not fork over A_0. Next choose an $\mathbf{F}^a_\lambda$-saturated M_1, such that $M_0 \cup A_1 \subseteq M_1$, $\operatorname{tp}_*(M_1, M_0 \cup A_2)$ does not fork over M_0, and an $\mathbf{F}^a_\lambda$-saturated M_2, such that $M_0 \cup A_2 \subseteq M_2$ and $\operatorname{tp}_*(M_2, M_1)$ does not fork over M_0 (this is easy). So clearly $\{M_1, M_2\}$ is independent over M_0, and by the latter part of $(\mathrm{d})_\chi$

$$\operatorname{stp}_*(\boldsymbol{I}, A_1 \cup A_2) \vdash \operatorname{stp}_*(\boldsymbol{I}, M_1 \cup M_2).$$

So $\operatorname{tp}_*(\boldsymbol{I}, M_1 \cup M_2)$ is $\mathbf{F}^a_\chi$-isolated, so there is M_4 $\mathbf{F}^a_\chi$-prime over $M_1 \cup M_2$, such that $\boldsymbol{I} \subseteq M_4$.

As M_4 contains an infinite indiscernible set $\boldsymbol{I}$ over $M_0 \cup M_1$, it is not $\mathbf{F}^a_\lambda$-minimal (by IV, 4.21) so we have proved $(\mathrm{b})_{\lambda,\chi}$.

(iv) $(\mathrm{b})_{\lambda,\chi} \Rightarrow (\mathrm{c})_{\lambda,\chi}$. Trivial (for χ singular, prove by induction on α that if $\langle M_1 \cup M_2, \langle a_i : i < \alpha \rangle \rangle$ is an $\mathbf{F}^a_\chi$-construction, then $M_1 \cup M_2 \cup \{a_i : i < \alpha\}$ is $\mathbf{F}^a_\kappa$-atomic over $M_1 \cup M_2$: for this use 2.3(1).

THEOREM 2.5: *Suppose T has the dop, T stable in λ and $\kappa \leqq \mu < \lambda$. Then T has 2^λ non-isomorphic $\mathbf{F}^a_\mu$-saturated models of cardinality λ.*

Proof. Let $A_0, A_1, A_2, \boldsymbol{I}$ be as $2.2(\mathrm{d})_\kappa$, $|\boldsymbol{I}| = \aleph_0$ and let J be any set of indices. W.l.o.g. we shall work in $\mathfrak{C}^{\mathrm{eq}}$ (see III, Section 6).

Let $A'_l = \operatorname{acl} A_l$; and we can define, for $l = 1,2, s \in J$, an elementary mapping f^l_s, such that:

(α) $\operatorname{Dom} f^l_s = A'_l$,

(β) $f^l_s \restriction A'_0 =$ the identity,

(γ) $\mathrm{tp}_*(f^l_s(A'_l)), \bigcup\{f^k_t(A'_k): (t,k) \neq (s,l), t \in I, k \in \{1,2\}\}$ does not fork over A'_0.

Next, for every $s, t \in J$ (not necessarily distinct) we choose an elementary mapping $f_{s,t}$, whose domain is $A'_1 \cup A'_2 \cup \boldsymbol{I}$, and which extends f^1_s, f^2_t. (This is possible as $f^1_s \cup f^2_t$ is an elementary mapping, which is true because $\mathrm{tp}(A'_l, A'_0)$ is stationary (by III, 6.9(1)) and the independence of $\{A'_1, A'_2\}$ and of $\{f^1_s(A_1), f^2_t(A_2)\}$ over A'_0.) Let $A^l_s = f^l_s(A_l)$ (so $|A^l_s| < \kappa$), $\boldsymbol{I}_{s,t} = f_{s,t}(\boldsymbol{I})$. Now:

$$\text{(st)}\quad \mathrm{stp}_*(\boldsymbol{I}_{s,t}, A^1_s \cup A^2_t) \vdash \mathrm{stp}_*\Big(\boldsymbol{I}_{s,t}, \bigcup_{v \in I} A^t_v \cup \bigcup_{v \in I} A^2_v \cup \bigcup_{(v,u) \neq (s,t)} \boldsymbol{I}_{v,u}\Big).$$

This holds by 2.4(d) for

$$B_0 = \bigcup\{A^l_v : (v,l) \neq (s,1), (t,2)\} \cup \{\boldsymbol{I}_{u,v} : \{u,v\} \cap \{s,t\} = \emptyset\},$$

$$B_1 = B_0 \cup A^1_s \cup \bigcup\{\boldsymbol{I}_{s,v} : v \neq t, v \in J\},$$

$$B_2 = B_0 \cup A^2_t \cup \bigcup(\boldsymbol{I}_{u,t} : u \neq s, u \in J\}.$$

Let $\boldsymbol{I}^*_{s,t}$ be any set indiscernible over $A^1_s \cup A^2_t$, extending $\boldsymbol{I}_{s,t}$, of power μ^+. Clearly (st) continues to hold for the $\boldsymbol{I}^*_{s,t}$'s.

Let R be any two place relation over J, and let $C_R = \bigcup_{l,s} A^l_s \cup \bigcup_{R(s,t)} \boldsymbol{I}^*_{s,t}$, and let M_R be an $\mathbf{F}^a_\mu$-prime model over C_R.

It is easy to check that $|J| = \lambda$ implies $\|M_R\| = \lambda$ (remember T is stable in λ, obviously $|C_R| \leqslant \lambda$, hence $\|M_R\| \leqslant \lambda$; on the other hand, clearly $A_1 \nsubseteq A_0$, hence $|C_R| \geqslant \lambda$, hence $\|M_R\| \geqslant \lambda$).

Now, using (st) we show:

(st 1) for $s, t \in J$, there is in M_R an $\boldsymbol{I}$ of power μ^+ realizing $\mathrm{tp}_*(\boldsymbol{I}^*_{s,t}, A^1_s \cup A^2_t)$ iff $R(s,t)$.

If $R(s,t)$ holds, clearly there is in M_R an $\boldsymbol{I}$ (of power μ^+) realizing $\mathrm{tp}_*(\boldsymbol{I}^*_{s,t}, A^1_s \cup A^2_t) : \boldsymbol{I}^*_{s,t}$ itself. Suppose $R(s,t)$ fail, $\boldsymbol{I} \subseteq M_R, |\boldsymbol{I}| = \mu^+$ realizes $\mathrm{tp}_*(\boldsymbol{I}^*_{s,t}, A^1_s \cup A^2_t)$. We shall work in $\mathfrak{C}^{\mathrm{eq}}$. Let $\boldsymbol{I}' \subseteq \boldsymbol{I}$, $|\boldsymbol{I}'| = \aleph_0$. If $\mathrm{stp}_*(\boldsymbol{I}', A^1_s \cup A^2_t) \equiv \mathrm{stp}_*(\boldsymbol{I}_{s,t}, A^1_s \cup A^2_t)$ we get an easy contradiction: by IV, 4.9, $\dim(\boldsymbol{I}', C_R, M) \leqq \mu$, and by (st), $\dim(\boldsymbol{I}', C_R, M) = \dim(\boldsymbol{I}', A^1_s \cup A^2_t, M)$ which should be $\geqq |\boldsymbol{I}| = \mu^+ > \mu$, a contradiction. However, if $\mathrm{stp}_*(\boldsymbol{I}', A^1_s \cup A^2_t) \neq \mathrm{stp}_*(\boldsymbol{I}^*_{s,t}, A^1_s \cup A^2_t)$, as $\mathrm{tp}(\boldsymbol{I}, A^1_s \cup A^2_t) = \mathrm{tp}(\boldsymbol{I}^*_{s,t}, A^1_s \cup A^2_t)$, still $A_0, A^1_s, A^2_t, \boldsymbol{I}$ satisfies $2.4(\mathrm{d})_\kappa$, so we can prove the assertion corresponding to (st).

So in M_R the relation R on $\langle A^1_t \cup A^2_t : t \in J\rangle$ is defined; and $|A^l_t| < \kappa$, $\lambda = \lambda^{<\kappa}$ (as T is stable in $\lambda, \kappa = \kappa_r(T)$). So as in VIII, 3.2, we can prove that there are 2^λ pairwise non-isomorphic $\mathbf{F}^a_\mu$-saturated models of T of cardinality λ. The case which most interests us, $\mu = \aleph_0$, T superstable (i.e. $\kappa(T) = \aleph_0$), has the same proof as in VIII, Section 2: we have just to choose the right Φ (see there). For more, see (remembering $\lambda = \lambda^{<\kappa}$!): [Sh 85e], III, §3 and (for $\kappa = \aleph_0$) [Sh 89], III, §7 and [Sh 88a].

Remarks. (1) For most λ we can also get 2^λ such models no one elementarily embeddable into the other, e.g. for λ regular $>|T|$; more generally, see also [Sh 83], [Sh 89], new version, III, §7.

(2) In 2.5, if T is not stable in λ we can still get similar results.

For the reader unsatisfied with this, we give two cases in more detail.

FACT 2.5A. *If T has the dop $\chi \geqq \lambda > \kappa, \lambda > \mu \geqq \kappa, \lambda$ regular, then there are sets A_i ($i < 2^\lambda$), each of power χ, such that, letting M_i be $\mathbf{F}^a_\mu$-prime over A_i, the following hold:*

(i) *The M_i's are pairwise not isomorphic.*

(ii) *For $i \neq j$, M_i is not elementarily embeddable into M_j.*

(iii) *For $i \neq j$, there is no elementary mapping from A_i into M_j.*

Proof. Let $S^* = \{\delta < \lambda : \mathrm{cf}\,\delta = \kappa\}$, and for each $\delta \in S$ let η_δ be an increasing sequence of successor ordinals of length κ converging to δ. For every $S \subseteq S^*$ let

$$J_S = \chi, \quad R_S = \{\langle \eta_\delta(i), \delta\rangle : \delta \in S, i < \kappa\}, \quad A_S = C_{R_S}, \quad M_S = M_{R_S},$$

$$A'_S = \bigcup_{\substack{l<2\\ i<\lambda}} A^l_i \cup \bigcup\{\boldsymbol{I}_{\langle i,\delta\rangle} : \langle i,\delta\rangle \in R\},$$

(note that $A'_S \subseteq A_S$).

Now suppose $S_1, S_2 \subseteq S, S_1 - S_2$ is stationary, $f: A'_{S_1} \to M_{S_2}$ an elementary mapping, and we shall eventually get a contradiction. This clearly suffices. By renaming, we can assume that f is into

$$B = B_0 \cup \{a_i : i < \lambda\}, \quad \text{where } B_0 = A_{S_2} \cup \bigcup_{\substack{i<\lambda\\ l<2}} A^l_i,$$

$\mathrm{tp}(a_i, B_i)$ is $\mathbf{F}^a_\mu$-isolated where $B_i = B_0 \cup \{a_j : j < i\}$.

Let $\lambda^* > 2^\chi$ be regular, and choose $N_i \prec (H(\lambda^*), \in)$, $\|N_i\| < \lambda$, $N_i \cap \lambda = \delta_i$, $\langle N_j : j \leqq i\rangle \in N_{i+1}$, and $f, A'_{s_1}, B, B_0, \langle a_i : i < \lambda\rangle, R_S, \langle A^l_i : l < 2, i < \chi\rangle, \langle \boldsymbol{I}^*_{s,t} : s, t \in J_{S_m}\rangle$ ($m = 1,2$) belongs to N_0. Choose $\zeta \in S_1 - S_2$, $\zeta = \delta_\zeta$.

Next choose $M \prec (H(\lambda^*), \in), \zeta \in M, M \cap \kappa$ an ordinal $\xi < \kappa$,

$\langle N_i : i < \lambda \rangle \in M$, and all the elements which we demand to be in N_0 will be in M too.

Now $f(A^0_\xi \cup A^1_\xi \cup \bigcup I_{\langle \eta_\zeta(\xi), \zeta \rangle})$ gives us the desired contradiction.

FACT 2.5B. *There are Φ, T_1 such that $T \subseteq T_1, |T_1| \leqq \lambda(T)$ and Φ is proper for (ω, T_1) (see Def.* VII, 2.6 *and Lemma* VIII, 2.3) *such that for every linear ordering I:*

(1) *$EM(I, \Phi)$ is $\mathbf{F}^a_{\aleph_0}$-saturated.*

(2) *There is a formula*

$$\psi(\bar{y}, \bar{z}) = (\exists x_0, \ldots, x_i, \ldots)_{i<\mu^+} \bigwedge_{\alpha<\mu^+} \varphi_i(\ldots, x_{i(\alpha,l)}, \ldots, \bar{y}, \bar{z}),$$

φ_i first order, $\bar{y}, \bar{z}$ of length $<\kappa$, such that, for $s, t \in I$,

$$EM(I, \Phi) \models \psi[\bar{b}_s, \bar{b}_t], \quad \textit{iff } s < t,$$

where

$$\bar{b}_t = \langle f_i(\bar{a}_t) : i < |A_1 \cup A_2| < \kappa \rangle, \quad \textit{and if } \kappa = \aleph_0, \bar{b}_t = \bar{a}_t.$$

Proof. Let J be $\{i : i < \beth^+_{\delta_0}\}$, where $\delta_0 = \beth_8(|T|)$, $I^*_{s,t}$ has power λ, M^0_R be $\mathbf{F}^a_\kappa$-prime over $\bigcup_{l,s} A^l_s \cup \bigcup \{I^*_{s,t} : s < t\}$ (so R is the natural ordering). Now expand M^0_R by $\lambda(T)$ functions and get M'_R such that:

(a) For every $\bar{b} \in \mathfrak{C}$, $\bar{a} \in M$, $\operatorname{stp}(\bar{b}, \bar{a})$ is realized in the closure of $\bar{a}$ by the functions of M'_R.

(b) $P = \{\bar{a}_s : s \in J\}$ where $\bar{a}_s \subseteq A^0_s \cup A^1_s$, and if $\kappa = \aleph_0$ then equality holds (i.e. the range of $\bar{a}_s$ is $A^0_s \cup A^1_s$) and always $A^0_s \cup A^1_s \subseteq \{f_i(\bar{a}_s) : i\}$.

(c) $f(-, \bar{a}_s, \bar{a}_t)$ $(s < t)$ is a one-to-one function from M_R into $I^*_{s,t}$.

(d) M'_R has Skolem functions.

(e) If $s < t$, $I' \subseteq$ Skolem hull of $\bar{a}_s \cup \bar{a}_t$ is countable, infinite and realized over $\bar{a}_s \cup \bar{a}_t$ the same type as $I_{t,s}$, then $Av(I'$, Skolem hull of $\bar{a}_s \cup \bar{a}_t)$ is omitted in M_R.

Now apply the proof of VII, 5.4 to get $EM^1(\omega, \Phi)$, a model of $T_1 = \operatorname{Th}(M'_R)$, which realizes only types M'_R, realizes.

2.5C *Corollary of* 2.5B: $I^a_{\aleph_0}(\lambda, T) = 2^\lambda$ *for* $\lambda \geqslant \lambda(T) + \aleph_1$.

Proof. From the Φ of 2.5B we can derive Φ_1 proper for $({}^{\omega\geqslant}\omega, T_1)$ such that for some (not first-order) L-formulas φ_n, $n < \omega$, $\nu \in {}^n\omega$, $\eta \in {}^\omega\omega$

$$EM(I, \Phi) \models \varphi_n(\bar{a}_\eta, \bar{a}_\nu), \quad \text{if and only if } \nu = \eta \restriction n$$

(just let $I = {}^{\omega\geqslant}\omega \times \{0, 1\}$ if η is a proper initial segment of ν, $I \models$

$(\eta,0) < (\nu,0) \leqslant (\nu,1) < (\eta,1)$, and if $\eta \in {}^n\omega$, $m < k$, then $(\eta^\frown\langle m\rangle,1) < (\eta^\frown\langle k\rangle,0)$ and if $\eta\in{}^\omega\omega$, $(\eta,0) = (\eta,1)$. $EM^1({}^{\omega\geqslant}\omega,\Phi_1)$ will be $EM^1(I,\Phi)$, choosing for $\eta\in{}^{\omega\geqslant}\omega$, $\bar{a}_\eta = \bar{a}_\eta = \bar{a}_{(\eta,0)}{}^\frown\bar{a}_{(\eta,1)}$. Now for $\lambda > |T_1|(+\lambda(T))$ the theorems of VIII, Section 2 work. For $\lambda = \lambda(T)$ by the proof in VIII, Section 3.1, Case I for almost every λ we get the result, and we can complete also the remaining case. Anyhow our main interest is $I(\lambda, T)$.

Remember that for T superstable, $|T|+\aleph_1 \leqq \lambda \leqq \lambda(T)$, $I(\lambda,T) = 2^\lambda$ was proved in IX, 1.20.

Now we shall mention a topic, not necessary for the rest of the book, but naturally connected to the dop. Just as we have looked at "hidden order" we can look for "hidden unstability", like the one caused by $\kappa(T) > \aleph_0$.

DEFINITION 2.6: T has the discontinuity dimensional property (didip for short) in the cardinal μ (μ regular) if there are $\mathbf{F}^a_\kappa$-saturated models M_α $(\alpha < \mu)$ such that $\alpha < \beta \Rightarrow M_\alpha \prec M_\beta$ and the $\mathbf{F}^a_\kappa$-prime model over $\bigcup_{i<\mu} M_i$ is not $\mathbf{F}^a_\kappa$-minimal over $\bigcup_{i<\alpha} M_i$.

Now if T has the didip for μ, then $\mu < \kappa(T)$ and there are 2^λ $\mathbf{F}^a_\chi$-saturated non-isomorphic models of power λ if $\lambda > \chi \geqq \kappa$, T stable in λ, at least when $(\forall\lambda_1 < \lambda)\,\lambda_1^{<\mu} < \lambda$, $\mathrm{cf}\,\lambda = \lambda$.

The proofs are parallel to the proofs in VIII on the number of $\mathbf{F}^a_\chi$-saturated models of power λ, when $\chi < \kappa(T)$ regular.

Also, the parallel of 2.3 holds, and if $\chi > \kappa$ is singular, T does not have the didip for cf χ, then the $\mathbf{F}^a_\chi$-prime model over any set A is minimal. Also, if each M_i is $\mathbf{F}^a_\lambda$-saturated $(i < \alpha) M \prec M_i$, $\{M_i : i < \alpha\}$ is independent over M, and N is $\mathbf{F}^a_\lambda$-prime over $\bigcup_{i<\alpha} M_i$, and in addition T does not have the dop nor any case of didip, then N is $\mathbf{F}^a_\lambda$-minimal over $\bigcup_{i<\alpha} M_i$.

X 3. The decomposition lemma

(Hyp) In this section T is superstable without the dimensional order property.

The main result of this section, the decomposition lemma, states that (for superstable T without the dimensional order property) every $\mathbf{F}^a_{\aleph_0}$-saturated model is $\mathbf{F}^a_{\aleph_0}$-prime over a non-forking tree $N_\eta(\eta\in I)$ (i.e., $I \subseteq {}^{\omega>}\|M\|$ is closed under initial segments, and

$\text{tp}(N_\eta, \bigcup\{N_\nu : \eta \not\leq \nu\})$ does not fork over $N_{\nu\restriction(l(\eta)-1)}$, when $l(\eta) > 0$, and $\eta \lessdot \nu \Rightarrow N_\eta \subseteq N_\nu$). This is a kind of structure theorem, so this division line (superstable + not dop) is significant. Though we shall eventually prove that some such theories (the deep ones, see Section 5) have many non-isomorphic models, all of them do not have a family of κ $\mathbf{F}^a_{\aleph_0}$-saturated models no one elementarily embedded into another, with κ of arbitrary cardinality (this is with the help of [Sh 82b]). Recall that for $A \subseteq B \subseteq C$, $B <_A C$ means that for each $\bar{c} \in C$, $\text{tp}(\bar{c}, B)$ is orthogonal to A.

THE ATOMIC DECOMPOSITION LEMMA 3.1: *Suppose $N_1 \prec M$ are $\mathbf{F}^a_{\aleph_0}$-saturated models. Then there are elements $a_i \in M$ $(i < \alpha)$ and models $N_{2,i}$, M_i such that:*

(a) $N_1 \prec N_{2,i} \prec M_i \prec M$,

(b) *M_i, $N_{2,i}$ are $\mathbf{F}^a_{\aleph_0}$-saturated,*

(c) *$\text{tp}(a_i, N_1)$ is regular, and for $i \neq j$, $\text{tp}(a_i, N_1)$, $\text{tp}(a_j, N_1)$ are orthogonal or equal,*

(d) *$a_i \in N_{2,i}$, and $N_{2,i}$ is $\mathbf{F}^a_{\aleph_0}$-prime over $N_1 \cup \{a_i\}$,*

(e) *$N_{2,i} <_{N_1} M_i$, and M_i is maximal with respect to this property (in fact, for no $a \in M - M_i$ is $\text{tp}_*(M_i \cup \{a\}, N_{2,i})$ orthogonal to N_1),*

(f) *M is $\mathbf{F}^a_{\aleph_0}$-prime and $\mathbf{F}^a_{\aleph_0}$-minimal over $\bigcup_{i<\alpha} M_i$,*

(g) *$\{M_i : i < \alpha\}$ is independent over N_1, and $\text{tp}_*(M_i, N_1 \cup \{a_i\})$ is almost orthogonal to N_1.*

Proof. Let $\boldsymbol{I} = \{a_i : i < \alpha\} \subseteq M$ be a maximal set, independent over N_1 of elements of M realizing over N_1 regular types and by V, 1.12 w.l.o.g. satisfying (c).

Let $N_{2,i} \subseteq M$ be $\mathbf{F}^a_{\aleph_0}$-prime over $N_1 \cup \{a_i\}$. So (c) and (d) hold trivially. By V, 3.2, we have:

FACT A. $\text{tp}(N_{2,i}, \bigcup_{i<\alpha, j\neq i} N_{2,j})$ *does not fork over N_1.*

Now for each $i < \alpha$, we define by induction on $j < \|M\|^+$ an element $b_{i,j} \in M - N_{2,j} \cup \{b_{i,\gamma} : \gamma < j\}$ such that $\text{tp}(b_{i,j}, N_{2,i} \cup \{b_{i,\gamma} : \gamma < j\})$ is $\mathbf{F}^a_{\aleph_0}$-isolated or is orthogonal to N_1 (we can assume that the second possibility occurs only if $N_{2,i} \cup \{b_{i,\gamma} : \gamma < j\}$ is the universe of an $\mathbf{F}^a_{\aleph_0}$-saturated model). There is a first $\beta(i) < \|M\|^+$ such that $b_{i,\beta(i)}$ is not defined. Obviously (by 1.6(5)):

FACT B. *$N_{2,i} \cup \{b_{i,\beta} : \beta < \beta(i)\}$ is the universe of an $\mathbf{F}^a_{\aleph_0}$-saturated model which we denote by M_i (clearly $M_i \prec M$).*

Now:

FACT C. $\mathrm{tp}_*(M_i, \bigcup_{j\neq i} M_j)$ *does not fork over* N_1.

To show Fact C, we just prove by induction on $\xi \leqq \sum_{i<\alpha}\beta(i)$ that if we let $A_i^\xi = N_{2,i} \cup \{b_{i,\beta}: \sum_{j<i}\beta(j)+\beta < \xi\}$, then for every $i<\alpha$, $\mathrm{tp}(A_i^\xi, \bigcup_{j\neq i} A_j^\xi)$ does not fork over N_1. The induction step is by V, 3.2, III, 0.1 (when we add an element realizing an $\mathbf{F}^a_{\aleph_0}$-isolated type) and V, 1.2(3), III, 0.1 (in the other case).

FACT D. $N_{2,i} <_{N_1} M_i$.

We prove by induction on $\beta \leqq \beta(i)$ that:

(*) $\qquad N_{2,i} <_{N_1} N_{2,i} \cup \{b_{i,\gamma}: \gamma < \beta\}$.

For $\beta = 0$ this is trivial, as well as for β limit. So let us prove it for $\beta+1$. Let $\mathrm{tp}(\bar{c}, N_{2,i})$ be a type which does not fork over N. Since N_1 is $\mathbf{F}^a_{\kappa(T)}$-saturated, by V, 1.2(3) it suffices to prove that $\mathrm{tp}(\bar{c}, N_{2,i})$, $\mathrm{tp}_*(\{b_{i,\gamma}: \gamma \leqq \beta\}, N_{2,i})$ are weakly orthogonal. This is equivalent to

$$\mathrm{tp}(\bar{c}, N_{2,i}) \vdash \mathrm{tp}(\bar{c}, N_{2,i} \cup \{b_{i,\gamma}: \gamma \leqq \beta\}).$$

By the induction hypothesis on β

$$\mathrm{tp}(\bar{c}, N_{2,i}) \vdash \mathrm{tp}(\bar{c}, N_{2,i} \cup \{b_{i,\gamma}: \gamma < \beta\}).$$

Hence $\mathrm{tp}(\bar{c}, N_{2,i} \cup \{b_{i,\gamma}: \gamma < \beta\})$ does not fork over $N_{2,i}$, hence (by transitivity, see III, 0.1) does not fork over N_1, and it suffices to prove it is weakly orthogonal to $\mathrm{tp}(b_{i,\beta}, N_{2,i} \cup \{b_{i,\gamma}: \gamma < \beta\})$. If the latter is $\mathbf{F}^a_{\aleph_0}$-isolated this holds by V, 3.2 and if the latter is orthogonal to N_1, this holds by V, 1.2(3).

FACT E. M *is* $\mathbf{F}^a_{\aleph_0}$*-minimal over* $\bigcup_{i<\alpha} M_i$.

PROOF OF FACT E. Let $M' \prec M$ be $\mathbf{F}^a_\kappa$-prime over $\bigcup_{i<\alpha} M_i$ (we know that there is such M'). Suppose $M' \neq M$ which holds if M is not $\mathbf{F}^a_\kappa$-prime over $\bigcup_{i<\alpha} M_i$, and M' can be chosen so that it holds if M is not $\mathbf{F}^a_{\aleph_0}$-prime and $\mathbf{F}^a_{\aleph_0}$-minimal over $\bigcup_{i<\alpha} M_i$; we shall eventually get a contradiction. Choose $b \in M - M'$ with $R[\mathrm{tp}(b, M'), L, \infty]$ minimal (it is $< \infty$ as T is superstable).

By V, 3.5, $\mathrm{tp}(b, M')$ is regular. Let us first assume that $\mathrm{tp}(b, M')$ is not orthogonal to N_1, then by 1.4 there is a regular type $p \in S^m(N_1)$ not orthogonal to it, hence by V, 1.12 some $b' \in M - M'$ realizes the stationarization of p over M'. But this contradicts the maximality of $I = \{a_i : i < \alpha\}$ as b' can serve as a_α.

So we can assume $\text{tp}(b, M')$ is orthogonal to N_1. Choose a set $B \subseteq M'$, $|B| < \kappa(T) = \aleph_0$ such that $\text{tp}(b, M')$ does not fork over B. Since M' is $\mathbf{F}^a_{\aleph_0}$-saturated we can also assume $\text{tp}(b, B)$ is stationary. Also there is a finite $S \subseteq \alpha$ such that $\text{tp}_*(B, \bigcup_{i \in S} M_i)$ is $\mathbf{F}^a_{\aleph_0}$-isolated [as $\text{tp}_*(B, \bigcup_{i<\alpha} M_i)$ is $\mathbf{F}^a_{\aleph_0}$-isolated by IV, 4.3 as $B \subseteq M'$, and M' is $\mathbf{F}^a_{\aleph_0}$-prime over $\bigcup_{i<\alpha} M_i$]. So there is $M^* \subseteq M'$ $\mathbf{F}^a_{\aleph_0}$-prime over $\bigcup_{i \in S} M_i$ and $B \in M$ so that $\text{tp}(b, M^*)$ is a regular type orthogonal to N_1. We prove that this is impossible by induction on $|S|$. If $|S| = 0$ then $\text{tp}(b, M^*)$ is not orthogonal to N_1 as $M^* = N_1$, a contradiction. If $|S| = 1, S = \{i\}$ we get a contradiction to the definition of M_i. Suppose $|S| = n+1$ and assume by induction that for any $U \subseteq \alpha$ with $|U| \leqq n$, if $N^* \prec M$ is $\mathbf{F}^a_{\aleph_0}$-prime over $\bigcup_{i \in U} M_i$ and $r \in S(N^*)$ is a regular type which is realized in M, then r is not orthogonal to N_1. Let $i \in S$ and choose $M^+ \subseteq M^*$, $\mathbf{F}^a_{\aleph_0}$-prime over $\bigcup_{j \in S, j \neq i} M_j$ such that M^* is $\mathbf{F}^a_{\aleph_0}$-prime over $M^+ \cup M_i$ (for some fixed $i \in S$). By V, 3.2 $\{M^+, M_i\}$ is independent over N_1. Since T does not have dop, by 2.3, $\text{tp}(b, M^*)$ is not orthogonal to one of M_i or M^+. Let N denote the model $\text{tp}(b, M^*)$ is not orthogonal to. By Lemma 1.4 there is a regular $q \in S^m(N)$ such that q is not orthogonal to $\text{tp}(b, M^*)$. But $\text{tp}(b, M^*)$ is orthogonal to every (regular) complete type over N_1. Since non-orthogonality is transitive on regular types (V, 1.13), it follows that q is orthogonal to every regular type in $S(N_1)$, i.e. by 1.4, q is orthogonal to N_1. But q is not orthogonal to $\text{tp}(b, M^*)$ and b is in the $\mathbf{F}^a_{\aleph_0}$-saturated model M, so by V, 1.12, the stationarization of q on M^* and hence q is realized in M. But then q and N contradict the hypothesis of induction.

So we prove Fact E and we can check that the only part of 3.1 to be proved is the last phrase of (e) which we leave to the reader. For (g) use 1.6(5), 1.6(1).

THE DECOMPOSITION LEMMA 3.2: *For any $\mathbf{F}^a_{\aleph_0}$-saturated model M we can find a set $I \subseteq {}^{\omega>}\|M\|$ (finite sequences of ordinals $< \|M\|$) closed under initial segments, and N_η, a_η for $\eta \in I$ and $p_\eta (\eta \in I - \{\langle\ \rangle\})$ such that:*

(1) *$N_\eta \prec M$ is $\mathbf{F}^a_{\aleph_0}$-saturated.*

(2) *$N_{\langle\rangle}$ is $\mathbf{F}^a_{\aleph_0}$-prime (over $\emptyset$).*

(3) *$p_{\eta\frown\langle i\rangle} = \text{tp}(a_{\eta\frown\langle i\rangle}, N_\eta)$ is regular, and for $\eta \frown \langle j\rangle \in I$, $p_{\eta\frown\langle i\rangle}$, $p_{\eta\frown\langle j\rangle}$ are orthogonal or equal.*

(4) *$N_{\eta\frown\langle i\rangle}$ is $\mathbf{F}^a_{\aleph_0}$-prime over $N_\eta \cup \{a_{\eta\frown\langle i\rangle}\}$.*

(5) *$\mathrm{tp}_*(N_{\eta\frown\langle i\rangle}, \bigcup\{N_\nu : \nu \in I$ but not $\eta\frown\langle i\rangle \trianglelefteq \nu\}$ does not fork over N_η (for $\eta\frown\langle i\rangle \in I$).*

(6) *M is $\mathbf{F}^a_{\aleph_0}$-prime, $\mathbf{F}^a_{\aleph_0}$-minimal over $\bigcup_{\eta\in I} N_\eta$.*

(7) *$\mathrm{tp}_*(\bigcup\{N_\nu : \eta \trianglelefteq \nu \in I\}, N_\eta)$ is orthogonal to $N_{\eta\restriction n}$ when $l(\eta) = n+1$.*

Proof. We define by induction on n, a set I_n of sequences of ordinals of length n, models N_η, M_η and elements a_η for each $\eta \in I_n$ such that:

(1) $\eta \in I_n, m < n$ implies $\eta\restriction m \in I_m$, and $I_0 = \{\langle\ \rangle\}$.

(2) N_η is $\mathbf{F}^a_{\aleph_0}$-saturated.

(3) $N_{\langle\rangle}$ is $\mathbf{F}^a_{\aleph_0}$-prime (over $\emptyset$) $M_{\langle\rangle} = M$.

(4) $N_{\eta\frown\langle i\rangle}$ is $\mathbf{F}^a_{\aleph_0}$-prime over $N_\eta \cup \{a_{\eta\frown\langle i\rangle}\}$.

(5) Let $p_{\eta\frown\langle i\rangle} = \mathrm{tp}(a_{\eta\frown\langle i\rangle}, N_\eta)$, then it is regular, and $p_{\eta\frown\langle i\rangle}, p_{\eta\frown\langle j\rangle}$ are equal or are orthogonal.

(6) M_η is $\mathbf{F}^a_{\aleph_0}$-saturated, $N_\eta \subseteq M_\eta \subseteq M$.

(7) $\eta \lessdot \nu$ implies $N_\eta \subseteq N_\nu \subseteq M_\nu \subseteq M_\eta$.

(8) M_η is $\mathbf{F}^a_{\aleph_0}$-prime over $\bigcup_i M_{\eta\frown\langle i\rangle}$.

(9) $N_{\eta\frown\langle i\rangle} <_{N_\eta} M_{\eta\frown\langle i\rangle}$.

(10) $\{M_{\eta\frown\langle i\rangle} : \eta\frown\langle i\rangle \in I\}$ is independent over N_η.

(11) $A_\eta = \{a_{\eta\frown\langle i\rangle} : \eta\frown\langle i\rangle \in I\}$ is a maximal subset of M_η (or even set of sequences from M_η) independent over N_η.

(12) A_η is a maximal subset of M which is independent over N_η and every element of it realizes over N_η a type orthogonal to $\bigcup_{k<l(\eta)} N_{\eta\restriction k}$ if $\eta \neq \langle\ \rangle$.

The definition is easy: for $n = 0$ trivial, for $n+1$, for each $\eta \in I_n$ we apply 3.1 with N_η, M_η standing for N_1, M and get $a_i, N_{2,i}, M_i$ and let $a_{\eta\frown\langle i\rangle} = a_i$, $N_{\eta\frown\langle i\rangle} = N_{2,i}$, $M_{\eta\frown\langle i\rangle} = M_i$ (so I_{n+1} is the set of ν's of length $n+1$ for which N_ν is defined).

Let $I = \bigcup_n I_n$. Now all the conditions of the lemma are obvious except "M is $\mathbf{F}^a_\kappa$-prime and $\mathbf{F}^a_\kappa$-minimal over $\bigcup_{\eta\in I} N_\eta$" and (11), (12).

Let us first prove (11), (12). If (11) or (12) fails for η, let $\bar a$ exemplify it; i.e. $\bar a \in M_\eta$, $\mathrm{tp}(\bar a, N_\eta)$ is orthogonal to $N_{\eta\restriction(k-1)}$ if $k = l(\eta) > 0$, and $\bar a \notin N_\eta$ and $\mathrm{tp}(\bar a, N_\eta \cup \{a_{\eta\frown\langle i\rangle} : \eta\frown\langle i\rangle \in I\})$ does not fork over N_η; w.l.o.g. $\mathrm{tp}(\bar a, N_\eta)$ is regular.

As in the proof of 3.1,

$$\mathrm{tp}(\bar a, \bigcup\{N_{\eta\frown\langle i\rangle} : \eta\frown\langle i\rangle \in I\}) \vdash \mathrm{tp}(\bar a, \bigcup(M_{\eta\frown\langle i\rangle} : \eta\frown\langle i\rangle \in I\}),$$

and as M_η is $\mathbf{F}^a_{\aleph_0}$-prime over $\bigcup\{M_{\eta\frown\langle i\rangle}:\eta\frown\langle i\rangle\in I\}$, $\operatorname{tp}(\bar{a},\bigcup_i M_{\eta\frown\langle i\rangle})$ does not fork over N_η; by V, 3.2,

$$\operatorname{tp}\left(\bar{a},\bigcup_i M_{\eta\frown\langle i\rangle}\right)\vdash\operatorname{tp}(\bar{a},M_\eta).$$

This proves (11) (i.e. when $\bar{a}\in M_\eta$); for (12) note that by the above $\operatorname{tp}(\bar{a},M_\eta)$ does not fork over N_η, hence is the stationarization of $\operatorname{tp}(\bar{a},N_\eta)$ which is orthogonal to $N_{\eta\restriction(l(\eta)-1)}$ when it is defined. Let $k=l(\eta)$, and we now prove by induction on $l\leqq k$ that $\operatorname{tp}(\bar{a},M_{\eta\restriction(k-l)})$ does not fork over N_η.

We have proved for $l=0$ and for $l+1$ notice that as $\operatorname{tp}(\bar{a},M_{\eta\restriction(k-l)})$ does not fork over N_η, it is parallel to $\operatorname{tp}(\bar{a},N_\eta)$, hence orthogonal to $N_{\eta\restriction(k-1)}$, hence to $N_{\eta\restriction(k-l-1)}$. So for $j\neq\eta(k-l-1)$, $\operatorname{tp}(\bar{a},M_{\eta\restriction(k-l)})$ is orthogonal to $\operatorname{tp}_*(M_{\eta\restriction(k-l-1)\frown\langle j\rangle},N_{\eta\restriction(k-l-1)})$.

As $\{M_{\eta\restriction(k-l-1)\frown\langle j\rangle}:j<\alpha\}$ is independent over $N_{\eta\restriction(k-l-1)}$, $\operatorname{tp}(\bar{a},M_{\eta\restriction(k-l)})$ is orthogonal to $\operatorname{tp}_*(\bigcup_{j\neq\eta(k-l-1)}M_{\eta\restriction(k-l-1)\frown\langle j\rangle},N_{\eta\restriction(k-l-1)})$, and

$$\operatorname{tp}(\bar{a},M_{\eta\restriction(k-l)})\vdash\operatorname{tp}\left(\bar{a},\bigcup_j M_{\eta\restriction(k-l-1)\frown\langle j\rangle}\right).$$

So as $M_{\eta\restriction(k-l-1)}$ is $\mathbf{F}^a_{\aleph_0}$-prime over $\bigcup_j M_{\eta\restriction(k-l-1)\frown\langle j\rangle}$, by IV, 4.10(2),

$$\operatorname{tp}(\bar{a},M_{\eta\restriction(k-l)})\vdash\operatorname{tp}(\bar{a},M_{\eta\restriction(k-l-1)}),$$

so also the latter does not fork over N_η. For $l=k$ we get a contradiction to $\bar{a}\in M=M_{\langle\,\rangle}$, as $\operatorname{tp}(\bar{a},M_{\langle\,\rangle})$ does not fork over N_η, and $\bar{a}\notin N_\eta$.

Now we shall prove that M is $\mathbf{F}^a_{\aleph_0}$-prime, $\mathbf{F}^a_{\aleph_0}$-minimal over $\bigcup_{\eta\in I}N_\eta$. So suppose $M'\subseteq M$, $M'\neq M$, M' is $\mathbf{F}^a_\kappa$-prime over $\bigcup_{\eta\in I}N_\eta$, and choose $b\in M-M'$, $R[\operatorname{tp}(b,M'),L,\infty]$ minimal, hence $\operatorname{tp}(b,M')$ is regular. Now $\operatorname{tp}(b,M')$ is orthogonal to each N_η. If not, choose η with minimal length, then by 1.4 for some regular $q\in S^m(N_\eta)$, $\operatorname{tp}(b,M')$, q are not orthogonal and $q=p_{\eta\frown\langle i\rangle}$ for some i, or q is orthogonal to every $p_{\eta\frown\langle i\rangle}$. Let $q'\in S^m(M')$ be the stationarization of q over M', so by V, 1.12 there is $\bar{a}'\in M$ which realizes q', so $\operatorname{tp}(\bar{a}',M')$ does not fork over N_η. By choice of η (being of minimal length) (if $\eta\neq\langle\,\rangle$) $\operatorname{tp}(b,M')$ is orthogonal to $\bigcup_{k<l(\eta)}N_{\eta\restriction k}$, hence by V, 1.13 also $\operatorname{tp}(\bar{a}',M')$ is, so we get a contradiction to (12). We conclude $\operatorname{tp}(b,M')$ is really orthogonal to each N_η.

Choose a finite $B\subseteq M'$ over which $\operatorname{tp}(b,M')$ does not fork, and a finite $I^*\subseteq I$ closed under initial segments such that $\operatorname{tp}_*(B,\bigcup_{\eta\in I}N_\eta)$ does not fork over $\bigcup_{\eta\in I^*}N_\eta$. As M' is $\mathbf{F}^a_{\aleph_0}$-prime over $\bigcup_{\eta\in I}N_\eta$, $\operatorname{tp}_*(B,\bigcup_{\eta\in I}N_\eta)$ is $\mathbf{F}^a_{\aleph_0}$-isolated, so $\operatorname{tp}_*(B,\bigcup_{\eta\in I^*}N_\eta)$ is $\mathbf{F}^a_{\aleph_0}$-isolated (see IV, 4.3).

Let $N \prec M'$ be $\mathbf{F}^a_{\aleph_0}$-prime over $\bigcup\{N_\eta : \eta \in I^*\}$ and $B \subseteq N$. We have found an $\mathbf{F}^a_{\aleph_0}$-prime model N over $\bigcup_{\eta \in I^*} N_\eta$, so that there is a regular type in $S^m(N)$, orthogonal to each N_η ($\eta \in I^*$), realized in M, and I^* is finite, closed under initial segments. We get a contradiction to the statement in the last sentence by induction on $|I^*|$. If $I^* = \{\eta \restriction l : l < \lg(\eta)\}$ this is trivial, and if not choose distinct η, ν in some $I^* \cap I_n$ with n minimal. Let $N^0 = N_{\eta\restriction(n-1)}$, $N' \prec N$ be $\mathbf{F}^a_{\aleph_0}$-prime over $\bigcup\{N_\sigma : \eta \triangleleft \sigma \in I^*\}$ and $N^2 \subseteq N$ be $\mathbf{F}^a_{\aleph_0}$-prime over $\bigcup\{N_\sigma : \text{not } \eta \triangleleft \sigma, \text{ but } \sigma \in I^*\}$ such that N is $\mathbf{F}^a_{\aleph_0}$-prime over $N^1 \cup N^2$. Now by 2.3 (and as T does not have the dop) for some $l \in \{1, 2\}$, there is a regular complete type over N^l orthogonal to N_0, realized in M, and we get a contradiction to the induction hypothesis on $|I^*|$.

In fact our proofs also prove:

LEMMA 3.3: *Suppose $I \subseteq {}^{\omega>}\lambda$ is closed under initial segments, $\{N_\eta : \eta \in I\}$ is a non-forking tree of $\mathbf{F}^a_\kappa$-saturated models* [*i.e.* $\eta \triangleleft \nu \Rightarrow N_\eta \prec N_\nu$, $\eta \in I - \{\langle\ \rangle\} \Rightarrow \operatorname{tp}(N_\eta, \bigcup\{N_\nu : \nu \in I,$ *not* $\eta \triangleleft \nu\})$ *does not fork over* $N_{\eta\restriction(l(\eta)-1)}$, *and each N_η is $\mathbf{F}^a_{\aleph_0}$-saturated*].

If T does not have the dop, M $\mathbf{F}^a_\kappa$-prime over $\bigcup_{\eta \in I} N_\eta$, then M is $\mathbf{F}^a_\kappa$-minimal over $\bigcup_{\eta \in I} N_\eta$ and every $q \in S^m(M)$ is not orthogonal to some N_η.

Proof. First note:

FACT 3.3A. *If S_1, S_2 are non-empty subsets of I which are downward closed, then $\{\bigcup_{\nu \in S_1} N_\nu, \bigcup_{\nu \in S_2} N_\nu\}$ is independent over $\bigcup_{\nu \in S_1 \cap S_2} N_\nu$.*

Proof. By the local properties of forking, it suffices to restrict ourselves first to the case $S_1 - S_2$ finite, then to the case $S_1 - S_2$, $S_2 - S_1$ finite, and finally S_1, S_2 finite (using III, 0.1).

Now we prove the statement by induction on $|S_1 \cup S_2|$; w.l.o.g. $S_1 \neq S_1 \cap S_2$, $S_2 \neq S_1 \cap S_2$, clearly $\langle\ \rangle \in S_1 \cap S_2$.

So choose $\eta \in S_1 - S_2$ of maximal length, and by the induction hypothesis and transitivity of forking (see III, 0.1) it is enough that $\operatorname{tp}_*(N_\eta, \bigcup\{N_\nu : \nu \in S_1 \cup S_2, \nu \neq \eta\})$ does not fork over $N_{\eta\restriction(l(\eta)-1)}$, but this follows from the hypothesis. Now return to 3.3.

If one of the conclusions fails we can reduce it to the case $I = I^*$ is finite; then, as in 2.2, the two conclusions are equivalent. If both fail, there is $M' \prec M$, M' $\mathbf{F}^a_{\aleph_0}$-prime over $\bigcup_{\eta \in I^*} N_\eta$, $b \in M - M'$, $\operatorname{tp}(b, M')$ orthogonal to every N_η; and continue as in the last part of the proof of 3.2: from the choice of B.

X.4. Deepness

(Hyp) In this section T is superstable without the dop.

As we know the number of non-isomorphic $\mathbf{F}^a_{\aleph_0}$-saturated models in every $\aleph_\alpha \geqq |T| + \aleph_1$ for unsuperstable T and for T with the dop, we concentrate on the case in the hypothesis. By the decomposition Lemma 3.2 we know every $\mathbf{F}^a_{\aleph_0}$-saturated model is $\mathbf{F}^a_{\aleph_0}$-prime over a non-forking tree of $\mathbf{F}^a_\kappa$-prime models. Clearly, if there are few trees then there are few models, but the converse is less clear. Anyhow, clearly the most important distinction is whether the tree (I in 3.2) is always well-founded. If it is always well-founded, naturally some rank is defined (called here the *depth*), and if the rank is small the number of models is small.

In this section we introduce the basic relevant notions and the simple facts about them. Notice that we could have chosen some other variants of the notions, but by later sections we can show that they would be equivalent.

We define the depth so that the results on the number of models can be stated smoothly (this is why 4.1(iii), 4.2(1) are such that the depth is not a limit ordinal).

Let us make some more specific remarks. Note that if the tree is not well-founded there are $N_l \prec N_{l+1} N_{l+1}$ $\mathbf{F}^a_{\aleph_0}$-prime over $N_l \cup \{a_l\}$, $\operatorname{tp}(a_l, N_l)$ regular orthogonal to N_{l-1}. So the rank is defined as an attempt to build such a sequence (i.e. the rank of (N, N', a) is ∞ iff there is such a sequence, $N = N_0, N' = N_1, a = a_0$).

In Definitions 4.1 and 4.3 we give some variants of this, in 4.2 we define the relevant property of a theory (deepness), in 4.4 we prove various facts, and in 4.5 the essential equivalence of some variants is given. Now 4.6 says that, looking for high depth, it is enough to look at types not orthogonal to $\emptyset$. For "canonical" examples see 4.9, 4.10.

DEFINITION 4.1: Let $K' = \{(N, N', \bar a) : \operatorname{tp}(\bar a, N)$ is regular, and N' is $\mathbf{F}^a_{\aleph_0}$-prime over $N \cup \{\bar a\}$ and N is $\mathbf{F}^a_{\aleph_0}$-saturated$\}$.

For every member of K' we define its depth, an ordinal (zero or successor but not limit) or infinity ∞, by:

(i) $\operatorname{Dp}(N, N', \bar a) \geqq 0$ iff $(N, N', \bar a) \in K'$;

(ii) $\mathrm{Dp}(N,N',\bar{a}) \geqq \alpha+1$ (α zero or successor) iff for some $N'',\bar{a}'$: $(N',N'',\bar{a}')\in K'$, $N' <_N N''$ and $\mathrm{Dp}(N',N'',\bar{a}') \geqq \alpha$;

(iii) $\mathrm{Dp}(N,N',\bar{a}) \geqq \delta+1$ (δ limit) iff $\mathrm{Dp}(N,N',\bar{a}) \geqq \beta$ for $\beta < \delta$;

(iv) $\mathrm{Dp}(N,N',\bar{a}) = \infty$ iff for every ordinal β $\mathrm{Dp}(N,N',\bar{a}) \geqq \beta$, $\mathrm{Dp}(N,N',\bar{a}) = \alpha$ iff $\mathrm{Dp}(N,N',\bar{a}) \geqq \alpha$ but not $\mathrm{Dp}(N,N',\bar{a}) \geqq \alpha+1$.

DEFINITION 4.2: (1) The depth of the theory $\mathrm{Dp}(T)$ is $\bigcup\{\mathrm{Dp}(N,N',a):(N,N',a)\in K'\}$ when this is finite and $\bigcup\{\mathrm{Dp}(N,N',a):(N,N',a)\in K'\}+1$ when this is an infinite ordinal.

(2) The theory T is *deep* if its depth is ∞; otherwise it is *shallow*.

DEFINITION 4.3: (1) $K_\lambda = \{(N,N',\bar{a}):N,N'$ are $\mathbf{F}^a_\lambda$-saturated, $\bar{a}\in N'$, $\bar{a}\notin N,N'$ $\mathbf{F}^a_\lambda$-atomic over $N\cup\{\bar{a}\}\}$. $\mathrm{Dp}((N,N',\bar{a}),K)$ is defined as in 4.1 for any set K of triples. If $(N,N',\bar{a})\notin K$ we interpret $\mathrm{Dp}((N,N',\bar{a}),K)$ as $\mathrm{Dp}(N,N',\bar{a}),K\cup\{(N,N',\bar{a})\}$ (not closing under isomorphism) and if $K=K'$ we omit it.

Let $K'_\lambda = \{(N,N',\bar{a})\in K_\lambda : \mathrm{tp}(\bar{a},N)$ is regular$\}$.

(2) For a tree I, we define $\mathrm{Dp}_\lambda(\eta,I)$ (λ a cardinal, $\eta\in I$): $\mathrm{Dp}_\lambda(\eta,I) \geqq \alpha+1$ iff for λ ν's, $\eta \lessdot \nu$, $\mathrm{Dp}(\nu,I) \geqq \alpha$, and for $\alpha=0$ or limit $\mathrm{Dp}_\lambda(\eta,I) \geqq \alpha$ iff $\mathrm{Dp}_\lambda(\eta,I) \geqq \beta$ for every $\beta<\alpha$. So $\mathrm{Dp}_\lambda(\eta,I)=\alpha$ if it is $\geqq\alpha$ but not $\geqq\alpha+1$. Let $\mathrm{Dp}_\lambda(I) = \sup\{\mathrm{Dp}_\lambda(\eta,I):\eta\in I\}$. For $\lambda=1$ we omit λ.

(3) $\mathrm{Dp}(T,K)$ is defined as in 4.2.

LEMMA 4.4: (1) *If* $(N,N',\bar{a})\in K$, $(N',N'',\bar{a}')\in K, N' <_N N''$, *then* $\mathrm{Dp}((N,N',\bar{a}),K) \geqq \mathrm{Dp}((N',N'',\bar{a}'),K)$ (*the inequality is strict except when both are* ∞) [*this holds for any class K of triples*].

(2) *If* $\alpha \leqslant \mathrm{Dp}((N,N',\bar{a}),K) < \infty$, $(N,N',a)\in K$, α *not limit, then some* $(N_0,N_0',\bar{a}_0)\in K$ *has depth* α (*in* K).

(3) $\mathrm{Dp}(N,N',\bar{a})=\infty$ *iff there are* $N_l,\bar{a}_l$ $(l<\omega)$, $N_{l+1} <_{N_l} N_{l+2}$, $N_0=N$, $N_1=N'$, $\bar{a}_0=\bar{a}$, $(N_{l+1},N_{l+2},\bar{a}_l)\in K'$. (*Similarly for any "reasonable" K*.)

(4) *If* $\alpha=\mathrm{Dp}(T)$ *or* $\alpha=\mathrm{Dp}(N,N',\bar{a})$, $\alpha<\infty$, *then* $\alpha<(2^{|T|})^+$ *and in fact* $\alpha<\delta(|T|)$ (*look at* VII, §5 *on the definition of* $\delta(|T|)$ *and information on it*).

(5) *The depth is preserved by automorphisms of* $\mathfrak{C}$.

(6) *If* $(N_l,N_l',\bar{a}_l)\in K'_{\aleph_0}$ $(l=0,1)$ *and* $\mathrm{tp}(\bar{a}_l,N_l)$ *are parallel or not orthogonal, then* $\mathrm{Dp}(N_0,N_0',\bar{a}_0)=\mathrm{Dp}(N_1,N_1',\bar{a}_1)$. *If only* $(N_1,N_1',\bar{a}_1)\in K_{\aleph_0}$, $\mathrm{tp}(\bar{a}_1,N_1)$ *regular, still* $\mathrm{Dp}(N_0,N',\bar{a}_0) \leqq \mathrm{Dp}(N_l,N_l',\bar{a}_l)$.

DEFINITION 4.4A: We let, for regular p, $\mathrm{Dp}(p)$ be the depth of $(N, N', \bar{a}) \in K'_{\aleph_0}$ when $\mathrm{tp}(\bar{a}, N)$, p *are parallel.*

PROOF OF LEMMA 4.4. Easy. Note for (3) we need (6).

(1) Trivial.

(2) We prove it by induction on $\beta = \mathrm{Dp}(N, N', \bar{a})$; for $\beta = \alpha$ there is nothing to prove; for $\beta > \alpha$, use 4.1(ii) applied to $\mathrm{Dp}(N, N', \bar{a}) \geqq \beta+1$, to get N'', a' with $\gamma = \mathrm{Dp}(N', N'', \bar{a}') \geqq \alpha$, but by 4.4(1), $\gamma < \beta$, and use the induction hypothesis.

(4) By (6), $\mathrm{Dp}(N, N', \bar{a})$ depends only on $\mathrm{tp}(\bar{a}, N)$ up to parallelism, and by (5) it is preserved by automorphisms of $\mathfrak{C}$, hence there are $\leqq 2^{|T|}$ possible depths. But by (2) the ordinals which are the depth of some triple form an initial segment of the set of the non-limit ordinals, hence

$$\alpha = \mathrm{Dp}(N, N', \bar{a}) < \infty \Rightarrow \alpha < (2^{|T|})^+.$$

The first phrase has been proved above. We work as in VII, Section 5. For the second phrase, let $\mathfrak{B} = (H(\lambda), \in, T, \delta(|T|))$, $H(\lambda)$– the family of sets of hereditary cardinality $<\lambda$, and w.l.o.g. $T \subseteq |T|$. Now consider $\mathfrak{B}'$ elementarily equivalent to $\mathfrak{B}$, with "T and $|T|$" standard: but non-well-ordered "ordinals" $<\delta(T)$. ($\mathfrak{B}'$ is known to exist.)

In this model we can consider various notions and check whether they are absolute, i.e. whether if $\mathfrak{B}'$ says something holds it really holds. Now this holds for

(a) being a model of T,

(b) $R(p, \Delta, \lambda) = n$ (Δ finite, $\lambda \leqq \aleph_0$),

(c) being orthogonal types,

(d) non-orthogonal types,

(e) $\mathrm{tp}_*(A, B)$ does not fork over C, and

(f) $A \subseteq B$, $p \in S^m(B)$; p has a unique extension in $S^m(N \cup B)$ when $\mathrm{tp}_*(N, B)$ does not fork over A.

We can conclude by 4.4(1) that also if $\mathfrak{B}' \vDash$ "$\mathrm{Dp}(p) \geqq a^*$, a^* an ordinal", then p has depth $\geqq$ the order type of $\{a \in \mathfrak{B}'' : a < a^*\}$. If $a^* \geqq \delta(|T|)$ this is not well-founded so $\mathrm{Dp}(p) = \infty$.

(5) Trivial.

(6) We prove by induction on γ that $[\mathrm{Dp}(N_0, N'_0, \bar{a}_0) \leqq \gamma$ or $\mathrm{Dp}(N_1, N'_1, \bar{a}_0) \leqq \gamma]$ implies the equality. We can choose $\mathbf{F}^a_\kappa$-saturated N_2, $N_0 \cup N_1 \subseteq N_2$, and by 4.4(5) w.l.o.g. $\mathrm{tp}(N'_l, N_2)$ does not fork over N_l (for $l = 0, 1$). By V, 3.2, N'_0 is $\mathbf{F}^a_\kappa$-constructible over $N_2 \cup \bar{a}_0$, and let

N'_2 be $\mathbf{F}^a_\kappa$-primary over $N_2 \cup N'_0 = N_2 \cup N'_0 \cup \bar{a}_0$. In N'_2 $\text{tp}(\bar{a}_1, N_2)$ is realized, so w.l.o.g. $\bar{a}_1 \in N'_2$, $N'_1 \subseteq N'_2$, N'_2 $\mathbf{F}^a_\kappa$-prime over $N_2 \cup \bar{a}_1$ and over $N_2 \cup N'_1$ (and over $N_2 \cup \bar{a}_0$ and over $N_2 \cup N'_0$).

By symmetry it is enough to prove $\text{Dp}(N, N'_0, \bar{a}_0) = \text{Dp}(N_2, N'_2, \bar{a}_0)$, because checking the definition we can observe that $\text{Dp}(N_2, N'_2, \bar{a}_0) = \text{Dp}(N_2, N'_2, \bar{a}_1)$. Now the inequality $\leqq$ is trivial (with the induction hypothesis for γ) and for the inequality $\geqq$ we have to replace parameters with others of the same type. For proving the second phrase act as before, observing that by the first phrase $\text{Dp}(N_1, N'_1, \bar{a}_1) = \text{Dp}(N_2, N'_2, \bar{a}_0)$.

(3) First suppose that there are N_l, $\bar{a}_l$, and let $\alpha_l = \text{Dp}(N_l, N_{l+1}, \bar{a}_l)$. By 4.4(1), $\alpha_l \geqq \alpha_{l+1}$, and if $\alpha_l \neq \infty$, $\alpha_l > \alpha_{l+1}$. So if α_0 is $\neq \infty$, then α_l $(l < \omega)$ is a strictly decreasing sequence of ordinals, a contradiction; so $\alpha_0 = \infty$, but $\alpha_0 = \text{Dp}(N_0, N_1, \bar{a}_0) = \text{Dp}(N, N', \bar{a})$, so we have proved the "if" part of 4.4(3).

Now suppose $\text{Dp}(N, N', \bar{a}) = \infty$. We define by induction on l, N_{l+1}, $\bar{a}_l$ such that N_{l+1} is $\mathbf{F}^a_\kappa$-saturated, $\text{Dp}(N_l, N_{l+1}, \bar{a}_l) = \infty$, $N_l <_{N_{l-1}} N_{l+1}$ (when $l > 0$). So let $N_0 = N$, $N_1 = N'$, $\bar{a}_0 = \bar{a}$. So the induction hypothesis holds. For $l+1$ as $\text{Dp}(N_l, N_{l+1}, \bar{a}_l) = \infty$, there are N_{l+2}, $\bar{a}_{l+1} = \langle a_{l+2} \rangle$, such that $(N_{l+1}, N_{l+2}, \bar{a}_{l+2}) \in K'$, $\text{Dp}(N_{l+1}, N_{l+2}, \bar{a}_{l+1}) \geqq (2^{|T|})^+$, hence by 4.4(4), $\text{Dp}(N_{l+1}, N_{l+2}, \bar{a}_{l+2}) = \infty$. As we can carry the induction we have proved the "only if" part of 4.4(3).

The following lemma shows that there is no real difference between the various $\text{Dp}(-, K)$'s, in particular, whether we use a or $\bar{a}$.

LEMMA 4.5: (1) *For any* $(N, N', \bar{a}) \in K_\lambda$

$$\text{Dp}((N, N', \bar{a}), K_\lambda) = \text{Dp}(N, N', \bar{a}).$$

(2) *If* $K'_1 = \{(N, N', a) \in K'_{\aleph_0} : N$ *is* $\mathbf{F}^a_{\aleph_0}$*-atomic over* $N \cup \{a\}\}$, *then on* K'_1, $\text{Dp}(-, K')$, $\text{Dp}(-, K'_1)$ *are equal.*

(3) *For any complete type* p, $\text{Dp}(p) = \sup\{\text{Dp}(r) : r$ *a complete regular type not orthogonal to* $p\}$.

(4) *Suppose* $N' <_N N''$, $\bar{a}' \in N'$, $\bar{a}'' \in N''$, $\text{tp}(\bar{a}', N)$ $\text{tp}(\bar{a}'', N')$ *are regular*, $\text{tp}(N', N \cup \bar{a}')$ *is almost orthogonal to* N *and* $\text{tp}(N'', N' \cup \bar{a}'')$ *is almost orthogonal to* N', *Then* $\text{Dp}(N, N', \bar{a}') > D(N', N'', \bar{a}'')$ (*or both are* ∞).

Proof. (1) Remember that by III, 4.22, if M is $\mathbf{F}^a_\lambda$-saturated, M' is $\mathbf{F}^a_\lambda$-prime over $M \cup \bar{a}$, $\kappa < \lambda$, then M is $\mathbf{F}^a_{\aleph_0}$-atomic over $M \cup \bar{a}$. So easily by the definitions and 4.4(6) $\mathrm{Dp}((N,N',\bar a),K_\lambda) \geqq \mathrm{Dp}(N,N',\bar a)$. So it suffices to prove by induction on α that:

(*) $\mathrm{Dp}((N,N',\bar a),K_\lambda) \geqq \alpha \Rightarrow \mathrm{Dp}(N,N',\bar a) \geqq \alpha$ for $(N,N',\bar a) \in K_\lambda$.

For $\alpha = 0$, α limit and successor of limit this is trivial; for $\alpha = \beta+1$, β not limit, there is $(N',N'',\bar a') \in K_\lambda$, $N' <_N N''$, and $\mathrm{Dp}((N',N'',\bar a'),K_\lambda) \geqq \beta$, hence by the induction hypothesis $\mathrm{Dp}(N,N'',\bar a') \geqq \beta$. Apply Lemma 3.1 for N,N'' standing for N_1,M and get $a_i,N_{2,i},M_i$ ($i < i(0)$). By V, Def. 3.2, Th. 3.2, w.l.o.g. $N_{2,i} = M_i$ and $i(0)$ is finite $(= w(\bar a',N))$, and let $\gamma = \mathrm{Max}_{i<i(0)} \mathrm{Dp}(N',N_{2,i},a_i)$ (so γ is not limit); clearly it suffices to prove $\gamma \geqq \beta$. Otherwise, as γ is not limit there is $(N'',N^*,\bar a^*) \in K'$, $N'' <_N N^*$, $\mathrm{Dp}(N'',N^*,\bar a^*) \geqq \gamma$. As $\mathrm{tp}(\bar a^*,N'')$ is regular, orthogonal to N', clearly as in the proof of 3.1 (see Fact E) for some $i < i(0)$ and regular $q \in S^1(N_{2,i})$, $\mathrm{tp}(\bar a^*,N'')$ and q are not orthogonal. Let c realize q, $N'_{2,i}$ be $\mathbf{F}^a_{\aleph_0}$-prime over $N_{2,i} \cup \{c\}$. So by 4.4(6)

(a) $\mathrm{Dp}(N'',N^*,\bar a^*) = \mathrm{Dp}(N_{2,i},N'_{2,i},c)$.

Now as q is not orthogonal to $\mathrm{tp}(\bar a^*,N'')$, it is orthogonal to any regular complete type over N' (as non-orthogonality is an equivalence relation among regular types, see V, 1.13) hence q is orthogonal to N'. Hence $N_{2,i} <_{N'} N'_{2,i}$, hence (by 4.4(1))

(b) $\mathrm{Dp}(N',N_{2,i},a_i) > \mathrm{Dp}(N_{2,i},N'_{2,i},\bar c)$.

As $\mathrm{Dp}(N'',N^*,\bar a^*) \geqq \gamma$ by (a), (b), $\mathrm{Dp}(N,N_{2,i},a_i) > \gamma$, contradicting γ's definition.

(3), (2), (4) Easy.

LEMMA 4.6: *If* $(N_0,N'_0,\bar a_0) \in K'_{\aleph_0}$, $\mathrm{tp}(\bar a_0,N_0)$ *is orthogonal to* $\emptyset$ *and has depth* $<\infty$, *then there is* $(N_1,N'_1,\bar a_1) \in K'_{\aleph_0}$ *such that* $\mathrm{tp}(\bar a_1,N_1)$ *is not orthogonal to* $\emptyset$ *and*

$$\mathrm{Dp}(N_0,N'_0,\bar a_0) < \mathrm{Dp}(N_1,N'_1,\bar a_1).$$

Proof. By 4.4(6), w.l.o.g. N_0 and N'_0 are $\mathbf{F}^a_{\aleph_0}$-prime over $\emptyset$. Let $B \subseteq N_0$ be finite, such that $\mathrm{tp}(\bar a_0,N_0)$ does not fork over B. There is B' realizing the stationarization of $\mathrm{stp}_*(B,\emptyset)$ over N_0, and let M' be $\mathbf{F}^a_{\aleph_0}$-prime over $N_0 \cup B'$. By 1.7, M' is $\mathbf{F}^a_{\aleph_0}$-prime over $\emptyset$. Hence by IV, 4.18, N_0,M' are $\mathbf{F}^a_{\aleph_0}$-prime over B,B', respectively, hence there is an isomorphism from M' onto N_0 taking B' to B. So there is a model $N \prec N_0$ such that $\mathrm{tp}_*(B,N)$ does not fork over $\emptyset$ and N_0 is $\mathbf{F}^a_{\aleph_0}$-prime over $N \cup B$. Hence by V, 3.9 there is a finite set $J = \{b_i : i < n\} \subseteq N_0$ independent over N, of elements realizing regular types, such that N_0 is $\mathbf{F}^a_{\aleph_0}$-prime over $N \cup J$. By V, 3.2 there are $N^*_m \prec N_0$ $\mathbf{F}^a_{\aleph_0}$-prime over $N \cup \{b_m\}$ such that N_0 is $\mathbf{F}^a_{\aleph_0}$-prime over $\bigcup_{l<n} N^*_l$. As $\operatorname{tp}(\bar{a}_0, N_0)$ is orthogonal to $\emptyset$, is parallel to $\operatorname{stp}(\bar{a}_0, B)$, and $\operatorname{tp}_*(B, N)$ does not fork over $\emptyset$, by 1.1 $\operatorname{tp}(\bar{a}_0, N_0)$ is orthogonal to N. On the other hand, by 3.3 $\operatorname{tp}(\bar{a}_0, N)$ is not orthogonal to some N^*_l, hence by 1.4 some regular $q \in S^m(N^*_l)$ is not orthogonal to $\operatorname{tp}(\bar{a}_0, N_0)$, so by V, 1.13 it is orthogonal to N.

Clearly $\operatorname{Dp}(N, N^*_l, b_l) > \operatorname{Dp}(q) = \operatorname{Dp}(N_0, N_1, \bar{a}_0)$ (see 4.4(1), 4.4(6), respectively) and $\operatorname{tp}(b_l, N)$ is not orthogonal to $\emptyset$ as it is not orthogonal to $\operatorname{tp}_*(B, N)$ (because $b_l \in N_0$, N_0 $\mathbf{F}^a_{\aleph_0}$-prime over $N \cup B$, $\operatorname{tp}(B, N)$ does not fork over $\emptyset$). Now put $(N, N'_1, \bar{a}_1) \stackrel{\text{def}}{=} (N, N^*_l, b_l)$.

THEOREM 4.7: *$I^a_{\aleph_0}(\aleph_\alpha, T)$ (the number of non-isomorphic $\mathbf{F}^a_{\aleph_0}$-saturated models of power $\aleph_\alpha$) is at most $\beth_{\operatorname{Dp}(T)}(|\alpha|^{2^{|T|}})$ for shallow T, so it is $< \beth_{\delta(|T|)}(|\alpha|^{2^{|T|}}) = \beth_{\delta(|T|)}(|\alpha|) < \beth_{(2^{|T|})^+}(|\alpha|)$ and if T is countable $< \beth_{\omega_1}(|\alpha|)$.*

Proof. Immediate by 3.2, and the bounds on $\delta(|T|)$ (see e.g. VII, 5.5 and 5.5(2)).

LEMMA 4.8: *If T is superstable without the dop, then $R(x = x, L, \infty) \geqslant \operatorname{Min}\{\omega, \operatorname{Dp}(T)\}$.*

Proof. Suppose $k < \omega$, $k \leqslant \operatorname{Dp}(T)$. Then we can find N_η, M_η as in 3.2's notation for $\eta \in {}^{k \geqslant}\lambda$ (any λ). Now we can prove by induction on $k - l(\eta)$ that for every $b \in M_\eta \backslash N_\eta$

$$k - l(\eta) + R(\operatorname{tp}(b, N_\eta), L, \infty) \leqslant R(x = x, L, \infty).$$

Example 4.9. A very natural example of a superstable T without the dop which is deep, is the following T: $T_{\text{dp}} = \operatorname{Th}({}^{\omega>}\omega, f)$ when $f(\eta)$ is η if $\eta = \langle\ \rangle$, and $\eta \restriction n$ if $\eta \in {}^{n+1}\omega$.

Notice that a model of T_{dp} consist of trees, exactly one with a root (i.e. $f(x) = x$), in which every element has infinitely many immediate predecessors (i.e. y's such that $f(y) = x$). A similar example is $T^*_{\text{dp}} = \operatorname{Th}({}^{\omega>}\omega, \ldots, P_n, f_n, \ldots)$ where $P_n = {}^n\omega, f_n$ is a partial function: $f \restriction P_n$.

Both theories are $\aleph_0$-stable, and by expanding a little we can get elimination of quantifiers.

Examples 4.9A. Examples of shallow theories can be obtained similarly to 4.9; we prefer to use the $T^{\alpha *}$ from II: the language consists just of the two-place relations E_i $(i < \alpha)$. The axioms of T state: each E_i is an equivalence relation, for $i < j$, E_i refines E_j, moreover each E_j-equivalence class is the union of infinitely many distinct E_i-equivalence classes. Also E_i has infinitely many equivalence classes and each E_i-equivalence class is infinite. It is not hard to check that $\mathrm{Dp}(N, N', a) = i$ *iff* $i = \gamma$ when $\gamma < \omega$, $i = \gamma + 1$ otherwise, where $\gamma = \min\{j\colon$ there is $b \in N, bE_j a\}$ (and $\gamma = \alpha$ if there is no such j).

CONCLUSION 4.10: *For every ordinal β, which is a natural number or a successor ordinal, for some $T = T^*_\beta$, $|T| = |\beta| + \aleph_0$, and $I^a_{\aleph_0}(\aleph_\alpha, T) = \beth_\beta(|\alpha| + \aleph_0)$.*

Proof. For $\beta < \omega$ or $\beta = \alpha + 2$ we use the previous example; for $\beta = \delta + 1$, δ limit, take the sum of models of T^{i*} $(i < \delta)$ with disjoint languages. The computation is easy, but it is a worthwhile exercise for the reader as we are proving in the chapter that every shallow theory T is in some sense similar to $T^*_{\mathrm{Dp}(T)}$.

X.5. Deep theories have many non-isomorphic models

(Hyp) In this section, T is superstable without the dop.

Clearly, if T is deep, we can construct trees like the one we get in 3.2, and try to prove that we get many models. The freedom we have is to determine various dimensions. So when $\alpha = \aleph_\alpha$ this is easy. Generally notice that we have much less freedom than, e.g., in 2.5 (when proving that the dop implies there are many models).

This section is dedicated to the proof of:

THEOREM 5.1: *If T is deep, $\lambda(T) \leqq \aleph_\alpha$, $\aleph_\beta < \aleph_\alpha$, then $I^a_{\aleph_\beta}(\aleph_\alpha, T) = 2^{\aleph_\alpha}$, i.e. T has $2^{\aleph_\alpha}$ non-isomorphic $\mathbf{F}^a_{\aleph_\beta}$-saturated models of cardinality $\aleph_\alpha$.*

Remark. But the lemmas will be used also for shallow theories. We shall concentrate on the case $\aleph_\alpha > \lambda(T)$, where $\lambda(T)$ is the first cardinal in which T is stable.

DEFINITION 5.2: We call $\langle N_\eta, a_\nu : \eta \in I, \nu \in I^+ \rangle$ a *representation if*:

(1) I is a tree with root $\langle\ \rangle$ of height $\leq \omega$, and we let η^- be the unique predecessor of η for $\eta \in I^+ \overset{\text{def}}{=} I - \{\langle\ \rangle\}$ and let $I^- \overset{\text{def}}{=} \{\eta^- : \eta \in I^+\}$.

(2) $N_{\langle\ \rangle}$ is $\mathbf{F}^a_{\aleph_0}$-prime over $\emptyset$.

(3) If $\eta = \nu_1^- = \nu_2^-$, then $p_{\nu_1} = \text{tp}(a_{\nu_1}, N_\eta)$ is regular, and p_{ν_1}, p_{ν_2} are equal or are orthogonal. If all p_ν with $\nu^- = \eta$ are equal, q_η will denote their common value.

(4) For $\eta \in I^+$, N_η is $\mathbf{F}^a_{\aleph_0}$-prime over $N_{(\eta^-)} \cup \{a_\eta\}$.

(5) For $\eta \in I^-$, $\{a_\nu : \nu^- = \eta\}$ is independent over N_η.

(6) If $\eta \in I, \eta^{--}$ is defined, then $\text{tp}(a_\eta, N_{\eta^-})$ is orthogonal to $N_{\eta^{--}}$.

Convention 5.2A: When we say "p is orthogonal to N_{η^-} but $\eta = \langle\ \rangle$", we interpret this as being always true.

Remark. We shall write in short $\langle N_\eta, a_\eta : \eta \in I \rangle$, though we do not need $a_{\langle\ \rangle}$, so $a_{\langle\ \rangle}$ is any element of $N_{\langle\ \rangle}$ or undefined.

DEFINITION 5.3: We say $\langle N_\eta, a_\eta : \eta \in I \rangle$ is an $\mathbf{F}$-representation of M if it is a representation and M is $\mathbf{F}$-primary over $\bigcup_{\eta \in I} N_\eta$. If $\mathbf{F} = \mathbf{F}^a_{\aleph_0}$, we omit it.

LEMMA 5.4: *Let $\langle N_\eta, a_\eta : \eta \in I \rangle$ be a representation, then:*

(1) $\text{tp}_*(\bigcup \{N_\nu : \eta \trianglelefteq \nu\}, \{N_\nu : \text{not } \eta \trianglelefteq \nu\})$ *does not fork over* $N_{(\eta^-)}$ (*for* $\eta \in I^+$).

(2) *For* $\eta \in I^+$, $\text{tp}_*(\bigcup_{\nu \geqslant \eta} N_\nu, N_\eta)$ *is orthogonal to* N_{η^-}.

(3) *For* $\eta, \nu \in I$, p_η, p_ν *are orthogonal or equal* (*and then* $\eta^- = \nu^-$).

(4) *Each* N_η *is* $\mathbf{F}^a_{\aleph_0}$*-prime over* $\emptyset$.

(5) *If* M *is* $\mathbf{F}^a_{\aleph_0}$*-prime over* $\bigcup_{\eta \in I} N_\eta$, $p \in S^m(M)$ *is regular, then* p *is not orthogonal to some* $p' \in S^{m'}(N_\eta)$ *for some* η.

(6) $\langle N_\eta, a_\eta : \eta \in I \rangle$ *represent some model of power* $\lambda(I) + |I|$.

(7) $\text{Dp}(N_{\eta^-}, N_\eta, a_\eta)$ *is at least* $\text{Dp}(\eta, I)$ (*see* Def. 4.3(2)) *even at least* $(-1) + \text{Dp}(\eta, I) + 1$.

Proof. As in the proof of 3.2, or easy using 3.3, 1.7. (1.7 is needed for (4).)

LEMMA 5.5: (1) *Every* $\mathbf{F}^a_{\aleph_0}$*-saturated model has a representation.*

(2) If $\langle N^l_\eta, a^l_\eta : \eta \in I_l \rangle$ $\mathbf{F}^a_{\aleph_0}$*-represents* M_l, $l = 0, 1$, $F : I_0 \to I_1$ *an isomorphism* (*for partially ordered sets so it preserves the level*), $F' : \bigcup_{\eta \in I_0} N^0_\eta \to \bigcup_{\eta \in I_1} N^1_\eta$ *is an elementary mapping, it maps* N^0_η *onto* $N^1_{F(\eta)}$

(equivalently, for each $\eta \in I_0$*,* $F' \upharpoonright N^0_\eta$ *is an elementary mapping onto* $N^1_{F(\eta)}$*,* $\mathrm{Dom}\, F' = \bigcup_{\eta \in I_0} N^0_\eta$*), then* M_0*,* M_1 *are isomorphic.*

(3) *If in (2)* F, F' *are not necessarily onto, then* M_0 *can be elementarily embedded into* M_1*. (We may demand* F' *maps* N^0_η *onto* N^1_η *or at least maps* N^0_η *into* $N^1_{F(\eta)}$ *and* $\mathrm{tp}_\lambda(F'(N^0_\eta), N^1_{F(\eta)^-})$ *does not fork over* $F'(N^0_{\eta^-})$*.*

Proof. (1) This is by 3.2 (and 5.4).

(2), (3) Trivial.

LEMMA 5.6: *For* $\beta > 0$*,* $\langle N_\eta, \bar{a}_\eta : \eta \in I\rangle$ $\mathbf{F}^a_{\aleph_0}$*-represents an* $\mathbf{F}^a_{\aleph_\beta}$*-saturated model* iff *for every* $\eta \in I$*, and regular* $p \in S^m(N_\eta)$ *orthogonal to* N_{η^-} *for at least* $\aleph_\beta$ ν*'s,* p_ν *is not orthogonal to* p*.*

Proof. Easy by 5.4(5), and usual arguments.

Let M be $\mathbf{F}^a_{\aleph_0}$-prime over $\bigcup_\eta N_\eta$, and suppose $A \subseteq M, |A| < \aleph_\beta$, $p \in S(A)$ is omitted. Let $\mathrm{tp}(a, M)$ be a stationarization of p over M; clearly w.l.o.g. p is stationary. By V, 3.9 there are $\{a_i : i < n\}$, independent over M, realizing over it regular types such that $\mathrm{tp}(a, M \cup \{a_i : i < n\})$ is $\mathbf{F}^a_{\aleph_0}$-isolated: w.l.o.g. $\mathrm{tp}(a, A \cup \{a_i : i < n\}$ is $\mathbf{F}^a_{\aleph_0}$-isolated, and $\mathrm{tp}(a_i, M)$ does not fork over A and $\mathrm{tp}(a_i, A)$ is stationary. We now try to define by induction on i, $b_i \in M$ realizing $\mathrm{stp}(a_i, A \cup \{b_j : j < i\})$. We have to fail for some i, so w.l.o.g. p is regular.

By 5.4(4) p is not orthogonal to some regular $q \in S(N_\eta)$, which is not orthogonal to some $p_{\eta\frown\langle i\rangle}$, so we know $p, p_{\eta\frown\langle i\rangle}$ are not orthogonal. By V, 2.3, 2.4, $\dim(p, M) = \dim(p_{\eta\frown\langle i\rangle}, M)$, which is $\geqq \aleph_\beta$ by hypothesis, so we finish.

LEMMA 5.7: *If* T *is deep, then for every tree* I *with root* $\langle\ \rangle$ *and height* $\leqq \omega$*, there is a representation* $\langle N_\eta, a_\eta : \eta \in I\rangle$ *(in fact* q_η *is well defined for every* $\eta \in I^-$*, i.e.* $p_{\eta\frown\langle i\rangle} = q_\eta$*).*

Proof. Easy.

By 4.4(3) there are $\mathbf{F}^a_{\aleph_0}$-saturated N_l, N_{l+1} $\mathbf{F}^a_{\aleph_0}$-prime over $N_l \cup \{a_l\}$, $\mathrm{tp}(a_l, N_l)$ regular, $N_{l+1} <_{N_l} N_{l+2}$. Complete the partial ordering of I to a well-ordering and let $\{\eta_i : i < i^*\}$ be a list of the members of I in increasing order. Let $n(i) = l(\eta_i)$ and $m(i)$ be maximal such that $\eta_i \upharpoonright m(i) \in \{\eta_j : j < i\}$, in fact $\eta_i \upharpoonright m(i) = \eta_{j(i)}$ (for $i > 0$). Now we define by induction on i an elementary mapping F_i. F_0 is the identity on N_0, F_i is an elementary mapping with domain $N_{n(i)}$, extending $F_{j(i)}$ such that $\mathrm{tp}_*(F_i(N_{n(i)}), \bigcup_{j<i} F_j(N_{n(i)}))$ does not fork over $F_{j(i)}(N_{m(i)})$. Let $N_{\eta_i} = F_i(N_{\eta_i})$, $a_{\eta_i} = F_i(a_{n(i)})$, and the checking is easy.

DEFINITION 5.8: For a representation $\langle N_\eta, \bar{a}_\eta : \eta \in I \rangle$ we let:

(1) E (E of the representation) is the equivalence relation on I^+ defined by:

$$\eta E \nu \text{ iff } p_\eta = p_\nu \text{ (iff } p_\eta, p_\nu \text{ are not orthogonal).}$$

We write $p_{\eta/E}$ instead of p_η and $N_{\eta/E}$ instead of N_{η^-}.

(2) The representation is standard if η/E is uncountable for every $\eta \in I$.

(3) We say J is a λ-large subtree of I (relative to E) if for $\eta \in J^+$ there are $s^x_{\eta/E} \subseteq \eta/E$ of cardinality $\leqslant \lambda$ such that $J = \{\eta \in I$: for no $l < l(\eta)$ does $\eta \restriction (l+1) \in s^x_{\eta \restriction (n+1)/E}\}$.

(4) We define similarly "J is a $(<\lambda)$-large subtree of I".

(5) J is a λ-big subtree of I if ($J \subseteq I$, $[\nu < \eta$ in I & $\eta \in J \Rightarrow \nu \in J]$ and) for $\eta \in J$ $|\{\alpha: \eta\frown\langle\alpha\rangle \in I \setminus J\}| \leqslant \lambda$. Similarly for $(<\lambda)$-big.

If we omit λ we mean $\aleph_0$.

CLAIM 5.9: *Suppose $\langle N_\eta, \bar{a}_\eta : \eta \in I \rangle$ is a (standard) representation of M. Then for any stationary type q over a finite subset of M:*

(a) *if q is not orthogonal to p_η, then* $\dim(q, M) = |\eta/E|$,

(b) *if q is orthogonal to every p_η, then* $\dim(q, M) = \aleph_0$.

Proof. Easy.

CLAIM 5.10: *Suppose $\langle N_\eta, \bar{a}_\eta : \eta \in I \rangle$ is a representation of M, $u \subseteq I$ finite and downward closed, $N \subseteq M$ is $\mathbf{F}^a_{\aleph_0}$-primary over $\bigcup_{\eta \in u} N_\eta$, $\rho \in I - u$, $\rho^- \in u$, $\bar{a}\frown\bar{b}\frown\bar{c} \in M$,* $\operatorname{tp}(\bar{a}, N)$ *regular, $\{a, \bar{a}_\rho\}$ depends on N,* $\operatorname{tp}(\bar{b}\frown\bar{c}, N \cup \bar{a})$ *is $\mathbf{F}^a_{\aleph_0}$-isolated, $r =$* $\operatorname{stp}(\bar{c}, \bar{b})$ *is regular orthogonal to N. Then r is orthogonal to p_ν when $\rho \not\leqslant \nu \in I$.*

Proof. Because we can find a representation $\langle N^1_\eta, \bar{a}^1_\eta : \eta \in I_1 \rangle$ of M with:

(i) $N^1_{\langle\rangle} = N$,

(ii) if $\eta \in I_1^+$, $\eta(0) > 0$, N^1_η is an $\mathbf{F}^a_{\aleph_0}$-prime model over $N_\nu \cup N$ for some $\nu \in I - u$, $\nu^- \in u$, $\nu \neq \rho$ (and $\bar{a}^1_\eta = \bar{a}_\nu$), and for every such ν there is such η,

(iii) $N^1_{\langle 0 \rangle}$ is $\mathbf{F}^a_{\aleph_0}$-prime over $N \cup \bar{a}\frown\bar{b}\frown\bar{c}$.

(iv) if $\eta \in I_1$, $\nu \in I$ are as in (ii) then for any sequence τ, $\eta\frown\tau \in I_1 \Leftrightarrow \nu\frown\tau \in I$, and if $\eta\frown\tau\frown\langle i \rangle \in I_1$, then $N^1_{\eta\frown\tau\frown\langle i \rangle}$ is $\mathbf{F}^a_{\aleph_0}$-prime over $N_{\nu\frown\tau\frown\langle i \rangle} \cup N^1_{\eta\frown\tau}$ and $\{N_{\nu\frown\tau\frown\langle i \rangle}, N^1_{\eta\frown\tau}\}$ is independent over $N_{\eta\frown\tau}$.

MAIN LEMMA 5.11: *Suppose for* $l = 1, 2$, $\langle N^l_\eta, \bar{a}^l_\eta; \eta \in I_l \rangle$ *is a standard representation of* M_l, *and* F *is an elementary embedding of* M_1 *into* M_2. *Let* E_l *be as in* 5.8. *Then there is a function* h *from* I_1 *into* I_2 *such that:*

(i) h *is one-to-one* $h(\langle\,\rangle) = \langle\,\rangle$ *and:* $\eta E_1 \rho \Rightarrow h(\eta) E_2 h(\rho)$; s^1_{η/E_1} *is a countable subset of* η/E_1, s^2_{η/E_1} *a countable subset of* $h(\eta)/E_2$,

(ii) *for* $\eta \in I^+_1$, $F(p^1_\eta)$, $p^2_{h(\eta)}$ *are not orthogonal,*

(iii) $\mathrm{Dp}(p^1_{\eta/E_1}) = \mathrm{Dp}(p^2_{h(\eta)/E_2})$,

(iv) *if* $\eta_0 \in I_1$, $\eta_0 \in v_0 \subseteq \eta_0/E_0$, v_0 *uncountable disjoint to* $s^1_{\eta_0/E_1}$, *for* $\eta \in v_0$ *we have* $\bar{d}_\eta {}^\frown \bar{c}_\eta \in N_\eta$, $\mathrm{tp}(\bar{d}_\eta {}^\frown \bar{c}_\eta {}^\frown \bar{a}^1_\eta, N^1_{\eta^-})$ *is the same (for all* $\eta \in v_0$) *and for some* $\alpha_\eta[\eta {}^\frown \langle \alpha_\eta \rangle \in I_1 \wedge p^1_{\eta \frown \langle \alpha_\eta \rangle}$ *is parallel* $\mathrm{tp}(\bar{d}_\eta, \bar{c}_\eta)]$ *then for all but finitely many* $\eta \in v_0$, $h(\eta) \lessdot h(\eta {}^\frown \langle \alpha_\eta \rangle)$.

(v) *If* F *is onto* M_2, h *is onto* I_2, *and* (iv) *holds for* h^{-1}, I_2, I_1 *too.*

Proof. Note that by 5.9, V, 1.16 (and as the representations are standard):

(1) $\dim(p^l_\eta, M_l) = |\eta/E_l|$ for $\eta \in I_l$.

First note that for any $\eta \in I^+_1$, $\dim(p^1_\eta, M_1) \geqslant \aleph_1$ (as the representation is standard). Hence $\dim(F(p^1_\eta), M_2) \geqslant \dim(F(p^1_\eta), F(M_1)) = \dim(p^1_\eta, M_1) \geqslant \dim(p^1_\eta, M_1) \geqslant \aleph_1$.

Hence by 5.9 for some $\nu = \nu_\eta \in I^+_2$, $F(p^1_\eta)$, $p^2_{\nu_\eta}$ are not orthogonal. Clearly, ν_η is not unique, but ν_η/E_2 is unique and depends on η/E_1 only, and

(2) $|\eta/E_1| = \dim(p^1_\eta, M_1) \leqslant \dim(F(p^1_\eta), M_2) = \dim(p_{\nu_\eta}, M_2) = |\nu_\eta/E_2|$.

We shall now define $h_{\eta/E_1} = h \restriction (\eta/E_1) : \eta/E_1 \to \nu_\eta/E_1$ for each E_1-equivalence class separately.

Class I. If $\mathrm{Dp}(p_{\eta/E_1}) = 0$, h_{η/E_1} is any one-to-one function from η/E_1 into ν_η/E_2.

Class II. $\mathrm{Dp}(p^1_{\eta/E_1}) > 0$.

By 7.2, p^1_{η/E_1} is trivial, and also $p^2_{\nu_\eta/E_2}$ is trivial by 7.3(4).

Let $\bar{b}_\eta = \bar{b}^1_{\eta/E_1} \subseteq N^1_{\eta^-}$ be such that p^1_{η/E_1} does not fork over $\bar{b}^1_{\eta/E_1}$, $p^1_{\eta/E_1} \restriction \bar{b}^1_{\eta/E_1}$ is stationary. Let $\bar{b}^2_{\eta/E_2} \subseteq N^2_{\nu_\eta/E_2}$ be such that $p^2_{\nu_\eta/E_2}$ does not fork over $\bar{b}^2_{\eta/E_2}$, $p^2_{\nu_\eta/E_2} \restriction \bar{b}^2_{\eta/E_2}$ is stationary.

Now $F(\bar{b}^1_\eta) \in M_2$, M_2 is $\mathbf{F}^a_{\aleph_0}$-atomic over $\bigcup_{\nu \in I_2} N^2_\nu$. So there is a finite $u_\eta = u_{\eta/E_2} \subseteq I_2$, $\nu^-_\eta \in u_\eta$, u_η closed under initial segments and $\mathrm{tp}_*(F(\bar{b}^1_\eta), \bigcup_{\nu \in u_\eta} N^2_\nu) \vdash \mathrm{tp}(F(\bar{b}^1_\eta), \bigcup_{\nu \in I_2} N^2_\nu)$. Hence $\mathrm{tp}(F(\bar{b}^1_\eta), \bigcup_{\nu \in I_2} N^2_\nu)$ is $\mathbf{F}^a_{\aleph_0}$-isolated over $\bigcup_{\nu \in u_\eta} N^2_\nu$. So there is N, $\mathbf{F}^a_{\aleph_0}$-primary over $\bigcup_{\nu \in u_\eta} N^2_\nu$, $N \subseteq M_2$, M_2 $\mathbf{F}^a_{\aleph_0}$-

primary over $\bigcup_{\nu\in I_2} N^2_\nu \cup N$ and $F(\bar b^1_\eta)\in N$. Let $\boldsymbol{I}^1_\eta = \boldsymbol{I}^1_{\eta/E_1}$ be a maximal set of sequences from $N^1_{\eta^-}$ realizing $p^1_{\eta/E_1}\restriction \bar b^1_\eta$, independent over $\bar b^1_\eta$.

By V, 1.16 it is countable. Easily $\boldsymbol{I}^1_\eta \cup \{\bar a^1_\rho : \rho\in\eta/E_1\}$ is a maximal set of sequences from M_1 realizing $p^1_{\eta/E_1}\restriction \bar b^1_\eta$ independent over $\bar b^1_\eta$. As $\dim(F(p^1_\eta), F(\bar b^1_\eta), N) = \aleph_0$, there is a countable subset $s^a_{\eta/E_1}\subseteq \eta/E_1$, s.t. $\{F(\bar a^1_\rho):\rho\in\eta/E_1 - s^a_\eta\}$ is independent over $(N, F(\bar b^1_{\eta/E_1}))$. Also $\{\bar a^2_\nu : \nu\in\nu_\eta/E_2, \nu\notin u_\eta$ hence $\mathrm{tp}(a_\nu, N)$ does not fork over $(N_{\nu_\eta^-})\}$ is a maximal subset of M_2 of sequences realizing p_{ν_η/E_2} independent over $(N, N_{\nu_\eta^-})$. Also $s^b_{\eta/E_2} = \{\nu : \nu\in\nu_\eta/E_2, \bar a_\nu\in N\}$ is countable. Now as $p^1_\eta, p^2_{\nu_\eta}$ are trivial, the relation "$\{F(\bar a^1_\rho), \bar a^2_\nu\}$ depends on $\bar b^1_\eta\cup\bar b^2_\eta$" for $\rho\in\eta/E_1 - s^a_{\eta/E_1}$, $\nu\in\nu_\eta/E_2 - s^b_{\eta/E_2}$, defines a one-to-one function from $\{\rho:\rho\in\eta/E_1 - s^a_\eta\}$ into $\{\nu:\nu\in\nu_\eta/E_2 - s^b_\eta\}$. Choose $s^1_{\eta/E_1}, s^2_{\eta/E_2}$ s.t. $s^a_{\eta/E_2}\subseteq s^1_{\eta/E_2}\subseteq \eta/E_2$, $s^b_{\eta/E_2}\subseteq s^2_{\eta/E_2}\subseteq \nu_\eta/E_2$, $|s^1_{\eta/E_1}| = |s^2_{\eta/E_2}| = \aleph_0$, and the function maps $\eta/E_1 - s^1_\eta$ onto $\nu_\eta/E_2 - s^2_\eta$. We define h_η on $\eta/E_1 - s^1_\eta$ according to the function mentioned above and $h_\eta\restriction s^1_\eta$ as any one-to-one function from s^1_η into s^2_η. If F is onto M_2 there is no problem making h_{η/E_1} be onto ν_η/E_2.

Now let us check that the demands on h are satisfied.

Demand (i). Holds by its definition (and $h/E_1 : I^+_1/E_1 \to I^+_2/E_2$ is one-to-one as the types p^2_{η/E_2} $(\eta/E_2\in I^+_2/E_2)$ are pairwise orthogonal).

Demand (ii). Trivially, by the choice of h.

Demand (iii). As $p^1_{\eta/E_1}, p^2_{h(\eta)/E_2}$ are regular, not orthogonal, by 4.4(6).

Demand (iv). So suppose $\eta_0\in I^+_1, \eta_0\in v_0\subseteq \eta_0/E_1, v_0\cap s^1_{\eta/E_1} = \emptyset$, v_0 uncountable, for $\eta\in v_0, \bar a_\eta\subseteq \bar d_\eta{}^\frown \bar c_\eta\in N^1_\eta, \mathrm{tp}(\bar a_\eta{}^\frown \bar d_\eta{}^\frown \bar c_\eta, N^1_{\eta^-}) = q$ (i.e. does not depend on $\eta\in v_0$), $\mathrm{tp}(d_\eta, \bar c_\eta)$ is stationary regular and parallel to some $p^1_{\eta^\frown\langle\alpha_\eta\rangle}$, by renaming w.l.o.g. to $p^1_{\eta^\frown\langle 0\rangle}$. Let $v_1 = h''(v_0)$.

We have to show that for all but finitely many $\eta\in v_0$, $h(\eta)\lessdot h(\eta^\frown\langle 0\rangle)$, i.e. (see 5.10) $F(p^1_{\eta^\frown\langle 0\rangle})$ is orthogonal to p^2_ρ when $h(\eta)\not\lessdot\rho$.

Now we proceed as in the definition of h. We choose $\bar b^*\in N^1_{\eta^-}$ such that for $\eta\in v_0$,

$$\mathrm{tp}(\bar a_\eta{}^\frown \bar d_\eta{}^\frown \bar c_\eta, \bar b^*\cup\bar a_\eta)\vdash \mathrm{tp}(\bar a_\eta{}^\frown \bar d_\eta{}^\frown \bar c_\eta, N_{\eta^-}\cup\bar a_\eta)$$

(possible as q exists).

Now we would like to have $F(\bar b^*)\in N$, this can be achieved by increasing u_{η_0/E_1}, to u'_{η_0/E_1} preserving its finiteness. This, naturally, costs us the omission of a finite number of members of v_0, v_1:

$|u'_{\eta_0/E_1} - u_{\eta_0/E_1}|$ by 5.10 we finish (note that $\mathrm{tp}(\bar{d}_\eta, \bar{c}_\eta)(\eta \in v_0)$ are pairwise orthogonal).

Demand (v). The same reasons for the existence (and uniqueness) of $h/E_1 : I_1/E_1 \to I_2/E_2$, show that it is onto.

In order to get demand (iv) for h^{-1}, note that if we allow $\aleph_0$ exceptions, the proof of (iv) dualizes. For the finite version note that the proof of demand (iv) for h^{-1} needs only that s^1_η, s^2_η are large enough in order to guarantee the existence of the "dual" of N_{η/E_1}.

COROLLARY 5.12: *Under the assumption of* 5.11, *the function h there satisfies:*

(1) *For some large (see Definition* 5.8(3)) *subset J_1 of I_1, $h \restriction J_1$ satisfies: it is (a one-to-one function from J_1 into I_2) preserving*

(∗) *for* $\eta, \rho \in J_1, \eta \lessdot \rho \Leftrightarrow h(\eta) \lessdot h(\rho)$.

(2) *If in addition F is onto M_2, for some large subsets J_1, J_2 of I_1, I_2, respectively, $h \restriction J_1$ is a one-to-one function from J_1 onto J_2 satisfying* (∗) *(and even h^{-1} satisfies the parallel demand).*

(3) *For any given cardinal μ, h maps* $I_1^{\geqslant\mu} = \{\eta \in I_1 : |\eta/E_1| \geqslant \mu\}$ *into* $I_2^{\geqslant\mu} = \{\nu \in I_2 : |\nu/E_2| \geqslant \mu\}$. *If in addition χ is regular* $> \aleph_0$, $(\leqslant \lambda(T)$, *naturally) and for every* $\eta \in I_1^{\geqslant\mu}, |\{\eta \frown \langle\alpha\rangle/E_1 : \eta \frown \langle\alpha\rangle \in I_1^{\geqslant\mu}\}| < \chi$, *then for some* $(<\chi)$*-large subset* $I_1^{\mu,\chi}$ *of* $I_1^{\geqslant\mu}$, $h \restriction I_1^{\mu,\chi}$ *satisfies* (∗).

(4) *If F is onto M_2, h maps $I_1^{\geqslant\mu}$ onto $I_2^{\geqslant\mu}$, and for χ as above w.l.o.g. maps $I_1^{\mu,\chi}$ onto a $(<\chi)$-large subset of $I_2^{\geqslant\mu}$.*

5.13 *Proof of 5.1.* So $\aleph_\alpha \geqslant \lambda(T), \aleph_\alpha > \aleph_\beta$.

By 5.7 (and 4.5) there is a representation $\langle N^0_\eta, a^0_\eta : \eta \in I^0 \rangle$ where $I^0 = {}^{\omega>}\aleph_\alpha$, q_η is well defined. By 4.6, w.l.o.g. $q_{\langle\rangle}$ does not fork over $\emptyset$.

As we want $\mathbf{F}^a_{\aleph_\beta}$-saturated models, by our system of notation (which is not convenient in this case) we have to increase I^0 to take care of this. So we can find $J, I \subseteq J$, and N_η, a_η for $\eta \in J - I^0$ such that:

(1) $\langle N_\eta, a_\eta : \eta \in J \rangle$ is a representation,

(2) for each $\eta \in J, \eta/E_J - I$ has cardinality $\aleph_\beta$ exactly,

(3) each regular $p \in S^m(N_\eta), \eta \in J$, is not orthogonal to some $p_{\eta \frown \langle\alpha\rangle}$.

For each $\xi < \aleph_\alpha$ choose I_ξ such that

(a) $\langle\xi\rangle \in I_\xi \subseteq I$,

(b) if $\eta \in I_\xi - \{\langle\xi\rangle, \langle\,\rangle\}$, then $\eta/E_I \subseteq I_\xi$,

(c) $\mathrm{Dp}_{\aleph_\alpha}(\langle\xi\rangle, I_\xi) = \xi$,
(d) $\eta \in I^+_\xi$ implies $\eta \upharpoonright 1 = \langle\xi\rangle$.
(e) if $\eta \in I^-_\xi$ then $|\{\alpha: \eta^1\langle\alpha\rangle \in I_\xi\}| = \aleph_\alpha$.

For any set $S \subseteq \aleph_\alpha$ let

$$I_S = \bigcup_{\xi \in S} I_\xi,$$

$$J_S = \{\eta \in J : \eta \in I_S \text{ or for some } m, \eta \upharpoonright m \in I_S, \eta \upharpoonright (m+1) \in J - I\},$$

$$M_S \text{ be a model represented by } \langle N_\eta, a_\eta : a \in J_S\rangle.$$

By (1), (2), (3) above, and 5.6, M_S is $\mathbf{F}^a_{\aleph_\beta}$-saturated and of cardinality $\aleph_\alpha$. Suppose $S(1), S(2) \subseteq \aleph_\alpha$, F is an isomorphism from $M_{S(1)}$ onto $M_{S(2)}$. Let $h: J_{S(1)} \to J_{S(2)}$ be as in 5.11.

Clearly any regular type over $M_{S(2)}$ of dimension $> \aleph_\beta$ is not orthogonal to some $p_\eta, \eta \in I_{S(2)}$. So necessarily h maps $I_{S(1)}$ into $I_{S(2)}$. Similarly, h^{-1} maps $I_{S(2)}$ into $I_{S(1)}$. Hence w.l.o.g. it maps $\{\langle\xi\rangle : \xi \in S(1)\}$ onto $\{\langle\xi\rangle : \xi \in S(2)\}$ (as $q_{\langle\rangle}$ does not fork over $\emptyset$). Now use 5.12.

THEOREM 5.14: *If T is deep and $\aleph_\alpha$ is $\geqslant \aleph_{\beta+1} + \lambda(T)$ but smaller than the first beautiful cardinal, then*

$$IE^a_{\aleph_\beta}(\aleph_\alpha, T) = 2^{\aleph_\alpha}.$$

Proof. We use the theory of κ-bqo as developed in [Sh 82]. Let $\mathcal{J}^{-2}(\omega, =)$ be the class of (I,f), I a well-founded tree (with root, no ω-branch and at most ω-levels, $f: I \to \omega$) ordered by $(I_1,f_1) \leqslant (I_2,f_2)$ *if and only if* the first is embeddable into the second, i.e. there is a function $G: I_1 \to I_2$, $I_1 \models s < t$ implies: $I_2 \models G(s) < G(t)$ and $f_1(s) = f_2(G(s))$.

By [Sh 82] there are $2^{\aleph_\alpha}$ pairs (I,f) each I of power $\aleph_\alpha$, no pair embeddable into another and w.l.o.g. each $\subseteq {}^{\omega>}(\aleph_\alpha)$ (use [Sh 82] 4.10, and translate it into trees as in the proof of [Sh 82, 5.7] for example). W.l.o.g. for each such I there are sequences in I of arbitrarily large (finite) length.

So let $\{(I^0_i, f^0_i) : i < 2^{\aleph_\alpha}\}$ be this family, so $I^0_i \subseteq {}^{\omega>}(\aleph_\alpha)$. Let $g: \aleph_\alpha \to \aleph_\alpha$ be such that $(\forall i < \aleph_\alpha)|\{j < \aleph_\alpha : g(j) = i\}| = \aleph_\alpha$, $I^1_i \stackrel{\text{def}}{=} \{\eta \in {}^{\omega>}(\aleph_\alpha) : g(\eta) \stackrel{\text{def}}{=} \langle g(\eta(l)) : l < l(\eta)\rangle \in I^0_i\}$, $f^1_i(\eta) = f^0_i(g(\eta))$. We leave the case $\alpha = 1$ to the reader, so by obvious monotonicity w.l.o.g. $\beta \geqslant 1$. Let for $i < 2^{\aleph_\alpha}$, $I_i = \{\eta : \eta \in I^1_i$, or for some m, $\eta \upharpoonright m \in I^1_i, \aleph_\alpha \leqslant \eta(m) < \aleph_\alpha + \aleph_\beta$, and for every $l > m$ if $(\exists k)[l = k^2 + (f^1_i(\eta \upharpoonright m) + m + 2)^2 + m < l(\eta)]$, then $\eta(l) < \aleph_\alpha$, otherwise $\eta(l) < \aleph_\beta\}$.

Let $I = {}^{\omega>}(\aleph_\alpha + \aleph_\beta)$, and by 5.7 there is a representation $\langle N_\eta, a_\eta : \eta \in I \rangle$ such that q_η is well defined. We choose $J, I \subseteq J, |J| = \aleph_\alpha$ and N_η, a_η for $\eta \in J - I$ such that

(1) $\langle N_\eta, a_\eta : \eta \in J \rangle$ is a representation,

(2) for each $\eta \in J - I$, η / E_J has cardinality $\aleph_\beta$ exactly,

(3) for each regular $p \in S^m(N_\eta), \eta \in J$, is not orthogonal to some $p_{\eta \frown \langle \alpha \rangle}$.

Let for $i < 2^{\aleph_\alpha}$, $J_i = \{\eta \in J : \eta \in I_i$ or for some $l, \eta \upharpoonright l \in I_i, \eta \upharpoonright (l+1) \notin I\}$. Let M_i be represented by $\langle N_\eta, a_\eta : \eta \in J_i \rangle$. Clearly each M_i is $\mathbf{F}^a_{\aleph_\beta}$-saturated and has cardinality $\aleph_\alpha$. So suppose $i \neq j < 2^{\aleph_\alpha}$, F is an elementary embedding of M_i into M_j. We shall use 5.11, so we have a function h as described there. Now applying (iv) on $h \upharpoonright I_i$ we can conclude that for some $\aleph_0$-large subset I_i^a of I_i, $h \upharpoonright I_i^a$ preserves $\lessdot$. By the choice of I_i (and $J_j - I_j$), necessarily h maps I_i^a into I_j.

We shall show later that w.l.o.g.

$\oplus$ for some increasing function $k : \omega \to \omega$, for $\eta \in I_i^a, l(h(\eta)) = k(l(\eta))$.

Now by the way we have defined I_i, I_j (and as I_i^1, I_j^1 are well founded)

$(*)_0$ k is the identity.

[Why? For any $\eta, \nu \in {}^\omega(\aleph_\alpha + \aleph_\beta)$ s.t. $\bigwedge_{l<\omega} \eta \upharpoonright l \in I_i^a, h(\eta \upharpoonright l) \in \{\nu \upharpoonright m : m < \omega\}$, then $\max\{m : \eta \upharpoonright m \in I_i^1\} = \max\{m : \nu \upharpoonright m \in I_j^1\}$ (by the choice of $I - I_i^1$). Hence for $\eta \in I_i^a \cap I_i^1, l(h(\eta)) = l(\eta)$. But with an assumption on the (I, f)'s we finish.]

Now as for every $\eta \in I_i^1 - \{\langle\ \rangle\}, |\{\nu \in I_i^1 : \nu^- = \eta^-\}| = \aleph_\alpha$ and similarly for I_j^1, clearly by the definition of I_i, I_j:

$(*)_1$ h maps $I_i^a \cap I_i^1$ into I_j^1, and

$(*)_2$ for $\eta \in I_i^a \cap I_i^1, f_i^1(\eta) = f_j^1(h(\eta))$.

Now as $I_i^a \cap I_i^1$ is an $\aleph_0$-large subtree of I_i^1, clearly there is a function g^* from I_i^0 into $I_i^a \cap I_i^1$ preserving level and $\lessdot$ and $=$, such that for $\eta \in I_i^0, g(g^*(\eta)) = \eta$. Now the composition $g^+ \stackrel{\text{def}}{=} (ghg^*)$ is a function from I_i^0 into I_j^0 preserving $\lessdot$ (but is not necessarily one-to-one) for $\eta \in I_i : f_j(g^+(\eta)) = (f_j g)\, h(g^*(\eta)) = f_j^1(h(g^*(\eta))) = f_i^1(g^*(\eta)) = f_i(g(g^*(\eta))) = f_i(\eta)$, contradicting the choice of the I_i^0's.

We still have to prove that w.l.o.g. $\oplus$ holds.

We shall prove later:

CLAIM 5.15: *If T is deep, then we can find $\langle N_l, a_l : l < \omega \rangle$ as in 4.4(3) such that:*

$(*)_1$ *for every $l > 0$ for some φ_l and $\bar{b}_l \in N_0$, $\models \varphi_l[a_l, \bar{b}_l]$ and for every*

$m > l$, $\operatorname{tp}(a_m, N_l)$ is orthogonal to every type to which $\varphi_l(x, \bar{b}_l)$ belong,

$(*)_2$ *$\operatorname{tp}(a_0, N_0)$ is not orthogonal to $\emptyset$, (if we waive regularity, it does not fork over $\emptyset$).*

Completion of the proof of 5.14. In the proof of 5.7, we can use the $\langle N_l, a_l : l < \omega\rangle$ from 5.15 (instead of 4.4(3)). Also in our present proof we could have used $\langle N_\eta : \eta \in {}^{\omega>}(\aleph_\alpha + \aleph_\beta)\rangle$ derived in this way. So there are formulas $\varphi_l(x, \bar{b}_l)$ $(l < \omega)$ $\bar{b}_l \in N_{\langle\rangle}$ such that for each $i < 2^{\aleph_\alpha}$, $\eta \in I_i^+$,

$$\varphi_{l(\eta)}(x, \bar{b}_{l(\eta)}) \in q_\eta,$$

$l(\eta) < m < \omega \Rightarrow q_\eta$ is orthogonal to every type to which $\varphi_{l(\eta)}(x, \bar{b}_{l(\eta)})$ belongs. W.l.o.g. $\bigcup_l \bar{b}_l \subseteq N_{\langle\rangle}$.

Now look at the proof of 5.11 (applies to $F : M_i \to M_j$).

So for some countable $u_1 \subseteq J_i$, $u_2 \in J_j$, closed under initial segments, and $N \subseteq M_j$, $\mathbf{F}^a_{\aleph_0}$-primary over $\bigcup_{\eta \in u_2} N_\eta$ the following hold:

(a) $F(\bigcup_{l<\omega} \bar{b}_l) \subseteq N$,

(b) $\operatorname{tp}_*(\langle \bar{b}_l : l < \omega\rangle, F(M_i))$ does not fork over $F(A) \subseteq N \subseteq M_j$, and $A \subseteq M_i$, $\bigcup_l \bar{b}_l \subseteq A$, A countable.

As $q_{\langle\rangle}$ does not fork over $\emptyset$, h maps $\{\eta \in J_i : l(\eta) = 1\}$ into $\{\eta \in J_j : l(\eta) = 1\}$. So w.l.o.g. $[\eta \in I_i^a \Rightarrow h(\eta) \restriction 1 \notin u_2$ and $\eta \restriction 1 \notin u_1]$. Let us prove that $h \restriction I_i^a$ preserves equality of the level. Let $\eta \in I_i^a$. If $l(\eta) = 1$, we have already noted this.

Now:

Observation. If $\eta_1, \eta_2 \in I_i^a$, $l(\eta_1) = l(\eta_2) > 1$, then for some automorphism H of $\mathfrak{C}$ over $N_{\langle\rangle}$, $H(p_{\eta_1})$, p_{η_2} are not orthogonal.

Hence,

Observation. If $\eta_1, \eta_2 \in I_i^a$, $l(\eta_1) = l(\eta_2) > 1$, then

(1) for some automorphism H of $\mathfrak{C}$ over A, $H(p_{\eta_1})$, p_{η_2} are not orthogonal,

(2) for some automorphism H of $\mathfrak{C}$ over $F(A) \cup \bigcup_{l<\omega} \bar{b}_l$, $H(F(p_{\eta_1}))$, $H(F(p_{\eta_2}))$ are not orthogonal.

However:

Observation. If $\nu_1, \nu_2 \in I_j$, $l(\nu_1) \neq l(\nu_2) > 1$, then for no automorphism H of $\mathfrak{C}$ over $\bigcup_{l<\omega} \bar{b}_l$, are $H(p_{\nu_1})$, p_{ν_2} not orthogonal.

As non-orthogonality is an equivalence relation, we finish.

Proof of 5.15. Choose $\varphi(x,\bar b)\langle N_l, a_l : l<\omega\rangle$ such that

(1) $\langle N_l, a_l : l<\omega\rangle$ is as in 4.4(3).

(2) $\models\varphi[a_l,\bar b]$ for $l<\omega$, and $\bar b\in N_0$.

(3) Under (1)+(2), $\alpha^*=R[\varphi(x,\bar b),L,\infty]$ is minimal.

By 4.4(3) we can find such $\varphi(x,\bar b), N_l, a_l$. We can prove that N_ω is $\mathbf{F}^a_{\aleph_0}$-primary over $N_0\cup a_0$ (see Exercises 5.1, 5.2), hence for $c\in N_\omega$, $\mathrm{tp}(c,N_0\cup a_0)$ forks over N_0, hence $t(c,N_1)$ forks over N_0. So for $l>0$, $R[\mathrm{tp}(a_l,N_1),l,\infty]<\alpha^*$, so for some $\bar b_l\in N_1$, and $\varphi_l:\models\varphi_l[a_l,\bar b_l]$ and $\alpha^*>R[\varphi_l(x,\bar b_l),L,\infty]$. Let $S_l=\{m: l<m<\omega$, and $\mathrm{tp}(a_m,N_m)$ is not orthogonal to some (regular) type to which $\varphi_l(x,\bar b_l)$ belongs$\}$. Suppose S_l is infinite. Let $S_l=\{n(l,i):i<\omega\}$, $n(l,i)<n(l,i+1)$. Now let $N^*_m=N_{n(l,m)}$, and for each m, as N^*_m is $\mathbf{F}^a_{\aleph_0}$-saturated, $\bar b_l\in N^*_m$ there is a regular $r_m\in S(N^*_m)$ to which $\varphi_l(x,\bar b_l)$ belongs and r_m is not orthogonal to $\mathrm{tp}(a_{n(l,m)},N^*_m)$. So r_m is realized by some $a^*_m\in N_{n(l,m)+1}$. Now $\varphi_l(x,\bar b_l),\langle N^*_m,a^*_m:m<\omega\rangle$ easily satisfies (1)+(2) (use Exercise 5.1, Claim V, 3.3), a contradiction to the choice of α^*.

So each S_l is finite. So we can find $\{l(k):k<\omega\}, 0<l(k)<l(k+1)<\omega, S_{l(k)}\subseteq\{i:i<l(k+1)\}$. Let $N^*_k=N_{l(k)}, a^*_k=a_{l(k)}$, and $\langle N^*_k,a^*_k: k<\omega\rangle,\varphi_{l(k)}(x,\bar b_{l(k)})$ are almost as required. For the requirement "$\mathrm{tp}(a_0,N_0)$ does not fork over N^*_0", prove as in 4.6 (and then use Exercise 5.1).

EXERCISE 5.1: If N_0 is $\mathbf{F}^a_{\aleph_0}$-prime over $\emptyset$, N_{l+1} $\mathbf{F}^a_{\aleph_0}$-prime over $N_l\cup\bar a_l$, for $l=0,1$, and $\mathrm{tp}(\bar a_1,N_1)$ is orthogonal to N_0, then N_2 is $\mathbf{F}^a_{\aleph_0}$-prime over $N_0\cup a_0$.

EXERCISE 5.2: If N_i is $\mathbf{F}^a_{\aleph_\alpha}$-prime over A for $i<i(*)$, N_i is increasing and $i(*)<\aleph_{\alpha+1}$, then $\bigcup_i N_i$ is $\mathbf{F}^a_{\aleph_\alpha}$-prime over A. (Hint: Use the characterization theorem IV, 4.21.)

EXERCISE 5.3: If N_0 is $\mathbf{F}^a_{\aleph_\alpha}$-prime over A, $\langle N_\alpha:\alpha\leqslant\alpha(*)\rangle$ increasing continuous, $N_{\alpha+1}$, $\mathbf{F}^a_{\aleph_\gamma}$-prime over $N_\alpha\cup\bar a_\alpha$, and for each α, $\{\beta:\mathrm{tp}(\bar a_\alpha,N_\alpha)$ not orthogonal to $\mathrm{tp}(\bar a_\beta,N_\beta)\}$ has cardinality $\leqslant\aleph_\gamma$, then $N_{\alpha(*)}$ is $\mathbf{F}^a_{\aleph_\gamma}$-prime over A. (Hint: see [Sh 88a].)

EXERCISE 5.3A: Show that in Lemma 5.5(1) we can add:

(8) *The representation is standard.*

(Hint: Use Exercise 5.3.)

THEOREM 5.16: (1) *If T is shallow, then $IE^a_{\aleph_\varepsilon}(\aleph_\alpha,T)\leqslant\beth_{\delta(|T|)}$ and even $IE^a_{\aleph_0}(T)<\beth_{\delta(|T|)}$, i.e.* $\sup\{|K|$: *K a family of $\mathbf{F}^a_{\aleph_0}$-saturated models, no one elementarily embeddable into another*$\}<\beth_{\delta(|T|)}$.

(2) *If T is deep, κ_1 the first beautiful cardinal $>|T|$, then $IE^a_{\aleph_\beta}(\aleph_\alpha, T) \leqslant \kappa_1^-$ and even $IE^a_{\aleph_0}(T) \leqslant \kappa_1^-$.*

(3) *If T is deep, and $|T|$ smaller than the first beautiful cardinal κ_0, $\aleph_\beta < \kappa_0$, then for $\beta < \alpha, IE^a_{\aleph_\beta}(\aleph_\alpha, T) = \mathrm{Min}\{\kappa_0^-, 2^{\aleph_\alpha}\}$*

$$IE^a_{\aleph_\beta}(T) = \kappa_0^-.$$

Proof. By 5.5(3) we can translate the problem of proving the upper bound to problems on embeddability of labelled trees. That is, if $\langle N_\eta, a_\eta : \eta \in I\rangle$ represents M, we can define model M_η such that

(a) M_η is a model of T with universe $\subseteq \{\alpha : \alpha < 2^{|T|}\} \times (1 + l(\eta))$,

(b) there are isomorphisms f_η from N_η onto $M_\eta, f_{\eta\restriction l} \subseteq f_\eta$ for $l < l(\eta)$.

So the function $g, g(\eta) \to M_\eta$ $(\eta \in I)$ has a range of cardinality $2^{2^{|T|}}$ (really, $2^{|T|}$ suffices). [Now if for $l = 1,2$, M_l has representation $\langle N^l_\eta, a_\eta : \eta \in I_l\rangle$ and above we have chosen $M^l_\eta, (\eta \in I_l), g_l$ as above, then

(*) if there is a function h from I_1 into I_2, one-to-one, preserving $\lessdot$, $\not\lessdot$, (and the level) $g_2(h(\eta)) = g_1(\eta)$, then M_1 can be elementarily embedded into M_2.]

For 5.16(1), see below (i.e. 5.16C implies the result by 5.5(3), 4.4(4)). For 5.16(2), (3) use what we have proved and 5.14 (for $\aleph_\beta \geqslant \kappa_0$, note that the proof of 5.14 still works).

Now the following problem on λ-well-ordering should have been dealt with in [Sh 82], but we forgot. We give below just what is needed here.

DEFINITION 5.16A: (1) A quasi-order Q (i.e. a pair $(|Q|, \leqslant), \leqslant$ a transitive, reflexive relation (but maybe $x \leqslant y \leqslant x, x \neq y$) is λ-well-ordered if for any $q_i \in Q$, $(i < \lambda)$ for some $i < j, q_i \leqslant q_j$.

(2) Q is λ-narrow if for any $q_i \in Q$ $(i < \lambda)$ for some $i \neq j, q_i \leqslant q_j$.

(3) $\mathscr{P}(Q)$ is the family of subsets of Q.

(4) For a given Q, $\leqslant^0$ is the following quasi-order on $\mathscr{P}(Q)$: $A \leqslant^0 B$ *if and only if* for some function f from A into B, $(\forall q \in A)\, q \leqslant f(q)$.

(5) For a given Q, $\leqslant^1$ is the following quasi-order on $\mathscr{P}(Q)$: $A \leqslant^0 B$ *if and only if for some one-to-one function f from A into B,* $(\forall q \in A)\, q \leqslant f(q)$.

DEFINITION 5.16B: Let $\mathscr{I}_{\leqslant\alpha}(Q)$ be the class of pairs (I, f), I a tree (with root and $\leqslant\omega$ levels) and depth $\mathrm{Dp}_1(I) \leqslant \alpha$, f a function from I into $\boldsymbol{Q}$, quasi-ordered by embeddability, where F is an embedding of (I_1, f_1) into (I_2, f_2) if F is a one-to-one function from I_1 into I_2, preserving level, $\not\lessdot$, $\lessdot$, depth and $f_1(\eta) \leqslant f_2(F(\eta))$ for $\eta \in I_1$.

THEOREM 5.16C: *If* $\|\boldsymbol{Q}\| = 2^\lambda, \alpha > 0$ *an ordinal, then* $\mathscr{I}_{<\alpha}(\boldsymbol{Q})$ *is* $\beth_\alpha(\lambda)^+$*-well-ordered.*

We shall prove 5.16C and 5.16D later.

LEMMA 5.16D: *For any cardinal* κ, *and ordinal* ζ, *letting* $\theta = 2^\kappa$, *there is a function* F *from* $\mathscr{P}({}^\kappa\zeta)$ *to* ${}^\theta\zeta$ *such that for* $A, B \in \mathscr{P}({}^\kappa\zeta)$, $A \leqslant^1 B$ *if* $F(A) \leqslant F(B)$.

Remark 5.16E. ${}^\theta\zeta, {}^\kappa\zeta$ are ordered coordinatewise.

DEFINITION 5.16F: We say $A \subseteq {}^\kappa\zeta$ is of *kind* $(w, \bar{\gamma}, \mu)$ if:

(i) $\bar{\gamma}$ is a sequence of ordinals $\leqslant\zeta$ of length κ,

(ii) w is a subset of κ,

(iii) for every $\bar{\alpha} = \langle \alpha_i : i < \kappa \rangle \in A$, $[i \notin w \wedge i < \kappa \Rightarrow \alpha_i = \gamma_i][i \in w \Rightarrow \alpha_i < \gamma_i]$,

(iv) if $\gamma_i^* < \gamma_i$ for $i \in w$, then $\{\bar{\alpha} \in A : \gamma_i^* \leqslant \alpha_i < \gamma_i \text{ for } i \in w\}$ has cardinality $|A|$,

(v) $|A| = \mu$.

DEFINITION 5.16G: We call $A \subseteq {}^\kappa\zeta$ simple if it is of some kind, we call it $\leqslant\lambda$-simple if it is the union of $\leqslant\lambda$ pairwise disjoint simple subsets.

Observation 5.16H. If $A_1, A_2 \in \mathscr{P}({}^\kappa\zeta)$, A_l of kind $(w^l, \bar{\gamma}^l, \mu_l)$ and $w^1 = w^2, \mu_1 \leqslant \mu_2, \bar{\gamma}^1 \leqslant \bar{\gamma}^2$, then $A_1 \leqslant^1 A_2$.

CLAIM 5.16I: *Any* $A \subseteq {}^\kappa\epsilon$ *is* $(\leqslant 2^\kappa)$*-simple.*

Proof. Let χ be regular large enough, and for $\zeta < \kappa^+, N_\zeta$ be an elementary submodel of $(H(\chi), \in)$ such that $\{\xi : \xi \leqslant 2^\kappa\} \subseteq N_\zeta$, $\|N_\zeta\| = 2^\kappa$, $A \in N_\zeta$, $\zeta \subseteq N_\zeta$, $\kappa \in N_\zeta$ and $[a \subseteq N \cap |a| \leqslant \kappa \Rightarrow \alpha \in N_\zeta]$, and $\xi < \zeta \Rightarrow N_\xi \in N_\zeta$ and let $N = N_{\kappa^+} \stackrel{\text{def}}{=} \bigcup_\zeta N_\zeta$.

Define an equivalence relation E_ζ on A

$$\bar{\alpha} E_\zeta \bar{\beta} \text{ if and only if } \bigwedge_{i<\kappa} (\forall \xi \in N_\zeta)(\alpha_i < \xi \equiv \beta_i < \xi \wedge \alpha_i = \xi \equiv \beta_i = \xi).$$

Clearly the number of E_ζ-equivalence classes is $\leqslant \|N\|^\kappa = 2^\kappa$, so for some $\bar{a}^* \in A$ for no $\zeta < \kappa^+$ is $\bar{a}^*/E_\zeta$ simple (otherwise $\{\bar{a}/E_\zeta : \bar{a} \in A, \bar{a}/E_\zeta$ is simple but for no $\xi < \zeta$ is $\bar{a}/E_\xi$ simple$\}$ a partition of A to $\leqslant 2^\kappa$ sets, each of them simple).

Let $\gamma_i = \text{Min}\{\xi \in N : \alpha_i^* \leqslant \xi\}$

$$\bar{\gamma} = \langle \gamma_i : i < \kappa \rangle,$$

$$w = \{i < \kappa : \alpha_i^* \notin N\}.$$

Let for $\bar{\beta} < \bar{\gamma}$, $A_{1,\zeta,\bar{\beta}} = \{\bar{a} \in \bar{a}^*/E_\zeta : \bar{\beta} \restriction w < \bar{a} \restriction w \leqslant \bar{\gamma} \restriction w\}$.

We shall prove that for some ζ $\bar{a}^*/E_\zeta$, is of kind $(w, \bar{\gamma}, |\bar{a}^*/E_\zeta|)$. The only problem is (iv), so let $\bar{\gamma}_\zeta^* \in {}^\kappa\epsilon$, $\bar{\gamma}_\zeta^* \restriction w < \bar{\gamma} \restriction w$ and suppose $B_{1,\zeta,\bar{\gamma}^*} = \{\bar{a} \in \bar{a}^*/E_\zeta : \bar{\gamma}_\zeta^* \restriction w < \bar{a} \restriction w\}$ has cardinality $< |\bar{a}^*/E_\zeta|$.

By the choice of N, as $\{\gamma_i : i < \kappa\} \subseteq N$, also for some $\zeta(0) < \kappa^+$, $\bar{\gamma} \in N_{\zeta(0)}$ and $A \in N_{\zeta(0)}$, hence $\{\bar{a}^*/E_{\zeta(0)+\zeta}, A, \bar{\gamma}\} \subseteq N_{\zeta(0)+\zeta+1} \subseteq N$. So there is $\bar{\gamma}^*$ with the above properties in N, hence in $N_{\zeta(1)}$ for some $\zeta(1), \zeta(0)+\zeta+1 < \zeta(1) < \kappa^+$ and $|\bar{a}^*/E_{\zeta(1)}| \leqslant |B_{1,\zeta,\gamma_\zeta}| < |\bar{a}^*/E_{\zeta(0)+\zeta}|$, so if we have chosen $\zeta < \kappa^+$ with $|\bar{a}^*/E_\zeta|$ minimal, also (iv) holds.

Proof of Lemma 5.16D. Straightforward.

Proof of Theorem 5.16C. By induction on α for α non-zero, we prove that there is a function $F : \mathscr{I}_{<\alpha}(Q) \to {}^{\lambda(\alpha)}\zeta$ (for some ordinal ζ) where $\lambda(\alpha) = \beth_{1+\alpha}(\lambda)$ such that

$$(I_1, f_1) \leqslant (I_2, f_2) \quad \text{if } F(I_1, f_1) \leqslant F(I_2, f_2).$$

For $\alpha = 1$, since in ${}^\lambda 2$ there are 2^λ pairwise incomparable elements, this is easy. For $\alpha = \beta + 1$, β non-zero use the induction hypothesis and Lemma 5.16D (the reduction to $(\mathscr{P}(\mathscr{I}_{\leqslant\beta}), \leqslant^1)$ is easy). For $\alpha = \beta$, β limit, note that to deal with $(\mathscr{P}(\bigcup_{\gamma<\beta} \mathscr{I}_{\leqslant\gamma}), \leqslant^1)$, we can reduce it to problems on $(\mathscr{P}(\mathscr{I}_{\leqslant\gamma}), \leqslant^1)$ for $\gamma < \beta$.

EXERCISE 5.4: (1) Suppose T is countable, superstable without the dop. Prove $\omega^2 \leqslant \text{Dp}(T) < \infty$ implies $IE^a_{\aleph_\beta}(T) = \beth_{\text{Dp}(T)}$.

(2) The natural upper bounds when T is not necessarily countable and/or $\text{Dp}(T) < \omega^2$.

EXERCISE 5.5: Show that for $\alpha < \omega^2$, there is a countable superstable T without the dop of depth $\omega + \alpha + 1$, and $IE^a_{\aleph_0}(T) = \beth_{\alpha+1}$.

5.17. Discussion on invariants. So the decomposition $\langle N_\eta, a_\eta : \eta \in I\rangle$ totally describes a model M which it represents; however, it is not read easily nor uniquely from (the isomorphism type of) M.

We can try to correct this somewhat. A technical point.

RE-DEFINITION: In the definition of a representation we replace (2), (4) by:

(2)′ $\|N_\eta\| \leqslant \lambda(T)$,

(4)′ $\mathrm{tp}(N_\eta, N_{\eta^-} \cup a_\eta)$ almost orthogonal to N_{η^-}.

DEFINITION 5.17A: (1) A representation $\langle N^0_\eta, a^0_\eta : \eta \in I\rangle$ is derived from the representation $\langle N^1_\eta, a^1_\eta : \eta \in J\rangle$, if $J \subseteq I$, $a^0_\eta = a^1_\eta$ for $\eta \in J$, there are for $\eta \in J$, $s_\eta \subseteq \{\nu : \eta \lessdot \nu \in I\}$, $[\nu^\frown\langle\alpha\rangle \in s_\eta \wedge \nu^\frown\langle\alpha\rangle \neq \eta \Rightarrow \nu \in s_\eta]$, $|s_\eta| \leqslant \lambda(T)$, $\langle\ \rangle \in J$, $N^0_{\langle\rangle}$ is $\mathbf{F}^a_{\aleph_0}$-prime over $\bigcup_{\rho \in s_{\langle\rangle}} N_\rho$; for $\eta \in J - \{\langle\ \rangle\}$, let ν be $\eta \restriction l_\eta \in J$ for a maximal $l_\eta < l(\eta)$, and N^0_η is $\mathbf{F}^a_{\aleph_0}$-prime over $N^0_\nu \cup \{N^1_\rho : \rho \in s_\eta\}$, and $I = \bigcup\{s_\eta : \eta \in J\}$. (We could rename the elements of J so that it is closed under initial segments. Of course, $\eta \in J \Rightarrow s_\eta \cap J = \{\eta\}$.)

(2) Two representations $\langle N^l_\eta, a^l_\eta : \eta \in I_l\rangle$ for $l = 1, 2$ are isomorphic if there is an isomorphism F from I_1 onto I_2 and an isomorphism F_η (for $\eta \in I_1$) from N^1_η onto $N^2_{F(\eta)}$, such that $I_1 \models \nu < \eta \Rightarrow F_\eta \subseteq F_\nu$.

DEFINITION 5.17B: Two representations will be called equivalent if they have isomorphic-derived representations.

We can prove

LEMMA 5.17C: *Two representations represent isomorphic models if and only if they are equivalent.*

However, is the equivalence class of representation of M a reasonable invariant?

We may try to replace $p_\eta(\eta \in I, l(\eta) = 1)$ which w.l.o.g. does not fork over $\emptyset$, by $\mathrm{cl}^3(p_\eta)$-simple types over $\emptyset$. Now for types of depth zero we need just their dimension, types of depth >0 are trivial, so we can represent a_η by an equivalence class for the natural dependency relation on the simple type (this is not so far from IX, 2.3's proof). However, the equivalence relation is not necessarily first order, so we do not know to continue this. *A priori*, we can use $\mathfrak{C}^{\mathrm{eq}}$ with a non-first-order equivalence relation, but we do not know enough corresponding theory to make this work, and get, as desired, instead $\langle N_\eta, a_\eta : \eta \in I\rangle$, a non-forking tree of imaginary elements such that the model is prime over it and is quite unique.

Note that in the representation $\langle N_\eta, a_\eta : \eta \in I\rangle$ not all types, e.g. $p_{\langle 8,t\rangle}$, are necessarily there in the tree. We could have replaced $N_{\langle\,\rangle}$ by $N_{\langle 8\rangle}$, and we could have replaced $N_{\langle 8,t\rangle}$ by any $\mathbf{F}^a_{\aleph_0}$-saturated submodel including $N_{\langle 8\rangle} \cup \bar{a}_{\langle 8,t\rangle}$. Moreover, if we try to continue to shrink $N_{\langle 8,t\rangle}$ the process is not necessarily well founded (when $\mathrm{Dp}(T) \geqslant \omega$). We may be tempted to take an inverse limit of the possible representation of a model, but it is not clear we get anywhere.

We can use standard representations (see Exercise 5.3A) and represent p_η's by a canonical base, but we still have non-orthogonal, non-parallel p_η's.

For a solution which is satisfying for me, see XIII, Section 1. By the way, 5.17C is true also for $(\subseteq^a_{|T|^+}, \boldsymbol{T}_t)$-decompositions.

X.6. Infinite depth

THEOREM 6.1: *If* $\aleph_\alpha \geqslant \lambda(T) + \aleph_{\beta+1}$, $\mathrm{Dp}(T) \geqslant \omega$, *then* $I^a_{\aleph_\beta}(\aleph_\alpha, T) \geqslant \beth_{\mathrm{Dp}(T)}(|\alpha| + \aleph_0)$.

Proof. Compared with the proofs of 5.1 and 5.14 we have one advantage: the clause in 5.11 saying $\mathrm{Dp}(p^1_{\eta/E_1}) = Dp(p^2_{h(\eta)/E_2})$ becomes meaningful. For finite depth in say 5.12, it would be many times enough to show that h usually preserves the level (i.e. when the level in the tree determines the depth). For infinite depth this is not true; however, $n + \mathrm{Dp}(T) = \mathrm{Dp}(T)$, so we can "dedicate" some low depth to "mark" the level.

Let $\chi = |\mathrm{Dp}(T)| + \aleph_0$. First assume $\alpha \geqslant \omega$.

We can find a representation $\langle N^*_\eta, a^*_\eta : \eta \in I^*\rangle$, s.t.

(i) $\mathrm{tp}(a^*_{\langle i\rangle}, N^*_{\langle\,\rangle})$ is not orthogonal to $\emptyset$ and has depth i,

(ii) $\mathrm{tp}(a^*_{\eta^\frown\langle i\rangle}, N_\eta)$ has depth i,

(iii) if p is a regular type of depth i not orthogonal to $\emptyset$, $i > 1$, then $\langle i\rangle \in I^*$,

(iv) if $\eta \in I^* - \{\langle\,\rangle\}$, and there is a regular type over N_η orthogonal to N_{η^-} of depth i, $i \geqslant 2$, then $\eta^\frown\langle i\rangle \in I^*$,

(v) if $\mathrm{Dp}(p^*_\eta) = 2$, then $\eta^\frown\langle i\rangle \in I^* \Leftrightarrow i = 1$.

(vi) $\eta^\frown\langle 0\rangle \notin I^*$ and $\eta^\frown\langle 1\rangle \in I^*$ implies $\mathrm{Dp}(p^*_\eta) = 2$.

Clearly there is such a representation, $|I^*| = |\mathrm{Dp}(T)| \leqslant \lambda(T) \leqslant \aleph_\alpha$ (or both are finite), and (by 4.6) $\mathrm{Dp}(T) = \bigcup\{i : \langle i \rangle \in I^*\} + 1$.

Now we define by induction on i, for $\eta \in (I^*)^-$, a set $H(\eta)$.

Case I: $\eta(l(\eta)-1) = 1$,

$H(\eta) = \{\aleph_\alpha\}$.

Case II: Not Case I and $\eta(l(\eta)-1) = 2$,

$H(\eta) = \{h;\ h$ a function with domain $H(\eta\frown\langle 1\rangle) = \{\aleph_\alpha\}$, and $h(\aleph_\alpha) = \{\aleph_{l(\eta)+1}\}$.

Case III: Not Cases I, II: $\eta(l(\eta)-1) \geqslant 3$,

$H(\eta) = \{g : g$ is a function with domain $\bigcup\{H(\eta\frown\langle i\rangle) : \eta\frown\langle i\rangle \in I^*\}$, and range $\subseteq \{\lambda : \aleph_0 < \lambda \leqslant \aleph_\alpha\} \cup \{0\}$, and $|\{Y : g(Y) \neq 0\}| \leqslant \aleph_\alpha\}$.

An exercise in cardinal arithmetic is to prove $|H(\langle\ \rangle)| = \beth_{\mathrm{Dp}(T)}(|\alpha| + \aleph_0)$ (using $\mathrm{Dp}(T) \geqslant \omega$) (check by cases). Now for every $Y \in H(\langle\ \rangle)$, we choose $I_Y, \langle N^Y_\nu, a_\nu : \nu \in I_Y\rangle, f^Y_\nu (\nu \in I_Y)$ and functions g^*_Y, g_Y, s.t.:

(a) $\langle N^Y_\nu, a_\nu : \nu \in I_Y\rangle$ is a representation,

(b) g_Y is a function from I_Y onto I^*, preserving level (equivalently length), $\lessdot$, mapping $I_Y - I^-_Y, I^-_Y$, into $I^* - (I^*)^-, (I^*)^-$, respectively,

(c) g^*_Y is a function with domain $I^-_Y, g^*_Y(\eta) \in H(g_Y(\eta))$,

(d) if $\eta\frown\langle\alpha\rangle \in I^-_Y$, then $g^*_Y(\eta\frown\langle\alpha\rangle)$ is in the domain of $g^*_Y(\eta)$,

(e) if $\eta \in I^{--}_Y$, Z is in the domain of $g_Y(\eta)$, then $|\{\eta\frown\langle\alpha\rangle \in I_Y : g^*_Y(\eta\frown\langle\alpha\rangle) = Z\}| = (g^*_Y(\eta))(Z)$; if $\eta \in I^-_Y - I^{--}_Y$ (so $\eta(l(\eta)-1) = 1$) then $|\{\eta\frown\langle\alpha\rangle : \eta\frown\langle\alpha\rangle \in I_Y\}| = \aleph_\alpha$,

(f) f^Y_ν is an isomorphism from N^Y_ν onto $N^*_{g_Y(\nu)}$, extending $f^Y_{\nu\restriction l}$ for $l < l(\nu)$.

There is no problem, and $|I_Y| = \aleph_\alpha + \lambda(T)$.

We now can find $J_Y, N^Y_\nu, a_\nu (\nu \in J_Y - I_Y)$ s.t.

(g) $\langle N^Y_\nu, a_\nu : \nu \in J_Y\rangle$ represent an $\mathbf{F}^a_{\aleph_\beta}$-saturated model M_Y of cardinality $\aleph_\alpha$ (see 5.6),

(h) $I_Y \subseteq J_Y$ and for $\eta \in J_Y$ of length m, if $\eta \notin I_y$ then

$$|\{\nu \in J_y : \mathrm{tp}(a_\nu, N_\nu) = \mathrm{tp}(a_\eta, N_{\eta-})\}| \leqslant \aleph_\beta,$$

(i) if $\beta = 0, J_Y = I_Y$.

For notational simplicity:

(j) for $Y \neq Z, J_Y \cap J_Z = \{\langle\ \rangle\}$.

So it is enough to prove:

(*) if $Y \neq Z$ are from $H(\langle\ \rangle)$, then M_Y, M_Z are not isomorphic.

So suppose F is an isomorphism from M_Y onto M_Z. We use the representations $\langle N^Y_\nu, a_\nu : \nu \in J_Y \rangle$, $\langle N^Z_\nu, \bar{a}_\nu : \nu \in J_Z \rangle$, and apply 5.11, obtaining a function h from J_Y onto J_Z. So there are $\aleph_0$-large $J^1_Y \subseteq J_Y$, $J^1_Z \subseteq J_Z$ such that $h \restriction J^1_Y$ is onto J^1_Z and preserves $\lessdot$ (by 5.12).

As $\eta \in J^1_Y \wedge |\{\nu : \eta \lessdot \nu \in J^1_Y\}| = \aleph_\alpha \Rightarrow |\{\nu : h(\eta) \lessdot \nu \in J_Z\}| = \aleph_\alpha$ clearly h maps $J^1_Y \cap (I_Y - I^-_Y)$ onto $J^1_Z \cap (I_Z - I^-_Z)$ (remember $\mathrm{Dp}(p^Y_\eta) = \mathrm{Dp}(p^Z_{h(\eta)})$). Similarly (looking for types of depth 2), $J^2_Y \cap (I^-_Y - I^{--}_Y)$ is mapped onto $J^1_Z \cap (I^-_Z - I^{--}_Z)$. However, looking at the definition of $H(\eta)$ when $\eta(l(\eta)-1) = 2$, we see that on $J^1_Y \cap (I^-_Y - I^{--}_Y)$ h preserves the level. From this we can easily deduce that $h \restriction J^1_Y \cap I_Y$ preserves the level. As it also preserves the depth, easily (by induction on $l(\nu)$):

$$\text{for } \nu \in J^1_Y \cap I_Y, g_Y(\nu) = g_Z(h(\nu)).$$

The rest is straightforward too. If $\alpha < \omega$ use Exercise 6.2.

EXERCISE 6.1: If T is multidimensional then $I^a_{\aleph_\beta}(\aleph_\alpha, T) \geqslant |\alpha + \omega|$ for $\aleph_\alpha \geqslant \lambda(T) + \aleph_{\beta+1}$; also $I(\aleph_\alpha, T) \geqslant |\alpha + \omega|$ if $\aleph_\alpha \geqslant |T| + \aleph_1$.

EXERCISE 6.2: Suppose $A_l (l \leqslant n+1)$ is increasing, $\bar{a}_l \in A_l (0 < l \leqslant n)$, $\mathrm{tp}(\bar{a}_{l+2}, A_{l+1})$ is orthogonal to A_l, $\mathrm{tp}_*(A_{l+1}, A_l \cup \bar{a}_{l+1})$ is almost orthogonal to A_l. Suppose also that there are μ pairwise orthogonal regular types, each orthogonal to A_n, but not to A_{n+1}, and $|A_l| = \aleph_\beta$. Prove that $n \leqslant \mathrm{Dp}(T)$ and for $\aleph_\alpha > \lambda(T) + \aleph_\beta, \alpha - \beta \geqslant \mu$

$$I^a_{\aleph_\beta}(\aleph_\alpha, T) \geqslant \beth_n(|\alpha - \beta|^\mu).$$

And even (if $n > 0$) $\geqslant \beth_{n-1}[|\alpha + \omega|^{(|\alpha - \beta + 1|^\mu)}]$.

EXERCISE 6.3: Suppose that for μ there are no $A_l, \bar{a}_l (l \leqslant n+1)$ as above for $n = \mathrm{Dp}(T)$. Prove $I^a_{\aleph_0}(\aleph_\alpha, T) \leqslant \beth_m(|\alpha|^{<\mu})$ for α large enough, and always $I^a_{\aleph_0}(\aleph_\alpha, T) \leqslant \beth_n(|\alpha|^{<\mu} + 2^{|T|})$ for every α. Phrase and prove the parallel for $I^a_{\aleph_\beta}(\aleph_\alpha, T)$ (Hint: Repeat the proof of 4.7.)

X.7. Trivial types

(Hyp) T stable, $\kappa = \kappa_r(T)$.

DEFINITION 7.1: We call $p \in S^m(B)$ trivial if it is regular stationary and the following holds:

If $\boldsymbol{I}$ *is an independent set of sequences realizing* p, $\bar{b}$ *realizes* p *but* $\mathrm{tp}(\bar{b}, B \cup \bigcup \boldsymbol{I})$ *forks over* B, *then for some* $\bar{c} \in \boldsymbol{I}$, $\mathrm{tp}(\bar{b}, B \cup \bar{c})$ *forks over* B.

LEMMA 7.2: *Suppose* $\mathrm{Dp}(N, N', \bar{a}) > 0, p \in S^m(B)$ *stationary and parallel to* $\mathrm{tp}(\bar{a}, N)$ *(which is regular). Then* p *is trivial provided* T *does not have the dimensional order property.*

Proof. Assume T is superstable. W.l.o.g. $B = |N|, N\ \mathbf{F}^a_{\aleph_0}$-saturated, $\boldsymbol{I}$ is finite and let $\boldsymbol{I} = \{\bar{a}_l : l < n\}$. Suppose there is no such $\bar{c}$, i.e. $\{\bar{b}, \bar{a}_l\}$ is independent over N for each l. Let M be $\mathbf{F}^a_{\aleph_0}$-prime over $N \cup \boldsymbol{I} \cup \bar{b}$. As $\boldsymbol{I} \cup \bar{b}$ is finite by Definition V, 3.2, Theorem V, 3.6(2), V, 3.9(1) there is a finite set $\boldsymbol{J} \subseteq M$ independent over N, of sequences realizing regular types, such that M is $\mathbf{F}^a_{\aleph_0}$-prime over $N \cup \boldsymbol{J}$.

FACT. *There is* $\bar{b}^* \in \boldsymbol{I} \cup \{\bar{b}\}$ *such that for no* $\bar{d} \in \boldsymbol{J}$, $\mathrm{tp}(\bar{b}^*, N \cup \bar{d})$ *forks over* N.

Otherwise there are $\bar{a}^0_l \in \boldsymbol{J}, \bar{b}^0 \in \boldsymbol{J}$, such that each of the pairs $\{\bar{a}^0_l, \bar{a}_l\}$ $(l < n)$ and $\{\bar{b}^0, \bar{b}\}$ is not independent over N.

So for each l, $\mathrm{tp}(\bar{a}_l, N \cup \{\bar{a}^0_l : l < n\})$ forks over N and $\mathrm{tp}(\bar{b}, N \cup \{\bar{a}_l : l < n\})$ forks over N, so by V, 1.14 $\mathrm{tp}(\bar{b}, N \cup \{a^0_l : l < n\})$ forks over N. By the choice of $\bar{b}^0$, $\mathrm{tp}(\bar{b}^0, N \cup \bar{b})$ forks over N, hence by V, 1.14 again $\mathrm{tp}(\bar{b}^0, N \cup \{\bar{a}^0_l : l < n\})$ forks over N. As $\boldsymbol{J}$ is independent $\bar{b}^0 \in \{\bar{a}^0_l : l < n\}$, so let $\bar{b}^0 = \bar{a}^0_l$. So, by our choice, $\mathrm{tp}(\bar{b}, N \cup \bar{a}^0_l)$ forks over N, and $\mathrm{tp}(\bar{a}^0_l, N \cup \bar{a}_l)$ forks over N, hence by V, 1.14 $\mathrm{tp}(\bar{b}, N \cup \bar{a}_l)$ forks over N; a contradiction to the assumption "$\boldsymbol{I}, \bar{b}$ is a counterexample" to 5.10. So we have proved the fact.

Let $\boldsymbol{J} = \{\bar{d}_l : l < m_0\}$. As M is prime over $N \cup \boldsymbol{J}$, there are N_l $\mathbf{F}^a_{\aleph_0}$-prime over $N \cup \bar{d}_l$, such that M is $\mathbf{F}^a_{\aleph_0}$-prime over $\bigcup_{l<m} N_l$ (see V, 3.2). Let $N^* \prec M$ be $\mathbf{F}^a_{\aleph_0}$-prime over $N \cup \bar{b}^*$. As $\bar{b}^*$ realizes p (as all members of $\boldsymbol{I} \cup \{\bar{b}\}$ do), by 4.4(6) $\mathrm{Dp}(N, N^*, \bar{b}^*) > 0$, hence there is a regular $q^* \in S^m(N^*)$ orthogonal to N, and let q be its stationarization over M. For each $l < m_0, \{\bar{b}^*, \bar{d}_l\}$ is independent over N (by the fact), hence by V, 3.2 $\mathrm{tp}_*(N_l, N^*)$ does not fork over N. Hence by 1.1 q^* is orthogonal to N_l. Hence q is orthogonal to each N_l. But as T lacks the dimensional order property, q is not orthogonal to some N_l. This contradiction proves the lemma.

What if T is not superstable? W.l.o.g. $B = |N|$, N is $|T|^+$-saturated, let $\boldsymbol{J}$ be a maximal subset of $\boldsymbol{I}$ independent over N, so $\mathrm{tp}(\bar{b}, N \cup \bigcup \boldsymbol{J})$ forks over M, hence is orthogonal over N, hence $\mathbf{F}^a_{|T|^+}$-isolated, so there is M $\mathbf{F}^a_{|T|^+}$-primary over $N \cup \bigcup \boldsymbol{J}$ to which $\bar{b}$ belongs. Continue as above.

LEMMA 7.3: *Let T be stable, $\kappa = \kappa_r(T)$.*

(1) *Suppose $r_l \in S^{m(l)}(A_l)$ for $l = 0, 1$, r_0 parallel to r_1, r_0 is stationary, regular and trivial, then so is r_1.*

(2) *Suppose $r \in S^m(A)$ is a (stationary) regular trivial type and $\{\bar{a}, \bar{b}\}$ is independent over A. Then for any $\bar{c}$ realizing r, $\mathrm{tp}(\bar{c}, A \cup \bar{a} \cup \bar{b})$ forks over A iff $\mathrm{tp}(\bar{c}, A \cup \bar{a})$ forks over A or $\mathrm{tp}(\bar{c}, A \cup \bar{b})$ forks over A.*

(3) *Suppose $r \in S^n(A)$ is stationary regular and trivial. For any $\bar{a}$ there are $\bar{d}_0, \ldots, \bar{d}_{n-1} \in r(\mathfrak{C})$ such that: $\{\bar{d}_0, \ldots, \bar{d}_{n-1}\}$ is independent over A, $\mathrm{tp}(\bar{d}_l, A \cup \bar{a})$ forks over A for $l < n$, and $n = \mathrm{low}_r(\bar{a}, A)$. So for any $\bar{d} \in r(\mathfrak{C})$, $\mathrm{tp}(\bar{d}, A \cup \bar{a})$ forks over A iff $\mathrm{tp}(\bar{d}, A \cup \bar{d}_l)$ forks over A for some $l < n$. Also $\mathrm{tp}(\bar{a}, A \cup \bigcup_{l<n} \bar{d}_l)$ is orthogonal to r.*

(4) *Suppose r_0, r_1 are stationary regular and not orthogonal. Then r_0 is trivial iff r_1 is trivial.*

(5) *Suppose $A \subseteq B$, $\mathrm{tp}(\bar{a}, B)$ does not fork over A, $r \in S^m(B)$ is stationary regular and trivial, not orthognal to $\mathrm{stp}(\bar{a}, A)$. Then for some $e \in \mathrm{acl}(A \cup \bar{a})$, $e \notin \mathrm{acl}\, A$, $\mathrm{stp}(e, A)$ is $\mathrm{cl}^3(r)$-simple of weight 1. In fact there are $e_0, \ldots, e_{n-1}$ as above, $\mathrm{tp}(\bar{a}, A \cup \{e_l : l < n\})$ orthogonal to r. If $\mathrm{stp}(\bar{a}, A)$ is semi-regular, $\mathrm{stp}(e, A)$ is regular.*

(6) *If for $l = 0, 1$, $p_l = \mathrm{stp}(\bar{a}_l, A)$ is not orthogonal to r, r a stationary regular trivial type, then p_0, p_1 are not weakly orthogonal; in fact p_1 has an extension over $A \cup \bar{a}_0$ which forks over A.*

Proof. (1) By the definition of parallel, r_1 is stationary and by V, 1.8(1), r_1 is regular. Let $A_0 \cup A_1 \subseteq M$, M $\mathbf{F}^a_\lambda$-saturated, $\lambda > (A_0 \cup A_1)$ and r_2 be the stationarization of r_0 (and r_1) over M. Clearly r_2 is regular and stationary.

Suppose r_1 is not trivial, then there are $b, \bar{a}_0, \ldots, \bar{a}_{n-1}$ realizing r_1, $\mathrm{tp}(\bar{b}, A_1 \cup \bar{a}_0 \cup \ldots \cup \bar{a}_{n-1})$ forks over A_1, but $\mathrm{tp}(\bar{b}, A_1 \cup \bar{a}_m)$ does not fork over A_1 for $m < n$. W.l.o.g. $\mathrm{tp}(\bar{b} {}^\frown \bar{a}_0 {}^\frown \ldots {}^\frown \bar{a}_{n-1}, M)$ does not fork over A_1, and then clearly $\bar{b}, \bar{a}_0, \ldots, \bar{a}_{n-1}$ exemplify r_2 is not trivial (use III, 0.1). So $\mathrm{tp}(\bar{b}, M \cup \bar{a}_0 \cup \ldots \cup \bar{a}_{n-1})$ forks over M, hence over A_0, and $\bar{b}, \bar{a}_0, \ldots, \bar{a}_{n-1}$ realizes r_2 and $r_0 \subseteq r_2$. So by V, 1.11, clearly $\mathrm{tp}(\bar{b}, A_0 \cup \boldsymbol{I})$ forks over A_0 where $\boldsymbol{I} = r_0(M) \cup \{\bar{a}_0, \ldots, \bar{a}_{n-1}\}$. Obviously for every $\bar{c} \in M$, $\mathrm{tp}(\bar{b}, A_0 \cup \bar{c})$ does not fork over A_0.

So $\bar{b}, \boldsymbol{I}$ exemplify r_0 is not trivial (except that $\boldsymbol{I}$ is not independent, but this can be discarded by r_0's regularity); a contradiction, hence r_1 is trivial as required.

(2) The implication $\Leftarrow$ is trivial. So suppose $\bar{c}$ is a counterexample to the other direction. Let M be an $\mathbf{F}^a_\kappa$-saturated model, $A \subseteq M$, $\text{tp}_*(M, A \cup \bar{a} \cup \bar{b} \cup \bar{c})$ does not fork over A. By some application of III, 0.1 clearly $\{\bar{a}, \bar{b}\}$ is independent over M, $\text{tp}(\bar{c}, M \cup \bar{a})$ and $\text{tp}(\bar{c}, M \cup \bar{b})$ do not fork over M, $\text{tp}(\bar{c}, M \cup \bar{a} \cup \bar{b})$ forks over M, and $\text{tp}(\bar{c}, M)$ is a stationarization of r, hence by (1) is a stationary regular trivial type. So w.l.o.g. $M = B$. Note: n^l may be infinite below.

Let $\{\bar{d}_m : m < n^0\}$ be a maximal set of sequences realizing $\text{tp}(\bar{c}, M)$ independent over M such that $\text{tp}(\bar{d}_m, M \cup \bar{a})$ forks over M, and similarly let $\{\bar{e}_m : m < n^1\}$ be a maximal set of sequences realizing $\text{tp}(\bar{c}, M)$, independent over M, such that $\text{tp}(\bar{e}_m, M \cup \bar{b})$ forks over M. By V, 3.16A, $n^0 = \text{low}_r(\bar{a}, M)$, $n^1 = \text{low}_r(\bar{b}, M)$ and $\text{tp}(\bar{d}_0 {}^\frown \ldots {}^\frown \bar{d}_{n_0-1}, M \cup \bar{a})$ and $\text{tp}(\bar{e}_0 {}^\frown \ldots {}^\frown \bar{e}_{n_1-1}, M \cup \bar{b})$ are $\mathbf{F}^a_\kappa$-isolated. So clearly $\{\bar{d}_0, \ldots, \bar{d}_{n_0-1}, \bar{e}_0, \ldots, \bar{e}_{n_1-1}\}$ is independent over M and $\text{tp}(\bar{d}_0 {}^\frown \ldots {}^\frown \bar{d}_{n_0-1} {}^\frown \bar{e}_0 {}^\frown \ldots {}^\frown \bar{e}_{n_1-1}, M \cup \bar{a} \cup \bar{b})$ is $\mathbf{F}^a_\kappa$-isolated (see V, 3.2).

So let N be $\mathbf{F}^a_\kappa$-prime over $M \cup \bar{a} \cup \bar{b}$, $\{\bar{d}_0, \ldots, \bar{d}_{n_0-1}, \bar{e}_0, \ldots, \bar{e}_{n_1-1}\} \subseteq N$. By V, 3.16(8), (10) $\text{low}_r(\bar{a} {}^\frown \bar{b}, M) = \text{low}_r(\bar{a}, M) + \text{low}_r(\bar{b}, M)$, and $\{\bar{d}_0, \ldots, \bar{d}_{n_0-1}, \bar{e}_0, \ldots, \bar{e}_{n-1}\}$ is a maximal subset of $r(N)$ independent over M.

Now what about $\bar{c}$? $\bar{c}$ realizes $r \in S^m(M)$, $\bar{c}$ depends on $\bar{a} {}^\frown \bar{b}$ (i.e., $\text{tp}(\bar{c}, M \cup \bar{a} \cup \bar{b})$ forks over M), hence by V, 1.16(3), (1) $\text{tp}(\bar{c}, M \cup r(N))$ forks over M, hence $\text{tp}(\bar{c}, M \cup \{\bar{d}_0 \cup \ldots \cup \bar{d}_{n_0-1} \cup \bar{e}_0 \cup \ldots \cup \bar{e}_{n_1-1}\})$ forks over M. By the definition of triviality, $\text{tp}(\bar{c}, M \cup \bar{d}_l)$ forks over M or $\text{tp}(\bar{c}, M \cup \bar{e}_l)$ forks over M for some l. By symmetry, suppose the former. Clearly by V, 1.9(2) $\{\bar{a}, \bar{c}\}$ is not independent over M, a contradiction.

(3) Assume, for simplicity, T is superstable (otherwise use V, 3.16, 3.16A). Let $A \subseteq M$, M $\mathbf{F}^a_\kappa$-saturated, $\text{tp}_*(M, A \cup \bar{a})$ does not fork over A. By V, 3.9 there are $n, \bar{d}_0, \ldots, \bar{d}_{n-1}$ such that: $\text{tp}(\bar{d}_0 {}^\frown \ldots {}^\frown \bar{d}_{n-1}, M \cup \bar{a})$ is $\mathbf{F}^a_\kappa$-isolated, $\bar{d}_l$ realizes the stationarization of r over M, $n = \text{low}_r(\bar{a}, M) = \text{low}_r(\bar{a}, A)$, and $\{\bar{d}_0, \ldots, \bar{d}_{n-1}\}$ is independent over (M, A). By the first part, $q_l = \text{tp}(\bar{d}_l, M \cup \bar{a})$ forks over M, hence q_l forks over A, hence for some $\bar{b}_l$, $\text{tp}(\bar{d}_l, A \cup \bar{a} \cup \bar{b}_l)$ forks over A, $\bar{b}_l \in M$.

The only property missing is $\text{tp}(\bar{d}_l, A \cup \bar{a})$ forks over A. If this fails we get a contradiction to part (2) (with $A, \bar{a}, \bar{b}_l, \bar{d}_l$ here standing for $A, \bar{a}, \bar{b}, \bar{c}$ there). The second sentence is by V, 3.12(a) and triviality.

We will still have to prove "$\text{tp}(\bar{a}, A \cup \bigcup_{l<n} \bar{d}_l)$ is orthogonal to r". Suppose not. Let N be $\mathbf{F}^a_\lambda$-saturated $\lambda > |A| + \kappa, A \cup \bigcup_{l<n} \bar{d}_l \subseteq N$, $\text{tp}(\bar{a}, N)$ does not fork over $A \cup \bigcup_{l<n} \bar{d}_l$.

As the conclusion fails, there is $\bar{d}_n$ realizing the stationarization of

r over N, $\text{tp}(\bar d_n, N \cup \bar a)$ forks over N, hence over A. As above we get $\text{tp}(\bar a, A \cup \bar d_n)$ forks over A. But this contradicts the second sentence in (3).

(4) Left to the reader (see V, 1.14).

(5) Let $\{\bar d_0, \ldots, \bar d_{n-1}\}$ be as in part (3) of the lemma (replacing A by B),

$$\boldsymbol{I}_l = \{\bar d \in r(\mathfrak{C}) : \text{tp}(\bar d, B \cup \bar d_l) \quad \text{forks over } B\},$$

and $\varphi_l, \bar b_l$ be such that $\bar b_l \in B, \varphi_l[\bar d_l, \bar a, \bar b_l]$ and $\varphi_l(\bar x, \bar a, \bar b_l)$ forks over B. By II, 2.2(8) there are $\bar c_l \in A \cup \bigcup \boldsymbol{I}_l$, and ψ_l such that for every $\bar d \in \boldsymbol{I}_l$, $\varphi[\bar d, \bar a, \bar b_l]$ iff $\psi_l[\bar d, \bar c_l, \bar b_l]$, and w.l.o.g. (by increasing $\bar c_l$) $\bar c_l$ is a concatanation of sequences from $\boldsymbol{I}_l$. Let $e_l \in \mathfrak{C}^{\text{eq}}$ be defined by $\bar c_l / E_l$, where

$$E_l(\bar y, \bar z) \stackrel{\text{def}}{=} (\forall \bar x)[\psi_l(\bar x, \bar y, \bar b_l) \equiv \psi_l(\bar x, \bar z, \bar b_l)].$$

By the second sentence in part (3), for every automorphism F of $\mathfrak{C}^{\text{eq}}$ which is the identity over $B \cup \bar a, F$ maps $\boldsymbol{I}_l$ into some $\boldsymbol{I}_m$ and if $m = l$, then $F(e) = e$. Hence e_l can have at most n possible images (varying F). Hence e_l is algebraic over $B \cup \bar a$.

Let, for $i < |T|^+$, f_i be an elementary mapping with domain $C = B \cup \bar a \cup \bigcup_l (\bar d_l \cup \{e_l\} \cup \bar c_l), f_i \restriction (A \cup \bar a) =$ the identity, $\text{stp}_*(f_i(C), \bigcup_{j \neq i} f_j(C))$ does not fork over $A \cup \bar a$, and extend $\text{stp}_*(C, A \cup \bar a)$, and f_0 is the identity.

SUBFACT: *For $i \neq j$, $\{f_i(\bar d_m), f_j(\bar d_m)\}$ is not independent over $\bigcup_\alpha f_\alpha(B)$.*

For every $i > 0$, $\text{tp}_*(f_i(B), \bigcup_{j<i} f_j(B) \cup \bigcup_{l<n} \bar d_l \cup \bar a)$ does not fork over $A \cup \bar a$, and $\text{tp}_*(f_i(B), A \cup \bar a)$ does not fork over A [as $\text{tp}_*(B, A \cup \bar a)$ does not fork over A by symmetry], hence by transitivity $\text{tp}_*(f_i(B), \bigcup_{j<i} f_j(B) \cup \bigcup_{l<n} \bar d_l \cup \bar a)$ does not over A. So by III, 0.1 $\text{tp}_*(\bigcup_{i>0} f_i(B), B \cup \bigcup_{l<n} \bar d_l)$ does not fork over A, hence over B, hence $\text{tp}_*(\bigcup_{l<n} \bar d_l, \bigcup_i f_i(B))$ does not fork over B. So $\{\bar d_l : l < n\}$ is independent over $(\bigcup_i f_i(B), B)$. Clearly $\langle f_i(C) : i < |T|^+ \rangle$ is indiscernible over $A \cup \bar a$, hence for every i $\{f_i(\bar d_l) : l < n\}$ is independent over $(\bigcup_j f_j(B), f_i(B))$. We can also conclude that $\text{tp}(\bar a, \bigcup_i f_i(B))$ does not fork over A. Now we know (for $i > 0$) that $\text{tp}(\bar a, f_i(B) \cup f_i(\bar d_m))$ forks over $f_i(B)$, hence $\text{tp}(\bar a, \bigcup_j f_j(B) \cup f_i(\bar d_m))$ forks over $f_i(B)$, so by the previous sentence $\text{tp}(\bar a, \bigcup_j f_j(B) \cup f_i(\bar d_m))$ forks over $\bigcup_j f_j(B)$. So by V, 3.12(a), $\{f_i(\bar d_m) : i < |T|^+\}$ cannot be independent over $\bigcup_j f_j(B)$. Hence, for some i, $\text{tp}(f_i(\bar d_m), \bigcup_j f_j(B) \cup \bigcup_{j \neq i} f_j(\bar d_m))$ forks over $f_i(B)$, and by the indiscernibility and the finite character of forking, for some n

$\mathrm{tp}(f_0(\bar{d}_m), \bigcup_j f_j(B) \cup_{j=1}^{n} f_j(\bar{d}_m))$ forks over $\bigcup_j f_j(B)$. We can choose a minimal n, so $\{f_j(\bar{d}_m): j = 1, n\}$ is independent over $\bigcup_j f_j(B)$. Applying 7.3(2) several times we find that for some $l, 1 \leqslant l \leqslant n$, $\mathrm{tp}(f_0(\bar{d}_m), \bigcup_j f_j(B) \cup f_l(\bar{d}_m))$ forks over $\bigcup_j f_j(B)$. But quite clearly

$$\mathrm{tp}(\bigcup\{f_j(B): j \neq 0, l\}, f_0(B) \cup f_0(\bar{d}_m) \cup f_l(B) \cup f_l(\bar{d}_m))$$

does not fork over A, hence $\{f_0(\bar{d}_m), f_l(\bar{d}_m)\}$ is not independent over $f_0(B) \cup f_l(B)$. By the indiscernibility we can replace $0, l$ by any $i \neq j$, thus proving the subfact.

As r is regular, for $l > 0$, $\mathrm{tp}(\bar{d}_l, B \cup \bigcup_{m \neq l} I_m)$ does not fork over B and easily $\mathrm{tp}(\bar{d}_l, B \cup \bigcup_{m \neq l}(\bar{d}_m \frown \bar{c}_m \frown \langle e_m \rangle))$ does not fork over B, hence $\mathrm{tp}(\bar{d}_l, B \cup \bigcup_{m \neq l}(\bar{d}_m \frown \bar{c}_m \frown \langle e_m \rangle) \cup \bigcup_{i>0} f_i(B \cup \bar{d}_0 \frown \bar{c}_0 \frown \langle e_0 \rangle))$ does not fork over B.

We can find an elementary mapping g:

$$\mathrm{Dom}(g) = \bar{a} \cup \bigcup_i f_i\Big(B \cup \bigcup_m \bar{d}_m \frown \bar{c}_m \frown \langle e_m \rangle\Big),$$

$g \restriction \bigcup_i f_i(B \cup \bar{d}_0 \frown \bar{c}_0 \frown \langle e_0 \rangle)$ is the identity and $\mathrm{stp}_*(\mathrm{range}(g), \bigcup_i f_i(B \cup \bar{a}_0 \frown \bar{c}_0 \frown \langle e_0 \rangle)$ extends $\mathrm{stp}_*(\mathrm{Dom}\, g, \bigcup_i f_i(B \cup \bar{d}_0 \frown \bar{c}_0 \frown \langle e_0 \rangle))$ and does not fork over $\bigcup_i f_i(B \cup \bar{d}_0 \frown \bar{c}_0 \frown \langle e_0 \rangle)$. Now $\{\bar{d}_m : m < n\} \cup \{g(\bar{d}_m): 0 < m < n\}$ is independent over $\bigcup_i f_i(B)$ by the previous paragraph, hence it is independent over $(\bigcup_i f_i(B), B)$.

As we have noted above, e_0 is algebraic over $B \cup \bar{a}$, hence $e_0 = g(e_0)$ is algebraic over $g(B \cup a) = B \cup g(\bar{a})$.

Checking closely, we see that for any automorphism F of $\mathfrak{C}^{\mathrm{eq}}$ which is the identity over $B \cup \bar{a} \cup g(\bar{a})$, $F(e_0) = e_0$ (as $F(e_0) \in \{e_l : l < n\}$ and $F(e_0) = F(g(e_0)) \in \{g(e_l): l < n\}$). So e_0 is definable over $B \cup \bar{a} \cup g(\bar{a})$, say by $\psi(x, \bar{a}, g(\bar{a}), \bar{b}^*)$ $(\bar{b}^* \in B)$. Let

$$E(\bar{y}_0 \frown \bar{z}_0, \bar{y}_1 \frown \bar{z}_1) \overset{\mathrm{def}}{=} \text{“}(\forall x)[\psi(x, \bar{y}_0, \bar{z}_0, f_i(\bar{b}^*)) \equiv \psi(x, \bar{y}_1, \bar{z}_1, f_i(\bar{b}^*))] \text{ holds for infinitely many } i\text{'s”}.$$

By III, 2.5, 1.7, E is a formula which is almost over A, and let $e^* = \bar{a} \frown g(\bar{a})/E$. Now exactly as in the proof of V, 4.11, $e^* \notin A$, $\mathrm{tp}(e^*, A)$ is $\mathrm{cl}^3(r)$-semi-simple not orthogonal to r and $e^* \in \mathrm{acl}(\bigcup_i f_i(B \cup \{e_0\}))$ but here, by the subfact, any two of $\{f_i(e_0): i\}$ depend on $\bigcup_i f_i(B)$, hence $\mathrm{low}_r(e^*, A) = 1$.

The only point left is “$e^* \in \mathrm{acl}(A \cup \bar{a})$”, but for any i we know that (k large enough) $e^* \in \mathrm{acl}(\bigcup_{j=i}^{i+k} f_j(B \cup \{e\}) \subseteq \mathrm{acl}(A \cup \bar{a} \cup \bigcup_{j=1}^{i+k} f_j(B))$ (because $e_0 \in \mathrm{acl}(B \cup \bar{a})$; see above). As this is true for every j and

$\{f_j(B):j\}$ is independent over $A \cup \bar{a}$, clearly $e^* \in \mathrm{acl}(A \cup \bar{a})$. The rest is obvious.

(6) Let $A \subseteq N_i$ for $i < |T|^+$, $|N_i| \leqslant |T|$, $\{N_i : i < |T|^+]$ independent over A, and r has a stationarization $r_i \in S^m(N_i)$. Let $\bigcup_i N_i \subseteq M$, M $\mathbf{F}^a_\kappa$-saturated, and w.l.o.g. $\mathrm{tp}(\bar{a}_0 {}^\frown \bar{a}_1, M)$ does not fork over A, and w.l.o.g. $r \in S^m(M)$. By the assumption there is $\bar{d}$ realizing r, $\bar{a}'_l$ realizing $\mathrm{tp}(\bar{a}_l, M)$ such that $\mathrm{tp}(\bar{a}'_l, M \cup \bar{d})$ forks over M. For each i there is $\bar{d}_i$ realizing the stationarization of r_i over M, $\mathrm{tp}(\bar{d}_i, M \cup \bar{d})$ forks over M. By V1.9(2), $\mathrm{tp}(\bar{d}_i, M \cup \bar{a}'_l)$ forks over M (for $i < |T|$, $l < 2$). As r_i does not fork over N_i by 7.3(2) for each l, i, $\mathrm{tp}(\bar{a}'_l, N_i \cup \bar{d})$ forks over N_i. Again by 7.2(2), $\{\bar{a}'_0, \bar{a}'_1\}$ independent over N_i is impossible. As this holds for each i as $A \subseteq N_i$, $\{N_i : i < |T|^+\}$ independent over A, easily $\mathrm{tp}(\bar{a}'_l, A \cup \bar{a}'_{1-l})$ forks over A.

Chapter XI

THE DECOMPOSITION THEOREM

XI.0. Introduction

(Hyp) T is stable.

This chapter is similar in spirit to Chapter IV, i.e. we give here an axiomatic treatment, not claiming for it any inherent meaning, just that it is a good way to organize the material. The notion we try to axiomatize is "a decomposition". The most elegant case was presented in Chapter X, Section 3, but in order to get the "main gap" for other cases we need other such theorems. In Section 2 we present the axioms, notions and prove the existence of decomposition (and the possibility to extend a partial one). $\mathbf{T}$ is a set of "small" types, $\subseteq^*$ a notion of "strong submodel". So we try to get decomposition theorems for models M realizing every $p \in \mathbf{T}$ over M, by $\subseteq^*$-submodels. In Section 3 we define various $\mathbf{T}$'s and $\subseteq^*$'s and prove that the axioms are satisfied for various pairs.

In Section 1 we prove some more facts on "p stationary inside A", (weak) orthogonality and conclusions of "$p \in S^m(M)$ is not regular" or "$p, q \in S^m(M)$ are not orthogonal" (for M not necessarily $\mathbf{F}^a_{\kappa_r(T)}$-saturated).

XI.1. Stationarization

Definition 1.1: (1) We say that $p \in S^m(B)$ is *stationary inside A if* $B \subseteq A$ and there is a unique extension of p in $S^m(A)$ which does not fork over B.

(2) We say that p is stationary inside (B, A) when p is an m-type which does not fork over $B, B \subseteq A$, and for every $q \in S^m(B)$, if $p \cup q$ does not fork over B, then q is stationary inside A.

FACT 1.2: *If $B \subseteq_t A$ (see Definitions 3.1(1) and 3.1(4)) and $p \in S^m(C)$ does not fork over B, then p is stationary inside (B, A).*

Proof. Suppose not. By Definition 1.1 w.l.o.g. $p \in S^m(B)$, so as p is a counterexample, for some $\bar{a} \in A$ and φ $p \cup \{\varphi(\bar{x}, \bar{a})\}$ and $p \cup \{\neg\varphi(\bar{x}, \bar{a})\}$ does not fork over B. By 1.4 below, for some $\bar{b} \in B$ and ϑ, $\models \vartheta[\bar{a}, \bar{b}]$ and every $\bar{a}'$ satisfying $\models \vartheta[\bar{a}', \bar{b}]$ satisfies this. As $B \subseteq_t A$ there is such $\bar{a}' \in B$, easy contradiction.

FACT 1.3: *If $p \in S^m(A), p \subseteq q, q$ forks over A, then for some $\varphi = \varphi(\bar{x}, \bar{b})$, $q \vdash \varphi(\bar{x}, \bar{b})$, and $p \cup \{\varphi(x, \bar{b}')\}$ forks over A for every $\bar{b}'$.*

Proof. It is easy from the next subfact.

SUBFACT 1.4: *For every $p \in S^m(A)$, and $\psi = \psi(\bar{x}, \bar{y})$ there is $\vartheta(\bar{y}, \bar{a})$, $\bar{a} \in A$ such that:*

for any $\bar{b}, p \cup \{\psi(\bar{x}, \bar{b})\}$ does not fork over A iff $\models \vartheta[\bar{b}, \bar{a}]$.

Proof. The set of $q \in S^m_\psi(|\mathfrak{C}|)$ such that $p \cup q$ does not fork over A is finite, in fact the number is $\leqslant \mathrm{Mlt}(p, \psi, \aleph_0)$.

Let $\{q_1, \ldots, q_n\}$ be an enumeration of those q's. We know (II, 2.2(1), (a), and II, 2.13(1), (3)) that for some $\vartheta_l(\bar{y}, \bar{a}_l)$,

$$\psi(\bar{x}, \bar{b}) \in q_l \text{ if and only if } \models \vartheta_l[\bar{b}, \bar{a}_l].$$

Let $\vartheta(\bar{y}, \bar{a}) = \bigvee_{l=1}^{n} \vartheta_l(\bar{y}, \bar{a}_l)$. So we can conclude (by III, 0.1, the extension property) that:

> $p \cup \{\psi(\bar{x}, \bar{b})\}$ does not fork over A iff for some $l, \psi(\bar{x}, \bar{b}) \in q_l$ iff for some $l \models \vartheta_l[\bar{b}, \bar{a}_l]$ iff $\models \vartheta[\bar{b}, \bar{a}]$.

The only point left is $\bar{a} \in A$. However, easily every automorphism of $\mathfrak{C}$ which is the identity on A maps $\vartheta(\bar{x}, \bar{a})$ to an equivalent formula. By III, 2.3(1) this implies that $\vartheta(\bar{y}, \bar{a})$ is equivalent to a formula with parameters from A.

Remark. Alternatively, for 1.4 we compute $R[-, \psi, k]$ for large enough k depending on p.

FACT 1.5: *Suppose $\{\bar{a}_i {}^\frown \bar{b}_i : i < \omega\}$ is indiscernible over $\bar{c}$, $\mathrm{tp}(\bar{b}_i, \bar{c} \cup \bigcup_{j \leqslant i} \bar{a}_j \cup \bigcup_{j < i} \bar{b}_j)$ does not fork over $\bar{c} \cup \bar{a}_i$, $\mathrm{tp}(\bar{d}, \bar{c} \cup \bigcup_{j < \omega} \bar{a}_i)$ does not fork* *over $\bar{c}$ and* $\mathrm{tp}(\bar{b}_0, \bar{c} \cup \bigcup_{j<\omega} \bar{a}_j)$, $\mathrm{tp}(\bar{d}, \bar{c} \cup \bigcup_{j<\omega} \bar{a}_j)$ *are not weakly orthogonal. Then for some n,* $\mathrm{tp}(\bar{b}_n, \bigcup_{i \leq n} \bar{a}_i)$ *and* $\mathrm{tp}(\bar{b}_0 {}^\frown \ldots {}^\frown \bar{b}_{n-1}, \bigcup_{i \leq n} \bar{a}_i)$ *are not weakly orthogonal.*

Proof. Clearly, $\mathrm{tp}(\bar{b}_i, \bar{c} \cup \bigcup_{j<\omega} \bar{a}_j \cup \bigcup_{j \neq i} \bar{b}_j)$ does not fork over $\bar{c} \cup \bar{a}_i$ (otherwise there is a formula witnessing it, and as $\{\bar{a}_j {}^\frown \bar{b}_j : j < \omega\}$ is indiscernible over $\bar{c}$, we may assume w.l.o.g. that the formula is over $\bar{c} \cup \bigcup_{j \leq i} \bar{a}_j \cup \bigcup_{j<i} \bar{b}_j$, a contradiction to one of the hypotheses). Also $\{\bar{a}_i : i < \omega\}$ is indiscernible over $\bar{c} {}^\frown \bar{d}$ (as $\{\bar{a}_i : i < \omega\}$ is indiscernible over $\bar{c}$, $\mathrm{tp}(\bar{d}, \bar{c} \cup \bigcup_{i<\omega} \bar{a}_i)$ does not fork over $\bar{c}$ and by III, 1.6(1) strong splitting implies forking).

As $\mathrm{tp}(\bar{b}_0, \bar{c} \cup \bigcup_{j<\omega} \bar{a}_j)$, $\mathrm{tp}(\bar{d}, \bar{c} \cup \bigcup_{j<\omega} \bar{a}_j)$ are not weakly orthogonal there are $\bar{b}'_0$ and φ such that $\bar{b}'_0$ realizes $\mathrm{tp}(\bar{b}_0, \bar{c} \cup \bigcup_{j<\omega} \bar{a}_j)$ and $\models \varphi[\bar{a}_0, \bar{b}_0, \bar{a}_1, \bar{a}_2, \ldots, \bar{a}_{k-1}, \bar{d}, \bar{c}]$, $\models \neg \varphi[\bar{a}_0, \bar{b}'_0, \bar{a}_1, \ldots, \bar{a}_{k-1}, \bar{d}, \bar{c}]$. By the indiscernibility of $\{\bar{a}_i : i < \omega\}$ over $\bar{c} \cup \bar{d}$, there are $\bar{b}'_i, \bar{b}''_i (i < \omega)$ such that $\bar{b}'_i$, $\bar{b}''_i$ realizes $\mathrm{tp}(\bar{b}_i, \bar{c} \cup \bigcup_{j<\omega} \bar{a}_j)$ and

$$\models \varphi[\bar{a}_i, \bar{b}'_i, \bar{a}_{i+1}, \ldots, \bar{a}_{i+k-1}, \bar{d}, \bar{c}], \models \neg \varphi[\bar{a}_i, \bar{b}''_i, \bar{a}_{i+1}, \ldots, \bar{a}_{i+k-1}, \bar{d}, \bar{c}].$$

Now the sequence $\langle \bar{a}_0 {}^\frown \bar{b}'_0, \bar{a}_1 {}^\frown \bar{b}''_1, \bar{a}_2 {}^\frown \bar{b}'_2, \bar{a}_3 {}^\frown \bar{b}''_3, \ldots, \bar{a}_{2i} {}^\frown \bar{b}'_{2i}, \bar{a}_{2i+1} {}^\frown \bar{b}''_{2i+1}, \ldots \rangle$ cannot be an indiscernible set (as the formula $\varphi(, \ldots, \bar{d}, \bar{c})$ contradicts II, 2.20). So it cannot realize the same type as $\langle \bar{a}_0 {}^\frown \bar{b}_0, \bar{a}_1 {}^\frown \bar{b}_1, \ldots \rangle$. The conclusion is immediate.

Remark 1.5A. (1) This is essentially like IV, 4.12, p. 190.

(2) Note that we can replace $\bar{a}_i$ by $\bar{a}'_i = \bar{a}_i {}^\frown \bar{c}$;

(3) We can let the sequences be infinite.

FACT 1.6: *Suppose in 1.5 that in addition there is $\bar{d}'$ realizing* $\mathrm{stp}(\bar{d}, \bar{c} \cup \bigcup_{i<\omega} \bar{a}_i)$ *such that* $\mathrm{tp}(\bar{d}', \bar{c} \cup \bigcup_{j<\omega} \bar{a}_j \cup \bar{b}_0)$ *fork over $\bar{c} \cup \bar{a}_0$. Then there are $n < \omega$ and $\bar{b}'_0$ realizing* $\mathrm{stp}(\bar{b}_0, \bar{c} \cup \bigcup_{i<n} \bar{a}_i)$ *such that* $\mathrm{tp}(\bar{b}'_0, \bigcup_{j=1}^{n} \bar{b}_j \cup \bar{c} \cup \bigcup_{i \leq n} \bar{a}_i)$ *fork over $\bar{c} \cup \bar{a}_0$.*

Proof. Work in $\mathfrak{C}^{\mathrm{eq}}$ and replace $\bar{a}_i, \bar{c}$ by an enumeration of $\mathrm{acl}(\bar{c} {}^\frown \bar{a}_i)$, $\mathrm{acl}(\bar{c})$, respectively. Now apply the previous fact.

FACT 1.7: *Suppose* $\mathrm{tp}(\bar{b}, M \cup \bar{a})$ *is not almost orthogonal to M,* $\models \psi(\bar{b}, \bar{a})$. *Then there are $\bar{a}_l, \bar{b}_l (l = 0, 1)$ such that:*

(i) $\mathrm{tp}(\bar{a}_0 {}^\frown \bar{b}_0, M) = \mathrm{tp}(\bar{a} {}^\frown \bar{b}, M)$, $\mathrm{tp}(\bar{a}_1, M) = \mathrm{tp}(\bar{a}, M)$, $\models \psi[\bar{b}_1, \bar{a}_1]$,

(ii) *the type* $\mathrm{tp}(\bar{a}_1, M \cup \bar{a}_0 \cup \bar{b}_0)$ *does not fork over M,*

(iii) $\models \varphi[\bar{a}_0, \bar{b}_0, \bar{a}_1, \bar{b}_1, \bar{c}]$ *and* $\{\varphi(\bar{a}_0, \bar{x}, \bar{a}^*_1, \bar{b}^*_1, \bar{c})\} \cup \mathrm{tp}(\bar{b}_0, \bar{a}_0 \cup \bar{c})$ *fork over $\bar{a}_0 \cup M$, for some φ and $\bar{c} \in M$, for every $\bar{a}^*_1, \bar{b}^*_1$.*

Proof. As $\operatorname{tp}(\bar{b}, M \cup \bar{a})$ is not almost orthogonal to M, we know that for some $\bar{d}$, $\operatorname{tp}(\bar{d}, M \cup \bar{a})$ does not fork over M but $\operatorname{tp}(\bar{d}, M \cup \bar{a} \cup \bar{b})$ forks over M. We can choose by induction on $n < \omega$, sequences $\bar{a}_n, \bar{b}_n$ such that $\bar{a}_n \frown \bar{b}_n$ realizes $\operatorname{stp}(\bar{a} \frown \bar{b}, M)$, and $\operatorname{tp}(\bar{a}_n \frown \bar{b}_n, M \cup \bar{d} \frown \bar{b} \frown \bar{a} \cup \bigcup_{l<n} \bar{a}_l \frown \bar{b}_l)$ does not fork over M. By 1.6 for some $n < \omega$ there is $\bar{b}'_0$ realizing $\operatorname{stp}(\bar{b}_0, M \cup \bigcup_{l<\omega} \bar{a}_l)$ such that $\operatorname{tp}(\bar{b}'_0, \bar{c} \cup \bigcup_{l<\omega} \bar{a}_l \cup \bigcup_{l=1}^{n} \bar{b}_l)$ forks over $M \cup \bar{a}_0$; w.l.o.g. n is minimal. By 1.3 there is a formula φ and $\bar{c} \in M$ such that

(i) $\vDash \varphi[\bar{a}_0, \bar{b}'_0, \bar{a}_1, \bar{b}_1, \ldots, \bar{a}_n, \bar{b}_n, \bar{a}_{n+1}, \bar{a}_{n+2}, \ldots, \bar{a}_k, \bar{c}]$,

(ii) for every $a_1^*, \ldots, \bar{a}_k^*, \bar{b}_1^*, \ldots, \bar{b}_n^*$,

$$\{\varphi(\bar{a}_0, \bar{x}, \bar{a}_1^*, \bar{b}_1^*, \bar{a}_2^*, \bar{b}_2^*, \ldots, \bar{a}_n^*, \bar{b}_n^*, \bar{a}_{n+1}^*, \ldots, \bar{a}_k^*, \bar{c})\} \cup$$
$$\operatorname{tp}(\bar{b}_0, M \cup \bar{a}_0) \text{ forks over } M \cup \bar{a}_0.$$

As $\{\bar{a}_0 \frown \bar{b}'_0, \bar{a}_1, \bar{a}_2 \frown \bar{b}_2, \ldots, \bar{a}_n \frown \bar{b}_n, \bar{a}_{n+1}, \ldots, \bar{a}_k\}$ is independent over M (by the minimality of n), $\operatorname{tp}(\bar{a}_2 \frown \bar{b}_2 \frown \ldots \frown \bar{a}_n \frown \bar{b}_n \frown \bar{a}_{n+1} \frown \ldots \frown \bar{a}_k, M \cup \bar{a}_0 \frown \bar{b}'_0 \frown \bar{a}_1)$ is finitely satisfiable in M, hence we can find in M a sequence $\bar{c}^* = \bar{a}'_2 \frown \bar{b}'_2 \frown \bar{a}'_3 \frown \bar{b}'_3 \frown \ldots \frown \bar{a}'_n \frown \bar{b}'_n \frown \bar{a}'_{n+1} \frown \ldots \frown \bar{a}'_k$ such that $\vDash (\exists \bar{y})[\varphi(\bar{a}_0, \bar{b}'_0, \bar{a}_1, \bar{y}, \bar{c}^*, \bar{c}) \wedge \psi(\bar{y}, \bar{a}_1)]$ (remember $\vDash \psi[\bar{b}, \bar{a}]$).

Let $\bar{b}'_1$ satisfy this formula, i.e.

$$\vDash \varphi[\bar{a}_0, \bar{b}'_0; \bar{a}_1, \bar{b}'_1, \bar{c}^*, \bar{c}] \wedge \psi[\bar{b}'_1, \bar{a}_1].$$

Clearly, $\bar{a}_0, \bar{b}'_0, \bar{a}_1, \bar{b}'_1$ are as required.

FACT 1.8: *Suppose $p \in S^m(M)$ is not regular, $\psi(\bar{x}, \bar{a}) \in p$. Then there are $\bar{c}, \bar{c}_0, \bar{c}_1$ such that*

(i) *$\bar{c}, \bar{c}_0$ realizes p, and $\vDash \psi[\bar{c}_1, \bar{a}]$,*

(ii) *$\{\bar{c}, \bar{c}_0\}$ is independent over M,*

(iii) *$\operatorname{tp}(\bar{c}, M \cup \bar{c}_1 \cup \bar{c}_0)$ and $\operatorname{tp}(\bar{c}_0, M \cup \bar{c}_1)$ fork over M.*

Proof. As p is not regular, by Theorem V, 1.9(1), (3) there are $\bar{c}_l (l \leqslant n)$, $\bar{c}$ and $\bar{c}^*$ realizing p such that $\{\bar{c}_l : l \leqslant n\} \cup \{\bar{c}\}$ is independent over M, $\operatorname{tp}(\bar{c}^*, M \cup \bigcup_{l \leqslant n} \bar{c}_l)$ forks over M and $\operatorname{tp}(\bar{c}, M \cup \bigcup_{l \leqslant n} \bar{c}_l \cup \bar{c}^*)$ forks over M. W.l.o.g. n is minimal. Still $n \geqslant 0$. If for every $l \leqslant n$, $\operatorname{tp}(\bar{c}_l, M \cup \bigcup_{m-l+1}^{n} \bar{c}_m \cup \bar{c}^*)$ does not fork over M, then $\operatorname{tp}(\bar{c}_0 \frown \ldots \frown \bar{c}_n, M \cup \bar{c}^*)$ does not fork over M, a contradiction. By renaming and monotonicity of forking, $\operatorname{tp}(\bar{c}_0, M \cup \bigcup \bigcup_{l=1}^{n} \bar{c}_l \cup \bar{c}^*)$ forks over M. By 1.3 there are formulas φ_1, φ_2 and $\bar{a}_1 \in M$ such that letting $\bar{d} = \bar{c}_1 \frown \ldots \frown \bar{c}_n$:

(a) $\vDash \varphi_1[\bar{c}^*, \bar{c}_0, \bar{d}, \bar{a}_1]$,

(b) $\vDash \varphi_2[\bar{c}, \bar{c}^*, \bar{c}_0, \bar{d}, \bar{a}_1]$,

(c) for every $\bar{c}^{**}, \bar{d}', \{\varphi_1(\bar{c}^{**}, \bar{x}, \bar{d}', \bar{a}_1)\} \cup \mathrm{tp}(\bar{c}_0, M)$ forks over M,

(d) for every $\bar{c}^{**}, \bar{d}', \bar{c}_0'\{\varphi_2(\bar{x}, \bar{c}^{**}, \bar{c}_0', \bar{d}', \bar{a}_1)\} \cup \mathrm{tp}(\bar{c}, M)$ forks over M.

So $\models (\exists \bar{y})[\varphi_1(\bar{y}, \bar{c}_0, \bar{d}, \bar{a}_1) \wedge \varphi_2(\bar{c}, \bar{y}, \bar{c}_0, \bar{d}, \bar{a}_1) \wedge \psi(\bar{y}, \bar{a})]$ (use $\bar{c}^*$ as a witness) and $\{\bar{d}, \bar{c}_0, \bar{c}\}$ is independent over M (as $\{\bar{c}_l : l \leqslant n\} \cup \{\bar{c}\}$ is independent over M). Hence for some $\bar{d}' \in M$

$$\models (\exists \bar{y})[\varphi_1(\bar{y}, \bar{c}_0, \bar{d}', \bar{a}_1) \wedge \varphi_2(\bar{c}, \bar{y}, \bar{c}_0, \bar{d}', \bar{a}_1) \wedge \psi(\bar{y}, \bar{a})].$$

So let $\bar{c}^+$ satisfy

$$\models \varphi_1[\bar{c}^+, \bar{c}_0, \bar{d}', a_1] \wedge \varphi_2[\bar{c}, \bar{c}^+, \bar{c}_0, \bar{d}', \bar{a}_1] \wedge \psi[\bar{c}^+, \bar{a}].$$

Clearly, $\bar{c}, \bar{c}_0, \bar{c}^+$ are as required (for $\bar{c}, \bar{c}_0, \bar{c}_1$) in the fact.

CLAIM 1.9: *Suppose $p \in S^n(N)$ is not orthogonal to some type q to which $\psi(\bar{x}, \bar{b})$ belongs, where $\bar{b} \in N$, $l(\bar{x}) = m$. Then*

(1) *for some r, $\psi(\bar{x}, \bar{b}) \in r \in S^m(N)$, r is not weakly orthogonal to p.*

(2) *If $N \subseteq N^*$ and N^* realizes p, then $\psi(N^*, \bar{b}) \neq \psi(N^*, \bar{b})$.*

Proof. (1) Let $N \subseteq M$, M $\mathbf{F}^a_{\kappa_r(T)}$-saturated, and $\mathrm{tp}(\bar{a}, M)$, $\mathrm{tp}(\bar{c}, M)$, are complete types over M which are non-orthogonal stationarization of p, q, respectively. By V, 1.2(3) $\mathrm{tp}(\bar{a}, M)$, $\mathrm{tp}(\bar{c}, M)$ are not weakly orthogonal, so w.l.o.g. $\mathrm{tp}(\bar{a}, M \cup \bar{c})$ forks over M, hence over N. So for some $\bar{d} \in M$, $\bar{e} \in N$, $\models \varphi[\bar{a}, \bar{c}, \bar{d}, \bar{e}]$ and for every $\bar{c}', \bar{d}', \{\varphi(\bar{x}, \bar{c}', \bar{d}', \bar{e})\} \cup p$ forks over N. Now as $\models (\exists \bar{x})[\varphi(\bar{a}, \bar{x}, \bar{d}, \bar{e}) \wedge \psi(\bar{x}, \bar{b})]$, $\{\bar{a}, \bar{d}\}$ independent over N for some $\bar{d}' \in N$, $\models (\exists \bar{x})[\varphi(\bar{a}, \bar{x}, \bar{d}', \bar{e}) \wedge \psi(\bar{x}, \bar{b})]$; hence for some $\bar{c}'$, $\models \varphi[\bar{a}, \bar{c}', \bar{d}', \bar{e}] \wedge \psi[\bar{c}', \bar{b}]$. Clearly, $r = \mathrm{tp}(\bar{c}', N)$ is as required.

(2) Follows by the proof of (1): if $N \subseteq N^1$, $\bar{a} \in N^1$ realizes p, obviously $\models (\exists \bar{x})[\varphi(\bar{a}, \bar{x}, \bar{d}', \bar{e}) \wedge \psi(\bar{x}, \bar{b})]$ and we can choose $\bar{c}' \in \psi(N^1, \bar{b})$, such that $\models \varphi[\bar{a}, \bar{c}', \bar{d}', \bar{e}] \wedge \psi[\bar{c}', \bar{b}]$. Now $\bar{c}' \notin N$ as $\mathrm{tp}(\bar{a}, N \cup \bar{c}')$ forks over N.

XI.2. The axiomatic treatment

We try here to axiomatize the various variants of the decomposition theorem. Notice that we have such theorems for $\mathbf{F}^a_{\aleph_0}$-saturation, $\prec$ (T superstable without the dop, in X, Section 3) for $\mathbf{F}^t_{\aleph_0}$-saturation, $\prec$ (T totally transcendental without the dop, in [Sh 82b]) for $\mathbf{F}^t_{\aleph_0}$-saturated, $\subseteq_a$ (in [Sh 86, Section 2]), T superstable satisfying the $(<\infty, 2)$-existence property; defined here in XII, 4.2) and we shall consider a few others.

We could have omitted the regularity in our treatment, but we can have it in all our applications. The most pleasant case is $\mathbf{T} = \mathbf{T}^t_{\aleph_0}$ (see Def. 3.1) $\subseteq^* = \subseteq^a_{\aleph_0}$ (see Def. 3.1), and a reader who does not like our generality can concentrate on it. The proof that our axioms are satisfied by some pairs is postponed to the next section.

CONTEXT 2.1: We shall consider a pair $(\mathbf{T}, \subseteq^*)$ where $\mathbf{T}$ is a set of types (i.e. m-types for $m < \omega$) and $\subseteq^*$ is a two-place relation on the family of sets $A \subseteq \mathfrak{C}$. If $\mathbf{T} = \mathbf{T}^t_{\aleph_0}$ we omit it (see Def. 3.2) and if $\subseteq^*$ is $\subseteq^t_{\aleph_0}$ (see Def. 3.1) we omit it. Let $\mathbf{T}(A) = \{p \in \mathbf{T};\ p$ a type over $A\}$.

Ax(A1): $\subseteq^*$ is preserved by automorphisms of $\mathfrak{C}$, $A \subseteq^* B$ implies $A \subseteq B$, if $A \not\subseteq B$ let $A \subseteq^* B$ means $A \subseteq^* B \cup A$, and assume $A \subseteq^* A$.

Ax(A2): $A \subseteq^* C$ and $A \subseteq B \subseteq C$ implies $A \subseteq^* B$.

Ax(A3): $\subseteq^*$ is transitive.

Ax(A4): If $A \subseteq^* B_i$, for $i < \alpha, B_i$ increasing, then $A \subseteq^* \bigcup_{i<\alpha} B_i$.

Ax($A4^-_{<\lambda}$): If $A \subseteq B$ and $A \subseteq^* C$ for every $C \subseteq B$ of power $<\lambda$, then $A \subseteq^* B$ (we write λ instead $<\lambda^+$).

Ax(A5): If $A_i \subseteq^* B$ for $i < \alpha, A_i$ increasing, then $\bigcup_{i<\alpha} A_i \subseteq^* B$.

Ax(A6): If $\{A_\eta : \eta \in I\}$ is a non-forking tree, $A_{\eta\restriction k} \subseteq^* A_\eta$, then $A_{\langle\rangle} \subseteq^* \bigcup_{\eta \in I} A_\eta$.

DEFINITION 2.2: (1) A set is $\mathbf{T}$-saturated if every type $p \in \mathbf{T}(A)$ is realized by some $\bar{c} \in A$.

(2) $\mathbf{F}(\mathbf{T}) = \{(p, B) : p \in S^m(A)$ for some $m < \omega, B \subseteq A$, and there is $q \subseteq p\restriction B, q \in \mathbf{T}, q \vdash p\}$.

(3) A is $\mathbf{T}$-atomic over B if A is $\mathbf{F}(\mathbf{T})$-atomic over B.

(4) A is $\mathbf{T}$-primitive over B, if $B \subseteq A$ and for every $\mathbf{T}$-saturated A^* extending B there is an elementary mapping $F, F(A) \subseteq A^*$, $F\restriction B =$ the identity.

(5) A is **T**-prime over B if A is **T**-saturated and **T**-primitive over B.

(6) A is **T**-minimal over B if A is **T**-saturated, $B \subseteq A$, and every **T**-saturated $A^*, B \subseteq A^* \subseteq A$, is equal to A.

Remark. Usually any **T**-saturated set is a model (follows by Ax(B2)). So usually we denote such sets by M, N even if this is not necessarily assumed.

Note that **T**-primitive is not necessarily equivalent to **F(T)**-primitive.

Ax(B1): **T** is preserved by automorphisms of $\mathfrak{C}$.

Ax(B2): Every finite type is in **T**.

Ax(B3): If $M \subseteq A, p \subseteq \mathrm{tp}(\bar{a}, A), p \in \mathbf{T}$, $\mathrm{tp}(\bar{a}, A)$ does not fork over M, and M is **T**-saturated, then p is realized in M.

Ax(B3⁻): Like Ax(B3) but $M \subseteq^* A$.

Ax(B3⁻⁻): Like Ax(B3) but $M \subseteq^* A \cup \bar{a}$.

Ax(B4): If B is **F(T)**-constructible over A, then B is **T**-atomic over A.

Ax(B4⁺): If B is **T**-atomic over $A, A \subseteq B$, and $\mathrm{tp}(\bar{b}, B)$ is **T**-atomic, then $\mathrm{tp}(\bar{b}, A)$ is **T**-atomic.

Ax(B5): If $p(\bar{x}, \bar{y})$ is a type, $p(\bar{x}, \bar{b}) \in \mathbf{T}$ and $p(\bar{x}, \bar{b}) \cup \{\varphi(\bar{x}, \bar{b})\}$ is consistent, then for some $q(\bar{y}) \in \mathbf{T}$ which $\bar{b}$ realizes, $\mathrm{Dom}\, q \subseteq \mathrm{Dom}\, p(\bar{x}, \bar{y})$ and for every $\bar{b}'$ realizing $q(\bar{y}), p(\bar{x}, \bar{b}') \in \mathbf{T}$ and $p(\bar{x}, \bar{b}') \cup \{\varphi(\bar{x}, \bar{b}')\}$ is consistent.

Ax(B6): If $p(\bar{x}, \bar{y}) \in \mathbf{T}$, then $p(\bar{x}, \bar{b}) \in \mathbf{T}$.

Ax(B7): If δ is a limit ordinal, $i < j < \delta \Rightarrow M_i \subseteq M_j$, each M_i is **T**-saturated, then $\bigcup_{i<\delta} M_i$ is **T**-saturated.

Ax(C1): If $N \subseteq N_1 \subseteq C, N \subseteq^* A, \{A, N_1\}$ is independent over N, C is **T**-atomic over $N_1 \cup A$, and N_1, N are **T**-saturated, then $N_1 \subseteq^* C$.

Ax(C1$^-$): Like Ax(C1), but A is **T**-saturated too and $N \subseteq^* N_1$.

Ax(C1^{--}): Like Ax(C1$^-$), but $C \subseteq \mathrm{acl}(N_1 \cup A)$.

Ax(C1$^+$): Like Ax(C1), but N not necessarily **T**-saturated.

Ax(C2): Suppose $M \subseteq^* N^1 \subseteq N$, all **T**-saturated and for some $\bar{c} \in N$, $\bar{c} \notin N^1$ $\mathrm{tp}(\bar{c}, N^1)$ is not orthogonal to M. Then for some $\bar{c} \in N, \bar{c} \notin N^1$ and $\mathrm{tp}(\bar{c}, N^1)$ does not fork over M.

Ax(C3): Suppose $M \subseteq^* A, A \subseteq N, A \neq N$ and M, N are **T**-saturated. If for every $\bar{b} \in N, \bar{b} \notin A$, $\mathrm{tp}(\bar{b}, A)$ is not almost orthogonal to M, then for some $\bar{b} \in N, \bar{b} \notin A, \mathrm{tp}(\bar{b}, A)$ does not fork over M.

Ax(C3$^+$): Like Ax(C3), but we strengthen the conclusion to: $\mathrm{tp}(\bar{b}, A)$ does not fork over M and is regular.

Ax(C3$^-$): Like Ax(C3), but we assume: $M \subseteq^* N$.

Ax(C3$^\pm$): Like Ax(C3$^+$), but we assume: $M \subseteq^* N$.

Ax(C4): If $M \subseteq^* N, M \neq N$, then for some $\bar{c} \in N, \bar{c} \notin M$ and $\mathrm{tp}(\bar{c}, M)$ is regular.

Ax(D1): Every $p \in \mathbf{T}(A)$ can be extended to a **T**-isolated complete type over A.

Ax(D2): Suppose $M \subseteq^* A, A \subseteq N, p \in \mathbf{T}(A)$ and M, N are **T**-saturated, then there is $\bar{c} \in N$ realizing p such that $\mathrm{tp}(\bar{c}, A)$ is almost orthogonal to M.

Ax(D2^{--}): Like Ax(D2) but $M \subseteq^* N$.

Ax(D2$^-$): Like Ax(D2), but suppose also $M_0 \subseteq M_1 \subseteq \ldots \subseteq M_n = M$, each M_l is **T**-saturated and $M_0 \subseteq^* N$ and for each $l < n (\forall B)[M_{l+1} \subseteq B \subseteq N \wedge \mathrm{tp}_*(B, M_{l+1})$ is almost orthogonal to $M_l \Rightarrow M_{l+1} \subseteq^* B]$ and $\mathrm{tp}_*(A, M_{l+1})$ is almost orthogonal to M_l.

Ax(E1): Suppose $M_0 \subseteq^* M_l (l = 1, 2) \{M_1, M_2\}$ is independent over M_0 and each $M_l (l < 3)$ is **T**-saturated. Then there is a **T**-prime **T**-atomic model M over $M_1 \cup M_2$.

Ax(E1$^+$): Like Ax(E1), but we omit the hypothesis $M_0 \subseteq^* M_l$ (but require that they are **T**-saturated).

Ax(E2): For M_l as in Ax(E1), every $p \in S^m(M_1 \cup M_2)$ which is **T**-isolated, is almost orthogonal to M_1 and to M_2.

Ax(E2$^+$): Like Ax(E2), for $M_0 \subseteq M_l$ **T**-saturated, $\{M_1, M_2\}$ independent over M_0.

DEFINITION 2.3: (1) Let $\lambda(\subseteq^*)$ be the first λ such that for every $A \subseteq B$, there is $A', A \subseteq A' \subseteq^* B, |A'| \leqslant \lambda + |A|$.

(2) Let $\lambda(\mathbf{T})$ be the first λ such that the following is impossible: $\mu \geqslant \lambda, A_i$ increasing continuous for $i < \mu^+, |A_i| \leqslant \mu$, and for every i some $p \in \mathbf{T}(A_i)$ is realized in A_{i+1}, but not in A_i.

(3) $\lambda(\mathbf{T}, \subseteq^*)$ is the first $\lambda \geqslant |T|$ such that:

(*) if $A_i (i < \lambda^+)$ is increasing continuous, $|A_i| \leqslant \lambda$, then for some $i: A_i \subseteq^* \bigcup_{j<\lambda^+} A_j$ and every $p \in \mathbf{T}(A_i)$ realized in $\bigcup_j A_j$ is realized in A_i.

(4) $\lambda_s(\mathbf{T}, \subseteq^*)$ is the first $\lambda \geqslant |T|$ such that: if $A_i (i < \lambda^+)$ is increasing continuous, $|A_i| \leqslant \lambda$, then for some closed unbounded subset S of λ^+, for every $i \in S$ of cofinality $\mathrm{cf}(\lambda), A_i \subseteq^* \bigcup_j A_j$ and every $p \in \mathbf{T}(A)$ realized in $\bigcup_j A_j$ is realized in A_i.

DEFINITION 2.4: (1) We say that $\langle N_\eta, \bar{a}_\eta : \eta \in I \rangle$ is a $(\mathbf{T}, \subseteq^*)$-decomposition if:

(a) I is a set of finite sequences closed under initial segments (but $\bar{a}_{\langle\rangle}$ is immaterial),

(b) each N_η is **T**-saturated, $\|N_\eta\| \leqslant \lambda(\mathbf{T}, \subseteq^*)$,

(c) $N_{\eta \restriction k} \subseteq^* N_\eta$,

(d) $\mathrm{tp}(\bar{a}_\eta, N_{\eta^-})$ is orthogonal to $N_{\eta^{--}}$ (when $l(\eta) > 1$),

(e) $\mathrm{tp}_*(N_\eta, N_{\eta \restriction k} \cup \bar{a}_{\eta \restriction (k+1)})$ is almost orthogonal to $N_{\eta \restriction k}$ (where $k < l(\eta), \eta \in I$),

(f) $\{\bar{a}_{\eta \frown \langle i \rangle} : \eta \frown \langle i \rangle \in I\}$ is independent over N_η,

(g) $\mathrm{tp}(\bar{a}_\eta, N_{\eta^-})$ is regular,

(2) we define a $(\mathbf{T}, \subseteq^*, \lambda)$-decomposition similarly, replacing $\lambda(\mathbf{T}, \subseteq^*)$ by λ.

Remark 2.4A. (1) Remember $\eta^- = \eta \restriction (l(\eta) - 1)$.

(2) All we shall prove for $(\mathbf{T}, \subseteq^*)$-decomposition, hold for $(\mathbf{T}, \subseteq^*, \lambda)$-decomposition if λ satisfies 2.3(3) (*).

DEFINITION 2.5: (1) We say that $\langle N_\eta, \bar{a}_\eta : \eta \in I\rangle$ is a $(\mathbf{T}, \subseteq^*)$-decomposition inside M if (in addition to (a)–(g)):

(h) $N_\eta \subseteq M$ for every $\eta \in I$,

(i) if $N_\eta \subseteq A \subseteq M$, $\mathrm{tp}_*(A, N_{\eta^-} \cup \bar{a}_\eta)$ is almost orthogonal to N_{η^-}, then $N_\eta \subseteq^* A$,

(j) M is $\mathbf{T}$-saturated,

(k) $N_{\langle\rangle} \subseteq^* M$.

(2) We say that $\langle N_\eta, \bar{a}_\eta : \eta \in I\rangle$ is a $(\mathbf{T}, \subseteq^*)$-decomposition of M if it is a maximal $(\mathbf{T}, \subseteq^*)$-decomposition inside M (maximal – under the natural ordering).

CLAIM 2.6: (1) *If* $\langle N_\eta, \bar{a}_\eta : \eta \in I\rangle$ *is a* $(\mathbf{T}, \subseteq^*)$*-decomposition, then* $\langle N_\eta : \eta \in I\rangle$ *is a non-forking tree.*

(2) *Any* $(\mathbf{T}, \subseteq^*)$*-decomposition inside* M *can be extended to a* $(\mathbf{T}, \subseteq^*)$*-decomposition of* M *(there is always a* $(\mathbf{T}, \subseteq^*)$*-decomposition inside* M*: the one with* I *empty).*

Proof. (1) Use (e) and (f) of Definition 2.4.

(2) Easy.

CLAIM 2.7: (1) [Ax(B2)] *Every* $\mathbf{T}$*-saturated set is a model.*

(2) Ax(B3) *implies* Ax(B3$^-$) *and* (B3^{--})*; and* Ax(A2), (B3$^-$) *imply* Ax(B3^{--}).

(3) Ax(C1$^+$) *implies* Ax(C1) *implies* Ax(C1$^-$) *implies* Ax(C1^{--}).

(4) Ax(C3$^+$) *implies* Ax(C3) *and* Ax(C3$^\pm$)*, each of which implies* Ax(C3$^-$). *Also,* Ax(C3$^\pm$) *implies* Ax(C4).

(5) Ax(D2) *implies* Ax(D2$^-$) *implies* Ax(D2^{--}).

(6) Ax(D1), (B4) *imply* Ax(E1$^+$) *implies* Ax(E1), *also* Ax(E2$^+$) *implies* Ax(E2).

(7) *Suppose that whenever* $M_l (l < 3)$ *are* $\mathbf{T}$*-saturated* $\{M_1, M_2\}$ *is independent over* M_0*, and* $M_0 \subseteq^* M_l$*, then* $\mathrm{acl}(M_1 \cup M_2)$ *is* $\mathbf{T}$*-saturated. Then* Ax(B2), (C1^{--}) *imply* Ax(C1$^-$).

(8) Ax(C1$^-$), (D2$^-$), (E1) *imply* Ax(E2).

(9) Ax(B3) *and* Ax(B5) *imply* Ax(E2)*; also* Ax(B3$^-$), Ax(B5) *and* Ax(C1^{--}) *imply* Ax(E2).

(10) Ax(B3$^-$), Ax(B5) *and* Ax(D1) *imply* Ax(D2).

(11) *For* $\lambda < \mu$, Ax(A4$^-_{<\lambda}$) *implies* Ax(A4$^-_{<\mu}$). *If* Ax(A2), *then* Ax(A4) *iff* Ax(A4$^-_{<\aleph_0}$).

(12) *Suppose* $\mathbf{T}$ *satisfies* Ax(B3). *If* M *is* $\mathbf{T}$*-saturated,* $p(\bar{x}, \bar{y}) \in \mathbf{T}(M)$*,* $\bar{a} {}^\frown \bar{b}$ *realizes* p *and* $\mathrm{tp}(\bar{b}, M \cup \bar{a})$ *is not almost orthogonal to* M*, then there are* $\bar{a}_l {}^\frown \bar{b}_l$ $(l = 0, 1)$ *satisfying* (i), (ii), (iii) *of* 1.7 *and* $\bar{a}_1 {}^\frown \bar{b}_1$ *realizes* p.

(13) *if T is superstable,* Ax(B3) *implies* Ax(B7).
(14) Ax($B4^+$) *implies* Ax(B4).

Proof. (8) Let M be **T**-prime and **T**-atomic over $M_1 \cup M_2$ (exists by Ax(E1)). Let us prove that $\mathrm{tp}_*(M, M_1 \cup M_2)$ is almost orthogonal to M_1. Let C be a maximal subset of M such that $\mathrm{tp}_*(C, M_1 \cup M_2)$ is almost orthogonal to M_1. Clearly, $M_1 \cup M_2 \subseteq C$. By Ax($C1^-$) $M_1 \subseteq^* M$, and $M_1 \subseteq^* C$ hence by Ax($D2^-$) (even ($D2^{--}$)) C is **T**-saturated. As M is **T**-prime over $M_1 \cup M_2$, M is isomorphic over $M_1 \cup M_2$ to some $M^-, M_1 \cup M_2 \subseteq M^- \subseteq C$. Now $\mathrm{tp}_*(C, M_1 \cup M_2)$ is almost orthogonal to M_1, hence $\mathrm{tp}_*(M^-, M_1 \cup M_2)$ is almost orthogonal to M_1, hence $\mathrm{tp}_*(M, M_1 \cup M_2)$ is almost orthogonal to M_1.

(9), (10) Use 1.3.

(12) Repeat the proof of 1.7.

THEOREM 2.8: [Ax(A1), (A2), (A3), (B2), (B3), ($B4^+$), (B5), (B6), (B7), ($C1^-$), (C2), (C4), ($D2^-$), (E1), (E2) and: (A4) or (A6), (C1)]. *Suppose T does not have the dop. If $\langle N_\eta, \bar{a}_\eta : \eta \in I \rangle$ is a* (**T**, $\subseteq^*$)*-decomposition of M, then M is* **T***-prime* **T***-atomic and* **T***-minimal over* $\bigcup_{\eta \in I} N_\eta$.

We break the proof to a series of claims.

CLAIM 2.9: (1) [Ax(A3), Ax($C1^-$)]. *Let $M_l, M (l < 3)$ be as in Ax*(E1). *Then $M_l \subseteq^* M$ for $l < 3$.*

(2) [Ax(B2), (B3), (B5), (B6)]. *Suppose $M_0 \subseteq^* M_1, M_0 \subseteq^* M_2$, $M_2 \subseteq M_2^*, \{M_1, M_2^*\}$ is independent over M_0 and each model is* **T***-saturated. Then every* **T***-isolated $p \in S^m(M_1 \cup M_2)$ has a unique extension in $S^m(M_1 \cup M_2^*)$.*

Proof. (1) Trivial.

(2) Easy too. Let p be a counterexample. By Ax(B2), (B3) (or use XII, 2.3) $M_1 \cup M_2 \subseteq_t M_1 \cup M_2^*$ (see Definition 3.1) hence by 1.2 p is stationary inside $M_1 \cup M_2^*$. So (as p is a counterexample) some extension of p in $S^m(M_1 \cup M_2^*)$ forks over $M_1 \cup M_2$. Let $\bar{a}$ realize such extension, so $\bar{a}$ realizes p and for some $\bar{b} \in M_2^*, \bar{c} \in M_1$, and $\varphi \vDash \varphi[\bar{a}, \bar{b}, \bar{c}]$ and $\varphi(\bar{x}, \bar{b}, \bar{c})$ fork over $M_1 \cup M_2$. By Fact 1.3 w.l.o.g. for every $\bar{b}', \bar{c}'$, $p \cup \varphi(\bar{x}, \bar{b}', \bar{c}')$ forks over $M_1 \cup M_2$.

As p is **T**-isolated there is $p_0 \subseteq p$, $p_0 \vdash p$ and $p_0 \in \mathbf{T}$ and w.l.o.g. $\bar{c} \subseteq \mathrm{Dom}\, p_0$. Clearly, $p_0 \cup \{\varphi(\bar{x}, \bar{b}, \bar{c})\}$ is consistent, hence by Ax(B5) there is $q(\bar{y},\bar{z})\in\mathbf{T}$ such that $q(\bar{y},\bar{z})\subseteq \mathrm{tp}(\bar{b}\frown\bar{c},\mathrm{Dom}\,p_0)$, and for every $\bar{b}'\frown\bar{c}'$ realizing $q(\bar{y},\bar{z})$, $p_0\cup\{\varphi(\bar{x},\bar{b}',\bar{c}')\}$ is consistent. Now $\mathrm{tp}(\bar{b}, M_1\cup M_2)$ does not fork over $M_2, \bar{c}\in M_1$ and $q(\bar{y},\bar{c})\in\mathbf{T}$ (by Ax(B6)), and $q(\bar{y},\bar{c})\subseteq \mathrm{tp}(\bar{b},M_1\cup M_2)$. So by Ax(B3) some $\bar{b}'\in M_2$ realizes $q(\bar{y},\bar{c})$, hence $\bar{b}'\frown\bar{c}$ realizes $q(\bar{y},\bar{z})$, hence $p_0\cup\{\varphi(\bar{x},\bar{b}',\bar{c})\}$ is consistent. As $p_0\vdash p$, also $p\cup\{\varphi(x,\bar{b}',\bar{c})\}$ is consistent, by the choice of φ it forks over $M_1\cup M_2$. As $p\in S^m(M_1\cup M_2), \bar{c},\bar{b}'\in M_1\cup M_2$, we get a contradiction.

CLAIM 2.10: [Ax(A1), (A2), (A3), (B2), (B3), (B4$^+$), (B5), (B6), (B7), (C1$^-$), (E1), (E2), and: (A4) or (A6), (C1)]. *Suppose $I\subseteq {}^{\omega>}\alpha$ is closed under initial segments, $\langle N_\eta:\eta\in I\rangle$ is a non-forking tree, each N_η is* **T**-*saturated and $N_{\eta\restriction k}\subseteq^* N_\eta$. Then there is a* **T**-*prime* **T**-*atomic model M over $\bigcup_{\eta\in I}N_\eta$, and $N_\eta\subseteq^* M$ for every $\eta\in I$ and $\mathrm{tp}_*(M,\bigcup_{\eta\in I}N_\eta)$ is almost orthogonal to each N_ν.*

Proof. Let $I=\{\eta_i: i<\beta\}$ be such that $\eta_i=\eta_j\restriction k\Rightarrow i\leqslant j$. We define by induction on i a model M_i such that:

(a) $\bigcup_{j<1+i}N_{\eta_j}\subseteq M_i$,
(b) M_i is **T**-prime and **T**-atomic over $\bigcup_{j<1+i}N_{\eta_j}\cup M_\gamma$ for each $\gamma<i$,
(c) $N_{\eta_j}\subseteq^* M_i$ for $j<1+i$,
(d) $M_0=N_{\eta_0}$,
(e) $M_\delta=\bigcup_{j<1+i}M_j$ for δ limit,
(f) $\mathrm{tp}_*(M_i,\bigcup_{j<1+i}N_{\eta_j})$ is almost orthogonal to each $N_{\eta_j}, j<1+i$,
(g) $\mathrm{tp}_*(M_i,\bigcup_{\eta\in I}N_\eta)$ does not fork over $\bigcup_{j<1+i}N_{\eta_j}$.

For $i=0$, no problem: (see (d)).

For $i=\delta$, limit.

We let $M_\delta=\bigcup_{j<\delta}M_j$.

Now (a), (d), (e) are trivial, (c) follows by Ax(A4) (or use Ax(A6), (C1): by Ax(A6) $N_{\eta_j}\subseteq^*\bigcup\{N_{\eta_\alpha}:\alpha<1+i, \eta_j\not\trianglelefteq\eta_\alpha\}$ and then by downward induction on $k\leqslant l(\eta_j)$, using Ax(C1), $N_{\eta_j}\subseteq^*\bigcup\{N_{\eta_\alpha}:\alpha<1+i, \eta_j\restriction k\leqslant\eta_\alpha\}$, so for $k=0$, $N_{\eta_j}\subseteq^*\bigcup_{\alpha<1+i}N_{\eta_\alpha}$; now use Ax(C1) with N_{η_j}, $N_{\eta_j}, \bigcup_{\alpha<1+i}N_{\eta_\alpha}, M_i$ standing for N, N_1, A, C, respectively, using M_δ is **T**-atomic over $\bigcup_{j<1+i}N_{\eta_j}$ proved below). Now (g) follows from (g) for $j<i$, by the finite character and monotonicity of forking. As for (f), let $\bar{c}\in M_\delta$, then for some $\alpha<\delta, \bar{c}\in M_\alpha$, hence $\mathrm{tp}(\bar{c},\bigcup_{j<1+\alpha}N_{\eta_j})$ is almost orthogonal to each $N_{\eta_j} j<1+\alpha$ by (g), $\mathrm{tp}(\bar{c},\bigcup_{j<\delta}N_{\eta_j})$ does not fork over $\bigcup_{j<1+\alpha}N_{\eta_j}$, so easily it is almost orthogonal to each N_{η_j} proving (f). Also, $\mathrm{tp}(\bar{c},\bigcup_{j<1+\alpha}N_{\eta_j})$ is **T**-isolated, hence for some $p\subseteq \mathrm{tp}(\bar{c},\bigcup_{j<1+\alpha}N_{\eta_j})$ $p\vdash \mathrm{tp}(\bar{c},\bigcup_{j<1+\alpha}N_{\eta_j})$ and $p\in\mathbf{T}$. But as $\mathrm{tp}(\bar{c},\bigcup_{j<1+\alpha}N_{\eta_j})$ is almost orthogonal to each $N_{\eta_j}(j<1+\alpha)$ and as $\langle N_\eta:\eta\in I\rangle$ is a non-forking tree, clearly (by XII, 2.3) $\mathrm{tp}(\bar{c}, \bigcup_{j<1+\alpha} N_{\eta_j}) \vdash \mathrm{tp}(\bar{c}, \bigcup_{j<\delta} N_{\eta_\delta})$. We can conclude that $p \vdash \mathrm{tp}(\bar{c}, \bigcup_{j<\delta} N_{\eta_j})$, hence $\mathrm{tp}(\bar{c}, \bigcup_{j<\delta} N_{\eta_j})$ is **T**-isolated.

The same argument proved half of (b); M_δ is **T**-atomic over $\bigcup_{j<\delta} N_{\eta_j} \cup M_\gamma$. For the other half (**T**-prime) let M^+ be **T**-saturated, $\bigcup_{j<1+\alpha} N_{\eta_j} \subseteq M^+$, and we shall define an elementary embedding of M into M^+ over $\bigcup_{j<1+\alpha} N_{\eta_j}$. For this we define by induction on $i \leqslant \delta$ an elementary embedding F_i of M_i into $M^+, F_i \restriction N_{\eta_j}$ = the identity for $j < 1+\alpha$ and $F_j \subseteq F_i$ for $j < 1+\alpha$. F_0 is the identity; for limit we take union, for successor we use (b) from the induction hypothesis. So F_δ is as required.

Really this proves (b) for $\gamma = 0$; the general case is proved similarly. M_i is **T**-saturated by Ax (B7).

For i a successor, let $1+i = j+1$, where $j > 0$ and let $l(\eta_j) = k_j$, $\eta_j \restriction (k_j - 1) = \eta_\alpha$. By (g) for $i-1, \{N_{\eta_j}, M_{i-1}\}$ is independent over N_{η_α}. Those three models are **T**-saturated (M_{i-1} by (b), the others by Def. 2.4), by (c) $N_{\eta_\alpha} \subseteq^* M_{i-1}$ and by Def. 2.4. $N_{\eta_\alpha} \subseteq^* N_{\eta_j}$.

So Ax(E1) applies and a **T**-prime **T**-atomic model M_i over $N_{\eta_j} \cup M_j$ exists. By Ax(E2) $\mathrm{tp}_*(M_i, N_{\eta_j} \cup M_j)$ is almost orthogonal to N_{η_j} and to M_j, and by 2.9(1) $N_{\eta_j} \subseteq^* M_i, M_j \subseteq^* M_i$. Now (a), (d), (e) hold trivially, (c) holds (by Ax(A3)). We can get (f) by 2.9(2), and it implies (g). Now (b) can be proved using (B4$^+$).

CLAIM 2.11: [Ax(A1), (A2), (A3), (B2), (B3), (B4$^+$), (B5), (B6), (B7), (C1$^-$), (E1), (E2) and: (A4) or (A6), (C1)]. *Suppose $\langle N_\eta : \eta \in I \rangle$ is as in the previous claim, and M is* **T**-*prime over $\bigcup_{\eta \in I} N_\eta$. Then $\mathrm{tp}_*(M, \bigcup_{\eta \in I} N_\eta)$ is almost orthogonal to N_ν and $N_\nu \subseteq^* M$ for each $\nu \in I$, and M is* **T**-*atomic over $\bigcup_{\eta \in I} N_\eta$.*

Proof. By 2.10 there is M^+ **T**-prime over $\bigcup_{\eta \in I} N_\eta$ which satisfies the conclusion of 2.11. As M is **T**-prime over $\bigcup_{\eta \in I} N_\eta$, there is an elementary embedding of M into M^+ over $\bigcup_{\eta \in I} N_\eta$, so w.l.o.g. $M \subseteq M^+$. Easily M inherits the required properties from M^+.

CLAIM 2.12: [Ax(A1), (A2), (A3), (B2), (B3), (B5), (B6), (C1$^-$), (C2), (E1), (E2) and: (A4) or (A6), (C1)]. *Suppose T does not have the dop. Suppose $\langle N_\eta, \bar{a}_\eta : \eta \in I \rangle$ is a* (**T**, $\subseteq^*$)-*decomposition $\bigcup_\eta N_\eta \subseteq M, M \neq M^+$, $\bigcup_{\eta \in I} N_\eta \subseteq M^+ \subseteq M, M^+$ is* **T**-*prime* **T**-*atomic over $\bigcup_{\eta \in I} N_\eta$, for each $\nu \in I$, $N_\nu \subseteq^* M^+$ and* $\mathrm{tp}_*(M^+, \bigcup_{\eta \in I} N_\eta)$ *is almost orthogonal to N_ν. Then for some $\bar{c} \in M, \bar{c} \notin M^+$ and $\eta \in I$* $\mathrm{tp}(\bar{c}, M^+)$ *does not fork over N_η and is orthogonal to N_{η^-} if $\eta \neq \langle\ \rangle$. [Actually we use only* (a), (b), (c) *of Definition* 2.4(1).]

Proof. Let $\lambda = \|M\|^+ + |T|^+$, and the following fact is true (under 2.12's hypothesis).

FACT 2.13: (A) *We can find* $\mathbf{F}^a_\lambda$*-saturated* $N^*_\eta (\eta \in I)$ *such that* $N_\eta \subseteq N^*_\eta$, $N^*_{\eta\restriction k} \subseteq N^*_\eta$, $\mathrm{tp}_*(N^*_\eta, M \cup \bigcup\{N^*_\nu : \eta \neq \nu$ *and not* $\eta \lessdot \nu\})$ *does not fork over* N_η.

(B) *For such* N^*_η, $\langle N^*_\eta : \eta \in I\rangle$ *is a non-forking tree and there is* M^*, $\mathbf{F}^a_\lambda$*-prime over* $\bigcup_{\eta\in I} N^*_\eta, M^+ \subseteq M^*$. (*We assume* M^+ *is* **T**-*atomic over* $\bigcup_{\eta\in I} N_\eta$ *and* Ax(B2), (B3), (B5), (B6).)

Proof of 2.13. Only the second phrase of (B) is problematic, and for this it suffices to prove that $\mathrm{tp}_*(M^+, \bigcup_{\eta\in I} N_\eta) \vdash \mathrm{tp}_*(M^+, \bigcup_{\eta\in I} N^*_\eta)$. This property of M^+ is inherited by any submodel, hence it suffices to prove that for some M^-, **T**-prime over $\bigcup_{\eta\in I} N_\eta$, $\mathrm{tp}_*(M^-, \bigcup_{\eta\in I} N_\eta) \vdash \mathrm{tp}_*(M^-, \bigcup_{\eta\in I} N^*_\eta)$. Let $\langle \eta_i : i < \beta\rangle$ be as in the proof of 2.10 and we define by induction on $i \leqslant \beta$ M_i as there and M^*_i such that in addition

(h) $M_i \subseteq M^*_i$, M^*_i is increasing (with i),

(i) M^*_i is $\mathbf{F}^a_\lambda$–prime over $\bigcup_{j<1+i} N^*_{\eta_j}$ (hence $\mathrm{tp}_*(M^*_i, \bigcup_{j<1+\alpha} N^*_{\eta_j}) \vdash \mathrm{tp}_*(M^*_i, \bigcup_{\eta\in I} N^*_\eta)$).

In the successor case $(1+i = j+1 > 1)$ we first define M_i, then prove $\mathrm{tp}_*(M_i, N_{\eta_j} \cup M_{i-1}) \vdash \mathrm{tp}_*(M_i, N_{\eta_j} \cup M^*_{i-1})$ (by 2.9(2)), then prove $\mathrm{tp}_*(M_i, N_{\eta_j} \cup M^*_{i-1}) \vdash \mathrm{tp}_*(M_i N^*_{\eta_j} \cup M^*_{i-1})$ (by 2.9(2)), and at last defining M^*_i.

End of proof of 2.12. So for any $\bar{c} \in M, \bar{c} \notin M^+$, $\mathrm{tp}(\bar{c}, M^*)$ is not orthogonal to some N^*_ν (as T does not have the dop and $\mathrm{tp}(\bar{c}, M^*)$ is parallel to some complete type over M^*, and use X 2.2). Choose such $\bar{c}, \nu$ with minimal $l(\nu)$. But $\{N^*_\nu, M^+ \cup \bar{c}\}$ is independent over N_ν, hence $\mathrm{tp}(\bar{c}, M^+)$ is not orthogonal to N_ν. Now apply Ax(C2).

CLAIM 2.14: [Ax(A2), (C4), (D2$^-$)]. (1) *Suppose* $\langle N_\eta, a_\eta : \eta \in I\rangle$ *is a* $(\mathbf{T}, \subseteq^*)$*-decomposition inside* $M, \nu \in I$, $\mathrm{tp}(\bar{c}, \bigcup_{\eta\in I} N_\eta)$ *does not fork over* N_ν *and is regular and orthogonal to* N_{ν^-} *if* $\nu \neq \langle\ \rangle$, *and* $\rho = \nu ^\frown \langle \alpha \rangle \notin I$. *Then we can define* $\bar{a}_\rho = \bar{c}$, *and* N_ρ *such that* $\langle N_\eta, \bar{a}_\eta : \eta \in I \cup \{\rho\}\rangle$ *is a decomposition inside* M.

(2) *If* $\mathrm{tp}(\bar{c}, \bigcup_{\eta\in I} N_\eta)$ *is not necessarily regular we can still define* $\bar{a}_\rho \notin \bigcup_{\eta\in I} N_\eta, N_\rho$.

Proof. We define by induction on $i < \lambda^+ = \lambda(\mathbf{T}, \subseteq^*)^+$ a set C_i, $N_\eta \cup \bar{c} \subseteq C_i \subseteq M$, $\mathrm{tp}_*(C_i, N_\eta \cup \bar{c})$ is almost orthogonal to N_η, C_i is increasing continuous, and for each i, $\neg[C_i \subseteq^* C_{i+1}]$ if possible and otherwise there is $p \in \mathbf{T}(C_i)$ realized in C_{i+1} but not in C_i if possible. By the definition of $\lambda(\mathbf{T}, \subseteq^*)$ for some i, every $p \in \mathbf{T}(C_i)$ realized in $C = \bigcup\{C_i : i < \lambda(\mathbf{T}, \subseteq^*)^+\}$ is realized in C_i and $C_i \subseteq^* C$. Clearly, by Ax(A2) $C_i \subseteq^* C_{i+1}$, hence

(*) if $C_i \subseteq B \subseteq M, |B| \leqslant \lambda(\mathbf{T}, \subseteq^*), \mathrm{tp}_*(B, N_\eta \cup \bar{c})$ is almost orthogonal to N_η, then $C_i^* \subseteq^* B$.

Also there is no $p \in \mathbf{T}(C_i)$ not realized in C_i, but realized by some $\bar{d} \in M$ such that $\mathrm{tp}_*(\bar{d} \cup C_i, N_\eta \cup \bar{c})$ is almost orthogonal to N_η.

Note that $N_\nu \subseteq^* C_i$ if $\nu = \langle\ \rangle$ as $N_\nu \subseteq^* M$, and if $\nu \neq \langle\ \rangle$ by (i) of Def. 2.5(1).

We want to let $|N_{\eta\frown\langle\alpha\rangle}| = C_i, a_{\eta\frown\langle\alpha\rangle} = \bar{c}$, but why is C_i $\mathbf{T}$-saturated? Ax(D2$^-$) is tailor-made for this (with $N_\nu, C_i, M, N_{\eta\restriction 0}, N_{\eta\restriction 1}, \dots, N_\eta$ here standing for $M, A, N, M_0, M_1, \dots, M_n$ there).

(2) Proved as above. However, $\mathrm{tp}(\bar{c}, N_\nu)$ is not necessarily regular. But by Ax(C4) there is $c' \in N_{\nu\frown\langle\alpha\rangle}, \bar{c}' \notin N_\eta$, such that $\mathrm{tp}(\bar{c}', N_\eta)$ is regular. Now $\mathrm{tp}(\bar{c}', \bigcup_{\eta \in I} N_\eta)$ does not fork over N_ν, hence $\bar{c}'$ satisfies the assumptions on $\bar{c}$ and we can repeat the proof with it.

Proof of 2.8. Easy.

CONCLUSION 2.15: [Ax(A1), (A2), (A3), (B2), (B3), (B4$^+$), (B5), (B6), (B7), (C1$^-$), (C2), (C4), (D2$^-$), (E1), (E2) and: (A4) or (A6), (C1)]. *Suppose T does not have the dop. If $\langle N_\eta, \bar{a}_\eta : \eta \in I\rangle$ is a $(\mathbf{T}, \subseteq^*)$-decomposition of M, then $N_\nu \subseteq^* M$, and $\mathrm{tp}_*(M, \bigcup_{\eta \in I} N_\eta)$ is almost orthogonal to N_ν for each $\nu \in I$.*

Proof. By 2.11 and 2.8.

THEOREM 2.16: [Ax(A1), (A2), (A3), Ax($4^-_{\lambda(\mathbf{T}, \subseteq^*)}$), (B2), (B3), (B4$^+$), (B5), (B6), (B7), (C1$^-$), (C2), (C4), (D2^{--}), (E1), (E2) and: (A4), or (A6), (C1); we can omit (C1$^-$) if we add (E1$^+$), (E2$^+$)]. *Suppose $M \subseteq^* N, M, N$ are $\mathbf{T}$-saturated, $\bar{a} \in N$, $\mathrm{Dp}(\bar{a}, M) = 1$, and $\mathrm{tp}_*(N, M \cup \bar{a})$ is almost orthogonal to M. Then we can find α and $N_\phi, N_{\{i\}}, \bar{a}_i$ for $i < \alpha$ such that:*

(a) $N_\phi \subseteq^* N$, $\mathrm{tp}(N_\phi, M)$ *does not fork over* $N_\phi \cap M$, $N_\phi \cap M \subseteq^* M$, and $\bar{a} \in N_\phi$,

(b) $N_\phi \cup \bar{a}_i \subseteq N_{\{i\}}$, $\mathrm{tp}(\bar{a}_i, N_\phi)$, *is regular*,

(c) $\{\bar{a}_{\{i\}} : i < \alpha\}$ *is independent over* $(M \cup N_\phi, N_\phi)$,

(d) $\mathrm{tp}_*(N_{\{i\}}, N_\phi \cup \bar{a}_{\{i\}})$ *is almost orthogonal to* N_ϕ, *hence* $\mathrm{tp}_*(N_{\{i\}}, N_\phi \cup \bar{a}_{\{i\}}) \vdash \mathrm{tp}_*(N_{\{i\}}, M \cup N_\phi \cup \bar{a}_{\{i\}})$,

(e) N *is* **T**-*prime and* **T**-*atomic over* $M \cup \bigcup_i N_{\{i\}}$,

(f) $\|N_{\{i\}}\|, \|N_\phi\|$ *are* $\leqslant \lambda_s(\mathbf{T}, \subseteq^*)$.

(g) $N_{\{i\}}, N_\phi$ *are* **T**-*saturated*.

Proof. First define N_ϕ to satisfy (a) and (f) (see Def. 2.3(4)), then $\bar{a}_i(i < \alpha)$ [just as a maximal family satisfying (c) and $\bar{a}_{\{i\}} \in N$] then $N_{\{i\}}$ [as a maximal subset of N satisfying (d)]. Now ${}^{-}N_{\{i\}}$ is **T**-saturated by Ax(D2)$^-$. We define by induction on $i \leqslant \alpha M_i$ **T**-prime **T**-atomic over $M \cup \bigcup_{j<i} N_{\{i\}}$ increasing continuous in i.

THEOREM 2.17: [Ax(A1), (A2), (A3), (B2), (B3), (B4$^+$), (B5), (B6), (B7), (C1$^-$), (C2), (C4), (D2^{--}), (E1), (E2) and: (A4) or (A6), (C1)]. *Suppose T does not have the dop and is superstable. Then every* **T**-*saturated model has a* $(\mathbf{T}, \subseteq^*)$-*decomposition* $\langle N_\eta, \bar{a}_\eta : \eta \in I\rangle$ *such that M is* **T**-*prime* **T**-*minimal* **T**-*atomic over* $\bigcup_{\eta\in I} N_\eta$, *and* $N_\nu \subseteq^* M$, $\mathrm{tp}_*(M, \bigcup_{\eta\in I} N_\eta)$ *is almost orthogonal to each* N_ν, *for* $\nu \in I$.

Remark. What is the difference between 2.8 and 2.17 ? The first has an extra hypothesis, Ax(D2$^-$), and stronger conclusion: every $(\mathbf{T}, \subseteq^*)$-decomposition will work, not just some.

Proof. Like the proof in X, Section 3 but using the claims above and their proofs.

DEFINITION 2.18: If $N \subseteq^* M$, then we define an equivalence relation $E = E_{(N,M)}$; its domain is $\{\bar{c} \in M : \mathrm{tp}(\bar{c}, N)$ is regular and trivial$\}$, $\bar{a} E \bar{b}$ iff there is an automorphism F of M over N such that $\{F(\bar{a}), \bar{b}\}$ is not independent over N (equivalently: for some $\bar{a}' \in M$, $\{\bar{a}', \bar{b}\}$ is not independent over N and there are $(N_0, \bar{a}^*), (N_1, a') \in \mathrm{CON}(N, M)$ (see Def. XIII, 3.1), N_0, N_1 isomorphic over N.

XI.3. Specifying the axiomatic treatment

DEFINITION 3.1: (1) Let $A \subseteq^t_\mu B$ means that $A \subseteq B$, and for every set of $<\mu$ formulas in $<\mu$ variables, all whose parameters are from A, if p is realized in B, then p is realized in A.

(2) Let $A \subseteq^s_\mu B$ means that $A \subseteq B$ and for every $a_i \in A$ $(i < \alpha < \mu)$ and $b_j \in B$ $(j < \beta < \mu)$ there are $b'_j \in A$ $(j < \beta < \mu)$ such that

$$\mathrm{tp}_*(\langle b_j : j < \beta\rangle, \{a_i : i < \alpha\}) = \mathrm{tp}_*(\langle b'_j : j < \beta\rangle, \{a_i : i < \alpha\}).$$

(3) Let $A \subseteq^a_\mu B$ means that $A \subseteq B$ and for every $a_i \in A$ $(i < \alpha < \mu)$ and $b_j \in B$ $(j < \beta < \mu)$ there are $b'_j \in A$ $(j < \beta < \mu)$ such that

$$\mathrm{stp}_*(\langle b_j : j < \beta \rangle, \{a_i : i < \alpha\}) \equiv \mathrm{stp}_*(\langle b'_j : j < \beta \rangle, \{a_i : i < \alpha\}).$$

(4) If $\mu = \aleph_0$ we omit it, but we may write $\subseteq_x$ instead of $\subseteq^x$.

(5) We say (A,B) satisfies the Tarski–Vaught condition if $A \subseteq_t A \cup B$.

Observation 3.1A: (5) For $x = t, s, a$, if $\mathrm{tp}(\bar{a}, A)$ is $\mathbf{F}^x_\kappa$-isolated $A \subseteq^x_\kappa B$, then $\mathrm{tp}(\bar{a}, B)$ is $\mathbf{F}^x_\kappa$-isolated, moreover $\mathrm{tp}(\bar{a}, A) \vdash \mathrm{tp}(\bar{a}, B)$.

Proof. Easy.

DEFINITION 3.2: $\mathbf{T}^x_\mu$ is the set of $\mathbf{F}^x_\mu$-types (see Def. IV, 2.1) for $x \in \{t, s, a\}$.

CLAIM 3.3: (1) *A is $\mathbf{T}^t_\mu$-saturated iff $A = |N|$, N is μ-compact.*

(2) *For $x = s, t, a$, A is $\mathbf{T}^x_\mu$-saturated iff $A = |N|$, N is $\mathbf{F}^x_\mu$-compact (see Def.* IV, 2.1(3)); *remember $\mathbf{F}^s_\mu$-compactness is the usual saturation, which, for $\mu > \aleph_0$, is equivalent to $\mathbf{F}^a_\mu$-compactness and to $\mathbf{F}^a_\mu$-saturation and that $\mathbf{F}^t_\mu$-saturation may be a very weak demand if there are few $\mathbf{F}^t_\mu$-isolated complete types.*

(3) *A is $\mathbf{T}^x_\mu$-saturated iff $A \subseteq^x_\mu B$ for every B extending A (for $x = t, s, a$). Similarly for $\mathbf{T}^x_{\mu,\kappa}$, $\subseteq^*_{\mu,\kappa}$. (See Definition* 3.4 *below.)*

(4) *A is $\mathbf{T}^t_{\aleph_0}$-saturated iff $A = |N|$, N model.*

DEFINITION 3.4: (1) For $x = s, t, a$ let $A \subseteq^x_{\mu,\kappa} B$ means that for ever reduct $\mathfrak{C}^-$ of $\mathfrak{C}$ with $<\kappa$ relations and functions, in $\mathfrak{C}^-$ $A \subseteq^x_\mu B$.

(2) For $x = s, t, a$ let $\mathbf{T}^x_{\mu,\kappa}$ means the set of types p which are in $\mathbf{T}^x_\mu$ for some reduct $\mathfrak{C}^-$ of $\mathfrak{C}$ with $<\kappa$ relations and functions.

Remark. Note that we shall use really only $\subseteq^a_{\aleph_0,\aleph_0}$, the other notions being defined for aesthetic reasons.

DEFINITION 3.5: (1) Let $A \subseteq^f_\lambda B$ (for $\lambda \geqslant 2$) means that $A \subseteq B$ and *for every $\alpha < \lambda$*, sequences $\bar{\boldsymbol{a}} \in A, \bar{\boldsymbol{b}} \in B$ (of length $<\lambda + \aleph_0$, so possibly infinite) and formulas $\varphi_i(\bar{\boldsymbol{y}}, \bar{\boldsymbol{x}})$, $\psi_i(\bar{z}_i, \bar{\boldsymbol{y}}, \bar{\boldsymbol{x}})$ for $i < \alpha$ (where $l(\bar{\boldsymbol{y}}) = l(\bar{\boldsymbol{b}}) < \lambda + \aleph_0$, $l(\bar{\boldsymbol{x}}) = l(\bar{\boldsymbol{a}}) < \lambda + \aleph_0$ and $\alpha < \lambda$, but φ_i, ψ_i are first order) *there is $\bar{\boldsymbol{b}}' \in A$* such that

(a) $\models \varphi_i[\bar{\boldsymbol{b}}, \bar{\boldsymbol{a}}]$ implies $\models \varphi_i[\bar{\boldsymbol{b}}', \bar{\boldsymbol{a}}]$,

(b) $R[\psi_i(\bar z_i, \bar b, \bar a), L, \infty] = \xi \geqslant 0$ implies $R[\psi_i(\bar z_i, \bar b', \bar a), L, \infty] = \xi \geqslant 0$.

(2) Let $A \subseteq^e_\lambda B$ (for $\lambda \geqslant 2$) means that $A \subseteq B$ and for $\alpha < \lambda$ and for every $\bar a \in A$ and $\bar b \in B$, of length $< (\lambda + \aleph_0), \varphi_i(\bar y, \bar x), \psi_i(\bar z_i, \bar y, \bar x)$ (for $i < \alpha$) there is $\bar b' \in A$ such that

(a) $\vDash \varphi_i[\bar b, \bar a]$ implies $\vDash \varphi_i[\bar b, \bar a]$,

(b) if $\psi_i[\bar z_i, \bar b, \bar a]$ forks over A, then $\psi_i(\bar z_i, \bar b', \bar a)$ fork over $\bar a$.

(3) Let $A \subseteq^{ef}_\lambda B$ means $A \subseteq^e_\lambda B$ and $A \subseteq^f_\lambda B$. If we omit λ we mean $\lambda = 2$, and write $\subseteq_e$, $\subseteq_f$ instead $\subseteq^e$, $\subseteq^f$.

FACT 3.6: (1) *We could waive* (a) *of* 3.5(1) (*hence the* φ_i*'s*).

(2) *If* $x \in \{f, e\}$ *and* $A \subseteq^x_\lambda C, \bar a \in A, \bar b \in C$, *and* $\operatorname{tp}(\bar c, C)$ *does not fork over* A *and* $\varphi_i(\bar y, \bar x, \bar u)$, $\psi_i(\bar z_i, \bar y, \bar x)$ *are formulas* ($i < \alpha < \lambda$), *then there is* $\bar b' \in A$ *satisfying*

(a) *if* $\vDash \varphi_i[\bar b, \bar a, \bar c]$, *then* $\vDash \varphi_i[\bar b', \bar a, \bar c]$ *for* $i < \alpha$,

(b) *if* $x = f$, *then* $R[\psi_i(\bar z_i, \bar b', \bar a), L, \infty] = R[\psi_i(\bar z_i, \bar b, \bar a), L, \infty]$ *for* $i < \alpha$,

(c) *suppose* $\aleph_0 + \lambda \geqslant \kappa_r(T)$. *If* $x = e$, $i < \alpha$ *and* $\psi_i(\bar z_i, \bar b, \bar a)$ *fork over* $\bar a$, *then* $\psi_i(\bar z_i, \bar b', \bar a)$ *fork over* $\bar a$.

(3) *Suppose* $\lambda + \aleph_0 \geqslant \kappa_r(T)$. *If* $A \subseteq^e_\lambda B$ ($\lambda \geqslant 2$) $A \subseteq C$, $\{B, C\}$ *independent over* A, *then* $C \subseteq^e_\lambda B \cup C$.

Proof. (1) Replace $\psi_i(\bar z_i, \bar y, \bar x)$ by $\psi_i(\bar z_i, \bar y, \bar x) \wedge \varphi_i(\bar y, \bar x)$. So if $\vDash \neg\varphi_i[\bar b', \bar a] \wedge \varphi_i[\bar b, \bar a]$, then $R[\psi_i(\bar z_i, \bar b', \bar a), L, \infty] = -1$, contradicting 3.5(1)(b).

(2) Let $\varphi_i^* = \varphi_i^*(\bar y, \bar c_i)$ be such that $\bar c_i$ is a finite sequence from A, $\vDash \varphi_i^*[\bar b, \bar c_i]$ and for every $\bar b' \in A$, $\vDash \varphi_i(\bar b', \bar a, \bar c)$ iff $\vDash \varphi_i^*(\bar b, \bar c_i)$ (see III, 0.1; note: $A \subseteq_t B$). Let $\bar d$ be a sequence of members of A of length $< \lambda + \aleph_0$ such that $\bar a \subseteq \bar d, \bar c_i \subseteq \bar d$ for $i < \lambda$. So there is $\varphi_i^+ = \varphi_i^+(\bar y, \bar d) \equiv \varphi_i^*(\bar y, \bar c_i)$. If $x = f$ apply 3.5(1) to $\bar d, \bar b, \varphi_i^+, \psi_i (i < \alpha)$ and get the desired conclusion. If $x = e$, as $\lambda \geqslant \kappa_r(T)$ [w.l.o.g. $\operatorname{tp}(\bar c, A)$ does not fork over $\bar d$], and apply 3.5(2) to $\bar d, \bar b, \varphi_i^+, \psi_i (i < \alpha)$.

(3) So suppose $\alpha < \lambda$, $\bar b \in B$, $\bar a \in A$, $\bar c \in C$ are sequences of length $< \lambda + \aleph_0$, $\varphi_i(\bar y, \bar x, \bar u)$, $\psi_i (i < \alpha)$ are formulas, and it suffices to find $\bar b' \in A$ such that

(a) $\vDash \varphi_i[\bar b, \bar a, \bar c]$ implies $\vDash \varphi_i[\bar b', \bar a, \bar c]$,

(b) if $\psi_i(\bar z_i, \bar b, \bar a, \bar c)$ forks over C, then $\psi_i(\bar z_i, \bar b', \bar a, \bar c)$ forks over $\bar a \cup \bar c$.

W.l.o.g. for each i, $\psi_i(\bar z_i, \bar b, \bar a, \bar c)$ forks over C (otherwise replace it by $z_0 \neq z_0$) and $\vDash \varphi_i[\bar b, \bar a, \bar c]$, and $\operatorname{tp}(\bar b {}^\frown \bar c, A)$ does not fork over $\bar a$. Now

(*) $\{\psi_i(\bar{z}_i, \bar{b}, \bar{a}, \bar{u})\} \cup \{\vartheta(\bar{u}, \bar{b}, \bar{a}') : \bar{a}' \in A, \models \vartheta[\bar{c}, \bar{b}, \bar{a}']\}$ fork over A ($\bar{z}_i \frown \bar{u}$ the variables).

Otherwise $\bar{e} \frown \bar{c}^1$ realizing it where $\bar{e}, \bar{c}^1$ correspond to $\bar{z}_i, \bar{u}$ such that $\text{tp}(\bar{e} \frown \bar{c}^1, A \cup \bar{b})$ does not fork over A. But clearly (by 1.2 as $A \subseteq_t B$ and $\text{tp}_*(B, A \cup C)$ does not fork over A), $\text{tp}(\bar{c}, A \cup \bar{b}) = \text{tp}(\bar{c}^1, A \cup \bar{b})$, hence w.l.o.g. $\bar{c}^1 = \bar{c}$; so $\models \psi_i[\bar{e}, \bar{b}, \bar{a}, \bar{c}]$ but $\text{tp}(\bar{e}, A \cup \bar{b} \cup \bar{c})$ does not fork over $A \cup \bar{c}$ (by III,0.1(3) as $\text{tp}(\bar{e} \frown \bar{c}, A \cup \bar{b})$ does not fork over A), a contradiction.

So there is $\vartheta_i(\bar{u}, \bar{b}, \bar{a}_i) \in \text{tp}(\bar{c}, A \cup \bar{b})$ such that $\psi_i(\bar{z}_i, \bar{b}, \bar{a}, \bar{u}) \wedge \vartheta_i(\bar{u}, \bar{b}, \bar{a}_i)$ forks over A. W.l.o.g. $\bar{a}_i \subseteq \bar{a}$ and even ϑ_i is $\vartheta_i(\bar{u}, \bar{b}, \bar{a})$.

Now we apply 3.6(2) for $x = e$ to $A \subseteq^e_\lambda B, \alpha, \bar{a}, \bar{b}, \bar{c}$ and the formulas

$$\varphi'_i(\bar{y}, \bar{x}, \bar{u}) = \varphi_i(\bar{y}, \bar{x}, \bar{u}) \wedge \vartheta_i(\bar{u}, \bar{y}, \bar{x}),$$

$$\psi'_i(\bar{z}_i \frown \bar{v}, \bar{y}, \bar{x}) = \psi_i(\bar{z}_i, \bar{y}, \bar{x}, \bar{v}) \wedge \vartheta_i(\bar{v}, \bar{y}, \bar{x})$$

(with $\bar{z}_i \frown \bar{v}$ standing for $\bar{z}_i \frown \bar{u}$, the infiniteness of the sequence does not bother us by the finite character of forking).

So we get $\bar{b}' \in A$ such that

(a)′ $\models \varphi_i[\bar{b}', \bar{a}, \bar{c}] \wedge \vartheta_i[\bar{c}, \bar{b}', \bar{a}]$,

(b)′ $\psi_i(\bar{z}_i, \bar{b}', \bar{a}, \bar{u}) \wedge \vartheta_i(\bar{u}, \bar{b}', \bar{a})$ forks over $\bar{a}$.

Now (a) follows from (a)′.

Why does $\psi_i(\bar{z}_i, \bar{b}', \bar{a}, \bar{c})$ fork over $\bar{a} \cup \bar{c}$? Otherwise for some $\bar{e}$, $\models \psi_i[\bar{e}, \bar{b}', \bar{a}, \bar{c}]$, and $\text{tp}(\bar{e}, \bar{b}' \cup \bar{a} \cup \bar{c})$ does not fork over $\bar{a} \cup \bar{c}$. As $\text{tp}(\bar{c}, \bar{b}' \cup \bar{a}) \subseteq \text{tp}(\bar{c}, A)$ does not fork over $\bar{a}$, by III, 0.1 $\text{tp}(\bar{e} \frown \bar{c}, \bar{a} \cup \bar{b}')$ does not fork over $\bar{a}$. However, $\psi_i(\bar{z}_i, \bar{b}', \bar{a}, \bar{u}) \wedge \vartheta_i(\bar{u}, \bar{b}', \bar{a})$ forks over $\bar{a}$ (by (b)′) and $\models \psi_i[\bar{e}, \bar{b}', \bar{a}, \bar{c}]$ (by the choice of $\bar{e}$) and $\models \vartheta_i[\bar{c}, \bar{b}', \bar{a}]$ (by (a)′) hence

$$\psi_i(\bar{z}_i, \bar{b}', \bar{a}, \bar{u}) \wedge \vartheta(\bar{u}, \bar{b}', \bar{a}) \in \text{tp}(\bar{e} \frown \bar{c}, \bar{a} \frown \bar{b}').$$

(When $\bar{z}_i, \bar{u}$ corresponds to $\bar{e}, \bar{c}$.) So $\text{tp}(\bar{e} \frown \bar{c}, \bar{a} \frown \bar{b}')$ forks over $\bar{a}$, a contradiction.

CLAIM 3.7: (1) *$A \subseteq^a_\lambda B$ implies $A \subseteq^s_\lambda B$ and $A \subseteq^a_{\lambda,\kappa} B$.*

(2) *$A \subseteq^s_\lambda B$ implies $A \subseteq^x_\lambda B$ implying $A \subseteq^t_\lambda B$ for $x = e, f$.*

(3) *$A \subseteq^x_\lambda$ implies $A \subseteq^x_{\lambda,\kappa} B$.*

Remark. Also, there are obvious monotonicity properties for the cardinals.

CLAIM 3.8: (1) *The axioms* (A1)–(A3) *are satisfied for* $\subseteq^*$ *being* $\subseteq^x_\lambda$, $\subseteq^x_{\lambda,\kappa}$ *($x = t, s, a$, and λ, κ cardinals) and* $\subseteq^f_\lambda$, $\subseteq^e_\lambda$.

(2) *When* $\lambda = \aleph_0$ *also* Ax(A4), (A5) *are satisfied and when* $\mu \geqslant \lambda$, $\mathrm{Ax}(4^-_{<\mu})$ *holds.*

(3) *If* $\subseteq^*$ *is* $\subseteq^x_{\aleph_0}$, $\subseteq^x_{\aleph_0,\kappa}$, $(x = t, a)$ *or* $\subseteq_e$, *also* Ax(A6) *is satisfied.*

(4) *If* **T** *is* $\mathbf{T}^x_\lambda, \mathbf{T}^x_{\lambda,\kappa} (x = t, s, a)$, *then* Ax(B1), (B2) *are satisfied.*

(5) *If* **T** *is* $\mathbf{T}^t_\lambda$ *or* $\mathbf{T}^t_{\lambda,\kappa}$, *or* $\mathbf{T}^a_\lambda$ $(\lambda \geqslant \kappa_r(T))$ *or* $\mathbf{T}^a_{\lambda,\kappa}$ (*where* $\lambda \geqslant \kappa_r(T)$ *or* $\lambda \geqslant \kappa + \aleph_1$ *or* $\lambda = \kappa = \aleph_0$ *and every reduct* $\mathfrak{C}^-$ *of* $\mathfrak{C}$ *with finitely many relations and functions is superstable*), *then* Ax(B3) *holds.*

(6) $\mathrm{Ax}(\mathrm{B4}^+)$ *holds for* **T** *being* $\mathbf{T}^t_\lambda, \mathbf{T}^a_\lambda$ *when* λ *is regular or* cf $\lambda \geqslant \kappa(T)$.

(7) Ax(B5), (B6) *hold for* **T** *being* $\mathbf{T}^x_\lambda, \mathbf{T}^x_{\lambda,\kappa} (x = t, s, a)$.

(8) *If* T *is superstable,* **T** *is* T^x_λ $(x = t, a$ *or* $x = s$, $\lambda > \aleph_0)$ *then* Ax(B7) *holds.*

Proof. Easy, e.g. for (3), $\subseteq_e$ use 3.6(3).

CLAIM 3.9: (1) *Suppose* $\kappa_r(T) \leqslant \lambda = \mathrm{cf}\,\lambda$ *or* $\kappa = \lambda = \mathrm{cf}\,\lambda > \aleph_0$ *or* $\kappa = \lambda = \kappa_r(T)$ *and* $(\mathbf{T}, \subseteq^*)$ *is one of* $(\mathbf{T}^x_\lambda, \subseteq^x_\lambda)$, $(\mathbf{T}^x_{\lambda,\kappa}, \subseteq^x_{\lambda,\kappa})$, *for* $x = t, s, a$. *Then* $\mathrm{Ax}(\mathrm{C1}^+)$ *holds.*

(2) *The pair* $(\mathbf{T}^t_{\aleph_0}, \subseteq_e)$ *satisfies* $\mathrm{Ax}(\mathrm{C1}^{--})$ *for* T *superstable.*

Proof. (1) Immediate by 3.3(3).

(2) By 3.6(3) $N_1 \subseteq^* N_1 \cup A$, and we know $C \subseteq \mathrm{acl}(N_1 \cup A)$. As $\subseteq_e$ satisfies Ax(A2) (see 3.8(1)) and Ax(B2) is satisfied it suffices to prove:

FACT 3.9A: *If* $N \prec \mathfrak{C}, N \subseteq_e B$, *then* $N \subseteq_e \mathrm{acl}\, B$.

Proof of the Fact. Suppose $\bar{a} \in N, \bar{b} \in \mathrm{acl}\, B$, and w.l.o.g. $\models \varphi[\bar{b}, \bar{a}]$, and $\psi(\bar{z}, \bar{b}, \bar{a})$ fork over $\bar{a}$. We know that $\mathrm{tp}(\bar{b}, B)$ is isolated, so there is $\bar{c} \in B$ and formula ϑ such that $\models \vartheta[\bar{b}, \bar{c}]$, $\vartheta(\bar{y}, \bar{c}) \vdash \mathrm{tp}(\bar{b}, N)$ and $\vartheta(\bar{y}, \bar{c})$ is algebraic. So let $\{b_0, \ldots, \bar{b}_{k-1}\}$ be the set of sequences realizing $\vartheta(\bar{y}, \bar{c})$ (note $k > 0$) w.l.o.g. $\bar{b} = \bar{b}_0$. So for each l, $\psi(\bar{z}, \bar{b}_l, \bar{a})$ forks over $\bar{a}$, hence $\bigvee_{l<k} \psi(\bar{z}, \bar{b}_l, \bar{a})$ forks over $\bar{a}$. But this formula is equivalent to $(\exists \bar{y})[\psi(\bar{z}, \bar{y}, \bar{a}) \wedge \vartheta(\bar{y}, \bar{c})]$, hence this formula too forks over A. Apply the definition of $N \subseteq_e A$, with

$$\bar{a}, \bar{c}, \bar{x}, \bar{u}, (\exists^{!k} \bar{y})\, \vartheta(\bar{y}, \bar{u}) \wedge (\forall \bar{y})[\vartheta(\bar{y}, \bar{u}) \rightarrow \varphi(\bar{y}, \bar{x})],$$

$$\psi'(\bar{z}, \bar{u}, \bar{x}) \stackrel{\text{def}}{=} (\exists \bar{y})[\psi(\bar{z}, \bar{y}, \bar{x}) \wedge \vartheta(\bar{y}, \bar{u})]$$

standing for $\bar{a}, \bar{b}, \bar{x}, \bar{y}, \varphi(\bar{y}, \bar{x}), \psi(\bar{z}, \bar{y}, \bar{x})$.

So for some $\bar{c}' \in N$, $\models (\exists^{!k}\bar{y})\vartheta(\bar{y},\bar{c}') \wedge (\forall \bar{y})[\vartheta(\bar{y},\bar{c}') \rightarrow \varphi(\bar{y},\bar{a})]$ and $\psi'(\bar{z},\bar{c}',\bar{a})$ fork over $\bar{a}$. So $\{\bar{y} : \vartheta(\bar{y},\bar{c}')\}$ has exactly k members; call them $\bar{b}'_0,\ldots,\bar{b}'_{k-1}$. As $N \prec \mathfrak{C}$, clearly $\bar{b}'_0,\ldots,\bar{b}'_{k-1} \in N$. Clearly, $\psi'(\bar{z},\bar{c}',\bar{a})$ is equivalent to $\bigvee_{l<k}\psi(\bar{z},\bar{b}'_l,\bar{a})$ but the former forks over $\bar{a}$, hence for some l, $\psi(\bar{z},\bar{b}'_l,\bar{a})$ forks over $\bar{a}$. Lastly, as $\models (\forall \bar{y})[\vartheta(\bar{y},\bar{c}') \rightarrow \varphi(\bar{y},\bar{a})]$ (by the choice of $\bar{c}'$), clearly $\models \varphi[\bar{b}'_l,\bar{a}]$. So $\bar{b}'_l$ is as required in Definition 3.5(2), so we finish the proof of the fact.

CLAIM 3.10: *Suppose T is superstable, and* $(\mathbf{T}, \subseteq^*)$ *is one of* $(\mathbf{T}^x_\lambda, \subseteq^y_\mu)$, $(\mathbf{T}^x_{\lambda,\kappa}, \subseteq^y_\mu)$ $(x = t, s, a; y = s, a)$ $(\mathbf{T}^t_\lambda, \subseteq^f_\mu)$, $(\mathbf{T}^t_\lambda, \subseteq^e_\mu)$. *Then* Ax(C3), (C3$^\pm$) *hold, hence* Ax(C4) (*by* 2.7(4)).

Proof. By the implications among the various $(\mathbf{T}, \subseteq^*)$ it suffices to deal with $(\mathbf{T}^t_{\aleph_0}, \subseteq_f)$, $(\mathbf{T}^t_{\aleph_0}, \subseteq_e)$. Let x be f or e, respectively. Among the sequences $\{\bar{b} \in N : \bar{b} \notin A\}$, choose one with minimal $R[\mathrm{tp}(\bar{b},M), L, \infty]$ and let

$$\bar{d} \in M, \vartheta(\bar{x},\bar{d}) \in \mathrm{tp}(\bar{b},M) \text{ be such that } \alpha \overset{\mathrm{def}}{=} R[\vartheta(\bar{x},\bar{d}), L, \infty] = R[\mathrm{tp}(\bar{b},M), L, \infty].$$

Among the $\{\bar{b} \in N : \bar{b} \notin A, \models \vartheta[\bar{b},\bar{d}]\}$ choose one with minimal $R[\mathrm{tp}(\bar{b}, A), L, \infty]$ and let $\bar{a} \in A, \psi(\bar{x},\bar{a}) \in \mathrm{tp}(\bar{b},A)$ be such that $\beta \overset{\mathrm{def}}{=} R[\psi(\bar{x},\bar{a}), L, \infty] = R[\mathrm{tp}(\bar{b},A), L, \infty]$.

W.l.o.g. $\mathrm{tp}(\bar{b},A)$, $\mathrm{tp}(\bar{a}{}^\frown\bar{b},M)$ does not fork over $\bar{a},\bar{d}$, respectively, and $\psi(\bar{x},\bar{a}) \vdash \vartheta(\bar{x},\bar{d})$ and $\bar{d} \subseteq \bar{a}$.

FACT 3.10A: $\alpha = \beta$.

Suppose not. So $\alpha > \beta$, and $\{\psi(\bar{x},\bar{a})\} \cup \mathrm{tp}(\bar{b},\bar{d})$ forks over $\bar{d}$, hence w.l.o.g. $\psi(\bar{x},\bar{a})$ forks over $\bar{d}$, and so it forks over M.

As $\mathrm{tp}(\bar{b},A)$ is not almost orthogonal to M, w.l.o.g. $\mathrm{tp}(\bar{b},M \cup \bar{a})$ is not almost orthogonal to M; hence, we can apply Fact 1.7 so there are $\varphi, \bar{a}_l, \bar{b}_l$ $(l = 0, 1)$ and $\bar{c} \in M$ as required there.

So clearly ($\bar{b}_1$ witnesses this)

$$\models (\exists \bar{y})[\varphi(\bar{a}_0,\bar{b}_0,\bar{a}_1,\bar{y},\bar{c}) \wedge \psi(\bar{y},\bar{a}_1)].$$

W.l.o.g. $\mathrm{tp}(\bar{a}_0{}^\frown\bar{b}_0{}^\frown\bar{a}_1{}^\frown\bar{b}_1, A)$ does not fork over M, hence

$$\mathrm{tp}(\bar{a}, M \cup \bar{a}_0{}^\frown\bar{b}_0) = \mathrm{tp}(\bar{a}_1, M \cup \bar{a}_0{}^\frown\bar{b}_0).$$

So

$$\models (\exists \bar{y})[\varphi(\bar{a}_0,\bar{b}_0,\bar{a},\bar{y},\bar{c}) \wedge \psi(\bar{y},\bar{a})] \wedge (\forall \bar{z})(\psi(\bar{z},\bar{a}) \rightarrow \vartheta(\bar{z},\bar{d})).$$

By 3.6(2) there is $\bar{a}^* \in M$ such that:

(a) $\models (\exists \bar{y})[\varphi(\bar{a}_0, \bar{b}_0, \bar{a}^*, \bar{y}, \bar{c}) \wedge \psi(\bar{y}, \bar{a}^*)]$ and $\psi(\bar{y}, \bar{a}^*) \vdash \vartheta(\bar{y}, \bar{d})$,

(b) if $x = f$, then $R[\psi(\bar{y}, \bar{a}^*), L, \infty] = R[\psi(\bar{y}, \bar{a}), L, \infty]$, and if $x = e$, then $\psi(\bar{y}, \bar{a}^*)$ forks over $\bar{d}$.

But if $M \not\subseteq_f A$, hence $M \subseteq_e A$, then as $\psi(\bar{y}, \bar{a}^*)$ forks over $\bar{d}$ and $\psi(\bar{y}, \bar{a}^*) \vdash \vartheta(\bar{y}, \bar{d})$, clearly

$$R[\psi(\bar{y}, \bar{a}^*), L, \infty] < R[\vartheta(\bar{y}, \bar{d}), L, \infty] = \alpha.$$

Also, if $M \subseteq_f A$, then

$$R[\psi(\bar{y}, \bar{a}^*), L, \infty] = R[(\psi(\bar{x}, \bar{a}), L, \infty] = \beta < \alpha.$$

So in any case

(b)′ $R[\psi(\bar{y}, \bar{a}^*), L, \infty] < \alpha$.

By (a), as $\mathrm{tp}(\bar{a}^\frown \bar{b}, M) = \mathrm{tp}(\bar{a}_0{}^\frown \bar{b}_0, M)$, clearly

$$\models (\exists \bar{y})[\varphi(\bar{a}, \bar{b}, \bar{a}^*, \bar{y}, \bar{c}) \wedge \psi(\bar{y}, \bar{a}^*)].$$

Choose $\bar{b}^* \in N$ such that

$$\models \varphi[\bar{a}, \bar{b}, \bar{a}^*, \bar{b}^*, \bar{c}] \wedge \psi[\bar{b}^*, \bar{a}^*].$$

Now as $\models \psi[\bar{b}^*, \bar{a}^*]$

$$R[\mathrm{tp}(\bar{b}^*, M), L, \infty] \leqslant R[\psi(\bar{y}, \bar{a}^*), L, \infty] < \alpha.$$

On the other hand, as $\models \varphi[\bar{a}, \bar{b}, \bar{a}^*, \bar{b}^*, \bar{c}]$ and $\mathrm{tp}(\bar{a}^\frown \bar{b}^\frown \bar{c}, \phi) = \mathrm{tp}(\bar{a}_0{}^\frown \bar{b}_0{}^\frown \bar{c}, \phi)$ and the choice of φ, necessarily $\mathrm{tp}(\bar{b}, \bar{c} \cup \bar{a} \cup \bar{a}^* \cup \bar{b}^*)$ forks over $\bar{c} \cup \bar{a}$. But as $\mathrm{tp}(\bar{b}, A)$ does not fork over $\bar{a}$, this implies $\mathrm{tp}(\bar{b}, A \cup \bar{b}^*)$ forks over A, hence $\bar{b}^* \notin A$. So $\bar{b}^*$ contradicts the choice of α.

So we have finished the proof of $\alpha = \beta$. We can conclude that $\mathrm{tp}(\bar{b}, M)$ does not fork over M, so w.l.o.g. $\bar{a} \in M$.

This proves Ax(C3). Now we continue for Ax(C3$^\pm$) by 3.10A w.l.o.g. $\psi = \theta$, $\bar{a} = \bar{d}$.

FACT 3.10B: $\mathrm{tp}(\bar{b}, M)$ *is regular*.

Suppose not, then apply Fact 1.8 to $p = \mathrm{tp}(\bar{b}, M)$, $\psi(\bar{x}, \bar{a}) \in p$, and get $\bar{c}, \bar{c}_0, \bar{c}_1$ satisfying 1.8(i), (ii), (iii). By the symmetry of non-forking (III, 0.1) $\mathrm{tp}(\bar{c}_1, M \cup \bar{c}_0)$ forks over M.

So there are formulas $\varphi_l (l = 1, 2)$ such that (after possibly increasing $\bar{a}$ but still $\bar{a} = \bar{d} \in M$):

(a) $\models \varphi_1[\bar{c}, \bar{c}_1, \bar{c}_0, \bar{a}]$,

(b) $\models \varphi_2[\bar{c}_1, \bar{c}_0, \bar{a}]$,
(c) $\varphi_1(\bar{x}, \bar{c}'_1, \bar{c}'_0, \bar{a})$ forks over $\bar{a}$ for any $\bar{c}'_1, \bar{c}'_0$,
(d) if $x = f$, $R[\varphi_2(\bar{x}, \bar{c}_0, \bar{a}), L, \infty] < R[\psi(\bar{x}, \bar{a}), L, \infty]$,
(e) if $x = e$, $\varphi_2(\bar{x}, \bar{y}, \bar{a}) \vdash \psi(\bar{x}, \bar{a}) \wedge \psi(\bar{y}, \bar{a})$, and $\varphi_2(\bar{x}, \bar{c}_0, \bar{a})$ forks over $\bar{a}$.

We can assume w.l.o.g. $\bar{c} = \bar{b}$. Clearly ($\bar{c}_1$ is a witness)

$$\models (\exists \bar{y})[\varphi_1(\bar{c}, \bar{y}, \bar{c}_0, \bar{a}) \wedge \varphi_2(\bar{y}, \bar{c}_0, \bar{a}) \wedge \psi(\bar{y}, \bar{a})].$$

Remember that $\bar{c}_0$ realizes p, hence $M \subseteq_x M \cup \bar{c}_0$ (we assume $M \subseteq_x N$ as we are proving Ax(C3$^\pm$)). As $\{\bar{c}, \bar{c}_0\}$ is independent over M, by 3.6(2) there is $\bar{c}'_0 \in M$ such that

$$\models (\exists \bar{y})[\varphi_1(\bar{c}, \bar{y}, \bar{c}'_0, \bar{a}) \wedge \varphi_2(\bar{y}, \bar{c}'_0, \bar{a}) \wedge \psi(\bar{y}, \bar{a})],$$
$$\models \psi[\bar{c}'_0, \bar{a}],$$

and $R[\varphi_2(\bar{y}, \bar{c}'_0, \bar{a}), L, \infty] < \alpha$ (a proof by the two cases, $x = f$, $x = e$, as in the previous Fact). Then we an find $\bar{c}'_1 \in N$ such that

$$\models \varphi_1[\bar{c}, \bar{c}'_1, \bar{c}'_0, \bar{a}] \wedge \varphi_2[\bar{c}'_1, \bar{c}'_0, \bar{a}] \wedge \psi[\bar{c}'_1, \bar{a}].$$

By (c) above, $\mathrm{tp}(\bar{c}, M \cup \bar{c}'_1)$ forks over $\bar{a}$, hence over M, hence (remember 3.10A) $\mathrm{tp}(\bar{c}, A \cup \bar{c}'_1)$ forks over A, so $\bar{c}'_1 \notin A$. On the other hand, as $\bar{c} = \bar{b}$:

$$R[\mathrm{tp}(\bar{c}'_1, A), L, \infty] \leqslant R[\mathrm{tp}(\bar{c}'_1, \bar{a} \cup \bar{c}'_0), L, \infty] \leqslant R[\varphi_2(\bar{y}, \bar{c}'_0, \bar{a}), L, \infty] < \alpha.$$

So $\bar{c}'_1$ contradicts the definition of α.

We have finished the proof of 3.10B too, and hence of 3.10.

CLAIM 3.11: *Suppose T is superstable. If* $(\mathbf{T}, \subseteq^*)$ *is one of* $(\mathbf{T}^x_\lambda, \subseteq^y_\mu)$, $(\mathbf{T}^x_{\lambda,\kappa}, \subseteq^y_\mu)$ $(x = t, s, a; y = s, a)$ *or* $(\mathbf{T}^t_\lambda, \subseteq^f_\mu)$, $(\mathbf{T}^t_\lambda, \subseteq^e_\mu)$, *then* Ax(C2) *holds; moreover we can get* $\mathrm{tp}(\bar{c}, N^1)$ *regular.*

Remark. For a simpler proof for $(\mathbf{t}^t_{\aleph_0}, \subseteq^a_{\aleph_0,\aleph_0})$, see [Sh 86].

Proof. The beginning is like that of 3.10. Again it is enough to deal with $(\mathbf{T}^t_{\aleph_0}, \subseteq_f)$ and $(\mathbf{T}^t_{\aleph_0}, \subseteq_e)$. Let x be e or f, respectively. We let M, N^1, N be as in Ax(C2), and $\bar{b}, \vartheta(\bar{x}, \bar{d}), \alpha, \bar{a}, \psi(\bar{x}, \bar{a}), \beta$ be chosen as in the proof of 3.10, restricting ourselves to $\bar{b}$'s such that $\mathrm{tp}(\bar{b}, N^1)$ is not orthogonal to M (with N^1 here standing for A there).

FACT 3.11A: $\alpha = \beta$.

We can find an $\mathbf{F}^a_{\aleph_0}$-saturated M^*, for $l = 0, 1$, $\bar{a}_l \in M^*$ and $\bar{b}_l$, such that $\text{tp}(\bar{a}_l{}^\frown\bar{b}_l, M) = \text{tp}(\bar{a}{}^\frown\bar{b}, M)$, $N_1 \subseteq M^*$, $\{\bar{a}_0, \bar{a}_1\}$ is independent over M, $\text{tp}(\bar{b}_l, M^*)$ does not fork over $\bar{a}_l$, and $\{\bar{b}_0, \bar{b}_1\}$ is independent over M^*. By V, 3.9 (and see Def. V, 3.2) there are $\bar{e}_0, \ldots, \bar{e}_{k-1}$ such that $\text{tp}(\bar{e}_m, M^*)$ is regular, $\{\bar{e}_m : m < k\}$ is independent over M^*, $\text{tp}(\bar{b}_0, M^* \cup \bigcup_{m<k} \bar{e}_m)$ and $\text{tp}(\bigcup_{m<k} \bar{e}_m, M^* \cup \bar{b}_0)$ are $\mathbf{F}^a_{\aleph_0}$-isolated, hence $k = w(p)$.

Suppose $\text{tp}(\bar{e}_m, M^*)$ is orthogonal to M iff $m \geqslant k(0)$. Now $k(0) > 0$ as $\text{tp}(\bar{b}, \bar{a})$ is not orthogonal to M, hence $\text{tp}(\bar{b}_l, \bar{a}_l)$ is not orthogonal to M. Choose $\bar{e}, \bar{a}_0 \subseteq \bar{e} \subseteq M^*$, such that:

(A) $\text{tp}(\bar{b}_0{}^\frown\bar{e}_0{}^\frown\bar{e}_1{}^\frown\ldots{}^\frown\bar{e}_{k-1}, M^*)$ does not fork over $\bar{e}$, and its restriction to $\bar{e}$ is stationary,

(B) $\text{stp}(\bar{e}_0{}^\frown\ldots{}^\frown\bar{e}_{k-1}, \bar{e} \cup \bar{b}_0) \vdash \text{stp}(\bar{e}_0{}^\frown\ldots{}^\frown\bar{e}_{k-1}, M^* \cup \bar{b}_0)$,

(C) $\text{stp}(\bar{b}_0, \bar{e} \cup \bigcup_{m<k} \bar{e}_m) \vdash \text{stp}(\bar{b}_0, M^* \cup \bigcup_{m<k} \bar{e}_m)$.

Now choose $\bar{e}^n (n < \omega)$ such that:

(D) $\{\bar{e}^n : n < \omega\} \subseteq M^*$ is an indiscernible set over $M \cup \bar{a}_0 \cup \bar{a}_1$ based on $\bar{a}_0$, each $\bar{e}^n$ realizing $\text{stp}(\bar{e}, \bar{a}_0)$.

Then choose $\bar{e}^n_0, \ldots, \bar{e}^n_{k-1}$, such that [letting $\bar{e}_{n,1} = \bar{e}^n_0{}^\frown\ldots{}^\frown\bar{e}^n_{k(0)-1}$, $\bar{e}_{n,2} = \bar{e}^n_{k(0)}{}^\frown\ldots{}^\frown\bar{e}^n_{k-1}$, $\bar{e}_{n,0} = \bar{e}_{n,1}{}^\frown\bar{e}_{n,2}$]:

(E) $\text{stp}(\bar{e}_{n,0}{}^\frown\bar{e}^n, \bar{a}_0 \cup \bar{b}_0) \equiv \text{stp}(\bar{e}_0{}^\frown\ldots{}^\frown\bar{e}_{k-1}{}^\frown\bar{e}, \bar{a}_0 \cup \bar{b}_0)$,

(F) $\text{tp}(\bar{e}_{n,0}, M^* \cup \bar{b}_0 \cup \bigcup_{m \neq n} \bar{e}_{m,0})$ does not fork over $\bar{e}^n \cup \bar{b}_0$.

So $\{\bar{e}_{n,0}{}^\frown\bar{e}^n : n < \omega\}$ is an indiscernible set over $\bar{a}_0 \cup \bar{b}_0$ based on $\bar{a}_0 \subseteq \bar{a}_0 \cup \bar{b}_0$. As $\text{tp}(\bar{b}, \bar{a})$ is not orthogonal to M, by V, 1.1, the types $\text{stp}(\bar{b}_0, \bar{a}_0)$, $\text{stp}(\bar{b}_1, \bar{a}_1)$ are not orthogonal, hence $\text{tp}(\bar{b}_0, M^*)$, $\text{tp}(\bar{b}_1, M^*)$ are not orthogonal, hence (V, 1.2(3)) not weakly orthogonal. So for some $\bar{b}_2$ realizing $\text{tp}(\bar{b}_1, M^*)$, $\text{tp}(\bar{b}_0, M^* \cup \bar{b}_2)$ forks over M^*, hence over $\bar{a}_0$. So by 1.4 for some $\bar{c} \in M^*$, and φ

(a) $\models \varphi[\bar{b}_0, \bar{b}_2, \bar{c}, \bar{a}_1, \bar{a}_0]$ and w.l.o.g. $\varphi(\bar{y}_0, y_2, \bar{z}, \bar{x}_1, x_0) \vdash \psi(y_0, \bar{x}_0) \wedge \psi(\bar{y}_2, \bar{x}_1)$,

(b) for any $\bar{b}'_2, \bar{c}', \bar{a}'_1$, $\{\varphi(\bar{x}, \bar{b}'_2, \bar{c}', \bar{a}'_1, \bar{a}_0)\} \cup \text{tp}(\bar{b}_0, \bar{a}_0)$ forks over $\bar{a}_0$, moreover for any $\bar{b}'_1, \bar{c}, \bar{a}'_1$ and $\bar{a}'_0$, if $\text{tp}(\bar{a}'_0, \phi) = \text{tp}(\bar{a}_0, \phi)$, then $\{\varphi(\bar{x}, b'_2, \bar{c}', \bar{a}'_1, \bar{a}'_0)\} \cup \{\vartheta'(\bar{x}, \bar{a}'_0) : \models \vartheta'(\bar{b}, \bar{a})\}$ forks over $\bar{a}'_0$;

Now we shall prove

(c) if $\bar{b}'_2$ realizes $\text{tp}(\bar{b}_2, M^*)$ and $\models \varphi[\bar{b}_0, \bar{b}'_2, \bar{c}, a_1, \bar{a}_0]$ and $n < \omega$, then $\text{tp}(\bar{e}_{n,1}, M^* \cup \bar{b}'_2)$ forks over $\bar{e}^n$.

If not, we shall prove by induction on $m, k(0) \leqslant m \leqslant k$, that $\text{tp}(\bar{e}^n_0{}^\frown\ldots{}^\frown\bar{e}^n_{m-1}, M^* \cup \bar{b}'_2)$ does not fork over $\bar{e}_n$. For $m = k(0)$ we have just assumed this.

Assume we have proved for $m < k$ and we shall prove for $m + 1$. By

(B) and (E) $\operatorname{tp}(\bar{e}_0^n\frown\ldots\frown\bar{e}_{k-1}^n, M^* \cup \bar{b}_0)$ does not fork over $\bar{e}^n \cup \bar{b}_0, \bar{e}^n \subseteq M^*$ and

$$\operatorname{stp}(\bar{e}_0^n\frown\ldots\frown\bar{e}_{k-1}^n\frown\bar{e}^n, \bar{a}_0 \cup \bar{b}_0) \equiv \operatorname{stp}(\bar{e}_0\frown\ldots\frown\bar{e}_{k-1}\frown\bar{e}, \bar{a}_0 \cup \bar{b}_0),$$

hence clearly for some automorphism F of M^* over $M \cup \bar{a}_0$

$$\operatorname{tp}(\bar{e}_0^n\frown\ldots\frown\bar{e}_{k-1}^n, M^*) = F[\operatorname{tp}(\bar{e}_0\frown\ldots\frown\bar{e}_{k-1}, M^*)].$$

Hence (as $\{\bar{e}_0, \ldots, \bar{e}_{k-1}\}$ independent over M^*) $\operatorname{tp}(\bar{e}_m^n, M^* \cup \bigcup_{l<m} \bar{e}_l^n)$ does not fork over M^* (see (A)), hence over $\bar{e}^n$, so it is parallel to $\operatorname{tp}(\bar{e}_m^n, \bar{e}^n)$, hence (as $\bar{e}^n\frown\bar{e}_m^n, \bar{e}\frown\bar{e}_m$ realize the same type over $M \cup \bar{a}_0$) it is orthogonal to M (but not to $M \cup \bar{a}_0$), hence (by (D) and X, Section 1) is orthogonal to $M \cup \bar{a}_1$, hence to $\operatorname{stp}(\bar{b}_1, \bar{a}_1)$, hence to $\operatorname{stp}(\bar{b}_2', M^* \cup \bigcup_{l<m} \bar{e}_l^n)$ (by the induction hypothesis). So $\operatorname{tp}(\bar{e}_m^n, M^* \cup \bar{b}_2' \cup \bigcup_{l<m} \bar{e}_l^n)$ does not fork over M^*. For $m = k$ we get a contradiction by V, 3.2.

Next we shall prove

(d) if $\bar{b}_2'$ realizes $\operatorname{stp}(\bar{b}_2, \bar{a}_0 \cup \bar{a}_1 \cup \bar{c})$ and $\models \varphi[\bar{b}_0, \bar{b}_2', \bar{c}, a_1, \bar{a}_0\}$, then for every n large enough $\operatorname{tp}(\bar{e}_{n,1}, \bar{a}_0 \cup \bar{a}_1 \cup \bar{c} \cup \bar{b}_2' \cup \bar{e}^n)$ forks over $\bar{e}^n$.

Suppose not. We know by III, 4.19(2), III, 4.21(11), III, 2.10 that for every n large enough, $\operatorname{stp}(\bar{b}_2', \bar{a}_0 \cup \bar{a}_1 \cup \bar{c} \cup \bar{e}^n \cup \bar{e}_{n,1}) \equiv \operatorname{stp}(\bar{b}_2, \bar{a}_0 \cup \bar{a}_1 \cup \bar{c} \cup \bar{e}^n \cup \bar{e}_{n,1})$. Choose such n for which the conclusion of (d) fails.

Note that $\operatorname{tp}(\bar{e}_{n,1}\frown\bar{b}_0, M^*)$ does not fork over $\bar{e}^n$. So we can find $\bar{b}_2''$ such that $\operatorname{stp}(\bar{e}_{n,1}\frown\bar{b}_0\frown\bar{b}_2'', \bar{a}_0 \cup \bar{a}_1 \cup \bar{c} \cup \bar{e}^n) \equiv \operatorname{stp}(\bar{e}_{n,1}\frown\bar{b}_0\frown\bar{b}_2', \bar{a}_0 \cup \bar{a}_1 \cup \bar{c} \cup \bar{e}^n)$, $\operatorname{stp}(\bar{e}_{n,1}\frown\bar{b}_0\frown\bar{b}_2'', M^*)$ does not fork over $\bar{a}_0 \cup \bar{a}_1 \cup \bar{c} \cup \bar{e}^n$.

So $\operatorname{tp}(\bar{b}_2'', M^*)$ does not fork over $\bar{a}_0 \cup \bar{a}_1 \cup \bar{c} \cup \bar{e}^n$, and also $\operatorname{tp}(\bar{b}_2, \bar{a}_0 \cup \bar{a}_1 \cup \bar{c} \cup \bar{e}^n)$ does not fork over $\bar{a}_1 \subseteq \bar{a}_0 \cup \bar{a}_1 \cup \bar{c} \cup \bar{e}^n$ and $\operatorname{stp}(\bar{b}_2'', \bar{a}_0 \cup \bar{a}_1 \cup \bar{c} \cup \bar{e}_n) \equiv \operatorname{stp}(\bar{b}_2', \bar{a}_0 \cup \bar{a}_1 \cup \bar{c} \cup \bar{e}^n) \equiv \operatorname{stp}(\bar{b}_2, \bar{a}_0 \cup \bar{a}_1 \cup \bar{c} \cup \bar{e}^n)$. We use the choice of $\bar{b}_2''$ and of n, respectively. We can conclude that $\operatorname{tp}(\bar{b}_2'', M^*) = \operatorname{tp}(\bar{b}_2, M^*)$ and also it is clear that $\models \varphi[\bar{b}_0, \bar{b}_2'', \bar{c}, \bar{a}_1, \bar{a}_0]$. So the hypothesis of (c) holds, hence its conclusion, i.e. $\operatorname{tp}(\bar{e}^{n,1}, M^* \cup \bar{b}_2'')$ forks over M^*.

But we have assumed (that for n the conclusion of (d) fails), i.e. $\operatorname{tp}(\bar{e}_{n,1}, \bar{a}_0 \cup \bar{a}_1 \cup \bar{c} \cup \bar{b}_2' \cup \bar{e}^n)$ does not fork over $\bar{e}^n$, hence (again by the choice of $\bar{b}_2''$ first demand) $\operatorname{tp}(\bar{e}_{n,1}, \bar{a}_0 \cup \bar{a}_1 \cup \bar{c} \cup \bar{b}_2'' \cup \bar{e}^n)$ does not fork over $\bar{e}_n$. As $\operatorname{tp}(\bar{e}_{n,1}\frown\bar{b}_0\frown\bar{b}_2'', M^*)$ does not fork over $\bar{a}_0 \cup \bar{a}_1 \cup \bar{c} \cup \bar{e}^n$, clearly $\operatorname{tp}(\bar{e}_{n,1}, M^* \cup \bar{b}_2'')$ does not fork over $\bar{a}_0 \cup \bar{a}_1 \cup \bar{c} \cup \bar{e}^n \cup \bar{b}_2''$; so together with the previous sentence $\operatorname{tp}(\bar{e}_{n,1}, M^* \cup \bar{b}_2'')$ does not fork over $\bar{a}_0 \cup \bar{a}_1 \cup \bar{c} \cup \bar{e}^n$, but $\bar{a}_0 \cup \bar{a}_1 \cup \bar{c} \cup \bar{e}^n \subseteq M^*$, so this contradicts the previous paragraph; thus we have proved (d). Now working in $\mathfrak{C}^{eq}$:

(e) There are formulas τ_1, τ_2 and $\bar{c}_1, \bar{c} \subseteq \bar{c}_1 \subseteq \mathrm{acl}(\bar{a}_0 \cup \bar{a}_1 \cup \bar{c})$, such that

(e1) $\models \tau_1[\bar{b}_0, \bar{b}_2, \bar{a}_0, \bar{a}_1, \bar{c}_1]$,

(e2) if $\models \tau_1[\bar{b}_0, \bar{b}'_2, \bar{a}_0, \bar{a}_1, \bar{c}_1]$ then for every n large enough

$$\models \tau_2[\bar{e}_{n,1}, \bar{a}_0, \bar{a}_1, \bar{c}_1, \bar{b}'_2, \bar{e}^n],$$

(e3) for any $\bar{a}'_0$, $\bar{a}'_1, \bar{c}'_1$, $\bar{b}'_2$ and $\bar{e}'$, such that $\mathrm{tp}(\bar{e}', \phi) = \mathrm{tp}(\bar{e}, \phi) = \mathrm{tp}(\bar{e}^n, \phi)$, $\{\tau_2(\bar{x}, \bar{a}'_0, \bar{a}'_1, \bar{c}'_1, \bar{b}'_2, \bar{e}')\} \cup \{\tau(\bar{x}, \bar{e}') : \models \tau[\bar{e}_{n,1}, \bar{e}^n]\}$ fork over $\bar{e}'$.

This is easy by 1.3, III, 2.5 and compactness (relying on (d), of course), and w.l.o.g. $c_1 = \bar{c}$ and we return to $\mathfrak{C}$.

(f) W.l.o.g. $\bar{a}_0 = \bar{a}$, $\bar{b}_0 = \bar{b}$, $N^1 \subseteq M^*$, $\mathrm{tp}_*(M^*, N)$ does not fork over N^1 and $\mathrm{tp}(\bar{a}_1, N^1)$ does not fork over M.

Define
$$\tau_3(\bar{x}_1, \bar{z}, \bar{y}_2) = (\exists^\infty n)\, \tau_2(\bar{e}_{n,1}, \bar{a}_0, \bar{x}_1, \bar{z}, \bar{y}_2, \bar{e}^n).$$

As $\{\bar{e}_{n,1} {}^\frown \bar{e}^n : n < \omega\}$ is an indiscernible set based on $\bar{a}_0 {}^\frown \bar{b}_0 = \bar{a} {}^\frown \bar{b}$. By III, 2.5 τ_3 is a first-order formula which is almost over $\bar{a}_0 {}^\frown \bar{b}_0$, hence $\tau_3 = \tau_3(\bar{x}_1, \bar{z}, \bar{y}, \bar{b}^*_0)$ for some $\bar{b}_0 \subseteq \bar{b}^*_0 \subseteq N$, and choose $\bar{a}^*_0, \bar{a}_0 \subseteq \bar{a}^*_0 \subseteq N^1$, $\mathrm{tp}(\bar{b}^*_0, N^1)$ does not fork over $\bar{a}^*_0$, and w.l.o.g. $\bar{a}^*_0 \subseteq \bar{b}^*_0$. Clearly by (e1) and (e2) above $\models \tau_3[\bar{a}_1, \bar{c}, \bar{b}_2, \bar{b}^*_0]$ and (by (e2)); for any $\bar{b}'_2$, if $\models \tau_1[\bar{b}_0, \bar{b}'_2, \bar{a}_0, \bar{a}_1, \bar{c}]$, then $\models \tau_3[\bar{a}_1, \bar{c}, \bar{b}'_2, \bar{b}^*]$. Let

$$\tau_4(\bar{x}_1, \bar{z}, \bar{b}^*_0, \bar{a}^*_0) = (\forall \bar{y}_2)[\tau_1(\bar{b}_0, \bar{y}_2, \bar{a}_0, \bar{x}_1, \bar{z}) \rightarrow \tau_3(\bar{x}_1, \bar{z}, \bar{y}_2, \bar{b}^*_0)].$$

So

(e4) $\models \tau_4[\bar{a}_1, \bar{c}, \bar{b}^*_0, \bar{a}^*_0]$.

Let
$$\varphi_1(\bar{y}_0, \bar{z}, \bar{x}_1, \bar{a}_0) = (\exists \bar{y}_2)[\varphi(\bar{y}_0, \bar{y}_2, \bar{z}, \bar{x}_1, \bar{a}_0) \wedge \tau_1(\bar{y}_0, \bar{y}_2, \bar{a}_0, \bar{x}_1, \bar{z})]$$

Clearly $\mathrm{tp}(\bar{b}^*_0, M^*)$ does not fork over $\bar{a}^*_0$, hence there is a formula $\varphi_2(\bar{z}, \bar{x}_1)$ with parameters from N^1 such that:

(*) for every $\bar{c}', \bar{a}'_1 \in M^*$
$$\models \varphi_2[\bar{c}', \bar{a}'_1] \quad \text{iff} \models \varphi_1[\bar{b}_0, \bar{c}', \bar{a}'_1, \bar{a}_0] \wedge \tau_4[\bar{a}'_1, \bar{c}', \bar{b}^*_0, \bar{a}^*_0].$$

So clearly $\models \varphi_2[\bar{c}, \bar{a}_1]$ [by (e4) $\models \tau_4(\bar{a}_1, \bar{c}, \bar{b}^*_0, \bar{a}^*_0)$ and using $\bar{b}_2$ for $\bar{y}_2$, by (a) and (e1) $\models \varphi_1[\bar{b}_0, \bar{c}, \bar{a}_1, \bar{a}_0]$, use (*) above] and $M \subseteq_x M \cup \bar{a}_1$ [as $M \subseteq_x N^1, \bar{a} \in N^1, \mathrm{tp}(\bar{a}, M) = \mathrm{tp}(\bar{a}_1, M)$] and $\mathrm{tp}(\bar{a}_1, N^1)$ does not fork over M [see (f)], so we can apply 3.6(2) and get $\bar{a}'_1 \in M$ such that

(i) $(\exists \bar{z})\varphi_2(\bar{z}, \bar{a}'_1)$ (remember φ_2 is over N^1),

(ii) $R[\psi(\bar{y}, \bar{a}'_1), L, \infty] < \alpha$ (we repeat the argument in 3.10A's proof).

By (i), for some $\bar{c}' \in N^1$,

(iii) $\models \varphi_2[\bar{c}', \bar{a}'_1]$.

Hence,

(iv) $\models \varphi_1[\bar{b}_0, \bar{c}', \bar{a}'_1, \bar{a}_0]$ and $\models \tau_4[\bar{a}'_1, \bar{c}', \bar{b}^*_0, \bar{a}^*_0]$.

By φ_1's definition for some $\bar{b}'_2 \in N$,

(v) $\models \varphi[\bar{b}_0, \bar{b}'_2, \bar{c}', \bar{a}'_1, \bar{a}_0]$ and $\models \tau_1[\bar{b}_0, \bar{b}'_2, \bar{a}_0, \bar{a}'_1, \bar{c}']$.

By (iv) and τ_4's definition, and (v),

(vi) $\models \tau_3[\bar{a}'_1, \bar{c}', \bar{b}'_2, \bar{b}^*_0]$.

By τ_3's definition,

(vii) $(\exists^\infty n)\, \tau_2[\bar{e}_{n,1}, \bar{a}_0, \bar{a}'_1, \bar{c}', \bar{b}'_2, \bar{e}^n]$.

By (e3),

(viii) $(\exists^\infty n)$ s.t. $\mathrm{tp}(\bar{e}_{n,1}, M^* \cup \bar{b}'_2)$ forks over M^*.

But this means $\mathrm{tp}(\bar{e}_{n,1}, M^*)$, $\mathrm{tp}(\bar{b}'_2, M^*)$ are not orthogonal, but the latter is parallel to $\mathrm{tp}(\bar{b}'_2, N^1)$ and the former is orthogonal to any regular type orthogonal to M. We conclude that

(ix) $\mathrm{tp}(\bar{b}'_2, N^1)$ is not orthogonal to M, hence $\bar{b}'_2 \notin N^1$.

But by (v) and the choice of φ (see (a)), $\models \psi[\bar{b}'_2, \bar{a}'_1]$, so by (ii),

$$R[\mathrm{tp}(\bar{b}'_2, M), L, \infty] \leqslant R[\psi(\bar{x}, \bar{a}'_1), L, \infty] < \alpha,$$

and by (ix) $\bar{b}'_2 \notin N^1$ and of course $\bar{b}'_2 \in N$. So $\bar{b}'_2$ contradicts the choice of α. So we prove $\alpha = \beta$ (i.e. Fact 3.11A), hence Ax(C2).

We can continue:

FACT 3.11B: *$\mathrm{tp}(\bar{b}, M)$ is regular.*

This is just like 3.10B [i.e. in the end, we assume $\bar{c} = \bar{b}$, get $\bar{c}'_1 \in N$, $\bar{c}'_1 \notin N^1$, such that $R[\mathrm{tp}(\bar{c}'_1, M), L, \infty] < \alpha$, $\mathrm{tp}(\bar{b}, N^1 \cup \bar{c}'_1)$ fork over M. So $\mathrm{tp}(\bar{c}'_1, N^1)$ is not orthogonal to $\mathrm{tp}(\bar{b}, N^1)$, hence to $\mathrm{tp}(\bar{b}, M)$, hence is not orthogonal to M. So $\bar{c}'_1$ contradicts the definition of α].

CLAIM 3.12: *Suppose T is superstable and $D(T)$ is scattered. Then $(\mathbf{T}^t_{\aleph_0}, \subseteq^t_{\aleph_0})$ satisfies*

(1) Ax(C2$^+$), (C3), (C3$^+$).

(2) Ax(D2).

Proof. (1) Like that of 3.10, 3.11; as the changes are similar, we concentrate on Ax(C3$^+$). For very finite B and m, $S^m(B)$ is a scattered topological space, hence the Cantor–Bendixon rank CB is defined. Among all pairs

$$\{(\bar{b}, \bar{d}) : \bar{b} \in N, \bar{b} \notin A, \bar{d} \in M\}$$

we choose one with minimal pair

$$(\alpha_0, \alpha_1) = (R[\mathrm{tp}(\bar{b}, \bar{d}), L, \infty], \mathrm{CB}(\mathrm{tp}(\bar{b}, \bar{d}))$$

(by the lexicographic order). Then choose $\vartheta(\bar{x}, \bar{d})$ which witness it, so

$$\vartheta(\bar{x}, \bar{d}) = \mathrm{tp}(\bar{b}, \bar{d}),$$

$$R[\mathrm{tp}(\bar{b}, \bar{d}), L, \infty] = R[\vartheta(\bar{x}, \bar{d}), L, \infty],$$

$$r \neq \mathrm{tp}(\bar{b}, \bar{d}), \vartheta(\bar{x}, \bar{d}) \in r \in S^m(\bar{d}) \Rightarrow \mathrm{CB}(r) < \mathrm{CB}(\mathrm{tp}(\bar{b}, \bar{d})).$$

We similarly choose among $\{(\bar{b}, \bar{a}) : \bar{b} \in N, \bar{b} \notin A, \bar{a} \in A, \models \vartheta[\bar{b}, \bar{d}]\, \bar{d} \subseteq \bar{a}$, $\mathrm{tp}(\bar{b}, M \cup \bar{a})$ not almost orthogonal to $M\}$ one with minimal

$$(\beta_0, \beta_1) = (R[\mathrm{tp}(\bar{b}, \bar{a}), L, \infty], \mathrm{CB}(\mathrm{tp}(\bar{b}, \bar{a}))$$

and let $\psi(\bar{x}, \bar{a})$ witness it.

FACT 3.12A: $\alpha_0 = \beta_0$.

This is proved like 3.10A, the only difference is the way we replace the use of (b) and (b)′ in showing that $\psi(\bar{x}, \bar{a}^*), \bar{b}^*$ contradicts the choice of (α_0, α_1). So in choosing $\bar{a}^*$ we replace (b) by

(b)′ $R[\psi(\bar{x}, \bar{a}^*), \psi(\bar{x}, \bar{y}), \aleph_0] \leqslant R[\psi(\bar{x}, \bar{a}), \psi(\bar{x}, \bar{y}), \aleph_0]$.

This is easy to do. W.l.o.g. (*) $R(\psi(\bar{x}, \bar{a}), \psi, \aleph_0) < R(\mathrm{tp}(\bar{b}, \bar{d}), \psi, \aleph_0)$ (by III, 4.1, II, 2.1). (If $n = R[\psi(\bar{x}, \bar{a}), \psi(\bar{x}, \bar{y}), \aleph_0]$, then by Exercise II, 3.7 (p. 55) for some $k < \omega$, $n = R[\psi(\bar{x}, \bar{a}), \psi(\bar{x}, \bar{y}), k]$. By II, 2.9(2) for some formula ϑ_1, for every $\bar{e}'$, $n = R[\psi(\bar{x}, \bar{e}), \psi(\bar{x}, \bar{y}), k]$ iff $\models \vartheta_1[\bar{e}]$. So for (b)′ to hold $\models \vartheta_1[\bar{b}^*]$ is enough, as always $R[\psi(\bar{x}, \bar{a}^*), \psi(\bar{x}, \bar{y}), k] \geqslant R[\psi(\bar{x}, \bar{a}^*), \psi(\bar{x}, \bar{y}), \aleph_0]$. We choose, as there, $\bar{b}^*$, clearly $\bar{b}^* \in N$, $\bar{b}^* \notin A$; now by the choice of (α_0, α_1), $\theta(x, \bar{d})$, necessarily $\mathrm{tp}(\bar{b}^*, \bar{d}) = \mathrm{tp}(\bar{b}, \bar{d})$, $\mathrm{tp}(\bar{b}^*, M)$, $\mathrm{tp}(\bar{b}, M)$ does not fork over M, hence $R(\mathrm{tp}(\bar{b}^*, M), \psi, \aleph_0) = R(\mathrm{tp}(\bar{b}, M), \psi, \aleph_0)$ (see III, 4.1), hence $R(\mathrm{tp}(\bar{b}, M), \psi, \aleph_0)$, $R(\mathrm{tp}(\bar{b}^*, M), \psi, \aleph_0) \leqslant R(\psi(\bar{x}, \bar{b}^*), \psi, \aleph_0) \leqslant R(\psi(\bar{x}, \bar{b}^*), \psi, \aleph_0)$, contradicting (*).)

FACT 3.12B: $\mathrm{tp}(\bar{b}, M)$ *is regular*.

A similar change made the proof of 3.10B work.

Proof of 3.12(2). Combine the proof of 3.14 below and the above.

Remark. Really we use:

FACT 3.12C: *Let T be superstable, $D(T)$ scattered.*

Suppose Θ is an ideal of m-formulas over A. Then for every m-formula $\varphi(\bar{x},\bar{a}) \notin \Theta$ over A there is an m-formula $\psi(\bar{x},\bar{b})$ over A such that:

(a) $\psi(\bar{x},\bar{b}) \vdash \varphi(\bar{x},\bar{b})$,

(b) $\bar{b} \in A$,

(c) $\psi(\bar{x},\bar{b}) \notin \Theta$,

(d) $\psi(\bar{x},\bar{b})$ has no extension in $S^m(A)$ disjoint to Θ which forks over $\bar{b}$,

(e) $\psi(\bar{x},\bar{b})$ has no two contradictory extensions over $\bar{b}$ which are not in Θ.

Proof. Clear.

CLAIM 3.13: *Suppose* $\mathbf{T}$ *is* $\mathbf{T}^a_\kappa$, $\kappa \geqslant \kappa(T)$, *or* $\mathbf{T}^t_\kappa$, *T totally transcendental, then* Ax(D1) *holds. If in addition* $(\mathbf{T}, \subseteq^*)$ *satisfies* Ax(B3$^-$), (B5), *then* Ax(D2), (D2$^-$), (D2^{--}) *hold.*

Proof. Immediate (see IV, 2.15–2.18 for those and other such theorems). As for Ax(D2), see 2.7(10), and as for Ax(D2$^-$), Ax(D2^{--}), see 2.7(5).

CLAIM 3.14: *Suppose* $(\mathbf{T}, \subseteq^*)$ *is* $(\mathbf{T}^t_{\kappa_r(T)}, \subseteq^a_\lambda)$ $\lambda \geqslant \kappa_r(T)$ *or* $(\mathbf{T}^t_{\kappa_r(T)}, \subseteq^a_{\lambda,\kappa})$. $\lambda \geqslant \kappa_r(T)$, $\kappa \geqslant \kappa_r(T)$ *or* $(\mathbf{T}^t_{\aleph_0}, \subseteq_e)$, $\kappa(T) = \aleph_0$. *Then* Ax(D2$^-$) *holds.*

Proof. Let $B = \operatorname{Dom} p$ and (as $\mathbf{T} = \mathbf{T}^t_{\kappa_r(T)}$) w.l.o.g. p has no extension over A which forks over B. Let $\bar{c} \in N$ realize p; so assume $\operatorname{tp}(\bar{c},A)$ is not almost orthogonal to M, and we shall eventually get a contradiction, thus finishing.

Let $l \leqslant n$ be minimal such that $\operatorname{tp}(\bar{c},A)$ is not almost orthogonal to M_l.

We can conclude that $M_l \subseteq^* A \cup \bar{c}$. [If $l = 0$, by a hypothesis $M_0 \subseteq^* N$, hence by Ax(A2) $M_0 \subseteq^* A \cup \bar{c}$. If $l = m+1$, then $\operatorname{tp}(\bar{c},A)$ is almost orthogonal to M_m by [choice of l], and $\operatorname{tp}_*(A,M_l)$ is almost orthogonal to M_m [by an assumption in Ax(D2$^-$)], so together $\operatorname{tp}_*(A \cup \bar{c},M_l)$ is almost orthogonal to M_m. By a hypothesis [of Ax(D2$^-$)] the last phrase implies $M_l \subseteq^* A \cup \bar{c}$.]

As we can replace p by any p', $p \subseteq p' \subseteq \operatorname{tp}(\bar{c},A)$ of power $< \kappa_r(T)$, we can assume that $\operatorname{tp}(\bar{c},M_l \cup B)$ is not almost orthogonal to $B \cap M_l$, and $\mathrm{tp}_*(B \cup \bar{c}, M_l)$ does not fork over $D \stackrel{\text{def}}{=} B \cap M_l$. By 2.7(12) there are $\bar{c}', B_1, \bar{c}_1$ such that $B_1 \cup \bar{c}_1$ realizes $\mathrm{stp}_*(B \cup \bar{c}, D)$, $\mathrm{tp}_*(B_1, D \cup B \cup \bar{c})$ does not fork over D, $\bar{c}'$ realizes p but $\mathrm{tp}(\bar{c}', D \cup B \cup B_1 \cup \bar{c}_1)$ forks over B; and let $\varphi(\bar{x}, \bar{a}) \in \mathrm{tp}(\bar{c}', D \cup B \cup B_1 \cup \bar{c}_1)$ fork over B. [Strictly speaking, in 2.7(12) B should be finite; this is anyhow the most important case, and we can imitate the proof generally.]

Case I. $(\mathbf{T}, \subseteq^*)$ is $(\mathbf{T}^t_{\kappa_r}(T), \subseteq^a_{\lambda,\kappa})$.

Let $\mathfrak{C}'$ be a reduct of $\mathfrak{C}$ to the predicates and function symbols appearing in p, or $\varphi(\bar{x}, \bar{d})$. This reduces Case I to Case II.

Case II. $(\mathbf{T}, \subseteq^*)$ is $(\mathbf{T}^t_{\kappa_r(T)}, \subseteq^a_{\lambda})$.

As $M_l \subseteq^* A \cup \bar{c}, B \cup \bar{c} \subseteq A \cup \bar{c}$, and $|B \cup \bar{c}| < \lambda$, $|D| \leqslant |B| < \lambda$ (and the definition of $\subseteq^a_\lambda$), clearly $\mathrm{stp}(B \cup \bar{c}, D)$ is realized in M_l. As $\mathrm{tp}(B, M_l)$ does not fork over D, w.l.o.g. $B_1 \cup \bar{c}_1 \subseteq M_l$. So $p \cup \{\varphi(\bar{x}, \bar{a})\}$ is realized by $\bar{c}'$, has power $\leqslant |p| + 1 < \kappa_r(T)$ and its parameters are from $A \subseteq N$; hence it is realized in N (N is $\mathbf{T}$-saturated by the hypothesis of Ax(D2$^-$)). So w.l.o.g. $\bar{c}' \in N$, $\bar{c}'$ realizes p and $\mathrm{tp}(\bar{c}', A)$ forks over B, a contradiction.

Case III. $(\mathbf{T}, \subseteq^*)$ is $(\mathbf{T}^t_{\aleph_0}, \subseteq_e)$, $\kappa(T) = \aleph_0$.

So w.l.o.g. $B = \bar{b}, B_1 = \bar{b}_1, D = \bar{d}$. As $\mathrm{tp}(\bar{c}', \bar{d} \cup \bar{b} \cup b_1 \cup \bar{c}_1)$ forks over $\bar{b}, \bar{d} \subseteq \bar{b}$ clearly $\mathrm{tp}(\bar{c}' {}^\frown \bar{b}, \bar{d} \cup \bar{b}_1 \cup \bar{c}^1)$ forks over $\bar{d}$. So for some $\varphi \models \varphi[\bar{c}', \bar{b}, \bar{c}_1, \bar{b}_1, \bar{d}], \varphi(\bar{x}, \bar{y}, \bar{c}_1, \bar{b}_1, \bar{d})$ forks over $\bar{d}$. We can find $\bar{d}^* \in M_l$ and ψ such that for every $\bar{c}'_1, \bar{b}'_1$ from M_l, $\models (\exists \bar{x})\varphi(\bar{x}, \bar{b}, \bar{c}'_1, \bar{b}'_1, \bar{d})$ iff $\models \psi[\bar{c}'_1, \bar{b}'_1, \bar{d}^*]$. W.l.o.g. $\varphi(\bar{x}, \bar{b}, \bar{y}, \bar{z}, \bar{d}) \vdash p(\bar{x})$, so as $\mathrm{tp}(\bar{b}_1 {}^\frown \bar{c}_1, M_l \cup \bar{b})$ does not fork over M_l, clearly $\models \psi[\bar{c}_1, \bar{b}_1, \bar{d}^*]$. Now $M_l \subseteq_e M_l \cup \bar{b}_1 \cup \bar{c}_1$ (as $\bar{b}_1 {}^\frown \bar{c}_1$ realizes $\mathrm{tp}(\bar{b} {}^\frown \bar{c}, M)$, and $M_l \subseteq_e A \cup \bar{c}, \bar{b} \subseteq A$), hence by the definition of $\subseteq_e$, there are $\bar{b}^*_1, \bar{c}^*_1 \in M_l$ such that $\models \psi[\bar{c}^*_1, \bar{b}^*_1, \bar{d}^*]$, and $\varphi(\bar{x}, \bar{y}, \bar{c}^*_1, \bar{b}^*_1, \bar{d})$ forks over $\bar{d}$. By the former and the choice of ψ, $\models (\exists \bar{x})\varphi(\bar{x}, \bar{b}, \bar{c}^*_1, \bar{b}^*_1, \bar{d})$, hence for some $\bar{c}'' \in N \models \varphi(\bar{c}'', \bar{b}, \bar{c}^*_1, \bar{b}^*_1, \bar{d})$. This, together with "$\varphi(\bar{x}, \bar{y}, \bar{c}'_1, \bar{b}'_1, \bar{d})$ forks over $\bar{d}$", implies $\mathrm{tp}(\bar{c}'' {}^\frown \bar{b}, \bar{c}^*_1 \cup \bar{b}^*_1 \cup \bar{d}) \subseteq \mathrm{tp}(\bar{c}'' {}^\frown \bar{b}, M_l)$ fork over $\bar{d}$. But $\mathrm{tp}(\bar{b}, M_l)$ does not fork over $\bar{d}$, hence (by III, 0.1) $\mathrm{tp}(\bar{c}'', M \cup \bar{b})$ forks over $\bar{d} \cup \bar{b}$, hence $\mathrm{tp}(\bar{c}'', A)$ forks over $\bar{b}$. But by the choice of φ, $\bar{c}''$ realizes p, so we get a contradiction to "p has no extensions over A forking over B".

CLAIM 3.15: *Suppose* $\mathbf{T}$ *is* $\mathbf{T}^t_{\aleph_0}$, T *superstable with the* $(<\infty, 2)$-*existence property (see Def.* XII, 4.2), *or* $\mathbf{T}^a_\lambda$ (cf $\lambda \geqslant \kappa_r(T)$), *or* $\mathbf{T}^t_\lambda$, T *totally transcendental. Then* Ax(E1), (E1$^+$) *hold.*

Proof. $\mathbf{T}^a_\lambda, \mathbf{T}^t_\lambda$ follows by 2.7(6), 3.13, 3.8(6). Then case $\mathbf{T}^t_{\aleph_0}$ is by Definition XII, 4.2.

CLAIM 3.16: (1) $\lambda(\mathbf{T}^x_{\lambda,\kappa}, \subseteq^y_{\mu,\kappa}) \leqslant \chi$ *hold if:* $|A| \leqslant \chi$ *implies* $|\mathbf{T}^x_{\lambda,\kappa}(A)| + |\mathbf{T}^y_{\mu,\kappa}(A)| \leqslant \chi$ *or* T *stable in* χ.

(2) $\lambda(\mathbf{T}^t_{\aleph_0}, \subseteq_{ef}) \leqslant |T|$ *if* T *is superstable*; $\lambda(\mathbf{T}^t_{\aleph_0}, \subseteq_{ef}) \leqslant |T|^+$; $\lambda(\mathbf{T}^t_{\aleph_0}, \subseteq_f) \leqslant |T|^+$, *and* $\lambda(\mathbf{T}^t_{\aleph_0}, \subseteq_e) \leqslant |T|$.

(3) *The same results hold for* λ_s.

Remark. (1) We can easily apply (1), and also generalize $\subseteq_{ef}$ to $\subseteq^{ef}_\lambda$, etc.

(2) In part (1) of 3.16 we can use $\mathbf{T}^x_{\lambda,\kappa_1}, \mathbf{T}^y_{\mu,\kappa_2}$.

Proof. Now part (2) (and the corresponding half of part (3)) is trivial. Also, part (1) follows from (3). So let us prove (1) for λ_s. Let $A_i (i < \chi^+)$ be increasing continuous, $|A_i| \leqslant \chi$. Let S be the set of $\delta < \chi^+$ such that:

(i) if T is stable in $\chi, i < \delta, p \in S^m(A_i)$, p realized in $\bigcup_j A_j$, then p is realized in A_δ,

(ii) if $i < \delta, |\mathbf{T}^x_{\lambda,\kappa}(A_i)| \leqslant \chi$, then every $p \in \mathbf{T}^x_{\lambda,\kappa}(A_i)$ realized in $\bigcup_j A_j$ is realized in A_δ,

(iii) similarly for $\mathbf{T}^y_{\mu,\kappa}, \mathbf{T}^t_{\aleph_0}$ (note that necessarily in all cases $|T| \leqslant \chi$),

(iv) if $|T|^{<\kappa} \leqslant \chi$, $L_1 \subseteq L(T), |L_1^*| < \kappa$, $\mathfrak{C} \restriction L_1^*$ is stable in χ, then every $p \in S^m_{L_1}(A_i)$ $(i < \delta)$ realized in $\bigcup_j A_j$ is realized in A_δ.

Clearly S is a closed unbounded set. We have still two points to clarify why S works, i.e. why for every $\delta \in S$ of cofinality cf χ (the problem for $\mathbf{T}^y_{\lambda,\kappa}$ are parallel):

(A) If cf $\chi < \lambda$ there may be $p \in \mathbf{T}^x_{\lambda,\kappa}(A_\delta)$ not in $\bigcup_{i<\lambda} \mathbf{T}^x_{\lambda,\kappa}(A_i)$.

However, if cf $\chi \geqslant \kappa(T)$ we can repeat the argument in III, 3.11. If cf $\chi < \kappa(T)$ and cf $\chi < \kappa$ or even $\chi^{<\kappa} > \chi$, we can contradict $|A| \leqslant \chi \Rightarrow |T^x_{\lambda,\kappa}(A)| \leqslant \chi]$ as in the proof of III, 3.6. So $\kappa \leqslant$ cf $\chi < \kappa(T)$, $\chi^{<\kappa} = \chi$ and by the latter it is enough to deal with each $\mathfrak{C} \restriction L_1$, $L_1 \subseteq L(T), |L_1| < \kappa$, and if cf $\xi \geqslant \kappa(\mathrm{Th}(\mathfrak{C} \restriction L_1))$ we can again finish and otherwise contradict $[|A| \leqslant \chi \Rightarrow |\mathbf{T}^x_{\lambda,\kappa}(A)| \leqslant \chi]$.

(B) If $x = a$, for $B \subseteq A_i$, stp$(\bar{b}, B)$ is not necessarily over A_i.

However, for $\delta \in S, A_\delta \subseteq_t \bigcup_{j<\chi^+} A_j$ (by (iii)), hence for $\bar{b} \in \bigcup_j A_j$, $E \in FE^m(A_\delta)$ there is $\bar{b}' \in A_j, \bar{b}E\bar{b}'$ [otherwise choose by induction on $n, \bar{b}_n \in A_\delta, \bigwedge_{l<n} \neg \bar{b}_l E \bar{b}_n$], so for the types we are interested in this does not occur.

CONCLUSION 3.17: (1) *If T is superstable, then* $(\mathbf{T}^t_{\aleph_0}, \subseteq^a_{\aleph_0})$ *and* $(\mathbf{T}^t_{\aleph_0}, \subseteq^t_{\aleph_0,\aleph_0})$ *satisfy* Ax(A1)–(A6), (B1)–(B7), (B4$^+$), (C1$^+$), (C2), (C3$^+$), (C4), (D2$^-$), (E2), *and* $\lambda(\mathbf{T}^t_{\aleph_0}, \subseteq^a_{\aleph_0}) \leqslant \lambda(T), \lambda(\mathbf{T}^t_{\aleph_0}, \subseteq^t_{\aleph_0,\aleph_0}) \leqslant |T| + 2^{\aleph_0}$. *If we assume the* $(<\infty, 2)$*-existence property, then* Ax(E1$^+$) *holds too.*

(2) *If T is superstable, $D(T)$ scattered, then* $(\mathbf{T}^t_{\aleph_0}, \subseteq^t_{\aleph_0})$ *satisfies* Ax(A1)–(A6), (B1)–(B7), (B4$^+$) (C1$^+$), (C2), (C3$^+$), (C4), (D2), (E2) *and* $\lambda(\mathbf{T}^t_{\aleph_0}, \subseteq^t_{\aleph_0}) \leqslant |T|$.

If we assume the $(<\infty, 2)$*-existence property, then* Ax(E1$^+$) *holds too and if T is totally transcendental,* Ax(E1$^+$) *and also* Ax(D1) *holds.*

(3) *If T is superstable, then* $(\mathbf{T}^t_{\aleph_0}, \subseteq_e)$ *satisfies* Ax(A1)–(A6), (B1)–(B7), (B4$^+$) (C1^{--}), (C2), (C3), (C3$^+$), (C4), (D2$^-$), (E2), *and* $\lambda(\mathbf{T}^t_{\aleph_0}, \subseteq_e) \leqslant |T|$.

If the hypothesis of 2.7(7) *holds, the* Ax(C1$^-$), (E1) *hold.*

(4) *If T is superstable, then* $(\mathbf{T}^a_{\aleph_0}, \subseteq^a_{\aleph_0})$ *satisfies* Ax(A1)–(A6), (B1)–(B7), (B4$^+$), (C1$^+$), (C2), (C3$^+$), (D1), (D2), (E1$^+$), (E2$^+$), *and* $\lambda(\mathbf{T}^a_{\aleph_0}, \subseteq^a_{\aleph_0}) = \lambda(T)$.

Proof. For Ax(A1), (A2), (A3) see 3.8(1), for Ax(A4), (A5) see 3.8(2), and for Ax(A6) see 3.8(3).

For Ax(B1), (B2) see 3.8(4), and for Ax(B3), (B4), (B5) see 3.8(5), (6), (7), respectively.

For Ax(C1$^+$) see 3.9(1) for (1), (2) and (4).

For Ax(C1^{--}) see 3.9(2) for (3).

For Ax(C2) see 3.11 for (1), (3), (4) and 3.12(2) for (2).

For Ax(C3$^+$) see 3.10 for (1), (3), (4) and 3.12(1) for (2).

For Ax(C4) note that it follows from Ax(C3$^\pm$) (by 2.7(4)).

For Ax(D1) see IV, 2 for (4).

For Ax(D2) see 3.12(2) for (2), and 2.7(6) for (4).

For Ax(D2$^-$) see 3.14 for (1) and (3), and see 2.7(5) for (2), (4).

For Ax(E1), (E1$^+$) see 3.15.

For Ax(E2) see 2.7(9).

For $\lambda(\mathbf{T}, \subseteq^*)$ see 3.16.

EXERCISE 3.1: Let T be countable $\aleph_0$-stable.

(A) If T has the dop or is deep, then $I(\lambda, T) = 2^\lambda$ for $\lambda > \aleph_0$.

(B) If T is shallow of infinite depth without the dop, then for $\alpha > 0$,

$$I(\aleph_\alpha, T) = \beth_{\mathrm{Dp}(T)}(|\alpha+\omega|).$$

(C) If T is shallow without the dop and of finite depth, then either for every $\alpha \geqslant \omega, I(\aleph_\alpha, T) = \beth_{\mathrm{Dp}(T)}(|\alpha|)$, or for every $\alpha \geqslant \omega, I(\aleph_\alpha, T) = \beth_{\mathrm{Dp}(T)}(|\alpha|^{\aleph_0})$ and for $\alpha < \omega$, $I(\aleph_0, T)$ is $\leqslant \beth_{\mathrm{Dp}(T)}(\aleph_0)$, or $\leqslant \beth_{\mathrm{Dp}(T)}(2^{\aleph_0})$, respectively.

[Hint for (C): Work as in Chapter X using $(\mathbf{T}^t_{\aleph_0}, \subseteq_t)$-decomposition, 3.17(2) and the decomposition theorems of Section 2; there is no new point.]

EXERCISE 3.2: Prove a parallel theorem for T totally transcendental not necessarily countable; in (B), (C): for α large enough in (A) for $\lambda \geqslant |T|+\aleph_1$.

EXERCISE 3.3: (A) If T is unsuperstable or with dop or deep, $I^a_{\aleph_0}(\lambda, T) = 2^\lambda$ for $\lambda \geqslant \aleph_1 + |D(T)|$.

(B) If T is shallow of infinite depth without dop, then for α large enough $I^a_{\aleph_0}(\aleph_\alpha, T) = \beth_{\mathrm{Dp}(T)}(|\alpha|)$ (really $\beth_{\mathrm{Dp}(T)}(|\alpha|) \leqslant I^a_{\aleph_0}(\aleph_\alpha, T) \leqslant \beth_{\mathrm{Dp}(T)}((|\alpha|+2)^{2^{|T|}})$.

(C) If T is superstable of finite depth without dop, then for some $\kappa \leqslant (2^{|T|})^+$, and $n, k, n+k = \mathrm{Dp}(T)$ for every α large enough

$$I^a_{\aleph_0}(\aleph_\alpha, T) = \beth\Big(\sum_{\mu<\kappa} \beth_k(|\alpha|^\mu)\Big).$$

(D) Determine the "large enough" in (C), (B).

[Hint: For (C) see XIII, Section 4.]

EXERCISE 3.4 [T superstable without dop, axioms as in 2.8]:

(1) If $\langle N_\eta, \bar a_\eta : \eta \in I\rangle$ is a $(\mathbf{T}, \subseteq^*, \lambda)$-decomposition of M, $J \subseteq I$ is non-empty closed under initial segments, $N \subseteq M^*$ is $\mathbf{T}$-prime, $\mathbf{T}$-atomic and $\mathbf{T}$-minimal over $\bigcup_{\eta\in I} N_\eta$, *then* M^* is $\mathbf{T}$-prime, $\mathbf{T}$-atomic and $\mathbf{T}$-minimal over $\bigcup_{\eta\in I} N_\eta \cup N$.

(2) If $\beta < \alpha$, $N \subseteq^* N_i \subseteq^* M^0 \subseteq^* M^1$ for $i < \alpha$, $\{N_i : i < \alpha\}$ independent over N, for $i < \beta$ N_i is an $(N, \bar b)$-component in M^0 and in M_1, $\bar b \in M^0$, N' is an $(N, \bar b)$-component in M^0, $\mathrm{tp}(\bar b, \bigcup_{\beta\leqslant i<\alpha} N_i)$ does not fork over N, *then* N' is an $(N, \bar b)$-component in M^1.

Chapter XII

THE MAIN GAP FOR COUNTABLE THEORIES

XII.0. Introduction

We shall concentrate on superstable T. The chapter ends with the proof of the main gap theorem for countable theories. The new missing part is, by Chapter XI, the existence of a $(T^t_{\aleph_0}, \subseteq^a_{\aleph_0,\aleph_0})$ decomposition, and for it the missing part is the $(\lambda, 2)$-existence property: if $M \prec M_1, M_2$, $\mathrm{tp}_*(M_1, M_2)$ does not fork over M, then over $M_1 \cup M_2$ there is a primary model. For this end we prove: if T does not have the $(\aleph_0, 2)$-existence property, then it has the otop (=omitting type order property). As the otop implies T has many non-isomorphic models, we can concentrate on T with the $(\aleph_0, 2)$-existence property. In Section 5 we prove that such T has the $(\lambda, 2)$-existence property. Though we are interested in diagrams consisting of three models, we need to consider for each $n < \omega$ a "stable system" $\langle M_s : s \in \mathscr{P}^-(n)\rangle$, $\|M_s\| = \lambda$, and want to prove by induction the existence of primary models over $\bigcup_s M_s$ (in fact there are more complications). For more on such diagrams see [Sh 83a].

In Sections 1–3 we do some preparatory work. In Section 1 we deal with two new $\mathbf{F}$'s: $\mathbf{F}^k_{\aleph_0}, \mathbf{F}^l_{\aleph_0}$, both strengthening $\mathbf{F}^l_{\aleph_0}$, and are useful for superstable countable T. Note that if $\langle A, \langle \alpha_i : i < \alpha\rangle\rangle$ is an $\mathbf{F}^l_{\aleph_0}$-construction, $A \subseteq_t B$, then $\langle B, \langle \alpha_i : i < \alpha\rangle\rangle$ is an $\mathbf{F}^l_{\aleph_0}$-construction. In Section 2 we prove all we need on stable systems. In Section 3 we deal with good sets: A is good if for every $p \in S^m(A)$ there is $B \subseteq A$, $|B| < \kappa_r(T)$ such that p does not fork over B and p is the unique extension of $p \restriction B$ in $S^m(A)$ not forking over B. So a $\mathbf{F}^a_{\kappa(T)}$-saturated model is a good set. We prove that for any system $\langle M_s : s \in \mathscr{P}^-(n)\rangle$ of $\mathbf{F}^a_{\kappa_r(T)}$-saturated models $\bigcup_s M_s$ is a good set. This helps in Section 5 to show how to go from the "non-$(\aleph_0, n)$-existence property" to the "non-$(\aleph_0, 2)$-existence property".

Notation. In notions like $\mathbf{F}^t_{\aleph_0}$-isolated, $\mathbf{F}^t_{\aleph_0}$-primary, $\mathbf{F}^t_{\aleph_0}$-prime, we omit the "$\mathbf{F}^t_{\aleph_0}$".

XII.1. On $\mathbf{F}^k_\lambda$ and $\mathbf{F}^j_\lambda$

(Hyp) T stable.

DEFINITION 1.1: Let (see IV§1) $\mathbf{F}^k_\lambda = \{(p,B): p \in S^m(A)$, B is a subset of A of power smaller than λ, and for each $\psi = \psi(\bar{x},\bar{y})$, for some formula $\varphi_\psi(\bar{x},\bar{a}_\psi)$ which is in p and is almost over B, $\varphi_\psi(\bar{x},\bar{a}_\psi) \vdash p \restriction \psi\}$.

LEMMA 1.2: (1) $(p,B) \in \mathbf{F}^k_\lambda$ *iff* $(p,B) \in \mathbf{F}^a_\lambda$ *and* $(p, \operatorname{Dom} p) \in \mathbf{F}^l_{\aleph_0}$.

(2) *Suppose T is countable and superstable, then for every $A, \bar{a} \in A$ and φ such that* $\models (\exists \bar{x})\varphi(\bar{x},\bar{a})$ *there is* $p \in S^m(A)$ *which includes* $\varphi(\bar{x},\bar{a})$ *and is* $\mathbf{F}^k_{\aleph_0}$*-isolated (and also* $\mathbf{F}^c_{\aleph_0}$*-isolated).*

(3) *If (A,C) satisfies the Tarski–Vaught condition,* $\operatorname{tp}(\bar{a},A)$ *is* $\mathbf{F}^k_{\aleph_0}$*-isolated, then so is* $\operatorname{tp}(\bar{a},C)$*. In fact, the same formulas witness it and* $\operatorname{tp}(\bar{a},A) \vdash \operatorname{tp}(\bar{a},C)$*. The same holds for* $\mathbf{F}^l_{\aleph_0}, \mathbf{F}^t_{\aleph_0}$.

(4) *In* (3) *if* $\operatorname{tp}(\bar{a},A)$ *is not isolated, then* $\operatorname{tp}(\bar{a},C)$ *too is not isolated.*

(5) $\lambda(\mathbf{F}^k_{\aleph_0}) = \aleph_0$ *and* $\mathbf{F}^k_{\aleph_0}$ *satisfies* Ax (I), (II, 1, 2, 3, 4), (III, 1, 2), (IV), (V, 1, 2), (VI), (VII), (IX).

Proof. (1) The implication $\Rightarrow$ is obvious: Let $(p,B) \in \mathbf{F}^k_{\aleph_0}$ be exemplified by $\varphi_\psi(\bar{x},\bar{a}_\psi)$ ($\psi = \psi(\bar{x},\bar{y} \in L(T))$, $\bar{c}$ realizes p, then

$$\operatorname{stp}(\bar{c},B) \vdash \{\varphi_\psi(\bar{x},\bar{a}_\psi): \psi \in L(T)\}$$

(see end of remark to Def. III, 2.1) and clearly $\{\varphi_\psi(\bar{x},\bar{a}_\psi\}: \psi \in L(T)\} \vdash \bigcup_\psi p \restriction \psi$ but

$$p = \bigcup_\psi p \restriction \psi,$$

hence $\operatorname{stp}(\bar{c},B) \vdash p$, so by IV, 2.1 $(p,B) \in \mathbf{F}^a_\lambda$, Now the $\varphi_\psi(\bar{x},\bar{a}_\psi)$'s exemplify $(p, \operatorname{Dom} p) \in \mathbf{F}^l_{\aleph_0}$.

Also, the other direction is easy. Let $(p,B) \in \mathbf{F}^a_\lambda$, $(p, \operatorname{Dom} p) \in \mathbf{F}^l_{\aleph_0}$, and $\bar{c}$ realizes p. We know by IV, 2.1 that $\operatorname{stp}(\bar{c},B) \vdash p$, and there are $\varphi_\psi(\bar{x},\bar{a}_\psi) \in p(\psi \in L)$ such that $\varphi_\psi(\bar{x},\bar{a}_\psi) \vdash \varphi_\psi(\bar{x},\bar{a}_\psi), \varphi_\psi(\bar{x},\bar{a}_\psi) \vdash p \restriction \psi$. As $\operatorname{stp}(\bar{c},B) \vdash p, \varphi_\psi(\bar{x},\bar{a}_\psi) \in p$ there is $E \in FE(B)$, $xE\bar{c} \vdash \varphi_\psi(\bar{x},\bar{a}_\psi)$, and let $\{\bar{d}_1,\ldots,\bar{d}_n\}$ be representatives of the E-equivalence classes and

$$\varphi^*_\psi(\bar{x},\bar{a}^*_\psi) = \bigvee\{\bar{x}E\bar{d}_l: xE\bar{d}_l \vdash \varphi_\psi(\bar{x},\bar{a}_\psi)\}.$$

Clearly, $\varphi^*_\psi(\bar{x},\bar{a}^*_\psi)$ is consistent with p (a formula equivalent to $\bar{x}E\bar{c}$ appears in the disjunction), is almost over B (as a disjunction of such formulas) and is equivalent to a formula $\varphi^+_\psi(\bar{x},\bar{a}^+_\psi)$ over A (as every automorphism of $\mathfrak{C}^{\mathrm{eq}}$ over A maps it into an equivalent formula). As $\varphi^+_\psi(\bar{x},a^+_\psi)\vdash\varphi^*_\psi(\bar{x},\bar{a}^*_\psi)$, $\varphi^*_\psi(\bar{x},\bar{a}^*_\psi)\vdash\varphi_\psi(\bar{x},\bar{a}_\psi)$, $\varphi_\psi(\bar{x},\bar{a}_\psi)\vdash p\restriction\psi$ (by their definitions), clearly $\varphi^+_\psi(\bar{x},\bar{a}^+_\psi)\vdash p\restriction\psi$. As $\varphi^*_\psi(\bar{x},\bar{a}^*_\psi)$ is consistent with p, also $\varphi^+_\psi(\bar{x},\bar{a}^+_\psi)$ is consistent with p, but as $\bar{a}^+_\psi\in A$, p is a complete type over A, clearly $\varphi^+_\psi(\bar{x},\bar{a}^+_\psi)\in p$.

(2) Choose $\psi(\bar{x},\bar{b})$ such that $\bar{b}\subseteq A$, $\vDash(\exists\bar{x})\psi(\bar{x},\bar{b})$, $\psi(x,\bar{a})\vdash\varphi(\bar{x},a)$ and (under those restrictions) $R^m[\psi(\bar{x},\bar{b}),L,\infty]$ is minimal. We know that every $p,\psi(\bar{x},\bar{b})\in p\in S^m(A)$, is $\mathbf{F}^a_{\aleph_0}$-isolated and (see IV, 2.18(4)) there is such p which is $\mathbf{F}^l_{\aleph_0}$-isolated. By part (1) we finish.

(3), (4) Also this part is easy.

(5) See IV, §2 (on $\mathbf{F}^a_{\aleph_0}$ and $\mathbf{F}^l_{\aleph_0}$), straightforward by (1).

CLAIM 1.3: (1) *If $A\subseteq B$, $p\in S^m(B)$, p is $\mathbf{F}^k_\lambda$-isolated and does not fork over A, then $p\restriction A$ is $\mathbf{F}^k_\lambda$-isolated.*

(2) *In fact, if $C\subseteq A\subseteq \mathrm{Dom}\, p$, $(p,C)\in\mathbf{F}^k_{\aleph_0}$, then $(p\restriction A,C)\vDash\mathbf{F}^k_{\aleph_0}$.*

Proof. (1) By Theorem IV, 4.3(2) $p\restriction A$ is $\mathbf{F}^a_\lambda$-isolated, and by IV, 4.3(5) $p\restriction A$ is $\mathbf{F}^l_{\aleph_0}$-isolated. Now use 1.2(1).

(2) Same proof.

Remark. The parallel of 1.3 holds for $\mathbf{F}^t_\lambda$, $\mathbf{F}^a_\lambda(\lambda\geqslant\kappa_r(T))$, $\mathbf{F}^c_\lambda(\lambda\geqslant\kappa_r(T))$, and $\mathbf{F}^l_\lambda$ (by IV, 4.3).

CLAIM 1.4: *Suppose $\langle N_\eta:\eta\in I\rangle$ is a non-forking tree of models, and A is $\mathbf{F}^l_{\aleph_0}$-constructible, or just $\mathbf{F}^l_{\aleph_0}$-atomic over $\bigcup_{\eta\in I}N_\eta$.*

(1) *If T does not have the dop, then every type not orthogonal to A is not orthogonal to some N_η.*

(2) *For every $\bar{a}\in A$, $\bar{a}\notin\bigcup_{\eta\in I}N_\eta$, and $\nu\in I$, $\mathrm{tp}(\bar{a},\bigcup_{\eta\in I}N_\eta)$ forks over N_ν,*

Proof. Like the proof of XI, 2.12, 13 and X, 2.2. Let $\lambda>|A|+\sum_\eta\|N_\eta\|$ be regular.

Choose $\{M_\eta:\eta\in I\}$ a non-forking tree of $\mathbf{F}^a_\lambda$-saturated models, $N_\eta\subseteq M_\eta$, $\mathrm{tp}_*(M_\eta,N_\eta\cup\bigcup\{M_\nu:\text{not }\eta\lessdot\nu\})$ does not fork over $M_{\eta^-}\cup N_\eta$ (for η of length zero omit M_{η^-}). By 2.10, 2.3 such M_η's exist and $(\bigcup_{\eta\in I}N_\eta$, $\bigcup_{\eta\in I}M_\eta)$ satisfies the Tarski–Vaught condition, hence (see 1.2(3)) A is $\mathbf{F}^l_{\aleph_0}$-atomic over $\bigcup_{\eta\in I}M_\eta$. So A is $\mathbf{F}^a_\lambda$-constructible over $\bigcup_{\eta\in I}M_\eta$. Now there is an $\mathbf{F}^a_{\aleph_0}$-primary M over $\bigcup_{\eta\in I}M_\eta$, $A\subseteq M$. By X, 3.3 we know that every type not orthogonal to M is not orthogonal to some M_η. So if r is not orthogonal to A it is not orthogonal to some $\mathrm{tp}(\bar{c}, M_\eta)$. As $\mathrm{tp}_*(M_\eta, A)$ does not fork over N_η, necessarily $\mathrm{tp}_*(\bar{c}, M_\eta)$ is not orthogonal to N_η (by X, 1.1). (w log $\mathrm{tp}(\mathrm{Dom}\ r, M)$ dnf over A.)

This completes the proof of (1). As for (2), a failure of it implies there is $\bar{a} \in M, \bar{a} \notin \bigcup_{\eta \in I} M_\eta$, $\mathrm{tp}(\bar{a}, \bigcup_{\eta \in I} M_\eta)$ does not fork over M_ν. But $\mathrm{tp}(\bar{a}, \bigcup_{\eta \in I} M_\eta)$ is $\mathbf{F}^a_\lambda$-isolated, hence $\mathrm{tp}(\bar{a}, M_\nu)$ is $\mathbf{F}^a_\lambda$-isolated (by IV, 4.3(2)). But $\bar{a} \notin M_\nu, M_\nu$ is $\mathbf{F}^a_{\aleph_0}$-saturated, a contradiction.

FACT 1.5: *Suppose* $\varphi(\bar{x}, \bar{a}) \vdash \mathrm{tp}(\bar{c}, \bar{a}), \bar{a} \subseteq A \subseteq B, (A, B)$ *satisfies the Tarski–Vaught condition. If* $\{\varphi(\bar{x}, \bar{a})\}$ *has no extension over* A *which forks over* $\bar{a}$, *then* $\{\varphi(\bar{x}, \bar{a})\}$ *has no extension over* B *which forks over* $\bar{a}$.

Proof. Suppose $\bar{d} \in B, \{\varphi(\bar{x}, \bar{a}), \vartheta(\bar{x}, \bar{d})\}$ is consistent and forks over $\bar{a}$. We can now define by induction on n, a distinct $p_n \in S^m_\vartheta(A)$ such that $\{\varphi(\bar{x}, \bar{a})\} \cup p_n$ is consistent. Hence for some n:

$$R^m[\mathrm{tp}(\bar{c}, \bar{a}), \vartheta, \aleph_0] = R^m[\{\varphi(\bar{x}, \bar{a})\}, \vartheta, \aleph_0]$$
$$> R^m[\{\varphi(\bar{x}, \bar{a})\} \cup p_n, \vartheta, \aleph_0] = R^m[\mathrm{tp}(\bar{c}, \bar{a}) \cup p_n, \vartheta, \aleph_0].$$

By III, 4.1, this implies that $\mathrm{tp}(\bar{c}, \bar{a}) \cup p_n$ forks over $\bar{a}$, a contradiction.

DEFINITION 1.6: $\mathbf{F}^l_{\aleph_0} = \{(p, B) : |B| < \aleph_0$ and letting $A = \mathrm{Dom}\, p$, $p \in S^m(A)$, $B \subseteq A$ and the following holds: $p \restriction B$ has no extension over A which forks over B, and for every ϑ for some $\psi_\vartheta(\bar{x}, \bar{b}_\vartheta) \in p \restriction B$, there are only finitely many $r \in S^m_\vartheta(A)$ which are consistent with $\{\psi_\vartheta(\bar{x}, \bar{b}_\vartheta)\}\}$.

FACT 1.7: *Suppose* $B \subseteq A$, p *has no extension in* $S^m(A)$ *forking over* B, *and* $p \in S^m(B)$. *Then for every* $\vartheta = \vartheta(\bar{x}, \bar{y})$ *the following conditions are equivalent:*

(α) *for some* $\psi(\bar{x}, \bar{b}) \in p$, *there are only finitely many* $r \in S^m_\vartheta(A)$ *consistent with* $\{\psi(\bar{x}, \bar{b})\}$,

(β) *for some* $\psi(\bar{x}, \bar{b}) \in p$, *for every* $r \in \{r \in S^m_\vartheta(A)$: *r consistent with* $\psi(\bar{x}, \bar{b})\}$ *the type* $p \cup r$ *does not fork over* B (*equivalently*, $p \cup r$ *is consistent*).

Remark. We shall use this fact freely when dealing with Definition 1.6.

Proof. $(\alpha) \Rightarrow (\beta)$. Let $\psi_0(\bar{x}, \bar{b}_0) \in p$ exemplify (α), and $\Gamma_0 = \{r \in S^m_\vartheta(A)$: r is consistent with $\psi_0(\bar{x}, \bar{b})\}$. For each $r \in \Gamma_0$ let p_r be a finite subset of p such that if $p \cup r$ is inconsistent, then $p_r \cup r$ is inconsistent. Let $\psi(\bar{x}, \bar{b}) = \psi_0(\bar{x}, \bar{b}_0) \wedge \bigwedge_{r \in \Gamma_0}(\bigwedge p_r)$. As $\bar{b}_0 \subseteq B$, each p_r is $\subseteq p$, hence is over B, clearly $\bar{b} \subseteq B$. As Γ_0 is finite as well as each p_r, clearly $\psi(\bar{x}, \bar{b})$ is a first-order formula. Clearly, $\psi(\bar{x}, \bar{b}) \in p$, ($p_r \subseteq p$, $\psi_0(\bar{x}, \bar{b}_0) \in p$, hence $p \vdash \psi(\bar{x}, \bar{b})$, hence $\psi(\bar{x}, \bar{b}) \in p$), We now show that $\psi(\bar{x}, \bar{b})$ exemplifies (β). So let $r \in \Gamma = \{r \in S^m_\vartheta(A)$: r consistent with $\psi(\bar{x}, \bar{b})\}$. Then $p_r \cup r$ is consistent (as $\psi(\bar{x}, \bar{b}) \vdash \bigwedge p_r$) and $r \in \Gamma_0$ [as $\psi(\bar{x}, \bar{b}) \vdash \psi_0(\bar{x}, \bar{b}_0)$], hence $p \cup r$ is consistent (by the choice of p_r), hence $p \cup r$ does not fork over B [as p has no extension over A which forks over B, by the hypothesis on p and the extension property for non-forking].

$(\beta) \Rightarrow (\alpha)$. Let $\psi(\bar{x}, \bar{b})$ exemplify (β), and $\Gamma = \{r \in S^m_\vartheta(A)$: r consistent with $\psi(\bar{x}, \bar{b})\}$. We shall show that Γ is finite, thus finishing.

As $\psi(\bar{x}, \bar{b})$ exemplifies (β), $\Gamma \subseteq \Gamma_1 \stackrel{\text{def}}{=} \{r \in S^m_\vartheta(A) : p \cup r$ does not fork over $B\}$. By Theorem III, 4.1 for every $r \in \Gamma$ (clearly $r \in \Gamma_1$ hence):

$$R^m[p, \vartheta, \aleph_0] = T^m[p \cup r, \vartheta, \aleph_0].$$

Hence $|\Gamma| \leqslant \mathrm{Mlt}[p, \vartheta] < \omega$. So we finish.

CLAIM 1.8: $\mathbf{F}^l_{\aleph_0} \subseteq \mathbf{F}^k_{\aleph_0}$.

Proof. Suppose $(p, B) \in \mathbf{F}^l_{\aleph_0}$. As $p \restriction B$ has no extension in $S^m(A)$ which forks over B, clearly $(p, B) \in \mathbf{F}^a_{\aleph_0}$, so by 1.2(1) it suffices to prove that p is $\mathbf{F}^l_{\aleph_0}$-isolated. So let $\vartheta = \vartheta(\bar{x}, \bar{y})$ and we have to find $\psi^*(\bar{x}, \bar{b}^*) \in p$, $\psi^*(\bar{x}, \bar{b}^*) \vdash (p \restriction \vartheta)$. By Definition 1.6 there is $\psi(\bar{x}, \bar{b}) \in p$ for which $\Gamma = \{r \in S^m_\vartheta(A) : r$ consistent with $\psi(\bar{x}, \bar{b})\}$ is finite. For each $r \in \Gamma - \{p \restriction \vartheta\}$ choose a formula $\varphi_r(\bar{x}, \bar{b}_r) \in r$, $\varphi_r(\bar{x}, \bar{b}_r) \notin p$ (hence $\neg\varphi_r(\bar{x}, \bar{b}_r) \in p$, and φ_r is ϑ or $\neg\vartheta$). Let $\psi^*(\bar{x}, \bar{b}^*) = \psi(\bar{x}, \bar{b}) \wedge \bigwedge\{\varphi_r(\bar{x}, \bar{b}_r) : r \in \Gamma - \{p \restriction \vartheta\}\}$. Clearly, $\psi(\bar{x}, \bar{b}) \in p$, $\psi(\bar{x}, \bar{b}) \vdash (p \restriction \vartheta)$.

CLAIM 1.9: *If* $(p, B) \in \mathbf{F}^l_{\aleph_0}$, $p \in S^m(A)$, (A, C) *satisfies the Tarski–Vaught condition, then:*

(a) *p has a unique extension $q \in S^m(C)$,*

(b) $(q, B) \in \mathbf{F}^l_{\aleph_0}$ *(the same witnesses $\psi_\vartheta(\bar{x}, \bar{b}_\vartheta)$ works),*

(c) *if p is not isolated, then q is not isolated.*

Proof. (a) By 1.8 $(p, B) \in \mathbf{F}^k_{\aleph_0}$, so apply 1.2(3).

(b) Trivial, as if there are exactly k types $r \in S^m_\vartheta(A)$ consistent with

$\psi(\bar{x},\bar{b}),(\bar{b}\in B)$, then there are exactly k types $r\in S^m_\vartheta(C)$ consistent with $\psi(\bar{x},\bar{b})$.

(c) Trivial.

FACT 1.10: *If* $(p,B)\in\mathbf{F}^l_{\aleph_0}, p\restriction B\subseteq q\in S^m(\mathrm{Dom}\,p)$, *then* $\mathrm{Dom}\,q=\mathrm{Dom}\,p$ *and* $(q,B)\in\mathbf{F}^l_{\aleph_0}$.

Proof. Trivial (check the definition).

LEMMA 1.11: $\mathbf{F}^l_{\aleph_0}$ *satisfies the following axioms:* $\lambda(\mathbf{F})=\aleph_0$, Ax (I), (II, 1, 2, 3, 4), (III, 1, 2), (IV), (V, 1, 2).

Proof. Let $\mathbf{F}=\mathbf{F}^l_{\aleph_0}, \lambda(\mathbf{F})=\aleph_0$. Trivial.

Ax (I), (II, 1, 2, 3, 4), (III, 1, 2), (IV). Trivially.

Ax (V, 1). If $q=\mathrm{tp}(\bar{a}\frown\bar{b},A), p=\mathrm{tp}(\bar{a},A\cup\bar{b}), B\subseteq A$ and $(q,B)\in\mathbf{F}$, then $(p,B\cup\bar{b})\in\mathbf{F}$.

First we have to show that $\mathrm{tp}(\bar{a},B\cup\bar{b})$ has no extension over $A\cup\bar{b}$ which forks over $B\cup\bar{b}$. Suppose $\mathrm{tp}(\bar{a}',A\cup\bar{b})$ extends $\mathrm{tp}(\bar{a},B\cup\bar{b})$ and forks over $B\cup\bar{b}$. As $\mathrm{tp}(\bar{b},A)$ does not fork over B, by III, 0.1(3), $\mathrm{tp}(\bar{a}'\frown\bar{b},A)$ forks over B, and clearly it extends $\mathrm{tp}(\bar{a}\frown\bar{b},B)$, a contradiction.

Now let $\vartheta=\vartheta(\bar{x};\bar{y})$ and we have to find a suitable $\psi(\bar{x},\bar{c})\in\mathrm{tp}(\bar{a},B\cup\bar{b})$ (which is consistent with just finitely many $r\in S^m_\vartheta(A\cup\bar{b})$). Now we can easily find $k<\omega$ and $\vartheta_l(\bar{x};\bar{y};\bar{z})$ (for $l<k$) such that $l(\bar{y})=l(\bar{b})$ and for every $\bar{c}\in(A\cup\bar{b})$ for some $\bar{c}'\in A$, and $l<k\,\vartheta(\bar{x},\bar{c})\equiv\vartheta_l(\bar{x},\bar{b},\bar{c}')$. For each ϑ_l there is $\psi_l(\bar{x},\bar{y},\bar{d}_l)\in\mathrm{tp}(\bar{a}\frown\bar{b},B)$ such that Γ_l is finite where $\Gamma_l=\{r:r\in S^{m+l(\bar{b})}_{\vartheta_l(\bar{x};\bar{y};\bar{z})}(A)$ is consistent with $\psi_l(\bar{x},\bar{y},\bar{d}_l)\}$.

Let $\psi(\bar{x},\bar{c})=\bigwedge_{l<k}\psi_l(\bar{x},\bar{b},\bar{d}_l)$ and $\Gamma=\{r\in S^m_{\vartheta(\bar{x};\bar{y})}(A\cup\bar{b})$: r is consistent with $\psi(\bar{x},\bar{c})\}$. Clearly, $|\Gamma|\leqslant\prod_{l<k}\Gamma_l$; by the choice of the ϑ_l's, k is finite, and each Γ_l is finite (by the choice of $\psi_l(\bar{x},\bar{y},\bar{d}_l)$). So we finish.

Ax (V, 2). As $\lambda(\mathbf{F})=\aleph_0$ it follows from (V, 1).

LEMMA 1.12: (1) *If T is countable and superstable, then* Ax(X, 1) *holds for* $\mathbf{F}^l_{\aleph_0}$.

(2) *If (A,B) satisfies the Tarski–Vaught condition, C is* $\mathbf{F}^l_{\aleph_0}$*-atomic over A, then* $\mathrm{tp}_*(C,A)\vdash\mathrm{tp}_*(C,B)$ *and* $(A\cup C,B\cup C)$ *satisfies the Tarski–Vaught condition.*

(3) *Suppose* $p\in S^m(A)$, $(p,B)\in\mathbf{F}^k_{\aleph_0}, p\restriction B$ *has a unique extension in*

$S^m(A)$ which does not fork over B. Then for every $\vartheta = \vartheta(\bar{x}, \bar{y})$ for some $\psi \in p \restriction B, \psi \vdash (p \restriction \vartheta)$, and $(p, B) \in \mathbf{F}^{\mathrm{i}}_{\aleph_0}$.

(4) *If $B \subseteq A, |B| < \aleph_0, p \in S^m(A), p \restriction B \vdash p$ and p is $\mathbf{F}^{l}_{\aleph_0}$-isolated, then $(p, B) \in \mathbf{F}^{\mathrm{i}}_{\aleph_0}$.*

(5) *Suppose $p \in S^m(A)$ is realized in some set $\mathbf{F}^{\mathrm{i}}_{\aleph_0}$-constructible over A. If for some finite $C \subseteq A$, p is the unique extension in $S^m(A)$ of $p \restriction C$ which does not fork over C, then for some finite $B \subseteq A, (p, B) \in \mathbf{F}^{\mathrm{i}}_{\aleph_0}$ and $p \restriction B \vdash p$.*

(6) *If $\langle A, \langle a_i : i < \alpha \rangle, \langle B_i : i < \alpha \rangle \rangle$ is an $\mathbf{F}^{l}_{\aleph_0}$-construction, $A \subseteq_t A_1$, then $\langle A_1, \langle a_i : i < \alpha \rangle, \langle B_i : i < \alpha \rangle \rangle$ is an $\mathbf{F}^{l}_{\aleph_0}$-contradiction too. Similarly for $(\mathbf{F}^{l}_{\lambda}, \subseteq^{t}_{\lambda}), (\mathbf{F}^{\mathrm{i}}_{\aleph_0}, \subseteq_t)$ and $(\mathbf{F}^{k}_{\aleph_0}, \subseteq_t)$.*

Proof. (1) Let $\bar{a} \in A, \models (\exists \bar{x}) \varphi(\bar{x}, \bar{a})$, and we have to find $p \in S^m(A)$ such that $\varphi(\bar{x}, \bar{a}) \in p$, p is $\mathbf{F}^{\mathrm{i}}_{\aleph_0}$-isolated. W.l.o.g. $\{\varphi(\bar{x}, \bar{a})\}$ has no extension over A which forks over $\bar{a}$. Let $\{\vartheta_l(\bar{x}, \bar{y}_l) : l < \omega\}$ list the formulas of L. We define by induction on $n < \omega, \varphi_n(\bar{x}, \bar{a})$ such that $\varphi_0(\bar{x}, \bar{a}) = \varphi(\bar{x}, \bar{a})$, $\varphi_{n+1}(\bar{x}, \bar{a}) \vdash \varphi_n(\bar{x}, \bar{a}), \models (\exists \bar{x}) \varphi_n(\bar{x}, \bar{a})$, and for each n there are only finitely many $r \in S^m_{\vartheta_n}(A)$ consistent with $\varphi_{n+1}(\bar{x}, \bar{a})$,

So $\varphi_0(\bar{x}, \bar{a})$ is defined. Suppose $\varphi_n(\bar{x}, \bar{a})$ is defined and we shall choose φ_{n+1}. Among $\Phi = \{\psi(\bar{x}, \bar{a}) : \models (\exists \bar{x}) \psi(\bar{x}, \bar{a}), \psi(\bar{x}, \bar{a}) \vdash \varphi_n(\bar{x}, \bar{a})\}$, choose one, $\varphi_{n+1}(\bar{x}, \bar{a})$ with minimal $R^m[\psi(\bar{x}, \bar{a}), \vartheta_n, \aleph_0]$.

Let $\Gamma = \{r \in S^m_{\vartheta_n}(A) : r$ be consistent with $\varphi_{n+1}(\bar{x}, \bar{a})\}$.

The only remaining point is "Γ is finite". We shall prove that for every $r \in \Gamma$,

$$R^m[\{\varphi_{n+1}(\bar{x}, \bar{a})\} \cup r, \vartheta_n, \aleph_0] = R^m[\varphi_{n+1}(\bar{x}, \bar{a}), \vartheta_n, \aleph_0],$$

and this will show that $|\Gamma| \leqslant \mathrm{Mlt}[\varphi_{n+1}(\bar{x}, \bar{a}), \vartheta_n]$, but the latter is finite so we shall finish.

For each $r \in \Gamma$, choose $q \in S^m(A), \{\varphi_{n+1}(\bar{x}, \bar{a})\} \cup r \subseteq q$. As $\varphi(\bar{x}, \bar{a})\} \in q$ [because $\varphi_{n+1}(\bar{x}, \bar{a}) \vdash \varphi_0(\bar{x}, \bar{a}), \varphi_0 = \varphi$], clearly q does not fork over $\bar{a}$ [as $\varphi(\bar{x}, \bar{a})$ has no extension over A which forks over A]. Hence,

$R^m[\{\varphi_{n+1}(\bar{x}, \bar{a})\} \cup r, \vartheta_n, \aleph_0] \leqslant$	by monotonicity,
$R^m[q, \vartheta_n, \aleph_0] =$	by III, 4.1 as q does not fork over $\bar{a}$,
$R^m[q \restriction \bar{a}, \vartheta_n, \aleph_0] =$	by the choice of $\varphi_{n+1}(\bar{x}, \bar{a})$,
$R^m[\varphi_{n+1}(\bar{x}, \bar{a}), \vartheta_n, \aleph_0] \leqslant$	by monotonicity,
$R^m[\{\varphi_{n+1}(\bar{x}, \bar{a})\} \cup r, \vartheta_n, \aleph_0]$.	

So the desired equality follows.

(2) Easy.

(3) Let $p^1 = \{\varphi \in p : \varphi$ is almost over $B\}$; as $(p,B) \in \mathbf{F}^k_{\aleph_0}, p^1 \vdash p$. If $p \restriction B \nvdash p^1$, then for some $\varphi \in p^1, p \restriction B \nvdash \varphi$, hence by III, 2.6(1) $(p \restriction B) \cup \{\neg\varphi\}$ does not fork over B, hence there is a $q \in S^m(A)$ extending $(p \restriction B) \cup \{\neg\varphi\}$ which does not fork over B. Now p, q contradict by a hypothesis of Fact 1.12(3), hence $p \restriction B \vdash p_1$. As $(p,B) \in \mathbf{F}^k_{\aleph_0}$, p is $\mathbf{F}^l_{\aleph_0}$-isolated, hence for every $\vartheta = \vartheta(\bar{x}, \bar{y})$ for some $\psi \in p, \psi \vdash p \restriction \vartheta$, but as $p \restriction B \vdash p^1 \vdash p$ we can choose $\psi \in p \restriction B$.

(4) For every ϑ for some $\psi \in p, \psi \vdash p \restriction \vartheta$, but $p \restriction B \vdash \psi$, hence some $\psi' \in p \restriction B, \psi' \vdash \psi$, hence $\psi' \vdash p \restriction \vartheta$.

(5) By IV, 1.4(2), as $\lambda(\mathbf{F}^l_{\aleph_0}) = \aleph_0$, there is an $\mathbf{F}^l_{\aleph_0}$-construction $\langle A, \langle c_i : i < n\rangle, \langle B_i : i < n\rangle\rangle$ such that p is realized by some $\bar{c} \subseteq \{c_i : i < n\}$, and $\operatorname{tp}(c_k, A \cup \{c_l : l < k\}) \in \mathbf{F}^l_{\aleph_0}(B_k)$ (so B_k is finite). By IV, 3.2 $\{c_i : i < n\}$ is $\mathbf{F}^l_{\aleph_0}$-atomic over A, hence p is $\mathbf{F}^l_{\aleph_0}$-isolated. Let $B = C \cup (\bigcup_{i<n} B_i \cap A)$, clearly it is a finite subset of A. We shall prove $p \restriction B \vdash p$; as p is $\mathbf{F}^l_{\aleph_0}$-isolated by 1.12(4) $(p,B) \in \mathbf{F}^l_{\aleph_0}$. As $\mathbf{F}^l_{\aleph_0} \subseteq \mathbf{F}^k_{\aleph_0} \subseteq \mathbf{F}^a_{\aleph_0}$ (see 1.8, 1.2(2)), $\langle A, \langle c_i : i < n\rangle, \langle B_i : i < n\rangle\rangle$ is an $\mathbf{F}^a_{\aleph_0}$-construction over A, hence $\{c_l : l < n\}$ is $\mathbf{F}^a_{\aleph_0}$-atomic over A, hence $\operatorname{stp}(\bar{c}, B) \vdash p$ (remember $p = \operatorname{tp}(\bar{c}, A), \bar{c} \in \{c_l : l < n\}$). So it suffices to prove (as $C \subseteq B$):

FACT 1.13: *If $B \subseteq A, p = \operatorname{tp}(\bar{c}, A) \in \mathbf{F}^a_{\aleph_0}(B), p \restriction B$ has a unique extension in $S^m(A)$ which does not fork over B, then $p \restriction B \vdash p$.*

Proof. We know that $\operatorname{stp}(\bar{c}, B) \vdash p$. Suppose $\vartheta(\bar{x}, \bar{d}) \in p$, and we shall prove that $p \restriction B \vdash \vartheta(\bar{x}, \bar{d})$. As $\operatorname{stp}(\bar{c}, B) \vdash p$ for some $E \in \operatorname{FE}(A)$, $xE\bar{c} \vdash \vartheta(\bar{x}, \bar{d})$. Clearly, $xE\bar{c}$ is almost over A, hence by IV, 4.1 there is a formula $\varphi(\bar{x}, \bar{b}), \bar{b} \in A$, which is almost over B, and

(*) for every formula $\psi(\bar{x}, \bar{a}), \bar{a} \in A, \vdash \bar{x}E\bar{c} \rightarrow \psi(\bar{x}, \bar{a})$ iff $\vdash \varphi(\bar{x}, \bar{b}) \rightarrow \psi(\bar{x}, \bar{a})$.

As $xE\bar{c}$ is consistent with p, necessarily $\varphi(\bar{x}, \bar{b})$ is consistent with p, hence $\varphi(\bar{x}, \bar{b}) \in p$. Also $p \restriction B \vdash \varphi(\bar{x}, \bar{b})$ as otherwise $(p \restriction B) \cup \{\neg\varphi(\bar{x}, \bar{b})\}$ being consistent does not fork over B (as $\varphi(\bar{x}, \bar{b})$ is almost over B, by III, 2.6(1)), hence there is $q \in S^m(A)$ extending $(p \restriction B) \cup \{\neg\varphi(\bar{x}, \bar{b})\}$ which does not fork over B, and clearly $q \neq p$. This contradicts the hypothesis.

So $p \restriction B \vdash \varphi(\bar{x}, \bar{b})$, and $\varphi(\bar{x}, \bar{b}) \vdash \vartheta(\bar{x}, \bar{d})$ (by (*) and as $xE\bar{c} \vdash \vartheta(\bar{x}, \bar{d})$). So $p \restriction B \vdash \vartheta(\bar{x}, \bar{d})$, and as $\vartheta(\bar{x}, \bar{d})$ was any formula in p, we have proved $p \restriction B \vdash p$, thus finishing.

(6) Left to the reader.

Remark. Note we have observed:

FACT 1.14: (1) *If* $p \in S^m(A)$ *is* $\mathbf{F}^a_{\aleph_0}$*-isolated over a set* $B \subseteq A$, *then there is* $q \subseteq p, q \equiv p$, *which is almost over* B.

(2) *Suppose* $\bar{a} \in A, \{\varphi(\bar{x}, \bar{a})\}$ *is consistent but has no extension over* A *which forks over* $\bar{a}, \bar{b} \in A, \{\varphi(\bar{x}, \bar{a}), \psi(\bar{x}, \bar{b})\}$ *is consistent. Then* $\varphi(\bar{x}, \bar{a}) \wedge \psi_1(\bar{x}, \bar{b}_1)$ *is almost over* $\bar{a}$ *for some* $\bar{b}_1 \in A$, $\psi_1(\bar{x}, \bar{b}_1) \equiv \psi(\bar{x}, \bar{b})$ *and* $\models (\exists \bar{x})[\varphi(\bar{x}, \bar{a}) \wedge \psi_1(\bar{x}, \bar{a}_1)]$.

XII.2. Stable systems

(Hyp) T is stable.

DEFINITION 2.1: We call $\mathbf{S} = \langle M_s : s \in I \rangle$ a stable system if $M_s \subseteq \mathfrak{C}$, I is a family of finite subsets of $\bigcup I$ closed under subsets, $s \subseteq t \Rightarrow M_s \subseteq M_t$, and for every $s \in I$, $\mathrm{tp}_*(M_s, \bigcup_{s \not\subseteq t} M_t)$ does not fork over $A^{\mathbf{S}}_s$ (see 2.2(3) below).

Notation 2.2. (1) $\mathscr{P}^-(s) = \{t : t \subseteq s, t \neq s\}$.

(2) For a stable system $\mathbf{S}$, let $\mathbf{S} = \langle M^{\mathbf{S}}_s : s \in I^{\mathbf{S}} \rangle$.

(3) If s is a finite subset of $\bigcup I, \mathscr{P}^-(s) \subseteq I$ let $A^{\mathbf{S}}_s = \bigcup_{s \subset t} M^{\mathbf{S}}_s$. $s \subseteq t \Rightarrow A^{\mathbf{S}}_s \subseteq A^{\mathbf{S}}_t$. Also, for $J \subseteq I$ let $A^{\mathbf{S}}_J = \bigcup_{t \in J} M^{\mathbf{S}}_t$.

(4) We omit the superscript $\mathbf{S}$ when its identity is clear.

LEMMA 2.3: (1) *If* $I = \{s_\alpha : \alpha < \alpha_0\}$, $s_\alpha \subseteq s_\beta \Rightarrow \alpha \leqslant \beta$; $M_s \prec \mathfrak{C}$ *and* $\mathrm{tp}_*(M_{s_\alpha}, \bigcup_{j<\alpha} M_{s_j})$ *does not fork over* A_{s_α}, *then* $\langle M_s : s \in I \rangle$ *is a stable system of models.*

(2) *If* $\langle M_s : s \in I \rangle$ *is a stable system,* $J \subseteq I$, *and* $s \in I \wedge s \subseteq \bigcup J \Rightarrow s \in J$, *then* $\bigcup_{s \in J} M_s \subseteq_t \bigcup_{s \in I} M_s$.

(3) *If* $\mathbf{S} = \langle M_s : s \in I \rangle$ *is a stable system,* $N_s \prec M_s$, $\mathrm{tp}(N_s, A^{\mathbf{S}}_s)$ *does not fork over* $\bigcup_{t \subset s} N_t$, *and* $s \subseteq t \Rightarrow N_s \subseteq N_t$, *then* $\langle N_s : s \in I \rangle$ *is a stable system and* $\bigcup_{s \in I} N_s \subseteq_t \bigcup_{s \in I} M_s$.

Proof. Essentially like [Sh 83a, 3.5], but we shall prove it. We first prove some facts.

FACT 2.4: *If* $\mathbf{S} = \langle M_s : s \in I \rangle$ *is a stable system, and for* $l = 0, 1$ $J_l \subseteq I$,

I_l is the closure under subsets of J_l, then $\mathrm{tp}_*(\bigcup_{s\in J_0} M_s, \bigcup_{s\in J_1} M_s)$ *does not fork over* $\bigcup\{\{M_s : s \in I_0 \cap I_1\}$.

Proof. W.l.o.g. J_l are closed under subsets, and let $J = J_1 \cap J_0$. We can find a list $\{s_\alpha : \alpha < \alpha^*\}$ of I such that $s_\alpha \subseteq s_\beta \Rightarrow \alpha \leqslant \beta$, $J = \{s_\alpha : \alpha < \alpha_1\}$, $J_0 = \{s_\alpha : \alpha < \alpha_2\}$, $J_1 = \{s_\alpha : \alpha < \alpha_1 \text{ or } \alpha_2 \leqslant \alpha < \alpha_3\}$. Clearly, for $\alpha < \alpha_3$, $\alpha \geqslant \alpha_2$, $\mathrm{tp}_*(M_{s_\alpha}, \bigcup_{\beta<\alpha} M_{s_\beta})$ is included in $\mathrm{tp}_*(M_{s_\alpha}, \bigcup\{M_t : s_\alpha \not\subseteq t \in I\})$, and hence does not fork over $A_{s_\alpha} \subseteq \bigcup\{M_{s_\beta} : \beta < \alpha_1 \text{ or } \alpha_2 \leqslant \beta < \alpha\}$. So $\mathrm{tp}_*(M_{s_\alpha}, \bigcup_{\beta<\alpha} M_{s_\beta})$ does not fork over $\bigcup\{M_{s_\beta} : \beta < \alpha_1, \text{ or } \alpha_2 \leqslant \beta < \alpha\}$. By IV, 3.2(1) (applied to $\mathbf{F}^f$) we can conclude that $\mathrm{tp}_*(\bigcup\{M_{s_\alpha} : \alpha_2 \leqslant \alpha < \alpha_3 \text{ or } \alpha < \alpha_1\}, \bigcup_{\beta<\alpha_2} M_{s_\beta})$ does not fork over $\bigcup_{\beta<\alpha_1} M_{s_\beta}$. But this is as required.

FACT 2.5: If $\mathbf{S} = \langle M_s : s \in I\rangle$ *is a stable system,* $\bar a_l \in M_{s(l)} (l < n)$, $t \subseteq \bigcup I$, *and* $\models \varphi[\bar a_0, \dots, \bar a_{n-1}]$, *then we can find* $\bar a'_l \in M_{s(l)\cap t}$ *such that* $\models \varphi[\bar a'_0, \dots, \bar a'_{n-1}]$ *and* $s(l) \subseteq t \Rightarrow \bar a'_l = \bar a_l$.

Proof. W.l.o.g. $s \subset s(l) \Rightarrow s \in \{s(m) : m < l\}$, the $s(l)\ (l < n)$ are distinct, $\{l : s(l) \subseteq t\}$ is an initial segment of $\{l : l < n\}$ and $s(n) \not\subseteq s(l)$ for $l < n$. We prove the assertion by induction on n. For $n = 0$ there is nothing to prove, and for $n = 1$ note $M_{s(l)\cap t}$ is an elementary submodel $M_{s(l)}$. So suppose we have proved for n and we shall prove for $n+1$, i.e. for given $\bar a_l \in M_{s(l)}\ (l < n+1), t \subseteq \bigcup I$ and φ satisfying the "w.l.o.g." above. If $s(l) \subseteq t$ for every l, let $\bar a'_l = \bar a_l$; so we assume (by the "w.l.o.g." above) that $s(n) \not\subseteq t$. As $\mathrm{tp}(\bar a_{s(n)}, \bigcup_{l<n} M_{s(l)})$ does not fork over $A^{\mathbf{S}}_{s(n)}$, clearly $\varphi(\bar a_0, \dots, \bar a_{n-1}, \bar x)$ does not fork over $A^{\mathbf{S}}_{s(n)}$, hence it is realized in every model which includes $A^{\mathbf{S}}_{s(n)}$. So there is type $p = p(\bar x_i)_{i<\alpha}$ over $A^{\mathbf{S}}_{s(n)}$ (in infinitely many variables) such that $p(\bar x_0, \dots, \bar x_i, \dots) \vdash \bigvee_{i<\alpha} \varphi(\bar a_0, \dots, \bar a_{n-1}, \bar x_i)$. So for some $\bar b \subseteq A^{\mathbf{S}}_{s(n)}$, and $\psi = \psi(\bar x_0, \bar x_1, \dots, \bar x_k, \bar b)$, and $k < \omega$

(i) $\models (\exists \bar x_0, \bar x_1, \dots, \bar x_k)\psi(\bar x_0, \dots, \bar x_k, \bar b)$,

(ii) $\psi(\bar x_0, \dots, \bar x_k, \bar b) \vdash \bigvee_{i\leqslant k} \varphi(\bar a_0, \dots, \bar a_{n-1}, \bar x_i)$.

As $\bar b \subseteq A^{\mathbf{S}}_{s(n)}$, and $(\forall s \subset s(n))[s \in \{s(l) : l < n\}]$, w.l.o.g. $\bar b = \bar b_0 {}^\frown \bar b_1 {}^\frown \dots {}^\frown \bar b_{n-1}$, $\bar b_l \subseteq M_{s(l)}$, and $[s(l) \not\subseteq s(n) \Rightarrow \bar b_l \text{ empty}]$. Now apply the induction hypothesis to $\bar a_l {}^\frown \bar b_l \in M_{s(l)}$ (for $l < n$) and the formula:

$$(\exists \bar x_0, \dots, \bar x_k)\psi(\bar x_0, \dots, \bar x_k, \bar b_0, \dots, \bar b_{n-1})$$
$$\wedge\ (\forall \bar x_0, \dots, \bar x_k)\left[\psi(\bar x_0, \dots, \bar x_k, \bar b_0, \dots, \bar b_{n-1}) \to \bigvee_{i\leqslant k} \varphi(\bar a_0, \dots, \bar a_{n-1}, \bar x_i)\right].$$

So there are $\bar a'_l {}^\frown \bar b'_l \in M_{s(l)\cap t} (l < n)$ satisfying the above formula and as in 2.5's conclusion; remembering that $s(l) \nsubseteq s(n) \Rightarrow \bar{b}_l = \langle\rangle$, clearly $\bar{b}'_0 {}^\frown \bar{b}'_1 {}^\frown \dots {}^\frown \bar{b}'_{n-1} \subseteq M^{\mathbf{S}}_{s(n)\cap t}$. Hence there are $\bar{c}_0, \dots, \bar{c}_k \in M_{s(n)\cap t}$ such that $\models \psi[\bar{c}_0, \dots, \bar{c}_k, \bar{b}'_0, \dots, \bar{b}'_{n-1}]$. So for some $i < k \models \varphi[\bar{a}'_0, \dots, \bar{a}'_{n-1}, \bar{c}_i]$. So $\bar{a}'_0, \dots, \bar{a}'_{n-2}, \bar{a}'_{n-1}, \bar{a}'_n \overset{\text{def}}{=} \bar{c}_i$ are as required.

Proof of 2.3. (1) An exercise in non-forking.

(2) Follows from Fact 2.5.

(3) First we prove that $\mathbf{S}^1 = \langle N_s : s \in I\rangle$ is a stable system. For every s $\mathrm{tp}_*(M_s, \bigcup\{M_t : t \in I, s \nsubseteq t\})$ does not fork over $\bigcup\{M_t : t \subset s\}$, hence (as $N_s \subseteq M_s$) also $\mathrm{tp}_*(N_s, \bigcup\{M_t : t \in I, s \nsubseteq t\})$ does not fork over $\bigcup\{M_t : t \subset s\}$. But $\mathrm{tp}_*(N_s, \bigcup\{M_t : t \subset s\})$ does not fork over $\bigcup\{N_t : t \subset s\}$, by the hypothesis. So by III, 0.1(2), $\mathrm{tp}_*(N_s, \bigcup\{M_t : t \in I, s \nsubseteq t\})$ does not fork over $\bigcup\{N_t : t \subset s\}$. As $N_t \subseteq M_t$, by monotonicity of non-forking we get the stability of the system ($s \subseteq t \Rightarrow N_s \subseteq N_t$ was assumed, and we know I is as required).

Then $A^{\mathbf{S}^1}_I \subseteq_t A^{\mathbf{S}}_I$ follows by Fact 2.5 and the following fact. Let $j \notin \bigcup I$, $J \overset{\text{def}}{=} I \cup \{s \cup \{j\} : s \in I\}$ and $N_{s\cup\{j\}} \overset{\text{def}}{=} M_s$.

FACT 2.6: *$\langle N_s : s \in J\rangle$ is a stable system (J, N_s as above).*

Proof. Let $s_\alpha (\alpha < \alpha_0)$ be as in 2.3(1) for $\mathbf{S}$, and define $t_\alpha (\alpha < 2\alpha_0)$ by: $t_{2\alpha} = s_\alpha$, $t_{2\alpha+1} = s_\alpha \cup \{j\}$. Clearly, $J = \{t_\alpha : \alpha < 2\alpha_0\}$ and $t_\alpha \subseteq t_\beta \Rightarrow \alpha \leqslant \beta$. Now use 2.3(1). For α even $(=2\beta)$ remember we have proved $\mathrm{tp}_*(N_{s_\beta}, \bigcup\{M_s : s \in I, s \nsubseteq s_\beta\})$ does not fork over $\bigcup\{N_s : s \subseteq s_\beta, s \neq s_\beta\})$ and this implies what we need. For α odd $(=2\beta+1)$ remember $\mathrm{tp}_*(M_{s_\beta}, \bigcup_{\gamma<\beta} M_{s_\gamma})$ does not fork over $\bigcup\{M_s : s \subset s_\beta\})$. As $N_{s_\beta} \subseteq M_{s_\beta}$, by III, 0.1(3) this gives $\mathrm{tp}_*(N_{t_\alpha}, \bigcup_{\gamma<\alpha} N_{t_\gamma}) = \mathrm{tp}_*(M_{s_\beta}, \bigcup_{\gamma<\beta} M_{s_\gamma} \cup N_{s_\beta})$ does not fork over $\bigcup\{M_s : s \subset s_\beta\} \cup N_{s_\beta} = \bigcup\{N_s : s \subset t_\alpha\}$, and this is what we need.

So we finish the proof of 2.3.

LEMMA 2.7: *Suppose $\langle M_s : s \in I\rangle$ is a stable system $|I| < \kappa$.*

(1) *If each M_s is κ-compact, p a type of cardinality $<\kappa$ in the variables $x_{t,i} (t \in I, i < i_t)$, and every finite subset of p is realized by an assignment sending each $x_{t,i}$ to a member of M_t, then p is realized by an assignment sending each $x_{t,i}$ to a member of M_t.*

(2) *If each M_s is $\mathbf{F}^a_\kappa$-saturated, $\kappa \geqslant \kappa_r(T)$, p a type which is almost over a set of power $<\kappa$ in the variables $x_{t,i} (t \in I, i < i_t < \kappa)$ and every finite subset of p is realized by an assignment sending each $x_{t,i}$ to a member*

of M_t, then p is realized by an assignment sending $x_{t,i}$ to a member of M_t.

Proof. (1) Clearly, $\kappa = \aleph_0$ is a trivial case; so w.l.o.g. $\kappa > \aleph_0$. Now we can replace $\mathfrak{C}$ (and all M_s) by their reduct to a language containing only the predicates (and function symbols) which appear in p, without affecting the hypothesis or the conclusion. So w.l.o.g. $|T| \leqslant |p| + \aleph_0 < \kappa$, hence $\kappa_r(T) \leqslant \kappa$, but then our desired conclusion follows from (2).

(2) So let p be almost over A, $|A| < \kappa$. As $\kappa_r(T) \leqslant \kappa$ there is $B \subseteq \bigcup_{s\in I} M_s$, $|B| < \kappa_r(T) + |A|^+ \leqslant \kappa$ such that $\mathrm{tp}_*(A, \bigcup_{s\in I} M_s)$ does not fork over B. Clearly, by IV, 4.2,

(*) for every $\bar{a} \in A$, and $\varphi = \varphi(\bar{x}, \bar{a})$ there are $\bar{b}_\varphi \in \bigcup_{s\in I} M_s$ and $\psi_\varphi = \psi_\varphi(\bar{x}, \bar{b}_\varphi)$ such that:
(i) ψ_φ is almost over B,
(ii) for every $\bar{c} \in \bigcup_s M_s, \varphi(\bar{c}, \bar{a}) \equiv \psi_\varphi(\bar{c}, \bar{b}_\varphi)$.

So w.l.o.g. p is over $\bigcup_{s\in I} M_s$, and almost over B. Checking the demands on κ ($\kappa \geqslant |B|^+ + \kappa_r(T) + |I|^+$) w.l.o.g. κ is regular. Looking at the demands on p it is clear that for every φ, $p \cup \{\varphi\}$ or $p \cup \{\neg\varphi\}$ satisfies the demands. Also we can allow more variables from $x_{t,i}$ ($i < i_t, t \in I$), and take increasing union of length $<\kappa$. So w.l.o.g.

$$p = \mathrm{stp}_*(\langle a_{t,i} : i < i_t, t \in I\rangle, B), \text{ and let } B_s = B \cap M_s, A_s \stackrel{\mathrm{def}}{=} \{a_{s,i} : i < i_t\}.$$

Now it suffices to prove that w.l.o.g.

(**) for each $t \in I$, $\mathrm{tp}_*(B_t \cup A_t, \bigcup\{B_s \cup A_s : t \not\subseteq s \in I\})$ does not fork over $\bigcup\{B_s \cup A_s : s \subset t\}$.

Because then we shall choose a list $\langle s_\alpha : \alpha < \alpha_0\rangle$ as in 2.3(1), and then define by induction on α, $a'_{t_\alpha, i} \in M_{s_\alpha}$ ($i < i_t$) such that $\mathrm{stp}_*(\langle a_{t_\beta, i} : i < i_{t_\beta}, \beta \leqslant \alpha\rangle, B) \equiv \mathrm{stp}(\langle a'_{t_\beta, i} : i < i_{t_\beta}, \beta \leqslant \alpha\rangle, B)$ using "M_{t_α} is $\mathbf{F}^a_\kappa$-saturated".

Why can we make this assumption? First, w.l.o.g. $\mathrm{tp}_*(B, A_t)$ does not fork over $A_t \cap B$, $\mathrm{tp}_*(B, M_t)$ does not fork over $M_t \cap B$ (just increase $B < \kappa$ times). Then we define by induction on $\alpha < \kappa$, $p_\alpha = \mathrm{stp}_*(\langle a^\alpha_{t,i} : i < i^\alpha_t, t \in I\rangle, B_\alpha)$ such that:

(a) p_0 is the p we have, p_α is increasing and continuous.

(b) Each finite subset of p_α is realized by some $\langle b^\alpha_{t,i} : i < i^\alpha_t, t \in I\rangle$, where $b^\alpha_{t,i} \in M_t$.

(c) $p_{\alpha+1}$ forks over B_α and $B_\alpha \subseteq \bigcup_{t\in I} M_t$.

(d) $\mathrm{tp}_*(B_\alpha, A_t)$ does not fork over $A_t \cap B$, $\mathrm{tp}_*(B_\alpha, M_t)$ does not fork over $M_t \cap B_\alpha$.

For $\alpha = 0$, α limit, there are no problems. For $\alpha = \beta+1$, we assume that p_β does not satisfy (**). Choose $t = t_\beta$ with minimal $|t|$, and finite $w_\beta \subseteq i^\beta_{t_\beta}$, and $\bar{b}_\beta \in B_\beta \cap M_{t_\beta}$ such that (letting $\bar{a}_\beta = \langle a^\beta_{t_\beta, i} : i \in w_\beta \rangle$, $B^\beta_t = B_\beta \cap M_t$, $A^\beta_t = \{a^\beta_{t,i} : i < i^\beta_t\}$), $\operatorname{tp}(\bar{a}_\beta {}^\frown \bar{b}_\beta, \bigcup\{B^\beta_s \cup A^\beta_s : t \not\subseteq s \in I\})$ forks over $\bigcup\{B^\beta_s \cup A^\beta_s : s \subset t\}$. Hence there are $\bar{d}_l = \langle d^\beta_{l,j} : j < j_{\beta,l} \rangle \in B^\beta_{s_l} \cup A^\beta_{s_l}$ for $l = 1, n, t \not\subseteq s_l \in I$, and $\varphi_\beta(\bar{x}, \bar{y}, \bar{d}_1, \ldots, \bar{d}_n)$, forks over $\bigcup\{B^\beta_s \cup A^\beta_s : s \subset t\}$ and is realized by $\bar{a}_\beta {}^\frown \bar{b}_\beta$.

Let $t(\beta, l) = s_{\beta,l} \cap t_\beta$, choose $j(\beta, l, j)\,(j < j_{\beta,l})$ so that $i_{t(\beta,l)} \leqslant j(\beta, l, j) < i_{t(\beta,l)} + \omega$, and $[t(\beta, l_1) = \tau(\beta, l_2) \wedge (l_1, j_1) \neq (l_2, j_2) \Rightarrow j(\beta, l_1, j_1) \neq j(\beta, l_2, j_2)]$. Now by 2.3(2) $p \cup \{\varphi(\langle x_{t_{\beta,i}} : i \in w_\beta \rangle, \bar{b}_\beta, x_{t(\beta,l), j(\beta,l)}, \ldots)\}$ satisfies (b). Then we can complete it to a $\operatorname{stp}(\langle a^\alpha_{t,i} : t \in I, i < i^\alpha_t \rangle, B_{\alpha+1})$ as required (notice $i^\beta_s = i^\alpha_s$ when $s \not\subseteq t_\beta$, and $i^\beta_s < i^\alpha_s + \omega$).

If $\kappa = \aleph_0$, choose t with maximal $|t|$ such that $t = t_\beta$ for unboundedly many β's. So $i^\alpha_t(\alpha < \omega)$ is eventually constant and we get a contradiction to $\kappa_r(T) \leqslant \kappa$. So let $\kappa > \aleph_0$, and remember that κ is regular. Note that for every pair $(t, w), t \in I$, w is a finite subset of $\bigcup_\alpha i^\alpha_t$, $S_{t,w} = \{\beta < \kappa : t_\beta = t, w_\beta = w\}$ is a bounded subset of κ, (by $\kappa_r(T) \leqslant \kappa$). Hence by the well-known properties of the closed unbounded filter, for some limit $\delta < \kappa$, $(\forall t \in I)(\forall w \subset \bigcup_{\alpha<\delta} i^\alpha_t)[S_{t,w} \subseteq \delta]$. So for this $\delta, t_\delta, w_\delta$, are not defined, a contradiction.

LEMMA 2.8: *In 2.7 we can replace (in assumption and conclusion) "$x_{t,i}$ assign to a member of M_t" by "each variable is assigned to a member of $\bigcup_{s\in I} M_s$", provided that I is finite.*

Proof. By the compactness theorem we can assign a model to each variable retaining the hypothesis and then use 2.7.

CONCLUSION 2.9: *Let $\langle M_s : s \in I \rangle$ be a stable system;*

$$J \subseteq I, (\forall t \in I)[t \subseteq \bigcup J \to t \in J].$$

(1) *If each M_s is κ-compact, then $\bigcup_{s\in J} M_s \subseteq^t_\kappa \bigcup_{s\in I} M_s$.*
(2) *If each M_s is $\mathbf{F}^a_\kappa$-saturated, $\kappa \geqslant \kappa_r(T)$, then $\bigcup_{s\in I} M_s \subseteq^a_\kappa \bigcup_{s\in I} M_s$.*

Proof. Immediate by 2.5 and 2.7(1) [for (1)] and 2.7(2) [for (2)].

FACT 2.10: *For the stable system $\langle M_t : t \in I \rangle$ and κ we can choose $j \notin \bigcup I$, let $J = I \cup \{t \cup \{j\} : t \in I\}$ and define $M_s(s \in J - I)$ such that $\langle M_s : s \in I \rangle$ is a stable system and for $s \in J - I$, M_s is $\mathbf{F}^a_\kappa$-saturated.*

Proof. Easy by 2.3(1) and the properties of forking.

CONCLUSION 2.11: *If* $\langle M_s : s \in I\rangle$ *is a stable system of* $\mathbf{F}^a_\kappa$*-saturated models,* $\mathrm{cf}\,\kappa \geqslant \kappa_r(T)$*, I finite, then for any* $\bar{c}$ *and an* $\mathbf{F}^a_\kappa$*-primary model M over* $\bigcup_{s\in I} M_s \cup \bar{c}$, $\mathrm{tp}_*(M \bigcup_{s\in I} M_s \cup \bar{c})$ *is* $\mathbf{F}^l_{\aleph_0}$*-atomic.*

Proof. Suppose $p = \mathrm{tp}(\bar{d}, \bigcup_{s\in I} M_s \cup \bar{c})$ is $\mathbf{F}^a_\kappa$-isolated, and we shall prove it is $\mathbf{F}^l_{\aleph_0}$-isolated. First observe that w.l.o.g. κ is regular $> |T|$. [Otherwise let $\mu = \kappa^+ + |T|^+$. Choose $j \notin \bigcup I$, let $J = I \cup \{t \cup \{j\} : t \in I\}$. By 2.10 we can find $\mathbf{F}^a_\kappa$-saturated M_s for $s \in J - I$ such that $\langle M_s : s \in J\rangle$ is a stable system. By 2.9(2) $\bigcup_{s\in I} M_s \subseteq^a_\kappa \bigcup_{s\in J} M_s$. W.l.o.g. $\mathrm{tp}(\bar{c}^\frown \bar{d}, \bigcup_{s\in J} M_s)$ does not fork over $\bigcup_{s\in I} M_s$, hence $\bigcup_{s\in I} M_s \cup \bar{c} \subseteq^a_\kappa \bigcup_{s\in J} M_s \cup \bar{c})$. So as $\mathrm{tp}(\bar{d}, \bigcup_{s\in I} M_s \cup \bar{c})$ is $\mathbf{F}^a_\kappa$-isolated, also $\mathrm{tp}(\bar{d}, \bigcup_{s\in J} M_s \cup \bar{c})$ is $\mathbf{F}^a_\kappa$-isolated, hence $\mathbf{F}^a_\mu$-isolated. By IV, 4.3(5) it suffices to prove that $\mathrm{tp}(\bar{a}, \bigcup_{s\in J} M_s \cup \bar{c})$ is $\mathbf{F}^l_{\aleph_0}$-isolated.]

Now let $\varphi = \varphi(\bar{x}, \bar{y})$, and suppose for no finite $q \subseteq p, q \vdash p \restriction^+ \varphi$ (remember $p = \mathrm{tp}(\bar{d}, \bigcup_{s\in I} M_s \cup \bar{c})$). Then w.l.o.g. for some $\varphi(\bar{x}, \bar{z}, \bar{c})$ and for no $\psi(\bar{x}, \bar{a}) \in p$ for every $\bar{b} \in \bigcup_{s\in I} M_s : \varphi(\bar{x}, \bar{b}, \bar{c}) \in p$ implies $\psi(\bar{x}, \bar{a}) \vdash \varphi(\bar{x}, \bar{b}, \bar{c})$. Then $r \overset{\mathrm{def}}{=} \{\neg(\forall \bar{x})[\psi(\bar{x}, \bar{a}) \to \varphi(\bar{x}, \bar{z}, \bar{c})] \wedge \varphi(\bar{d}, \bar{z}, \bar{c})] : \psi(\bar{x}, \bar{a}) \in p\}$ is finitely satisfiable in $\bigcup_{s\in I} M_s$ [if $n < \omega, \psi_l(\bar{x}, \bar{a}_l) \in p$ for $l < n$ let $\psi(\bar{x}, \bar{a}) \overset{\mathrm{def}}{=} \bigwedge_{l<n} \psi_l(\bar{x}, \bar{a}_l)$, so $\psi(\bar{x}, \bar{a}) \in p$ hence for some $\bar{b} \in \bigcup_{s\in I} M_s$, $\varphi(\bar{x}, \bar{b}, \bar{c}) \in p$ but $\psi(\bar{x}, \bar{a}) \nvdash \varphi(\bar{x}, \bar{b}, \bar{c})$; now substituting $\bar{b}$ for $\bar{z}$ is as required]. Hence every subset of r of power $< \kappa$ is realized in $\bigcup_{s\in I} M_s$. But as κ is regular $> |T|$, for some complete $p_1 \subseteq p, |p_1| < \kappa, p_1 \vdash p$, so $\{\neg(\forall \bar{x})[\psi(\bar{x}, \bar{a}) \to \varphi(\bar{x}, \bar{z}, \bar{c})] \wedge \varphi(\bar{d}, \bar{z}, \bar{c}) : \psi(\bar{x}, \bar{a}) \in p_1\}$ is realized in $\bigcup_{s\in I} M_s$, say by $\bar{e}$. But as p_1 is complete, this means that $p_1 \nvdash \varphi(\bar{x}, \bar{e}, \bar{c})$ and $\models \varphi[\bar{d}, \bar{e}, \bar{c}]$, hence $\varphi(x, \bar{e}, \bar{c}) \in p$. This contradicts $p_1 \vdash p$.

CONCLUSION 2.12: *For a stable system* $\langle M_s : s \in I\rangle$ *and* $s \in I$, $\mathrm{tp}_*(M_s, \bigcup_{t\subset s} M_t)$ *is stationary inside* $\bigcup\{M_s : s \nsubseteq t, t \in I\}$.

Proof. By 2.3(2), Def. 2.1 and XI, 1.2.

XII.3. On good sets

LEMMA 3.1: *Let T be stable,* $\kappa = \kappa_r(T)$,

If $M_l\,(l < 3)$ are $\mathbf{F}^a_\kappa$-saturated, $M_0 \prec M_l, \{M_0, M_1\}$ independent over M_0, then $M_1 \cup M_2$ is a good set, where

DEFINITION 3.2: A set A is called *good* if for every $p \in S^m(A)$ there is $B \subseteq A, |B| < \kappa_r(T)$, such that $p \restriction B$ is stationary inside B (see Def. XI, 1.1).

We shall prove 3.1 later (in 3.5).

FACT 3.3: (1) *Every $\mathbf{F}^a_\kappa$-saturated $M(\kappa = \kappa_r(T))$ is a good set.*

(2) *If $A \subseteq^a_\kappa C, \kappa = \kappa_r(T)$, C is good, then A is a good set.*

(3) *If A is a good set in $\mathfrak{C}$, then A is a good set in $\mathfrak{C}^{eq}$.*

(4) *If A is good, $|B| < \kappa_r(T)$, then $A \cup B$ is good.*

Proof. (1) If $p \in S^m(M)$ let $C \subseteq M, |C| < \kappa$, be such that p does not fork over C, let $\bar{b}$ realize p and let $\bar{b}'$ realize $\mathrm{stp}(\bar{b}, C)$ and belong to M (exists by the definition of $\mathbf{F}^a_\kappa$-saturated). Now $B = C \cup \bar{b}'$ is as required in Def. 3.2.

(2) Let $p \in S^m(A)$, and we shall find a set $B \subseteq A$ as required in Def. 3.2. Let $q \in S^m(C)$ be an extension of p which does not fork over A. As C is a good set there is $B_1 \subseteq C, |B_1| < \kappa$, such that q does not fork over B_1, and $q \restriction B_1$ is stationary inside C. Choose $B_0 \subseteq A, |B_0| < \kappa$, such that $\mathrm{tp}_*(B_1, A)$ does not fork over B_0, and w.l.o.g. $B_1 \cap A \subseteq B_0$ and p (which is in $S^m(A)$) does not fork over B_0. Let f be an elementary mapping from $B_1 \cup B_0$ into $A, f \restriction B_0 =$ the identity, $\mathrm{stp}_*(f(B_1), B_0) \equiv \mathrm{stp}(B_1, B_0)$. Let B be the range of f, and we shall show that it is as required. Clearly, $|B| < \kappa$. Also p does not fork over B (as $B_0 \subseteq B$), and as $B_0 \subseteq A \subseteq C$, q extends p and does not fork over A, clearly q does not fork over B_0.

So suppose $p' \in S^m(A), p' \neq p$, p' extends $p \restriction B$ and does not fork over B, and let $q' \in S^m(A)$ extend p' and does not fork over B. Now as $p \neq p'$, clearly $q' \neq q$, hence $q' \restriction B_1 \neq q \restriction B_1$ so for some $\bar{b} \in B_1$, $\vartheta(\bar{x}, \bar{b}) \in q, \neg\vartheta(\bar{x}, \bar{b}) \in q'$. But $\bar{b}, f(\bar{b})$ realizes the same strong type over B_0, q, q' does not fork over B_0, hence (see III, 4.8(1)) $\vartheta(\bar{x}, f(\bar{b})) \in q$, $\neg\vartheta(\bar{x}, f(\bar{b})) \in q'$. But q, q' extend $p \restriction B, B \supseteq f(B_1)$, a contradiction.

(3) Suppose $p = \mathrm{tp}(\bar{b}, A, \mathfrak{C}^{eq})$, w.l.o.g. $\bar{b} = b = \bar{a}/E$, E an equivalence relation in $\mathfrak{C}$ defined by a first-order formula without parameters; now use the goodness of A for $\mathrm{tp}(\bar{a}, A)$.

(4) Suppose $p = \mathrm{tp}(\bar{c}, A \cup B)$; for some $C \subseteq A$, $|C| < \kappa$, we have: $\mathrm{tp}(\bar{b}', C)$ is stationary inside A for every $\bar{b}' \in B \cup \bar{c}$ (C exists since A is good). Suppose $\bar{c}'$ realizes $\mathrm{tp}(\bar{c}, C \cup B)$, and $\mathrm{tp}(\bar{c}, A \cup B)$ does not fork over C; then $\mathrm{tp}*(\bar{c} \cup B, A)$ does not fork over C, hence $\mathrm{tp}_*(\bar{c}' \cup B, A) \equiv \mathrm{tp}(\bar{c} \cup B, A)$, hence $\mathrm{tp}(\bar{c}', A \cup B) \equiv \mathrm{tp}(\bar{c}, A \cup B)$, as required.

A more general lemma than 3.1 is

LEMMA 3.4: *Let T be stable, $\kappa = \kappa_r(T)$.*

Suppose $D_0 \subseteq D_l (l = 1, 2), D_2 = |M_2|$, M_2 is $\mathbf{F}^a_\kappa$-saturated, each D_l is good, $\{D_1, D_2\}$ independent over D_0 and $D_0 \subseteq^a_\kappa D_1$. Then $D_1 \cup D_2$ is good.

Proof. Let $q \in S^m(D_1 \cup M_2)$. Let $\bar{c}$ realize q, q does not fork over $C, C \subseteq D_1 \cup M_2$, $\|C\| < \kappa$. Also we can define $C_l = D_l \cap C$, w.l.o.g. $\text{tp}_*(C_l, D_0)$ does not fork over C_0.

We shall work in $\mathfrak{C}^{\text{eq}}$ (see III, §6, and easily we can transform the hypothesis to $\mathfrak{C}^{\text{eq}}$, and the conclusion back to $\mathfrak{C}$).

Stage A: W.l.o.g. for some regular $\lambda > (2^{|T|})^+$, M_2 is $\mathbf{F}^a_\lambda$-saturated, $D_0 \subseteq^a_\lambda D_1$. Straightforward; choose elementary mapping $f_i (i < \lambda)$ such that $\text{Dom} f_i = D_1, f_i \restriction D_0 =$ the identity, $\text{tp}_*(f_i(D_1), \bigcup_{j<i} f_j(D_1) \cup M_2)$, does not fork over D_0, and f_0 is the identity on D_1. Next choose M_2^*, an $\mathbf{F}^a_\lambda$-saturated model, $M_2 \cup \bigcup_{0<j<\lambda} f_j(D_1) \subseteq M_2^*$, and $\text{tp}_*(M_2^*, \bigcup_{j<\lambda} f_j(D_1))$ does not fork over $\bigcup_{0<j<\lambda} f_j(D_1)$. Now the triple $D_0^* = \bigcup_{j<\lambda/j\neq 0} f_j(D_1), D_1^* = D_0^* \cup D_1, D_2^* = |M_2^*|$ is as required and easily $D_1 \cup D_2 \subseteq^a_\kappa D_1^* \cup D_2^*$. So by 3.3(2) it is enough to prove $D_1^* \cup D_2^*$ is a good set.

Stage B: $q \in S^m(D_1 \cup M_2)$ is the unique extension of $\text{tp}(\bar{c}, A_2 \cup D_1)$ (in $S^m(D_1 \cup M_2)$) which does not fork over C, where $A_2 = \text{acl}(D_0 \cup C_2)$ (of course $A_2 \subseteq M_2$). Clearly, q does not fork over $A_2 \cup D_1$ (by monotonicity of forking). It is enough to prove that q is the unique extension of $\text{tp}(\bar{c}, A_2 \cup D_1)$ in $S^m(D_1 \cup M_2)$ which does not fork over $A_2 \cup D_1$ (by III, 0.1(2)). So, by symmetry of forking (see III, 0.1(1)) it is enough to prove that $\text{tp}_*(M_2, A_2 \cup D_1 \cup \bar{c})$ is the unique extension (complete over $A_2 \cup D_1 \cup \bar{c}$) of $\text{tp}_*(M_2, A_2 \cup D_1)$ which does not fork over $A_2 \cup D_1$. As $\text{tp}_*(M_2, A_2 \cup D_1 \cup \bar{c})$ does not fork over $A_2 \cup D_1$ (by symmetry of non-forking, see III, 0.1(1)) it is enough to show that $\text{tp}_*(M_2, A_2 \cup D_1)$ is stationary. But as $\{D_1, M_2\}$ is independent over $D_0, D_0 \subseteq A_2 \subseteq M_2$, clearly $\text{tp}_*(M_2, A_2 \cup D_1)$ does not fork over A_2 (see III, 0.1(3)), hence it is enough to prove that $\text{tp}_*(M_2, A_2)$ is stationary. But we are working in $\mathfrak{C}^{\text{eq}}$ and $A_2 = \text{acl}(D_0 \cup C_2)$, hence $A_2 = \text{acl} A_2$ (by III, 6.2(4)), hence every complete type over A_2 is stationary (by III, 6.9(1)).

Stage C: Let $q_1 = q \restriction (A_2 \cup D_1)$. We can choose $B_0 \subseteq A_2 \cup D_1, |B_0| \leqslant |T|$ such that for every finite set Δ of formulas, $R^n(q_1, \Delta, \aleph_0) = R^n(q_1 \restriction B_0, \Delta, \aleph_0)$, and $\mathrm{Mlt}(q, \Delta, \aleph_0) = \mathrm{Mlt}(q_1 \restriction B_0, \Delta, \aleph_0)$, and $C \subseteq B_0$.

Choose B_1, such that $B_0 \subseteq B_1 \subseteq D_1 \cup A_2, |B_1| \leqslant 2^{|T|} < \lambda$ and $B_1 \cup \bar{c}$ is (the universe of) an $L_{|T|^+,|T|^+}$-elementary submodel of $\mathfrak{A} = (D_1 \cup A_2 \cup \bar{c}, D_0, D_1, A_2, \bar{c}, \ldots, d, \ldots, \varphi(-), \ldots)_{d \in C, \varphi(-) \in L(T)}$. That is, D_l, A_2 are monadic relations, the members of $\bar{c}$ and each $d \in C$ an individual constant and each $\varphi = \varphi(x_0, \ldots, x_{n-1})$ an n-place relation, i.e. $\varphi = \{\langle a_0, \ldots, a_{n-1}\rangle : a_l \in |\mathfrak{A}| = |D_1| \cup |A_2| \cup \bar{c},$ and $\mathfrak{C} \models \varphi[a_0, \ldots, a_{n-1}]\}$. Remember also that $L_{|T|^+,|T|^+}$ is an infinitary logic. Clearly, $\mathrm{tp}_*(B_1 \cap A_2, B_1 \cap D_1)$ does not fork over $B_1 \cap D_0$.

Now choose an elementary mapping f, $\mathrm{Dom} f = B_1 \cup \bar{c}, f \restriction (B_1 \cap A_2) =$ the identity, f maps $B_1 \cap D_1$ into D_0 and it maps $\bar{c}$ into M_2, and

$$\mathrm{stp}_*(B_1 \cup \bar{c}, B_1 \cap A_2) \equiv \mathrm{stp}_*(f(B_1 \cup \bar{c}), B_1 \cap A_2).$$

There is no problem in doing so: first define $f \restriction (B_1 \cap D_1)$ by the hypothesis on D_0, D_1 from Stage A. As $\mathrm{tp}(B_1 \cap A_2, B_1 \cap D_1)$ does not fork over $D_0 \cap B_1$, clearly $id_{(B_1 \cap A_2)} \cup (f \restriction (B_1 \cap D_1))$ is an elementary mapping. As M_2 is $\mathbf{F}^a_\lambda$-saturated we can define $f(\bar{c})$ suitably.

Stage D: For every extension $q' \in S^m(D_1 \cup M_2)$ of $q \restriction (D_1 \cup C_2 \cup f(\bar{c}))$ which does not fork over $C, q' \restriction B_1 = q \restriction B_1$. So suppose $\varphi(\bar{x}, \bar{b}, \bar{d}) \in q \restriction B_1 = \mathrm{tp}(\bar{c}, B_1), \bar{b} \subseteq B_1 \cap A_2, \bar{d} \subseteq B_1 \cap D_1$. As $\bar{b} \subseteq B_1 \cap A_2$ and we have defined $A_2 = \mathrm{acl}(D_0 \cup C_2)$, clearly for some $\bar{c}^* \in C_2$ and $\bar{e} \in D_0$ and $\vartheta \in L(T), k < \omega$, the following holds:

$$\models \vartheta[\bar{b}, \bar{c}^*, \bar{e}] \wedge (\exists^{\leqslant k} \bar{y}) \vartheta(\bar{y}, \bar{c}^*, \bar{e}).$$

As $\bar{c}^*, \bar{b} \subseteq B_1 \cap A_2$, by B_1's choice there is $\bar{e}$ as required which belongs to $D_0 \cap B_1$. It is also clear that $\vartheta(\mathfrak{C}^{\mathrm{eq}}, \bar{c}^*, \bar{e}) \subseteq B_1 \cap A_2$ (as $A_2 = \mathrm{acl}\, A_2$). As f is the identity on $B_1 \cap A_2$, clearly

$$\models (\forall \bar{y}) [\vartheta(\bar{y}, \bar{c}^*, \bar{e}) \to \varphi(\bar{c}, \bar{y}, \bar{d}) \equiv \varphi(f(\bar{c}), \bar{y}, f(\bar{d}))]$$

(just try every $\bar{y} \in B_1 \cap A_2$ as this is enough). Hence the formula,

$$\psi(\bar{x}) \stackrel{\mathrm{def}}{=} (\forall \bar{y}) [\vartheta(\bar{y}, \bar{c}^*, \bar{e}) \to \varphi(\bar{x}, \bar{y}, \bar{d}) \equiv \varphi(f(\bar{c}), \bar{y}, f(\bar{d})],$$

belongs to q (as $\bar{c}$ satisfies it); and its set of parameters is

$$\bar{c}^* \cup \bar{e} \cup \bar{d} \cup f(\bar{c}) \cup f(\bar{d}) \subseteq C_2 \cup (B_1 \cap D_0) \cup (B_1 \cap D_1) \cup f(\bar{c}) \cup f(B_1 \cap D_1) \subseteq D_1 \cup C_2 \cup f(\bar{c})$$

(remember that f maps $B_1 \cap D_1$ into D_0).

So $\psi(\bar{x}) \in q \upharpoonright (D_1 \cup C_2 \cup f(\bar{c}))$. Clearly, $\psi(\bar{x}) \vdash (\varphi(\bar{x}, \bar{b}, \bar{d}) \equiv \varphi(f(\bar{c}), \bar{b}, f(\bar{d}))$; [as $\varphi(\bar{x}, \bar{b}, \bar{d}) \in \mathrm{tp}(\bar{c}, B_1)$, $\models \varphi[\bar{c}, \bar{b}, \bar{d}]$, hence $\models \varphi[f(\bar{c}), \bar{b}, \bar{d}]$ so $\psi(\bar{x}) \vdash \varphi(\bar{x}, \bar{b}, \bar{d})$] so we finish Stage D.

Stage E: For some $C_1^* \subseteq D_1, |C_1^*| < \kappa, C_1 \subseteq C_1^*$ and $q \upharpoonright (D_1 \cup C_2 \cup f(\bar{c}))$ is the unique extension of $q \upharpoonright (C_1^* \cup C_2 \cup f(\bar{c}))$ in $S^m(D_1 \cup C_2 \cup f(\bar{c}))$ which does not fork over $C_2 \cup C_1^*$. We choose $C_1^* \subseteq D_1, C_1 \subseteq C_1^*, |C_1^*| < \kappa$, such that $\mathrm{tp}_*(C_2 \cup f(\bar{c}) {}^\frown \bar{c}, D_1)$ does not fork over C_1^*, and every extension of $\mathrm{tp}_*(C_2 \cup f(\bar{c}) {}^\frown \bar{c}, C_1^*)$ which does not fork over C_1^* and is complete over D_1 *is equal* to $\mathrm{tp}_*(C_2 \cup f(\bar{c}) {}^\frown \bar{c}, D_1)$. This is possible as D_1 is good and $\kappa = \kappa_r(T)$ is regular.

Suppose C_1^* is not as required, then some $\bar{c}^*$ realizes some $q^* \in S^m(D_1 \cup C_2 \cup f(\bar{c}))$, where q^* extends $q \upharpoonright (C_1^* \cup C_2 \cup f(\bar{c}))$, does not fork over $C_2 \cup C_1^*$, but $q \neq q^*$. Then by III, 0.1(3) $\mathrm{tp}_*(C_2 \cup f(\bar{c}) {}^\frown \bar{c}^*, D_1)$ does not fork over C_1^*; it extends $\mathrm{tp}_*(C_2 \cup f(\bar{c}) {}^\frown \bar{c}, C_1^*)$, but is different from $\mathrm{tp}_*(C_2 \cup f(\bar{c}) {}^\frown \bar{c}, D_1)$. This contradicts the choice of C_1^*.

Stage F: Conclusion of the proof of Lemma 3.4. We shall show that $C_2 \cup f(\bar{c}) \cup C_1^*$ satisfies the requirements, i.e. $q \upharpoonright (C_2 \cup f(\bar{c}) \cup C_1^*)$ is stationary inside $M_2 \cup D_1$. Clearly, the power of this set is $< \kappa$ (C_2 – see the beginning; C_1^* – see Stage E). So suppose $q' \in S^m(D_1 \cup M_2)$ extends $q \upharpoonright (C_2 \cup f(\bar{c}) \cup C_1^*)$, does not fork over $C_2 \cup f(\bar{c}) \cup C_1^*$. Clearly, q' does not fork over C, hence over $C_2 \cup C_1^*$. By Stage E, $q' \upharpoonright (D_1 \cup C_2 \cup f(\bar{c})) = q \upharpoonright (D_1 \cup C_2 \cup f(\bar{c}))$. By Stage D, $q' \upharpoonright B_1 = q \upharpoonright B_1$. But by the construction in Stage C, $B_0 \subseteq B_1$, and so $q' \upharpoonright B_0 = q \upharpoonright B_0$, and remember q, q' does not fork over C, and $C \subseteq B_0$.

We want now to prove that $q' \upharpoonright (A_2 \cup D_1) = q \upharpoonright (A_2 \cup D_1)$, if this fails, for some $\varphi(\bar{x}, \bar{e}), \varphi(\bar{x}, \bar{e}) \in q \upharpoonright (A_2 \cup D_1), \neg\varphi(\bar{x}, \bar{e}) \in q' \upharpoonright (A_2 \cup D_1)$. By the choice of B_0,

$$R^m[q \upharpoonright (A_2 \cup D_1), \varphi, \aleph_0] = R^m[q \upharpoonright B_0, \varphi, \aleph_0],$$

$$\mathrm{Mlt}[q \upharpoonright (A_2 \cup D_1), \varphi, \aleph_0] = \mathrm{Mlt}[q \upharpoonright B_0, \varphi, \aleph_0].$$

As q' does not fork over B_0,

$$R[q' \upharpoonright (A_2 \cup D_1), \varphi, \aleph_0] = R[q' \upharpoonright B_0, \varphi, \aleph_0].$$

Let $\Gamma_0 = \{r_l : l < \alpha\}$ be the set of $r \in S_\varphi^m(|\mathfrak{C}^{eq}|)$ such that $(q \upharpoonright B_0) \cup r$ does not fork over B_0. It is known that $\alpha = \mathrm{Mlt}[q \upharpoonright B_0, \varphi, \aleph_0] < \omega$. For at least one $l = l_0$, $(q' \upharpoonright (A_2 \cup D_1)) \cup r_l$ does not fork over B_0, so, $\neg\varphi(\bar{x}, \bar{c}) \in r_{l_0}$, hence $r_{l_0} \notin \Gamma_1$, where $\Gamma_1 = \{r \in S_\varphi^m(|\mathfrak{C}^{eq}|) : q \upharpoonright (A_2 \cup D_1) \cup r$ does not fork over $B_0\}$. So Γ_1 is a proper subset of $\{r_l : l < \alpha\}$, hence

$$\mathrm{Mlt}[q \restriction (A_2 \cup D_1), \varphi, \aleph_0] = |\Gamma_1| < |\Gamma| = \mathrm{Mlt}[q \restriction B_0, \varphi, \aleph_0],$$

Contradicting the choice of B_0.

So we have to conclude that $q \restriction (A_2 \cup D_1) = q' \restriction (A_2 \cup D_1)$ so by Stage B, $q = q'$, so we finish the proof of 3.4.

CONCLUSION 3.5: *Suppose* $\langle N_t : t \in \mathscr{P}^-(n) \rangle$ *is a stable system of* $\mathbf{F}^a_\kappa$*-saturated models, where* $\kappa = \kappa_r(T)$ ($\mathscr{P}^-(n)$ *is* $\{t : t \subseteq n = \{0, 1, \ldots, n-1\}, |t| < n\}$, *and on a "stable system"*, see Def. 2.1). *Then* $A = \bigcup_t N_t$ *is a good set.*

Proof. We prove this by induction on n.

For $n = 0, A = \phi$, so A is trivially good.

For $n = 1, A = N_\phi$, and this is Fact 3.3(1).

For $n = 2$, use Lemma 3.4 with

$$D_0 = N_\phi, D_1 = M_{\{0\}}, M_2 = M_{\{1\}}.$$

For $n = k+1$, let $D_0 = \bigcup\{N_t : t \in \mathscr{P}^-(k)\}$, $M_2 = N_k = N_{\{0,1,\ldots,k-1\}}$, and $D_1 = \bigcup\{N_t : t \in \mathscr{P}^-(n), t \neq k\}$.

Now $D_2 = |M_2|$ is a good set by 3.3(1), D_0, D_1 are good sets by the induction hypothesis (for D_0 – clear: for D_1 – note $D_1 = \bigcup\{N_{t \cup \{k\}} : t \in \mathscr{P}^-(k)\}$ and $\langle N_{t \cup \{k\}} : t \in \mathscr{P}^-(k) \rangle$ is a stable system of $\mathbf{F}^a_\kappa$-saturated models). The remaining hypothesis of 3.4 (on D_0, D_1) follows by 2.9(2). The conclusion of 3.4 says that $A = D_1 \cup D_2$ is a good set.

XII.4. The otop/existence dichotomy

(Hyp) T is stable.

DEFINITION 4.1. T has the omitting type order property (otop) if there is a type $p(\bar{x}, \bar{y}, \bar{z})$ such that for every λ and a two-place relation R on λ, there is a model M of T, and $\bar{a}_\alpha \in M$ for $\alpha < \lambda$ such that:

for any $\alpha < \beta < \lambda : \alpha R \beta$ iff the type $p(\bar{a}_\alpha, \bar{a}_\beta, \bar{z})$ is realized in M.

Remark. We can think of various variants; for T countable superstable without the dop they will be equivalent. For T stable, for example, it is natural to make $\bar{x}, \bar{y}$ into infinite sequences.

DEFINITION 4.2. The theory T has the $(\lambda, 2)$-existence property if, whenever $M_0 \prec M_l (l = 1, 2)$, $\Sigma_{l<3} \|M_l\| = \lambda\ (l < 3)$ and $\{M_1, M_2\}$ is independent over M_0, there is a primary model over $M_1 \cup M_2$.

The main theorem of this section is

THEOREM 4.3: *A countable superstable* T *has the otop if* (a) *or at least* (b) *below hold:*

(a) T *does not have the* $(\aleph_0, 2)$*-existence property,*

(b) *there are* $M_0 \prec M_l (l = 1, 2)$, $\{M_1, M_2\}$ *independent over* M_0 *and* B, $\mathbf{F}^l_{\aleph_0}$*-constructible over* $M_1 \cup M_2$ *but not atomic over* $M_1 \cup M_2$.

Proof of 4.3 is broken down into a series of facts.

FACT 4.4: *Suppose condition* 3.3(a) *holds. Then* 3.3(b) *holds.*

Proof. As the $(\aleph_0, 2)$-existence property fails, there are countable $M_l (l < 3)$, $M_0 \prec M_l (l = 1, 2)$, $\{M_1, M_2\}$ independent over M_0 such that there is no primary model over $M_1 \cup M_2$. Hence there are $\bar{a} \subseteq M_1 \cup M_2$, and φ such that $\vDash (\exists \bar{x}) \varphi(\bar{x}, \bar{a})$ but $\{\varphi(\bar{x}, \bar{a})\}$ has no isolated extension in $S^m(M_1 \cup M_2)$ (remember that T and the M_l's are countable). As T is superstable there is $\bar{a}^* \in M_1 \cup M_2$ and $\varphi^*, \varphi^*(\bar{x}, \bar{a}^*) \vdash \varphi(\bar{x}, \bar{a})$, $\vDash (\exists \bar{x}) \varphi^*(\bar{x}, \bar{a}^*)$, and $\{\varphi^*(\bar{x}, \bar{a}^*)\}$ has no extension over $M_1 \cup M_2$ which forks over $\bar{a}^*$. By 1.12(1) there is an $\mathbf{F}^l_{\aleph_0}$-isolated $p \in S^m(M_1 \cup M_2)$, such that $\varphi^*(\bar{x}, \bar{a}^*) \in p$. So (b) of 4.3 holds.

DEFINITION 4.5. Let ζ be the minimal ordinal such that there are M_l, $(l < 3)$ $\bar{a}^*, \bar{c} = \langle c_l : l < n \rangle, \bar{c}^* \subseteq \bar{c}$ and $\psi^+(y, \bar{b})$ such that:

(1) $M_0 \prec M_1, M_0 \prec M_2$,

(2) $\{M_1, M_2\}$ is independent over M_0,

(3) $\mathrm{tp}(c_k, M_1 \cup M_2 \cup \{c_l : l < k\})$ is in $\mathbf{F}^l_{\aleph_0}(\bar{a}^* \cup \{c_l : ; < k\})$,

(4) $\bar{c}^* \subseteq \psi^+(\mathfrak{C}, \bar{b})$,

(5) $\mathrm{tp}(\bar{c}^*, M_1 \cup M_2)$ is not isolated,

(6) ζ is $R^1[\psi^+(x, \bar{b}), L, \infty]$,

(7) $\bar{a}^*, \bar{b}$ are from $M_1 \cup M_2$.

FACT 4.6: ζ *is well defined (and* $< \infty$*), and w.l.o.g.* $\bar{b} = \bar{a}^*$.

Proof. Clearly, by 4.4, condition 4.3(b) holds, and the derivation of an example for 4.5 is easy by 1.11, IV, 1.3(1) and IV, 1.4(2). We can define $\bar{a}^*$ by Ax(III.2) from IV, §1. W.l.o.g. $\bar{b} = \bar{a}^*$ as we can enlarge both.

FACT 4.7: *We can find* $\psi^+, M_l (l < 3)\, p, \bar{a}^*, \bar{c}, \bar{c}^*$ *as in* 4.6 *such that*

(a) $\bar{c} = \bar{c}^*, p \restriction \bar{a}^* \vdash p, (p, \bar{a}^*) \in \mathbf{F}^l_{\aleph_0}$, *letting* $p = \mathrm{tp}(\bar{c}, M_1 \cup M_2)$,

(b) *each* M_l *is* $\aleph_1$*-saturated. Moreover, we can have for some* $\lambda = \lambda^{\aleph_0}$, M_0 *is a saturated of power* λ, M_1 *and* M_2 *saturated of power* λ^+.

Proof. Let $M_l (l < 3)\, p, \bar{a}^*, \bar{c}, \bar{c}^*$ be as in 4.5. Choose $\lambda = \lambda^{\aleph_0} \geqslant \sum_{l<3} \|M_l\|$. We can choose saturated $M^*_l\ (l < 3), M_l \prec M^*_l M^*_0 \prec M^*_l$, $\{M^*_1, M^*_2\}$ independent over M^*_0, $\|M^*_0\| = \lambda$, $\|M^*_1\| = \|M^*_2\| = \lambda^+$, and for $l = 1, 2$ $\mathrm{tp}(M^*_l, M_l \cup M^*_{3-l})$ does not fork over $M_l \cup M^*_0$ and $\mathrm{tp}(M^*_0, M_1 \cup M_2)$ does not fork over M_0 (choose M^*_0, then M^*_1, and lastly M^*_2 and use III, 0.1). By 2.3(3) $(M_1 \cup M_2, M^*_1 \cup M^*_2)$ satisfies the Tarski–Vaught condition.

Clearly, $\mathrm{tp}(\langle c_l : l < k\rangle, M_1 \cup M_2)$ is $\mathbf{F}^l_{\aleph_0}$-isolated (by IV, 3.2), hence by 1.12(2), for each $k < n\, (M_1 \cup M_2 \cup \{c_l : l < k\}, M^*_1 \cup M^*_2 \cup \{c_l : l < k\})$ satisfies the Tarski–Vaught condition, hence by 1.9(b) $\mathrm{tp}(c_k, M^*_1 \cup M^*_2 \cup \{c_l ; l < k\})$ is $\mathbf{F}^l_{\aleph_0}$-isolated. So there is $\bar{a}^1 \subseteq M^*_1 \cup M^*_2$ such that for every k, $\mathrm{tp}_*(c_k, M^*_1 \cup M^*_2 \cup \{c_l : l < k\}) \in \mathbf{F}^l_{\aleph_0}(\bar{a}^1 \cup \{c_l : l < k\})$. By 3.1, w.l.o.g. $\mathrm{tp}(\bar{c}^*, M^*_1 \cup M^*_2)$ is the unique extension of $\mathrm{tp}(\bar{c}, \bar{a}^1)$ which does not fork over $\bar{a}^*$. So by 1.12(5),

$$\mathrm{tp}(\bar{c}^*, \bar{a}^1) \vdash \mathrm{tp}(\bar{c}^*, M^*_1 \cup M^*_2),$$

$$(p, \bar{a}^*) \in \mathbf{F}^l_{\aleph_0}.$$

So we can replace $\bar{a}^*$ by $\bar{a}^1$ and $\bar{c}^*$ by $\bar{c}$.

FACT 4.8: *There are countable* $M^*_l, p, \bar{a}^*, \psi^+$ *as in Fact* 4.7(a) *such that* $M^*_0 \subseteq_a M^*_l (l = 1, 2)$ *and* M^*_1, M^*_2 *are isomorphic over* M^*_0. *Moreover, for some countable* $M_3, M^*_l \subseteq_a M_3 (l = 1, 2)$, *and* M_3 *realizes* p *and* M_3 *is* $\mathbf{F}^l_{\aleph_0}$*-constructible over* $M^*_1 \cup M^*_2$.

Proof. Let $M^*_l, p, \bar{a}^*$ be as in 4.7. Let M_3 be $\mathbf{F}^l_{\aleph_0}$-primary over $M^*_1 \cup M^*_2$, by a construction starting with a sequence $\bar{c}$ realizing p (see IV, Def. 1.2). It exists by 1.12(1)) and IV, 3.1(5)). By the saturation of the M^*_l's, $M^*_0 \subseteq_a M^*_1 \subseteq_a M^*_3, M^*_0 \subseteq_a M^*_2 \subseteq_a M^*_3$, and M^*_1, M^*_2 are isomorphic over M^*_0 by the saturativity conditions in 4.7. Everything is as required except the countability, but by the Lowenheim–Skolem argument we can find $M^*_l, p, \bar{a}^*$, as required.

FACT 4.9: *Let* $M_l (l \leqslant 3), p, \bar{a}^*, \psi^+$ *be as in* 4.8 (*renaming* M^*_l *as* M_l). *There is* M^*_3 *such that* M^*_3 *is countable*, $M_1 \cup M_2 \subseteq M^*_3, M^*_3$ *is* $\mathbf{F}^l_{\aleph_0}$*-constructible over* $M_1 \cup M_2$ *and* M^*_3 *omits* p.

Remark. Really for our purpose, "M_3 $\mathbf{F}^l_{\aleph_0}$-atomic over $M_1 \cup M_2$" also suffices, but it is aesthetically nicer to have the $\mathbf{F}^l_{\aleph_0}$-constructible.

Proof. First note that there is no consistent $\varphi(\bar{x}, \bar{b}), \bar{b} \in M_1 \cup M_2 \varphi(\bar{x}, \bar{b}) \vdash p \restriction \bar{a}^*$. So by the omitting type theorem there is M_3^* omitting $p, M_1 \cup M_2 \subseteq M_3^*$. However, we need a special kind of M_3^* so we have to be a little more careful. Clearly, it suffices to prove

SUBFACT 4.10: *Suppose A is countable, p an m-type over some (finite) $\bar{c} \subseteq A$ with no support over A [i.e. there is no consistent $\psi(\bar{x}, \bar{a}, \bar{c}), \bar{a} \in A$, $\psi(\bar{x}, \bar{a}, \bar{c}) \vdash p$] and $\models (\exists y)\varphi(y, \bar{a}), \bar{a} \in A$. Then some b satisfies $\varphi(y, \bar{a})$, $\mathrm{tp}(b, A)$ is $\mathbf{F}^l_{\aleph_0}$-isolated and p has no support over $A \cup \{b\}$.*

Proof. As T is superstable, w.l.o.g. $\{\varphi(y, \bar{a})\}$ has no extension over A which forks over $\bar{a}$ and $\bar{c} \subseteq \bar{a}$. We shall define a type $q(y)$ over $\bar{a}$, s.t.

(a) for every $\vartheta(y, \bar{z})$ for some $\psi(y) \in q(y), \{r \in S^1_\vartheta(A): r$ consistent with $\psi(y)\}$ is finite,

(b) there is b realizing q s.t. over $A \cup \{b\}$ there is no support for p.

(c) $\varphi(y, \bar{a}) \in q(y)$ and q is over $\bar{a}$.

This suffices. A sufficient (and necessary) condition for (b) is:

(b)′ there is a model $M, A \subseteq M$, M realizes q but omits p;

and a sufficient condition for (b)′ is:

(b)″ for every $\bar{d} \in A$ and formula $\vartheta(\bar{x}, y, \bar{d})$, if $q(y) \cup \{\vartheta(\bar{x}, y, \bar{d})\}$ is consistent, then so is $q(y) \cup \{\vartheta(\bar{x}, y, \bar{d}), \neg\psi(\bar{x})\}$ for some $\psi(\bar{x}) \in p$.

Let $\{\vartheta^0_m(\bar{x}, y, \bar{d}_n): n < \omega\}$ list all formulas with $\vartheta(\bar{x}, y, \bar{d}), \bar{d} \in A$, and $\{\vartheta^1_n(y, \bar{z}_n): n < \omega\}$ list all formulas $\vartheta(y, \bar{z})$. We defined by induction on $n < \omega$ a formula $\varphi_n(y, \bar{a})$ s.t.

(i) $\models (\exists y)\varphi_n(y, \bar{a})$,

(ii) $\varphi_0(y, \bar{a}) = \varphi(y, \bar{a})$,

(iii) $\vdash (\forall y)[\varphi_{n+1}(y, \bar{a}) \to \varphi_n(y, \bar{a})]$,

(iv) $\varphi_{2n+1}(y, \bar{a})$ is consistent with only finitely many $r \in S^1_{\vartheta^1_n}(A)$,

(v) either $\{\varphi_{2n+2}(y, \bar{a}), \vartheta^0_n(\bar{x}, y, \bar{d}_n)\}$ is inconsistent or for some $\psi(\bar{x}, \bar{c}) \in p$, for every $\varphi'(y, \bar{a})$

$$\models (\exists y)[\varphi_{2n+1}(y, \bar{a}) \wedge \varphi'(y, \bar{a})] \to (\exists y, \bar{x})$$

$$[\varphi_{2n+2}(y, \bar{a}) \wedge \varphi'(y, \bar{a}) \wedge \vartheta^0_n(\bar{x}, y, \bar{d}_n) \wedge \neg\psi(\bar{x}, \bar{c})].$$

Clearly, if we succeed, then $q(y) = \{\varphi_n(y, \bar{a}): n < \omega\}$ is as required: (iv) takes care of (a), and (v) takes care of (b)″. For $n = 0$ there is no problem. For $n = 2m+1$ act as in the proof of 1.12(1). So we are left with $n = 2m+2$. Let $\vartheta = \vartheta^0_n$.

If there is $\varphi'(y,\bar{a})$ s.t. $\{\varphi'(y,\bar{a}),\varphi_{n-1}(y,\bar{a})\}$ is consistent but $\{\varphi'(y,\bar{a}),\varphi_{n-1}(y,\bar{a}),\vartheta(\bar{x},y,\bar{d}_m)\}$ is inconsistent, then choose $\varphi_n(y,\bar{a}) = \varphi'(y,\bar{a}) \wedge \varphi_{n-1}(y,\bar{a})$ (easily (v) holds, as well as (i) and (iii)). So we assume there is no such φ'.

So, in particular $\{\varphi_{n-1}(y,\bar{a}) \wedge \vartheta(\bar{x},y,\bar{d}_n)\}$ is consistent. As $p(\bar{x})$ has no support over A and $\bar{a},\bar{d}_n \in A$, clearly for some $\psi(\bar{x},\bar{c}) \in p$, $\{\varphi_{n-1}(y,\bar{a}),\vartheta(\bar{x},y,\bar{d}_n),\neg\psi(\bar{x},\bar{c})\}$ is consistent. So $\varphi^*(y) \overset{\text{def}}{=} (\exists\bar{x})[\varphi_{n-1}(y,\bar{a}) \wedge \vartheta(\bar{x},y,\bar{d}_n) \wedge \neg\psi(\bar{x},\bar{c})\}$ is consistent and it implies $\varphi_{n-1}(y,\bar{a})$, and its parameters are from A. By the choice of φ and by (iii), clearly $\{\varphi^*(y)\}$ does not fork over $\bar{a}$. Now apply IV, 4.2 with $\varphi^*(y),\bar{a},\bar{a}$ here standing for $\varphi(\bar{x},\bar{a}),C,A$ there and get $\varphi^+(y,\bar{a})$ (standing for $\psi(\bar{x},\bar{b})$) s.t.

(A) $\{\varphi^*(y),\neg\varphi^+(y,\bar{a})\}$ forks over $\bar{a}$, hence as $\varphi^*(y)\vdash\varphi(y,\bar{a})$ and $\varphi(y,\bar{a})$'s choice, $\{\varphi^*(y),\neg\varphi^+(y,\bar{a})\}$ is inconsistent so $\varphi^*(y)\vdash\varphi^+(y,\bar{a})$,

(B) for every $\vartheta'(y,\bar{a}')(\bar{a}' \subseteq \bar{a}),\varphi^*(y)\vdash\vartheta'(y,\bar{a}')$ implies $\varphi^+(y,\bar{a})\vdash \vartheta'(y,\bar{a}')$.

By (A) we can replace "implies" by "iff". We let $\varphi_n(y,\bar{a}) = \varphi^+(y,\bar{a})$, and as (iii) holds by (B) (as $\varphi^*(y)\vdash\varphi_{n-1}(y)$ by its definition) and (i) is immediate by (A) (as $\varphi^*(y)$ does not fork over A) we have just to prove that the second possibility in (v) holds. Suppose $\{\varphi_n(y,\bar{a}),\varphi'(y,\bar{a})\}$ is consistent, and it is enough to prove $\{\varphi_n(y,\bar{a}), \varphi'(y,\bar{a}),\vartheta(\bar{x},y,\bar{d}_n),\neg\psi(\bar{x},\bar{c})\}$ is consistent. If not, then

$$\varphi_n(y,\bar{a}) \wedge \varphi'(y,\bar{a})\vdash\neg(\exists\bar{x})[\vartheta(\bar{x},y,\bar{d}_n) \wedge \neg\psi(\bar{x},\bar{c})],$$

hence

$$\varphi_n(y,\bar{a}) \wedge \varphi'(y,\bar{a})\vdash\neg\varphi^*(y),$$

hence

$$\varphi^*(y)\vdash\neg(\varphi_n(y,\bar{a}) \wedge \varphi'(y,\bar{a})),$$

hence (by (B))

$$\varphi^+(y,\bar{a})\vdash\neg(\varphi_n(y,\bar{a}) \wedge \varphi'(y,\bar{a})),$$

i.e.

$$\varphi^+(y,\bar{a})\vdash\neg(\varphi^+(y,\bar{a}) \wedge \varphi'(y,\bar{a})),$$

hence

$$\{\varphi^+(y,\bar{a}) \wedge \varphi'(y,\bar{a})\} \quad \text{is consistent},$$

contradicting the choice of $\varphi'(y,\bar{a})$.

THE CONSTRUCTION 4.11: Let $M_l(l \leqslant 3)M_3^*,\psi^+(x,\bar{a}^*),p\in S^m(M_1 \cup M_2)$ be as in 4.9, and let f be an isomorphism from M_1 onto M_2 over M_0. Let $\bar{a}_l^* = \bar{a}^* \cap M_l$ and w.l.o.g. $f(\bar{a}_1^*) = \bar{a}_2^*$, $\text{tp}(\bar{a}_l^*,M_0)$ does not fork over $\bar{a}_0^*$.

Now let λ and a two-place relation R on λ be given. We shall define by induction on n, for every $t \subseteq \lambda, |t| = n$, a countable model N_t.

For $n = 0$ we let $N_\phi = M_0$.

For $n = 1$ let us choose for each $i < \lambda$ a model $N_{\{i\}}$ isomorphic to M_1 over M_0, f_i an isomorphism from M_1 onto $N_{\{i\}}$ such that $f_i \restriction M_0 =$ the identity, and $\{N_{\{i\}} : i < \lambda\}$ is independent over M_0. We let $\bar{a}_i = f_i(\bar{a}_1^*)$, $f_i' = f_i \circ f^{-1} : M_2 \to M_{\{i\}}, p_{\alpha,\beta} = (f_\alpha \cup f_\beta')(p)$ for $\alpha < \beta$.

For $n = 2$ we have to define $N_{\{\alpha,\beta\}}$ for $\alpha < \beta$.

We separate two cases:

Case a: $\alpha R \beta$. We can extend $(f_\alpha \cup f_\beta')$ to an elementary mapping $f_{\alpha,\beta}$ from M_3, and we denote its range by $N_{\{\alpha,\beta\}}$.

Case b: Not $\alpha R \beta$. We can extend $(f_\alpha \cup f_\beta')$ to an elementary mapping $f_{\alpha,\beta}$ from M_3^* and we denote its range by $N_{\{\alpha,\beta\}}$.

For $n > 2$: We let N_t be $\mathbf{F}^l_{\aleph_0}$-constructible over $\bigcup\{N_s : s \subseteq t, s \neq t\}$. We call $\bigcup\{N_t : t \subseteq \lambda \text{ finite}\}$ by M_R.

FACT 4.12: *For every finite* $t \subseteq \lambda, |t| > 1$, *and* $\bar{c} \in N_t$, $\operatorname{tp}(\bar{c}, \bigcup\{N_t : s \subseteq t, s \neq t\})$ *is* $\mathbf{F}^l_{\aleph_0}$*-isolated.*

Proof. Quite clear.

FACT 4.13: $\langle N_t : t \subseteq \lambda, |t| < \aleph_0 \rangle$ *is a stable system* (*see Def.* 2.1).

Proof. Let $\{t_i : i < \alpha\}$ be an enumeration of $\{t \subseteq \lambda : t \text{ finite}\}$, such that $t_i \subseteq t_j$ implies $i \leqslant j$ and for $i < \lambda, t_{1+i} = \{i\}$, and $t_0 = \emptyset$. We prove by induction on $\beta \leqslant \alpha$ that $\langle N_t : t \in \{t_i : i < \beta\} \rangle$ is a stable system.

There are no problems for $\beta = 0$ or β limit or $\beta < \lambda$. So let $\beta = \gamma + 1, \gamma \geqslant \lambda$; by the induction hypothesis and 2.3(2) $(\bigcup\{N_s : s \subseteq t_\gamma, s \neq t_\gamma\}, \bigcup\{N_{t_i} : i < \gamma\})$ satisfies the Tarski–Vaught condition. Also by 4.12 for every $\bar{c} \in N_t$, $\operatorname{tp}(\bar{c}, \bigcup\{N_s : s \subseteq t_\gamma, s \neq t_\gamma\})$ is $\mathbf{F}^l_{\aleph_0}$-isolated. So by 1.2(3) this type has a unique extension over $\bigcup_{i<\gamma} N_{t_i}$, hence this extension necessarily does not fork over $\bigcup\{N_s : s \subseteq t_\gamma, s \neq t_\gamma\}$. So we finish 4.13 by 2.3(1).

Remark. Observe we really have proved:

CLAIM 4.14: *If I is a family of finite subsets of* $\lambda, s \subseteq t \in I \Rightarrow s \in I$, $N_s (s \in I)$ *models*, $\{N_{\{\alpha\}} : \{\alpha\} \in I\}$ *independent over* $N_\emptyset, s \subseteq t \Rightarrow N_s \subseteq N_t$, *and* $t \in I \wedge |t| > 1 \Rightarrow \operatorname{tp}_*(N_t, \bigcup\{N_s : s \subseteq t, s \neq t\})$ *is* $\mathbf{F}^l_{\aleph_0}$*-atomic, then* $\{N_s : s \in I\}$ *is a stable system and* $\operatorname{tp}_*(N_t, \bigcup\{N_s : s \subseteq t, s \neq t\}) \vdash \operatorname{tp}_*(N_t, \bigcup\{N_s : t \nsubseteq s \in I\})$.

FACT 4.15: *For* $\alpha < \beta, p_{\alpha,\beta} \restriction (\bar{a}^*_\alpha \cup \bar{a}^*_\beta) \vdash p_{\alpha,\beta}$ *(see* 4.11 *for notations).*

Obvious, as $f_\alpha \cup f'_\beta$ is an elementary mapping, and 4.7(a) (which is inherited by 4.8, 4.9).

THE MAIN FACT 4.16: M_R *realizes* $p_{\alpha,\beta} \restriction (\bar{a}^*_\alpha \cup \bar{a}^*_\beta)$ *iff* $\alpha R \beta$.

Proof. If $\alpha R\beta$, then $p_{\alpha,\beta} \restriction (\bar{a}^*_\alpha \cup \bar{a}^*_\beta)$ is realized by $f_{\alpha,\beta}(\bar{c})$ for any $\bar{c} \in M_3$ realizing p ($\bar{c}$ exists by the choice of M_3).

We now deal with the difficult half. Suppose "not $\alpha R\beta$". Then $N_{\{\alpha,\beta\}}$ omits $p_{\alpha,\beta} \restriction (\bar{a}^*_\alpha \cup \bar{a}^*_\beta)$ as M^*_3 omits $p \restriction \bar{a}^*$.

We now prove by induction on $n, 2 \leqslant n < \omega$, that if $\{\alpha,\beta\} \subseteq t \subseteq \lambda$, $|t| = n$, then N_t omits $p_{\alpha,\beta}$ (this suffices by 4.15). For $n = 2$ we know it. Suppose we have proved for n and we shall prove for $n+1$. Let $\{\alpha,\beta\} \subseteq t \subseteq \lambda, |t| = n+1$. Choose $\gamma \in t - \{\alpha,\beta\}$ and let $s = t - \{\gamma\}$. So $p_{\alpha,\beta}$ is over N_s and N_s omits it. We also can easily show that N_t is $\mathbf{F}^l_{\aleph_0}$-constructible over $N_s \cup N_{\{\gamma\}}$ (by 1.9(b), 1.12(2) and as any $\mathbf{F}^l_{\aleph_0}$-isolated type is $\mathbf{F}^l_{\aleph_0}$-isolated). We also know that $N_\emptyset \prec N_s, N_{\{\gamma\}}$ and $\{N_s, N_{\{\gamma\}}\}$ is independent over $N_\emptyset$. Let $A = \{d \in N_t : R[\mathrm{tp}(d, N_s \cup N_{\{\gamma\}}), L, \infty] < \zeta\}$. However, if $\bar{d} \in A$, $\mathrm{tp}(\bar{d}, N_s \cup N_{\{\gamma\}})$ is isolated (for otherwise we get a counterexample to the choice of ζ in Def. 4.5). So A is atomic over $N_s \cup N_{\{\gamma\}}$. As 4.8 says $N_\emptyset \subseteq_a N_{\{\gamma\}}$ by XI, 3.9(1) $N_s \subseteq_a A$.

Also, if $d \in N_t - N_s$, $\models \psi^+(d, \bar{a}^*_\alpha {}^\frown \bar{a}^*_\beta)$ (ψ^+ – from 4.7 (and 4.5)), then $r = \mathrm{tp}(d, N_s \cup N_{\{\gamma\}})$, being realized in an $\mathbf{F}^l_{\aleph_0}$-constructible set over $N_s \cup N_{\{\gamma\}}$, is also $\mathbf{F}^l_{\aleph_0}$-isolated; hence for some formula ϑ^* in it, $\vartheta^*(x) \vdash \mathrm{tp}_\vartheta(d, N_s \cup N_{\{\gamma\}})$ for $\vartheta(x,y) = [x = y]$. Now r forks over N_s [otherwise it is finitely satisfiable in N_s, hence ϑ^* is satisfiable in N_s, say by $d' \in N_s$, hence $\vartheta^*(x) \nvdash x \neq d'$, $d' \in N_s$ contradiction]; hence,

$$R^1[\mathrm{tp}(d, N_s \cup N_{\{\gamma\}}), L, \infty] < R^1[\mathrm{tp}(d, N_s), L, \infty] \leqslant R^1[\psi^+(x, \bar{a}^*_\alpha {}^\frown \bar{a}^*_\beta), L, \infty] = \zeta.$$

We can conclude that (as $N_s \subseteq A$)

$$\psi^+(N_t, \bar{a}^*_\alpha {}^\frown \bar{a}^*_\beta) \subseteq A.$$

So if $p_{\alpha,\beta}$ is realized in N_t, any sequence realizing it is included in A. But by XI, 3.9(1) $N_s \subseteq_a A$, so if in A there is a sequence realizing $p_{\alpha,\beta} \restriction (\bar{a}_\alpha \cup \bar{a}_\beta)$, then (remembering $\bar{a}_\alpha \cup \bar{a}_\beta \subseteq N_{\{\alpha\}} \cup N_{\{\beta\}} \subseteq N_s$) there is such a sequence in N_s; contradicting the induction hypothesis. So we have proved 4.16 and hence Theorem 4.3.

LEMMA 4.17: *Suppose that* (b) *of* 4.3 *fails, T is a countable superstable without the dop* (*for* (3) *superstable without dop suffices*).

(1) *If $M_0 \subseteq M_1, M_2, \{M_1, M_2\}$ is independent over M_0, and A is $\mathbf{F}^l_{\aleph_0}$-constructible over $M_1 \cup M_2$, then M is atomic over $M_1 \cup M_2$.*

(2) *If M_l are as above, M is $\mathbf{F}^l_{\aleph_0}$-constructible over $M_1 \cup M_2$ (and $M_1 \cup M_2 \subseteq M$), then M is minimal over $M_1 \cup M_2$ (i.e. there is no M^*, $M_1 \cup M_2 \subseteq M^* \subset M$).*

(3) *If M_l are as above and are $\mathbf{F}^a_\lambda$-saturated and M is $\mathbf{F}^a_{\aleph_0}$-constructible over $M_1 \cup M_2 (M_1 \cup M_2 \subseteq M)$, then M is $\mathbf{F}^a_\lambda$-saturated.*

Proof. (1) We can find M $\mathbf{F}^l_{\aleph_0}$-constructible over $M_1 \cup M_2, A \subseteq M$; so w.l.o.g. $A = |M|$. Also if there is a counterexample, there is a countable one (by the Lowenheim–Skolem argument). By 4.3(b) (and the countability) there is $M^* \subseteq M$ primary over $M_1 \cup M_2$, hence atomic over $M_1 \cup M_2$. So if M is not atomic over $M_1 \cup M_2$, necessarily $M \neq M^*$. So it suffices to prove (2).

However we first prove (3).

(3) We can find an $\mathbf{F}^a_\lambda$-saturated model M^+ $\mathbf{F}^a_\lambda$-constructible over $M_1 \cup M_2, M \subseteq M^+$. If $M = M^+$, we can finish, so assume $M \neq M^+$ and choose $c \in M^+ - M$ with $R[\mathrm{tp}(c, M), L, \infty]$ minimal. By X, 3.3 $\mathrm{tp}(c, M)$ is not orthogonal to some M_l, but $M_l \subseteq_a M^+$ trivially, so by XI, 3.11 there is $c \in M^+ - M$ such that $\mathrm{tp}(c, M)$ does not fork over M_l, a contradiction (say to V, 3.2).

(2) Suppose $M_l (l < 3) M$ are a counterexample, $M_1 \cup M_2 \subseteq M^* \subset M$, and let $\lambda > \|M\| + |T|$ be a regular cardinal. We can find $\mathbf{F}^a_\lambda$-saturated ($\equiv \lambda$-saturated) $N_l (l < 3)$ such that $M_l \subseteq N_l$, $\mathrm{tp}_*(N_0, M)$ does not fork over M_0, $\mathrm{tp}_*(N_l, M \cup N_{3-l})$ does not fork over $N_0 \cup M_l$, for $l = 1, 2$. Now by 2.3 $(M_1 \cup M_2, N_1 \cup N_2)$ satisfies the Tarski–Vaught condition, hence by 1.2(3) M is $\mathbf{F}^l_{\aleph_0}$-constructible over $N_1 \cup N_2$, hence there is N, $\mathbf{F}^l_{\aleph_0}$-constructible over $N_1 \cup N_2, M \cup N_1 \cup N_2 \subseteq N$. By (3) N is $\mathbf{F}^a_\lambda$-saturated. Now let $\langle c_i : i < \alpha \rangle$ be a maximal sequence such that $c_i \in N, c_i \notin \{c_j : j < i\}$, and $\mathrm{tp}(c_i, M^* \cup N_1 \cup N_2 \cup \{c_j : j < i\})$ is $\mathbf{F}^l_{\aleph_0}$-isolated. Clearly, $\alpha < \|N\|^+$. Also $B = M^* \cup N_1 \cup N_2 \cup \{c_i : i < \alpha\}$ is the universe of some $N^* \prec \mathfrak{C}$, because every $\{\varphi(x, \bar{b})\} (\bar{b} \subseteq B)$ can be extended to an $\mathbf{F}^l_{\aleph_0}$-isolated complete type over B, and this type is realized in N (as N is $\mathbf{F}^a_\lambda$-saturated). Now for every $\bar{b} \in B$, $\mathrm{tp}(\bar{b}, M^* \cup N_1 \cup N_2)$ is $\mathbf{F}^l_{\aleph_0}$-isolated, whereas $(N_1 \cup N_2 \cup M^*, N_1 \cup N_2 \cup M)$ satisfies the Tarski–Vaught condition [because $\mathrm{tp}_*(N_1 \cup N_2, M)$ does not fork over M^*]. So $N^* \cap M = M^* \cap M = M^*$, hence $N^* \neq N$ and by (3) N^* is $\mathbf{F}^a_\lambda$-saturated. So N_0, N_1, N_2, N, N^* contradict "T does not have the dop" (see Def. X, 2.1).

Remark. Why have we not assumed in 4.3 "T does not have the dop", thereby simplifying the proof? As the present 4.3 says more, e.g. it may help to prove "if T is superstable with the otop, then $I(\aleph_0, T) = 2^{\aleph_0}$".

XII.5. From the $(\aleph_0, 2)$-existence property to the $(\lambda, 2)$-existence property

(Hyp) T countable without 4.3(b), T stable.

DEFINITION 5.1: We call $\mathscr{S} = \langle M_s : s \in I \rangle$ ($I \subseteq \mathscr{P}(n)$ is closed under subsets, $n \geqslant 2$) a sp. stable system if

(a) $\langle M_s : s \in I \rangle$ is a stable system (see Def. 2.1),

(b) $\bigcup I - \{0\} \in I, \bigcup I - \{1\} \in I$,

(c) $\bigcup_{s \in I} M_s$ is $\mathbf{F}^t_{\aleph_0}$-atomic over $M_{\bigcup I - \{0\}} \cup M_{\bigcup I - \{1\}}$,

(d) if $s \in I, \{0, 1\} \subseteq s$, then M_s is $\mathbf{F}^l_{\aleph_0}$-constructible over $A^{\mathscr{S}}_s$, where we let $A^{\mathscr{S}}_s = \bigcup \{M_t : t \in \mathscr{P}^-(s)\}$.

Remark. Note that for $I = \mathscr{P}^-(2)$, a sp. stable system is exactly a stable system.

DEFINITION 5.2: T has the (λ, n)-existence property (for sp. stable systems) if for every sp. stable system $\mathscr{S} = \langle M_s : s \in \mathscr{P}^-(n) \rangle, \lambda = \sum_{s \in I} \|M_s\|$, there is over $A^{\mathscr{S}}_n$ an $\mathbf{F}^t_{\aleph_0}$-primary model.

DEFINITION 5.3: A true sp. stable system $\mathscr{S} = \langle M_s : s \in I \rangle$ is a sp. stable system which satisfies

(e) if $s \in I, \{0, 1\} \subseteq s$, then M_s is $\mathbf{F}^t_{\aleph_0}$-constructible over $\bigcup_{t \subset s} M_t$ (hence M_s is prime over $A^{\mathscr{S}}_s$).

DEFINITION 5.4: (1) T has the true (λ, n)-existence property if for every true sp. stable system $\mathscr{S} = \langle M_s : s \in \mathscr{P}^-(n) \rangle, \lambda = \sum_{s \in I} \|M_s\|$, there is over $\bigcup \{M_t : t \in \mathscr{P}^-(n)\}$ an $\mathbf{F}^t_{\aleph_0}$-primary model.

(2) T has the strong $(\aleph_0, n)$-existence property, if for every true sp. stable system $\mathscr{S} = \langle M_s : s \in \mathscr{P}^-(n) \rangle$, in which each M_s is countable every $\mathbf{F}^l_{\aleph_0}$-constructible set over $A^{\mathscr{S}}_s$ is atomic over $A^{\mathscr{S}}_s$.

FACT 5.5: (1) *Every true sp. stable system is a sp. stable system.*

(2) *The (λ, n)-existence property implies the true (λ, n)-existence property.*

(3) *In Def.* 5.1, 5.3(e) *implies* (d) *and* (d) *implies* (c).

(4) *If* $\mathscr{S} = \langle M_s : s \in I \rangle$ *is a sp. stable system,* $\{0,1\} \subseteq s \in I$, $J \subseteq I$ *is downward closed,* $s - \{0\}, s - \{1\} \in J$, *then* M_s *is* $\mathbf{F}^i_{\aleph_0}$*-constructible over* $\bigcup_{t \in J} M_t$. *If* $\mathscr{S}$ *is a true sp. stable system* M_s *is* $\mathbf{F}^t_{\aleph_0}$*-constructible over* $\bigcup_{t \in J} M_t$. *Also,* $\bigcup_{s \in I} M_s$ *is* $\mathbf{F}^i_{\aleph_0}$*-constructible over* $M_{\bigcup I - \{0\}} \cup M_{\bigcup I - \{1\}}$.

(5) *The strong* $(\aleph_0, n)$*-existence property implies the true* $(\aleph_0, n)$*-existence property.*

Proof. Easy (in (3), (d) implies (c) as 4.3(b) fails and use (4)).

CLAIM 5.6: (1) *T has the* $(\aleph_0, n)$*-existence property iff for every sp. countable stable system* $\mathscr{S} = \langle M_s : s \in \mathscr{P}^-(n) \rangle$, *every type* $\{\varphi(\bar{x}, \bar{a})\}$ *over* $A^{\mathscr{S}}_s$ *can be completed to an isolated complete type.*

(2) *Similarly for the true* $(\aleph_0, n)$*-existence property.*

Proof. Immediate.

CLAIM 5.7: *Suppose* $\mathscr{S} = \langle M_s : s \in \mathscr{P}^-(n) \rangle$ *is a true sp. stable system* $\lambda = \sum \|M_s\| > \aleph_0$, *Then we can define* $M^\alpha_s, \alpha < \lambda$, *such that*

(a) $\|M^\alpha_s\| \leqslant \aleph_0 + |\alpha|, M^\alpha_s$ *is increasing continuous with* α,

(b) $M^\alpha_s \cap M_t = M^\alpha_{s \cap t}$,

(c) $\mathrm{tp}_*(M^\alpha_s, \bigcup_{t \in J} M_t)$ *does not fork over* $\bigcup_{t \in J} M^\alpha_t$ *for every* $J \subseteq \mathscr{P}^-(n)$,

(d) $\mathscr{S}^\alpha = \langle M^\alpha_s : s \in \mathscr{P}^-(n) \rangle$ *is a true sp. stable system,*

(e) $\mathscr{S}^{\alpha,\beta} = \langle M^{\alpha,\beta}_s : s \in \mathscr{P}^-(n+1) - \{n\} \rangle$ *is a true sp. stable system, where* $M^{\alpha,\beta}_s$ *is* M^α_s *if* $n \notin s$ *and* $M^\beta_{s-\{n\}}$ *if* $n \in s$, *if* $\beta = \lambda$ *let* $M^\beta_{s-\{n\}} = M_{s-\{n\}}$.

Proof. Easy: for (d) check the definition of a stable system and for (e) use 2.3(1) for an enumeration $\langle s(l) : l < 2^{n+1} - 2 \rangle$ such that $s(2l) \in \mathscr{P}^-(n)$, $s(2l+1) = s(2l) \cup \{n\}$. For checking Def. 5.3(e) (hence Def. 5.1(d)) use Theorem IV, 3.3. For 5.1(c) use IV, 4.3.

LEMMA 5.8: *Suppose* $n < \omega, \lambda > \aleph_0$ *and T has the true* $(\aleph_0, n)$*-existence and the true* $(\mu, n+1)$*-existence property for every* $\mu < \lambda$, *then T has the true* (λ, n)*-existence property.*

Remark 5.8A. Lemmas 5.7 and 5.8 remain true if we omit the "true".

Proof. Let $\langle M_s : s \in \mathscr{P}^-(n) \rangle$ be a true sp. stable system, $\|M_s\| \subseteq \lambda$. We should find a primary model over $\bigcup M_s$. We use 5.7 and its notation.

We define by induction on $\alpha < \lambda$ an ordinal ζ_α and $a_i(\bigcup_{\beta<\alpha}\zeta_\beta \leqslant i < \zeta_\alpha)$ such that

(1) for $i < \zeta_\alpha$, $\mathrm{tp}(a_i, A_n^{\mathscr{S}} \cup \{a_j : j < i\})$ is isolated; moreover, some formula over $A_n^{\mathscr{S}^\alpha} \cup \{a_j : j < i\}$ isolates it,

(2) $A_n^{\mathscr{S}^\alpha} \cup \{a_j : j < \zeta_\alpha\}$ is the universe of a model M^α.

For $\alpha = 0$, we know that $\mathscr{S}^0$ is a true sp. stable system. As $\|M_s^0\| = \aleph_0$ there is a primary model over $A_n^{\mathscr{S}^0}, M_0$ and let $|M_0| = A_s^{\mathscr{S}^0} \cup \{a_i : i < \zeta_0\}$ be such that $\mathrm{tp}(a_i, A_n^{\mathscr{S}^0} \cup \{a_j : j < i\})$ is isolated (e.g. $\zeta_0 = \omega$ is necessarily alright). This gives (2) and somewhat less than (1), but by 5.7(e), and 2.3(2) $(A_n^{\mathscr{S}^0}, A_n^{\mathscr{S}})$ satisfies the Tarski–Vaught condition, so (1) holds. For α limit let $\zeta_\alpha = \bigcup_{\beta<\alpha}\zeta_\beta$ and there is nothing to do.

So let $\alpha = \beta + 1$, it is easy to check that we can extend the system $\mathscr{S}^{\alpha,\beta}$ by $M_s = M^\alpha$ for $s = \{0, \ldots, n-1\}$. We get a stable system by 2.3(1) [let $\{0, \ldots, n-1\}$ be the last s] and it is a true sp. stable system – use 5.5(4).

Apply the true $(|\alpha| + \aleph_0, n+1)$-existence property (as above (1) holds by 5.7(e), 2.3(2)).

Finally, $\bigcup_{\alpha<\lambda} M^\alpha$ is a primary model over $A_n^{\mathscr{S}}$.

LEMMA 5.9: *Suppose T has the strong $(\aleph_0, m)$-existence property for $m \leqslant n$. If $\mathscr{S} = \langle M_t : t \in I\rangle$ is a sp. stable system, $s \in I$, $\{0, 1\} \subseteq s$, $|s| \leqslant n$, then M_s is atomic over $A_s^{\mathscr{S}}$.*

Proof. If there is a counterexample, w.l.o.g. for some $m \leqslant n$, $I = \mathscr{P}(m)$, $s = m$, m is minimal and $\|M_s\| = \aleph_0$ (as in the proof of 5.7.) As m is minimal, for every $t \subseteq s, t \neq s, \{0, 1\} \subseteq t$ implies M_t is atomic over $A_t^{\mathscr{S}}$, hence by the countability M_s is primary over $A_t^{\mathscr{S}}$. So $\langle M_t : t \in \mathscr{P}^-(n)\rangle$ is a true stable system. As M_s is $\mathbf{F}^i_{\aleph_0}$-primary over $A_s^{\mathscr{S}}$, and the strong $(\aleph_0, m)$-existence property holds, clearly M_s is atomic over $A_s^{\mathscr{S}}$, a contradiction.

Notation 5.10: Let n^* be the minimal n for which the strong $(\aleph_0, n)$-existence property fails if there is such n and ω otherwise.

FACT 5.11: $n^* > 2$.

By the hypothesis of the section "T countable stable and condition 4.3(b) fails".

CLAIM 5.12: *Suppose T is superstable without the dop. Then $n^* = \omega$.*

Remark. Really "without the dop" is not necessary, this will be shown in a subsequent paper.

Proof. Suppose $n = n^* < \omega$, and let $\mathbf{S} = \langle M_s : s \in \mathscr{P}^-(n) \rangle$ be a counter-example, i.e. $\langle M_s : s \in \mathscr{P}^-(n) \rangle$ a true sp. stable system, $\|M_s\| = \aleph_0, M_n$ is $\mathbf{F}^l_{\aleph_0}$-constructible but not atomic over A^s_n.

Let $I = \mathscr{P}^-(n+1) - \{n\}$. Suppose $\lambda = (2^{\aleph_0})^+$.

FACT 5.13: *We can define* $M_s(s \in I - \mathscr{P}^-(n))$ *such that*

(a) $\langle M_s : s \in I \rangle$ *is a sp. stable system,*

(b) *if* $s \in I - \mathscr{P}^-(n)$, *then* M_s *is* λ*-saturated (in fact* $\mathbf{F}^a_\lambda$*-prime over* $\bigcup \{M_t : n \in t \subseteq s\}$ *when* $s \neq \{n\}$*).*

Proof. Let $I = \{s(l) : l < 2^{n+1} - 2\}$ be such that $s(l) \subseteq s(m) \Rightarrow l \leqslant m$, and $\mathscr{P}^-(n) = \{s(l) : l < 2^n - 1\}$. So $M_{s(l)}(l < 2^n - 1)$ are defined. We define the rest by induction on l, such that the relevant parts of (a) and (b) hold.

Suppose we have defined for every $m < l$. If $0 \notin s(l)$ or $1 \notin s(l)$ there is no problem, so assume $\{0, 1\} \subseteq s(l)$. As $\langle M_t : t \in \mathscr{P}^-(s(l)) \rangle$ is a sp. stable system, $|s(l)| \leqslant n^*$, $M_{s(l)-\{n\}}$ is atomic over $\bigcup_{t \subset s(l)-\{n\}} M_t$ (by 5.9) and $(\bigcup_{t \subset s(l)-\{n\}} M_t, \bigcup_{t \subset s(l)-\{n\}} M_{t \cup \{n\}})$ satisfies the Tarski–Vaught condition. Hence by 1.2(3) $\operatorname{tp}_*(M_{s(l)-\{n\}}, \bigcup_{t \subset s(l)-\{n\}} M_t) \vdash \operatorname{tp}_*(M_{s(l)-\{n\}}, \bigcup_{t \subset s(l)-\{n\}} M_{t \cup \{n\}})$ and so (remembering $\|M_{s(l)-\{n\}}\| = \aleph_0$) $M_{s(l)-\{n\}}$ is $\mathbf{F}^l_{\aleph_0}$-constructible over $\bigcup_{t \subset s(l)} M_t$ and so we can find $M_{s(l)}$ which is $\mathbf{F}^l_{\aleph_0}$-constructible over $\bigcup_{t \subset s(l)} M_t$, hence over $\bigcup \{M_t : t \subset s(l), n \in t\}$. By 4.17(3) it is λ-saturated and by 5.9 atomic over $\bigcup \{M_t : t \subset s(l), n \in t\}$. So the relevant part of (b) holds, and also (a) is easy, as w.l.o.g. $\operatorname{tp}_*(M_{s(l)}, \bigcup_{m<l} M_{s(m)})$ does not fork over $\bigcup_{t \subset s(l)} M_t$.

So we have proved 5.13. Let us continue with 5.12.

Let for $s \in \mathscr{P}^-(n), M^*_s = M_{s \cup \{n\}}$, so $\langle M^*_s : s \in \mathscr{P}^-(n) \rangle$ is a stable system, each M^*_s is λ-saturated, and by 2.3 $(\bigcup_s M_s, \bigcup_s M^*_s)$ satisfies the Tarski–Vaught condition. Hence by 1.9, 1.12(2) for some $\bar{c} \in M_n$, $\operatorname{tp}(\bar{c}, \bigcup_s M^*_s)$ is not isolated, $\operatorname{tp}(c_k, \bigcup_s M^*_s \cup \{c_l : l < k\})$ is $\mathbf{F}^l_{\aleph_0}$-isolated. By 3.5 $\operatorname{tp}(\bar{c}, \bigcup_s M^*_s)$ is the unique extension in $S^m(\bigcup_s M^*_s)$ of $\operatorname{tp}(\bar{c}, \bar{b})$ which does not fork over $\bar{b}$ for some $\bar{b} \subseteq \bigcup_s M^*_s$. By 1.12(5) for some $\bar{d} \in \bigcup_s M^*_s : \operatorname{tp}(\bar{c}, \bar{d}) \vdash \operatorname{tp}(\bar{c}, \bigcup_s M^*_s)$, $(\operatorname{tp}(\bar{c}, \bigcup_s M^*_s), \bar{d}) \in \mathbf{F}^l_{\aleph_0}$, hence $\operatorname{tp}(\bar{c}, \bar{d})$ is not isolated, $\operatorname{tp}(\bar{c}, \bigcup_s M^*_s)$ does not fork over $\bar{d}$.

Let for $s \in \mathscr{P}^-(n-1), M^{**}_s = M^*_{s \cup \{n-1\}}$. Then $\langle M^{**}_s : s \in \mathscr{P}^-(n-1) \rangle$ is a sp. stable system. Let $t(*) = n-1 (= \{0, 1, \ldots, n-2\})$. By 5.9, $M^*_{t(*)}$

is atomic over $\bigcup\{M^*_s : s \in \mathscr{P}^-(n-1)\}$, hence even $\bigcup\{M^*_{s\cup\{n-1\}} : s \in \mathscr{P}^-(n-1)\}$ by 2.3, 1.2(3) hence $M^*_{t(*)}$ is atomic over $\bigcup\{M^{**}_s : s \in \mathscr{P}^-(n-1)\}$. So, clearly $\bigcup\{M^{**}_s : s \in \mathscr{P}^-(n-1)\} \cup \bar{d}$ is $\mathbf{F}^l_{\aleph_0}$-constructible over $\bigcup\{M^{**}_s : s \in \mathscr{P}^-(n-1)\}$, and by the previous paragraph $\operatorname{tp}(\bar{c}, \bigcup\{M^{**}_s : s \in \mathscr{P}^-(n-1)\} \cup \bar{d}\})$ is $\mathbf{F}^l_{\aleph_0}$-isolated. Together, $\bigcup\{M^{**}_s : s \in \mathscr{P}^-(n-1)\} \cup \bar{c}\frown\bar{d}$ is $\mathbf{F}^l_{\aleph_0}$-constructible over $\bigcup\{M^{**}_s : s \in \mathscr{P}^-(n-1)\}$. By 5.9, $\operatorname{tp}(\bar{c}\frown\bar{d}, \bigcup\{M^{**}_s : s \in \mathscr{P}^-(n-1)\})$ is isolated; hence (by Ax(V.1) for $\mathbf{F}^t_{\aleph_0}$) $\operatorname{tp}(\bar{c}, \bigcup\{M^{**}_s : s \in \mathscr{P}^-(n-1)\} \cup \bar{d})$ is isolated; as it does not fork over $\bar{d}$, also $\operatorname{tp}(\bar{c}, \bar{d})$ is isolated, contradicting the previous paragraph.

CONCLUSION 5.14: *Suppose T is countable superstable and without the dop and the otop. Then T has the true sp. (λ, n)-existence property for every λ, n; hence it has the $(\lambda, 2)$-existence property for every λ.*

XII.6. The book's main theorem

THE MAIN GAP THEOREM 6.1: *Let T be countable.*

(1) *If T is not superstable or (is superstable) deep or with the dop or the otop, then for every uncountable $\lambda, I(\lambda, T) = 2^\lambda$.*

(2) *If T is shallow superstable without the dop and without the otop, then for every $\alpha > 0$ $I(\aleph_\alpha, T) < \beth_{\omega_1}(|\alpha|)$.*

In fact, if $\mathrm{Dp}(T) \geqslant \omega$ $I(\aleph_\alpha, T) = \beth_{\mathrm{Dp}(T)}(|\alpha|)$, *and if* $\mathrm{Dp}(T)$ *is finite, then* $\beth_{\mathrm{Dp}(T)}(|\alpha|) \leqslant I(\aleph_\alpha, T) \leqslant \beth_{\mathrm{Dp}(T)}(|\alpha|^{(\beth_1)})$.

Remark 6.1A. In (2) also every model of T is prime and even primary over some non-forking tree of models $N_\eta (\eta \in I)$, $\|N_\eta\| \leqslant 2^{\aleph_0}$.

Proof. (1) If T is not superstable, this was proved in VIII, §2. If T is superstable but with the dop this (and more) was proved in X, 2.5, IX, 1.20, and if T is superstable with the otop the same proof works. If T is superstable and deep, more was proved in X, 5.1.

(2) By 4.3, condition 4.3(b) fails, so the hypothesis of Section 5 holds. By 5.14 the $(<\infty, 2)$-existence property holds.

By XI, 2.4 for $(\mathbf{T}^t_{\aleph_0}, \subseteq_a)$ for every model M there a non-forking tree $\{N_\eta : \eta \in I\}$ of models, $\|N_\eta\| \leqslant 2^{\aleph_0}$ so that M is prime over $\bigcup_{\eta \in I} N_\eta$. So exactly as in X, 4.7 we get the upper bound (by X, 4.5(4) the depth is the same) and by X, 6.1, we get the lower bounds (for finite depth by the same proof, without marking levels).

This way for Dp(T) finite we get only $I(\aleph_\alpha, T) \leqslant \beth_{\mathrm{Dp}(T)}(|\alpha|^{\beth_2})$. Using also XI, 2.16, we can in (2) have $|\alpha|^{\beth_1}$ instead of $|\alpha|^{\beth_2}$.

LEMMA 6.2: *If T is superstable without the dop and with the existence property, then the otop fails.*

Proof. For such T we have a structure theory, contradicting the otop.

CHAPTER XIII

FOR THOMAS THE DOUBTER

XIII.0. Introduction

Let T be countable in the introduction.

The aim of this chapter is to justify the assertion that XII, 6.1 is "the book's main theorem". For this we deal with some related problems and show that we can solve them too.

In Section 1 we show that if T is superstable, shallow without dop and otop, then any model can be characterized by up isomorphism by generalized cardinal invariants of countable depth. If T is superstable deep without dop and otop, then we show a weaker result is the best possible. For other theories, strong negative results appear in [Sh 87].

In the second section we show that we can essentially compute $IE(\lambda, T)$ for countable T. The main difficulty is the case T superstable without the dop but with the otop. Why is there a difficulty which does not occur when T has the dop? There the embedding preserves "$\dim(I,M) \geqslant \lambda$" (but not necessarily "$\dim I,M) \leqslant \lambda$"); however, we know quite accurately which indiscernible sets have large dimension (the $\boldsymbol{I}_{s,t}$ and some types not orthogonal to $\emptyset$). Here realizing a type is preserved but not necessarily omitting a type; and working just with $EM(I, \Phi)$, where $EM(I, \Phi) \models [(\exists \bar{x}) \wedge p(\bar{x}, \bar{a}_s, \bar{a}_t]^{\text{if}\,(s<t)}$, is not enough. So we repeat an analysis like the one in Section 4 of Chapter XII. So we have countable models $M_{\{i\}}(i < \lambda)$ independent over $M_\emptyset$, and we use an $\mathbf{F}^l_{\aleph_0}$-primary model M on their union. We look for $i < j$, only at types realized in $\{\bar{c}\colon \text{tp}(\bar{c}, M_{\{i\}} \cup M_{\{j\}})$ has rank $\leqslant \xi\}$, with ξ minimal so that continuum many types may occur. The result is that we have $2^{\aleph_0}$ possible types, but in M only countably many types are realized. So it makes sense that if for each pair $\{i\}, \{j\}$ we make a random choice we shall get many pairwise non-elementarily embeddable models, and this we do (so the section has a combinatorial part).

In Section 3 we prove the Morley conjecture: the function $I(\lambda, T)$ $\lambda \geqslant |T|$ is non-decreasing in λ except for T categorical in $\aleph_1$ but not in $\aleph_0, \lambda > \mu = \aleph_0$. Our strategy is that for many kinds of T we can compute $I(\lambda, T)$ for $\lambda > \aleph_0$ and then see that it is non-decreasing (as $2^{\aleph_0} \leqslant I(\aleph_1, T)$). The remaining case is T superstable shallow of finite depth, without the dop and otop. Then in the main case we can decompose any model M of power μ: i.e. it is prime over a non-forking tree, find a large splitting and then add more copies to get a larger tree so that the prime model over it has power λ. So we found a function from $\{M: \|M\| = \mu\}$ to $\{M: \|M\| = \lambda\}$, which is near enough to being one-to-one, i.e. to preserving non-isomorphism to prove our conclusion.

If we cannot find a good enough splitting in the tree, we prove $I(\lambda, T) \geqslant 2^{\|M\|}$ for $\lambda \geqslant \|M\|$.

In Section 4 we compute $I(\aleph_\alpha, T)$ for α large enough; this involves some finer analysis of the trees (mainly 4.4) and the result is 4.11. We finish in 4.15 proving that, e.g. $I(\aleph_\alpha, T) > \beth_2$ for some α implies $I(\aleph_\alpha, T) \geqslant |\alpha + 1|$ for every α.

XIII.1. Can the models be characterized by invariants?

In this section we try to show that in some sense a "set of invariants for models of T, which characterize each model up to isomorphism" is possible for suitable countable T. The point is of course that for countable theories which are not "suitable" there are complementary results.

One such invariant is the $L_{\infty,\lambda}$-theory of a model of power $\lambda > 2^{\aleph_0}$ which is sufficient for T superstable without the dop and otop. For a model of power λ, its $L_{\infty,\lambda}$-theory looks a very strong invariant, in particular the quantification may look too strong. So the negative complementary results (see [Sh 87]) seem more convincing. However, we then strengthen the positive results by replacing $L_{\infty,\lambda}$ by $L_{\infty,\beth_1^+}(Q^B_{(\mu^+)})_{\mu<\lambda}$, where Q^B_μ is a simple generalized quantifier, similar to a cardinality quantifier (saying a dimension is $\geqslant \mu$).

We may still complain on the use of sentences with arbitrary depth, but we can prove that this is necessary if T is deep. On the other hand, if T is shallow, then we can consider only sentences of small depth (a bound is determined by the depth of T).

THEOREM 1.1: *Suppose* $\lambda > |T| + 2^{\aleph_0}$, *and* T *is superstable without the dop, with the* $(<\infty, 2)$*-existence property. Then any two* $L_{\infty,\lambda}$*-equivalent models of* T *of power* λ *are isomorphic.*

This will follow from the stronger results but for this we have to define Q^B_μ.

DEFINITION 1.2: Syntactically, Q^B_μ operates in the following way for $\alpha \leqslant \omega$: $(Q^B_\mu x_0, x_1, \ldots, x_n \ldots)_{n<\alpha} \langle \varphi_n(x_0, \ldots, x_{n-1}, \bar{y}_n) : n < \alpha \rangle$ is a formula if each $\varphi_n(x_0, \ldots, x_{n-1}, \bar{y}_n)$ is the quantifier bounded by the variables $x_0, x_1, \ldots$, and

$$M \models (Q^B_\mu, x_0, x_1, \ldots, x_n, \ldots)_{n<\alpha} \langle \varphi_n(x_0, \ldots, x_{n-1}, \bar{c}_n) : n < \alpha \rangle$$

iff for every set $A \subseteq M$, $|A| < \mu$ for some $b \in M - A$ for every $n < \alpha$, $a_1, \ldots, a_{n-1} \in A$, $M \models \neg\varphi_n[b, a_1, \ldots, a_{n-1}, \bar{c}_n]$.

Remark. The thing we have in mind is: a vector space over a field has dimension $\geqslant \mu$. In fact, this is the only way we use it, so we can demand in the definition that the φ_n's define a dependence relation (as in V, 1.14). In this case if $\chi > \aleph_0$, $L_{\chi,\lambda}(Q^B_\mu) \subseteq \Delta(L_{\chi,\lambda}(\exists^{\geqslant\mu} x))$ where $\exists^{\geqslant} x$ "means" there are $\geqslant \mu$ x's such that..., and on the Δ-closure see [MSS 76].

CLAIM 1.3: $L_{\infty,\kappa}(Q^B_{\mu^+})_{\mu<\lambda} \subseteq L_{\infty,\lambda}$ *when* $\kappa \leqslant \lambda$, *i.e. every formula in the first logic is equivalent to one in the second.*

Proof. Trivial.

THEOREM 1.4: *Suppose* $\lambda > |T| + 2^{\aleph_0}$ *and* T *is superstable without the dop and with the* $(<\infty, 2)$*-existence property. Then any two* $L_{\infty,\beth_1^+ + |T|^+}(Q^B_{\mu^+})_{\mu<\lambda})$*-equivalent models of* T *of power* λ *are isomorphic.*

QUESTION: Can we demand only $L_{\infty,|T|^+}(Q^B_{\mu^+})_{\mu<\lambda}$-equivalence?

Proof. We shall use $(\mathbf{T}^t_{\aleph_0}, \subseteq^a_{\aleph_0,\aleph_0})$-decomposition inside the models. Let $\|M_1\| = \|M_2\| = \lambda$, M_1, M_2 are $L_{\infty,\beth_1^+ + |T|^+}(Q^B_{\mu^+})_{\mu<\lambda}$-equivalent models. We shall define by induction on n a set $I_n \subseteq {}^{n\geqslant}\lambda$, models N^l_η, $\bar{a}^l_\eta$ $(\eta \in I_n - \bigcup_{m<n} I_m$, $l = 1, 2)$ and a function F_n such that:

(1) $N^l_\eta \subseteq M_l$.

(2) $\langle N^l_\eta, \bar{a}^l_\eta : \eta \in I_n \rangle$ is a $(\mathbf{T}^t_{\aleph_0}, \subseteq^a_{\aleph_0,\aleph_0})$-decomposition inside M_l (see Def. XI, 2.4, 2.5).

(3) $\langle N^l_\eta, \bar{a}^l_\eta : \eta \in I_n \rangle$ is maximal in the following restricted sense: there is no $\eta \in I_n, l(\eta) < n$ and $\alpha, \nu \overset{\text{def}}{=} \eta ^\frown \langle \alpha \rangle \notin I$, and $N^l_\nu, \bar{a}^l_\nu$, such that $\langle N^l_\rho, \bar{a}^l_\rho : \rho \in I_n \cup \{\nu\} \rangle$ is a $(\mathbf{T}^t_{\aleph_0}, \subseteq^a_{\aleph_0,\aleph_0})$-decomposition inside M_l.

(4) F_n is an elementary mapping with domain $\bigcup_{\eta \in I_n} N^1_\eta$, mapping N^1_η onto N^2_η.

(5) $(M_1, \dots, c, \dots,)_{c \in N^1_\eta}$ is $L_{\infty, \beth_1^+ + |T|^+}(Q_{\mu^+})_{\mu<\lambda}$-equivalent to $(M_2, \dots, F_n(c), .)_{c \in N^1_\eta}$.

(6) $F_m \subseteq F_n$ for $m < n$.

Clearly, if we succeed, then $\langle N^l_\eta : \eta \in \bigcup_{n<\omega} I_n \rangle$ is a $(\mathbf{T}^t_{\aleph_0}, \subseteq^a_{\aleph_0,\aleph_0})$-decomposition of M_l, so by XI, 2.8 M_l is prime over $\bigcup\{N^l_\eta : \eta \in \bigcup_{n<\omega} I_n\}$, and $\bigcup_{n<\omega} F_n$ is an elementary mapping from $\bigcup\{N^1_\eta : \eta \in \bigcup_n I_n\}$ onto $\bigcup\{N^2_\eta : \eta \in \bigcup_n I_n\}$.

So we can find an elementary embedding F of M_1 into M_2 extending $\bigcup_{n<\omega} F_n$. But XI, 2.8 also says that M_2 is minimal over $\bigcup\{N^2_\eta : \eta \in \bigcup_{n<\omega} I_n\}$, hence F is onto M_2. So M_1, M_2 are isomorphic (alternatively we use the uniqueness of prime models).

For $n = 0$ there is no problem. So suppose we have defined for n (and all $m \leqslant n$) and shall define for $n+1$. Clearly, it suffices to prove

FACT 1.4A: *Suppose $\eta \in I_n, l(\eta) = n, p \in S^m(N^1_\eta)$ is a regular type orthogonal to N_{η^-} (if it exists). Then we can find α and $N^1_\beta, a^1_\beta, N^2_\beta, \bar{a}^2_\beta$ ($\beta < \alpha$ such that:*

(A) $\operatorname{tp}(\bar{a}^l_\beta, N^l_\eta)$ *is regular not orthogonal to* p.

(B) $\{\bar{a}^l_\beta : \beta < \alpha\} \subseteq M_l$ *is independent over* N^l_η.

(C) $\{\bar{a}^l_\beta : \beta < \alpha\}$ *is maximal (under conditions* (A), (B)).

(D) $N^l_\eta \cup \bar{a}^l_\beta \subseteq N^l_\beta \subseteq M$, $\operatorname{tp}_*(N^l_\beta, N^l_\eta \cup \bar{a}^l_\beta)$ *is almost orthogonal to* N^l_η.

(E) $N^l_\beta \subseteq A \subseteq M$, $\operatorname{tp}_*(A, N^l_\eta \cup \bar{a}^a_\beta)$ *almost orthogonal to* N^l_η *implies* $N^l_\beta \subseteq^a_{\aleph_0,\aleph_0} A$.

(F) $\|N^l_\eta\| \leqslant |T| + 2^{\aleph_0}$.

(G) *For each $\beta < \alpha, F_n \restriction N^1_\eta$ can be extended to an isomorphism F^β from N^1_β onto N^2_β.*

(H) $(M_1, \dots, c, \dots,)_{c \in N^1_\beta} (M_2, \dots, F^\beta(c), \dots,)_{c \in N^1_\beta}$ *are* $L_{\infty, \beth_1^+ + |T|^+}(Q^B_{\mu^+})_{\mu<\lambda}$-*equivalent.*

Proof. Let $\boldsymbol{I}_l = \{\bar{a} \in M_l : \operatorname{tp}(\bar{a}, N^l_\eta)$ is regular not orthogonal to $p\}$. We define by induction on $\gamma < \gamma(*)$ $\boldsymbol{I}^\gamma_1$ so that

(i) $\boldsymbol{I}^\gamma_1$ is the set of $\bar{a} \in \boldsymbol{I}_1$ realizing some complete $L_{\infty, \beth_1^+ + |T|^+}(Q^B_{\mu^+})_{\mu<\lambda}$-type over N^1_η in M_1.

(ii) The dimension of $\boldsymbol{I}^\gamma_1$ over $(\bigcup_{j<\gamma} \boldsymbol{I}^j_1, N^1_\eta)$ is $< \lambda$.

There is no problem to do this and to define corresponding I_2^η. For each γ we can find maximal subsets of I_l^η, independent over $(\bigcup_{j<\gamma} I_l^j, N_\eta^l)$, $\{\bar{a}_{\gamma,\beta}^l : \beta < \alpha_\gamma^l\}$. We can assume $\alpha_\gamma^1 = \alpha_\gamma^2$ (all call it α_γ) by (H) for n.

By the definition of the I_l^η's, the dimension of I_l over $(\bigcup_{j<\gamma(*)} I_l^j, N_\eta^l)$ is λ or zero. In the first case let $\{\bar{b}_\beta^l : \beta < \lambda\}$ be maximal independent (over $(\bigcup_{j<\gamma(*)} I_l^j, N_\eta^l)$) subset of I_l.

We now define by induction on $\beta < \lambda$ $\bar{a}_\beta^l (l = 1, 2)$ such that

(i) $\bar{a}_\beta^l \in I_l$, $\bar{a}_\beta^l$ does not depend on $\bigcup_{\gamma<\gamma(*)} I_l^\gamma \cup \{\bar{a}_\zeta^l : \zeta < \beta\}$ over N_η^l.

(ii) $(M_1, \ldots, c, \ldots, \bar{a}_\beta^1)_{c\in N_\eta^1}$ and $(M_2, \ldots, F_n(c), \ldots, \bar{a}_\beta^2)_{c\in N_\eta^1}$ are $L_{\infty, \beth_1^+ + |T|^+}(Q_{\mu^+}^B)_{\mu<\lambda}$-equivalent.

(iii) $\bar{b}_\beta^1$ depends on $\{a_\zeta^1 : \zeta \leqslant 2\beta\} \cup \bigcup_{\gamma<\gamma(*)} I_1^\gamma$.

(iv) $\bar{b}_\beta^2$ depends on $\{\bar{a}_\zeta^2 : \zeta \leqslant 2\beta + 1\} \cup \bigcup_{\gamma<\gamma(*)} I_2^\gamma$.

If we cannot continue, we contradict the maximality of $\gamma(*)$. If the dimension of I_l over $(\bigcup_{j<\gamma(*)} I_l^j, N_\eta^l)$ is zero, we do not define any $\bar{a}_\beta^l$.

After doing it we have defined $\bar{a}_\beta^l (\beta < \alpha)$ as required (with a different indexing) and there is no problem defining N_η^l.

THEOREM 1.5: *If in 1.4 we assume in addition that T is shallow, $\gamma = 2\mathrm{Dp}(T)$, then any two $L_{\infty, (\beth_1)^+ + |T|^+, \gamma}(Q_{\mu^+}^B)_{\mu<\gamma}$-equivalent models of T of power λ are isomorphic, where:*

DEFINITION 1.6: $L_{\infty, \lambda, \gamma}(Q_\mu^B)_{\mu\in I}$ is the set of sentences of $L_{\infty, \lambda}(Q_\mu^B)_{\mu\in I}$ of quantifier depth $<\gamma$.

Proof. The same as the proof of 1.4 but for each η we demand only: $(M_0, c)_{c\in N_\eta^0}, (M_1, f_\eta(c))_{c\in N_\eta^1}$ are $L_{\infty, \beth_1^+ + |T|^+, \gamma(\eta)}$ equivalent where $\gamma_\eta = 2\mathrm{Dp}(\bar{a}_\eta, N_{\eta^-})$ if $l(\eta) > 0$, $\gamma_\eta = 2\mathrm{Dp}(T)$ if $l(\eta) = 0$.

THEOREM 1.7: *In 1.4 if we restrict ourselves to $\mathbf{F}_{\aleph_0}^a$-saturated models, then we can omit the requirement that "the $(<\infty, 2)$-existence property" holds.*

Proof. Just use X, 2.1 instead of XI, 2.8.

THEOREM 1.8: *Suppose T is deep and superstable, $\lambda > \aleph_0$. Then the $L_{\infty,\lambda}$-Scott height of models of T of power $\mu \geqslant \lambda + \lambda(T)$ can be any ordinal $<\mu^+$.*

Remark. If $\mu \geqslant \lambda^+ + \lambda(T)$ of course there are $L_{\infty,\lambda}$-equivalent non-isomorphic models of power μ, in fact $\mathbf{F}^a_\lambda$-saturated (except, of course, when T is unidimensional).

Proof. Let $\alpha < \mu^+$ be limit; it is known that there are $\xi < \zeta < \mu^+$, $\xi \geqslant \mu$, such that $(\xi, <), (\zeta, <)$ are $L_{\infty,\lambda,\alpha}$-equivalent (but of course not $L_{\infty,\lambda}$-equivalent) (by Kino [K 66]).

Let $I = \{\eta : \eta$ a strictly decreasing sequence of ordinals and $\eta(0)$, if defined, is ξ or $\zeta\}$. So in $(I, \lessdot)$, $\langle\xi\rangle, \langle\zeta\rangle$ satisfies the same $L_{\infty,\lambda,\alpha}$-formulas. As T is deep there are A_n, a_n, s.t. $|A_n| = \aleph_0, A_n \subseteq A_{n+1}, a_n \in A_{n+1}$ $\operatorname{tp}(a_n, A_n)$ is orthogonal to A_{n-1} stipulating, $A_{-1} = \emptyset$, and $\operatorname{tp}_*(A_{n+1}, A_n \cup \{a_n\})$ is almost orthogonal to A_n for $n > 0$. We can find elementary mapping $f_\eta(\eta \in I)$, $\operatorname{Dom} f_\eta = A_{l(\eta)}, f_{\eta \restriction k} \subseteq f_\eta$, such that letting $A_\eta = \operatorname{Rang}(f_\eta)$, $\langle A_\eta : \eta \in I\rangle$ is a non-forking tree. Let M be $\mathbf{F}^a_{\aleph_0}$-prime over $\bigcup_{\eta\in I} A_\eta$. By [Sh 87], letting $A_1 = \{b_n : n < \omega\}$, the sequences $\langle f_{\langle\xi\rangle}(b_n) : n < \omega\rangle$ and $\langle f_{\langle\xi\rangle}(b_n) : n < \omega\rangle$ realize the same $L_{\infty,\lambda,\alpha}$-type in M; but as in X, 5.1's proof, not the same $L_{\infty,\lambda}$-type.

XIII.2. On having many models, no one elementarily embeddable into another

THEOREM 2.1: *Let T be countable.*

(1) *If T is not superstable or superstable with the dop or otop, then for every regular $\lambda \geqslant 2^{\aleph_0}$ (as in 2.9) $IE(\lambda, T) = 2^\lambda$.*

(2) *If T is superstable without the dop and otop, but is deep, then for every $\lambda, \lambda < \kappa_0$ (=smaller than the first beautiful cardinal (see Silver* [Si]; *the relevant facts are summed up in* [Sh 82, *Theorem* 2.4, *p.* 194]) *$IE(\lambda, T) = 2^\lambda$, and for every $\lambda \geqslant \kappa_0 IE(\lambda, T) = \kappa_0^-$.*

(3) *If T is superstable without the dop and otop, and is shallow, then for every $\lambda, IE(\lambda, T) \leqslant \beth_{\mathrm{Dp}(T)} < \beth_{\omega_1}$.*

Proof. Most parts appear in Chapter X or have similar proofs. The only missing part is (1) when T is superstable without the dop but with the otop. As the dop fails by XII, 6.1 the $(<\infty, 2)$-existence property and even the $(\aleph_0, 2)$-existence property fail.

Reflection will explain the difference with the non-dop: there we know exactly which types have uncountable dimension, whereas here we do not know when $p(\bar{x}, \bar{a}', a'')$ is realized. So we have to rework somewhat XII, §4.

We assume the following combinatorial principle (and prove the relevant cases later).

DEFINITION 2.2: ($CP_{\lambda,\chi,\mu,\kappa,\vartheta}$). There are ordered sets $I_\alpha \cong (\lambda, <)$ $(\alpha < \chi)$ each of power λ, and g_α a two-place function from I_α to ϑ, such that:

Suppose $\alpha \neq \beta < \chi$ (see [Sh 83]), I_α is $\psi(\langle x_0, y_0\rangle, \ldots, \langle x_i, y_i\rangle, \ldots)_{i<\vartheta}$-unembeddable into I_β, where $\psi(\ldots\langle x_i, y_i\rangle, \ldots)_{i<\vartheta}$ = "the $g_\alpha(x_i, y_i)$ are distinct for $i < \vartheta$". This means: let $L^*_{\mu,\kappa}$ be a vocabulary $\{F_{\alpha,i}: \alpha < \mu, i < \kappa\}$ $F_{\alpha,i}$ is i-place $M_{\mu,\kappa}(I_\beta)$ is the free $L^*_{\mu,\kappa}$-algebra which $\{t: i \in I_\beta\}$ generates, so we demand:

(*) if $f: I_\alpha \to M_{\mu,\kappa}$, then we can find $\langle a^1_\zeta, a^2_\zeta : \zeta < \vartheta\rangle$ in I_α s.t.

(a) $f(a^l_\zeta) = \tau^l(\bar{\gamma}^l_\zeta)$, τ^l an $L^*_{\mu,\kappa}$-term, and for $\zeta < \xi < \theta$, $\bar{\gamma}^1_\zeta{}^\frown\bar{\gamma}^2_\zeta, \gamma^1_\xi{}^\frown\bar{\gamma}^2_\xi$ realize the same atomic type over $\{i : i < \vartheta\}$ in I_β,

(b) $\langle g_\alpha(a^1_\zeta, a^2_\zeta) : \zeta < \vartheta\rangle$ is with no repetition.

LEMMA 2.3: *Suppose* ($CP_{\lambda,\chi,\beth_1,\aleph_1,\aleph_1}$). *Then* $IE(\lambda, T) \geqslant \chi$ (*assuming T is countable superstable without the dop, but with the otop*).

Remark. We can get the conclusion also for T with the independence property or with the dop (T stable).

The following is closely related to XII, 4.4.

DEFINITION 2.4: Let ξ be the minimal ordinal such that there are $M_l (l < 3)$ $\bar{a}^* \in M_1 \cup M_2$, $p_\eta (\eta \in {}^\omega 2)$ and $\psi^+(y, \bar{a}^*)$ such that

(1) $M_0 \prec M_l (l = 1, 2)$ and $\{M_1, M_2\}$ is independent over M_0.

(2) $p_\eta \in S^m(M_1 \cup M_2)$ are all $\mathbf{F}^l_{\aleph_0}$-isolated, not isolated and distinct.

(3) If $\bar{c}$ realizes p_η, then $\bar{c} \subseteq \psi^+(\mathfrak{C}, \bar{a}^*)$.

(4) $\xi = R[\psi^+(y, \bar{a}^*), L, \infty]$.

(5) Each p_η does not fork over $\bar{a}^*$.

FACT 2.5: (1) *ξ is well defined (when T is countable, superstable without the dop but with the otop).*

(2) *In the definition, w.l.o.g. each M_l is countable and $M_0 \subseteq_a M_l$* and *M_1 is isomorphic to M_2 over M_0. (Of course ξ is well defined and* $< \infty$.)

Proof. (1) By XII, 6.2 the $(\aleph_0, 2)$-existence property fails, so for some $M_l (l < 3)$ satisfying 2.4(2) and $\bar{a} \in M_1 \cup M_2$ and $\varphi \models (\exists \bar{x})\varphi(\bar{x}, \bar{a})$, but $\varphi(\bar{x}, \bar{a})$ has no isolated extension in $S^m(M_1 \cup M_2)$. Now we can easily find the p_η's, using $\psi^+ = (y = y)$.

(2) Just like XII, 4.7, 4.8.

FACT 2.6: *If* $M_\emptyset \prec M_{\{i\}}(i < \alpha)$ *independent over* $M_\emptyset$ *and* M *is* $\mathbf{F}^l_{\aleph_0}$-*constructible over* $\bigcup_{i<j} M_{\{i\}}$ *and* $A = \{b \in M : R[\mathrm{tp}(b, \bigcup M_{\{i\}}, L, \infty] < \xi\}$, *then* A *is atomic over* $\bigcup_{i<j} M_{\{i\}}$.

We shall later prove it for $\alpha = 2$.

Suppose not, then, as in the proof of XII, 4.7 w.l.o.g. each $M_{\{i\}}$, $M_\emptyset$ are $\aleph_1$-saturated. By XII, 4.17(3) M is $\aleph_1$-saturated too. So we can define by induction on $\beta < \alpha$ a model N_β, such that $N_0 = M_\emptyset, N_\delta = \bigcup_{i<\delta} N_i$ and $N_{\gamma+1}$ is $\mathbf{F}^a_{\aleph_0}$-prime over $N_\gamma \cup M_{\{\gamma\}}, N_{\gamma+1} \subseteq M$. As T does not have the dop, $M = N_\alpha$, and so there is a minimal β such that

$$A_\beta = \left\{ b \in N_\beta : R\left[\mathrm{tp}\left(b, \bigcup_{i<\beta} M_{\{i\}}\right), L, \infty\right] < \xi \right\}$$

is not atomic over $\bigcup_{i<\beta} M_{\{i\}}$.

It is easy to see that for β limit $A_\beta = \bigcup_{\gamma<\beta} A_\gamma$, and that for $\bar{c} \in A_\gamma$ or even $\bar{c} \in N_\gamma$, $\mathrm{tp}(\bar{c}, \bigcup_{i<\gamma} M_{\{i\}}) \vdash \mathrm{tp}(\bar{c}, \bigcup_{i<\beta} M_{\{i\}})$. Hence our β is not limit, and trivially not zero, hence $\beta = \gamma + 1$. Let $\bar{c} \in A_\beta$, $\mathrm{tp}(\bar{c}, \bigcup_{i<\beta} M_{\{i\}})$ not isolated. But (as we have assumed the fact for $\alpha = 2$) $\mathrm{tp}(\bar{c}, M_{\{\gamma\}} \cup N_\gamma)$ is atomic, and A_γ is atomic over $\bigcup_{i<\gamma} M_{\{i\}}$.

We now prove

(*) $\quad \mathrm{tp}_*(N_\gamma, A_\gamma \cup M_{\{\gamma\}} \cup \bar{c})$ does not fork over A_γ.

This suffices, since by symmetry this proves $\mathrm{tp}(\bar{c}, N_\gamma \cup M_\gamma)$ does not fork over $A_\gamma \cup M_\gamma$, hence by IV, 4.3. $\mathrm{tp}(\bar{c}, A_\gamma \cup M_{\{\gamma\}})$ is atomic, and as A_γ is atomic over $\bigcup_{i<\gamma} M_{\{i\}}$, hence over $\bigcup_{i<\beta} M_{\{i\}}$, we shall finish by IV, 3.2(4).

Let us prove (*).

So let $\bar{d} \in N_\gamma$, choose $\bar{b} \in A_\gamma, \bar{a} \in M_\emptyset$, $\mathrm{tp}(\bar{d}, A_\gamma)$ does not fork over $\bar{b}$, $\mathrm{tp}(\bar{b}, M_\emptyset)$ does not fork over $\bar{a}, \bar{a} \subseteq \bar{b}$.

Let $\bar{e} \in M_{\{\gamma\}}$ and we shall prove that $\mathrm{tp}(\bar{d}, \bar{b} \cup \bar{c} \cup \bar{e})$ does not fork over $\bar{b}$. W.l.o.g. for some ψ $\bar{c} \subseteq \psi(\mathfrak{C}, \bar{b}_1, \bar{e})$, where $\bar{b}_1 \subseteq \bar{b} \cap (\bigcup_{i<\gamma} M_{\{i\}})$, $R[\psi(x, \bar{b}_1, \bar{e}), L, \infty] < \xi$. Suppose $\mathrm{tp}(\bar{d}, \bar{b} \cup \bar{c} \cup \bar{e})$ forks over $\bar{b}$, then for some $\varphi \models \varphi[\bar{d}, \bar{e}, \bar{c}, \bar{b}]$, and for every $\bar{e}', \bar{c}'$, $\mathrm{tp}(\bar{d}, \bar{b}) \cup \varphi(\bar{y}, \bar{e}', \bar{c}', \bar{b})$ forks over $\bar{b}$ (see XI, 1.3). We can find $\bar{e}^* \in M_\emptyset$, realizing $\mathrm{stp}(\bar{e}, \bar{a})$ [because $M_\emptyset$ is $\aleph_1$-saturated]. Now $N_\gamma, M_{\{\gamma\}}$ are independent over $M_\emptyset$, hence $\bar{d}^\frown\bar{e}^\frown\bar{b}, \bar{d}^\frown\bar{e}^{*\frown}\bar{b}$ realizes the same type. So $R[\psi(x, \bar{b}_1, \bar{e}^*), L, \infty] < \xi$, and there is $\bar{c}^* \in N_\gamma$ such that $\wedge \bigwedge_i \psi[c_i^*, \bar{b}_1, \bar{e}^*] \models \varphi[\bar{d}, \bar{e}^*, \bar{c}^*, \bar{b}]$. Clearly, $\bar{c}^* \subseteq A_\gamma$ and $\mathrm{tp}(\bar{d}, \bar{b} \cup \bar{c}^* \cup \bar{e}^*)$ forks over $\bar{b}$, hence $\mathrm{tp}(\bar{d}, A_\gamma)$ forks over $\bar{b}$, a contradiction. As we may increase $\bar{b}$, we proved (*).

We are left with the case $\alpha = 2$, and w.l.o.g. the M_l are countable, and by the choice of ξ we can find B atomic over $M_1 \cup M_2$ (hence $B \neq A$), $M_1 \cup M_2 \subseteq B \subseteq A$ such that $B \subseteq^t_{\aleph_0} A$. We then can find a model N which is $\mathbf{F}^l_{\aleph_0}$-constructible over B, now $A \not\subseteq N$. But as $A \subseteq_t B$, N is $\mathbf{F}^l_{\aleph_0}$-constructible over A too, hence we can find M $\mathbf{F}^l_{\aleph_0}$-constructible over $N \cup A$, hence over A, hence over $M_1 \cup M_2$, and as $N \cap A = B \neq A$, N a proper submodel of M. This is a contradiction to "T does not have the dop" by XII, 2.8.

2.7. *Proof of Lemma* 2.3: Let $M_l, p_\eta (l < 3, \eta \in {}^\omega 2)$ be as in 2.4, 2.5, so by 2.5 we can assume that there is an isomorphism f from M_1 onto M_2 over M_0. Let $I_\alpha, g_\alpha (\alpha < \chi)$ exemplify $CP_{\lambda, \chi, \beth_1, \aleph_1, \aleph_1}$ and for $\alpha < \chi$ we shall define a model M^α. Let $M^\alpha_\emptyset = M_0$, for $i \in I_\alpha f^\alpha_i$ an isomorphism from M_1 onto $M^\alpha_{\{i\}}$, $f^\alpha_i \restriction M_0$ = the identity, $\{M^\alpha_{\{t\}} : t \in I_\alpha\}$ independent over $M^\alpha_\emptyset$. Let $f^\alpha_{i,j} = f^\alpha_i \cup (f^\alpha_j \circ f^{-1}) : M_1 \cup M_2 \to M^\alpha_{\{i\}} \cup M^\alpha_{\{j\}}$, $p^\eta_{i,j} = f^\alpha_{i,j}(p_\eta)$ so $p^\eta_{i,j} \in S^m(M_{\{i\}} \cup M_{\{j\}})$.

Let $\{\eta_\alpha : \alpha < \omega_1\}$ be distinct members of ${}^\omega 2$. We define by induction on n for every finite subsets t of I_α of power n a model M^α_t.

For $n = 0, 1$: we have defined.

If $t = \{i,j\}\ i < j$, let M^α_t be $\mathbf{F}^l_{\aleph_0}$-constructible over $M_i \cup M_j$ realizing $p^{\eta(i,j)}_{i,j}$, where $\eta(i,j) = \eta^\alpha(i,j) \overset{\text{def}}{=} \eta_{g_\alpha(i,j)}$.

If $n > 2$, M^α_t is $\mathbf{F}^l_{\aleph_0}$-constructible over $\bigcup\{M_s : s \subseteq t, s \neq t\}$. We make M_t depend on no more than is necessary, i.e. it depends on $g^\alpha \restriction t$, not on α.

Let $M^\alpha = \bigcup_{t \subseteq I_\alpha} M^\alpha_t$. Clearly, $\langle M_t : t \subseteq I_\alpha \rangle$ is a stable system, and

FACT 2.8: *For every countable $A \subseteq M^\alpha$, only countably many types $p \in S^m(A)$ such that $p \vdash \bigwedge_{l<m} \psi(x_l, \bar{b})$, where $\bar{b} \in A$, $R[\psi(y,\bar{b}), L, \infty] \leqslant \xi$, are realized in M^α [this is because for some countable $t \subseteq \lambda$, (letting $M^\alpha_t = \bigcup\{M^\alpha_s : s \subseteq t, s$ finite$\})$ $A \subseteq M^\alpha_t$ and for every such type realized by $\bar{c} \in M$, $\mathrm{tp}(\bar{c}, M^\alpha_t \cup \bigcup_{\gamma < \lambda / \gamma \notin t} M_\gamma)$ is atomic (by Fact 2.6 as in the proof of* XII, 4.16)].

So we get an easy contradiction to the combinatorial principle we have assumed.

LEMMA 2.9: *Suppose λ is $> \mu^{<\kappa}, \mu^+ \geqslant \kappa, \kappa \leqslant \vartheta < \lambda$, $\vartheta^{<\kappa} < \lambda$, ϑ, κ regular and λ is not a strong limit of cofinality $\leqslant \kappa$, and $\neg(\exists \sigma < \lambda)$ [σ strong limit $\wedge$ cf $\sigma < \vartheta \wedge \sigma < \lambda < 2^\sigma$]. Then $CP_{\lambda, 2^\lambda, \mu, \kappa, \vartheta}$ holds.*

Remark. We can get the "full strong" properties as in [Sh 83, §2].

Proof. Our proof is split into some cases.

Case A. λ is regular and $(\forall\alpha<\kappa)(\forall i<\lambda)[|i|^{|\alpha|}<\lambda]$.

Let $S_i \subseteq \{\delta<\lambda : \mathrm{cf}\,\delta=\vartheta\}$ $(i<\lambda)$ be pairwise disjoint stationary sets and for each $\delta\in\bigcup_{i<\lambda}S_i$ let η_δ be an increasing continuous sequence of ordinals $<\delta$, of length ϑ, each of cofinality $<\vartheta$, $\delta=\bigcup_{\alpha<\vartheta}\eta_\delta(\alpha)$. For any $A\subseteq\lambda$ we define a two-place function $g_A:\lambda\to\vartheta$ as follows:

$g_A(\delta,j)$ is $\mathrm{Min}\{\alpha<\vartheta:\eta_\delta(\alpha)\geqslant j\}$, if $\delta\in\bigcup_{i\in A}S_i, j$ an ordinal $<\delta$,

$g_A(\delta,j)$ is 0, otherwise.

Now if $\mathscr{P}$ is a family of pairwise incomparable subsets of λ, $|\mathscr{P}|=2^\lambda$, then $\{(\lambda,g_A):A\in\mathscr{P}\}$ exemplifying the desired conclusion.

For suppose $A,B\subseteq\lambda$, $A-B\neq\emptyset$, f a function from I_A into $M(I_B)$ ($I_A=(\lambda,g_A)$, $I_B=(\lambda,g_B)$; for the definition of $M(I_B)$ see [Sh 83] 1.1). Let $i(*)\in A-B$. For each $i<\lambda$ let $f(i)=\tau_i(\bar a_i)$, $\bar a_i$ a sequence of length $<\kappa$ from I_B, τ_i a term in $L^*_{\mu,\kappa}$.

We can find a stationary $S\subseteq S_{i(*)}$ such that (remember $\mu^{<\kappa}<\lambda$):

(1) for every $i\in S$, $\tau_i=\tau^*$, $\bar a_i=\langle a_{i,\alpha}:\alpha<\alpha_0\rangle$, $\alpha_0<\kappa$, and w.l.o.g. $a_{i,0}=i$.

Similarly, w.l.o.g.:

(2) the truth value of "$a_{i,\beta}<a_{i,\gamma}$", "$a_{i,\beta}\in\bigcup_{j\in B}S_j$", "$\eta_{a_{i,\gamma}}(j)<a_{i,\beta}$", "$\eta_{a_{i,\gamma}}(j)=a_{i,\beta}$", "$\eta_{a_{i,\gamma}}(j)>a_{i,\beta}$" and also $\zeta=\mathrm{Min}\{\xi:\eta_{a_{i,\beta}}(\xi)\geqslant a_{i,\gamma}\}$ for $\beta,\gamma<\alpha_0$, $\zeta,j<\vartheta$, is the same for all $i\in S$ (as there are $\leqslant\vartheta^{|\alpha_0|}<\lambda$ possibilities).

By Fodor's lemma, as $(\forall i<\lambda)[|i|^{|\alpha_0|}<\lambda]$ w.l.o.g.:

(3) for every $i\in S$, $\beta<\alpha_0$ if $a_{i,\beta}<i$, then $a_{i,\beta}=a_{i_0,\beta}$, where $i_0=\mathrm{Min}\,S$ and also $\eta_{a_{i,\beta}}(\xi_{i,\beta}-1)$ is constant, where $\xi_{i,\beta}=\mathrm{Min}\{\xi:\eta_{a_{i,\beta}}(\xi)\geqslant i\}$ (if it is well defined).

Also, w.l.o.g.:

(4) if $i<j$ are in S, then $a_{i,\beta}<j$ for every $\beta<\alpha_0$.

Now choose δ a limit point of S of cofinality ϑ, $\delta\in S$ (really $\delta\in S_{i(*)}$ suffices). Then choose $\{j_\xi:\xi<\vartheta\}\subseteq S\cap\delta$ such that $\mathrm{Min}\{\zeta:\eta_\delta(\zeta)\geqslant j_\xi\}$ is strictly increasing; then $x_\xi=\delta$, $y_\xi=j_\xi$, fulfil the requirement because: $\delta\notin\bigcup_{j\in B}S_J$, hence $\eta_{a_{\delta,\gamma}}(\zeta)$ is $=\eta_{a_{\mathrm{Min}\,S,\gamma}}(\zeta)$ or is $\geqslant\delta$ for $\zeta<\vartheta$, $\gamma<\alpha_0$.

Case B. For some σ, $\sigma<\lambda\leqslant 2^\sigma$, $\sigma^{<\kappa}=\sigma\geqslant\mu^{<\vartheta}$.

The proof is similar to [Sh 83, 2.7]. Let $S_i\subseteq\{\delta<\sigma^+:\mathrm{cf}\,\delta=\vartheta\}$ $(i\leqslant\sigma)$ be stationary, pairwise disjoint. For every $\delta\in\bigcup_{i<\lambda}S_i$ choose an increasing continuous sequence η_δ of length ϑ, $\delta=\bigcup_{\xi<\vartheta}\eta_\delta(\xi)$. For any $A\subseteq\sigma$ let I_A be defined as in Case A.

Let $\{A_\xi : \xi < \lambda\}$ be a family of subsets of σ, any non-trivial Boolean combination of $<\kappa$ of them has power σ (they exist; see AP 1.5(2)). Now for any $B \subseteq \lambda$ let $J_B = \sum_{\varepsilon \in B} I_{A_\varepsilon}$, and it suffices to prove that if $B_1 \nsubseteq B_2, f: J_{B_1} \to M(J_{B_2})$, then there are x_ξ, y_ξ as required in Def. 2.2. We can choose $\epsilon(*) \in B_1 - B_2$ and consider $f \restriction (I_{A_{\epsilon(*)}} \times \{\epsilon(*)\})$. Let $f(\langle i, \epsilon(*)\rangle) = \tau_i(\langle \ldots, \langle a_{i,\alpha}, \epsilon_{i,\alpha}\rangle, \ldots\rangle_{\alpha<\alpha_i})$ and again for some $S \subseteq S_\sigma$, S stationary, and

(1) $\tau_i = \tau^*, \alpha_i = \alpha^*$ for $i \in S$,

(2) $\{\langle \ldots \langle a_{i,\alpha}, \epsilon_{i,\alpha}\rangle_{\alpha<\alpha^*} : i \in S\}$ forms a Δ-system, i.e. for distinct $i_1, i_2, i_3, i_4 \in S$,

$$[a_{i_1,\alpha_1} = a_{i_2,\alpha_2} \Rightarrow a_{i_3,\alpha_1} = a_{i_4,\alpha_1} = a_{i_3,\alpha_2} = a_{i_4,\alpha_2}],$$

$$[\epsilon_{i_1,\alpha_1} = \epsilon_{i_2,\alpha_2} \Rightarrow \epsilon_{i_3,\alpha_1} = \epsilon_{i_4,\alpha_1} = \epsilon_{i_3,\alpha_2} = \epsilon_{i_4,\alpha_2}],$$

(3) the truth values of $\epsilon_{i,\alpha} = \epsilon_{i,\beta}$, $a_{i,\alpha} = \eta_{a_{i,\beta}}(\xi)$, "$a_{i,\alpha} \in \bigcup_{j \in A_{\epsilon_{i,\beta}}} S_j$", "$a_{i,\alpha} = a_{i,\beta}$"; and the values of $\mathrm{Min}\{\xi : \eta_{a_{i,\alpha}}(\xi) \geqslant a_{i,\beta}\}$ are the same for all $i \in S, \alpha < \alpha^*, \beta < \alpha^*$,

(4) if $i_1 < i_2$ are in S, then $a_{i_1,\alpha} < i_2$,

(5) if for some $\gamma^*, i \in S \;\&\; a_{\alpha,i} \neq i \;\&\; \zeta < \vartheta \Rightarrow \eta_{a_{i,\alpha}}(\zeta) \geqslant i \vee \eta_{a_{i,\alpha}}(\zeta) = \eta_{a_{\mathrm{Min}\, S,\alpha}}(\zeta) < \gamma^*$.

Now choose $j(i) \in A_{\epsilon(*)} - \bigcup_{\alpha<\alpha^*} A_{\epsilon_{i,\alpha}}$ for $i \in S$ [exists as $\epsilon(*) \in B_1 - B_2$, $\epsilon_{i,\alpha} \in B_2$ and the choice of $\{A_\xi : \xi < \lambda\}$]. As $j(i) < \sigma$ w.l.o.g. $j(i) = j(*)$ for every $i \in S$. Now choose $\delta \in S_{j(*)}$ δ a limit point of S; for appropriate $\{j_\xi : \xi < \vartheta\} \subseteq S$, $x_\xi = \langle \delta, \epsilon(*)\rangle$, $y_\xi = \langle j_\epsilon, \epsilon(*)\rangle$ we get the desired conclusion.

Case C. λ is a strong limit of cofinality $>\kappa$ (or $>\kappa^+$ if κ is singular).

Similar to [Sh 83, 2.6, Case B]. We use (λ, g) such that for every $\delta, i < \lambda$ if $\{j : g(i,j) \neq 0\} \neq \emptyset$ then it is a set of successor ordinals, i is limit, the set has order type $\vartheta \times \kappa$, $g(i, -)$ is a one-to-one function over every interval of order type ϑ, its limit is a strong cardinal $\lambda^i, \lambda^i < i < 2^{\lambda^i}$, $\mathrm{cf}\, \lambda^i = \kappa$.

FACT 2.10: *Let $\vartheta < \kappa < \lambda$ be regular cardinals and let*

$\mathscr{P}^+ = \{S : S \subseteq \lambda, (\forall \delta \in S)\, \mathrm{cf}\, \delta = \vartheta$, and for every closed unbounded $C \subseteq \lambda$ for some $\xi \in C$, $\mathrm{cf}\, \xi = \kappa$ and $S \cap \xi$ is a stationary subset of $\xi\}$.

Then any $S \in \mathscr{P}^+$ can be partitioned to λ subsets each in $\mathscr{P}^+$.

Proof. For every $\delta \in S$ let η_δ be an increasing sequence of length ϑ of ordinals $<\delta$ whose limit is δ. Let for each $\alpha < \vartheta, i < j$

$$A^\alpha_{i,j} = \{\delta \in S : i \leqslant \eta_\delta(\alpha) < j\}.$$

We shall prove

SUBFACT 2.10A: *For some* $\alpha < \vartheta$ *for every* $i < \lambda$ *for some* $j, i < j < \lambda$ *and* $A^{\alpha}_{i,j} \in \mathscr{P}^+$.

This suffices as then we can define by induction on $j < \lambda$ an ordinal $i(j)$ such that $j_1 < j \Rightarrow i(j_1) < i(j)$ and for j-successor, $A^{\alpha}_{i(j-1),i(j)} \in \mathscr{P}^+$. This is clearly possible and $\{A^{\alpha}_{i(j),i(j+1)} : j < \lambda\}$ is a family of λ pairwise disjoint subsets of S which are in $\mathscr{P}^+$. So we can get a partition as required.

Proof of the Subfact. If not, then for every $\alpha < \vartheta$ there is $i_\alpha < \lambda$ such that for no $j, i_\alpha < j < \lambda$, there is a closed unbounded C^{α}_{j} subset of λ such that

$$[\xi \in C^{\alpha}_{j} \wedge \operatorname{cf}\xi = \kappa] \Rightarrow [A^{\alpha}_{i_\alpha,j} \cap \xi \text{ is not a stationary subset of } \xi].$$

Let $i(*) = \bigcup_{\alpha<\vartheta} i_\alpha < \lambda$, and

$$C^* = \{\delta < \lambda : \delta \text{ is limit, } \delta > i(*), \text{ and } \delta \in C^{\alpha}_{j} \text{ for each } \alpha < \vartheta, j < \delta\}.$$

It is well known that C^* is a closed unbounded subset of λ. But $S \in \mathscr{P}^+$, so there is $\xi \in C^*$ such that $\operatorname{cf}\xi = \kappa$ and $S \cap \xi$ is a stationary subset of ξ. For each $\delta \in S \cap \xi$ there is $\alpha_\delta < \vartheta$ such that $\eta_\delta(\alpha_\delta) > i(*)$. So $S \cap \xi = \bigcup_{\alpha<\vartheta}\{\delta \in S \cap \xi : \alpha_\delta = \alpha\}$ and as $\vartheta < \kappa = \operatorname{cf}\xi$, for some β,

$$S_1 = \{\delta \in S \cap \xi : \alpha_\delta = \beta\}$$

is a stationary subset of ξ. Now the function $\delta \to \eta_\delta(\beta)$ is regressive on S_1 (i.e. $\eta_\delta(\beta) < \delta$) so by a variant of the Fodor lemma, for some $j < \delta$,

$$S_2 = \{\delta \in S_1 : \eta_\delta(\beta) < j\}$$

is a stationary subset of ξ. So by the definition of $\alpha_\delta, \beta, S_1, S_2$ for every $\delta \in S_2$, $i_{\beta^*} \leqslant i(*) < \eta_\delta(\alpha_\delta) = \eta_\delta(\beta) < j$, hence $\delta \in A^{\beta^*}_{i_\alpha,j}$. So $S_2 \subseteq A^{\beta}_{i_\beta,j}$.

But $\xi \in C^*$, hence $\xi \in C^{\beta^*}_{i_\alpha,j}$, hence $A^{\beta^*}_{i_\alpha,j} \cap \xi$ is not a stationary subset of ξ, but $S_2 \subseteq A^{\beta}_{i_\beta,j}$ and S_2 is a stationary subset of ξ, a contradiction.

Remark. We would like to have in 1.1(2) that $IE(\lambda, T) = 2^\lambda$ for every $\lambda > \aleph_0$. We think that it is true, but having spent considerable time on proving "$I(\lambda, T) = 2^\lambda$ for T unsuperstable, $\lambda \geqslant |T| + \aleph_1$" in all cases we have not found it challenging to do this again.

XIII.3. On the Morley conjecture

DEFINITION 3.1: We say that M_1 is an $(N, \bar{a})$-component of M if $M_1 \subseteq M, N \cup \bar{a} \subseteq M_1$, $\mathrm{tp}_*(M_1, N \cup \bar{a})$ is almost orthogonal to N and $|M_1|$ is a maximal set under the previous conditions.

CLAIM 3.2: *If* $\{M_1, M_2\}$ *is independent over* $M_0, M_0 \subseteq M_1, M_0 \subseteq M_2, M$ *primary over* $M_1 \cup M_2, \bar{a} \in M_1$, *and* M^* *is an* $(M_0, \bar{a})$*-component of* M_1, *then* M^* *is an* $(M_0, \bar{a})$*-component of* M *provided that every regular stationary type not orthogonal to* $\mathrm{tp}(M_1, M^*)$ *is not orthogonal to* M_0.

Proof. Left to the reader.

We now work in the context of Section 2 of Chapter XI. The following lemma shows that under suitable circumstances, we can "blow up a model M" by increasing a suitable dimension, without losing control.

LEMMA 3.3 [The axioms used in XI, 2.8]: *We assume* T *is superstable without the dop and* λ *satisfies* $(*)$ *of* XI, 2.3(3). *Suppose* $N \subseteq^* M^1, \{\bar{a}_i : i < \beta\}$ *independent over* N *(not necessarily maximal!)* $\mathrm{tp}(\bar{a}_i, N)$ *regular,* $\|N_i\| \leqslant \lambda$, $\lambda \leqslant \alpha < \beta$, *for* $i < \alpha, \bar{a}_i \in M^1$ *and* N_i *an* $(N, \bar{a}_i)$*-component of* M^1; *for* $i \geqslant \alpha$ *(but* $< \beta$*)* $\mathrm{tp}_*(N_i, M^1 \cup \bigcup_{j \neq i} N_j)$ *does not fork over* N, *the* N_i*'s are pairwise isomorphic over* N, *and* M^2 *is* **T**-*primary over* $M^1 \cup \bigcup_{i<\beta} N_i$. *Suppose further that all the models are* **T**-*saturated. Then for every* $\bar{b} \in M^1$, $\mathrm{tp}(\bar{b}, N)$ *regular, and* $(N, \bar{b})$*-component* M_2 *of* M^2, *there is an* $(N, \bar{b})$ *component* M_1 *of* M^1 *isomorphic to* M_2 *over* $N \cup \bar{b}$ *provided that* $\lambda^+ \leqslant \beta$.

LEMMA 3.3A: (1) *Really for some automorphism of* M^2, $F \restriction (N \cup \bar{b}) =$ *the identity,* $F(M_2) \subseteq M^1$.

(2) *Moreover, if* $N \cup \bar{b} \subseteq M \subseteq M^1 \cap M_2$, $F(M_2)$ *an* $(N, \bar{b})$*-component of* M^1 *and* M^2, *then w.l.o.g.* $F \restriction M =$ *the identity provided that* $\lambda^+ \leqslant \alpha$.

Proof. We should note:

FACT 3.4: *If* $\alpha \geqslant \lambda^+$ $A \subseteq M^2$ *of power* $\leqslant \lambda$, *then there is an automorphism* F *of* M^2, *mapping* A *into* M^1 *such that* $F \restriction (M^1 \cap A) =$ *the identity.* [*Because, letting* $N^0_{\langle\rangle} = N$, $N^0_{\langle i \rangle} = N_i$, $\langle N^0_\eta : \eta \in \{\langle i \rangle : i < \alpha\} \cup$

$\{\langle\rangle\}\rangle$ is a $(\mathbf{T}, \subseteq^*, \lambda)$-decomposition inside M^1, hence we can extend it to $\langle N^0_\eta : \eta \in J^1\rangle$, a $(\mathbf{T}, \subseteq^*, \lambda)$-decomposition of M^1 with $\|N^0_\eta\| \leqslant \lambda$, w.l.o.g. $\langle i\rangle \notin J^1$ when $\alpha \leqslant i < \beta$. As $N^0_{\langle i\rangle}$ is a $(N, \bar{a}_i)$-component of M^1, $\langle i\rangle \lessdot \eta \in J^1 \Rightarrow \langle i\rangle = \eta$ and easily $\langle N^0_\eta : \eta \in J^2\rangle$, where $J^2 \stackrel{\text{def}}{=} J^1 \cup \{\langle i\rangle; \alpha \leqslant i < \beta\}$ is a $(\mathbf{T}, \subseteq^*, \lambda)$-decomposition of M^2. W.l.o.g. $A = |M|$, M prime over $\bigcup_{\eta \in J} N^0_\eta$, $J \subseteq J^2$, $|J| \leqslant \lambda$, and the rest should be clear using the uniqueness of the **T***-prime,* **T***-minimal model over A).]*

FACT 3.4A: *If $A \subseteq M^2$ has power $\leqslant |\alpha|$, $B \subseteq A \cap M^1$ and $|B| < |\alpha|$, then there is an automorphism F of M^2 mapping A into M^1 such that $F \restriction (B \cup N) =$ the identity.*

Proof. Same as the proof of 3.4.

FACT 3.4B: *If $N \cup \bar{b} \subseteq B$, $A \subseteq M^2$, $B \subseteq M^1$, $|A| \leqslant |\alpha|$ and* $\text{tp}_*(B, N \cup \bar{b})$ *is almost orthogonal to N, then for some automorphism F of M^2, F maps A into M^1 and $F \restriction (N \cup B) =$ the identity.*

Remark. Instead of $\bar{b}$ finite, $\bar{b}$ of length $< \lambda$ suffices.

Proof. As we can replace $\langle N_i : i < \alpha\rangle$ by $\langle N_i : i \in \alpha \backslash w\rangle$ for any subset w of α of cardinality $< |\alpha|$, w.l.o.g. $\bar{b}$ does not depend over $(N, \bigcup_{i<\alpha} N_i)$, hence over $(N, \bigcup_{i<\beta} N_i)$. Now we can find $\langle N^0_\nu : \nu \in J^1\rangle$ be as there, such that: $B \subseteq N^*$, N^* **T**-prime, **T**-atomic over $\bigcup\{N^0_\nu : \nu \in J^3\}$, where $J^3 = \{\nu : \nu = \langle\,\rangle$ or $\nu(i) \geqslant \alpha$ (hence $\geqslant \beta$)$\}$. The rest is the same.

Remark 3.4C. In 3.4A, 3.4B, $\alpha \geqslant \omega$ suffices.

Continuation of the proof of 3.3. Let $\langle N^2_\eta, \bar{a}^2_\eta : \eta \in I^2\rangle$ be a $(\mathbf{T}, \subseteq^*, \lambda)$-decomposition of M_2 with $\|N^2_\eta\| \leqslant \lambda$, and $N \cup \bar{b} \subseteq N^2_{\langle\rangle}$. Let $I^2_n = \{\eta \in I^2 : l(n) \leqslant n\}$. We can define by induction on n, $N^1_\eta, \bar{a}^1_\eta, F_\eta$ for $\eta \in I^2_n$ such that

(1) $\langle N^1_\eta, \bar{a}^1_\eta : \eta \in I^2_n\rangle$ is a $(\mathbf{T}, \subseteq^*, \lambda)$-decomposition inside M^1,

(2) F_η is an automorphism of M^2 mapping N^2_η onto N^1_η, $F_\eta(\bar{a}^2_\eta) = \bar{a}^1_\eta$, $F_{\eta\restriction k} \subseteq F_\eta$ for $k < l(\eta)$, and $F_{\langle\rangle} \restriction (N \cup \bar{b}) =$ the identity.

(3) For $\eta \in I_{n-1}$ $\{\bar{a}^1_{\eta\frown\langle i\rangle} : \eta \frown \langle i\rangle \in I^2_n\}$ is independent over N^1_η, each $\text{tp}(\bar{a}_{\eta\frown\langle i\rangle}, N^1_\eta)$ is regular orthogonal to N_{η^-} if $\eta \neq \langle\rangle$ and to N otherwise.

For $n = 0$: Let $F_{\langle\rangle}$ be an automorphism of M^1, $F^{\langle\rangle} \restriction (N \cup \bar{b}) =$ the identity, and $F_{\langle\rangle}$ maps $N^2_{\langle\rangle}$ into M^1 (exists by 3.4A). We let $N^1_{\langle\rangle} = F_{\langle\rangle}(N^2_{\langle\rangle})$ (and $\bar{a}^1_{\langle\rangle}$ are meaningless).

For $n = m+1$: We deal with a specific $\eta \in I_m$ of length m, $\{F_\eta(\bar{a}^2_{\eta\frown\langle i\rangle}) : \eta\frown\langle i\rangle \in I_n\}$ is as required in (3); however, $F_\eta(\bar{a}^2_{\eta\frown\langle i\rangle})$ does not necessarily belong to M^1. Let r be regular and we deal with $\boldsymbol{I} = \{F_\eta(\bar{a}^2_{\eta\frown\langle i\rangle}) : \eta\frown\langle i\rangle \in I_n,\ \mathrm{tp}(F_\eta(\bar{a}^2_{\eta\frown\langle i\rangle}, N^1_\eta)$ not orthogonal to $r\}$. For each $\bar{a} \in \boldsymbol{I}$, $p = \mathrm{tp}(\bar{a}, N^1_\eta)$ is a regular type orthogonal to N [as $N \cup \bar{b} \subseteq N^2_{\langle\rangle} \subseteq^* N^2_\eta \subseteq^* M_2$, for every $\bar{c} \in M_2$, $\mathrm{tp}(\bar{c}, N^2_\eta)$ is almost orthogonal to N (as $\mathrm{tp}_*(M_2, N \cup \bar{b})$ is) hence is orthogonal to N (by Ax(C2))]. But $\mathrm{tp}_*(\bigcup_{\alpha \leqslant i < \beta} N_i, M^1)$ does not fork over N, hence it is orthogonal to $\mathrm{tp}(\bar{a}, N^1_\eta)$. So clearly, $\mathrm{tp}(\bar{a}, M^1)$ forks over N^1_η, and it does not fork over some $\bar{c} \subseteq M^1$.

We can (using 3.4A) define by induction on $k < \omega$, $\bar{d}_l \in M^1$ realizing $\mathrm{tp}(\bar{a}, N^1_\eta \cup \bar{c} \cup \bigcup_{k<l} \bar{d}_k)$, so for some l $\{\bar{d}_k : k < l\}$ is independent over N^1_η and $\bar{d}_l$ (hence $\bar{a}$) depends on them. [Otherwise, $\{\bar{d}_k : k < \omega\}$ is an indiscernible set based on $\mathrm{tp}(\bar{a}, N^1_\eta)$, but $\mathrm{tp}(\bar{a}, N^1_\eta \cup \bar{c}) = \mathrm{Av}(\{\bar{d}_k : k < \omega\}, N^1_\eta \cup \bar{c})$, a contradiction.] We can conclude that each $\bar{a} \in \boldsymbol{I}$ depends on $\boldsymbol{J} = \boldsymbol{J}_r = \{\bar{b} \in M^1 : \mathrm{tp}(\bar{b}, N^1_\eta)$ is regular not orthogonal to $r\}$ over N^1_η.

Now it suffices to find for each $\bar{a}_{\eta\frown\langle i\rangle} \in \boldsymbol{I}$ an automorphism F^i of M^2, $F^i \restriction N^1_\eta =$ the identity, $F^i(\bar{a}_{\eta\frown\langle i\rangle}) \in M^1$ and $\{F^i(\bar{a}_{\eta\frown\langle i\rangle}) : \bar{a}_{\eta\frown\langle i\rangle} \in \boldsymbol{I}\}$ is a maximal subset of $\boldsymbol{J}$ independent over N^1_η. [As then (by 3.4B) for some automorphism F^i_1 of M^2, $F^i_1 \restriction (N^1_\eta \cup F^i(\bar{a}_{\eta\frown\langle i\rangle})) =$ the identity and F^i_1 maps $F^i(F_\eta(N^2_{\eta\frown\langle i\rangle}))$ into M^1, and let $F_{\eta\frown\langle i\rangle} = F^i_1 \circ F^i \circ F_{\eta}$.]

If $|\boldsymbol{I}| \leqslant \lambda$, there is an automorphism F of M^2, $F \restriction N^1_\eta =$ the identity, F maps $\boldsymbol{I}$ into M^1 (hence into $\boldsymbol{J}$), hence the maximality of $\{a^2_{\eta\frown\langle i\rangle} : \eta\frown\langle i\rangle \in \boldsymbol{I}^2\}$ implies the maximality of $\{F(\bar{a}) : \bar{a} \in \boldsymbol{I}\}$, hence $F^i = F$ is as required.

If $|\boldsymbol{I}| > \lambda$, we still know that $\boldsymbol{J}$ has a maximal subset $\boldsymbol{J}^0$ independent over N^1_η, and that $|\boldsymbol{I}| = |\boldsymbol{J}^0|$. We then define by induction on $\xi < |\boldsymbol{I}|$, countable $\boldsymbol{I}^1_\xi \subseteq \boldsymbol{I} - \bigcup_{\zeta<\xi} \boldsymbol{I}^1_\zeta$, $\boldsymbol{I}^2_\xi \subseteq \bigcup_{\zeta<\xi} \boldsymbol{I}^1_\zeta$, $\boldsymbol{J}^1_\xi \subseteq \boldsymbol{J}^0 - \bigcup_{\zeta<\xi} \boldsymbol{J}^1_\zeta$, $\boldsymbol{J}^2_\xi \subseteq \bigcup_{\zeta<\xi} \boldsymbol{J}^1_\zeta$ such that $\boldsymbol{I} = \bigcup_\xi \boldsymbol{I}^1_\xi$, $\boldsymbol{J}^0 = \bigcup_i \boldsymbol{J}^1_i$, $\mathrm{tp}_*(\boldsymbol{I}^1_\xi, \bigcup_{\zeta<\xi} \boldsymbol{J}^1_\zeta \cup \boldsymbol{J}^0 \cup N^1_\eta)$ does not fork over $N^1_\eta \cup \bigcup \boldsymbol{J}^1_\xi \cup \boldsymbol{J}^2_\xi \cup \boldsymbol{I}^2_\xi$ and $\mathrm{tp}(\boldsymbol{J}^1_\xi, \boldsymbol{I} \cup \bigcup_{\zeta<\xi} \boldsymbol{J}^1_\zeta \cup N^1_\eta)$ does not fork over $N^1_\eta \cup \bigcup \boldsymbol{I}^1_\xi \cup \boldsymbol{I}^2_\xi \cup \boldsymbol{J}^2_\xi$ and $|\boldsymbol{I}^l_\xi| = |\boldsymbol{J}^l_\xi| \leqslant \aleph_0$ and $[\bar{c} \in \boldsymbol{I}^2_\xi \cap \boldsymbol{I}^1_\zeta$ implies $\boldsymbol{J}^1_\zeta \cup \boldsymbol{J}^2_\zeta \subseteq \boldsymbol{J}^2_\xi]$. Now for each i let F^ξ_1 be an automorphism of M^2, $F^i_1 \restriction (N^1_\eta \cup \boldsymbol{J}^1_\xi \cup \boldsymbol{J}^2_\xi) =$ the identity, $F^\xi_1(\boldsymbol{I}^1_\xi \cup \boldsymbol{I}^2_\xi) \subseteq M^1$ exists by 3.4B with $\bar{b}$, $N^1_\eta \cup \boldsymbol{J}^1_\xi \cup \boldsymbol{J}^2_\xi$, $N^1_\eta \cup \boldsymbol{J}^1_\xi \cup \boldsymbol{J}^2_\xi \cup \boldsymbol{I}^1_\xi \cup \boldsymbol{I}^2_\xi$ here corresponding to b, B, A there). Lastly, for $\bar{a}_{\eta\frown\langle i\rangle} \in \boldsymbol{I}$ let F^i be F^ξ_1 when $\bar{a}_{\eta\frown\langle i\rangle} \in \boldsymbol{I}^1_\xi$.

If r is trivial (e.g. when it has depth > 0) the proof is easier. *Note*: We can find a one-to-one function h from $\boldsymbol{I}$ onto $\boldsymbol{J}_0$ s.t. $\{\bar{a}, h(\bar{a})\}$ is not independent over N^1_η; now use 3.4 for $A = A \cup \bar{a} \cup h(\bar{a})$ for each $\bar{a} \in \boldsymbol{I}$.

As was said above, this suffices to define $F_{\eta\frown\langle i\rangle}$ (for $\eta\frown\langle i\rangle\in I^2$).

So we have carried the definition of $\langle F_\eta:\eta\in I^2\rangle$. As $\langle N^1_\eta:\eta\in I^2\rangle$ is a non-forking tree of models, $F\stackrel{\text{def}}{=}\bigcup\{F_\eta:\eta\in I^2\}$ is an elementary map from $\bigcup_{\eta\in I^2}N^1_\eta$, we can extend F to an elementary mapping F^+ from M_2 into M^1, thus finishing.

One point is still missing: Why is $M_1=\text{Range}(F^+)$ an $(N,\bar b)$-component of M^1 (and of M^2)? The reason is that for each $\eta\in I^2$: $(*)_\eta$ $\{F_{\eta\frown\langle i\rangle}(\bar a^2_{\eta\frown\langle i\rangle}):\eta\frown\langle i\rangle\in I^2\}$ is a maximal independent subset of $J_\eta=\{\bar a\in M^2:\text{tp}(\bar a,N_\eta)$ is regular not orthogonal to N_{η^-} if $\eta\neq\langle\,\rangle$, to N if $\eta=\langle\,\rangle\}$ (or use the proof of 3.3A(1)).

Proof of 3.3A(1). We can find a $(\mathbf{T},\subseteq^*,\lambda)$-decomposition $\langle N^3_\eta,\bar a^3_\eta:\eta\in I^3\rangle$ of M^2 such that $I^2=\{\eta:\langle 0\rangle\frown\eta\in I^3\}$, and for $\eta\in I^2$, $\bar a^2_\eta=\bar a^3_{\langle 0\rangle\frown\eta}$, $N^2_\eta=N^3_{\langle 0\rangle\frown\eta}$, $N^3_{\langle\rangle}=N$. Now easily $F^+\cup\bigcup\{\text{id}_{N^3_\eta}:\eta\in I^3,\ \eta(0)>0\}$ is an elementary mapping and (by XI, §2) M^2 is $\mathbf{T}$-prime over its domain, so it can be extended to an embedding F^* of M^2 to itself (extending F^+). But we can check that $\langle F(N^3_\eta),(F(\bar a^3_\eta):\eta\in I^3\rangle$ is a $(\mathbf{T},\subseteq^*,\lambda)$-decomposition of M^2 (see $(*)_\eta$ above). As T does not have dop, M^2 is $\mathbf{T}$-minimal over $\bigcup_{\eta\in I^3}F^*(N^3_\eta)$ hence the mapping is onto.

Proof of 3.3A(2). It is similar, but choosing a $(\mathbf{T},\subseteq^*,\lambda)$-decomposition $\langle N^2_\eta,\bar a^2_\eta:\eta\in I^2\rangle$ of M_2 we take care that M is $\mathbf{T}$-prime, $\mathbf{T}$-atomic over $\bigcup_{\eta\in I}N^2_\eta, I\subseteq I^2$ close under initial segments (see XI, §2); when we choose F_η add the condition:

(4) if $\eta\in I^2_n\cap I$, then F_η is the identity.

Note that in the induction step, for each $\eta\in I^2_n\cap I$ our commitment on $F(\bar a_{\eta\frown\langle i\rangle})='\bar a_{\eta\frown\langle i\rangle}$ when $\eta\frown\langle i\rangle\in I$ is not an obstacle.

In the end, instead of extending $\bigcup_{\eta\in I^2}F_\eta$ to an elementary embedding of M_2 into M^1, we note that also $\bigcup_{\eta\in I^2}F_\eta\,\text{id}_M$ is also an elementary mapping and M_2 is $\mathbf{T}$-prime over its domain, and extend it to an embedding F^+ of M_2 into M_1. Then continue as in the proof of 3.3A(1).

LEMMA 3.5 [The axioms used in XI, 2.8]: *Assume T is superstable without the dop. Suppose M^* is a model of T of power $\mu>\lambda\stackrel{\text{def}}{=}\lambda(\mathbf{T},\subseteq^*)$, $2^\mu>2^\lambda$, and M^* is $\mathbf{T}$-saturated and $(*)$ fails but $(**)$ holds where*

$(*)$ *there are $\chi\geqslant\lambda$ and a $\mathbf{T}$-saturated $N\subseteq^*M^*$, $\bar a_i\in M^*$ $(i<\chi^+)$, $\{\bar a_i:i<\chi^+\}$ independent over N, N_i an $(N,\bar a_i)$-component of M^*, the N_i pairwise isomorphic over N, $\|N_i\|\leqslant\chi$, $\|N\|\leqslant\lambda$;*

(**) *there are* $N \subseteq N_i \subseteq M^*, N_i \neq N$, $\|N\| \leqslant \lambda$, $\|N_i\| \leqslant \lambda$ $(i < \lambda^+)$, $\{N_i : i < \lambda^+\}$ *independent over* N *and the* N_i*'s are pairwise isomorphic over* N.

Then for every $\mu(1) \geqslant \mu$, T *has (at least)* 2^μ *pairwise non-isomorphic* **T**-*saturated models of power* $\mu(1)$.

Proof. Let $N, N_i\,(i < \lambda^+)$ exemplify (**). Let $\mu(*) = \mathrm{Min}\{\vartheta : 2^\vartheta = 2^\mu\}$, hence $\lambda < \mu(*) \leqslant \mu$. Let $\langle N_\eta, \bar{a}_\eta : \eta \in I\rangle$ be a $(\mathbf{T}, \subseteq^*)$-decomposition of M^* (so $\|N_\eta\| \leqslant \lambda$) and w.l.o.g. $N \subseteq N_{\langle\rangle}$, and for each η $\{i : \eta ^\frown \langle i\rangle \in I\}$ is an ordinal $\alpha_I(\eta)$. We concentrate on the case $\mu(*)$ is not strong limit (as for T countable we use only the case $\mathrm{Dp}(T) < \omega, \mu < \beth_\omega$). If $\mu(*)$ is regular, then for some $\eta \in I$, for $\mu(*)$ ordinals i, $\eta ^\frown \langle i\rangle \in I$, but $\{\nu : \eta ^\frown \langle i\rangle \lessdot \nu\}$ has power $< \mu(*)$. If there is such η, w.l.o.g. $\eta = \langle\rangle$, and $(\forall i < \mu(*))[\langle i\rangle \in I \wedge |\{\nu : \langle i\rangle \lessdot \nu \in I\}| < \mu(*)]$. If there is no such η, then necessarily $\mu(*)$ is singular, and let $\mu(*) = \sum\{\mu_i : i < \mathrm{cf}\,\mu(*)\}$ and w.l.o.g. μ_i is increasing, each μ_i regular, 2^{μ_i} (strictly) increasing and $2^{\mu_i} > 2^\lambda$. For each i we can find $\eta_i \in I$ such that η_i is $\lessdot$-incomparable with η_j for $j < i$, and for μ_i ordinals $j, \eta ^\frown \langle j\rangle \in I$, $|\{\nu : \eta ^\frown \langle j\rangle \lessdot \nu\}| < \mu_i$. W.l.o.g. $\eta_i = \langle i\rangle$ for $i < \mathrm{cf}\,\mu(*)$, and for $j < \mu_i$: $\langle i,j\rangle \in I$, $|[\nu : \langle i,j\rangle \lessdot \nu \in I\}| < \mu_i$. If $\mathrm{cf}\,\mu(*) \leqslant \lambda$ we can again assume w.l.o.g. that $\alpha_I(\langle\rangle) \geqslant \mu(*)$ and for $i < \mu(*)$, $|\{\nu : \langle i\rangle \lessdot \nu \in I\}| < \mu(*)$.

So we have exactly one of the following cases.

Case A. For $i < \mu(*)$, $\langle i\rangle \in I$ and $|\{\nu \in I : \eta ^\frown \langle i\rangle \lessdot \nu\}| < \mu(*)$ and even, $|\{\nu \in I : \nu(0) \leqslant i\}| < \mu(*)$.

Case B. Not Case A, $\mu(*) > \mathrm{cf}\,\mu(*) > \lambda$, for $i < \mathrm{cf}\,\mu(*) : \langle i\rangle \in I$, $(\forall j < \mu_i)[\langle i,j\rangle \in I \wedge |\{\nu : \langle i,j\rangle \lessdot \nu \in I\}| < \mu_i]$ and $\mu = \sum\{\mu_i : i < \mathrm{cf}\,\mu(*)\}$, $\mu_i, 2^{\mu_i}$ strictly increasing.

As $\{N_i : i < \lambda^+\}$ is independent over $N, N \subseteq N_{\langle\rangle}$ for some $i_0 < \lambda^+$, $\{N_i : i_0 \leqslant i < \lambda^+\}$ is independent over $(N, N_{\langle\rangle})$, and w.l.o.g. $i_0 = 0$. Now in Case A, w.l.o.g. $\{N_i : i < \lambda^+\}$ is independent over $(N, N \cup \bigcup_{i<\mu(*)} \bar{a}_{\langle i\rangle})$. [If $\mu(*) > \lambda^+$ is regular, for some $\alpha < \mu(*)$, $\{\bar{a}_{\langle i\rangle} : \alpha < i < \mu(*)\}$ is independent over $(N_{\langle\rangle}, N_{\langle\rangle} \cup \bigcup_{i<\lambda^+} N_i)$. Now rename $\bar{a}_{\langle\alpha+i\rangle}$ as $\bar{a}_{\langle i\rangle}$ (for $i < \mu(*)$), $\bar{a}_{\langle i\rangle}$ as $\bar{a}_{\langle \alpha_I(\langle\rangle)+i\rangle}$ for $i < \alpha$ (and make the corresponding renaming for $N_\nu, \bar{a}_\nu$). If $\mu(*) > \lambda^+$ is singular, the proof is similar. If $\mu(*) = \lambda^+$, define by induction on $i < \lambda^+$, $\beta(i) < \lambda^+$, $\gamma(i) < \lambda^+$ such that $\mathrm{tp}_*(N_{\beta(i)}, N_{\langle\rangle} \cup \bigcup_{j<i} N_{\langle\gamma(j)\rangle} \cup \bigcup_{j<i} N_{\beta(j)})$ does not fork over N and $\mathrm{tp}_*(N_{\langle\gamma(i)\rangle}), N_{\langle\rangle} \cup \bigcup_{j<i} N_{\langle\gamma(j)\rangle} \cup \bigcup_{j\leqslant i} N_{\beta(j)})$ does not fork over $N_{\langle\rangle}$ and then rename $\bar{a}_{\langle\gamma(i)\rangle}$ as $\bar{a}_{\langle i\rangle}, N_{\langle\gamma(i)\rangle}$ as $N_{\langle i\rangle}$, and $\mathrm{N}_{\beta(i)}$ as N_i, etc.]

Similarly in Case B, w.l.o.g. $\{N_i : i < \lambda^+\}$ is independent over $(N, N_{\langle\rangle} \cup \bigcup_{i<\mathrm{cf}\mu(*)} a_{\langle i\rangle})$ (remember $\mathrm{cf}\,\mu(*) > \lambda$).

Now we define M, M_i, $\bar{a}_i$ (for $i < \mu(*)$). If Case A, we let $M = N_{\langle\rangle}$, $\bar{a}_i = \bar{a}_{\langle i\rangle}$ and M_i be **T**-prime over $\bigcup\{N_\nu : \langle i\rangle \lessdot \nu\} \in I$. If Case B, we let $M \subseteq M^*$ be **T**-prime over $\bigcup_{i<\mathrm{cf}\,\mu(*)} N_{\langle i\rangle}$, for $i < \mu(*)$ let $j(i) < \mathrm{cf}\,\mu(*)$ be minimal such that $\mu_{j(i)} > i$, $\bar{a}_i = \bar{a}_{\langle i, j(i)\rangle}$ and M_i be **T**-prime over $M \cup \bigcup\{N_\nu : \langle j(i), i\rangle \lessdot \nu \in I\}$. Clearly, $\|M\|, \|M_i\|$ are $<\mu(*)$, and really $\|M\| \leqslant \lambda$ if Case A, $\|M\| = \lambda + \mathrm{cf}\,\mu(*)$, otherwise. Note that for $\gamma < \mu(*)$, $\sum_{i<\gamma} \|M_i\|$ is $<\mu(*)$. *Note*: $\{\bar{a}_i : i < \mu(*)\} \subseteq M^*$ is independent over M, M_i an $(M, \bar{a}_i)$-component of M^*.

Now we define by induction on $\gamma < \mu(*)$ an ordinal $i(\gamma)$ and a set $\boldsymbol{J}_\gamma \subseteq \boldsymbol{J} \stackrel{\mathrm{def}}{=} \{\bar{a} \in M^* : \mathrm{tp}(\bar{a}, M)$ is regular$\}$ such that (letting $\chi = \|M\|$)

(1) $i(\gamma) < (|\gamma| + \chi)^+$,

(2) $\bar{a}_{i(\gamma)} \in \boldsymbol{J}_\gamma$, $\|\boldsymbol{J}'_\gamma\| \leqslant \|M\| + \lambda + |\gamma|$ for every $\boldsymbol{J}'_\gamma \subseteq \boldsymbol{J}_\gamma$ independent over μ,

(3) $\bar{a}_{i(\gamma)}$ does not depend on $\bigcup_{\beta<\gamma} \boldsymbol{J}_\beta \cup B$ over M, where $B \stackrel{\mathrm{def}}{=} \bigcup_{i<\lambda} N_i$.

(4) $\boldsymbol{J}_\beta$ is a minimal subset of $\boldsymbol{J}$ such that:

(a) $\{\bar{a}_{i(\beta)}\} \cup \bigcup_{\gamma<\beta} \boldsymbol{J}_\gamma \subseteq \boldsymbol{J}_\beta$, and

(b) if $\bar{a}', \bar{a}'' \in \boldsymbol{J}$, and there is an automorphism F of M^* over M, $F(\bar{a}') = \bar{a}''$, then $\bar{a}' \in \boldsymbol{J}_\beta$ iff $\bar{a}'' \in \boldsymbol{J}_\beta$.

We shall show we can carry the induction.

Remark 3.5A. Note that there is an automorphism F of M^* over $M, F(\bar{a}') = \bar{a}''$ iff there are M', an $(M, \bar{a}')$-component of M^*, and M'', an $(M, \bar{a}'')$-component of M^* and an isomorphism F' from M' onto M''. $F' \restriction M =$ the identity and $F'(\bar{a}') = \bar{a}''$. [Why? The implication $\Rightarrow$ is trivial; so let us prove $\Leftarrow$. We can find $\alpha, M'_i, \bar{a}'_i (i < \alpha)$ such that $\{M'_i : i < \alpha\}$ is independent over $M, M \subseteq^* M'_i \subseteq^* M^*$, M prime over $\bigcup_{i<\alpha} M'_i$, M'_i an $(M, \bar{a}'_i)$-component of M^*, $M'_0 = M'$, and if $\{\bar{a}', \bar{a}''\}$ is independent, $M'_1 = M''$. If $\{\bar{a}', \bar{a}''\}$ is independent, let F be a mapping of $\bigcup_{i<\alpha} M'_i$ on itself, $F \restriction M'_i =$ the identity if $i > 1$, $F \restriction M'_0 = F'$, $F \restriction M'_1 = (F')^{-1}$. Clearly, F is well defined, elementary, hence can be extended to an automorphism of M^*. If $\{\bar{a}', \bar{a}''\}$ is not independent, define F by $F \restriction M'_i =$ the identity for $i > 0$, $F \restriction M'_0 = F'$, and when $\mathrm{tp}(\bar{a}', M)$ is regular we finish easily and this is the case which interests us. (For the general case, decompose M' and prove by induction on $w(\bar{a}, M)$.).] Hence, by $\neg(*)$: for every $\bar{a} \in \boldsymbol{J}$ and $\boldsymbol{J}' \subseteq \{\bar{a}' \in \boldsymbol{J}$; for some automorphism F of M^* over $M, F(\bar{a}) = \bar{a}'\}$ if $\boldsymbol{J}'$ is independent over M, then $|\boldsymbol{J}'| \leqslant \chi$. Let us do the induction step.

As $\{\bar{a}_\beta : \beta < (|\gamma| + \chi)^+\}$ is an independent subset of $\boldsymbol{J}$ over M and $|\boldsymbol{J}_\beta| \leqslant \|M\| + \chi^+ |\beta|$, there is no problem to choose $i(\gamma)$ for which (1), (3) hold. Then we can choose J_β satisfying (4) to which $\bar{a}_{i(\gamma)}$ belongs. The other half of (2) holds as (*) fails and 3.5A. Now for $i < \lambda^+$ let $M_{\mu(*)+i}$ be **T**-prime over $M \cup N_i$ and $\subseteq M^*$. Clearly, $M_{\mu(*)+i}$ $(i < \lambda^+)$ are pairwise isomorphic over M. We define M_i $(\mu(*)+\lambda^+ \leqslant i < \mu(*)+\mu(1))$ such that M_i is isomorphic to $M_{\mu(*)}$ over M and $\mathrm{tp}(M_i, M^* \cup \bigcup_{j<i} M_j)$ does not fork over M. So $\{M_i : i < \mu(*)+\mu(1)\}$ is independent over M, $\|M_i\| < \mu(*)$, $M_i \neq M$, each M_i (and M) is **T**-saturated; and $M \subseteq^* M_i$ (as $\langle N_\eta, \bar{a}_\eta : \eta \in I\rangle$ is a $(\mathbf{T}, \subseteq^*)$-decomposition of $M^*, N_{\langle\rangle} \subseteq^* M^*$, etc.). For each $S \subseteq \mu(*)$ let M_S^0 be **T**-prime over $\bigcup_{i\in S} M_{i(\gamma)} \cup \bigcup_{i<\lambda} M_{\mu(*)+i}$, $M_S^0 \subseteq M^*$, and then let M_S^1 be **T**-prime over $M_S^0 \cup \bigcup_{i<\mu(1)} M_{\mu(*)+\lambda+i}$. Clearly, each M_S^1 is **T**-saturated and has power $\mu(1)$.

It suffices to prove $\{S : M_S^1 \cong M_{S(0)}^1\}$ is $\leqslant \mu(*)^{\|M\|}$ (as $\mu(*)^{\|M\|} < 2^{\mu(*)}$ because $\mu(*)$ is not a strong limit and $(\forall\vartheta < \mu(*))2^\vartheta < 2^{\mu(*)}$). Let M_S^2 be **T**-prime over $M_S^0 \cup \bigcup_{i<\mu(*)} M_{\mu(*)+\lambda^++i}, M_S^2 \subseteq M_S^1$. Clearly, if $M_S^1 \cong M_{S(0)}^1$, then also $M_S^2 \cong M_S^2$. By VIII, 1.2 it suffices to prove that $|S(0)\backslash S(1)| > \chi \Rightarrow M_{S(0)}^2, M_{S(1)}^2$ are not isomorphic over M. So for some $\gamma, \gamma \in S(0)$, $\gamma \notin S(1)$, and F an isomorphism from $M_{S(0)}^2$ onto $M_{S(1)}^2$ over M, we have: $F(\bar{a}_{\langle i(\gamma)\rangle})$ belongs to the model M_S^0. So $F(M_{i(\gamma)})$ is an $(M, F(\bar{a}_{i(\gamma)}))$-component of $M_{S(1)}^2$. By 3.3A(1), w.l.o.g. F maps $M_{i(\gamma)}$ into $M_{S(1)}^0$, hence $F(M_{i(\gamma)})$ is an $(M, F(\bar{a}_{\langle i(\gamma)\rangle}))$-component of $M_{S(1)}^2$, hence of $M_{S(1)}^0$. Now, using Exercise XI, 3.4(2), $F(M_{i(\gamma)})$ is an $(M, F(\bar{a}_{i(\gamma)}))$-component of M^*. As $F(\bar{a}_{i(\gamma)}) \in M_{S(1)}^0$ there are $\gamma(1) < \gamma(2) < \ldots < \gamma(k) < \mu(*)$, such that $F(\bar{a}_{i(\gamma)})$ depends on $\{\bar{a}_{i(\gamma(1))}, \ldots, \bar{a}_{i(\gamma(k))}\} \cup B$ over M and $\{\gamma(1), \ldots, \gamma(k)\}$ is minimal with this property. By (4) above $\gamma > \gamma(k)$ (and $k = 0$) are impossible and by the choice of $\gamma, \gamma \neq \gamma(k)$, so $\gamma < \gamma(k)$. Now $\bar{a}_{i(\gamma)}$ does not depend on $\{\bar{a}_{i(\gamma(1))}, \ldots, \bar{a}_{i(\gamma(k-1))}\} \cup B$ over M. Hence $\bar{a}_{i(\gamma(k))}$ depends on $\{a_{i(\gamma(1))}, \ldots, \bar{a}_{i(\gamma(k-1))}\} \cup \{F(\bar{a}_{i(\gamma)}) \cup B$ over M. But $a_{i(\gamma(l))} \in J_{\gamma(l)}$, and $\bar{a}_{i(\gamma)} \in J_\gamma$ (check the definition of J_γ). By V, 1.14 $\bar{a}_{i(\gamma(k))}$ depends on $J_{\gamma(1)} \cup \ldots \cup J_{\gamma(k-1)} \cup J_\gamma \cup B \subseteq \bigcup_{\beta<\gamma(k)} J_\beta \cup B$ over M, a contradiction. Hence we finish.

Remark 3.5A. (1) In the case $\mu(*)$ a strong limit, use $S \subseteq \mu(*)$ of cardinality $\|M\|^+$.

(2) If $N \subseteq^* M^*$ (i.e. this is added to the assumption (**)), then it was enough to have N_i for $i < \lambda$.

LEMMA 3.6 [The axioms used in IX, 2.8]: *Assume T is superstable without the dop. Suppose $\lambda = \lambda(\mathbf{T}, \subseteq^*), 2^\mu > 2^\lambda, \mu > 2^{|T|}, \mu(1) > \mu$. Then $I(\mu, \mathbf{T}, T) \leqslant I(\mu(1), \mathbf{T}, T) \times |\{\langle \chi, \chi_1\rangle : \lambda \leqslant \chi < \chi_1 \leqslant \mu\}|$.*

Remark 3.6A. (1) $I(\mu, \mathbf{T}, T)$ is the number of **T**-saturated models of T of power μ, up to isomorphism.

(2) "$\mu > 2^{|T|}$" is needed only to get that (**) of 3.5 holds.

(3) For simplicity we assume that the demand in Definition IX, 2.3(3) holds for every $\lambda' \geqslant \lambda$ (as this is true in the cases we shall use 3.6).

Proof. As $\mu > 2^{|T|}$, every model of T of power μ satisfies (**) of 3.5. If some **T**-saturated model of T of power μ fails to satisfy (*) of 3.5, then by 3.5, $I(\mu(1), \mathbf{T}, T) \geqslant 2^{\mu}$, but trivially $2^{\mu} \geqslant I(\mu, T) \geqslant I(\mu, \mathbf{T}, T)$, so we finish. So we can assume that each **T**-saturated model of T of power μ satisfies (*). In (*), we can assume that $N_i, \bar{a}_i$ are defined for $i < \chi_1$ (for some $\chi_1, \chi < \chi_1 \leqslant \mu$), such that $\{(N_i, \bar{a}_i) : i < \chi_1\}$ is maximal (under the conditions there). It suffices to prove that for each pair (χ, χ_1), the number of **T**-saturated models of T of power μ for which we can get $\{(N_i, \bar{a}_i) : i < \chi_1\}$ as above (up to isomorphism) is $\leqslant I(\mu(1), \mathbf{T}, T)$. So let (χ, χ_1) be fixed, so $\lambda \leqslant \chi < \chi_1 \leqslant \mu$.

Let $\{\langle M^{\alpha}, N^{\alpha}, \langle N_i^{\alpha}, \bar{a}_i^{\alpha} : i < \chi_1 \rangle\rangle : \alpha < \alpha(*)\}$ be a maximal list such that M^{α}, M^{β} are not isomorphic for each $\alpha \neq \beta$, each M^{α} a **T**-saturated model of T of power μ, $N^{\alpha}, \langle N_i^{\alpha}, \bar{a}_i^{\alpha} : i < \chi_1 \rangle$ is as above (in 3.5(*)). It suffices to prove that $|\alpha(*)| \leqslant I(\mu(1), \mathbf{T}, T)$.

We can define for each α, for $\chi_1 < i < \mu(1)$ a model N_i^{α} isomorphic over N^{α} to N_0^{α}, such that $\mathrm{tp}(N_i^{\alpha}, M^{\alpha} \cup \bigcup_{j<i} N_j^{\alpha})$ does not fork over N^{α}. Lastly, let M_*^{α} be **T**-prime over $M^{\alpha} \cup \bigcup_{i<\mu(1)} N_i^{\alpha}$. Clearly, it is enough to prove that for $\alpha < \beta < \alpha(*)$, $M_*^{\alpha}, M_*^{\beta}$ are not isomorphic. For notational simplicity we let $\alpha = 0, \beta = 1$ so we assume from F is an isomorphism from M_*^0 onto M_*^1. Clearly, if F_l is an automorphism of M_*^l (for $l = 0, 1$), then we can replace F by $F_1 \circ F \circ F_0$. So, for example, w.l.o.g. there are $M_a^l \subseteq^* M_*^l, M_a^l$ **T**-prime over $M^l \cup \bigcup\{N_j^l : j < \mu + \mu\}$ such that F maps M_a^0 onto M_a^1.

By the Lowenheim–Skolem argument there are, for $l = 0, 1$ $W_l \subseteq W^l \subseteq \mu(1)$, $|W_l| = |W^l - W_l| = \chi$ and models $N_{\langle\rangle}^l$ such that: $N_{\langle\rangle}^l$ is **T**-saturated, $\|N_{\langle\rangle}^l\| = \chi$, $N_{\langle\rangle}^l \subseteq^* M_*^l, N_i^l \subseteq N_{\langle\rangle}^l$ for $i \in W_l$, $\{N_i^l : i < \mu(1), i \notin W_l\}$ is independent over $(N^l, N_{\langle\rangle}^l)$, F maps $N_{\langle\rangle}^0$ onto $N_{\langle\rangle}^1$. Moreover, by some manipulation (and $F(M_a^0) = M_a^1$) there are $N_{\langle\rangle}^{l,*} \subseteq M_*^l$ **T**-primary over $N_{\langle\rangle}^l \cup \bigcup_{i \in W^l} N_i^l$ such that F maps $N_{\langle\rangle}^{0,*}$ onto $N_{\langle\rangle}^{1,*}$. By Fact 3.4, w.l.o.g. $N_{\langle\rangle}^l \subseteq M^l$ and even $N_{\langle\rangle}^{l,*} \subseteq M^l$, and by its proof we can preserve even the last sentence. Let $\mathbf{I}_l^a \overset{\mathrm{def}}{=} \{\bar{b} \in M_*^l : \mathrm{tp}(\bar{b}, N_{\langle\rangle}^l)$ is regular$\}$. Clearly, F maps $\mathbf{I}_0^a$ onto $\mathbf{I}_1^a$. As in the proof of 3.3 it suffices to find $\mathbf{I}_l \subseteq \mathbf{I}_l^a \cap M^l$ independent over $N_{\langle\rangle}^l$, maximal, and for each

$\bar{b} \in \boldsymbol{I}_l$ an $(N^l_{\langle\rangle}, \bar{b})$-component $M^l_{\bar{b}}$ of M^l, and a one-to-one function h from $\boldsymbol{I}_0$ onto $\boldsymbol{I}_1$, such that for $\bar{b} \in \boldsymbol{I}_0$, $F \restriction N^0_{\langle\rangle}$ can be extended to an isomorphism from $M^0_{\bar{b}}$ onto $M^1_{\bar{b}}$. By 3.3 it suffices to find such $M^l_{\bar{b}}$ which are $(N^l_{\langle\rangle}, \bar{b})$-components of M^l_*; hence it is enough that some isomorphism $F_{\bar{b}}$ from M^0_* onto M^1_* extend $F \restriction N^0_{\langle\rangle}$ and maps $\bar{b}$ to $h(\bar{b})$. Let $\boldsymbol{I}^b_l = \{\bar{b} \in M^l_* : \mathrm{tp}(\bar{b}, N^l_{\langle\rangle}))$ be regular not orthogonal to $\mathrm{tp}(\bar{a}^l_i, N^l)$ for some (every) $i \in \mu(1) \setminus W^l\}$. Clearly, $\mathrm{tp}(\bar{a}^1_i, N^1)$, $\mathrm{tp}(F(\bar{a}^0_i), F(N^1))$ are not orthogonal [as every indiscernible set $\subseteq M^1_*$ orthogonal to $\mathrm{tp}(\bar{a}^1_0, N^1)$ has dimension $\leqslant \mu$], hence F maps $\boldsymbol{I}^b_0$ onto $\boldsymbol{I}^b_1$.

Now when we restrict ourselves to $\boldsymbol{I}^a_l \setminus \boldsymbol{I}^b_l$ $(l = 0, 1)$ we can as in the proof of 3.3 define appropriately $\boldsymbol{I}_l \cap (\boldsymbol{I}^a_l - \boldsymbol{I}^b_l) \subseteq M^l$ and $h \restriction (\boldsymbol{I}_l \cap (\boldsymbol{I}^a_l - \boldsymbol{I}^b_l))$. So we can concentrate on $\boldsymbol{I}^b_l (l = 0, 1)$. Let

$\boldsymbol{I}^c_l = \{\bar{b} \in M^l_* : \bar{b}$ realizes $\mathrm{tp}(\bar{a}^l_i, N^l_{\langle\rangle})$ for $i \notin W^l$ and there is an $(N^l_{\langle\rangle}, \bar{b})$-component of M^l_* isomorphic to the $\mathbf{T}$-prime model over $N^l_i \cup N^l_{\langle\rangle}$ (for $i \notin W^l$) over $N^l_{\langle\rangle}\}$.

Clearly, $\boldsymbol{I}^c_l \subseteq \boldsymbol{I}^b_l$ and let

$$\boldsymbol{I}^d_l = \{\bar{b} \in \boldsymbol{I}^b_l : \bar{b} \text{ depends on } \boldsymbol{I}^c_l \text{ over } N^l_{\langle\rangle}\}.$$

Now F maps $\boldsymbol{I}^d_0$ onto $\boldsymbol{I}^d_1$ [for this it suffices to prove that for every $\bar{b} \in \boldsymbol{I}^c_0, F(\bar{b})$ depends on $\boldsymbol{I}^c_1$ over $N^1_{\langle\rangle}$ and also that every $\bar{c} \in \boldsymbol{I}^c_1$ depends on $F(\boldsymbol{I}^c_0)$ over $N^1_{\langle\rangle}$. By the automorphisms M^l_* have over $N^l_{\langle\rangle}$ (see proof of 3.4) it suffices to prove this for at least one $\bar{b} \in \boldsymbol{I}^c_0, \bar{c} \in \boldsymbol{I}^c_1$. But the choice of $N^l_{\langle\rangle}{}^*$, W^l ensure this.]

Again by the use of $N^l_{\langle\rangle}{}^*$, it is clear how to define $\boldsymbol{I}_l \cap \boldsymbol{I}^d_l \subseteq M^l$ and $h \restriction (\boldsymbol{I}^a_l \cap \boldsymbol{I}^d_l)$. Lastly, we let $\boldsymbol{I}^e_l \subseteq \boldsymbol{I}^c_l \cap M^l$ be maximal subset, independent over $(N^l_{\langle\rangle}, N^l_{\langle\rangle} \cup \bigcup \boldsymbol{I}^d_l)$ and we can deal with them too.

THEOREM 3.7 (Morley conjecture): *Let T be a countable complete first-order theory. Then for $\lambda > \mu \geqslant \aleph_0$, $I(\lambda, T) \geqslant I(\mu, T)$ except when $\lambda > \mu = \aleph_0$, T is complete, $\aleph_1$-categorical, not $\aleph_0$-categorical.*

Remark. For not necessarily complete T, the conclusion holds except when $\lambda > \mu = \aleph_0$, every completion of T is $\aleph_1$-categorical (or have finite models only), there are only finitely many completions of T and at least one such completion is not $\aleph_0$-categorical (but have infinite models).

Proof. If T is not superstable, or superstable but with the dop or with the otop or is deep, then by XII, 6.1 we know that for $\lambda >$

$\aleph_0 I(\lambda,T) = 2^\lambda$ but $2^\lambda \geqslant 2^\mu \geqslant I(\mu,T)$, hence we finish. So we assume T is superstable without the dop without the otop and is shallow. We concentrate on non-$\aleph_0$-stable T. Let $\mu_0 = \mathrm{Min}\{\chi : 2^\chi > 2^{2^{\aleph_0}}\}$ (hence $(\mu_0 > 2^{\aleph_0})$). By IX, 1.20 for $\mu < \mu_0$, $I(\lambda,T) \geqslant \mathrm{Min}\{2^{2^{\aleph_0}}, 2^\lambda\} \geqslant \mathrm{Min}\{2^\mu, 2^\mu\} \geqslant I(\mu,T)$. So we assume $\mu \geqslant \mu_0$, and apply 3.5, 3.6 for $(\mathbf{T}, \subseteq^*) = (\mathbf{T}^t_{\aleph_0}, \subseteq^a_{\aleph_0})$. We see that together they give the conclusion for $\lambda > \mu \geqslant \mu_0$, provided that

$$(*)\qquad I(\lambda,T) \geqslant \aleph_0 + |\{(\chi,\chi_1) : 2^{\aleph_0} \leqslant \chi < \chi_1 \leqslant \mu\}|.$$

As T is $\aleph_0$-unstable, $I(\lambda,T) \geqslant \mathrm{Min}\{2^{2^{\aleph_0}}, 2^\mu\} \geqslant \aleph_0$. If T is not unidimensional, $I(\lambda,T) \geqslant |\{\chi : \chi \leqslant \mu\}|$ easily, hence by cardinal arithmetic $(*)$ holds. We deal with unidimensional T in detail in 4.15.

What about $\aleph_0$-stable T? For $\mu = \aleph_0$, if T is categorical in $\aleph_0$, this is trivial, otherwise by IX, 2.1 $I(\lambda,T) \geqslant \aleph_0$ for every λ, so the only non-trivial case is $I(\aleph_0,T) > \aleph_0$; but then by [SHM 84] $I(\lambda,T) \geqslant 2^{\aleph_0} = I(\aleph_0,T)$. For $2^\mu > 2^{\aleph_0}$, apply 3.5, 3.6, 3.6A(2) for $(\mathbf{T}, \subseteq^*) = (\mathbf{T}^t_{\aleph_0}, \subseteq_t)$, and again it suffices to prove

$$(*)'\qquad I(\lambda,T) \geqslant \aleph_0 + \{|(\chi,\chi_1) : \aleph_0 \leqslant \chi + \chi_1 \leqslant \mu\}|.$$

If T is multidimensional, this holds by Ex.X, 6.1; if T is not multidimensional, use IX, 2.4.

We are left with the case $2^\mu = 2^{\aleph_0}$. We can assume T not multidimensional (by IX, 2.4)). If $\lambda \geqslant \aleph_\omega$ use Ex.XI, 3.1 (which gives the exact number for λ and a corresponding upper bound for μ). If $\lambda < \aleph_\omega$, $\mathrm{Dp}(T) > 2$, by X, 6.2, $I(\lambda,T) \geqslant 2^{\aleph_0} = 2^\mu \geqslant I(\mu,T)$. If $\lambda < \aleph_\omega$, $\mathrm{Dp}(T) \leqslant 2$, repeat the analysis of [SHM 84] considering all regular types and get either $I(\mu,T) = \aleph_0 \leqslant I(\lambda,T)$ or $I(\mu,T) = 2^{\aleph_0} \leqslant I(\lambda,T)$.]

XIII.4. $I(\aleph_\alpha, T)$ for α large enough

(Hyp) T is superstable.

DEFINITION 4.1: (1) For any pair (N_0,N_1) of models of T such that $N_0 \subseteq N_1$ we define a cardinal $SND_\lambda(N_0,N_1)$ (=special number of dimensions) to be the number of models $N \in SE_\lambda(N_0,N_1)$ up to isomorphism over N_1, where $SE_\lambda(N_0,N_1)$ is the class of N satisfying

(*) $(N_1,b)_{b\in N_0} \subseteq^a_\lambda (N,b)_{b\in N_0}$, and for some $\bar a \in N$ $\mathrm{tp}(\bar a,N_1)$ is a regular type of depth zero orthogonal to N_0, and $\mathrm{tp}_*(N,N_1 \cup \bar a)$ is almost orthogonal to N_1.

(2) If $\bar a$ exemplifies $N \in SE_\lambda(N_0,N_1)$ we shall write also $(N,\bar a) \in SE_\lambda(N_0,N_1)$.

(3) We write $N \in SE_\lambda(N_1)$, $(N,\bar a) \in SE_\lambda(N_1)$ when (*) above holds omitting N_0 (and the orthogonality of $\mathrm{tp}(\bar a,N_1)$ to N_0). If $\lambda = |T|^+$, we omit it.

Remark. (1) Clearly, if λ increases, $SE_\lambda(N_0,N_1)$ decreases.

(2) For $\lambda = 0$, $(N_1,b)_{b\in N_0} \subseteq^a_\lambda (N,b)_{b\in N_0}$ becomes $N_1 \subseteq N$.

(3) We may consider $\subseteq^x_\lambda$ and then write SND^x_λ, SE^x_λ.

(4) Why are we restricting ourselves to "$\mathrm{Dp}(\mathrm{tp}(\bar a,N_1)) = 0$"? Because by 4.5, X, 7.2 the other cases are not so interesting.

LEMMA 4.2: (1) *If $N \in SE_{\aleph_0}(N_1)$, then N is a minimal extension of N_1.*

(2) *$N^* \in SE_{\aleph_0}(N_1)$, N_1 is $\mathbf{F}^a_\lambda$-saturated, then N^* is $\mathbf{F}^a_\lambda$-saturated.*

(3) *If $(N,\bar a) \in SE_0(N_1)$, then N is $\mathbf{F}^a_{\aleph_0}$-constructible over $N_1 \cup \bar a$ and for every $\bar b \in N, \bar b \notin N$, $\mathrm{tp}_*(N,N_1 \cup \bar b)$ is almost orthogonal to N_1 and* $\mathrm{Dp}(\bar b,N_1) = 0$. *If T is countable N is even $\mathbf{F}^k_{\aleph_0}$-constructible over $N_1 \cup \bar a$.*

(4) *If $N_1 \subseteq^a_{\aleph_0} N, \bar a \in N$, $\mathrm{tp}_*(N,N_1 \cup \bar a)$ is almost orthogonal to N, then N is $\mathbf{F}^a_{\aleph_0}$-atomic over $N_1 \cup \bar a$.*

Remark 4.2A. If T is countable, we can replace $SE_{\aleph_0}$ by SE_0 in (1), (2) (e.g. in (1) use $N^{**} \subseteq M$, $\mathbf{F}^l_{\aleph_0}$-constructible over $M_1 \cup N^*$). We can omit countability in (3) by III.

Proof. (1) Like XII, 4.17. Suppose N is not minimal over N_1, so there is $N^*, N_1 \subseteq N^* \subseteq N, N_1 \neq N^* \neq N$. Let $\lambda > \|N\| + |T|$ be regular, M_1 be an $\mathbf{F}^a_\lambda$-saturated extension of N_1, such that $\{M_1, N\}$ is independent over N_1. Now $\mathrm{tp}_*(N,N_1 \cup \bar a)$ is almost orthogonal to N_1, hence N is $\mathbf{F}^a_\lambda$-constructible over $M_1 \cup \bar a$ (remember $\lambda > \|N\|$). Let M be $\mathbf{F}^a_\lambda$-primary over $M_1 \cup N$, hence also over $M_1 \cup \bar a$. As $\mathrm{Dp}(\bar a,N_1) = 0$ every non-algebraic $p \in \bigcup_{m<\omega} S^m(M)$ is not orthogonal to M_1, hence for every $\bar c \in N, \bar c \notin N^*$, $\mathrm{tp}(\bar c,N^*)$ is not orthogonal to N_1. Remember $N_1 \subseteq^a_{\aleph_0} N$, hence by XI, 3.11 for some such $\bar c$ $\mathrm{tp}(\bar c,N^*)$ does not fork over N_1, and choose $\bar b \in N^*, \bar b \notin N$. As M is $\mathbf{F}^a_\lambda$-primary over $M_1 \cup \bar a$, $\mathrm{tp}(\bar a,M_1)$ is regular by V, 3.3, M is $\mathbf{F}^a_\lambda$-primary over $M_1 \cup \bar b$. By III, 4.22 it is also $\mathbf{F}^a_{\aleph_0}$-atomic over $M_1 \cup \bar b$, hence $\mathrm{tp}(\bar c,M_1 \cup \bar b)$ is $\mathbf{F}^a_{\aleph_0}$-isolated. But $\mathrm{tp}(\bar c,M_1 \cup N^*)$ does not fork over M_1, hence over N_1. So by IV, 4.3 $\mathrm{tp}(\bar c,N_1)$ is $\mathbf{F}^a_{\aleph_0}$-isolated. But this contradicts $N_1 \subseteq^a_{\aleph_0} N, \bar c \in N, \bar c \notin N_1$.

(2) Let $\mu > \|N^*\| + |T|^+$, $(N^*, \bar{a}) \in SE_{\aleph_0}(N_1)$ and let N be $\mathbf{F}^a_\lambda$-primary over $|N^*|$, so we assume $N \neq N^*$ and essentially get a contradiction. Let M_1 be an $\mathbf{F}^a_\mu$-saturated extension of N_1, $\text{tp}_*(M_1, N)$ does not fork over N_1. Easily N^* is $\mathbf{F}^a_\mu$-constructible over $M_1 \cup \bar{a}$, N is $\mathbf{F}^a_\mu$-constructible over $M_1 \cup N^*$, and there is an $\mathbf{F}^a_\mu$-primary model M over $M_1 \cup N$, hence over $M_1 \cup \bar{a}$. From this we see that for every $\bar{c} \in N$, $\bar{c} \notin N^*$, $\text{tp}(\bar{c}, N^*)$ is not orthogonal to N_1, hence by XI, 3.11 for some such $\bar{c} \in N$, $\bar{c} \notin N^*$ and $\text{tp}(\bar{c}, N^*)$ does not fork over N_1. As N is $\mathbf{F}^a_\lambda$-primary over N^*, $\text{tp}(\bar{c}, N^*)$ is $\mathbf{F}^a_\lambda$-isolated, and as N_1 is $\mathbf{F}^a_\lambda$-saturated, we get a contradiction to IV, 4.3.

(3) Let A be a maximal subset of N, which includes $N_1 \cup \bar{a}$ and is $\mathbf{F}^a_{\aleph_0}$-constructible over $N_1 \cup \bar{a}$. We assume $A \neq N$ and get a contradiction. Let μ, M_1, M be as in the proof of 4.2(2). By 4.2(2) M is $\mathbf{F}^a_{\aleph_0}$-saturated; over $M_2 \cup A$ there is an $\mathbf{F}^a_{\aleph_0}$-primary model M^2, hence w.l.o.g. $M^2 \subseteq M$. Clearly, $M \in SE_{\aleph_0}(M_1)$, hence by 4.2(1) $M^2 = M$. But M^2 is $\mathbf{F}^a_{\aleph_n}$-atomic over $M_1 \cup A$. Hence for every $\bar{c} \in N$, $\text{tp}(\bar{c}, M_1 \cup A)$ is $\mathbf{F}^a_{\aleph_0}$-isolated. This is a contradiction to $A \neq N$ (and the maximality of A).

So we have proved "N is $\mathbf{F}^a_{\aleph_0}$-constructible over $N_1 \cup \bar{a}$". If T is countable, we can find a maximal $A \subseteq N$ including $N_1 \cup \bar{a}$ and $\mathbf{F}^l_{\aleph_0}$-constructible over $N_1 \cup \bar{a}$. We can find $M^2 \subseteq M$ which is $\mathbf{F}^l_{\aleph_0}$-constructible over $M_1 \cup \bar{a}$, hence $\mathbf{F}^k_{\aleph_0}$-atomic over A (see XII, 1.2) and continue as above.

The second conclusion to 4.2(3) can be similarly proved using V, 3.3.

(4) Similar to the above.

CONCLUSION 4.3: (1) *Suppose T has the $(<\infty, 2)$-existence property. If $N \in SE_{\aleph_0}(N_1)$, $M_1 \subseteq N_1$, $M_1 \subseteq M$, and $M \subseteq N$, $\{M, N_1\}$ independent over M_1, $M \neq M_1$, then N is primary atomic and minimal over $M \cup N_1$.*

(2) *If also $M \in SE_0(M_1)$, N atomic over $M \cup N_1$, then $N \in SE_0(N_1)$, if in addition $M \in SE_{\aleph_0}(M_1)$, then $N \in SE_{\aleph_0}(N_1)$.*

(3) *If in addition $M \in SE_\lambda(M_1)$, N primary over $M \cup N_1$, then $N \in SE_\lambda(N_1)$.*

Proof. (1) By the hypothesis there is N^* primary and atomic over $M \cup N_1$, so w.l.o.g. $N^* \subseteq N$. But by 4.2(1) $N^* = N$, so N is primary and atomic over $M \cup N_1$. Again by 4.2(1) N is minimal over $M \cup N_1$.

(2), (3) Easy.

LEMMA 4.4: *Suppose T has the $(<\infty, 2)$-existence property but not the dop. Suppose, further, $N_1 \subseteq N^l \subseteq N (l = 1,2), N_1 \subseteq^a_{|T|^+} N$, and $(N^l, \bar{a}^l) \in SE(N_1)$ (for $l = 1,2$). If $\{\bar{a}^1, \bar{a}^2\}$ is not independent over N_1 or even $N_1 \subseteq A \subseteq N$, $\mathrm{tp}(\bar{a}^l, A)$ does not fork over $N_1 (l = 1,2)$ but $\{\bar{a}_1, \bar{a}_2\}$ is not independent over A, then N^1, N^2 are isomorphic over N_1.*

Proof. Clearly, $\mathrm{tp}_*(A, N^l)$ does not fork over N_1. By XI, §2 there is $M \subseteq N, N_1 \subseteq A \subseteq M$, $\mathrm{tp}_*(M, N^1)$ does not fork over N_1 and N is primary over $M \cup N^1$. Clearly, $\mathrm{tp}(\bar{a}^2, M)$ does not fork over N_1 (by V, 1.14): otherwise as $\mathrm{tp}(\bar{a}^1, A \cup \bar{a}^2) \subseteq \mathrm{tp}(\bar{a}^1, M \cup \bar{a}^2)$ forks over N_1, also $\mathrm{tp}(\bar{a}^1, M)$ forks over N_1, a contradiction. Hence $\mathrm{tp}_*(N^2, M)$ does not fork over N_1, and by XI, §2, $M \subseteq^a_{|T|^+} N$, N is primary over $M \cup N^2$ too. Let λ be large enough regular cardinal so that $T, N, M, N_1, N^1, N^2 \in H(\lambda)$. Let $\mathfrak{A}$ be an elementary submodel of $(H(\lambda), \in)$ of power $|T|$ which includes $\{i : i < |T|\} \cup \{|T|, N, M, N_1, N^1, N^2, \bar{a}^1, \bar{a}^2\}$. Clearly by 4.3, N^l is primary (and atomic) over $N_1 \cup (N^l \cap \mathfrak{A})$ (for $l = 1,2$). Also $N \cap \mathfrak{A}$ is primary over $(M \cap \mathfrak{A}) \cup (N^1 \cap \mathfrak{A})$: this is proved as follows: as there is a construction $\langle M \cup N_1, \langle a_i : i < \alpha \rangle\rangle$ of N over $M \cup N_1$, there is such a construction in $\mathfrak{A}$. Clearly, $\langle (M \cap \mathfrak{A}) \cup (N^1 \cap \mathfrak{A}), \langle a_i : i \in \alpha \cap \mathfrak{A} \rangle\rangle$ is a construction of $N \cap \mathfrak{A}$. Clearly, $\{M \cap \mathfrak{A}, N^1 \cap \mathfrak{A}\}$ is independent over $N_1 \cap \mathfrak{A}$, and $N_1 \cap \mathfrak{A} \prec M \cap \mathfrak{A}, N^1 \cap \mathfrak{A} \prec N$.

So $N \cap \mathfrak{A}$ is primary over $(M \cap \mathfrak{A}) \cup (N_1 \cap \mathfrak{A})$.

Similarly $N \cap \mathfrak{A}$ is primary over $(M \cap \mathfrak{A}) \cup (N^2 \cap \mathfrak{A})$.

Now there is an elementary mapping f from $M \cap \mathfrak{A}$ into N_1 which is the identity over $N_1 \cap \mathfrak{A}$ (as $N_1 \subseteq^a_{|T|^+} N$, $\|\mathfrak{A}\| \leqslant |T|$) and let $M^* = \mathrm{Rang}\, f$. We can extend f to f_l by the identity on $N^l \cap \mathfrak{A}$ (and it is still elementary, by the non-forking) and then extend it to an elementary mapping g_l from $N \cap \mathfrak{A}$ into N^l [as $N \cap \mathfrak{A}$ is primary over $(N^l \cup M) \cap \mathfrak{A}$]. We know that $M^l \stackrel{\mathrm{def}}{=} \mathrm{Rang}(g_l)$ is atomic over $M^* \cup (N^l \cap \mathfrak{A})$, and easily $M^* \cup (N^l \cap \mathfrak{A}) \subseteq_t N_1 \cup (N^l \cap \mathfrak{A})$, hence $\mathrm{tp}_*(\mathrm{Rang}(g_l), N_1)$ does not fork over M^*. Clearly, $\{M^l, N_1\}$ is independent over M^*, and by 4.3 N^l is primary over $M^l \cup N_1$. As $g_2 \circ g_1^{-1}$ is an isomorphism from M^1 into M^2, over M^*, it can be extended to an isomorphism from N^1 into N^2 over N_1. By 4.2 it is onto N^2, so N^1, N^2 are isomorphic over N_1.

LEMMA 4.5: *Suppose $(N^l, \bar{a}^l) \in SE(N_1) (l = 1,2)$ and $\mathrm{tp}(\bar{a}^1, N_1)$, $\mathrm{tp}(\bar{a}^2, N_1)$ are not orthogonal, and they are not trivial. Then N^1, N^2 are isomorphic over N_1.*

Proof. By 4.2(3), w.l.o.g. $\bar{a}^l = \langle b^l \rangle$, and for every $c^l \in N^l - N_1$, $R[\mathrm{tp}(c^l,$

$N_1), L, \infty] \geqslant R[\text{tp}(b^l, N_1), L, \infty]$. We also can find $\varphi_l(x; \bar{c}^l) \in \text{tp}(b^l, N_1)$, such that $\alpha_l \overset{\text{def}}{=} R[\varphi_l(x, \bar{c}^l), L, \infty] = R[\text{tp}(b^l, N_1), L, \infty]$. W.l.o.g. $\alpha_1 \leqslant \alpha_2$.

Next notice that $\text{tp}(b^1, N_1), \text{tp}(b^2, N_1)$ are not weakly orthogonal. This is because we can define by induction on $i < \omega$ $b_i^1, b_i^2 \in N_1$, such that b_i^1 realizes $\text{stp}(b^1, \bar{c}^1 \cup \bar{c}^2 \cup \bigcup_{j<i} b_j^1 \cup \bigcup_{j<i} b_j^2)$ and b_i^2 realizes $\text{stp}(b^2, \bar{c}^1 \cup \bar{c}^2 \cup \bigcup_{j\leqslant i} b_j^1 \cup \bigcup_{j<i} b_j^2)$. Now apply, V, 2.7.

Now note that by XI, 1.9 there is $b' \in N^2$ such that $\models \varphi_1[b', \bar{c}^1]$ and $b' \notin N_1$, hence $\alpha_2 \leqslant R[\text{tp}(b', N_1), L, \infty] \leqslant R[\varphi_1(x, \bar{c}^1), L, \infty] = \alpha_1$, so necessarily $\alpha_1 = \alpha_2$ and w.l.o.g. $\bar{c}^1 = \bar{c}^2, \varphi_1 = \varphi_2$, so let $\bar{c} = \bar{c}^l, \varphi = \varphi_l$.

W.l.o.g. $\text{tp}(N^1, N^2)$ does not fork over N_1, and let N be primary and atomic over $N^1 \cup N^2$ (exists by the $(<\infty, 2)$-existence property). By XI, §§2, 3, $N_1 \subseteq^a_{|T|^+} N$. By Fact XI, 1.4 (and X, 7.1, like XI, 1.8) there is $b^3 \in N, \models \varphi[b^3, \bar{c}]$ such that $\{b^1, b^3\}$, and $\{b^2, b^3\}$ are independent over N_1 but $\{b^1, b^2, b^3\}$ is not. By XI, §2 (applied to $\subseteq^a_{\aleph_0}$) there is $N^3 \subseteq N$ such that $N_1 \cup \{b^3\} \subseteq N^3$ and $\text{tp}_*(N^3, N_1 \cup \{b^3\})$ is almost orthogonal to N_1. As $N_1 \subseteq^a_{|T|^+} N$ also $N_1 \subseteq^a_{|T|^+} N^3$. We want to apply 4.4 twice: first to prove that N^1, N^3 are isomorphic over N_1 and then that N^2, N^3 are isomorphic over N_1. The only missing point is "$\text{tp}(b^3, N^1)$ is regular". If this fails, then by XI 3.10 (formally, its proof) there is $b^4 \in \varphi(N, \bar{c}) - N_1$, such that $\text{tp}(b^4, N_1)$ forks over $\bar{c}$, hence $R[\text{tp}(b^4, N_1), L, \infty] < \alpha_1$. There are $\bar{d}^1 \in N^1, \bar{d}^2 \in N^2$ such that $\text{tp}(b^4, N^1 \cup N^2) \in \mathbf{F}^t_{\aleph_0}(\bar{d}_1 \cup \bar{d}_2)$ and so necessarily $\text{tp}(\bar{d}_1, N_1 \cup \bar{d}_2 \cup \{b^4\})$ forks over N_1. W.l.o.g. $\text{tp}(\bar{d}_1 {}^\frown \bar{d}_2 {}^\frown \langle b^4 \rangle, N_1)$ does not fork over $\bar{c}$, and we can easily find $\bar{d}' \in N_1$ realizing $\text{stp}(\bar{d}_2, \bar{c})$ hence $\text{tp}(\bar{d}', \bar{c} \cup \bar{d}_1) = \text{tp}(\bar{d}_2, \bar{c} \cup \bar{d}_1)$; then find (see XI, 1.4) $b^5 \in N^1$ such that $\models \varphi[b^5, \bar{c}]$ and $\text{tp}(\bar{d}_1, \bar{c} \cup \bar{d}' \cup \{b^5\})$ forks over $\bar{c}$, hence $R[\text{tp}(b^5, \bar{c} \cup \bar{d}_1 \cup \bar{d}'), L, \infty] < \alpha_1$. So b^5 contradicts the choice of α_1.

DEFINITION 4.6: Assume T has finite depth. Let $N_0 \subseteq^a_{|T|^+} N_1$, and we define

(1) $n(N_0, N_1)$ is the maximal n for which there are $N_l (2 \leqslant l \leqslant n+1)$ such that for $l \geqslant 1$, $(N_l, b)_{b \in N_{l-1}} \subseteq^a_{|T|^+} (N_{l+1}, b)_{b \in N_{l-1}}$, and there is $a_l \in N_{l+1} - N_l$, $\text{tp}(a_l, N_l)$ is regular orthogonal to N_{l-1} and $\text{tp}_*(N_{l+1}, N_l \cup a_l)$ is almost orthogonal to N_l.

(2) $snd(N_0, N_1)$ is the first $\kappa \geqslant 1$ such that for every $N_l (2 \leqslant l \leqslant n(N_0, N_1)+1)$ as in (1), $SND(N_{n-1}, N_n) < \kappa$.

(3) $k(N_0, N_1)$ is the maximal k such that $k < n(N_0, N_1)$ and there are $N_l (2 \leqslant l \leqslant k+1)$ as in (1) such that for every $\lambda < \kappa$ there are $N_l (k+1 < l \leqslant n(N_0, N_1)+1))$ such that $\langle N_l : 2 \leqslant l \leqslant n(N_0, N_1)+1 \rangle$ is as in (1) and $SND(N_{n-1}, N_n) \geqslant \lambda$.

(4) We define $l(N_0, N_1)$ as $n(N_0, N_1) - k(N_0, N_1)$.

DEFINITION 4.7: (1) We define $n(N_1)$, $snd(N_1)$, $k(N_1)$, $l(N_1)$ as in 4.6, just omitting the requirements connected with N_0.

(2) We define $n(T_1)$ as the maximal $n(N_1)$, we define $snd(T)$ as the supremum of $\{snd(N_1) : n(N_1) = n(T)\}$. If this supremum is obtained, $k(T)$ is the maximal

$$\{k(N_1) : n(N_1) = n(T_1), snd(N_1) = snd(T)\}$$

and $l(T) = n(T) - k(T)$.

If the supremum is not obtained, $k(T) = 0, l(T) = n(T)$.

MAIN THEOREM 4.8: *Suppose T is superstable without the dop but with the $(\infty, 2)$-existence property and of finite depth. Then for any $\aleph_\alpha$ such that $\alpha \geqslant 2^{2^{|T|}}$,*

$$I(\aleph_\alpha, T) = \beth_{k(T)}\left[\sum_{\lambda < snd(T)} \beth_{l(T)}(|\alpha + 1|^\lambda)\right].$$

This section up to 4.13 is devoted to the proof of this theorem. So we assume its hypothesis, and 4.11 and 4.12 implies it easily.

FACT 4.9: (1) *In Def. 4.6, 4.7 let $N_0 \subseteq^a_{|T|^+} N_1$; if $n(N_0, N_1) = 0$, then $snd(N_0, N_1)$, $k(N_0, N_1)$, $l(N_0, N_1)$ are not defined, if $n(N_0, N_1) > 0$, then $snd(N_0, N_1)$, $k(N_0, N_1)$, $l(N_0, N_1)$ are defined, $0 \leqslant k(N_0, N_1) < n(N_0, N_1)$. If $n(N_1) = 0$, $snd(N_0)$, $k(N_0)$, $l(N_0)$ are not defined; if $n(N_1) > 0$, then they are defined and $0 \leqslant k(N_1) < n(N_1)$. Lastly, $n(T) = \mathrm{Dp}(T) \geqslant 0$.*

(2) $1 \leqslant snd(N_0, N_1) \leqslant (2^{|T|})^+$, $1 \leqslant snd(N_1) \leqslant (2^{|T|})^+$ *[but if N is $\mathbf{F}^a_\kappa$-saturated or just for some $N_2, N_1 \subseteq^a_{|T|^+} N_2$, $N_1 \neq N_2$, then $2 \leqslant snd(N_1)$] and $2 \leqslant snd(T) \leqslant (2^{|T|})^+$, if $n(N_0, N_1) = 1$, then $snd(N_0, N_1) = SND(N_0, N_1)^+$.*

(3) *In Def. 4.6, 4.7 we can restrict ourselves to models of power $\leqslant 2^{|T|}$, without changing the results (for those models and for T).*

Of course in 4.6, 4.6(2) we should assume $\|N_1\| \leqslant 2^{|T|}$.

(4) *If $N_0 \subseteq^a_{|T|^+} N_1$, $\bar a \in N_1$, $\mathrm{tp}_*(N_1, N_0 \cup \bar a)$ is almost orthogonal to N_0, then $n(N_0, N_1) \leqslant \mathrm{Dp}(N_0, N_1)$. (If N_0, N_1 are $\mathbf{F}^a_\kappa$-saturated, equality holds.)*

LEMMA 4.10: (1) *Suppose $N_0 \subseteq^a_{|T|^+} N_1$, $\|N_1\| \leqslant 2^{|T|}$, $n(N_0, N_1) = \mathrm{Dp}(N_0, N_1) \geqslant 1$ and $\aleph_\alpha \geqslant 2^{|T|}$. Then the number of models in $K = K^{\aleph_\alpha}_{(N_0, N_1)} = \{N : N_1 \subseteq^a_{|T|^+} N, \|N\| = \aleph_\alpha$, and $\mathrm{tp}_*(N, N_1)$ orthogonal to $N_0\}$ up to isomorphism over N_1 is at most $\beth_{\mathrm{Dp}(N_0, N_1)}(|\omega + \alpha|^{2^{|T|}})$.*

(2) *If in addition* $n(N_0, N_1) = \mathrm{Dp}(N_0, N_1)$, *then the number is at most*

$$\beth_{k(N_0,N_1)}\Big[\sum_{\substack{\lambda < snd(N_0,N_1)\\ \text{or } \lambda = 1}} \beth_{l(N_0,N_1)}(|\omega+\alpha|^\lambda + 2^{2^{|T|}})\Big].$$

Proof. We prove this by induction on $n = \mathrm{Dp}(N_0, N_1)$ (simultaneously for (1) and (2)). We concentrate on the proof of (2) ((1) is easier). Let $N \in K$. If $n(N_0, N_1) = 1$, then $k(N_0, N_1) = 0$, $l(N_0, N_1) = 1$ by XI, 2.6(2), 2.8 there are $(N^i, \bar{a}^i) \in SE(N_0, N_1)$ for $i < \alpha$, such that $\{N^i : i < \alpha\}$ is independent over N_1, $\mathrm{tp}(\bar{a}^i, N_1)$ is regular orthogonal to N_0, $\mathrm{tp}_*(N^i, N_1 \cup \bar{a}^i)$ is almost orthogonal to N_1, $N_1 \subseteq N^i$ and N is primary over $\bigcup_i N^i$. So N is determined by the function giving for each isomorphism type of $N \in SE(N_0, N_1)$ over N, how many times it appears among the N_i's. This function has domain of power $SND(N_0, N_1)$ and range $\{\lambda : 0 \leqslant \lambda \leqslant \aleph_\alpha\}$ which has power $|\alpha + \omega|$. As by 4.9(3) $snd(N_0, N_1) = SND(N_0, N_1)^+$, we finish.

Next let $n(N_0, N_1) = 2$.

By XI, 2.6(2), 2.8 there are $(N^i, \bar{a}^i) \in SE(N_0, N_1)$ for $i < \alpha$ such that $\{N^i : i < \alpha\}$ is independent over N_1, $\mathrm{tp}(\bar{a}^i, N_1)$ is regular orthogonal to N_0, $\mathrm{tp}_*(N^i, N_1 \cup \bar{a})$ is almost orthogonal to N_1, $N_1 \subseteq^a_{|T|^+} N^i$ and N is primary over $\bigcup_{i<\alpha} N^i$. Clearly $\mathrm{Dp}(\bar{a}^i, N_1) \leqslant 1$ and w.l.o.g. $\mathrm{Dp}(\bar{a}^i, N_1) = 1$ iff $i < \beta$. The number of isomorphism types of N^i over N_1 ($\beta \leqslant i < \alpha$) is $\leqslant 2^{|T|}$ (see 4.3(1)) and the number of times each appears belongs to $\{\chi : 0 \leqslant \chi \leqslant \aleph_\alpha\}$, so the number of possible isomorphism types of $\bigcup_{\beta \leqslant i < \alpha} N^i$ over N_1 is at most

$$|\omega + \alpha|^{2^{|T|}}.$$

Now let $i < \beta$ and apply XI, 2.16 for $(\mathbf{T}^t_{\aleph_0}, \subseteq_a)$ (so $N_1 \subseteq N_{\langle\rangle} \subseteq_a N_i$). There are $\leqslant 2^{|T|}$ possible isomorphism types for $N_{\langle\rangle}$, hence $\leqslant \beth_1(|\alpha+\omega|^{\lambda(i)} + 2^{|T|})$ possible isomorphism types of N^i over N_1, and $\lambda(i) = SND(N_1, N^i_2)$ for suitable N^i_2. So the number of isomorphism types of $\bigcup_{i<\beta} N^i$ over N_1 is the number of functions to $\{\chi : 0 \leqslant \chi \leqslant \aleph_0\}$ from the set of isomorphism types of N^i over N_1, which we have bound above, a bound $\geqslant |\{\chi : 0 \leqslant \chi \leqslant \aleph_0\}|$, hence the number is

$$\leqslant \beth_1\Big[\sum_{\lambda < snd(N_0,N_1)} \beth_1(|\alpha+\omega|^\lambda + 2^{|T|})\Big].$$

Together with the bound on the number of possible isomorphism types of $\bigcup_{\beta \leqslant i < \alpha} N^i$ we get the required bound, when $k(N_0, N_1) = 1$. If

$k(N_0, N_1) = 0$, then $\sup_{i<\alpha} \lambda_i < snd(N_0, N_1)$ and we get $\sum_{\lambda < snd(N_0, N_1)} \beth_2(|\alpha+\omega|^\lambda + 2^{|T|})$ implying the condition.

For $n > 2$ choose $N^i, \bar{a}^i$ as above and w.l.o.g. $\mathrm{Dp}(\bar{a}^i, N_1) = n-1$ iff $i < \beta$. Choose $N_2^i \subseteq^a_{|T|^+} N^i, N_1 \cup \bar{a}_i \subseteq N_2^i, \|N_2^i\| \leqslant 2^{|T|}$. The number of possible isomorphism types of $N_2^i, \beta \leqslant i < \alpha$, is $\leqslant \beth_2(|T|)$ and for each the number of possible isomorphism types of N^i over it is $\leqslant \beth_{n-1}(|\omega+\alpha|^{2^{|T|}})$ (by the induction hypothesis and 4.9(2)). The number of possible isomorphism types of N_2^i over N_1 is $\leqslant \beth_2(|T|)$ and the number of possible isomorphism types of N^i over N_2^i for $i < \alpha$ is

$$\leqslant \beth_{k(N_1, N_2^i)}\left[\sum_{\lambda < snd(N_1, N_2^i)} \beth_{l(N_1, N_2^i)}(|\omega+\alpha|^\lambda + 2^{|T|})\right].$$

The rest is easy, too.

LEMMA 4.11: (1) *Suppose* $\|N_1\| \leqslant 2^{|T|}, \aleph_\alpha \geqslant 2^{|T|}$. *Then the number of models in*

$$K = \{N, N_1 \subseteq^a_{|T|^+} N, \|N\| = \aleph_\alpha\}$$

up to isomorphism over N_1 *is at most* $\beth_{\mathrm{Dp}(T)}(|\omega+\alpha|^{2^{|T|}})$, when $n(N_1) = \mathrm{Dp}(T)$ is $\leqslant$

$$\beth_{k(N_1)}\left[\sum_{\lambda < snd(N_1)} \beth_{l(N_1)}(|\omega+\alpha|^\lambda + 2^{|T|})\right].$$

(2) $$I(\aleph_\alpha, T) \leqslant \beth_{k(T)}\left[\sum_{\lambda < snd(T)} \beth_{l(T)}(|\omega+\alpha|^\lambda + 2^{|T|})\right].$$

Proof. Just like 4.10.

LEMMA 4.12: *Suppose* $|T| = \aleph_\beta, \beta < \alpha$. *Then*

(1) $I(\aleph_\alpha, T) \geqslant \beth_{k(T)}\left[\sum_{\lambda < snd(T)} \beth_{l(T)}(|\alpha-\beta|^\lambda)\right]$ if $|\alpha-\beta|^+ \geqslant snd(T)$.

(2) $I(\aleph_\alpha, T) \geqslant \beth_{n(T)-2}(|\alpha| + \aleph_0)$ *if* $n(T) \geqslant 2$.

(3) $I(\aleph_\alpha, T) \geqslant \beth_{n(T)-1}(|\alpha-\beta|^\lambda)$ *when* $\lambda < snd(T)$.

Proof. Straightforward (remembering X, §6, Exercise XI, 3.2 (we shall not use in (2) the "$+\aleph_0$")).

EXAMPLE 4.13: *For every* $k, l < \omega$ *and cardinal* $\kappa \geqslant \aleph_0$ *there is a totally transcendental theory* $T, |T| = \kappa$ *such that for* $\aleph_\alpha \geqslant \kappa^+ + \aleph_\alpha$

$$I(\aleph_\alpha, T) = \beth_l\left[\sum_{\lambda<\kappa} \beth_k(|\alpha|^\lambda)\right].$$

Proof. There is a theory T^0_λ such that $|T^0_\lambda| = \lambda, I(\aleph_\alpha, T^0_\lambda) = |\alpha|^\lambda$ (just λ pairwise disjoint infinite monadic predicates). So it is easy to find $T^l_\lambda\, I(\aleph_\alpha, T^l_\lambda) = \beth_l(|\alpha|)^\lambda)$. By disjoint sums of $|\{\chi : \chi < \kappa\}|$ such theories we find $T^k_{<\kappa}, |T^k_{<\kappa}| \leqslant \bigcup_{\mu<\kappa} \mu^+ + \aleph_0, I(\aleph_\alpha, T^k_{<\lambda} = \sum_{\lambda<\kappa} \beth_k(|\alpha|^\lambda)$. Now it is easy to define $T^{k,l}_{<\kappa}$ as required.

DISCUSSION 4.14: (1) So in 4.12(1) equality holds when T (is superstable and) with $(<\infty, 2)$-existence without dop otop and of finite depth, and $\alpha \geqslant \beta + 2^{|T|}$.

So for T countable $\alpha \geqslant 2^{\aleph_0}$, we can compute $I(\aleph_\alpha, T)$ from few invariants of T. But what can $snd(T)$ be? So we are stuck with the continuum hypothesis type problem, i.e. whether $snd(T) > \aleph_1 \Rightarrow snd(T) = (2^{\aleph_0})^+$. Of course, if $2^{\aleph_0} = \aleph_1$ this holds, and if we start with $V \models G.C.H.$ and add κ, $(\mathrm{cf}\,\kappa > \aleph_0)$ generic reals, still $snd(T) > \aleph_2 \Rightarrow snd(T) = (2^{\aleph_0})^+$. If we translate our problem to descriptive set theory we get as a sufficient condition:

> Assume there is a set A of reals $|A| > \aleph_0$, and $(\forall r_0 r_1 \ldots \in A([\bigwedge_l r_0 \notin B_l(r_0, r_1, \ldots)]$ where B_l is an analytic function. Is there such $A, |A| = 2^{\aleph_0}$?

THEOREM 4.15: *For every countable theory exactly one of the following occurs:*

(i) $I(\aleph_\alpha, T) = 1$ *for every* $\alpha > 0$,

(ii) $I(\aleph_\alpha, T) = \beth_2$, (*i.e.* Min$\{2^{\aleph_\alpha}, \beth_2\}$) *for every* $\alpha > 0$,

(iii) $I(\aleph_\alpha, T) \geqslant |\alpha + 1|$, *for every* α.

Proof. By XII, 6.1 we can assume that T is superstable, without the dop and otop, and shallow. If T is totally transcendental, then if T is unidimensional by IX, 1.8(2), Case (i) holds, and if T is not unidimensional by IX, 2.1, Case (iii) holds. If T is not unidimensional (not necessarily totally transcendental), Case (iii) holds by V, 2.10(2) (and Def. V, 2.2).

So we can assume that T is not $\aleph_0$-stable, is unidimensional, hence has depth 1. By IX, 1.20 $I(\aleph_\alpha, T) \geqslant$ Min$\{2^{\aleph_\alpha}, \beth_2\}$ for $\alpha > 0$. Now if $snd(T) > 2$, by 4.12 $I(\aleph_\alpha, T) \geqslant |\alpha|$ when $\alpha \geqslant 2^{\aleph_0}$, hence together with the previous sentence implies (iii). However, if $snd(T) \leqslant 2$, it is 2, and by 4.11 $I(\aleph_\alpha, T) \leqslant 2^{2^{|T|}} = \beth_2$ so (ii) necessarily holds. [As an alternative to the use of 4.12 we can note that if there is (for T) a regular non-trivial type, then every regular type is not orthogonal to it (as T is unidimensional), hence is non-trivial (by X, 7.3(4)), so 4.5

is always $SND(N_0) \leqslant 1$, and it is easy to prove (ii). On the other hand, if every regular type is trivial the computation of the dimension becomes easy (see X, §7).]

A personal remark. This theorem has a personal significance for me. In 1969, just after reading Morley's paper, I conjectured it. As this has been, in a sense, the first step toward this book, it seems appropriate to close the book with an affirmation of the conjecture.

APPENDIX

A.0. Introduction

This is a technical chapter containing the combinatorial theorems needed in the book. Of some self-interest may be 1.6, in which we find when a family of sets has independent subfamilies; 2.4 which is a weakened version of Halperin–Lauchli's theorem, proved by a different proof; and 2.6 which generalizes the Erdös–Rado theorem to trees.

In Section 1 we introduce filter and ultrafilters, with some basic lemmas (e.g., every non-trivial ultrafilter can be extended to an ultrafilter). Then we deal with the filter $D(\alpha)$ (generated by the closed unbounded subsets of α) and prove that we can usually split a stationary set to $\mathrm{cf}\,\alpha$ stationary subsets. We prove some theorems saying that families of sets have some nicely looking subfamilies, and the existence of families of 2^λ functions from λ to λ (or subsets of λ) which are very independent.

In Section 2 we deal with the partition calculus: we prove the theorems of Ramsey, a weakening of Halperin and Lauchli, Erdös–Rado and a generalization to trees; and that if $|S| > \lambda$, $x \in S \Rightarrow |f(x)| < \lambda$, there is an independent $S^* \subseteq S$, $|S^*| = |S|$.

Section 3 deals with κ-contradictory orders and connected anti-symmetric relations.

A.1. Filters, stationary sets and families of sets

Definition 1.1: (1) D is a filter over I, if D is a non-empty family of subsets of I and

(i) $A, B \in D$ implies $A \cap B \in D$,

(ii) if $A \in D$, $A \subseteq B \subseteq I$ then $B \in D$.

(2) The filter is *trivial* if $\emptyset \in D$, we sometimes "forget" to say our filter is non-trivial.

(3) D is an ultrafilter over I, if D is a non-trivial filter over I, and for every $A \subseteq I$, $A \in D$ or $I - A \in D$.

(4) Let S be a family of subsets of I, the filter $[S]$ generated by the family S is $\{A: A \subseteq I$, and $\bigcap_{i<n} B_i \subseteq A$ for some $B_i \in S\}$.

(5) The filter D over I is principal if for some A, $D = \{B \subseteq I: A \subseteq B\}$.

(6) The filter D is λ-complete if $A_i \in D(i < \alpha < \lambda)$ implies $\bigcap_{i<\alpha} A_i \in D$.

THEOREM 1.1: (1) *For any family E of subsets of I, $[E]$ is a filter over I.*

(2) *The filter $[E]$ is non-trivial iff every finite intersection of members of S is non-empty.*

(3) *Any non-trivial filter can be extended to an ultrafilter.*

(4) *Any filter is $\aleph_0$-complete.*

(5) *An ultrafilter D over I is principal iff for some $t \in I$, $\{t\} \in D$, so $D = \{A \subseteq I: t \in A\}$.*

Proof. Part (3) follows by Zorn's Lemma. The other parts are easy and well known.

DEFINITION 1.2: (1) A set S is *unbounded* under α if for every $\beta < \alpha$, there is $\gamma \in S$, $\beta \leq \gamma < \alpha$. (We omit "under α" when it is clear.)

(2) A subset S of α is closed, if for every limit $\delta < \alpha$ when S is unbounded under δ, $\delta \in S$.

(3) $D(\alpha)$ is the filter generated by the family of closed unbounded subsets of α.

THEOREM 1.2: *If* $\mathrm{cf}\,\alpha > \aleph_0$, *then $D(\alpha)$ is a* $(\mathrm{cf}\,\alpha)$*-complete non-trivial filter over α.*

Proof. As any unbounded subset of α is not empty; it suffices to prove that if A_i $(i < \beta < \mathrm{cf}\,\alpha)$ are closed unbounded subsets of α then $A = \bigcap_{i<\beta} A_i$ is too. Clearly A is closed; if $\gamma < \alpha$ we shall define by induction on $n < \omega$ ordinals β_n so that $\beta_0 = \gamma$, $\beta_n < \beta_{n+1}$, $\beta_n < \alpha$ and $(\beta_n, \beta_{n+1}) \cap A_i \neq \emptyset$ for $i < \beta$. $[(\beta_n, \beta_{n+1}) = \{\gamma: \beta_n < \gamma < \beta_{n+1}\}.]$ As $\mathrm{cf}\,\alpha > \aleph_0$, $\bigcup_{n<\omega} \beta_n < \alpha$ and as each A_i is closed and unbounded under $\bigcup_{n<\omega} \beta_n$, $\bigcup_{n<\omega} \beta_n \in \bigcap_{i<\beta} A_i$, hence $\bigcap_{i<\beta} A_i$ is unbounded. Let $\beta_0 = \gamma$ and if β_n is defined, define inductively β_n^i $(i < \beta)$ so that $\beta_n^0 = \beta_n$, $\beta_n^{1+i} \in A_i$, $\bigcup_{j<i} \beta_n^j < \beta_n^i < \alpha$. This is possible as each A_j is unbounded under α, and $\beta < \mathrm{cf}\,\alpha$. Clearly $\beta_{n+1} = \bigcup_{i<\beta} \beta_n^i < \alpha$ will be suitable.

For simplicity, we shall now concentrate on $D(\lambda)$ for regular λ.

DEFINITION 1.3: A subset A of λ is *stationary* if $\lambda - A \notin D(\lambda)$.

THEOREM 1.3: (1) *If A is a stationary subset of λ, λ regular $> \aleph_0$, f a function from A into λ, $0 \neq \alpha \in A \rightarrow f(\alpha) < \alpha$, then for some β $\{\gamma \in A: f(\gamma) = \beta\}$ is stationary.*

(2) *If A is a stationary subset of λ, λ regular $> \aleph_0$, then A can be partitioned into λ pairwise disjoint, stationary subsets, provided that for some $\mu < \lambda$, $(\forall \alpha \in A)[\mathrm{cf}\,\alpha \leq \mu]$.*

(3) *If $\kappa < \lambda$ are regular, then $\{\alpha < \lambda: \mathrm{cf}\,\alpha = \kappa\}$ is a stationary subset of λ.*

Remark. The additional condition in (2) involving μ is not, in fact, necessary.

Proof. (1) Suppose not, then for every $\beta < \lambda$ there is $S_\beta \in D(\lambda)$ such that $\gamma \in S_\beta \Rightarrow f(\gamma) \neq \beta$. Let $R = \{\langle \gamma, \beta \rangle: \gamma \in S_\beta\}$ and $M = \langle \lambda, <, f, R \rangle$ and M_α the submodel with universe α. By VII, 1.4 $S = \{\alpha < \lambda: M_\alpha \prec M\} \in D(\lambda)$ hence A is not disjoint to S.

So let $\delta \in S \cap A$, then clearly for every $\beta < \delta$, S_β is unbounded under λ hence under δ (using that $M \models (\forall x)(\exists y)[x < y \wedge R(y, \beta)]$ and $M_\delta \prec M$) hence $\delta \in S_\beta$. But $f(\delta) < \delta$, contradiction.

(2) For every $\alpha \in A$ choose an increasing sequence $\beta_\alpha(i)$ ($i < \mathrm{cf}\,\alpha \leq \mu$) whose limit is α. Let $S(\beta, i) = \{\alpha \in A: \beta_\alpha(i) = \beta\}$ ($\beta < \lambda, i < \mu$), so clearly $\beta_1 \neq \beta_2$, $i < \mu \Rightarrow S(\beta_1, i) \cap S(\beta_2, i) = \emptyset$. If for some $i < \mu$, $\{\beta < \lambda: S(\beta, i)$ is stationary$\}$ has cardinality λ, clearly we finish. Otherwise for every $i < \mu$ there is $\beta[i] < \lambda$ such that for $\beta \geq \beta[i]$, $S(\beta, i)$ is not stationary, so let it be disjoint to some closed unbounded $S^*(\beta, i)$. Clearly $\beta^* = \sup_{i<\mu} \beta[i] < \lambda$ as $\mu < \lambda$. Let

$$R_1 = \{\langle \alpha, i, \beta_\alpha(i) \rangle: \alpha \in A, i < \mathrm{cf}\,\alpha\},$$

$$R_2 = \{\langle \beta, i, \gamma \rangle: \gamma \in S^*(\beta, i), \beta \geq \beta^*, i < \mu\},$$

$M = \langle \lambda, <, R_1, R_2, \beta^* \rangle$ and M_α the submodel of M with universe α.

By VII, 1.4 $S = \{\alpha < \lambda: M_\alpha \prec M\} \in D(\lambda)$, so there is $\delta > \beta^*$, $\delta \in S \cap A$. As in the proof of (1) it is easy to show that for $i < \mu$, $\beta \geq \beta^*$, $\beta < \delta$, $\delta \in S^*(\beta, i)$, hence $\beta_\delta(i) \neq \beta$. But as $\beta^* < \delta$, and the sequence $\beta_\delta(i)$ ($i < \mathrm{cf}\,\delta$) converges to δ, we easily get a contradiction.

(3) We leave it to the reader.

THEOREM 1.4: (1) *If λ is regular and $\chi < \lambda \Rightarrow \chi^{<\mu} < \lambda$, and $|S_t| < \mu$ for $t \in W$, where $|W| = \lambda$, then for some $W' \subseteq W$, $|W'| = \lambda$ and S, for any $s \neq t \in W'$, $S_s \cap S_t = S$.*

(2) *Moreover if* $S_t = \{s^t_i: i < \alpha(t)\}$ *we can assume that* $t \in W' \Rightarrow \alpha(t) = \alpha^0$ *and for some* $U \subseteq \alpha^0$; $i \in U, t \in W' \Rightarrow s^t_i = s^i$ *and* $S = \{s^i: i \in U\}$ *and* $s^i_{t(1)} = s^j_{t(1)} \Leftrightarrow s^i_{t(2)} = s^j_{t(2)}$ for $t(1), t(2) \in W'$, and $s^i_{t(1)} = s^j_{t(2)}, i \neq j \in W' \Rightarrow t(1), t(2) \in U$.

Proof. W.l.o.g. $W = \lambda$, $S_\alpha \subseteq \lambda$ and let $\kappa = \mu$ when μ is regular, and $\kappa = \mu^+$ otherwise. Clearly $\kappa < \lambda$ and κ is regular. For any $\alpha < \lambda$, $\mathrm{cf}\,\alpha = \kappa$, clearly $S_\alpha \cap \alpha$ is a bounded subset of α, so let $h(\alpha) < \alpha$ be a bound of it. By 1.3(3) $\{\alpha < \lambda: \mathrm{cf}\,\alpha = \kappa\}$ is stationary, and by 1.3(1) on some stationary $W^1 \subseteq \{\alpha < \lambda: \mathrm{cf}\,\alpha = \kappa\}$ h has constant value β. As $|\beta|^{<\mu} < \lambda$, by 1.3(1) there are a stationary $W^2 \subseteq W^1$, and $S \subseteq \beta$ such that for every $\alpha \in W^2$, $S_\alpha \cap \alpha = S_\alpha \cap \beta = S$. We can easily find a stationary $W' \subseteq W^2$ such that $\alpha < \gamma \in W'$ implies $S_\alpha \subseteq \gamma$. Clearly W', S satisfy our requirements.

(2) The same proof, essentially.

THEOREM 1.5: (1) *If* $\lambda^{<\kappa} = \lambda$, *then there is a family* $\mathscr{G}$ *of* 2^λ *functions from* λ *into* λ, *such that for any distinct* $f_i \in \mathscr{G}$ $(i < \alpha < \kappa)$ *and any ordinals* $\gamma_i < \lambda$ $(i < \alpha)$, $\{\zeta < \lambda$: *for every* $i < \alpha, f_i(\zeta) = \gamma_i\} \neq \emptyset$.

(2) *If* $\lambda^{<\kappa} = \lambda$ *there are subsets* S_i $(i < 2^\lambda)$ *of* λ, *such that for any disjoint non-empty* $U, V \subseteq \lambda$, $|U| + |V| < \kappa$, *the set* $\bigcup_{i\in U} S_i - \bigcup_{i\in V} S_i$ *has cardinality* λ.

Proof. (1) Let $\{(A_i, \langle C^i_\zeta: \zeta < \alpha_i\rangle, \langle j^i_\zeta: \zeta < \alpha_i\rangle): i < \lambda\}$ be an enumeration of all triples $(A, \langle C_\zeta: \zeta < \alpha\rangle, \langle j_\zeta: \zeta < \alpha\rangle)$ such that:

(i) A is a subset of λ, $|A| < \kappa$.

(ii) C_ζ is a subset of A and $\zeta(1) \neq \zeta(2)$ implies $C_{\zeta(1)} \neq C_{\zeta(2)}$.

(iii) $\alpha < \kappa$ and $j_\zeta < \lambda$.

(Clearly the number of triples is λ, as $\lambda^{<\kappa} = \lambda$.) Now for every set $B \subseteq \lambda$ we define a function $f_B: \lambda \to \lambda$. We define $f_B(i)$ as follows: if $B \cap A_i = C^i_\zeta$, $f_B(i) = j^i_\zeta$. Otherwise $f_B(i) = 0$.

If $B_{\alpha(\zeta)}$ $(\zeta < \alpha < \kappa)$ are distinct and $j_\zeta < \lambda$ $(\zeta < \alpha)$ we can find $A \subseteq \lambda$, $|A| < \kappa$ so that $B_{\alpha(\zeta)} \cap A$ are distinct; so for some i, $A_i = A$, $\alpha_i = \alpha$, $C^i_\zeta = B_{\alpha(\zeta)} \cap A$, $j^i_\zeta = j_\zeta$, so $f_{B_{\alpha(\zeta)}}(i) = j_\zeta$. Hence $\{f_B: B \subseteq \lambda\}$ is a family satisfying our requirements.

(2) Immediate by (1), for if $\mathscr{G} = \{f_i: i < 2^\lambda\}$, let $S_i = \{\alpha < \lambda: f_i(\alpha) = 0\}$.

THEOREM 1.6: (1) *If* S *is an infinite family of subsets of* I, *then we can find* $t_n \in I$, $A_n \in S$ $(n < \omega)$ *such that*

(i) $t_n \in A_m$ *iff* $n = m$, *or*

(ii) $t_n \in A_m$ *iff* $n \neq m$, *or*

(iii) $t_n \in A_m$ *iff* $n < m$, *or*

(iv) $t_n \in A_m$ *iff* $n \geq m$.

(2) *For every* $n < \omega$ *there is* $k < \omega$ *such that if* $|S| \geq k$, S *a family of sets then we can find* t_i, A_i $(i < n)$ *as above.*

Proof. (1) We define by induction on n, t_n, A_n, S_n such that:

(α) $S_0 = S$, $S_{n+1} \subseteq S_n$, S_n is infinite.

(β) $S_n^1 = \{A \in S_n: t_n \in A\}$ and $S_n^2 = \{A \in S_n: t_n \notin A\}$ are non-empty.

(γ) $A_n \in S_n^1 \Leftrightarrow A_n \notin S_n^2 \Leftrightarrow S_{n+1} \neq S_n^1 \Leftrightarrow S_{n+1} = S_n^2$.

This is easy as we can always find a suitable t_n, since S_n is infinite, and then it is easy to choose A_n and S_{n+1}. By Ramsey's theorem for a 2-place function with range of cardinality 4 (see 2.1) the conclusion is easy.

(2) Essentially the same proof.

DEFINITION 1.4: (1) Ded λ is the first cardinal μ, such that no tree with λ nodes has $\geq \mu$ branches (a tree is a partially ordered set T such that for every $x \in T$ $\{y \in T: y < x\}$ is well-ordered. A branch is a maximal linearly ordered subset).

(2) $\mathrm{Ded}_r\, \lambda$ is the first regular cardinal which is $\geq$ Ded λ.

THEOREM 1.7: (1) *Suppose* S *is a family of subsets of* I, I *infinite,* $|S| \geq \mathrm{Ded}_r|I|$. *Then there are, for every* n, *elements* $t_0, \ldots, t_{n-1}$ *of* I, *such that for every* $w \subseteq n$ *there is* $A_w \in S$ *such that:* $t_i \in A_w \Leftrightarrow i \in w$ *for* $i < n$.

(2) *Suppose* S *is a family of subsets of* I, I *is finite,* $|S| > \sum_{i \leq n} \binom{|I|}{i}$. *Then there are elements* $t_0, \ldots, t_{n-1} \in I$ *such that for every* $w \subseteq n$ *there is* $A_w \in S$ *such that* $t_i \in A_w \Leftrightarrow i \in w$.

Proof. (1) Let $\lambda = |I|$, $\mu = \mathrm{Ded}_r\, \lambda$, and w.l.o.g. $I = \lambda$, and $J \subseteq I$, $|J| < \lambda$ implies $|\{A \cap J: A \in S\}| < \mu$. Let for $\alpha \leq \lambda$ $S_\alpha^0 = \{\langle A \cap \alpha, \alpha\rangle: A \in S\}$. So clearly $|S_\alpha^0| < \mu$ for $\alpha < \lambda$. Let $\langle A, \alpha\rangle \restriction \beta = \langle A \cap \beta, \min\{\alpha, \beta\}\rangle$. On $\bigcup_{\alpha \leq \lambda} S_\alpha^0$ we define the following partial order: $\langle A_1, \alpha_1\rangle \leq \langle A_2, \alpha_2\rangle$ iff $\alpha_1 \leq \alpha_2$, $A_1 = A_2 \cap \alpha_1$. Clearly under this order $\bigcup_{\alpha \leq \lambda} S_\lambda^0$ is a tree. Let for $\alpha < \lambda$

$$S_\alpha^1 = \{s \in S_\alpha^0: |\{t \in S_\lambda^0: s \leq t\}| \geq \mu\},$$

$$S_\lambda^1 = \{s \in S_\lambda^0: \text{for every } \alpha < \lambda, s \restriction \alpha \in S_\alpha^1\}.$$

Clearly

$$|S^0_\lambda - S^1_\lambda| = \left| \bigcup_{\substack{s \in S^0_\alpha - S^1_\alpha \\ \alpha < \lambda}} \{t \in S^0_\lambda : s \le t\} \right| < \mu$$

as μ is regular and $|\bigcup_{\alpha<\lambda} S^0_\alpha| < \mu$.

We will now prove by induction on n that:

$(*)_n$ For every $s \in S^1_\alpha$, $\alpha < \lambda$ there are $\alpha_0, \alpha_1, \ldots, \alpha_{n-1} > \alpha$ such that for every $w \subseteq n$ there is $\langle A_w, \lambda\rangle \in S^1_\lambda$ such that $s \le \langle A_w, \lambda\rangle$ and $\alpha_i \in A_w \Leftrightarrow i \in w$.

For $n = 0$ there is nothing to prove. Suppose we have proved for n and we shall prove for $n + 1$, and let $s \in S^1_\alpha$, $\alpha < \lambda$.

Clearly $T_s = \{t \in \bigcup_{\beta<\lambda} S^1_\beta : s \le t\}$ is a tree, and for every $t \in S^1_\lambda$, $s \le t$, $B_t = \{t^* : s \le t^* < t\}$ is a branch of T_s, and if $t_1 \ne t_2 \in S^1_\lambda$, $s \le t_1$, $s \le t_2$ then $B_{t_1} \ne B_{t_2}$. As $|S^0_\lambda - S^1_\lambda| < \mu$, the number of branches of T_s is $\ge \mu$; hence by the definition of $\mathrm{Ded}_r\, \lambda$, $|T_s| > \lambda$, hence for some $\alpha(*) < \lambda$ $|T_s \cap S^1_{\alpha(*)}| > \lambda$. So let $s_i \in T_s \cap S^1_{\alpha(*)}$ ($i < \lambda^+$) be distinct. For each of them by $(*)_n$ we can find $\alpha(*) < \alpha^i_0, \ldots, \alpha^i_{n-1} < \lambda$ as mentioned there. As the number of n-tuples of ordinals $< \lambda$ is λ, we can assume $\alpha_k = \alpha^0_k = \alpha^1_k$ for $k < n$. As $s_0 \ne s_1$ there is an $\alpha_n < \alpha(*)$, $\alpha_n > \alpha$, such that $s_0 \restriction \alpha_n = s_1 \restriction \alpha_n$ but $s_0 \restriction (\alpha_n + 1) \ne s_1 \restriction (\alpha_n + 1)$. Now it is easy to check that $\alpha_0, \ldots, \alpha_n$ satisfy our requirements.

(2) Essentially the same proof.

EXERCISE 1.1: Show that if $\mu < \mathrm{Ded}_r\, \lambda$, $|I| = \lambda$, then there is a family S of subsets of I such that the conclusion of 1.7(1) fails even for $n = 2$.

EXERCISE 1.2: Show that the bound in 1.7(2) is the best possible.

PROBLEM 1.3: Can we replace $\mathrm{Ded}_r|I|$ by $\mathrm{Ded}|I|$ in 1.7(1)?

EXERCISE 1.4: Prove $\chi < \mathrm{Ded}\,\lambda$ iff there is an ordered set J, $|J| \ge \chi$, and a dense subset $I \subseteq J$, $|I| \le \lambda$. (Hint: For $\Rightarrow$ use a lexicographic order on the tree plus its branches. For $\Leftarrow$ the nodes will be intervals of I.)

LEMMA 1.8: (1) *Suppose $\aleph_\beta$ is regular, $|S_t| < \aleph_\beta$ for $t \in W$, $|W| = \aleph_{\beta+n}$ then for some $W' \subseteq W$, $|W'| = \aleph_{\beta+n}$ and V, $|V| < \aleph_\beta$ and for every distinct $s \ne t \in W'$, $S_t \cap S_s \subseteq V$. If $W \subseteq \{\delta < \aleph_{\delta+1} : \mathrm{cf}\,\delta \geqslant \aleph_\beta\}$ is stationary, we can have W' stationary too.*

(2) *Instead "$\aleph_\beta$ is regular" we can demand* $|S_t| < \mathrm{cf}\, \aleph_\beta$ *for* $t \in W$. *If* $W = \aleph_{\beta+n}$, $S_\alpha \subseteq \aleph_{\beta+n}$, *we can assume* $\gamma \in S_\alpha - V$ *implies* $\gamma \geq \alpha$ *for* $\gamma \in W'$.

Proof. (1) Similar to 1.4: we can assume for some μ $|S_t| = \mu$ for every t; $W = \lambda, S_\alpha \subseteq \lambda$. Then for some stationary $W^1 \subseteq \{\alpha < \aleph_{\beta+n}: \mathrm{cf}\,\alpha = \mu^+\}$, and $\gamma < \aleph_\alpha$, for every $\alpha \in W^1$, $S_\alpha \cap \alpha \subseteq \gamma$; and $S_\xi \subseteq \alpha$ for $\xi < \alpha$. Now we prove by downward induction on $l \leq n$ that there is $V_l \subseteq \gamma$, $|V| < \aleph_{\beta+l}$, and $|\{\alpha \in W^1: S_\alpha \cap \gamma \subseteq V_l\}| = \aleph_{\alpha+n}$. For $l = n$ $V_n = \gamma$; for $l = 0$ we get our conclusion, and if V_{l+1} is defined, let $V_{l+1} = \bigcup_i V_l^i$, $|V_l^i| < \aleph_{\beta+l}$, V_l^i increasing ($i < \aleph_{\beta+l}$); one of the V_l^i is as required on V_l.

(2) Similarly.

A.2. Partition theorems

THEOREM 2.1 (Ramsey's Theorem): (1) *For any infinite ordered set I, and n-place function f from I, with range of cardinality* $< \aleph_0$ *there is an infinite set* $J \subseteq I$ *such that if* $\bar{s}, \bar{t} \in {}^nJ$, $\bar{s}, \bar{t}$ *increasing, then* $f(\bar{s}) = f(\bar{t})$.

(2) *For any* $n, k, l < \omega$ *there is* $m = m_r(n, k, l)$ *such that if I is an ordered set of cardinality* $\geq m$, *and f an n-place function from I with range of cardinality* $\leq l$, *then for some* $J \subseteq I$, $|J| = k$; $\bar{s}, \bar{t} \in {}^nJ$, $\bar{s}, \bar{t}$ *increasing* $\Rightarrow f(\bar{s}) = f(\bar{t})$.

Proof. (1) For simplicity we assume $I = \omega$. We prove the assertion by induction on n. For $n = 0$ or $n = 1$ there is nothing to prove, so assume we have proved for n, and we shall prove for $n + 1$.

We now define by induction on $k < \omega$, natural numbers $l(k)$ and infinite sets $S_k \subseteq \omega$ such that

(i) $l(k) < l(k + 1)$, $S_k \subseteq S_{k-1}$,

(ii) for any $m \in S_k$, $l(k) < m$,

(iii) for any $m_0, \ldots, m_{n-1} < k$, $m \in S_k$

$$f(l(m_0), \ldots, l(m_{n-1}), l(k)) = f(l(m_0), \ldots, l(m_{n-1}), m).$$

We choose $l(0) = 0$, $S_0 = \{i: 0 < i < \omega\}$, and if $l(k)$, S_k are defined, let E_k be the equivalence relation on S_k defined by: $iE_kj \Leftrightarrow$ for any $m_0, \ldots, m_{n-1} \leq k$, $f(l(m_0), \ldots, l(m_{n-1}), i) = f(l(m_0), \ldots, l(m_{n-1}), j)$. Clearly E_k has only finitely many equivalence classes, so at least one of them is infinite; choose $l(k + 1)$ from such a class and let $S_{k+1} = \{i: i \in S_k, l(k + 1) < i, iE_kl(k + 1)\}$.

Now define on $I' = \{l(k): k < \omega\}$ a function f', $f'(l(k_0), \ldots, l(k_{n-1})) = f(l(k_0), \ldots, l(k_{n-1}), l(k))$ for all large enough k.

Now by the induction hypothesis (applied to I', f') we get our conclusion easily.

(2) We leave it to the reader.

THEOREM 2.2: *For any n, $k < \omega$ we can define an increasing sequence $l(i)$ $(i < \omega)$ such that the following holds: Suppose f is an n-place function from ${}^{l(i)}2$ into k, then there are $h_m: {}^{i \geq}2 \to {}^{l(i) \geq}2$, $m < n$, and $k^* < k$ such that*

(i) *$h_m(\eta)$ has length $l(l(\eta))$.*

(ii) *$\eta \lessdot \nu \Leftrightarrow h_m(\eta) \lessdot h_m(\nu)$.*

(iii) *h_m is one-to-one.*

(iv) *If $\eta_m \in {}^{i}2$ $(m < n)$ are distinct, then $k^* = f(h_0(\eta_0), \ldots, h_{n-1}(\eta_{n-1}))$.*

Proof. Let $l(0) = 0$ and if $l(i)$ is defined, we want $l(i + 1)$ to be such that:

(*) If $|S_i| \geq l(i + 1)$ for $i \in I$, $|I| = n2^{l(i)}$, and g is an n-place function from $\bigcup_{\alpha \in I} S_\alpha$ into k then there are distinct $a_\alpha^0, a_\alpha^1 \in S_\alpha$ $(\alpha \in I)$ and an n-place function g' from I into k such that $g(a_{\alpha(1)}^{j(1)}, \ldots, a_{\alpha(n)}^{j(n)}) = g'(\alpha(1), \ldots, \alpha(n))$ for any distinct $\alpha(1), \ldots, \alpha(n) \in I$ and any $j(1), \ldots, j(n) \in \{0, 1\}$.

We can do this, for let $l_i^*(0) = k$, $l_i^*(\alpha + 1)$ be 2 to the power $n[l_i^*(\alpha)n2^{l(i)}]^n$, and $2^{l(i+1)-l(i)} \geq l_i^*(n2^{l(i)})$. Now for any suitable I, S_α, g let $I = n2^{l(i)}$ for simplicity, and now define by induction on $\alpha < n2^{l(i)}$ sets W_α and elements a_α^0, a_α^1 such that:

(i) $W_\alpha \subseteq \bigcup\{S_j: \alpha \leq j < n2^{l(i)}\}$ and $W_{\alpha+1} \subseteq W_\alpha$,

(ii) $|W_\alpha \cap S_j| = l_i^*(n2^{l(i)} - \alpha)$ for $j > \alpha$,

(iii) $a_\alpha^0 \neq a_\alpha^1 \in S_\alpha \cap W_\alpha$,

(iv) if $b_0, \ldots, b_{n-2} \in \{a_\beta^\gamma: \gamma < 2, \beta < \alpha\} \cup \bigcup\{S_j: \alpha < j < n2^{l(i)}\}$ and $m \leq n - 1$, then

$$g(b_0, \ldots, b_{m-1}, a_\alpha^0, b_m, \ldots, b_{n-2}) = g(b_0, \ldots, b_{m-1}, a_\alpha^1, b_m, \ldots, b_{n-2}).$$

If we have defined for $\beta < \alpha$, we first choose W_α to satisfy (i) and (ii) and then we can easily satisfy (iii) and (iv). Clearly a_α^0, a_α^1 are as required in (*).

Now we prove the theorem by induction on i. For $i = 0$ there is nothing to prove. Suppose we have proved for i and we shall prove for

$i+1$. We apply (*) for $I = {}^{l(i)}2 \times n$, f (or more exactly, some suitable extension of it) and $S_{\langle \eta,m\rangle} = \{\nu : \nu \in {}^{l(i+1)}2,\ \eta \lessdot \nu\}$, and get distinct $\rho^{\alpha}_{\langle \eta,m\rangle} \in S_{\langle \eta,m\rangle}$, $\alpha = 0, 1$ and an n-place function f' from ${}^{l(i)}2$ into k, so that $f(\rho^{\alpha(0)}_{\langle \eta(0),0\rangle}, \rho^{\alpha(1)}_{\langle \eta(1),1\rangle}, \ldots, \rho^{\alpha(n-1)}_{\langle \eta(n-1),n-1\rangle} = f'(\eta(0), \eta(1), \ldots, \eta(n-1))$. Now we use the induction hypothesis on i for f' and get suitable h'_m $(m < n)$, k^*. Now define h_m so that for $\eta \in {}^{i\geq}2$, $h_m(\eta) = h'_m(\eta)$ and for $\eta \in {}^{i}2$, $h_m(\eta^\frown\langle\alpha\rangle) = \rho^{\alpha}_{\langle \eta,m\rangle}$.

THEOREM 2.3: *Suppose $n, k < \omega$ and f is an n-place function from ${}^{\omega>}2$ into k. Then we can find functions h^0_m, h^1_m from ${}^{\omega>}2$ into ${}^{\omega>}2$ (for $m < n$) and $k^* < k$ and natural numbers $l^0(i)$, $l^1(i)$ such that*

(i) *$l^0(i)$, $l^1(i)$ increase with i,*

(ii) *$h^0_m(\eta)$ has length $l^0(l(\eta))$, $h^1_m(\eta)$ has length $l^1(l(\eta))$; h^0_m and h^1_m are one-to-one,*

(iii) *$\eta \lessdot \nu$ iff $h^1_m(\eta) \lessdot h^1_m(\nu)$,*

(iv) *$h^1_m(\eta) \lessdot h^0_m(\eta)$,*

(v) *for any i and $\eta_0, \ldots, \eta_{n-1} \in {}^{i}2$, distinct,*

$$k^* = f(h^0_0(\eta_0), \ldots, h^1_{n-1}(\eta_{n-1})).$$

Proof of 2.3. Define $l(i)$ as in 2.2; and apply 2.2 to get for each $i < \omega$, $m < n$, functions $h_{m,i}$ $(m < n)$ and $k^*_i < k$.

For every j the number of possible $\langle h_{m,i} \restriction {}^{j\geq}2 : m < n\rangle$ is finite, hence by Konig's lemma there is an increasing sequence $i(\alpha) < \omega$ $(\alpha < \omega)$ and k^* so that $h_{m,i(\alpha)} \restriction {}^{j\geq}2 = h_{m,i(j)} \restriction {}^{j\geq}2$ for $\alpha \geq j$ and $k^*_{i(\alpha)} = k^*$. Now for $\eta \in {}^{j}2$ let $\eta' \in {}^{i(j)}2$ be defined by $\eta'[\alpha] = \eta[\alpha]$, $\alpha < j$, $\eta'[\alpha] = 0$, $j \leq \alpha < i(j)$ and let $h^0_m(\eta) = h_{m,i(j)}(\eta')$, $h^1_m(\eta) = h_{m,i(j)}(\eta)$ and clearly they prove 2.3.

THEOREM 2.4: *Suppose f_m is an $n(m)$-place function from ${}^{\omega>}2$ into a finite set. Then there are functions $h^0, h^1 : {}^{\omega>}2 \to {}^{\omega>}2$ such that:*

(i) *$l(\eta) = l(\nu)$ implies $l(h^0(\eta)) = l(h^0(\nu))$, $l(h^1(\eta)) = l(h^1(\nu))$; h^0 and h^1 are one-to-one.*

(ii) *$\eta \lessdot \nu$ iff $h^1(\nu) \lessdot h^1(\nu)$,*

(iii) *$h^1(\eta) \lessdot h^0(\eta)$,*

(iv) *if $m \leq i \leq \min\{\alpha, \beta\}$, $\eta_l \in {}^{\alpha}2$, $\nu_l \in {}^{\beta}2$ (for $l < n(m)$), $\eta_l \restriction i = \nu_l \restriction i$, $l(1) \neq l(2) \Rightarrow \eta_{l(1)} \restriction i \neq \eta_{l(2)} \restriction i$, then*

$$f_m(h^0(\eta_0), \ldots, h^0(\eta_{n(m)-1})) = f_m(h^0(\nu_0), \ldots, h^0(\nu_{n(m)-1})).$$

Proof. Immediate by repeated use of 2.3.

DEFINITION 2.1: $\lambda \to (\kappa)^n_\mu$ if for every n-place function f from λ into μ there are $S \subseteq \lambda$, $|S| = \kappa$ and $\alpha < \mu$ such that for any increasing sequence $\bar{s} \in {}^nS$, $f(\bar{s}) = \alpha$.

THEOREM 2.5: $\beth_n(\lambda)^+ \to (\lambda^+)^{n+1}_\lambda$.

Proof. By induction on n. For $n = 0$ it is easy, so suppose we have proved it for n, and we shall prove it for $n + 1$. Let $\chi = \beth_{n+1}(\lambda)$, $\mu = \beth_n(\lambda)$; and define an increasing sequence β_i ($i < \mu^+$) of ordinals $<\chi^+$ such that (*) if $S \subseteq \beta_i$, $|S| \leq \mu$, then for every $\gamma < \chi^+$ there is an ordinal $\gamma' < \beta_{i+1}$ such that for any $s_0, \ldots, s_{n-1} \in S$ $f(s_0, \ldots, s_{n-1}, \gamma) = f(s_0, \ldots, s_{n-1}, \gamma')$ and $\gamma' \in S \Leftrightarrow \gamma \in S$. This is easily done by induction, by cardinality considerations; let $\beta^* = \bigcup_{i<\mu^+} \beta_i$. Now define by induction on $i < \mu^+$, $\gamma_i < \beta_{i+1}$, such that for any $s_0, \ldots, s_{n-1} \in \{\gamma_j: j < i\}$,

$$f(s_0, \ldots, s_{n-1}, \gamma_i) = f(s_0, \ldots, s_{n-1}, \beta^*), \qquad j < i \Rightarrow \gamma_j \neq \gamma_i.$$

Define an n-place function f' on μ^+:

$$f'(i(0), \ldots, i(n-1)) = f(\gamma_{i(0)}, \ldots, \gamma_{i(n-1)}, \beta^*).$$

By the induction hypothesis there is $S' \subseteq \mu^+$, $|S'| = \lambda^+$ so that for any $i(0) < \cdots < i(n-1) \in S'$, $f'(i(0), \ldots, i(n-1)) = \alpha_0$, for some fixed α_0. Now $S = \{\gamma_i: i \in S'\}$ proves the theorem.

THEOREM 2.6: *For every $n, m < \omega$ there is $k = k(n, m) < \omega$ (and $k(n, 1) = 0$) such that whenever $\lambda = \beth_k(\chi)^+$ the following holds: If f is an m-place function from ${}^{n\geq}\lambda$ into χ (or any set of cardinality $\leq \chi$), then there is $I \subseteq {}^{n\geq}\lambda$ such that:*

(i)$_\chi$ $\langle\,\rangle \in I$ *and if* $\eta \in I \cap {}^{n>}\lambda$, *then* $|\{\alpha: \alpha < \lambda, \eta^\frown\langle\alpha\rangle \in I\}| = \chi^+$,

(ii)$_f$ *if* $\eta_0, \ldots, \eta_{m-1}, \nu_0, \ldots, \nu_{m-1} \in I$ *and* $\langle\eta_0 \cdots\rangle \sim \langle\nu_0, \ldots\rangle$ (*see* VII, *Definitions* 2.3 *and* 3.1), *then* $f(\eta_0, \ldots) = f(\nu_0, \ldots)$.

Remark. We do not try to get the best $k(n, m)$.

Proof. We first prove for $m = 1$ that we can choose $k(n, 1) = 0$, and then prove by induction on n. Clearly we can assume $n, m > 0$.

Case 1. $m = 1$. Let $k = 0$, so $\lambda = \chi^+$. We define, by induction on $j \leq n$, for any $\eta \in {}^{n-j}\lambda$ a set $I_\eta \subseteq \{\nu: \nu \in {}^{n\geq}\lambda, \eta \lessdot \nu$ or $\eta = \nu\}$ and ordinals $\alpha_n(\eta), \ldots, \alpha_{n-j}(\eta) < \chi$ such that:

(i) $\eta \in I_\eta$ and if $\nu \in I_\eta \cap {}^{n>}\lambda$, then $|\{\alpha < \lambda: \nu^\frown\langle\alpha\rangle \in I_\eta\}| = \chi^+$,

(ii) for any $\nu \in I_\eta \cap {}^i\lambda$, $n - j \le i \le n$, $f(\nu) = \alpha_i(\eta)$.

For $j = 0$, $\eta \in {}^n\lambda$, $I_\eta = \{\eta\}$. Suppose we have defined I_η for any $\eta \in {}^{n-j+1}\lambda$, and let $\eta \in {}^{n-j}\lambda$. For any $\beta < \lambda$, the sequence $\langle \alpha_n(\eta^\frown\langle\beta\rangle), \ldots, \alpha_{n-j+1}(\eta^\frown\langle\beta\rangle)\rangle$ is well defined, and there are only $\le \chi^j = \chi$ such sequences, hence there is a sequence $\langle \alpha_n, \ldots, \alpha_{n-j+1}\rangle$ which we get for $\lambda = \chi^+$ β's. Let $\alpha_n(\eta) = \alpha_n, \ldots, \alpha_{n-j+1}(\eta) = \alpha_{n-j+1}$ and $\alpha_{n-j}(\eta) = f(\eta)$ and

$$I_\eta = \{\eta\} \cup \bigcup \{I_{\eta^\frown\langle\beta\rangle} : \langle \alpha_n(\eta^\frown\langle\beta\rangle), \ldots, \alpha_{n-j+1}(\eta^\frown\langle\beta\rangle)\rangle = \langle \alpha_n, \ldots, \alpha_{n-j+1}\rangle\}.$$

Clearly this satisfies our demands.

Case 2. $n = 1$. This follows by 2.5, taking $k(n, m) = m - 1$.

Case 3. Suppose we have proved for n, m, and we shall prove for $n + 1$, m.

Let $\mu = \beth_{m^2+m+1}(\chi)$, $k(n + 1, m) = k(n, m) + m^2 + m + 2$. Define an m-place function g on ${}^{n\ge}\lambda$:

$$\begin{aligned} g(\eta_0, \ldots, \eta_{m-1}) = \{\langle h, \beta_0, \ldots, \beta_{m-1}, \alpha\rangle &: h \text{ a function, } w \subseteq m,\ h: w \to m, \\ &\beta_l < \mu\ (l < m) \text{ and } \alpha = f(\nu_0, \ldots, \nu_{n-1}) \text{ where} \\ &l \in w \Rightarrow \nu_l = \eta_{h(l)}{}^\frown\langle\beta_l\rangle,\ l \notin w \Rightarrow \nu_l = \eta_l\}. \end{aligned}$$

Clearly the range of g has cardinality $\le 2^\mu$. By the hypothesis of Case (3) there is a set $I_0 \subseteq {}^{n\ge}\lambda$ which satisfies:

(i)′ $\langle\ \rangle \in I_0$ and $\eta \in I_0 \cap {}^{n>}\lambda$ implies $|\{\alpha < \lambda : \eta^\frown\langle\alpha\rangle \in I_0\}| = (2^\mu)^+$,

(ii)′ if $\bar\eta, \bar\nu \in I_0$, $\bar\eta \sim \bar\nu$, then $g(\bar\eta) = g(\bar\nu)$.

We can find $I_1 \subseteq I_0$ such that

(i)″ $\langle\ \rangle \in I_1$, and $\eta \in I_1 \cap {}^{n>}\lambda$ implies $|\{\alpha < \lambda: \eta^\frown\langle\alpha\rangle \in I_1\}| = \chi^+$ (define $I_1 \cap {}^i\lambda$ by induction on i).

In view of (ii)′ it suffices to find for each $\eta \in {}^n\lambda \cap I_1$ a set $I_\eta \subseteq \{\eta^\frown\langle\alpha\rangle : \alpha < \mu\}$, $|I_\eta| = \chi^+$ such that:

(*) If $\bar\nu, \bar\eta$ are similar sequences of length m from $I^* = \bigcup\{I_\eta : \eta \in I_1 \cap {}^n\lambda\} \cup I_1$ and $\bar\nu[l] \restriction n = \bar\eta[l] \restriction n$ *then* $f(\bar\eta) = f(\bar\nu)$.

For this it suffices to prove Theorem 2.7 below.

THEOREM 2.7: *Suppose* $|X_i| = \beth_{m(m+1)}(\chi)^+$ *for* $i < \chi$ (X_i *well ordered by* $\le$) *and* f *is an* m*-place function from* $\bigcup_{i<\chi} X_i$ *into a set of cardinality*

$\leq \chi$. *Then there are sets* $Y_i \subseteq X_i$, $|Y_i| = \chi^+$ *such that: if* $s(l), t(l) \in Y_{i(l)}$ $(l < m)$ *and* $i(l_1) = i(l_2) \Rightarrow [s(l_1) < s(l_2) \Leftrightarrow t(l_1) < t(l_2)]$ *then*

$$f(s(0), \ldots, s(m-1)) = f(t(0), \ldots, t(m-1)).$$

Proof. We define for $n \leq m$, $i < \chi$ sets X_i^n such that

(1) $X_i^0 = X_i$, $X_i^{n+1} \subseteq X_i^n$, $|X_i^n| = \beth_{m(m-n+1)}(\chi)^+$,

(2) if $s(l), t(l) \in X_{i(l)}^n$, $l < m$, $|\{i(l): i(l) \geq i(l_0)\}| \leq n$, and

$$i(l_1) = i(l_2) \Rightarrow [s(l_1) < s(l_2) \Leftrightarrow t(l_1) < t(l_2)],$$
$$i(l) \neq i(l_0) \Rightarrow s(l) = t(l),$$

then $f(s(0), \ldots, s(m-1)) = f(t(0), \ldots, t(m-1))$.

We define X_i^n by induction on n. For $n = 0$, $X_i^0 = X_i$ and clearly (1) and (2) hold. Suppose we have defined for n, and we shall define X_i^{n+1} by induction on i. Suppose we have defined X_j^{n+1} for $j < i$; choose $S_\alpha^n \subseteq X_\alpha^n$, $|S_\alpha^n| = m$ for $i < \alpha < \chi$.

By 2.5 we can easily find $X_i^{n+1} \subseteq X_i^n$, $|X_i^{n+1}| = \beth_{m(m-n)}(\chi)^+$ such that (2) holds for $n + 1$, $i(l_0) = i$ when $i(l) \leq i \Rightarrow s(l), t(l) \in X_{i(l)}^{n+1}$ and $i(l) > i \Rightarrow s(l), t(l) \in S_{i(l)}^n$. By the induction hypothesis on n clearly (2) holds. Now let $Y_i = X_i^m$ and clearly the Y_i's are as required.

THEOREM 2.8: *Suppose* $|S| > \lambda$, S *is the domain of* f, *and for* $x \in S$, $|f(x)| < \lambda$ ($f(x)$ *is a set*). *Then there is* $S^* \in S$, $|S^*| = |S|$ *such that* $s \neq t \in S^*$ *implies* $s \notin f(t)$.

Proof. Case 1. $|S| = \mu$ is regular. W.l.o.g. $S = \mu$; $x \in S \Rightarrow f(x) \subseteq \mu$ and define S_i by induction on $i \leq \lambda$, so that S_i is a maximal subset of $\mu - \alpha_i^2$ [where $\alpha_i^1 = \sup \bigcup_{j<i} S_j$, $\alpha_i^2 = \sup \bigcup \{f(\beta) \cup \{\beta\}: \beta < \alpha_i^1\}$] which satisfies $s \neq t \in S_i \Rightarrow s \notin f(t)$.

If for some i, $|S_i| = \mu$ we finish. Otherwise, $i \leq \lambda \Rightarrow \alpha_i^2 < \mu$, and notice that $j < i$, $s \in \mu - \alpha_i^2$ implies $f(s) \cap S_j \neq 0$. Now clearly the S_i's are pairwise disjoint and for any $s \in \mu - \alpha_\lambda^2$; $f(s) \cap S_i \neq 0$ for $i < \lambda$, hence $|f(s)| \geq \lambda$, contradiction.

Case 2. $|S| = \mu$ is singular. Let $\mu = \sum_{i<\kappa} \mu(i)$, $\kappa = \mathrm{cf}\, \mu$, $\mu(i)$, $i < \kappa$, is a strictly increasing sequence, $\lambda, \kappa < \mu(0)$. W.l.o.g. we can assume that $S = \mu$, and $x \in S \Rightarrow f(x) \subseteq S$; and $\alpha < \beta \Rightarrow \beta \notin f(\alpha)$ (as we can replace S by any $S' \subseteq S$, $|S'| = \mu$) and $\beta \neq \gamma \in (\mu(i), \mu(i)^+) = \{\alpha: \mu(i) < \alpha < \mu(i)^+\}$ implies $\beta \notin f(\gamma)$ (by Case 1). Now we define by induction on $\zeta < \chi = \max\{\kappa^+, \lambda^+\}$ sets $S^i_\zeta \subseteq (\mu(i), \mu(i)^+)$ (for $i < \kappa$) such that

(i) $\alpha \in S^i_\zeta$ implies $f(\alpha)$ is disjoint to $\bigcup_{j<i} S^j_\zeta$,

(ii) $\alpha \in S^i_\zeta$, $\beta \in \bigcup_{\xi<\zeta} S^i_\xi$ implies $\beta < \alpha$,

(iii) $|S^i_\zeta| = \mu(i)$ *or* $|S^i_\zeta| < \mu(i)$ and S^i_ζ is a maximal subset of $(\mu(i), \mu(i)^+)$ satisfying (i) and (ii). Clearly for any $\zeta < \chi$, $\alpha \neq \beta \in \bigcup_{i<\kappa} S^i_\zeta \Rightarrow \alpha \notin f(\beta)$ hence if $|\bigcup_{i<\kappa} S^i_\zeta| = \mu$ we finish. Otherwise, for every $\zeta < \chi$ there is $j(\zeta) < \kappa$ such that $|\bigcup_{i<\kappa} S^i_\zeta| < \mu(j(\zeta))$. As χ is regular and $> \kappa$ there is $j^0 < \kappa$ such that $|U| = \chi$ where $U = \{\zeta < \chi: j(\zeta) \leq j^0\}$; so for all $\zeta \in U$, $|S^{j^0}_\zeta| < \mu(j^0)$. There is α such that $\mu(j^0)$, $\sup \bigcup_{\zeta \in U} S^{j^0}_\zeta < \alpha < \mu(j^0)^+$ so by (iii) for all $\zeta \in U$, $f(\alpha) \cap \bigcup_{j<j^0} S^j_\zeta \neq \emptyset$ hence $|f(\alpha)| \geq \chi > \lambda$, contradiction.

EXERCISE 2.1: Suppose $I \subseteq {}^{\alpha\geq}\lambda$, and $\nu \lessdot \eta$, $\eta \in I \Rightarrow \nu \in I$ and for $\eta \in {}^{\alpha>}\lambda \cap I$ define $\lambda_\eta = |\{i < \lambda: \eta^\frown\langle i\rangle \in I\}|$.

(1) If $\alpha = n < \omega, f: I \rightarrow \chi$ and cf $\lambda_\eta > \chi$ for every $\eta \in {}^{n>}\lambda \cap I$ then there are $J \subseteq I$ and α_l, $l \leq n$ such that $\langle\ \rangle \in J$ and $\eta \in J \cap {}^{n>}\lambda$ implies $\lambda_\eta = |\{i < \lambda: \eta^\frown\langle i\rangle \in J\}|$ and $\eta \in J$ implies $f(\eta) = \alpha_{l(\eta)}$.

(2) If $\alpha = \omega$ and for every $\eta \in I$, cf $\lambda_\eta > \chi^{\aleph_0}$ then there are $J \subseteq I$ and α_l as above.

EXERCISE 2.2: Reprove 2.7 as follows: assume $|X_i| = \lambda = \beth_m(\chi^+)^+$, $m > 1$, $X_i = \lambda \times \{i\}$ (for $i < \mu$) and $\chi^\mu \leq \chi^+$ and define

$$g(\alpha_0, \ldots, \alpha_{m-1}) = \{(c, i_0, \ldots, i_{m-1}): c = f(\langle\alpha_0, i_0\rangle, \ldots, (\alpha_{m-1}, i_{m-1}))\}$$

and use 2.5; and get suitable $S \subseteq \lambda$, $|S| = \chi^+$ w.l.o.g. $S = \chi^+$ and let $Y_i = (\chi i, \chi(i + 1))$ (or χi, $\chi(i + 1)$-ordinal products).

EXERCISE 2.3: Reprove 2.6 similarly, assuming $\lambda = \beth_{mn}(\chi^+)^+$, $\mu = |\text{Range} f|$, $2^\mu \leq \chi^+$ using

$$\begin{aligned} g(\alpha_0, \ldots, \alpha_{mn-1}) &= \\ &= \{(c, \langle i^0_0, \ldots, i^0_{l(0)}\rangle, \ldots, \langle i^{m-1}_0, \ldots, i^{m-1}_{l(m-1)}\rangle): \\ &\qquad c = f(\langle\alpha_{i^0_0}, \ldots\rangle, \ldots, \langle\alpha_{i^{m-1}_0}, \ldots\rangle), l(k) < n, i^k_l < mn\} \end{aligned}$$

get $S \subseteq \lambda$, $|S| = \chi^+$ from 2.5, w.l.o.g. $S = \chi^+$ and let

$I = \{\langle\alpha_0, \ldots, \alpha_l\rangle: l \leq n$, and for each $k \leq l$

$$\lambda^{k+1}\alpha_k < \alpha_{k+1} < \lambda^{k+1}\alpha_k + \lambda \text{ and } 0 < \alpha_0 < \lambda\}.$$

A.3. Various results

LEMMA 3.1: *If $\kappa < \lambda \leq \chi$, χ singular and for every $\mu < \chi$* AD$(\mu, \lambda, \lambda, \kappa)$ *holds, then* AD$(\chi, \lambda, \lambda, \kappa)$ *holds (see Definition* VII, 1.11).

Proof. Clearly if AD$(\mu, \lambda, \lambda, \kappa)$ holds, S is a set of cardinality λ, there is a (λ, κ)-family of μ subsets of S. Let $\chi = \sum_{i<\mu_0} \mu_i$, $\mu_i < \chi$; and let S' be a (λ, κ)-family of subsets of λ, $S' = \{A_i : i < \mu_0\}$. As $|A_i| = \lambda$, there is a (λ, κ)-family, S_i, of μ_i subsets of A_i for $i < \mu_0$. Clearly $\bigcup_{i<\mu_0} S_i$ is a (λ, κ)-family of χ subsets of λ.

EXERCISE 3.1: Prove AD$(\lambda, \lambda, \lambda, 1)$ and natural implications.

LEMMA 3.2: *Suppose* AD$(\chi, \lambda, \mu, \kappa)$ holds.

(1) *Then $\lambda^\kappa \geq \chi$; so $\chi = 2^\lambda$, $\lambda \geq \kappa$ implies $\lambda^\kappa = 2^\lambda$.*

(2) *If $\lambda > 2^\kappa$, $\aleph_\alpha = \min\{\lambda_1 : \lambda_1^\kappa \geq \lambda\}$ and $\lambda \leq \aleph_\mu$, then for some β, $\lambda \geq \aleph_\beta = \beta > \aleph_\alpha > 2^\kappa$.*

Proof. (1) Let $S = \{A_i : i < \chi\}$ be a (μ, κ)-family of subsets of λ. Choose $B_i \subseteq A_i$, $|B_i| = \kappa$; so $i \neq j \rightarrow B_i \neq B_j$, hence $\lambda^\kappa = |\{B \subseteq \lambda : |B| = \kappa\}| \geq |\{B_i : i < \chi\}| = \chi$.

(2) Immediate by VII, 1.9 (λ, κ, μ correspond to $\aleph_\gamma$, μ, χ).

On κ-contradictory orders and κ-skeleton like sequences see Definitions VIII, 3.1 and 3.2 (VIII, Section 3 is the only place they are used).

THEOREM 3.3: *For every $\lambda = \lambda + \kappa^+$, κ regular, there are 2^λ pairwise κ-contradictory orders of cardinality λ.*

We prove 3.3 by a series of claims.

DEFINITION 3.1: The orders I, J will be called strongly κ-contradictory if they have cofinalities $\geq \kappa$ and there are *no* orders I_1, J_1 with cofinality $\geq \kappa$ such that there is a model M with an anti-symmetric relation $<$ and κ-skeleton like sequences $\langle a_s : s \in I_1 + I^* \rangle$ and $\langle b_s : s \in J_1 + J^* \rangle$ such that:

(1) For every $t \in J^*$ and $s^1 \in I_1$ there is $s \in I_1$, $s^1 < s$, $M \models a_s < b_t$.

(2) For every $t \in I^*$ and $s^1 \in J_1$ there is $s \in J_1$, $s^1 < s$, $M \models b_s < a_t$.

(I^* is I with order the inverse of the order of I.)

Remark. By the definition of κ-skeleton like sequences (Definition VIII, 3.1) (1) implies that for every $t \in J^*$ there are $s^0 \in I_1$, $s^1 \in I^*$ such that

$I_1 + I^* \vDash s^0 \leq s \leq s^1$ implies $M \vDash a_s < b_t$. Similarly for (2). We shall use this many times.

CLAIM 3.4: *Any two strongly κ-contradictory orders are κ-contradictory.*

Proof. Immediate.

QUESTION 3.2: Is the converse true?

CLAIM 3.5: *If I, J have distinct cofinalities $\geq \kappa$, then I, J are strongly κ-contradictory.*

Proof. W.l.o.g. $\mathrm{cf}\, J = \mu > \lambda = \mathrm{cf}\, I$, $\lambda \geq \kappa$. Suppose I, J are not strongly κ-contradictory, and we shall get a contradiction. By Definition 3.1, there are $M, I_1, J_1, \langle a_s: s \in I_1 + I^*\rangle, \langle b_t: t \in J_1 + J^*\rangle$ satisfying the conditions mentioned there. Let $\langle s(\alpha): \alpha < \lambda\rangle$ be an increasing unbounded sequence in I, and for each $\alpha < \lambda$ choose $t(\alpha) \in J$ and $t'(\alpha) \in J_1$ such that $J_1 + J^* \vDash t'(\alpha) \leq t \leq t(\alpha)$ implies $b_t < a_{s(\alpha)}$. As J has cofinality $> \lambda$, there is a bound $t(*) \in J$ to $\{t(\alpha): \alpha < \lambda\}$.

So for every $\alpha < \lambda$, $J_1 + J^* \vDash t'(\alpha) \leq t(*) \leq t(\alpha)$, hence $b_{t(*)} < a_{s(\alpha)}$. This contradicts (1) from Definition 3.1 (by the remark to it).

CLAIM 3.6: *If λ is a regular cardinal $> \kappa$, κ regular, $I = \sum_{\alpha<\lambda} I^*_\alpha$, $J = \sum_{\alpha<\lambda} J^*_\alpha$ and $\{\alpha < \lambda$: $\mathrm{cf}\,\alpha \geq \kappa$ and I_α, J_α are strongly κ-contradictory$\}$ $\neq \emptyset \bmod D(\lambda)$ ($A = \emptyset \bmod D(\lambda)$ iff $\lambda - A \in D(\lambda)$, $A \subseteq \lambda$), then I, J are strongly κ-contradictory.*

Proof. Suppose there are a model M, orders I^1, J^1 and sequences $\langle a_s: s \in I^1 + I^*\rangle$, $\langle b_t: t \in J^1 + J^*\rangle$ satisfying the conditions from Definition 3.1, and we shall get a contradiction.

As I_α, J_α are non-empty, we can choose $s_\alpha = s(\alpha) \in I_\alpha$, $t_\alpha = t(\alpha) \in J_\alpha$. Now we define ordinals $\alpha_i = \alpha(i)$ for $i < \lambda$ such that:

(i) $j < i < \lambda$ implies $\alpha_j < \alpha_i < \lambda$,

(ii) for a limit ordinal δ, $\alpha_\delta = \sup\{\alpha_i: i < \delta\}$,

(iii) if $s \in I^*$, $I^1 + I^* \vDash s \leq s_{\alpha(i+1)}$ then $M \vDash a_s < b_{t(\alpha(i))}$,

(iv) if $t \in J^*$, $J^1 + J^* \vDash t \leq t_{\alpha(i+1)}$ then $M \vDash b_t < a_{s(\alpha(i))}$.

Let us define by induction:

Case 1. $i = 0$; then $\alpha_0 = 0$.

Case 2. $i = \delta$ is limit; then $\alpha_\delta = \bigcup_{j<i} \alpha_j$ (it exists as $i < \lambda$ and λ is regular).

Case 3. α_i is defined and we shall define α_{i+1}.

By the remark to Definition 3.1, there is $s^1 \in I^*$ such that $s \in I^*$, $I^1 + I^* \models s \leq s^1$ implies $M \models a_s < b_{t(\alpha(i))}$. Similarly, there is $t^1 \in J^*$ such that $t \in J^*$, $J^1 + J^* \models t \leq t^1$ implies $b_t < a_{s(\alpha(i))}$. Let α_{i+1} be the first ordinal $> \alpha_i$ such that $I^1 + I^* \models s_{\alpha(i+1)} < s^1, J^1 + J^* \models t_{\alpha(i+1)} < t^1$.

Clearly $\{\alpha_i: i < \lambda, i \text{ limit}\}$ is a closed unbounded subset of λ, so it $\in D(\lambda)$. Hence by the hypothesis there is a limit $\delta < \lambda$ such that $\text{cf}\, \alpha_\delta \geq \kappa$ (so clearly $\text{cf}\, \delta = \text{cf}\, \alpha_\delta \geq \kappa$) and $I_{\alpha(\delta)}, J_{\alpha(\delta)}$ are strongly κ-contradictory.

Let us define $I^+ = I_{n(\delta)}$, $J^+ = J_{n(\delta)}$, $I_1^+ = I \restriction \{s_{\alpha(i)}: i < \delta\}$, $J_1^+ = J \restriction \{t_{\alpha(i)}: i < \delta\}$. We define the model N such that $|N| = |M|$, $<^N = \{\langle a, b\rangle: \langle b, a\rangle \in <^M\}$ and $N = (|N|, <^N)$. Now the sequences $\langle a_s: s \in I_1^+ + (I^+)^*\rangle\langle b_t: t \in J_1^+ + (J^+)^*\rangle$ in the model N, and the orders I^+, J^+, I_1^+, J_1^+ satisfy the conditions mentioned in Definition 3.1, hence $I^+ = I_{\alpha(\delta)}$, $J^+ = J_{\alpha(\delta)}$ are not strongly κ-contradictory, contradiction.

CLAIM 3.7: *If λ is regular and $> \kappa$, κ regular, then there is a family of 2^λ pairwise strongly κ-contradictory orders of cardinality λ.*

Proof. By 1.3(3) $A = \{\alpha: \alpha < \lambda, \text{cf}\, \alpha = \kappa\}$ is a stationary subset of λ, and by 1.3(2) there are pairwise disjoint, stationary $A_i \subseteq A$ $(i < \lambda)$. For any set $W \subseteq \lambda$ and $\alpha < \lambda$ let $I_{W,\alpha}$ be κ if $\alpha \in \bigcup_{\beta \in W} A_\beta$ and κ^+ otherwise; and let $I_W = \sum_{\alpha<\lambda} I_{W,\alpha}^*$. By claim 3.5 κ, κ^+ are strongly κ-contradictory. If W, U are distinct subsets of λ, for some γ, $\gamma \in W \Leftrightarrow \gamma \notin U$, so

$$\{\alpha: \text{cf}\, \alpha \geq \kappa; I_{W,\alpha}, I_{U,\alpha} \text{ are strongly } \kappa\text{-contradictory}\}$$

includes A_γ, hence by claim 3.6, I_W, I_U are strongly κ-contradictory. Our conclusion follows immediately.

CLAIM 3.8: *If λ is singular and $> \kappa$, κ regular, then there is a family of 2^λ pairwise strongly κ-contradictory orders of cardinality λ.*

Proof. Let $\lambda = \sum_{\alpha<\mu} \lambda_\alpha$ where $\mu = \text{cf}\, \lambda$, $\kappa < \lambda_\alpha$, λ_α increasing, λ_α regular and choose a regular χ, $\mu + \kappa^+ \leq \chi < \lambda$, and let A_i $(i < \mu)$ be disjoint stationary subsets of $\{\alpha: \alpha < \chi, \text{cf}\, \alpha = \kappa\}$. Let K_α be a family of 2^{λ_α} pairwise strongly κ-contradictory orders of cardinality λ_α. For any $f \in \prod_{\alpha<\mu} K_\alpha$ (i.e., a function with domain μ, $f(\alpha) \in K_\alpha$) and $\beta < \chi$

let $I_{f,\beta}$ be $f(\alpha)$ when $\beta \in A_\alpha$, and κ when $\beta \notin \bigcup_{\alpha<\mu} A_\alpha$. Let $I_f = \sum_{\beta<\chi} I^*_{f,\beta}$; as in 3.7 we can prove (by 3.5, 3.6) that I_f, I_g are strongly κ-contradictory when $f \neq g \in \prod_{\alpha<\mu} K_\alpha$. As $|I_f| = \lambda$ and $|\prod_{\alpha<\mu} K_\alpha| = \prod_{\alpha<\mu} 2^{\lambda_\alpha} = 2^\lambda$ we finish.

Proof of Theorem 3.3. Immediate by Claims 3.7, 3.8, and 3.4.

Definition 3.2: (1) An m-place relation R over a set S is connected [antisymmetric] iff for every distinct $s_0, \ldots, s_{m-1} \in S$ there is a permutation σ of m such that $R(s_{\sigma(0)}, \ldots, s_{\sigma(m-1)})$ holds [does not hold].

(2) If R is an m-place relation over S, σ a permutation over S, then

$$R^\sigma = \{\langle s_0, \ldots, s_{m-1}\rangle : \langle s_{\sigma(0)}, \ldots, s_{\sigma(m-1)}\rangle \in R\}.$$

LEMMA 3.9: (1) *Suppose R is an m-place, connected and antisymmetric relation over ω, and ω is a Δ_R-m-indiscernible sequence, $\Delta_R = \{R(x_0, \ldots, x_m)\}$.*

Then there is a relation $R^*(x_0, \ldots, x_{3m-1})$ which is a Boolean combination of instances of R, such that when $i < k \neq l$

$$R^*(mi, mi+1, \ldots, mi+m-1, mk, mk+1, \ldots, mk+m-1, ml, ml+1, \ldots, ml+m-1)$$

holds iff $k < l$.

(2) *Suppose R is an m-place, connected and antisymmetric relation over $m = \{0, \ldots, m-1\}$. Then for some $n < m-1$ and permutation σ of m*

$$R^\sigma(0, \ldots, n-1, n, n+1, n+2, \ldots, m-1) \equiv \neg R^\sigma(0, \ldots, n-1, n+1, n, n+2, \ldots, m-1).$$

Proof. (1) Let R^* be

$$\bigwedge \{R(x_{h(0)}, \ldots, x_{h(m-1)})^t : h \text{ a function from } m \text{ into } 3m, h \text{ one-to-one, and } t = 0 \Leftrightarrow t \neq 1 \Leftrightarrow R(h(0), \ldots, h(m-1)) \text{ holds}\},$$

and let

$$P(i, k, l) = R^*(im, im+1, \ldots, im+m-1, km, \ldots, lm, \ldots)$$

(so P is a three-place relation over ω).

By the R-m-indiscernibility of ω, $0 < k < l$ implies $P(0, k, l)$ holds. Suppose it holds for $0 < l < k$, and we shall get a contradiction. By the R-m-indiscernibility of ω we can assume $l = 1$, $k = 2$.

Let σ be a permutation of m, $n < m - 1$, then

$$\begin{aligned}
&R^{\sigma}(0, 1, \ldots, m - 1)\\
&\quad \text{iff}\quad R^{\sigma}(0, 1, \ldots, n, m + n + 1, 2m + n + 2, \ldots, 2m + m - 1)\\
&\qquad\qquad\qquad\qquad\qquad\qquad \text{(as } \omega \text{ is } R\text{-}m\text{-indiscernible)}\\
&\quad \text{iff}\quad R^{\sigma}(0, \ldots, n, 2m + n + 1, m + n + 2, \ldots, m + m - 1)\\
&\qquad\qquad\qquad\qquad\qquad\qquad \text{(as } P(0, 2, 1) \text{ holds)}\\
&\quad \text{iff}\quad R^{\sigma}(0, \ldots, n - 1, m + n, 2m + n + 1,\\
&\qquad\qquad\qquad\qquad m + n + 2, \ldots, m + m - 1)\\
&\qquad\qquad\qquad\qquad\qquad\qquad \text{(as } \omega \text{ is } R\text{-}m\text{-indiscernible)}\\
&\quad \text{iff}\quad R^{\sigma}(0, \ldots, n - 1, 2m + n, m + n + 1,\\
&\qquad\qquad\qquad\qquad 2m + n + 2, \ldots, 2m + m - 1)\\
&\qquad\qquad\qquad\qquad\qquad\qquad \text{(as } P(0, 2, 1) \text{ holds)}\\
&\quad \text{iff}\quad R^{\sigma}(0, \ldots, n - 1, n + 1, n, n + 2, \ldots, m - 1)\\
&\qquad\qquad\qquad\qquad\qquad\qquad \text{(as } \omega \text{ is } R\text{-}m\text{-indiscernible).}
\end{aligned}$$

Hence it suffices to prove (2).

(2) Let $n_\sigma = \min\{i \le m\colon \sigma(i) \ne i \text{ or } i = m\}$, assume there are no such σ and n; and we shall prove by downward induction on n_σ that $\sigma \in \Sigma = \{\sigma'\colon R^{\sigma'}(0, \ldots, m - 1) \equiv R(0, \ldots, m - 1)\}$. Clearly our assumption means $\sigma \in \Sigma$, $n < m - 1$ implies $\sigma(n, n + 1) \in \Sigma$ ($(n, n + 1)$ is the permutation interchanging n and $n + 1$). If $n_\sigma = m$ this is trivial, $n_\sigma = m - 1$ is impossible, and $n_\sigma = m - 2$ follows by the assumption.

Suppose we have proved for $n + 1$, and $n_\sigma = n$, and let $\sigma(i) = n$ (so clearly $n < i < m$) and let $\sigma_0 = \sigma(i, i - 1)(i - 1, i - 2)\cdots(n + 1, n)$; clearly $j < n_\sigma$ implies $\sigma_0(j) = \sigma(j) = j$, and $\sigma_0(n) = \sigma(i) = n$, hence $n_{\sigma_0} > n_\sigma$, hence $\sigma_0 \in \Sigma$. But clearly $\sigma = \sigma_0(n, n + 1)(n + 1, n + 2)\cdots(i - 1, i)$ so clearly $\sigma \in \Sigma$.

EXERCISE 3.3: Suppose $|I| \ge 2m$, R an m-place connected and antisymmetric relation over a, $<^1$, $<^2$ orders on I, and $(I, <^1)$, $(I, <^2)$ are R-m-indiscernible sequences.

Then (1) or (2) holds:

(1) $I = I_1 \cup I_2$, (I_1, I_2 disjoint); and $<^1$, $<^2$ are identical on each I_l ($l = 1, 2$) and $s_l \in I_l$ implies $s_1 <^1 s_2$, $s_2 <^2 s_1$.

(2) $I = I_1 \cup I_2 \cup I_3$ (the I_l's pairwise disjoint); $<^1$, $<^2$ are identical on I_2, $|I_1 \cup I_3| \le m - 2$ and $s_l \in I_l$ implies $s_1 <^1 s_2 <^1 s_3$, $s_1 <^2 s_2 <^2 s_3$.

DEFINITION 3.3: A *dependency relation* on a set W, is a relation R between members of W and subsets of W (x depends on w) satisfying the following conditions (where w is called independent if, for no $x \in w$, does x depend on $w-\{x\}$):

(0) x depends on $\{x\}$.

(1) Exchange principle: if $\{x_i : i < \alpha\}$ is not independent, then for some $\beta < \alpha, x_\beta$ depends on $\{x_i : i < \beta\}$.

(2) Finite character: x depends on w iff x depends on some finite $u \subseteq w$ (so we have monotonicity: if x depends on $u, u \subseteq w$, then x depends on w).

(3) (Weak) transitivity: if w, u are independent, x depends on u, and every $y \in u$ depends on w, then x depends on w.

DEFINITION 3.4: A *nice dependency relation* on W is defined similarly, strengthening (3) to

(3)′ Full transitivity: if x depends on u, and every $y \in u$ depends on w (where $u, v \subseteq W$), then x depends on w.

LEMMA 3.10: *For a dependency relation on W, any $w \subseteq W$, any two maximal independent subsets of w have the same power. Also, every independent subset of w can be extended to a maximal independent subset of w.*

Proof. The second sentence follows from the finite character (2). For the first, suppose $u_0 \subseteq w$ is a maximal independent subset of w, and $u \subseteq w$ is an independent subset of w. It suffices to prove $|u_0| \geqslant |u|$ so assume this fails. If $|u|$ is infinite, for every $x \in u$ there is a finite $w_x \subseteq u_0$ on which x depends. By cardinality consideration for some finite $u_0' \subseteq u_0, u' = \{x : w_x = u_0'\}$ is infinite, hence $|u'| > |u_0'|$ and clearly every $x \in u'$ depends on u_0'. So $u_0' \cup u', u_0', u'$ satisfy the assumptions on w, u_0, u and u_0' is finite. So w.l.o.g. u_0 is finite. We choose a counterexample with minimal $|u_0 - u|$. Let $u_0 = \{x_i : i = 1, n\}$ with $u_0 \cap u = \{x_i : i = 1, m\}$. Choose $x_0 \in u - u_0$ (exists as $|u_0| < |u|$). So $\{x_i : i \leqslant m\}$ is independent (being $\subseteq u$), but $\{x_i : i \leqslant n\}$ is not (as $x_0 \in u$ depends on $u_0 = \{x_i : i = 1, n\}$ and $x_i (i \leqslant n)$ are distinct as $x_0 \notin u_0$). So let I, $\{0, \ldots, m\} \subseteq I \subseteq \{0, \ldots, n\}$ be maximal s.t. $\{x_i : i \in I\}$ is independent and w.l.o.g. $I = \{0, \ldots, k\}$ so $m \leqslant k < n$.

Clearly, $u_0' \stackrel{\text{def}}{=} \{x_i : i \in I\}$ is independent, and every member x_i of u_0 depends on it (if $i \in I$ by (0) and monotonicity, if $i \notin I$, by the exchange principle for $\{x_0, x_1, \ldots, x_k, x_i\}$ x_i depends on u_0'). Now for every $x \in u$, x depends on u_0 by assumption and every $y \in u_0$ depends on u_0'. As u_0, u_0' are independent, x depends on u_0'. So u_0', u satisfy the assumptions, but $|u_0' - u| = |\{i : m < i \leqslant k\}| < |\{i : m < i \leqslant n\}| = |u_0 - u|$ as $k < n$ and u_0', u form a counterexample (as $|u_0'| \leqslant |u_0| \leqslant |u_0|$). This is a contradiction to the choice of u_0, u.

HISTORICAL REMARKS

Unfortunately, though this will surely be the most carefully read part of the book, the author had become allergic to writing by the time he reached this section. So he apologizes in advance for any inaccuracy. Theorems and notions not credited are due to the author and/or are trivial. Chapters II, III, VII and VIII were distributed in Spring, '74; most of Chapter V in the Fall. It is still valuable to look at [Sh 71], Section 0.

I.1

The theorems are classical, for credits see, e.g. [CK 73] (in 1.3 the use of VIII, 4.7 can be easily avoided by constructing the expansion by approximations of cardinality $< \lambda$).

I.2

The notion "T totally transcendental" ($\neq \aleph_0$-stable, by our terminology, as totally transcendental here doesn't imply countable), was suggested by Morley [Mo 65], "T stable in λ", by Rowbottom [Ro 64], "stable", "superstable" by [Sh 69] and [Sh 69a] "indiscernible sequence" by Ehrenfeucht–Mostowski [EM 56], "indiscernible set over A", by Morley [Mo 65]. Lemmas 2.1, 2.12, 2.13 and 2.9 are trivial, 2.4 was deduced by Ehrenfeucht–Mostowski [EM 56], from Ramsey theorem [Ra 29]. Theorem 2.8 proved through 2.5, 6 and 7 is from Shelah [Sh 72b]. Morley proved 2.8 for models of countable $\aleph_0$-stable theories. His proof was by choosing a type $p \in S^m(A)$ with minimal rank $\alpha = R(p, \mathrm{L}, \aleph_0)$, realized by $> \lambda$ member of I, choose inductively $a_i \in I$ realizing p such that $R[\mathrm{tp}(\bar{a}_i, A \cup \bigcup_{j<i} \bar{a}_j] = \alpha$, and prove it is an indiscernible sequence like 2.5. See II, 2.17, 8 and 9 for a similar proof. Notice that for stable T, there is no corresponding to minimal. Rowbottom has a weaker unpublished result (a direct generalization of

[Mo 65]). In [Sh 69] and [Sh 70] there were approximations to this. See III, 4.23 for another way to prove. Theorems 2.10 and 2.11 are from [Sh 72b]. (Erdös and Makkai [ErM 66] is the case $\lambda = \aleph_0$ of 2.11 for the parallel combinatorial theorem.)

II.1

Morley [Mo 65], influenced by the Cantor–Bendixon rank, defines $R^m(p, \mathrm{L}, \aleph_0)$ and $\mathrm{Mlt}^1(p, \mathrm{L}, \aleph_0)$ for complete p (in different notation). He proves for this version the parallels of 1.1, 1.2, 1.4(3), 1.6, 1.9(1), 1.10(1) and (2). In [Sh 69] and [Sh 69a] $R^m(p, \Delta, 2)$ was defined for complete Δ-types, the interesting case was Δ finite or singleton whose research was continued in [Sh 71], where the completeness was dropped; $R^m(p, \mathrm{L}, \infty)$ was used in [Sh 70a]. In [Sh 72a] $R^m(p, \mathrm{L}, \aleph_0)$ for p not necessarily a Δ-type, and [Sh 71d] introduce $R^m(p, \Delta, \lambda)$. The "(Δ, n)-indiscernibility" was introduced in [Sh 71].

Exercises 1.5 and 1.6 are due to L. Sfard and A. Hinkis, resp., in their Master thesis.

II.2

Everything is from [Sh 71]. Concerning the stability spectrum (2.13) see III, Section 3. Baldwin [Bl 70a] proved independently, a weaker version of 2.12(1) (for T totally transcendental A a definable set). 2.17, 8 and 9 generalize the proof of Morley [Mo 65] on the existence of indiscernible sets. Concerning 2.20, Harnik and Ressayre [HR] independently prove that if T stable in λ, $2^\mu > \lambda$, I indiscernible over A, then for every $\bar{a}$ there is $J \subseteq I$, $|J| < \lambda$, $I - J$ is indiscernible over $A \cup \bar{a}$, and Shelah [Sh 70] proved there is $J \subseteq I$, $|J| < \kappa(T)$, such that $I - J$ is indiscernible over $A \cup \bar{a} \cup \bigcup J$ (see III, Section 3 for definition, but $\kappa(T)$ is always $\leq \mu$, and sometimes $< \mu$) and Shelah [Sh 71] proved 2.20 which is the finite version.

Morley [Mo 65] proved that if there is an indiscernible sequence which is not an indiscernible set, T is not $\aleph_0$-stable, exhibiting in the proof order (see 2.16).

II.3

Morley [Mo 65] proved that for countable $\aleph_0$-stable T, $\sup_p R^m(p,\mathrm{L},\aleph_0) < \omega_1$; and Lachlan [La 71] solving a question from [Mo 65] remove the "$\aleph_0$-stable" from the hypothesis. Theorem 3.1 is a slight generalization. Morley [Mo 65] proves 3.3 and 3.2 in different terminology. $D^m(p, \Delta, \lambda)$

was defined in [Sh 71], Section 6, some of its properties and its connection to superstability 3.9 and 3.14 were established.

More on ranks, see III, Section 4, V, Section 7, and Baldwin and Blass [BB 74].

II.4

The finite cover property was suggested by Keisler [Ke 67] (for his order, see VI), and prove the property (E) implies it (Ehrenfeucht property from [Eh 57]—there is a formula $\varphi(x_1, \ldots, x_n)$ defining on an infinite set a connected antisymmetric relation; (E) is related to but weaker than the order property); so this is related to 4.2.

Most of the section is from [Sh 71], but 4.16 (if some $\varphi(\bar{x}; \bar{y})$ has the strict order property, then some $\varphi(x; y)$ has the strict order property) conjectured in [Sh 71], is of Lachlan [La 75a], and 4.4 was added lately.

III

Sections 1, 2, 3 and 5 are from [Sh 71a] (with some additions; 5.17, the proof of 2.8), some parts are exposed in [Sh 75c] with different proofs.

III.1–3

The stability spectrum theorem evolves as follows. Morley [Mo 65] proves that if T is countable, stable in $\aleph_0$ then it is stable in every λ. Successively and independently Rowbottom [Ro 64] (using GCH) and Ressayre [Re 69] prove that if T is stable in λ, $\lambda < \lambda^{\aleph_0}$ then T is stable in every $\mu \geq \lambda$, and Shelah [Sh 68a] [Sh 69] proves that if T is stable in one λ, it is stable in every $\lambda = \lambda^{|T|}$, and if T is stable in one λ, $\lambda < \lambda^{\aleph_0}$, T is stable in every $\mu \geq 2^{|T|}$. In [Sh 69b] and [Sh 70] 3.8(1) is proved and in [Sh 71d], 5.15 was proved.

Theorem 3.12 for $\lambda > |T|$ was proved by Harnik [Ha 75], seeing a proof of VI, 5.3 (then III, 3.10 was essentially included in it) and then 3.11 for cf $\lambda > |T|$ was clear.

III.4

A result based on this was announced in [Sh 71], p. 275–6. Concerning 4.6(3), Lachlan [La 72], confirming a conjecture of L. Blum, proved $\mathrm{Mlt}^1[\mathrm{tp}(\bar{a}, |M|), L, \aleph_0] = 1$. In different terms, some of this, mainly the symmetry lemma 4.13, was developed later and independently by Lascar [Ls 73], [Ls 75], who concentrates on superstable T and complete types over models (see also on V, 7.12).

III.5

Theorem 5.18 was proved independently by Lachlan [La 74] (using a version of 2.8) and Shelah (it is immediate from [Sh 71a], Lemmas 38, 40 (p.106, 108), as noticed somewhat later). Theorem 5.14 is from [Ke 71a].

III.7

Theorems 7.4 and 7.2 were proven for countable T by Keisler [Ke 76] (so $\kappa = \aleph_0$) partially confirming a conjecture 4E, p. 330 [Sh 71] (see (4) there). He proved a version of 7.3 by induction.

IV.1,2

Vaught [Va 61] investigates $\mathbf{F}^t_{\aleph_0}$-prime models over $\emptyset$, Morley uses $\mathbf{F}^t_{\aleph_0}$-prime model (mainly for $\aleph_0$-stable T), Ressayre [Re 69] and Shelah [Sh 69], [Sh 69a] use, independently, $\mathbf{F}^t_\lambda$-prime, isolated. In [Sh 70] $\mathbf{F}^s_\lambda$-prime, isolated, were used. About the same time and independently Lachlan [La 72] (ad hoc) and Shelah (see [Sh 71] 7) p. 275) use $\mathbf{F}^l_{\aleph_0}$. Fragments of lemmas here were parts of proofs. In Rowbottom [Ro 64], Ressayre [Re 69] and Shelah [Sh 69], [Sh 69a], 2.18(1) appeared for $\mathbf{F}^t_\lambda$, $2^\lambda > \mu$, T stable in μ.

Morley [Mo 65] proved 2.18(1) for $\mathbf{F}^t_{\aleph_0}$, T $\aleph_0$-stable.

IV.3

For the appropriate specific $\mathbf{F}$'s, 3.1, 3.10, 3.12 and 3.17 appear in Morley [Mo 65], Ressayre [Re 69], Shelah [Sh 69], [Sh 69a]; 3.2 appear in [Mo 65] and [Re 69]; 3.3–3.8, 3.11, 3.16, 3.18 were not traced exactly by the author; 3.9 for $\mathbf{F}^t_\lambda$ is due to Ressayre (unpublished) and 3.15 is essentially from [Sh 72a].

IV.4

Most results appeared in [Sh 71d] and [Sh 72a], and were announced in [Sh 71].

IV.5

Theorems 5.1 and 5.2 for $\lambda = \aleph_0$ are from Vaught [Va 61], and generally from Harnik and Ressayre [HR 71]. For 5.3 see credits in [CK 73], 5.4 is of Vaught [Va 61]. Theorem 5.6 was announced in [Sh 74b]. Theorem 5.13 (without maximality) is due to Silver, answering a question from [Mo 65]. Theorem 5.17 for countable T, is due to Grilliot [Gr 72], and the main case was announced in [Sh 76].

V.1

Minimal formulas were introduced by Marsh [Mr 66], who proved the existence of dimension for it. Harnik and Ressayre [HR 71] generalize this to minimal types, weakly minimal formulas were introduced in [Sh 74].

V.2

The part of Theorem 2.10 asserting the equivalence of (1), $(3)_I$ and $(4)_\lambda$ is from [Sh 70], for regular λ. Harnik [Ha 76] proved, independently of this, the following weaker result (GCH): every T satisfies exactly one of the following; (i) for every $\mu > \lambda \geq |T|$, λ regular, T has a maximally λ-saturated model of cardinality $\geq \mu$, (ii) there is λ_0 such that every λ_0-saturated model of T of cardinality $> \mu^{|T|}$ is μ^+-saturated. Harnik [Ha 75a] and the author independently observe that the regularity is not needed. Most of the theorem was proved in [Sh 71d].

V.3

Theorems 3.11, 3.19, and Exercises 3.7–3.16 onward were added after the author saw Lascar [Ls 76], and was motivated by problems there.

V.4

Exercise 4.4 was asked in the first version and was answered independently by Rosental and Shelah on the one hand and Lascar on the other hand (using algebraically closed fields).

V.6

On n-cardinals theorems generally see [CK 73]. In [Sh 69] (see [Sh 71]) the 2-cardinal theorem for stable theories was proved (6.14(2)) and then the n-cardinal theorem was announced in [Sh 72d]. Later Forrest [Fo 7x] proved those theorems. In [Sh 69] models of stable theories with the Chang quantifier Q^{ec} were also discussed, and the transfer theorem proved (starting from a regular cardinal).

Lachlan [La 72] proved a much stronger theorem: 6.14(1) (if T countable and stable, $P(M) \subseteq N \prec M$, then there is M_1, $M \prec M_1$, $M \neq M_1$, $P(M_1) = P(M)$). Baldwin [Bl 75] gives another proof. Harnik [Ha 75a] generalizes this to λ-compact models, with P replaced by a type.

Theorem 6.16 was proved by Lascar [Ls 76] for T $\aleph_0$-stable, λ regular.

V.7

Gaifman [Ga 73] proved 7.4(2) (without stability assumptions). Baldwin [Bl 73], proves $\sup_p R^m(p, \mathrm{L}, \aleph_0) < \omega$ for T $\aleph_1$-categorical; essentially he proved 7.9 for $\varphi(x, \bar{b})$ minimal; he solved by this one of the problems from Morley [Mo 65]. Erimbetov [Er 75] and the author proved 7.11(2) independently.

The rank $L(p)$ and Theorem 1.12 on it are due to Lascar [Ls 76].

VI.1

Ultraproducts were introduced by Łoš and revived in Frayne, Morel and Scott [FMS 62]. See [CK 73] for references.

VI.2

For most results see [CK 73] for references, Theorems 2.6 (and 2.7, 2.8), 2.9 and 2.10 are due to Shelah, and announced in [Sh 75b] (a weaker form of 2.6 was announced in [Sh 71g]). Exercise 2.10 for ultrafilters was asked in Keisler [Ke 67a], proved in [Sh 70b]. Koppelberg [Kp 75] proved this for filters (which is the exercise) by another proof and in the exercise we hint how the proof in [Sh 70b] works for this too.

VI.3

Keisler [Ke 64] proved the existence of a good $\aleph_1$-incomplete ultrafilter over λ when $\lambda^+ = 2^\lambda$, Kunen [Ku 72] eliminates this hypothesis, and we represent this result in 3.1–3.4; 3.5 is Keisler's theorem on isomorphic ultrapower (see [Ke 64a]), 3.6 is of Frayne, Morel and Scott [FMR 62]; 3.7 and 3.8 is of Keisler [Ke 65]. Exercise 3.1–3.5 represent [Sh 71c] (with a slight improvement), Exercises 3.6 and 3.7 are from [Sh 72c], and Exercise 3.11(2) was announced in [Sh 70]. Exercises 3.17, 3.18 and 3.19 are of Benda [Be 72]. For uniform filters and Exercises 3.20–3.23 see [CK 73]. Exercise 3.26(1) and Lemma 3.9 for $M_n = P^{M_n}$ is from Ellentuck and Rucker [ER 72]. The results of the rest of the section were announced in [Sh 76b]. Keisler [Ke 67a] shows (assuming GCH) that for "almost all sets S of successor cardinals, $\leq \lambda^+$, for some ultrafilter D over λ, $S = \omega \cup \{\prod n_i/D : n_i < \omega\}$; he uses products of ultrafilters, and the operation from Exercise 3.20. On Boolean ultrapower (Definition 3.8, Exercise 3.32) see Mansfield [Mn 71]. Theorem 3.12 answers a question of Keisler (see [CK 73]).

VI.4

Most of the results in Sections 4, 5 and 6 appear in [Sh 72], and announced in [Sh 71]. Keisler's order, and 4.1, 4.2, 4.5 and 4.6 are of Keisler [Ke 67]. Independently Keisler and the author prove 4.7 for the T_{ord}, the theory of infinite linear order. On limit ultrapower (Definition 4.2, Exercise 4.10) see Keisler [Ke 63]. Concerning 4.3, Keisler [Ke 67] gives a similar theorem with a much stronger condition on T. By Benda [Be 72] the condition could be weakened to: T is $\otimes$-maximal if for some $\varphi(x, y)$ for every n, and $S \subseteq \{w: w \subseteq n\}$,

$$(\exists y_1, \ldots, y_n)\left(\bigwedge_{w \subseteq n} \left[(\exists x) \bigwedge_{i=1}^{n} \varphi(x, y_i) \right]^{\mathrm{if}(w \in S)} \right).$$

VI.5

Conjecture 5.1 was made in [Sh 72], later and independently Eklof asks the same question (in a little different terminology). Keisler and Prikry [KP 74] prove a positive answer for good ultrafilters; see also Exercise 5.10, 5.11 and 5.12 (which is the result of [Kp 74]).

Concerning 5.8, Keisler [Ke 67] proved that $\otimes$-minimal theories are not $\otimes$-maximal, and does not have the f.c.p. Exercise 5.8 is of Keisler [Ke 63], and Exercise 5.7 was proved by Morley independently of Theorem 5.3.

VI.6

Ultralimits (Definition 6.1) are from Kochen [Ko 61] and Keisler [Ke 63], 6.1 should have appeared there; on μ-descending complete filters see [CK 73]; 6.3 appear in [Sh 72], but it was not traced.

VII.1

The main results of the section were announced in [Sh 71e]. On 1.1–1.4 see [CK 73] for references. On AD (see Definition 1.11) see Baumgartner [Ba 70] [Ba 76].

Exercise 1.7 was not traced.

VII.2

Ehrenfeucht–Mostowski [EM 56] introduce the models (in our notation) $EM(I)$, I an order, and prove their existence, i.e., 2.3. We give here a generalization. Ehrenfeucht [Eh 57] proves a variant of 2.4.

[EM 56] proved $EM(I)$ realized few types, and Morley [Mo 65] proved that for well ordered I, the model is stable in appropriate cardinalities (i.e., 2.7(2) and 2.9(1)).

VII.3

Theorem 3.6 is from [Sh 74]. Theorem 3.7 was first proved assuming there is a measurable cardinal. The proof is hinted in Exercise 3.3. (The main consequence was announced in [Sh 72e] with measurable, and in [Sh 73] without measurable.)

VII.4

The main result was announced in [Sh 73].

VII.5

Most references are well known (starting with Morley's omitting type theorem). Now 5.5(8) seems new; $\mu(\lambda) = \beth_{\delta(\lambda)}$ was by Barwise and Kunen [BK 71] (and see Shelah [Sh 72d]).

VIII.1

1.1 is trivial, 1.2(1) is essentially due to Ehrenfeucht [Eh 58]; it seemed 1.3 was first mentioned in [Sh 71]. Concerning 1.7(2), Keisler [Ke 70] proved that when $|D(T)| = 2^{\aleph_0}$ there are $2^{\aleph_1}$ models for which the sets of types realized in distinct models are distinct. In [Sh 72e] (with measurable) and [Sh 73] (without) 1.7(2) was announced.

Concerning 1.8, the simplest example ($T = \mathrm{Th}(M)$, $M = ({}^{\omega}2, E_0, F_0, \ldots)$, $\eta E_n v \Leftrightarrow \eta \restriction n = v \restriction n$ and $(F_n(\eta))[k] = \eta[k]$ iff $n \neq k$) was proved by Baumgartner and Laver.

VIII.2,3

Ehrenfeucht [Eh 57] proved that if T has property (E) (which implies unstability) $\lambda = 2^{\mu}$ then it has at least two non-isomorphic models in λ. Scott replaces $\lambda = 2^{\mu}$ by $\lambda = \mu^{\kappa} > \mu$. In [Sh 69a] it is proved: if T is unstable, $|T_1| = \aleph_\alpha < \aleph_\beta$, then $I(\aleph_\beta, T_1 T) \geq |\beta - \alpha|$ and similar results for unsuperstable T). In [Sh 71b], 3.1 and 3.2 for $\kappa = \aleph_0$, $\lambda > |T|$ are proved (using only Case 1 and a proof similar to 3.1) and in [Sh 73] the whole result (3.3, 3.1) was announced.

In [Sh 74a] the proofs are sketched and they are discussed also in the introduction of [Sh 75c].

VIII.4

In 4.1 the fact that T is stable in μ, when $|T| \leq \mu < \lambda$, was proved by Morley for $|T| = \aleph_0$, $T = T_1$, and it was observed that the proof is given generally by Keisler [Ke 71], Cudnovski [Cu 70] (for countable), Rowbottom [Ro 64], Ressayre [Re 69] and Shelah [Sh 68] [Sh 69a] (independently). Keisler [Ke 67] proved for countable T, T does not have the f.c.p.

In Theorem 4.3, for countable T, (1) $\Rightarrow$ (3) is essentially proved in Keisler [Ke 66]; and (2) $\Rightarrow$ (3) was proved, independently by Cudnovskii [Cu 70], Keisler [Ke 71] and Shelah [Sh 68] [Sh 69] [Sh 69a]. The same holds for 4.4. Most cases of 4.7 which are proved here are from [Sh 70]. Keisler [Ke 66] proved that 4.3(1) (hence (2), (3)) holds when $\mu = \aleph_1 > |T|$.

IX.1

Theorem 1.4 answered a question of Sabbagh [Sb 75] who gave modules as a counterexample (so its theory is stable but not superstable, see Exercise 1.2 for another example). Theorem 1.6 is due independently to Engeler [En 59], Ryll–Nardzewski [Ry 59] and Svenonius [Sv 59]. Theorem 1.8(1)–(4) and 1.9 is Morley categoricity theorem confirming Łŏs' conjecture from [Mo 65], and (5) was added in Morley [Mo 67]. Lemma 1.10 and many others are essentially from [Sh 74]. Concerning 1.11(1) Baldwin [Bl 73] proved for $\aleph_1$-categorical countable T that $\sup_p R^m(p, L, \aleph_0) < \omega$ and also 1.1(2) in this context.

Theorem 1.15, first part, answers a question in [Mo 65]. Rowbottom [Ro 64], Ressayre [Re 69] and Shelah [Sh 68a] [Sh 69] [Sh 69a] were successive and independent approximations. Rowbottom proves (GCH) that categoricity in $\lambda_0 > \chi_0 = \min\{\chi: \chi^{\aleph_0} > \chi \geq |T|\}$, implies T is categorical in each $\lambda \geq \lambda_0$. Ressayre eliminates GCH, and proves it also for $|T|^+ < \lambda_0 < \chi_0$, and mostly for $\lambda_0 = |T|^+$, and shows T is categorical in some $\lambda < \beth_{(2^{|T|})^+}$, Shelah shows it for $\lambda_0 > |T|$, $\lambda_0 \neq \chi_0$, and that T is categorical in some $\lambda < \mu(|T|)$. The final answer is [Sh 74]. The second part of 1.15 answers a question of Keisler and so is 1.16(i); they were announced in [Sh 70c], [Sh 71f]. 1.16(ii) too answers a question of Keisler, and appears in [Sh 70].

On countable T categorical in $\aleph_1$ see also Baldwin [Bl 72] [Bl 72a], Dickmann [Di 73] and Makowski [Ma 74]; on uncountable T categorical in $|T|^+$ see also Andler [An 75].

Concerning 1.19, it was asked in Morley [Mo 65], Keisler [Ke 71a] proves it when $\aleph_0 < |T| < 2^{\aleph_0}$, $|T|$ regular, and Shelah [Sh 70a] when $|T|^{\aleph_0} = |T|$.

Theorem 1.20 was announced in [Sh 71].

IX.2

In Theorem 2.2(1) one inequality ($\le \aleph_0$) is of Morley [Mo 67], the other inequality and 2.2(2) and (3) are of Baldwin and Lachlan [BL 71], thus answering a question from [Mo 65]. Theorem 2.1(1) is of Lachlan [La 73], Lascar [Ls 76] gave an alternative proof, and here we give another (but in Lachlan's proof we can weaken a little the superstability demand). Morley [Mo 70] proves $I(\aleph_0, T_0) > \aleph_1 \Rightarrow I(\aleph_0, T_0) = 2^{\aleph_0}$ (thus partially confirming Vaught conjecture). The first part of 2.1(2) is from [Sh 70] and confirms a conjecture of Harnik, the second, for T countable is of Lachlan [La 75]. Theorem 2.3 was essentially announced in [Sh 71], and much more than 2.4 in [Sh 74] and [Sh 75c]. Latter and independently Lachlan proves 2.4 (for countable totally transcendental theories). Exercise 2.2 is from Harnik and Ressayre [HR 71].

X

This is based on [Sh 82a], [Sh 82b] (and we thank the *Israel Journal of Mathematics* for permission to use part of it). It was done in 1972 (with the parallel for $\aleph_0$-stable T), and announced, for example, in [Sh 74], and distributed in 1979 and 1980, respectively. The main change here is in writing down fully the proofs for the *IE* case.

It was planned in this series to compute $I(\lambda, T)$ for T totally transcendental (the complication being as in Example XIII, 4.13), hence the trivial cases of computing $I(\lambda, T)$ for λ large enough were not written down. (See Ex. XI, 3.1.) However, Saffe computed and proved the hard cases for T totally transcendental.

XI

This chapter, as well as Chapters XII and XIII, were written in 1982, circulated in September 1983. It continues on the appropriate parts of [Sh 82a], [Sh 86] (but is more general).

XII

Done and announced in early 1982.

XIII

Completed, done and announced in the Fall of 1982.

A.1

See [CK 73] for references for 1.1–1.6; Theorem 1.7 appears in [Sh 71] (in different notation) (the finite case was completed together with Perles) and Sauer [Su 72] proved later and independently 1.7(2) (the finite case).

A.2

Theorem 1.1 is of Ramsey [Ra 29]; Theorems 2.2, 2.3 and 2.4 form a weak version of the Halperin and Lauchli [HL 60] theorem (more exactly, a variant of it which Laver proved, and Pincus showed inessential changes in their proof gives); 2.5 is the Erdös–Rado Theorem (see [EHR 65]; 2.6 was mentioned in [Sh 71]; 2.7 is from Erdös, Hajnal and Rado [EHR 65]. Theorem 2.8 is of Hajnal [Hj 61].

A.3

For 3.1 see [Sh 71b] (inside the proof) and [Ba 76], for 3.2 and 3.3 see [Sh 71b]. Theorem 3.9 is essentially from Morley [Mo. 65].

REFERENCES

[An 75] D. Andler, Semi-minimal theories and categoricity, *J. Symb. Logic* **40** (1975) 419–440.

[BSh 85] J. T. Baldwin and S. Shelah, Classification of theories by second order quantifiers, *Notre Dame J. of Formal Logic* **26** (1985) 229–303.

[Ba 70] J. E. Baumgartner, Results and independence proofs in combinatorial set theory, Ph.D. Thesis, University of California, 1970.

[Ba 76] J. E. Baumgartner, Almost-disjoint sets, the dense set problem and partition calculus, *Annals of Math. Logic* **9** (1976) 401.

[Ba 76a] J. E. Baumgartner, A new class of order types, *Annals of Math. Logic* **9** (1976) 187.

[BB 73] J. T. Baldwin and A. Blass, Skolemization and categoricity, *Notices A.M.S.* **20** (1973) A-588.

[BB 74] J. T. Baldwin and A. Blass, An axiomatic approach to rank in model theory, *Annals Math. Logic* **7** (1974) 295–324.

[Be 72] M. Benda, On reduced products and filters, *Annals of Math. Logic* **4** (1972) 1–29.

[BK 71] J. Barwise and K. Kunen, Hanf numbers for fragments of $L_{\infty,\omega}$, *Israel J. Math.* **10** (1971) 306–320.

[Bl 70] J. T. Baldwin, Countable theories categorical in uncountable power, Ph.D. Thesis, Simon Fraser University, 1970.

[Bl 70a] J. T. Baldwin, A note on definability in totally transcendental theories, *Notices A.M.S.* **17** (1970, Nov.) 1087.

[Bl 72] J. T. Baldwin, Almost strongly minimal theories, I, *J. Symb. Logic* **37** (1972) 481–493.

[Bl 72a] J. T. Baldwin, Almost strongly minimal theories, II, *J. Symb. Logic* **37** (1972) 657–666.

[Bl 73] J. T. Baldwin, α_T is finite for $\aleph_1$-categorical T, *Trans. A.M.S.* **181** (1973) 37–51.

[Bl 73a] J. T. Baldwin, The number of automorphisms of a model of an $\aleph_1$-categorical theory, *Fund. Math.* (1)**83** (1973), 1–6.

[Bl 75] J. T. Baldwin, Conservative extensions and the two cardinal theorem for stable theories, *Fund. Math.* **88** (1975) 7–9.

[BL 71] J. T. Baldwin and A. H. Lachlan, On strongly minimal sets, *J. Symb. Logic* **36** (1971) 79–96.

[BuSh 89] S. Shelah and S. Buechler, On the existence of regular types, *Annals of Pure and Applied Logic* **45** (1989) 277–308.

[CHL 86] G. Cherlin, L. Harrington and A. Lachlan, $\aleph_0$-categorical, $\aleph_0$-stable structures, *Annals of Pure and Applied Logic* **28** (1986) 103–136.

[CK 73] C. C. Chang and H. J. Keisler, *Model Theory*, North-Holland Publ. Co., Amsterdam, 1973.

[Cu 70] G. V. Čudnovskiĭ, Questions of the theory of models that are connected with categoricity, *Algebra i Logika* **9** (1970) 80–120.

[Di 73] M. Dickmann, The problem of a non-finite axiomatizability of $\aleph_1$-categorical theories, *Proc. of Russell Memorial logic conference*, Leeds (1971), 1973, 141–216.

[Eh 57] A. Ehrenfeucht, On theories categorical in power, *Fund. Math.* **44** (1957) 241–248.

[Eh 58] A. Ehrenfeucht, Theories having at least continuum many non-isomorphic models in each infinite power, *Notices A.M.S.* **5** (1968) 680.

[EM 56] A. Ehrenfeucht and A. Mostowski, Models of axiomatic theories admitting automorphism, *Fund. Math.* **43** (1956) 50–68.

[En 59] E. Engeler, A characterization of theories with isomorphic denumerable models, *Notices A.M.S.* **6** (1959) 161.

[Er 75] M. M. Erimbetov, Complete theories with 1-cardinal formulas, *Algebraica i Logika* **14** (1975) 255–257.

[ER 72] E. Ellentuck and V. B. Rucker, Martin's axiom and saturated models, *Proc. A.M.S.* **34** (1972) 243–249.

[ErM 66] P. Erdös and M. Makkai, Some remarks on set theory X, *Studia Scientiarum Mathematirum Hungarica* **1** (1966) 157–159.

[EHR 65] P. Erdös, A. Hajnal and R. Rado, Partition relations for cardinal numbers, *Acta Math.* **16** (1965) 93–196.

[Fl 71] U. Felgner, Comparisons of the axiom of local and universal choice, *Fund. Math.* **71** (1971) 43–62.

[Fo 7x] Forrest, Some results for ω-stable theories.

[FMS 62] T. Frayne, A. Morel and D. Scott, Reduced direct products, *Fund. Math.* **51** (1962) 195–228.

[Ga 73] H. Gaifman, Intuitive syntactical characterizations of some semantical properties, mimeograph, Stanford, 1973.

[Ga 75] H. Gaifman, Global and local choice function, *Israel J. Math.* **22** (1975) 257–265. (Announced *Notices A.M.S.* (1968) 947.)

[Gr 72] T. J. Grilliot, Omitting types; application to recursion theory, *J. Symb. Logic* **37** (1972) 81–89.

[GS 73] F. Galvin and S. Shelah, Some counterexamples in partition calculus, *J. Comb. Theory, Series A* **15** (1973) 167–174.

[GSh 83] R. Grossberg and S. Shelah, On universal locally finite groups, *Israel J. Math.* **44** (1983) 289–302.

[Ha 71] V. Harnik, Stable theories and related concepts, Ph.D. Thesis, The Hebrew University, Jerusalem, Israel, 1971.

[Ha 75] V. Harnik, On the existence of saturated models of stable theories, *Proc. A.M.S.* **52** (1975) 361–367.

[Ha 75a] V. Harnik, A two cardinal theorem for sets of formulas in a stable theory, *Israel J. Math.* **21** (1975) 7–23.

[HR 71] V. Harnik and J. P. Ressayre, Prime extensions and categoricity in power, *Israel J. Math.* **10** (1971) 172–185.

[He 73] S. H. Hechler, Independence results concerning the number of nowhere dense sets necessary to cover the real line, *Acta Math. Acad. Sci. Hungar.* **22** (1973) 27–32.

[He 74] S. H. Hechler, On the existence of certain cofinal subsets of ${}^{\omega}\omega$, *Proc. Symp. in Pure Math.*, XIII, Part II, A.M.S., Providence, R.I. (1974) 155–173.

[Hj 61] A. Hajnal, Proof of a conjecture of S. Ruziewicz, *Fund. Math.* **50** (1961) 123–128.

[HL 60] J. D. Halperin and H. Lauchli, A partition theorem, *Trans. A.M.S.* **124** (1960) 360–367.

[HSh 89] E. Hrushovski and S. Shelah, Dichotomy theorem for regular types, *Annals of Pure and Applied Logic* **45** (1989) 157–169.

[Je 71] T. Jech, *Lectures in set theory*, Lecture Notes 217, Springer Verlag, Berlin (1971).

[K 66] A. Kino, On definability of ordinals in logic with infinitely long expressions, *J. Symb. Logic* **31** (1966) 365–375.

[Ke 63] H. J. Keisler, Limit ultrapowers, *Trans. A.M.S.* **107** (1963) 382–408.

[Ke 64] H. J. Keisler, Good ideals in field of sets, *Annals of Math.* **79** (1964) 338–359.

[Ke 64a] H. J. Keisler, Ultraproducts and saturated models, *Indag. Math.* **26** (1964) 178–186.

[Ke 65] H. J. Keisler, Ideals with prescribed degree of goodness, *Annals of Math.* **81** (1965) 112–116.

[Ke 66] H. J. Keisler, Some model theoretic results for ω-logic, *Israel. J. Math.* **4** (1966) 249–261.

[Ke 67] H. J. Keisler, Ultraproducts which are not saturated, *J. Symb. Logic* **32** (1967) 23–46.

[Ke 67a] H. J. Keisler, Ultraproducts of finite sets, *J. Symb. Logic* **32** (1967) 47–57.

[Ke 70] H. J. Keisler, Logic with the quantifier "there exists uncountably many", *Annals of Math. Logic* **1** (1970) 1–94.

[Ke 71] H. J. Keisler, *Model theory for infinitary logic*, North-Holland Publ. Co., Amsterdam, 1971.

[Ke 71a] H. J. Keisler, On theories T categorical in their own power, *J. Symb. Logic* **36** (1971) 240–244.

[Ke 76] H. J. Keisler, Six classes of theories, *J. Australian Math. Soc.* **21** (1976) 257–265.

[Kk 77] D. Kueker, *Annals Math. Logic* **11(1)** (1977) 57–103.

[Ko 61] S. Kochen, Ultraproducts in the theory of models, *Annals of Math.* **72** (1961) 221–262.

[Kp 75] S. Koppelberg, Homomorphic images of a σ-complete Boolean algebra, *Proc. A.M.S.* **51** (1975) 171–175.

[KP 74] H. J. Keisler and K. Prikry, A result concerning cardinalities of ultraproducts, *J. Symb. Logic* **39** (1974) 43–48.

[Kr 66] K. Kuratowski, *Topology*, Academic Press, New York, 1966.

[Ku 72] K. Kunen, Ultrafilters and independent sets, *Trans. A.M.S.* **172** (1972) 299–306.

[La 71] A. H. Lachlan, The transcendental rank of a theory, *Pacific J. Math.* **37** (1971) 119–122.

[La 72] A. H. Lachlan, A property of stable theories, *Fund. Math.* **77** (1972) 9–20.

[La 73] A. H. Lachlan, The number of countable models of a countable superstable theory, *Proc. of the Inter. Congress on Logic, Methodology and the Philosophy of Science*, Rumania (Aug.–Sept. 1971), North-Holland Publ. Co., Amsterdam, 1973, pp. 45–56.

[La 74] A. H. Lachlan, Two conjectures on the stability of ω-categorical theories, *Fund. Math.* **81** (1974) 133–145.

[La 75] A. H. Lachlan, Theories with a finite number of models in an uncountable power are categorical, *Pacific J. Math.* **61** (1975) 465–481.

[La 75a] A. H. Lachlan, A remark on the strict order property, *Z., Math. Logic. Grundl. Math.* **21** (1975) 69–76.

[Lo 54] J. Łoš, On the categoricity in power of elementary deductive systems and related problems, *Colloq. Math.* **3** (1954) 58–62.

[Ls 73] D. Lascar, Types definissables et produit de types, *C.R.A.S. Paris* t. 276 (9 mai 1973) 1253–1256.

[Ls 76] D. Lascar, Ranks and definability in superstable theories, *Israel. J. Math.* **23** (1976) 53–87.

[Ma 74] J. A. Makowsky, On some conjectures connected with complete sentences, *Fund. Math.* **81** (1974) 193–202.

[MSS 76] J. A. Makowsky, S. Shelah and J. Stavi, Δ-logics and generalized quantifiers, *Annals of Math. Logic* **10** (1976) 155–192.

[Mo 65] M. D. Morley, Categoricity in power, *Trans. A.M.S.* **114** (1965) 514–538.

[Mo 67] M. D. Morley, Countable models of $\aleph_1$-categorical theories, *Israel. J. Math.* **5** (1967) 65–72.

[Mo 70] M. D. Morley, The number of countable models, *J. Symb. Logic* **35** (1970) 14–18.

[Mn 71] R. Mansfield, The theory of Boolean ultrapowers, *Annals Math. Logic* **2** (1971) 279.

[Mr 66] W. E. Marsh, On ω_1-categorical and not ω-categorical theories, Ph.D. Thesis, Dartmouth College, 1966.

[MS 70] D. Martin and R. M. Solovay, Internal Cohen extensions, *Annals Math. Logic* **2** (1970) 143–178.

[PSh 85] Pillay and S. Shelah, Classification over a predicate I, *Notre Dame J. of Formal Logic* **26** (1985) 361–376.

[Ra 29] F. D. Ramsey, On a problem of formal logic, *Proc. London Math. Soc.* **30** (1929) 338–384.

[Re 69] J. P. Ressayre, Sur les théories du premier ordre catégorique en un cardinal, *Trans. A.M.S.* **142** (1969) 481–505.

[Ro 64] F. Rowbottom, The Łoš conjecture for uncountable theories, *Notices A.M.S.* **11** (1964) 248.

[Rs 72] J. Rosenthal, A new proof of a theorem of Shelah, *J. Symb. Logic* **37** (1972) 133–134.

[Ry 59] C. Ryll-Nardzewski, On categoricity in power $\leq\aleph_0$, *Bull. Acad. Polon. Sci. Ser. Sci. Math. Astronom. Phys.* **7** (1959) 545–548.

[Sa 72] G. Sacks, *Saturated Model Theory*, Benjamin, Reading, Mass., 1972.

[Sb 75] G. Sabbagh, Sur les groupes qui ne sont pas réunion d'une suite croissante de sous-groupes propres, *C.R. Acad. Sci. Paris, Ser. A-B* **280** (1975) 763–766.

[Sf] J. Saffe, The number of uncountable models of ω-stable theories, *Annals of Pure and Applied Logic* **24** (1983) 231–261.

[So 71] R. M. Solovay, Real-valued measurable cardinals, *Proc. Symp. in Pure Math.*, XIII, Part I, ed. D. Scott, A.M.S., Providence, R.I., 1971, pp. 397–428.

[Su 72] M. Sauer, On the density of families of sets, *J. Combinatorial Theory, Series A* **13** (1972).

[Sh 68] S. Shelah, Class with homogeneous models only, *Notices A.M.S.* **15** (1968, Aug.) 803.

[Sh 68a] S. Shelah, Categoricity in power, *Notices A.M.S.* **15** (1968, Oct.) 930.

[Sh 69] S. Shelah, Categoricity of classes of models, Ph.D. Thesis, The Hebrew University, Jerusalem, Israel, 1969.

[Sh 69a] S. Shelah, Stable theories, *Israel. J. Math.* **7** (1969) 187–202.

[Sh 69b] S. Shelah, On stable homogeneity, *Notices A.M.S.* **16** (1969, Feb.) 426.

[Sh 69c] S. Shelah, On generalization of categoricity, *Notices A.M.S.* **16** (1969, June) 683.

[Sh 70] S. Shelah, Finite diagrams stable in power, *Annals Math. Logic* **2** (1970) 69–118.

[Sh 70a] S. Shelah, On theories T-categorical in $|T|$, *J. Symb. Logic* **35** (1970), 73–82,

[Sh 70b] S. Shelah, On the cardinality of ultraproducts of finite sets, *J. Symb. Logic* **35** (1970) 83–84.

[Sh 70c] S. Shelah, Solution of Łoš conjecture for uncountable languages, *Notices A.M.S.* **17** (1970, Oct.) 968.

[Sh 70d] S. Shelah, When every reduced product is saturated, *Notices A.M.S.* **17** (1970, Jan.) 453.

[Sh 71] S. Shelah, Stability, the f.c.p.; and superstability: model theoretic properties of formulas in the first order theory, *Annals Math. Logic* **3** (1971) 271–362.

[Sh 71a] S. Shelah, The number of non-almost isomorphic models of theory T in a power, *Pacific J. Math.* **36** (1971) 811–818.

[Sh 71b] S. Shelah, The number of non-isomorphic models of an unstable first-order theory, *Israel. J. Math.* **9** (1971) 473–487.

[Sh 71c] S. Shelah, Every two elementary equivalent models have isomorphic ultrapowers, *Israel. J. Math.* **10** (1971) 224–233.

[Sh 71d] S. Shelah, Lecture notes, Spring 1971, by R. Gail, U.C.L.A., mimeograph.

[Sh 71e] S. Shelah, Generalizations of saturativity, *Notices A.M.S.* **18** (1971, Jan.) 258.

[Sh 71f] S. Shelah, Some unconnected results in model theory, *Notices A.M.S.* **18** (1971, April) 563.

[Sh 71g] S. Shelah, Isomorphism of ultrapowers, *Notices A.M.S.* **18** (1971, July) 666.

[Sh 72] S. Shelah, Saturation of ultrapowers and Keisler's order, *Annals Math. Logic* **4** (1972) 75–114.

[Sh 72a] S. Shelah, Uniqueness and characterization of prime models over sets for totally transcendental first-order theories, *J. Symb. Logic* **37** (1972) 107–113.

[Sh 72b] S. Shelah, A combinatorial problem stability and order for models and theories in infinitary languages, *Pacific J. Math.* **41** (1972) 247–261.

[Sh 72c] S. Shelah, For what filters every reduced product is saturated, *Israel. J. Math.* **12** (1972) 23–31.

[Sh 72d] S. Shelah, On models with power-like ordering, *J. Symb. Logic* **37** (1972) 247–267.

[Sh 72e] S. Shelah, Various results in model theory, *Notices A.M.S.* **19** (1972, Nov.) A-764.

[Sh 73] S. Shelah, On the numbers of non-isomorphic models, *Notices A.M.S.* **20** (1973, Aug.) A-498.

[Sh 74] S. Shelah, Categoricity of uncountable theories (first version was lecture notes by M. Brown, U.C.L.A. distributed in Fall '70) Proc. of the Symp. in Honour of Tarski's seventieth birthday in Berkeley 1971, ed. Henkin, *Symp. Pure Math.* Vol. XXV (1974) 187–204.

[Sh 74a] S. Shelah, Why there are many non-isomorphic models for unsuperstable theories, *Proc. of the International Congress of Math.*, Vancouver 1974, pp. 553–557.

[Sh 74b] S. Shelah, Uniqueness of prime models, with an application to differential fields, *Notices A.M.S.* **21** (1974, Feb.) A-318.

[Sh 75] S. Shelah, A two cardinal theorem, *Proc. A.M.S.* **48** (1975) 207–213.

[Sh 75a] S. Shelah, Categoricity in $\aleph_1$ of sentences of $L_{\omega_1,\omega}(Q)$, *Irael. J. Math.* **20** (1975) 127–148.

[Sh 75b] S. Shelah, Various results in mathematical logic, *Notices A.M.S.* **22** (1975, Jan.) A-23.

[Sh 75c] S. Shelah, The lazy model theorist guide to stability, Proc. of a Symp. in Louvain, March 1975, ed. P. Henrard, *Logique et Analyse*, 18 année, **71–72** (1975) 241–308.

[Sh 76] S. Shelah, Some remarks in model theory, *Notices A.M.S.* **23** (1976, Feb.) A-289.

[Sh 76a] S. Shelah, On powers of singular cardinals, compactness of second order logic, *Notices A.M.S.* **23** (1976, June) A-449.

[Sh 76b] S. Shelah, Ultraproduct of finite cardinalities and Keisler order, *Notices A.M.S.* **23** (1976, Aug.) A-494.

[Sh 78] S. Shelah, End extensions and number of non-isomorphic models, *J. Symb. Logic* **43** (1978) 550–562.

[Sh 78a] S. Shelah, On the numbers of minimal models, *J. Symb. Logic* **43** (1978) 475–486.

[Sh 79] S. Shelah, Hanf number of omitting types for simple first-order theories, *J. Symb. Logic* **44** (1979) 319–324.

[Sh 79a] S. Shelah, On uniqueness of prime models, *J. Symb. Logic* **44** (1979) 215–226.

[Sh 80] S. Shelah, Simple unstable theories, *Annals of Math. Logic* **19** (1980) 177–204.

[Sh 80a] S. Shelah, Independence results, *J. Symb. Logic* **45** (1980) 563–573.

[Sh 81] S. Shelah, On saturation for a predicate, *Notre Dame J. of Formal Logic* **22** (1981) 239–248.

[Sh 82] S. Shelah, Better quasi-orders for uncountable cardinals, *Israel J. Math* **42** (1982) 177–226.

[Sh 82a] S. Shelah, The spectrum problem I, $\aleph_\epsilon$-saturated models, the main gap, *Israel J. Math* **43** (1982) 324–356.

[Sh 82b] S. Shelah, The spectrum problem II, totally transcendental theories and the infinite depth case, *Israel J. Math* **43** (1982) 357–364.

[Sh 83] S. Shelah, Construction of many complicated uncountable structures and Boolean algebras, *Israel J. Math.* **45** (1983) 100–146.

[Sh 83a] S. Shelah, Classification theory for non-elementary classes I, the number of uncountable models of $\psi \in L_{\omega_1,\omega}$, *Israel J. Math*, Part A, **46** (1983) 212–240; Part B, **46** (1983) 241–273.

[Sh 83b] S. Shelah, Models with second order properties IV, A general method and eliminating diamonds, *Annals of Math. Logic* **25** (1983) 183–212.

[Sh 84] S. Shelah, On universal graphs without instances of CH, *Annals of Pure and Applied Logic* **26** (1984) 75–87; Universal graphs without instances of CH revisited, *Israel J. Math.*, in press.

[Sh 84a] S. Shelah, A combinatorial principle and endomorphism rings of abelian groups II, *Proc. of the Conference on Abelian groups*, ed. R. Gobel, C. Metteli, A. Orsatti and L. Salce, International Centre Mechanical Sciences, Abelian Groups and Modules, pp. 37–86.

[Sh 84b] S. Shelah, On co-κ-Soulsin relations, *Israel J. Math.* **47** (1984) 139–153.

[Sh 85] S. Shelah, *A Classification of Generalized Quantifiers*, Springer Verlag Lecture Notes **1182** (1985) 1–46.

[Sh 85a] S. Shelah, A Classification theory for non-elementary classes II, Abstract elementary classes, *Proc. of the USA–Israel Symp. in Classification Theories* (1985), ed. J. Baldwin, Springer Verlag Lecture Notes **1292** (1987) 419–497.

[Sh 85b] S. Shelah, A Classification of first order theories which have a structure theory, *Bulletin of AMS* **12** (1985) 227–232.

[Sh 85c] S. Shelah, Monadic logic: Lowenheim numbers, *Annals of Pure and Applied Logic* **28** (1985) 203–216.

[Sh 85d] S. Shelah, Classification theory over a predicate III, Notes from Lectures in Simon Fraser, Summer (1985).

[Sh 85e] S. Shelah, Universal classes, *Proc. of the USA–Israel Symp. in Classification Theory* (1985), ed. J. Baldwin, Springer–Verlag Lecture Notes **1292** (1987) 264–418.

[Sh 86] S. Shelah, The spectrum problem III, universal theories, *Israel J. Math.* **55** (1986) 229–250.

[Sh 86a] S. Shelah, *Monadic Logic: Hanf Numbers*, Springer–Verlag Lecture Notes **1182** (1986) 203–223.

[Sh 86b] S. Shelah, *Classification Theory Over a Predicate* II, Springer–Verlag Lecture Notes **1182** (1986) 47–90.

[Sh 87] S. Shelah, Existence of many $L_{\infty,\lambda}$-equivalent non-isomorphic models of T of power λ, *Proc. of the Classification Theory Conference*, organized by G. Cherlin, P. Mangani and A. Marcja, *Annals of Pure and Applied Logic* **34** (1987) 291–310.

[Sh 88a] S. Shelah, On the number of strongly $\aleph_\varepsilon$-saturated models of power λ, *Annals of Pure and Applied Logic*, **36** (1987) 279–288; an addition: **40** (1988) 89–91.

[Sh 88b] S. Shelah, Number of pairwise non-elementarily embeddable models, *J. Symb. Logic* **54** (1989) 1431–1455.

[Sh 89] S. Shelah, Universal classes, second version, preprint V (1987), VI (1987), III §6 §7 (1988); IV and III §1–§5 were revised.

[Sh 89a] S. Shelah, Multidimensionality.

[SHM 84] S. Shelah, L. Harrington and M. Makkai, A proof of Vaught conjecture for w-stable theories, *Israel J. Math.* **49** (1984) 259–286.

[Si] J. Silver, A large cardinal in the constructible universe, *Fund. Math.* **69**, 93–100.

[Sv 59] L. Svenonius, $\aleph_0$-categoricity in first-order calculus, *Theoria* (Lund) **25** (1959) 82–94.

[Va 61] R. L. Vaught, Denumerable models of complete theories, infinistic methods, *Proc. of Symp. Foundation of Math.* (Warsaw 1959), Pergamon Press, London and PWN, Warsaw, 1961, pp. 303–321.

INDEX OF DEFINITIONS AND ABBREVIATIONS*

* Notice that for e.g., Av(I, A), $D^m(p, \Delta, \lambda)$, strongly λ-homogeneous, you should look at Av, D, homogeneous, resp. So look at the main word, ignoring strongly, explicitly, etc., but semi-, uni-, multi-, bi-, are considered part of the word. Note that the same word may have different meaning depending on the text (e.g., regular cardinal and regular type; Av(I, A) and Av(D, I)), whereas some variations are only shortening (e.g. $D^m(p, \Delta, \lambda)$, $D(p, \Delta, \lambda)$). So we number the distinct meanings. Some too well-known notions were omitted.

independence	(1) I is independent over A, [over (B, A)] [over p]	III, Definition 4.4(1), (2), p. 118
	(2) Π is independent over Φ and mod Ψ	III, Definition 5.4, p. 125
	(3) $\mathcal{G}$ is independent mod D	VI, Definition 3.1, p. 345; VI, Definition 3.6, p. 358
	(4) $(\mathcal{G}_1, \mathcal{G}_2, D)$- is κ-independent	VI, Definition 3.3, p. 350
indiscernible	(1) (Δ, n) indiscernible (set of sequences), Δ-n-indiscernible, n-indiscernible, etc.	I, Definition 2.4, p. 10
	(2) (Δ, n)-indiscernible sequences, Δ-n-indiscernible, n-indiscernible	I, Definition 2.3, p. 10
	(3) maximal indiscernible set	III, Definition 3.2, p. 106
	(4) absolutely indiscernible	IV, Definition 5.1, p. 212
	(5) (Δ, n)-indiscernible indexed set, etc.	VII, Definition 2.4, p. 413
inevitable	$p \in S^m(A)$ is inevitable	IV, Definition 5.9A, p. 210
inp (independent partitions)	$\kappa^m_{\text{inp}}(T)$, $\kappa_{\text{inp}}(T)$, κr^m_{inp}, $\kappa r_{\text{inp}}(T)$	III, Definition 7.3, p. 145
ird (independent orders)	$\kappa^m_{\text{ird}}(T)$, $\kappa_{\text{ird}}(T)$	III, Definition 7.1, p. 137
isolated	p is $\mathbf{F}$-isolated over A	IV, Section 1, p. 153
isomorphic	M, N are isomorphic	I, Definition 1.4, p. 5
k	(1) $k(N_0, N_1)$	XIII, Definition 4.6(3), p. 647
	(2) $k(N_1)$	XIII, Definition 4.7(1), p. 648
	(3) $k(T)$	XIII, Definition 4.7(2), p. 648
K	(1) $K = (\mathbf{F}, W, \lambda, \mu)$	V, Section 6, p. 289
	(2) K^h	V, Definition 6.2, p. 296
	(3) $K^m_\Delta(\lambda, T)$, $K^m_\Delta(\lambda)$, $K^m(\lambda)$, $K(\lambda)$, etc.	II, Definition 4.4(1), (3), p. 75
	(4) $K(I)$	VII, Definition 2.5, p. 413
	(5) K'	X, Definition 4.1, pp. 527–528
	(6) K_λ	X, Definition 4.3(1), p. 528
	(7) K'_λ	X, Definition 4.3(1), p. 528
Keisler	Keisler order $\otimes$, $\otimes_\lambda$	VI, Definition 4.1, pp. 370–371
kind	(1) $\theta_{\bar{a}}$ is the kind of $\bar{a}$	III, Definition 6.4, p. 132
	(2) (used only in X, §5)	X, Definition 5.16F, p. 545
Kr	$\mathrm{Kr}^m(\lambda, T)$, $\mathrm{Kr}^m(\lambda)$, etc.	II, Definition 4.4(2), (3), p. 75
L (Lascar rank)	$L(p)$	V, Definition 7.5, p. 316
ℓ	(1) $\ell(N_0, N_1)$	XIII, Definition 4.6(4), p. 648
	(2) $\ell(N_1)$	XIII, Definition 4.7(2), p. 648
	(3) $\ell(T)$	XIII, Definition 4.7(2), p. 648
large	J is a λ-large ($< \lambda$-large) subtree	X, Definition 5.8, p. 536

regular	(1) regular type	V, Definition 1.2, p. 233
	(2) (p, φ) is regular	V, Definition 3.1, p. 251
	(3) (p, φ) is strongly regular	V, Definition 3.6, p. 265
	(4) $\mathscr{P}$ is regular	V, Definition 4.4(1), p. 277
	(5) $\mathscr{P}$-regular type	V, Definition 4.5(4), (5), p. 278
	(6) regular family of sets	VI, Definition 1.3(1), p. 326
	(7) D is λ-regular [regular] filter	VI, Definition 1.3(3), (4), p. 326
	(8) λ is a regular cardinality	see Notation
regularize	family of sets regularize a filter	VI, Definition 1.3(2), p. 326
representation	(1) $\langle N_\eta, a_\nu : \eta \in I \rangle$ is a representation	X, Definition 5.2, p. 534; redefined X, 5.17, p. 547
	(2) an **F**-representation	X, Definition 5.3, p. 534
	(3) E of the representation	X, Definition 5.8, p. 536
	(4) equivalent representation	X, Definition 5.17B, p. 547
	(5) standard representation	X, Definition 5.8(2), p. 536
s (strongly)	$I \leq_s J$, $I \leq_w J$	V, Definition 2.1(2), (3), p. 240
S	(1) $S_\lambda(A)$, $S_\lambda^*(A)$, $S^\lambda(A)$, $S_*^\lambda(A)$	VII, Definition 1.2, p. 402
	(2) $S_\Delta^m(A, M)$, $S_\Delta(A, M), \ldots, S(A)$ (all eight possibilities)	I, Definition 2.1(2), (3), p. 9
saturated	(1) saturated, λ-saturated	I, Definition 1.2(1), (3), p. 3
	(2) $(\mathbf{F}, \mu)$-saturated	IV, Definition 1.1(1), p. 155
	(3) **F**-semi-saturated	IV, Definition 1.1(3), p. 155
	(4) **F**-saturated	IV, Definition 1.1(4), p. 155
	(5) (λ, H_1, H_2)-saturated model	VII, Definition 1.4(1), p. 402; VII, Definition 1.9, p. 410
	(6) (λ, κ)-saturated model	VII, Definition 1.5, p. 403; see VII, Section 1
	(7) strongly κ-saturated	I, Definition 1.5, p. 5
	(8) M is **T**-saturated	XI, Definition 2.2(1), p. 562
sct	$\kappa_{\mathrm{sct}}^m(t)$, $\kappa_{\mathrm{sct}}(T)$, $\kappa r_{\mathrm{sct}}^m(T)$, $\kappa r_{\mathrm{sct}}(T)$	III, Definition 7.5, p. 149
SE	(1) $N \in SE_\lambda(N_0, N_1)$	XIII, Definition 4.1(1), pp. 643–644
	(2) $(N, \bar{a}) \in SE_\lambda(N_0, N_1)$	XIII, Definition 4.1(2), p. 644
	(3) $N \in SE_\lambda(N_1)$	XIII, Definition 4.1(3), p. 644
	(4) $(N, \bar{a}) \in SE_\lambda(N_1)$	XIII, Definition 4.1(3), p. 644
semi-definable	p is semi-definable over I (over A)	VII, Definition 4.1(2), (3), p. 426

INDEX OF SYMBOLS

Latin letters

a, b, c, d	elements of $\mathfrak{C}$
f, g	function
h	function, or see VII, Definition 3.1, p. 422
i, j	ordinals
k, l, m, n	natural numbers
p, q, r	types
s, t	elements of an ordered set
u, v	sets
w	finite set
x, y, z	variables
A, B, C	subsets of $\mathfrak{C}$
D	filter
E	equivalence relation or such a formula, or rarely a family of sets generating a filter
F, G	functions
H	function, or see beginning of V, Section 6
I, J	ordered sets or trees or index symbols (in Chapters I–IV and IX–XIII they are interchanged with $\boldsymbol{I}, \boldsymbol{J}$)
	on I^+, I^- see X, Section 2
K	class of models, or see beginning of V, Section 6
L	first-order language
M, N	models
P	predicate or relation, usually unary
Q, R	predicates or relations
	on Q^{β}_{μ} see XIII, Definition 1.2
S	set
T	theory (usually fixed)
W	set, or see beginning of V, Section 6
X, Y	sets

Greek letters

α, β, γ	ordinals
δ	limit ordinal; on $\delta(\lambda)$, $\delta(\lambda, \kappa)$ see VII, Definition 5.1, p. 432
	on η^- see X, Section 2
ζ	ordinal
η	sequence, usually of ordinals

θ **formula**

κ **cardinal (usually infinite)**
on $\kappa^m(T)$, $\kappa(T)$; see III, Definition 3.1, p. 102

λ **cardinal (usually infinite);**
on $\lambda_l(T)$ see IV, Definition 2.4, p. 165
on $\lambda^l(T)$ see IV, Definition 2.5, p. 165
on $\lambda(T)$ see IX, Conclusion 1.14(4), p. 488
on $\lambda(\mathbf{F})$, $\lambda_r(\mathbf{F})$ see IV, p. 153
on $\lambda(K)$ see V, p. 289
on $\lambda^*(p)$ see III, Definition 5.1, p. 124
on $\lambda(\subseteq^*)$, $\lambda(\bar{\mathbf{T}})$, $\lambda(\mathbf{T}, \subseteq^*)$, $\lambda_s(\mathbf{T}, \subseteq^*)$ see XI, 2.3

μ **cardinal (usually infinite);**
on $\mu(\lambda)$, $\mu(\lambda, \kappa)$ see VII, Definition 5.1, p. 432
on $\mu(D)$ see VI, Definition 6.2, pp. 391–392
on $\mu(\mathbf{F})$ see IV, Definition 1.1(2), p. 155
on $\mu(K)$ see V, p. 289

ν **sequence, usually of ordinals**

ξ **ordinal**

ρ **sequence, usually of ordinals**

σ **permutation, or term, or sequence of ordinals**

τ **term**

φ **formula**

χ **cardinal, usually infinite**

ψ **formula**

ω **first infinite ordinal**

ω_α **α^{th} infinite ordinal**

Γ **set of formulas; see II, Definition 3.3, p. 44; see II, Theorem 2.2(4), p.31**

Δ, Π, Ψ **sets of formulas**

Φ **set of formulas or types**

Other letters

$\mathbf{a}$ **infinite sequence of elements**

$\mathbf{B}$ **Boolean algebra**

$\mathscr{A}$ **construction; see IV, Section 1**

$\mathfrak{B}$ **model**

$\mathfrak{C}$ **see p. 7**

$\mathbf{F}$ **see IV, Section 1, p.** 153;
on $\mathbf{F}^{\mathfrak{x}}_{\chi}$ see IV, Definition 2.1, IV, Definition 2.2, IV, Definition 2.3, pp. 157–158, IV, Definition 2.6, IV, Definition 2.7, p. 168, VII, Definition 4.2, p. 428, XII, Definition 1.1, p. 591, XII, Definition 1.6, p. 593
on $\mathbf{F}^{\mathfrak{x}}_{\chi,k}$ see IV, Definition 4.1, p. 185
on $\mathbf{F}(\mathbf{T})$ see XI, 2.2
on $\mathbf{F}^k_\lambda$ see XII, Definition 1.1
on $\mathbf{F}^l_\lambda$ see XII, Definition 1.6

$\mathscr{G}$ **family of functions**

I, J **sets of sequences from $\mathfrak{C}$, usually of fixed length, mostly indiscernible**
(in chapters I–V and IX–XIII they are interchanged with I, J)

$\mathscr{P}$ power set see X, Section 5, XII
on $\mathscr{P}^{-}(s)$ see XII, 2.1

T	$T(A)$ a family of "small types" over A, see XI, 2.1, on T^{x}_{μ} see XI, 2.2
t	**truth value (0, 1)**

Other symbols

$\exists$	**existential quantifier;** **on** $\exists^{<\chi}$ see V, Section 6, p. 289
$\emptyset$	**empty set**
$\prec$	**elementary submodel; see I, Definition 1.1, p. 2;** **on** $\prec^{\lambda}_{\kappa}$ see V, Section 6, p. 289
$\approx$	**equivalence relation;** **on** $\approx_D$ **see VI, Definition 1.1(1), p.** 324
$\prod$	**product**
$\sum$	**sum**
$\oplus$	**natural sum; see V, Definition 7.6, p.** 316
$\otimes$, $\otimes_\lambda$	**Keisler order; see VI, Definition 4.1(2), p.** 370
$\otimes^*$	see VI, Definition 4.3, p. 375
$\triangleleft$	**initial segment; see Notation**
$\restriction$	**restriction; see Notation and VI, Notation, p.** 331
∞	**infinity sign**
$\subseteq$, $\subset$	**subsets (not necessarily proper)** on $\subseteq^{\chi}_{\mu,\kappa}$, $\subseteq_x$ see XI, Definitions 3.4, 3.5
$\models$	**satisfaction; see Notation**
$\vdash$	**see II, Definition 1.3, p. 23**
$<_A$	on $B <_A C$ see X, Definition 1.2